Martin Hanson
Ann Hanson

Hodder & Stoughton
A MEMBER OF THE HODDER HEADLINE GROUP

The publishers would like to thank the following individuals, institutions and companies for permission to reproduce photographs in this book. Every effort has been made to trace ownership of copyright. The publishers would be happy to make arrangements with any copyright holder whom it has not been possible to contact:

Chapter openers: Part one – mitochondrion courtesy of Science Photo Library/Bill Longcore; Part two – culture of nerve cells from human brain courtesy of Science Photo Library/Eye of Science; Part three – stomata in surface view courtesy of Science Photo Library/Dr Jeremy Burgess.

Biophoto Associates (47 all, 49, 56 bottom, 57, 299, 301, 306, 350 top, 494, 499 both, 506, 569, 577, 643 bottom, 661); Bruce Coleman Limited (144, 234, 237, 262 both, 369, 513, 550, 631, 689); Holt Studios/Nigel Cattlin (589, 596); Science Photo Library/A B Dowsett (272), /Andrew Syred (111 left, 361, 397, 535, 554), /Astrid & Hanns-Frieder Michler (385), /Bill Longcore (54), /Biophoto Associates (171, 469), /Bruce Iverson (42, 575 top), /Claude Nuridsany & Marie Perennou (232, 448), /CNRI (114 all, 199), /David Parker (218), /D Phillips (514), /Dr David Patterson (350 bottom), /Dr Don Fawcett (48, 50), /Dr Jeremy Burgess (40, 58, 540, 643 top, 658, 707), /Dr Kari Lounatmaa (55), /Dr Kenneth R Miller (618), /Dr Tony Brain (504), /Eric Grave (277), /Eye of Science (28, 61, 162), /J C Revy (647), /John Heseltine (579), /John Paul Key, Peter Arnold Inc. (380), /John Reader (779), /Juergen Berger, Max-Planck Institute (270), /K R Porter (115), /L Willatt, East Anglian Regional Genetics Service (111 right), /Manfred Kage (454, 479), /Michael Abbey (143), /National Library of Medicine (7, 122), /Omikron (56 top), /Patrick Lynch (575 bottom), /Profs P M Motta & S Correr (391), /Pr S Cinti/CNRI (300, 317), /Quest (298), /Science Pictures Ltd (119), /Science Source (165), /Simon Fraser (211), /Sinclair Stammers (342)

The publishers would also like to thank Blackwell Science for the use of the graph on page 808, showing the change in relative thickness of eggshells of peregrine falcons over 60 years, which was originally included in an article by D A Ratcliffe in *Journal of Applied Ecology* (1970) 7:67–115.

Finally, we are very grateful to the following examining bodies for permission to reproduce questions: AEB, NEAB, WJEC, UCLES and Edexcel.

Orders: please contact Bookpoint Ltd, 39 Milton Park, Abingdon, Oxon OX14 4TD. Telephone: (44) 01235 400414, Fax: (44) 01235 400454. Lines are open from 9.00–6.00, Monday to Saturday, with a 24 hour message answering service. Email address: orders@bookpoint.co.uk

A catalogue record for this title is available from The British Library

ISBN 0 340 66443 6

First published 1999
Impression number 10 9 8 7 6 5 4 3 2 1
Year 2004 2003 2002 2001 2000 1999



Cover photo from Sharpshooters.
Illustrated by Martin Hanson, Ann Hanson and Ian Foulis Associates.

Typeset by Wearset, Unit 1, Boldon Business Park, Boldon, Tyne and Wear, NE35 9PE.
Printed in Great Britain for Hodder & Stoughton Educational, a division of Hodder Headline Plc, 338 Euston Road, London NW1 3BH by Redwood Books, Trowbridge, Wilts.

# CONTENTS

## Cells, Heredity and Evolution


CHAPTER 1 Characteristics of living things and the nature of biology . . . . . . . . 3

CHAPTER 2 The chemicals of life . . . . . . . . . . . . . . . . . . . . . . . . . . . . . . . . . . . . 14

CHAPTER 3 Units of life – the structure of cells . . . . . . . . . . . . . . . . . . . . . . . . . 40

CHAPTER 4 Biological catalysts – enzymes . . . . . . . . . . . . . . . . . . . . . . . . . . . . 64

CHAPTER 5 Energy from molecules – respiration and fermentation . . . . . . . . . 76

CHAPTER 6 How things enter and leave cells . . . . . . . . . . . . . . . . . . . . . . . . . . 94

CHAPTER 7 Chromosomes and cell division . . . . . . . . . . . . . . . . . . . . . . . . . . 109

CHAPTER 8 Mendel and heredity . . . . . . . . . . . . . . . . . . . . . . . . . . . . . . . . . . 122

CHAPTER 9 What do genes do? . . . . . . . . . . . . . . . . . . . . . . . . . . . . . . . . . . . 153

CHAPTER 10 The molecule of heredity – DNA . . . . . . . . . . . . . . . . . . . . . . . . . 160

CHAPTER 11 Making proteins . . . . . . . . . . . . . . . . . . . . . . . . . . . . . . . . . . . . . 173

CHAPTER 12 Changing the message – mutation . . . . . . . . . . . . . . . . . . . . . . . 190

CHAPTER 13 Putting genes to work: gene technology . . . . . . . . . . . . . . . . . . 205

CHAPTER 14 Genes in populations: evolution . . . . . . . . . . . . . . . . . . . . . . . . . 222


## The Functioning Animal


CHAPTER 15 The variety of life . . . . . . . . . . . . . . . . . . . . . . . . . . . . . . . . . . . . 257

CHAPTER 16 Animal organisation – size and complexity . . . . . . . . . . . . . . . . . 295

CHAPTER 17 The nervous system . . . . . . . . . . . . . . . . . . . . . . . . . . . . . . . . . . 303

CHAPTER 18 Detecting change – receptors . . . . . . . . . . . . . . . . . . . . . . . . . . . 328

CHAPTER 19 Movement and support in animals . . . . . . . . . . . . . . . . . . . . . . . 348

CHAPTER 20 Chemical communication – the endocrine system . . . . . . . . . . . . 372

CHAPTER 21 Homeostasis . . . . . . . . . . . . . . . . . . . . . . . . . . . . . . . . . . . . . . . . 382

CHAPTER 22 Transport in mammals – the blood system . . . . . . . . . . . . . . . . . 388

CHAPTER 23 Gas exchange and breathing . . . . . . . . . . . . . . . . . . . . . . . . . . . 412

CHAPTER 24 Nitrogenous excretion and osmotic regulation . . . . . . . . . . . . . . 431

## The Functioning Animal *continued*


CHAPTER 25 Regulation of body temperature . . . . . 447

CHAPTER 26 Nutrition in mammals . . . . . 461

CHAPTER 27 Defence against micro-organisms . . . . . 486

CHAPTER 28 Reproduction . . . . . 501


## The Functioning Plant


CHAPTER 29 Plant organisation . . . . . 530

CHAPTER 30 Plant life cycles . . . . . 537

CHAPTER 31 Growth in plants . . . . . 562

CHAPTER 32 The control of plant growth . . . . . 585

CHAPTER 33 Gathering energy and raw materials . . . . . 606

CHAPTER 34 Transport in plants . . . . . 641


## Organisms and Their Environment


CHAPTER 35 Introduction to ecology . . . . . 670

CHAPTER 36 The physical environment . . . . . 679

CHAPTER 37 The biotic environment . . . . . 687

CHAPTER 38 Populations . . . . . 710

CHAPTER 39 Communities and ecosystems . . . . . 725


## The Human Impact


CHAPTER 40 Primate legacy . . . . . 752

CHAPTER 41 The evolution of humans . . . . . 775

CHAPTER 42 The human predicament . . . . . 801

**Appendix I** . . . . . 821

**Appendix II** . . . . . 827

**Answers** . . . . . 843

**Index** . . . . . 854

# ACKNOWLEDGEMENTS

I owe a great debt to the many people who, directly and indirectly, have had a hand in this book. I have always felt that teaching a subject to young people with critical minds is the best way of learning it, and for that reason I am grateful to the students I have taught since 1965.

It is a special pleasure to pay tribute to Arthur Ellis, whose lively and stimulating teaching at Cheadle Hulme School was such a source of inspiration to colleagues and students alike. As an influential figure in the development of Nuffield Biology, Arthur always leaned heavily towards inquiry rather than didactic transfer of facts and terminology.

I should also like to express my thanks to Michael Thain and Stephen Winrow-Campbell, who each read the entire manuscript and whose critical comments were of great value. Dr Don Love of Auckland University read Chapter 13 and made numerous helpful suggestions. Any remaining errors are, of course, solely my responsibility, and I would be grateful to anyone who takes the trouble to point out any outstanding mistakes or other defects.

Finally, I wish to record my thanks to Ann Hanson, whose artistic flair has added so much to the visual appeal of the book.

Martin Hanson
King's College
Auckland
New Zealand

m.hanson@kings.ak.school.nz

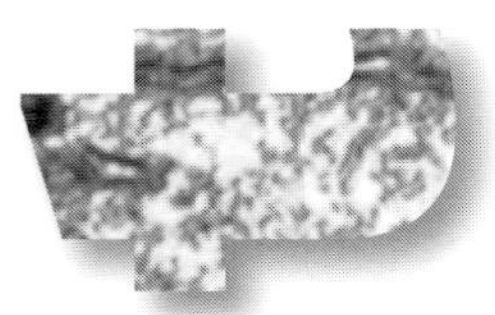

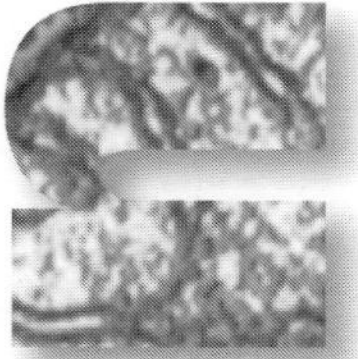
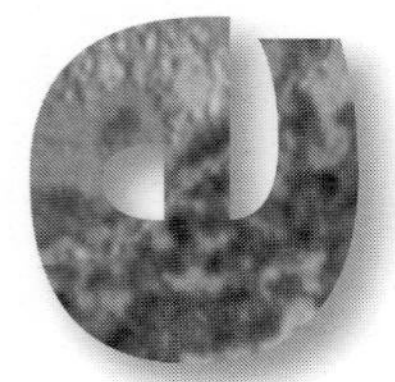


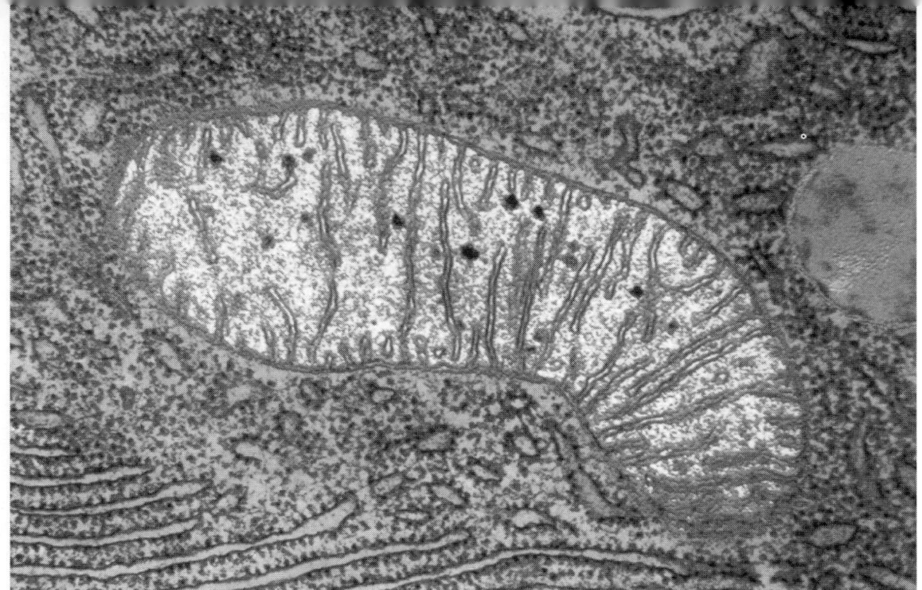

# CELLS, HEREDITY and EVOLUTION


CHAPTER 1
CHARACTERISTICS of LIVING THINGS and the NATURE of BIOLOGY . . . . . 3

CHAPTER 2
The CHEMICALS of LIFE . . . . . 14

CHAPTER 3
UNITS of LIFE – the STRUCTURE of CELLS . . . . . 40

CHAPTER 4
BIOLOGICAL CATALYSTS – ENZYMES . . . . . 64

CHAPTER 5
ENERGY from MOLECULES – RESPIRATION and FERMENTATION . . . . . 76

CHAPTER 6
HOW THINGS ENTER and LEAVE CELLS . . . . . 94

CHAPTER 7
CHROMOSOMES and CELL DIVISION . . . . . 109

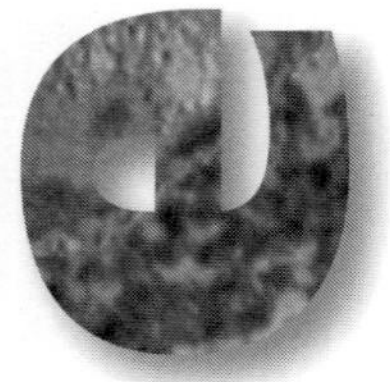
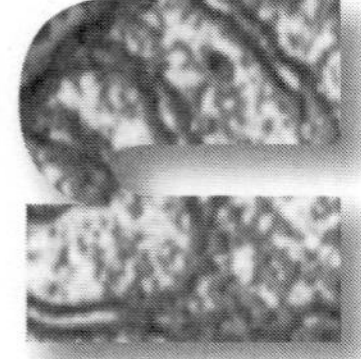

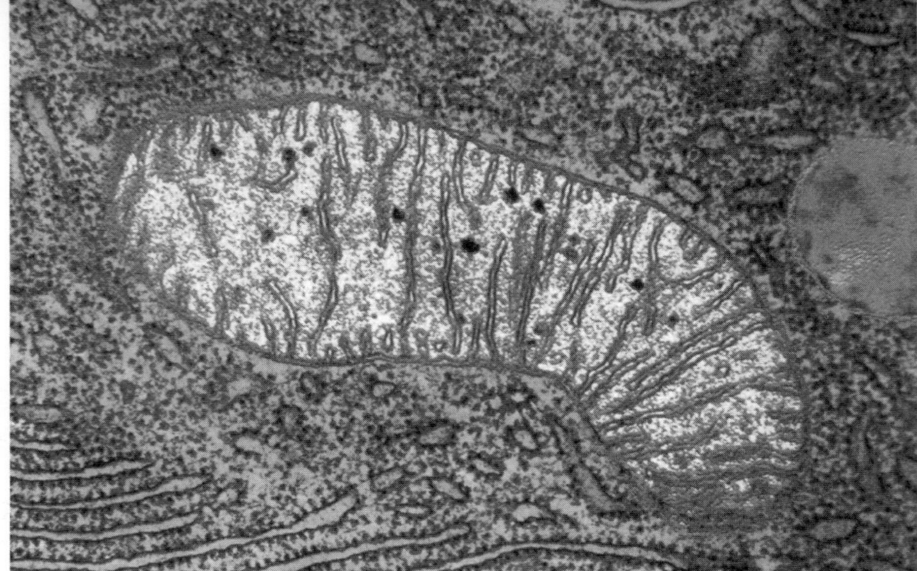

# CELLS, HEREDITY and EVOLUTION


**CHAPTER 8**
MENDEL and HEREDITY .................................. 122

**CHAPTER 9**
WHAT DO GENES DO? .................................. 153

**CHAPTER 10**
The MOLECULE of HEREDITY – DNA .................................. 160

**CHAPTER 11**
MAKING PROTEINS .................................. 173

**CHAPTER 12**
CHANGING the MESSAGE – MUTATION .................................. 190

**CHAPTER 13**
PUTTING GENES to WORK: GENE TECHNOLOGY .................................. 205

**CHAPTER 14**
GENES in POPULATIONS: EVOLUTION .................................. 222

CHAPTER 1

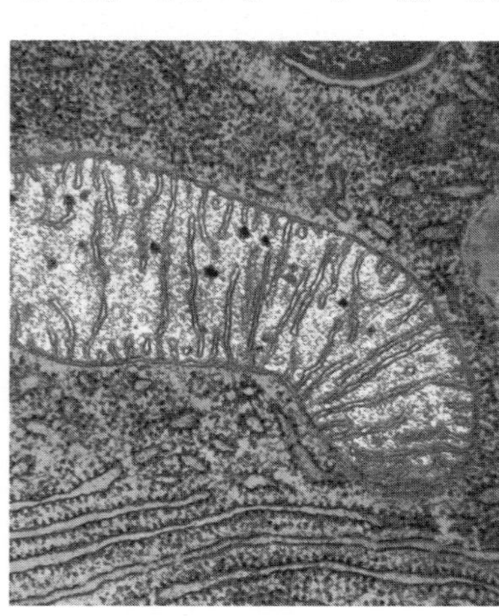

# CHARACTERISTICS of LIVING THINGS and the NATURE of BIOLOGY

**LEARNING OBJECTIVES**

By the time you have completed your study of this chapter you should be able to:

- explain how living things differ from non-living matter.
- explain how autotrophs differ from heterotrophs.
- explain what is meant by a hypothesis and the importance of controls in designing a scientific experiment.
- describe Darwin's ideas on natural selection and explain how they differ from Lamarck's ideas on the inheritance of acquired characters.
- Understand what is meant by teleology and explain why it is so contrary to modern evolutionary ideas.

## What is life?

Biology is the study of the most complex things known – living things. Though it is difficult to give a simple, water-tight definition of life, most people have little difficulty in distinguishing living things from non-living ones. Living things, or **organisms**, have certain characteristics that, taken together, distinguish them from non-living matter. These include:

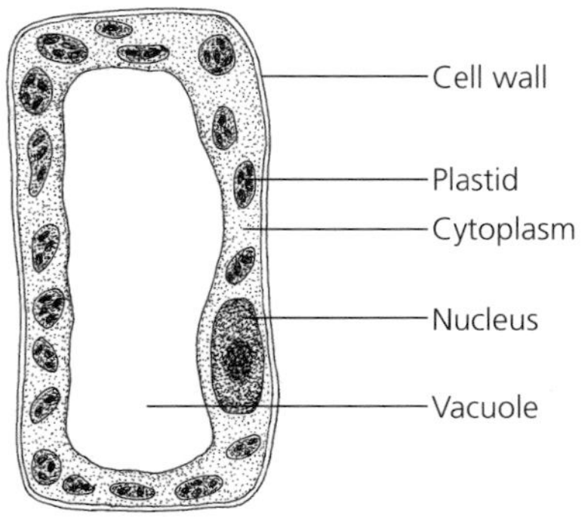


**Figure 1.1**
Light microscope view of a plant cell

- **Complexity** Unlike non-living matter, living things have a highly organised structure. Most consist of many small units called **cells** (Figure 1.1), but some consist of only one cell (Figure 1.2). A cell has its own complex internal organisation, consisting of many smaller structures called **organelles** (Figure 1.3).
- **Metabolism** Building up a complex structure requires matter and energy from the surroundings. Inside cells, substances undergo many chemical changes, speeded up by complex catalysts called **enzymes** (Chapter 4). The chemical reactions of life are collectively called **metabolism**. Some substances are built up into more complex molecules – a process called **anabolism**. Others are broken down into simpler substances in **catabolism**. The most important catabolic process is **respiration**, in which the energy released is used to drive

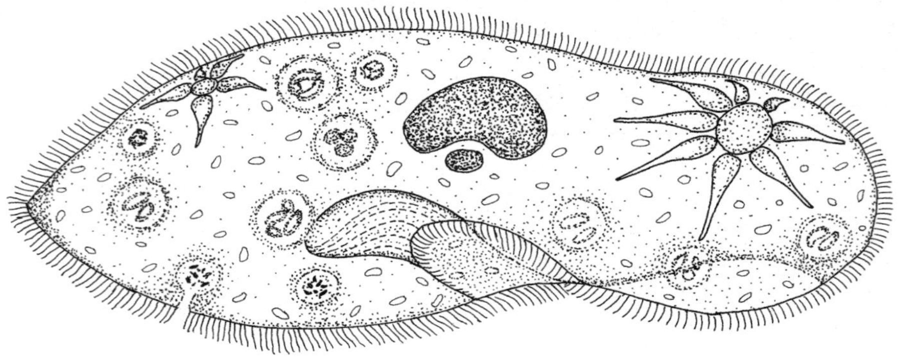

**Figure 1.2▶**
*Paramecium*, an organism consisting of a single cell

energy-requiring processes. In being used, some of the energy is transformed into heat.

- **Feeding** Living things take in materials and energy from their surroundings. This process is called feeding. The raw materials for growth are called **nutrients**, and the processes by which these are obtained and incorporated into the body are called **nutrition**.

According to the kind of nutrients they use, living things can be divided into two major groups (Figure 1.4). Some, such as green plants and some bacteria, are called **autotrophs** ('self-feeders'). Autotrophs take in their nutrients in the form of simple inorganic raw materials, namely carbon dioxide, water and mineral ions. These are built up into the complex organic substances that form the major constituents of cells.

In contrast to autotrophs, **heterotrophs** – which include all animals, all fungi, and many bacteria – cannot synthesise organic substances from inorganic raw materials. Heterotrophs use organic materials made by other living things (*heteros* = other). Though some heterotrophs feed on other heterotrophs, all heterotrophs are ultimately dependent on autotrophs for their supplies of organic materials.

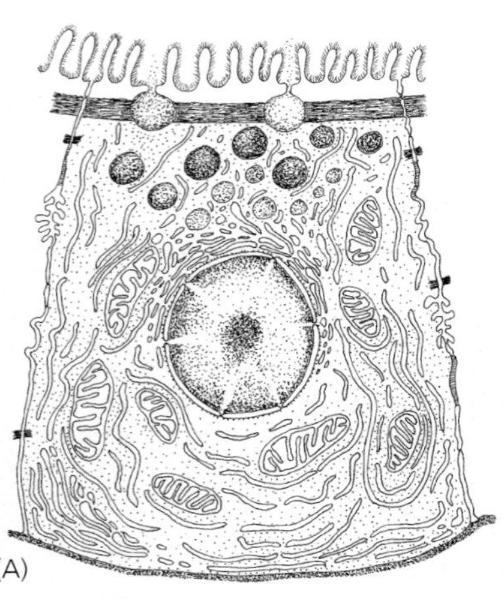

(A)

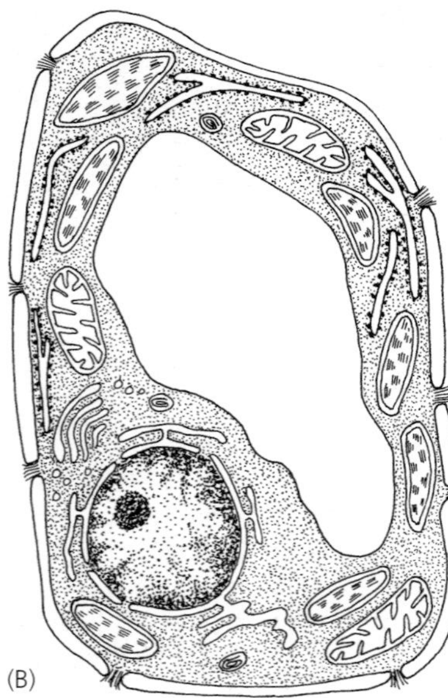

(B)

**Figure 1.3▲**
Electron microscope views of (A) an animal cell and (B) a plant cell

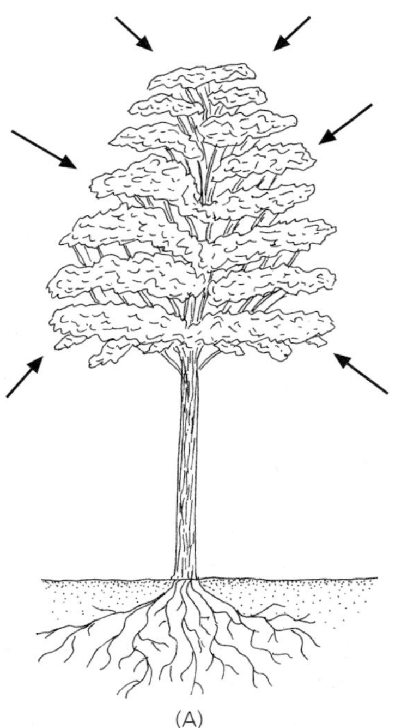

(A)

(B)

**Figure 1.4▶**
(A) Autotrophic and (B) heterotrophic nutrition. Autotrophs take in thinly-dispersed, inorganic nutrients. Heterotrophs take in organic nutrients produced by other organisms

## Life and the Second Law of Thermodynamics

Only two centuries ago many biologists thought that living things did not obey physical and chemical laws, but were governed by some mysterious 'vital force'. We now know that life obeys the same laws of nature as non-living matter. With regard to energy, all changes in the universe obey two laws: the First and Second Laws of Thermodynamics.

The First Law, otherwise known as the Law of Conservation of Energy, states that matter cannot be created or destroyed, but can only be changed from one form to another. Electrical energy can be changed into kinetic (movement) energy as in an electric motor, or vice versa as in a dynamo. In photosynthesis, light energy is changed into chemical energy.

The Second Law describes the *direction* in which changes occur. It can be stated in various ways, but perhaps one that relates most to everyday experience is:

> "*All changes result in a net increase in randomness or disorder in the universe.*"

Another name for disorder is **entropy**. All changes therefore result in a net increase in entropy. For example:

- ▶ Substances tend to diffuse from a higher to a lower concentration.
- ▶ Heat tends to be conducted through a material from higher to lower temperature.
- ▶ Complex structures tend to break down into more random arrangements.
- ▶ Systems rich in stored energy (e.g. chemical, elastic, gravitational, nuclear) tend to change to a state in which some of this stored energy is converted to heat, which then becomes dispersed and thus randomised.

As highly complex structures, organisms are the essence of non-randomness. In fact, they increase their organisation as they grow, decreasing their entropy. Are they breaking the Second Law?

Actually, no. To see why, consider the growth of an animal. During its lifetime, it takes in far more than its own mass of high energy materials in its food. Most of this concentrated energy is converted into heat, only a small proportion being retained as new growth. To produce a gram of new, highly organised flesh, it has to convert many times that amount of equally highly organised food into carbon dioxide, water and heat. Overall, there is a *net increase* in entropy in the universe as a whole.

Plants are subject to the same restrictions. As a plant grows, the amount of energy in the new tissue produced is much less than the light that it absorbed; over 95% of the energy is radiated out into space as heat, and thus randomised.

- **Generation of wastes** Some metabolic processes generate waste products. These may leave the body passively by *diffusion* – for example when ammonia diffuses out of an *Amoeba*. Alternatively – as in most animals – the organism expends energy in actively getting rid of wastes. This process is called **excretion**. Metabolic wastes are quite different from *faeces*, which consist largely of undigested food and are not actually produced by the body.
- **Active transport** All cells are constantly transporting substances across membranes from regions of lower concentration to regions of higher concentration. This phenomenon is called **active transport**, and because substances are moved up a concentration gradient, in the opposite direction to the 'natural' one, it requires energy (Chapter 6).
- **Homeostasis** All cells and organisms devote much of their energy to actively maintaining stable internal conditions. For example, when a leaf wilts it reduces further water loss by closing the tiny pores in its

outer layer (epidermis). This regulation of internal conditions is called **homeostasis**. A familiar human example is what happens when you are short of water – urine output (and hence water loss) decreases and at the same time you become thirsty, which normally leads to increased water intake. Many other examples of homeostasis are described in Chapter 21.

- **Sensitivity** All organisms have the ability to respond to both internal and external changes. Such responses are *adaptive* because they promote survival and reproduction.

  Responses may involve purely internal adjustments. For instance, the heart rate may increase in response to chemical changes in the blood. Alternatively, they may involve changes in behaviour of the entire organism, for example when woodlice run more quickly in response to lower humidity.

  Adaptive responses by plants are less dramatic, but they are just as crucial to survival. An example is the ability of plants to orientate their leaves to the light, thus enabling them to maximise light absorption.
- **Movement** Although movement occurs in many non-living things, only living organisms move using energy obtained by the organism itself. Movement in organisms may be obvious – as when an animal moves around – or more subtle, for instance when a plant moves by growth. Some kinds of movement can only be detected under the microscope, and may be very slow, for example when a cell divides.
- **Growth** All living things grow. The growth of an organism is quite different from the increase in size of non-living things such as snowballs and crystals. A snowball rolling down a snowy slope simply adds material to its *outer* surface, without changing it. Growth in a living thing involves taking in material and converting it to new substances *inside* the body.
- **Reproduction** Closely linked to growth is **reproduction** – the production of more individuals like the parents. Like growth, reproduction involves the production of new cells. This requires not only a supply of raw materials and energy from the surroundings – it also requires *information*. This information is stored in a chemical called **DNA**. It is carried in thread-like bodies called **chromosomes** which (except in bacteria) are located in the cell nucleus. Every time a cell divides, its DNA is copied, one copy passing to the nucleus of each daughter cell.

## The growth of biological ideas

It has been said that scientific knowledge is doubling every seven years or so. Only half a century ago biologists had little idea of the fine structure of cells, still less did they understand how cells work. Yet nowadays it is possible to transplant human genes into bacteria, which then make human hormones!

### Early beginnings – scientific method

Only a few centuries ago, science as we now know it hardly existed. Ideas that would be regarded by us as incredible were accepted without question. One such idea was that of **spontaneous generation**, according to which living things were thought to develop from non-living matter.

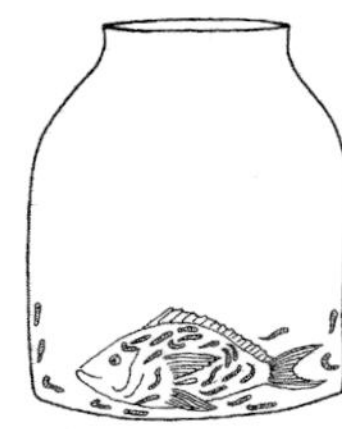

**Figure 1.5**
Redi's experiment disproving spontaneous generation. Only in the jars not covered by muslin did maggots appear

Maggots, for example, were said to develop from rotting meat, and this was accepted as fact by educated and intelligent people.

One of the few people who did not accept the doctrine of spontaneous generation merely because other people said that it was so was an Italian, Francesco Redi. He suspected that the maggots that appeared in meat came from eggs laid by flies, and in 1668 he reported an experiment in which he put this to the test (Figure 1.5). He placed a dead snake, some veal and some fish in each of four jars, and left them open to the air. He set up another four in the same way except that he prevented flies from entering by covering them with gauze. He observed flies entering and leaving the open jars, and after a few days he observed that maggots had appeared in the uncovered meat but not in the meat protected from flies.

One reason why Redi's discovery was so important was that it severely weakened the idea that life could develop from non-life. However, it had a deeper significance than this, for what Redi had done was help bring about a new way of thinking. Instead of uncritically parroting what others before him had said, he was sceptical. He put forward a testable explanation or **hypothesis** that could explain why maggots appear in rotting meat.

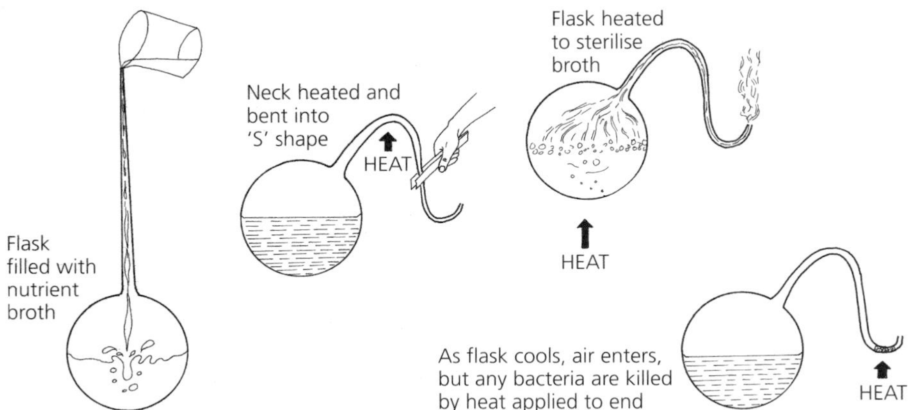


**Figure 1.6►**
Pasteur's experimental disproof of spontaneous generation in bacteria

Redi had put his hypothesis to the test by carefully designing an **experiment**. If he had set up only jars covered with gauze, it would have been possible to argue that maggots did not appear because the meat was unsuitable in some way, or perhaps for some other reason unconnected with the ability of flies to reach the meat. Redi had excluded such alternative explanations by including **controls**, consisting of jars open to the air. Since the only way in which the experimental jars differed from the control jars was that they were covered, it must have been the cover that was responsible for the failure of maggots to appear.

Convincing though Redi's experiment was, the idea that microscopic organisms could be generated spontaneously lingered on for another two centuries. It was Louis Pasteur who finally proved that even bacteria have parents. He boiled some nutrient broth in a flask with a 'swan' neck, and allowed it to slowly cool. Though the broth was in contact with the air, it remained clear even after several months. When he broke the neck, it turned cloudy and smelly within days (Figure 1.6).

**Figure 1.7**
Charles Darwin

## The idea that shook the world – evolution

Long before Darwin published his *Origin of Species*, naturalists were aware that there is a pattern in the similarities and differences between organisms. All animals with feathers, for instance, also have a beak, scaly

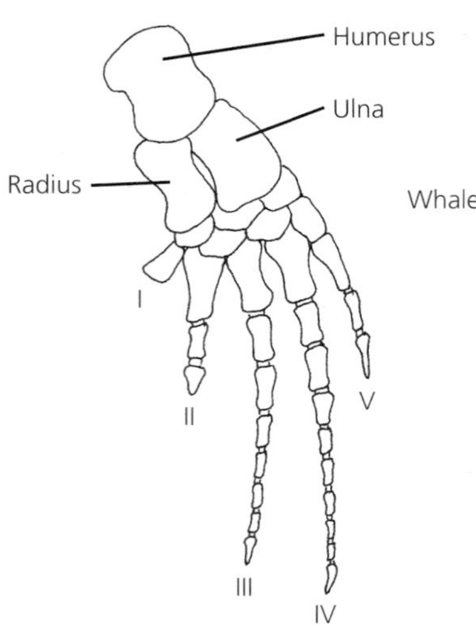


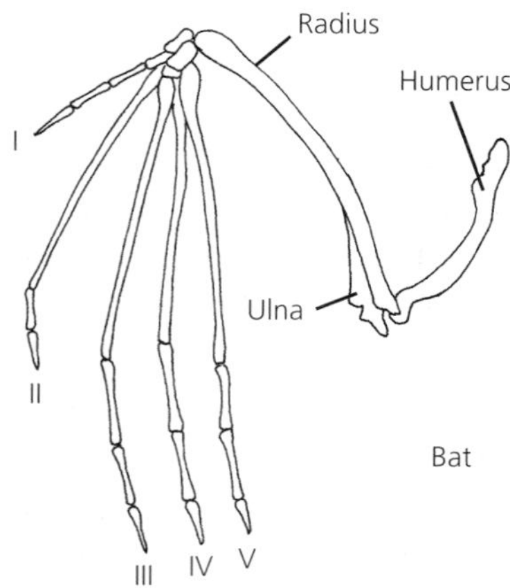


**Figure 1.8▲**
Similarities between limbs of bat and whale

legs and claws, and lay large yolky eggs with shells of lime. All animals with a diaphragm also have a four-chambered heart, three bones in the middle ear, a large brain, and suckle their young on milk. Why should such features occur in characteristic combinations?

One possibility is that it is all a gigantic coincidence. An alternative view was that it reflected the sense of order in the mind of the Creator. Darwin and a few other audacious biologists, on the other hand, preferred an explanation to a belief. Darwin maintained that the reason rats and mice, for example, have so much in common is that they are *related*. In other words, they are descended from a common ancestor.

Biblical fundamentalists reply that similar lifestyles impose similar demands, requiring that such creatures be given similar body structures by a Creator. Whilst this might seem reasonable to some, there are problems when we consider animals such as bats and whales. These have totally different lifestyles but share a fundamentally similar internal anatomy (Figure 1.8). Structures that are built on a fundamentally similar plan and have similar embryonic development, even though they may have different functions, are said to be **homologous**. **Analogous** structures, on the other hand, have similar functions but are built on quite different plans, for example the wings of insects and birds.

Why should the anatomy of a bat resemble that of a whale more closely than that of a sparrow? Why should the anatomy of a whale differ from that of a trout far more than it differs from that of a rabbit?

Creationists find it even harder to explain why an intelligent Creator should go to the trouble of including a vestigial left lung in snakes, or the centimetre-long wings of a kiwi. And why should the embryos of baleen whales have teeth, only for these to disappear before birth? Dandelions expend considerable energy in making nectar – yet insects are not needed, since seeds are produced without fertilisation. Why do they do this unless their ancestors needed insects for fertilisation? There are many more examples of functionless structures, a human example being shown in Figure 1.9.

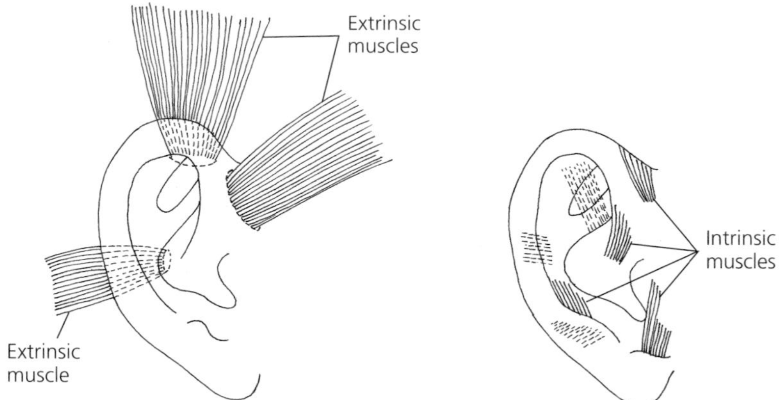


**Figure 1.9▶**
Vestigial human ear muscles (courtesy of Professor J. Carman, Auckland School of Medicine)

To biologists these oddities of nature are only to be expected. There is now a colossal body of evidence, from many branches of biology to support the view that fundamental similarities between organisms are the result of descent from a common ancestral form. The classification of organisms thus takes on a far greater significance than merely putting them into groups for the sake of orderliness. When biologists classify organisms, they are discovering relationships that already existed long before there were any biologists to study them. The groups into which biologists put animals and plants represent clusters of distant relatives. Classification is thus inseparable from the study of evolution.

## Lamarck and the inheritance of acquired characters

Though Darwin was the first to present substantial evidence for evolution, he was by no means the first to suggest the idea. In 1809, the year of Darwin's birth, the French zoologist Jean Baptiste Lamarck published *Philosophie Zoologique*, in which he suggested that species have changed over long periods of time. It was on the question of *how* evolution occurred that Lamarck differed so fundamentally from Darwin half a century later.

Lamarck's ideas contained two core statements:

- evolutionary changes occur in response to *need*.
- body parts that are put to greater use become more highly developed, and vice versa, and *these effects are inherited.*

Lamarck's second statement was built on the well-established fact that many bodily features can adapt to change. For example in response to ultraviolet radiation, the skin produces more of the pigment melanin, and it develops thickened calluses in response to abrasion. However, there is no evidence that lifetime experience is inherited – children of farm workers are not born with thicker calluses on their skin because of the experience of their parents. It seems difficult to imagine a mechanism by which signals from suntanned skin could reach the developing eggs and sperm. It is even more difficult to see how such changes could be brought about in just those genes that affect melanin production in the next generation.

## Darwin and natural selection

The catalyst that set Darwin questioning the doctrine of Divine Creation was his voyage round the world in the *Beagle*. Most influential of all was his visit to the Galápagos islands, a small group of islands about 1000 km off the coast of Ecuador in South America (Figure 1.10). As ship's naturalist, Darwin spent several weeks on the islands in 1835 and was greatly impressed by the peculiarities of the animal and plant life.

**Figure 1.10**
The Galápagos islands

There are very few species on the Galápagos islands, entire taxonomic groups being absent. There are only six species of reptile (including the giant tortoises, after which the islands are named), two species of mammals, 23 species of bird and no amphibians (as they cannot tolerate salt and would be most unlikely to be dispersed from the mainland on floating logs). Numerous plant groups are also absent (e.g. conifers, lilies and palms).

Darwin collected specimens of the birds of the islands, and on his return home he had them examined by John Gould, a specialist on birds. The Galápagos cuckoo was indistinguishable from the mainland species. The Galápagos warbler, on the other hand, was sufficiently different from the mainland type to be regarded as a local race. The Galápagos swallow is considered to be a subspecies of the mainland species, and one of the flycatchers belongs to a different species. The Galápagos mockingbirds belong to a genus different from, but clearly related to, that of the mainland. There is thus every degree of difference between Galápagos and mainland species.

Even more interesting than the differences between Galápagos and mainland animals were the differences between the inhabitants of different islands. The mocking birds Darwin had collected were actually of three distinct kinds, each peculiar to a particular island. Some time later he realised that some of the finches were also peculiar to particular islands. He was later to recall another interesting fact: the Governor had told him that each island had its own variety of giant tortoise, which could easily be distinguished by the carapace (shell).

It was these observations that led Darwin to begin thinking the unthinkable. The general similarity of the animals to those of the mainland, together with the volcanic origin of the islands, led him to the conclusion that the animals and plants had not been created *in situ* but were the descendants of colonists from the mainland. The fact that some of the animals were restricted to particular islands, even though the islands were very similar in climate and vegetation to their neighbours, led him to believe that they had changed since their arrival.

Darwin opened his first notebook on 'The transmutation of species' in 1837. Though convinced of the *fact* of evolution, Darwin had no idea of its *mechanism*. Then, in 1838 he happened to read 'for amusement' *An Essay on the Principle of Population, as It Affects the Future Improvement of Society* by Thomas Malthus. In it Malthus pointed out that the potential for increase in population is far greater than the potential for increase in food production. Accordingly, Malthus said, starvation was inevitable.

This led Darwin to the idea that for all species there is a 'struggle for existence', and that this could be the key to the mechanism of evolution. He maintained that evolution occurs by **natural selection**, sometimes known as the 'survival of the fittest'. Reduced to its essentials, Darwin's ideas are extremely simple, and can be summed up in a number of observations (O) and deductions (D):

1 All species produce far more offspring than are needed to replace the parents (O).

2 Though populations fluctuate, they do so about a mean, which does not change much over long periods of time (O).

3 Therefore most offspring die without reproducing (D).

4 Within a population, individuals *vary*, no two being exactly alike (O).

5 Some ('fitter') individuals therefore have a better chance of surviving than others (D).

6 Variation is, to some extent at least, hereditary (O).

7 The survivors will tend to hand on their favourable characteristics to their offspring (D).

8 Over successive generations, the population will slowly change (D).

Darwin's ideas could not have been more different from those of Lamarck (Figure 1.11). Lamarck would have explained the evolution of features in terms of their use to *existing* generations. Darwin, on the other hand, would have explained them in terms of their advantage to *previous* generations.

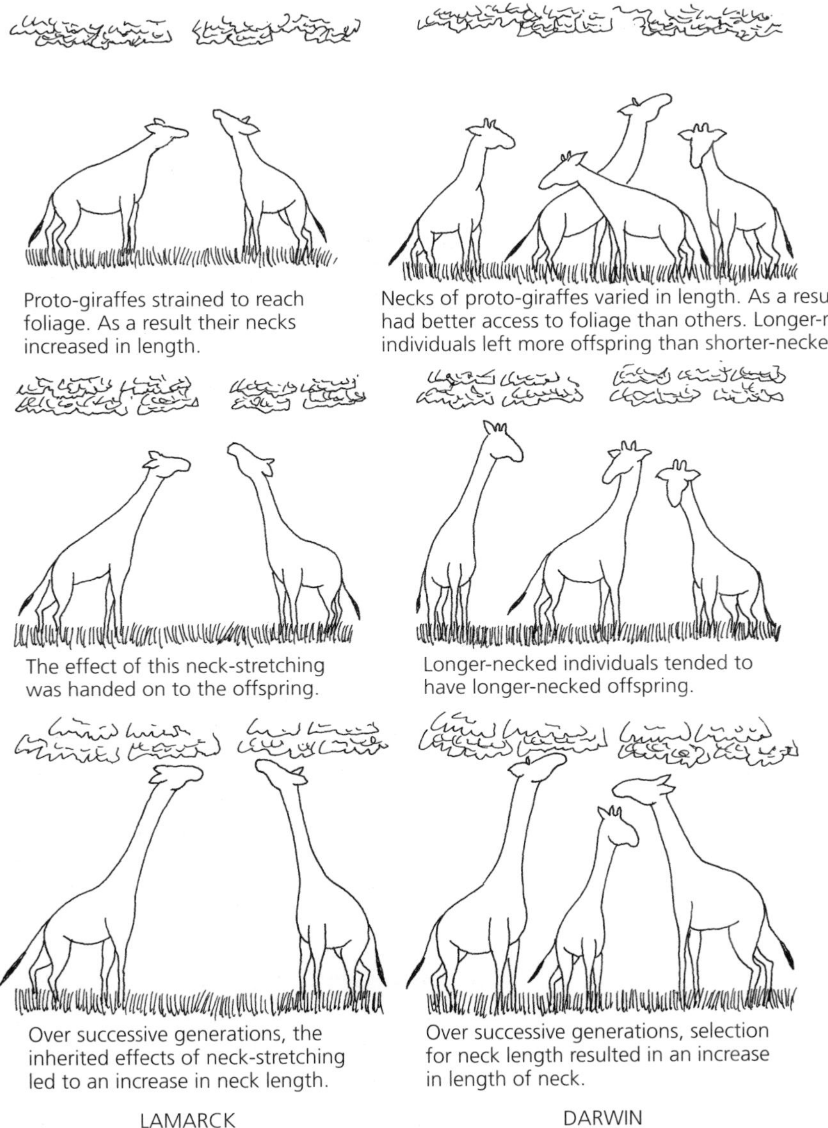

Proto-giraffes strained to reach foliage. As a result their necks increased in length.
Necks of proto-giraffes varied in length. As a result some had better access to foliage than others. Longer-necked individuals left more offspring than shorter-necked ones.
The effect of this neck-stretching was handed on to the offspring.
Longer-necked individuals tended to have longer-necked offspring.
Over successive generations, the inherited effects of neck-stretching led to an increase in neck length.
Over successive generations, selection for neck length resulted in an increase in length of neck.
LAMARCK
DARWIN


**Figure 1.11**
How Darwin and Lamarck would have explained the evolution of the giraffe's neck

The appeal of natural selection lay not only in its simplicity, but also in the fact that, as animal and plant breeders had long known, artificial selection by humans can produce hereditary change, such as higher yielding rice.

Yet the theory had one major weakness – no one at that time understood how heredity worked. Until this was understood, natural selection could not be regarded as a fully satisfactory explanation of how evolution occurred.

## Mortal bodies and immortal germs

Lamarck's ideas were based on a fundamental misunderstanding regarding what can and what cannot be inherited. The first person to appreciate the distinction between 'heritable' and 'non-heritable' was August Weismann in the late 19th Century. He realised that, so far as heredity is concerned, an organism consists of two parts (Figure 1.12):

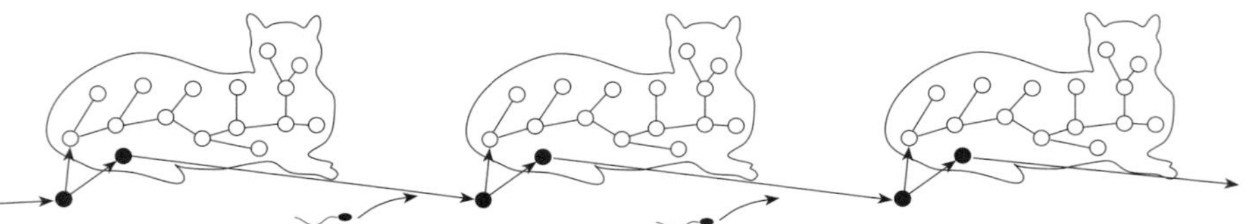

**Figure 1.12**
What can be inherited and what cannot. Successive generations are connected by an unbroken line of descent of cells – the germ line (black). Each zygote produces a body composed of somatic cells (unshaded) which eventually die, but it also produces germ cells (gametes)

- the **germ**, which consists of the gametes and the cells ancestral to them. These cells have the potential to leave a (theoretically) unlimited number of descendants through future generations. It follows that only changes in these cells can be transmitted to the offspring.
- the body, or **soma**, which consists of all the cells except the germ. Since the body eventually dies, all somatic cells are ultimately doomed. Any adaptive change in the body therefore cannot be transmitted to the next generation – it is not a suntan that is inherited, but the *capacity to develop* one.

The realisation that only the 'germ' can be passed from one generation to the next was but the first step towards an understanding of how heredity works. It was not until three-quarters of a century after Darwin's death that biologists understood the nature of the hereditary material and how it works. The early steps towards this understanding are described in Chapters 8 and 9.

### CLEAR THINKING

**Despite the lack of supporting evidence, the idea that evolution occurs in response to need still finds appeal even today. We can see this in many wildlife programmes and in popular books, in statements like this:**

> ***"Whales have a layer of blubber in order to reduce heat loss."***

**The phrase *'in order to'* implies that the whales had some kind of view of their future needs. In reality, whales vary randomly with regard to the amount of fat under the skin. Those individuals better endowed with fat were better at conserving heat, and consequently had more energy available for growth and reproduction. These individuals thus had an advantage and were more likely to become parents. Since offspring tend to resemble their parents, the next generation had, on average, a thicker layer of blubber than the previous one.**

**To attribute purpose to a structure or process is called teleology. Suppose, for example, we ask 'why do we have eyes?'. The teleological answer would be 'to see with', which would imply that eyes somehow develop in anticipation of future needs.**

**But genes cannot see into the future. A more correct answer to the question would be that eyes develop under the control of genes that were advantageous to *previous* generations.**

**Though 'purpose' is a term with which biologists do not feel comfortable, the word 'function' is a different matter. The function of a structure or process is simply a statement of how it contributes to survival. Thus the function of the heart is to pump blood round the body, and the function of leaves is photosynthesis.**

## Biology in modern society

Though Redi and Darwin relied more on their own mental powers than on sophisticated equipment, most branches of modern biology are utterly dependent on sophisticated tools and techniques the like of which Darwin and his contemporaries could never have imagined.

The supporting knowledge upon which biology rests has also increased greatly. In Darwin's time, mathematics was considered to have little role in biology. Now, however, it is indispensable, as is a knowledge of chemistry and, in many cases, physics. The traditional boundaries between some sciences have largely disappeared, to be replaced by 'hybrid' disciplines such as biophysics and biochemistry.

With the explosion of biological knowledge, specialisation has become inevitable. This can mean that a person in one discipline may have virtually no idea of many developments in other areas of biology. Biology has become so large that it now has common boundaries with medicine, agriculture, psychology, chemistry, physics, nutrition, and many other disciplines.

The greater understanding of life processes has also brought an increased potential to control and exploit them. Genes are now routinely transplanted from one species to another and, at the time of writing, the complete analysis of human genetic material is close to completion. The time may not be far off when potential parents can choose the characteristics of their children.

As for discoveries in other sciences, there is a danger that new biological knowledge may be misused. The more we understand life's mechanisms, the more some people see danger in the new knowledge being used for harmful purposes. As in all sciences, then, questions of how to use new knowledge almost invariably accompany new discoveries.

### QUESTIONS

**1** Give an example of a functionless structure in
**a)** humans
**b)** a plant.

**2** Why is it that houseflies have become immune to DDT while humans have not?

**3** Which of the following statements is *untrue*? Evolutionary change always
**A** has a genetic basis
**B** involves changes in structure
**C** involves populations
**D** involves more than one generation

**4** Gifted musicians often have musically talented children. How might this be explained by
**a)** Lamarck,
**b)** a modern biologist?

CHAPTER 2

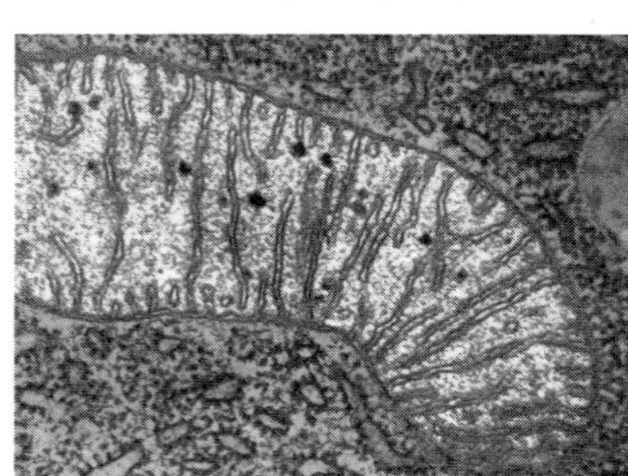

# The CHEMICALS of LIFE

## LEARNING OBJECTIVES

By the time you have completed your study of this chapter you should be able to:

- explain how the special properties of carbon enable it to form the skeletons of complex molecules.
- distinguish between condensation and hydrolysis and give examples.
- explain how the unique properties of water relate to its biological importance.
- give examples of the diverse roles of minerals in living things.
- describe the distinctive chemical features of amino acids and explain how they are linked to form polypeptides and proteins.
- explain the chemical basis for the diversity of proteins.
- explain the significance of molecular shape in protein function, and how the shape of globular proteins is maintained.
- explain the chemical basis behind the sensitivity of globular proteins to heat.
- describe the distinguishing chemical features of fats and explain the chemical basis for their richness in energy.
- explain the relationship between the chemistry of phospholipids and their role in cell membranes.
- recognise a steroid and be able to give examples of the biological roles of steroids.
- know the basic chemical features of the common monosaccharides, disaccharides and polysaccharides, and understand how different kinds of linkage can account for the different properties of starch and cellulose.

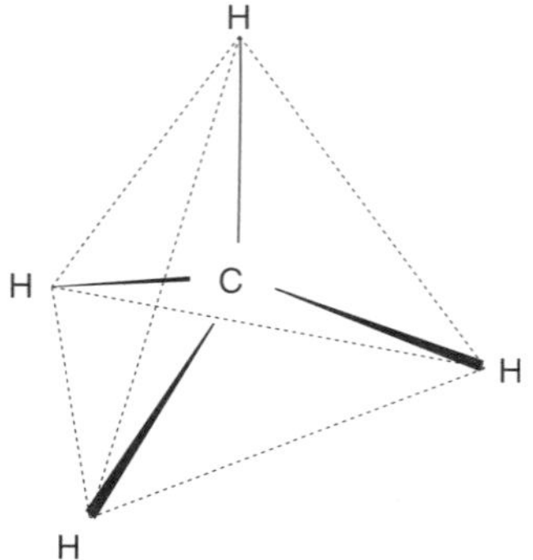


**Figure 2.1**
Tetrahedral carbon atom. The four numbered atoms lie at the points of a tetrahedron

## Organic molecules

Organic compounds are those containing **carbon**, in which at least one carbon atom is bonded to hydrogen (i.e. reduced). Of all the elements essential for life, carbon is perhaps the most important because of the ease with which it forms covalent bonds with other atoms, especially other carbon atoms.

When a carbon atom is linked to four other atoms, the latter occupy the points of a tetrahedron (Figure 2.1). This means that, except when double bonds are formed, a carbon atom can be linked to another atom in any of four directions. Since each additional carbon atom provides up to four more potential branching points, there is enormous scope for building elaborate chains and rings (Figure 2.2).

Though carbon forms the 'skeleton' of organic molecules, other atoms are also present. Hydrogen has a valency of 1 (see Appendix 1), meaning that it can only combine with one other atom and can therefore only form

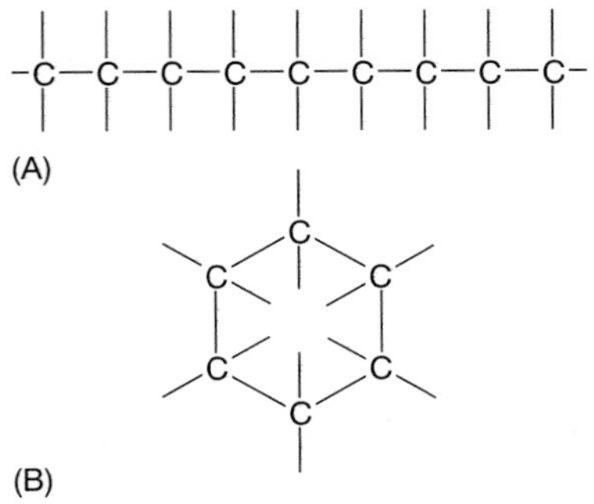

(A)

(B)

**Figure 2.2▲**
Carbon atoms can be built into chains (A) and rings (B). Other atoms attached to the carbons are not shown

the tips of chains and branches. The majority of organic molecules also contain **oxygen**. With a valency of 2, oxygen can link two other atoms together. Nitrogen, with a valency of 3, is an invariable constituent of proteins and nucleic acids. Sulphur and phosphorus are other common constituents of organic molecules.

Because the four other atoms that may be bonded to carbon form a tetrahedron, carbon chains are not straight but form a zigzag (Figure 2.3). In chemical formulae this is normally ignored and chains are usually drawn straight. There are some cases, however, where the three dimensional appearance of a molecule has to be shown in order to distinguish between different forms called **optical isomers** (see below).

## Functional groups

Many of the chemical properties of an organic compound are due not to the structure of the whole molecule, but to a localised group of atoms called a **functional group**. Thus a carbon atom joined to three hydrogen atoms constitutes a **methyl** (—$CH_3$) group, and a hydrogen atom joined to an oxygen atom forms an **alcohol** (—OH) group. Some of the more commonly encountered functional groups are shown in Figure 2.4.

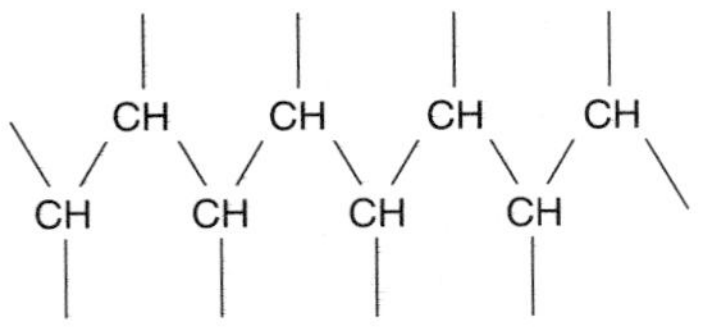

**Figure 2.3▲**
The carbon backbone of organic molecules forms a zigzag. In this case each carbon is shown bonded to a hydrogen

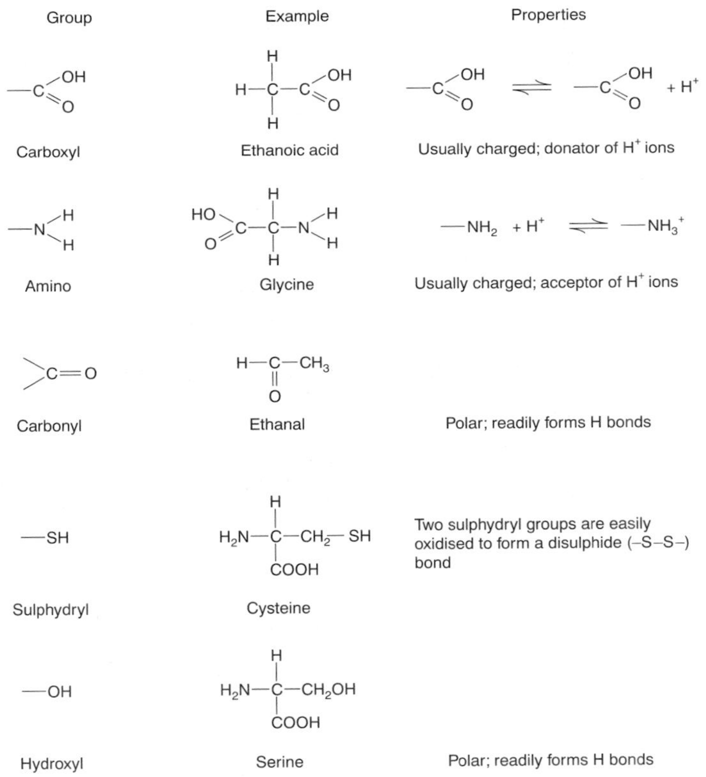

| Group | Example | Properties |
|---|---|---|
| —COOH<br>Carboxyl | $CH_3COOH$<br>Ethanoic acid | —COOH ⇌ —$COO^-$ + $H^+$<br>Usually charged; donator of $H^+$ ions |
| —$NH_2$<br>Amino | HOOC—$CH_2$—$NH_2$<br>Glycine | —$NH_2$ + $H^+$ ⇌ —$NH_3^+$<br>Usually charged; acceptor of $H^+$ ions |
| >C=O<br>Carbonyl | H—C(=O)—$CH_3$<br>Ethanal | Polar; readily forms H bonds |
| —SH<br>Sulphydryl | $H_2N$—CH(COOH)—$CH_2$—SH<br>Cysteine | Two sulphydryl groups are easily oxidised to form a disulphide (–S–S–) bond |
| —OH<br>Hydroxyl | $H_2N$—CH(COOH)—$CH_2OH$<br>Serine | Polar; readily forms H bonds |

**Figure 2.4►**
Some of the more important chemical groups and their properties

## Chemical shorthand

Many organic molecules are complex and it would be tedious to write out a full formula every time. Fortunately this is not usually necessary because in general we are only interested in certain functional groups, so the rest of the molecule can often be represented in an abbreviated or skeletal form. The use of abbreviated formulae is particularly convenient when representing large molecules built from many repeating subunits, such as starch and cellulose.

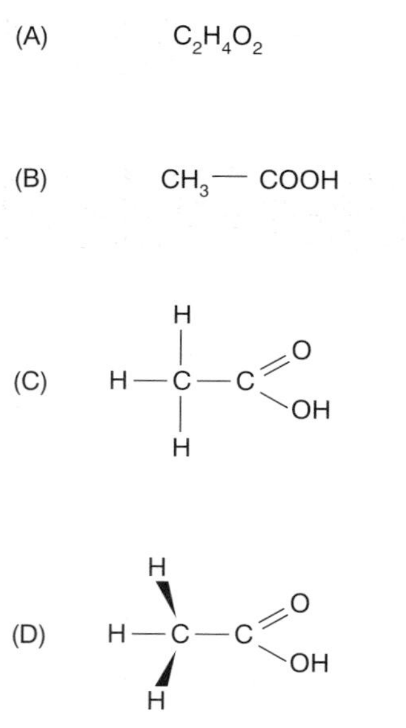

(A) $C_2H_4O_2$

(B) $CH_3—COOH$

(C) H, H—C—C=O, OH, H

(D) H, H—C—C=O, OH, H

**Figure 2.5▲**
Four different ways of representing ethanoic acid. In (D) two of the hydrogen atoms are above the plane of the paper, as indicated by the 'wedged' bonds

The distinctive features of organic compounds depend on their functional groups, so it is often necessary to give some idea of the arrangement of atoms in an organic molecule. Ethanoic acid, for example, can be represented in four different ways, as shown in Figure 2.5.

## Isomers

Some compounds have the same numbers of constituent atoms, but they are arranged in different ways and have different properties. The amino acids leucine and isoleucine are an example (Figure 2.6). Chemicals which contain the same constituent atoms linked in different ways are called **isomers**, and more examples will be encountered later in this chapter.

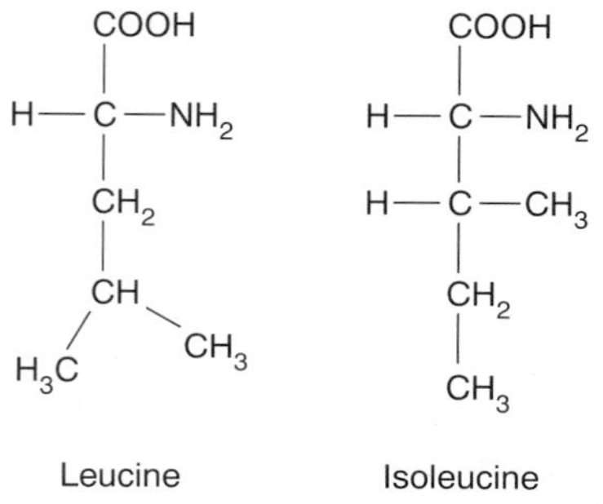

COOH, H—C—$NH_2$, $CH_2$, CH, $H_3C$, $CH_3$

Leucine

COOH, H—C—$NH_2$, H—C—$CH_3$, $CH_2$, $CH_3$

Isoleucine

**Figure 2.6◄**
Two isomeric amino acids, leucine and isoleucine

# The composition of living things

Of the 92 naturally-occurring elements, only about 25 **bioelements** are known to be essential to life. Table 2.1 compares the proportions of some of the bioelements in living matter and in the Earth's crust. Since there are considerable differences between them, it is evident that life did not evolve by random use of the available elements. Clearly, bioelements must have some special properties making them suitable for life.

**Table 2.1**
Relative abundance of the main elements of the human body

| Human body Element | % | Earth's crust Element | % |
|---|---|---|---|
| Hydrogen | 63.00 | Oxygen | 47.00 |
| Oxygen | 25.50 | Silicon | 28.00 |
| Carbon | 9.50 | Aluminium | 7.90 |
| Nitrogen | 1.40 | Iron | 4.50 |
| Calcium | 0.31 | Calcium | 2.50 |
| Phosphorus | 0.22 | Sodium | 2.50 |
| Sulphur | 0.05 | Potassium | 2.50 |
| Magnesium | 0.01 | Magnesium | 2.20 |

Six of the lighter, non-metallic elements – carbon, hydrogen, oxygen, nitrogen, phosphorus and sulphur – account for the greatest proportion of bioelements in organisms as diverse as bacteria and humans. Mineral salts account for a very small proportion, though they are nonetheless vital.

Just as organisms use a relatively small number of elements, they also contain a small number of different kinds of chemical compound (Table

2.2). The main ones are water and a handful of 'families' of organic compounds – proteins, fats and related substances, carbohydrates, nucleic acids and steroids.

**Table 2.2**
Chemical composition of two very different kinds of cell

| Constituent | Percentage of total live body mass | |
|---|---|---|
| | Human liver cell | Bacterium |
| Water | 70 | 70 |
| Proteins | 16 | 15 |
| Nucleic acids | 1.5 | 7 |
| Lipids | 5 | 2 |
| Carbohydrates | 3 | 3 |
| Other organic molecules | 2 | 2 |
| Inorganic ions | 2.5 | 1 |

## Water

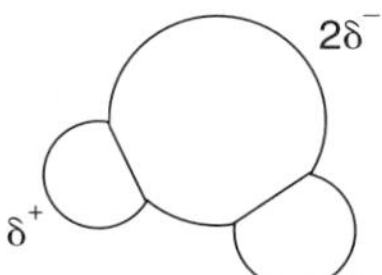


**Figure 2.7**
The polar water molecule

Directly or indirectly, all cell processes involve water. So important is water to life that most cells cannot tolerate much change in their water content. The central role of water in life processes arises from the unequal sharing of electrons in the water molecule. The oxygen atom takes more than its 'fair' share of the two shared electrons in each covalent bond. This gives it a slight negative charge ($\delta^-$), and leaves each hydrogen atom with a slight positive charge ($\delta^+$). Water molecules are thus said to be **polar** (Figure 2.7).

As a result of this polarity, the negative end of each water molecule is attracted to the positive end of a neighbouring water molecule (Figure 2.8). Water molecules are thus held together much more strongly than would otherwise be expected. Were it not for this, water would not be a liquid at room temperature (methane, with similar molecular mass, has a boiling point of $-164$°C).

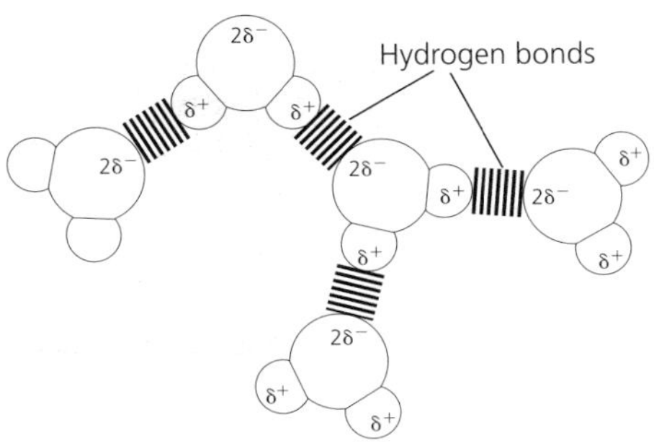


**Figure 2.8►**
Hydrogen bonding between water molecules

The attraction between the oxygen and hydrogen of neighbouring molecules is an example of a **hydrogen bond**. Hydrogen bonds can exist between hydrogen and nitrogen which, to a lesser extent than oxygen, also tends to draw electrons from hydrogen.

Individual hydrogen bonds are weak (about 20 times weaker than covalent bonds) but, like the hooks in a Velcro fastener, they can be very effective in large numbers. Moreover, although they are easily broken by the energy of molecular vibrations and collisions, they are continually reforming as fast as they are broken. As we shall see, the weakness of hydrogen bonds plays a fundamental role in the working of enzymes (Chapter 4).

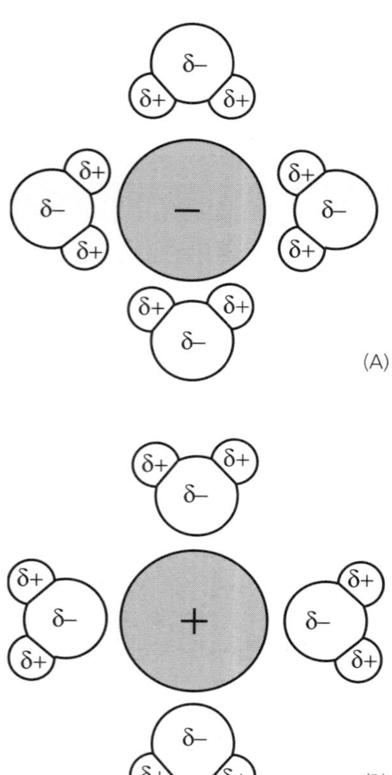


**Figure 2.9**
Attraction between water molecules and (A) negative ions and (B) positive ions

## Water as a solvent

Because water molecules are **polar** they are attracted to other polar molecules and to ions (Figure 2.9). As a result water is an extremely good solvent. This is important because molecules can only react if they are able to freely intermingle, as they do in a solution. Substances that are attracted to water are said to be **hydrophilic** ('water-loving'), whilst those that are not attracted to water are **hydrophobic** ('water-fearing').

It is interesting to note that whilst life depends on water being a good solvent, life could not function if it dissolved all substances. As we shall see later in this chapter, the fact that some chemical groupings are hydrophobic actually helps to keep many protein molecules in their correct shapes. Cell membranes also can only exist because one end of the constituent phospholipid molecules is hydrophobic (Chapter 3).

## Water as a heat buffer

By virtue of its unique chemical properties, some of which are given below, water helps to prevent organisms becoming too hot or too cold.

- **Density** Ice is less dense than water, so it floats. As a result of this, lakes and ponds freeze from the top downwards. The layer of ice helps to insulate the water below, protecting the organisms in it from freezing. If water froze from the bottom upwards, even the tropical oceans would be mostly ice, with only a thin layer of liquid at the surface.
- **High latent heat of fusion** Compared with other liquids, water requires a lot of heat to melt, and conversely, much heat is given up when it freezes. Because of this high latent heat of fusion, even small bodies of water do not freeze as readily as they would otherwise.
- **High heat capacity** When water is warmed, much of the heat supplied is absorbed in breaking hydrogen bonds rather than increasing the speed of movement of the water molecules. As a result it takes more heat to warm up a gram of water through a degree Celsius than it does for most other substances, so seas and lakes warm up more slowly in summer and cool down more slowly in winter than does the land.
- **High latent heat of evaporation** Because of the attraction between molecules in liquid water, large amounts of energy are needed to evaporate water. This high latent heat of evaporation enables many land organisms to keep cool through the loss of water by evaporation.

## Surface tension

Wherever water meets air it behaves as if it has an elastic 'skin' because at the surface, the net pull of neighbouring water molecules is *inwards*. A number of organisms make use of this **surface tension**. Pond skaters use their waxy feet to walk on the surface of the water, and the larvae of mosquitoes and some other insects are able to breathe at the surface by hanging from it (Figure 2.10).

Another example of the importance of surface tension is in the leaves of most land plants, which are protected by a waxy cuticle perforated by tiny pores or **stomata**. If water entered the stomata, it would flood the internal air space system, severely reducing the rate of diffusion of gases. This does not happen because in order to squeeze through a pore, a water drop would have to change from its spherical shape, but this is resisted by surface tension (Figure 2.11).

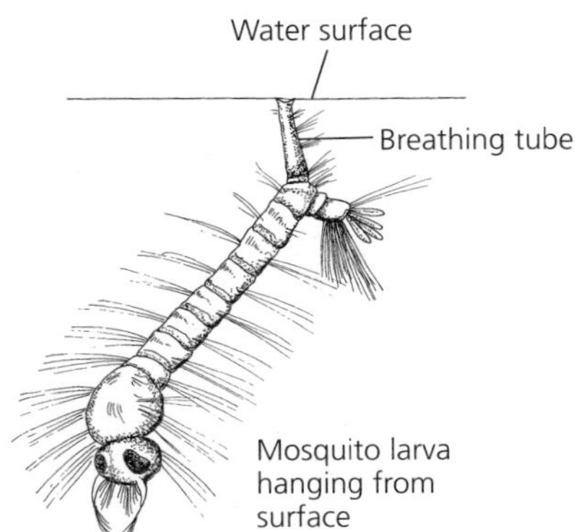


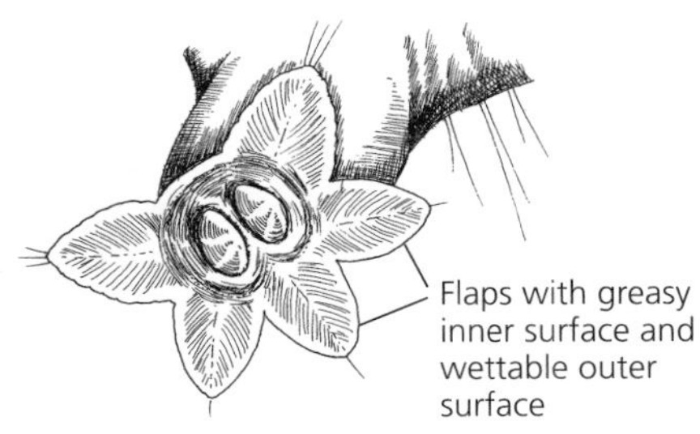


**Figure 2.10**
Role of surface tension in air-breathing by mosquito larva

## Capillarity

An important biological phenomenon that involves surface tension is **capillarity**, which is the tendency of water to invade narrow spaces in porous, wettable materials. It can be seen when a narrow glass tube is placed in water (Figure 2.12). At the edge of the tube the water is attracted to the glass and creeps up the sides, causing the water surface to become concave, pulling the water up behind it. The narrower the tube, the higher the water climbs. This is because the upward pull is proportional to the amount of contact between glass and water i.e. the *circumference* of the tube, but the weight of the water column is proportional to its cross-sectional *area*. Doubling the circumference of a tube quadruples its cross-sectional area.

Capillarity results from the interaction of two phenomena:

- **adhesion**, the attraction of water for the molecules of a wettable surface;
- **cohesion**, the attraction of water molecules for each other.

Capillarity plays an important part in the movement of water through plants. The surfaces of plant cells contain submicroscopic water-filled spaces and, as explained in Chapter 34, it is the tendency of these spaces to hold water that sets up the forces which pull water up the plant.

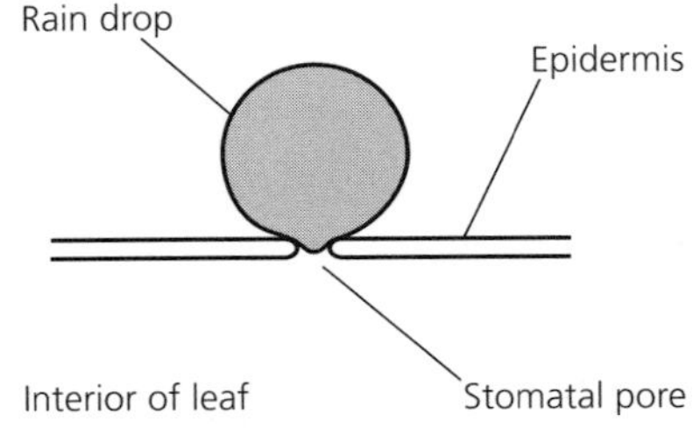


**Figure 2.11**
How surface tension protects a leaf from 'drowning'

# Inorganic ions (mineral salts)

Although bioelements present in the largest quantities are non-metals (e.g. carbon, hydrogen, oxygen, nitrogen, phosphorus and sulphur), most of the others are metals and are present in living matter in minute amounts – in some cases in parts per *billion*. Some, such as copper, are poisonous at higher concentrations.

Inorganic ions play diverse roles, for instance:

- some are constituents of larger molecules; for example iron forms part of the oxygen-carrying blood pigment haemoglobin.
- some have a structural function, e.g. calcium and phosphate in bones and teeth.
- metals such as magnesium, cobalt, zinc and manganese are essential for the action of certain enzymes (Chapter 4).
- some are important in maintaining the water balance between the inside and outside of cells, for example sodium and chloride ions in the blood.

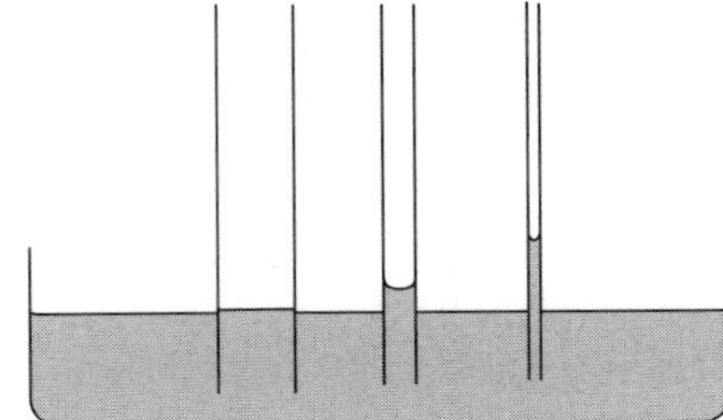

**Figure 2.12**
Capillarity in glass tubes of various diameters

# Proteins

Proteins are amongst the most complex substances known and they account for about 50% of the dry matter in cells. Some of their diverse functions are listed below.

- **Catalysis** Almost all **enzymes** are proteins. Without enzymes, the chemical processes of life could not occur.
- **Transport** Many substances are transported into and out of cells by specific proteins in the plasma (outer cell) membrane. Proteins may also transport substances over much longer distances; for example haemoglobin transports oxygen round the body. Over intermolecular distances, some proteins are specialised for the transport of electrons (Chapter 5).
- **Mechanical support** Some proteins help maintain body shape, such as collagen in tendons and ligaments, and keratin in the outer layer of the skin.
- **Movement** Some proteins interact to produce movement, such as tubulin which is involved in the movement of chromosomes in cell division, and actin and myosin of muscle.
- **Communication** Some proteins carry information between cells as **hormones**. Others are situated in the plasma membrane and act as **receptors**, detecting changes outside the cell.
- **Defence** All **antibodies** are proteins. Antibodies are substances which help defend against invading micro-organisms.
- **Storage** A number of proteins play a storage role. Ferritin, for example, is an iron-containing protein that stores iron in the liver. Ovalbumin is the protein of egg white, which contains large quantities of water for the developing chick.
- **Regulation of gene expression** (Chapter 11).

## Amino acids – the building units of proteins

Proteins are **polymers**, consisting of long chains of **amino acids**. There are many naturally-occurring amino acids but only 20 are used as raw materials for building proteins. A few other amino acids do occur in proteins, but these are produced by modification of one of the 20 'standard' amino acids after the protein has been produced. For example, hydroxyproline is produced from proline after its incorporation into collagen.

All the protein amino acids can be represented by the general formula shown in Figure 2.13. At the centre of this part of the molecule is a carbon atom known as the **alpha (α) carbon atom**. The **R group**, one of the four groups linked to the α carbon atom, differs from one amino acid to another and is responsible for the individual properties of each amino acid.

The other three groups are the same for all 20 amino acids and are shaded in Figure 2.14. The **amino** (—$NH_2$) group is basic because it can accept a proton to form an —$NH_3^+$ group. The **carboxyl** (—COOH) group is acidic because it can dissociate to form a —$COO^-$ group and an $H^+$ ion, or proton. Some examples of amino acids are shown in Figure 2.14.

One of the most important characteristics of the R group is whether or not it is attracted to water. Charged and polar R groups are hydrophilic;

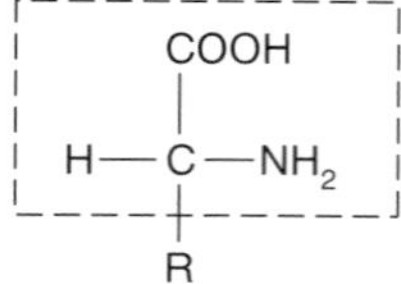


**Figure 2.13**
General formula for an amino acid. The enclosed part is common to all amino acids

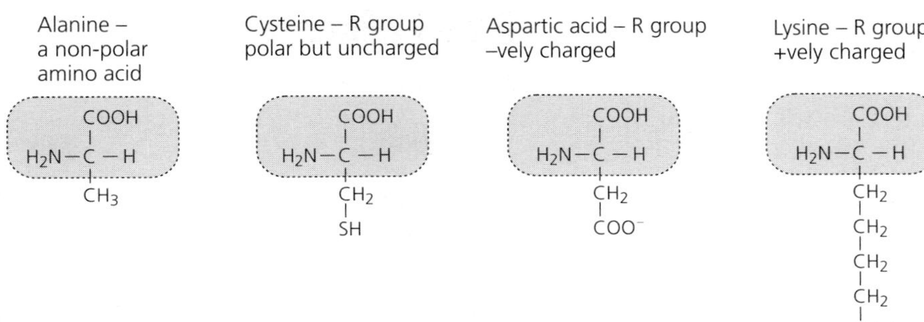

**Figure 2.14**
Four of the 20 amino acids. The part common to all amino acids is enclosed. Amino acids with polar R groups are attracted to water, while those with non-polar R groups are not

non-polar R groups are hydrophobic. As we shall see later, the attraction or non-attraction of R groups for water is the single most important factor influencing the shape of a protein molecule, and is crucial to its function. Also playing a part in maintaining the shape of protein molecules is cysteine, one of two amino acids that contain sulphur.

Another important property of amino acids depends on whether the R group is charged or not. In two of the 20 amino acids it has a carboxyl group which, as explained above, makes it acidic. In three of them it contains an amino group and is basic. Since R groups are not 'used up' in linking amino acids, they can contribute an electrical charge to the protein.

## Linking amino acids

Amino acids are linked by the reaction of the amino group of one amino acid with the carboxyl group of another with the production of water (Figure 2.15). The joining together of two molecules to form a larger one with the production of water is an example of a **condensation reaction**. The process does not occur spontaneously – it requires a source of energy and the presence of enzymes and other factors.

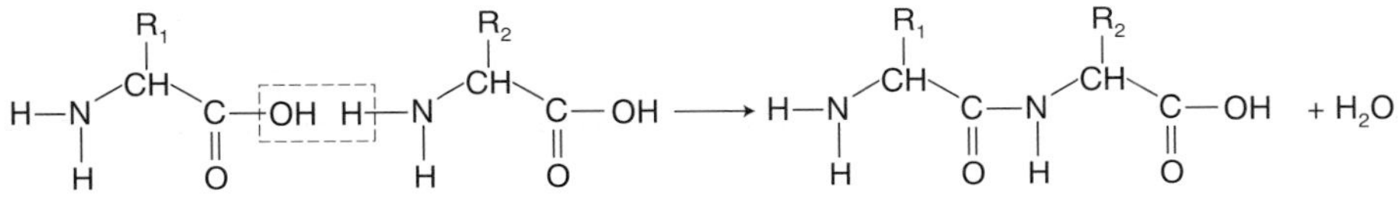

**Figure 2.15**
Formation of a dipeptide

The link between two amino acids is called a **peptide bond**. Two amino acids linked together form a **dipeptide**, three form a **tripeptide** and larger numbers are called **polypeptides**. Figure 2.16 shows part of a polypeptide chain, the dotted lines passing through the peptide bonds. Polypeptides longer than a few dozen amino acids are usually considered to be proteins. Since they have had water removed, the units of a polypeptide chain are no longer amino acids as such, and are more appropriately called amino acid **residues**.

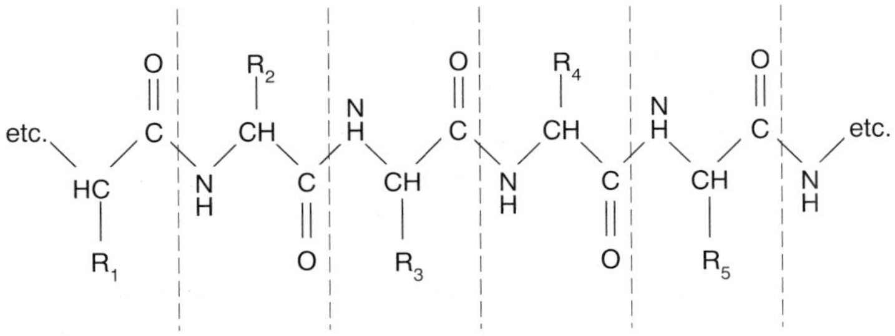

**Figure 2.16**
Part of a polypeptide chain showing five amino acid residues. The dotted lines pass through the peptide bonds

## Condensation and hydrolysis

Proteins are formed by a kind of reaction called **condensation**, in which large molecules are made by joining smaller ones together with the production of water. As we shall see, other large molecules such as nucleic acids, fats and the more complex carbohydrates, are also formed by condensation. For every link that is formed one molecule of water is produced, and this is always one less than the number of units in the chain. Thus when a polypeptide consisting of 100 amino acids is produced, 99 molecules of water are formed.

The reverse of condensation (in which large molecules are broken down into smaller ones with the consumption of water) is **hydrolysis**. Digestion is essentially a series of hydrolytic reactions.

Hydrolytic reactions are 'downhill' energetically in the sense that there is a net release of energy. They therefore occur spontaneously (though usually only at a high enough temperature to provide the necessary activation energy, see Chapter 4). In contrast, condensation reactions are 'uphill', meaning that they cannot occur without the input of energy and cannot occur by a reversal of hydrolysis. Instead, condensation reactions occur by different pathways. This energy requirement for making large molecules is why growth – which involves building large molecules from small ones – is such an energy-demanding process.

Although the amino and carboxyl groups of the α carbon atom are 'used up' in the formation of peptide links, there is always a free amino group at one end of the chain and a free carboxyl group at the other – no matter how long the chain. The two ends of a polypeptide chain are thus distinguishable and are referred to as the amino end and the carboxyl end, respectively. By convention, the amino acids in a polypeptide chain are numbered from the amino end.

## Endless variety

When you consider that for every amino acid in a polypeptide chain there are 20 possibilities, the scope for different sequences is beyond human comprehension. To keep things simple, lets ignore the fact that the two ends of an amino acid are actually different (with carboxyl and amino groups, respectively). The number of different dipeptides that could be produced would be $20 \times 20 = 400$ (20 possibilities for the first amino acid, and for each of these, 20 possibilities for the second). There would be $20^3 = 8000$ possible kinds of tripeptide. For an average-sized protein of 300 amino acids, the number of possible sequences would be $20^{300}$, or about $10^{390}$. To put this in perspective, the number of electrons in the universe is estimated to be a 'mere' $10^{75}$! It follows that the millions of different proteins that occur in nature are only an infinitesimal fraction of those that could, in principle, exist.

## What determines the order of amino acids?

In any particular protein the amino acids are linked in a specific sequence. It follows that to make a protein a cell must have the *information* to join the amino acids up in the right order. This information is encoded in the **genes**, with each gene specifying the amino acid sequence of a polypeptide. Genes consist of a chemical called **deoxyribonucleic acid**, or **DNA**, which forms a major part of chromosomes.

## Proteins and pH

An important property of most proteins is that they are sensitive to pH. This is due to the fact that a number of amino acids have basic R groups, containing an amino group (e.g. lysine) or acidic R groups, containing a carboxyl group (e.g. glutamic acid). Both kinds of R group are ionised to an extent which depends on the pH (Figure 2.17). When dissolved in water, a protein with equal numbers of acidic and basic amino acids will have equal numbers of negative and positive charges, so there is no net charge.

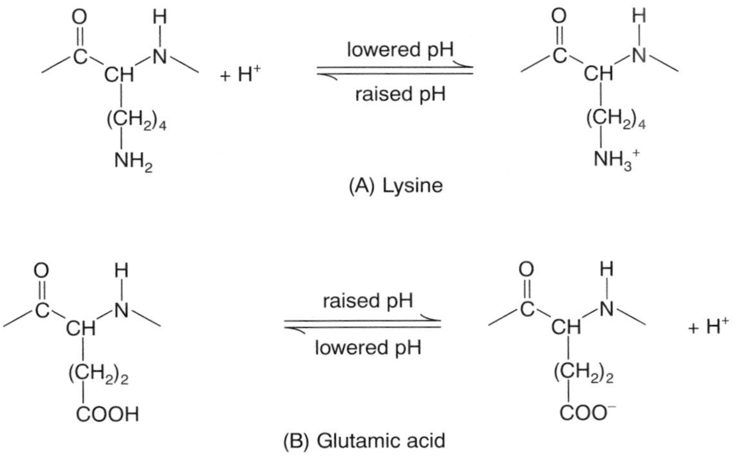

**Figure 2.17**
Effect of pH on dissociation of (A) lysine, a basic amino acid, and (B) glutamic acid, an acidic amino acid

Most soluble proteins have an excess of acidic or basic amino acids and are thus acidic or basic. As a result, they have a net charge when dissolved in water. The **histones** (constituents of chromosomes) are basic proteins, and in water have a positive charge because of the excess of —$NH_3^+$ groups. Acidic proteins have a net negative charge when dissolved in water.

The net charge on a protein also depends on the pH. Imagine an acidic protein dissolved in water and suppose that acid is added. The additional protons ($H^+$) react with the —$COO^-$ groups on the protein, reducing the number of negative charges. Protons also react with $OH^-$ ions, the removal of which encourages the formation of more —$NH_3$ groups. The opposite changes occur when a base is added to a soluble protein (Figure 2.18).

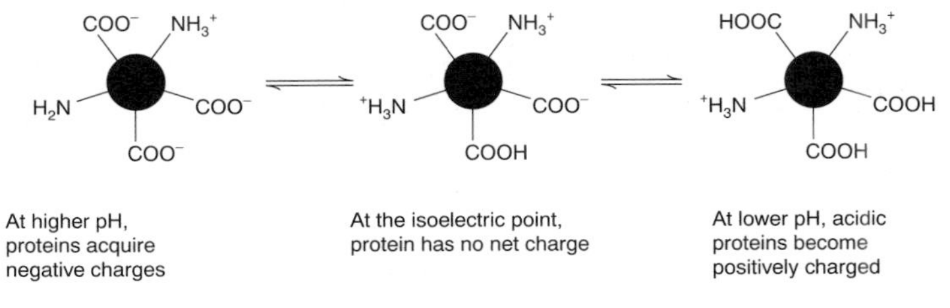

**Figure 2.18**
Effect of pH on charge of a protein

The net charge on a protein thus depends on the pH of the solution. At a particular pH there is no net charge. At this pH, called the **isoelectric point**, there is no net repulsion, and the protein may be precipitated. Acidic proteins have isoelectric points below pH 7 and basic proteins have isoelectric points above pH 7.

The ability of many proteins to 'soak up' $H^+$ ions or $OH^-$ ions enables them to act as **buffers**. A buffer is a substance which resists changes in pH. If, for example, you add a drop of hydrochloric acid to a beaker of water, there is a sharp fall in pH because all the $H^+$ ions remain in solution. If, on the other hand, a similar drop of acid is added to a solution containing a protein (such as milk), the fall in pH is very much smaller. This is because most of the added $H^+$ ions either combine with $—COO^-$ groups on the protein to form —COOH groups, or with $OH^-$ ions provided by the ionisation of $—NH_2$ groups.

Buffering by proteins is important during intense muscular exercise. Despite the large quantities of lactic acid being generated, the blood pH only falls slightly because of the presence of proteins in the blood plasma (other buffering agents such as hydrogencarbonate are even more important as buffers).

At the pH inside cells (about 7.2) most proteins have a slight negative charge, which is balanced by $K^+$ ions. This is important in the establishment of the potential difference across the plasma membrane (Chapter 17).

### Separating proteins by electrophoresis

The fact that proteins are electrically charged provides a simple way of separating them. Except at the isoelectric pH, a protein has either a net positive or a net negative charge. In an electric field, negatively-charged proteins thus migrate towards the anode (positive pole), and positively-charged proteins move towards the cathode (negative pole). The rate of movement depends on the ratio of its charge to its mass; the smaller the mass and the larger the charge, the faster it moves. By adjusting the pH the charge on the proteins can be varied, and hence the rate and direction of migration change accordingly.

The use of an electrical field to separate molecules is called **electrophoresis.** The diagram shows the kind of apparatus used. Although filter paper can be used as a medium for holding the proteins, modern methods use a polyacrylamide gel (PAG). This acts as a kind of molecular sieve because, as its pores are so small, the movement of large molecules is hindered to a greater extent than that of smaller ones.

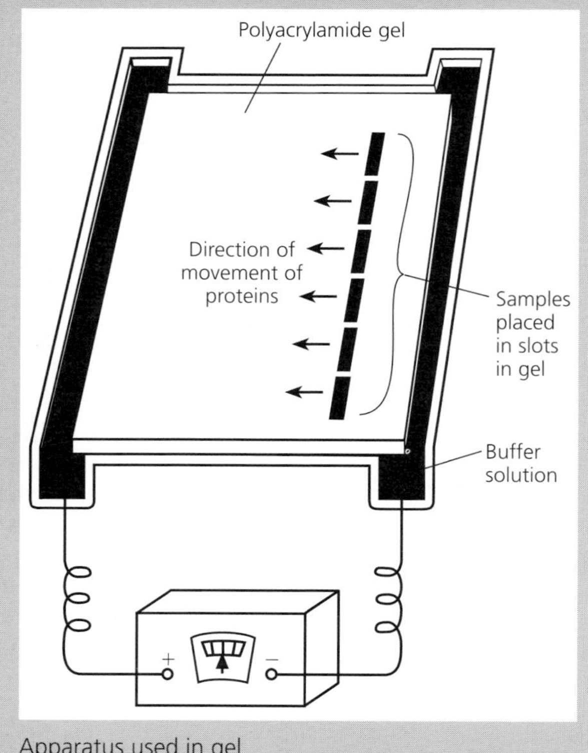


Apparatus used in gel electrophoresis (after Strickberger)

## Protein shapes

Protein molecules can be divided into two groups according to their shapes:

- **fibrous proteins** have long, rope-like molecules, and most have a *mechanical* function, providing strength. Examples are **collagen**, the main constituent of tendon and ligament (Figure 2.19); **keratin**, which occurs in nails, hair and the outer layer of the skin; and **fibroin** in silk.
- **globular proteins** have ball-shaped molecules, the polypeptide chain being coiled and folded (Figure 2.20). Most globular proteins have a *chemical* function, for example enzymes (Chapter 4).

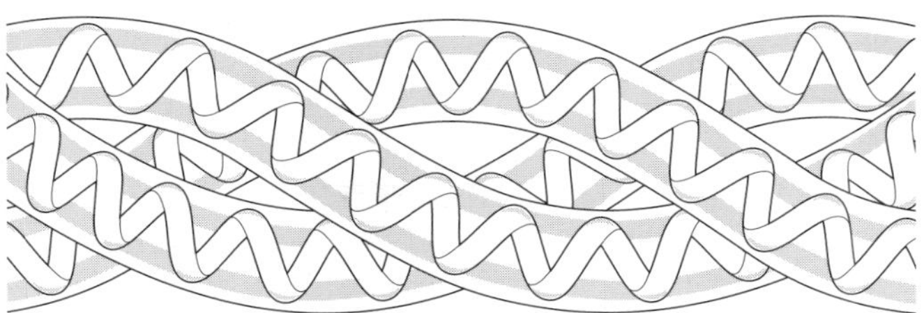

**Figure 2.19▶**
Collagen, a fibrous protein, consists of three polypeptide chains wrapped round each other

**Figure 2.20▲**
Myoglobin, a globular protein. The disc represents the non-protein part, haem, which contains iron

The shape, or **conformation**, of a protein molecule ultimately depends on the sequence of amino acids along the polypeptide chain. This is the **primary structure** and is under the control of DNA (Chapter 10).

In all proteins the polypeptide chain is further organised into some kind of geometrically regular **secondary structure**, of which there are two main kinds. In the **α helix** the chain is coiled like a spiral staircase, held in shape by hydrogen bonds between N—H and C=O groups of successive turns (Figure 2.21). The α helix occurs in many globular proteins and also in the keratin of hair.

Another kind of secondary structure is the **pleated sheet** (or **β structure**) which occurs in fibroin. Here, different polypeptide chains run in parallel bundles, cross-linked by hydrogen bonding. In some globular proteins a pleated sheet structure is formed by the polypeptide chain doubling back on itself (Figure 2.22).

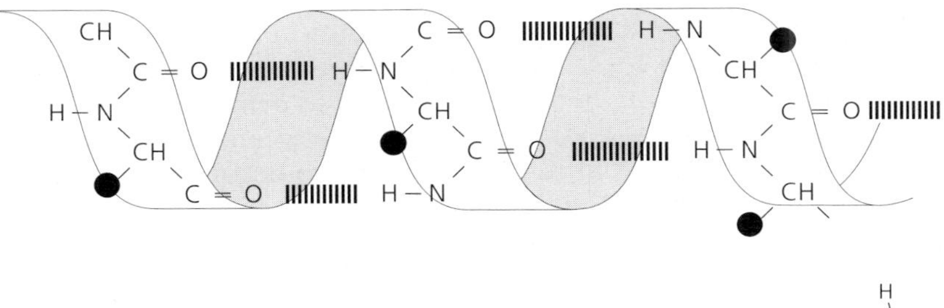


**Figure 2.21▶**
The α helix secondary structure, found in hair protein. The hydrogen bonds holding adjacent turns of the helix together are indicated by dotted lines

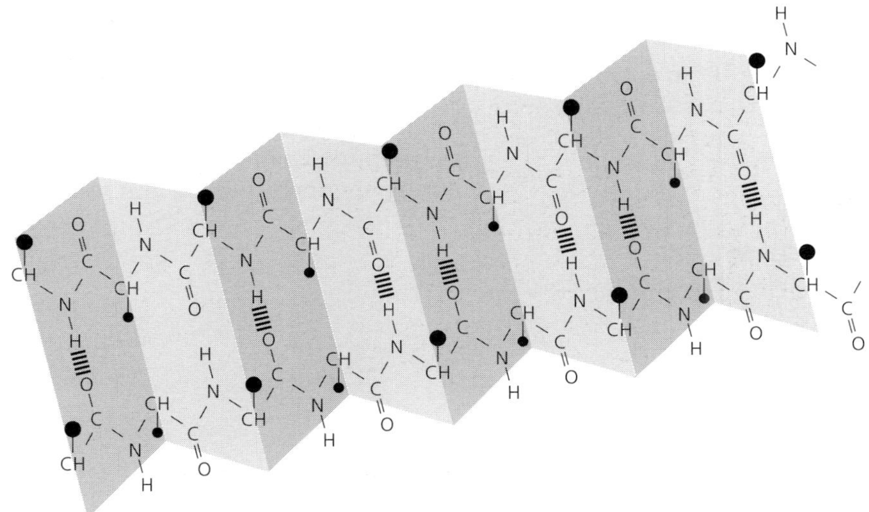

**Figure 2.22▶**
The pleated sheet secondary structure, found in the keratin of nails and hooves

In all globular proteins, the polypeptide chain is folded on itself to form a **tertiary structure**. In some globular proteins the polypeptide chain consists mainly of α helix, for example myoglobin, while others contain sections of both α and β structures (Figure 2.23).

While secondary structures are regular and of a few distinct kinds, a protein's tertiary structure is irregular and is unique to that protein. Moreover, whilst secondary structures are held in place by hydrogen bonds between N—H and —C=O groups of the polypeptide backbone, tertiary structures are stabilised by forces involving the R groups.

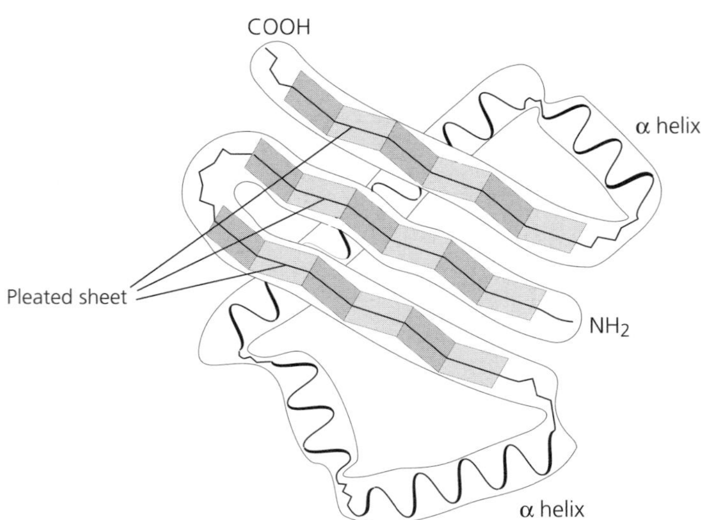


**Figure 2.23►**
A globular protein containing sections of α and β structure

**Figure 2.24▲**
Haemoglobin, a protein with a quaternary structure

Whilst many proteins consist of a single polypeptide chain, some globular proteins consist of two or more polypeptide chains loosely held together to form a **quaternary structure**. By itself, each polypeptide chain is non-functional; only in combination do they make a functional protein. An example of this is haemoglobin, in which there are two α chains and two β chains (Figure 2.24). When haemoglobin combines with an oxygen molecule, the loose binding between the polypeptide chains allows them to move slightly relative to one another, causing the other haem groups to combine with oxygen more readily (Chapter 22). Similar movements between polypeptide chains (or between different regions of the same chain) occur in many other globular proteins, especially enzymes (see Chapter 4).

## How globular proteins keep their shape

Although a globular protein typically consists of several hundred or more amino acids, only a few R groups are actually involved in its biological function. These special R groups may lie a long way apart along the polypeptide chain, but are brought close together into specific geometrical arrangements by coiling and folding (Figure 2.25).

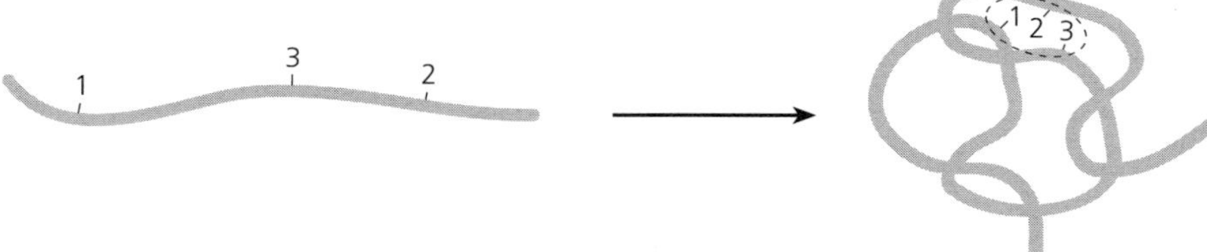


**Figure 2.25►**
Catalytically-active R groups brought together by 2° and 3° folding

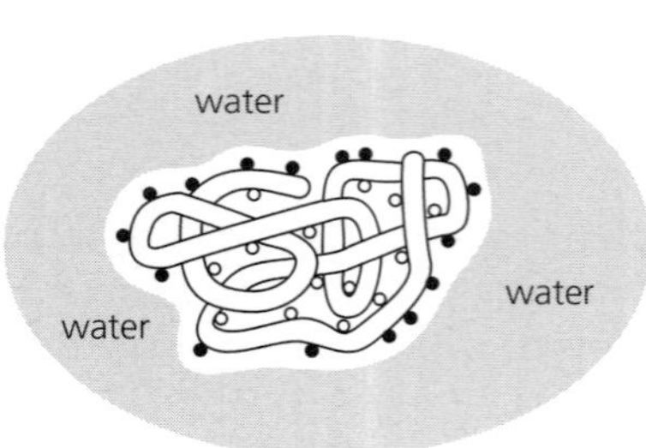


**Figure 2.26▲**
How water helps to hold a globular protein molecule in shape. Dark circles are water-seeking R groups

The most important mechanism holding a polypeptide chain in shape is dependent on the fact that some R groups are hydrophilic and others are hydrophobic. The polypeptide chain is 'floppy' enough for it to adopt the conformation which allows most of the water-attracting R groups to form hydrogen bonds with the surrounding water, causing them to face outwards. This leaves the hydrophobic R groups buried inside (Figure 2.26).

In addition to the interactions with water, protein shape is also stabilised by various kinds of attractive forces between different parts of the polypeptide chain (Figure 2.27).

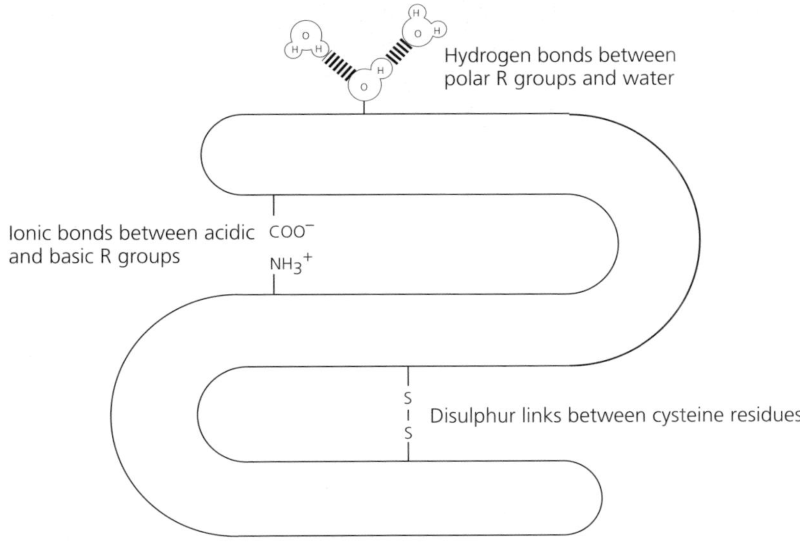


**Figure 2.27**
Forces that maintain the tertiary structure of globular proteins

- **Hydrogen bonds** between different parts of the polypeptide chain.
- **Disulphur links** are covalent bonds formed by the removal of hydrogen from adjacent cysteine residues after the protein has been synthesised (Figure 2.28).

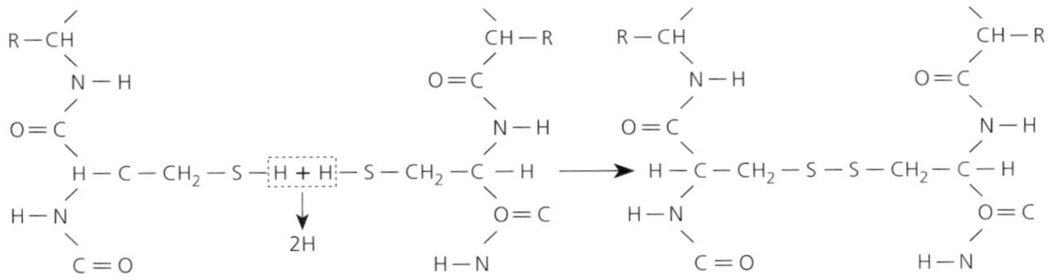

```
  R—CH                                        CH—R          R—CH                        CH—R
      N—H                                 O=C                   N—H                 O=C
  O=C                                         N—H           O=C                         N—H
      H—C—CH2—S—[H + H]—S—CH2—C—H   ──►   H—C—CH2—S—S—CH2—C—H
  H—N                ↓                    O=C           H—N                         O=C
      C=O            2H               H—N                   C=O                 H—N
```

**Figure 2.28**
Formation of a disulphur link between two cysteine units

- **Salt linkages.** Basic amino acids have an R group containing an —$NH_2$ group, and acidic amino acids have an R group containing a —COOH group. Depending on the pH, these may be ionised as —$NH_3^+$ and —$COO^-$ groups. By attraction, oppositely charged R groups can hold neighbouring regions of the chain together.

In proteins with a quaternary structure, such as haemoglobin, the polypeptide chains are held together by hydrogen bonding between water and polar groups, non-polar R groups being concentrated in the inside of the molecule (Figure 2.29).

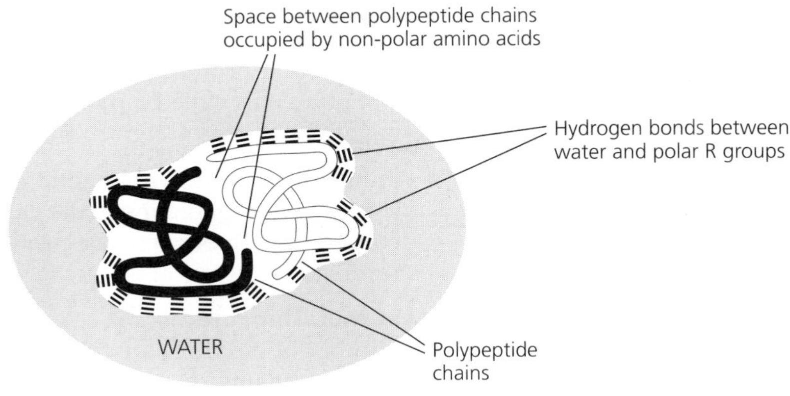


**Figure 2.29**
The role of hydrogen bonding with water in maintaining quaternary structure. For clarity, hydrogen bonds have been shown much longer than in reality

## A molecular 'mistake'

The importance of joining amino acids in the right sequence is illustrated by an inherited condition called **sickle cell anaemia**. In homozygotes (individuals with two copies of the defective gene) the red blood cells become sickle-shaped at the lower oxygen concentrations encountered in the tissues. One effect of this is to impede the free flow of blood through the capillaries. Another result is that the red cells live for a much shorter time, causing severe anaemia. When only one of the two genes is affected (i.e. in heterozygotes) there is little harm done.

Sickle cell anaemia is caused by a 'mistake' in the gene coding for the production of the β chain of haemoglobin. The result is that at position number 6, glutamic acid is replaced by valine. Whereas glutamic acid is strongly attracted to water because of a charged R group, valine is non-polar. The β chains, instead of being water-soluble, are more attracted to each other and tend to crystallise, forming the typical sickle-shaped red cells.

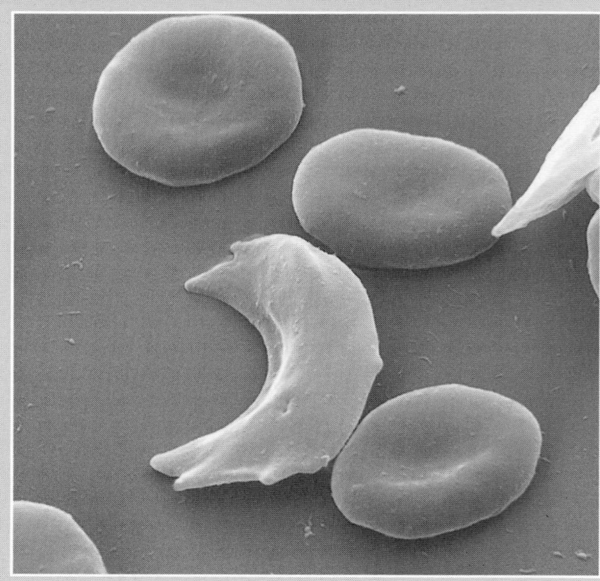

SEM of sickled red corpuscles

Of these forces, only disulphur links are strong, because they are covalent. Hydrogen bonds are individually weak, but there are so many that at moderate temperatures they are strong enough to hold the molecule in shape. At higher temperatures molecular collisions and vibrations become violent enough to break up the specific folding pattern. The R groups responsible for its activity no longer have the same geometrical arrangement, and as a result the protein is inactivated and usually precipitated, as for instance when egg white is cooked. The inactivation of a protein by changing the folding of the polypeptide chain is called **denaturation**, and is usually permanent. There are, however, some cases in which a protein can be renatured, restoring full activity. This shows that the information for creating a specific folding pattern must lie in the polypeptide chain itself – that is, in its primary structure.

## Conjugated proteins

Many globular proteins are only fully functional when combined with a non-peptide part called a **prosthetic group**. Some examples of conjugated proteins are given below:

- in haemoglobin the prosthetic group is itself a fairly complex, iron-containing, organic molecule called **haem**. It is the haem which combines with oxygen, the protein part (globin) rendering the reaction reversible.
- in **metalloproteins** it is a metal such as copper, iron or zinc. A number of enzymes are of this type, for instance carbonic anhydrase, which contains zinc, and catalase, which contains iron.

- in **glycoproteins** it is a polysaccharide, for example the mucin that gives saliva its sliminess.
- in **phosphoproteins**, where the polypeptide is bonded to phosphate groups, for instance milk protein (casein) and the protein of egg yolk (vitellin).

# Lipids

Fats, oils, phospholipids and steroids are collectively called **lipids**. Their common feature is that they dissolve in organic solvents such as alcohol. They play a variety of roles, some of which are essential to life.

## Functions of lipids

- Phospholipids form the structural framework of *cell membranes*.
- Fats and oils are an important *energy store* in animals and many seeds.
- The water-resistant *cuticle* of plants consists of a kind of lipid called wax.
- Some *hormones*, for example the sex hormones, are steroids.
- In mammals, fat is stored under the skin in a layer of fat-storage cells which acts as a *heat insulator*. It also helps protect against bruising.
- In some single-celled organisms fat droplets give *buoyancy*.

## Fatty acids

Fatty acids are major components of fats. A fatty acid molecule consists of two chemically distinct parts – a carboxyl group and a hydrocarbon chain typically 13 to 21 carbon atoms long. In most fatty acids the effect of the non-polar hydrocarbon chain is enough to overcome the polarity of the carboxyl group, so most fatty acids have low solubility in water.

Besides the length of the hydrocarbon chain, fatty acids differ as to whether they are **saturated** or **unsaturated** (Figure 2.30). An unsaturated fatty acid is free to bond with more hydrogen atoms because it contains one or more double bonds (the carbon atoms on either side of a double bond are linked to only one hydrogen atom each). A saturated fatty acid is unable to react with hydrogen because the hydrocarbon chain has no double bonds, so every carbon atom is already linked to two carbon atoms.

A double bond in a hydrocarbon chain imparts a 'kink' to it. Fats containing unsaturated fatty acids therefore do not pack together so readily to form a solid, and consequently, they melt at a lower temperature. Since fats containing unsaturated fatty acids are usually liquid at room temperature they are called **oils**. Most plant fats (e.g. palm oil, sunflower oil), and also fish oils, contain mainly unsaturated fatty acids. Many animal fats (e.g. lard, butter) contain mainly saturated fatty acids.

## Fats

Fats are more properly known as **triglycerides** because they are formed by condensation between a molecule of **glycerol** and three **fatty acid** molecules (Figure 2.31). Since there are about 70 common kinds of fatty acid, many different triglycerides are possible.

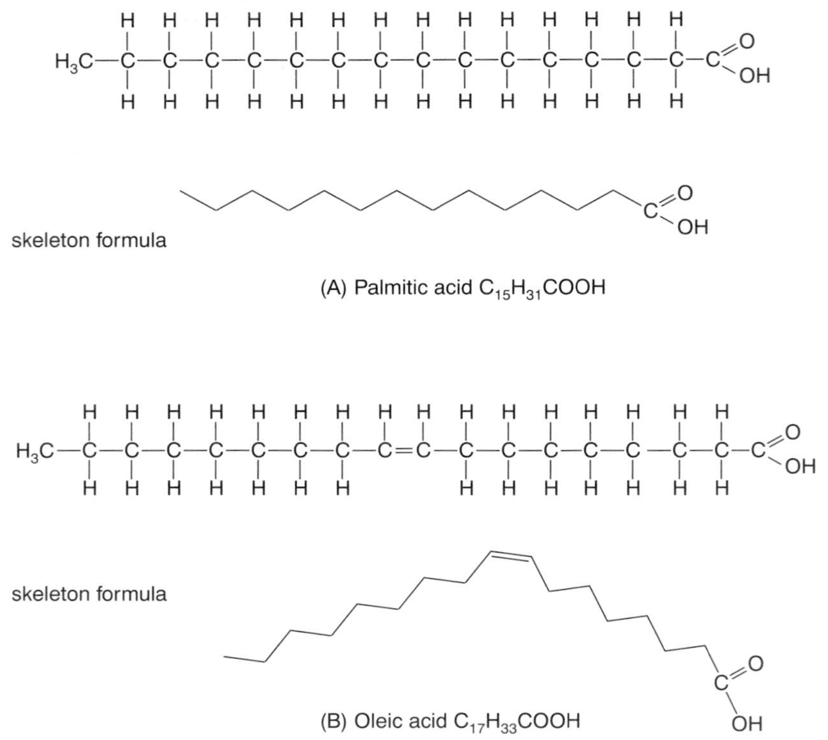

**Figure 2.30▶**
Saturated (A) and unsaturated (B) fatty acids

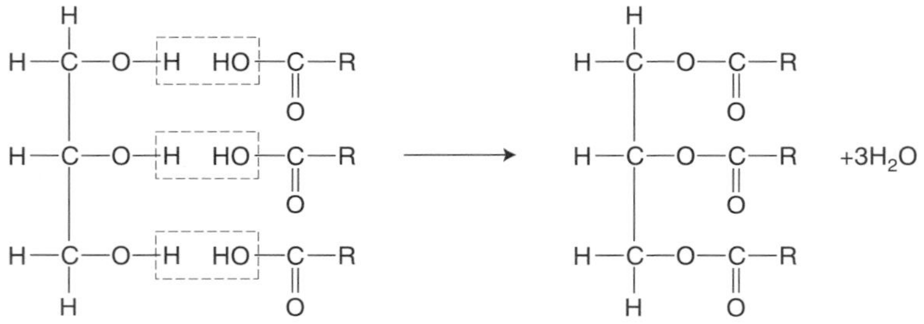

**Figure 2.31▶**
Formation of a triglyceride from fatty acids and glycerol. The R groups represent the hydrocarbon chains of the fatty acids

Each —OH group of the glycerol combines with a —COOH group of a fatty acid forming a water molecule, so three water molecules are produced in the formation of each triglyceride. During the reaction, the polar parts of both fatty acid and glycerol are 'used up', so the resultant triglyceride is quite insoluble in water.

In organic chemistry, the condensation reaction between an alcohol group and a carboxyl group results in the formation of an **ester**. Triglycerides are thus esters of fatty acids and glycerol.

## Fats as fuels

Fats are more reduced than carbohydrates, meaning that they contain proportionately less oxygen. Consequently, they have to undergo more oxidation to convert them to carbon dioxide and water. As a result, the oxidation of a gram of fat yields nearly two and a half times as much energy as a gram of dry carbohydrate. The difference is actually greater than this because glycogen and starch contain a lot of water, which adds further weight. When energy has to be carried around, for example in birds, insects and wind-dispersed seeds, fat is an economical way of storing energy. It is no coincidence that migrating birds store large quantities of fat before setting out on their journeys.

## Phospholipids

Phospholipids differ from fats in that one of the three fatty acids is replaced by a group containing a phosphate and usually a nitrogenous base. Because these are electrically charged, this part of the molecule is attracted to water and is **hydrophilic**. The rest of the molecule, however, is not and is **hydrophobic** (Figure 2.32). (Though hydrophobic literally means 'water-fearing', it does not actually repel water; it is simply not attracted to it.)

A molecule with hydrophilic and hydrophobic ends is said to be **amphipathic**. As a result of this, phospholipid molecules readily arrange themselves into double sheets called **bilayers**, with the hydrophilic ends facing out into the water and the hydrophobic ends facing inwards towards each other (Figure 2.33). All cell membranes consist partly of a lipid bilayer.

**Figure 2.32▲**
A phospholipid

**Figure 2.33►**
A lipid bilayer

## Waxes

Waxes are esters of fatty acids and long-chain alcohols. They are the main component of the cuticle of plants, and have been a major factor in the colonisation of the land by plants.

## Steroids

Though grouped with fats for convenience, steroids are quite unrelated to them chemically and cannot be hydrolysed into smaller units. All steroids consist of four joined rings, the 'corners' of each being occupied by a carbon atom.

Steroids have diverse functions. One of the most important steroids is **cholesterol**, which is a major constituent of the plasma membranes of all cells except bacteria (Figure 2.34). The four rings are non-polar, but the —OH group is hydrophilic and faces out of the membrane towards the water.

Some hormones are steroids, e.g. the *sex hormones* and the hormones secreted by the adrenal cortex. Vitamin D is also a steroid, as are the bile salts that emulsify fats in the small intestine.

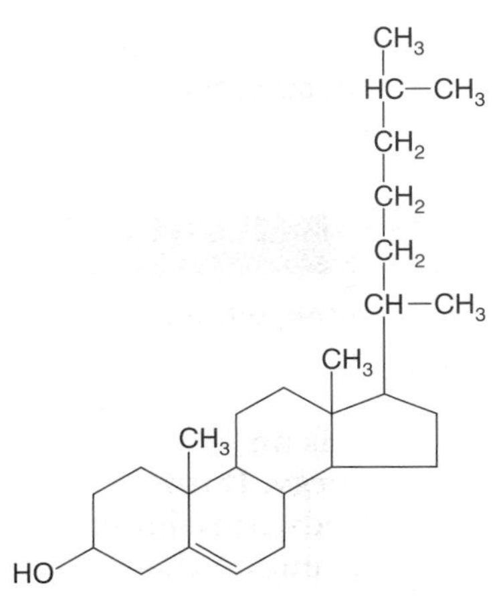


**Figure 2.34**
Cholesterol, an important steroid

# Carbohydrates

Carbohydrates contain the elements carbon, hydrogen and oxygen. The hydrogen and oxygen are present in the same ratio as in water. They can be represented by the general formula $CH_2O$.

Carbohydrates can serve any of three biological roles:

- **fuels** – for example glucose, or an energy-storage carbohydrate such as starch.

- **support** – for example cellulose in plant cell walls.
- **information** – by acting as specific markers on the outer surfaces of cell membranes, some complex carbohydrates enable cells to 'recognise' one another.

The common carbohydrates can be divided into the following groups:

- **monosaccharides**, or *simple sugars*, in which the molecule consists of a single unit.
- **disaccharides**, or *complex sugars*, consisting of two monosaccharide units joined together. Disaccharides are actually the main subcategory of a larger group, the **oligosaccharides**, which consist of three or four simple sugar units.
- **polysaccharides**, the most complex carbohydrates, consisting of many monosaccharide units linked together.

## Monosaccharides

Monosaccharides have the general formula $C_nH_{2n}O_n$. All are crystalline, soluble, sweet-tasting and reduce Benedict's solution. They are classified according to the number of carbon atoms in the molecule. Thus **trioses** (three carbon atoms) are represented by the general formula $C_3H_6O_3$, **tetroses** by $C_4H_8O_4$, **pentoses** by $C_5H_{10}O_5$ and **hexoses** by $C_6H_{12}O_6$.

### Benedict's test for reducing sugars

**All simple sugars and most complex sugars give a red precipitate when heated with Benedict's solution. In the reaction, the blue $Cu^{2+}$ ions in Benedict's solution are reduced to the red $Cu^{+}$ of copper (I) oxide (CuO), which appears as a red precipitate. If there is insufficient sugar to reduce all the $Cu^{2+}$ ions, the granules of CuO are seen through a blue solution. A trace of precipitate thus appears green and a slightly larger amount appears yellow. All monosaccharides and all common disaccharides (except sucrose) are reducing sugars.**

**The reducing properties of glucose are due to the aldehyde group of carbon number 1, which only exists in the open chain form. Although neither of the two ring forms ($\alpha$ and $\beta$) can reduce Benedict's solution, the removal of the chain form by the Benedict's solution causes more of the ring form to be converted to the reducing, chain form, so all the sugar is eventually oxidised.**

### *Pentoses*

Biologically, the most important pentose is **ribose**, since it forms a constituent of ATP and RNA (Chapters 5 and 11). An important derivative of ribose is **deoxyribose**, a constituent of DNA. Like all sugars with five or more carbon atoms, ribose can exist in both chain and ring forms (Figure 2.35).

### *Hexoses*

The most widely occurring hexose is **glucose**. It is particularly important, not only because it is one of the main energy sources used by cells, but also because it forms the building units of polysaccharides such as starch and cellulose.

A glucose molecule can exist in a number of alternative forms: an **open chain form** and a number of **ring structures**, one of which is shown in

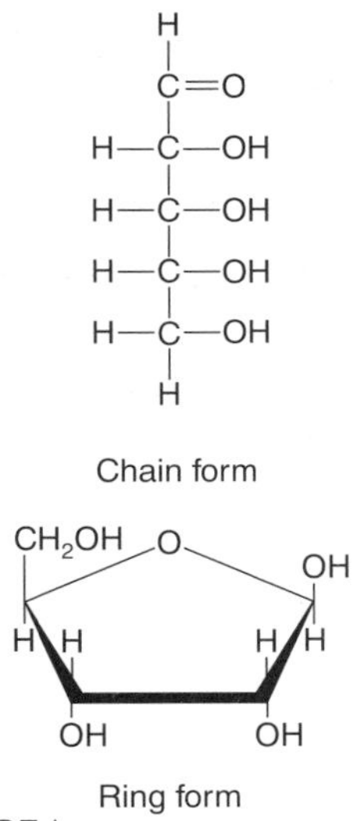

Chain form

Ring form

**Figure 2.35▲**
Ribose, one of the most biologically-important sugars

Figure 2.36. A ring form is shown tilted on its side, the thicker bonds representing those above the plane of the paper.

For most purposes it is enough to show the ring forms of sugars as if they are flat, with the —OH groups directed vertically. In reality, because carbon chains form a zigzag, hexose molecules are chair-shaped with the —OH groups directed laterally. This is important when we come to consider the structure of cellulose.

In the formation of a ring, the hydrogen of the —OH of carbon number 5 links with the C=O of carbon number 1 to form an —OH group, and at the same time the oxygen attached to carbon number 5 joins with carbon number 1 (Figure 2.37). The reason why *two* alternative ring structures can form is that the new —OH group can form either 'above' or 'below' the plane of the ring. Since the two ring forms are both in equilibrium with the open chain form, they are interconvertible.

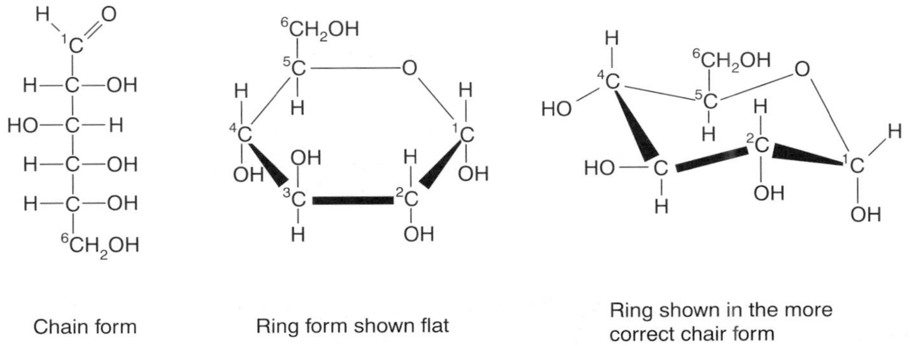

**Figure 2.36►**
Chain and ring forms of glucose

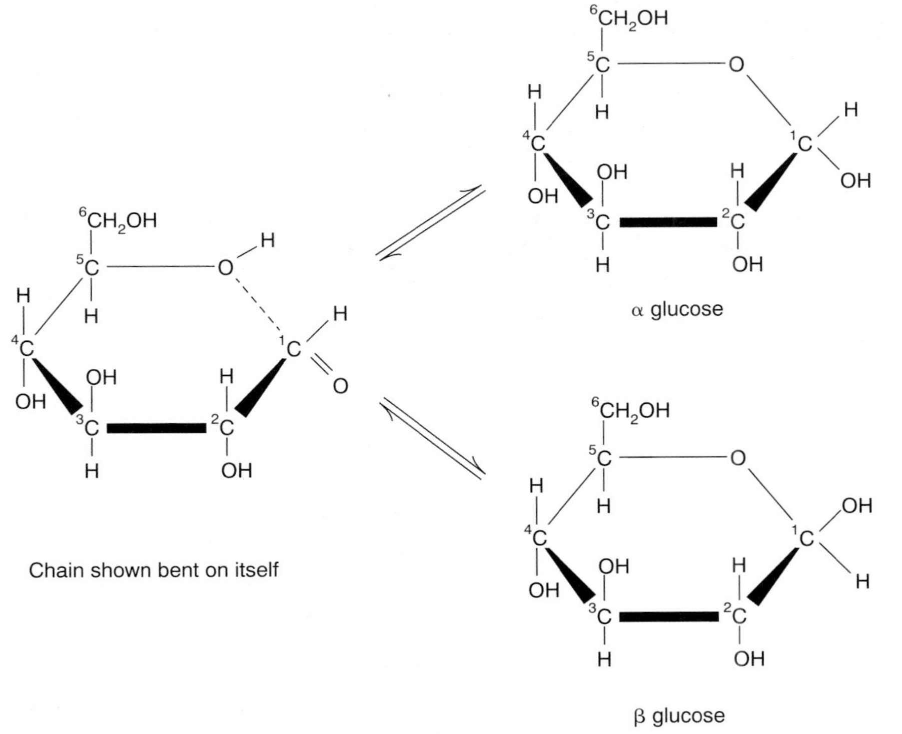

Chain form

Ring form shown flat

Ring shown in the more correct chair form

Chain shown bent on itself

α glucose

β glucose

**Figure 2.37►**
Ring formation in glucose

The two ring structures are known as **α glucose** and **β glucose**. Their importance lies in the fact that α glucose is the building unit of starch and glycogen, whilst β glucose is the building unit of cellulose. All other hexoses, and all pentoses, also exist as α and β forms.

**Fructose** is another hexose which, like glucose, exists in chain and ring forms (Figure 2.38). It differs from glucose in a number of ways: first, when combined with other sugars it exists in a five-membered rather than a six-membered ring; second, the reducing properties of the open-chain form are due to a keto (C=O) group rather than an aldehyde group (—CHO).

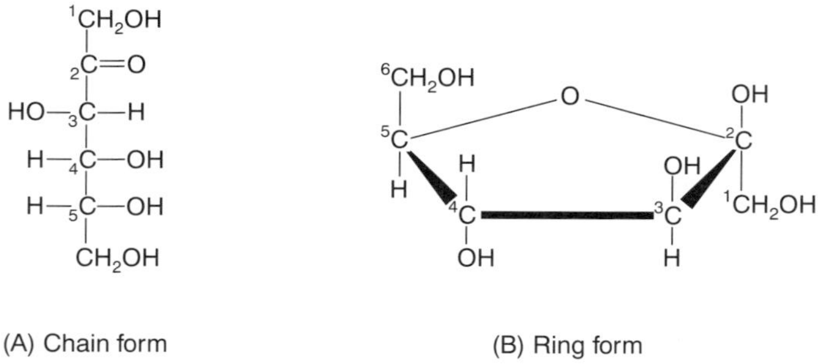

(A) Chain form

(B) Ring form

**Figure 2.38**
Chain and ring forms of fructose, a ketose sugar

## Sugar 'families'

All monosaccharides can be divided into two 'families' – **aldoses and ketoses.** Aldose sugars can all be considered to be related to the triose sugar **glyceraldehyde** because they have an aldehyde (—CHO) group in the open chain form. Ketoses have a ketone (C=O) group, as does the triose sugar dihydroxyacetone. Of the more common sugars, only fructose and ribulose are ketoses.

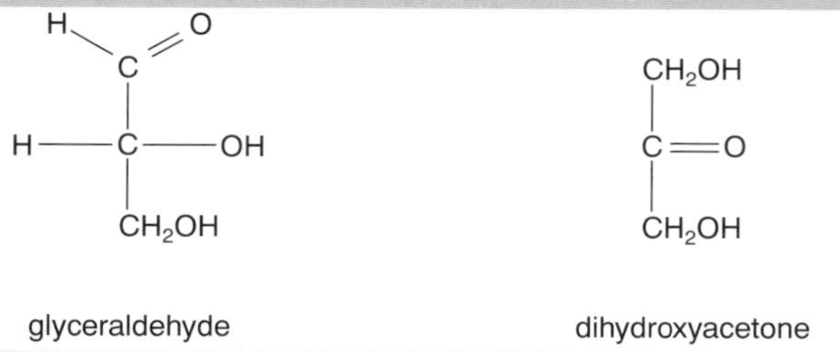

glyceraldehyde

dihydroxyacetone

Glyceraldehyde and dihydroxyacetone

## Left-handed and right-handed molecules

All sugars can exist in contrasting forms which, like left and right hands, are mirror images of each other. The two **stereoisomers** (or optical isomers) can be distinguished by their effect on polarised light. (This is light in which all the waves are in the same plane, unlike normal light in which the waves are vibrating in all planes.) Optical isomers differ from each other in that one (the '+' form) rotates polarised light clockwise, and the other (the '−' form) rotates it anticlockwise.

Whether a molecule is optically active depends on whether or not it has one or more *asymmetric carbon atoms*. An asymmetric carbon atom has four *different* groups attached to it. This is most easily seen in the triose sugar (+) glyceraldehyde, which is optically active because all four groups attached to carbon number 2 are different, and the ketose sugar dihydroxyacetone, which has only three different groups attached to carbon number 2.

The diagram shows the left- and right-handed forms of the chain form of glucose, called L-glucose and D-glucose, respectively. For comparison, their respective stereochemical relatives (+) glyceraldehyde and (−) glyceraldehyde are also shown. The prefix D- indicates that the shaded part of the molecule is the same as in the triose sugar (+) glyceraldehyde. Likewise the prefix L- denotes a relationship with (−) glyceraldehyde.

## Left-handed and right-handed molecules *continued*

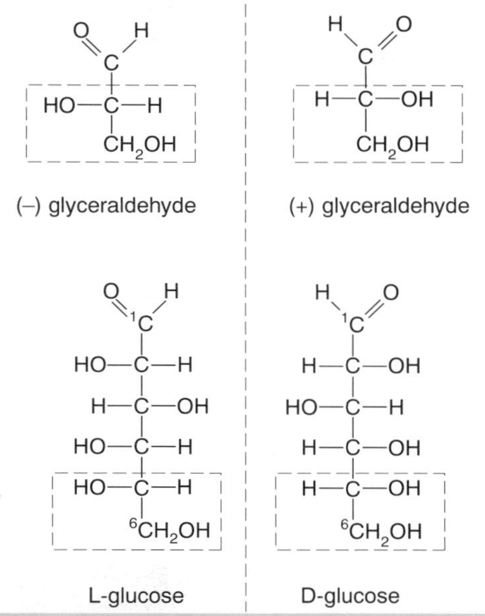

(–) glyceraldehyde

(+) glyceraldehyde

L-glucose

D-glucose

When a sugar is made in a test tube, D- and L-forms are produced in equal amounts so the solution is not optically active. However, with rare exceptions, living things produce only D-sugars and cannot use the L-forms.

Sugars are not the only biological molecules to be optically active. Amino acids also exist as left- and right-handed forms, but in this case almost all naturally-occurring ones belong to the L- series.

Left- and right-handed sugars



## Disaccharides

Disaccharides are actually the simplest structures of a group called **oligosaccharides**, which consist of a small number of sugar units strung together by **glycosidic bonds**. Disaccharides consist of two sugar units, trisaccharides have three and tetrasaccharides have four. All can be broken down into their constituent monosaccharides, either by boiling with acid, or by treatment at room temperature with a suitable enzyme. Disaccharides are sweet-tasting, crystalline solids, and all the common ones (except sucrose) reduce Benedict's solution.

**Maltose** is a disaccharide which has no biological function in its own right, since it is produced solely as an intermediate product during the breakdown of starch. It consists of two glucose units linked by carbon atoms number 1 and number 4, respectively (Figure 2.39). The link is thus called an **α 1,4 glycosidic bond**.

Maltose is a reducing sugar because, although the reducing carbon atom of one glucose residue (carbon number 1) is permanently locked in the ring form, the other can open out into the reducing chain form. By contributing its carbon number 1 to the glycosidic bond, one of the two glucose units is unable to open out into the chain form and is thus non-reducing. Since the other unit can still form the open chain, maltose is a reducing sugar.

**Sucrose** (cane or beet sugar) is the main form in which energy is transported through a plant (Figure 2.40). It also acts as an energy storage compound, for example in beet, carrots and onions. It consists of glucose and fructose linked by a 1,2 glycosidic bond (carbon number 1 of α-glucose linked to carbon number 2 of β-fructose). Since these are the reducing carbon atoms, neither ring can open out to form the open chain form, so sucrose is a non-reducing sugar.

**Lactose** (milk sugar) is another important disaccharide consisting of β-galactose and glucose.

glucose + glucose
maltose + $H_2O$
$H_2O$
α 1,4 link

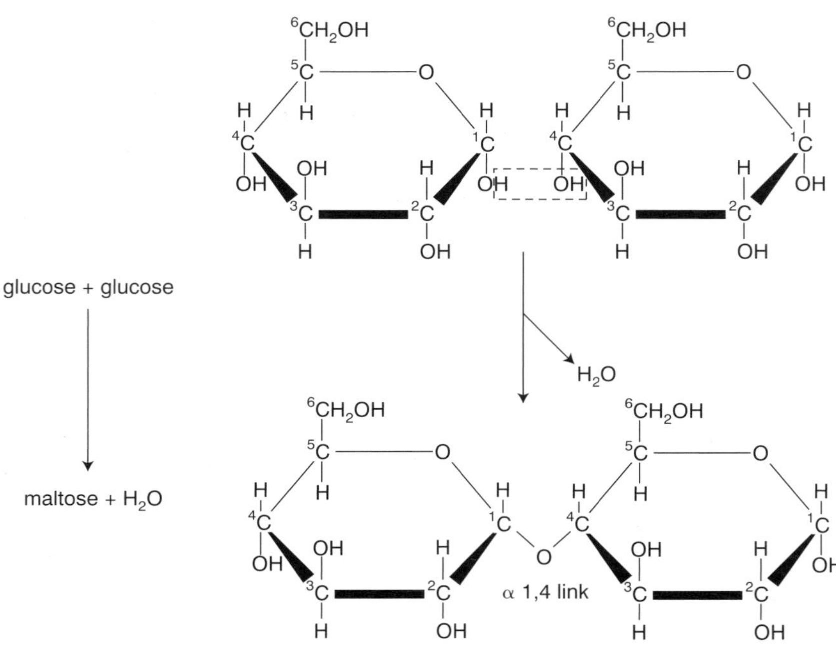

**Figure 2.39**
Maltose consists of two glucose units linked together

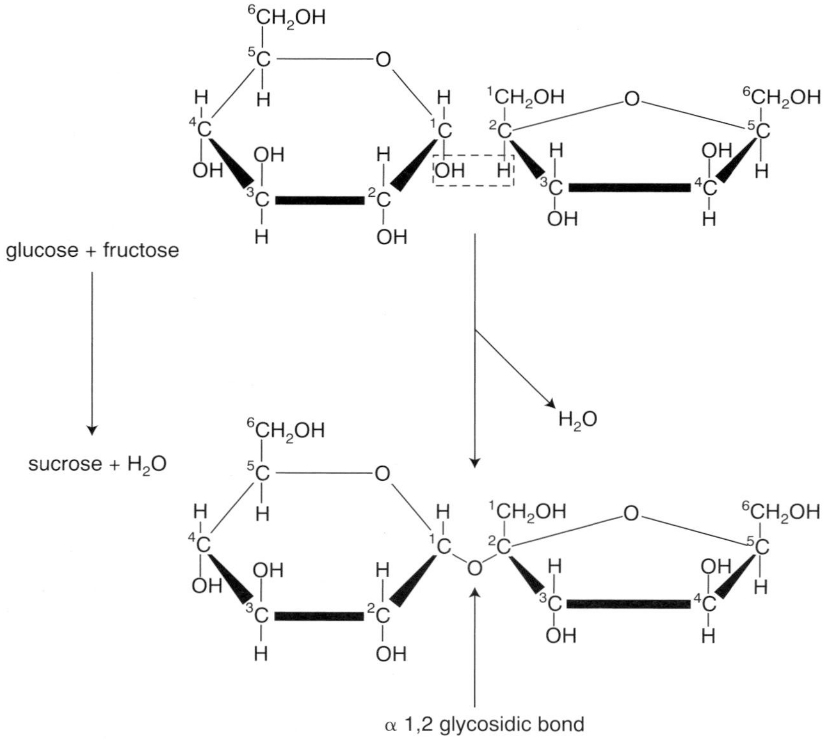

glucose + fructose
sucrose + $H_2O$
$H_2O$
α 1,2 glycosidic bond

**Figure 2.40**
Sucrose

## Polysaccharides

The most complex carbohydrates consist of chains of up to several thousand sugar units and are called **polysaccharides** ('poly' means many). Polysaccharides are insoluble in cold water, have no sweet taste and do not reduce Benedict's solution.

Most of the more important polysaccharides are similar in that they consist of glucose units, yet their physical properties can be very different because of the various ways in which the monosaccharide subunits are linked together.

**Starch** is the main energy-storage compound of plants and is stored in many tubers and seeds. It actually consists of a mixture of two polysaccharides, **amylose** and a higher proportion of **amylopectin**.

Amylose consists of a chain of α glucose units, linked by α 1,4 glycosidic bonds (Figure 2.41).

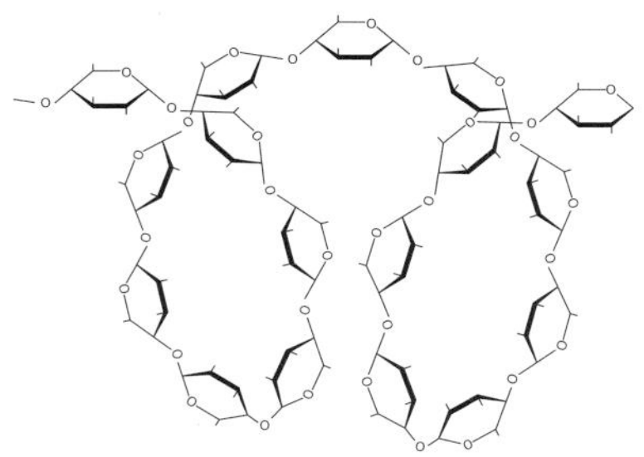

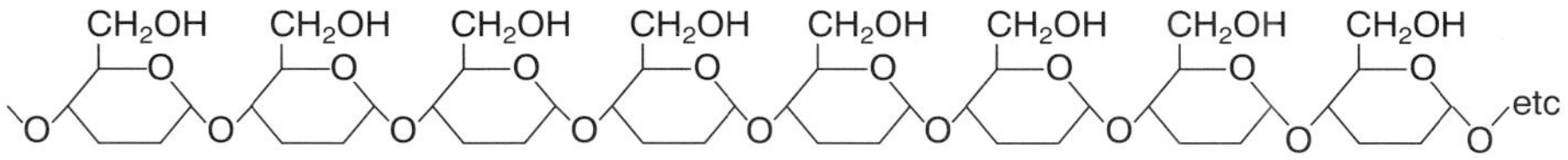

**Figure 2.41▲►**
Amylose

The chain is held in a spiral by hydrogen bonds and is just wide enough for iodine molecules to fit inside, forming a complex with an intense blue colour. Heating causes the hydrogen bonds to break so the blue colour disappears, but it reappears on cooling. Amylopectin also consists of glucose units, but there are frequent branches due to α 1,6 linkages (Figure 2.42). It gives a brown colour with iodine solution.

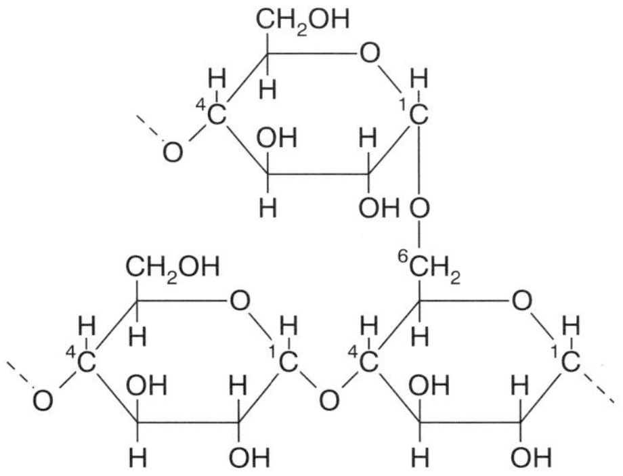

**Figure 2.42**
Amylopectin is a branched chain of glucose units (A), the branches form α1,6 links (B)

**Glycogen** is the energy storage carbohydrate in animals and is similar to amylopectin except that it is even more highly branched (approximately every eight residues). It occurs in the liver and muscle of vertebrates.

**Cellulose** forms the main constituent of plant cell walls. It consists of unbranched chains of several thousand glucose molecules, but instead of α glucose as in starch, the units are β glucose (Figure 2.43).

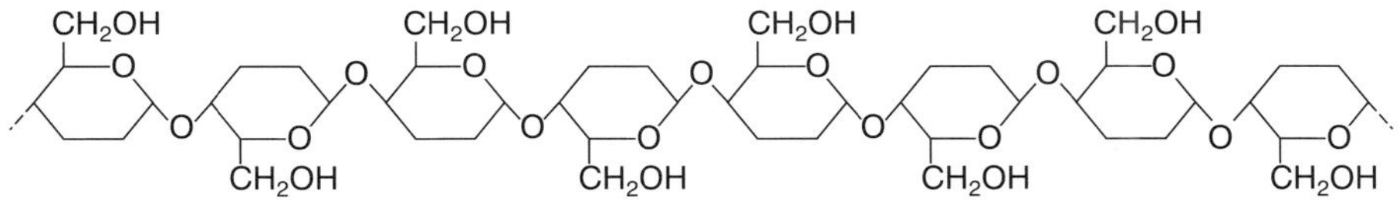

**Figure 2.43**
Part of a cellulose molecule, consisting of β-linked glucose units

As a result of this slight difference in bonding, alternate glucose residues are 'upside down'. This has the important effect of bringing the oxygen bridge and the —OH of carbon number 3 of the next glucose residue close enough together to enable hydrogen bonds to form between them. (The orientation of this —OH group is only clear when the glucose residues are shown in the more truly representational chair form, as shown in Figure 2.44). These hydrogen bonds stabilise a cellulose molecule in a straight, cable-like configuration – ideal for resisting tension.

Besides these *intrachain* hydrogen bonds, adjacent chains are cross-linked by *interchain* hydrogen bonds to form rope-like bundles or **microfibrils**. Each microfibril consists of about 300 cellulose molecules.

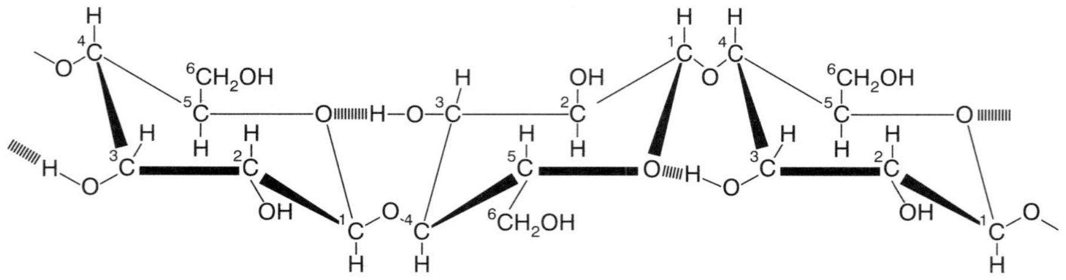

**Figure 2.44**
Intrachain H-bonding in cellulose stabilises the molecules in a cable-like form

Because of the difference in bonding, starch-digesting enzymes have no effect on cellulose, and thus plant-eating animals are unable to digest cellulose for themselves (they rely on micro-organisms in their gut to do it for them).

**Chitin** is similar to cellulose except that the units are a nitrogen-containing derivative of glucose (N-acetyl glucosamine). It forms the main constituent of the cuticle of insects and other arthropods, and also the walls of many fungal hyphae.

**Inulin** is a storage polysaccharide produced by dahlias and other members of the daisy family, and is polyfructose.

Like disaccharides, polysaccharides are built up by condensation reactions. They can also be hydrolysed into simple sugars by boiling with water using a little acid as a catalyst or by reaction with a suitable enzyme. For each link broken, a water molecule is added. Since the number of links in a polysaccharide is always one less than the number of units, hydrolysis of a starch molecule consisting of $n$ glucose units uses up $n - 1$ water molecules.

## Carbohydrates as cell markers

In all cells some of the globular proteins in the plasma membrane are **glycoproteins** – conjugated proteins in which the prosthetic group is a carbohydrate. The protein anchors the molecule in the membrane leaving the carbohydrate part protruding. Under the electron microscope these cell surface glycoproteins appear as a fringe called the **glycocalyx.** The function of some of these proteins is to serve as 'recognition' sites by which cells can distinguish one another, for example when a sperm enters the egg of the same species but does not enter an egg of a different species.

To act as a marker molecule a polysaccharide must have *specificity* – that is, it must differ from all others. In this respect carbohydrates have far greater potential than proteins. Although there are 20 kinds of amino acid, they are linked in the same way and are not branched. Sugars, on the other hand, not only come in many kinds, but can be joined in more ways since any of several carbon atoms can be involved in a link. For example, carbon atom number 1 of glucose can be linked in 11 different ways to another glucose molecule – and that is when only one kind of sugar is used! Moreover, polysaccharide chains can branch since each sugar can be linked to more than two others.

Communication by cell surface markers is important in defence. During infection, cells lining blood vessels in the affected region are stimulated to produce specific carbohydrates on their surfaces. The presence of these carbohydrates stimulates certain types of white corpuscles to leave the blood vessels and enter the tissues (Chapter 27). Many disease-causing bacteria use these carbohydrate cell markers to distinguish one kind of host cell from another. This is why some bacteria attack only certain parts of the body – i.e. those in which the host cells carry the marker.

The role of many cell surface carbohydrates is still mysterious. For example, although the medical importance of the oligosaccharides that make up the ABO blood group antigens has been known for many years, their natural function is still unknown.

## Nucleotides

Nucleotides are the subunits of **nucleic acids**. These are complex polymers concerned with the storage and translation of genetic information and are dealt with in more detail in Chapters 10 and 11.

A nucleotide consists of three parts – a phosphate group, a pentose sugar and a nitrogenous base. Depending on the kind of sugar, nucleotides fall into two groups. The **deoxyribonucleotides** are subunits of DNA (Chapter 10) and the **ribonucleotides** are constituents of RNA. Besides playing a major role in protein synthesis (Chapter 11), RNA is also the genetic material in some viruses, e.g. HIV. Ribonucleotides also have a number of functions in their own right, for example:

- some are *energy carriers*, such as ATP (Chapter 5).
- some are constituents of *respiratory coenzymes* (see Chapter 5).
- cyclic AMP (adenosine monophosphate) is a *'second messenger'*, amplifying and relaying a hormonal signal after it has been received by a cell.

## QUESTIONS

**1** Which of the following is the most fundamental reason why water is vital to life?
**A** Cells die very quickly when they lose too much water.
**B** Water acts as a solvent for chemical reactions in cells.
**C** Water forms over 80% of most cells.
**D** The water content of many cells is regulated within narrow limits.

**2** Each of the substances **A–D** is a constituent of one of the substances **(i)–(iv)**, but not necessarily in the right order. For each of the substances **(i)–(iv)**, write the letter of the appropriate constituent.

| | |
|---|---|
| **i)** cellulose | **A** glycerol |
| **ii)** fat | **B** amino acids |
| **iii)** RNA | **C** glucose |
| **iv)** protein | **D** ribose |

**3** How many water molecules would be needed to convert an amylose molecule consisting of 1000 glucose units to maltose?

**4** Write the simplest formula for a trisaccharide consisting of a hexose, a pentose and a tetrose.

**5** Which of the following *could* represent **(a)** a monosaccharide, **(b)** a disaccharide **(c)** a trisaccharide?
**i)** $C_{11}H_{20}O_{10}$ **ii)** $C_9H_{14}O_7$ **iii)** $C_7H_{14}O_7$

**6** Which of the alternatives **(a)–(e)** could represent
**i)** a fat,
**ii)** a saturated fatty acid,
**iii)** an unsaturated fatty acid,
**iv)** a carbohydrate,
**v)** an amino acid?
**a)** $C_{18}H_{36}O_2$ **b)** $C_6H_{13}O_2N$ **c)** $C_{18}H_{34}O_2$
**d)** $C_{57}H_{104}O_6$ **e)** $C_5H_{10}O_5$

**7** How many different *dipeptides* could be formed using any combination of the amino acids glycine and alanine?

**8** A haemoglobin molecule consists of two $\alpha$ chains, each consisting of 141 amino acids, and two $\beta$ chains, each consisting of 146 amino acids. How many water molecules are produced in the manufacture of a haemoglobin molecule?

**9** In the amino acid alanine the R group is $—CH_3$. In glycine the R group is a hydrogen atom. Write a structural formula for a dipeptide in which the peptide bond is formed between the amino group of alanine and the carboxyl group of glycine.

CHAPTER 3

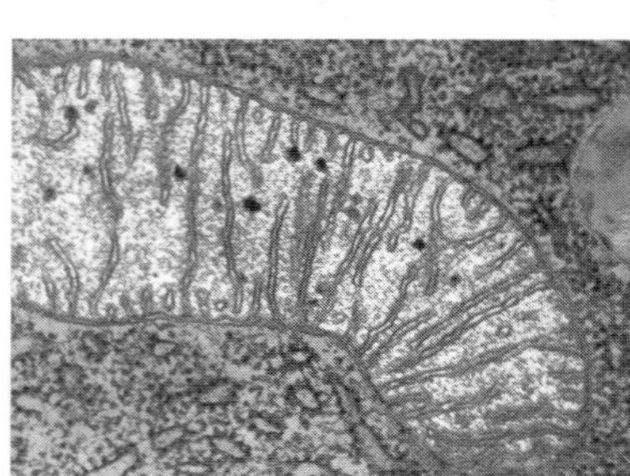

# UNITS of LIFE – the STRUCTURE of CELLS

## LEARNING OBJECTIVES

By the time you have completed your study of this chapter you should be able to:

- appreciate the limitations of the light microscope, and the special strengths and limitations of the electron microscope.
- give the names, locations, functions and relative sizes of the parts of a cell.
- describe some of the methods used to investigate the functioning of cells.
- describe the similarities and differences between prokaryotic and eukaryotic cells, and between animal and plant cells.
- know the approximate size of viruses and describe their structural features.
- describe how viruses reproduce.

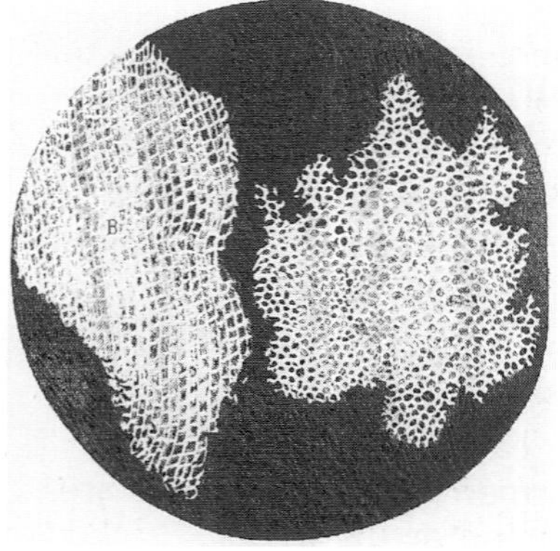

**Figure 3.1**
Hooke's drawing of a slice of cork

The magnifying power of curved glass surfaces has been known since ancient times, but it was not until the 17th Century that lenses were used to study living things. One of the first microscopes was used by Robert Hooke, who is generally credited with the discovery of cells. He studied a thin slice of cork under the microscope and saw that it resembled a honeycomb with many compartments separated by walls (Figure 3.1). He called the compartments **cells** (a cell was a term used for a small room).

Hooke did not realise that his 'little boxes' had been produced by a transparent, jelly-like material that had died by the time the cork had fully developed. The importance of this jelly, or 'protoplasm' as it was later called, was not realised until biologists developed staining techniques many years later.

## Parts of a light microscope

A compound light microscope has two kinds of lens:

- one or more **objectives**, mounted on a rotating **nosepiece**. The objective produces an *inverted image* (upside down and back to front) just below the eyepiece.
- the **eyepiece** at the top of the microscope, which magnifies the image produced by the objective. Because the rays of light leaving the eyepiece are *diverging*, the eyepiece does not form an image. This is left to the lens system of the eye, which forms an image on the retina.

## Parts of a light microscope *continued*

Since the image produced by the objective is inverted, to move the image upwards and to the right you have to move the slide downwards to the left.

The specimen is normally mounted on a glass **slide** which is placed on the **stage** and held by two clips. Underneath the stage is an inbuilt light source, or alternatively a **mirror**, which directs light from a lamp onto the specimen. In higher quality microscopes there is also a **condenser** lens between the mirror and the stage. Its function is to focus light on the specimen. If the microscope has a condenser, the *flat* side of the mirror is used, otherwise the *concave* side of the mirror can substitute for a condenser.

There are two focusing knobs: a large, **coarse adjustment** which is used on low and middle power only, and a smaller, **fine adjustment**, which is used with high power objectives. Beneath the stage there is usually an **iris diaphragm**, which is used to adjust the brightness and the contrast.

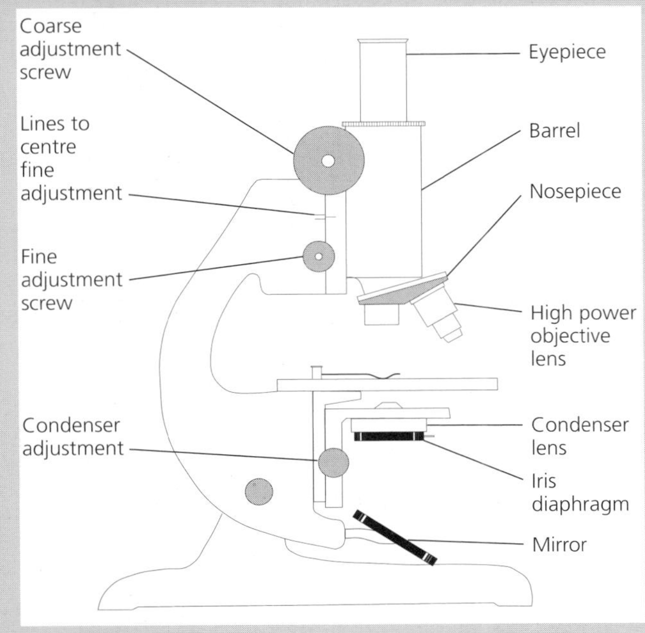


Parts of a school microscope

## The cellular basis of life

During the 18th and 19th Centuries microscopists gradually realised that all living things are composed of cells or of only one cell. The idea that cells are the units of life became known as the **Cell Theory**, first formally put forward by the botanist Mattias Schleiden and the zoologist Theodor Schwann. It contained three essential principles:

- *all living things consist of cells*. Some organisms are said to be **unicellular** because they consist of only one cell. Even gametes (eggs and sperms) are cells. There are a few exceptions to this, for example, some fungi and algae are **coenocytes** (Chapter 15), in which nuclei divide without division of the cytoplasm (Figure 3.2). A superficially similar state of affairs (arising in quite a different way) is seen in the **syncytial** structure of skeletal muscle fibres, which are formed by fusion of many cells during development (Chapter 19).

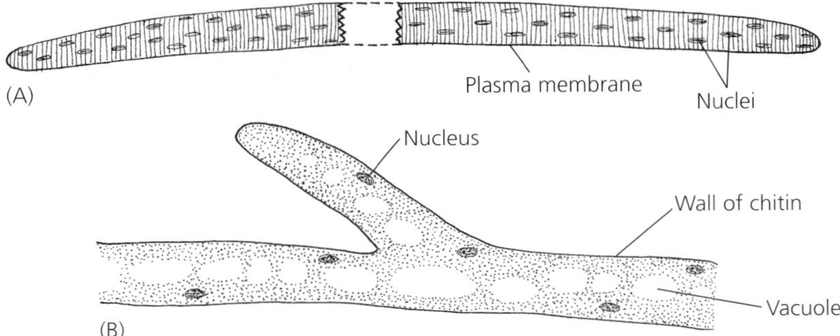


**Figure 3.2**
Syncytium of skeletal muscle fibre (A) and coenocytic hypha of fungus such as *Rhizopus* (B)

- *the activities of living things are the outward signs of processes occurring in their cells.* For example, the saliva that is poured into the mouth is made in the cells of the salivary glands, and the pumping of the heart is due to the contraction and relaxation of its muscle cells.
- *all cells arise from pre-existing cells.* This was simply an extension of the notion of the continuity of life – that all living things have parents, and that life cannot arise from non-life.

## Animal cells under the light microscope

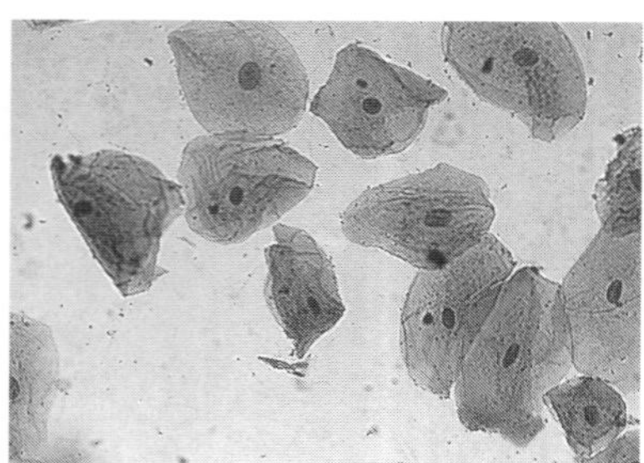

**Figure 3.3**
Human cheek cells as seen under the light microscope

Any multicellular organism consists of many different kinds of cell, each specialised for carrying out a particular function. There is, therefore, no such thing as a 'typical' animal or plant cell – each kind has one or more features that make it different from others. Figure 3.3 shows a cell from the lining of the cheek which is specialised for *protection*.

Notice that the cells are flattened, so their appearance depends on the angle of viewing. With a school microscope you should be able to see the following parts:

- the **cytoplasm** – a transparent, jelly-like material which forms the bulk of the cell. Its outermost layer is the **plasma membrane** or **plasmalemma**, often just called the **cell membrane**. Its function is to regulate the movement of materials into and out of the cell (Chapter 6). It is too thin to see even with the most powerful light microscope – what you actually see is not the plasma membrane itself but the *boundary* between the cytoplasm and the outside of the cell.
- the **nucleus** – containing the **chromosomes** – thread-like bodies containing **DNA**, the genetic material. With other stains, one or more darkly-staining **nucleoli** can be seen. These are concerned with making certain kinds of **RNA**, a close chemical relative of DNA and which plays an essential role in protein synthesis (Chapter 11).

With the most powerful light microscopes certain other structures can just be made out, for example mitochondria, Golgi bodies, and centrosomes. Other structures such as cilia and microvilli (which are absent in cheek cells) occur in certain other kinds of cell. Since no details of these can be seen under the light microscope, they will be dealt with later in this chapter.

## Plant cells under the light microscope

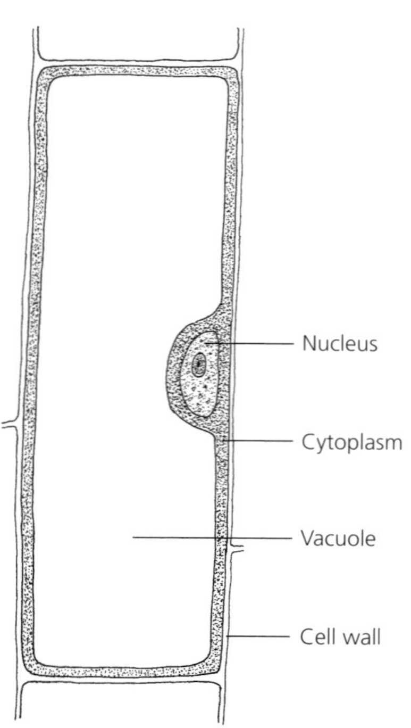


**Figure 3.4**
Section through a plant cell

Figure 3.4 shows a thin slice through a plant cell.

In addition to the nucleus and cytoplasm, a number of other parts can be seen:

- the **cell wall**, made largely of **cellulose**. This acts like the outer layer of a football – it resists stretching, enabling the cell to withstand internal pressure. A cell that is under pressure is firm and helps to support the plant. Though strong, the unmodified cell wall is permeable to most substances.
- most plant cells have small structures called **plastids** in the cytoplasm. These can be of several kinds. **Chloroplasts** contain the green, light-trapping pigment **chlorophyll** and carry out **photosynthesis**, in which light energy is used to convert carbon dioxide and water to sugar and oxygen (Chapter 33). Plant cells that do not carry out photosynthesis lack chloroplasts and often have starch-storing **amyloplasts** or brightly-coloured **chromoplasts**. Under certain

conditions, one kind of plastid can give rise to another. For example, when potatoes slowly turn green after exposure to light, it is because amyloplasts develop into chloroplasts, and in a ripening tomato, chloroplasts change into chromoplasts. All plastids develop from small proplastids present in dividing cells.

- one or more large spaces called **vacuoles**, which make up most of the cell. Vacuoles contain **cell sap**, a watery solution of various substances such as salts and sugars. The vacuole is separated from the cytoplasm by a membrane called the **tonoplast**. In many leaf cells the vacuoles contain **tannins**, which render plant cells less palatable to herbivorous animals and less easily digestible. In some cells the vacuoles contain crystals of various kinds, such as calcium oxalate, which is believed to have an excretory function.

The vacuole plays two important roles in the life of the cell. Firstly, because it is so large, it displaces the cytoplasm to the edge of the cell. In photosynthetic cells the chloroplasts are therefore as close as possible to the source of $CO_2$ (which diffuses very slowly in solution). Secondly, the salts and other dissolved substances in the sap produce a strong tendency for water to enter the cell by osmosis (Chapter 6), causing the cell to become tightly inflated and firm. This plays a very important role in the support of plants.

## Units for microscopic objects

**Most of the things you see under the light microscope are so small that it is best to measure their sizes in micrometres (µm).** 'µ' is the Greek letter *mu* (pronounced 'mew'). An even smaller unit, the **nanometre (nm)**, is needed when dealing with the very small things seen under the electron microscope.

1 µm = 0.001 mm ($10^{-6}$ m)
1 nm = 0.001 µm ($10^{-9}$ m)

Many microscopic objects are so small that it is difficult to appreciate their sizes in relation to each other. The diagram shows a range of objects in relation to each other on a logarithmic scale, each point on the scale being 10 times smaller than the previous one.

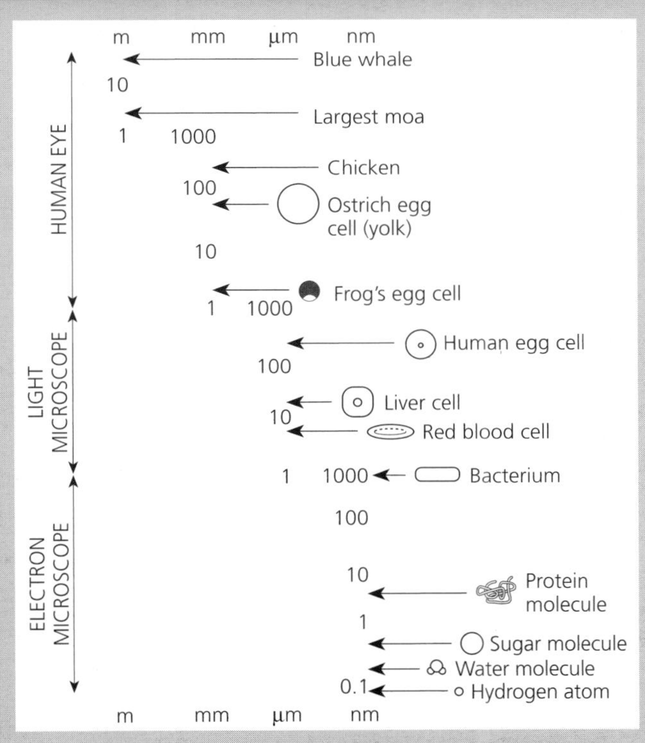


Relative sizes of biological objects

## Cells in three dimensions

It is very important to remember that when you look down a microscope you see only those things that are at the same horizontal level. The picture you see is thus two-dimensional, like a page of a book. To get a mental picture of what the specimen is really like you need to visualise it in three dimensions.

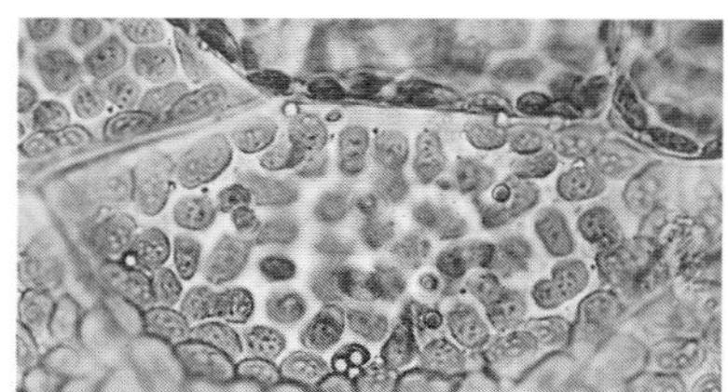

**Figure 3.5**
Plant cell under the light microscope

Figure 3.4 shows a view of a plant cell that you are unlikely to see in a specimen you have prepared yourself. Rather than a thin section through a cell, you are much more likely to see an entire one, which looks quite different (Figure 3.5). Though the chloroplasts seem to fill the entire cell, in reality they don't. The illusion occurs because you are actually looking through two thin 'windows' of cytoplasm separated by a thick layer of vacuolar sap. There also seem to be more chloroplasts at the sides of the cell because you are looking through a greater thickness of cytoplasm (i.e. edge on).

## Mounting a specimen

Living specimens are mounted in water, which helps to keep them alive. A thin glass **coverslip** is carefully lowered onto a drop of water containing the specimen. If the specimen needs to be viewed for more than a few minutes, it should be mounted in 10% glycerine, which slows down evaporation.

The coverslip has several important functions:

- it holds the specimen steady.
- it keeps the specimen flat so that you do not have to vary the focusing as you move the slide around.
- it allows the stage to be tilted without water running off the slide.
- it stops the high power objective lens getting wet, and prevents it misting up due to condensation.
- a drop of water has a curved surface which acts as a lens, distorting the image.

It is important to use the right amount of water. Too little and a delicate animal may be crushed by the surface tension pulling the coverslip down onto the slide; too much and the specimen cannot be held steady.

## Temporary stains

Living cells are transparent, so it often makes it easier to see details if they are first **stained**. To be of any use a stain must give contrast, so it must be taken up to different extents by different parts of the cell. A considerable number of stains exist, each being taken up preferentially by a particular part of the cell. Iodine dissolved in potassium iodide solution, for example, stains starch deep-blue to black, and stains nuclei brown. Methylene blue stains nuclei blue and, unlike most stains, does not kill cells. Lignin in plant cell walls is stained yellow by aniline sulphate and red by phloroglucinol and concentrated HCl.

## Making a permanent preparation

A normal wet mount is quick to make but only lasts a few hours at most. Permanently stained preparations are more durable but take much longer to make, involving the following steps:

- **fixing**, in which the tissue is placed in a liquid such as formalin. This penetrates rapidly and preserves the material in as natural a state as possible, rendering it more stable to subsequent treatment.
- **dehydration**, in which water is progressively removed by passing the material through a series of increasingly concentrated ethanol solutions.
- **embedding**, in which the specimen is placed in hot wax and then cooled. The specimen is now effectively part of a solid block of wax, making it firm enough to be sectioned.

## Making a permanent preparation *continued*

- **sectioning**, or the cutting of the material into thin slices using a microtome.
- **mounting**, or placing the section on a glass slide.
- **clearing**, in which the wax is removed from the section by immersing it in an organic solvent such as xylene.
- **rehydration** (by a reversal of dehydration).
- **staining**, followed by another round of dehydration.
- **mounting** in some material such as DPX, that sets solid and preserves the material indefinitely.

## Estimating the size of a specimen

When you draw a specimen you should, if possible, include the magnification of your drawing. *This is not the magnification of the microscope, but the number of times the drawing is 'larger than life'.* You therefore need to know the size of the specimen and the size of your drawing. Provided that you know the diameter of the field of view, you can estimate the size of an object by estimating how many times the cell would fit across it. In other words, you use the field of view as a kind of crude ruler. If the field of view is 300 μm wide and you think that the cell would fit across it ten times, the cell must be about 300/10 = 30 μm wide.

How do you find the diameter of the high power field of view? Although you can measure the low power field of view by placing a ruler across it, you can't do this with high power because the field diameter is less than 1 mm. Instead, you can calculate it in the following way: suppose the low power objective lens magnifies 4 times and the high power objective magnifies 40 times. The high power magnification is thus ten times as powerful as low power. The high power field of view is therefore a tenth of the low power field of view (which you can measure with a ruler).

## Limits to the light microscope – resolving power

The most powerful light microscopes magnify about 1500 times. Though higher magnifications are possible, you don't see any more *detail* – the picture just becomes more and more blurred. You see similar 'empty' magnification when you look at a newspaper photograph with a lens. The photograph is made up of thousands of dots – the more dots per square centimetre, the more detailed the picture. Rather than revealing more dots, a lens just enlarges the dots you can already see.

The amount of detail that can be revealed by a microscope is termed its **resolving power**. This is expressed as *the minimum distance between two points at which they are still visible as two separate points.* Resolving power depends on the objective lens, because the eyepiece cannot add detail that the objective has failed to pick up. The resolving power of the human eye is about 0.1 mm or 100 μm. Thus two points 90 μm apart are seen as one, whilst two points 100 μm apart appear distinct. Most school microscopes have a resolving power of about 0.5 μm (about half the width of an average bacterium), giving a useful magnification of about 400. The best light microscopes can resolve points about 0.2 μm (200 nm) apart.

The reason why things smaller than about 200 nm cannot be resolved with a light microscope is related to the fact that light travels in waves, the distance between successive wave peaks being called the **wavelength**. Objects closer than about half the wavelength of the light cannot be distinguished. Of visible light, violet has the shortest wavelength – about 400 nm – so with violet light objects of about 200 nm can be distinguished.

## Oil immersion microscopy

Since the eyepiece can only enlarge the image produced by the objective, the performance of the objective lens places limits on the quality of the microscope as a whole. The resolving power of an objective depends not only on the wavelength of the light used but also on the angle of the cone of light picked up by the objective. To accept a wide-angled cone of light the objective needs to have a short focal length, which is why the high power objective is so close to the coverslip.

The most useful rays of light – those carrying the most information – are those at the edge of the cone (Figure 3.10). The trouble is that these rays are refracted as they leave the coverslip and are not collected by the objective. These peripheral rays can be 'rescued' by placing a drop of cedar wood oil between the coverslip and the objective. Because the oil has the same refractive index as glass, the rays are not refracted as they leave the coverslip and enter the objective. Most oil immersion objectives have magnifications of about 90.

## Differences between light and electron microscopes

The electron microscope differs in several important ways from the light microscope:

- instead of using light rays focused by glass lenses, it uses a beam of electrons focused by powerful electromagnets.
- electrons do not pass easily through air, so the interior of the microscope has to be a vacuum.
- because organisms cannot survive in a vacuum, only dead specimens can be studied.
- the specimens must be about 50–100 nm thick, which is about a hundredth of the thickness of sections used in light microscopy, and about 1/200 the thickness of an average-sized animal cell. To cut such thin sections the specimen is embedded in resin. The microtome blade used must be especially sharp and is made from broken glass.
- after fixing with glutaraldehyde, the specimen is 'stained' with a substance that absorbs electrons, such as lead or uranium salts.
- since the eye cannot see electrons, the final image is produced on a fluorescent screen by a **projector lens**.

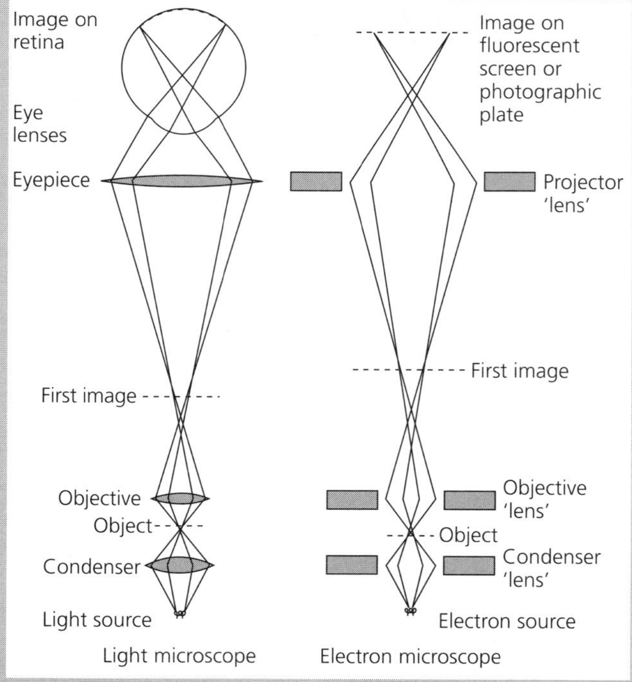


Light and electron microscopes compared

# Cells under the electron microscope

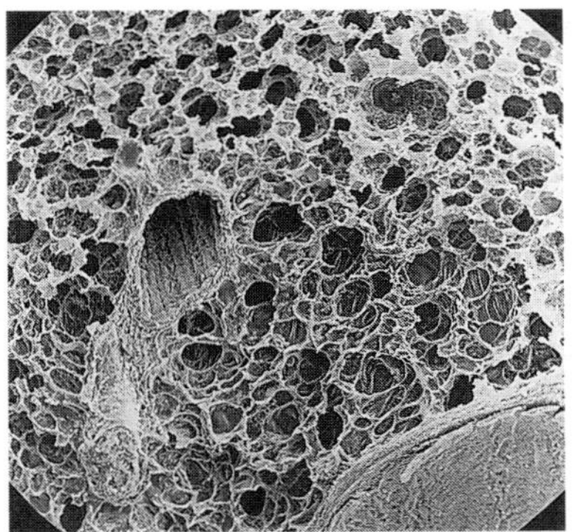

**Figure 3.6**
Scanning electron micrograph of lung capillary network and alveoli

The second half of this century saw the invention of a new kind of microscope, the **electron microscope**. It is based on the principle that a beam of electrons has wave-like properties. The higher the voltage, the shorter the wavelength of the electrons. A 100 000 Volt electron beam has a wavelength of 0.004 nm, which could theoretically give a resolution of 0.002 nm. In practice, engineering problems make the actual resolving power about 0.5 nm, with useful magnifications of up to 500 000 times.

The most powerful electron microscopes are called **transmission electron microscopes (TEM)**, because the electron beam passes through the specimen. Another kind of electron microscope is the **scanning electron microscope (SEM)**. It uses electrons scattered by the surface of the specimen, giving an image with a 3-dimensional perspective (Figure 3.6). The resolving power is less than that of the TEM, giving an effective magnification of up to 20 000 times.

## Two levels of cellular organisation

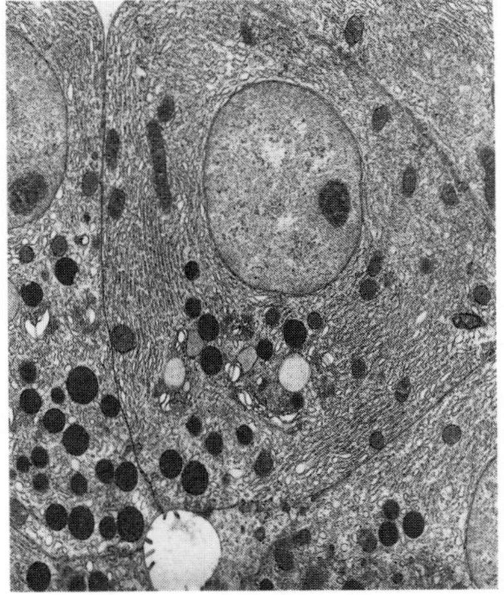

**Figure 3.7**
Transmission electron micrograph of an animal cell

The invention of the electron microscope enabled biologists to probe more deeply than ever before into the structure of cells. It became clear that cells could be divided into two fundamentally different groups. One group, the bacteria, have no clearly defined nucleus and are called **prokaryotes**. The other group are the **eukaryotes** and comprise all other living things. Eukaryotes have their chromosomes packaged inside a **nuclear envelope**, so that there is a distinct nucleus. The section immediately following applies to the eukaryotes; prokaryotes are dealt with at the end of this chapter.

Eukaryotes have a complex organisation, with distinct structures called **organelles**, which are specialised for different functions. Some organelles can be seen under the light microscope (e.g. the nucleus and chloroplasts), but most can only be resolved with the electron microscope. An electron microscope view of an animal cell is shown in Figure 3.7 and a plant cell is shown in Figure 3.8.

## Cell membranes

Long before the invention of the electron microscope it was evident that cells had to be separated from their environment by some kind of barrier. Animal cells have a clear boundary between the inside and the outside of the cell. Plant cells have a cell wall, but after immersion in strong solutions, the cytoplasm pulls away from the wall, showing that the cytoplasm is also surrounded by some kind of membrane (Chapter 6). Whatever its nature, it is too thin to be seen under the light microscope, so indirect methods had to be used to deduce its structure.

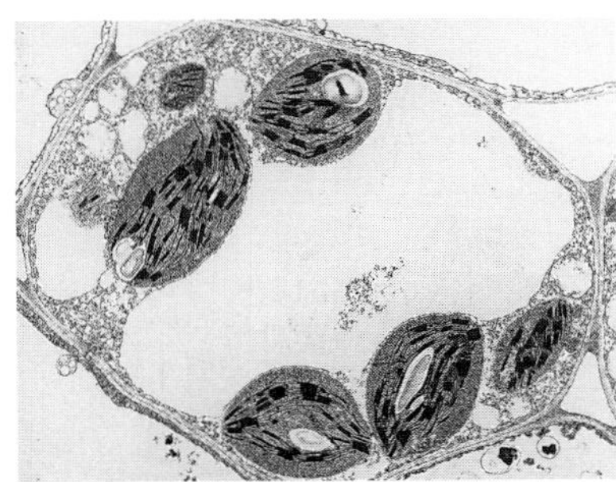

**Figure 3.8**
Transmission electron micrograph of a plant cell

A very early clue came from a discovery by Overton at the end of the last century. He found that the speed with which a substance penetrates a cell depends on how fat-soluble it is. On the strength of this observation, he concluded that the plasma membrane must contain lipid.

A direct analysis of the composition of the plasma membrane was made by Gorter and Grendel in 1925. Using red blood cells they found that the membranes consisted of roughly equal amounts of **lipid** and **protein**, with small amounts of **carbohydrate**. Moreover, they calculated that the amount of lipid was just enough to form a layer two molecules thick.

Using this and other information, in 1935 Davson and Danielli proposed a model for membrane structure shown in Figure 3.9.

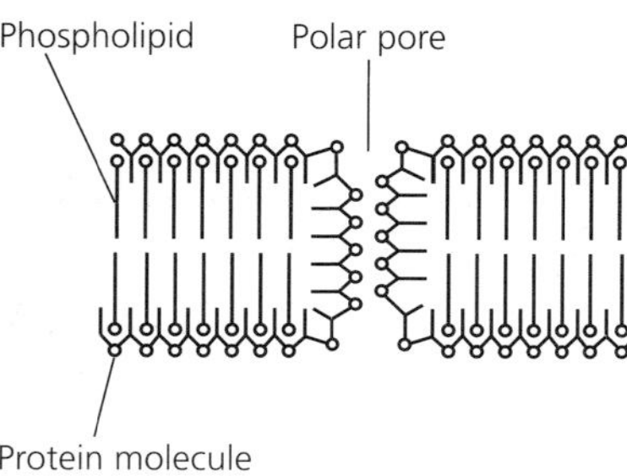


**Figure 3.9▲**
Davson–Danielli model of membrane structure

According to this model the outer layers of the membrane consist of protein, with the two layers of phospholipid facing head-to-tail.

The Davson–Danielli model seemed all the more plausible when the first electron micrographs of plasma membranes were taken. They showed two dark lines with a paler layer between, so it was tempting to think that the dark lines represented the protein and the pale area the lipid.

## The fluid mosaic model of membrane structure

The development of the technique of **freeze fracture** threw serious doubt on the Davson–Danielli model. In this technique a tissue is rapidly frozen and then broken along a plane of weakness with a sharp blow. The cleavage planes often coincide with the middle of a cell membrane, separating it into two complementary halves (Figure 3.10). The membrane is then viewed from the inside, so to speak. Instead of appearing as a smooth surface, as would be expected from the Davson–Danielli model, it had a cobbled appearance. If each 'cobble' corresponded to a protein molecule, then the proteins must be buried within the membrane itself.

In the light of this discovery, in 1972 Singer and Nicholson put forward their **fluid mosaic model** of membrane structure. The proteins were envisaged as floating like icebergs in the lipid bilayer, with some proteins spanning the entire thickness of the bilayer, some confined to the outer half and others to the inner half (Figure 3.11). Membrane proteins anchored in the lipid bilayer are called **integral proteins**. Others, called **peripheral proteins**, are attached to the integral proteins.

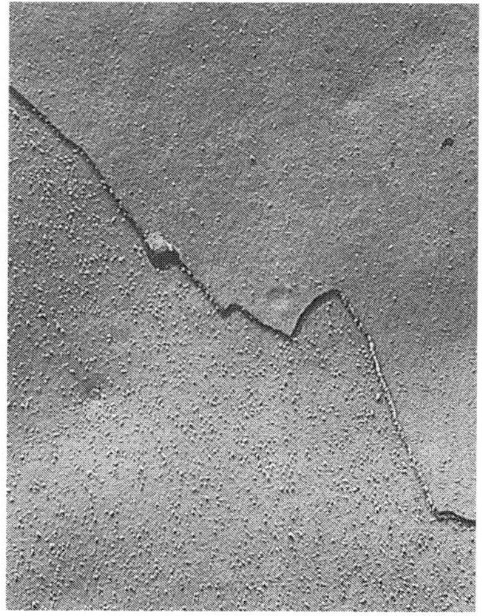

**Figure 3.10▲**
TEM view of freeze fracture through middle of cell membrane showing membrane proteins

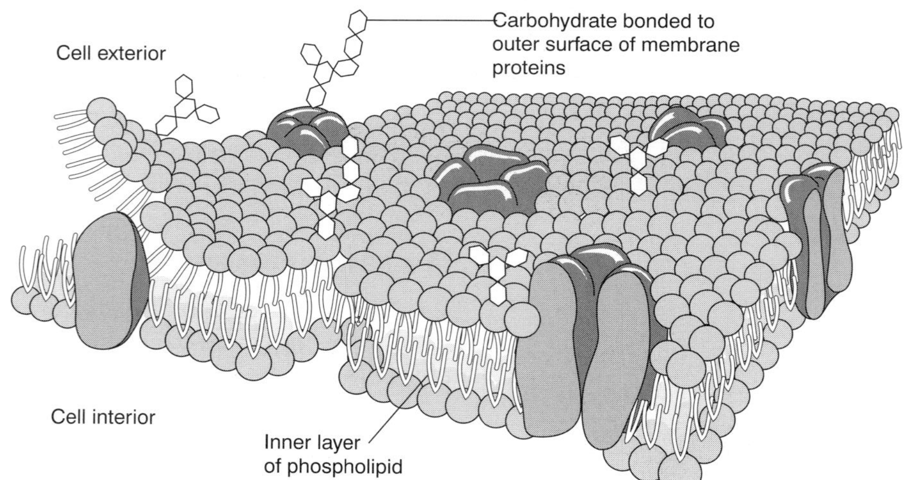


**Figure 3.11►**
Fluid mosaic model of membrane structure. Part of the membrane is shown as if the two layers of phospholipid have been pulled apart (after Satir)

An elegant experiment that lent support to the fluid mosaic model involved the fusion of mouse and human cells (Figure 3.12). Firstly, the proteins on the surface of each kind of cell were tagged with an antibody coupled to a dye that fluoresced when illuminated. The mouse proteins fluoresced green and the human proteins fluoresced red, so that they could be distinguished from each other. The mouse and human cells were then artificially fused together. After about 40 minutes the two kinds of protein had intermingled, showing that they must have been free-floating in the plasma membrane rather than forming a continuous layer as implied by the Davson–Danielli model. Evidently the two dark lines seen in electron micrographs must have been an **artefact** – something produced by human action during the preparation of the material.

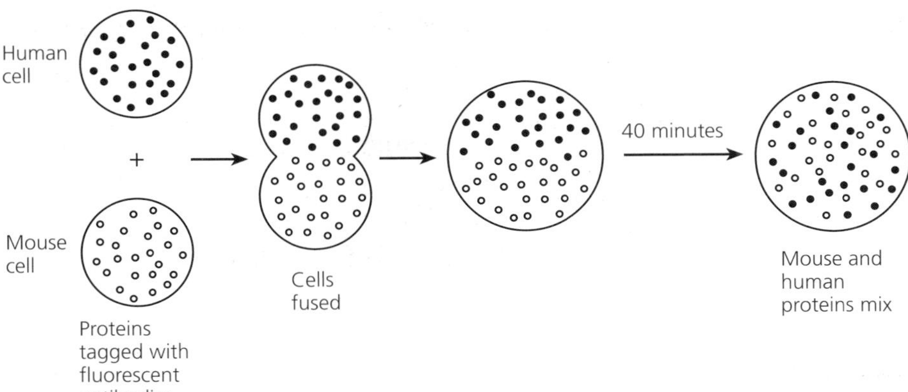


**Figure 3.12**
Evidence for fluid mosaic model – fusion of mouse and human cells

Although red blood cells have only one membrane (the plasma membrane) virtually all other eukaryotic cells also have a complex system of internal membranes. It is now accepted that all cell membranes, both external and internal, conform broadly to the fluid mosaic model.

This is not to say that all cell membranes are identical. For one thing, membrane thickness varies slightly from 5–10 nm, depending on the location. Membranes also vary in composition. The plasma membrane for instance contains significantly more cholesterol than other membranes, and the inner mitochondrial membrane has a higher proportion of proteins than other membranes.

## The plasma membrane

The plasma membrane forms the boundary between the cell and its surroundings. It has two important functions:

- it controls the movement of substances into and out of the cell (see Chapter 6). It does this both passively and actively. Firstly, it is partially or differentially permeable, allowing certain molecules to enter and leave the cell (e.g. oxygen and $CO_2$) whilst restricting the movement of others. Secondly, it actively transports substances into and out of the cell.
- it enables the cells of multicellular organisms to 'recognise' each other and, in more complex animals, to distinguish the body's own cells ('self') from foreign cells ('non-self').

Like other cell membranes the plasma membrane consists of a lipid bilayer with protein molecules floating in it. In addition, many of the protein molecules are linked to branched, oligosaccharide carbohydrates that protrude outwards from the cell surface forming a fringe called the **glycocalyx**. These cell surface carbohydrates are believed to play an important part in contact interactions between cells.

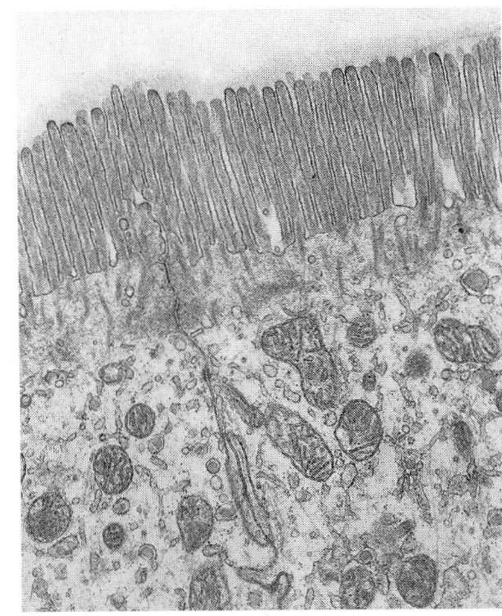

**Figure 3.13**
TEM of microvilli on epithelial cells of convoluted tubule of kidney nephron

## Microvilli

In the cells lining the small intestine, kidney tubules and some other cells, parts of the plasma membrane are extended into tiny finger-like processes called **microvilli** (Figure 3.13). These greatly increase the surface area for active uptake of substances such as glucose, amino acids and mineral salts. A microvillus is held in shape by a longitudinal bundle of actin filaments, held to the plasma membrane by myosin molecules and cross-linked by other proteins.

## Separating organelles

You may have wondered how we know the function of each type of organelle in the life of the cell. To study an organelle reasonable quantities must be obtained, uncontaminated by other cell constituents. To separate cell constituents biologists use a **centrifuge**, which is simply a device for generating artificial G-forces in a tube containing a suspension of particles. Particles settle out according to their size, shape and density.

The first step is to use a blender or homogeniser to break the cells up into a soup-like **homogenate**. The cells are suspended in a salt solution with the same concentration as the cytoplasm, thus ensuring that organelles do not take up or lose water by osmosis. By running the centrifuge at successively higher speeds, particles can be separated off. At relatively low speeds, nuclei and unbroken cells settle. Higher speeds bring down mitochondria, and to collect ribosomes the highest speeds (up to 500 000 times gravity) must be used.

A refinement of the technique is **density-gradient centrifugation**. The centrifuge tube is first filled with a solution of sucrose or caesium chloride with decreasing concentration (and thus decreasing density) from bottom to top. The suspension is then placed on the top and the tube spun in the centrifuge. Particles move down the tube until they reach a level at which the surrounding liquid has the same density.

Besides being used to separate particles, high speed centrifuges are also used to obtain information about the relative densities and sizes of particles. In fact the rate of sedimentation of a particle is a useful diagnostic character in its own right and is measured in **Svedberg units** (symbol S).

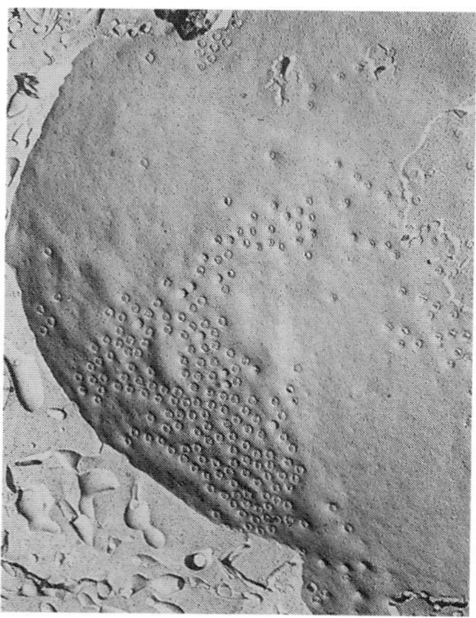

**Figure 3.14**
Freeze fracture of nuclear envelope showing nuclear pores

## The nucleus

The nucleus is the 'control centre' of the cell. It contains the chromosomes which consist of DNA and protein. The layer separating the nucleus from the cytoplasm was called the 'nuclear membrane' by light microscopists, but the electron microscope has since shown that it is actually two-layered and is thus more appropriately called the **nuclear envelope** (Figure 3.14). The nucleus and cytoplasm are not isolated from each other, as the nuclear envelope is penetrated by many pores through which a constant two-way traffic of molecules passes. For example, 'working copies' of the genes (in the form of messenger RNA) pass from nucleus to cytoplasm, whilst raw materials pass in the reverse direction.

The pores in the nuclear envelope are not simply holes, but are lined by specialised proteins concerned with the transport of nucleic acids and other macromolecules between nucleus and cytoplasm.

## The endoplasmic reticulum

With the invention of the electron microscope it became clear that the cytoplasm is far from being a formless jelly (as had previously been believed) but has an extremely complex organisation. One of the largest and most complex structures in eukaryotic cells is the **endoplasmic reticulum**. Invisible under the light microscope, it appears under the electron microscope as an extensive system of membrane-bound spaces. The total area of the endoplasmic reticulum may be very large – in liver cells it has been estimated to be 25 times larger than that of the plasma membrane.

It is easier to understand the structure of the endoplasmic reticulum if you think of it as a greatly folded outgrowth of the outer layer of the nuclear envelope (Figure 3.15).

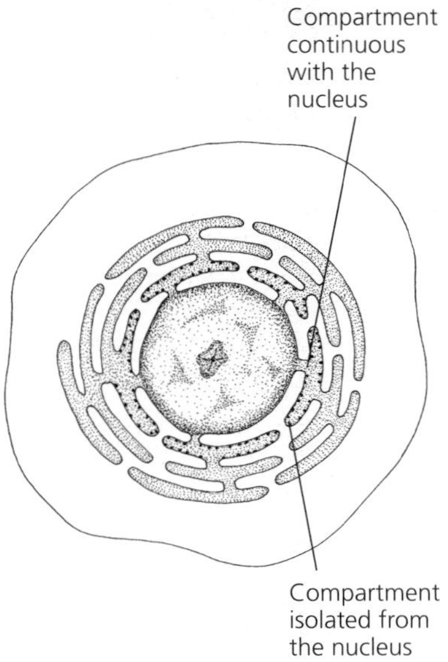


**Figure 3.15**
Endoplasmic reticulum as an outgrowth of nuclear envelope

In the two-dimensional sections with which electron microscopists have to work, the endoplasmic reticulum seems to consist of many separate spaces. However, it is believed that there are really only two compartments, separated from each other by a single greatly-folded membrane. One compartment, the **cytosol**, communicates with the nucleus via the pores in the nuclear envelope. The other, comprising the **cisternae**, is continuous with the cavity within the nuclear envelope but not with the nucleus itself.

Parts of the endoplasmic reticulum contain large numbers of **ribosomes** – small granules approximately 25 nm across. The ribosomes are involved in the manufacture of proteins and consist of protein and a kind of RNA called **ribosomal RNA** or **rRNA** (see Chapter 11). They are assembled in the nucleus and reach the cytoplasm via the pores in the nuclear envelope. Since ribosomes cannot pass through membranes they can only reach the cytosol, so the cisternae do not contain ribosomes (Figure 3.16).

Two regions of the endoplasmic reticulum can usually be distinguished:

- the **rough endoplasmic reticulum**, so-called because its flattened cavities are studded with ribosomes giving it a granular appearance. Cells which actively make proteins have an extensive rough ER.
- the **smooth endoplasmic reticulum** with tubular cavities. It has few ribosomes and is concerned with production of lipids and steroids, and is the site of membrane synthesis in eukaryotes.

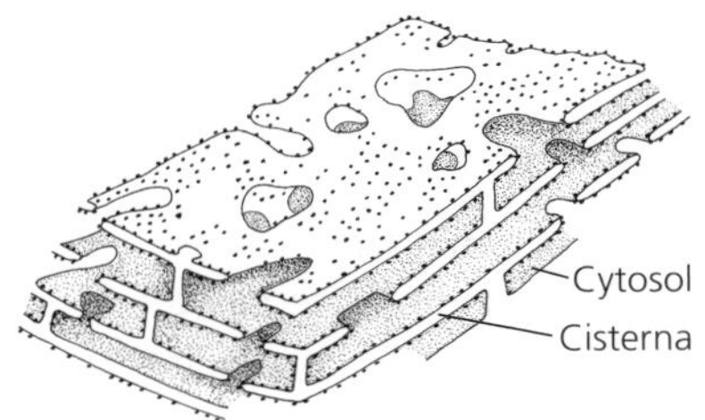


**Figure 3.16**
Cisternae and cytosol, showing ribosomes present only in the latter

## The Golgi apparatus

Most cells make several hundred different kinds of protein. Some are for 'export' and are secreted through the plasma membrane, but the majority function within the cell. Of these, some become components of membranes and others operate within the cytosol. How does the cell ensure that each protein reaches its correct destination?

It seems that this is one of the main functions of the **Golgi bodies**, or **dictyosomes** as they are sometimes called in plants. For many years after their discovery by Camillo Golgi at the end of the last century their existence was doubted, but all that changed with the invention of the electron microscope. This revealed that a Golgi body consisted of a stack of flattened sacs, with small sacs round the edges (Figure 3.17). At least one Golgi body appears to be characteristic of all eukaryotic cells. Most animal cells have only one Golgi body but in plant cells there may be many. When more than one are present they are collectively known as the **Golgi apparatus**.

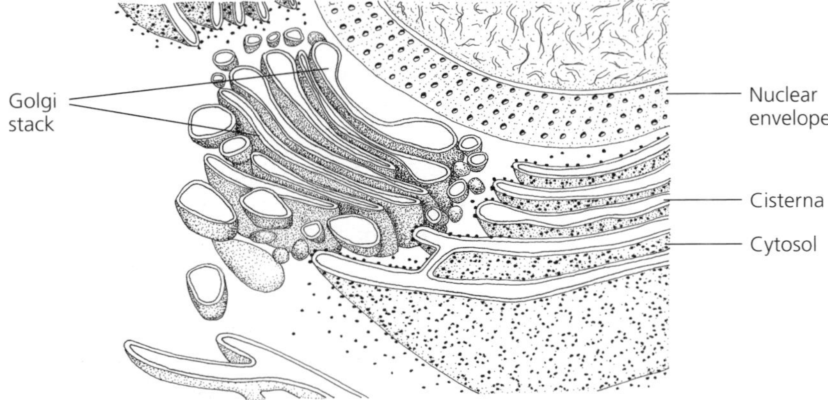


**Figure 3.17**
A Golgi stack after Du Praw

Golgi bodies typically lie just outside the nucleus and are closely associated with portions of smooth endoplasmic reticulum. Electron micrographs show vesicles – apparently just budded off from the rough endoplasmic reticulum – in the process of fusing with Golgi sacs. This suggests that newly-manufactured proteins go to the Golgi bodies before they reach their final destination.

In recent years it has been possible to mark specific proteins with antibodies and follow their progress through the Golgi bodies. As a result of these and other experiments it is clear that a Golgi stack has two functionally-different sides. At the convex side of the stack – the so-called **cis** or **forming face** – vesicles bud off from the rough endoplasmic reticulum and fuse with Golgi sacs. Proteins move from one Golgi sac to the next by vesicles being budded off sacs at the forming face and fusing with the next vesicle in the stack. At the concave side (the **trans** or **maturing face**) vesicles are budded off to be sent to their various destinations.

Within the Golgi body, vesicles containing proteins receive oligosaccharide 'labels' which direct them to their destination. Of those vesicles destined for internal use, some become lysosomes (see below), whilst others contain proteins destined to form components of internal membranes.

Another function of the Golgi apparatus is to modify certain proteins after their production in the rough endoplasmic reticulum. For example, it is in the Golgi apparatus that glycoproteins (such as the mucin of saliva) receive their carbohydrate component. It is in the Golgi sacs that proinsulin is converted to the active hormone insulin by 'pruning' off part of the polypeptide chain.

In addition to these major functions the Golgi apparatus has a number of other subsidiary roles that will be dealt with in their appropriate contexts.

## Lysosomes

Lysosomes are tiny vacuoles about 0.5 μm in diameter which occur in almost all eukaryotic cells. They contain a wide variety of hydrolytic enzymes which can carry out a variety of roles, for example:

- in most cells they are used to break down worn-out organelles such as mitochondria.
- they may be used to digest ingested particles, for example when white blood corpuscles digest bacteria, or when a single-celled protoctistan digests a food particle (Figure 3.18).
- the head of a sperm contains a modified lysosome which, at the moment of contact with the ovum, releases its enzymes. These enzymes break down the plasma membrane of the ovum, enabling the head of the sperm to penetrate it.

Lysosomal enzymes work best in acidic conditions, at a pH of about 5. This is maintained by the lysosome membrane which actively pumps in $H^+$ ions. Thus, if a lysosome did burst and release its deadly contents, the pH of the cytosol (about 7.2) would be too high for the enzymes to do any damage.

It is said that lysosomes are also involved in **programmed cell death**, or **apoptosis**. This happens, for example, when a tadpole's tail shrinks as it changes into a frog. A similar process occurs in a pupa as it changes into a butterfly or moth, when many of the muscles of the caterpillar are

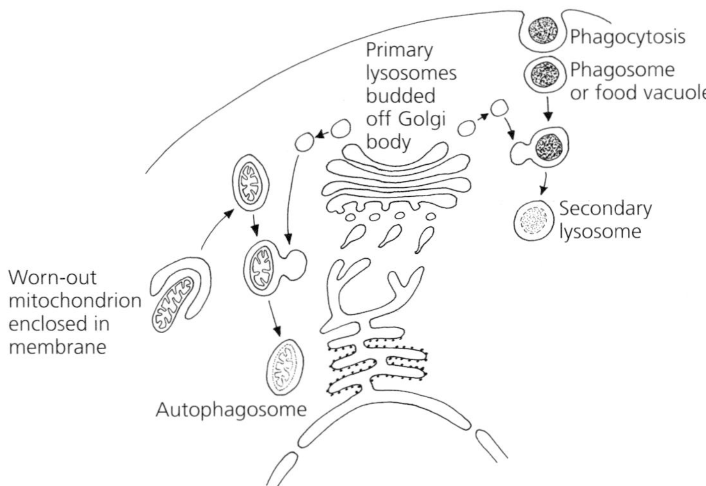


**Figure 3.18**
Role of lysosomes in intracellular digestion

broken down to form a soupy liquid which provides the raw materials to build the adult organs. This raises an interesting point: if lysosomes are involved in self-digestion or autolysis, their enzymes must be able to work at pH values closer to 7.

Lysosomes originate by being budded off from the Golgi bodies, and in this state are called **primary lysosomes**. During autolysis the membrane of the primary lysosome breaks down to release its enzymes, with the result that the entire cell is digested. How then does a cell digest an ingested food particle whilst leaving the cell as a whole intact? As Figure 3.18 shows, the food particle is initially isolated from the rest of the cell by the membrane of the food vacuole. The primary lysosome fuses with the vacuole forming a **secondary lysosome**, within which digestion can occur safely. A worn-out mitochondrion is first surrounded by a membrane derived from the endoplasmic reticulum. This forms a vacuole with which the primary lysosome then fuses, forming an **autophagosome**.

## Tracing the fate of chemicals in cells – autoradiography

An important question cell biologists ask is: what happens to the raw materials absorbed by cells? How are they used, and where? Problems such as these can be solved by the use of radioactive tracers. In one experiment the utilisation of amino acids by insulin-secreting pancreas cells was traced. Living pancreas cells were first 'fed' with amino acids labelled with tritium, the radioactive isotope of hydrogen ($^3H$). After five minutes the radioactive amino acids were replaced with normal amino acids. After a series of progressively longer intervals, sections of the tissue were cut for viewing under an electron microscope. Although the electron microscope itself cannot detect radioactivity, it can be used in conjunction with **autoradiography**, which can.

The sections of tissue were overlaid with a thin layer of photographic emulsion. The preparations were left in darkness for a few days, during which time radiation emitted by the tissue caused grains of silver to be deposited in the emulsion. The distribution of silver was then compared with an electron micrograph of the section. In sections cut shortly after removal of the tracer, radioactivity was concentrated in the rough endoplasmic reticulum. In sections cut after ten minutes the radioactivity had shifted to the Golgi bodies and after 45 minutes it was concentrated in secretory granules. A reasonable interpretation is that the amino acids are assembled into proteins in the rough endoplasmic reticulum. Then the proteins are transported to the Golgi bodies before being 'packaged' in vacuoles for secretion.

## Peroxisomes

Peroxisomes were called microsomes before their function was understood. They resemble lysosomes in size and appearance and in their occurrence in almost all eukaryotic cells. Peroxisomes are particularly prominent in liver and kidney cells, in which they oxidise various toxic chemicals such as ethanol. A by-product of such oxidative reactions is **hydrogen peroxide**. This reactive and potentially toxic chemical is broken down in peroxisomes to water and oxygen by the enzyme **catalase**.

In seeds that store fat, peroxisomes convert fatty acids into carbohydrate – a chemical feat that mammalian cells cannot do. In leaf cells the peroxisomes are also the site of **photorespiration** (Chapter 33).

## Mitochondria

Mitochondria (singular – mitochondrion) are sausage-shaped organelles which are the sites of respiration, and occur in almost all cells except bacteria. They are similar in size to bacteria, typically about 0.5–1.0 μm thick and several μm long, and are just visible under the best light microscopes. Their function is to provide chemical energy for the cell in the form of **adenosine triphosphate (ATP)** (Chapter 5). For this reason, they have been termed the 'powerhouses of the cell'. Most cells have many mitochondria and some can have more than a thousand.

A mitochondrion consists of two membranes enclosing an inner space (Figure 3.19). The inner membrane contains proteins that play a key part in respiration. Its area is extended by folds called **cristae**, whose inner surface is studded with thousands of tiny **stalked particles**. The relationship between the structure of mitochondria and the production of ATP is dealt with in Chapter 5. Mitochondria contain their own DNA and ribosomes of the prokaryotic type, and reproduce by dividing into two.

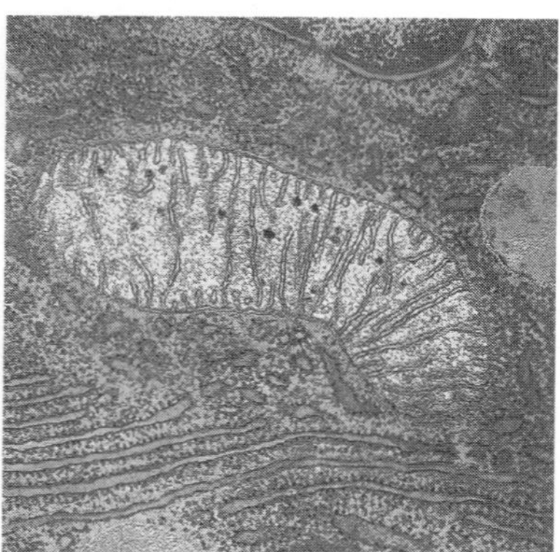

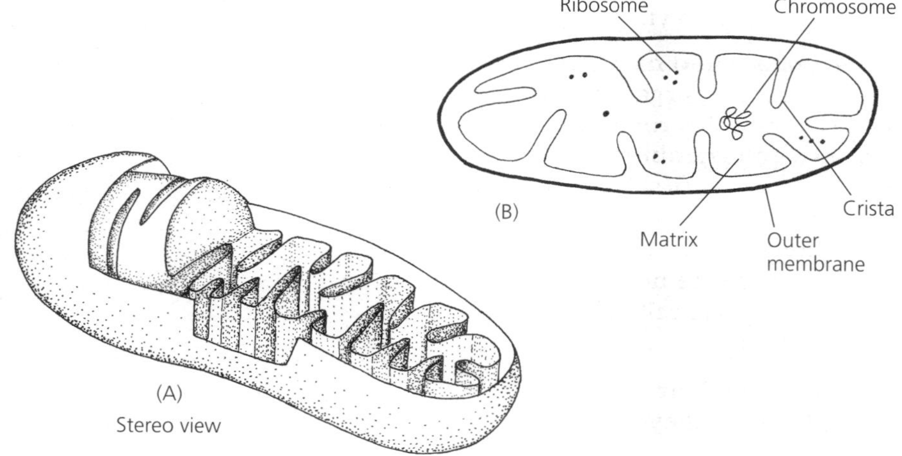


**Figure 3.19**
Diagrammatic views of a mitochondrion in 3-D (A) and in section (B). The photo above is a TEM of a mitochondrion

## Plastids

Plastids are a family of self-replicating organelles found in plant cells. They contain DNA and are able to make some of their own proteins. The most familiar plastids are **chloroplasts**, the site of photosynthesis (Figure 3.20). These lozenge-shaped structures are 4–8 μm long and 2–3 μm thick. Their detailed structure and its functional significance are dealt with in Chapter 33.

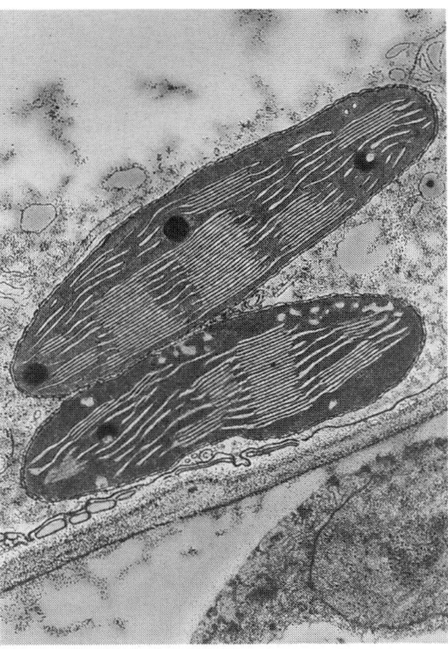

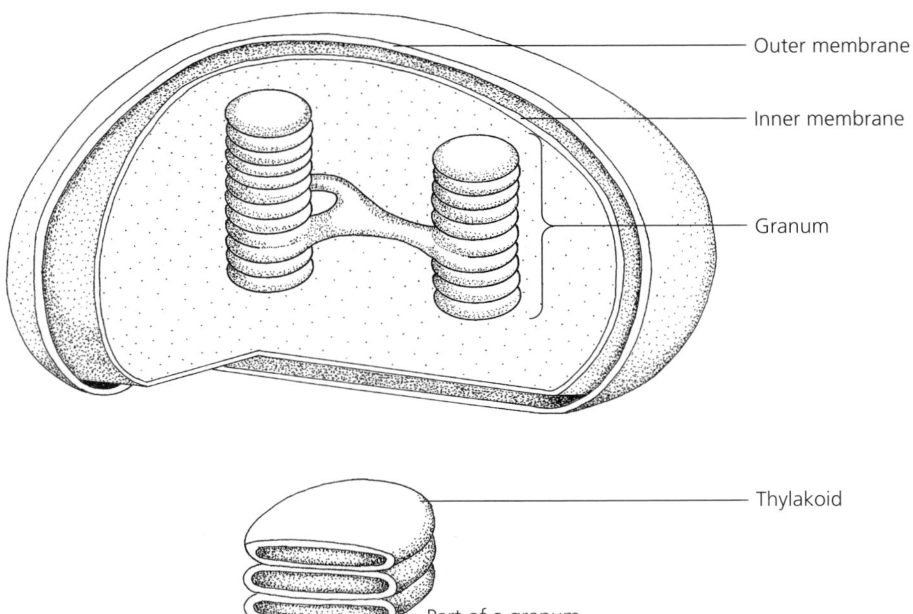


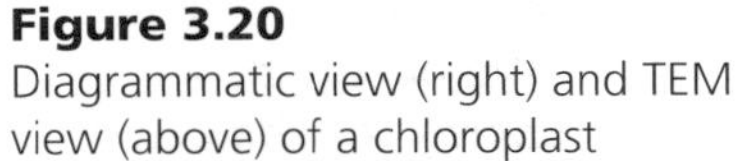

**Figure 3.20**
Diagrammatic view (right) and TEM view (above) of a chloroplast

## Are plastids and mitochondria bacterial 'guests'?

There are many instances of a unicellular organism living inside another in a state of **mutualism** – a relationship in which both partners benefit. The endodermal cells of the fresh water *Hydra*, for instance, contain single-celled photosynthetic organisms, and many other examples of this **endosymbiosis** (one organism living inside the cells of another) are known.

Chloroplasts and mitochondria are both self-replicating organelles with their own DNA. Could this mean that these organelles are also endosymbionts? This daring hypothesis was first put forward about twenty years ago by Lynn Margulis, who suggested that both were once free-living bacteria that are now living in partnership with eukaryotic cells. Some of the evidence is given below:

- both have their own DNA and ribosomes with which they make some of their own proteins.
- like the DNA of bacteria, chloroplast DNA and mitochondrial DNA form a closed loop.
- the ribosomes of chloroplasts and mitochondria are more similar to those of prokaryotes than eukaryotes in a number of ways: they are smaller, with a sedimentation rate of 70 S, compared with 80 S for eukaryotic ribosomes; and their action is inhibited by the same antibiotics (such as chloramphenicol and streptomycin) that inhibit protein synthesis in bacteria.
- both chloroplasts and mitochondria have a double membrane. This is what would be expected if an organism were to live inside a vacuole of a eukaryotic cell. The inner membrane would be the plasma membrane of the bacterium and the outer membrane would be the vacuolar membrane of the eukaryote.

## Cilia and flagella (undulipodia)

Cilia and flagella are hair-like outgrowths of the cytoplasm which beat against fluid outside the cell. They are extremely widespread amongst animals, algae and simpler plants. The flagella of bacteria are fundamentally different in structure and work in a completely different

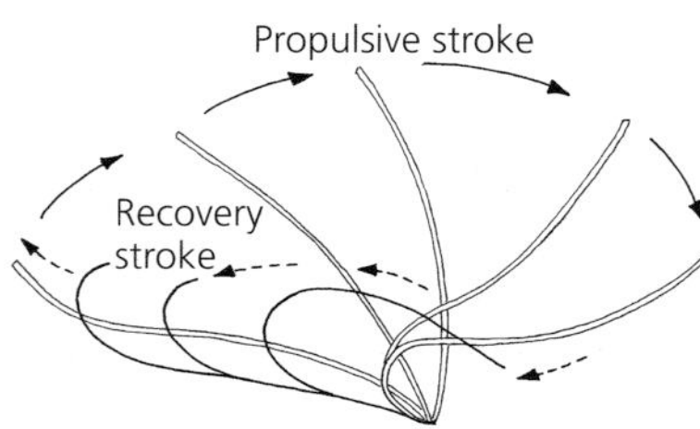


**Figure 3.21**
Drawing of cilium to show pattern of beat

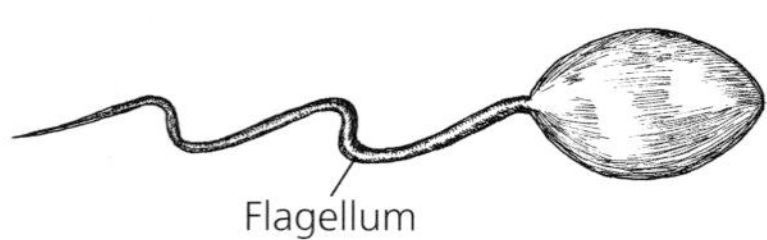


**Figure 3.22**
A flagellated unicellular organism showing several waves present simultaneously

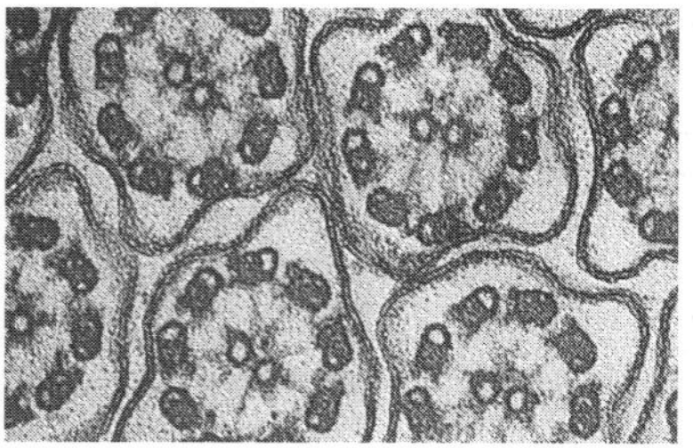

**Figure 3.23**
TEM of section through cilia showing 9:2 structure

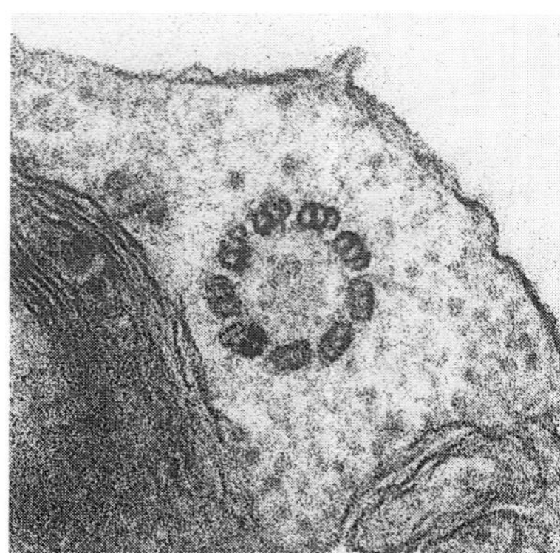

**Figure 3.24**
TEM of basal body of cilium in transverse section

way, and for this reason the cilia and flagella of eukaryotes are sometimes called **undulipodia**. With a width of about 200 nm they are just visible under the light microscope.

Cilia (singular, cilium) are relatively short with distinct propulsive and recovery phases to their beat, pushing the water parallel to the plasma membrane (Figure 3.21). There are usually large numbers per cell, each beating slightly out of phase with its neighbour in a **metachronal rhythm**. The co-ordination of the beat of neighbouring cilia implies the existence of some form of communication system between them.

In very small organisms such as *Paramecium* (a common inhabitant of ponds and ditches) cilia are used for swimming and for creating feeding currents in the water. In larger animals they are used to push fluid over stationary surfaces, such as the gills of bivalve molluscs and the lining of the bronchial tubes of mammals.

Flagella (singular, flagellum) are longer than cilia and are typically present in much smaller numbers (Figure 3.22). They usually develop several waves simultaneously, pushing against the water along the axis of the flagellum. They most often occur as propulsive organelles of single-celled organisms, and in the sperm tails of animals.

One of the most striking things about cilia and flagella is the fact that in the cells of organisms as diverse as lizards and seaweeds, they all have a ring of nine pairs of **microtubules** surrounding two central ones (Figure 3.23). The mechanism of bending is complicated, but involves the two members of each outer pair sliding past each other in a co-ordinated sequence.

At the base of a cilium or a flagellum is a cylindrical **basal body** (Figure 3.24). In cross section this resembles a cilium, except that the peripheral tubules are in triplets rather than pairs and there are no central tubules.

The basal body appears to be a kind of organising centre for the formation of microtubules; if a cilium or flagellum is severed, new microtubules grow out from the basal body.

## The centrosome

Just outside the nucleus is a small, clear area of cytoplasm called the **centrosome**, containing two tiny cylindrical granules called **centrioles**. These always lie at right angles to each other and are similar in structure to the basal bodies of cilia (Figure 3.25). The centrosome seems to be involved in the organisation of microtubules (see below), such as the spindle fibres which are responsible for chromosome movement during cell division. However, centrioles are absent in organisms lacking a flagellated stage in the life cycle, so in these organisms microtubule organisation is independent of centrioles.

Centrosomes occur in animal cells and in plants with flagellated gametes such as mosses and ferns. They are absent from the cells of seed-bearing plants in which the male gametes lack flagella.

## The cytoskeleton

Changing and maintaining cell shape are the twin functions of a complex network of proteins called the **cytoskeleton** (Figure 3.26). If the term ‘cytoskeleton’ conveys an impression of something static like the girders of a building, nothing could be further from the truth. The cytoskeleton is actually in a highly dynamic state, its fibres being simultaneously assembled and broken down in different parts of the cell.

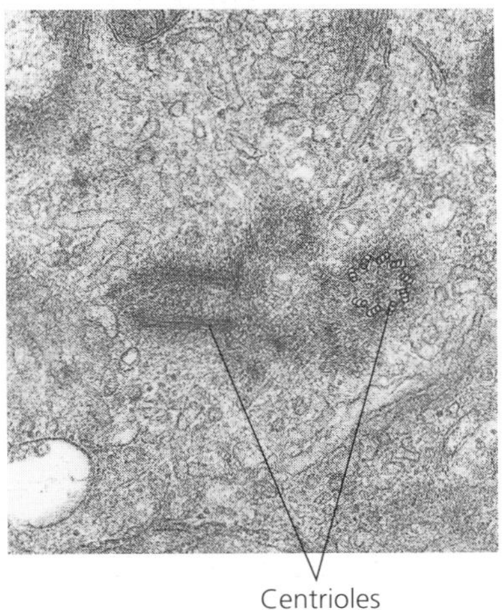


**Figure 3.25▲**
TEM of centrioles

**Microtubules** form the threads of the spindle during cell division, and are an important component of cilia and flagella. They are also involved in the movement of mitochondria and chloroplasts. In plant cells they direct the formation of the intercellular material (middle lamella) between newly-divided cells. They also determine the orientation of new microfibrils as a cell wall grows, thus indirectly determining the shape of the cell (Chapter 29).

Microtubules consist of tubes made up of chains of a globular protein called **tubulin** (Figure 3.26). Each tubulin molecule is actually a *dimer*, consisting of two slightly differing subunits (α and β tubulin). Microtubules are continually assembling and reassembling; on average half the microtubules in a cell are assembled and reassembled every ten minutes. The growth of microtubules is organised by the centrosome from which microtubules appear to radiate. Under appropriate conditions (including the presence of $Mg^{2+}$ ions and GTP, a chemical relative of ATP), microtubules assemble spontaneously from their tubulin subunits. Microtubule formation is inhibited by the drug colchicine (extracted from *Colchicum*, the autumn crocus).

**Figure 3.26►**
Part of a microtubule showing tubulin subunits

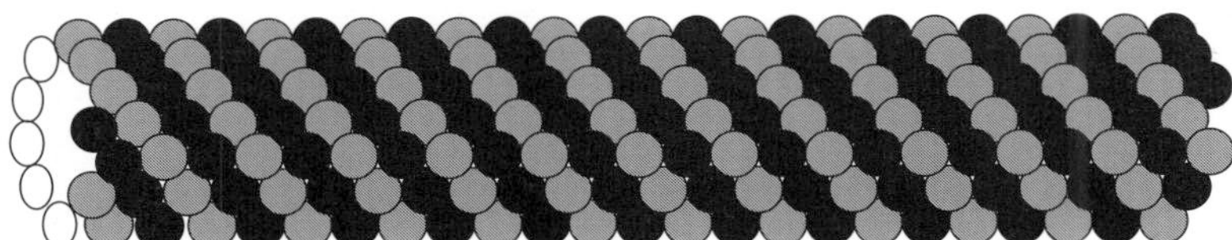

**Microfilaments** occur in all eukaryotic cells as a network just beneath the plasma membrane, forming the **cell cortex** (Figure 3.27). They consist of the protein **actin** which is best known for its role in muscle (Chapter 19). Microfilaments are involved in amoeboid movement, and in the formation of a cleavage furrow in dividing animal cells. They also help maintain the shape of microvilli.

**Figure 3.27►**
Microfilaments consist of two-strands of actin wound round each other in a helix (A). They form a network just below the plasma membrane (B), and also form the support for microvilli (C). In conjunction with myosin, they form contractile bundles responsible for cytoplasmic cleavage in animal cells (D)

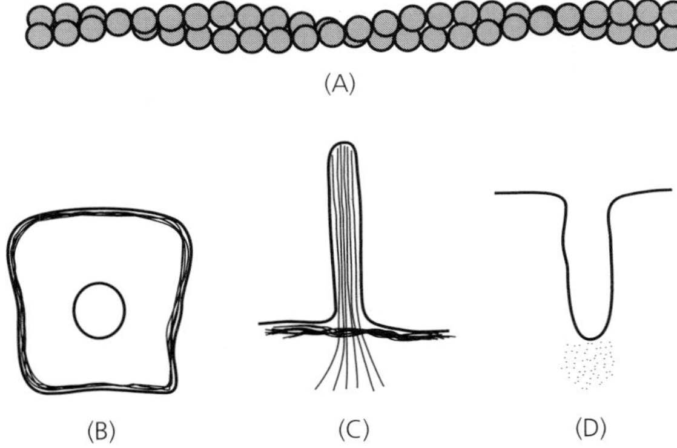


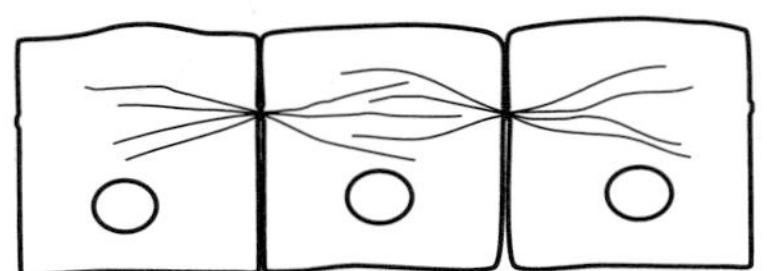

**Figure 3.28▲**
Intermediate filaments. These are formed from a variety of types of fibrous proteins. On the kind shown, keratin fibres extend between intercellular junctions and serve to bind neighbouring epithelial cells together

**Intermediate filaments** are rope-like fibres which occur in most animal cells, but not in plants (Figure 3.28). Their chief function seems to be to help resist distortion. The cells of animal epithelial tissues, for example, are held together by bundles of intermediate filaments made of **keratin** which link neighbouring cells.

## The plant cell wall

Plant cells have a **cell wall** immediately outside the plasma membrane. Its framework consists of a mesh of cellulose **microfibrils**, which can be seen under the electron microscope (Figure 3.29). In unmodified cell walls the spaces between the microfibrils are normally wide enough for most molecules to pass through.

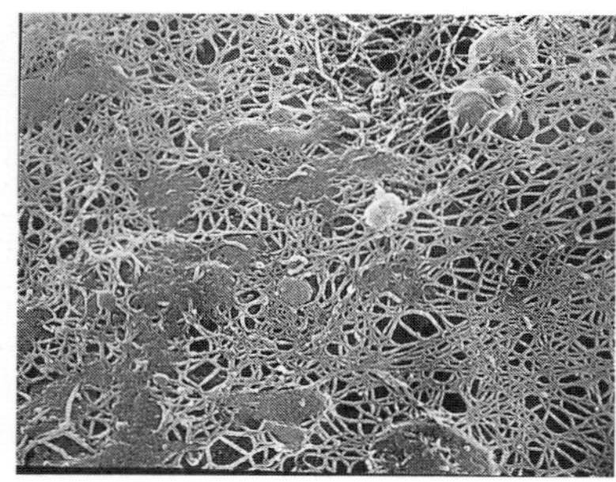

**Figure 3.29▲**
SEM of the surface of a protoplast (cell with wall removed) showing cellulose microfibrils growing over plasma membrane

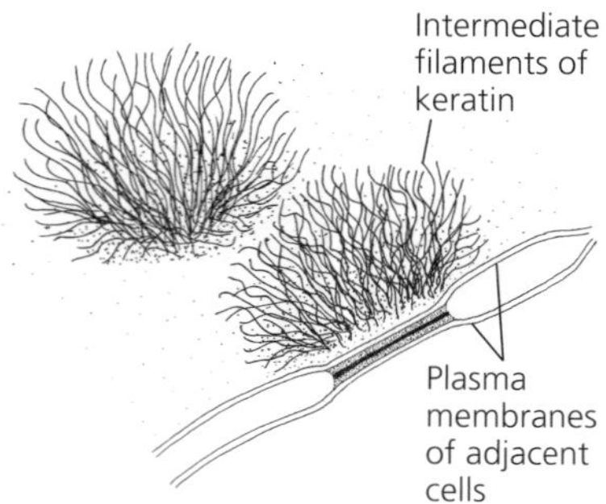


**Figure 3.30▲**
Drawing of a desmosome which acts like a 'rivet' connecting adjacent cells together via clusters of intermediate filaments (not shown in the lower cell)

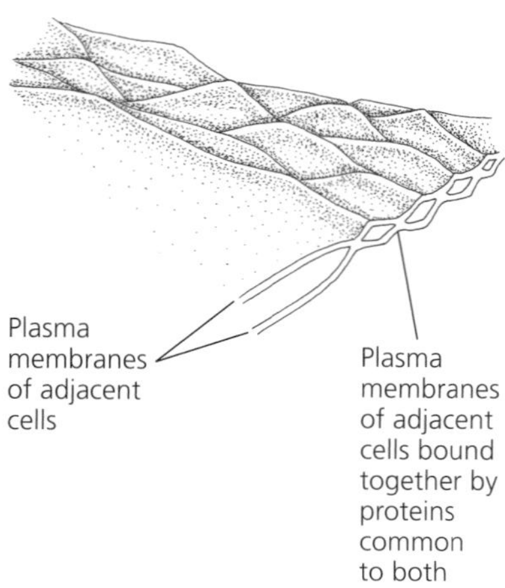


**Figure 3.31▲**
Drawing of a tight junction

Besides cellulose, two other carbohydrates and also some protein form significant components of the cell wall. **Hemicelluloses** are a heterogeneous group of polysaccharides consisting of a variety of β-linked sugars with occasional short side branches. Hemicelluloses form a network of cross links between the microfibrils. **Pectins** are a group of branched polysaccharides that are particularly characteristic of the **middle lamella** (the intercellular cement that binds adjacent cells together) and also the primary wall. Pectins are acidic polysaccharides that are normally present as calcium pectate (the main constituent of jelly).

Whilst the cell is still growing and the wall is expanding, the cell wall is relatively extensible and is called the **primary wall**. Microfibrils cannot stretch, and the extension of the primary wall of a growing plant cell is believed to involve microfibrils sliding past each other. Once the cell has finished enlarging, a **secondary wall** is laid down inside the primary wall. The microfibrils of the secondary wall are thought to be more extensively cross-linked than those of the primary wall and are thus less able to slide past each other, so considerable **turgor pressure** may be built up.

In many plant cells, other substances are deposited between the cellulose microfibrils. One of these substances is **lignin** which binds the cellulose microfibrils together. By preventing the microfibrils from slipping past each other, lignin makes cell walls stiff and resistant to buckling. As a result the cell does not cave in easily when its contents are under tension – an important property in the water-conducting cells of land plants. By filling the spaces between microfibrils, lignin also greatly reduces the permeability of the cell wall. Another substance that may be deposited in cell walls is **cutin**, a water-resistant material that is deposited in the outer walls of the epidermis of leaves and stems rendering them relatively impermeable to water.

## No cell is an island

With the exception of single-celled organisms, cells are only independent units in a limited sense. Not only is each cell in contact with several others, but many cells are actually attached to their neighbours by various kinds of connection, briefly discussed below.

**Desmosomes** are points connecting the cytoskeleton of adjacent animal cells (Figure 3.30). This is particularly important in epithelial tissues in which adjacent cells are bound together to form sheets.

**Tight junctions** play an important part in the active transport of materials across animal cells (Figure 3.31). By sealing the narrow cleft between cells they prevent substances that have been transported across the cell from diffusing back between cells (Chapter 6). Tight junctions also restrict the free movement of membrane proteins in the lipid bilayer. This is important in the absorption of nutrients from the small intestine and kidney tubules (Chapters 24 and 26). In the case of the small intestine, glucose is transported into the cell on the side facing the intestinal cavity and out of the cell on the side facing the blood. These two processes involve different transport proteins, each of which must be restricted to the appropriate side of the cell and thus must be prevented from moving freely.

**Gap junctions** are channels that connect the interiors of adjacent animal cells, thus allowing small molecules to move between neighbouring cells (Figure 3.32). They are particularly important in cardiac muscle fibres in which they form junctions which allow electrical signals to pass directly from one muscle cell to its neighbours (Chapter 22).

**Plasmodesmata** are fine strands of cytoplasm passing through the walls of adjacent plant cells. Each plasmodesma is about 20–40 nm thick

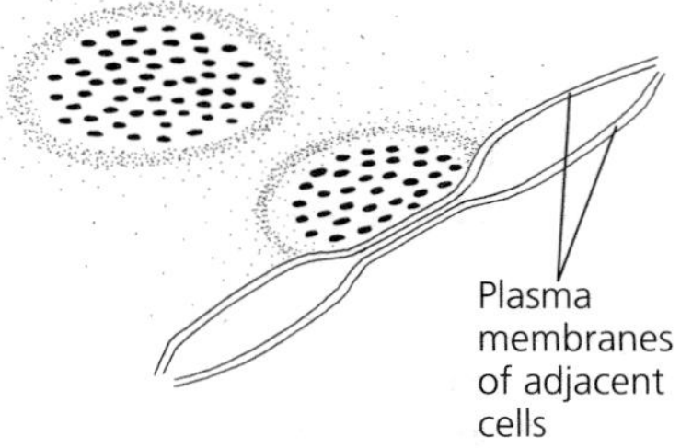


**Figure 3.32▲**
Drawing of a gap junction, consisting of a cluster of pores, each pore being formed from an aggregation of proteins

and consists of a cylinder of plasma membrane surrounding an extension of the smooth endoplasmic reticulum (Figure 3.33). Plasmodesmata are present from the time a newly-formed cell is separated from its 'sister', and originate as thin channels where new cell wall is not laid down. Like gap junctions, plasmodesmata allow free passage of materials between adjacent cells.

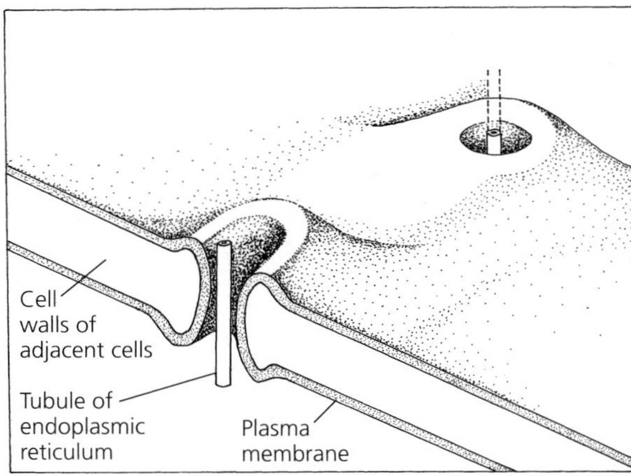


**Figure 3.33►**
Drawing of two plasmodesmata, one shown in section

## Animal and plant cells compared

Table 3.1 summarises the similarities and differences between animal and plant cells. You will notice that most of the differences are visible under the light microscope, and that most of the features that can only be seen under the electron microscope are present in both animals and plants.

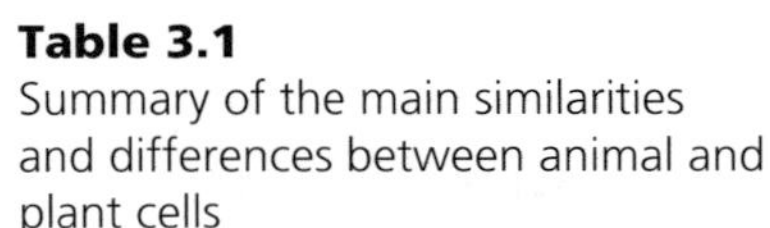

**Table 3.1**
Summary of the main similarities and differences between animal and plant cells

| Structure | Visible using | Animals | Plants |
|---|---|---|---|
| Cell wall | LM | – | + |
| Plastids | LM | – | + |
| Vacuole | LM | – | + |
| Plasma membrane | EM only | + | + |
| Endoplasmic reticulum | EM only | + | + |
| Golgi bodies | LM (just) | + | + |
| Lysosomes | EM only | + | + |
| Mitochondria | LM (just) | + | + |
| Flagella | LM (just) | + | some |
| Centrosome | LM (just) | + | some |

LM, light microscope; EM, electron microscope; –, absent; +, present

## Prokaryotic organisation

As mentioned earlier, bacteria have a much simpler organisation than eukaryotes and also have some special features of their own (Figure 3.34(a)).

The most important features of prokaryotes are set out below.

- They are much smaller, rarely exceeding 2 μm in width, whereas few eukaryotic cells are less than 5 μm in size.

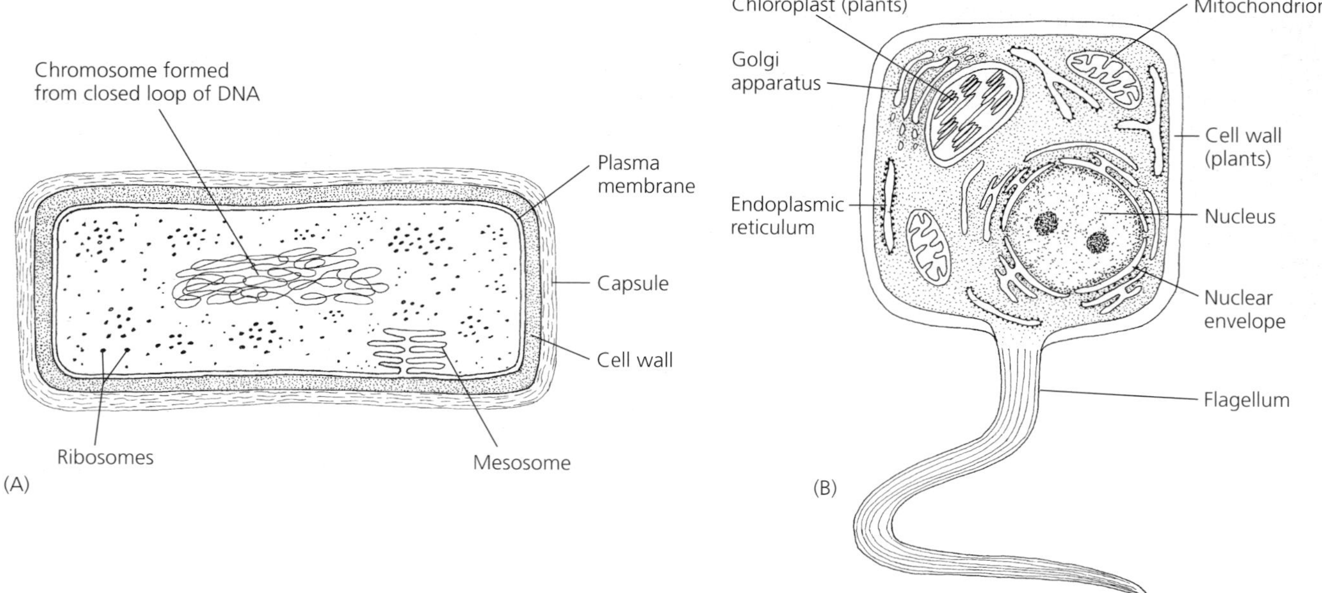


**Figure 3.34**
Comparison between prokaryotic cell (A) and eukaryotic cell (B)

- They are much simpler internally, lacking endoplasmic reticulum, Golgi bodies, lysosomes, mitochondria and plastids (the absence of the latter two would be expected if they are themselves prokaryotes).
- The DNA is restricted to a region called the **nucleoid**, but there is no nuclear envelope and hence no clearly defined nucleus.
- Their chromosomes form a closed loop and lack the substantial protein component of eukaryotic chromosomes. Many bacteria also have small accessory chromosome loops called **plasmids**. These carry genes for – amongst other things – resistance to antibiotics.
- The ribosomes of prokaryotes are slightly smaller than those of eukaryotes.
- Part of the plasma membrane is invaginated to form a structure called the **mesosome**, the function of which is uncertain.
- Though many bacteria have flagella, these are quite different from those of eukaryotes.
- Bacteria have a **cell wall** outside the plasma membrane, but it is very different from the cell wall of plants. It consists of a very complex polymer called **peptidoglycan**, the subunits of which contain sugars and amino acids, some of which are D-amino acids (which are otherwise rare in nature).
- Many bacteria have a **capsule** outside the cell wall. In pathogenic (disease-causing) bacteria the capsule helps resist attack by white blood corpuscles.

## Viruses: non-living or living exceptions to the Cell Theory?

Viruses are particularly interesting because they seem to straddle the boundary between living and non-living matter. People still argue about whether they are living or not.

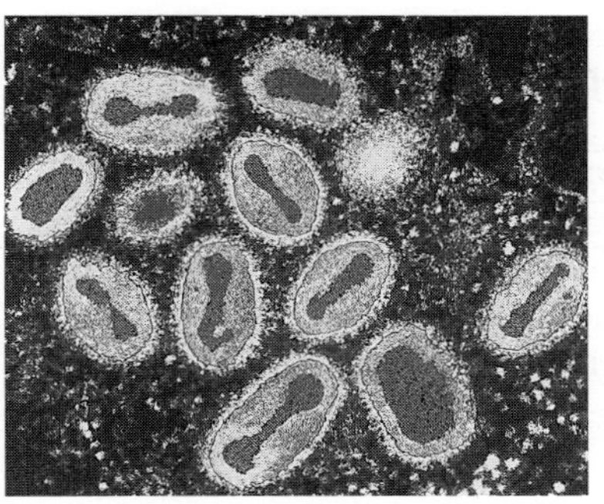

**Figure 3.35**
Electron micrograph of smallpox

## Discovery

Our knowledge of the nature of viruses can be traced to 1982 when Dimitri Ivanowski demonstrated that when an extract of leaves infected by tobacco mosaic disease was passed through a porcelain filter, the filtrate could induce the disease in healthy tobacco plants. Ivanowski believed that the cause of the disease was a toxic substance. A few years later another Russian, Martinus Beijerinck, showed that the agent could be passed from one plant to another, and then to another and so on, apparently without limit. The conclusion was inevitable – whatever the agent was, it could reproduce.

Since the porcelain filter was fine enough to retain all bacteria, the infective agent was clearly smaller than any bacterium. This new kind of pathogen was called a **filterable virus** (*virus* is Latin for poison).

Many other diseases of both plants and animals have been shown to be caused by viruses. Some viruses, called **bacteriophages** (often shortened to **phage**), even attack bacteria.

## Chemical composition

An early discovery that set viruses apart from living things was the fact that they could be *crystallised*. When analysed chemically the crystals are found to contain protein and nucleic acid (for a fuller treatment of DNA and RNA see Chapters 10 and 12). The nucleic acid can be either **DNA** or a close chemical relative, **RNA**, but never both (unlike cells). Many of the more common viruses, such as the influenza virus, contain double-stranded RNA.

## Size

Virus particles vary in diameter from 10 nm to 300 nm (1 nanometre is a millionth of a mm, or $10^{-9}$ metres). The largest viruses, such as the smallpox virus, are just visible under the light microscope (Figure 3.35). With the invention of the electron microscope viruses could be seen and their sizes measured.

## Structure

A virus particle, or **virion**, consists of a core of nucleic acid surrounded by a protein coat called a **capsid** (Figure 3.36). The individual protein molecules of the capsid are called **capsomeres**. It is the regular stacking of the capsomeres that allows viruses to be crystallised. All four possible types of nucleic acid occur in viruses: double-stranded DNA, single-stranded DNA, double-stranded RNA and single-stranded RNA.

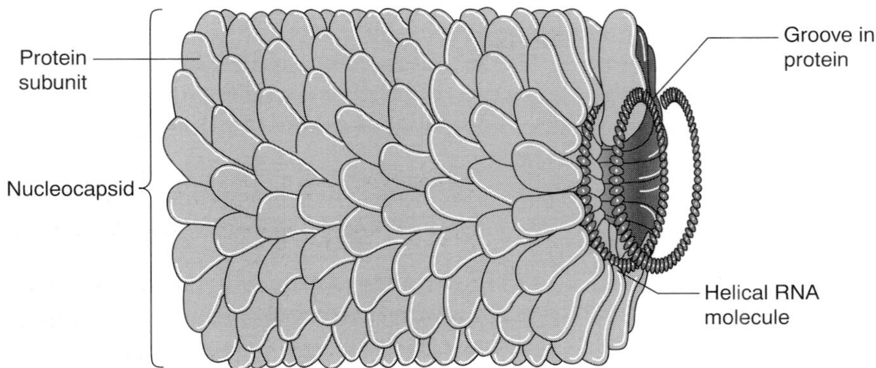


**Figure 3.36**
Drawing of a virion

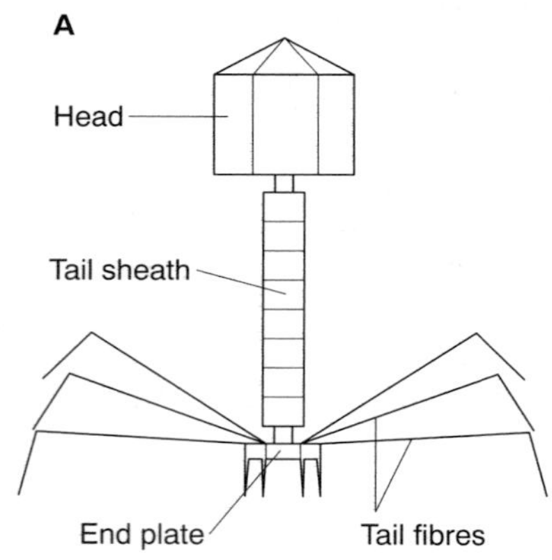


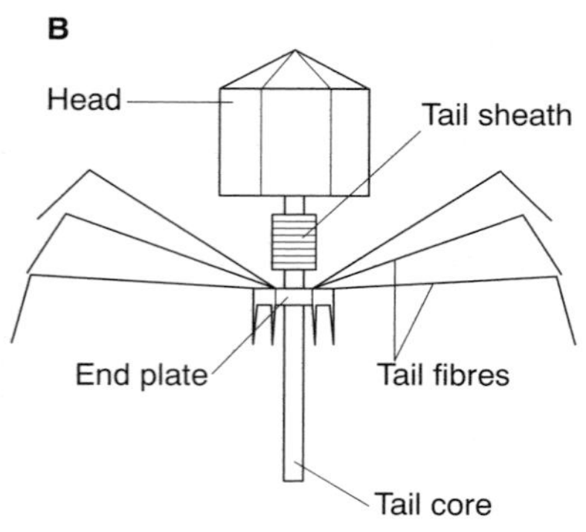


**Figure 3.37▲**
Diagram of a $T_2$ or $T_4$ phage

**Figure 3.38►**
Reproductive cycle of a bacteriophage

Besides nucleic acid and protein, viruses such as pox, herpes and influenza are surrounded by a lipid **envelope** derived from membranes of the host cell.

Bacteriophage particles have an even more complex structure. The $T_4$ phage, which attacks the human colon bacterium *Escherichia coli*, consists of a **head** containing DNA, a **contractile sheath**, and a tail attached to six **tail fibres** (Figure 3.37). The role of the tail fibres is to attach the phage particle to the surface of the host cell.

## Reproduction

A virus can only reproduce inside a host cell, which provides raw materials, energy and catalysts – in fact everything except the information for making the virus. Viral reproduction takes place in six stages (Figure 3.38):

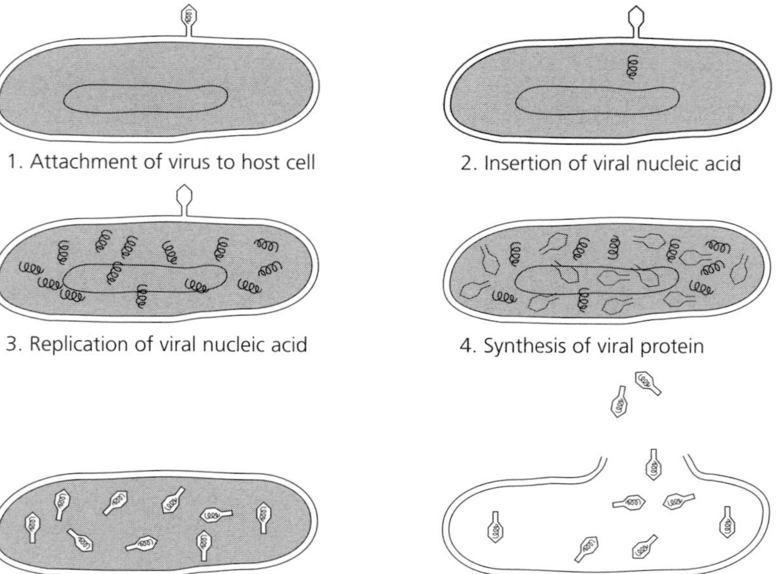


- *contact with the host cell.* This involves combination between the protein exterior of the virus and specific protein molecules in the host plasma membrane. These carry out normal cell functions, but have been used by the virus as a means of 'recognising' the host cell. Most receptors are limited to certain cells which explains why, for example, polio virus can only attack certain cells in the spinal cord and cells lining the gut.
- *entry into the host cell.* In animal viruses this occurs by endocytosis. The bacterial cell wall makes this impossible, and the nucleic acid is injected by contraction of the viral sheath.
- *replication of the nucleic acid.* The precise manner in which this occurs depends on whether the nucleic acid is RNA or DNA, and whether it is single-stranded or double-stranded.
- *production of the capsomeres of the protein coat.*
- *assembly of the capsomeres round the nucleic acid to form new virus particles.* The combination between capsomeres and nucleic acid occurs spontaneously. This is one of many examples of **self-assembly**, in which molecules spontaneously aggregate to form larger structures.
- *release of virus particles by lysis (disintegration) of the host cell.*

## Retroviruses

Viruses such as HIV and tumour viruses deliver single-stranded RNA into the host cell, together with the enzyme **reverse transcriptase**. Using the RNA as a template, the enzyme then makes a complementary strand of DNA which then is replicated to form double-stranded DNA. This then follows either of two pathways: it may make new viral RNA and viral particles, or it may be incorporated into the host DNA. In the latter case it behaves like a normal part of the host genome, replicating with it when the host cell divides. At some later stage the cell may become transformed into a cancer cell.

These viruses are particularly interesting because viral RNA is used as a template to make DNA – the reverse of the normal process – which is why they are called **retroviruses**.

## QUESTIONS

**1** Which of the following can be seen using:
**a)** the best light microscopes **b)** an electron microscope?
chloroplast, nucleus, mitochondrion, Golgi body, skin cell, lysosome, haemoglobin molecule, cell membrane, cell wall.

**2** A microscope has a ×4 low power objective and a ×40 high power objective. If the low power field of view is 3 mm across, what is the diameter of the high power field of view?

**3** Arrange the following in *decreasing* order of size:

**A** haemoglobin molecule
**B** mitochondrion
**C** liver cell
**D** ribosome
**E** nucleus
**F** glucose molecule

**4** Below is a list of structures found in cells.
**i)** plasma membrane
**ii)** endoplasmic reticulum
**iii)** mitochondrion
**iv)** nuclear envelope
**v)** cellulose wall
**vi)** large vacuole
**vii)** Golgi body
**viii)** ribosome
**ix)** cilium
**x)** plastid
**xi)** centriole
**xii)** lysosome

For each of the above, write
**A** if it is present in all cells,
**B** if it is absent in eukaryotes,
**C** if it is absent in prokaryotes,
**D** if it is present in flowering plants but not in animals,
**E** if it is present in animals but not in flowering plants.

**5** Match the structures **(i)–(x)** to the functions **A–L**.
**i)** mitochondrion
**ii)** chloroplast
**iii)** plasma membrane
**iv)** nucleus
**v)** lysosome
**vi)** cilium
**vii)** ribosome
**viii)** centriole
**ix)** Golgi body
**x)** cell wall
**xi)** nucleolus
**xii)** vacuole

**A** manufactures ribosomal RNA
**B** gives mechanical support to cell
**C** converts chemical energy to mechanical energy
**D** makes ATP using energy from oxidation of organic matter
**E** links amino acids into long chains
**F** contains the genetic material
**G** reduces distance between chloroplasts and air spaces
**H** contains digestive enzymes used in intracellular digestion
**I** organises the fibres of the spindle in animal cells
**J** modifies proteins after their production
**K** regulates movement of substances into and out of cells
**L** converts light to chemical energy

CHAPTER

# 4

# BIOLOGICAL CATALYSTS – ENZYMES

## LEARNING OBJECTIVES

By the time you have completed your study of this chapter you should be able to:

- explain the conditions that are necessary for chemical reactions to occur.
- understand what catalysts do and what they cannot do.
- explain how enzymes differ from inorganic catalysts.
- describe how enzymes interact with the molecules on which they act.
- explain the effect of temperature and pH on enzyme activity in terms of the properties of globular proteins.
- interpret the effect of substrate concentration on enzyme activity in terms of the interaction between enzyme and substrate.
- explain how the induced fit mechanism depends on the weakness of forces holding globular proteins in shape.
- give examples of the various kinds of enzyme cofactor.
- explain the various ways in which enzymes can be inhibited.
- explain how enzymes are increasingly being used in industry.

All life processes such as the contraction of a muscle, the opening of a flower and the emission of light by a glow worm, are the direct or indirect result of chemical reactions occurring within cells. Most cells are a hive of chemical activity, with hundreds or even thousands of different chemical reactions going on at any given moment. Collectively these chemical processes are called **metabolism**. The chemicals produced in metabolic reactions are called **metabolites**.

Metabolic reactions require the presence of **enzymes**. Enzymes are **catalysts** – that is, they speed up chemical reactions without being used up.[1] The reason why they are not used up is that although they take part in the reaction, they are unchanged at the end. Since each enzyme molecule can be recycled millions of times over, enzymes are effective in minute concentrations.

Enzymes differ from other catalysts in several important ways:

- they work thousands of times faster than other catalysts. For example, to break down starch into sugar using acid as a catalyst, it has to be boiled for many hours. The enzyme in saliva (salivary amylase) does a similar job in a few minutes at room temperature.

[1] *Enzyme* means 'in yeast', and was coined because much of the early research into biological catalysis was done using yeast.

- they are *specific*, meaning that they only catalyse one kind of reaction.
- they are very sensitive to changes in pH.
- they are easily inactivated by heat.
- they are very sensitive to inhibition by certain chemicals.

These peculiarities of enzymes can all be explained by the fact that (with the exception of certain recently-discovered RNA enzymes or 'ribozymes') all enzymes are globular proteins (see below).

### From vitalism to mechanism

For thousands of years it has been known that after crushing, grapes can be used to make wine. How this process of **fermentation** occurred was, of course, a mystery, but it was known that the sediment (yeast) that accumulated on the bottom played an important part, for it could be used to accelerate a fresh cycle of fermentation.

With the development of the microscope biologists could watch yeast divide, and it became clear that yeast was living. Just how living cells performed the conversion of sugar to ethanol became the subject of heated argument.

In one camp were the chemists, led by von Liebig, who maintained that fermentation was the result of chemicals produced within the yeast cells. On the other side were the biologists, led by Louis Pasteur. In their view, fermentation was not explicable in chemical terms, but could only occur within intact living cells, and was in some way bound up with a mysterious 'vital force'.

The problem was resolved in 1897 by Eduard Buchner. He succeeded in obtaining a juice from ground up yeast cells, and showed that this liquid – which contained no cells – could convert sugar to ethanol. Whatever the mechanism, it could evidently occur outside living cells.

The difference between biological and chemical processes became further blurred when, in 1926, James Sumner extracted the enzyme urease from jack beans and succeeded in purifying it in crystalline form. On analysis, he pronounced that the enzyme consisted of pure protein. Although this was too much for some biologists, other enzymes were also purified and found to be protein. Some of the mystery was thus removed – enzymes had been shown to be mere molecules that could be manipulated in a test tube.

## Enzymes are 'fussy'

Although specificity distinguishes enzymes from other catalysts, the degree of specificity varies considerably. One of the least specific enzymes is lipase, which will bring about the hydrolysis not only of fats but of other esters too. At the other extreme are enzymes which are so specific that they will only act on one substance. Urease, for example, will only act on urea. Enzymes that act on optically-active substances (such as sugars and amino acids) act only on the natural isomer (Chapter 3).

## Naming enzymes

When enzymes were first discovered and identified, many were named according to the **substrate**, or substance they act on, the name ending with the suffix *-ase*. Thus the enzyme that hydrolyses maltose to glucose was called **maltase**, and lipid-hydrolysing enzymes were called **lipases**, and so on.

About 2000 enzymes are now known, and it has become necessary to devise a systematic way of naming and classifying them. Some of these technical names are rather a mouthful. For instance, the enzyme that digests lactose (milk sugar) was first called lactase, but is now more properly called β-galactosidase.

## Types of enzyme

Though thousands of enzymes are known, they all fall into six general categories:

- **hydrolases** catalyse hydrolytic reactions. All digestive enzymes are hydrolases.
- **oxidoreductases** catalyse the transfer of oxygen, hydrogen or electrons. Some of the most important of these enzymes are involved in respiration and in photosynthesis (Chapters 5 and 33).
- **isomerases** catalyse the rearrangement of atoms within the substrate to produce an isomer. For example, phosphoglucomutase catalyses the conversion of glucose-1-phosphate to glucose-6-phosphate.
- **ligases** catalyse the linking together of two molecules with the hydrolysis of ATP. An example of this is the linking of an amino acid to transfer RNA during protein synthesis (Chapter 11).
- **transferases** catalyse the transfer of a group from one molecule to another, such as the transfer of an amino group from an amino acid to a keto acid.
- **lyases** catalyse the partial breakdown of a molecule by non-hydrolytic removal of a group, for example the removal of a carboxyl group from pyruvate to form ethanal in the fermentation of sugar (Chapter 5).

## What makes a chemical reaction go?

Though enzymes are special catalysts, they are not magicians. Enzymic reactions obey the same chemical laws as do other chemical reactions. For any reaction to occur, two conditions must be satisfied whether or not a catalyst is present:

- the molecules must collide with a certain minimum energy, called the **activation energy**.
- the energy of the products must be less than the energy of the raw materials – in other words, the reaction must be **exergonic**, with a *net release of energy*.

These two ideas can be illustrated by the 'ball and hill' model shown in Figure 4.1(a), which shows the hydrolysis of sucrose to glucose and fructose. The raw materials and products are represented by the balls on the two sides of the hill, which represents the activation energy. In this scheme the ball representing the raw materials can take three routes to the other side. In the uncatalysed reaction the activation energy is high, in the presence of acid as catalyst the activation energy is lower, and with the enzyme sucrase it is lower still. Whichever route is taken, the *net* energy change is the same.

Because catalysts do not affect the overall energetics of reactions they cannot affect the direction a reaction takes, as shown in Figure 4.1(b).

When the difference in energy between the raw materials and end products is very small, an equilibrium is set up which is not changed by a

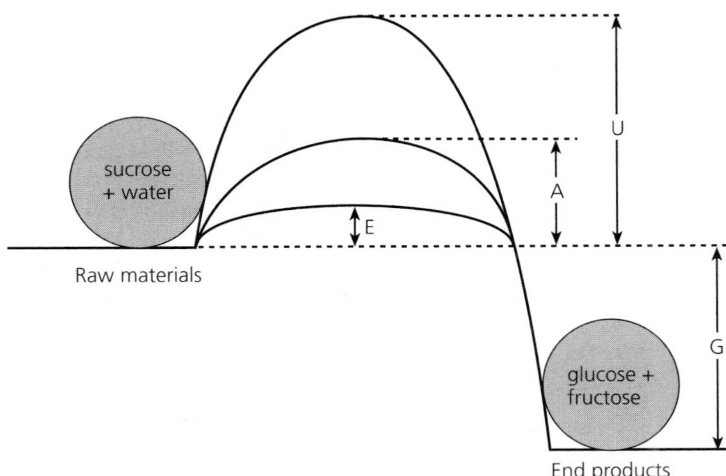


**Figure 4.1**
'Ball and hill' model showing energy changes in a chemical reaction such as the hydrolysis of sucrose. The ball on the left represents the reactants (sucrose and water) and that on the right represents the products (glucose and fructose). The hill represents the activation energy with no catalyst (U), with acid as catalyst (A), and with enzyme as catalyst (E). G represents the overall energy yield of the reaction. A reaction will only occur if the products have lower energy than the reactants

catalyst. Adding a catalyst makes no difference to the composition of the final equilibrium mixture – by lowering the energy barrier it simply *speeds up the attainment of equilibrium.*

The important thing to notice about all this is that catalysts *do not add energy to a system*. They do not make impossible reactions possible – *they simply make possible reactions easier*. An analogy is to liken the action of a catalyst to the addition of oil to a rusty nut – it may help you to turn it, but it won't provide the necessary force.

## How enzymes work

Several kinds of evidence (see below) point to the fact that an enzyme temporarily combines with its substrate to form an **enzyme–substrate complex** (Figure 4.2). Since most enzyme molecules are much larger than their substrate molecules, only a small part of an enzyme can combine with its substrate. This special part of the enzyme molecule is called the **active site**, and it forms a cleft or depression in the surface of the enzyme. The surface of the active site is quite small, usually involving only 3–12 amino acids.

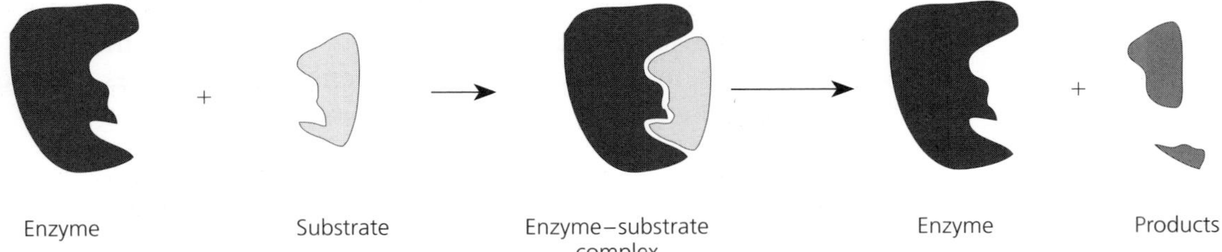


**Figure 4.2▲**
Formation of enzyme–substrate complex

For many years it was thought that the substrate molecules fitted into the active site like a key fits into a lock. This would neatly explain why enzymes are so specific – only substrate molecules with the right shape could fit into the active site. In recent years however it has become apparent that the shape of the active site is not rigid, but holds the substrate molecule in a flexible, **induced fit** (Figure 4.3).

**Figure 4.3►**
Induced-fit mechanism of enzyme action

Evidently the secondary and tertiary (and in some cases quaternary) structures of an enzyme need to be able to move slightly. The weakness of the forces holding the enzyme molecule in shape are thus necessary in order that it may undergo the change of shape necessary for it to work. Unfortunately it is this very weakness that renders enzymes easily inactivated by heat (see below).

The effect of the active site is not merely to cradle the substrate molecule – it also makes it more reactive. The enzyme–substrate complex is then converted to an **enzyme–product complex**. This then breaks away from the active site and is free to combine with another substrate molecule.

The maximum rate at which an enzyme can work is expressed as the **turnover number**. It is equal to the maximum number of molecules of substrate that each enzyme molecule can convert into product each second, under optimum conditions. It is a slow enzyme that converts only a hundred substrate molecules per second. Carbonic anhydrase, which catalyses the combination between carbon dioxide and water, has a turnover number of 600 000 per second!

## Factors that affect the activity of enzymes

Enzymes are extremely sensitive to conditions. Some of the most important of these are dealt with in the following sections.

### Substrate concentration

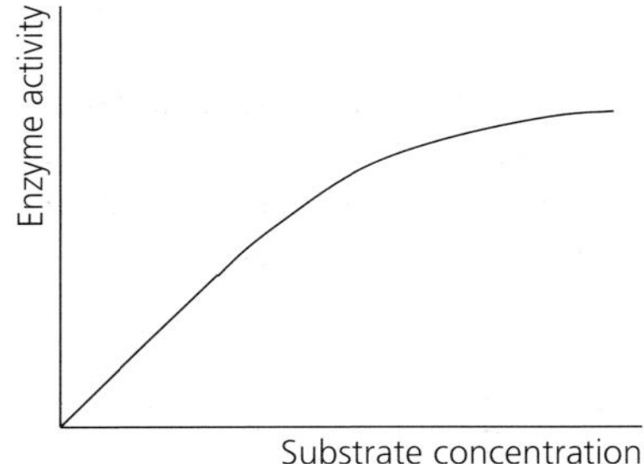


**Figure 4.4**
Effect of substrate concentration on enzyme activity

One might expect that the more concentrated the substrate, the faster the enzyme would work. So it does, but only at lower substrate concentrations (Figure 4.4). At higher concentrations a change in substrate concentration makes no difference. Clearly there is a limit to how fast an enzyme molecule can work.

The fact that an enzyme becomes 'saturated' at higher substrate concentrations provided some of the earliest evidence that enzymes combine with their substrates. Just as only one key can fit into a lock at one time, the active site can only process one substrate molecule at a time. Nevertheless, under favourable conditions, most enzymes can process thousands of substrate molecules every second.

**CLEAR THINKING**

**In an essay on enzymes, a student drew the graph shown, with the caption: 'The effect of enzyme concentration on enzyme activity'. In the essay, the student made two statements. Firstly, as long as there is excess substrate, enzyme activity is directly proportional to enzyme concentration (straight, solid line). Secondly, if substrate concentration is limited, the graph tails off (curved, dotted line).**

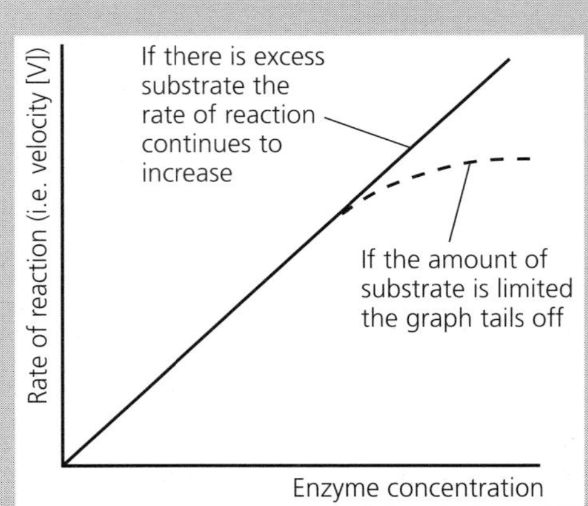


Enzyme 'activity'

**CLEAR THINKING**

**These statements and the graph raise two criticisms:**

- **we need to be clear about what we mean by 'activity'. If we talk about the activity of typists, we normally mean the output per typist – doubling the number of typists would not normally be considered to be an increase in activity. In the same way, enzyme activity means the number of product molecules produced per unit time *per unit amount of enzyme*. Thus, whilst substrate concentration does affect enzyme activity, enzyme concentration does not. We therefore need to distinguish between enzyme activity, which is a measure of the 'productivity' of each enzyme molecule, and the reaction rate, which is the total quantity of product formed per unit time.**
- **the graph could only level off in the way shown if enzyme molecules were in excess of the substrate – in other words, if enzyme molecules were *competing* for substrate. In this case, each enzyme molecule would only have to react once for all the substrate to be converted into product. Since the time taken for this to happen would be a minute fraction of a second, the reaction rate would be too fast to measure. Such a state of affairs could not happen under physiological conditions.**

## Temperature

Chemical reactions go faster at higher temperatures because molecules move faster and they also collide more violently. The effect of temperature on a reaction is expressed as the **temperature coefficient** or $\mathbf{Q_{10}}$. This is the number of times the rate of reaction increases for a 10°C rise in temperature, i.e.:

$$Q_{10} = \frac{\text{Rate of process at T + 10°C}}{\text{Rate of process at T°C}}$$

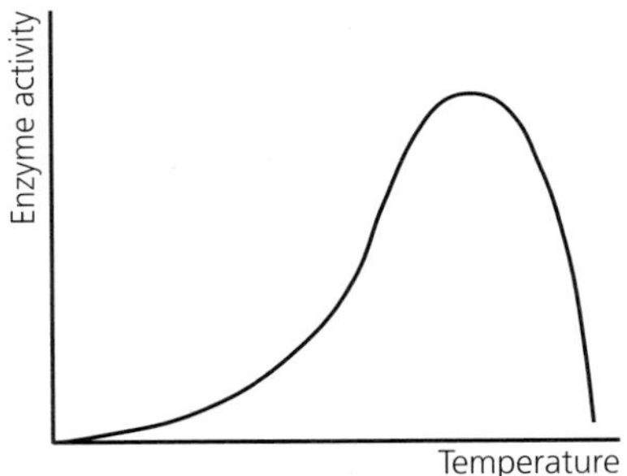


**Figure 4.5**
Effect of temperature on activity of salivary amylase

The $Q_{10}$ for most chemical reactions is about 2, which means that the rate of the reaction roughly doubles for every 10°C rise in temperature. At moderate temperatures enzymic reactions follow the same rule, but at higher temperatures the situation becomes more complicated.

Figure 4.5 shows the effect of temperature on the activity of salivary amylase which catalyses the hydrolysis of starch to maltose. The increase in activity with temperature only occurs at moderate temperatures. There is therefore an **optimum temperature** for enzyme activity (although the situation is more complicated than Figure 4.5 would suggest – see below).

Why are enzymes so unstable to heat? Firstly, we need to remember that the arrangement of the amino acid R groups at the active site depends on the precise way the polypeptide chain is coiled and folded on itself. The forces holding the molecule in this specific shape are *weak*. At higher temperatures molecular collisions and vibrations become violent enough to change the shape of the enzyme permanently. Most important of all, heat destroys the specific shape of the active site. As a result the enzyme is said to have been **denatured**.

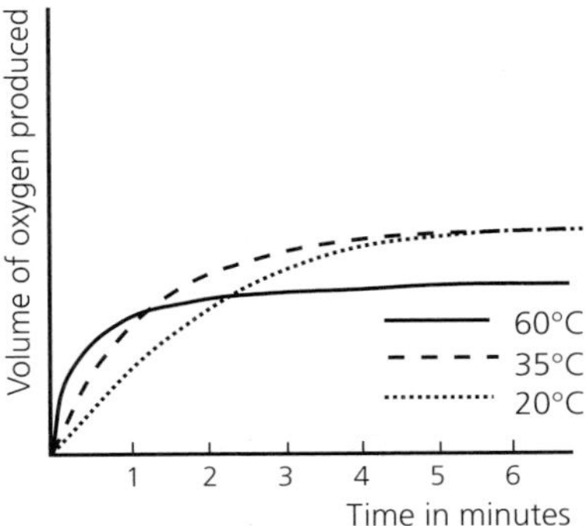


**Figure 4.6**
Effect of time on optimum temperature for enzyme activity

The temperature at which an enzyme works fastest is the optimum temperature. This sounds simple enough, but in a test tube the optimum temperature depends on the time over which the rate is measured, as Figure 4.6 shows.

Over the first minute more product is formed at 50°C than at 35°C. Hence if the rate of reaction is determined by measuring the amount of product formed during the minute after mixing, the optimum would be nearer to 50°C. At 50°C, the 'life expectancy' of a given enzyme molecule is shorter than at 35°C, but while it is functional, it is actually working faster.

In living cells the rate of denaturation of an enzyme must be balanced by its rate of synthesis, so stability is more important than rate of working. Organisms adapted to cold environments have enzymes with optimum temperatures as low as 5–10°C. Those adapted to hot climates have enzymes with much higher optima – in some hot springs bacteria flourish at 90°C or more. Endothermic ('warm-blooded') animals regulate their body temperatures at between 37°C and 40°C, and their enzymes work best at these temperatures. A rise of only a few degrees above the optimum may cause heat stroke or death.

These chemical effects explain why insects and other ectothermic ('cold-blooded') animals are active on warm sunny days and sluggish in cooler weather. Mammals and birds on the other hand have body temperatures which are maintained at, or very near, the optimum, allowing them to remain active in widely-varying temperatures.

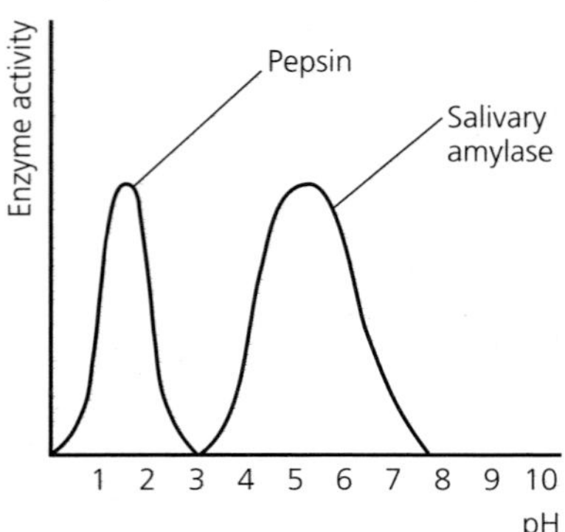


**Figure 4.7**
Effect of pH on enzyme activity

## pH

Most enzymes work inside the cells where they were produced. Since the pH inside the cytosol is normally about 7, it is not surprising that most enzymes work best at or near this pH value. Enzymes secreted into the gut operate outside the cells that make them, and may have different pH optima. Some work best under acid conditions; pepsin in gastric juice has an optimum pH of about 1.5, as do most lysosomal enzymes (Figure 4.7).

Why does pH affect enzymes? As with temperature, the answer lies in the forces holding the molecule in shape. Most globular proteins contain some amino acids with acidic (—COOH) or basic (—$NH_2$) R groups. Changing the pH alters the electrical charges on these R groups. This in turn changes the forces between them, and thus the shape of the molecule (see Chapter 3).

## Cofactors

Although some enzymes consist entirely of protein (e.g. pancreatic ribonuclease), in others the active site contains a non-protein substance called a **cofactor** (Figure 4.8).

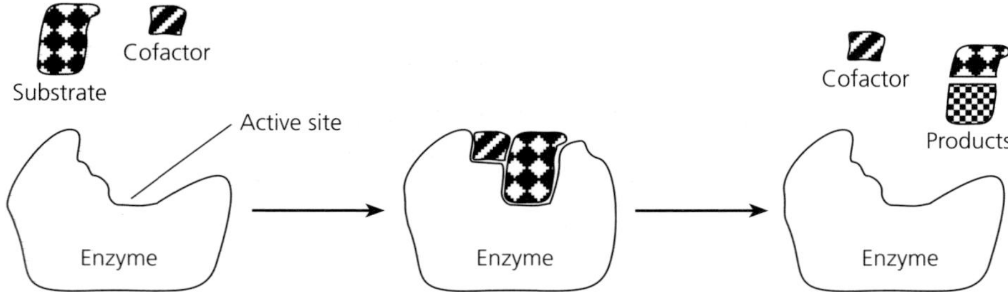


**Figure 4.8**
Role of a cofactor in enzyme activity

Cofactors may be simple metal ions or quite complex organic molecules. Some cofactors are loosely associated with the protein part of the enzyme, only combining with it whilst it is bound to the substrate. Other cofactors are permanently bound to the rest of the enzyme and are called **prosthetic groups**. Organic cofactors that form a temporary association with the rest of the enzyme are called **coenzymes**.

Many vitamins are enzyme cofactors (or are used to make them). Some of the vitamins of the B-group are used to make coenzymes which are important in the reactions of respiration. For example, the coenzyme nicotinamide adenine dinucleotide (NAD) is derived from nicotinic acid (Vitamin B3).

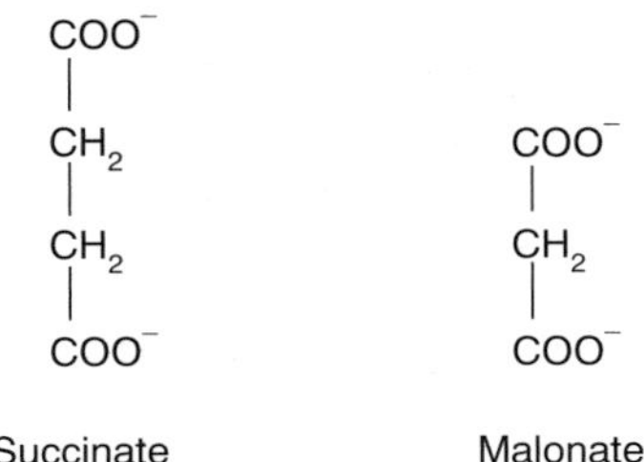


**Figure 4.9**
Succinate and its competitive inhibitor, malonate

A number of inorganic **trace elements** are enzyme cofactors. Zinc, for example, is a constituent of carbonic anhydrase. Numerous other metal ions such as manganese and magnesium are also enzyme cofactors.

## Enzyme poisons

A considerable number of chemicals are poisonous because they interfere with the action of enzymes. These enzyme poisons or **inhibitors** can work in two quite different ways.

**Competitive inhibitors** resemble the normal substrate sufficiently to temporarily occupy the active site. The best known is malonate which inhibits the action of succinate dehydrogenase (Figure 4.9).

Malonate is able to combine with the active site and, during this brief period, the enzyme cannot combine with its normal substrate, succinate. Succinate and malonate are thus 'competing' for the active site of the enzyme. If however the ratio of succinate to malonate is increased, the inhibitory effect is reduced. This is because it becomes more likely that a malonate ion is denied access to the active site because a succinate ion is already occupying it (Figure 4.10).

One particularly important example of competitive inhibition (and one that occurs quite naturally) is the competition between oxygen and carbon dioxide for the active site of ribulose bisphosphate carboxylase (Chapter 33).

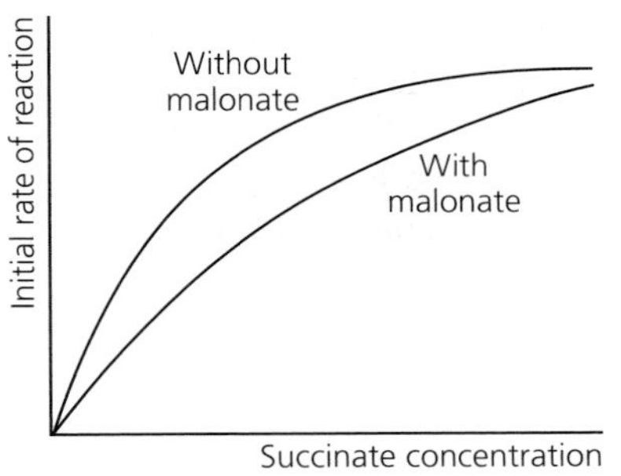


**Figure 4.10**
Graph showing competition between succinate and malonate

**Non-competitive inhibitors**, on the other hand, have no resemblance to the normal substrate. Some non-competitive inhibitors combine temporarily with enzymes; cyanide, for example, combines with the copper ion in the prosthetic group of cytochrome oxidase, one of the enzymes involved in respiration (Chapter 5).

Some of the most deadly enzyme poisons combine with a part of the enzyme molecule that may be some way from the active site. As a result the folding pattern of the polypeptide chain is changed, distorting the active site. Heavy metals such as lead and mercury act in this way, poisoning enzymes in the nervous system. Another example is di-isopropylfluorophosphate (DFP), a nerve gas. DFP inactivates acetylcholinesterase, an enzyme that plays an essential part in the transmission of information from one nerve cell to another (Chapter 17).

## Enzymes in teams: metabolic pathways

In most metabolic processes the product of one reaction becomes the raw material for another. Such a sequence of reactions is called a **metabolic pathway** (Figure 4.11). Different metabolic pathways are linked into a network so that one substance may be converted into a great many others.

**Figure 4.11**
A metabolic pathway

Reactions in cells differ in one important respect from the same reactions in test tubes. *In vitro* (literally, 'in glass') the conditions are constantly changing because the substrate concentration is decreasing as it is used up, and the product concentration is increasing. In a living cell the product of one reaction is the substrate for another and is normally used up as quickly as it is formed. As a result the substrate and product concentrations tend to remain steady.

## Why so many steps?

It might seem unnecessary to convert one chemical into another via many steps rather than a single step. Why this complexity? There seem to be several possible reasons:

- when a metabolic pathway releases energy, as in respiration, it does so in several small amounts rather than one big one. This state of affairs allows some of the energy to be conserved in a chemically-useful form (see Chapter 5).
- if all the energy were to be released in a single step, most would be released as heat which would damage the cell.
- a number of the intermediate compounds on a pathway may be of importance in themselves as raw materials for other pathways.

## Controlling catalysis

Although enzymes obey chemical laws they are subject to considerable biological control. Some of the most important are considered below.

### *Keeping enzymes and substrates apart*

Some enzymes – especially digestive enzymes – are potentially hazardous because they could attack the cells that produce them. Cells can protect themselves from these deadly products in two ways depending on whether the enzymes are for use within the cell or outside it.

1 Enzymes that are used to digest worn out organelles are packaged inside lysosomes which isolate the enzymes from the rest of the cell (Chapter 3).

2 Many protein-digesting enzymes are secreted as harmless inactive forms called **zymogens** or **proenzymes**. After secretion a small section of polypeptide is chemically clipped off, 'unmasking' the active site and producing the active enzyme. For example, the stomach enzyme pepsin is secreted as the inactive precursor **pepsinogen** which is converted into its active form in the stomach cavity.

## Bringing enzyme and substrate together

Some enzymes are organised so as to increase the chances of a molecule 'finding' the next enzyme in the pathway. One way of achieving this is to concentrate enzymes inside compartments, such as mitochondria or the cytoplasmic cisternae.

A more effective mechanism depends on the fact that some enzyme molecules are permanently located within membranes, floating like icebergs in the lipid bilayer. Since they can move in only two dimensions enzymes are more likely to collide with, and hand on their products to, the next enzyme in the sequence. This kind of organisation is a feature of some of the mitochondrial enzymes.

Some enzymes don't have to wait for collisions with other enzymes because they are permanently aggregated into **multienzyme complexes**. This enables a product molecule to be 'handed on' directly to the next enzyme without having to 'wait' for a chance collision.

## Regulating the activity of enzymes

Some metabolic pathways are subject to **end-product inhibition**, in which the activity of the first enzyme in a sequence is inhibited by the final product of the pathway. It is dependent on the fact that the enzyme has *two* binding sites which may be some way apart. Besides the active site, it also has a site to which the end product of the pathway can bind reversibly. When this happens the enzyme changes its shape and becomes inactive. When the end product breaks away from the enzyme the latter resumes its catalytic shape.

Most enzymes of this type consist of more than one polypeptide chain, held together so loosely that they can shift their position relative to each other. Enzymes that can do this are said to be **allosteric**. Allosteric means 'other shape', referring to the fact that the enzyme can alternate between two shapes.

End-product inhibition is an example of **negative feedback** – the more concentrated the end product, the more the first enzyme is inhibited and vice versa. As a result, if there is a change in the rate of use of the end product, the output of the pathway tends to adjust accordingly.

## Controlling enzyme production

Another way in which enzyme action can be regulated is by controlling the rate of enzyme production (Chapter 11).

## Enzymes in the test tube

Most experiments on enzymes are performed under conditions that are significantly different from those present in nature. In a living cell the concentration of the substrate doesn't usually change very much over time because other reactions are continuously supplying more raw materials and removing end products.

In a test tube one usually starts off with a fixed amount of substrate and no end product. The concentration of the substrate thus falls and the end product accumulates (unless it is given off as a gas). This can be important in an experiment in which you want to change *only one condition at a time*. The problem can be partly solved by using *saturating* concentrations of substrate, so that initial decreases in substrate concentration have no effect on the activity of the enzyme.

## Building up and breaking down

Metabolic pathways are of two general kinds:

- **catabolic** processes (catabolism), in which large molecules are broken down into smaller ones, for example during digestion and respiration. Most catabolic reactions release energy (though this energy is only put to useful purposes in respiration).
- **anabolic** processes (anabolism), in which small molecules are built up into larger ones, as in photosynthesis and protein synthesis. Anabolic reactions require the input of energy from respiration.

Anabolism and catabolism are not independent processes. As shown in Figure 4.12 (and explained more fully in Chapter 5) the energy released in certain catabolic reactions is used to drive anabolic processes.

It might seem wasteful, but all cells are continuously breaking down proteins and other large molecules (except their own DNA), as well as making them. Cell components are said to be in a state of **turnover**. Apart from the DNA of certain cells that cease to divide early in life (such as neurones), not a single molecule that was in your body when you were born is still there now! Even a cell that has finished growing and will never

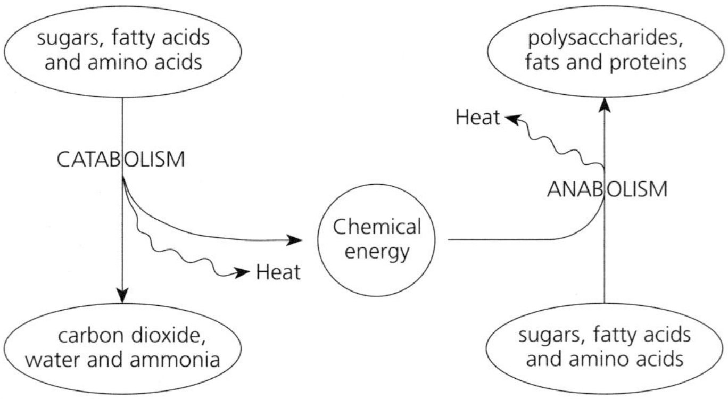

**Figure 4.12**
Relationship between anabolism and catabolism

divide again (such as a nerve cell) continues to make and break down proteins and other large molecules.

The rate of turnover varies from one substance to another and also from one tissue to another. In a week about half your liver proteins are replaced, but the 'half life' of muscle proteins is about a month. Brain phospholipids are replaced much more slowly and, as mentioned, DNA is not turned over at all.

## Enzymes as tools

Making chemicals by conventional industrial processes has always had certain drawbacks. Firstly, side reactions may result in a low percentage yield. Secondly, some industrial processes require the input of a good deal of heat, which is costly.

Enzymes offer a solution to both these problems. With their great specificity they convert virtually all the raw material into end product. Moreover their ability to work at ordinary temperatures results in energy saving.

The simplest way of putting enzymes to work is to use micro-organisms, allowing the enzymes to work inside the cells that made them. This method has been used more or less unknowingly for thousands of years in brewing and wine making. The micro-organisms are incubated with the material to be converted in a large vat, and after a suitable time the product is recovered.

One of the disadvantages of this method is that, although each individual enzyme converts almost all of its substrate to product, different metabolic pathways may compete for the raw material supplied. For example, much of the sugar supplied to yeast in the brewing industry is used as raw material for growth instead of being converted into ethanol and $CO_2$.

To an increasing extent enzymes are being used outside the cells that produce them. Most are hydrolases that are secreted into the environment, such as the proteases used in enzyme washing powders. These enzymes have the advantage that they have relatively simple requirements. They do not require metabolic energy in the form of ATP (Chapter 5) and when cofactors are involved these are usually inorganic ions rather than complex coenzymes. An example of an intracellular enzyme that is used industrially is glucose isomerase, used in the production of fructose from glucose.

There are two ways in which enzymes can be used extracellularly. The simplest is to mix enzyme and substrate in a large vessel. The disadvantage of this is that before the product can be recovered, it has to be separated from the enzyme, which is then discarded.

A much more efficient way is to attach the enzyme to a solid surface and to allow the substrate to flow over the enzyme (Figure 4.13). This allows continuous production, the enzyme being (in theory) capable of infinite reuse.

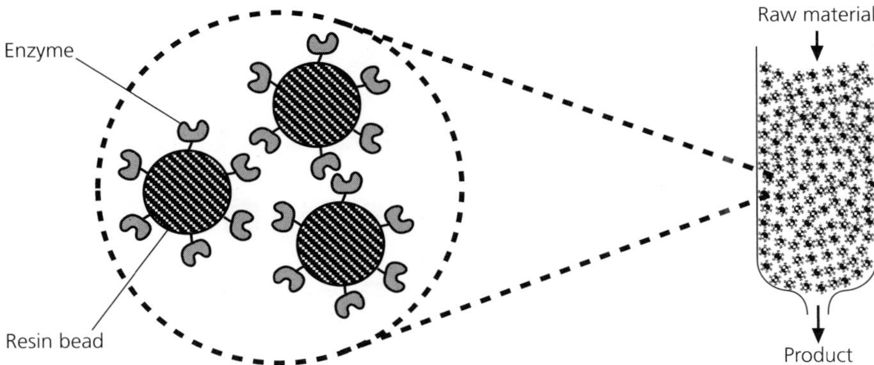


**Figure 4.13**
Chemical conversion using an immobilised enzyme

Another advantage is that immobilised enzymes are usually more stable than those in free solution. Immobilised glucose isomerase, for instance, remains active at temperatures of up to 60°C for hundreds of hours.

There are various ways of immobilising enzymes. They may be attached to a solid surface such as beads of plastic or resin; they may be trapped in a gel or inside tiny beads permeable to both substrate and products or they may be held behind a selectively-permeable membrane. Whatever method is used it must involve binding to non-catalytic parts of the enzyme molecule.

Despite the fact that immobilised enzymes are more stable they are still sensitive to pH changes and to inhibition by chemical impurities, so apparatus must be kept scrupulously clean.

## QUESTIONS

**1** True or false? All enzymes
- **a)** have large molecules.
- **b)** are inactivated at 90°C.
- **c)** are concerned with the digestion of food.
- **d)** supply energy for the reactions they catalyse.
- **e)** do not take part in the reactions they catalyse.
- **f)** contain nitrogen.
- **g)** have an optimum temperature of about 37°C.

**2** What effect does an enzyme have on the **(a)** speed, **(b)** direction of a reaction?

**3** Why are enzymes effective at such low concentrations?

**4** Give *three* ways in which enzymes differ from other catalysts.

**5** Consider the following statements:
- **A** When substrates can exist as 'left- and right-handed' forms, enzymes can only act on one of them.
- **B** Enzymes are very sensitive to pH.
- **C** Enzymes become saturated at high substrate concentrations.
- **D** The effect of low temperatures on enzyme activity is reversible.
- **E** Some enzymes change their light absorption properties when the substrate is added.
- **F** Some enzymes are inhibited by substances that have a similar molecular structure to the normal substrate.

- **a)** Which statement(s) supports the view that enzymes combine with their substrates?
- **b)** Which statement(s) supports the view that the substrate fits into part of the enzyme surface and that this depends at least partly on shape?

CHAPTER 

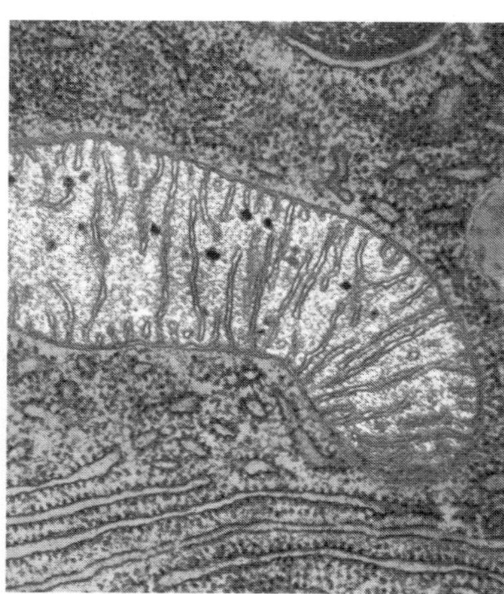

# ENERGY from MOLECULES – RESPIRATION and FERMENTATION

## LEARNING OBJECTIVES

By the time you have completed your study of this chapter you should be able to:

- explain how respiration differs from gas exchange and breathing.
- explain the theory behind simple investigations into the nature of respiratory gas exchange, and understand how respiratory rates of small animals are measured.
- give the approximate relative energy values of different biological fuels.
- distinguish between exergonic and endergonic reactions and give examples of endergonic processes in cells.
- explain the role of ATP in linking exergonic and endergonic reactions in cells.
- outline the three stages of carbohydrate breakdown in terms of their raw materials and end products, their relative ATP yields and the location of each stage within the cell.
- explain why the Krebs cycle is strictly aerobic, and the basis behind the inhibitory effect of cyanide in respiration.
- outline how fats and amino acids are used as energy sources.
- distinguish between respiration and fermentation, and explain the advantages and disadvantages of each.
- distinguish between fermentation and anaerobic respiration.

## The need for energy

All cells are constantly absorbing energy from their surroundings and using it to drive cell processes. Though organisms perform a seemingly endless variety of activities, the need for energy can be reduced to five types of process, the first three being characteristic of all life:

- *anabolism* – the building up of large molecules from smaller ones e.g. proteins from amino acids and polysaccharides from sugars.
- *active transport* – the movement of a substance across a cell membrane from a low to a high concentration (Chapter 6).
- *movement* – all cells show some form of movement, whether it be moving from place to place, or the less obvious movement of their organelles such as vacuoles, flagella or chromosomes.

- *bioluminescence* – some organisms are able to convert chemical energy to light, for example 'glow worms', some planktonic protoctists and some bacteria.
- *maintenance of body temperature* – in birds and mammals.

## Sources of energy

**Table 5.1**
Energy values of different energy substrates

| 'Fuel' | Energy value (kJ) per gram |
|---|---|
| Carbohydrate | 17 |
| Fat | 39 |
| Protein | 17 |

Cells obtain their energy by breaking down organic molecules into simpler substances. The energy yields of the three main classes of 'fuel' or **energy substrates** are shown in Table 5.1. As explained in Chapter 2, fat is much richer in energy than either protein or carbohydrate. In animals in which weight saving is important – especially in flying animals – fat forms the principal energy reserve.

The overall equation for the complete oxidation of glucose is as follows:

$$C_6H_{12}O_6 + 6O_2 \rightarrow 6CO_2 + 6H_2O - 2868 \text{ kJ}$$

The figure '−2868' is the energy yield for every mole of glucose used. The minus sign indicates the *release* of energy, since the end products have less energy than the raw materials.

The above equation is only an overall summary; many reactions lie between the raw materials and the end products. When carbohydrate is being used there are two distinct phases in the breakdown process: glycolysis and respiration.

- **Glycolysis**, in which glucose is converted to **pyruvate** and yields only a small amount of useful energy. Oxygen is not used in the process, but the final product is dependent on whether oxygen is present. In the presence of oxygen, pyruvate is fully oxidised in respiration. When oxygen is absent, pyruvate undergoes further conversion. In plants and in yeast it is converted to **ethanol** and $CO_2$ and in animal tissues such as muscle it is converted to **lactate**. Anaerobic glycolysis is called **fermentation**, and is often incorrectly referred to as 'anaerobic respiration'. As we shall see later in this chapter anaerobic respiration and fermentation are fundamentally different processes and should not be confused.
- **Respiration**, which occurs in the mitochondria and yields a large amount of useful energy. Pyruvate, fatty acids and amino acids can all serve as **respiratory substrates** and are completely oxidised to $CO_2$. In most organisms respiration is **aerobic**, meaning that oxygen is used up, producing water.

The release of energy occurs continuously in all living cells, but in plants $CO_2$ production is normally 'masked' in the daytime by photosynthesis. Two other processes, **respiratory gas exchange** and **breathing**, are closely linked to respiration but are actually quite different processes. They are often confused, particularly by the medical profession!

Respiratory gas exchange is the process by which $CO_2$ diffuses out of the organism (or an individual cell) and oxygen diffuses in. The concentration gradients down which these gases move are set up as a result of respiration. The area over which gas exchange occurs is called the **respiratory gas exchange surface**. In smaller animals the gas exchange surface covers the entire body. In most larger and more complex animals the gas exchange surface is concentrated in specialised **respiratory gas exchange organs** such as gills or lungs.

Breathing is the pumping of air or water to and from the gas exchange organs. By **ventilating** the gas exchange surface, the water or air next to it is kept fresh.

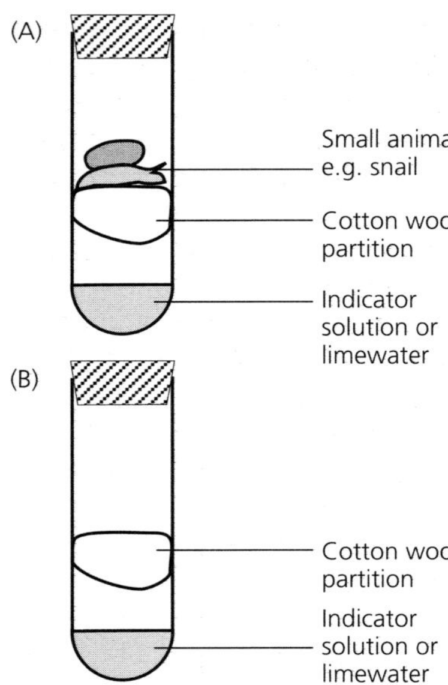


**Figure 5.1**
Simple demonstration of $CO_2$ production by a small animal

## Demonstrating respiration

The simplest way of showing that an organism is respiring is by detecting the $CO_2$ produced. Figure 5.1 shows a simple method using hydrogencarbonate indicator solution (limewater, which turns milky in the presence of $CO_2$, is an alternative).

Carbon dioxide dissolves in water and then reacts with it to form carbonic acid, so the concentration of $CO_2$ in the gas above the indicator affects its colour. If the indicator is in equilibrium with normal air (0.035% $CO_2$) it has a rose colour. If the air is richer in $CO_2$ the indicator turns yellow. If the air contains significantly less than 0.035% $CO_2$ it turns purple. If the liquid in tube A becomes yellow but the liquid in tube B (the control) does not, then the change *must be due to the animal.* With small animals you will need to leave the apparatus for several hours to see any change.

During the daytime plants do not appear to be respiring because of photosynthesis, but the uptake of oxygen can be detected using a heavy isotope of oxygen, $^{18}O$.

## Measuring the rate of respiration

The experiment described above is a **qualitative** one – it only tells us *whether* an organism is using oxygen. To make it into a **quantitative** experiment would mean measuring the *rate* of respiration. The rate of respiration can be expressed as:

- the amount of oxygen absorbed in a given time;
- the amount of $CO_2$ produced in a given time;
- the amount of organic matter used up in a given time. In practice this is much more difficult to measure because it involves estimating changes in dry mass (mass after drying). Since dry mass can only be measured once (a dry organism is a dead one), changes in dry mass have to be estimated by comparing the dry masses of samples of different organisms at successive time intervals.

The easiest method is to measure oxygen uptake using a **respirometer**. Figure 5.2 shows a simple one. Although the animal cannot be prevented from producing $CO_2$, the soda lime absorbs it as fast as it is produced. As a control, a similar apparatus is set up, but without an animal. To minimise any warming by body heat (which would cause the gas to expand), the barrel of the syringe should not be handled when setting up the apparatus.

After the apparatus has been assembled it should be left in a constant temperature water bath for about 15 minutes to ensure that both are at the same temperature. Next, a small drop of coloured water is introduced

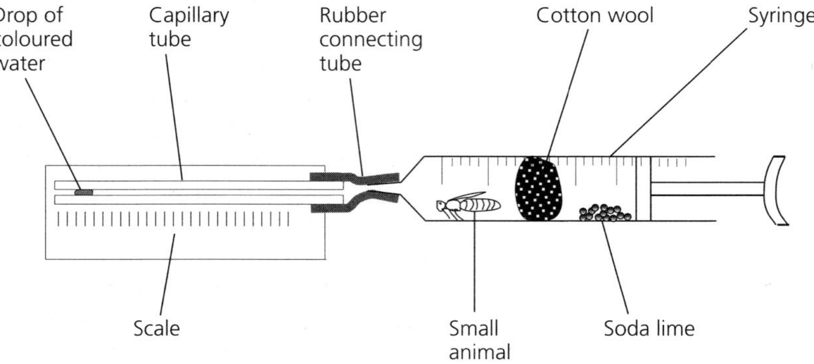


**Figure 5.2**
A simple respirometer

into the end of the capillary and drawn a centimetre or so up the tube by gently pulling out the plunger of the syringe. Any uptake of oxygen by the animal would result in the water drop moving slowly toward the syringe, which is measured using the scale. When the water reaches the end of the scale you can return it to the beginning using the plunger (taking care not to touch the barrel of the syringe).

If the liquid in the control moves slightly, then there has been a small change in either room temperature or atmospheric pressure, and a small correction must be made. Suppose, for example, that over the period of measurement the drop moves 2 $mm^3$ towards the syringe in the control and 22 $mm^3$ towards the syringe in the apparatus with the animal. The volume change due to the animal would thus be $22 - 2 = 20$ $mm^3$. If, on the other hand, the control drop had moved 3 $mm^3$ *away* from the syringe, the change due to the animal would be $22 + 3 = 25$ $mm^3$.

Of course, a big animal uses more oxygen in a given time than a small one. To compare the rate of oxygen uptake by a snail and a centipede you must, therefore, take account of the different body sizes. You would do this by expressing the rate of respiration as the volume of oxygen absorbed per minute *per gram of animal*, calculated as follows:

$$\frac{\text{volume of oxygen absorbed}}{\text{time in minutes} \times \text{mass of animal}} \text{ cm}^3 \text{ min}^{-1} \text{ g}^{-1}$$

Suppose for example a mouse with a mass of 20 grams uses 2 $cm^3$ oxygen in 2 minutes. In 1 minute the animal uses 1 $cm^3$, but each gram of mouse must be using 1/20 of 1 $cm^3$ (0.05 $cm^3$).

## Exergonic and endergonic reactions

Many of the processes that occur in cells are **endergonic** because they cannot occur without an input of energy. One example is the conversion of water to hydrogen and oxygen – this cannot occur spontaneously but must be 'driven' by energy supplied from an outside source such as mains electricity. Since energy is gained *from* the surroundings, the products have more energy than the raw materials; in the example just given, hydrogen and oxygen have more energy than water. The overall energy change ($\Delta G$) in the splitting of water is, therefore, positive.

**Exergonic** reactions, such as the formation of water from hydrogen and oxygen, give out energy to the surroundings. Because the products have less energy than the raw materials there is a net loss of energy and $\Delta G$ is negative. Provided that the necessary activation energy is supplied, exergonic reactions occur spontaneously.

How, then, do cells make endergonic reactions go? The problem is akin to getting a ball to the top of a hill – energy must be supplied from outside. In cells, the energy that drives endergonic reactions is supplied by glycolysis and/or respiration.

This energy is not used directly, however, because the breakdown of a glucose molecule releases far more energy than is needed to 'drive' any single endergonic reaction that occurs in cells. If a cell were to 'spend' the energy of an entire molecule of glucose to 'drive' an endergonic reaction (such as joining together two amino acids), most of the energy would be 'left over' as heat, which would be extremely damaging to the cell.

A useful way of envisaging the situation is to compare the 'spending' of energy to the spending of money. A glucose molecule would be the equivalent of a £10 note and a starch molecule the equivalent of a £100 note. There is, however, an important difference between spending energy and spending money. When a cell 'spends' energy, any that is not used is wasted as heat. Unlike financial transactions, no change is given! Obviously, it is advantageous for cells to spend energy in small 'packets', so that little is wasted. These 'packets' are molecules of a compound called **adenosine triphosphate** or **ATP**.

## The role of ATP – the universal energy carrier in cells

ATP is the 'small change' of the cell's energy economy and is the chief energy carrier in all living cells (Figure 5.3).

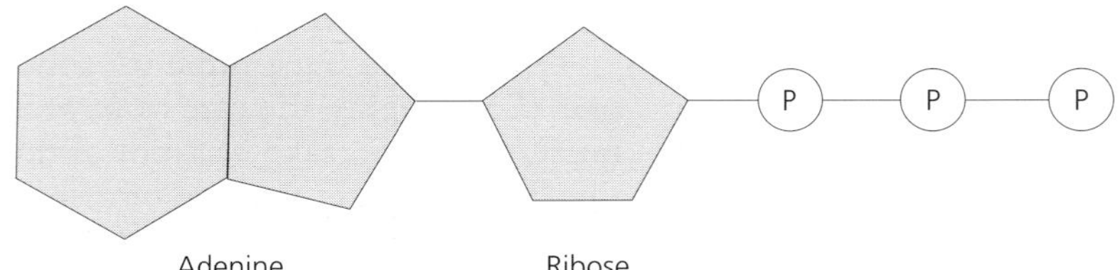


**Figure 5.3**
Structure of ATP

ATP can be hydrolysed to **adenosine diphosphate (ADP)** and inorganic phosphate (shortened to $P_i$) with the release of about 50 kJ per mol of energy:[1]

$$ATP + H_2O \rightarrow ADP + P_i - 50\ kJ$$

In some cell processes ATP is hydrolysed to **adenosine monophosphate (AMP)** with the release of even more energy:

$$ATP + 2H_2O \rightarrow AMP + 2P_i - 100\ kJ$$

When ATP is hydrolysed to ADP, the 50 kJ of energy released is not a great deal more than is required to drive many of the endergonic reactions that occur in cells. Take for instance the conversion of glucose to glucose-6-phosphate. If this were to occur in the most direct way using inorganic phosphate, it would require the input of 13.8 kJ per mole of energy:

$$\text{glucose} + P_i \rightarrow \text{glucose-6-phosphate} + H_2O + 14\ kJ$$

In the form shown of course, this reaction cannot occur. What actually happens is that the phosphate and the energy are supplied by ATP:

$$\text{glucose} + ATP \rightarrow \text{glucose-6-phosphate} + ADP - 36\ kJ$$

Under the conditions in the cell the hydrolysis of ATP yields 50 kJ per mole of energy. If 14 kJ of this is 'spent' to make a mole of glucose-6-phosphate, the net energy change is −36 kJ, so the overall process is exergonic (Figure 5.4).

### *Energy for making ATP*

Since hydrolysis of ATP is exergonic its production must be endergonic, so it cannot occur spontaneously. The energy to drive ATP production is supplied by the exergonic processes of glycolysis and/or respiration, summarised as follows:

$$\text{glucose} + \text{oxygen} + \text{many } ADP + \text{many } P_i \rightarrow CO_2 + \text{water} + \text{many } ATP$$

For reasons to be explained later, it is not possible to give a precise figure

**CLEAR THINKING**

**It is often stated that two of the phosphate groups of ATP are linked to the rest of the molecule by 'energy-rich bonds' (sometimes represented by curly lines), and that when such bonds are split, energy is released. Since a chemical bond is an *attraction* between atoms, it follows that to break a chemical bond must *require* energy. The stronger a bond, the more energy is needed to break it and, conversely, the more energy is released when it is formed.**

**A more accurate way of putting it would be to say that an exergonic reaction releases energy because the formation of new, stronger bonds releases more energy than is absorbed in the breaking of the old, weaker ones. Overall, the bonds in ADP and inorganic phosphate are stronger than those in ATP and water.**

[1]Most books give a figure of 30 kJ per mole. However, this applies to 'standard' conditions in which all reactants are initially at equimolar concentrations. In cells, ATP is at a much higher concentration and under these conditions $\Delta G$ is thought to be about 50 kJ per mol.

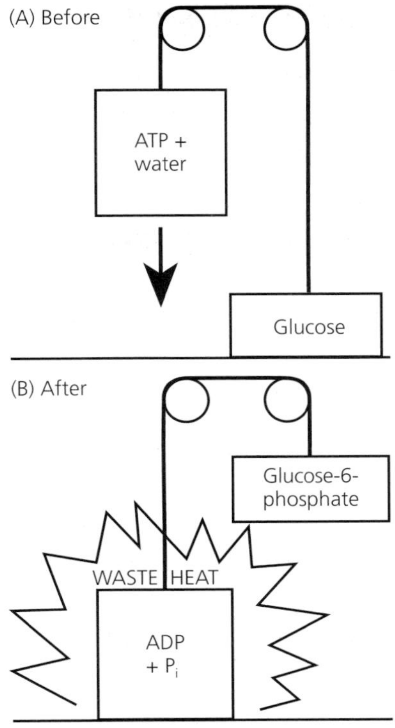


**Figure 5.4▲**
Model to show how ATP hydrolysis is used to drive an endergonic reaction

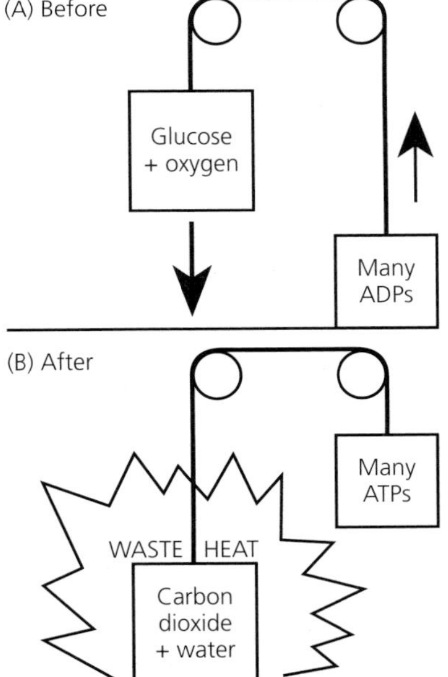


**Figure 5.5▲**
Model showing conversion of large 'packets of energy' in glucose to smaller 'packets of energy' in ATP

for the number of ATP molecules produced for each molecule of glucose used. The important thing is that the large 'packet of energy' in each glucose molecule is converted to many smaller 'packets of energy' in the form of ATP. Overall roughly half the energy in the glucose is conserved as ATP energy, the rest being wasted as heat (Figure 5.5).

In driving endergonic reactions, the hydrolysis of ATP produces ADP and inorganic phosphate. These are then reconverted to ATP using energy obtained in respiration. ATP thus acts as a kind of energy carrier – rather like a rechargeable battery (Figure 5.6).

The actual amount of ATP present in an average human body at any one time is only about 50 g, yet an average person uses about 140 kg of ATP daily. Each molecule of ATP must therefore, on average, be recycled about 2800 times a day – or about once every 2 minutes. Of course this figure will vary greatly with the individual and with the kind of tissue. An average cell has enough ATP to last about 6 seconds during normal activity.

## Biological oxidation

At various stages in its breakdown the energy substrate undergoes oxidation (those who are unsure of oxidation are advised to consult Appendix I). Only the final reaction involves the addition of oxygen; all other oxidations involve the removal of hydrogen or of electrons. Hydrogen is not removed as a gas but is picked up by a **hydrogen carrier** under the influence of a **dehydrogenase** enzyme. Likewise, electrons are not removed in a free state but are transferred from one electron carrier to another.

Some of the hydrogen and electron carriers are important in respiration as coenzymes. One of these is **NAD$^+$** (the nicotinamide adenine dinucleotide ion). This is derived from one of the B-group vitamins, **nicotinic acid**. When NAD$^+$ accepts hydrogen it actually picks up one hydrogen atom plus the electron of another leaving a proton in solution:

$$NAD^+ + 2H \rightarrow NADH + H^+$$

This is sometimes represented as the simpler but slightly less correct:

$$NAD + 2H \rightarrow NADH_2$$

Except when it is necessary to balance chemical equations, 'NADH + H$^+$' is shortened to the less cumbersome 'NADH'.

Other important hydrogen carriers are the **flavoproteins**. These have a prosthetic group consisting of either **FAD** (flavine adenine dinucleotide), or **FMN** (flavine adenine mononucleotide). Both FAD and FMN contain **riboflavine** (Vitamin $B_2$). Although respiratory coenzymes are reused and should theoretically never run out, they are very slowly lost from the body and must be replenished from vitamins in the diet.

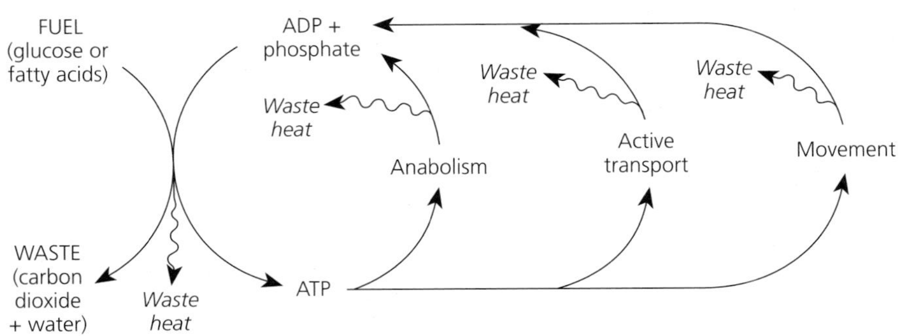


**Figure 5.6▶**
ATP as an energy carrier

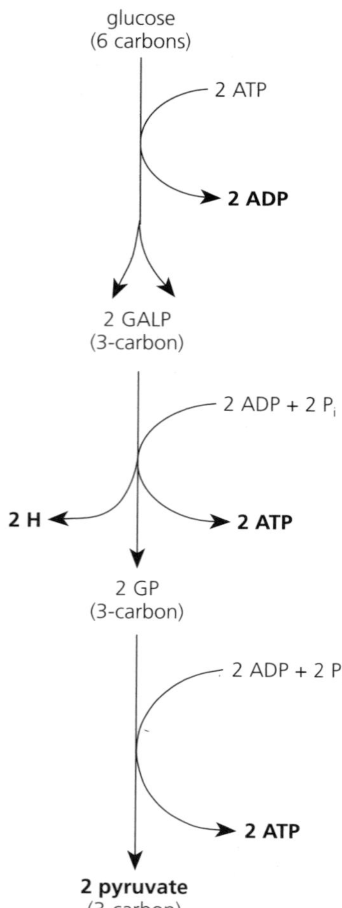


**Figure 5.7**
Simplified scheme of glycolysis

# The first phase of carbohydrate breakdown: glycolysis

Glycolysis (*glyco* = sugar and *lysis* = breakdown) occurs in the cytosol and does not use oxygen. For every glucose molecule used glycolysis yields only a small amount of ATP. It is thus an inefficient process, with most of the energy remaining in the pyruvate end product. Figure 5.7 shows a simplified scheme of the process, represented by the following stages:

- it actually begins with the *consumption* of ATP in the formation of fructose bisphosphate.[2] This means that the *net* ATP gain is actually less than the total produced.
- the fructose bisphosphate is split into two molecules of a triose phosphate (glyceraldehyde-3-phosphate or GALP). This means that in the 'balance sheet' for sugar breakdown *all numbers have to be doubled after this point.*
- the triose phosphate is then oxidised to glycerate-3-phosphate (GP). This involves the removal of two hydrogen atoms, which are picked up by $NAD^+$ to form NADH. At the same time a molecule of ATP is produced (two for each glucose used). How the NADH gets rids of its two electrons to regenerate $NAD^+$ depends on the availability of oxygen and is dealt with later.
- glycerate-3-phosphate is converted to pyruvate with the production of another molecule of ATP (two for each original glucose).

Overall, four molecules of ATP are produced for each glucose molecule used. Since two are used up at the beginning this gives a net gain of two ATPs.

# Respiration

Respiration occurs in the mitochondria, described in Chapter 3. Like chloroplasts they are bounded by two membranes. The surface of the inner membrane is increased by folds or **cristae** which extend into the fluid interior or **matrix**. Projecting into the matrix from the inner membrane are thousands of tiny **stalked particles** where most of the ATP is produced.

With carbohydrate as substrate, respiration consists of four stages:

1 Conversion of pyruvate to acetyl coenzyme A.

2 The **Krebs cycle,** in which acetyl CoA is broken down into $CO_2$ and hydrogen in the form of a reduced hydrogen carrier.

3 **Proton pumping**, in which hydrogen removed in the Krebs cycle is passed along a series of electron and hydrogen carriers to oxygen, forming water. The energy released is used to pump protons out of the mitochondrion, setting up a **proton concentration gradient**.

4 ATP production. The tendency of protons to diffuse back into the mitochondrion is used to make ATP.

[2] In a *bisphosphate*, two phosphate groups are attached to different parts of a molecule, whereas in a *diphosphate* such as ADP, a phosphate group is attached via another phosphate group to the same part of a molecule.

## Production of acetyl CoA

Pyruvate produced in glycolysis enters the mitochondria. Here it combines with Coenzyme A to form a two-carbon derivative of acetic acid, **acetyl coenzyme A** (abbreviated to acetyl CoA). This is a very complex process involving removal of both $CO_2$ and of hydrogen by $NAD^+$. One of the coenzymes involved is derived from thiamine (Vitamin $B_1$).

## The Krebs cycle

Named after its Nobel Prize-winning discoverer Sir Hans Krebs, the Krebs cycle is also known as the **citric acid cycle** and the **tricarboxylic acid cycle** (Figure 5.8). It results in the complete breakdown of acetyl CoA to $CO_2$ and hydrogen in the form of NADH and reduced flavoprotein ($FPH_2$). The Krebs cycle occurs in the inner compartment of the mitochondrion (the matrix) and although oxygen is not used in the cycle itself, it is strictly aerobic.

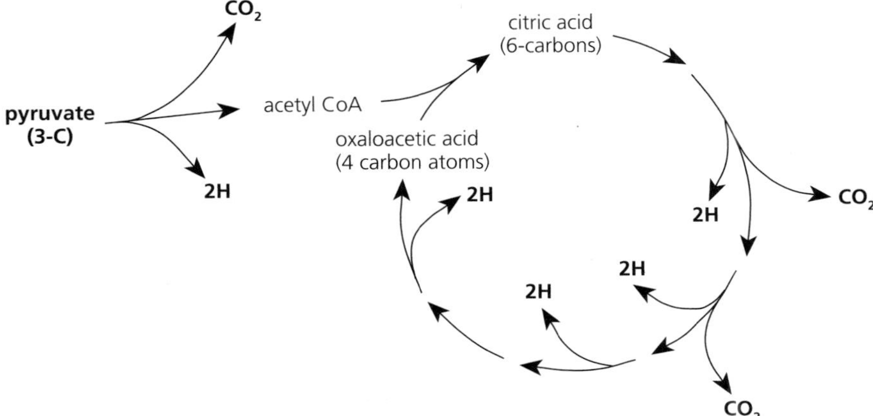


**Figure 5.8**
Outline summary of the Krebs cycle. Hydrogen is removed from the substrate at five points. In one of these dehydrogenation reactions the hydrogen acceptor is flavoprotein, in the other four it is $NAD^+$

The Krebs cycle can be summarised as follows:

$$\text{acetyl CoA} + 3NAD^+ + FP \rightarrow 2CO_2 + 3NADH + 3H^+ + FPH_2$$

There are eight steps in the cycle, but we only need consider the following:

- acetyl CoA combines with the 4-carbon compound **oxaloacetate** (oxaloacetic acid) forming the 6-carbon **citrate** (citric acid). At the same time coenzyme A is regenerated to combine with more pyruvate.
- during subsequent reactions two $CO_2$ molecules and three pairs of hydrogen molecules are removed (three as NADH, one as $FPH_2$). A molecule of ATP is also produced in one of these reactions (two for each original glucose molecule). The overall result is the regeneration of oxaloacetate; this picks up another 2-carbon molecule to start another turn of the cycle. Oxaloacetate thus acts as a kind of carrier for the breakdown of the 2-carbon compound entering the cycle.

Though the Krebs cycle itself makes very little ATP, it provides the 'fuel' for the third stage of respiration in the form of NADH and $FPH_2$.

## Proton pumping

The raw materials for this stage are the NADH and $FPH_2$ produced in the Krebs cycle. Proton pumping occurs in the inner membrane of the mitochondrion. In a way to be explained shortly, it results in the pumping of protons out of the matrix of the mitochondrion, making it alkaline. The energy for proton pumping is derived from the oxidation of NADH and $FPH_2$. The overall equation for NADH oxidation is:

$$NADH + H^+ + \tfrac{1}{2}O_2 \rightarrow NAD^+ + H_2O - 220\ kJ$$

Rather than occurring in a single reaction (as implied by the above equation) the transfer of electrons from NADH to oxygen is actually a multistep process. Electron transfer occurs via a series of proteins, collectively called the **respiratory chain** because they work together rather like a 'bucket chain' (Figure 5.9). Each carrier protein alternately oxidises the carrier 'upstream' and reduces the one 'downstream'. For this to happen, each carrier has to be a slightly stronger reducing agent (and weaker oxidising agent) than the one 'upstream'. By dividing the total energy release into small steps a greater proportion of the total energy is trapped.

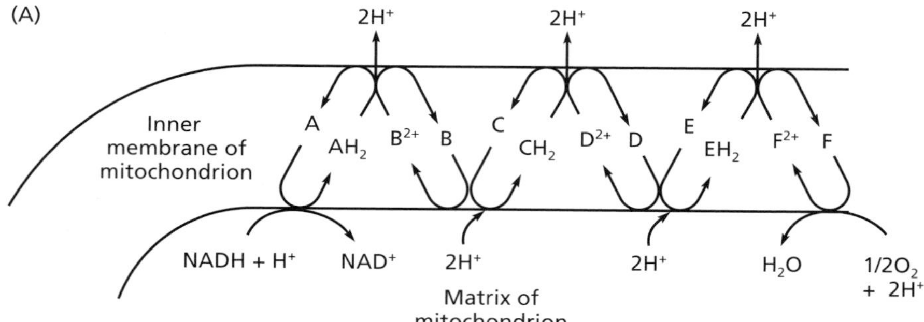


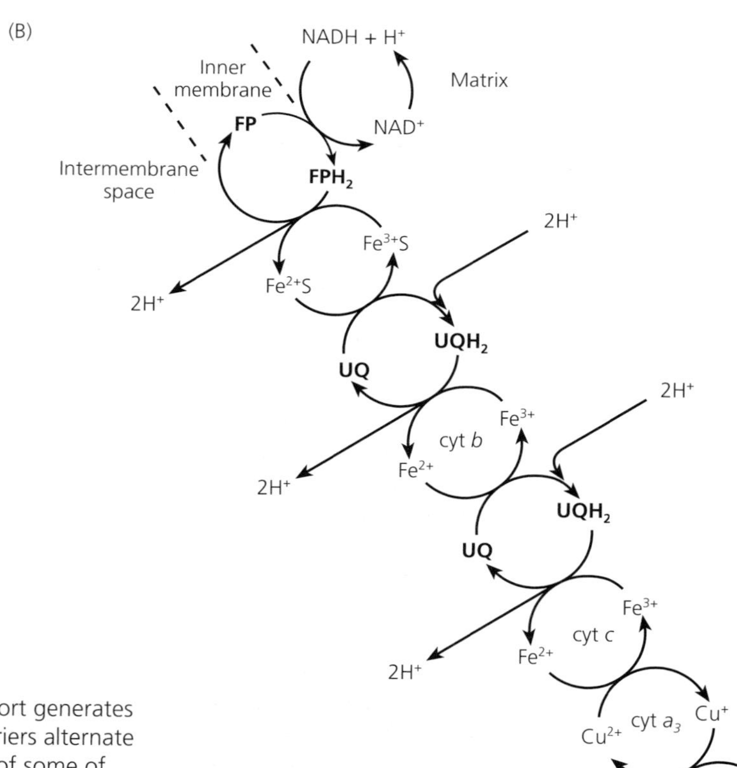


**Figure 5.9**
(A) General idea of how hydrogen transport generates a proton gradient. Note that electron carriers alternate with hydrogen carriers. (B) Arrangement of some of the carriers in the hydrogen/electron 'bucket chain'. Hydrogen carriers are flavoprotein and ubiquinone (coenzyme Q) and are shown in bold. Electron carriers include iron-sulphur proteins and cytochromes

Among the electron carriers are the iron-containing proteins called the **cytochromes**. The iron can be in either the ferrous ($Fe^{2+}$) or ferric ($Fe^{3+}$) state, depending on whether it has picked up or released an electron. The last of the cytochromes is called **cytochrome oxidase** or **cytochrome $a_3$** and contains copper.

Quite how the energy released in these electron transfers is used to make ATP remained unclear until Peter Mitchell proposed his **chemiosmotic hypothesis** in 1967. Mitchell proposed that the energy released in electron transport was used to pump protons out of the mitochondrion, producing a **proton gradient** across the inner mitochondrial membrane. It is the tendency for protons to diffuse back into the mitochondrion that powers ATP synthesis.

Though his idea was slow to gain support, evidence for it is now so great that it is generally accepted, and Mitchell was awarded the Nobel Prize for Medicine in 1978. One of the attractions of chemiosmosis is that it also accounts for ATP production in chloroplasts, except that the direction of proton pumping is reversed (Chapter 33).

The key to proton pumping is the fact that some of the carriers transfer entire hydrogen atoms whilst others carry only electrons. When a hydrogen carrier gives up a hydrogen atom to a carrier that can only accept an electron, the proton is released across the inner membrane of the mitochondrion and into the intermembrane space. Conversely, when a hydrogen carrier picks up an electron from an electron carrier it also picks up a proton from the matrix. Proton pumping is believed to occur at three sites along the chain, three pairs of protons being pumped for each pair of electrons passing down the chain.

The result of proton pumping is both a **pH gradient** and a **potential difference** across the inner mitochondrial membrane. These two factors contribute to a **proton motive force** causing protons to diffuse back into the mitochondrion.

Overall, 12 pairs of electrons pass down the respiratory chain for each molecule of glucose originally used. When these electrons recombine with protons and oxygen at the end of the chain, 12 molecules of water are formed, yet the net yield is only six molecules of water. The reason for the discrepancy is that at three stages in the Krebs cycle, a molecule of water is used up. This means that for each glucose molecule used, $2 \times 3 = 6$ molecules of water are used, resulting in a net gain of $12 - 6 = 6$.

## ATP production

The chemiosmotic hypothesis proposes that the proton motive force provides the energy for ATP synthesis. The inward leakage of protons is confined to channels through the stalked particles – the inner membrane as a whole is impermeable to protons. The catalyst for ATP synthesis is an enzyme called **ATP synthase** which is located in the head of the particles. The precise mechanism is uncertain, though it is known to depend on the establishment of a high local $H^+$ concentration inside the 'head' of each stalked particle (Figure 5.10). Each ATP molecule synthesised involves the inward movement of three protons.

The formation of ATP using the energy released in oxidation is called **oxidative phosphorylation**. An important consequence of chemiosmosis is that the link between oxidation (electron transfer) and phosphorylation (ATP production) is *indirect*, energy being transferred from the former to the latter by the proton gradient. It is thus inappropriate to show ATP formation occurring at specific sites along the electron transport chain, since this would imply a direct link between the two processes (Figure 5.11).

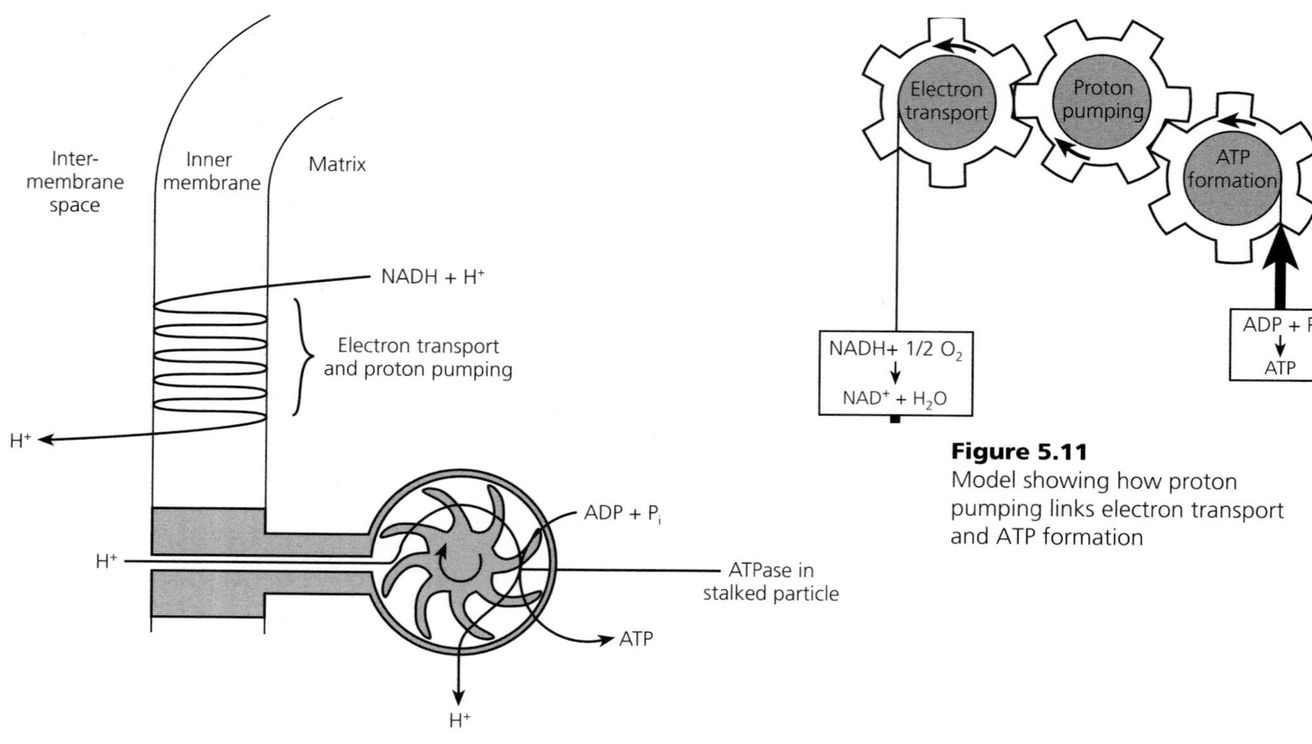


**Figure 5.10**
Diagram showing ATP formation in stalked particle

**Figure 5.11**
Model showing how proton pumping links electron transport and ATP formation

The proton gradient in mitochondria is not only used to make ATP. It is also the immediate source of energy for the transport of $Ca^{2+}$ in mitochondria and for the movement of flagella in bacteria, neither of which uses ATP.

### *Interfering with ATP synthesis*

ATP production in mitochondria can be prevented in various ways:

- lack of oxygen (anoxia). Although oxygen only interacts directly with the last of the electron carriers, absence of oxygen causes *all* the carriers to cease working, and thus stops proton pumping. This is because each one can only pick up hydrogen or an electron if it can pass it on to the next carrier. Absence of oxygen thus causes the supply of $NAD^+$ to dry up, bringing the Krebs cycle (and as we shall see, most of the ATP production) to a halt.
- **cyanide** inhibits cytochrome oxidase (the last of the electron carriers in the respiratory chain) preventing oxygen from accepting electrons. The result is the same as that of anoxia.
- **dinitrophenol** prevents ATP production by making the inner mitochondrial membrane permeable to protons, enabling them to take the ‘easy’ way back into the matrix. Because of this ‘short circuit’ the proton gradient collapses. Electron transport and oxygen consumption continue at an increased pace but ATP synthesis stops, the energy released in electron transport being entirely dissipated as heat.

An interesting naturally-occurring short circuit exists in the mitochondria of a tissue called **brown fat**, in which the inner mitochondrial membranes are naturally leaky to protons. Brown fat is present in human babies, which need to maintain a high rate of heat production. It is so-called because of the colour produced by its abundant mitochondria.

## Structure and function in the mitochondrion

Mitochondria are adapted to their function in a number of ways:

- because the carrier proteins are located in a lipid bilayer they can move in only two dimensions, making it easier for them to transfer electrons.
- proton pumping is dependent on the fact that the electron carriers are orientated the same way, some facing into the matrix and others towards the intermembrane space (can you explain this in terms of polar and non-polar R groups?). This is important since, as we shall see, some of the carriers pump protons outward through the inner membrane while others accept them from the matrix.
- the area of the inner membrane (and therefore the number of stalked particles) is greatly enlarged by the cristae.
- except at the stalked particles, the inner membrane is impermeable to protons.
- although less extensive than the inner membrane, the outer surface of a mitochondrion has a large external surface for the absorption of oxygen, pyruvate, ADP and inorganic phosphate (Figure 5.12).

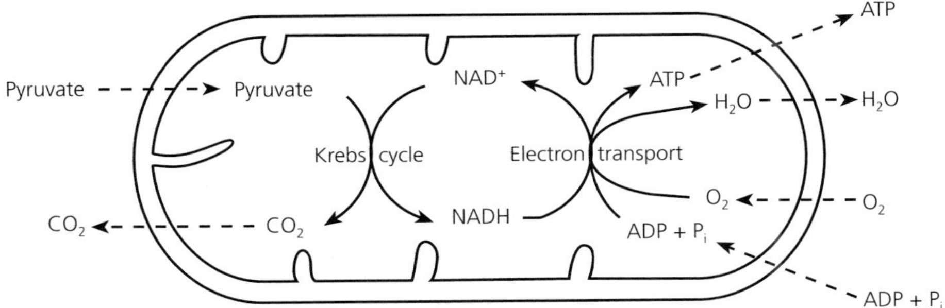


**Figure 5.12**
Raw materials and end products of mitochondrial activity

Over 90% of a cell's ATP production occurs in the mitochondria, which have aptly been called the 'powerhouses of the cell'. Entering a mitochondrion are: pyruvate, oxygen, NADH (produced in glycolysis), ADP and inorganic phosphate. Leaving it are ATP, water, $CO_2$ and $NAD^+$.

## Energy accounting

How efficiently do cells convert the energy in the energy substrate to ATP energy? In other words, what proportion of the 2886 kJ of energy released when a mole of glucose is oxidised is actually trapped and stored as ATP? To answer this we need to know two things – the energy content per mole of ATP, and the number of moles of ATP produced for each mole of glucose used.

The first of these is known approximately. Although the energy yield for ATP hydrolysis is often quoted as 30.6 kJ per mole, this is under specified laboratory conditions in which all reactants are present at molar quantities. In cells ATP is present at higher concentrations, and under these conditions the actual energy yield of ATP hydrolysis is thought to be 50.2 kJ per mol.

The ATP yield in glucose breakdown is less certain. Before the general acceptance of Mitchell's chemiosmotic hypothesis it was believed that ATP was made as electrons passed down the respiratory chain, and that for every NADH molecule oxidised, three ATP molecules were produced.

The chemiosmotic hypothesis makes energy accounting difficult because not all the energy of the proton gradient is used to make ATP. Some is used to pump other particles across the membrane, such as $Ca^{2+}$ ions, pyruvate and ADP (see Chapter 6).

It follows that there is no reason why there should be a whole number relationship between the number of electrons passing down the respiratory chain and the number of ATP molecules produced. For every pair of electrons carried by NADH it is thought that about 2.5 ATP molecules are produced, and for every pair of hydrogen atoms carried by $FPH_2$ about 1.5 ATP molecules are produced. The total ATP yield for each glucose molecule used can be estimated thus:

| | |
|---|---:|
| Glycolysis | 2 |
| Krebs cycle | 2 |
| NADH | $10 \times 2.5 = 25$ |
| $FPH_2$ | $2 \times 1.5 = 3$ |
| **Total** | **32** |

If 32 moles of ATP are produced for each glucose used, then with a yield of 50 kJ per mole of ATP, this gives $50.2 \times 32 = 1606$ kJ per mole of glucose used. This represents an efficiency of 1606/2886, or about 56% – far better than the 25% efficiency of a motor car engine.

Note that the 32 molecules of ATP quoted here is somewhat lower than the yield of 38 quoted in many older texts.

Since 30/32 of the ATP molecules produced in the breakdown of a molecule of glucose are formed in the mitochondria, the description of them as the 'powerhouses of the cell' is appropriate. Mitochondria are particularly numerous in those parts of a cell in which energy is rapidly being expended. For example, they are frequently clustered close to cell membranes in which active transport is occurring.

## Energy without oxygen

Many animal and plant tissues, and also yeast, can make ATP from chemical processes in which oxygen plays no part. There are two fundamentally different mechanisms: **anaerobic respiration** and **fermentation**.

## Anaerobic respiration

As in aerobic respiration, hydrogen passes from organic matter down a chain of carriers to a hydrogen acceptor. In this case the acceptor is not oxygen but another inorganic substance such as nitrate, sulphate or $CO_2$. Organisms that obtain their energy in this way include certain soil bacteria, for example *Pseudomonas denitrificans*. Under anaerobic conditions – for example after prolonged flooding – it is able to respire anaerobically. Instead of using oxygen as a hydrogen acceptor it uses nitrate, which is eventually converted to nitrogen gas. The result is **denitrification** – the loss of nitrate, with consequent loss of soil fertility.

Anaerobic respiration generates less ATP than aerobic respiration. This is because these alternative hydrogen acceptors are much less powerful oxidising agents than oxygen, so the electrons lose less energy as they flow down the respiratory chain.

In well-aerated soils *Pseudomonas denitrificans* takes advantage of the greater efficiency of aerobic respiration and uses oxygen. Since it is able to respire with or without oxygen it is said to be a **facultative anaerobe**. Bacteria that use sulphate or $CO_2$ as hydrogen acceptors can only respire anaerobically and are said to be **obligate anaerobes**.

## Fermentation

An alternative way of using organic matter to make ATP without using oxygen is **fermentation.** It occurs in yeasts and in many other micro-organisms and for limited periods in many plant and animal tissues. Some organisms are **obligate fermenters**, such as some bacteria. Others resort to fermentation when the occasion demands and are **facultative fermenters**.

Though anaerobic respiration is often confused with fermentation, the two processes are fundamentally different (Figure 5.13). In respiration, whether it be aerobic or anaerobic, hydrogen is removed from the substrate and passes down a chain of carriers to a hydrogen acceptor obtained from the environment. In aerobic respiration the hydrogen acceptor is oxygen and in anaerobic respiration it is nitrate, sulphate or carbonate. In fermentation, as we shall see, the organic fuel itself acts as hydrogen acceptor, and there is no transport of electrons along a chain of carriers.

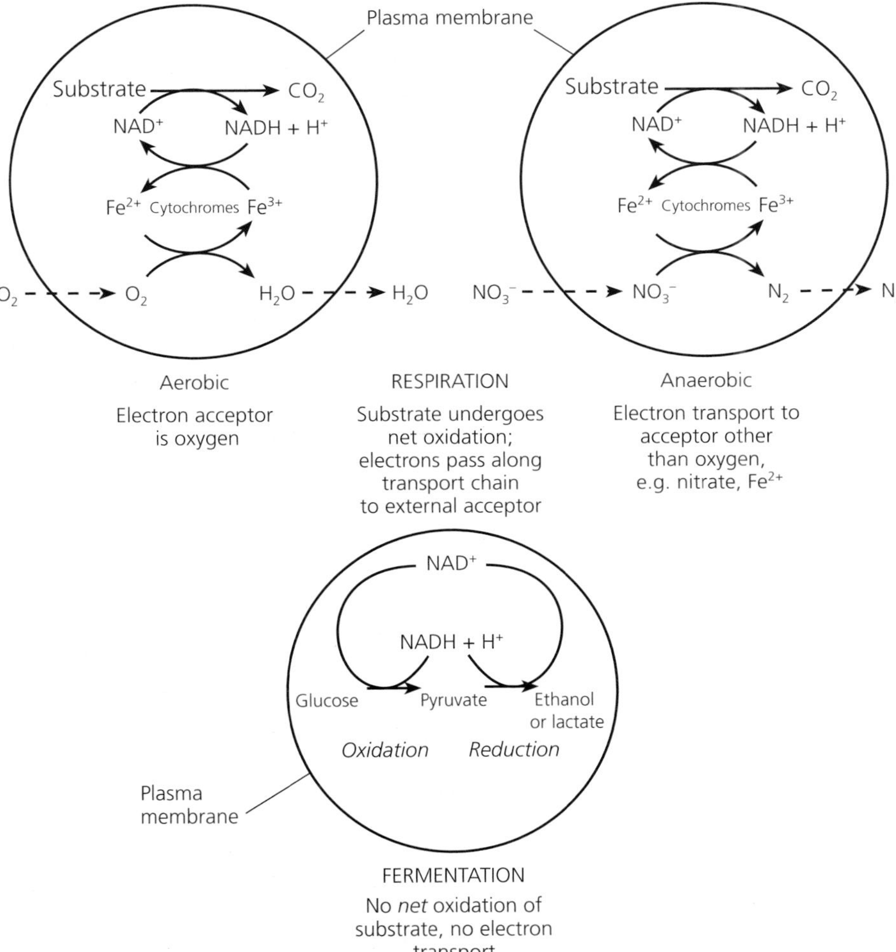


**Figure 5.13**
Distinction between respiration and fermentation

To understand how this happens, think back to glycolysis. For it to occur there must be a continuous supply of $NAD^+$ as well as glucose, ADP and phosphate. Though $NAD^+$ is continuously being used up in the oxidation of GALP to GA, it is regenerated because the NADH and $H^+$ donate their hydrogen to another substance. In the presence of oxygen, the hydrogen is given up to the respiratory chain. If oxygen is scarce or absent the hydrogen is removed using an alternative acceptor, which is the fuel itself.

In animals and many bacteria (such as those that cause milk to go sour), the hydrogen acceptor is pyruvate, which is reduced to lactate:

$$\text{pyruvate} + \text{NADH} + \text{H}^+ \rightarrow \text{lactate} + \text{NAD}^+$$

In plants and in fungi such as yeast, the pyruvate is first decarboxylated to ethanal, which then accepts hydrogen from $NADH^+ + H^+$ to form ethanol:

$$\text{pyruvate} \rightarrow \text{ethanal} + \text{CO}_2$$

$$\text{ethanal} + \text{NADH} + \text{H}^+ \rightarrow \text{ethanol} + \text{NAD}^+$$

Both lactate and ethanol thus serve as hydrogen 'dustbins'. Their function is simply to allow $NAD^+$ to be regenerated, so permitting glycolysis to continue.

Fermentation has two disadvantages:

- only two ATP molecules are made for each glucose molecule used, instead of the 32 made in aerobic respiration. This is only 1/16 as much – most of the energy is still locked up in the lactate or ethanol (both can be burned). Like charcoal, ethanol and lactate are partially burnt fuels. Fermentation is only 1/16 as efficient as aerobic respiration, so to make ATP at the same rate as it could aerobically a cell would have to use glucose 16 times as fast (Figure 5.14).

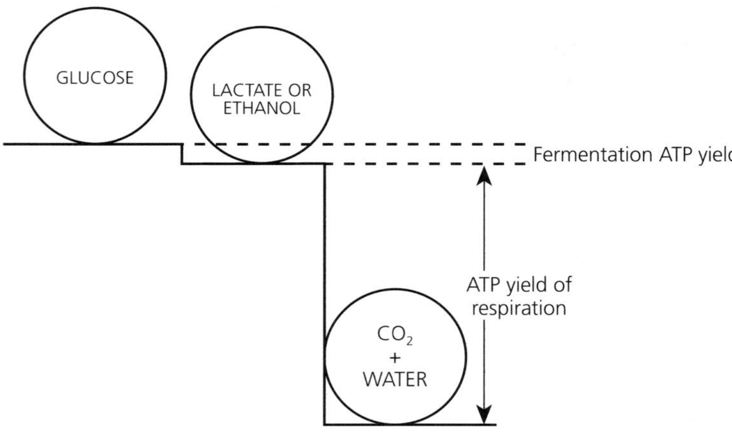


**Figure 5.14**
Relative energy yields of fermentation and respiration

- since lactate and ethanol are slightly toxic, many organisms can only sustain fermentation for a limited period. During intense exercise fermentation is very rapid and lactate accumulates. This is one reason why muscles cannot work very hard for long.

Despite its inefficiency, fermentation has its advantages. It enables some organisms to live in environments that are uninhabitable to other organisms, such as the cavity of the small intestine in which tapeworms live. Here there is little problem with inefficiency because food is so abundant that waste does not matter and the lactate can readily be got rid of into the surroundings. Fermentation is particularly useful in skeletal muscle in which energy demands can far outstrip the ability of the heart to cope (Chapter 19).

Fermentation is also put to commercial use. Alcoholic fermentation is the basis for brewing, in which both the $CO_2$ and ethanol are used, and in bread making, where the $CO_2$ is used to make dough rise. In the cheese industry, lactate, produced by bacteria, causes the pH of the milk to fall sufficiently to precipitate the milk protein as **curd**, after which the water is slowly released as **whey**.

## Respiration using fat or protein

Although glucose is the major fuel in most human tissues, fat and protein are also important sources of energy. In carnivores protein is the main respiratory fuel.

Fat is first broken down to fatty acids and glycerol. Glycerol is converted into pyruvate, which then enters the Krebs cycle. Fatty acids undergo a complex process called **β-oxidation**. Here 2-carbon fragments are repeatedly pruned off as acetyl coenzyme A which enters the Krebs cycle. In animals β-oxidation occurs in the mitochondria, and in plants it occurs in the peroxisomes. Since fatty acids enter the Krebs cycle they cannot be broken down anaerobically.

Proteins are first broken down to amino acids which are then relieved of their amino groups by a process called **deamination**, which occurs in the liver in vertebrates. The resulting **keto acids** then enter the Krebs cycle, either as pyruvate or one of the Krebs cycle intermediate compounds.

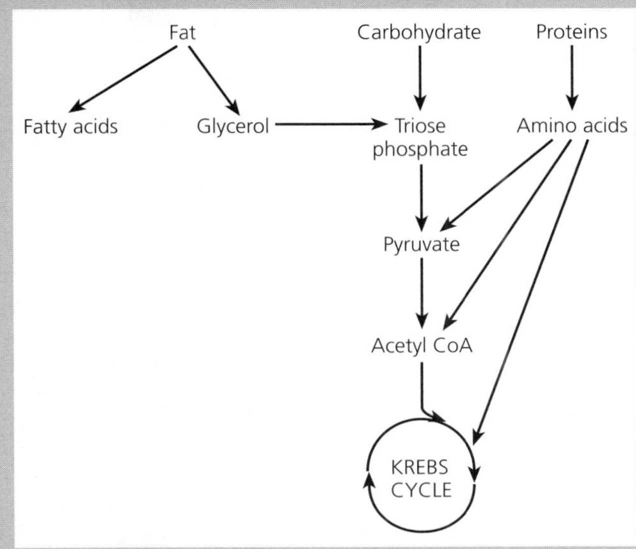


Respiration using fat and protein

## Respiratory quotients

It is often possible to get an idea of what kind of fuel an organism is using by measuring its **respiratory quotient**, or **RQ**. This is defined as:

$$\frac{\text{volume of } CO_2 \text{ given off}}{\text{volume of } O_2 \text{ taken in}} \text{ per unit time}$$

Since equal volumes of all gases at the same temperature and pressure contain the same number of molecules, the RQ is also the ratio of the number of moles of the two gases. For carbohydrate the RQ = 1, since equal numbers of molecules of $CO_2$ and oxygen (and hence equal volumes) are exchanged:

$$C_6H_{12}O_6 + 6O_2 \rightarrow 6CO_2 + 6H_2O$$

For a fat, the RQ is less than 1 because more oxygen is taken in than $CO_2$ given out. The equation for the fat tripalmitin is:

$$2C_{51}H_{96}O_6 + 145O_2 \rightarrow 102CO_2 + 98H_2O$$

This gives an RQ of 102/145 or about 0.7. The RQ for protein is about 0.9. Fermentation would, of course, give an RQ of infinity (since no oxygen is used). An RQ of over 1 could thus indicate that some fermentation is occurring – or it could imply that organic acids (which have a higher oxygen content than carbohydrate) are being respired. It should also be remembered that in many cases a mixture of substrates may be used.

RQ values are also apt to be distorted if one kind of fuel is being changed into another. For example, when a germinating seed converts fat into sugar (which contains more oxygen than fat), oxygen must be added to the fat in addition to that being used in respiration, so the RQ may be as low as 0.3.

## Control of ATP production

When a cell becomes more active ATP is used up more rapidly so the supply of ADP increases. Since ADP is a raw material for ATP production we would expect respiration to speed up, and so it normally does.

However cells have more sophisticated controls than this. A fall in the concentration of ATP or a rise in the concentration of ADP has a direct influence on the activity of certain key enzymes. One example is phosphofructokinase, a glycolytic enzyme. This catalyses the addition of a second phosphate group to fructose-6-phosphate to form fructose-1,6-bisphosphate. If the ATP concentration falls, or if the ADP concentration rises, the activity of the enzyme increases, speeding up glycolysis and boosting the supply of ATP.

The inhibitory effect of ATP on the activity of phosphofructokinase is an example of negative feedback – the more active the enzyme, the more ATP is produced and the more the enzyme is inhibited. Negative feedback is a characteristic of many self-regulating processes.

How does feedback inhibition of enzymes work? In other words, how do ATP and ADP regulate the activity of the active site of phosphofructokinase? The answer is that enzymes subject to feedback control are *allosteric*. This means that in addition to the active site they have another site at which a molecule other than the substrate can bind to the enzyme. The effect of binding with ADP is to increase the activity of the active site; ATP has the reverse effect.

It is worth pointing out that ATP and ADP exert their effects on an enzyme situated at an early stage of the respiratory pathway – precisely the point at which control would be expected to be most sensitive.

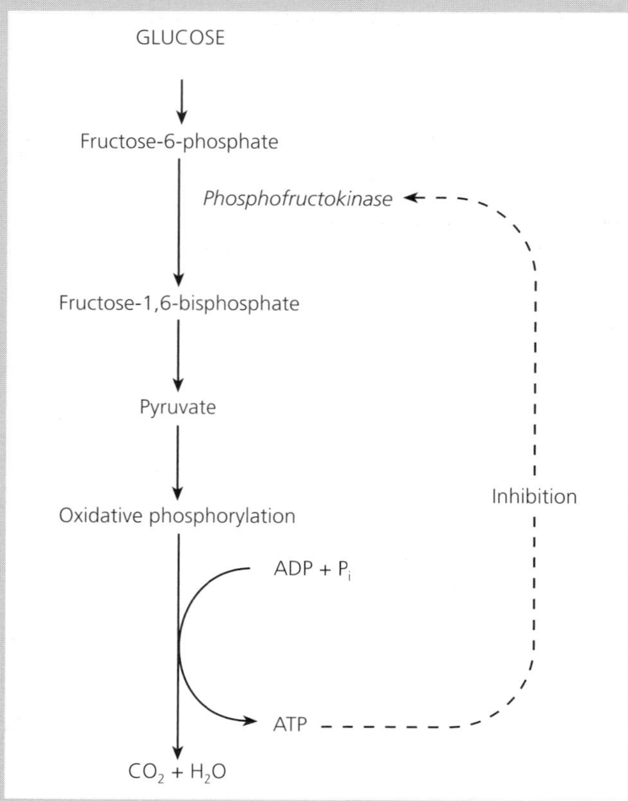


Control of ATP production

## QUESTIONS

**1** The same chemical equation summarises both the combustion (burning) of glucose and its breakdown in respiration. There are, however, some fundamental differences between them. Give *two*.

**2** Which of the following provides **(a)** the most and **(b)** the least rapidly available source of energy for cells?

starch, ATP, glucose, pyruvate.

**3** The following substances yield different amounts of useful energy:

glucose, starch, ATP, pyruvate.

Write them down in *decreasing* order of the amount of useful energy *per molecule*.

**4** Why does fat give more than twice as much energy per gram as carbohydrate on oxidation?

**5** All of the following can be used by cells to give energy. Which would you expect to yield the greatest amount of useful energy *per gram* if used in respiration, and why?

| $C_3H_8O_3$ | $C_3H_4O_3$ | $C_3H_6O_3$ | $C_2H_6O$ |
|---|---|---|---|
| glycerol | pyruvate | lactate | ethanol |

**6** In the presence of oxygen a molecule of lactate can yield more ATP than a molecule of pyruvate. Why?

**7** It is easy to demonstrate that in respiration oxygen is used up and $CO_2$ is produced. It is much more difficult to show that water is produced. One way is to use 'heavy' oxygen, or $^{18}O$. This is chemically identical to ordinary oxygen but has slightly heavier atoms, which can be distinguished from the ordinary form, $^{16}O$, by sensitive instruments. To show that water is being produced by a mouse, which of the following procedures would be the best? Supply the mouse with:

**A** sugar containing $^{18}O$, and then analyse the air it breathes out for the presence of water containing $^{18}O$.
**B** sugar containing $^{18}O$, and then analyse the urine for the presence of water containing $^{18}O$.
**C** 'heavy' oxygen gas, and then analyse the urine for water containing 'heavy' oxygen.
**D** water containing 'heavy' oxygen, and then analyse water vapour in the exhaled air for the presence of 'heavy' oxygen.

**8** A mouse was given glucose containing the radioactive isotope $^{14}C$ and air containing the 'heavy' but non-radioactive isotope, $^{18}O$. The water in the animal's urine and the $CO_2$ breathed out by the animal were subsequently analysed for the presence of these isotopes. Which of the following would you expect to be the result?

**A** The $CO_2$ would be radioactive and 'heavy', and the water would be normal.
**B** The $CO_2$ would be radioactive and the water would be 'heavy'.
**C** The $CO_2$ would be non-radioactive and the water would be 'heavy'.
**D** The $CO_2$ and water would both be radioactive and the water would be 'heavy'.

CHAPTER

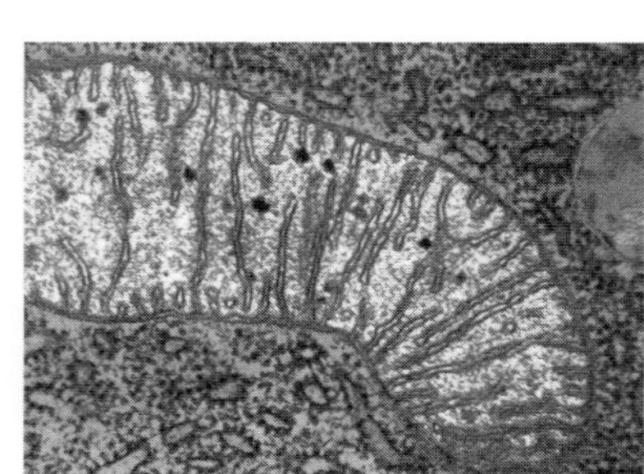

# 6 HOW THINGS ENTER and LEAVE CELLS

## LEARNING OBJECTIVES

By the time you have completed your study of this chapter you should be able to:

- explain what is meant by an open system and be able to give examples of substances entering and leaving cells.
- understand the nature of diffusion and the biological significance of the factors affecting its rate.
- explain why osmosis is a special case of diffusion.
- explain the effects of placing animal cells in different kinds of solution.
- explain how the cellulose wall of plant cells affects the water relations of plant cells.
- use the concept of water potential to explain the movement of water into and out of plant cells.
- distinguish between active and passive transport.
- explain how a concentration gradient of one substance can be used to transport another substance across a membrane.
- be aware of the similarities and differences between facilitated diffusion and active transport.
- be able to distinguish the various forms of cytosis and to give an example of each.

Living cells are **open systems**, meaning that they are constantly exchanging materials and energy with their surroundings. All this coming and going occurs through the **plasma membrane**, a 5–10 nm-thick layer separating a cell from its environment. To some extent the kind of substances moving across the plasma membrane reflect the function of the cell and its circumstances (Figure 6.1). For example oxygen may leave a leaf cell in the day and enter it at night, whilst the reverse holds for $CO_2$.

Substances enter and leave cells in a number of different ways: **diffusion** (of which osmosis is a special case), **active transport**, **facilitated diffusion** and **cytosis**.

**Figure 6.1**
Some examples of chemical exchanges in cells

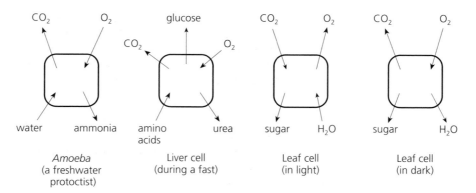

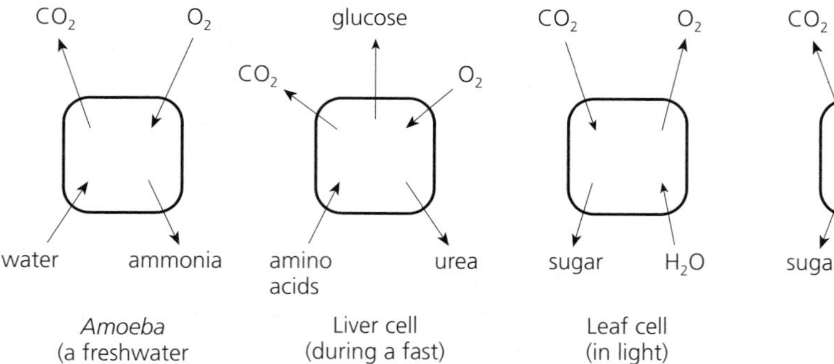

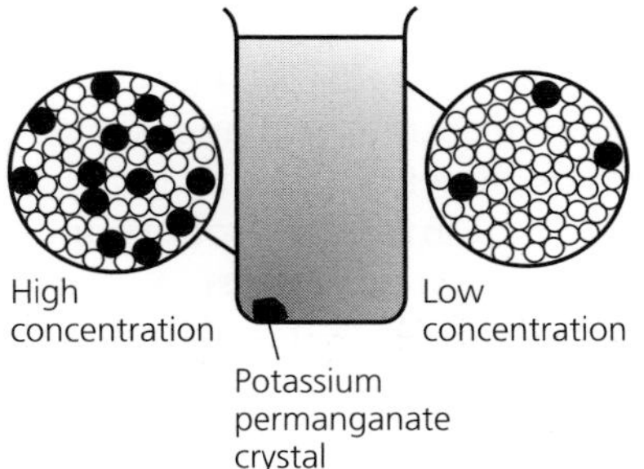


**Figure 6.2**
Diffusion of coloured material (black) in water (white)

# Diffusion

Diffusion is *the movement of a substance from an area of high concentration to an area of lower concentration as a result of the random movement of its particles* (Figure 6.2).

Diffusion depends on the fact that in gases, liquids and dissolved solids, molecules are in constant, random motion. You can see diffusion in action if you drop a crystal of potassium permanganate into a beaker of water that has been allowed to stand for a couple of hours to ensure complete stillness. Over the next few hours, the purple colour very slowly spreads upwards, and by the following day it has usually become evenly spread (much of the overnight spreading is actually due to minute convection currents associated with cooling). Once the pigment is evenly spread, all *net* movement ceases – although molecular and ionic movement continues in all directions.

The energy for diffusion comes from the kinetic energy of particles. Diffusion is not something that a cell 'does' or 'carries out', but is a purely passive process. The organism or cell creates the concentration difference and diffusion is the result of this. An amoeba does not 'excrete' ammonia in the sense that it actively propels the ammonia out of its body – the ammonia leaves quite passively with the energy coming from the thermal movement of the ammonia molecules. Similarly oxygen diffuses into a cell because it is being used in respiration and thus the concentration of oxygen within the cell is lower than that outside. The cell is not 'doing' the transporting.

It is important to realise that diffusion is a *net* movement of molecules – at any given moment, some oxygen molecules are moving out of a cell, but more are moving into it.

## Factors that affect diffusion rates

Organisms must be able to obtain their raw materials and get rid of wastes quickly enough for their needs. The speed of diffusion of a substance is therefore very important and depends on five factors:

1 **the nature of the material through which diffusion occurs** On the cellular scale distances are short and diffusion is generally rapid. Over distances more than a millimetre or so, diffusion through liquids is too slow to support an organism's need for gas exchange. On land a considerable proportion of the diffusion pathway is through air, in which diffusion is much faster. Oxygen, for instance, diffuses over 300 000 times more quickly in air than it does in solution. This is because molecules in a gas are much further apart, so they collide and change direction far less often.

   The significance of this can be seen when soil becomes flooded. Oxygen normally diffuses quite rapidly down to plant roots through the honeycomb of air spaces. In flooded soils it has to diffuse in solution and, since it is only sparingly soluble, the oxygen content of the soil water falls quickly. If flooding is prolonged, roots may die. Earthworms incidentally can take evasive action and usually come up to the surface after heavy rain.

   Diffusion through membranes is slower than through water. The resistance to diffusion depends on lipid solubility, with lipid-soluble particles penetrating faster than polar molecules and ions. It also depends on particle size, small particles passing through membranes faster than larger ones.

2 **concentration gradient** The rate of diffusion is directly proportional to the steepness of the concentration gradient. The concentration gradient is the rate at which concentration changes with distance (Figure 6.3). The rate of diffusion is therefore directly proportional to the concentration difference and inversely proportional to the length of the diffusion pathway. Thus the rate of diffusion would be doubled if the concentration difference were to be doubled, or if the length of the diffusion pathway were to be halved.

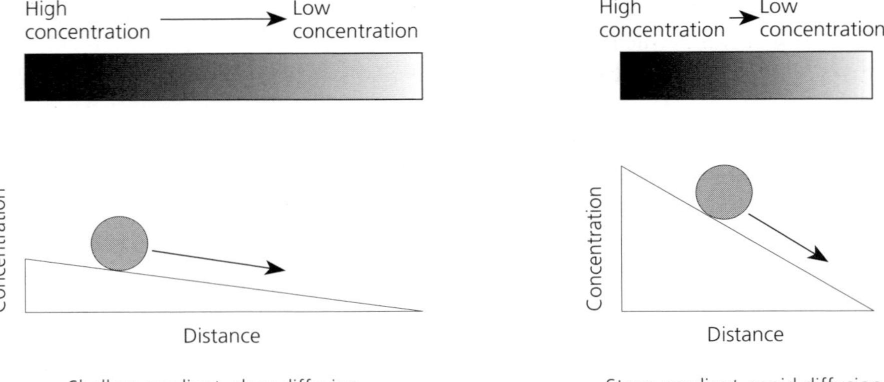


**Figure 6.3**
Model showing effect of concentration gradient on diffusion rate

Because cells are very small, substances diffuse in and out quickly, even in solution. Though plant cells tend to be larger than animal cells, the large vacuole displaces the cytoplasm to the outside of the cell. As a result the chloroplasts are very close to the intercellular air spaces (a distance of about 2 μm).

3 **particle size** At any given temperature the molecules of all kinds of substances have the same average energy, so small particles must be moving faster than big ones. In a solution of sucrose and glucose therefore the glucose molecules are, on average, travelling about 1.4 times faster than those of sucrose.

4 **temperature** Diffusion is slightly quicker at higher temperatures because molecules move faster. The effect is actually very small – a rise from 10°C to 20°C increases the average molecular speed by about 1.7%.

5 **surface area** The larger the area available for diffusion the greater the quantity of material diffusing in a given time. Many organisms increase the surface area of their bodies by folding.

## CLEAR THINKING

**The rate of any process is the amount of change divided by the time taken for that change to take place. Where diffusion is concerned, 'rate' can have two very different meanings, as illustrated by the following two examples:**

- **if we are talking about the rate of diffusion of oxygen across the lining of the lungs, we mean the *quantity* of oxygen diffusing from air to blood per second per unit area of membrane.**
- **when talking about the rate of diffusion of a chemical signal from one nerve cell to another, we mean the *time* taken for the chemical to cross the gap between one nerve cell and another.**

**In the first situation, there is a more or less steady concentration gradient, oxygen being continuously delivered to one side of the membrane by breathing movements and removed from the other side by the blood. The quantity of oxygen crossing the membrane**

**CLEAR THINKING**

**per second is inversely proportional to the thickness of the membrane. Other things being equal, doubling the length of the diffusion pathway halves the rate of diffusion.**

**The rate of diffusion across a membrane also depends on the permeability of the membrane, and on the area of the membrane. This is summed up by Fick's First Law of Diffusion, which states that the rate of diffusion across a membrane is proportional to:**

$$\frac{\text{difference in concentration}}{\text{thickness of membrane}} \times \text{surface area of membrane}$$

**In the second case we are dealing with non-steady state conditions which apply when a substance is suddenly released from a membrane. As it spreads from its point of release its concentration gradient decreases, and hence so does its rate of travel. The speed of diffusion is proportional to the square of the distance between the two points, summed up by Fick's Second Law of Diffusion.**

**For example, suppose it takes an oxygen molecule an average time of 0.0001 seconds to diffuse a distance of 1 µm. To diffuse 1 metre (a million times as far) it would take about a million million times as long, or about 3 years!**

## Osmosis

Osmosis is a special case of diffusion in which water moves through a **selectively-** or **differentially-permeable membrane** such as the dialysis tubing shown in Figure 6.4. This is like a very fine sieve in that it allows small molecules (such as water) to pass through, but not larger ones like sugar. The level in the tube slowly rises and eventually overflows.

To begin with it may be helpful to recall that there is a universal tendency for energy to become evenly spread. Thus heat is conducted from the hotter to the cooler end of a metal bar until its temperature is the same throughout. Similarly, when an object rolls downhill it moves from a higher to a lower gravitational energy.

So too with the water and sugar solution. By entering the sugar solution, water is moving from an area of higher energy to one of lower energy. In doing so it could be made to do work, which is what happens in the generation of hydroelectricity.

In a solution, the energy of the water that is available to do work (**free energy**) is called the **water potential** of that solution (denoted by $\psi$, the Greek letter psi). When two solutions are separated by a selectively-permeable membrane, water always tends to move from a higher to a lower water potential. Thus in Figure 6.4 the sugar solution inside the dialysis tubing has a lower water potential than the water in the beaker.

We can now define water potential. The water potential of a solution is the tendency of water to leave that solution if it is separated from distilled water by a selectively-permeable membrane. (The fact that water enters the sugar solution in Figure 6.4 simply means that it has a *negative* tendency to leave it.)

The movement of water from a higher to a lower water potential through a selectively-permeable membrane is called **osmosis**. We should remember, however, that as in other cases of diffusion, water actually moves across the membrane in *both directions simultaneously*, but at a given moment more molecules are moving into the bag than out of it.

There are two reasons why, in Figure 6.4, the water in the beaker has more energy than the water in the sugar solution. Firstly, the water

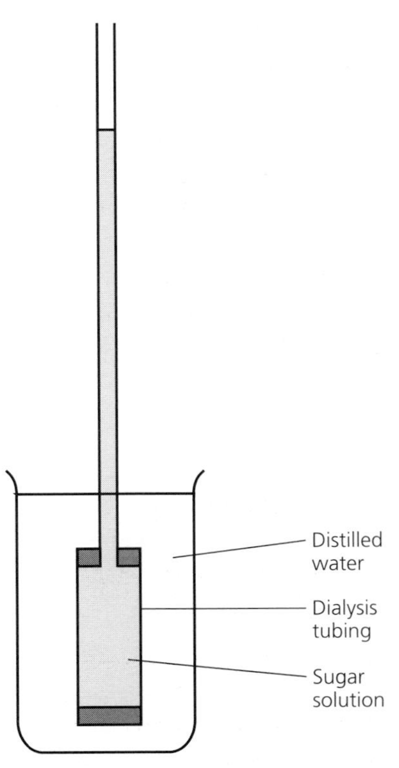


**Figure 6.4**
Simple demonstration of osmosis

concentration is lowered slightly (in effect, the sugar 'dilutes' the water), so there are fewer water molecules per unit volume. Secondly, water molecules are attracted to solute molecules so each sugar molecule is surrounded by an 'overcoat' of water molecules. A water molecule that is attracted to a sugar molecule is less free to move, and so has less kinetic (movement) energy than one which can move freely. The effect of the sugar (or any other solute) is thus to lower the average kinetic energy of the water molecules.

How is water potential measured? As in the case of electrical energy we can only measure *differences* in potential, which means that we need a standard against which other solutions can be compared. The standard used is pure water at atmospheric pressure. The water potential of a solution ($\psi$) is the sum of two factors:

- its **solute potential** ($\psi_s$). This is a measure of the lowering of the water potential due to the presence of solutes. The solute potential of pure water at atmospheric pressure is arbitrarily fixed at zero, so *all solutions have a negative solute potential*. The greater the concentration of solutes, the lower – that is the more negative – the water potential.
- its **pressure potential** ($\psi_p$), or simply 'pressure'. This can be either positive or negative. A solution at atmospheric pressure has a pressure potential of zero. In living plant cells pressure potential is usually positive, but in the xylem vessels of plants (through which water is pulled) it is nearly always negative. When the pressure in a solution is negative it is under **tension**. (In animal cells the pressure inside and outside the cell is always the same, so pressure potential does not enter into the matter.)

The water potential of a solution is thus the sum of the pressure potential (positive or negative) and the solute potential (always negative):

$$\psi = \psi_s + \psi_p$$

Water potential, solute potential and pressure potential are all measured in pressure units. The Pascal (Pa) is 1 Newton per $m^2$ and is too small for convenience, so kiloPascals (kPa) or megaPascals (MPa) are used (1 kPa = 1000 Pa, and 1 MPa = 1000 kPa).

## Measuring solute potential

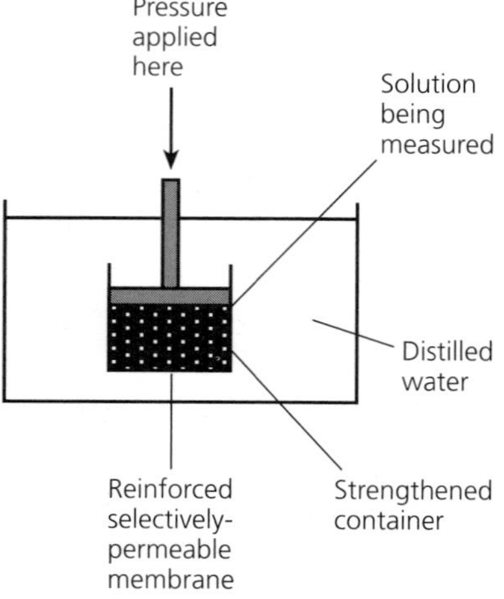


**Figure 6.5**
Measuring solute potential by applying back pressure

Imagine a solution separated from distilled water by a selectively-permeable membrane, as in Figure 6.5. Water tends to enter the solution by osmosis but, if enough pressure is applied, osmosis can be prevented or even reversed. The solute potential of the solution is defined as the pressure that would have to be exerted on it to just prevent the water entering it by osmosis.

If, in Figure 6.5, a pressure of +3 MPa is needed to prevent water entering the solution, then its solute potential is −3 MPa. Under a pressure of 3 MPa its water potential is zero ($-3 + 3 = 0$). On the other hand, if the pressure in the solution is +1 MPa, its water potential is therefore $-3 + 1 = -2$ MPa.

## Measuring osmotic concentrations

Where osmosis is concerned, concentrations of dissolved substances are best expressed in terms of **molarity** rather than as a percentage. A **molar** solution is one in which 1 $dm^3$ (a litre) of solution contains the molecular mass (one mole) of the substance in grams. Thus a litre of molar glucose solution contains 180 grams of glucose (molecular mass of glucose = $(6 \times 12) + (12 \times 1) + (6 \times 16) = 180$). Similarly a litre of molar sucrose contains 342 grams of sucrose.

Expressing concentrations in terms of molarity is very useful for one simple reason: for substances whose particles do not dissociate into ions, *solutions of equal molarity have the same number of solute particles per unit volume of solution and have the same solute potential.* In contrast, a given volume of 10% sucrose solution contains only just over half as many solute molecules as the same volume of 10% glucose solution (since sucrose molecules are almost twice as big as glucose molecules). This is important because the solute potential of a solution depends on the *number* of solute particles per unit volume rather than their total mass.

## Osmosis in animal cells

In mammals and other vertebrates the cells are bathed by **tissue fluid**. This is **isotonic** with the cells, meaning that it has the same solute potential as the cells. Under these conditions the cells have no need to regulate their water content (and indeed are quite unable to do so). The solute potential of the tissue fluid is regulated by the kidneys, which thus have a similar function to the contractile vacuole of a protozoan.

The inability of mammal cells to regulate their water content is evident when blood is diluted with water. The surrounding solution is now **hypotonic** to the cells (meaning that the solute concentration is lower). Water enters by osmosis, causing the cells to swell and burst, releasing the haemoglobin and leaving the empty cell membranes (Figure 6.6). If a salt solution that is **hypertonic** to the cells (having a higher solute concentration) is added to blood, the cells lose water by osmosis and shrink.

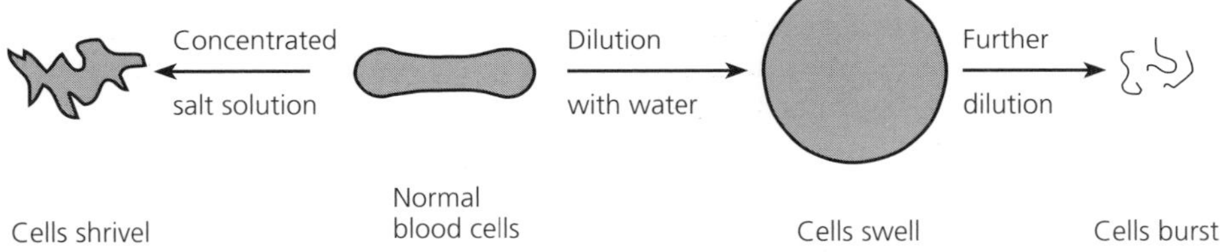


**Figure 6.6**
Effect of different solutions on red blood cells

## Osmosis in plant cells

As far as osmosis is concerned, a plant cell consists of three parts (Figure 6.7):

- the **cell wall**, which allows all molecules to pass through – i.e. is fully permeable. Because the cell wall is strong enough to resist tension the cell can develop a high internal pressure.
- the **vacuole**, which contains a watery **cell sap**, a solution of salts and other dissolved substances. The vacuole is separated from the cytoplasm by another selectively-permeable membrane, the **tonoplast**.
- the **cytoplasm**, which is bounded by two selectively-permeable membranes – externally by the plasma membrane and internally by the tonoplast.

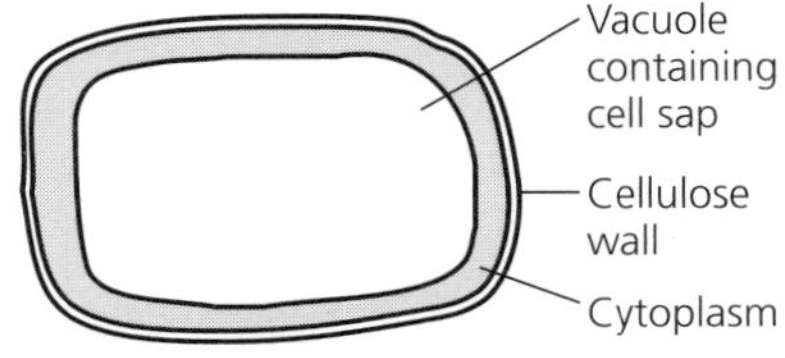


**Figure 6.7**
Osmotically important parts of a plant cell

Osmosis can be readily observed in plant cells by mounting a tissue, such as a strip of rhubarb or onion epidermis, in molar sucrose solution. After a few minutes the cytoplasm can be seen to have pulled away from the cell wall, a process called **plasmolysis** (Figure 6.8A).

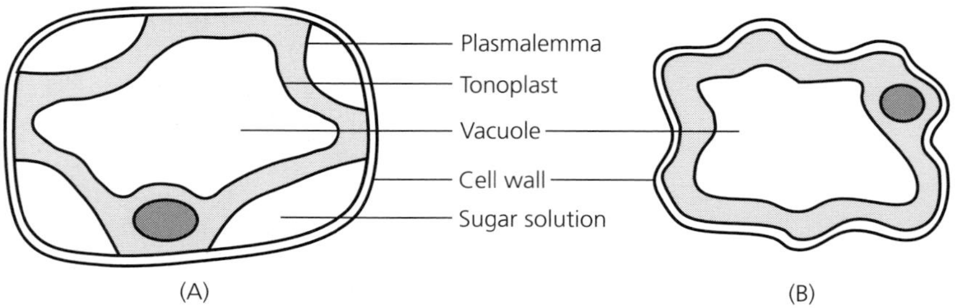


**Figure 6.8**
Effect of water loss in plant cells in concentrated solutions (A) and air (B)

The explanation depends on the fact that the sugar solution has a lower (more negative) solute potential than the cell sap. Water leaves the cell by osmosis, and continues to do so even after the wall has shrunk as much as it can. As a result, the cytoplasm separates from the wall like a deflated football bladder.

The fact that it is the cytoplasm that appears to shrink rather than the wall itself shows that the wall must be permeable to the sugar solution, and that the selectively-permeable layer must be the cytoplasm. In a plasmolysed cell, the liquid occupying the space between the wall and the cytoplasm must, therefore, be sugar solution.

Obviously, the degree of plasmolysis depends on the concentration of the surrounding solution. A solution causing *just detectable* plasmolysis has the same solute potential as that of the cell sap, and a cell in this state is said to be at **incipient plasmolysis**.

*Point to note:* Plasmolysis only occurs when cells are surrounded by strong solutions, and rarely, if ever, occurs under natural conditions. When a plant cell loses water by **evaporation** the result is quite different. Instead of the cytoplasm contracting away from the wall, the water at the surface of the cell retreats into the submicroscopic spaces between the cellulose microfibrils in the wall. The curved water surfaces (menisci) act like tiny trampolines, setting up a negative pressure and the cell collapses (Figure 6.8B).

So much for the behaviour of plant cells in hypertonic solutions. What happens in hypotonic solutions? Imagine that a completely limp or **flaccid** cell is placed in distilled water. To begin with, $\psi_p = 0$, so $\psi = \psi_s$. As water enters by osmosis, the cell swells, stretching the wall and causing the pressure potential to rise. At the same time, the cell sap becomes slightly diluted, causing its solute potential to rise (become less negative). The higher the pressure potential inside the cell, the more it opposes further water entry. When the pressure potential is equal in magnitude (but of course opposite in sign to the water potential) the water potential is zero, and the cell can absorb no more water. A cell in this state is like a tightly inflated football and is said to be **turgid**, and its internal pressure is called **turgor pressure**. These changes are illustrated in Figure 6.9 in which some imaginary values for $\psi$, $\psi_p$ and $\psi_s$ have been used.

## The plasma membrane as a barrier

A cell membrane consists of a lipid bilayer with globular proteins floating in it (Chapter 3). This fluid mosaic structure (Figure 3.11) looks solid enough, yet clearly a wide variety of molecules are able to cross it. Oxygen, $CO_2$, water, glucose, amino acids, ions, proteins and steroids are just some of the substances making up this traffic. It is important to realise however that only certain molecules such as oxygen, $CO_2$ and water, cross by diffusion, most substances crossing by other mechanisms.

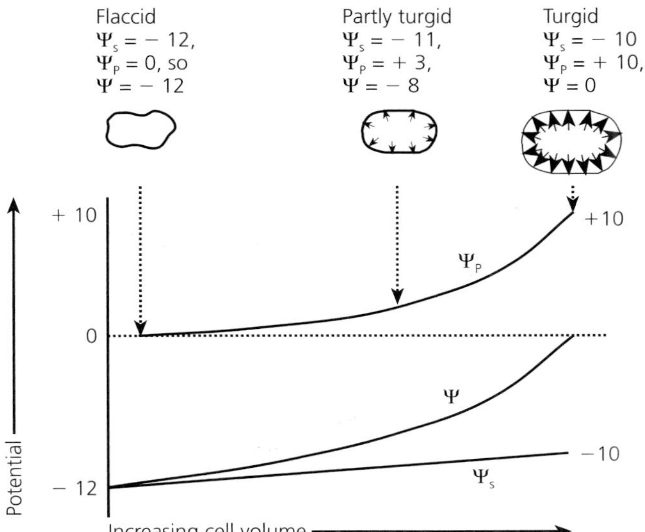


**Figure 6.9**
Changes in solute and pressure potential in a plant cell as its water content changes (values of $\psi_p$, $\psi_s$ and $\psi$ are imaginary)

The lipid and protein components of the plasma membrane offer two different routes into and out of the cell and involve quite different mechanisms:

- very small molecules such as water, oxygen and $CO_2$, can cross an artificial lipid bilayer (containing no protein) by simple diffusion, squeezing between the lipid molecules. Even so, cell membranes impose considerable resistance, even to quite small molecules. Water, for example, moves 10 000 times more slowly across a cell membrane than it does in the absence of a membrane. Larger molecules can only do so if they are fat soluble, such as steroid hormones and ethanol (which is why alcohol is absorbed so quickly by the stomach).
- ions, and molecules the size of glucose, cross an artificial lipid bilayer very slowly, and proteins cannot do so at all. These substances cannot cross a cell membrane by simple diffusion, but by chemical mechanisms involving the membrane proteins.

## Comparing the permeability of membranes to different solutes

When a drop of blood is added to a test tube containing about 5 $cm^3$ of urea solution that is isotonic with the plasma (0.3 molar), the resulting suspension of red blood corpuscles initially looks cloudy. After about 5 seconds the red cells burst, releasing haemoglobin and forming a clear red solution.

Why do the cells burst when put in an isotonic solution? Although the total solute concentration is initially the same on either side of the red cell membranes, the urea is much more concentrated outside, so it diffuses into the red cells. This lowers their water potential to below that of the external solution and water diffuses in, causing them to swell and burst.

The time taken for the cells to burst is a measure of their permeability to urea. When 0.3 molar glycerol solution is used instead of urea the solution takes about half a minute to go clear. Evidently glycerol penetrates red cells more slowly than urea. When 0.3 molar glucose solution is used the liquid remains cloudy, suggesting that red cell membranes have very low permeability to glucose.

These observations make sense when we compare the molecular masses of the three solutes: 60 for urea, 90 for glycerol and 180 for glucose, suggesting that small molecules penetrate membranes more quickly than larger ones.

Another factor affecting how quickly a substance enters a cell is its solubility in lipid. For a given molecular size, the more lipid-soluble a substance the faster it penetrates.

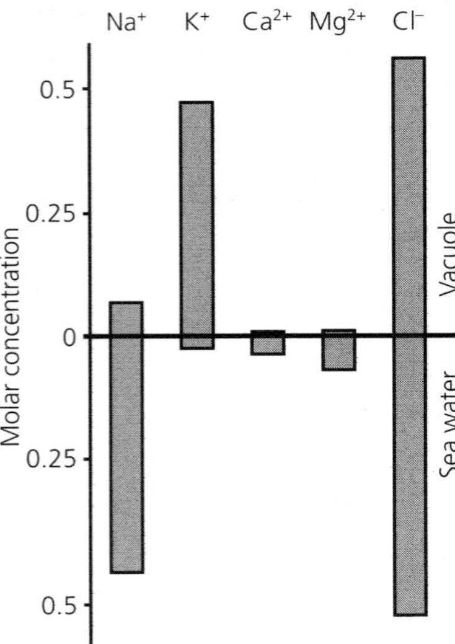


**Figure 6.10**
Comparison between external and internal ion concentrations in *Valonia*, a marine protoctist

# Active transport

As we have seen, diffusion tends to result in molecules and ions becoming randomly dispersed so that concentrations become equal. Yet the ionic composition of living cells is always very different from that of the surroundings. Some minerals are more concentrated inside than outside, whilst for others the reverse is true (Figure 6.10).

In some organisms the ability to concentrate minerals is spectacular. Sea squirts, for instance, are able to concentrate vanadium, a metal that occurs in extremely low concentrations in sea water, and many seaweeds can accumulate iodine.

At first sight this ability to concentrate chemicals might appear to be contrary to the Second Law of Thermodynamics, which states that all systems tend to become more random or disordered. Actually they are not an exception to the Second Law. It is possible for 'unnatural' changes to occur, *provided energy is supplied.* A refrigerator can remain cooler than its surroundings by expending energy to pump heat from its cold interior to the warm environment. As soon as it is switched off its interior warms up to the same temperature as the environment. Similarly, living things can transport substances from low to high concentrations by expending energy. Cells are thus **open systems**, constantly exchanging materials and energy with their surroundings.

Movement of a substance from a low concentration to a higher concentration ('uphill') is called **active transport**. In some form or another active transport is universal in living cells. For example, in all cells there is a slow outward leakage of potassium ions which is counteracted by the active uptake of potassium. In most animals the active accumulation of potassium is linked to the pumping out of sodium, which slowly leaks in from the higher concentration outside.

Active transport is not confined to the plasma membrane; it also occurs across the membranes of organelles. Lysosomes and the thylakoids of chloroplasts, for instance, actively accumulate $H^+$ ions. Small wonder that active transport accounts for about a third of the energy budget of most cells – in nerve cells it may be two-thirds.

### Detecting transport

When investigating rates of solute uptake, it must be possible to distinguish between solute that is absorbed during the experimental period, and solute that was already inside the cells. Radioactive tracers are particularly useful here. The tissue is placed in a solution containing the tracer at a known concentration and left for a specified period under the conditions being investigated. The tissue is then washed and its level of radioactivity determined using a **Geiger counter.** Knowing the concentration of the tracer and the period of time the tissue was immersed in it, the rate of uptake of the solute can be calculated.

## Characteristics of active transport

As an 'uphill' process, active transport is dependent upon a supply of metabolic energy. It is therefore not surprising that cells heavily engaged in transport work such as those of the proximal tubules of the kidney – have many mitochondria.

Active transport differs from diffusion in a number of important ways:

1 it is impeded by anything that interferes with ATP production, such as a lack of oxygen or the presence of cyanide (Figure 6.11A).

2 whereas diffusion is only slightly temperature sensitive, active transport is highly responsive to temperature (Figure 6.11B).

3 whereas the rate of diffusion of a solute into a cell is proportional to its external concentration, active transport shows a 'saturation' effect (Figure 6.12). As the external potassium concentration increases, so does its initial rate of uptake, *but only up to a point.* Beyond this, further increases in concentration have no effect. This is readily explained by supposing that at higher potassium concentrations the carrier becomes fully occupied, or 'saturated', just as the active site of an enzyme becomes saturated.

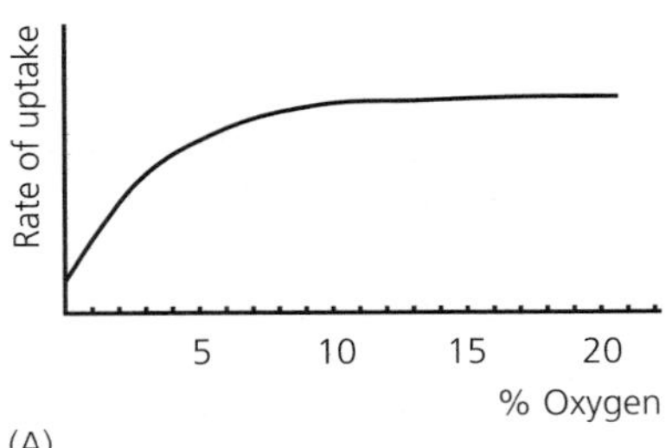


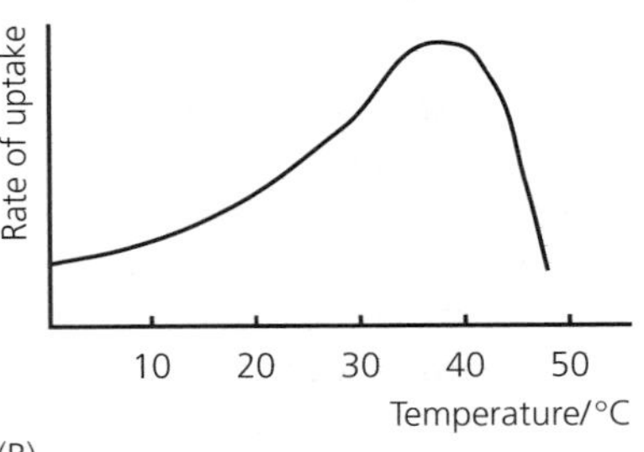


**Figure 6.11▲**
Effect of oxygen supply on rate of active uptake of bromide ions by barley roots (A), and effect of temperature on rate of uptake of potassium ions by carrot discs (B)

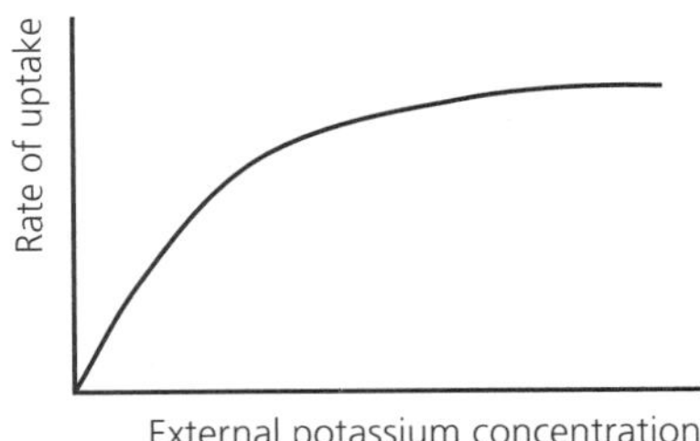


**Figure 6.12▲**
Effect of solute concentration on its rate of active uptake by carrot tissue

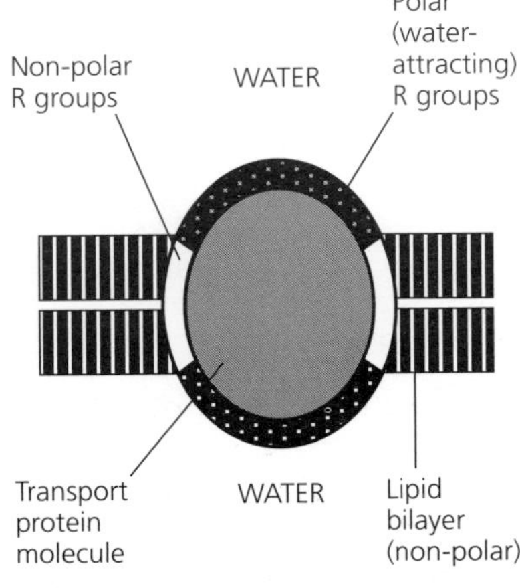


**Figure 6.13▲**
How the orientation of membrane transport proteins is maintained

4 like enzymes, active transport is susceptible to the action of inhibitors. Phlorizin, for example, inhibits the uptake of glucose by kidney tubule cells. In some cases an inhibitor is chemically similar to the normally transported solute. For example, the uptake of potassium by roots may be inhibited by rubidium, and strontium may compete with calcium. In nerve cells the uptake of glucose is greatly reduced by the presence of deoxyglucose. This suggests that carriers in the membranes have sites that fit the substance, but that other substances with similar structures can compete with it.

## How does active transport work?

The above facts all suggest that active transport is a chemical process involving protein carriers located in cell membranes, and research has confirmed this. To transport a substance across a membrane a protein must span its entire thickness, one side combining with the chemical and the other releasing it.

Transport proteins are orientated within the membrane in the following way. Their polar R groups tend to be concentrated on the sides of the molecule that face the water on either side of the membrane. The R groups in contact with the lipid bilayer are predominantly non-polar (Figure 6.13). The polar R groups form hydrogen bonds with the water and these hydrogen bonds keep these surfaces facing towards the water. Hence, although transport proteins are free-floating within the bilayer and can spin round like a top, they cannot easily flip over and 'do a somersault'.

A carrier protein must therefore contain a channel through which the particle is transported. Some carrier proteins consist of more than one polypeptide, loosely held together into a quaternary structure, with a channel running between them.

It is probable that the carrier alternates between two different conformations or shapes. Having picked up its load, the protein undergoes a change in its conformation, releasing the particle on the other side (Figure 6.14). To transport another molecule the protein must first regain its original shape, a process requiring energy in the form of ATP.

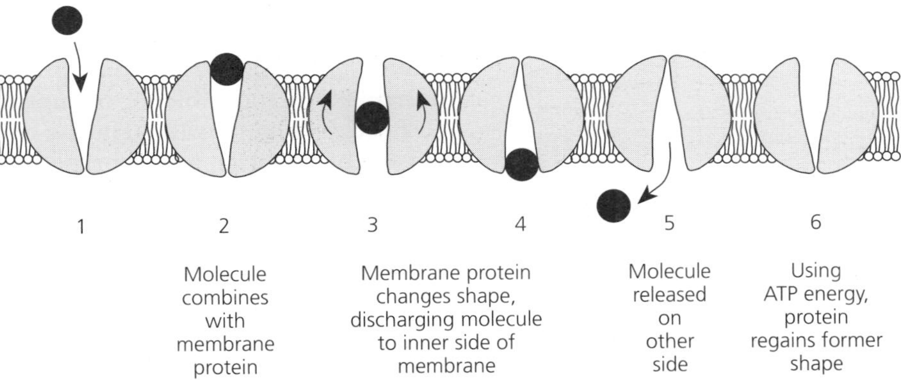


**Figure 6.14▲**
Reversible change in conformation of carrier protein molecule

## Cotransport

Many active transport mechanisms are more complicated than that just described, the transport of one solute being linked with the transport of another.

An example of this **cotransport** is the regulation of the balance between sodium and potassium ions in animal cells. The outward leakage of potassium and inward leakage of sodium are offset by a carrier protein in the plasma membrane called **sodium–potassium ATPase**. This actively pumps out three sodium ions in exchange for two potassium ions, accompanied by the hydrolysis of one ATP molecule (Figure 6.15).

In the above example sodium and potassium are transported in opposite directions across the membrane. Solutes may be transported in the same direction. For example, in the small intestine glucose is absorbed with sodium (see below).

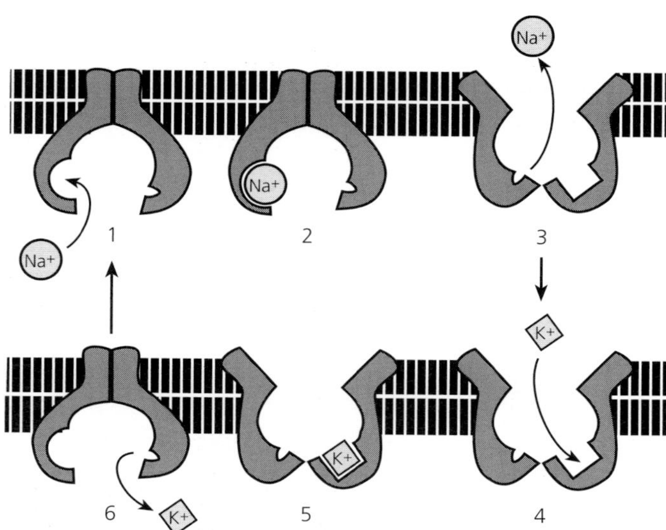


**Figure 6.15**
Cotransport of sodium and potassium (after Alberts, *et al.*)

## Facilitated diffusion

Facilitated diffusion is a form of transport which shares features with both diffusion and active transport. There are two quite different mechanisms involving permeases and ion channels, respectively.

1 **Permeases** differ from active transporters in that they only transport solutes *down* a concentration gradient. One example is **glucose permease**, responsible for the uptake of glucose by red blood corpuscles (Figure 6.16). Having combined with glucose on the outside of the plasma membrane, the permease undergoes a conformational change and discharges the glucose molecule into the cell.

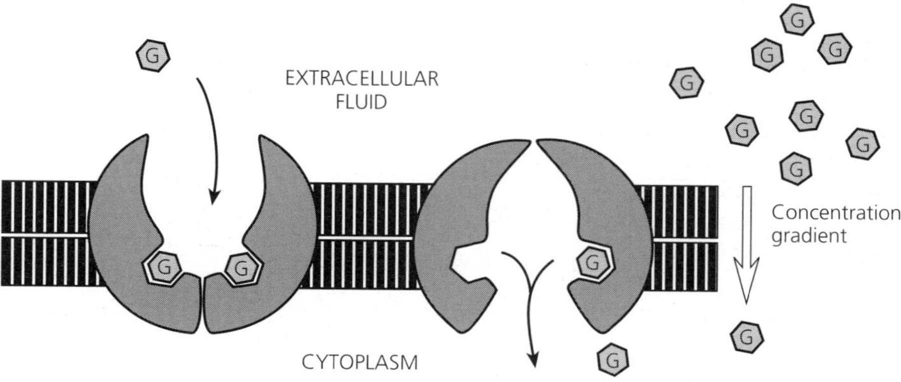


**Figure 6.16**
Facilitated diffusion of glucose by glucose permease

Without any 'push' from ATP it then reverts to its original conformation, ready to pick up another glucose molecule. On release inside the corpuscle, glucose is then phosphorylated to glucose-6-phosphate. This lowers the glucose concentration inside the cell and helps to maintain a steep concentration gradient.

Since ATP is not required to convert one conformational state into the other, permeases are more like turnstiles than pumps, the direction of transport always being 'downhill'.

In other respects, however, permeases resemble active transporters:

- they have the same high degree of specificity with regard to the solute carried.
- they may be competitively inhibited. Glucose permease, for instance, is competitively inhibited by deoxyglucose.
- permeases show a saturation effect.

2 **Ion channels** operate in a quite different way. Each forms a tube lined by polar R groups which thus allow the passage of ions down a concentration gradient. Ion channels play an important part in the propagation of nerve impulses. There are several kinds, each specific for one kind of ion (Figure 6.17).

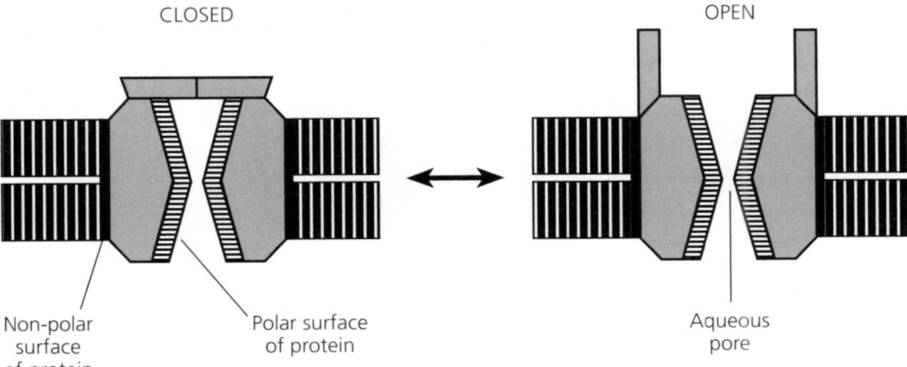


**Figure 6.17**
Diffusion via ion channels (after Alberts, *et al.*)

One of the most distinctive features of ion channels – and one that plays a key role in the propagation of nerve impulses – is that they can open or close within a millisecond in response to stimuli such as a change in membrane potential or a chemical transmitter substance (Chapter 17).

## Transport across entire cells

Active transport is not limited to serving the needs of individual cells. As an example consider the absorption of glucose by the small intestine. In travelling from gut cavity to blood, glucose must cross the cells of the gut lining. It must therefore enter the cells on the side facing the gut cavity and leave them on the side nearest the blood. This **transcellular transport** involves the co-operative action of active transport, facilitated diffusion, microvilli and tight junctions.

The diagram illustrates what is involved. Using energy from ATP hydrolysis, the sodium–potassium pump maintains a low sodium concentration inside the cell. On the side of the cell facing the intestinal cavity sodium ions and glucose enter the cell, carried by a cotransport protein. Note that although sodium is moving 'downhill', the gradient is maintained by active pumping on the other side of the cell. In effect glucose 'hitches a ride', using the energy of the sodium gradient. There is no need for

## Transport across entire cells *continued*

glucose to move down a gradient; so long as the glucose gradient is less steep than the sodium gradient, glucose will enter the cell.

In this way a high concentration of glucose is built up inside the cells lining the intestine, enabling it to leave the other side of the cell by facilitated diffusion. For this to work the glucose permease molecules (which, remember, float freely in the lipid bilayer) must be prevented from moving along the plasma membrane from the side of the cell farthest from the intestine to the side nearest it – otherwise glucose could leak back into the intestine. This is the role of the **tight junctions**, which also prevent glucose from slipping between cells back into the intestine.

The uptake of glucose by intestinal cells is an example of a **secondary pump**. Here the immediate source of energy is a concentration gradient built up by a primary pump, in this case the sodium–potassium pump. Part of the energy expended in pumping out sodium is thus temporarily stored in the form of the sodium concentration gradient, and is used in the uptake of glucose.

Another example of a secondary pump is the transport of pyruvate acid into mitochondria. Pyruvate is transported with $H^+$ ions across the inner mitochondrial membrane, the $H^+$ ions moving down the gradient established by electron transport (Chapter 5). This process is not powered by ATP hydrolysis, but by the same energy source that powers ATP synthesis – the proton gradient developed by electron transport.

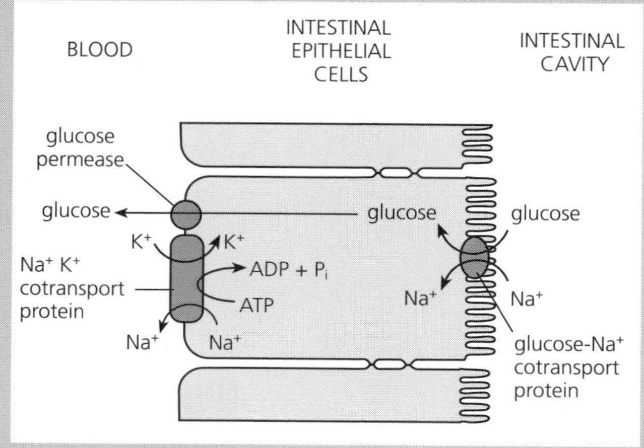


Transport across an intestinal cell

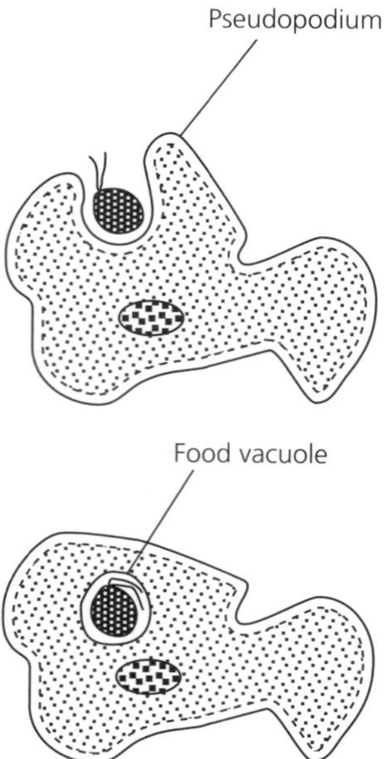


**Figure 6.18**
Phagocytosis by an amoeba engulfing a unicellular alga

# Cytosis

Cytosis differs from previously mentioned mechanisms of membrane transport in that a relatively large 'package' of material crosses a membrane without actually passing through it. It involves the folding and fusion of membranes and, since this is a form of movement, it must involve energy expenditure. Depending on its direction, cytosis can be outward (**exocytosis**) or inward (**endocytosis**).

## Endocytosis

Endocytosis is the taking in of bulk material from the environment by an intucking of the plasma membrane, forming small 'bags' or vesicles. It occurs in two forms – phagocytosis and pinocytosis.

**Phagocytosis** occurs when, for example, a phagocyte (a kind of white blood corpuscle) engulfs a bacterium or when an amoeba engulfs a food particle such as a diatom (Chapter 15). The cytoplasm flows round the food particle by forming outgrowths called **pseudopodia**. Eventually the particle is enclosed in a **food vacuole** which then fuses with a lysosome containing digestive enzymes (Figure 6.18).

How does phagocytosis work? There is evidence that some form of specific receptor in the plasma membrane must be involved. For example,

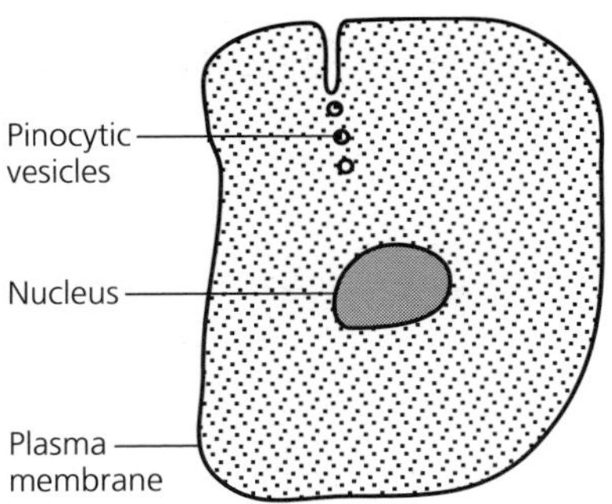


**Figure 6.19**
Pinocytosis

white blood corpuscles called phagocytes show far more 'appetite' for bacteria which have been coated with antibody. Once activated, the antibody receptors presumably cause the underlying cytoskeleton to fold inwards to engulf the bacterium.

**Pinocytosis** ('cell drinking') is superficially like phagocytosis on a much smaller scale. The plasma membrane tucks in repeatedly, enclosing tiny vacuoles containing extracellular fluid (Figure 6.19). Pinocytosis occurs in most, if not all, eukaryotic cells – even in plant cells, where it obviously must occur beneath the cell wall.

The formation of a pinocytic vesicle involves small depressions in the plasma membrane called **clathrin-coated pits** (Figure 6.20). Beneath each coated pit is a layer of a protein called **clathrin**. This forms a kind of basket round the pit, causing it to tuck in to form a **clathrin-coated vesicle** enclosing extracellular fluid. The vesicles then shed their clathrin coats and return to the plasma membrane.

In many cells pinocytosis results in the indiscriminate uptake of extracellular material without any specific stimulus. However, in some cases it involves the uptake of specific proteins. One example is the absorption of antibody from milk by young mammals. The membranes of the cells lining the gut contain receptors which are transmembrane proteins. These bind specifically to antibodies and form clusters in coated pits before being tucked in as coated vesicles. The transfer of antibodies (made in lymph nodes) from maternal blood to the milk ducts also involves pinocytosis.

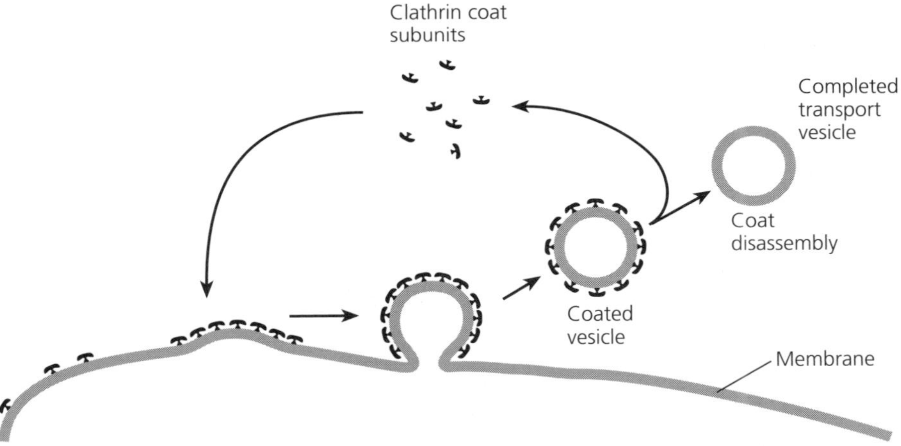


**Figure 6.20**
Formation of clathrin-coated pits in pinocytosis (after Alberts, *et al.*)

## Exocytosis

The reverse of endocytosis is **exocytosis** which occurs when a cell secretes a substance for use elsewhere (Figure 6.21). The secretory product is packaged in vesicles budded off the Golgi sacs. The secretory vesicles are then directed by microtubules to the plasma membrane and – in cells which secrete into a duct – to a specific region of it.

How do secretory vesicles find their way to the correct part of the plasma membrane? In nerve cells, synaptic vesicles have to travel from the cell body to the axon terminals – a distance of up to a metre or more. The vesicles are propelled by special motor proteins, and guided along the axon by microtubules.

Once at its destination, a secretory vesicle must await the appropriate signal for release. In nerve cells the release of a neurotransmitter (such as acetylcholine) is triggered by a sudden increase in intracellular calcium concentration following the arrival of a nerve impulse.

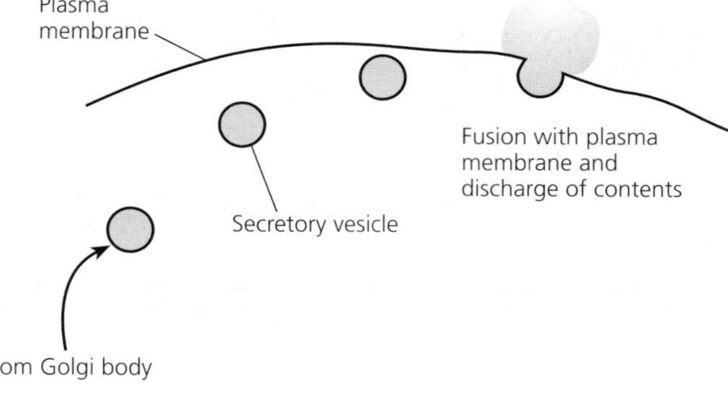


**Figure 6.21**
Exocytosis

### *Endocytic–exocytic cycling*

With every pinocytic vesicle formed the area of plasma membrane would be reduced were it not for the fact that exocytosis occurs at the same rate. The two processes are thus interdependent and form complementary parts of a **membrane cycle**.

## QUESTIONS

**1** True or false:

**a)** If water is neither entering nor leaving a plant cell, then the water concentration inside and outside the cell must be equal.

**b)** $CO_2$ diffuses out of cells because it is poisonous.

**c)** If oxygen is diffusing into a cell, then no oxygen molecules are leaving the cell.

**d)** All other things being equal, substances with large molecules diffuse faster than substances with small molecules.

**e)** Oxygen diffuses faster in air because molecules of a gas are moving faster than those in a liquid.

**2** Each of the following sentences consists of two parts. The second part may or may not be a correct explanation for the first.

**a)** Lactic acid diffuses out of active muscle fibres **because** they need to get rid of it.

**b)** When the temperature of the surrounding pond water is raised from 5°C to 15°C, oxygen diffuses more quickly into a *Paramecium* **because** diffusion is faster at higher temperatures.

For each sentence, write the letter **A**, **B**, **C** or **D**, depending on whether:

**A** Both statements are true, and the second is a correct explanation for the first.

**B** Both statements are true, but the second is not the reason for the first.

**C** The first statement is true and the second is false.

**D** Both statements are false.

**3** If a raw potato chip neither gains nor loses mass when placed in a sugar solution, then the potato cells must have the same

**A** water concentration as the sugar solution.

**B** water potential as the sugar solution.

**C** pressure as the sugar solution.

**D** pressure and solute potential as the sugar solution.

**4** Which two of the following solutions have identical solute potentials?

**A** 10% sucrose and 10% glucose

**B** 10% sucrose and 5% glucose

**C** 0.2 molar sucrose and 0.2 molar glucose

**D** 0.2 molar sucrose and 0.1 molar glucose

**5** 0.3 M glycerol solution has the same water concentration as blood plasma. When such a solution is added to blood, the red cells slowly swell and burst after about 30 seconds. When the experiment is repeated with 0.3 M urea solution, the red cells burst after about five seconds. When glucose solution is used, the red cells show no change even after an hour. Explain these observations, using the fact that the relative masses of the molecules of urea, glycerol and glucose are 60, 92 and 180, respectively.

CHAPTER

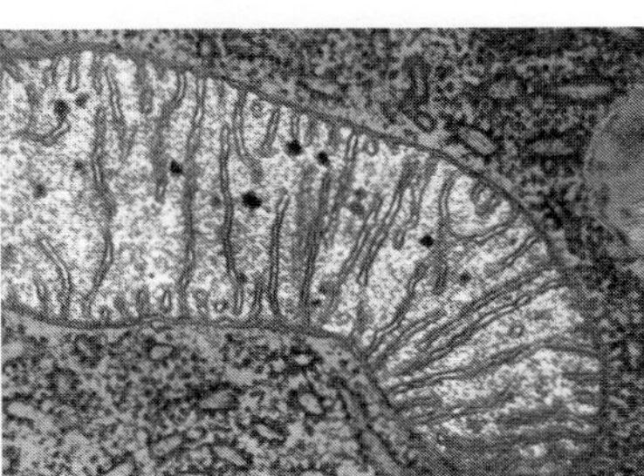

# 7 CHROMOSOMES and CELL DIVISION

## LEARNING OBJECTIVES

By the time you have completed your study of this chapter you should be able to:

- understand and use the terms diploid, haploid, karyotype, homologues.
- describe the events of the cell cycle.
- explain how mitosis generates genetically identical daughter nuclei.
- compare the mechanisms of cytoplasmic cleavage in animal and plant cells.
- explain the complementary roles of fertilisation and meiosis in maintaining chromosome number.
- know the stages of meiosis, and understand how independent segregation produces new combinations of chromosomes.
- explain how crossing over recombines material within homologous pairs of chromosomes.

## All cells have parents

All living things originate from pre-existing organisms – in other words, all organisms have parents. Moreover, offspring are very similar (though never identical) to their parents. Clearly the information for building each generation must somehow be transmitted from the previous generation. How is this information carried and how is it transmitted from parents to offspring?

Since the gametes (sex cells) are the only link between generations the information must be contained inside the eggs and sperms. Like all cells, gametes arise by division of pre-existing cells. The information carried by the eggs and sperms must have come from the cells that gave rise to them. These cells in turn must have got the information from the cells that gave rise to *them*, and so on over all the cell generations leading back to the fertilised egg. And of course it doesn't stop there. A fertilised egg is not the beginning of a living cell, it is simply a cell produced by the joining of two other cells, each of which is the product of a long line of cellular ancestors. Every cell thus has a line of ancestors stretching back to the time of the first cells.

### The importance of the nucleus

Though the nucleus was first observed by Robert Brown in 1833, its importance in heredity was not immediately appreciated. When, in 1875, Oscar Hertwig observed that in fertilisation, it is the nucleus of the egg and sperm that join together, attention began to focus on the nucleus.

## The importance of the nucleus continued

The first experimental evidence that the nucleus carries hereditary material was obtained by Theodor Boveri in 1889. He found that sea urchin eggs from which he had removed the nucleus could sometimes be fertilised when mixed with sperms. Apart from their small size, the resulting larvae were of normal appearance. In one dramatic experiment he managed to fertilise artificially an egg that lacked a nucleus with the sperm of another species of sea urchin. The embryo that developed resembled the species that provided the sperm, rather than the species that supplied the egg. Since the only part of a sperm to enter a sea urchin egg is the head, which consists almost entirely of nucleus, this was strong evidence for the role of the nucleus in heredity.

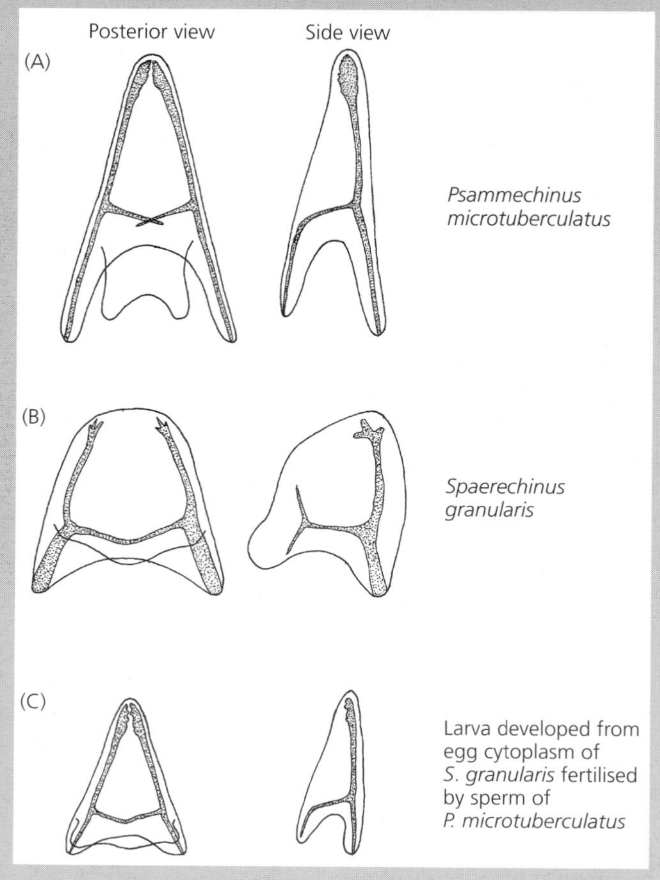


Boveri's demonstration of the importance of the nucleus in heredity. Posterior and side views of larval sea urchins of two species (A and B), and one produced from nucleus of one species and cytoplasm of another (C)

## The vehicles of inheritance: chromosomes

When cells divide the information for building each generation must also divide and be distributed to the daughter cells. With this fact in mind, many 19th Century biologists focused their attention on cell division.

As fixing and staining techniques were developed, a number of 19th Century microscopists began to notice the presence of thread-like bodies in dividing cells. Because they readily took up dyes they were called **chromosomes** (*chromo* = colour, *soma* = body).

By the end of the 19th Century biologists had discovered a number of interesting facts about chromosomes:

- their number is constant for all the cells of the body and is characteristic of a species (Table 7.1).[1] Chromosome numbers vary considerably: the horse roundworm has only 2, whilst some ferns have over 500.
- in any given cell, chromosomes vary in length.[2] They also differ in the position of the **centromere** (a small structure that plays a key role in the movement of chromosomes during cell division).

**Table 7.1**
Chromosome numbers of some common organisms

| Organism | Chromosome number |
|---|---|
| Human | 46 |
| Chimpanzee | 48 |
| Dog | 78 |
| Rat | 42 |
| Mouse | 40 |
| Fruit fly | 8 |
| Mosquito | 6 |
| Garden pea | 14 |
| Maize | 20 |
| Crocus | 6 |
| Tomato | 24 |
| Onion | 16 |

[1] There are some exceptions to this, even in a single organism. For example, liver cells are **polyploid**, with more than the usual two chromosome sets.

[2] Human chromosomes are numbered from largest (#1) down to the smallest.

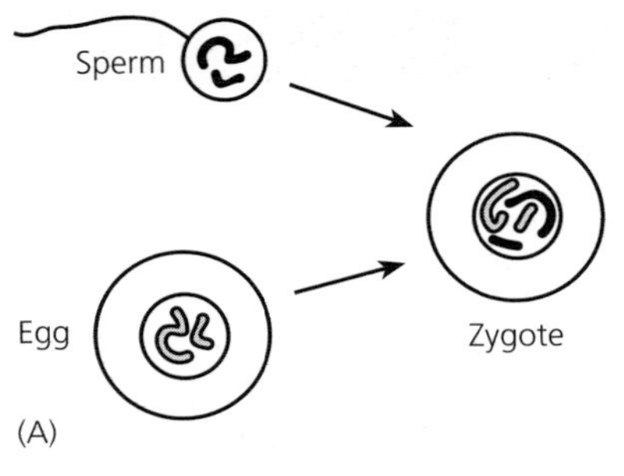


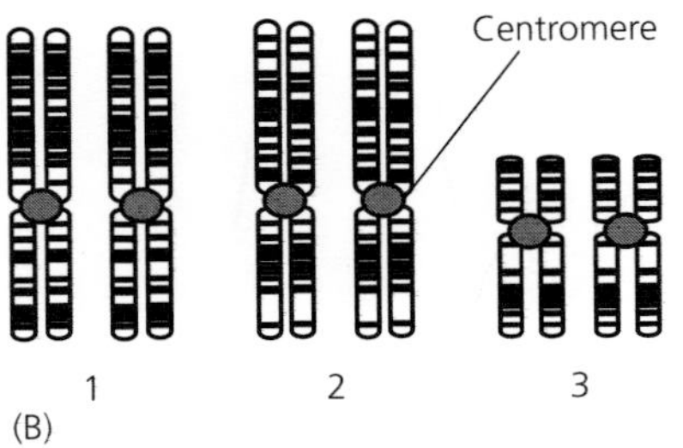


**Figure 7.1▲**
Homologous chromosomes. (A) Gametes have one of each pair of chromosomes; the zygotes have both members of each pair. (B) With the exception of sex chromosomes, the two members of a pair are visibly indistinguishable, with identical length, centromere position and banding pattern. Chromosomes 1 and 2, though similar in length and centromere position, have different banding patterns and are not homologous

- in multicellular animals and in the more complex land plants, body cells contain two of each kind of chromosome and are said to have the **diploid** number. Chromosomes thus exist in pairs, the two members of a pair being called **homologues**. This is why (in animals and land plants) the number of chromosomes is normally an even number. One member of each pair has been inherited from each parent, so we can speak of *maternal* and *paternal* chromosomes (Figure 7.1).
- gametes (eggs and sperms) have only one of each homologous pair, and are said to have the **haploid** number (Figure 7.1A).

Organisms differ not only in their bodily features, but also in their chromosomal characteristics. In recent decades it has been found that the various regions of each chromosome differ in their staining properties, resulting in a pattern of bands rather like a supermarket bar code (Figure 7.2). When these banding patterns are taken into account, each chromosome pair can be distinguished from every other one.

To display a chromosome set, a photograph of the stained chromosomes is taken and enlarged. The individual chromosomes are then cut out of the photograph and arranged in matching pairs in decreasing order of size to produce a **karyogram**. The sum total of an organism's chromosomal characteristics constitute its **karyotype** (Figure 7.3).

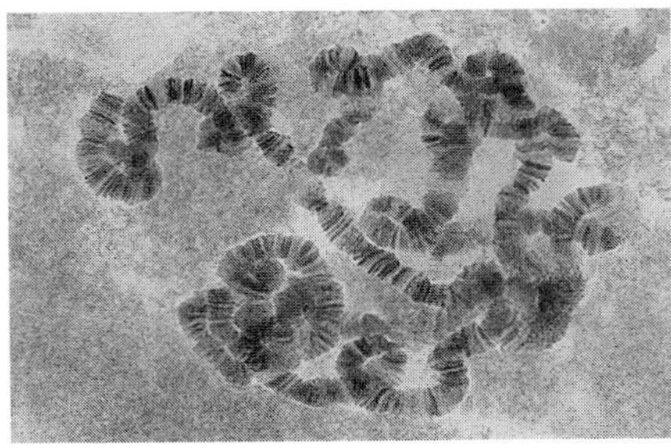

**Figure 7.2▲**
Banding pattern in a giant chromosome from the fruit fly, *Drosophila funebris*, as revealed by staining

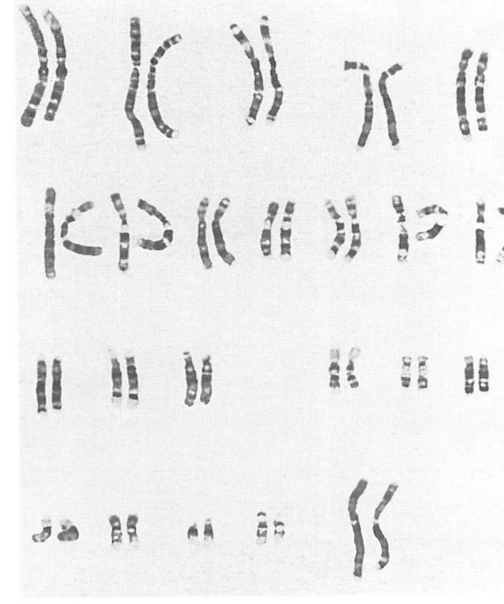

**Figure 7.3▲**
Female human karyotype

# Cell division

Cell division involves the division of the nucleus followed by the division of the cytoplasm. There are two kinds of nuclear division:

- **mitosis**, which produces daughter nuclei genetically identical to the parent.
- **meiosis**, in which a diploid nucleus gives rise to four genetically different haploid nuclei through two consecutive divisions.

As far as their effects on chromosome number are concerned, meiosis and fertilisation have opposite effects. In every sexual life cycle both processes occur, so that there is no change in chromosome number over successive generations. The relationship between the two processes in a mammal is shown in Figure 7.4.

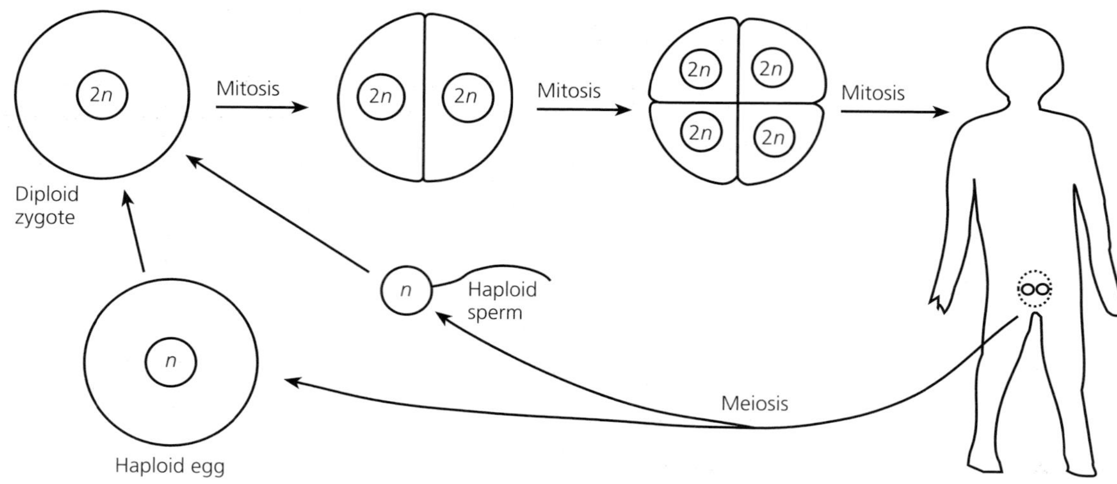


**Figure 7.4**
Relationship between fertilisation, meiosis and mitosis

## The cell cycle

Actively dividing cells go through a process called the **cell cycle**. Each cycle consists of mitosis (the **M phase**), alternating with **interphase** (the period between divisions) which is dominated by growth and synthesis. The duration of the cell cycle varies considerably. In rapidly dividing mammalian tissues it takes 12–18 hours, with mitosis accounting for an hour or less. At least 90% of each cycle is accounted for by interphase. We know this because when actively dividing cells are viewed under the microscope, only about 10% are 'caught in the act' of division.

A nucleus in interphase used to be called a 'resting nucleus' because it doesn't appear to be doing anything. Even after staining all that can be seen of the chromosomes is a seemingly structureless material called **chromatin**. In reality the nucleus is extremely active, for it is preparing for the next division. The drama of mitosis arises from the fact that it involves mechanical events that can be seen under the microscope; the processes occurring in interphase are chemical, and so escape the eye.

One such process is the duplication of chromosomes. The timing of this process was investigated by supplying actively dividing cells with radioactively-labelled thymine for about 30 minutes. Since thymine is a constituent of DNA, but not of RNA, any cells that became radioactive must have been making DNA. When the cells were examined for radioactivity, about 40% of them were found to have taken up the thymine. Evidently DNA synthesis does not occupy the whole of interphase. Further experiments along these lines showed that the period of DNA synthesis (the **S phase**) occupies just under half of interphase (Figure 7.5). It is separated from successive divisions by two periods called $\mathbf{G_1}$ and $\mathbf{G_2}$ (G standing for gap).

$G_1$ is a period of intense activity, as the various organelles and other cytoplasmic components increase in number and bulk. A cell that is not going to divide again goes no further than $G_1$, but if it is going to divide again it enters the S phase, during which each chromosome splits lengthways into two identical strands called sister **chromatids**, held together at the centromere. The various chromosomal proteins such as the histones are also produced during S phase, as are the centrioles of the centrosome in animal cells.

$G_2$ phase is concerned with the production of proteins which are directly involved in mitosis, such as the **tubulins** that polymerise to form the microtubules of the spindle.

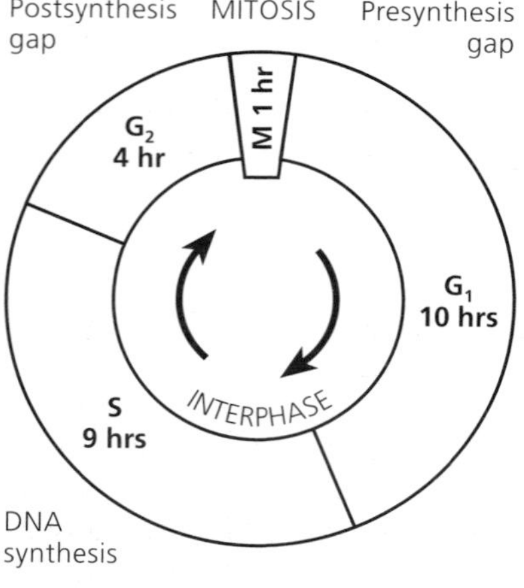


**Figure 7.5**
The cell cycle; M = mitosis, S = DNA synthesis (after Alberts, *et al.*)

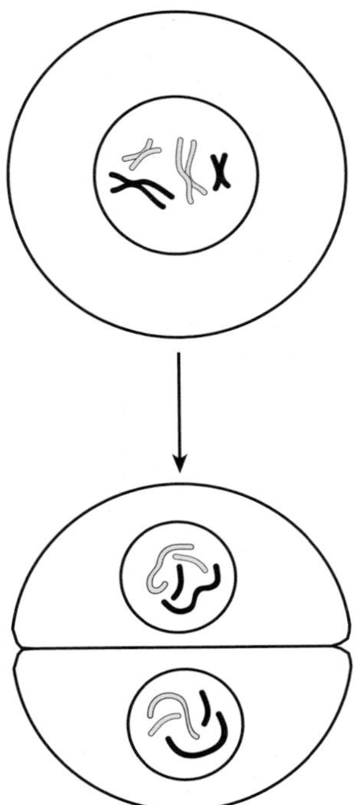

**Figure 7.6**
Essential result of mitosis. The parent cell divides into two daughter cells, each with identical chromosome sets

## The control of cell division

Why do some cells never divide whilst others do so continuously, and some do so only in response to injury or other external event? Clearly, cell division must be under some kind of control. Take, for instance, a cell in the dividing layer of the skin. After a number of cycles of division and growth it fails to replicate its DNA and begins to differentiate into a skin cell. At some stage in the cell cycle dividing cells must make some kind of 'decision' as to whether to enter another cycle of division or begin to specialise. This decision is taken during $G_1$ phase.

Ultimately the control of cell division lies with genes. When a gene controlling cell division changes – that is, when it **mutates** – cells may begin to divide uncontrollably producing a tumour or cancer. An understanding of the genetic control of cell division is, therefore, closely linked to the prevention and cure of cancer.

# Mitosis

The function of mitosis is to increase the *number* of genetically identical copies of the parent nucleus (Figure 7.6).

Mitosis occurs during growth, repair (which is really localised growth) and asexual reproduction. Mitotic divisions occur throughout a developing embryo, but in adult multicellular organisms they tend to be restricted to certain regions. In mammals these include such places as the skin, gut lining, bone marrow and in healing wounds. Some cells in a mature mammal never divide, such as nerve cells, receptor cells of the eye and ear, and heart muscle cells. Others only divide in response to injury, such as liver cells. In flowering plants mitosis occurs in regions called **meristems**, such as the tips of shoots and roots and the cambium. In single-celled organisms, it occurs as part of asexual reproduction.

Since most complex organisms are diploid we tend to associate mitosis with diploidy. Mitosis does however occur in haploid cells, for instance during the growth of moss and fern gametophytes (Chapter 30).

The events of mitosis centre round the chromosomes – in fact the name comes from the Greek *mitos*, meaning 'thread', referring to the appearance of the chromosomes during mitosis. Except when using phase-contrast microscopy, chromosomes are only visible after treatment with a dye such as aceto-orcein or Fuelgen stain. Since staining kills the cells, each appears as a 'snapshot'. Reconstructing the events of mitosis is thus like doing a jigsaw, except that the pieces fit together in time rather than in space.

Mitosis is a continuous process but for convenience it is divided into five stages: **prophase**, **prometaphase**, **metaphase**, **anaphase** and **telophase** (Figure 7.7).

### *Prophase*

Prophase is by far the longest stage of mitosis and, in actively dividing cells, the majority are seen to be in this stage. It begins with the chromosomes coiling up and slowly becoming distinct. The need for this is obvious when you consider that during interphase, an average human chromosome is about 4 cm long – about 10 000 times the diameter of the nucleus. In this highly-extended state the chromosomes are far too tangled up to be packaged into two identical groups for distribution to daughter cells.

In animal cells prophase is also marked by the division of the centrosome, with the two pairs of daughter centrioles migrating to

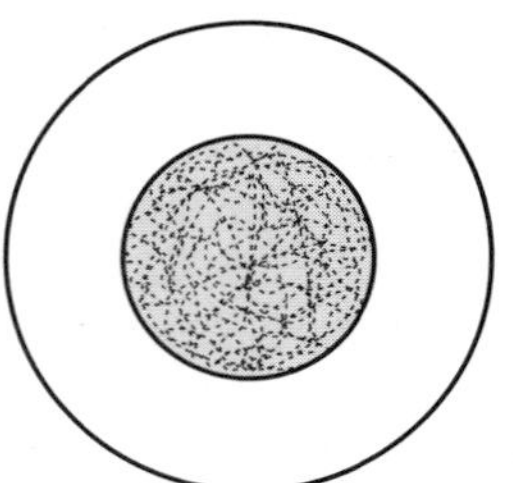

1. **Interphase**: chromosomes replicating but not visible as separate structures

2. **Prophase**: chromosomes shorten and become visible

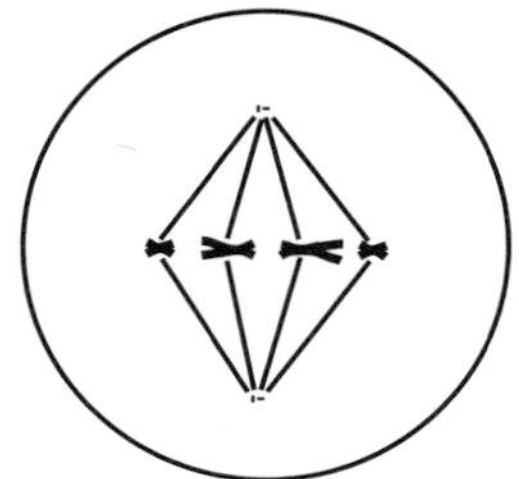

3. **Metaphase**: chromosomes become arranged on equator of spindle

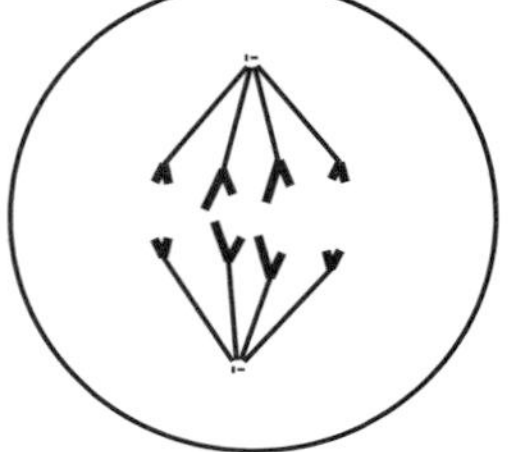

4. **Anaphase**: chromatids move apart towards opposite poles of spindle

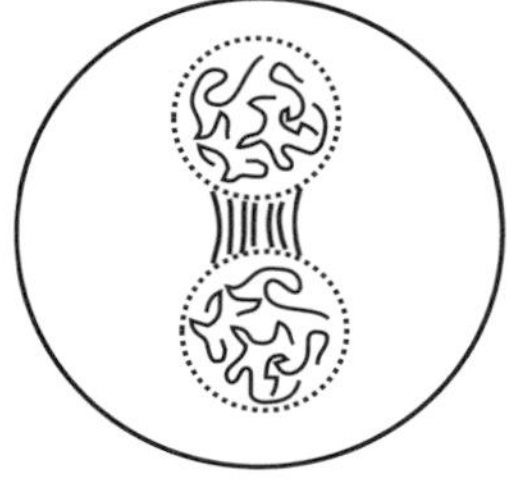

5. **Telophase**: daughter chromosomes lengthen and become indistinct

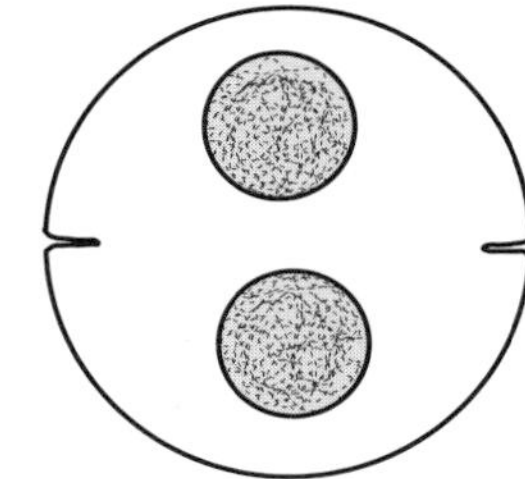

6. **Cytoplasmic cleavage**

**Figure 7.7**
The stages of mitosis in an animal cell. Though not part of mitosis, interphase and cytoplasmic cleavage are shown for convenience

opposite sides of the nucleus. From each centriole many microtubules radiate in all directions, forming an **aster**. Some of the microtubules make contact with the chromosomes via a specialised region of the centromere called the **kinetochore**. The nucleoli disappear, the chromosomes continue to shorten and chromatids are now distinct from each other.

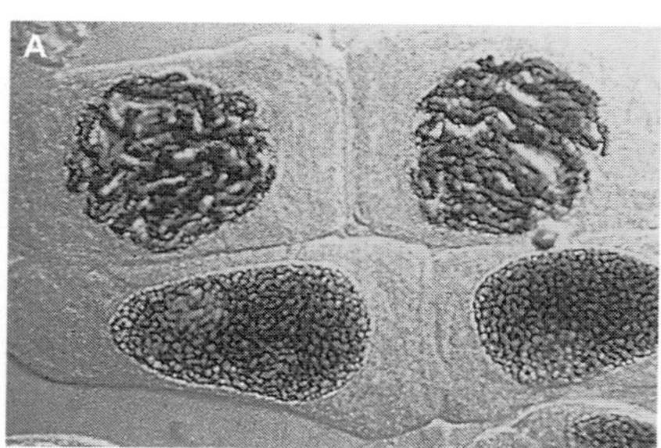


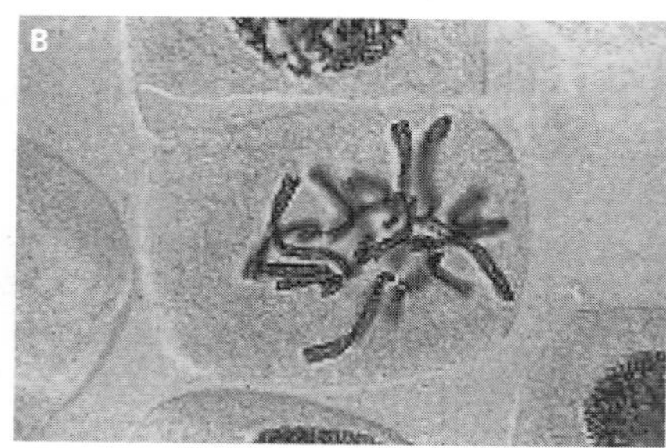


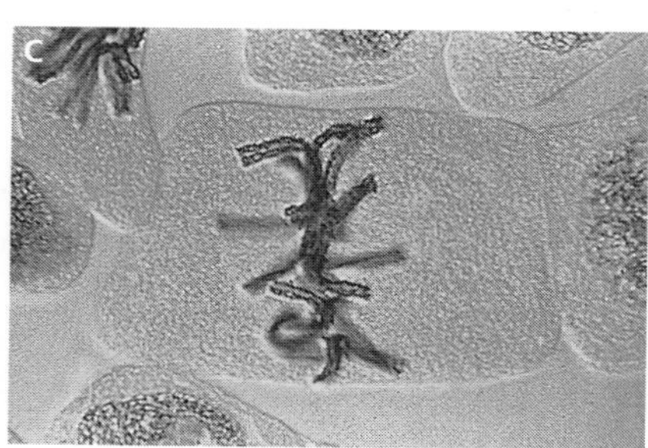


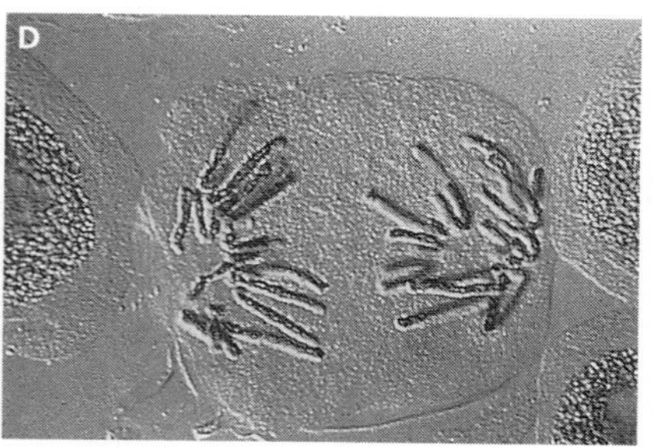


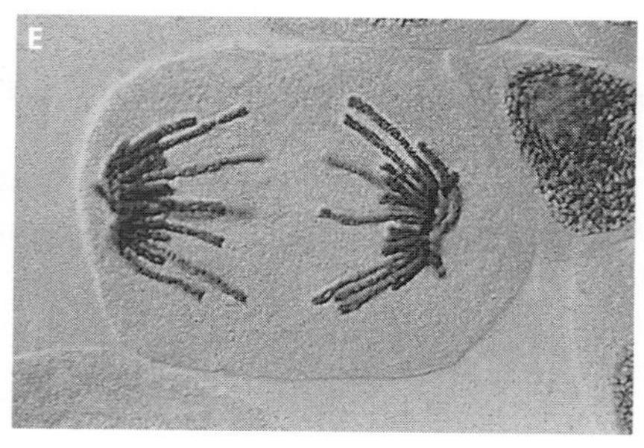


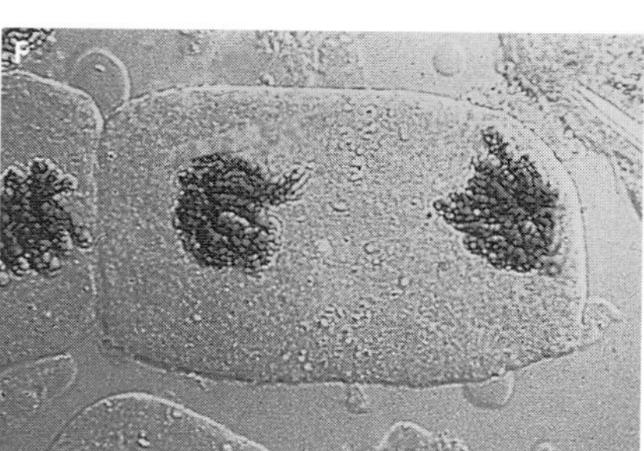


**Figure 7.8**
Photomicrographs showing the stages of mitosis in a bluebell. (A) interphase (lower two cells); (B) prophase; (C) metaphase; (D) anaphase; (E) late anaphase; (F) telophase

### *Prometaphase*

Prometaphase (often regarded as the last part of prophase) is marked by the breakdown of the nuclear envelope into vesicles resembling those of the endoplasmic reticulum. Nucleus and cytoplasm are no longer distinguishable. In animal cells, microtubules from the two centrosomes extend towards each other and begin to form the **spindle**. The centrosomes occupy the **poles** of the spindle. In plant cells there are no centrioles and the spindle develops independently.

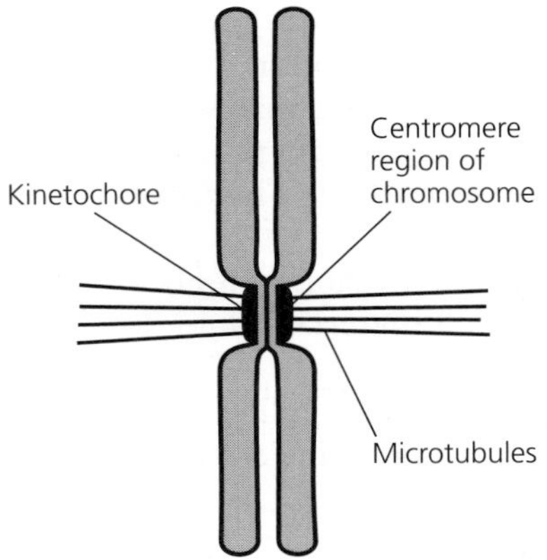


**Figure 7.9**
Spindle microtubules attached to kinetochore

By this time each chromosome has become 'caught' by microtubules extending from both spindle poles. The other microtubules meet up and overlap with microtubules extending from the opposite pole (Figure 7.9).

### Metaphase

Held by microtubules the centromeres with their chromatids move to the equator of the spindle. This is a relatively short stage, with only a few cells in a root tip preparation being likely to be 'caught' in it.

### Anaphase

Anaphase is also a relatively short stage, and in it the microtubules play two quite different roles. Those attached to the kinetochores pull the chromatids apart, towards the spindle poles. They move at about 1 μm per minute and eventually collect at the two poles. Each chromatid is now considered to be a chromosome in its own right. The microtubules which extend from pole to pole push against each other in the region where they overlap, causing the spindle poles to move further apart.

### Telophase

The final stage of mitosis can be compared to prophase in reverse. The chromosomes become extended and indistinct. The nucleoli reappear and a nuclear envelope develops round each cluster of daughter chromosomes forming a nucleus. Each daughter nucleus now has one of the two identical chromatids of each chromosome of the parent nucleus. The daughter nuclei are thus genetically identical.

The changes undergone by a chromosome during one cell cycle are diagrammatically summarised in Figure 7.10.

**Figure 7.10▶**
Chromosomal events of the cell cycle

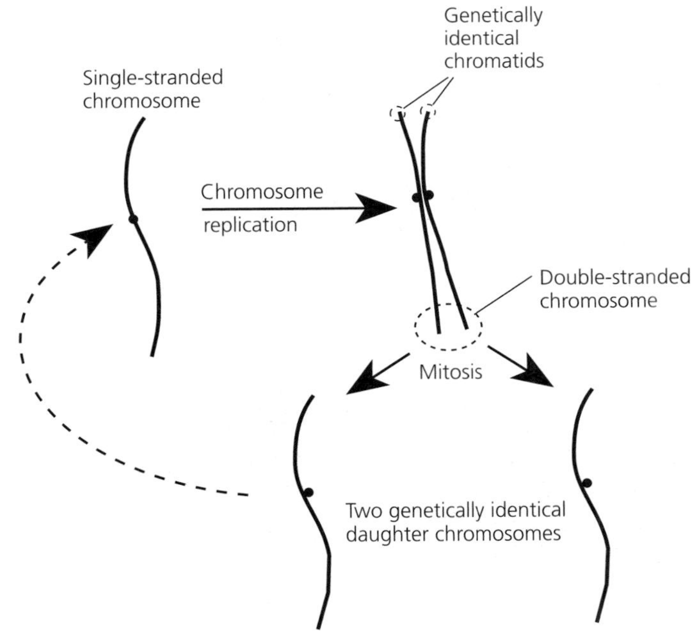


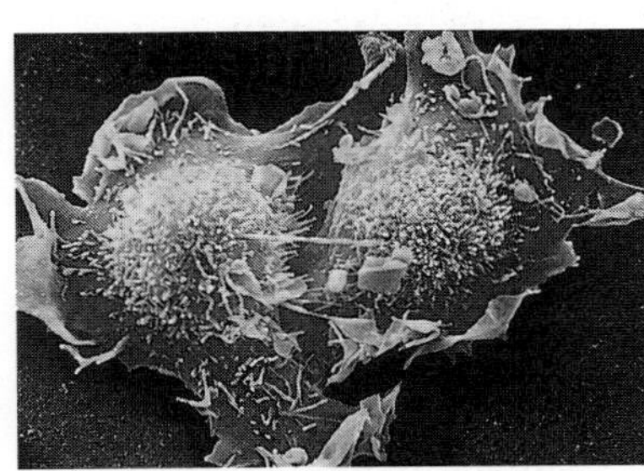

**Figure 7.11**
Cytoplasmic cleavage in an animal cell

## Cytoplasmic division

Mitosis is usually followed by division of the cytoplasm, though this is not always so. In the coenocytic hyphae of many fungi, for instance, the nuclei divide without division of the cytoplasm, so that many nuclei share a common envelope of cytoplasm.

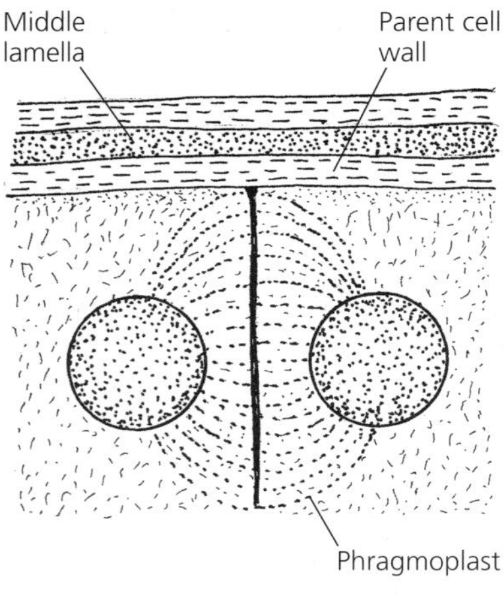


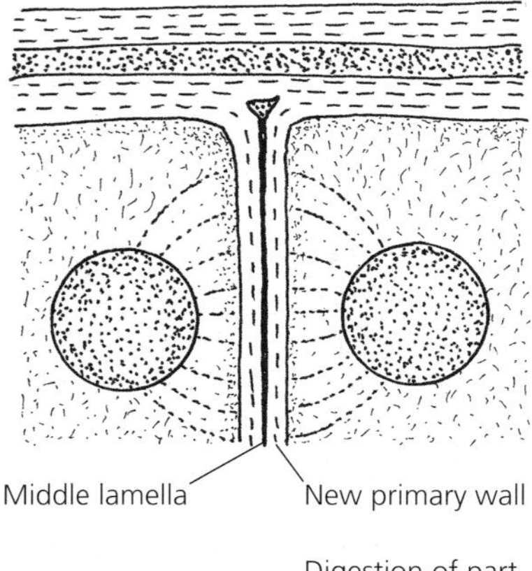


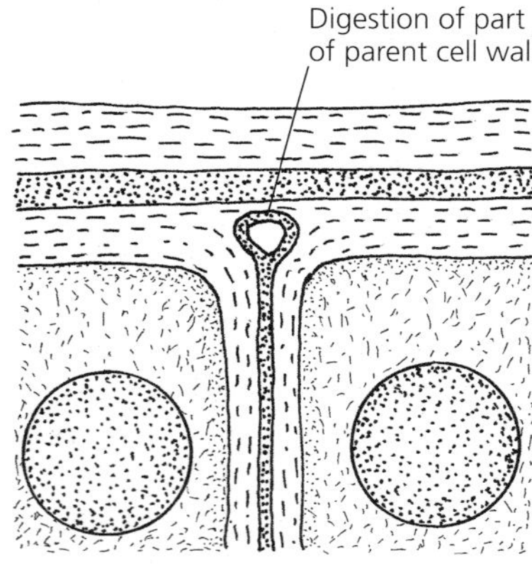


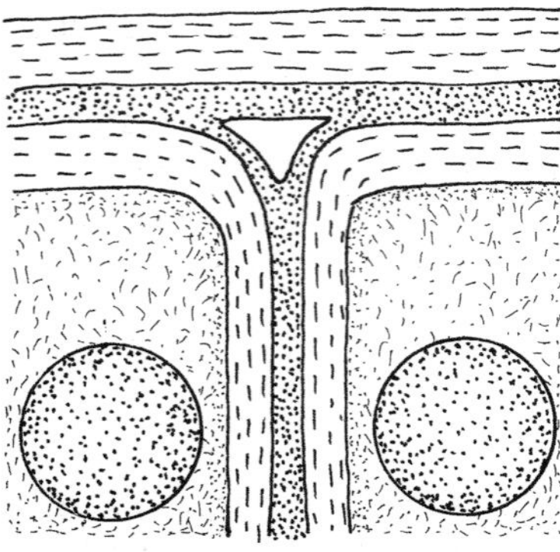

**Figure 7.12**
Cytoplasmic cleavage in a plant cell

Cytoplasmic cleavage (or **cytokinesis** as it is also called) often begins during telophase. In animal cells it begins with the formation of a **cleavage furrow** in the same plane as the equator of the spindle (Figure 7.11). The furrow is due to the contraction of a ring of two proteins, **actin** and **myosin**, better known for their role in muscle. The furrow deepens and eventually divides the cell into two.

Straitjacketed by its cellulose wall, cytoplasmic division in a plant cell follows quite different lines. Instead, new cell wall material is laid down within the existing cell dividing it into two (Figure 7.12).

During telophase a barrel-shaped set of microtubules called the **phragmoplast** develops with its equator in the same plane as that of the spindle. Next, vesicles produced by the Golgi bodies collect on the equator of the phragmoplast. The vesicles contain pectins and hemicelluloses and fuse together to form the **cell plate**. New cell wall material is then deposited on either side of the cell plate, which becomes the **middle lamella** (the material between adjacent cells). On each side of the middle lamella a new cell wall is laid down.

One final act remains to be performed. Since each new cell wall is laid down inside the parent cell wall, the two daughter cells are still bound together, imprisoned inside the wall of the parent cell. They become freed from each other by the digestion of a girdle of the parent cell wall.

# Meiosis

Meiosis is a key component of sexual reproduction and occurs only in the reproductive organs. In animals it occurs in ovaries and testes and results in the production of gametes. In plants the products are not gametes but **spores**, and in flowering plants these are produced in the anthers and ovules of the flower.

Meiosis is more complicated and prolonged than mitosis. This is partly because it consists of two divisions, known as meiosis I and meiosis II respectively. Since it is easy to lose sight of essentials among the details, a full description will be delayed until after some of the basics have been dealt with.

The most important thing about meiosis is its role in the creation of new genetic types. It achieves this by two quite different mechanisms:

- independent segregation of chromosome pairs.
- crossing over.

## Segregation of chromosomes

A key feature in meiosis is the reduction in chromosome number, with the diploid parent nucleus dividing twice to form four haploid nuclei. The reduction in chromosome number occurs during meiosis I which is thus referred to as a **reduction division** (meiosis comes from the Greek *meion*, meaning 'smaller').

Figure 7.13 shows how this happens in a cell with only one pair of chromosomes. In the first division both members of the pair move to opposite poles of the spindle – that is, they **segregate**. Each daughter nucleus thus receives only *one* member of the chromosome pair. Moreover, the chromosomes remain two-stranded, both chromatids of each chromosome finishing up in the same daughter nucleus. Only in the second division do the chromatids go their separate ways.

Figure 7.14 shows meiosis in a cell with two pairs of chromosomes. As we have just seen, in the first division each daughter nucleus receives only

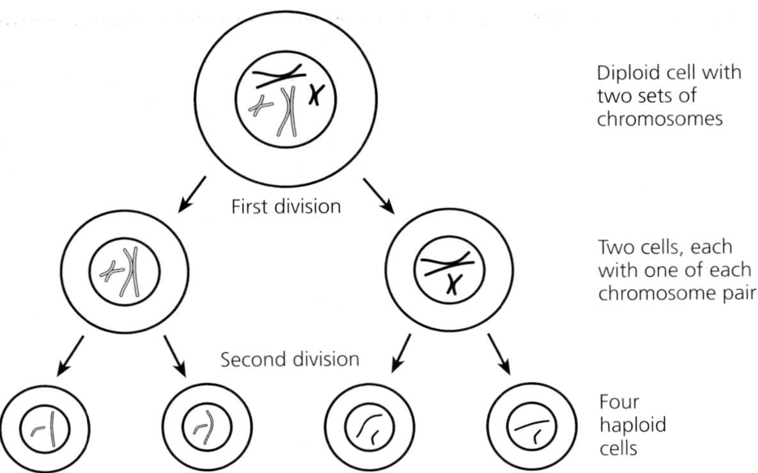


**Figure 7.13►**
How meiosis reduces chromosome number from diploid to haploid

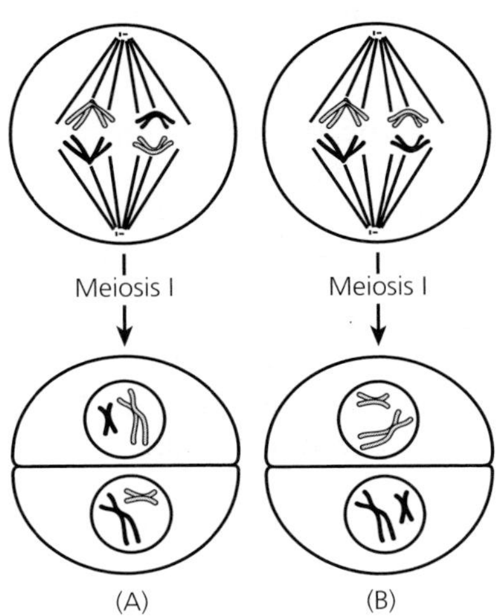


**Figure 7.14▲**
How meiosis produces genetically different daughter nuclei. For a cell with two pairs of chromosomes there are two ways in which they can arrange themselves on the equator of the spindle, with equal probabilities

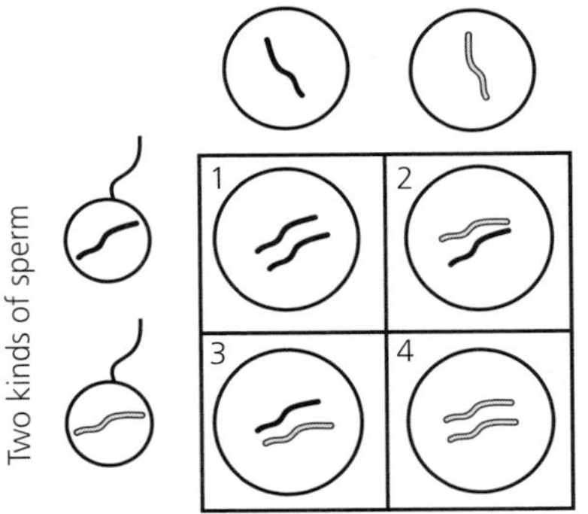


**Figure 7.15▲**
How random fertilisation involving two kinds of gamete produces three kinds of zygote. Although 2 and 3 are different events, they produce the same results. Each additional pair of chromosomes doubles the number of kinds of gamete and triples the number of kinds of zygote

one of each pair of chromosomes. It is a matter of chance which nucleus receives which member of a pair, the behaviour of one pair having no effect on the behaviour of any other pair. In other words chromosome pairs segregate **independently** of each other.

The upshot of all this is that in a nucleus with two pairs of chromosomes segregation can occur in two ways, each producing two genetically different nuclei [(A) and (B) in Figure 7.14]. In a *population* of cells about half would segregate as in (A) and half would behave like (B). A population of cells would thus produce four different chromosomal combinations. Since each additional chromosome pair doubles the number of kinds of segregation, a population of cells with three pairs of chromosomes would produce eight different genetic types, and with four pairs there would be 16 different products. A human with 23 pairs of chromosomes could produce $2^{23}$ (or over 8 million) different kinds of gamete.

Large though this number is, it pales into insignificance when we go on to consider how many kinds of human **zygote** could be obtained. As we have just seen, for each chromosome pair two kinds of gamete can be produced in meiosis. A species with one pair of chromosomes would produce two kinds of gamete, and random fertilisation would result in three different kinds of zygote (not four, since two kinds of fertilisation give the same result as shown in Figure 7.15). Fruit flies (with four pairs of chromosomes) could produce $2^4$ kinds of gamete and $3^4$ kinds of zygote. Humans could produce $3^{23}$, or about 94 thousand million kinds of zygote! Moreover, this calculation ignores crossing over, which vastly increases the number of possibilities.

The reshuffling of chromosomes is thus a two-stage process – a halving of the chromosome number in meiosis, followed by a doubling in fertilisation. Neither is more 'sexual' than the other – the two processes are complementary. This is sometimes the subject of confusion. Spore production by ferns, for instance, is sometimes incorrectly described as asexual reproduction. It is merely one half of a sexual process that straddles two generations (Chapter 30).

## Crossing over

Independent segregation reshuffles entire chromosomes but leaves each individual chromosome unchanged. Crossing over is a kind of cut-and-paste mechanism in which parts of homologous chromosomes are exchanged.

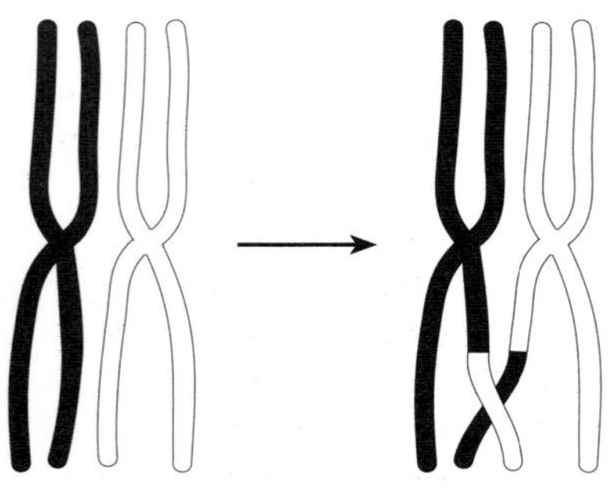

**Figure 7.16**
Crossing over

Crossing over occurs during the early stages of meiosis I when homologous chromosomes come together in pairs. Each consists of two sister chromatids, produced by replication during the previous interphase. Each chromosome pair thus consists of four chromatids, two belonging to each chromosome. Chromatids belonging to different members of a pair are called **non-sister** chromatids.

During crossing over, breaks occur in non-sister chromatids at points opposite each other (Figure 7.16). Non-sister strands then join so that each chromosome now consists of non-identical strands, each consisting of both maternal and paternal segments. Figure 7.16 shows only one crossover, but there are usually more than one. As a result of this reshuffling each has lost its original identity.

As a result of crossing over each chromosome you inherited from your mother actually consists of some bits from your mother's mother's chromosomes and some from your mother's father's chromosomes. Each chromosome you inherited from your father similarly consists of segments from each of his parents. On average, half of each of your chromosomes came from each of your parents, a quarter from each of their parents, an eighth from each of their parents, and so on.

## Meiosis I

As with mitosis, each of the two divisions of meiosis is a continuous process, but for convenience they are subdivided into stages. They are given the same names as those of mitosis, but the Roman numeral I or II is added to indicate the first or second division (Figure 7.17).

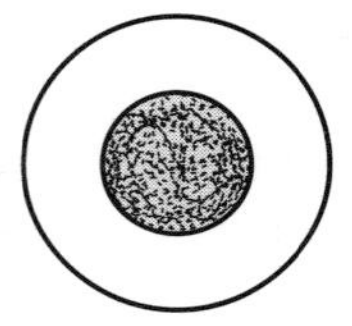

1. Chromosomes are replicating but not yet visible

2. Chromosomes shorten and become visible

3. Homologous chromosomes come together in pairs

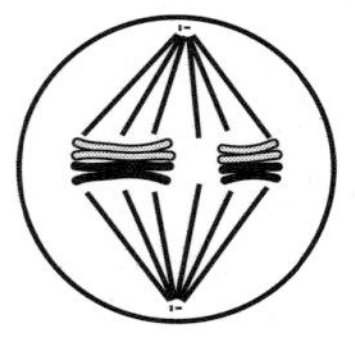

4. Chromosomes line up on spindle equator

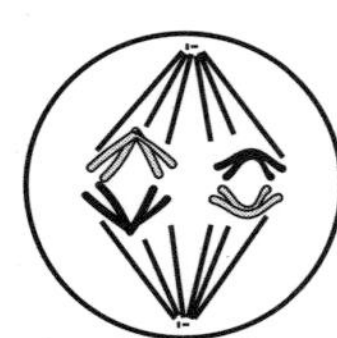

5. The two chromosomes of each pair move apart

6. Two daughter nuclei are formed, each containing one of each chromosome pair

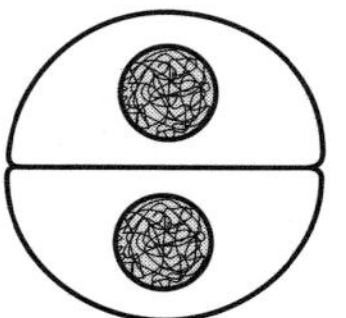

7. The cytoplasm divides

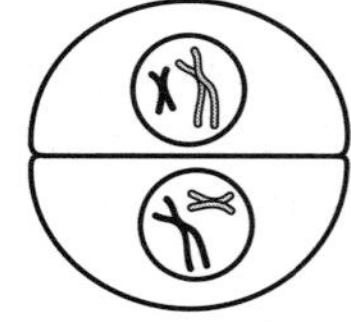

8. Chromosomes again become visible

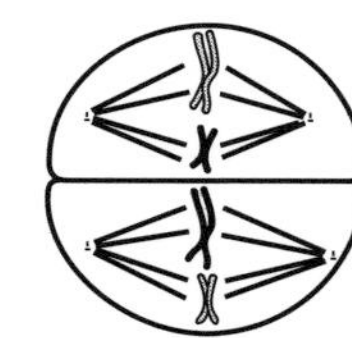

9. A new spindle forms and chromosomes line up on the equator

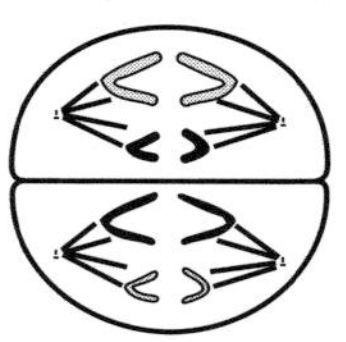

10. Chromatids move to opposite poles

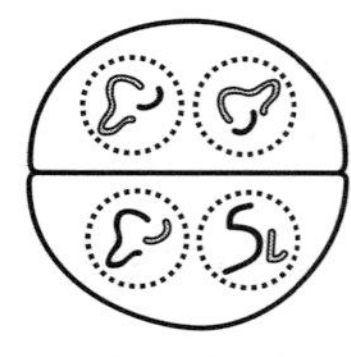

11. Four haploid nuclei are formed

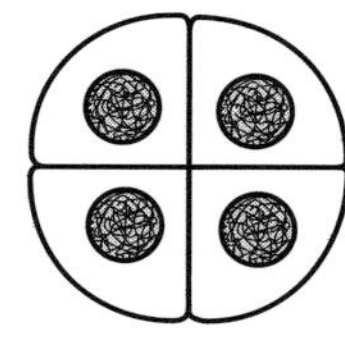

12. Cytoplasm cleaves again, producing four haploid cells

**Figure 7.17**
The stages of meiosis in an animal cell (crossing over not shown)

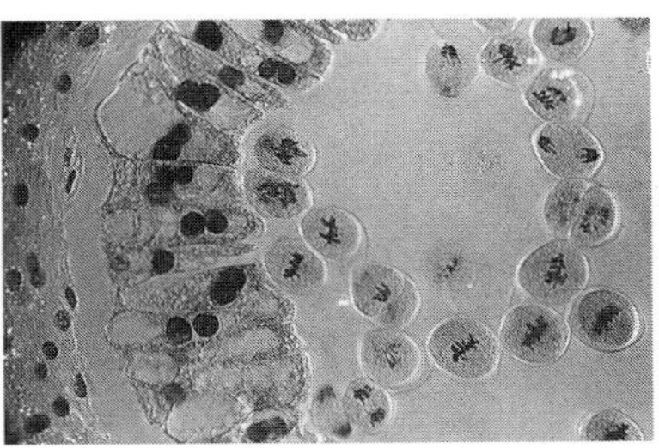

**Figure 7.18**
Photomicrograph of meiosis in a bluebell

### Prophase I

The chromosomes are two-stranded because they have already replicated in the preceding interphase. As in prophase of mitosis the chromosomes shorten by coiling, but there are important differences. Firstly, for a considerable time the sister chromatids remain too close to each other to be distinct so each chromosome seems to be single-stranded. Secondly (and most important of all) chromosomes come together in homologous pairs (the *only* time they are physically associated). The pairing of chromosomes is called **synapsis** and each chromosome pair forms a **bivalent**. The pairing is so intimate that the two appear as one, so for a time only the haploid number of chromosomes can be distinguished.

During pairing chromosomes continue to shorten, and each now begins to appear separate from its partner. Each chromosome is also seen to consist of two chromatids, so each bivalent is now visibly four-stranded. At one or more points along a bivalent the chromosomes appear interlocked for a time by one or more **chiasmata**. A chiasma is the visible evidence that a crossover has occurred. Towards the end of prophase I the nucleoli disappear and the nuclear envelope breaks down.

Prophase I is by far the longest and most complex stage of meiosis. In the male mouse both meiotic divisions are completed in about 12 days, 11.5 days being taken up by prophase I. In the human female it begins at about birth and then stops, not to be resumed until after the egg is shed from the ovary, which may be 45 years later!

### Metaphase I

As in mitosis, metaphase begins with the appearance of the spindle and the arrangement of the chromosomes on its equator. Unlike mitotic metaphase, the chromosomes in metaphase I are associated in bivalents.

### Anaphase I

This differs from anaphase of mitosis in that spindle fibres pull entire, two-stranded chromosomes apart. Each pole of the spindle thus receives only *one* of each pair of homologous chromosomes and thus has the haploid number of centromeres. Note that unlike anaphase of mitosis, in anaphase I the centromeres do not divide.

It is worth noting at this stage that, as far as the products of meiosis I are concerned, it is best to avoid reference to the haploid number of chromosomes for the following reason. As will become apparent in Chapter 8, crossing over can lead to two non-sister chromatids carrying different alleles of a given gene. It is clearly inappropriate to refer to a heterozygous nucleus as being haploid. The problem is avoided by referring to the haploid number of *centromeres*.

### Telophase I

The chromosomes lengthen and become indistinct, and the nucleoli and nuclear envelope reappear.

### Interphase

In many plants there is no interphase, telophase I merging with prophase II. In animal cells and in some plants there is a distinct interphase, but there is no DNA synthesis (chromosomes are double-stranded at the end of meiosis I).

## Meiosis II

Though similar to mitosis *mechanically*, the genetic consequences of meiosis II differ fundamentally from those of mitosis because the daughter nuclei are genetically different. This is because of crossing over, as a result of which the two chromatids of each chromosome are no longer identical.

### Prophase II

The chromosomes shorten again and become distinct. At the end of this stage the nuclear envelope breaks down and the spindle develops.

### Metaphase II

The centromeres become arranged on the equator of the spindle, attached to the spindle microtubules by their kinetochores.

### Anaphase II

The chromatids are pulled apart and collect at the poles of the spindle. Each group contains the haploid number of chromosomes.

### Telophase II

The chromosomes lengthen and become indistinct and the spindle fibres disappear. A nuclear envelope reforms round each daughter nucleus and nucleoli also become re-established.

Telophase is followed by cytoplasmic cleavage which results in the formation of four haploid cells.

## Chromosomes and DNA

Though the structure and replication of DNA are not dealt with in detail until Chapter 10, it is appropriate at this point to consider briefly the relationship between changes in DNA content and the number of chromosomes (Figure 7.19).

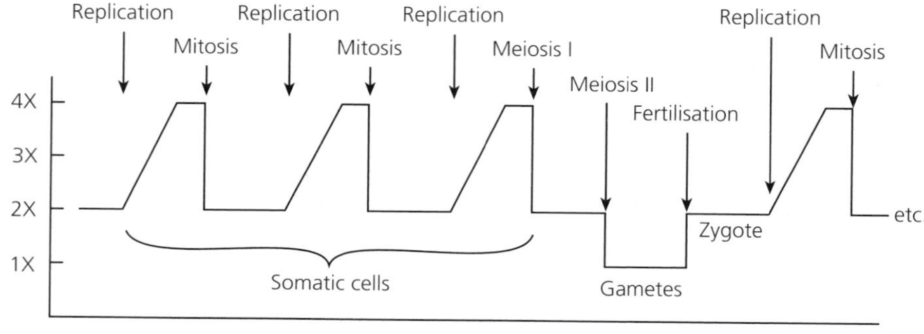


**Figure 7.19**
Changes in DNA content at different stages of the animal life cycle

A chromatid is a single molecule of DNA combined with proteins. One might therefore expect that the DNA content of nuclei at different stages in a life cycle would bear a direct relationship to the number of chromosomes.

Actually, this is not always the case because chromosomes can be single- or double-stranded. Thus a cell produced in meiosis I has the same amount of DNA as a zygote immediately after fertilisation; the former has half as many chromosomes as the zygote but they are double-stranded.

Ignoring the fact that certain tissues (such as liver) can be polyploid, it follows that for a given species, cells can have three different amounts of DNA:

- gametes, with the haploid number of single-stranded chromosomes, have X picograms of DNA (a pg is $10^{-12}$ grams).
- diploid cells just after mitosis have 2X pg of DNA.
- diploid cells entering mitosis have 4X pg of DNA.

**Table 7.2**
Differences between mitosis and meiosis

| Mitosis | Meiosis |
|---|---|
| One division | Two divisions |
| May occur in diploid or haploid cells | Only occurs in diploid cells |
| Products genetically identical | Products genetically different |
| No change in chromosome number | Chromosome number reduced from diploid to haploid |
| Chromosomes do not associate in pairs during the process | Chromosomes associate in pairs during the process |

## QUESTIONS

**1** Name the enclosed structures in the diagram below.

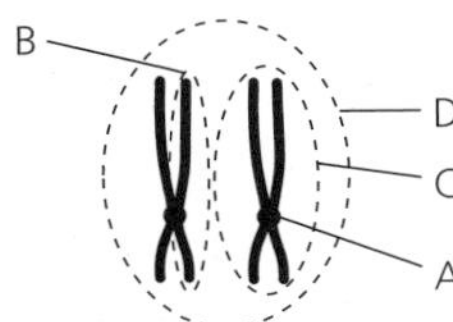


**2** A student was asked to draw an animal cell with six chromosomes undergoing mitosis, and produced the diagram shown on the below. Redraw it, correcting *three* errors.

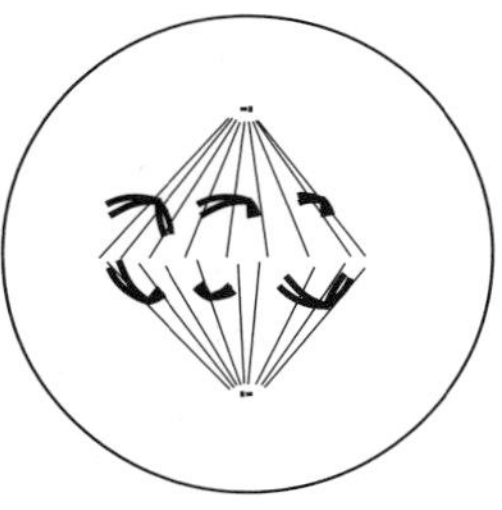

**3** The horse roundworm (*Parascaris equorum*) has one pair of chromosomes in each body cell. Which method of generating genetic variety *cannot* occur in this species?

**4** At what stage in the cell cycle does a
**a)** chromosome come to consist of two chromatids,
**b)** chromatid become a chromosome?

**5** Which of the diagrams **A–F** at the bottom of the page could represent the chromosomes of a
**a)** sperm,
**b)** skin cell,
**c)** cell that is about to divide to produce two pollen grains,
**d)** zygote that is about to divide?

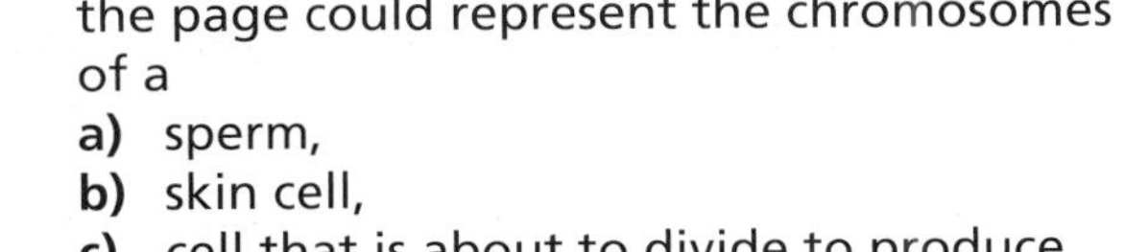

**6** In what sense is the second division of meiosis genetically different from mitosis?

**7** Which of the following drawings showing crossing over is incorrect, and why?

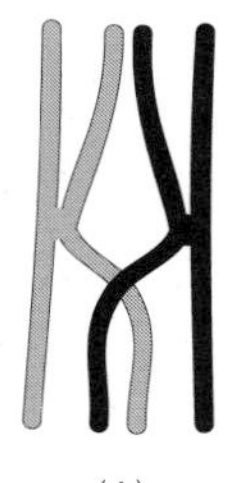
(A)

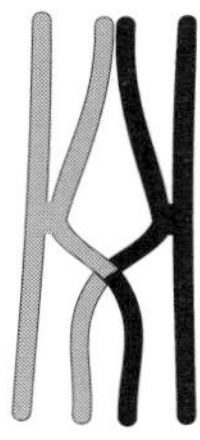
(B)

**8** A certain species of animal has eight chromosomes in each of its body cells.
**a)** Ignoring crossing over, how many chromosomal combinations would such a cell produce in its gametes?
**b)** How many chromosomal combinations could be produced in the zygote by random fertilisation?

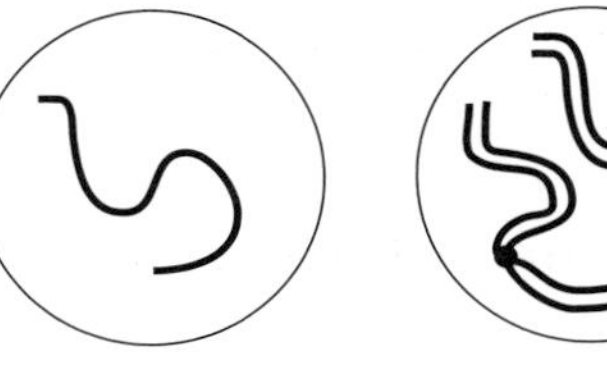
(A)

(B)

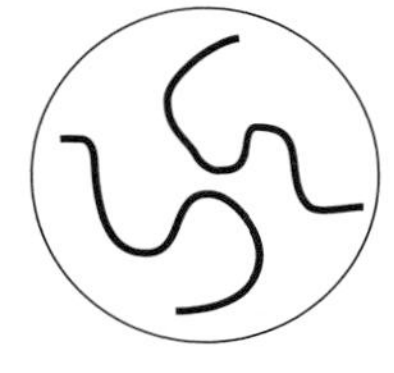
(C)

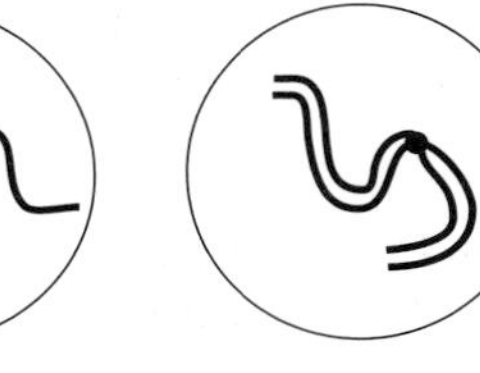
(D)

(E)

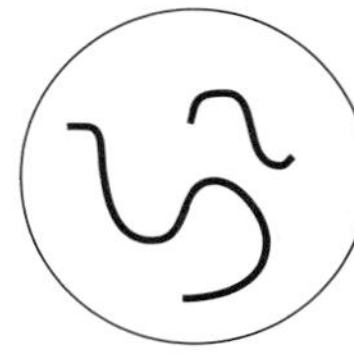

(F)

CHAPTER 

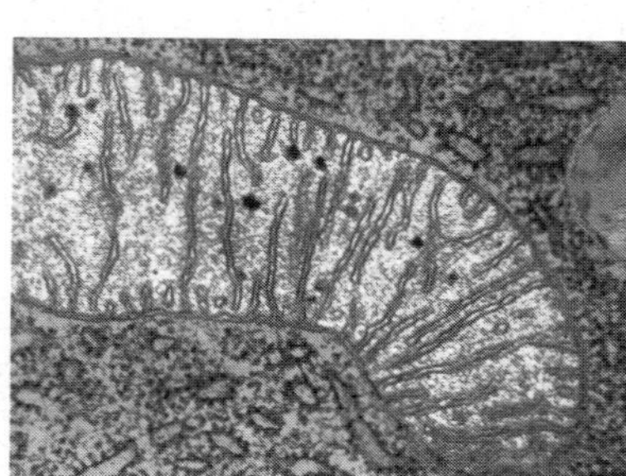

# MENDEL and HEREDITY

## LEARNING OBJECTIVES

By the time you have completed your study of this chapter you should be able to:

- distinguish between blending and particulate inheritance.
- distinguish between monohybrid and dihybrid crosses.
- explain what is meant by segregation of genes and how it accounts for the production of 3:1 and 1:2:1 monohybrid ratios.
- understand the use of a test cross to determine an organism's genetic makeup.
- explain what is meant by independent assortment and how this accounts for the production of a 9:3:3:1 and a 1:1:1:1 ratio.
- explain how the behaviour of linked genes accounts for the failure of genes to assort independently.
- calculate recombination frequencies between genes and use them to compare the relative distances between genes.
- explain how chromosomes determine sex in mammals and fruit flies.
- explain what is meant by sex-linked inheritance and show how it differs from autosomal inheritance.
- analyse pedigrees to determine whether a trait is dominant or recessive, and whether it is autosomal or sex-linked.
- solve simple problems involving incomplete dominance, codominance, and multiple alleles.
- explain the production of 9:3:4 and 9:7 ratios.

**Figure 8.1**
Gregor Mendel

## Particles or fluids?

One of Darwin's biggest problems in presenting his case for evolution was that Victorian ideas about inheritance seemed incompatible with natural selection. It was commonly thought at the time that the hereditary material was a kind of fluid. In sexual reproduction the fluid from the two parents was thought to mix, producing offspring with the average of the parents' characteristics. The trouble with this idea of **blending inheritance** is that with each successive generation, the population would become more and more uniform so there would be no variation upon which natural selection could act.

The solution came in 1900 when three biologists independently came across an obscure scientific paper called *Experiments in Plant Hybridisation*. The paper had been published 34 years earlier in *The Proceedings of the Natural History Society of Brünn* and had been gathering dust ever since.

Brünn is a small town in what is now the Czech Republic, and the author of the paper was Gregor Mendel (Figure 8.1), an obscure monk

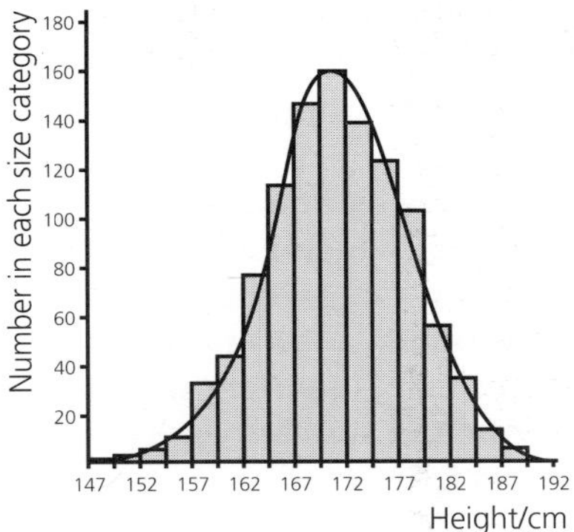


**Figure 8.2**
Bell-shaped curve for a continuously varying character – height in adult males

who was interested in science. In his paper he described the results of 8 years of breeding experiments, as a result of which he had unravelled the basic laws of heredity.

Like others before him Mendel used the garden pea. Besides being easy to grow it had some important biological advantages. It is normally self-fertilising so each variety **breeds true** for its particular characteristics. This means that generation after generation the same characters are shown. Each variety is said to be a **pure line**. This was important to Mendel because he had to be sure that any new character appearing in his plants was due to the crosses he had performed rather than something that had been 'hidden' in the parents. A second advantage of self-fertilisation was that when Mendel wanted to inbreed the peas by self-pollination, all he had to do was to leave them alone without the need to protect them from 'foreign' pollen.

Another useful feature of the garden pea was that it was available in a number of clearly contrasting varieties such as purple or white flowers, round or wrinkled seed, tall or short stem. Characters like these are said to be **qualitative**, being of the 'either/or' kind. Since each individual organism can be placed in one of two or more distinct categories, qualitative characters are said to vary **discontinuously**. In contrast, characters such as body mass or skin colour in humans vary **continuously** because each individual can be anywhere on a continuous spectrum between two extremes. A frequency distribution of a continuously varying character approximates a normal curve (Figure 8.2).

Discontinuously varying characters have two great advantages in the study of heredity. Firstly, they tend to be insensitive to environmental factors. This means that variation in the offspring is less likely to be due to variation in their environment than to differences in their hereditary makeup. Secondly, since offspring can be placed in distinct groups, they can be counted and expressed as ratios.

Like most flowers the garden pea is **hermaphrodite**, having both male and female organs. This means that each plant could be used as a male (pollen donor) or a female (pollen recipient). Mendel transferred pollen using a fine brush, after first removing the stamens from the flowers of the plant to be used as female. To protect them from foreign pollen, he covered the flowers in small bags (Figure 8.3).

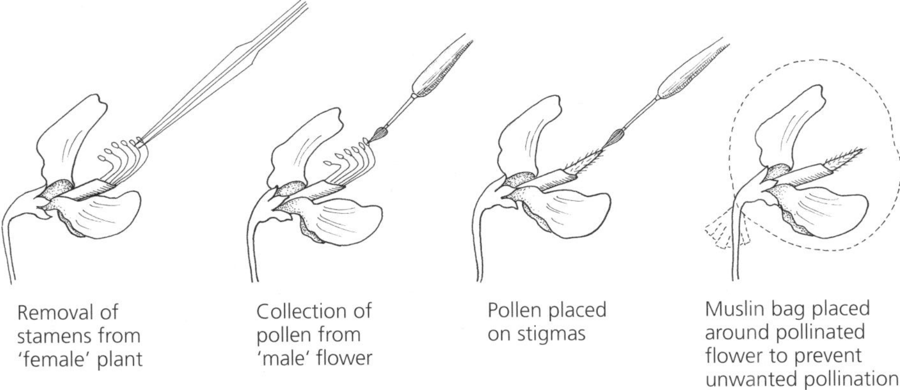


**Figure 8.3**
Artificial cross-pollination of the garden pea

# Monohybrid crosses – 3:1 and 1:1 ratios

Mendel's first experiments were **monohybrid** crosses, in which he paid attention to only one pair of contrasting characters at a time.

In one experiment he crossed a true-breeding tall plant with a true-breeding short plant. He did this in both ways (called **reciprocal** crosses) – in one cross he used the tall parent as a male (pollen donor), and in the other the tall parent as a female (pollen recipient).

Regardless of which way round he crossed the two varieties he got similar results (Figure 8.4). All the hybrid offspring (which he called the 'first filial generation' or $\mathbf{F_1}$) were just as tall as their tall parent. He then allowed the $F_1$ plants to fertilise themselves to produce a 'second filial generation' or $\mathbf{F_2}$. Of these 'grandchildren' about three-quarters were tall and a quarter were short – a ratio of 3:1. He obtained similar results with crosses involving other pairs of characters (Table 8.1).

**Figure 8.4►**
Results of Mendel's monohybrid experiment

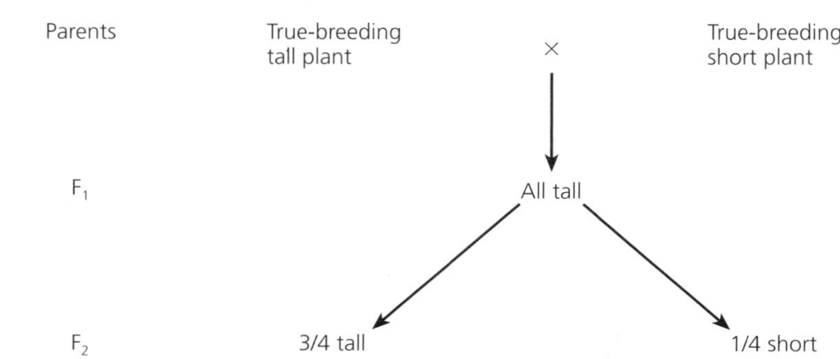


**Table 8.1**
Summary of Mendel's monohybrid results

| Character | Contrasting traits: Dominant | Recessive | $F_2$ phenotypes | | Ratio |
|---|---|---|---|---|---|
| Seed shape | round | wrinkled | 5474 | 1850 | 2.96:1 |
| Seed colour | yellow | green | 6 022 | 2 001 | 3.01:1 |
| Flower colour | red | white | 705 | 224 | 3.14:1 |
| Pod shape | inflated | constricted | 882 | 299 | 2.95:1 |
| Pod colour | green | yellow | 428 | 152 | 2.95:1 |
| Flower position | axillary | terminal | 651 | 207 | 2.82:1 |
| Stem height | tall | short | 787 | 277 | 2.84:1 |
| **TOTAL** | | | **14 949** | **5 010** | **2.98:1** |

From Mendel's results it is possible to draw the following conclusions:

- since reciprocal crosses produced the same results, male and female parents must make equal contributions to the characteristics of the offspring.
- despite receiving factors for both tall stems and short stems, the factor for short stems was not expressed in the $F_1$ plants. Mendel concluded that tall is **dominant** to short, which is **recessive**.
- the factors for tall and short stems had not 'diluted' each other in the $F_1$ generation because the two heights of the $F_2$ plants were indistinguishable from those of their grandparents. Rather than being a fluid, the hereditary material must therefore behave like *particles*, each retaining its separate identity. This is implied by **Mendel's First Law** (see information box 'Mendel's Laws' on page 133).
- prior to the formation of the gametes the two members of each pair of factors must split up or **segregate**.

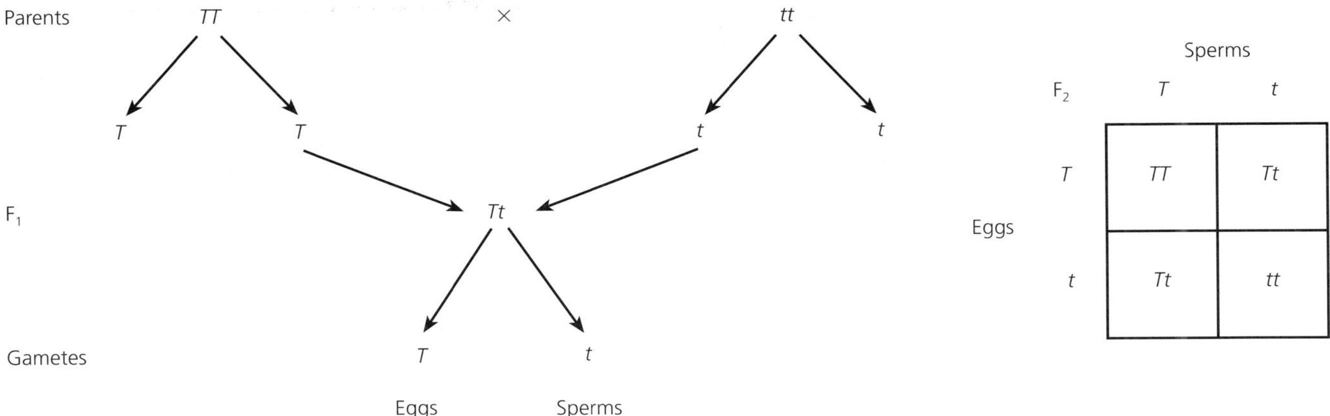


**Figure 8.5**
Explanation for Mendel's monohybrid experiment

That the two factors must have segregated can be proven in the following way. A short $F_2$ plant must have been produced by fusion of gametes carrying only the factor for short stems. Similarly those $F_2$ plants that were true-breeding for tallness must have been produced by fusion of gametes carrying only the factor for tallness.

The gametes produced by the $F_1$ plants must therefore have carried only *one* factor for height. If half the $F_1$ gametes carried the factor for tall and half carried the factor for short (i.e. a 1:1 ratio), and if fertilisation were random, then about $\frac{1}{2} \times \frac{1}{2} = \frac{1}{4}$ of the offspring would carry two factors for tall stems, and similarly $\frac{1}{4}$ would carry two factors for short stems. The remainder would carry one of each and would have tall stems. The result would be a ratio of 3 tall:1 short.

Figure 8.5 shows how this result was obtained using the symbols *T* for the factor for tall stems and *t* for the factor for short.

The various ways in which the gametes can combine at fertilisation are shown in a **Punnett square**, after the Cambridge geneticist Reginald Punnett who first used it. It is important to realise that the four separate compartments do not represent an individual offspring but a particular kind of fertilisation event. Notice that the joining of a *T* egg with a *t* sperm is a different event from the joining of a *t* egg with a *T* sperm, although it produces the same result *Tt*. Since *Tt* can be produced in two different ways, it is produced with twice the frequency of *TT* or *tt*.

The Punnett square also shows that as far as their hereditary makeup is concerned, there are actually *three* kinds of plant in the $F_2$ – true-breeding tall, non-true-breeding tall and short, in a ratio of 1:2:1. Mendel demonstrated this by allowing a selection of the $F_2$ plants to self-fertilise. All the short plants bred true to type, meaning that they only produced short offspring. Of the tall plants only about a third were true-breeding. This is because, of the three kinds of fertilisation that produce tall plants, only one gives rise to true-breeding offspring.

Since Mendel's time new terms have been introduced and some of his conclusions have been modified. The hereditary factors are called **genes**. The alternative forms of a gene are called **alleles**. Mendel used an upper case letter to denote dominant allele, and a lower case letter for the recessive allele (*T* and *t* in the above example). An organism that is true-breeding for a particular trait carries two copies of the same allele and is said to be **homozygous** for that trait. Thus a true-breeding tall plant is represented by *TT* and a true-breeding short plant by *tt*. The $F_1$ plants, on the other hand, carry both alleles for height and are **heterozygous** (*Tt*). Although the $F_1$ plants had the same bodily characteristics or **phenotype** as their tall parents, they differed in their hereditary makeup or **genotype**. (Phenotype is not limited to external appearance. A person's blood group, intelligence, reflex time and cholesterol level are all aspects of the phenotype.)

## Determining an organism's genotype – the testcross

In order to show that his $F_1$ plants were heterozygous for height, Mendel performed a **testcross** (also known as a backcross) by mating them with short plants (Figure 8.6). In a testcross the organism under test is mated with one with the recessive phenotype. Since the latter only produces gametes carrying the recessive allele, any variation in the offspring must be due to variation in the gametes produced by the $F_1$ parent. Moreover the ratio of the two kinds of offspring will be the same as the ratio of the different kinds of gamete produced by the $F_1$ plants.

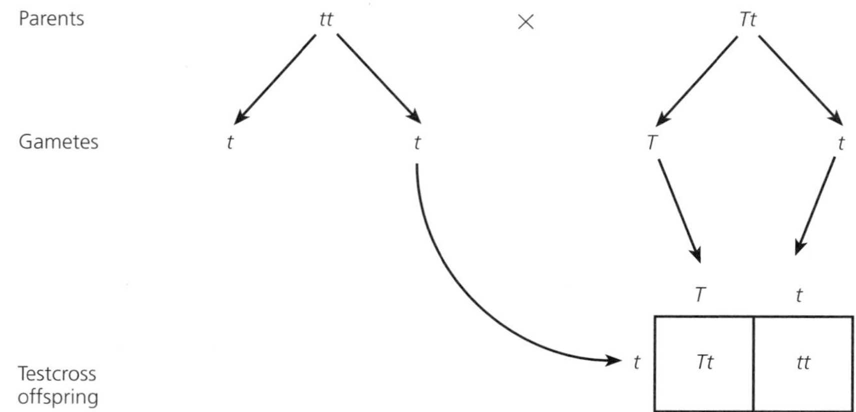


**Figure 8.6**
Mendel's monohybrid testcross

Of the 166 offspring, 85 had tall stems and 81 had short stems. This is approximately a 1:1 ratio, showing that the $F_1$ plants were producing two kinds of gamete in roughly equal numbers. Had the $F_1$ plants been genetically similar to their tall parents all the testcross offspring would have been tall.

### Asking the right question

People sometimes ask: 'which is the more important influence in human intelligence – genes or environment?' The question is meaningless for it is like saying: 'which is more important, the recipe or the cook?' Just as a cake requires both recipe and cook, any phenotypic characteristic requires both information (the genes) *and* the necessary raw materials, energy and other environmental influences. Both are indispensable.

A more sensible question would be: 'which plays the more important part in explaining differences in phenotype – differences in genes or differences in environment?' Some characteristics develop in the same way in any environment that supports life, such as blood groups. In this case, differences in environment have no influence on differences in phenotype. Other characters such as skin pigmentation are very sensitive to environmental factors like sunbathing. Here, differences in environment have much more influence than differences in phenotype.

# Dihybrid crosses: the 9:3:3:1 and 1:1:1:1 ratios

Figure 8.7 illustrates a **dihybrid** experiment in which Mendel considered the inheritance of two traits *simultaneously*. He crossed a round yellow-seeded variety (*RRYY*) with a wrinkled green-seeded variety (*rryy*).

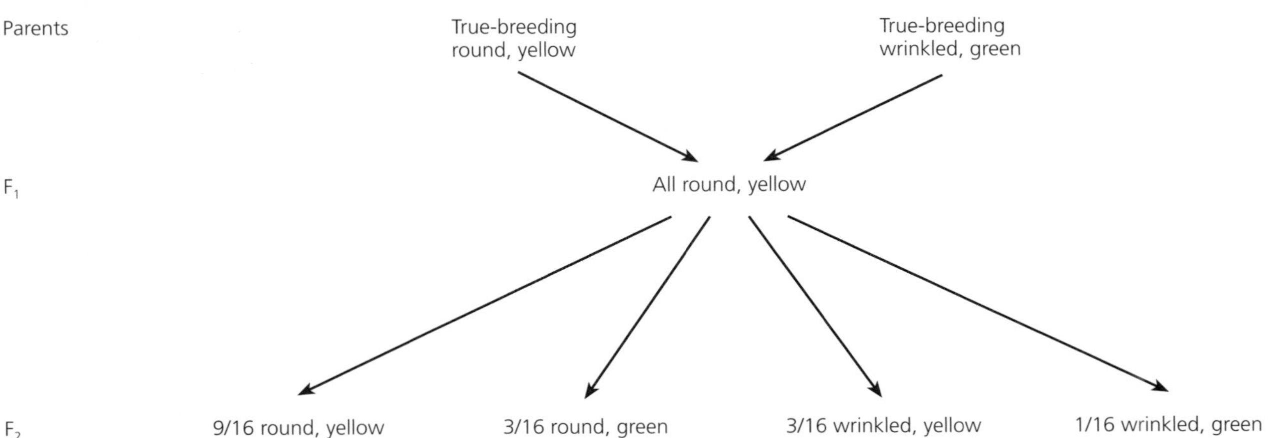


**Figure 8.7**
Results of Mendel's dihybrid cross

From his monohybrid experiments Mendel knew that round is dominant to wrinkled and yellow is dominant to green. He would therefore have expected that all the $F_1$ seeds (*RrYy*) would be round and yellow. He would also have expected that $\frac{3}{4}$ of the $F_2$ seeds would be round and $\frac{1}{4}$ wrinkled, and that $\frac{3}{4}$ would be yellow and $\frac{1}{4}$ green.

What Mendel could not predict was how seed shape would behave with respect to seed colour. For example, what proportion of the round seeds would also be yellow, and what proportion would be green? Likewise what proportion of the green seeds would be round and what proportion would be wrinkled?

Mendel's results were as follows:

- 315 round and yellow – about $\frac{9}{16}$ of the total
- 108 round and green – about $\frac{3}{16}$ of the total
- 101 wrinkled and yellow – about $\frac{3}{16}$ of the total
- 32 wrinkled and green – about $\frac{1}{16}$ of the total

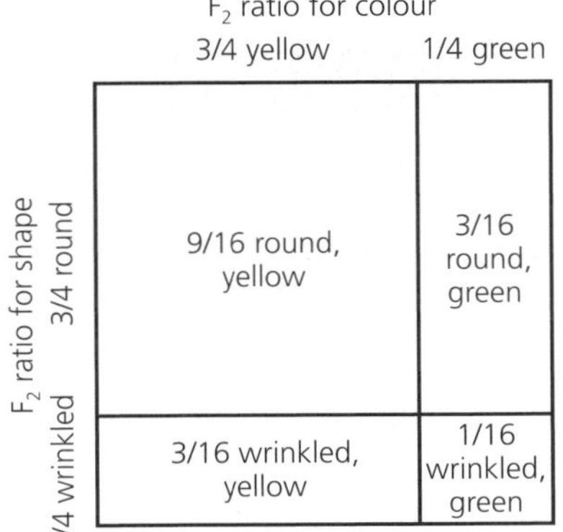


**Figure 8.8**
How a 9:3:3:1 ratio is two independent 3:1 ratios

This is a ratio of approximately 9:3:3:1. The most obvious feature of the $F_2$ is that two of the phenotypes – wrinkled, yellow and round, green – have combinations of characteristics not shown by either parent. Because they show new combinations of characters they are called **recombinants**.

Not quite so obvious is the fact that *seed shape is inherited independently of seed colour*. To see what this means, consider seed colour and shape separately. Of the 423 round seeds, approximately $\frac{3}{4}$ were yellow and $\frac{1}{4}$ were green. Likewise of the 140 green seeds about $\frac{3}{4}$ were round and $\frac{1}{4}$ were wrinkled. When we say that shape and colour are inherited independently we mean that the probability of a seed being round does not affect the probability of it being yellow. What this means is that an $F_2$ seed was three times more likely to be yellow than green *irrespective of whether it is round or wrinkled.*

Another way of putting this is to say that the 3:1 ratio for seed colour is independent of the 3:1 ratio for seed shape. Figure 8.8 shows how a 9:3:3:1 ratio is the product of two independent 3:1 ratios.

If *R* and *r* segregate independently of *Y* and *y*, four kinds of pollen and four kinds of egg (*RY*, *Ry*, *rY* and *ry*) would be produced in approximately equal numbers, i.e. a ratio of 1:1:1:1. This is a reflection of the fact that the 1:1 ratio for *R* and *r* gametes and the 1:1 ratio for *Y* and *y* gametes are independent of each other.

As Figure 8.9 shows, there are 16 different ways in which four kinds of egg and four kinds of sperm could combine. If fertilisation is random, these will occur with equal frequency. Notice that although there are 16 different kinds of fertilisation event, only nine different genotypes are produced. This is because some fertilisations give the same result (e.g. an *RY* egg and an *ry* sperm give the same product as an *Ry* egg and an *rY* sperm).

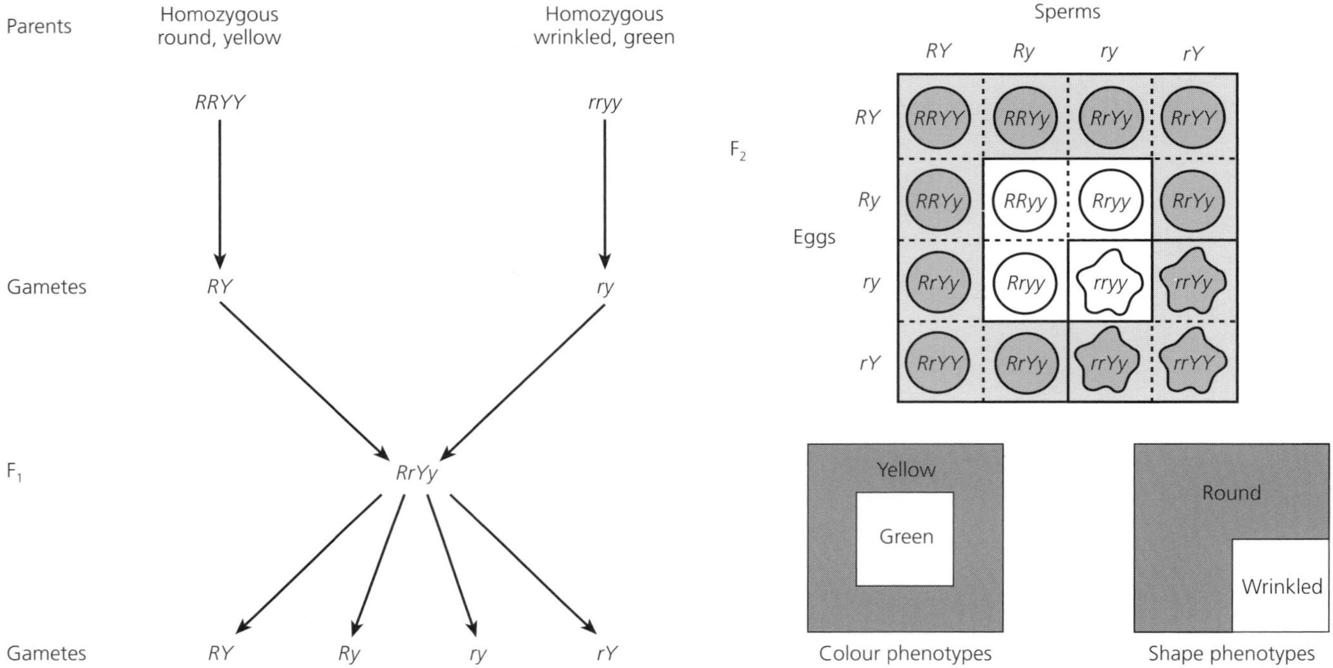


**Figure 8.9**
Explanation for Mendel's dihybrid cross

A simple way to see why there are nine genotypes in a dihybrid $F_2$ is as follows. There are three genotypes for seed shape (*RR*, *Rr* and *rr*). For each of these there are three genotypes for colour (*YY*, *Yy* and *yy*), giving $3 \times 3 = 9$ possible genotypes (Figure 8.10).

Three genotypes for colour

| Three genotypes for shape | *YY* | *Yy* | *yy* |
|---|---|---|---|
| *RR* | *RRYY* | *RRYy* | *RRyy* |
| *Rr* | *RrYY* | *RrYy* | *Rryy* |
| *rr* | *rrYY* | *rrYy* | *rryy* |

**Figure 8.10**
Punnett square showing why there are nine $F_2$ genotypes in a dihybrid cross

## Correct use of a Punnett square

A Punnett square is used to show the various ways in which gametes can combine at fertilisation. Each compartment of the Punnett represents a different kind of fertilisation event. More than one kind of fertilisation is only possible if at least one of the parents is heterozygous. Despite this, Punnett squares are sometimes used to illustrate matings between **homozygotes**, such as *AA* × *aa*, or *AABB* × *aabb*, in which only one kind of fertilisation is possible. In such cases the different cells of a Punnett square have no meaning, so the use of a Punnett square in this situation is inappropriate.

## A dihybrid testcross

To check that the $F_1$ plants were indeed producing four kinds of egg and four kinds of pollen Mendel testcrossed the $F_1$ plants by mating them with plants grown from wrinkled, green seeds. His results were as shown below:

- Round, yellow 57
- Round, green 51
- Wrinkled, yellow 49
- Wrinkled, green 53

This result is fairly close to the theoretical 1:1:1:1 ratio which would be expected if the two pairs of alleles were segregating independently, and is shown diagramatically in Figure 8.11 (see information box below for Chi-squared analysis).

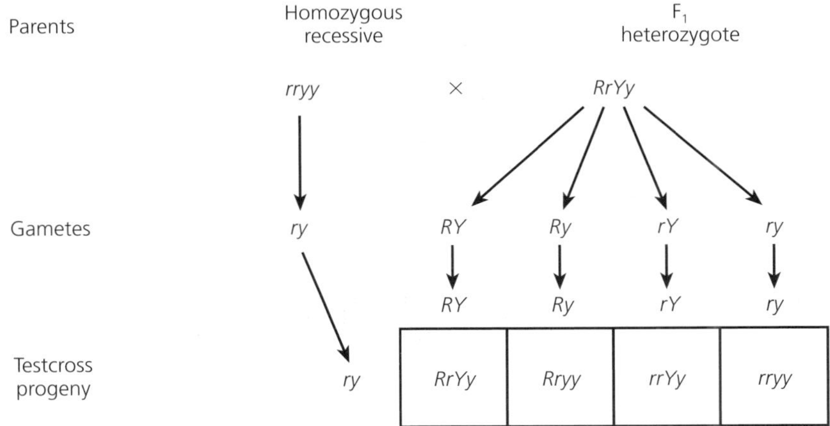


**Figure 8.11►**
Mendel's dihybrid testcross

## Chi-squared analysis of Mendel's dihybrid testcross

$\chi^2 = \Sigma$

| Phenotype | O | E | O–E | (O–E)$^2$ | $\frac{(O-E)^2}{E}$ |
|---|---|---|---|---|---|
| Round, yellow | 57 | 52.5 | 4.5 | 20.25 | 0.38 |
| Round, green | 51 | 52.5 | −1.5 | 2.25 | 0.043 |
| Wrinkled, yellow | 49 | 52.5 | −3.5 | 12.25 | 0.233 |
| Wrinkled, green | 53 | 52.5 | 0.5 | 0.25 | 0.00048 |
| | | | | | $\Sigma$ = **0.66** |

Since there are four phenotypic categories there are three degrees of freedom (see Appendix II).

Values of chi-squared for three degrees of freedom

| | Probability | | | | | | | | |
|---|---|---|---|---|---|---|---|---|---|
| | 0.9 | 0.8 | 0.7 | 0.5 | 0.3 | 0.2 | 0.1 | 0.05 | 0.01 |
| df: 3 | 0.58 | 1.0 | 1.42 | 2.37 | 3.66 | 4.64 | 6.25 | 7.82 | 11.34 |

The observed result of 0.66 corresponds to a probability of between 0.9 and 0.8. We therefore conclude that the deviation from the expected 1:1:1:1 ratio is not significant.

## Too good to be true?

Mendel's results were very close to the theoretical values – so close in fact that the statistician R.A. Fisher took a keen interest in them. He calculated that if the deviations from the theoretical values were due to chance alone, the most probable chi-squared value for Mendel's experiments taken as a whole is 41. He then worked out the chi-squared value for Mendel's actual results and obtained a value of 15.54.

For one degree of freedom, the most probable chi-squared value is 0, so a small value would not be suspicious. However some of Mendel's experiments involved three pairs of characters, giving more degrees of freedom. In these circumstances a very low chi-squared value is actually much less likely than a larger one. Fisher calculated that the probability that Mendel's ratios could be as 'good' as they were by chance alone was about 1 in 2000.

It seems then that Mendel's results were just too good to be true. Various explanations have been offered. If his assistant knew what Mendel expected to happen he may have tried to please his superior by 'adjusting' the results. Alternatively, Mendel himself may have discarded results that did not fit so well in the belief that the parent plants had somehow been contaminated by foreign pollen. We will never know, since all the documents containing his results were burned.

## Why Mendel succeeded

Mendel was by no means the first to perform breeding experiments with the garden pea. In 1799, T.A. Knight published the results of an experiment in which he had crossed a purple-flowered variety with a white-flowered variety. He found that all the offspring had purple flowers, but that some of the next generation had purple flowers and some had white flowers. Knight had thus discovered dominance. Moreover, he also anticipated Mendel in that he showed that reciprocal crosses yielded similar results.

Twenty-five years later, J. Goss crossed a yellow-seeded variety of garden pea with a green-seeded variety and inbred the $F_2$ plants. He found that the $F_2$ green seeds all bred true to type but that the $F_2$ yellow seeds yielded a mixture of both types of offspring. He had thus shown that $F_2$ plants were of two genetic types, albeit of the same appearance.

Unfortunately, neither Knight nor Goss *counted* the offspring and so did not realise that $F_2$ plants were produced in definite ratios. Mendel was far ahead of his contemporaries in that he appreciated the place of mathematics in breeding experiments. It was left to Mendel to discover that $F_1$ plants produce two types of pollen and two types of eggs each in a 1:1 ratio and, moreover, that different pairs of factors segregate independently.

## Chance and probability in genetics

As far as each individual egg is concerned it is a matter of chance which kind of sperm happens to fertilise it. Similarly chance determines which kind of egg a sperm fertilises. Consequently we rarely encounter exact whole number ratios such as 1:1 or 9:3:3:1.

Chance deviations from the most probable results are called **sampling errors.** Any individual group of offspring can be thought of as a sample of an infinitely large number of offspring that – had there been time – could theoretically have been obtained.

The larger the sample the more closely a ratio approaches the theoretical – in other words the smaller the sampling error. The most probable composition of a family of two is one boy and one girl, but two of the same sex would not be in the least surprising. With larger families, deviations due to chance get proportionately smaller. We might be a little surprised to find that a mouse gives birth to eight females and two males, but a family of 80 male and 20 female fruit flies would be very surprising – in fact we would strongly suspect that something other than chance was responsible.

What degree of difference from an 'expected' ratio can we accept as due to chance? Out of 20 offspring we would surely accept nine and 11 to be a chance deviation from a 1:1 ratio. But what about eight and 12, or five and 15? At what point are we entitled to suspect that something other than chance is at work? This is where we need a statistical test such as the chi-squared test, discussed in Appendix II.

In humans and other species with small families, ratios do not mean a lot. It is more sensible to speak of the **probability** that an individual or group of individuals will be of a particular kind.

The idea of probability flows naturally from sampling error. If we toss a coin a thousand times the proportion of heads would be very close to half. A million tosses would give a ratio almost exactly half heads and half tails – extremely close to 1:1. The larger the number of trials, the more closely the proportion of heads would approach 0.5. Given an *infinitely* large number of tosses, the proportion of heads would be exactly 0.5. *The probability of an event is the proportion of occasions it would occur out of an infinite number of trials.*

There are two laws of probability:

- **product rule** If two events are independent of each other (i.e. neither influences the probability of the other), the probability that they will *both* occur is the product of their individual probabilities. For example, if a cell with the genotype *RrYy* undergoes meiosis the probability that a gamete receives *R* is 0.5 and the probability that it receives *Y* is also 0.5. The probability that it receives *both R* and *Y* is $0.5 \times 0.5 = 0.25$.
- **sum rule** The probability that *either* of two independent events occurs is the sum of their individual probabilities. Thus the probability that a gamete will receive *R* or *r* is $0.5 + 0.5 = 1$ (i.e. certainty).

Example: In mice, black coat (*B*) is dominant to brown (*b*). A brown mouse was mated with a mouse heterozygous for coat colour, and a litter of two was produced. What is the probability that one will be black and the other brown?

There are two possible ways in which this could occur – black followed by brown, or brown followed by black.

Probability of black followed by brown $= \frac{1}{2} \times \frac{1}{2} = \frac{1}{4}$

Probability of brown followed by black $= \frac{1}{2} \times \frac{1}{2} = \frac{1}{4}$

Hence the probability of *either* of these occurring is $\frac{1}{4} + \frac{1}{4} = \frac{1}{2}$

## Genetics short cuts

An organism with the genotype *AaBbCc* is mated with one with the genotype *aaBbCc*. Assuming that these genes are inherited independently, what proportion of the offspring would have the genotype *aabbcc*?

To do this using a Punnett square would take time (it would have 32 compartments!). Provided that you know your monohybrid ratios (3:1 for $Aa \times Aa$ and 1:1 for $Aa \times aa$) it is much quicker to work it out for each gene pair separately and then multiply the results:

For $Aa \times aa$, $\frac{1}{2}$ the offspring will have the genotype *aa*.

For $Bb \times Bb$, $\frac{1}{4}$ of the offspring will have the genotype *bb*.

For $Cc \times Cc$, $\frac{1}{4}$ of the offspring will have the genotype *cc*.

Hence the proportion that would be homozygous recessive for *all three* gene pairs is $\frac{1}{2} \times \frac{1}{4} \times \frac{1}{4} = \frac{1}{32}$.

## Genes and chromosomes

After the rediscovery of Mendel's work the close similarity between the behaviour of genes and chromosomes was striking. Some are listed below; others are dealt with later in this chapter.

- Mendel had deduced that of the two genes for each character, only *one* was carried by each gamete. In parallel to this the microscopists later showed that chromosomes also exist in pairs but each gamete receives only one of each pair.
- The resemblance between parents and offspring implies an accurate mechanism for distributing copies of the hereditary material to daughter cells. The regular way in which chromatids are distributed to daughter cells during mitosis is consistent with this.
- The independent segregation of chromosomes can explain how the offspring of sexual reproduction differ in minor ways both from their parents and their siblings.

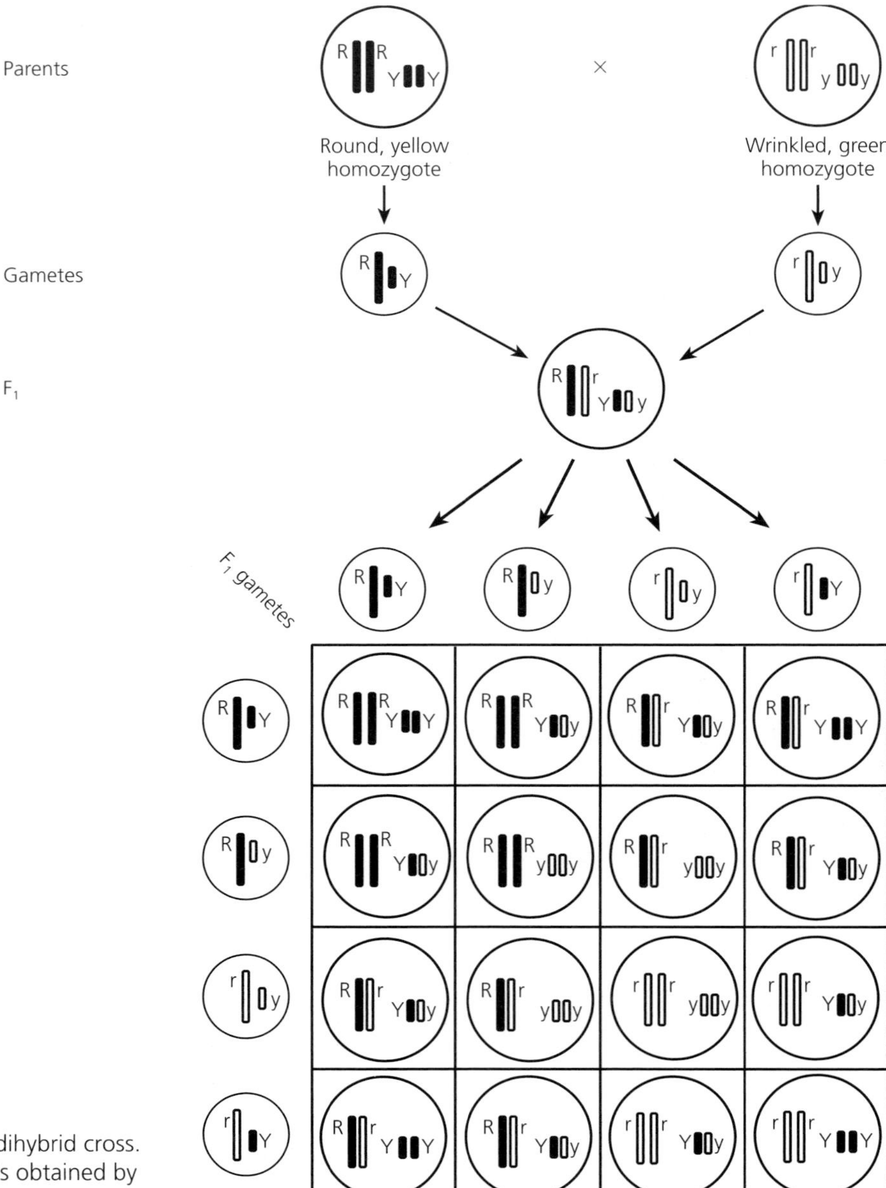


**Figure 8.12**
Chromosomal explanation for Mendel's dihybrid cross. The genotypic ratio of 1:2:1:2:4:2:1:2:1 is obtained by multiplying the two 1:2:1 monohybrid genotypic ratios

## Mendel's Laws

Carl Correns (one of the people who rediscovered Mendel's work) summed up the essence of Mendel's findings in two laws:

1 First Law, the Law of Segregation: *Each inherited character is determined by two genes, only one of which can be carried by a gamete.*
2 Second Law, the Law of Independent Segregation (or Law of Independent Assortment):

*The segregation of each member of a pair of alleles is not affected by the segregation of any other pair.* In other words *either member of one pair of alleles may associate* with equal probability *with either member of another pair.*

Soon after Mendel's work was rediscovered some genes were found to behave in a way that was contrary to Mendel's Second Law. In the event, far from undermining the idea that genes are borne on chromosomes, these cases of **linkage** provided further evidence for this theory (see later in this chapter).

- In many animals male and female phenotypes are associated with chromosomal differences (see below).
- Certain abnormal phenotypes are associated with chromosomal abnormalities, for example Down's syndrome (see Chapter 12).

The second and third points are particularly important because they help to explain the two key features of heredity – the great *similarities* between parents and offspring and the *variation* with regard to detail.

These observations (together with other findings) led some biologists to the view that genes are borne on chromosomes. Figure 8.12 shows how the behaviour of chromosomes at meiosis and fertilisation accounts for Mendel's dihybrid results described earlier.

The segregation of the genes (Mendel's First Law) is simply the result of the segregation of homologous chromosomes in meiosis. Moreover, Mendel's Second Law is explicable if each chromosome pair segregates independently of the other pairs (Figure 8.13).

# The exception that proved the rule – linkage

Living things have far more genes than chromosomes – in the case of humans over 4300 times as many (23 pairs of chromosomes and about 100 000 genes). Each chromosome must therefore contain many genes. Genes that are on the same chromosome are said to be **linked** (Figure 8.14).

Since they are literally 'stuck together', linked genes cannot segregate independently and are thus an exception to Mendel's Second Law. Linked genes do not give the Mendelian dihybrid 1:1:1:1 testcross ratio and consequently do not give a 9:3:3:1 $F_2$ ratio.

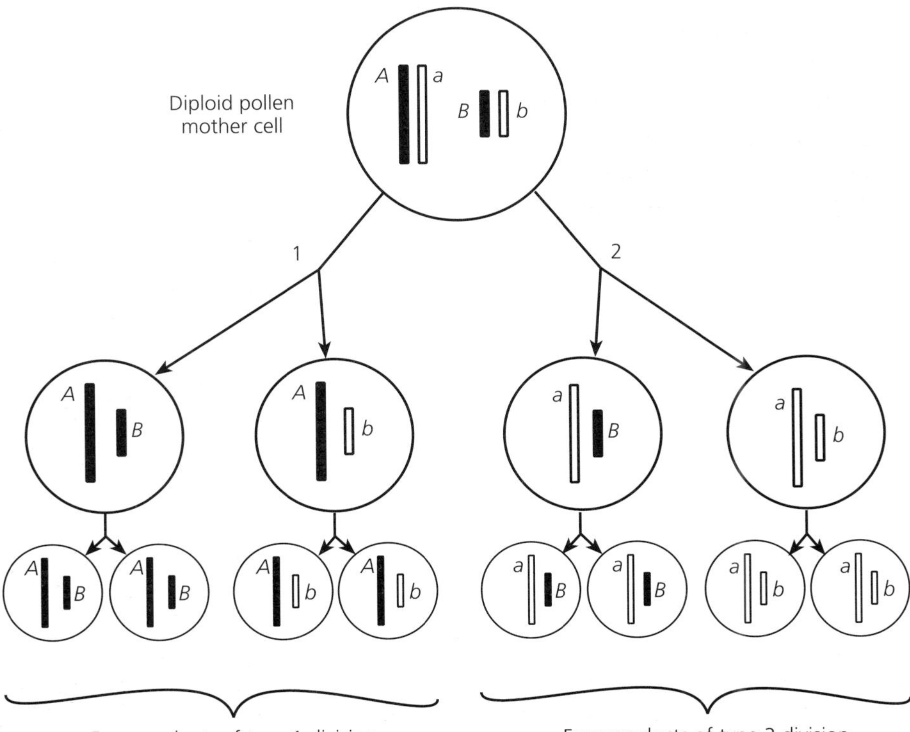


**Figure 8.13**
How a cell with two pairs of chromosomes can produce four kinds of pollen grains. The diploid pollen cell may divide in either of two ways (1 and 2), each producing two different chromosomal combinations

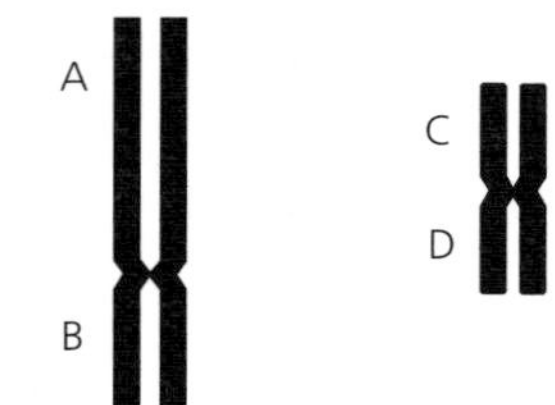


**Figure 8.14**
Linked genes. Genes A and B are linked, as are genes C and D. A is not linked to C or D

Inheritance of linked genes is complicated by **crossing over** (see Chapter 7). For this reason it will be helpful if we consider first a hypothetical case in which crossing over does not occur. Note that when dealing with linked genes each chromosome is represented by a single line (even though it actually consists of two chromatids during the first division of meiosis).

In the fruit fly (*Drosophila melanogaster*) vestigial wings (*vg*) and a black body (*b*) are recessive to their normal (wild-type) alleles for long wings (*Vg*) and a grey body (*B*). A fly homozygous for grey body and long wings was mated with a vestigial-winged, black-bodied fly, and the $F_1$ females were then testcrossed with vestigial-winged, black-bodied males. Without crossing over we would expect the following results:

- 50% vestigial-winged, black-bodied
- 50% long-winged, grey bodied

In this imaginary example, there are only two phenotypes in a 1:1 ratio, just as we would obtain in a monohybrid testcross. There are neither vestigial-winged, grey-bodied, nor long-winged, black-bodied offspring. The two pairs of traits are present in the same combinations as in the original parents. As Figure 8.15 shows, this is just what would be expected of genes linked on the same chromosome pairs.

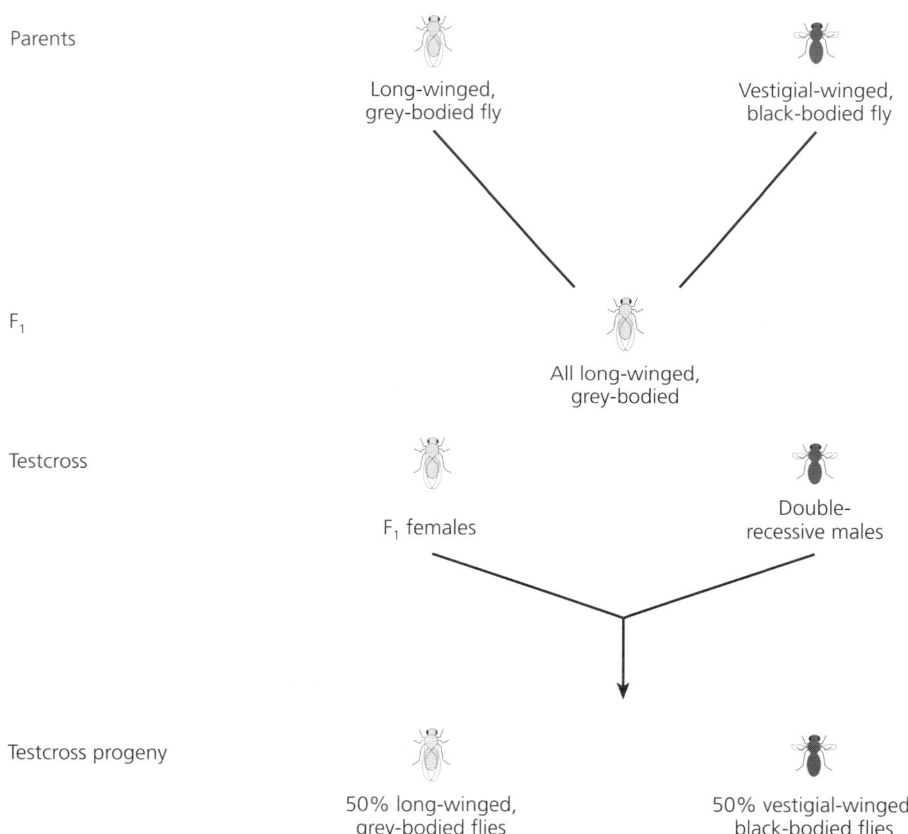


**Figure 8.15**
Behaviour of linked genes in a dihybrid cross without crossing over

Now consider what actually happens *with* crossing over. The proportions of the offspring (with percentages slightly rounded off) were actually as follows:

- 40% long-winged, grey-bodied
- 40% vestigial-winged, black-bodied
- 10% long-winged, black-bodied
- 10% vestigial-winged, grey-bodied

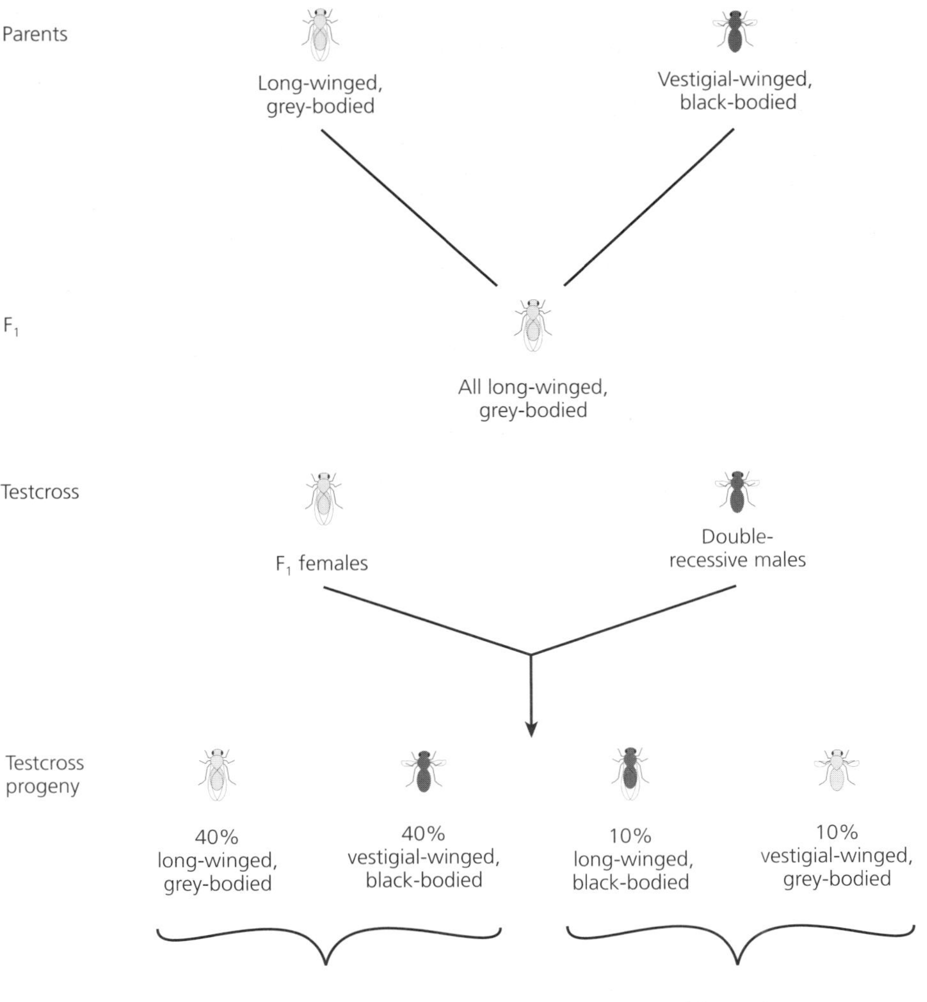


**Figure 8.16**
Behaviour of linked genes in a dihybrid cross with crossing over

Two of the phenotypes (long wings, black body, and vestigial wings, grey body) show combinations of characters not present in either of the original parents. These are **recombinants**. The long-winged, grey-bodied flies and the vestigial-winged, black-bodied flies are called **parental types** (so-called because they have the same combinations of characters as the original parental stocks – actually the *grandparents* of the testcross progeny).

Notice the following:

- the recombinants are fewer than the parental phenotypes. If the genes were segregating independently the four phenotypes would be produced in roughly equal numbers.
- the two recombinant classes are in approximately equal numbers, as are the two parental types.

Clearly, linked genes are not permanently stuck together, but can recombine. This is due to crossing over, when homologous chromosomes pair up during the first division of meiosis (Figure 8.17 and Chapter 7). You may recall that a crossover is a kind of cut-and-paste process in which breaks occur at points opposite each other in non-sister chromatids, followed by rejoining of non-sister strands.

The result of this is that each of the chromatids involved in crossing over now consists of maternal and paternal segments. Before crossing

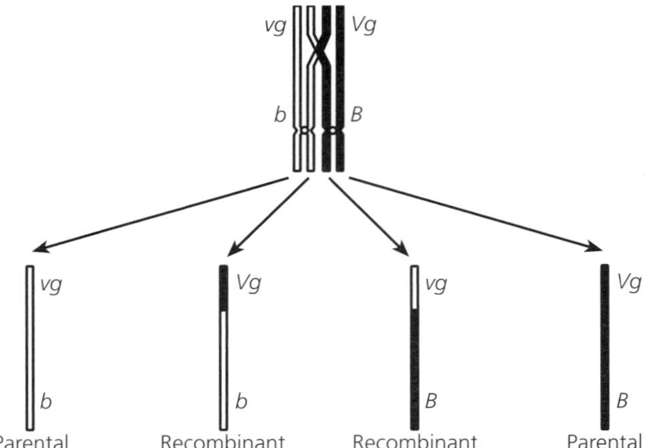


**Figure 8.17**
Crossing over between linked genes

over, the two chromatids of each chromosome are identical, but afterwards all four strands are different. The two that were not involved in the crossover are unchanged whilst each of the two that were involved now carries a new combination of genes. Each crossover thus generates *two* recombinant chromatids.

Notice that in a testcross the homozygous-recessive parent can only hand on alleles governing the recessive trait, so the $F_1$ flies are the only source of variation in the testcross progeny. Consequently the proportions of the four phenotypes in the progeny are the same as the proportions of the four kinds of gamete produced by the $F_1$ flies. This is why linkage is usually studied by testcrossing the $F_1$ generation rather than inbreeding them to produce an $F_2$.

Crossing over explains how linked genes can recombine, but we have not yet answered the question of why recombinants are less common than parental types. Firstly, we need to appreciate that although at least one crossover usually occurs on every chromosome pair, it can occur at *any* point on a chromosome. It may occur between two particular genes, or somewhere else. As far as the *Vg*/*vg* and *B*/*b* genes are concerned each ovary cell undergoing meiosis can behave in one of two ways (Figure 8.18):

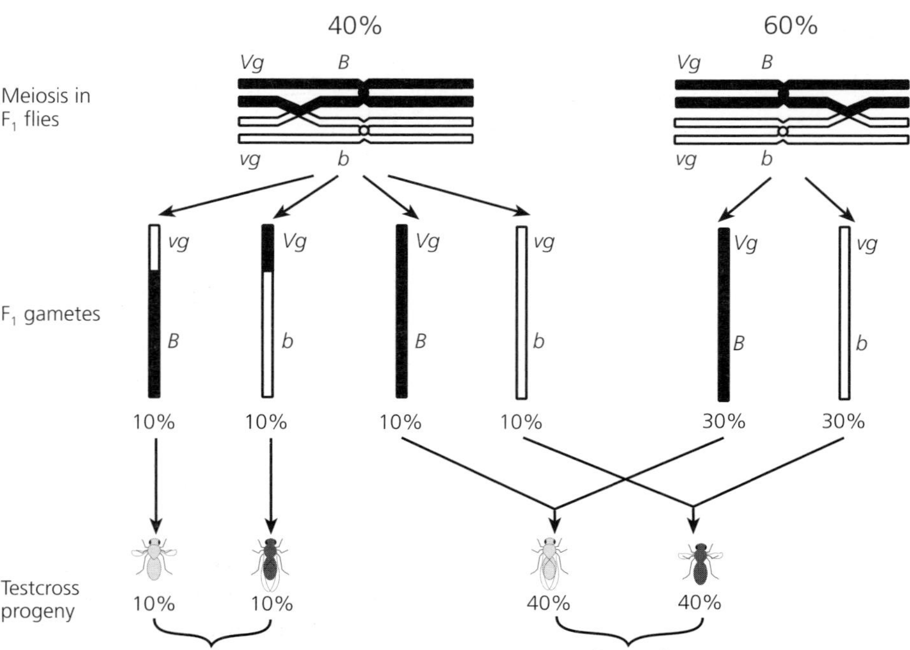


**Figure 8.18**
Chromosomal explanation for the results of the cross depicted in Figure 8.16. 60% of the cells in the ovaries of the $F_1$ females underwent meiosis in the manner shown on the right, while the remaining 40% behaved in the way shown on the left

- Case 1 – In some cells crossing over occurs between *Vg/vg* and *B/b*. This gives *four* different meiotic products in a 1:1:1:1 ratio.
- Case 2 – In the remainder, crossing over occurs *outside* the region between these two genes. These cells give equal numbers of only *two* kinds of meiotic product in the monohybrid ratio of 1:1.

What actually happens of course is that some cells yield four kinds of meiotic product and some only two kinds. The eggs produced by an $F_1$ female are therefore of four kinds but with an excess of parental types.

Notice an important thing – if a crossover were to occur between *Vg/vg* and *B/b* in *all* the cells, *the percentage recombination would only be 50%*. This is because for every crossover that occurs, two of the chromatids are unchanged (and are thus parental types). The percentage recombination is thus only *half* the proportion of cells in which a crossover occurs between the two genes in question. If 20% of the cells undergo crossing over between two genes, the percentage recombination is only 10%.

You may have wondered why, in the above testcross, $F_1$ *females* were used. The reason is that in *Drosophila*, and some other insects, crossing over does not occur in males – even in the autosomes. If the $F_1$ males were to be used in a testcross only two kinds of offspring would be produced, both of which would be parental types.

## Double crossovers

**The number of crossovers between two loci (gene positions) is not limited to one – in fact large chromosomes usually show two or more chiasmata at meiosis I. The diagram shows how a double crossover between two loci *A/a* and *B/b* affects their recombination. You will see that as far as *A/a* and *B/b* are concerned, each crossover cancels out the effect of the other. Only if a third locus (*C/c*) between the two crossovers is taken into account can the double crossover be detected.**

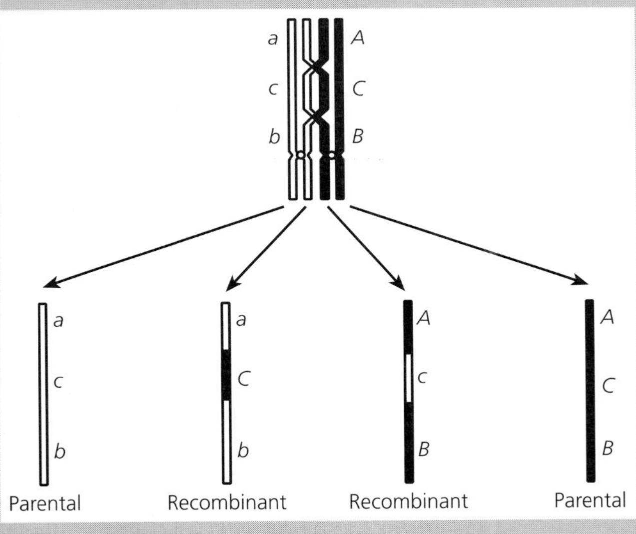


Effect of a double crossover

## Chromosome mapping

For any two gene loci the frequency of recombination has a characteristic value called the **crossover value** (C.O.V.). It is normally expressed as a percentage, given by the formula:

$$\text{C.O.V.} = \frac{\text{number of recombinant offspring}}{\text{total number of testcross offspring}} \times 100\%$$

In 1911 (a decade after Mendel's work was rediscovered) Thomas Morgan realised that the closer two genes are on a chromosome the less likely they are to recombine. Thus if two genes have a C.O.V. of 2%, they are closer together than two other genes with a C.O.V. of 5%.

In 1913 this idea was taken a step further by Alfred Sturtevant, who suggested that crossover values could be used to map the relative positions of the genes on a chromosome. If two genes have a C.O.V. of 1%, they are said to be 1 **map unit** apart. Such distances are of course purely relative, having no relation to micrometres or other standard units of distance.

The following imaginary example illustrates the basic idea. Suppose that for the genes *A*/*a*, *B*/*b* and *C*/*c* we have the following crossover values:

*A*/*a* – *B*/*b* 6%
*B*/*b* – *C*/*c* 4%
*A*/*a* – *C*/*c* 2%

Since *A*/*a* and *B*/*b* genes have the highest crossover values they must be the farthest apart. Accordingly we can place them at opposite ends of the map (Figure 8.19A). We know that the *C*/*c* gene is 2 map units away from *A*/*a*, but *C*/*c* could be on either side of *A*/*a* (Figure 8.19B). The 4% recombination between *B*/*b* and *C*/*c* shows that it must be as indicated in Figure 8.19C.

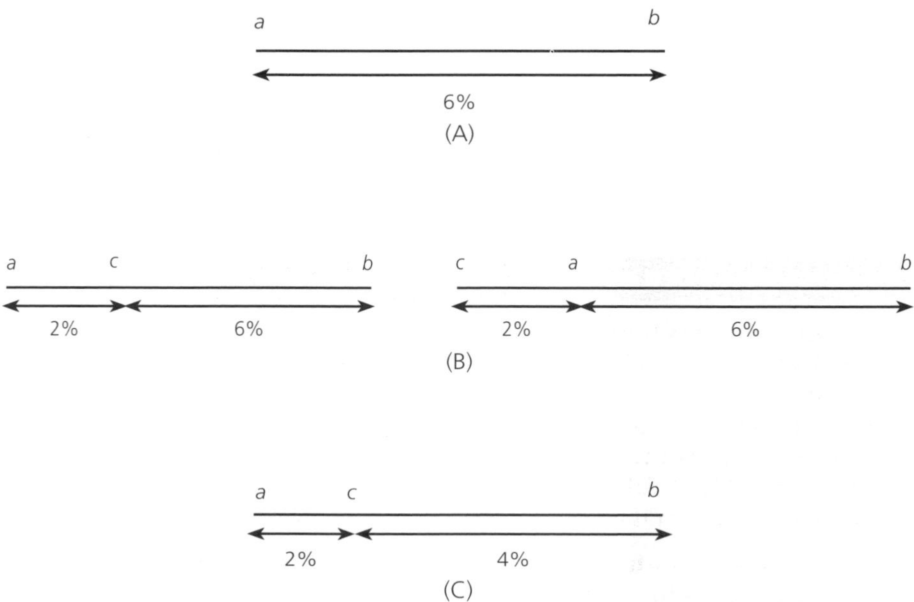


**Figure 8.19►**
Working out the positions of genes on a chromosome from recombination data

This simple chromosome map illustrates an important point – for genes that are close together, crossover values are *additive*. In this case the crossover value of 6% for *A*/*a* and *B*/*b* is equal to the 2% for *A*/*a* and *C*/*c* added to the 4% for *C*/*c* and *B*/*b*. This is consistent with the idea that genes are arranged in *linear* fashion on the chromosomes, like beads on a necklace.

As a result of these studies the idea was developed that each gene occupies a fixed position or **locus** (plural, **loci**), which is the same for both homologues.[1] Figure 8.20 shows the position of some of the genes on a human chromosome.

As we have just seen, the further apart two genes are along a chromosome the higher the probability that at least one crossover will occur between them. When genes are a long way apart they recombine with a frequency approaching 50% – the value obtained when genes are

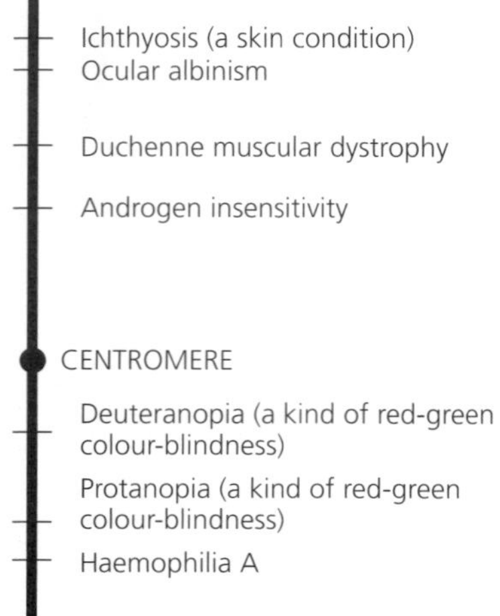


**Figure 8.20**
Some of the genes on the human X chromosome in their approximate relative positions

[1]It is now known that some genes, called **transposons**, can change position.

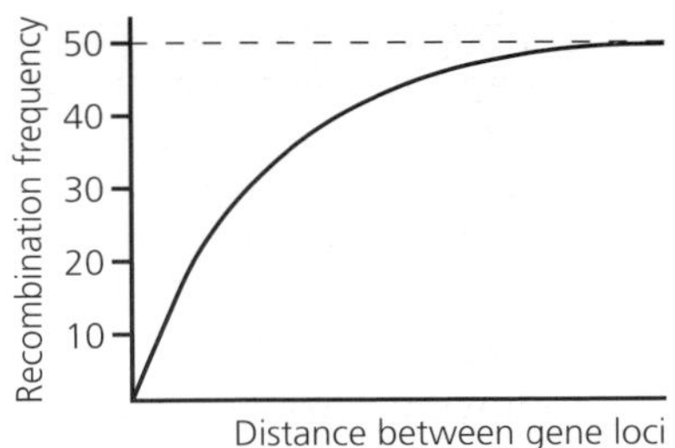


**Figure 8.21**
Graph showing how frequency of recombination between genes is affected by their distance apart

on different chromosomes. Genes a very long way apart on the same chromosome thus recombine almost as frequently as genes on different chromosomes. Yet maps of large chromosomes show that genes at opposite ends are separated by distances of much more than 50 map units – 160 units in the case of chromosome #1 of maize.

The reason for the apparent contradiction is that crossover values between genes are only proportional to their distance apart if they are close together (Figure 8.21). Distances between genes that are a long way apart have to be worked out in small steps by adding the distances between intervening genes that are much closer together.

## More evidence that genes are borne on chromosomes

For any given species, genes can be arranged in groups. The genes in one group do not assort independently with respect to each other, but do assort independently with respect to members of any other group.

The conclusion that each of these **linkage groups** corresponds to a particular chromosome pair is supported by the fact that, for a given species, there is a close correspondence between the number and size of its linkage groups. *Drosophila melanogaster*, for example, has four pairs of chromosomes, one of which is very small. It also has three large linkage groups and one very small one.

### The significance of linkage and crossing over

Crossing over has been of enormous help to geneticists in unravelling the mechanism of heredity and has provided endless raw material for genetics problems. But what about the importance of crossing over to living things? The following imaginary example should make this clear.

Suppose that all individuals in a population are homozygous for two linked genes *A* and *B*. Suppose that in one individual, *A* mutates to an allele *a*, and that in another individual *B* mutates to *b*. One individual now has the genotype *AABb* and the other has the genotype *AaBB*. If these two individuals mate, some of the offspring will have the genotype *AaBb*.

Now suppose that the alleles *a* and *b* work particularly well together as a combination. Offspring with both *a* and *b* will be more likely to survive and become parents. But – and here is the crux – though *a* and *b* are in the same organism *they are on different members of a homologous pair*. As a result they segregate at meiosis, so the advantageous combination *ab* is not preserved.

Crossing over enables advantageous alleles originating in different individuals to be brought together into the same chromosome strand, thus keeping them together long enough for natural selection to act.

Crossing over vastly increases the possibilities for recombination. It is estimated that humans have about 100 000 gene loci. If only one per cent of these are heterozygous, and each locus exists as two alleles, then the number of possible genotypes is $3^{1000}$, or about $10^{477}$. To put this in perspective, the number of electrons in the universe is estimated to be about $10^{75}$. Imagine this number raised to the power 6 and you have $10^{450}$. Multiply this by $10^{27}$ and you have $10^{477}$.

## Did Mendel discover linkage?

It is an interesting fact that, of the 34 different varieties of pea Mendel had at his disposal, his paper described experiments involving only seven. It could be coincidence, but this happens to be the same as the haploid number of chromosomes for the garden pea. Seven is therefore the maximum number of gene pairs that could segregate independently of each other. It is possible that Mendel worked with more than seven varieties but ignored those results that he could not explain. If by any chance he did stumble on the phenomenon of linkage, we will never know because, as mentioned earlier, his written records were burned.

## The origin of alleles

Independent assortment and crossing over simply reshuffle existing alleles into new combinations. The creation of new alleles occurs by a completely different process called **mutation** in which one allele changes into another. Newly-produced alleles (and their corresponding phenotypes) are called **mutants.** More often than not mutant alleles are harmful so they remain uncommon. 'Normal' or 'wild-type' alleles are those that become established in the population because they have passed the test of natural selection.

# What determines gender?

In many animals (and also in some plants) one pair of chromosomes, called **sex chromosomes**, determines whether the organism is male or female. The other chromosomes have no influence on sex and are called **autosomes**. Humans have 22 pairs of autosomes and one pair of sex chromosomes.

In many animals the sex chromosomes are visibly different in one of the sexes. In mammals and in fruit flies the males have a long **X chromosome** and a shorter **Y chromosome**, whilst females have two X chromosomes. Though they look different, the X and Y chromosomes behave like a pair because they segregate during meiosis – each sperm receiving either an X chromosome or a Y chromosome.

It follows that, whilst all the eggs carry an X chromosome, the sperms are of two kinds, half carrying an X chromosome and half a Y chromosome. The male is thus said to be the **heterogametic** sex and the female is the **homogametic** sex. It is the males that determine the gender of the offspring; Y-carrying sperms producing males and X-carrying sperms producing females (Figure 8.22).

On this basis one would expect equal numbers of male and female offspring. In fact the human sex ratio at birth is not quite 1:1; for every 100 girls born there are about 106 boys. This does not mean that more Y-carrying sperms are produced than X-carrying sperms. Y-carrying sperms may swim faster, or it may be that XY embryos are more successful at implanting into the uterine wall.

Whatever the explanation the initial advantage conferred by the Y chromosome later becomes a liability. Males die more frequently than females at all ages, and the sex ratio reaches about 1:1 by reproductive age. By old age females heavily outnumber males.

Interestingly, the higher male mortality begins before birth. Amongst early miscarriages the sex ratio is about 130 males:100 females.

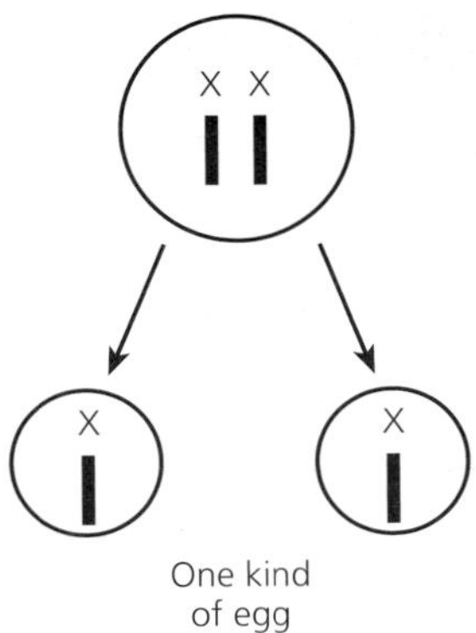


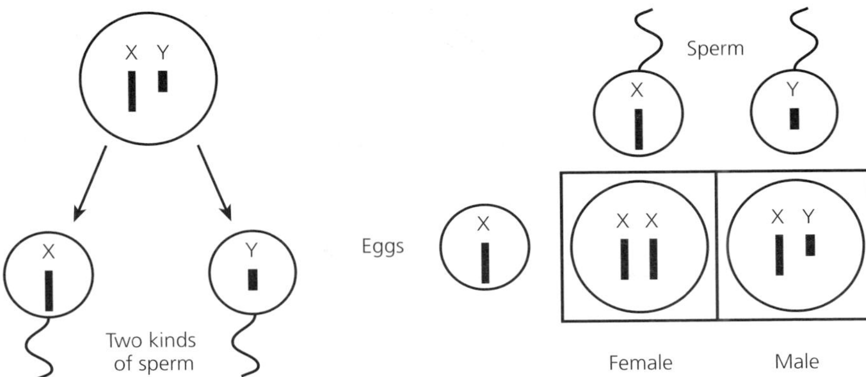


**Figure 8.22**
How gender is determined at fertilisation

In butterflies, moths and birds, the female is heterogametic and determines the gender of the offspring. Ants, bees and wasps have a completely different mechanism – fertilised (diploid) eggs develop into females and unfertilised (haploid) eggs develop into males.

## Sex-linked inheritance

You may remember that in his experiments on the garden pea Mendel found that reciprocal crosses gave similar results, showing that male and female gametes make equal genetic contributions to the offspring. Though this is true for characters controlled by loci on the autosomes, it is not the case for loci on the X chromosome.

As far as inheritance is concerned the sex chromosomes of mammals and fruit flies have two important characteristics. Firstly, since the Y chromosome is shorter than the X, it must carry fewer genes. There must therefore be many loci on the X chromosome that have no counterpart on the Y chromosome. As far as these genes are concerned, males carry only one copy and females have the normal complement of two. Gene loci that are on the X chromosome but not on the Y chromosome are said to be **X-linked**. Apart from the gene that determines maleness, the **testis determining factor**, there are no genes definitely known to be on the human Y chromosome, although 'hairy ears' is often quoted as a possible example of **Y-linked inheritance**. Since few, if any, Y-linked genes are known, the term 'sex-linkage' normally refers to genes on the X chromosome.

Secondly, females have two X chromosomes, one of which they hand on to each offspring. Males have only one, which they only hand on to their daughters. Females can therefore be homozygous or heterozygous for an X-linked allele. Males, with only one X chromosome, are said to be **hemizygous**. As far as X-linked alleles are concerned there is no such thing as dominance in males. Alleles on the X chromosome that govern characters that are recessive in females are thus always expressed in males. Heterozygous females are thus **carriers** for X-linked recessive alleles.

We can now see why X-linked recessive traits are so much more common in males than in females. A male only has to have one parent (his mother) carrying the allele to show the condition. Females on the other hand only show such traits if they inherit the allele from *both* parents. In birds and butterflies the situation is reversed.

Figure 8.23 shows the inheritance of red-green colour-blindness, an X-linked recessive trait in which either the red-sensitive or the green-sensitive cone pigments are defective (see Chapter 18). Consequently the person is unable to distinguish between the two colours.

Another X-linked recessive trait is **haemophilia** in which the blood

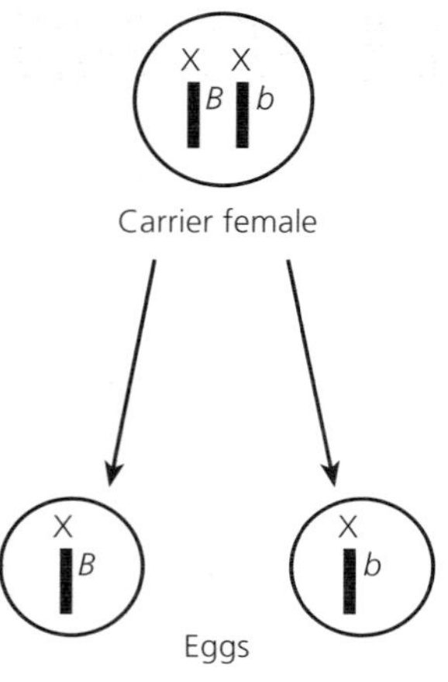


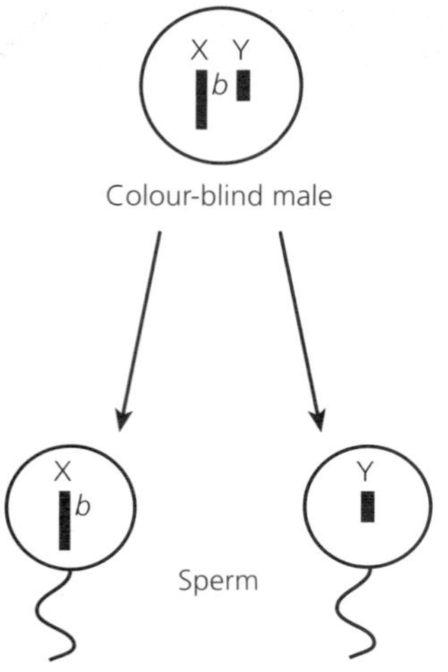


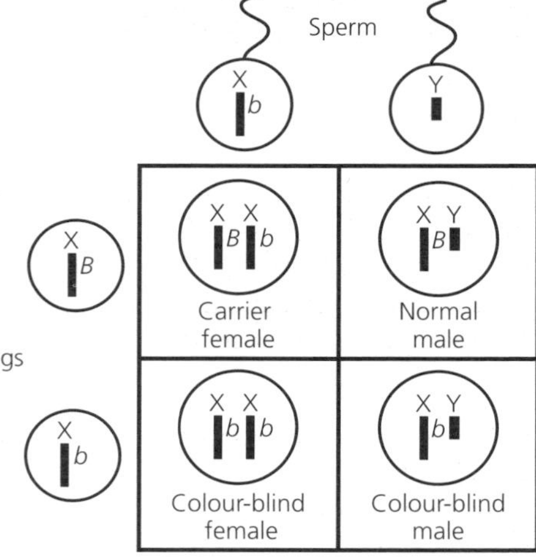


**Figure 8.23**
Results of a mating between a red-green colour-blind male and a carrier female, where *B* is the allele for normal vision and *b* the allele for colour-blindness

cannot clot because one of the protein-clotting factors is deficient. In the days before the clotting factor could be extracted and given by transfusion, haemophilia was often fatal. The condition is much rarer in females since both parents must carry the allele.

## The discovery of sex linkage

In his experiments on fruit flies, the American geneticist Thomas Morgan found an interesting exception to the general rule that reciprocal crosses give the same result. In one of his cultures, he found a fly with white eyes, unlike the normal red-eyed type. He mated the white-eyed male with a true-breeding red-eyed female, and all the $F_1$ had red eyes. It was the $F_2$ that provided the surprise. Although as expected there was a ratio of approximately 3:1 of red-eyed to white-eyed flies, the white-eyed flies were all males!

By mating the $F_1$ females with their white-eyed father, Morgan also obtained white-eyed females. This enabled him to perform the reciprocal of the original cross – a white-eyed female with a red-eyed male. All the $F_1$ males had white eyes like their mother and all the females had red eyes like their father (see diagram).

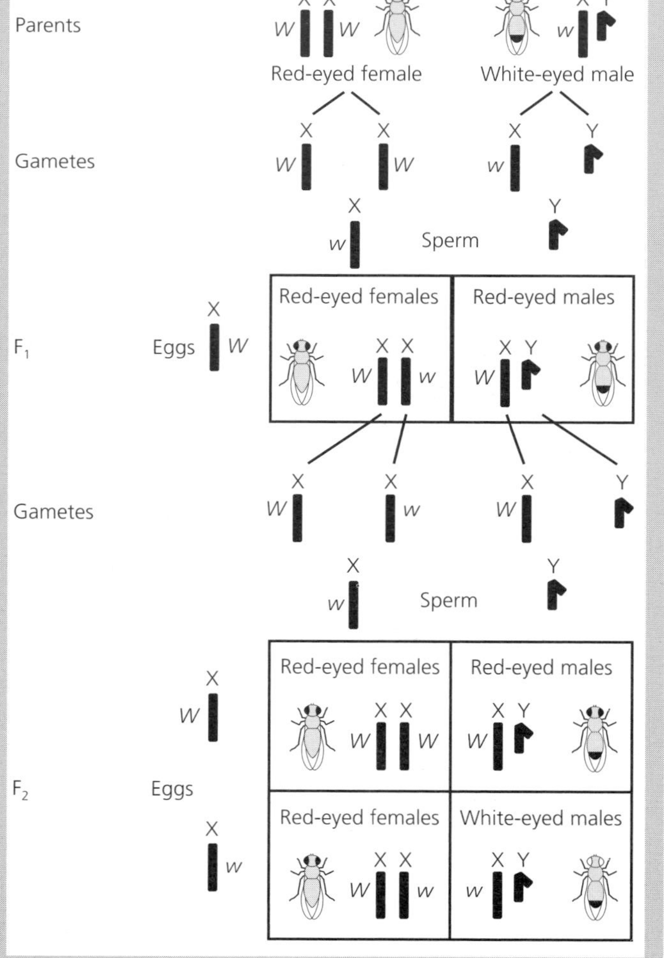


Inheritance of white-eye in *Drosophila*, where *W* is the allele for red eyes and *w* the allele for white eyes

## The discovery of sex-linkage *continued*

On inbreeding the $F_1$, Morgan found that the difference between the two reciprocal crosses persisted into the $F_2$. When the original white-eyed parent was male only a quarter of the $F_2$ flies had white eyes, and these were all male. When the original white-eyed parent had been female, *half* the $F_2$ flies (both male and female) had white eyes.

In some way the female parent was having a greater influence on the eye colour of the offspring than the male parent. To explain this, Morgan drew upon the (then) recent discovery that *Drosophila* females have two X chromosomes and males have only one. Morgan suggested that the allele for white-eye was carried on the X chromosome but not on the Y chromosome. This would explain why the female parent had a greater influence on the eye colour of the offspring than the male did, because whilst the male gives his X chromosome only to his daughters, the female gives an X chromosome to each of her children.

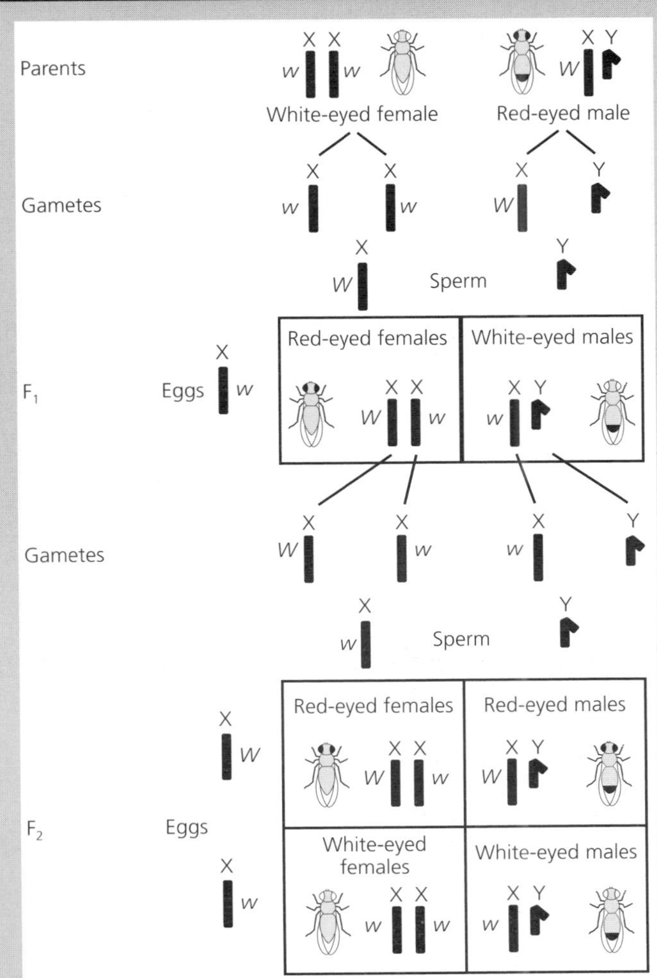


## Barr bodies and tortoiseshell cats

It is a peculiar fact that although female mammals have two X chromosomes, only one of them is active. The other exists as a **Barr body** (after Murray Barr, who first observed it). A Barr body is visible in non-dividing cells as a dark blob just under the nuclear envelope. The Barr body is formed by condensation of one of the X chromosomes early in development. Which X chromosome is inactivated is a matter of chance. In roughly half the cells the paternal X chromosome is inactivated, and in the other half the maternal chromosome is inactivated. In its highly condensed state a Barr body cannot interact with its surroundings.

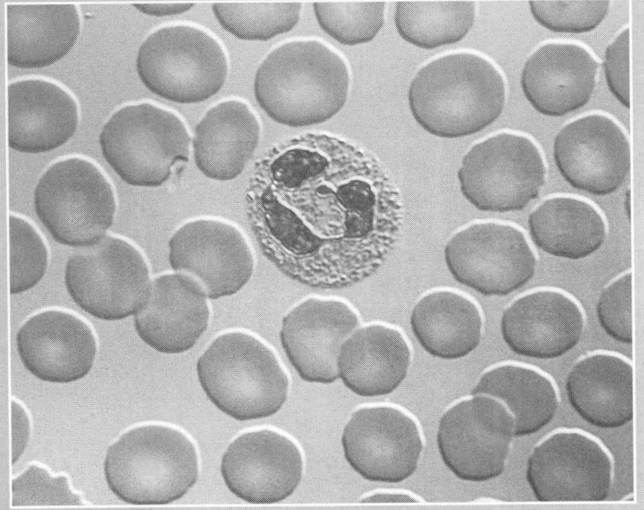

A Barr body, visible as a 'drumstick-shaped' appendage in the neutrophil in the centre of the photo

## Barr bodies and tortoiseshell cats *continued*

Though otherwise inactive, a Barr body continues to replicate with every subsequent cell division. Each cell in which the X chromosome is inactivated therefore leaves a line of descendants all having the same non-functional X chromosome. The number of descendent cells depends on the stage of development at which X-inactivation occurs – the earlier this is, the more cells are derived from it.

Each of the two X chromosomes is functional in roughly half the cells of the body, so that the female body is thus a kind of mosaic. In a female heterozygous for the allele for haemophilia only about half her liver cells have the allele for making the clotting protein involved, but it seems that this is enough. This is why females do not express X-linked recessive alleles as often as males.

The random inactivation of X chromosomes can have some interesting effects. In cats the **tortoiseshell** coat is a mosaic of black and yellow. Tortoiseshell cats are all females which are heterozygous for the X-linked alleles for black and yellow fur. Each patch represents a clone of skin cells.

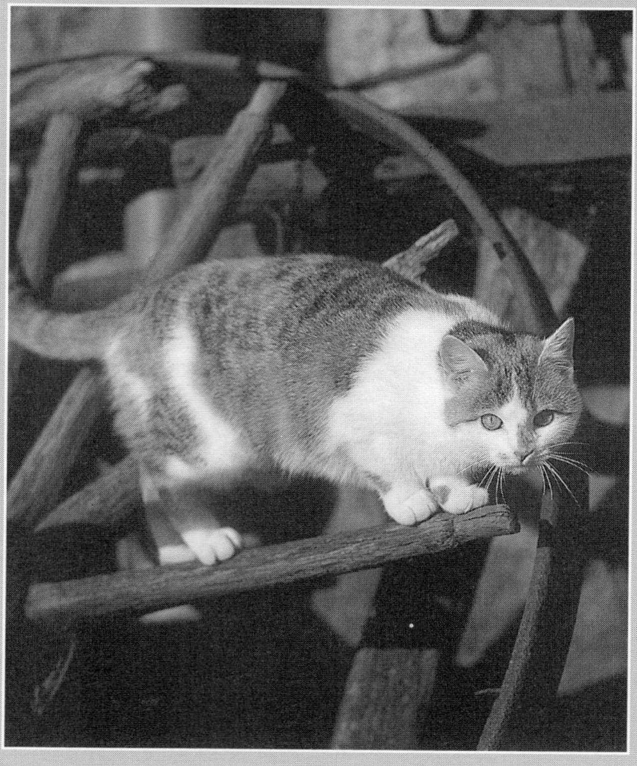

A tortoiseshell cat displaying the characteristic mosaic fur

## Lethal alleles

Some genes have alleles that are serious enough to be fatal. Manx cats are heterozygous for an allele which in homozygotes is fatal in early development. On average the litters resulting from Manx × Manx matings contain only about three-quarters as many kittens as normal, the other quarter having died early in development. Moreover Manx kittens make up about two-thirds of the total of those born – just what would be expected if those homozygous for one of the alleles had all died. Although heterozygotes (Manx cats) account for half the total zygotes formed, they constitute two-thirds of those carrying the normal allele.

Some alleles take longer to exert their lethal effects. In Duchenne muscular dystrophy – an X-linked recessive condition in which the muscles progressively waste away – death occurs by the age of 20. Huntingdon's disease – an autosomal recessive trait in which the brain undergoes progressive deterioration – does not begin to manifest itself until middle age.

## Pedigree studies

Humans are not suitable subjects for breeding experiments because they prefer to choose their mating partners. Under these circumstances genetic studies have to be **retrospective**, tracing the histories of families that already exist. Family histories are called **pedigrees**.

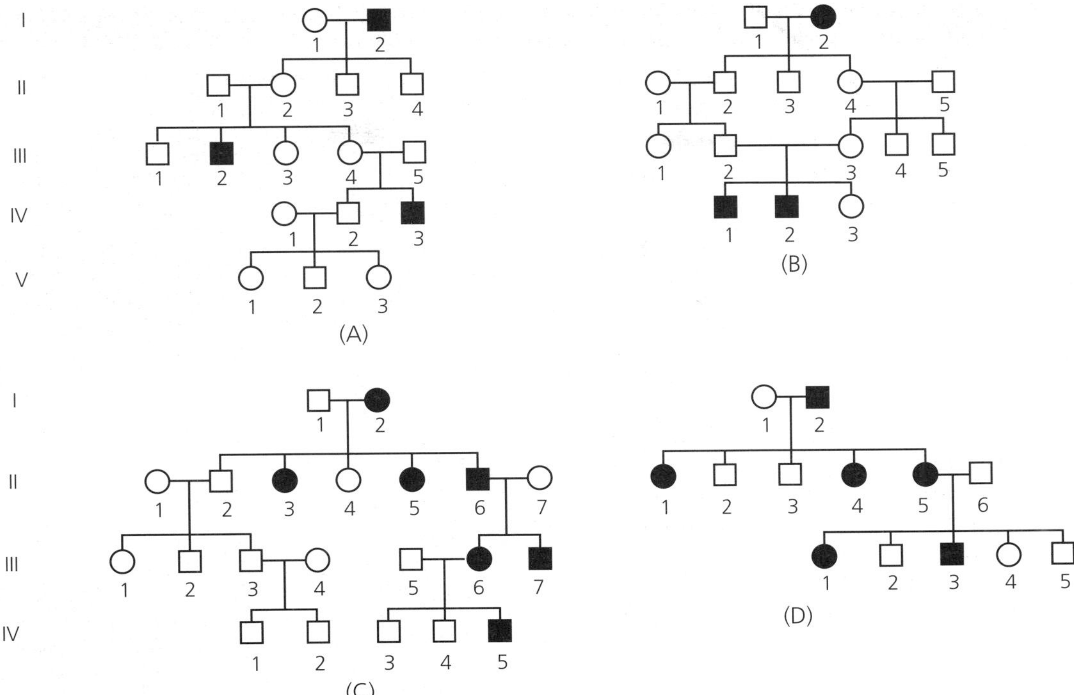


**Figure 8.24**
Four different pedigrees to show different modes of inheritance

Even though human families are small it is often possible to deduce from a pedigree whether a trait is dominant or recessive and whether it is autosomal or X-linked. Figure 8.24 shows four pedigrees to illustrate these possibilities.

In the diagram females are represented by circles and males by squares. Mating partners are connected by horizontal lines. Each generation is identified by Roman numerals on the left. Affected individuals are indicated by shaded shapes. In these examples, assume that the allele for the shaded condition is rare in the general population and that none of the alleles changes by mutation. In analysing pedigrees it is helpful to remember the following rules:

- any individual differing from *both* parents must show the *recessive* character.
- X-linked recessive traits are more common in males than in females.
- a male can only receive an X-linked allele from his mother and can only pass it on to his daughters.
- carrier females can hand on X-linked alleles to sons *or* daughters.
- all the sons of a woman homozygous for an X-linked recessive trait will show the condition.
- all the daughters of a man with an X-linked dominant trait will show the trait but, unless his partner also shows the condition, none of his sons will.

The following paragraphs relate to Figure 8.24.

**A** Since the condition appears to 'skip' generation II, individual III2 must be showing the recessive character (he could only have got the gene from his parents, but neither expresses it). As the allele is rare in the general population we can assume that II1 does not carry the gene. Hence III2 could only have inherited the allele from his mother, suggesting that it is X-linked. This is consistent with the fact that III2 and IV3 both inherit the allele from their mothers.

**B** Since the trait 'skips' generations II and III it must be recessive. It cannot be X-linked otherwise both the sons of I2 would show the condition. It must therefore be autosomal recessive.

**C** Since the allele for the shaded condition is rare we can take it that I1 is homozygous. Because the children are of both types, I2 must be heterozygous so the allele must be dominant. If it were X-linked dominant all the daughters of I2 would show the condition. It must therefore be autosomal dominant.

**D** Since all the daughters of I1 show the condition but none of the sons does, it is probably X-linked dominant.

## Incomplete dominance

In some cases a heterozygote is phenotypically intermediate between both homozygotes. Flower colour in snapdragons is an example (Figure 8.25). Plants with the genotype *RR* have red flowers, those of genotype *rr* have white flowers, whilst heterozygotes have pink flowers. Red is said to be **incompletely dominant** to white. Since however it would be just as true to say that white is incompletely dominant to red, **non-dominance** would be a more appropriate term.

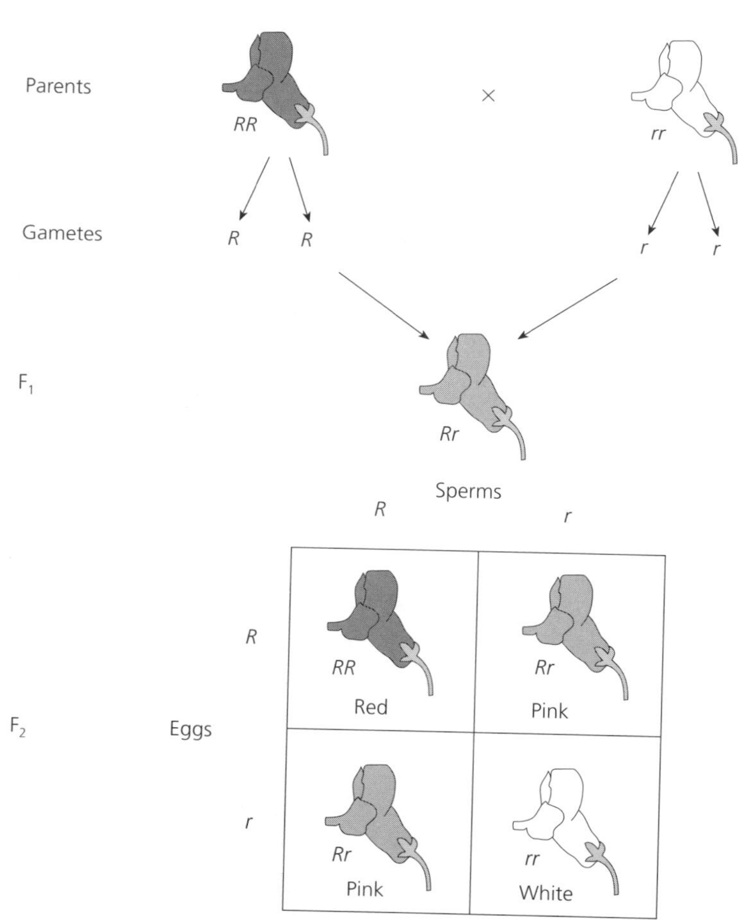


**Figure 8.25**
Incomplete dominance in snapdragons

## Codominance

In codominance two qualitatively different characters (characters that differ in *kind*) are expressed simultaneously. An example is the M and N blood group antigens on the surface of red blood cells, determined by the alleles $L^M$ and $L^N$, respectively. Neither M nor N blood group is dominant to the other so the three genotypes ($L^M L^M$, $L^M L^N$ and $L^N L^N$) are all distinguishable phenotypically.

## Multiple alleles

Though many genes exist in two allelic forms, this is not always so. Blood groups of the ABO series for example are under the control of a gene with three alleles, $I^A$, $I^B$ and $I^O$. $I^A$ and $I^B$ are codominant, and both are dominant to $I^O$. The various genotypes associated with each phenotype are shown below:

- Group O $I^O I^O$
- Group A $I^A I^A$, $I^A I^O$
- Group B $I^B I^B$, $I^B I^O$
- Group AB $I^A I^B$

Figure 8.26 shows the possible outcomes of a mating between a Group A heterozygote and a Group B heterozygote.

Parents

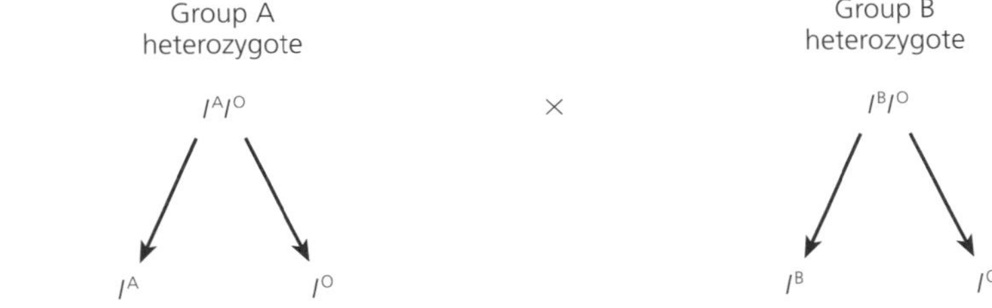


Gametes

Offspring

| | $I^B$ | $I^O$ |
|---|---|---|
| $I^A$ | $I^A I^B$ Group AB | $I^A I^O$ Group A |
| $I^O$ | $I^B I^O$ Group B | $I^O I^O$ Group O |

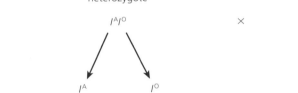

**Figure 8.26**
Possible outcomes of a mating between a Group A heterozygote and a Group B heterozygote

When a gene exists as multiple alleles the number of possible genotypes increases – for example, with three alleles there are six possible genotypes. An individual however cannot carry more than two alleles.

## Interaction between gene products

So far we have considered inheritance involving a one-to-one relationship between gene and characteristic, with each character being the result of a particular gene. In reality things are not usually so simple, as the following examples show.

## Epistasis

In some cases a gene at one locus can only be expressed in the presence of a specific allele at another locus. In mice (as in all vertebrates) there is a dominant allele (*C*) which is necessary for the development of the pigment melanin. The homozygous recessive (*cc*) is albino. In the presence of *C*, two alleles of another gene determine what *kind* of colour, with black (*B*) being dominant to brown (*b*). Since the *C*/*c* locus permits the expression of the *B*/*b* locus, *C*/*c* is said to be **epistatic** to *B*/*b* (*epi* = above).

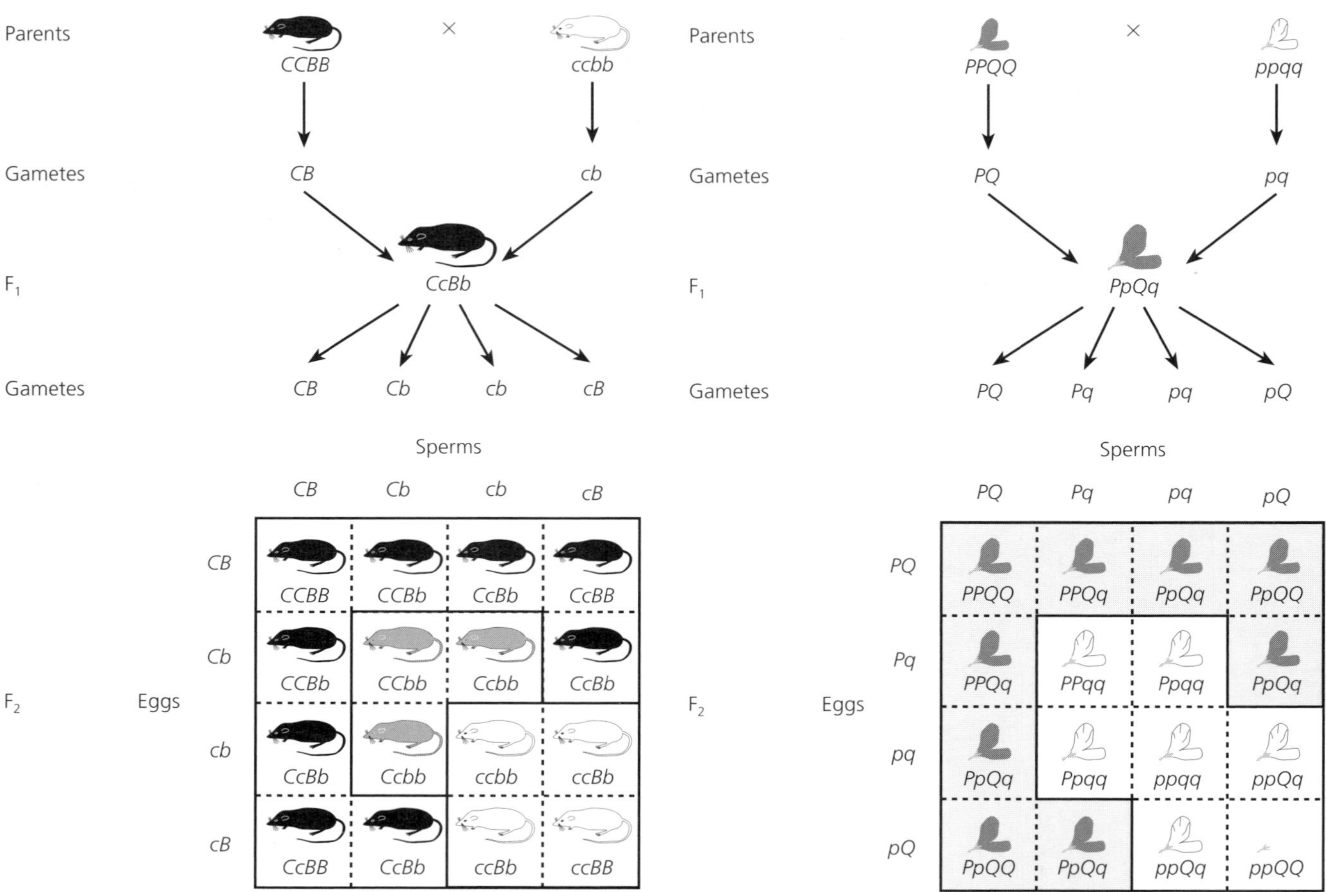


**Figure 8.27▲**
The 9:3:4 ratio in mice, illustrating epistasis

**Figure 8.28▲**
The 9:7 ratio in sweet peas, illustrating the effect of complementary genes

For example if a mouse with the genotype *CCBB* is mated with a mouse of genotype *ccbb* and the $F_1$ inbred, the $F_2$ mice are in the ratio of $\frac{9}{16}$ black:$\frac{3}{16}$ brown:$\frac{4}{16}$ albino (Figure 8.27). Three-quarters (i.e. $\frac{12}{16}$) are coloured and one-quarter (i.e. $\frac{4}{16}$) are albino. Of the albinos, three-quarters (i.e. $\frac{3}{16}$ of the total) are 'cryptoblack', meaning that they would be black if they had the *C* allele, and one-quarter ($\frac{1}{16}$ of the total) are 'cryptobrown'.

## Complementary genes

Another modified dihybrid $F_2$ ratio is the 9:7 ratio which occurs when a trait is only developed in the presence of dominant alleles at *two* loci. One example is the development of purple pigment in sweet pea flowers. Plants of genotypes *ppQQ* and *PPqq*, respectively, are both white-flowered. When they are crossed, the $F_1$ are all purple-flowered and the $F_2$ consist of purple and white flowers in the ratio $\frac{9}{16}$ purple:$\frac{7}{16}$ white (Figure 8.28). Since each of the parents provides what the other lacks, the two loci are **complementary**. (For the explanation of this result see Chapter 9.)

## Continuously varying characters

In his breeding experiments Mendel used discontinuously varying characters – i.e. those of the 'either/or' type. But of course most characteristics are not like this. Body height and weight, nose width, skin colour, reflex time, IQ, blood pressure – and, indeed, the majority of distinguishing characteristics – vary continuously between two extremes. Individuals are not tall *or* short, dark- *or* fair-haired, and so cannot be put into discrete groups.

There are two reasons why such characters vary continuously. Firstly, many of them are at least partly susceptible to differences in environment. Although different genotypes are distinct from each other, environments intergrade with one another with any two extremes being connected by a smooth continuum of intermediates.

The second reason is genetic. With a character that is under the control of one gene pair, there are only three possible genotypes (*AA*, *Aa* and *aa*), and if dominance is incomplete, only three phenotypes are possible. However, many continuously varying characters are known to be under the control of several, perhaps many, gene loci, each gene making a small contribution to the overall phenotype.

Suppose, for example, that skin pigment is under the control of two gene loci, each existing as two alleles, *A*/*a* and *B*/*b*. Suppose also that *A* and *B* each produce two units of pigmentation and *a* and *b* each produce one unit. There will be five phenotypes, having four, three, two, one or no 'upper case' alleles, respectively.

Punnett square showing how incomplete dominance at two loci could produce five shades of skin colour

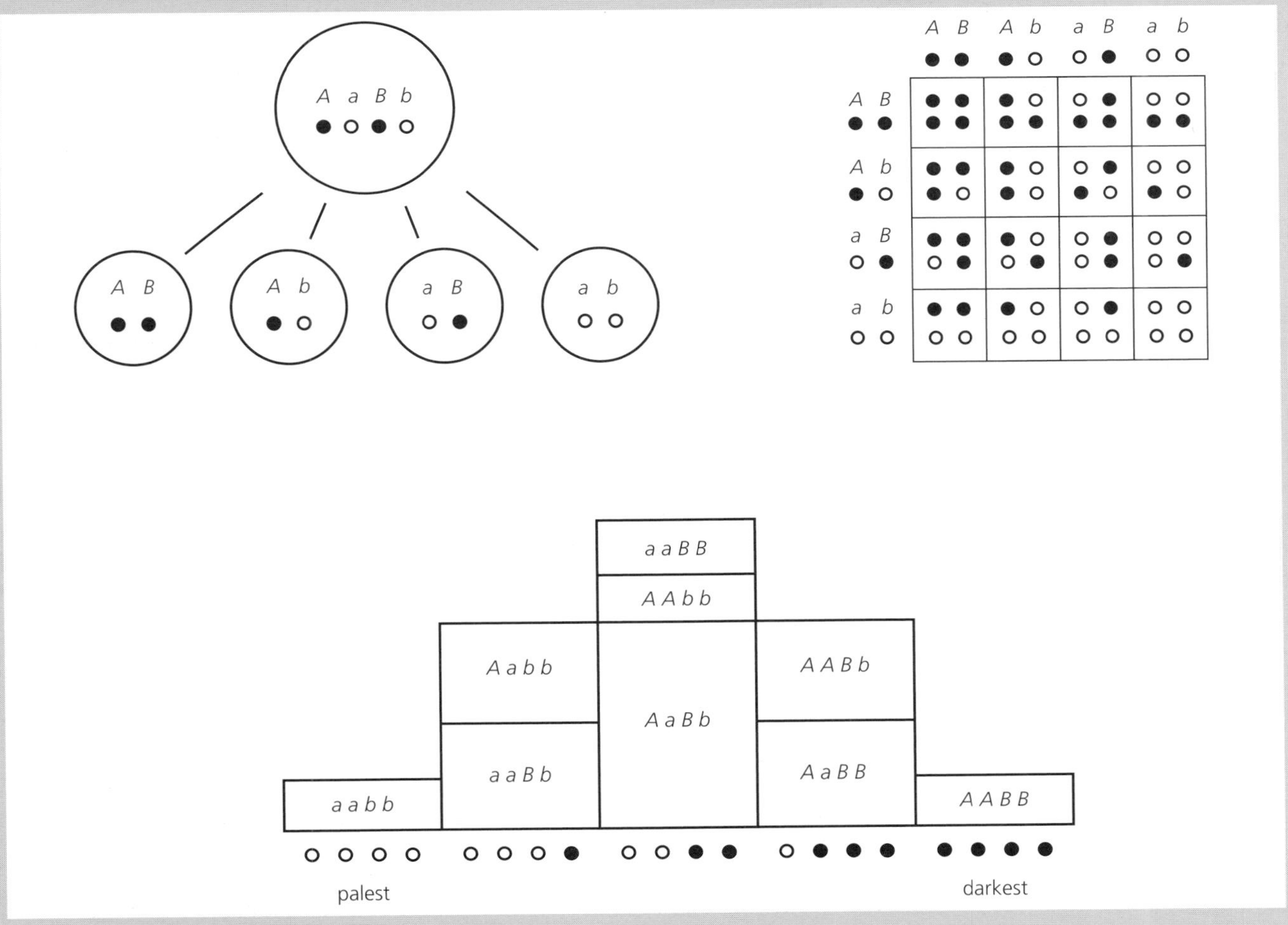

## Continuously varying characters *continued*

| | |
|---|---|
| *AABB* | very dark ($\frac{1}{4}$) |
| *AABb*<br>*AaBB* | dark ($\frac{4}{16}$) |
| *AaBb*<br>*AAbb*<br>*aaBB* | medium ($\frac{6}{16}$) |
| *aaBb*<br>*Aabb* | pale ($\frac{4}{16}$) |
| *aabb* | very pale ($\frac{1}{16}$) |

The five phenotypes would not be present with equal frequency. This is because organisms homozygous for both loci (*AABB* and *aabb*) can only be produced in one way, but individuals heterozygous for one locus can be produced in two ways and so are produced twice as often. Double heterozygotes can be produced in four ways and occur four times as often. If large numbers of double heterozygotes have children, the variation in skin colour (assuming equal exposure to sunlight) would be expected to have the distribution shown in the diagram.

With three loci (six genes) there would be seven possible phenotypes, and with four loci (eight genes) there would be nine genotypes. As the number of loci increases, the steps in the distribution become smaller and the curve begins to approach more closely the normal curve. When the effects of environmental differences are taken into account, even a character under control of quite a small number of loci would be expected to have a smooth distribution.

## Pleiotropy

A situation opposite to that where a character is influenced by more than one gene is when a gene affects the development of more than one character. This phenomenon is called **pleiotropy**. An interesting example is the fact that people with Group O blood are more likely to develop a duodenal ulcer than those of Groups A, B or AB, and Group A people are more prone to developing stomach cancer. Albinos have slightly impaired vision because the optic nerve fibres do not follow the normal pathways. There are many other examples – in fact it is likely that most (perhaps all) genes have pleiotropic effects.

## Glossary of terms used in this chapter

**Allele** One of a number of alternative forms that the gene at a given locus can take. Allelic genes occupy the same position (locus) on a chromosome. Hence genes that are allelic cannot exist in the same homologue. All the genes in a given gamete are therefore non-allelic.

**Diploid** Having two sets of chromosomes.

**Dominant** Describes a character or trait whose expression is the same in homozygotes and heterozygotes.

**$F_1$ generation** The offspring of a mating between two different true-breeding varieties.

**$F_2$ generation** The offspring produced by inbreeding an $F_1$ generation (the term should not be used for the offspring of a testcross or backcross).

**Gene** A unit of hereditary material that determines the sequence of amino acids in a polypeptide (see Chapter 11 for an elaboration).

## Glossary of terms used in this chapter *continued*

**Genotype** The hereditary makeup of an organism, unchanging throughout the life of the individual.

**Haploid** Having one set of chromosomes.

**Heterozygote** An organism carrying two different allelic forms at a locus.

**Homologue** One of a pair of chromosomes.

**Homozygote** An organism carrying two copies of an allele.

**Linkage group** All the genes on a given chromosome pair. In humans, for example, there are 23 linkage groups because there are 23 pairs of chromosomes.

**Locus** The position that alternative alleles occupy on a chromosome.

**Phenotype** The physical characteristics of an organism, which change during life.

**Recessive** Describes a character or trait that is only expressed in homozygotes.

**Testcross** A mating performed to determine frequencies of the kinds of gamete produced by an organism. It is often used to determine whether an organism is homozygous or heterozygous.

## QUESTIONS

**1** A long-eared mouse was mated with a short-eared mouse and they produced a litter consisting of six long-eared mice and two short-eared mice. Which of the following statements is justified?

A Both the parents must have been heterozygous.
B Both the parents must have been homozygous.
C One of the parents must have been homozygous and the other heterozygous.
D No firm deduction can be made.

**2** In humans, the ability to taste phenylthiocarbamide (PTC) is inherited and is due to a single pair of genes. Bill and Pam can both taste PTC, but their first child cannot.

**a)** What is Bill's genotype?
**b)** What is the probability that their second child will be able to taste PTC?
**c)** Bill's mother cannot taste PTC, and neither could his father's father. What can you say about the genotype of Bill's father?

**3** How many kinds of gamete could be produced by organisms with the following genotypes:

**a)** *AABBCCddEEff*
**b)** *AaBB*
**c)** *AaBbCc*?

**4** True or false?

**a)** If a plant shows a recessive trait, then we know its genotype.
**b)** A gamete carries one allele at each locus.
**c)** In a mating between two mice, each of which has the genotype *Bb*, four kinds of fertilisation are possible.
**d)** Only a diploid nucleus can undergo meiosis.
**e)** A sperm has the same number of chromosomes as the cell which produced it.
**f)** If both parents are heterozygous for tongue-rolling, then the probability that their first two children will be tongue-rolling boys will be $\frac{9}{64}$ (tongue-rolling is dominant).

## QUESTIONS

**5** How many cells (compartments) should the Punnett square have when working out the progeny of each of the following matings:

**a)** *Aa* × *aa* **b)** *Aabb* × *AaBb*
**c)** *AABB* × *AaBb* **d)** *AaBbCc* × *AaBbcc*

**6** In genetic notation, allelic genes are usually represented by variants of the same symbol (e.g. upper and lower case). The symbols thus 'give away' which genes are allelic. In the following question alleles are represented by *numbers*. An organism is heterozygous for three pairs of genes, indicated by the numbers 1–6, respectively. It produces gametes with the following genotypes:

*1 3 4, 1 5 4, 2 3 4, 1 5 6, 2 5 4, 2 3 6, 2 5 6, 1 3 6.*

Which genes are allelic to which? (Give the pairs of numbers).

**7** Organisms with the genotype *EEFF* and *eeff* are crossed, and the $F_1$ testcrossed. The following are four different testcross results, the four phenotypes being represented by *EF*, *Ef*, *eF* and *ef*. Numbers are percentages.

| | *EF* | *Ef* | *eF* | *ef* |
|---|---|---|---|---|
| A | 50 | 0 | 0 | 50 |
| B | 20 | 30 | 30 | 20 |
| C | 25 | 25 | 25 | 25 |
| D | 40 | 10 | 10 | 40 |
| E | 35 | 15 | 5 | 45 |

**i)** Which one of the testcross results **A–E** could be obtained if the genes are not linked?
**ii)** Which result could be obtained if the genes are linked but there is no crossing over?
**iii)** Which result could be obtained if the genes are linked with some crossing over?
**iv)** Which two results could not be accounted for by normal chromosome behaviour, and why?

**8** The diagram shows the arrangement of certain genes on the chromosomes of a particular animal. Consider the gametes produced by this animal.

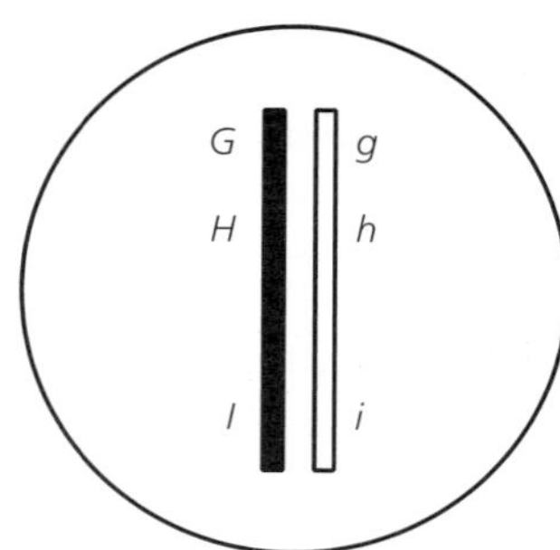


**a)** Ignoring locus *H/h*, write the genotype(s) of the
i) least frequent
ii) most frequent kind(s) of gamete produced by this organism.

**b)** Now consider locus *H/h* as well. The following are the gene combinations of some of the kinds of gamete the animal produces:

**A)** *GHI* **B)** *GhI* **C)** *GHi* **D)** *Ghi*

**i)** Write the letters of these different kinds of gamete in *decreasing* order of frequency.
**ii)** Give the genotype of the kind of gamete that would be produced with equal frequency as **(A)** *GHI*; **(B)** *GhI*.

**9** Consider a cell in the bone marrow of an albino. If it is just about to enter prophase, how many copies of the gene for albinism does the cell have?

**10** What is the probability that a man's Y chromosome has been inherited from his
**a)** father's mother's father
**b)** father's father's father?

**11** True or false?
**a)** Mitosis only occurs in diploid cells.
**b)** In a homozygote the products of meiosis are genetically the same.
**c)** Homologous chromosomes are usually not genetically identical.
**d)** Human X-linked recessive traits only occur in males.

CHAPTER 9

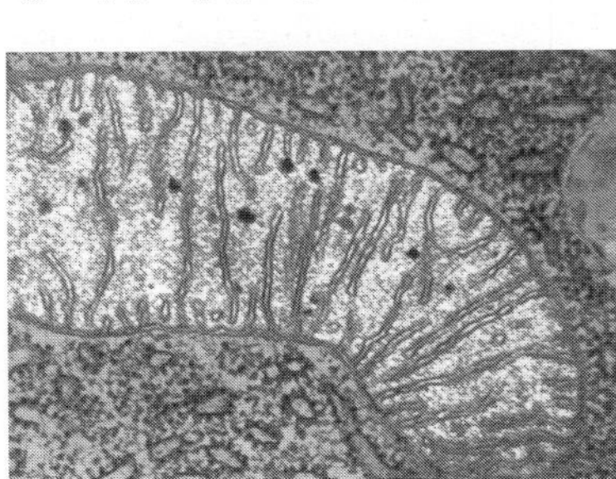

# WHAT DO GENES DO?

## LEARNING OBJECTIVES

By the time you have completed your study of this chapter you should be able to:

- explain what is meant by the one gene–one enzyme hypothesis, and present evidence to support it.
- explain in biochemical terms why some alleles are dominant and others recessive.
- explain in chemical terms how a 9:3:3:1 ratio can be modified to produce ratios such as 9:7 or 9:3:4.
- explain in chemical terms how it is possible for a gene to have multiple effects.
- give examples of how the environment can influence the phenotypic expression of a given genotype.

## One gene–one polypeptide

By early this century, biologists knew how genes were passed from parents to offspring. The more intriguing question of what genes *are* and how they work, remained a mystery.

One of the first people to shed any light on this question was a physician, Archibald Garrod. He took a particular interest in **alkaptonuria**, a rare condition characterised by the excretion of dark-brown urine. This colour is due to a substance called **homogentisic acid** or **alkapton**. From studies of the pedigrees of affected people, Garrod concluded that alkaptonurics are homozygous for a recessive allele. He further suggested that alkaptonurics lack the enzyme that normally converts homogentisic acid into another substance. What he was implying therefore was that the function of a gene is to control the production of an enzyme.

Garrod made this daring intellectual leap in 1908 but, like Mendel before him, he was ahead of his time. It was not until the early 1940s that George Beadle and Edward Tatum tested Garrod's hypothesis experimentally (see information box on 'Moulds, genes and enzymes' below).

### Moulds, genes and enzymes

Beadle and Tatum used the red bread mould *Neurospora crassa*, the life cycle of which is shown in Figure 15.21. Besides being easy to keep and having a life cycle of only a few days it has two key advantages over more complex organisms:

- since it is haploid for most of its life cycle, all its genes are expressed – there is no dominance.

## Moulds, genes and enzymes *continued*

- besides reproducing sexually it also produces vast numbers of asexual spores. Since these are genetically identical to the parent, many copies of a given strain can be quickly produced.

*Neurospora* can be grown on agar jelly containing only glucose, mineral salts and one vitamin. This is called **minimal medium**, and from these simple raw materials the fungus is able to synthesise all the chemicals that make up its body.

The ability to manufacture its body chemicals must be under the control of its genes. By damaging these genes Beadle and Tatum hoped to learn something about how they exerted their control. First, they irradiated large numbers of *Neurospora* spores with X-rays (which were known to cause mutations). They hoped that this would produce fungi that would be unable to carry out all the chemical processes of the normal fungus. They succeeded in producing a number of strains that were unable to grow unless certain organic substances were added to the minimal medium. Amongst these **nutritional mutants** were strains that needed the amino acid arginine. Since amino acids are the building units of proteins, any protein containing arginine could not be produced.

Beadle and Tatum found that these 'arginine mutants' were of different kinds which differed in their 'fussiness'. For convenience, three of these mutant strains will be called 1, 2 and 3.

- **Strain 1** could grow provided that arginine, citrulline or ornithine was provided.
- **Strain 2** could grow if arginine or citrulline was present.
- **Strain 3** could grow only if arginine was present – no other substance would do.

The simplest explanation for these observations is that arginine is produced in a sequence of reactions as follows:

precursor substance $\rightarrow$ ornithine $\rightarrow$ citrulline $\rightarrow$ arginine

Beadle and Tatum suggested that Strain 1 lacked the enzyme necessary to convert the precursor substance to ornithine, although the fungus would still be able to make arginine so long as ornithine or citrulline were provided. Strain 2 presumably lacked the enzyme that converted ornithine to citrulline, in which case arginine could be made from citrulline if this were available. Strain 3 would have lacked the enzyme that converted citrulline to arginine, so only arginine would support growth.

The next step was to determine the nature of the genetic differences between the three strains. The key question was this: by how many gene loci did the three strains differ? There were two possibilities:

- in each strain, a gene at a different locus was defective – in which case the genes would be **non-allelic**.
- in each strain, a gene at the same locus had gone wrong but in a different way in each strain. In this case the genes would be **allelic**.

To resolve the issue they mated the different mutant strains together. If the *same* gene was defective in three different ways, then all the offspring would require arginine, half being of each parental type. If, on the other hand, a *different* gene was defective in each strain, then each mutant parent would be normal for the gene for which the other mutant parent was defective. Each parent would thus make up for the deficiency of the other. If the genes were on different chromosomes (not linked), a quarter of the offspring would be normal, a quarter would resemble one parent, a quarter would resemble the other, and a quarter would be mutant for both genes. If the genes were linked, some normal offspring would be produced by crossing over.

In the event, some of the offspring were of wild type. This showed that the different strains *complemented* each other – in other words each parent provided what the other lacked. The genes must therefore have occupied different loci – i.e. they were non-allelic.

## Moulds, genes and enzymes *continued*

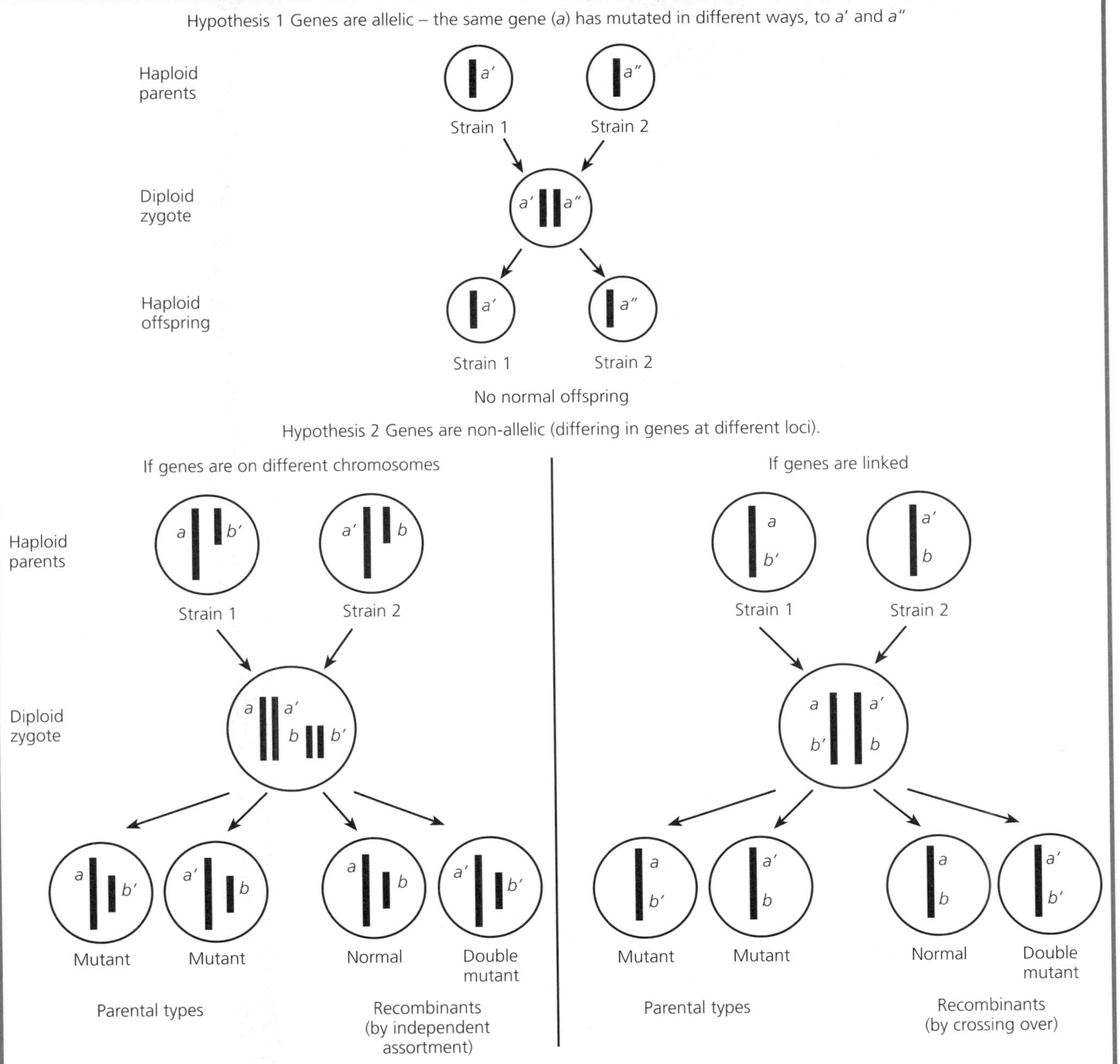

Hypothesis 1 Genes are allelic – the same gene (*a*) has mutated in different ways, to *a*′ and *a*″

Haploid parents

*a*′ Strain 1

*a*″ Strain 2

Diploid zygote

*a*′ *a*″

Haploid offspring

*a*′ Strain 1

*a*″ Strain 2

No normal offspring

Hypothesis 2 Genes are non-allelic (differing in genes at different loci).

If genes are on different chromosomes

Haploid parents

*a* *b*′ Strain 1

*a*′ *b* Strain 2

Diploid zygote

*a* *a*′ *b* *b*′

*a* *b*′ Mutant

*a*′ *b* Mutant

*a* *b* Normal

*a*′ *b*′ Double mutant

Parental types

Recombinants (by independent assortment)

If genes are linked

*a* *b*′ Strain 1

*a*′ *b* Strain 2

*a* *b*′ *a*′ *b*

*a* *b*′ Mutant

*a*′ *b* Mutant

*a* *b* Normal

*a*′ *b*′ Double mutant

Parental types

Recombinants (by crossing over)

How a breeding experiment can establish whether two mutant genes of *Neurospora* are allelic

In the light of their experiments Beadle and Tatum formally proposed what Garrod had hinted at, that *the function of a gene is to control the production of an enzyme*. Their '**one gene–one enzyme**' proposal has since been modified to take account of new knowledge:

- many proteins are not enzymes, for example insulin and keratin.
- many proteins consist of more than one polypeptide chain. It is thus more appropriate to speak of 'one gene–one polypeptide'.

- some genes do not code for the production of proteins at all, but for ribosomal RNA and transfer RNA (see Chapter 11).

## Explaining $F_2$ ratios

The one gene–one polypeptide hypothesis explains some of the modified $F_2$ ratios described in Chapter 8. The 9:7 ratio in sweet peas would be expected if purple pigment is produced in two sequential reactions, with each of the enzymes concerned being under the control of a gene at a different locus.

Only if both functional enzymes (and therefore both dominant alleles) are present is purple colour produced.

The 9:3:4 ratio for coat colour in mice is explained if the product of one gene is an enzyme controlling the production of pigment, and the product of a second gene is an enzyme responsible for controlling the dispersion of the pigment:

Another modified $F_2$ ratio, the 15:1, can be explained if the substance needed to produce a character is produced in two pathways, *either* of which can substitute for the other:

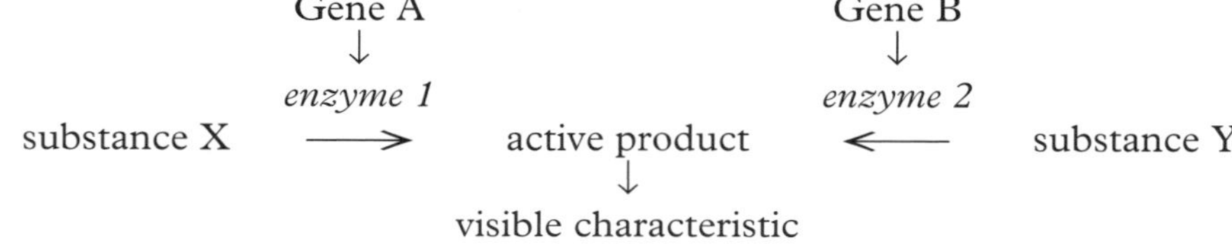


## Metabolic blocks in humans

A large number of genetic disorders in humans are caused by an inability to produce a particular enzyme. The following examples are all due to autosomal recessive alleles which fail to produce the necessary enzyme.

- **Albinism** The body cannot make melanin, the pigment that gives skin and hair its colour.
- **Alkaptonuria** Homogentisic acid accumulates in the blood and is excreted in the urine, which slowly darkens on exposure to air.
- **Phenylketonuria (PKU)** Phenylalanine cannot be converted into tyrosine, and accumulates in the blood. In young children this causes mental retardation. Another result is the shortage of tyrosine (from which melanin is produced), so phenylketonurics tend to be very blond. If diagnosed early enough the condition can be treated with a special diet containing proteins low in phenylalanine.

Figure 9.1 shows where some of these metabolic blocks occur.

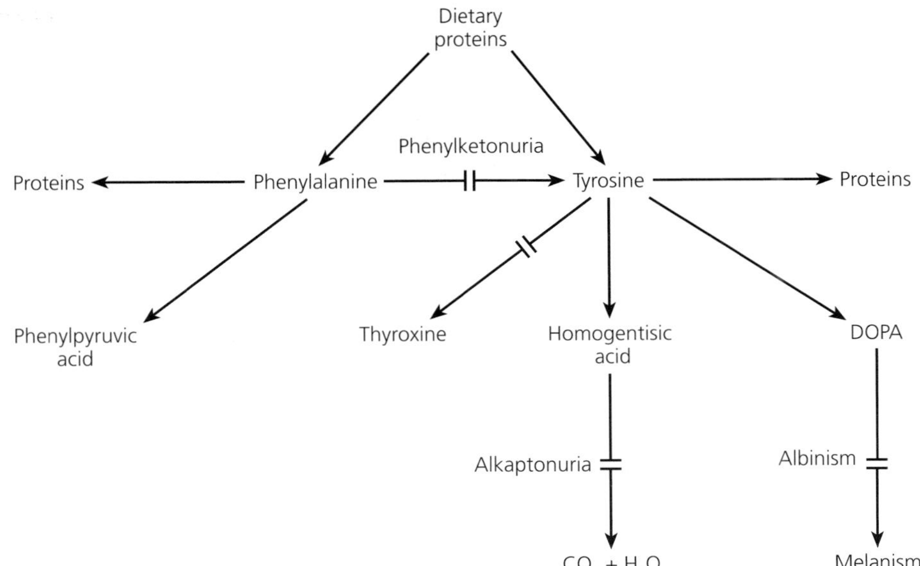


**Figure 9.1**
Some examples of metabolic blocks in humans

## Dominant and recessive traits

What makes some alleles dominant and others recessive? Figure 9.2 shows one possible mechanism. Suppose a mutant allele makes a non-functional protein, and that a certain minimum amount of protein must be made to produce the normal phenotype. With only one working allele a heterozygote would be expected to produce only half as much functional protein as would a homozygous-normal individual. If this is sufficient then a heterozygote would have the normal phenotype, and the mutant allele would be recessive. A person who is heterozygous for albinism, for instance, is able to produce normal quantities of melanin with only half as much enzyme. If, on the other hand, having only half as much gene product was insufficient to produce a normal phenotype, then the heterozygote would be phenotypically different from the normal homozygote so the mutant allele would be dominant.

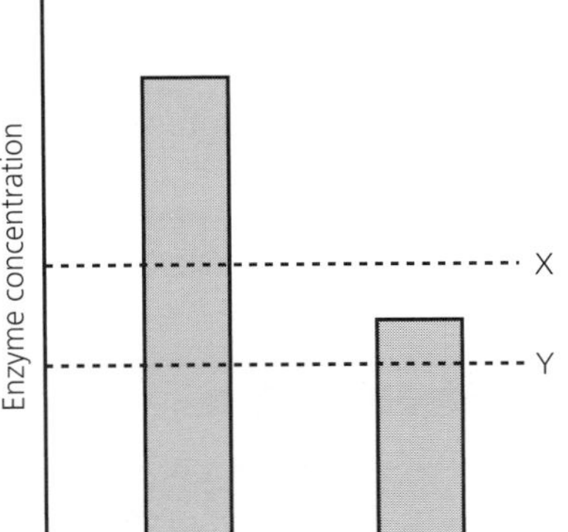


**Figure 9.2**
One possible mechanism of dominance. If the minimum concentration of enzyme needed to produce a functional phenotype is X, then *a'a'* and *a'a* will be phenotypically different, and *a'* will be recessive. If the critical enzyme concentration is Y, then *a'a'* and *a'a* will be phenotypically similar and *a'* will be dominant

## Pleiotropy

Though the primary effect of a gene is the production of a polypeptide, the effects of a gene rarely end there because gene products have their own effects, for example:

- an enzyme may take part in several metabolic pathways.
- the product of a reaction may subsequently be used in several other metabolic pathways.

Indirect effects of a gene can go much further, as Figure 9.3 shows. Many apparently trivial characteristics – such as the ability to roll the tongue or to taste phenylthiocarbamide (PTC) – are probably pleiotropic effects of alleles whose main function is quite different. In fact it is likely that pleiotropy is the rule rather than the exception.

## Genotype and phenotype

An understanding of protein synthesis has done much to demystify the workings of cells. An organism is however far more complex than a system

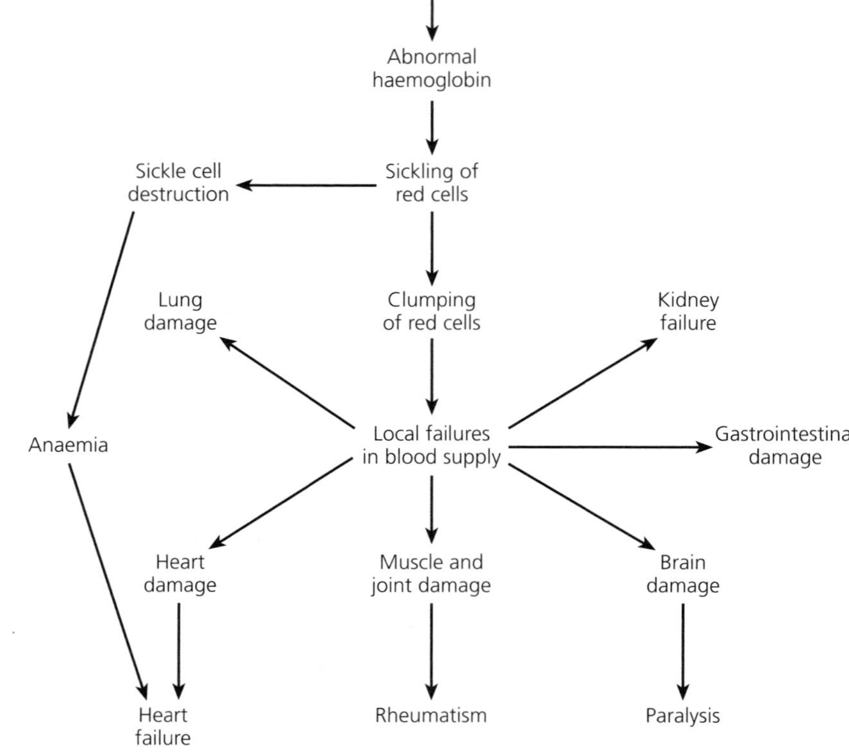


**Figure 9.3**
Pleiotropic effects of the sickle-cell allele

of individual chemical processes. Proteins are only the first link in a chain of effects a gene has on the total phenotype. Multicellular organisms are so complex that even if we knew every molecular detail of an organism's genetic makeup, it would not be possible to predict all the details of the body that develop. There are a number of reasons for this:

- genes usually have pleiotropic effects.
- a phenotypic character may be affected by many genes.
- differences in the environment may give rise to differences in phenotype.
- the activity of a gene may be affected by the products of other genes (Chapter 11).

## Environmental effects on the phenotype

Building a body requires two essential things: information – encoded in the *genes* – and a suitable *environment*. Just as baking a cake requires ingredients *and* heat, neither genes nor environment alone is sufficient to build a body.

The environment can influence growth and development in two quite different ways. The most obvious way is when some adverse environmental factor prevents the genetic potential of the organism from being fully realised. For example, when denied sufficient energy or raw materials organisms grow more slowly. The increase in height of children of Japanese immigrants to the United States shows that some environmental factor – probably diet – had been preventing previous generations from reaching their potential height.

A quite different kind of effect occurs when some environmental signal triggers an adaptive response, for example:

- whether a shoot produces leaves or flowers may depend on photoperiod (day length) or, in some cases, on previous exposure to a period of cold.
- the white winter coat of most Arctic mammals develops in response to changes in day length.
- the enzymes produced by a bacterium depend on the chemicals in its environment. For instance, the human gut bacterium *Escherichia coli* only produces lactase when lactose is present (Chapter 11).
- the anatomy of the leaves of many forest trees varies according to the light intensity during early development (Chapter 33).
- in mammals, a fall in body temperature triggers an increase in output of the hormone thyroxine which stimulates an increase in metabolic rate and therefore heat production, so helping to maintain body temperature.

In all of these cases the response by the organism is **adaptive** – that is, it increases the likelihood of the organism passing its genes on to the next generation.

Genes therefore do not rigidly specify every detail of the body. Rather, they determine *the limits of a range of possible phenotypes*. Moreover, even in a given organism (whose genotype of course does not change), the phenotype is constantly changing – not only with age, but also in response to changing environmental conditions.

## QUESTIONS

**1** What kind of children would you expect to be produced by a marriage between a man with alkaptonuria and a woman with phenylketonuria?

**2** Albinism is an inherited condition in which the pigment melanin cannot be produced, so the skin is pale, the hair white and the iris pink. It has generally been assumed that albinism is due to the recessive allele of a single gene, but there is at least one known case of two albino parents having a normal child.

**a)** Assuming that there was no error over parentage, suggest a possible explanation based on Figure 9.2 (your answer to Question 1 should provide a clue).

**b)** Suggest one other possible explanation.

**3** A haemoglobin molecule consists of four polypeptide chains, two α chains and two β chains. In the hereditary disease sickle cell anaemia the β chain is defective, causing the haemoglobin to crystallise at the lower oxygen concentrations which occur in the capillaries. As a result the red cells become sickle-shaped. In heterozygotes, sickling only occurs under abnormally low oxygen concentrations.

**a)** Assuming that in a heterozygote, normal and abnormal β chains are produced in equal amounts, how many kinds of haemoglobin molecule could be produced, and in what relative quantities?

**b)** Explain why crystallisation occurs much less readily in heterozygotes.

**4** Suppose that an enzyme consists of four identical polypeptide chains, and that the enzyme is only functional if all four are normal. What proportion of the enzyme molecules would be functional in a heterozygote, assuming that equal quantities of normal and abnormal polypeptide are produced?

CHAPTER

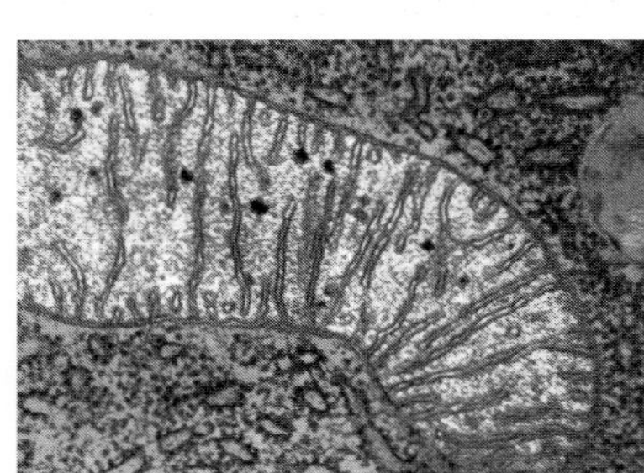

# 10

# The MOLECULE of HEREDITY – DNA

**LEARNING OBJECTIVES**

By the time you have completed your study of this chapter you should be able to:

- summarise the evidence for the view that DNA is the genetic material.
- describe the structure of DNA in sufficient detail to explain how it can carry information and replicate itself.
- summarise the evidence that DNA replication is semiconservative.
- describe the mechanics of DNA replication.
- describe the organisation of DNA and histones in a eukaryotic chromosome.

## What are genes made of?

By the end of the last century, biologists were fairly sure that the key to genetics lay somewhere in the nucleus. With the rediscovery of Mendel's work, attention focused on the chromosomes. These were known to consist mainly of two substances: **deoxyribonucleic acid** (or **DNA** for short) and protein. It seemed likely that one of these was the genetic material – but which?

DNA was first isolated in 1869 by the Swiss biochemist Friedrich Miescher – only four years after Mendel published his results. Miescher had extracted it from white corpuscles in pus and later in salmon sperm – in both of which the nuclei are particularly prominent. He found it to be a phosphorus-rich substance and called it 'nuclein'. Because of its acidic properties it was later rechristened **nucleic acid**. It was later discovered that there were two such substances – DNA (which is almost confined to the nucleus) and **ribonucleic acid**, or **RNA**. Although RNA occurs mainly in the cytoplasm, the term 'nucleic acid' had become too well-established to change.

In the 1920s it became possible to further pinpoint the location of DNA in the cell when R. Feulgen developed a stain that was taken up by DNA, but not RNA. Treatment of dividing cells with Feulgen stain showed that it was taken up specifically by the chromosomes.

To carry enough information to build anything as complex as an organism, the genetic material must be able to exist in an almost infinite variety of forms. DNA can be broken down into subunits called **nucleotides**. These are of four kinds and, for a long time after the discovery of DNA, it was believed that they were present in the same repetitive sequence in all DNA, which has a structural rather than an informational role.

Proteins, on the other hand, consist of 20 kinds of subunit and, for this reason, they were thought more likely to be the genetic material. By 1952 however the conclusion that genes consist of DNA had become inescapable because of the results of experiments involving bacteria and viruses (see information box below).

## Evidence that DNA is the genetic material

### Bacterial transformation

In 1928 Frederick Griffith, a bacteriologist, made an astounding discovery. He was working on two strains of *Streptococcus pneumoniae* – the normal, pneumonia-causing (or virulent) strain, and a mutant, harmless strain. The normal strain is called *smooth* (*S*) because of the smooth, glistening colonies it forms when grown on agar. The other, *rough* (*R*) strain is harmless and is so-called because of the rough appearance of its colonies. The ability of the *S* strain to cause disease lies in the protective polysaccharide capsule round the cells, which is lacking in the *R* strain.

Griffith found that when he injected either living *R* bacteria or boiled *S* bacteria into mice, the mice remained healthy as would be expected. Each of these treatments *by itself* was harmless, but when he injected a *mixture* of heat-killed *S* bacteria *and* living *R* bacteria into mice they died of pneumonia. Moreover, Griffith was able to recover large numbers of living *S* bacteria from the corpses of the mice (Figure 10.1).

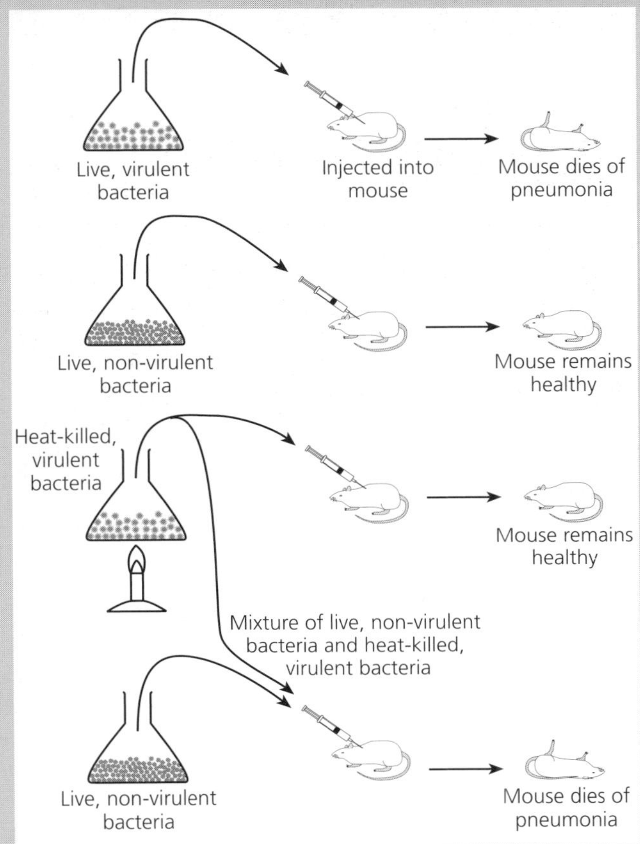


Transformation in bacteria

What was going on? Somehow the live, harmless bacteria had been **transformed** into virulent ones. The parents of the *S* bacteria recovered from the bodies of the mice must have been *R* bacteria, but the *information* for virulence must have somehow come from the *S* bacteria, which were dead!

The identity of the **transforming principle** (as Griffith had called it) was discovered in 1944 by Avery, MacLeod and McCarty. First, they found that a chemical extract of *S* bacteria was sufficient to transform *R* bacteria. However, when an extract of *S* bacteria was treated with deoxyribonuclease (DNase) – an enzyme that digests only DNA – it lost all transforming ability. In contrast, treatment with ribonuclease (which attacks RNA) or proteases had no effect on transforming ability. There was only one conclusion – the information for making the polysaccharide capsule must have been carried in DNA.

### Evidence from viruses

Though bacterial transformation provided strong evidence that genes are made of DNA, most biologists remained unconvinced. In 1952 however conclusive proof was provided by Alfred Hershey and Martha Chase. They used a **bacteriophage** (abbreviated to **phage**), a virus that attacks bacteria. The phage they used, called T2 (T for 'type'), attacks *Escherichia coli*, the human colon bacterium, and consists of two chemicals only – DNA and protein. One of

## Evidence that DNA is the genetic material *continued*

these – the genetic material – was known to be injected into the bacterium, the other remaining outside as an empty case or 'ghost'.

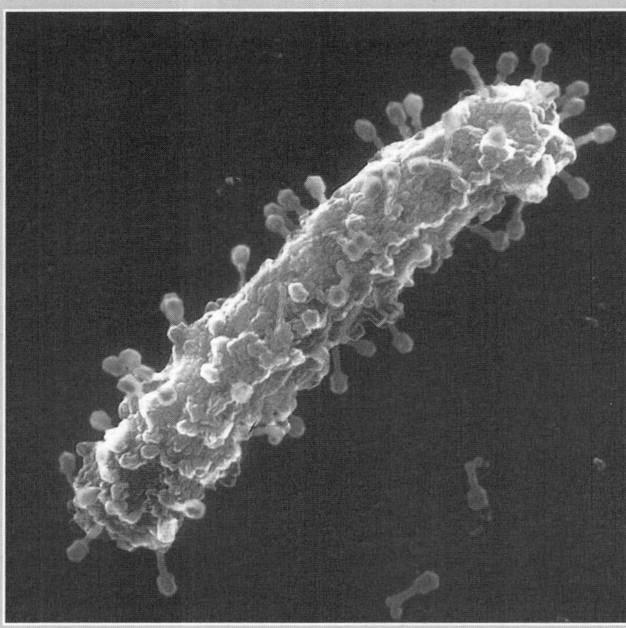

SEM of bacteriophages attacking *E. coli*

The problem was to identify which of the two chemicals was injected. Hershey and Chase exploited the fact that DNA contains phosphorus but not sulphur, and protein contains sulphur but not phosphorus.

They grew two separate phage cultures. One was supplied with nutrients containing the radioactive isotope of phosphorus, $^{32}P$, but normal sulphur, $^{32}S$. As a result the viral DNA became radioactive but the protein remained normal. The other culture was supplied with nutrients containing normal phosphorus, $^{31}P$, and radioactive sulphur, $^{35}S$. In this culture the viral protein became radioactive but the DNA was normal.

The two phage cultures were added to separate cultures of *E. coli*, and left just long enough for the phage particles to attach themselves to the bacteria and inject their genes. The cultures were then violently agitated in a blender, causing the empty viral shells ('ghosts') to become separated from the bacteria. The infected bacteria and empty viral casings were then separated by centrifugation, and each was then tested for radioactivity.

The results were dramatic and conclusive. In the culture that had been infected with DNA-labelled phage, the radioactivity was concentrated in the bacteria. In the complementary experiment using protein-labelled phage, the radioactivity was concentrated in the phage ghosts. Clearly then, it must have been the DNA that was injected into the bacteria, the protein remaining outside.

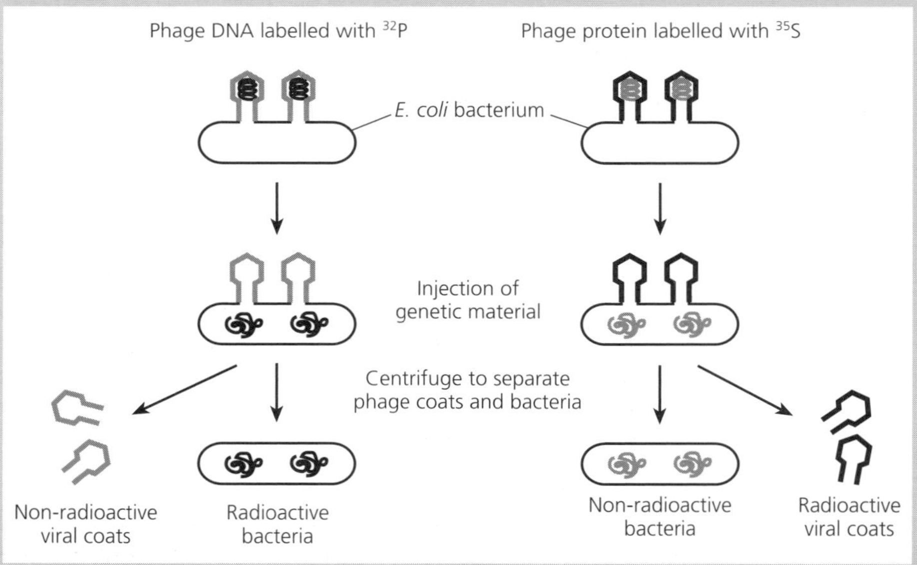


Hershey and Chase's experiment using bacteriophage

## Evidence that DNA is the genetic material *continued*

### Ultraviolet absorption

Even before Avery and coworkers showed that DNA could transform bacteria, other experiments had pointed to DNA as the genetic material. In 1939 it had been shown that DNA absorbed ultraviolet radiation most strongly at just those wavelengths that were most effective in inducing mutations.

### DNA content of nuclei

In the late 1940s it was found that cells from a variety of tissues (e.g. thymus, pancreas, kidney) from domestic cattle were found to contain the same amount of DNA. Most significantly, the sperms contained only half as much DNA as body cells, and similar results were obtained for a variety of other species.

The DNA content of the cells thus depends on the number of chromosomes, as would be expected of the genetic material. However, chromosomes also contain protein, so this finding could only be regarded as being *consistent* with the view that DNA is the genetic material, rather than proof. A further point to bear in mind is that small quantities of DNA are also present in mitochondria and chloroplasts. Since the number of these organelles varies independently of the number of chromosomes, cytoplasmic DNA would be expected to slightly distort the exact whole number ratios resulting from differences in chromosome number.

# The structure of DNA

As genetic material, DNA must have four essential properties:

1 it must carry information, which must be different for different organisms. There must therefore be great scope for DNA to vary from one organism to another, and even more, from one species to another.

2 because offspring resemble their parents there must be a mechanism by which DNA can copy itself, or **replicate**.

3 genetic information must either be very stable chemically, or at least be able to be repaired.

4 since life has changed over long periods of time, resistance to change cannot be absolute; there must be some mechanism by which DNA can change over time.

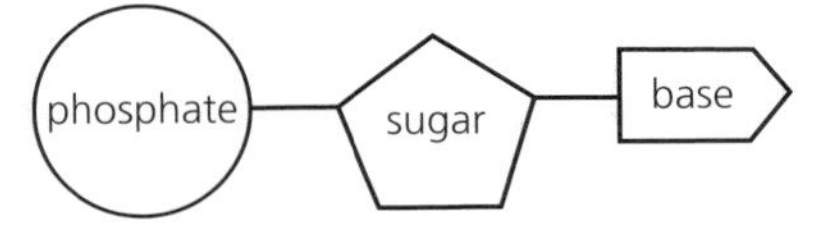


**Figure 10.1**
Structure of a nucleotide

Once it was accepted that the genetic material was DNA, there was intense interest in its structure. As we have seen, it was known that nucleic acids could be hydrolysed into subunits called nucleotides (Figure 10.1).

Each nucleotide consists of three parts:

- a **pentose sugar**. In DNA this is **deoxyribose**, and in RNA it is **ribose**.
- a **nitrogenous base**.
- a **phosphate group**.

When the phosphate group is removed from a nucleotide, the remainder, consisting of sugar and base, is called a **nucleoside**. An alternative name for a nucleotide is therefore a **nucleoside monophosphate**. ATP (adenosine triphosphate) is a nucleoside triphosphate.

In DNA the bases are of four kinds: **adenine**, **guanine**, **thymine** and **cytosine** (Figure 10.2). Adenine and guanine have double-ringed

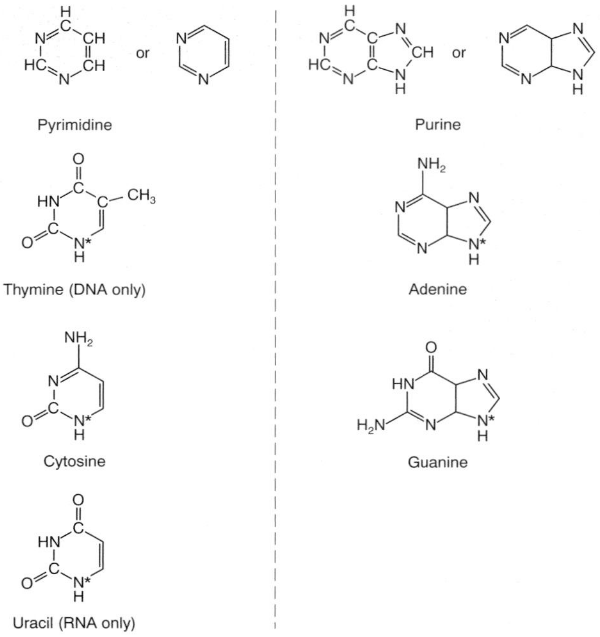

Pyrimidine

Purine

Thymine (DNA only)

Adenine

Cytosine

Guanine

Uracil (RNA only)

**Figure 10.2**
The nucleotide bases. Thymine and cytosine are derivatives of pyrimidine, while adenine and cytosine are derived from purine. The asterisks mark the nitrogen atoms which are bonded to pentose sugar in nucleotides

structures and are called **purines**. Thymine and cytosine have single-ringed structures and are called **pyrimidines**. In RNA, thymine is replaced by **uracil**.

In DNA nucleotides are linked in the manner shown in Figure 10.3. An important feature of all polynucleotide chains is the fact that the two ends are different, one being referred to as the **3'** (spoken as '3 prime') end and the other as the **5'** end. The basis for this lies in the numbering of the carbon atoms in the sugar. One end is nearest carbon #3 of the sugar, and the other is nearest carbon #5.

The backbone of all polynucleotide chains consists of the sequence: phosphate – sugar – phosphate – sugar – and so on. Since this is the same in all nucleic acids, the information content of DNA cannot lie within the sugar or phosphate; it must reside in the *sequence of bases* along the chain.

But what about the other property of DNA – its ability to replicate? This question could not be answered without detailed information about its structure – in particular, its three-dimensional organisation.

Crucial to the eventual solution to the problem was a curious discovery by Erwin Chargaff in 1950. Chargaff had found that hydrolysis of DNA always yielded equal numbers of moles of adenine and thymine, and equal numbers of moles of guanine and cytosine. Moreover, this was true regardless of the species from which the DNA had been extracted (Table 10.1).

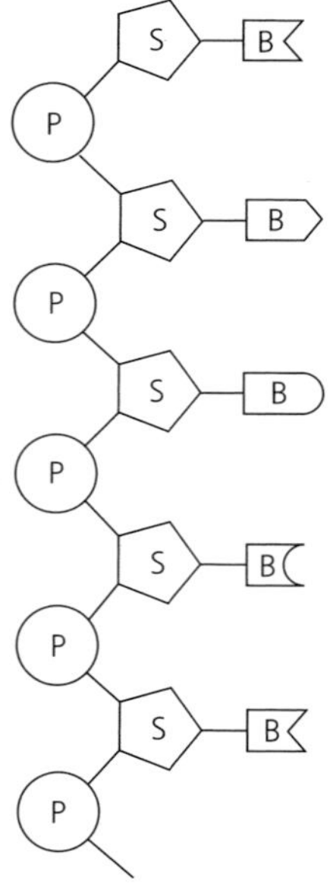


**Figure 10.3**
A polynucleotide chain. Each of the four kinds of base is represented by a different shape (P = phosphate, S = sugar, B = base)

## The double helix

The significance of 'Chargaff's rules' remained obscure until 1953 when James Watson and Francis Crick, working at the Cavendish Laboratory at Cambridge University, worked out the structure of DNA.

Watson and Crick's success depended on the results of X-ray crystallographic work by Rosalind Franklin and Maurice Wilkins at King's College, London. The technique is dependent on the fact that when a beam of X-rays is directed through a material in which the atoms are regularly arranged in space – as they are in a crystal – they are scattered in a regular manner. By very careful preparation, DNA molecules can be oriented in parallel bundles in which the atoms are regularly spaced. The

**Table 10.1▶**
Base composition of DNA from various species (molar %)

| Species | Adenine | Guanine | Thymine | Cytosine | A + G/ C + T | A + T/ G + C |
|---|---|---|---|---|---|---|
| *E. coli* | 24.7 | 26.0 | 23.6 | 25.7 | 1.03 | 0.93 |
| Yeast | 31.3 | 18.7 | 32.9 | 17.1 | 1.0 | 1.79 |
| Wheat | 27.3 | 22.7 | 27.1 | 22.8 | 1.0 | 1.2 |
| Locust | 29.3 | 20.5 | 29.3 | 20.7 | 1.0 | 1.42 |
| Sea urchin | 32.8 | 17.7 | 32.1 | 17.3 | 1.02 | 1.85 |
| Salmon | 29.7 | 20.8 | 29.1 | 20.4 | 1.02 | 1.43 |
| Turtle | 29.7 | 22.0 | 27.9 | 21.3 | 1.07 | 1.33 |
| Chicken | 28.8 | 20.5 | 29.2 | 21.5 | 0.97 | 1.38 |
| Human | 30.9 | 19.9 | 29.4 | 19.8 | 1.03 | 1.60 |

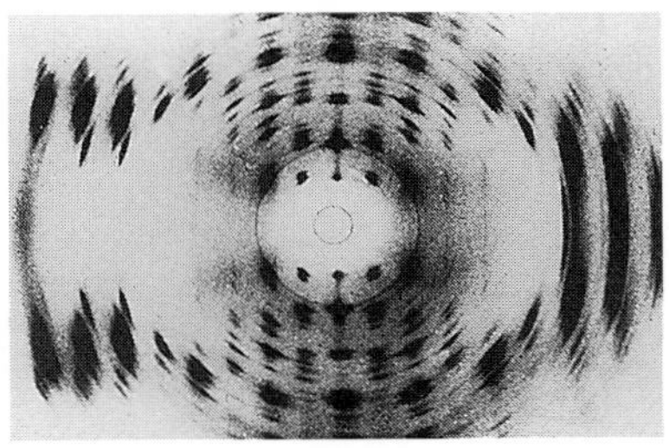

**Figure 10.4▲**
X-ray diffraction pattern produced by DNA

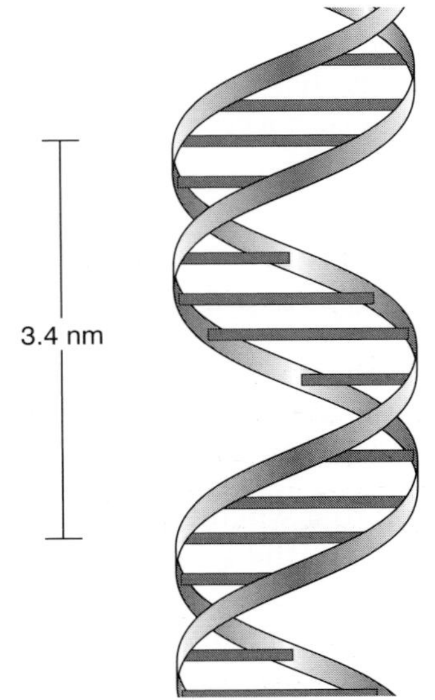


**Figure 10.5**
The double helix

scattered rays can be detected by a photographic plate, and from the pattern produced it is possible to work out the spacing of the atoms (Figure 10.4).

Franklin's X-ray photographs indicated that the DNA molecule is a helix or spiral about 2 nm wide and with one complete turn every 3.4 nm (Figure 10.5). A width of 2 nm strongly indicated that the molecule must be a **double helix**, consisting of two polynucleotide strands running side by side. The phosphate groups (being negatively charged) had to be on the outside – had they been on the inside they would have repelled each other.

That left the question of the bases. Watson and Crick reasoned that the only way that they could be fitted into the helix was by forming pairs, linking the two polynucleotide strands. The DNA molecule could thus be compared to a twisted ladder, with the sides being represented by the phosphate–sugar backbones and the rungs by pairs of bases. Moreover, the only way the rungs could all be the same length would be if the larger purines were paired with the smaller pyrimidines.

But which purine was paired with which pyrimidine? Chargaff's rules strongly suggested that adenine would be paired with thymine and guanine with cytosine. Watson and Crick proposed that the bases were loosely attracted to each other by hydrogen bonds – two between adenine and thymine, and three between guanine and cytosine (Figure 10.6).

Thymine Adenine Cytosine Guanine

**Figure 10.6▶**
Pairing between DNA bases. There are two hydrogen bonds between adenine and thymine and three between guanine and cytosine

Another feature of the double helix is the fact that the two polynucleotide chains are **antiparallel** or 'head to tail', with the 5' end of one chain adjacent to the 3' end of the other. The importance of this will become apparent when we consider DNA replication.

The significance of the specific pairing between the bases in the two strands cannot be overemphasised. Since the base sequences in the two strands are complementary to each other, *each strand contains the information to build the other*. The double helix thus explained how DNA could be replicated.

Because of complementary base pairing, the ratios of adenine/thymine and guanine/cytosine are both equal to 1, as is the ratio of total purines/total pyrimidines. There are, however, no restrictions on the ratio adenine/guanine or thymine/cytosine. In other words, as one moves along either of the two DNA strands there are no restrictions on possible sequences.

If DNA is to carry information that is unique to each species there must potentially be an almost infinite variety of base sequences. One would therefore expect considerable variation in the ratio of A + T/G + C, an expectation that is consistent with Chargaff's data (right hand column in Table 10.1).

### The impact of the double helix

Five years after their work was published, James Watson, Francis Crick and Maurice Wilkins were awarded the Nobel Prize for Medicine. Rosalind Franklin had died by this time, and Nobel prizes are not awarded posthumously. If its effect on future research is anything to go by, the discovery of the structure of DNA must rate as one of the most momentous in the history of science. It is fair to say that a new branch of science – molecular biology – was spawned by that discovery.

## How DNA is replicated

In their letter to the journal *Nature*, in which they reported their findings, Watson and Crick wrote:

> *It has not escaped our notice that the specific pairing we have postulated immediately suggests a possible copying mechanism for the genetic material*"

Behind this coy remark lay the implied suggestion that when DNA is replicated, each strand acts as a template or mould for the synthesis of a complementary strand.

The mechanism of DNA replication envisaged by Watson and Crick was as follows. Firstly, the two strands separate, exposing the bases. Next, by complementary pairing, bases of free nucleotides combine with the bases of the existing strands. Finally, the new nucleotides become longitudinally linked by covalent bonds.

A consequence of this proposed mechanism would be that each daughter DNA strand would consist of one new and one old – or 'parental' – strand. Because half the original molecule would be represented in each 'daughter' molecule, this was termed **semiconservative** replication.

### Evidence that replication is semiconservative

So beautifully did the double helix explain how DNA is copied, that most molecular biologists felt intuitively that replication *had* to be semiconservative. However, plausibility is not proof.

The semiconservative hypothesis was put to the test in a classic experiment by Matthew Meselson and Franklin Stahl in 1958. They grew *E. coli* cells in a medium in which nitrogen was supplied (as ammonium ions) in the form of the heavy, but non-radioactive isotope, $^{15}N$. The bacteria were grown in this medium for a sufficient number of generations to ensure that virtually all the DNA was 'heavy'. They then replaced the heavy nitrogen with the normal isotope, $^{14}N$. Immediately before changing the medium, and then at intervals corresponding to successive generations, they removed samples of the bacteria and extracted the DNA from each sample.

## Evidence that replication is semiconservative *continued*

Each DNA sample was then centrifuged in a solution of caesium chloride. At the very high speeds used, the caesium chloride formed a gradient of concentration, so the further down the tube, the more concentrated (and hence the more dense) the solution. The DNA settled out at a level at which the caesium chloride solution had the same density. The position of each DNA band was made visible by viewing the tubes in ultraviolet light, which is strongly absorbed by DNA.

The results were exactly as predicted by the semiconservative hypothesis. Immediately before the change of medium, all the DNA occupied a single band corresponding to 'heavy' ($^{15}N^{15}N$) DNA. After one generation all the DNA was again concentrated in a single band, but a little higher up. This is just as would be expected if all the DNA were 'hybrid' (i.e. $^{15}N^{14}N$), consisting of one heavy and one normal strand. After two generations, there were two bands, one consisting of $^{15}N^{14}N$ DNA and the other of normal DNA ($^{14}N^{14}N$) a little higher up.

As a further test, Meselson and Stahl took the DNA obtained after one generation and heated it, rupturing the hydrogen bonds holding the two strands together. After centrifugation the DNA separated as two bands, corresponding to heavy and normal (but single-stranded) DNA, respectively.

Meselson and Stahl's experiment. Note that when spinning, centrifuge tubes are horizontal

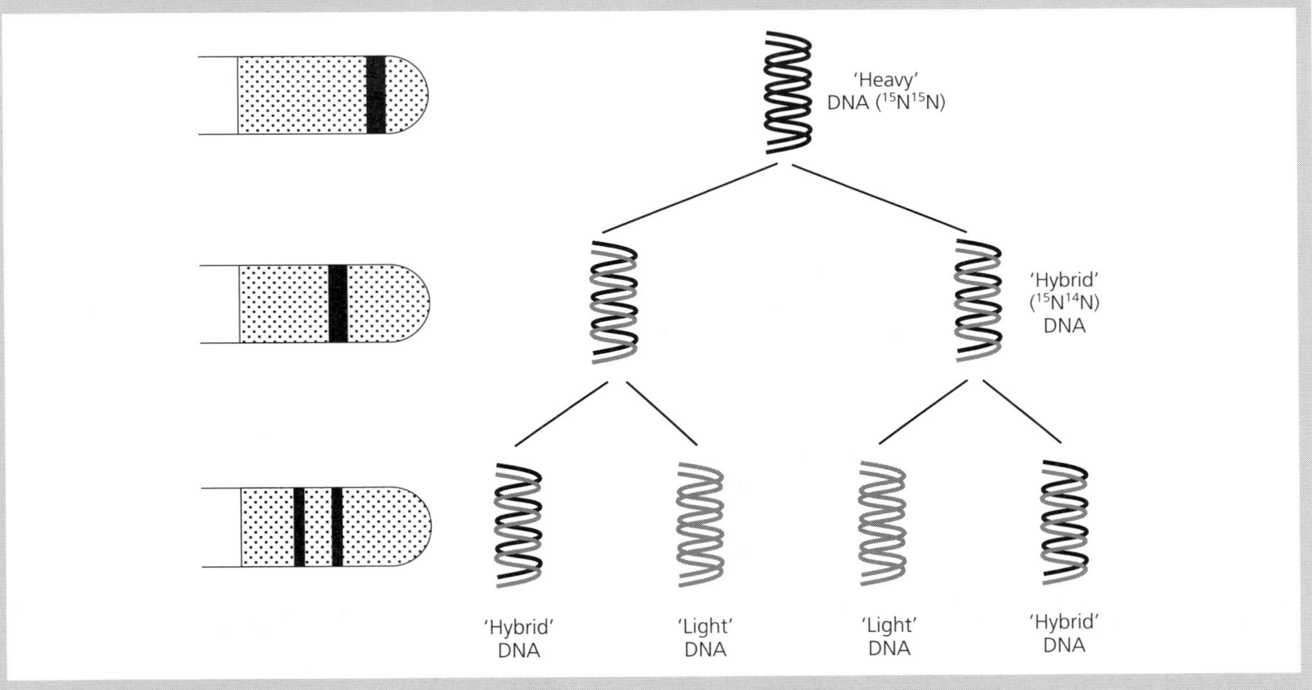


## The mechanism of DNA replication

DNA synthesis in eukaryotes occurs during interphase. First, an enzyme called **DNA helicase** separates the two DNA strands, exposing the bases. Next, enzymes called **DNA polymerases** catalyse the formation of two new strands by the addition of nucleotides complementary to those of the existing strands (Figure 10.7). Since the linking of nucleotides occurs by condensation, a molecule of water is produced for every new nucleotide added. Like other condensation reactions the process requires energy.

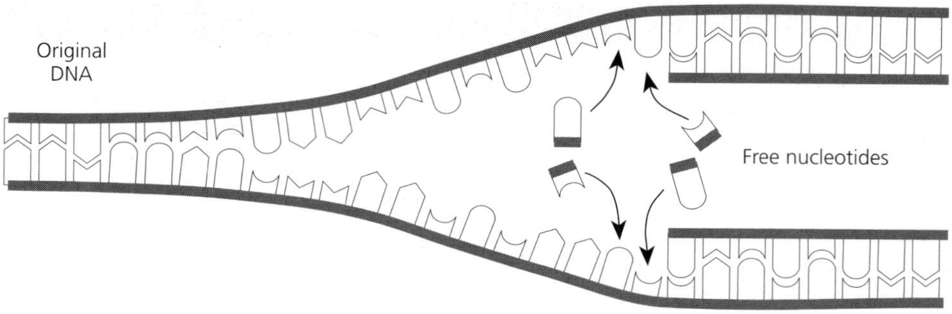

Original DNA

'Daughter' DNA

Free nucleotides

'Daughter' DNA

**Figure 10.7**
Simple representation of DNA replication

Actually, DNA replication is more complex than this in a number of ways. For example, the raw materials are *nucleoside triphosphates* rather than nucleotides (nucleoside monophosphates). As each nucleoside triphosphate pairs up with its complementary nucleotide, its two terminal phosphate groups are split off by hydrolysis. As in the hydrolysis of ATP, this releases energy, which in this case drives the polymerisation process (Figure 10.8). Other factors complicating the process are described in the information box, on 'Okazaki fragments' on page 169.

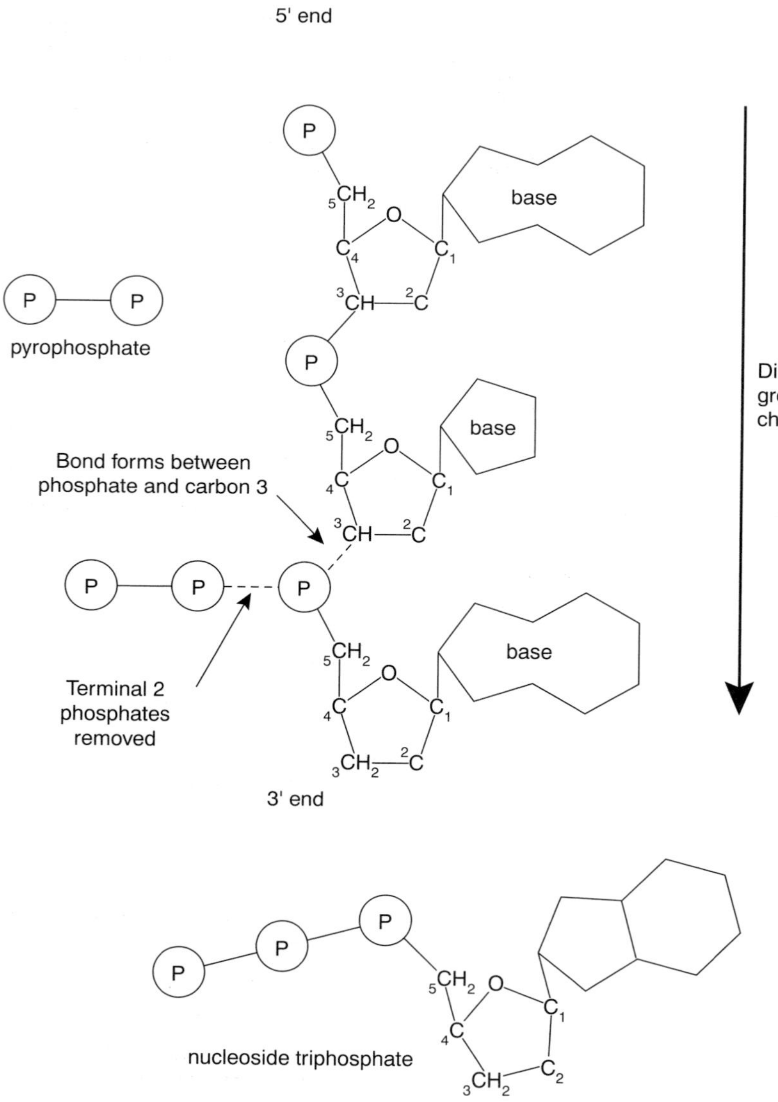

5' end

pyrophosphate

Bond forms between phosphate and carbon 3

Terminal 2 phosphates removed

Direction of growth of chain

3' end

nucleoside triphosphate

**Figure 10.8**
Formation of a polynucleotide

## Okazaki fragments

Polynucleotide chains can only grow by adding nucleotides to the 3' end. A DNA strand can, therefore, only grow in the 5' → 3' direction. Since the two DNA strands are **antiparallel**, only one of the two new strands – the **leading strand** – can be synthesised continuously. The other **lagging strand** is produced discontinuously in separate segments called **Okazaki fragments** which are subsequently joined together by an enzyme called **DNA ligase.**

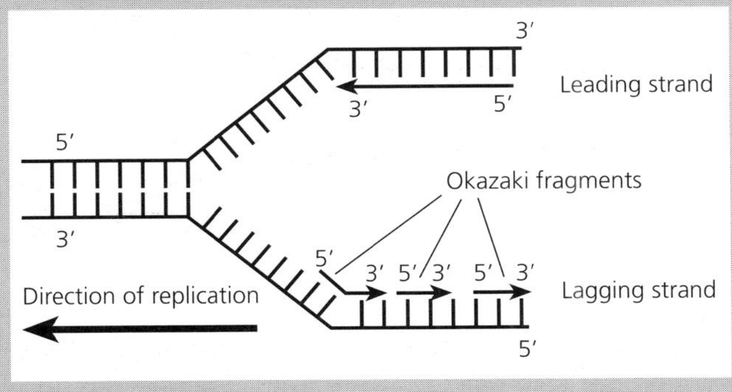


Discontinuous DNA replication

## DNA replication in eukaryotes and prokaryotes

Each chromosome contains a single continuous molecule of DNA. If replication were to begin at one end of the chromosome and work towards the other, it would take several weeks to replicate the DNA in a chromosome of average length. In eukaryotes, DNA replication begins simultaneously at several hundred points along a chromosome and proceeds in both directions from each point of origin (Figure 10.9). In prokaryotes (which have far less DNA to replicate) replication begins at one point.

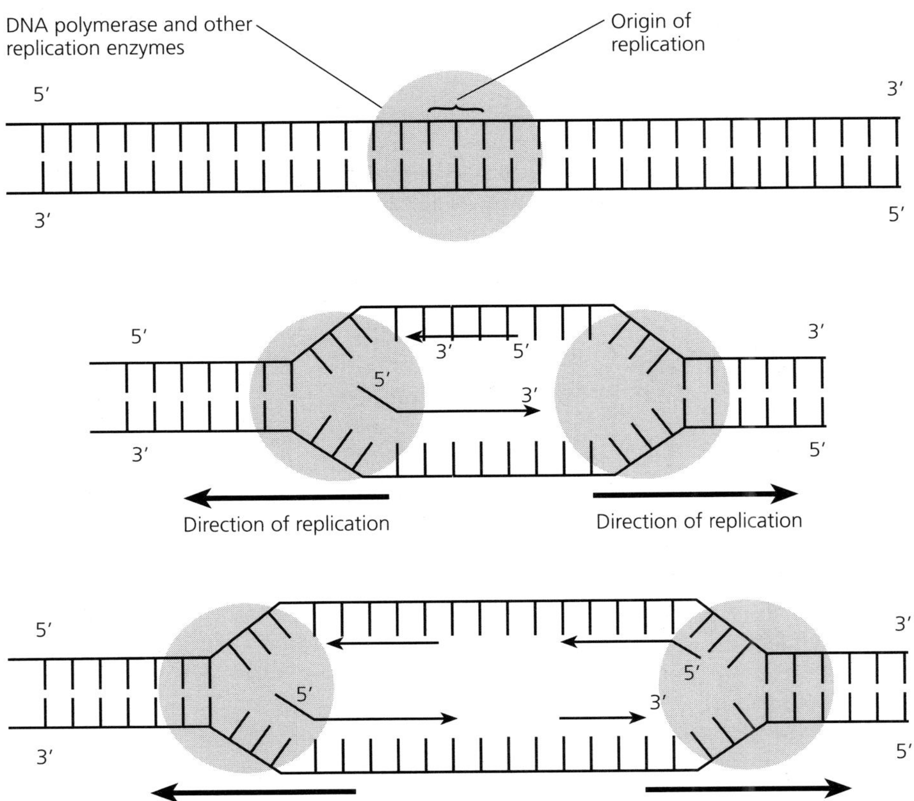


**Figure 10.9**
Replication of DNA spreading in both directions from point of origin

## DNA hybridisation

When DNA is heated to 70–90°C the hydrogen bonds holding the bases together break and the two strands come apart. The DNA is said to 'melt' or become **denatured.** DNA melting can be detected by a sudden decrease in viscosity and an increase in its absorption of ultraviolet light. DNA can also be denatured by a strong alkali, a fact that is put to good use in using gene probes (Chapter 13).

The melting point ($T_m$) of DNA is the temperature at which half the DNA is single-stranded and varies with the proportion of the two kinds of base pair. Since guanine is linked to cytosine by three hydrogen bonds and adenine is linked to thymine by two hydrogen bonds, G—C base pairs require more energy to break than A—T pairs. DNA that has a high proportion of G—C pairs 'melts' at a higher temperature than DNA that is richer in adenine and thymine. The melting point thus provides a measure of the G—C content.

Unlike thermal denaturation of proteins, denaturation of DNA can be reversed if the temperature is lowered and maintained at 65°C for several hours. This process of **reannealing** is dependent on the complementary sequences of bases in the single strands coming back together by random collision. If DNA from two closely-related species is mixed, melted and then reannealed, some of the DNA forms **hybrid duplexes** composed of one strand of each species.

DNA hybridisation

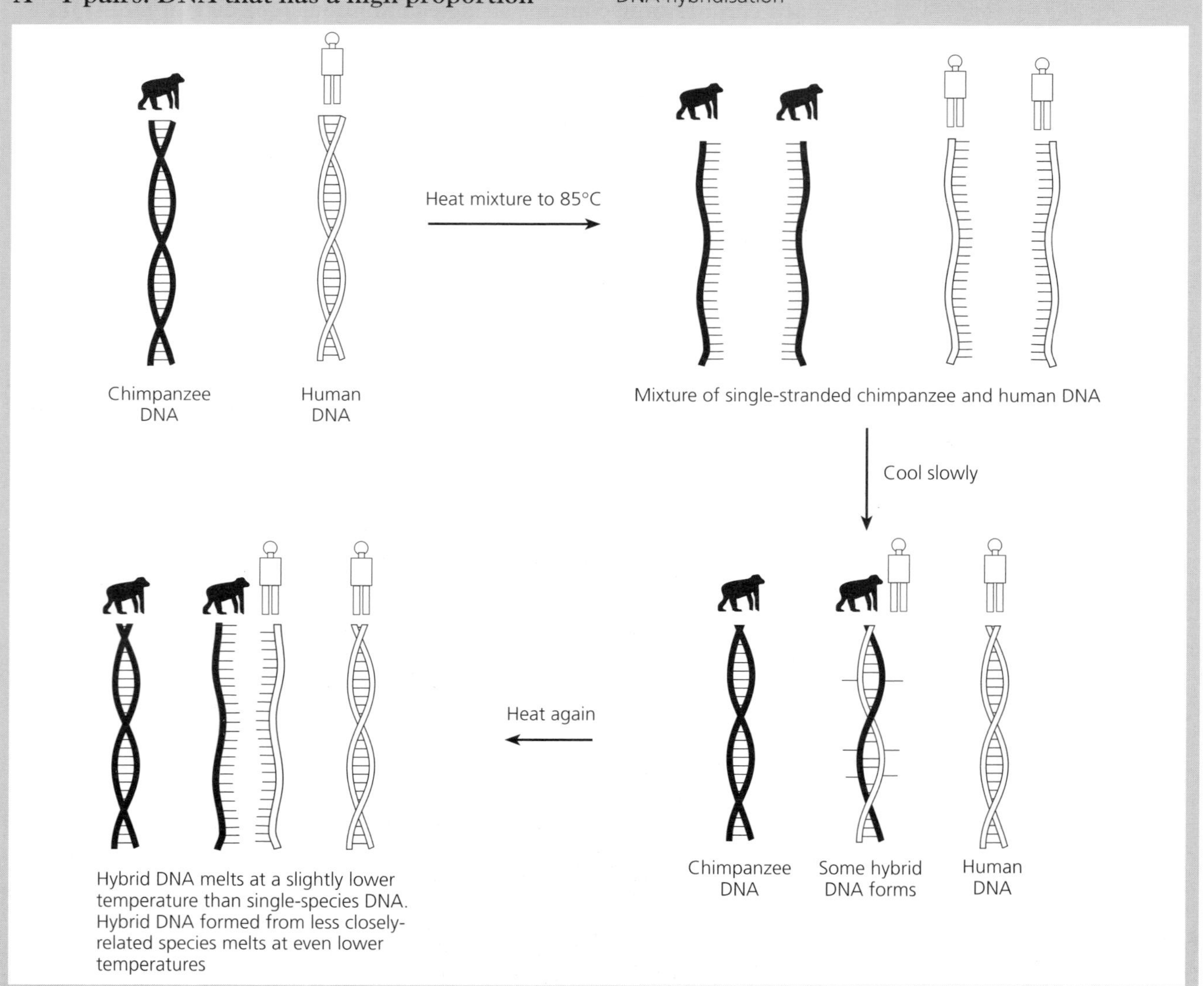
Chimpanzee DNA
Human DNA
Heat mixture to 85°C
Mixture of single-stranded chimpanzee and human DNA
Cool slowly
Chimpanzee DNA
Some hybrid DNA forms
Human DNA
Heat again
Hybrid DNA melts at a slightly lower temperature than single-species DNA. Hybrid DNA formed from less closely-related species melts at even lower temperatures

## Repair of DNA

One of the key properties of DNA is its resistance to change. It used to be assumed that this was due to inherent chemical stability. In fact DNA is vulnerable to a variety of spontaneous chemical changes, such as oxidation and hydrolysis. As a result, the real rate of mutation is actually higher than it would appear – it is just that most mutations are repaired before the DNA replicates.

Although any damage to the DNA can result in mutation, only about one in a thousand actually does so because cells can repair their genetic material. The repair of DNA depends on the same key property of DNA as does replication – the fact that each strand contains the information to build the other.

If a base is chemically changed to an abnormal base, this is 'recognised' and removed by a DNA repair enzyme. The normal base on the other strand is used as a template for the replacement of the abnormal base with the correct one.

The enzymes involved in DNA repair are themselves under the control of genes which, like any other genes, can mutate. Affected individuals cannot repair their DNA so they suffer from unusually high rates of cancer (a result of mutation).

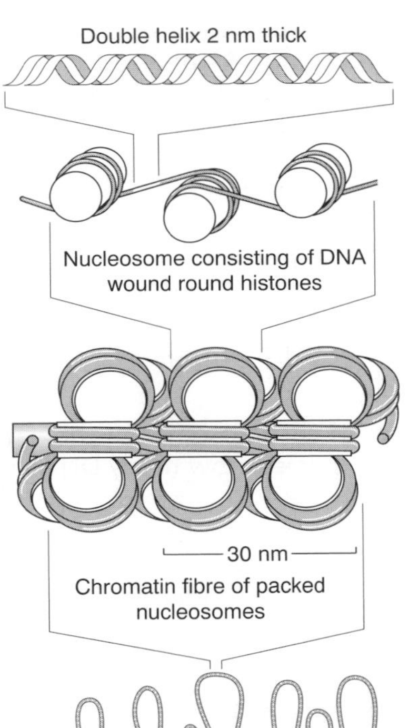

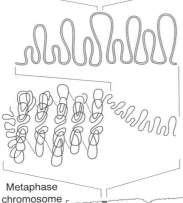

**Figure 10.10**
Organisation of DNA and histones into nucleosomes

## Organisation of DNA in eukaryotic chromosomes

Besides DNA, eukaryotic chromosomes contain considerable quantities of basic proteins called **histones**, the DNA–histone complex being called **chromatin**. It seems that one of the major functions of histones is the packaging of DNA into a more compact form.

The DNA in the nucleus of a human cell consists of $6 \times 10^9$ base pairs, which would be about 2 metres long if fully extended. This is distributed amongst 46 chromosomes, so the DNA in an average human chromosome would be just over 4 cm long. The longest human chromosome (#1) would be nearly 9 cm long if fully extended.

The DNA and histones are not in the form of a disorganised tangle, but are organised into packaging units called **nucleosomes** (Figure 10.11).

Each nucleosome consists of two turns of the double helix wrapped round a group of histone molecules. These proteins are basic because they contain a high proportion of basic amino acids. These give histones a net positive charge, enabling them to be attracted to the negatively-charged phosphate groups of DNA.

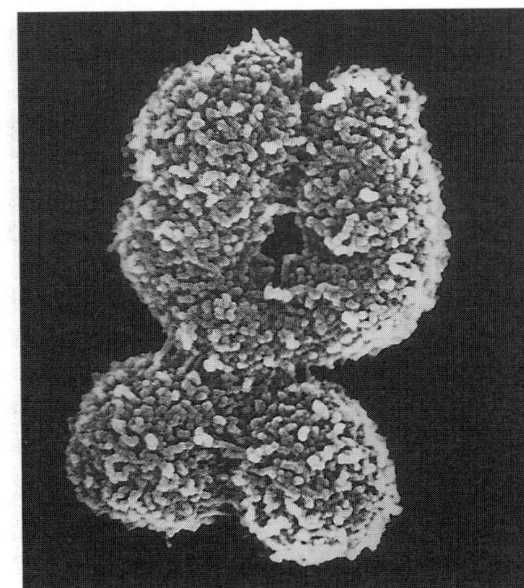

**Figure 10.11**
Electron micrograph of single chromosome to show coiling of DNA

## QUESTIONS

**1** If one strand of DNA has the sequence G A T C A T A A T, what is the sequence of bases in the other strand?

**2** Of the three kinds of constituents of a polynucleotide, which could be removed without breaking the chain?

**3** Which of the following statements is incorrect?

A Before a cell divides, its DNA content doubles.
B The proportions of total purines to total pyrimidines is the same for all species.
C The proportions of adenine to guanine and of cytosine to thymine are the same for all species.
D For a diploid species, the DNA content of the gametes is half that of the body cells.

**4** The ability of DNA to be replicated depends on which of the following properties?

A When broken down, it yields an equal number of adenine and thymine molecules, and an equal number of guanine and cytosine molecules.
B It is a giant molecule.
C It contains two repeating phosphate–sugar backbones.
D It has a helical structure.

**5** Which of the following is the best evidence that DNA is the genetic material?

A It is found mainly in the nucleus.
B It contains four kinds of base in definite proportions.
C Ultraviolet light is maximally absorbed by DNA at the wavelengths which most readily cause mutations.
D The two strands are linked by hydrogen bonds which allow them to separate easily.

**6** Which of the following features of DNA best explains why organisms resemble their parents?

A It consists of a double helix.
B The sequence of bases in one strand defines the sequence in the other.
C The bases form pairs.
D The proportions of bases differ from one species to another.

**7** In the DNA of a certain species 20% of the bases are thymine. What proportion of the bases are guanine?

**8** In 1963 Herbert Taylor grew roots of bean plants in thymidine in which the hydrogen was radioactive isotope, tritium ($^3H$). Thymidine is the nucleoside formed from thymine and deoxyribose. He then placed the roots in a non-radioactive medium containing colchicine. This inhibits spindle formation and so prevents separation of chromatids. After just enough time for one chromosomal duplication, he removed some of the cells and placed them over a photographic plate to determine the location of the radioactivity. After another cell cycle he removed a further sample, and determined the location of radioactivity in these chromosomes.

**a)** After one cell cycle in the non-radioactive medium, every chromatid was still radioactive. Explain.
**b)** How would you expect the radioactivity to be distributed after one more replication cycle in the non-radioactive medium?

**9** The diploid number of chromosomes of the mouse is 40. For this species, how many DNA molecules are there in

**a)** a sperm;
**b)** a zygote in early prophase;
**c)** a cell that is just about to undergo meiosis II?

**10** The table shows the % base composition in two viruses, cowpox and the M13 bacteriophage.

| | A | T | G | C |
|---|---|---|---|---|
| Cowpox | 29.5 | 29.9 | 20.6 | 20.0 |
| M13 phage | 23.3 | 32.8 | 21.1 | 19.8 |

**a)** Which of these two viruses has an unusual base composition?
**b)** Suggest an explanation.
**c)** How would you test your explanation?

CHAPTER

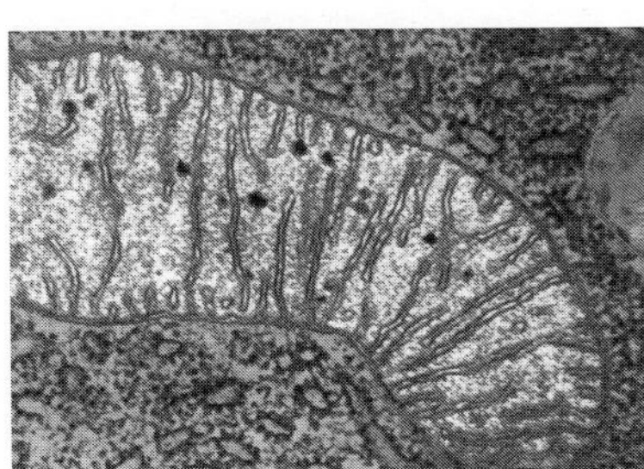

# 11

# MAKING PROTEINS

## LEARNING OBJECTIVES

By the time you have completed your study of this chapter you should be able to:

- give some of the evidence that protein synthesis occurs in the cytoplasm, and that information flows from nucleus to cytoplasm.
- describe the process by which DNA is copied into messenger RNA.
- explain why, in the genetic code, at least three bases must represent each amino acid.
- explain what is meant by saying that the code is 'degenerate'.
- use the table of mRNA codons to work out the sequence of amino acids corresponding to a given sequence of mRNA bases.
- explain what is meant by 'translation' in protein synthesis, and describe the roles of transfer RNA and ribosomes.
- appreciate the difference between 'code' and 'message', and use the terms correctly.
- briefly explain the differences between protein synthesis in prokaryotes and eukaryotes.
- explain why multicellular organisms must be able to switch genes on and off, and how gene action is regulated in bacteria.
- explain what is meant by the 'central dogma', and its evolutionary significance.

## Genes and proteins

In Chapter 10 we saw that the function of many genes is to provide information for making a particular polypeptide. Since the properties of a polypeptide depend on the sequence of its amino acids, the function of a gene must be to provide the information for joining the amino acids in a specific sequence. In some way then, the base sequence in a gene must specify the amino acid sequence in a protein.

Experiments have shown that proteins are made in the cytoplasm (see information box on page 174). Yet we know that, with the exception of small amounts in the mitochondria and plastids, DNA is restricted to the nucleus. It follows that DNA cannot be directly involved in making proteins; its role must be indirect.

The near absence of DNA from the cytoplasm is in marked contrast to the abundance of another nucleic acid, **RNA** (**ribonucleic acid**). RNA is particularly abundant in cells that are active in protein synthesis.

Somehow, therefore, the information needed to make proteins must travel from the nucleus to the cytoplasm. Evidence for the existence of such a messenger was obtained in experiments beginning in the late 1930s (see information box on page 174). Later experiments showed that the substance involved was a form of ribonucleic acid (RNA).

## Where are proteins made?

As we saw in earlier chapters, genes are present on the chromosomes, yet the majority of the proteins they code for are in the cytoplasm. This raises the question of whether proteins are produced in the nucleus (where the information is) or in the cytoplasm, where most proteins carry out their functions.

In one experiment bacterial cells were given a 15 second 'pulse' of sulphate labelled with $^{35}S$, after which the tracer was removed from the nutrient medium. Since sulphur is a constituent of two amino acids, newly produced proteins would thus be expected to become radioactive. A sample of the bacterial culture was then removed, and another sample was removed two minutes later. In each of the two samples the cells were broken up and the various constituents separated by centrifugation. The levels of radioactivity of the various fractions were then compared.

In the first sample, most of the radioactivity was located with the ribosomes. In the second sample, the radioactivity was concentrated in a lighter fraction, consisting of free protein molecules. The conclusion was clear – ribosomes are involved in protein synthesis and, after completion, proteins become detached from the ribosomes. In eukaryotes, ribosomes are restricted to the cytoplasm which, therefore, must be where proteins are made.

## Evidence for the messenger

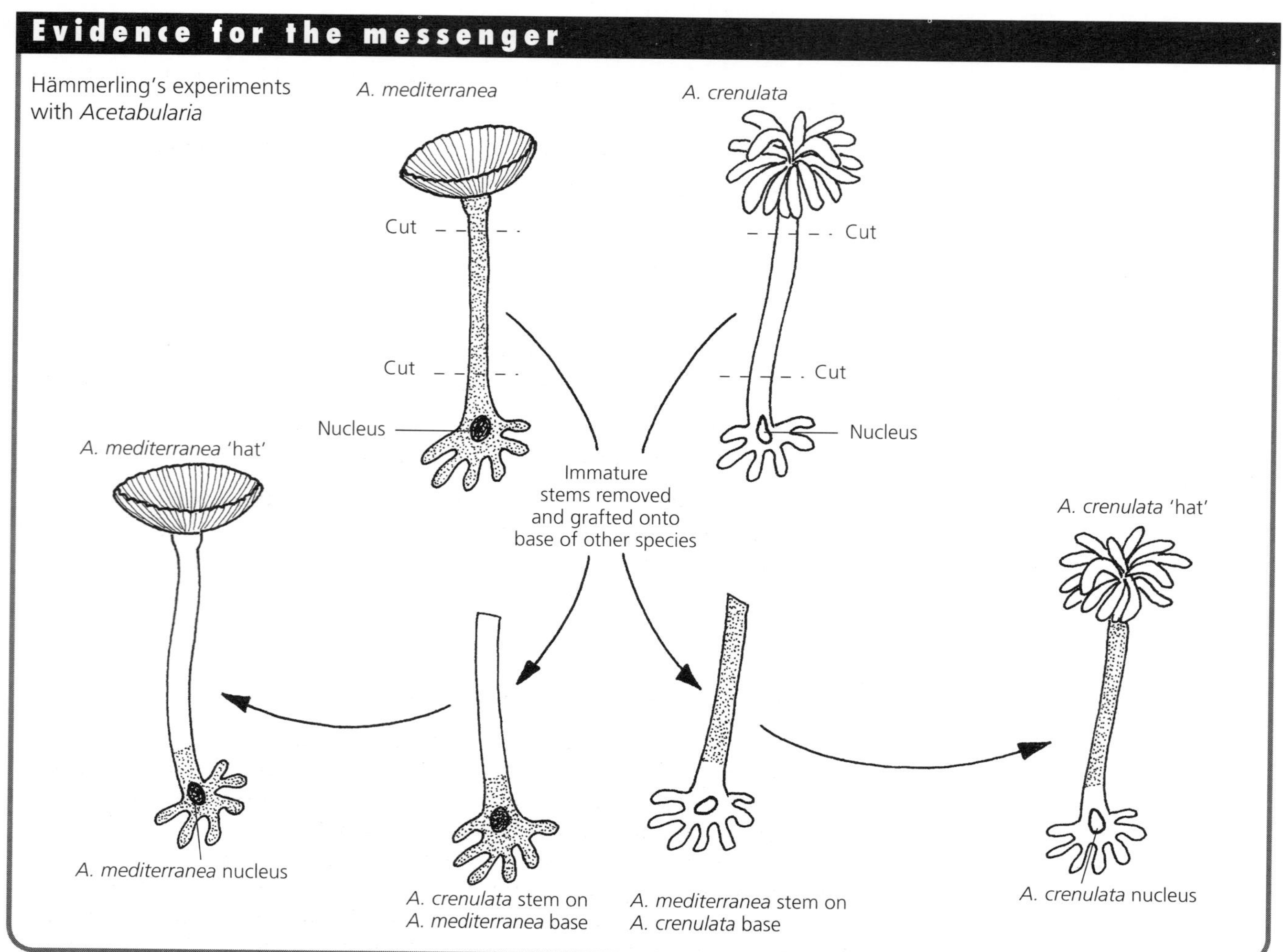

### Evidence for the messenger *continued*

The existence of some kind of messenger was demonstrated by Joachim Hämmerling in experiments using the marine protoctist *Acetabularia*. This extraordinary organism resembles a miniature umbrella, and though it grows up to 6 cm high, it is actually a single giant cell. The nucleus is located in the rhizoid which anchors it to the rock.

In one experiment Hämmerling took a young specimen that had not yet developed its 'hat', and cut it into three parts. The tip developed a hat and the rhizoid regenerated an entire individual, but the middle piece developed no further. However, when he removed the tip and waited several days before severing the middle piece from the rhizoid, the middle piece *did* develop a new hat. Hämmerling concluded that some kind of 'hat-forming substance' was produced in the base and moved into the middle piece.

Further experiments showed that the origin of this 'hat-forming substance' must be the nucleus. If the stalks of *A. mediterranea* and *A. crenulata* were removed and the nuclei exchanged, each rhizoid regenerated a new stalk and hat. However the hat was of the type characteristic of the species from which the *nucleus* had been taken.

## Ribonucleic acid (RNA)

RNA differs from DNA in four ways:

1 in RNA the sugar is **ribose** (Figure 2.35). This is similar to deoxyribose, except that it has the full sugar complement of oxygen atoms.

2 instead of thymine, RNA contains another pyrimidine, **uracil**.

3 except in certain viruses, RNA is **single-stranded**.

4 RNA molecules are always much shorter than DNA, which extends the entire length of a chromosome.

In prokaryotes there are two stages in the manufacture of a protein:

- **transcription**, or the making of a copy or **transcript** of the DNA. This is in the form of **messenger RNA**, usually shortened to **mRNA**.
- **translation**, or the actual making of a polypeptide, i.e. the linking of amino acids in the specific sequence encoded in the mRNA.

In eukaryotes protein synthesis is more complicated because there are *three* stages. After its formation, the RNA transcript has to undergo further 'editing' before it is ready to be translated. This additional stage is called **RNA processing** and occurs in the nucleus.

# Stage I: transcription

Transcription results in the production of 'working copies' (or transcripts) of the gene in the form of RNA. In prokaryotes, transcription produces mRNA, but in eukaryotes it is more properly called a **primary RNA transcript**. This is later processed to produce mRNA.

As in DNA replication, transcription involves the principle of complementary base pairing, but there are some important differences. First, a DNA molecule contains thousands of genes, so an RNA molecule is a copy of only a very small part of a DNA strand. Second, only one of the two DNA strands is copied. This raises the twin questions of how does the cell 'know' where to begin transcription, and which strand to copy? As we shall see in a moment, both questions have the same answer.

Transcription begins when the enzyme **RNA polymerase** 'recognises' and binds to a **promoter** (Figure 11.1) – a particular sequence of bases on one of the DNA strands (the sequence on the other strand is, of course, complementary, and therefore different).

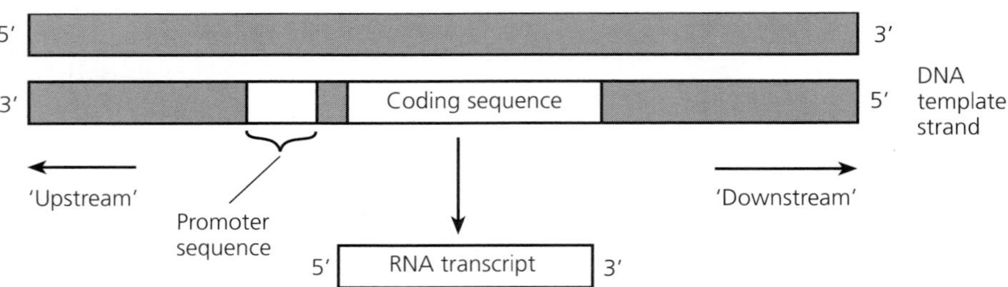


**Figure 11.1**
Promoter and coding regions of a eukaryotic gene

The RNA polymerase then begins to 'unzip' the two DNA strands, and moves along one of them, adding nucleotides to the 3' end of the RNA. As it does so, it links up nucleotides to form a complementary strand of RNA. Just as the two strands in DNA are antiparallel, the RNA strand runs antiparallel to the DNA template strand (Figure 11.2). The pairing of the bases is the same as in DNA, except that adenine of DNA pairs with uracil of RNA.

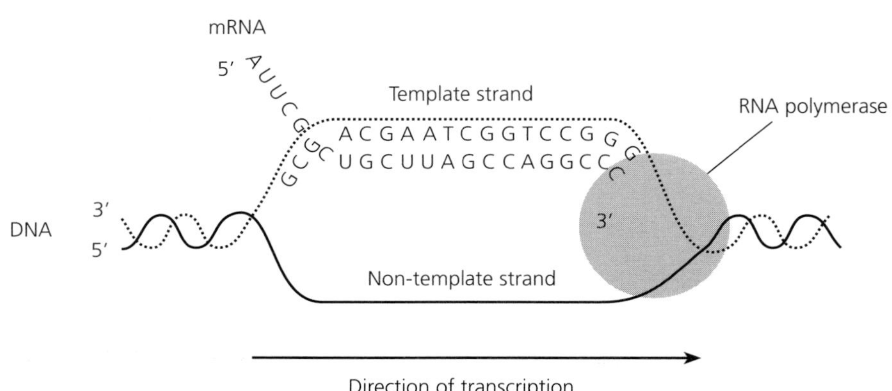


**Figure 11.2**
How an RNA transcript is produced on a DNA template

Like all polynucleotide chains, the two ends of an RNA molecule are different (the 5' and 3' ends). Without this feature the sequences AUG and GUA would be the same. By convention, the order of nucleotides is written with the 5' end as the beginning (i.e. on the left). The 3' end is said to be 'downstream' of the 5' end.

Since the production of the RNA transcript uses one of the DNA strands as a template, the latter is called the **template strand**. The use of the terms 'sense strand' for the template strand and 'antisense strand' for the non-template strand are ambiguous and best avoided, since the RNA (the message) is actually a copy of the 'antisense' strand.

The RNA polymerase continues to move along the DNA until it reaches a **terminator**, a sequence of bases signifying 'end of message'.

One important result of transcription is that the number of copies of the gene is greatly increased. From a single 'master copy' in the form of DNA, many RNA copies can be produced. This enables a protein to be produced much faster than would be possible if the DNA were to be used directly.

## Stage II: RNA processing

The information in the RNA transcript is not yet ready to be translated. This is because, in a eukaryotic gene, coding regions are separated by sections of DNA that are not translated into proteins. The segments carrying information that is translated are called **exons** because they are *ex*pressed. The *in*tervening sequences seem to have no function and are called **introns**.

Initially, the RNA copy of a gene includes both introns and exons. Before the message can be used to make a protein, the introns are removed and the exons are then spliced together to make a continuous strand of mRNA (Figure 11.3).

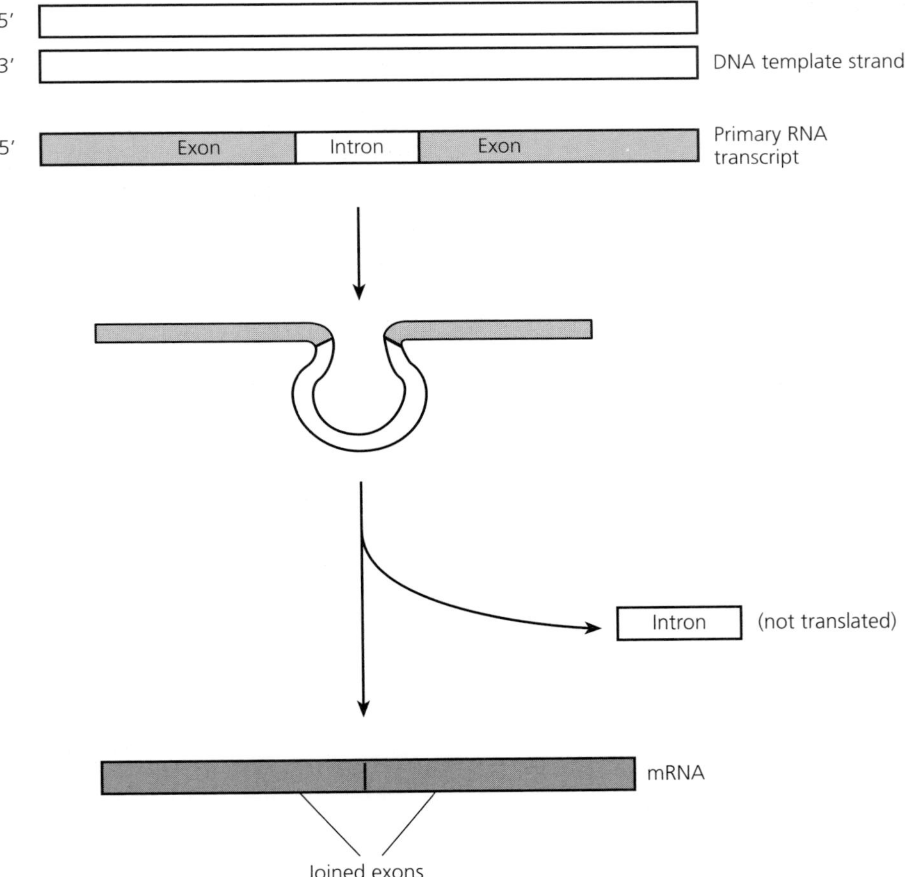


**Figure 11.3**
How the primary transcript is 'edited' by removal of introns

In many genes there are a dozen or more introns, which may in total be of considerably greater length than the exons. Why should cells waste energy producing DNA that seems to carry no useful information? At the moment no firm answer seems to be available.

Perhaps even more surprising was the discovery that in some cases the introns catalyse their own removal. RNA molecules that can act as catalysts are called **ribozymes**.

After its production, the mRNA leaves the nucleus via one of the pores in the nuclear envelope. These pores are not simply holes through which mRNA molecules diffuse. They have a complex structure and seem to actively 'feed' mRNA (and a variety of other large molecules) through them by a process as yet incompletely understood.

**Table 11.1**
The 16 possible codons based on a doublet code

| UU | UC | UA | UG |
|---|---|---|---|
| CU | CC | CA | CG |
| AU | AC | AA | AG |
| GU | GC | GA | GG |

## The genetic code

Having arrived in the cytoplasm, the information in the mRNA can be used to make a polypeptide. How, then, is this information encoded? Clearly, each individual base cannot specify an amino acid, because there are five times as many kinds of amino acids as there are bases.

What about two bases – i.e. a doublet code? For each of the four possibilities for the first base there would be four for the second, making a total of 16 different 'words', which would still be insufficient to code for the 20 amino acids that are used to make proteins (Table 11.1).

A triplet code, with a sequence of three bases specifying each amino acid, would provide four times as many possibilities as a doublet code (four for each of the 16 possible doublets). In a triplet code there would be $4 \times 4 \times 4 = 64$ possible 'words' – more than enough to specify the 20 kinds of amino acid. Theory demands, then, that a sequence of at least three bases must specify each amino acid. Experiments briefly outlined in the information box 'Cracking the code' on page 180 show that the code is indeed a **triplet code**. Thus, a sequence of 300 nucleotides would be needed to specify a polypeptide 100 amino acids long.

Although a triplet code would be the most economical, it doesn't follow that it *must* be a triplet code. That DNA 'words' actually *do* have three letters was shown in the 1960s, as was the identity of the triplets coding for each amino acid.

Another feature of the code is that it is **non-overlapping**. This means that each base is part of only one triplet. In an overlapping code, each base would be part of three triplets. Thus in the base sequence

CATCATCATCAT

the third letter (T) would be not only the last letter of the first triplet, but also the second letter of the second triplet, and the first letter of the third triplet. Thus the above sequence would be read as:

CAT ATC TCA CAT ATC TCA etc.

If the code were overlapping, it would be impossible to change just one triplet, thus imposing great restrictions on the number of possible amino acid sequences. Strong support for the non-overlapping nature of the code is provided by the case of sickle cell anaemia, in which the β chain of haemoglobin differs from normal by only *one* amino acid.

The genetic code is shown in Figure 11.4. The triplets specifying each amino acid are shown as mRNA triplets, and the names for the amino acids are given as abbreviations (aspN and gluN stand for asparagine and glutamine, respectively). If the genetic code were written as DNA triplets, there would be the problem of which of two complementary triplets should represent each amino acid. mRNA triplets are called **codons**. Except for the fact that uracil replaces thymine, mRNA codons are the same as those of the non-template strand of DNA.

Remember that the standard convention is to write each codon with the base at the 5' end first. AUG could thus be written more fully as (5') AUG (3'). The full names of the amino acids are not given – only their standard abbreviations (e.g. trp is short for tryptophan). The first base of a codon is indicated down the left side of the table, and the second base along the top. The four possibilities for the third base are given down the right hand side of the table. Thus CUG specifies arginine (arg).

The code has a number of important features:

- there are more than three times as many triplets as are needed, but all 64 mean something. All but two of the 20 amino acids are represented

| | | Second base of codon | | | | | |
|---|---|---|---|---|---|---|---|
| | | U | C | A | G | | |
| First base of codon (5' end) | U | phe<br>phe<br>leu<br>leu | ser<br>ser<br>ser<br>ser | tyr<br>tyr<br>STOP<br>STOP | cys<br>cys<br>STOP<br>trp | U<br>C<br>A<br>G | Third base of codon (3' end) |
| | C | leu<br>leu<br>leu<br>leu | pro<br>pro<br>pro<br>pro | his<br>his<br>gluN<br>gluN | arg<br>arg<br>arg<br>arg | U<br>C<br>A<br>G | |
| | A | ile<br>ile<br>ile<br>met + START | thr<br>thr<br>thr<br>thr | aspN<br>aspN<br>lys<br>lys | ser<br>ser<br>arg<br>arg | U<br>C<br>A<br>G | |
| | G | val<br>val<br>val<br>val | ala<br>ala<br>ala<br>ala | asp<br>asp<br>glu<br>glu | gly<br>gly<br>gly<br>gly | U<br>C<br>A<br>G | |

**Figure 11.4**
The genetic code 'dictionary'

by more than one triplet. A code in which several 'words' can have the same meaning is said to be **degenerate**.

- the degeneracy is not random – where several codons represent the same amino acid, the alternative codons are nearly always in the same cell of the table indicating that only the last base is different. Substitution of a base in the third position would, therefore, usually have no effect.
- four of the codons are 'punctuation marks'. One (AUG) means 'start' and also represents the amino acid methionine. Three are 'stop' codons, meaning 'end translation here'.
- the code is almost universal. With the exception of mitochondrial DNA, a given codon specifies the same amino acid in all organisms. Even in mitochondria, the differences are minor. For example, UGA is a stop codon in nuclear DNA, but is read as tryptophan in mitochondrial DNA.

Though the numerical relationship between nucleotides and amino acids is simple, the relationship between proteins and DNA is not quite so straightforward because some proteins consist of more than one polypeptide – for instance haemoglobin. A length of DNA coding for one polypeptide is a **gene**. It therefore needs two genes to code for the protein haemoglobin (one for the two $\alpha$ chains and one for the two $\beta$ chains).

**CLEAR THINKING**

**In an essay, Bill wrote: 'An mRNA molecule contains a sequence of triplet codes.' In another essay, Mary wrote 'a mutation is a change in the genetic code'.**

**Bill and Mary were using the term 'code' as if it was interchangeable with 'message'. In fact they have quite different meanings. Think of the Morse code.**

**Using this one code it is possible to send an infinite number of different messages. Likewise, most cells are busy translating thousands of different genetic messages (genes), using the *same* genetic code. A mutation is a change in the genetic information; the code remains the same.**

## Cracking the code

Within a few years of Watson and Crick's discovery of the structure of DNA, intensive efforts were being made to break the genetic code.

The first problem was to find out how many bases represented each amino acid. A triplet code seemed the most obvious, but this hypothesis had to be tested experimentally. The experiments were performed by Francis Crick using a bacteriophage. He used frame-shift mutations (Chapter 12). These are of two kinds, depending on whether a base is deleted from the DNA (a '−' mutation) or added (a '+' mutation). Either a '+' or a '−' mutation would alter the reading frame, so that every triplet 'downstream' of the mutation would be changed. If a base were added, and another deleted not too many bases 'downstream' of the addition, the deletion would restore the reading frame to normal. As a result only the amino acids specified by the region between the two mutations would be abnormal. If the abnormal region were not too long the protein could still be functional. Crick found this to be the case; in some instances a '+' mutation reversed a '−' mutation.

To cut a long, complicated story short, Crick found that by either adding or deleting two bases, a mutant phenotype was invariably produced. However, deletion or addition of *three* bases did, in some cases, produce a normal phenotype. Moreover, addition or deletion of four or five bases always produced a mutant phenotype, but addition or removal of six bases did, in some cases, restore the phenotype to normal. This is precisely what would be expected if the code were a triplet code.

The next problem was to find out which base sequence specifies which amino acid. The first step towards a solution was taken when it was learned how to make proteins in test tubes when supplied with ribosomes, transfer RNAs, amino acids and other necessary chemicals.

The next discovery was how to make artificial mRNAs of known base sequences. By using these artificial mRNAs and determining the amino acid sequence of the polypeptides produced, it became possible to work out which triplet specified which amino acid, and by 1966 the code had been cracked.

# Stage III: Translating the message

After transcription the genetic information is relocated from nucleus to cytoplasm, but it is still in coded form. In the next stage the information in the mRNA is used to join amino acids together into the correct sequence. In this process the information in the 4-base nucleic acid 'alphabet' is translated into the 20-amino acid 'alphabet' of proteins. It is a very complex process, involving two more kinds of RNA: **transfer RNA** (**tRNA**) and **ribosomal RNA** (**rRNA**). First, however, the amino acids have to be activated.

## Activating the amino acids

The linking of amino acids to form polypeptides requires energy and therefore does not occur spontaneously. To enable it to react, each amino acid has to be made more reactive using energy provided by ATP. Two phosphates are split off from ATP leaving adenosine monophosphate (AMP) linked to the amino acid:

$$\text{amino acid} + \text{ATP} \rightarrow \text{amino acid–AMP complex} + 2P_i$$

## Transfer RNA

Once activated, amino acids have enough energy to link up into polypeptide chains. But making a protein requires information besides raw

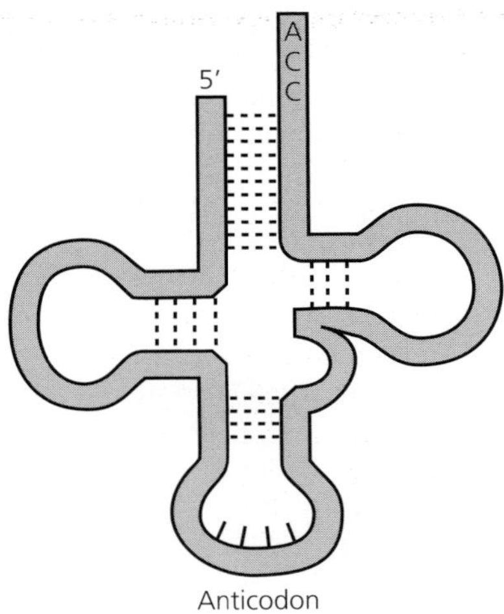


**Figure 11.5**
A transfer RNA molecule

materials and energy. As we have seen, this information is encoded in mRNA.

If each amino acid could combine with the codon specifying it, amino acids could simply line up alongside mRNA in the correct sequence. The trouble is, amino acids have no tendency to combine directly with mRNA – still less with the codons specifying them.

This is where tRNA comes in. It is the smallest of the three kinds of RNA, approximately 70 to 90 nucleotides in length. Though single-stranded, it is folded on itself into a shape resembling a twisted clover leaf. The folds are held in place by complementary base pairing (Figure 11.5).

The function of tRNA is to act as an **adapter** – one end of the molecule combines with a particular kind of amino acid and the other links with the mRNA codon specifying that amino acid. It follows that there must be at least 20 kinds of tRNA, one for each amino acid (there are actually more, some amino acids having more than one tRNA).

Combination with the amino acid occurs at the 3' end of the molecule. Since this part of the molecule is the same for all tRNAs (it has the sequence CCA), how does each amino acid 'recognise' its corresponding tRNA?

The answer lies in the fact that although all tRNAs have the same general shape, their folding patterns are probably slightly different (as a result of differences in the base sequences).

The combination between an activated amino acid and a tRNA is catalysed by an enzyme, which is different for each amino acid. Each of the 20 enzymes, therefore, must be specific both for one particular amino acid and also for the appropriate kind of tRNA.

$$\text{amino acid–AMP complex} + \text{tRNA} \rightarrow \text{amino acid–tRNA} + \text{AMP}$$

We have seen how a tRNA combines with 'its' amino acid, but how does a tRNA molecule combine with mRNA? At another part of the tRNA molecule are three unpaired bases called an **anticodon**. The anticodon is so-called because it is complementary to an mRNA codon, so can combine with it by base pairing. In this way, a tRNA–amino acid complex can link up with its corresponding mRNA codon.

## One base sequence – three possible messages

Although the base sequence in mRNA must be 'read' in threes, the bases are not grouped in threes. In other words, there are no spaces between triplets as there are between words on a page. Imagine the following sequence of letters:

THECATSAWTHEBIGDOGEATTHEFATRAT

There are three different ways (called **reading frames**) in which this message could be read three letters at a time, depending on whether one starts at the first, second or third base along. These are shown below with spaces inserted to make them easier to read:

THE CAT SAW THE BIG DOG EAT THE FAT RAT
T HEC ATS AWT HEB IGD OGE ATT HEF ATR AT
TH ECA TSA WTH EBI GDO GEA TTH EFA TRA T

Only one reading frame makes any sense. To ensure that the message is read using the correct reading frame *it must be read from the first base, one triplet at a time*. This is the function of the **ribosomes**. Only in the presence of a ribosome can a codon form complementary base pairs with an anticodon of tRNA.

## The role of the ribosomes

Ribosomes are the structures that actually 'read' the message in the mRNA. They are tiny granules about 20 nm in diameter, and consist of two smaller subunits (Figure 11.6). Ribosomes consist of about two-thirds ribosomal RNA (rRNA), and one third protein. Like tRNA, rRNA is folded on itself by base pairing.

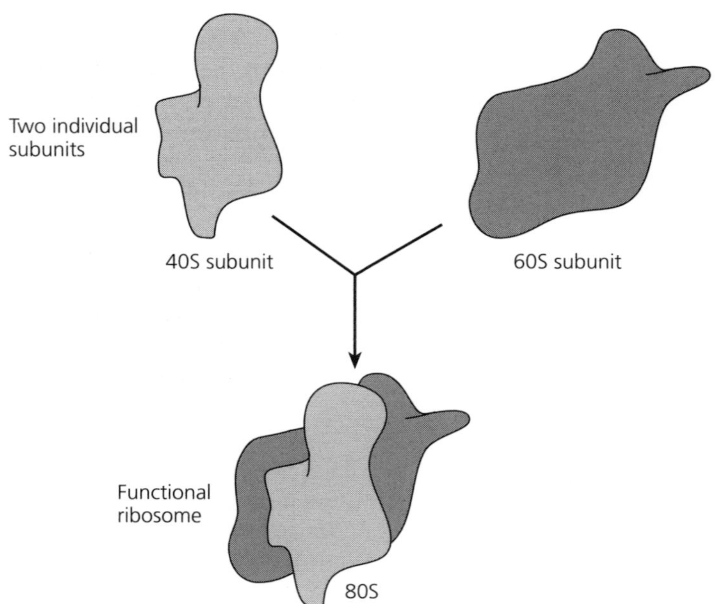


**Figure 11.6**
Subunits of a eukaryotic ribosome. 'S' stands for Svedberg, a unit of sedimentation rate. This is not directly proportional to mass, so a 40 S unit can combine with a 60 S unit to form an 80 S unit rather than a 100 S unit

The genes responsible for the production of rRNA are located in chromosome segments which collectively form the **nucleolus**. Both ribosomal subunits are assembled in the nucleus before being exported, still as separate subunits, to the cytoplasm.

The details of how ribosomes make proteins are complicated, but in essence what happens is as follows. When the ribosomal subunits collide with the 5' end of an mRNA molecule they link up with it. This allows the anticodon of a tRNA carrying the first amino acid to combine with the first codon of the mRNA. In eukaryotes, the first amino acid is always methionine, but this is later removed.

Next, the ribosome moves along one codon, allowing the appropriate tRNA to combine with the next codon. At the same time a peptide bond forms between the carboxyl group of the first amino acid and the amino group of the next (Figure 11.7).

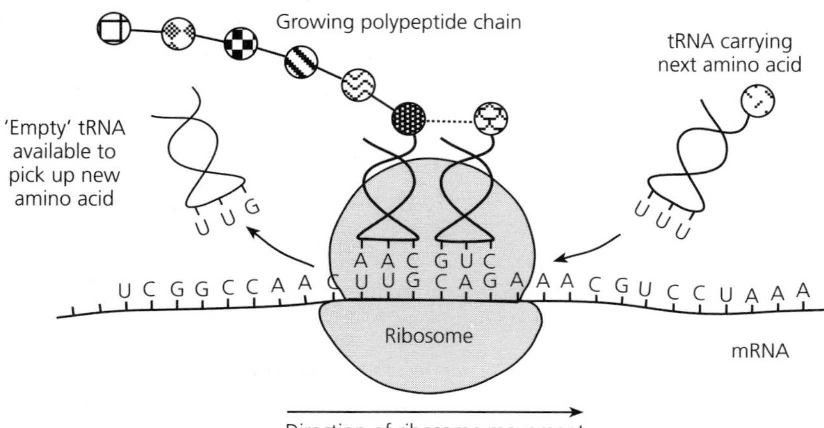


**Figure 11.7**
How a ribosome 'reads' mRNA

Meanwhile the tRNA that had been linked to the first amino acid breaks away and is now free to pick up another amino acid.

The ribosome then moves along another triplet, and a new tRNA with its appropriate amino acid attaches to the next mRNA codon. Apart from the time when the ribosome is moving, two tRNA molecules are attached to the ribosome – one attached to the amino acid most recently added to the polypeptide, and one carrying the amino acid that is just about to be added.

Each time the ribosome moves along one codon, one more amino acid is added to the polypeptide chain. When the ribosome encounters a 'stop' codon it detaches itself and is free to begin another round of translation. The polypeptide chain assumes its secondary and tertiary structure. In proteins with a quaternary structure, the individual polypeptides spontaneously clump together to form the complete protein.

The end result is that amino acids are linked up in the sequence corresponding to the sequence of codons in the mRNA. It normally takes about half a minute for a ribosome to produce a polypeptide 150 amino acids long. Normally, up to a dozen or so ribosomes are reading the same mRNA molecule at any one time, the mRNA and the attached ribosomes forming **polyribosomes** (Figure 11.8).

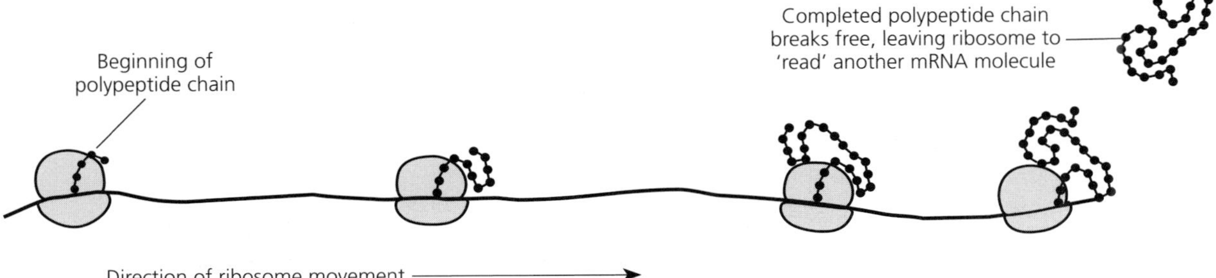


**Figure 11.8**
Polyribosomes

Although each mRNA molecule is 'read' many times, mRNA molecules are continuously being broken down by the enzyme **ribonuclease** (**RNase**). In eukaryotes, the time for half the mRNA molecules to be broken down is a few hours (in prokaryotes the halflife of mRNA is only a few minutes).

Because mRNA is so short-lived, a given protein can only continue to be produced if there is a constant supply of new mRNA molecules from the nucleus. If transcription ceases, translation soon stops. Without this continuous breakdown of mRNA a cell would not be able to decrease production of a protein in response to a fall in demand.

The rapid turnover of mRNA is in marked contrast to the ribosomal RNA and tRNA, both of which last much longer.

Protein synthesis uses a lot of energy; for every amino acid incorporated into a protein, the equivalent of four ATP molecules are used up. To make a haemoglobin molecule a cell in the bone marrow needs to 'spend' about 2300 ATP molecules. To produce these the cell has to use 76 glucose molecules.

An interesting tailpiece to protein synthesis has been the discovery that the formation of peptide bonds is catalysed not by the protein part of the ribosome but by the RNA component. This is another example of RNA acting as an enzyme: a **ribozyme**. (See earlier in this chapter, the section on RNA processing.)

## *Post-translational processing*

After their production on the ribosomes, many proteins have to undergo further processing in the Golgi bodies before they are functional. This

**post-translational processing** may take the form of the removal of a section of the polypeptide. Insulin, for instance, undergoes two such cleavages with a single polypeptide being converted into two chains held together by disulphur links. Conjugated proteins (such as haemoglobin) must have the haem group added to the globin chains.

## Protein synthesis in prokaryotes

Protein synthesis in bacteria is fundamentally similar to that in eukaryotes, but there are some quite important differences. For example:

- in prokaryotes, genes are not 'split'. That is, there are no introns to be removed before translation can begin.
- in prokaryotes, translation and transcription occur simultaneously. The absence of a nuclear envelope means that a ribosome can begin moving along mRNA before the other end has been completed. This is possible because ribosomes move along the mRNA in the same (5' → 3') direction as mRNA is synthesised. In other words, ribosomes move 'downstream'.
- in prokaryotes, a group of adjacent genes is transcribed together, so the resulting mRNA molecule is a copy of several genes strung together. Though the mRNA copies remain together, they are translated individually.
- the ribosomes of prokaryotes (and also mitochondria and plastids) are slightly smaller than those of eukaryotes.

# Controlling the protein factory – regulating gene action

In a multicellular organism different kinds of cell are specialised for different functions – in a human, for example, there are up to 200 phenotypically different kinds of cell. Yet all these cells are produced by mitosis, which produces daughter cells with identical chromosome sets.

It could be argued that just because all body cells have the same chromosomes, it does not mean they have identical genes. Experiments have shown, however, that individual body cells contain the entire genetic programme of the organism. It is possible, for instance, to grow an entire carrot plant from a single root cell. Similar experiments have been performed on frogs, in which the nucleus of a fertilised egg was replaced by the nucleus of an intestinal cell from a tadpole. In some cases, eggs developed into frogs (Figure 11.9).

Experiments such as these prove that mitosis distributes identical sets of genes to the daughter cells. How, then, do cells become phenotypically different if they are genetically identical?

The answer must be that each cell only uses the small part of the genome which relates to its particular role. Thus connective tissue cells translate the gene for collagen but not the gene for insulin, and developing red blood corpuscles ignore the gene for making keratin but concentrate on translating the genes for haemoglobin.

It follows that during development, as cells differentiate to take on different functions, some genes must be active and others inactive. Development must, therefore, involve the orchestrated switching on and off of genes in an orderly sequence. For this to happen, cells must 'know' not only which proteins to make but when to make them.

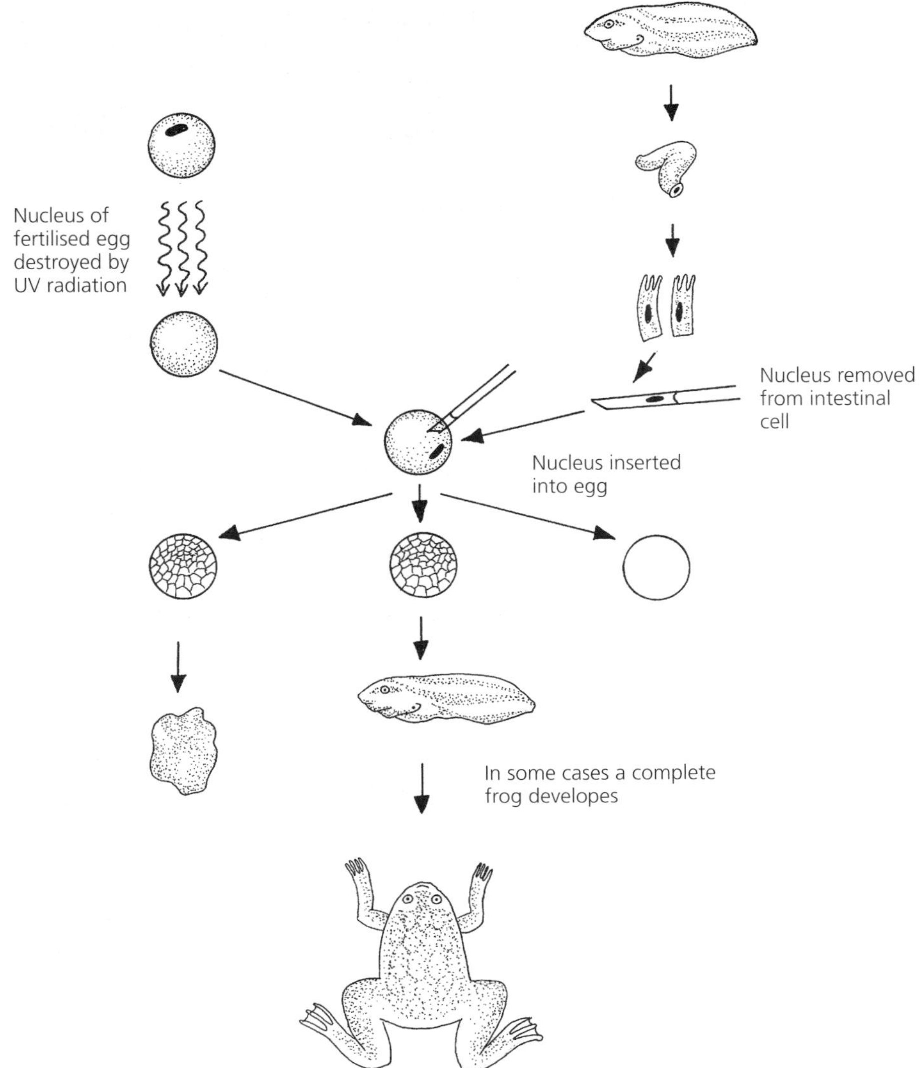


**Figure 11.9**
Evidence that body cells carry a complete set of genetic information

## Which came first – DNA or protein?

Proteins cannot be synthesised without the information in nucleic acid. In both prokaryotes and eukaryotes this is DNA. DNA cannot synthesise itself because it needs enzymes to link the nucleotides together. Thus DNA needs enzymes for its production and enzymes need DNA for their production.

Until recently it was believed that all enzymes are proteins. This leads us to a classic 'chicken and egg' problem: how could either of two interdependent chemicals have originated first? This seemingly insurmountable problem of how life originated has been solved – at least in principle – with two discoveries:

- in some viruses the genetic material, though consisting of RNA, can replicate itself.
- it has been found that some RNA molecules (called ribozymes) can act as enzymes. Examples are the catalysis of peptide bond formation by the RNA of ribosomes, and the removal of introns from mRNA by the mRNA itself.

RNA can thus carry out the two key functions of life: catalysis and self-replication. The first biomolecule may thus have been neither DNA nor protein, but RNA.

## How are genes regulated?

The ability of cells to switch genes on and off is clear enough, but the question of how it is achieved is another matter. The first insight into how it works in bacteria was provided by François Jacob and Jacques Monod in 1960, working with the bacterium *Escherichia coli*. Gene regulation in bacteria is quite different from that in eukaryotes, a factor that has to be taken into account when genetic engineers place eukaryotic genes into bacteria.

If *E. coli* is grown in a medium containing lactose, it produces the enzyme lactase, more properly known as **β-galactosidase**. This enzyme hydrolyses lactose to glucose and galactose which can then be used as fuel to supply energy. In a lactose-free medium containing glucose the bacteria produce almost no β-galactosidase, but within minutes of adding lactose, production of the enzyme begins. The advantage of this is obvious: by tailoring enzyme production to suit the circumstances, the bacteria save energy.

We can thus distinguish between two categories of enzymes:

- enzymes that are produced continuously under all conditions favourable to growth. Enzymes of this type are said to be **constitutive**, for example those catalysing the reactions of glycolysis.
- enzymes whose production depends on requirements – such as β-galactosidase.

Enzymes whose production can be switched on or off according to circumstances are of two general kinds:

- **inducible enzymes** – for example β-galactosidase. These are only produced in the presence of an **inducer**. Inducible enzymes are usually the raw materials for catabolic pathways, such as respiration.
- **repressible enzymes** – these are only produced in the *absence* of a particular substance, which is usually the end product of an anabolic pathway. For example, in an arginine-free environment, *E. coli* produces the amino acid for itself, but if arginine is added to the medium it stops making it.

For lactose to be used it must first enter the cell. Besides being able to produce β-galactosidase, *E. coli* can also make **lactose permease**, a membrane protein that enables lactose to enter the cell rapidly. Lactose permease production is also induced by lactose; when lactose is supplied to lactose-deficient *E. coli* it penetrates the cells very slowly at first, but within a few minutes it begins to enter rapidly.

Jacob and Monod set out to investigate the mechanism of enzyme induction. Crucial to this work were mutants which produced β-galactosidase **constitutively** – in other words, even in the absence of lactose.

An important feature of these constitutive mutants was that the β-galactosidase was indistinguishable from the enzyme produced by normal *E. coli* cells. This showed that the gene specifying the structure of the enzyme was quite different from the gene or genes controlling its rate of production. On this basis Jacob and Monod distinguished between two kinds of gene:

- **structural genes**, which specify the amino acid sequence of polypeptides. These are the ordinary genes encountered in traditional genetics. The genes coding for β-galactosidase and lactose permease are structural genes.
- **regulator genes**, which control the activity of structural genes.

One of the most interesting things about these mutants was that they produced both lactose permease and β-galactosidase constitutively. Evidently the production of β-galactosidase and lactose permease is controlled by the *same* switch mechanism (a third protein, transacetylase, of uncertain function, is also under the control of the same switch).

Another piece in the puzzle fell into place when Jacob and Monod mapped the position of the genes for β-galactosidase (the *z* gene) and lactose permease (the *y* gene). The genes turned out to be *next to each other* on the bacterial chromosome. Here, it seemed, was the physical basis for the common switch: the *z* and *y* genes and their switch mechanism were acting as a single unit. Jacob and Monod called this unit an **operon**.

## How are genes regulated? *continued*

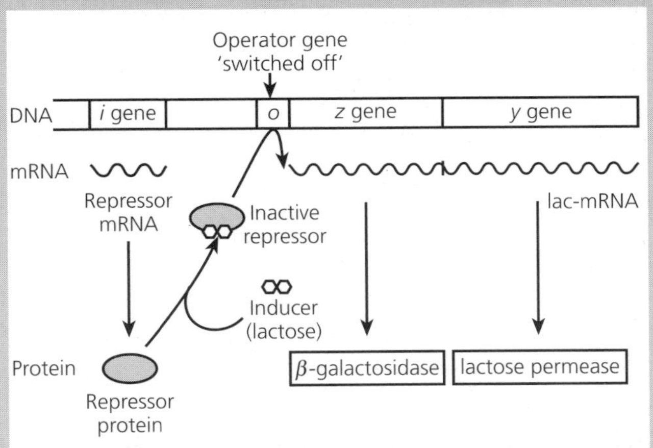


Lactose operon

As a result of these and other experiments, Jacob and Monod proposed that the switch mechanism controlling the lactose operon has two components:

- a section of DNA called the **operator** (symbol *o*). This is immediately upstream of the *z* gene, next to the promoter.
- a regulator gene (symbol *i*) which produces an allosteric protein called the **repressor.**

In the absence of lactose the repressor binds to the operator. This prevents RNA polymerase from binding to the promoter, preventing transcription. If lactose is present, it combines with the allosteric site of the repressor. This changes its shape sufficiently to prevent it from combining with the operator, leaving it in the 'on' position.

With one small modification, Jacob and Monod's scheme can account for enzymic repression in bacteria. Take the case of arginine production by *E. coli*, mentioned earlier. If arginine is present it combines with the repressor protein, just as lactose does. In this case, however, combination with arginine *enables* the repressor to combine with the operator (rather than *preventing* it as in the case of enzyme induction).

The operon hypothesis accounts very nicely for the regulation of enzyme production in prokaryotes. Eukaryotes are another matter. Unlike prokaryotes, in which genes involved in a metabolic pathway may be adjacent on the chromosome, in eukaryotes they may be scattered on different chromosomes, and thus cannot be controlled in the same way as in bacteria. Progress is being made in this area but the mechanisms are complex and beyond the scope of this book.

## Genetic switches in humans

Human babies normally produce the enzyme lactase which digests lactose, or milk sugar. In most human populations production of the enzyme usually ceases before adolescence, with the result that lactose cannot be digested. When it reaches the large intestine it is fermented by the intestinal bacteria causing diarrhoea and vomiting, the symptoms of **lactose intolerance**. Amongst Europeans (and some other populations with a long history of milk consumption) a high percentage of individuals are lactose-tolerant because lactase is produced throughout life.

What happens is that in most humans the gene for lactase production is switched off during childhood. The switching off of the lactase gene is brought about by an autosomal recessive gene. In populations with a long history of association with cattle (and therefore milk-drinking), the switching off of the lactase gene is disadvantageous, and in most individuals the genetic switch does not operate.

Another example of a genetic switch in humans is the change in haemoglobin production which occurs at birth. Before birth the baby produces foetal haemoglobin, which consists of two α chains and two γ chains. At birth the developing red cells begin to produce adult haemoglobin, in which each γ chain is replaced by a β chain. In some way, the gene for γ globin must be switched off and that for β globin switched on.

## The central dogma

In outline, the mechanism by which the information encoded in DNA is used to make proteins can be summed up as 'DNA makes RNA and RNA makes proteins', or

$$\text{DNA} \rightarrow \text{RNA} \rightarrow \text{protein}$$

The DNA constitutes the genotype and the proteins make up the phenotype. This relationship between DNA, RNA and protein was called the **central dogma** of molecular biology by Crick. He uses the word 'dogma' because at that time there was no concrete evidence to support it.

By stating that information can flow from DNA to protein but not in the reverse direction, Crick was saying that genes provided the information to build the body, but that the body cannot directly influence the plan that helped to build it. This assertion goes to the heart of the way modern biologists envisage the mechanism of evolution. The neo-Darwinian view is that, although changes in DNA are random, a small proportion of such changes are, by chance, advantageous. These 'better' genotypes help build bodies which are more successful because they produce more offspring and therefore more copies of the changed DNA.

A modern follower of Lamarck, on the other hand, would say that when the body (and therefore proteins) adjusts to changing conditions, these changes are inherited. In other words, changes in the body result in changes in the genes. This would be equivalent to saying that changes in the amino acid sequences of proteins can lead to changes in the base sequence of DNA. Not only is there no evidence of this, it is difficult to imagine how it could happen.

Since Crick stated the central dogma, it has had to be modified as a result of two discoveries:

- in some viruses RNA is used as a template for making more RNA, so the RNA must contain the information for its own replication, just like DNA.
- in retroviruses, viral RNA is injected into the host cell and is used as a template to make *DNA*. Later, this viral DNA is used to make viral RNA. In these viruses, then, information in RNA is used to make DNA.

The possible pathways of information flow can thus be written:

$$\text{DNA} \underset{\textit{information}}{\overset{\textit{information}}{\rightleftharpoons}} \text{RNA} \xrightarrow{\textit{information}} \text{protein}$$

The essential feature of the central dogma can thus be restated:

> "*Information can flow from nucleic acid to protein, or from one kind of nucleic acid to another, but not from protein to nucleic acid or from protein to protein*".

*Point to note:* The control of gene activity, as outlined for the lactose operon, is not a breach of the central dogma, nor does it necessitate its modification. This is because although repressor proteins affect DNA, they only affect the rate at which it is transcribed. They have no influence on its information content (i.e. its base sequence).

## QUESTIONS

**1** A molecule of human haemoglobin consists of two $\alpha$ chains, each of 141 amino acids, and two $\beta$ chains, each of 146 amino acids.

**a)** How many kinds of mRNA are needed for a cell to make haemoglobin?

**b)** Including the start and stop codons, how many nucleotides would be needed to specify the structure of human haemoglobin?

**2** If there were only two kinds of base in DNA, one purine and one pyrimidine, how many bases would be needed to specify each of the 20 amino acids found in proteins?

**3** Use the table of the genetic code (Figure 11.4) to determine the sequence of amino acids that is specified by the following base sequence in mRNA:

(5′ end) U U G A A C C C A G G U (3′ end)

## QUESTIONS

**4** In some viruses, the rule that DNA is double-stranded and RNA is single-stranded does not apply. The table shows the relative numbers of molecules of each kind of base extracted from the nucleic acids from five different sources, including viruses, bacteria and eukaryotes. For each, state whether the nucleic acid is DNA or RNA, and whether it is single-stranded or double-stranded.

| Source | A | G | T | C | U |
|---|---|---|---|---|---|
| A | 32 | 18 | 32 | 18 | |
| B | 34 | 16 | | 16 | 34 |
| C | 19 | 31 | 19 | 31 | |
| D | 30 | 26 | 22 | 22 | |
| E | 22 | 24 | | 24 | 30 |

**5** Each of the following is a property of one or more of the following: DNA, mRNA, tRNA, rRNA

Excluding bacteria and viruses, name the nucleic acid(s) to which each of the following applies:

**a)** single-stranded
**b)** contains uracil
**c)** combined with protein
**d)** contains paired bases
**e)** has a short life-span
**f)** produced in the nucleus

**6** True or false?

**a)** A mutation is a change in the genetic code.
**b)** A change in a DNA triplet may not result in a change in protein structure.
**c)** An mRNA molecule can be involved in the production of only one kind of polypeptide, but tRNA molecules and ribosomes are involved in the production of more than one kind of polypeptide.

**7** When we say that the genetic code is 'universal', we mean that:

**A** It is the same for all organisms
**B** It is the same for all proteins made by an organism
**C** It does not change throughout the life of the organism
**D** DNA contains the same four bases in all species
**E** In all species, proteins are built of different sequences of the same 20 amino acids

**8** The pores in the nuclear envelope carry a constant stream of molecular traffic. For each of the substances listed below write 'O' if you expect it to move *out* of the nucleus, 'I' if you expect it to move *into* the nucleus from the cytoplasm, and 'N' if you expect it to do *neither*.

**a)** DNA
**b)** mRNA
**c)** tRNA
**d)** rRNA alone
**e)** ATP
**f)** completed ribosomal subunits
**g)** ribosomal proteins
**h)** histones
**i)** DNA polymerase
**j)** RNase
**k)** deoxyribonucleoside triphosphates
**l)** inorganic phosphate

**9** Provided they are given the necessary raw materials and other essential conditions, bacterial cells that have been broken up can still synthesise DNA, RNA and proteins. For each of the treatments listed in the table at the bottom of the page, write the letter of the most probable consequence from the list **A–F** below. Note that amanitin (the toxin of the death cap fungus) binds irreversibly with RNA polymerase, and diphtheria toxin prevents the movement of ribosomes along mRNA.

| | Treatment | | Symptom |
|---|---|---|---|
| 1 | Uracil deprivation | A | Protein synthesis stops immediately, DNA synthesis continues |
| 2 | Thymine deprivation | B | DNA synthesis stops, RNA and protein synthesis continue |
| 3 | RNase added | C | Synthesis of DNA, RNA and proteins stops immediately |
| 4 | Diphtheria toxin added | D | RNA synthesis stops, protein synthesis continues for a few hours |
| 5 | Amanitin added | E | Protein synthesis stops, RNA synthesis continues |
| 6 | ATP deprivation | F | No effect |

CHAPTER 12

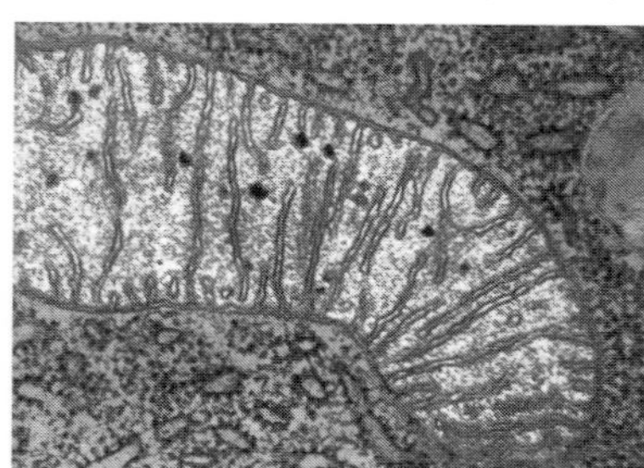

# CHANGING the MESSAGE – MUTATION

## LEARNING OBJECTIVES

By the time you have completed your study of this chapter you should be able to:

- explain what is meant by mutation, and how it differs from recombination.
- explain what is meant by the statement that gene mutations are random, and explain the evolutionary significance of this.
- explain why most gene mutations are harmful in most environments.
- list three other characteristics of gene mutations.
- distinguish between base substitutions and frame shifts, and compare their possible effects on the structure of the protein product.
- give examples of various kinds of mutagen.
- distinguish between chromosomal translocations, inversions, deletions and duplications, and compare their biological effects.
- define aneuploidy and give a human example.
- explain how the various kinds of aneuploidy can arise.
- explain the difference between autopolyploidy and allopolyploidy and give an example of each.
- explain why autotriploids are highly infertile.
- explain why hybrids between different species are usually sterile.
- explain what is meant by an amphidiploid and how it can be produced.

In Chapter 8 we saw how independent assortment and crossing over lead to genetic variation within a species. When it comes to differences between species it is another matter. There is a limit to the range of form that can occur within a species; it seems that no combination of chimpanzee genes can produce a gorilla, and vice versa. Since the two species retain their distinctness, gorillas must have at least some alleles that chimpanzees do not have. It needs more than the reshuffling of existing alleles to explain how one species can give rise to two or more species. Evolution must, therefore, involve the production of new alleles as well as the forging of new combinations of existing ones.

A similar argument applies to entire chromosomes and to chromosome sets. Closely-related species often have different chromosome numbers. Even when they have the same number, individual chromosomes often differ in size and in the position of the centromere. This means that chromosomes must also be able to change during evolution in both structure and number.

These genetic changes are called **mutations** (from the Latin *mutare* = to change), and the organisms showing them are called **mutants**. Mutations resemble recombination in one key respect – they are *random* processes. There is, however, an important difference – mutation is relatively rare, so the products tend to persist over many generations. Changes by recombination, on the other hand, occur every sexual generation.

A mutation, then, is a sudden and persistent change in the hereditary material. Mutations are of three general kinds:

- **gene mutations**, which affect single genes.
- **block mutations**, which affect the distribution and arrangement of genes on chromosomes.
- **changes in chromosome number.**

## Gene mutations

Otherwise known as **point mutations**, gene mutations involve changes to individual genes, as a result of which one allele changes to another.

### Characteristics of gene mutations

Gene mutations have a number of biologically important characteristics, discussed below.

#### *Mutations are random*

Randomness in this sense means that mutations have no relation to need. Although environmental agents can influence the *rate* of mutation, the environment has no influence on which genes mutate or in what way they mutate. Thus ultraviolet radiation increases mutation rate, but it does not preferentially cause changes to those genes controlling the amount of melanin in the skin. All the environment does is to speed up the rate of a random process – it has no effect on its direction.

### Evidence that mutation is not purposeful – replica plating

**When an antibiotic is added to a pure culture of bacteria, a tiny proportion of mutants survive because they are resistant to the antibiotic. The question is, does the mutation occur in *response* to the antibiotic, or does the antibiotic simply reveal a characteristic that was already present in some individuals?**

**This is an important question to which modern supporters of Darwin and Lamarck would give quite different answers. A Lamarckian would opt for the first explanation; a Darwinian would prefer the second.**

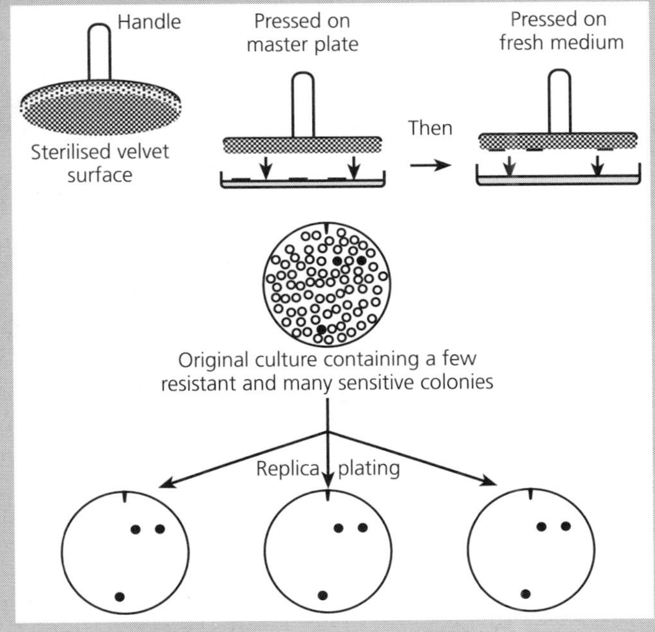


Use of replica plating to show that advantageous mutations occur independently of the need for them (after Hayes)

## Evidence that mutation is not purposeful – replica plating *continued*

**The difficulty in distinguishing between these two hypotheses is this: we don't know whether a bacterium is resistant until *after* it has been treated with the antibiotic.**

**Joshua and Esther Lederberg found an ingenious solution to this problem, using a technique called replica plating. Instead of using resistance to an antibiotic, they used resistance of *E. coli* to a bacteriophage, but the principle is the same. They mounted a piece of sterile velvet onto a wooden disc the same size as a Petri dish. They gently placed the velvet onto an agar plate containing millions of bacteria, pressing just hard enough for some of them to adhere to the velvet. Next, they pressed the velvet onto a series of sterile agar plates. Finally, they added a suspension of the bacteriophage to the replicas.**

**On each replica, only a few colonies of resistant bacteria developed. The important thing was that the distribution of these colonies was identical in all the replicas. If resistance had developed *after* addition of the bacteriophage, one would not expect it to occur in bacteria occupying identical positions on the agar. Their positions were the same and this was because the velvet had picked up bacteria that were *already resistant*.**

### *Most are harmful*

Under most environmental conditions mutations are disadvantageous. This is to be expected. Since most existing alleles have 'passed the test' of natural selection, almost any random change is more likely to be harmful than beneficial (imagine a random change to a clockwork motor).

It is important to realise, however, that a mutation can be harmful in some circumstances but beneficial in others. For example, many animals living in deep, underground caves lack the skin pigment melanin. Above ground albinism is disadvantageous, but in constant darkness energy is saved by not producing a chemical that has no protective value.

Besides depending on the external environment, the effect of an allele may also depend on the 'genetic environment' – in other words, the other genes present. For example, in malarial areas the sickle cell allele is beneficial in heterozygotes (Chapter 14).

### *Most are recessive*

In diploid organisms most mutant alleles are **recessive**. As long as a mutant allele is rare it will be present almost entirely in the heterozygous state and will not be expressed. In haploid organisms, of course, all alleles are expressed because there is no such thing as dominance.

### *Mutations are rare*

The statement that mutations are rare needs to be qualified. If we consider a particular gene locus, then in any given generation it is extremely unlikely to mutate. In *Drosophila*, a new mutation at a particular locus is typically present in about 1 in 30 000 to 1 in 50 000 gametes. However, if we consider the tens of thousands of genes present in an organism, a mutation at *some* locus is quite likely. What this means is that *most* gametes will carry one or more new mutations.

Mutation rates differ from one locus to another; some loci mutate hundreds of times more frequently than others.

### *They can occur in any cell at any time*

Only **germ-line mutations** (those that occur in the gametes or cells that give rise to them) can be handed on to the next generation. Mutations in body cells are called **somatic mutations** and cannot be passed on to future generations (though they can cause cancer).

### *Mutations are reversible*

Mutations can occur in both directions (though usually with different frequencies).

## Types of gene mutation

A gene mutation is a change in the base sequence in the DNA. There are two fundamentally different kinds: **base substitutions** and **frame shifts**.

### *Base substitutions*

In a base substitution, one base *replaces* another. The effects range from zero to lethal, as Figure 12.1 shows in summary form.

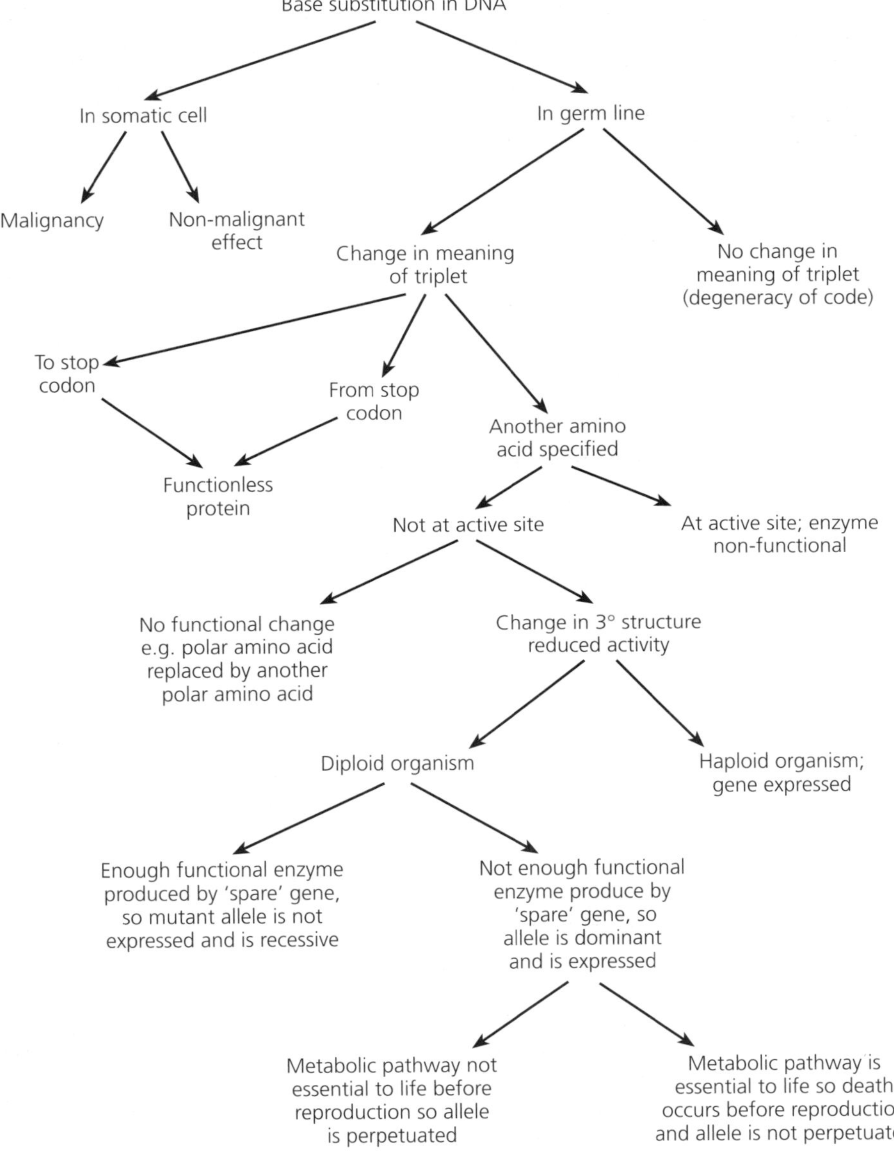


**Figure 12.1**
Possible effects of a base substitution in a gene coding for an enzyme

A well-known example of a base substitution is the mutation responsible for sickle cell haemoglobin. As a result of this mutation, amino acid #6 in the β chain of haemoglobin is changed from glutamic acid to valine. The mRNA codons for glutamic acid are GAA and GAG, and the codons for valine are GUU, GUC, GUA and GUG. The mutation must, therefore, have resulted in a change from GAA to GUA or GAG to GUG.

## Himalayan albinism and Siamese cats

Himalayan rabbits have white fur except for the ears and feet, which are black. In these areas the temperature is several degrees lower than that of the rest of the body. If the animals are kept in a very warm environment the new hair that grows is white. If, on the other hand, hair is shaved from the back and the skin kept cool by an ice pack, the new hair growth is black. Siamese cats provide another example of temperature-dependent albinism. Evidently, one of the enzymes involved in melanin production is less stable to heat but functions normally at lower temperatures.

The fact that the enzyme is normal at lower temperatures suggests that the amino acids occupying the active site are unchanged by the mutation. The effect must lie with those amino acids that help to maintain the specific shape of the enzyme. Evidently the forces holding different parts of the peptide chain together are weaker, so that even at body temperature the enzyme is denatured. At lower temperatures these weaker forces are sufficient to keep the enzyme in shape.

Temperature-sensitive mutant

### *Frame shifts*

**Frame shifts** are caused by the insertion or deletion of a base. The result is a change in the **reading frame**, so every triplet 'downstream' of the mutation is changed. Consider, for example, the following sentence (with spaces inserted between triplets to make easier reading):

THE CAT ATE THE BIG FAT RAT

By adding 'S' after CAT it destroys the sense thereafter:

THE CAT SAT ETH EBI GFA TRA T

The deletion of a base would be equally catastrophic:

THC ATA TET HEB IGF ATR AT

Unless a frame shift occurs near the 3' ('downstream') end of a gene, it will almost certainly lead to the production of a completely functionless protein. However, a second frame shift of the opposite kind to the first will restore the reading frame. Combining the above two 'mutations' we have:

THC ATS ATE THE BIG FAT RAT

The greater the distance between an insertion and a deletion, the greater the chance that the change in the reading frame will produce a 'stop' codon. Since three out of the 64 triplets signal 'stop', there is a $\frac{3}{64}$ chance that any given triplet will be transformed into a stop codon. On the other hand, a stop codon could change to a codon specifying an amino acid (what effect would that have on the protein?)

If an insertion and a deletion are close enough, the number of changed amino acids may be small enough for the protein to be functional. Most back-mutations are probably of this kind (though an exact reversal of the original mutation will be extremely rare, as will be the exact reversal of a base substitution).

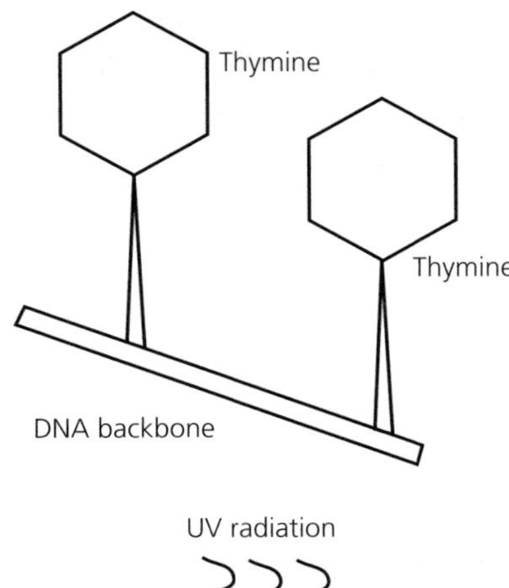


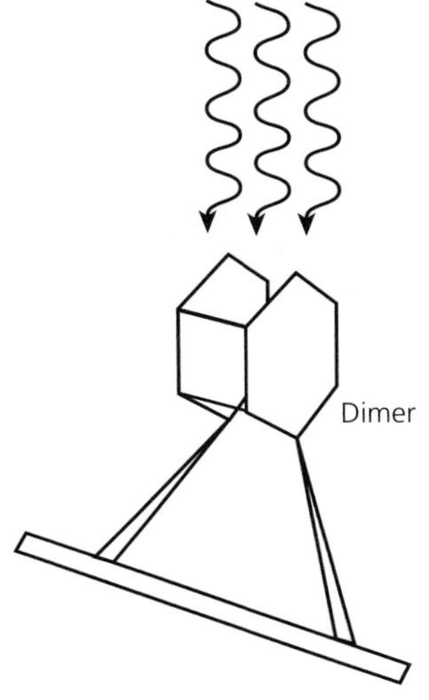


**Figure 12.2**
Formation of a thymine dimer by UV radiation

## How gene mutations are produced

Though gene mutations may be spontaneous (without apparent external cause), most are induced by external agents called **mutagens**. These are of three kinds: radiation, chemicals and viruses.

### *Radiation*

X-rays, γ-rays, α-rays, β-rays and neutrons are called **ionising radiation**. This is because they contain so much energy they can break covalent bonds to produce ions and other highly-reactive particles. These are so reactive that they combine with almost any organic molecule with which they come into contact, changing it chemically. If this happens to be DNA the result could be a mutation. A particle of ionising radiation, therefore, does not have to collide with DNA – it may damage it indirectly by way of the reactive particles it creates.

Ultraviolet light, on the other hand, specifically damages DNA. Where two thymine bases occur next to each other on one of the two DNA strands, they become linked to form a **thymine dimer** (Figure 12.2). This may interfere with the replication in the next division causing a mutation that may result in skin cancer.

### *Chemicals*

A large number of chemicals are known to be mutagenic, for example:

- **nitrous acid** deaminates adenine to hypoxanthine (which pairs like guanine), and cytosine to uracil (which pairs like thymine).
- **base analogues** such as 5-bromouracil and 2-amino purine are so-called because they are sufficiently like normal DNA bases to be incorporated into replicating DNA, but have slightly different pairing properties.
- **acridine dyes** become lodged in the DNA helix and interfere with replication, causing an extra base to be inserted – i.e. a frame shift.

Many other mutagens are known, for example formaldehyde and mustard gas.

### *Viruses*

Some viruses are known to cause cancer, which results from a somatic mutation. The hepatitis B virus, for instance, produces liver cancer in some people many years after the original infection.

## No safe dose

The probability of a gene mutation occurring is directly proportional to the size of the dose of mutagen, as each mutation results from a single chemical event. Even at 'background' or natural radiation levels some mutations still occur. A medical X-ray is 'safe' only because the risks are trivial in comparison with the risks of not diagnosing what could be a serious condition.

## Why uracil in RNA?

Very occasionally cytosine spontaneously changes to uracil. In RNA this does not matter much because each RNA molecule is only one of many copies. Such a change would be much more serious in DNA, since there is only one copy (two in homozygotes). This kind of mutation is prevented by an enzyme which regularly moves along the DNA 'checking' for the

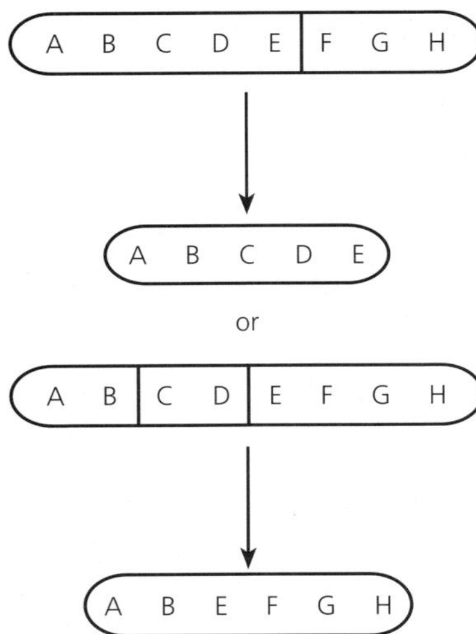


**Figure 12.3▲**
A chromosomal deletion

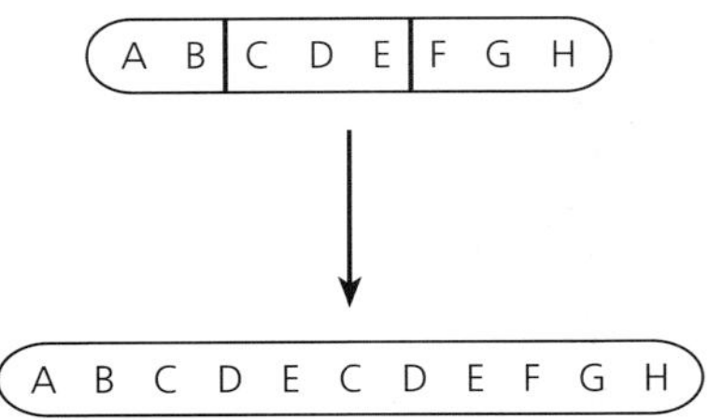


**Figure 12.4▲**
A chromosomal duplication

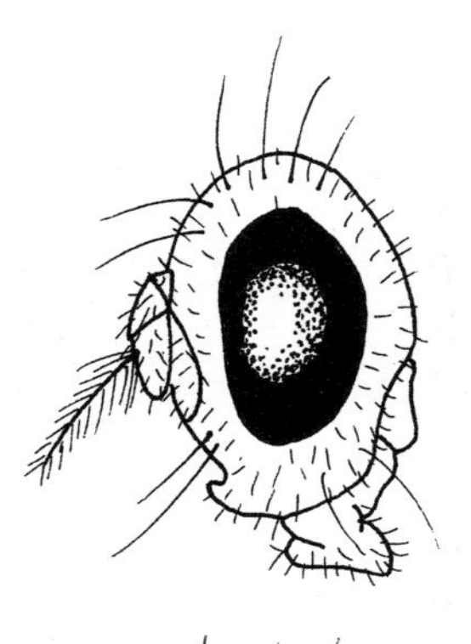

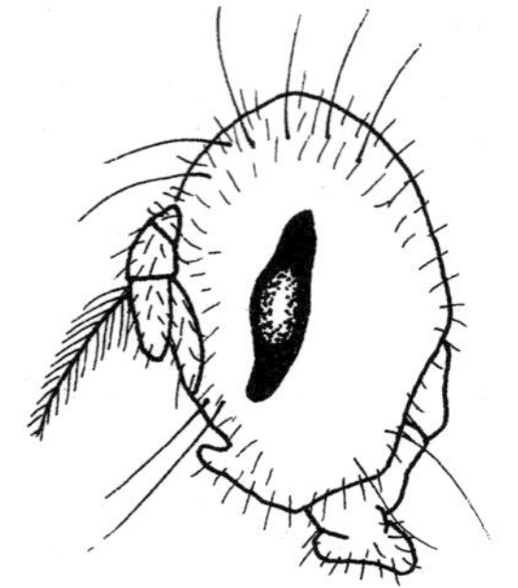

**Figure 12.5▲**
The bar-eye condition in *Drosophila*

presence of uracil. Any uracil encountered is removed and cytosine is substituted. If uracil was found naturally in DNA, the cell would not be able to distinguish between 'bona fide' uracil and uracil produced by mutation.

### Cancer as a mutation

In some cases mutation can lead to cancer. This is a group of diseases in which the normal controls of cell division break down. As a result cells multiply in unrestrained fashion and may spread to other parts of the body. Cell division and the ability of cells to retain their positions in the body are under genetic control. A cell becomes a cancer cell when these controls break down. A key event in the formation of a cancer cell is the mutation of a **proto-oncogene** to form an **oncogene**. Over 20 different oncogenes have been discovered. Proto-oncogenes are normal genes that play an essential part in regulating cell behaviour such as division.

## Changes in chromosome structure

Otherwise known as **block mutations**, these affect many genes and produce changes that may be visible during meiosis. As a result of radiation and chemicals such as mustard gas, chromosomes may be broken. Although cells have enzymes that can repair such breaks, chromosomes can still undergo permanent change, for two reasons. First, a break is not always repaired, and second, if two breaks occur, the wrong ends may be rejoined.

As a result of a structural change, a chromosome will no longer be able to pair with its partner along its entire length during meiosis. Such mispairings can sometimes be seen down the microscope (in contrast to gene mutations, which can only be detected by their effects on the phenotype).

Structural changes in chromosomes fall into five categories: **deletions**, **duplications**, **inversions**, **translocations** and **fusions**.

### *Deletions*

Here, a section of a chromosome becomes broken off (Figure 12.3). Deletions are often fatal, even when heterozygous, since they usually involve the loss of many genes. This can lead to the expression of recessive and possibly harmful traits. Deletions can sometimes be detected when homologous chromosomes pair up in meiosis. Since the chromosome without the deletion is longer than its homologue, it may show up as a loop in the bivalent.

A human example of a deletion is the 'cri-du-chat' syndrome, caused by the loss of the short arm of chromosome #5. It is so-called because of the cat-like cry of the affected child, who is not only severely mentally retarded but also suffers physical deformities and has a short life expectancy.

### *Duplications*

A duplication is a repeated section of a chromosome (Figure 12.4). Duplications are generally less serious than deletions since there is no loss of genes, but some duplications result in abnormal phenotypes.

An example of a duplication is the bar-eye condition in *Drosophila*, caused by a duplicated section of the X-chromosome (Figure 12.5). The quantitative balance between genes (or rather the relative amounts of their protein products) must, therefore, be an important factor influencing the phenotype.

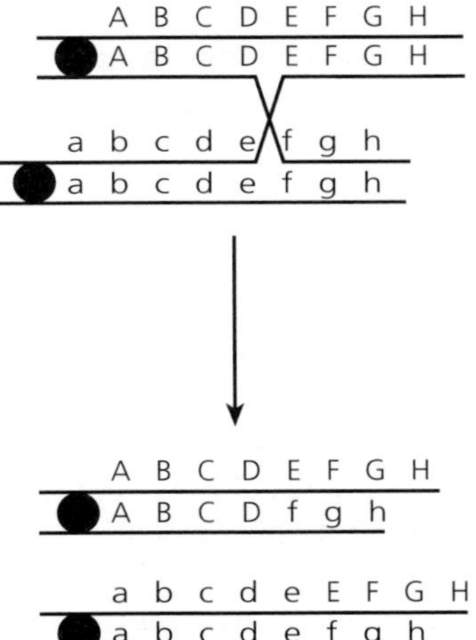


**Figure 12.6▲**
Formation of a deletion and a duplication by unequal crossing over

Duplications and deletions probably arise when the two breaks that occur in crossing over are not quite opposite each other (Figure 12.6). For every duplication produced in this way there will be a deletion in one of the chromatids of the homologous chromosome.

Duplications are probably important in evolution because the two loci can evolve independently of each other and take on different functions. There are several proteins whose amino acid sequences show that they are descended from a common ancestor and have probably originated by duplication. For example, myoglobin and the α and β chains of haemoglobin can be traced to a common ancestral protein.

## Inversions

If a chromosome is broken in two places and repaired 'the wrong way round', an inversion occurs (Figure 12.7). Since two breaks must occur, two radiation 'hits' are needed, and the second must occur before the first break has been repaired.

Although there is no loss of DNA, inversions can cause problems during meiosis in organisms heterozygous for the inversion. When chromosomes pair during meiosis I, the only way that loci can match up is by forming loops (Figure 12.8).

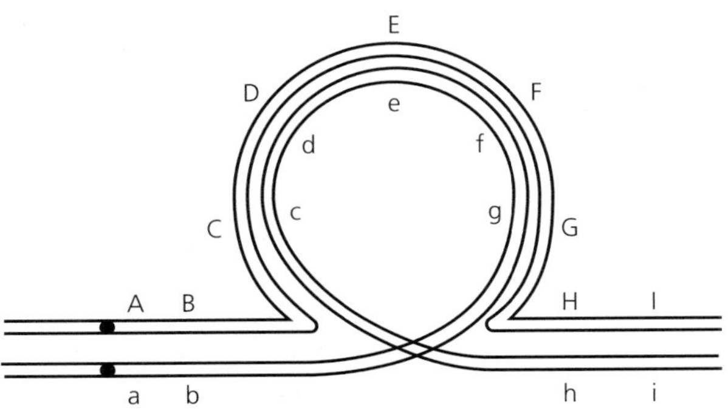


**Figure 12.8►**
An inversion loop with a crossover

This is not, however, the problem. The difficulties arise if a crossover occurs within the inverted segment. When the chromosomes separate at anaphase I, only two of the four chromatids have the full complement of genes. In Figure 12.8 one chromatid will have a 'double dose' of the *A*/*a* and *B*/*b* loci and two centromeres, and will lack *F*/*f*, *G*/*g* and *H*/*h*. The other chromatid will lack a centromere and will lack *A*/*a* and *B*/*b*, and will have a double dose of *F*/*f*, *G*/*g* and *H*/*h* (see if you can work out why).

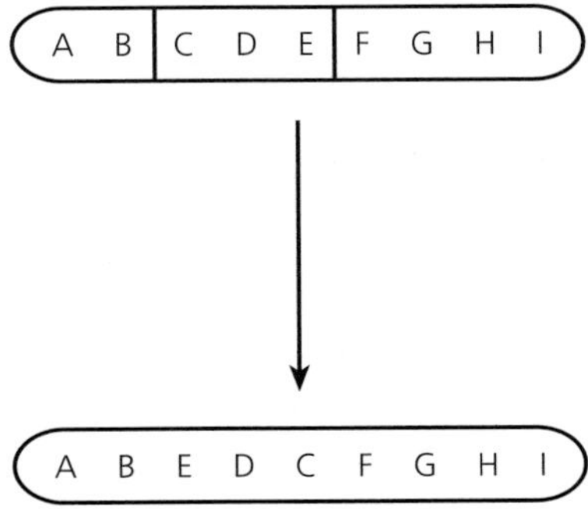


**Figure 12.7▲**
Formation of a chromosomal inversion

## Translocations

When breaks occur in non-homologous chromosomes the ends of the non-partner chromosomes may rejoin (Figure 12.9).

As in the case of inversions, translocations result in a relocation of the genetic material rather than a change in quantity. The problems arise in translocation heterozygotes, in which only one member of each of the two chromosome pairs has the translocation. As with inversions, some of the

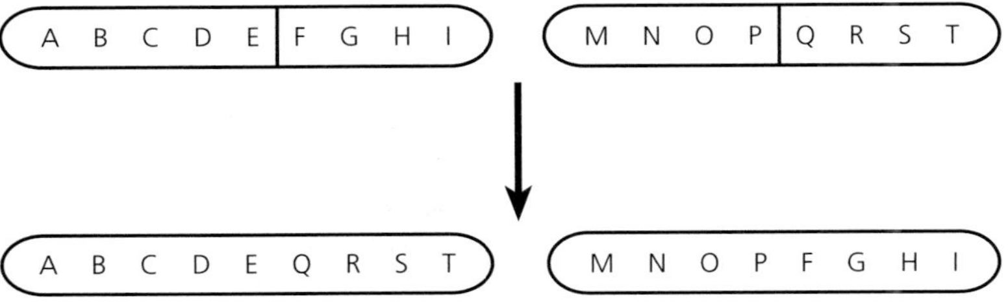


**Figure 12.9►**
Formation of a chromosomal translocation

products of meiosis are inviable. Despite this, translocations are widespread in nature and have played a part in the evolution of many species, including humans. Clearly, there must be advantages as well as disadvantages. There are two possibilities:

- genes may be brought into more favourable positions relative to each other.
- new linkage groups will be established, thus restricting recombination within these gene combinations.

### Fusions

If two non-homologous chromosomes become broken, they may rejoin to form a single, larger chromosome. If breaks occur near centromeres close to the ends of the chromosomes, few genes may be lost (Figure 12.10).

Chromosome fusions have been important in our own evolution. Humans have 23 pairs of chromosomes, whilst chimpanzees, gorillas and orangutans have 24 pairs. Comparison of banding patterns shows that two of the chromosomes of the great apes have fused to form chromosome #2 of humans.

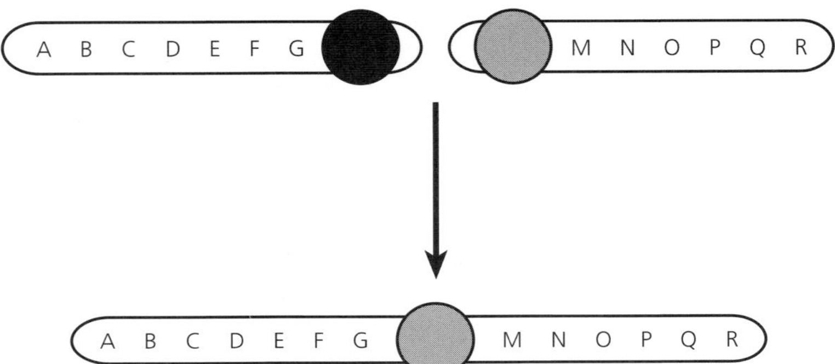


**Figure 12.10**
Fusion of non-homologous chromosomes

## Changes in chromosome number

Although uncommon in animals, numerical changes in chromosomes have been important in the evolution of over a third of flowering plant species. To begin with, some new terms will be needed, and the meaning of some established ones extended slightly:

Variations in chromosome number are of two kinds:

- **aneuploidy**, which involves only part of the chromosome set, with certain chromosomes being represented an unusual number of times.
- **euploidy**, in which every chromosome is present an unusual number of times.

### Aneuploidy

The simplest kind of aneuploidy occurs when only one type of chromosome is involved. The various possibilities are named according to the number of times that chromosome is represented. Remembering that the normal diploid set is $2n$, we have:

- **monosomy** – only one is present instead of the usual two, giving a total of $2n - 1$ chromosomes.
- **trisomy** – three copies ($2n + 1$)
- **tetrasomy** – four copies ($2n + 2$)

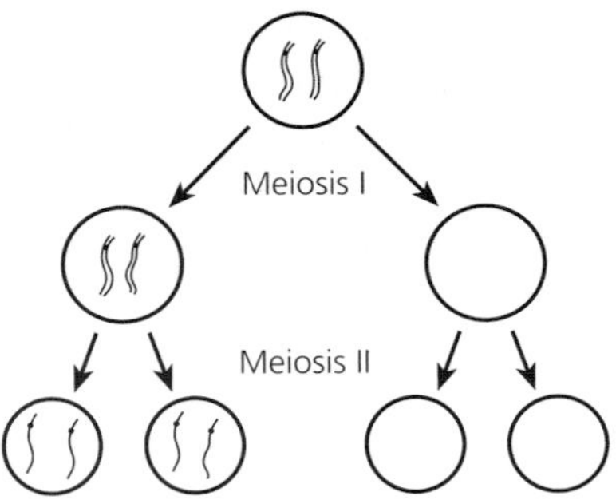


**Figure 12.11**
Origin of aneuploidy by non-disjunction in meiosis I. Only one pair of chromosomes is shown

Aneuploidy is more common in plants than in animals. In the latter the effects are almost always harmful. Since it involves loss of an entire chromosome, monosomy is more serious than trisomy. The severity of the effects also depends on which chromosome is involved; loss of a large chromosome (carrying more genes) would be expected to be more serious than the loss of a small one.

Aneuploidy results from the failure of chromosomes to separate or **disjoin** during cell division. This is called **non-disjunction** and can occur during meiosis or mitosis. In meiosis I, instead of moving to opposite poles of the spindle, homologous chromosomes move to the *same* pole. In the second meiotic division and in mitosis, it is the chromatids which fail to disjoin. As shown in Figure 12.11, one daughter cell receives both chromosomes and can be represented by $(n + 1)$ whilst the other receives neither and can be represented as $(n - 1)$.

When a normal gamete $(n)$ joins with a gamete with $(n + 1)$ chromosomes, the result is a trisomic. If a normal gamete joins with an $(n - 1)$ gamete, a monosomic results. Fertilisation between two $(n + 1)$ gametes would produce a tetrasomic $(2n + 2)$. Joining of two $(n - 1)$ gametes would produce a nullisomic $(2n - 2)$, but these are very rare.

A monosomic carries only one allele for all the loci on the affected chromosome. As a result any recessive alleles will be expressed. Harmful recessive alleles will not be protected by dominant alleles. For this reason, monosomy would be expected to be more serious in the larger chromosomes, since more genes are involved. Some human examples of aneuploidy are given below.

### Turner's syndrome

Turner's syndrome is monosomy for the X chromosome, and occurs in 1 in 2500 births. The person is phenotypically female but with undeveloped sex organs, small stature and a characteristically 'webbed' neck. The person tends to score below average in spatial and mathematical tests, though normally in verbal tests.

### Klinefelter's syndrome

Klinefelter's syndrome occurs in about 1 in 400 births, and originates by fusion of an XX egg and a Y sperm, or an X egg and an XY sperm. Affected individuals have male genitalia but tend to have a somewhat female body shape which can be treated with testosterone. IQ and life expectancy are normal.

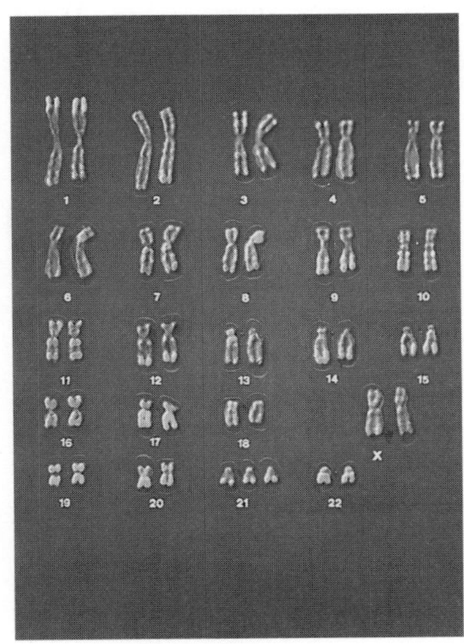

**Figure 12.12**
The karyotype of a female suffering from Down's syndrome showing trisomy of chromosome number 21

### Down's syndrome

Down's syndrome is trisomy of chromosome 21 (Figure 12.12), and is associated with certain facial features, reduced IQ and increased susceptibility to disease. The overall frequency of Down's syndrome is about 1 in 650. One of its most striking features is the fact that its frequency rises as the age of the mother rises, particularly after age 40. Evidently, the probability of non-disjunction increases with the age of the potential egg. The cause of this maternal effect is not known, but it may have something to do with the very long time the egg spends in prophase I in mammals (Chapter 7).

Non-disjunction probably occurs with equal frequency in all chromosomes. The reason why it only seems to affect certain chromosomes may be that these are the only cases that survive. It may be no coincidence that chromosome 21 is one of the smallest chromosomes (containing the fewest genes), and that other cases of trisomy in humans also involve smaller chromosomes (Edwards' syndrome, #18, and Patau's syndrome, #13).

In humans most cases of aneuploidy involve the sex chromosomes. This may be connected with the fact that in males, all X-linked alleles are expressed and are subjected to selection. As a result the frequency of harmful X-linked recessive alleles is probably lower than that of autosomal recessive alleles. Harmful autosomal recessive alleles are protected from selection by dominant alleles.

## Euploidy

In euploidy, the genome (the entire complement of genetic material) is represented three or more times (polyploidy) or, in the case of monoploidy, once. Since monoploids are rare, almost all variation in ploidy is due to polyploidy. There are two kinds of polyploidy: **autopolyploidy** and **allopolyploidy**.

### *Autopolyploidy*

Autopolyploidy results from the multiplication of the entire genome *within a single species*. Thus, an autotriploid has three sets of chromosomes and an autotetraploid has four sets.

Autopolyploidy is caused by the failure of the spindle to function properly. As a result, all the chromosomes finish up in the same nucleus, which will have twice as many chromosomes as it should have. It can happen in either of two ways:

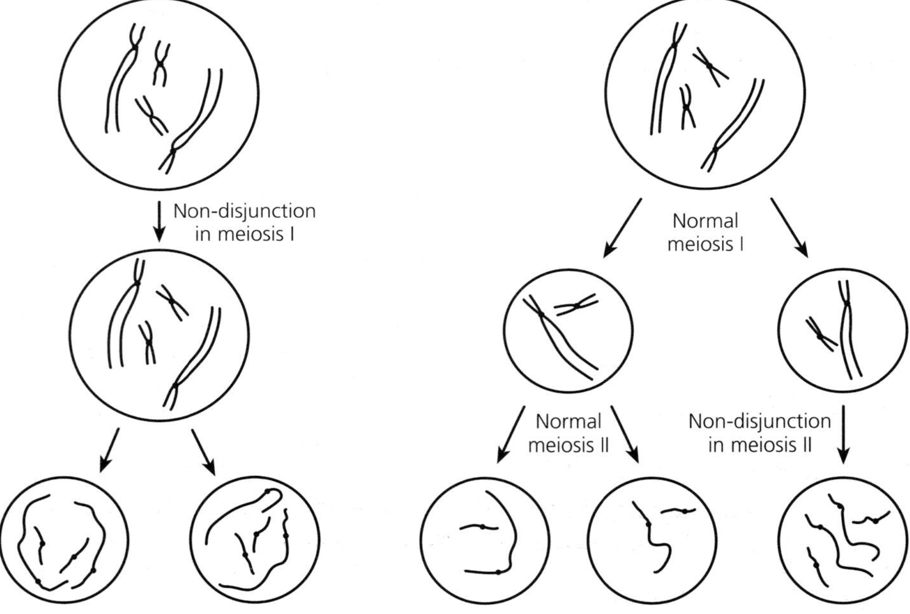


**Figure 12.13►**
Chromosome doubling as a result of non-disjunction in meiosis

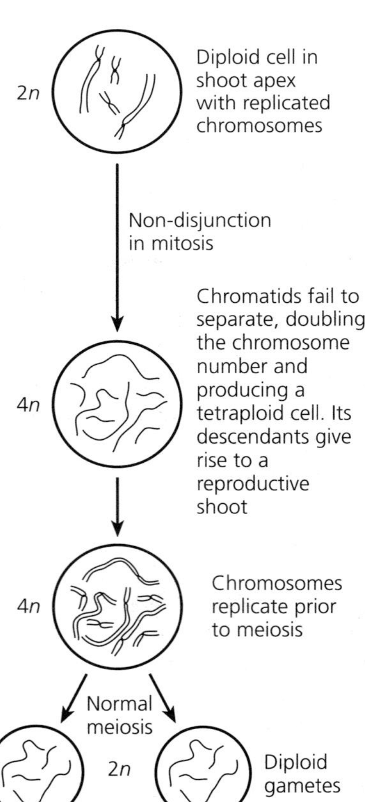


**Figure 12.14▲**
Chromosome doubling as a result of abnormal mitosis

- in meiosis. Whether it occurs in the first or the second division, the normal halving of the chromosome number fails to occur (Figure 12.13). The result is two diploid nuclei, which later give rise to diploid gametes. If a diploid gamete fuses with a normal haploid gamete, a triploid zygote is produced. If it fuses with another diploid gamete, the resulting zygote is tetraploid.
- in mitosis, resulting in a tetraploid cell (Figure 12.14). The descendants of this cell will also be tetraploid. This is called **somatic doubling**, and if it occurs in a shoot tip, it may give rise to a tetraploid shoot. If the shoot later produces flowers, tetraploid cells in the stamens and ovules undergo meiosis to produce diploid pollen and eggs.

Although chromosome doubling can occur quite naturally, it can also be induced artificially by **colchicine** (extracted from *Colchicum*, the autumn crocus), which prevents the formation of the spindle. When the drug is removed, normal cell division resumes.

### Why are autotriploids infertile?

Although autopolyploids are often vigorous and may be larger than diploids, they are usually less fertile. Autotriploids are especially infertile because of the very low probability that meiosis will produce cells with balanced chromosome sets. This is why cultivated bananas (which are triploid) are infertile and therefore seedless. Dandelions are triploid, with $3n = 24$. With three homologues, meiosis I results in each daughter nucleus receiving either one or two of each chromosome trio. The probability that a daughter nucleus will receive eight chromosomes (one of each kind) is $(\frac{1}{2})^8$, or $\frac{1}{256}$. Similarly, the probability that it will receive 16 (two of each kind) is also $\frac{1}{256}$. Only these two kinds of pollen grain have balanced chromosomes sets, so only $\frac{2}{256} = \frac{1}{128}$ of the pollen grains and eggs would be viable. But, of course, to produce a viable zygote, both pollen and egg must be viable, so the probability of producing a zygote with two of each kind of chromosome would be $(\frac{1}{128})^2$, or about 1 in 16 000. No wonder dandelions are virtually sterile. It is just as well, then, that they are **parthenogenetic**, the eggs developing without fertilisation.

Autotetraploids tend to be more fertile than autotriploids. With four homologues, each daughter nucleus can receive two chromosomes. Such 2 + 2 segregation often occurs, so a much higher proportion of pollen and eggs receive balanced chromosome sets.

### *Allopolyploidy*

Allopolyploidy has probably been more important in evolution than autopolyploidy. It results from hybridisation between species. The first direct proof that polyploids can arise in this way was obtained by Karpechenko in the 1920s. He crossed radish (*Raphanus sativus*, $2n = 18$) and cabbage (*Brassica oleracea*, $2n = 18$). The hybrid was sterile, as expected, since each chromosome had no homologue with which it could pair at meiosis. Normal chromosome segregation could not, therefore, occur and no viable pollen or eggs could be formed (Figure 12.15).

Eventually, however, some viable seeds were obtained. Chromosome doubling had occurred within the sterile hybrid producing somatic cells with 36 chromosomes. Each chromosome now had a partner which was not only homologous, but identical, since it had arisen by chromosome duplication.

The fully fertile hybrid (called *Raphanobrassica*) had two sets of cabbage chromosomes and two sets of radish chromosomes and is called an **amphidiploid** (*amphi* = Greek for 'both'). Numerous other amphidiploids have been produced artificially. The method involves treating shoots with colchicine to induce chromosome doubling, thus converting a sterile hybrid into a fertile amphidiploid.

The best known example of an allopolyploid is cord grass (*Spartina townsendii*). First observed in 1870 in Southampton water, it has since spread rapidly and is now widely distributed round British and nearby European coasts. It originated by hybridisation between a British species, *S. maritima* ($2n = 60$), and an introduced American species, *S. alterniflora* ($2n = 62$). Some of the hybrids have 61 chromosomes and are sterile, whilst others – the amphidiploids – have 122 and are fertile.

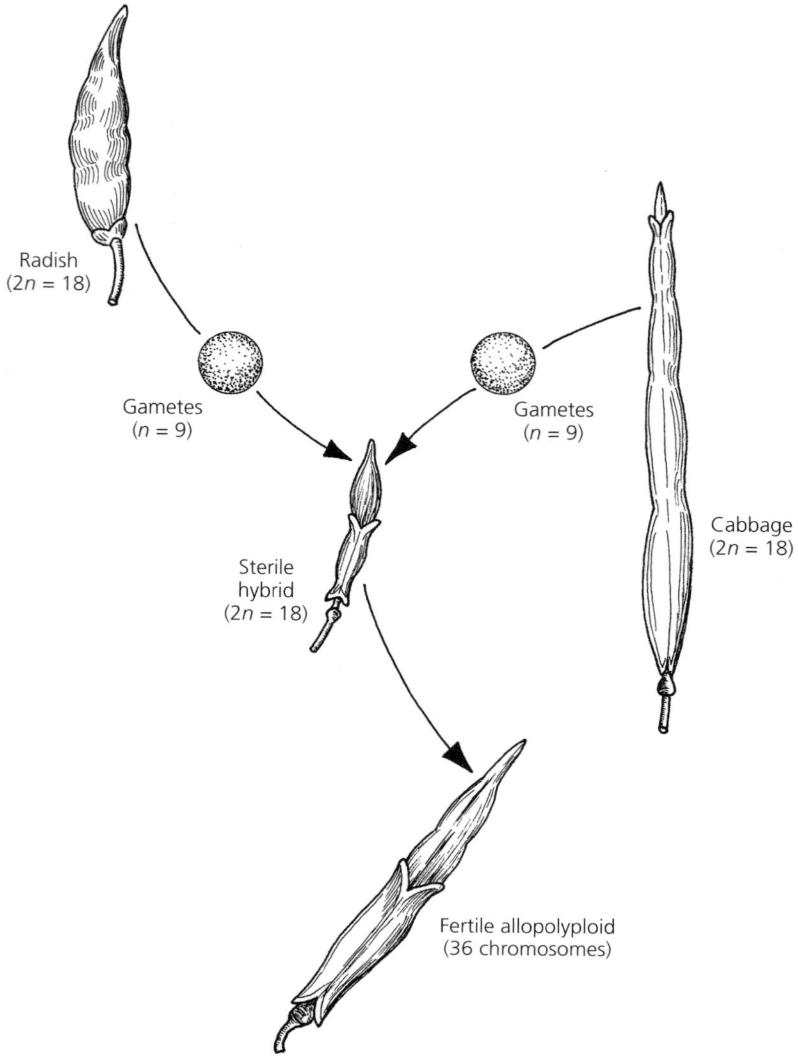

Radish ($2n = 18$)

Gametes ($n = 9$)

Gametes ($n = 9$)

Cabbage ($2n = 18$)

Sterile hybrid ($2n = 18$)

Fertile allopolyploid (36 chromosomes)

**Figure 12.15**
Formation of *Raphanobrassica*

Many other plants are suspected to have had an allopolyploid history, but this often has to be deduced by comparison with present day species. One example is the modern wheat *Triticum aestivum* ($2n = 42$). Its probable ancestry is shown in Figure 12.16.

Modern wheat is an allohexaploid with three ancestors, each of which has contributed 14 chromosomes. Thus the haploid number ($n = 21$) is made up of three groups of 7 (indicated by $x$, the *monoploid* number). In wheat, therefore, haploid and monoploid are not the same. Many other cultivated plants have an allopolyploid ancestry, for example:

Cotton, $2n = 4x = 52$, an allotetraploid
Loganberry, $2n = 6x = 42$, an allohexaploid
Strawberry, $2n = 8x = 56$, an allo-octaploid

## *The effects of polyploidy in plants*

Besides fertility, polyploidy has other effects on plants. With more chromosomes to accommodate, the cell nuclei are obviously larger, as are the cells as a whole. As a result, the plants themselves are often larger and sometimes more vigorous.

The frequency of polyploid plant species increases as one travels from the equator towards the poles. For example, 37% of flowering plant species are polyploid in Sicily, while in southern Greenland over 70% are

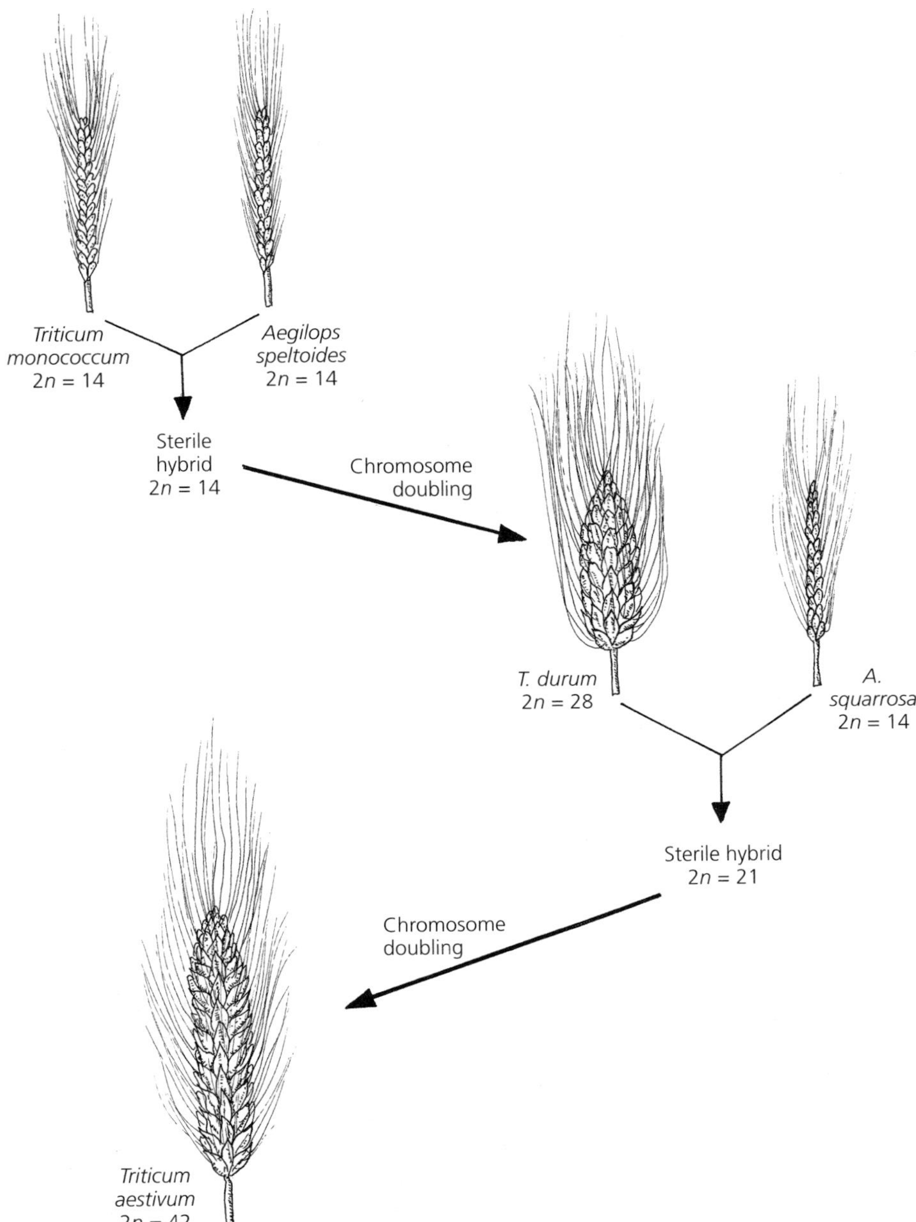


**Figure 12.16**
Possible origin of modern wheat

polyploid. A similar gradient exists with altitude. Like other aspects of genetic variation, the advantages or disadvantages of polyploidy depend on the environment.

### Polyploidy in animals

In animals polyploidy is far less common than in plants. It occurs in a number of animals that reproduce parthenogenetically (without fertilisation) such as some crustaceans and some leeches. The evidence for its harmfulness (in humans at any rate) is clear from the high proportion of polyploids among miscarried foetuses, about 10% of these being triploids.

The explanation for the relative rarity of polyploidy in animals is uncertain. It may have something to do with the mechanism of sex-determination, which, in many animals, is under the control of special sex-chromosomes.

## QUESTIONS

**1** Match each of the terms **(i)–(x)** to one of the descriptions **A–J**.

**i)** karyotype
**ii)** homologues
**iii)** allopolyploid
**iv)** autopolyploid
**v)** aneuploidy
**vi)** non-disjunction
**vii)** euploidy
**viii)** amphidiploid
**ix)** genome
**x)** colchicine

**A** produced by increasing chromosome number within a species
**B** variation in chromosome number involving the entire genome
**C** failure of chromosomes to separate during cell division
**D** an organism's chromosomal characteristics
**E** produced by chromosome doubling of sterile hybrid
**F** has chromosomes derived from more than one ancestral species
**G** used to artificially increase chromosome number
**H** the entire genetic complement of an organism
**I** members of a pair of chromosomes
**J** change in chromosome number involving only part of a chromosome set

**2** 'Most mutations are harmful'. Rewrite this statement in a more acceptable form, explaining why it should be criticised.

**3** What is meant by saying that mutations are *random*?

**4** In what kind of organisms is mutation the *only* source of genetic variation?

**5** *Chlamydomonas* is a single-celled organism which spends the active part of its life cycle in the haploid state. Why would you expect mutant alleles to be rarer in a population of *Chlamydomonas* than in a diploid species?

**6** Normal individuals of a species of plant have a diploid number of four. Which of the diagrams **A–E** could show cells of individuals that are

**a)** diploid
**b)** triploid
**c)** monosomic
**d)** trisomic
**e)** nullisomic?

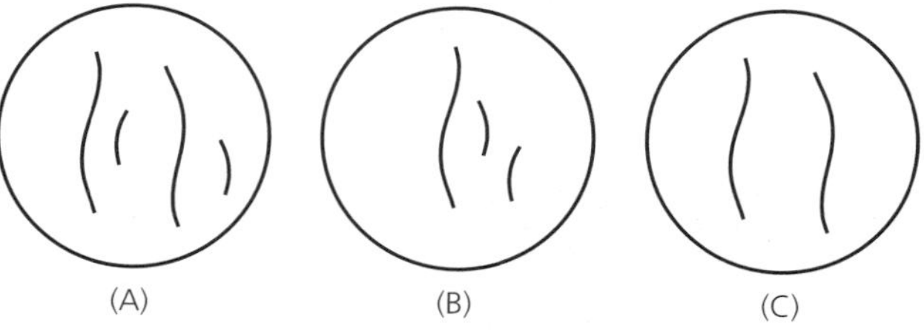


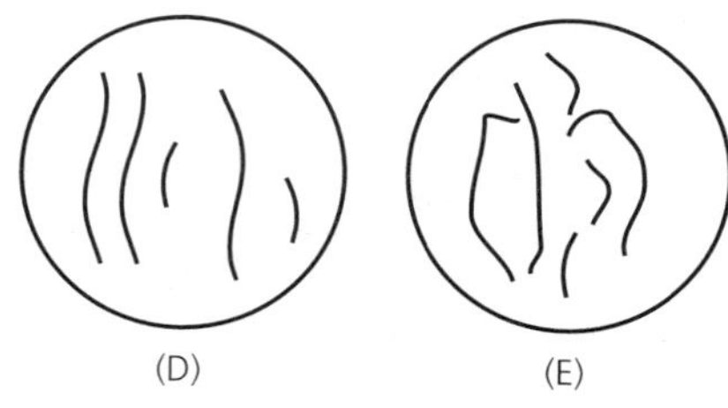


**7** The diagram shows how one kind of polyploidy may originate. *A* and *B* each represent a chromosome *set*.

**a)** Why is the hybrid *AB* sterile?
**b)** If the two original parent plants have six and eight chromosomes, respectively, how many chromosomes would the *AABB* fertile hybrid have?

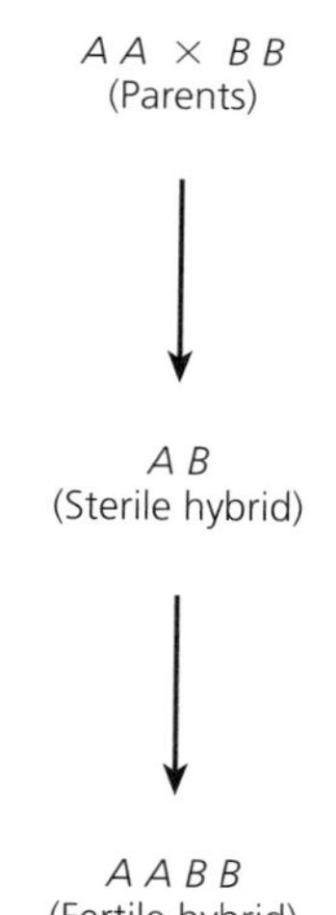

CHAPTER

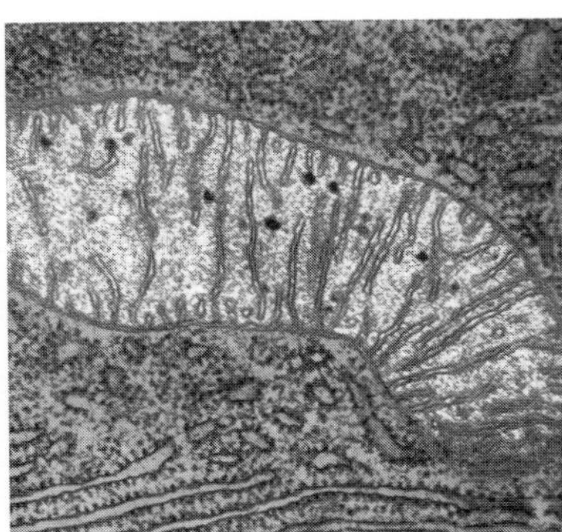

# 13

# PUTTING GENES to WORK: GENE TECHNOLOGY

## LEARNING OBJECTIVES

By the time you have completed your study of this chapter you should be able to:

- explain the use of restriction endonucleases, DNA ligase, reverse transcriptase and terminal transferase in genetic engineering.
- explain the use of plasmids and bacteriophage as 'carriers' of DNA.
- explain how fragments of DNA can be separated.
- explain how specific fragments of DNA can be inserted into a plasmid or bacteriophage.
- describe how a gene can be cloned.
- explain the use of a gene probe and Southern blotting technique to find a specific gene.
- briefly discuss some of the problems in expressing eukaryotic genes in prokaryotes.
- describe the use of *Agrobacterium* to insert genes into cells of flowering plants.
- describe ways of inserting genes into animal cells.
- describe the uses to which gene technology can be put.
- discuss some of the commonly perceived drawbacks in gene technology.

## What is gene technology?

Until the origin of agriculture all genetic change was the result of two processes: the production of random variation by recombination and mutation, and selection of the 'best' genes by the environment. About 10 000 years ago, humans began to bring about genetic change in the animals and plants they domesticated.

At first, the process of selection would probably have been subconscious. For example, less docile animals would have been more troublesome and so would have been more likely to be used for food than kept for breeding purposes. The most easily harvested grain would come from those plants in which the seed heads were most strongly attached to the stem. Seeds that became detached easily would have escaped the harvest and so would not have been used for next year's harvest.

At some stage plants and animals would have been deliberately chosen for breeding because of their desirable qualities. By selective breeding, the genetic make-up of many species has been progressively altered under domestication. These traditional methods of plant and animal breeding had three major limitations:

1 the production of 'better' varieties was largely a matter of luck, since recombination and mutation are both random processes.

2 gene pools of different species are normally isolated from each other, so geneticists were limited to crossing varieties of the same species, or, in some cases, closely-related species.

3 it takes much longer to produce new varieties by selective breeding.

The explosion of molecular genetics has changed all that. Ever since it was discovered that the genetic code is universal, the transplantation of genes between unrelated organisms has been a possibility. Within the last two decades this has become a reality.

This entirely new technology is called **genetic engineering**, and is becoming a major branch of **biotechnology**, the industrial use of biological processes. DNA produced by artificially combining genes from different organisms is called **recombinant DNA**, and organisms carrying genes that have been transplanted from other species are said to be **transgenic**.

The applications of gene technology go far beyond the improvement of agricultural yields. Here are a few examples to illustrate the diversity of uses to which DNA technology is now being put.

- Medically-important human proteins that are normally produced in minute quantities can be grown on an industrial scale by transplanting the respective genes into bacteria. Until recently, for example, diabetics had to rely on insulin extracted from cattle and pigs. Amino acid sequences of beef, pig and human insulin are not quite the same, and some diabetics develop antibodies to the foreign insulin, so much of the benefit is lost. Blood products, such as the clotting protein factor VIII, can be produced by bacteria, eliminating the danger of contamination by viruses such as HIV and hepatitis through blood transfusions. Other proteins that have been produced by genetically-engineered bacteria include human growth hormone, epidermal growth factor, interferons (antiviral proteins) and erythropoietin (a protein that stimulates red blood cell production).

- Antiviral vaccines can now be produced without risk. Traditionally these have been prepared using a live, but weakened (attenuated), virus. This has always carried a slight danger if the virus has not been completely inactivated. The part of the virus that stimulates antibody production is the protein coat, so a vaccine consisting only of the protein is just as effective and carries no risk. Viral coat proteins can be made by bacteria into which the genes for viral coat proteins have been transplanted. A vaccine against hepatitis B has already been produced.

- Transgenic animals and plants. Genes can be transplanted into domesticated animals and plants in order to boost their performance.

  Transgenic animals and plants can be used to make products quite alien to them. For example, melanin, serum albumin and antibodies have all been produced in transgenic plants. Transgenic rabbits have been used to produce the protein interleukin-2 in their milk.

- It is now possible to identify, with an extremely high level of certainty, a person from whom a small blood (or other tissue) sample is available. This application, called **genetic fingerprinting**, is of enormous importance to forensic science and has played a key part in a number of murder trials.

- **gene therapy**. The base sequences of individual genes can be determined, making it possible to diagnose exactly why a given gene, such as the gene causing cystic fibrosis, is faulty. This paves the way for the eventual replacement of faulty genes by normal ones.
- by comparing DNA sequences in present day populations it is possible to learn much about their history.

## A genetic engineer's mini-dictionary

| | |
|---|---|
| **Agarose gel** | jelly-like material used to separate DNA fragments of different sizes. |
| ***Agrobacterium*** | common soil bacterium used to insert plasmids into flowering plant cells. |
| **Anneal** | joining of complementary nucleotide sequences by controlled lowering of temperature. |
| **Bacterial transformation** | process by which bacteria can take up 'foreign' naked DNA; promoted by addition of calcium ions. |
| **Bacteriophage** | virus that attacks bacteria; used as a vehicle for replicating DNA inside bacteria, i.e. a **vector**. |
| **Biolistics** | insertion of DNA by firing minute DNA-coated particles of tungsten or gold into cells. |
| **cDNA** | DNA produced by copying RNA. |
| **Clone** | a large number of genetically-identical offspring produced by the copying of an original type. |
| **Cloning vector** | plasmid or bacteriophage DNA, into which foreign DNA is inserted for replication. |
| **DNA ligase** | enzyme that joins fragments of DNA. |
| **Electroporation** | technique of inserting DNA into animal cells or plant protoplasts by using electric pulses to create minute, transient holes in the plasma membrane. |
| **Gene cloning** | process by which multiple copies of a gene are produced in a bacterium or bacteriophage. |
| **Gene library** | the entire genome of an organism, distributed as DNA fragments in a large number of bacteria or bacteriophage. |
| **Gene probe** | a segment of single-stranded nucleic acid, used to find a complementary sequence of DNA. |
| **Genetic fingerprinting** | technique by which fragments of DNA in individuals are compared to establish relationships between them. |
| **Genome** | collective name for all the genes of an organism. |
| **Linker** | short, artificially synthesised section of single-stranded DNA used to make complementary ('sticky') ends to DNA fragments so that they can be joined up. |
| **Plaque** | a clear area in a dense population of bacteria, resulting from lysis (breakdown) of bacteria after infection by bacteriophage. |
| **Plasmid** | a small, independently-replicating bacterial mini-chromosome bearing genes for resistance to antibiotics and used as a carrier for replicating DNA. |
| **Polymerase chain reaction** | technique whereby a minute quantity of DNA can be replicated many times in the test tube. |
| **Recombinant DNA** | produced by *in vitro* (in the test tube) splicing of DNA from two different sources. |
| **Restriction endonuclease** | bacterial enzyme that cuts DNA at specific sites. |
| **Reverse transcriptase** | enzyme used to make DNA from RNA template. |

**A genetic engineer's mini-dictionary** ***continued***

| | | | |
|---|---|---|---|
| **Southern blotting** | technique for hybridising DNA fragments with a probe. | **Transgenic organism** | organism that contains artificially-introduced genes from another species. |
| **Terminal transferase** | enzyme used to add short nucleotide sequences to DNA. | **Vector** | a virus or a bacterium into which DNA is inserted for replication. |

## Steps in transplanting a gene

Transplanting a gene involves the following basic steps:

- the DNA responsible for a particular characteristic is identified and isolated from other DNA.
- large numbers of copies of the gene or genes are prepared by a process called **gene cloning**.
- the gene is inserted into the organism in which the gene is to be expressed.

A number of key discoveries during recent decades have made it possible to extract, isolate and transplant genes from one organism to another. These tools include enzymes that can be used for cutting, splicing and copying.

### *Restriction endonucleases*

A **restriction endonuclease** is part of an enzymic defence system by which a bacterium defends itself against bacteriophage. For this to work, the restriction endonuclease must be able to distinguish 'self' DNA from 'alien' DNA. In a way to be described later, the bacterium does this by 'labelling' its own DNA.

When a bacteriophage injects its DNA into a resistant bacterium, the viral DNA is recognised as foreign because it is unlabelled. The restriction enzyme cuts the phage DNA whenever it encounters a specific base sequence four, six or eight base pairs in length. The longer the recognition sequence, the less often it will occur by chance (a particular sequence of six bases, for example, would be encountered by chance every $4^6 = 4096$ bases).

Hundreds of restriction enzymes have been identified and extracted. Each is named according to the species and strain of bacterium from which it is extracted. Thus *Eco*RI denotes an endonuclease extracted from a strain of *E. coli*.

Although the recognition sequences vary from one restriction enzyme to another, most of them are **palindromic**, meaning that they read the same in both directions. An oft quoted example is 'Able was I ere I saw Elba.' Figure 13.1 shows the sequences recognised by some commonly used restriction enzymes, the arrows indicating the points at which the DNA is cut.

Restriction enzymes can cut DNA in two ways:

- some produce staggered cuts when the cleavage points in the two strands are not opposite each other, such as *Eco*RI (Figure 13.2). The two ends are thus **sticky** because they can base-pair with complementary ends. Any two pieces of DNA cut with the same

| Micro-organism | Enzyme abbreviation | Recognition sequence |
|---|---|---|
| *Escherichia coli* | *Eco*RI | 5′. . . G↑A A T T C . . . 3′<br>3′. . . C T T A A G . . . 5′ |
| *Hemophilus influenzae* | *Hin*dIII | 5′. . . A↑A G C T T . . . 3′<br>3′. . . T T C G A A . . . 5′ |
| *Hemophilus aegyptius* | *Hae*III | 5′. . . G G↑C C . . . 3′<br>3′. . . C C G G . . . 5′ |
| *Bacillus amyloliquefaciens* | *Bam*I | 5′. . . G↑G A T C C . . . 3′<br>3′. . . C C T A G G . . . 5′ |
| *Serratia marcescens* | *Sma*I | 5′. . . C C C↑G G G . . . 3′<br>3′. . . G G G C C C . . . 5′ |

**Figure 13.1**
Base sequences recognised by some commonly used restriction enzymes

restriction enzyme will, therefore, have sticky ends that are complementary to each other. This is the basis of how human DNA can be spliced into bacterial DNA; the same restriction enzyme is used to make the cut in the human and the bacterial DNA.

The process of sticking complementary ends together by base-pairing is called **annealing**. The adhesion is, however, weak since it is due to hydrogen bonding only. To convert it into a strong, covalent bond requires another enzyme, **DNA ligase**.

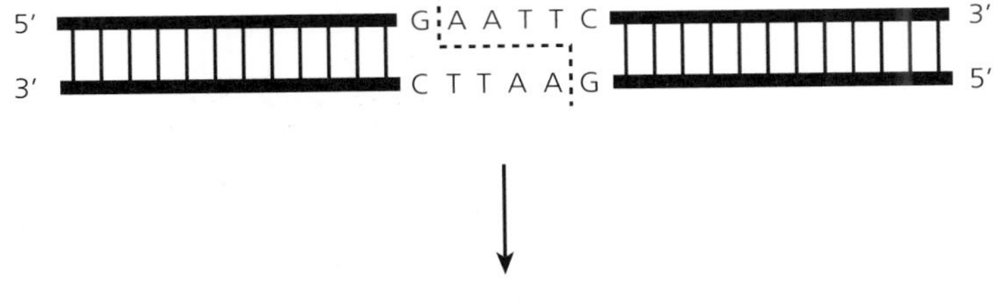

**Figure 13.2**
Production of sticky-ended DNA fragments by *Eco*RI

- some produce blunt ends, such as *Hae*III produced by *Hemophilus aegyptius* (Figure 13.3). These can be made 'sticky' by adding 'linkers' in a way to be described later.

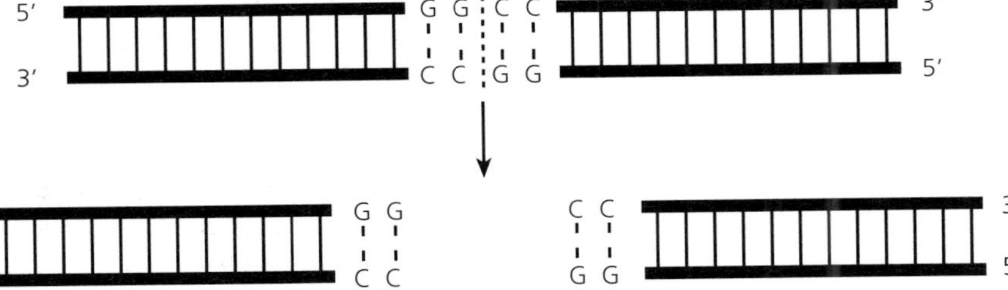


**Figure 13.3**
Production of flush-ended DNA fragments by *Hae*III

How do bacteria label their own DNA? This is where another enzyme, called a **modification enzyme**, comes in. Every time the bacterial DNA replicates, the modification enzyme labels it by adding methyl groups to certain bases. The enzyme only does this within a specific base sequence – *the same sequence, in fact, that the restriction enzyme recognises* (Figure 13.4). Since it is unlabelled, phage DNA is cut up by the restriction enzyme.

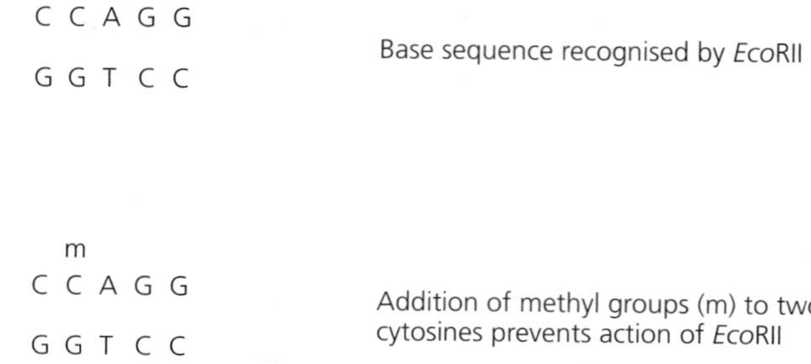


**Figure 13.4**
Protection of bacterial DNA by methylation

The fewer the bases in the recognition sequence, the more often that sequence will occur. Hence, the more bases that have to be methylated by the modification enzyme and the more fragments into which any alien DNA will be cut. With a recognition sequence of more than eight bases there would be a danger that an alien viral genome might not contain that sequence at all and so would not be dealt with.

## DNA ligase

This enzyme is used to splice together pieces of DNA from different sources (Figure 13.5). Its normal role is to join Okazaki fragments together during DNA replication (see Chapter 10), and also the repair of DNA.

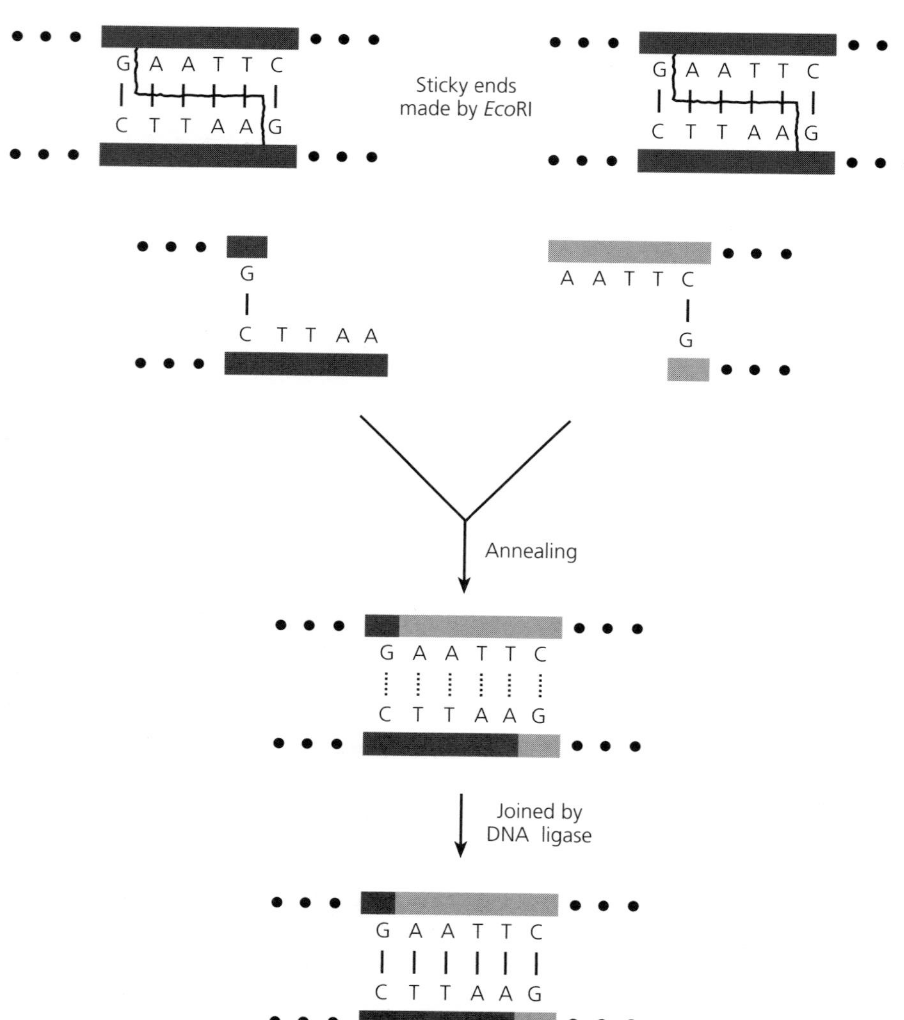


**Figure 13.5**
Use of DNA ligase to join sections of DNA with sticky ends (after Watson, *et al.*)

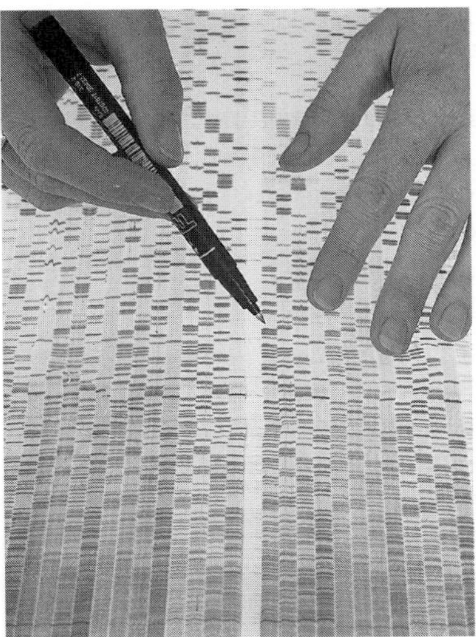

**Figure 13.6**
Separation of DNA fragments using an agarose gel

### *Reverse transcriptase*

Reverse transcriptase is obtained from **retroviruses** (Chapter 11). It is so called because it catalyses the copying of RNA into DNA, with the RNA acting as a template – the reverse of what normally happens. DNA made on an RNA template is called **complementary DNA** or **cDNA**. Reverse transcriptase makes it possible to synthesise a gene from an mRNA copy.

Why do retroviruses need reverse transcriptase? Unlike other RNA viruses, the RNA of retroviruses is single-stranded. Its replication obviously poses a problem since the host cell lacks an enzyme to catalyse RNA synthesis using RNA as a template. This is the role of reverse transcriptase, which is injected alongside the viral RNA.

## Separating DNA fragments

DNA molecules are so long that they can easily be broken up into small fragments, either mechanically or chemically using restriction enzymes. The resulting fragments (of which there may be up to a thousand or more) can be separated by **gel electrophoresis** (Chapter 2). The DNA mixture is placed on agarose gel and subjected to an electric field. DNA fragments are negatively charged (due to the phosphate groups), and move towards the anode (+ve electrode). The larger the fragments, the more slowly it moves through the pores in the gel (Figure 13.6).

## Cloning a gene

One of the most fundamental procedures in genetic engineering is **gene cloning**. A clone is the collective name for a series of genetically-identical copies produced from a single 'parent'. In genetic engineering, then, cloning is the process of producing many copies from an original DNA fragment. Cloning a gene involves several operations:

1 obtaining DNA from the source organism. This can be done in various ways, for example:
   - by obtaining an extract of the entire genome of the organism. The DNA extract is first digested by restriction enzymes to yield many different fragments, only a tiny proportion of which contain the gene of interest.
   - a cDNA copy of a gene's mRNA transcript can be made using reverse transcriptase.

2 the DNA fragments are inserted into either a bacterial plasmid or the DNA of a bacteriophage (described in more detail below). The plasmid or bacteriophage acts as a **vector**. Since they also serve to copy not only the carrier DNA but also the 'passenger' DNA, they are also called **cloning vectors**. If the source DNA and the cloning vector are cut with the same restriction enzyme, their sticky ends can be joined by annealing, followed by incubation with DNA ligase.

3 insertion of the recombinant DNA into a 'host' organism within which it may replicate. The host is usually a bacterium, but eukaryotic cells are sometimes used, particularly yeasts. If the cloning vector is a plasmid, it is introduced into the host by **bacterial transformation** (Chapter 10). If the vector is a bacteriophage, the recombinant DNA in the phage is replicated by infecting a host bacterium. Whatever kind of cloning vector is used, the result is a large number of colonies of the host organism, each colony being a clone.

If the source DNA consists of fragments of the entire genome (as in 1 above), then many different clones will be produced. Between them these contain the entire genome of the donor organism and constitute a **gene library** or **DNA library**.

4 Since the individual clones have no obvious distinguishing features, a gene library has no 'directory'. The next stage is, therefore, to find the clone containing the desired gene. Before discussing the various ways in which this can be done we need to take a closer look at the two main kinds of cloning vector – plasmids and bacteriophage.

### *Plasmids*

Plasmids are present in most bacteria, and are small circular sections of DNA. Unlike the ordinary bacterial chromosome, plasmids are present in 20 or more copies. Though small, they are vital to the bacterium's survival, for they carry genes for resistance to antibiotics. Their small size makes them easily separable from the main chromosome.

Plasmids have two properties that are crucial to their role as cloning vectors:

- they can be taken up by bacteria of the same (or even another) species. The uptake of DNA by bacteria is the basis for transformation, and is the mechanism by which resistance to antibiotics can spread from one species of bacterium to another.
- because they carry genes for resistance to antibiotics, they can be used as markers. A gene is incorporated into a plasmid carrying a gene for resistance to, say, streptomycin, and the plasmid transplanted into a streptomycin-sensitive strain of bacteria. If the recipient bacteria are treated with streptomycin, any survivors must have received the plasmid and hence the 'passenger' gene.

If the 'passenger' DNA has been cut from the donor species by use of a restriction enzyme generating sticky ends, it can be joined to other pieces of DNA that have been produced using the same restriction enzyme, as both will have the same kind of sticky ends (Figure 13.7).

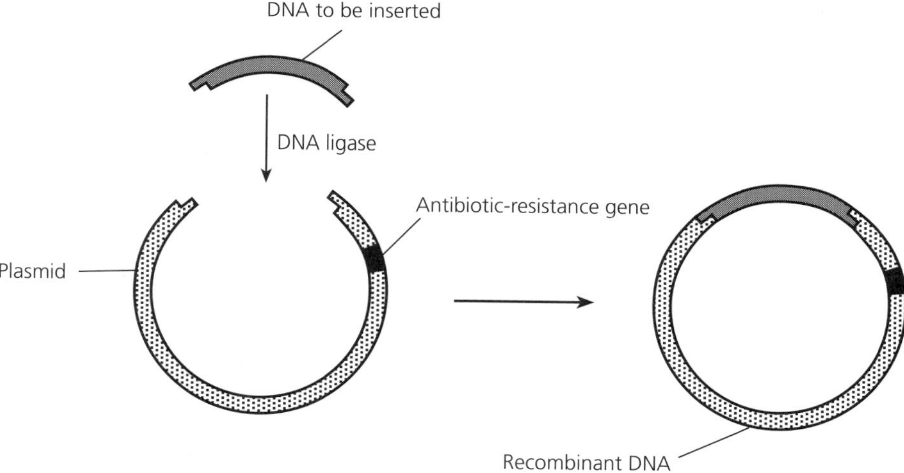


**Figure 13.7**
Inserting DNA into a plasmid

If the DNA to be inserted has been synthesised *in vitro* (i.e. in the test tube) using reverse transcriptase, its ends will be blunt. These must first be made sticky by addition of linkers that are complementary to the ends that will be created when the plasmid is cut. A common method is to add **poly-A tails** using the enzyme **terminal transferase** (Figure 13.8).

The bacterial plasmids are then opened up using a restriction enzyme that generates blunt ends. These are then 'tailed' by addition of poly-T to

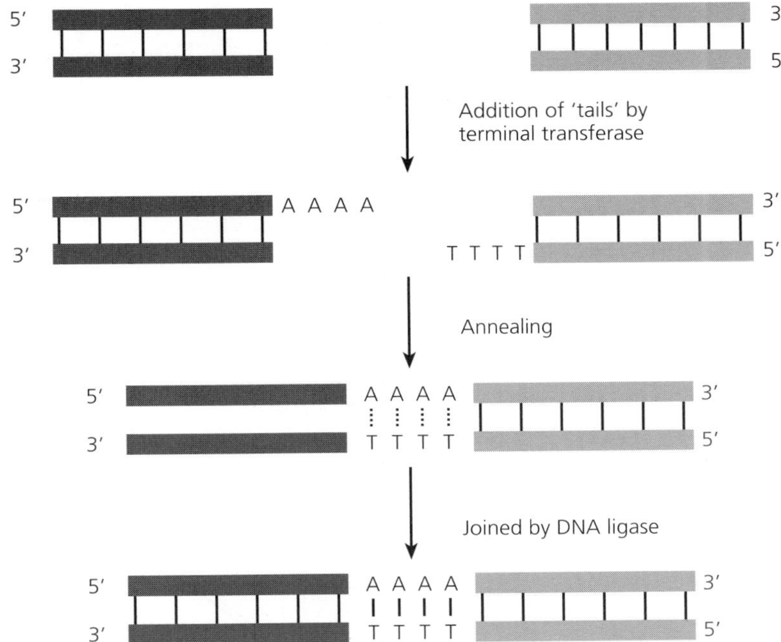


**Figure 13.8**
Joining flush-ended DNA using terminal transferase to create 'linkers'

make them complementary to the poly-A tails of the cDNA. After mixing the plasmids with the cDNA, they are covalently joined using DNA ligase.

The plasmids are then added to a suspension of the bacteria in the presence of calcium ions (which promote transformation). A small proportion of the bacteria are transformed by taking up the DNA; the plasmids are then maintained in the cells as self-replicating units.

After treatment with the vector plasmids, the bacteria are allowed to multiply. A suspension of the bacteria is then treated with the appropriate antibiotic and plated out onto agar. Those bacteria that grow are the ones that received the 'passenger' DNA.

### *Bacteriophage*

The usual result of a phage attack on a bacterium is the **lysis** (break-up) of the host cell and the release of hundreds of new phage particles. Each of these can then infect neighbouring cells, and so on. When this happens in a dense population of bacteria on an agar plate (called a 'lawn' of bacteria), a clear area or **plaque** is produced on the agar. A plaque thus marks the position of millions of phage particles.

The phage most commonly used is phage lambda. To insert a piece of DNA into a phage, the phage particles are first broken up into their protein and DNA components. A non-essential part of the phage DNA is then removed to make room for the 'passenger' DNA (Figure 13.9).

The DNA of the phage and the 'passenger' DNA are then cut to give either blunt or compatible cohesive ends. In the former case, the ends can then be tailed with compatible single-stranded DNAs, such as poly-A and poly-T. The viral and 'passenger' DNA are then mixed and, after the ends have been annealed, they are covalently joined using DNA ligase. The protein coat is then added back and the phage then used to infect the cells of the recipient bacteria.

## Finding the right gene

The result of cloning a gene is an agar plate containing a large number of clones. If a plasmid is used as a cloning vector, the clones take the form of

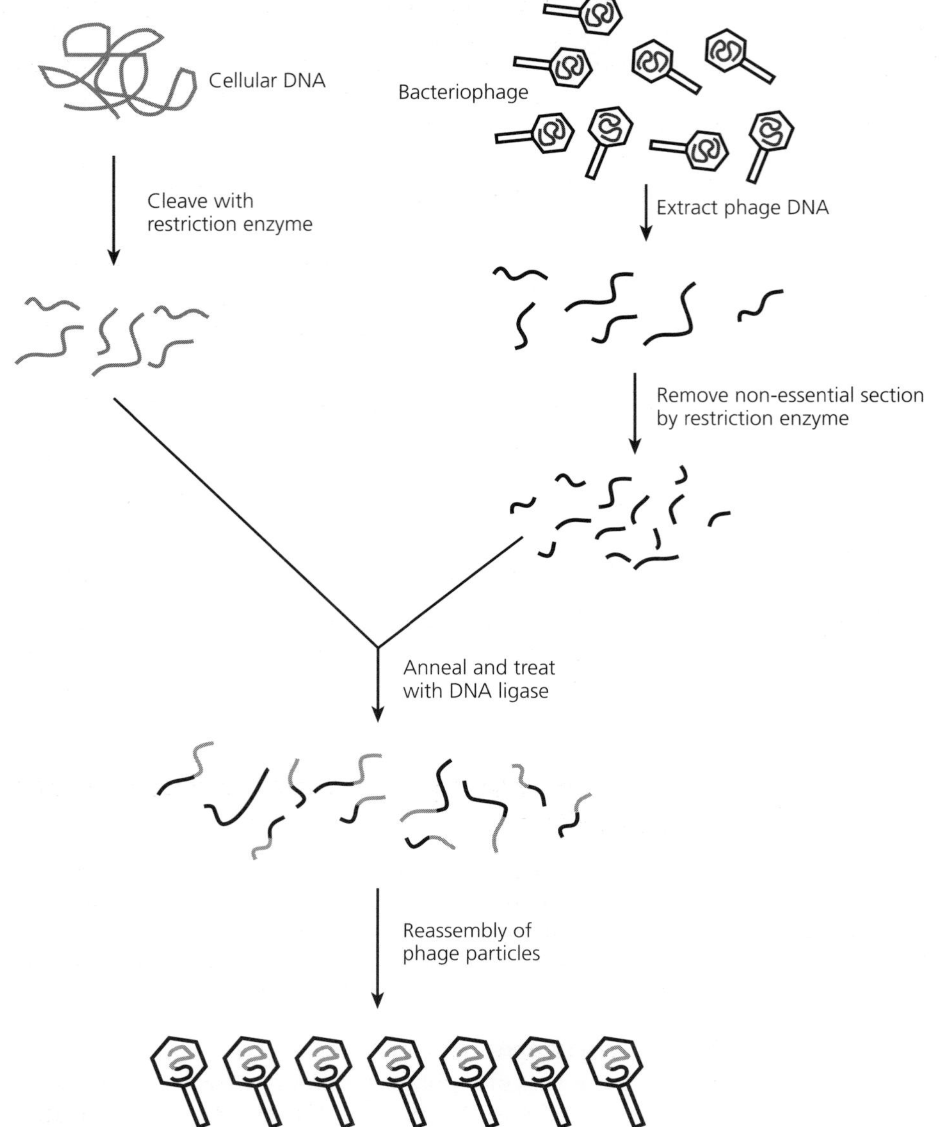


**Figure 13.9**
Insertion of 'passenger' DNA into phage lambda

bacterial colonies. If a bacteriophage is used, the clones will be plaques on a 'lawn' of host bacteria.

Only a tiny proportion of these clones will contain the desired gene. There are two ways of finding this gene:

- when the gene is expressed in the host cell, it is possible to detect the protein product of the gene using an antibody to that protein (Figure 13.10).
- by looking for the gene itself using a **gene probe**.

## *Gene probes*

All genes are chemically very similar, differing only in base sequence. The sequence is specific for each gene, and it is this fact which is exploited in the use of gene probes.

A gene probe is a length of nucleic acid with a base sequence that is complementary to at least part of the gene being sought. As a result, each probe can only base-pair with DNA containing a complementary base sequence. The probe has to be long enough to contain a unique base sequence that will only pair with the DNA of interest (a 20-base sequence

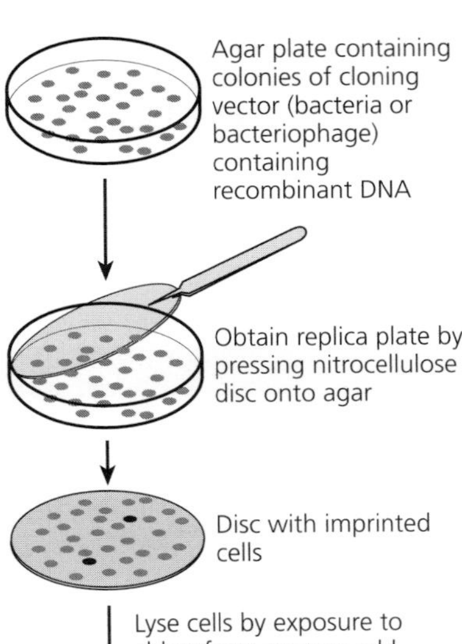
Agar plate containing colonies of cloning vector (bacteria or bacteriophage) containing recombinant DNA
Obtain replica plate by pressing nitrocellulose disc onto agar
Disc with imprinted cells
Lyse cells by exposure to chloroform vapour; add antibody; radioactive label binds to antibody
Detect labelled colonies using autoradiography
From the position of the radioactivity on the replica, the position of the desired colonies on the original can be determined

**Figure 13.10▲**
Using antibody to locate a gene

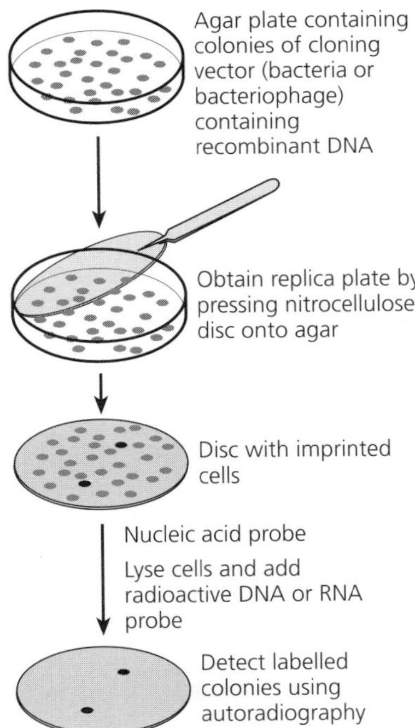
Agar plate containing colonies of cloning vector (bacteria or bacteriophage) containing recombinant DNA
Obtain replica plate by pressing nitrocellulose disc onto agar
Disc with imprinted cells
Nucleic acid probe
Lyse cells and add radioactive DNA or RNA probe
Detect labelled colonies using autoradiography
From the position of the radioactivity on the replica, the position of the desired colonies on the original can be determined

**Figure 13.11▲**
Using a gene probe to find the desired gene

is sufficient). The simplest method of preparing a probe is to use an mRNA copy of the gene. This is most easily extracted from cells specialised for making a very small number of types of protein, in which there are correspondingly few types of mRNA. The probe is made radioactive by labelling the 5' end with $^{32}P$.

The procedure for using the probe is described below and shown in Figure 13.11. If bacteriophage is used as the vector, the essentials of the technique are similar.

1 A piece of nitrocellulose filter or nylon is gently pressed on the agar containing the colonies and carefully lifted off. Some of the bacteria adhere to the filter, producing an exact replica of the pattern of colonies.
2 The bacteria are then lysed (broken open) chemically to free the DNA.
3 The DNA is then denatured by alkali to make it single-stranded.
4 The probe is added and allowed to hybridise with the DNA on the filter.
5 The filter is then washed to remove any unbound probe.
6 The positions of the bound probe are then revealed by leaving the filter in contact with X-ray film. After development, the position of the hybridised probe is revealed and the colonies on the agar therefore identified.

## Transplanting genes between prokaryotes and eukaryotes

Though many of the genes that are manipulated by genetic engineers are those of eukaryotes, they are usually cloned in prokaryotes. This presents a problem that we have conveniently ignored so far – that mechanisms of gene expression differ in important ways in prokaryotes and eukaryotes. Some of these problems are described briefly below.

### *The problem of introns*

Unlike genes of eukaryotes, genes of prokaryotes are not composed of introns and exons (Chapter 11). Prokaryotes lack the ability to remove introns from RNA, so eukaryotic genes cannot be expressed in prokaryotes unless the introns are first removed.

A simple way round this problem is to use an mRNA copy of the gene. Although most cells make many proteins and thus have many different kinds of mRNA, many cells make predominantly one kind of protein. Immature red blood cells, for example, make mainly the α and β chains of haemoglobin, and certain β cells of the pancreas produce mainly insulin. The mRNA is then used to make a cDNA copy using reverse transcriptase.

To obtain the desired mRNA, the cells are first broken up and the ribosome–mRNA complexes collected by centrifugation. Each ribosome–mRNA complex will have a polypeptide nearing completion at one end. The problem is how to separate the desired protein–mRNA complex from all the other chemicals present. Genetic engineers use the fact that when an antibody is added to the antigen that stimulated its production, the antibody combines with the antigen and forms a precipitate (Chapter 27). By adding the antibody to the cell extract, the desired protein–mRNA complex is precipitated and can be isolated. The mRNA is then used as a template for the synthesis of cDNA using reverse transcriptase.

Though the cDNA is a copy of the required gene, it lacks the sticky ends needed to splice it into the DNA of a bacterium, and also the start and stop codons necessary for its transcription; both must be added before the gene can be transcribed.

## The polymerase chain reaction

**One of the most useful techniques in the genetic engineer's tool kit is the polymerase chain reaction or PCR. By this technique a minute quantity of DNA (as little as one molecule) can be increased a billion-fold or more. In essence, it involves the repeated replication of DNA in the test tube, with the amount of DNA doubling each replication cycle. After 20–30 replication cycles the amount has increased between a million ($10^6$) and a billion ($10^9$) times. The procedure is very quick, each cycle taking only about five minutes.**

**In principle, all that is needed to replicate DNA in the test tube is a parent DNA molecule, the enzyme DNA polymerase and a supply of the four kinds of nucleotide (actually nucleoside triphosphates, which are more reactive). By heating the DNA to denature it, two single-strands of DNA are produced. Each of these can be used as a template to build the other strand.**

**There is, however, a complicating factor – part of the base sequence of the DNA which is to be amplified *has to be known in advance*. The reason for this lies in the fact that DNA polymerase cannot work on a template that is single-stranded throughout its entire length. Two single-stranded, oligonucleotide DNA primers have to be added first. Each primer is only about 20 bases long and hybridises with a section of polynucleotide at either end of the region of DNA to be replicated (see diagram). Since the primers have to be synthesised artificially, the base sequences of the complementary parts of the DNA must be known in advance.**

**A further complicating factor is that after each cycle of DNA replication, the DNA has to be denatured by heat, which would also denature most DNA polymerases. To avoid this, DNA polymerase from a bacterium that is adapted to high temperatures is used.**

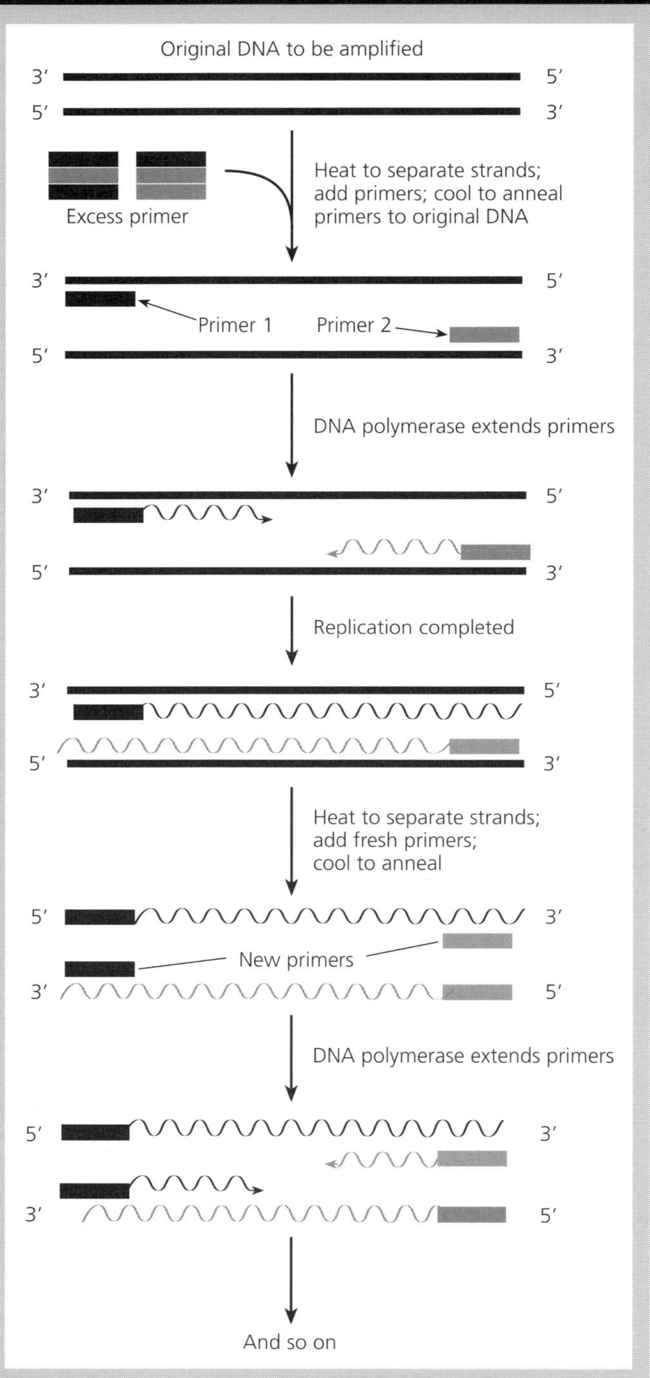


Polymerase chain reaction

## The polymerase chain reaction *continued*

The PCR consists of the following steps:

1 The DNA is heated to 95°C to denature it.
2 DNA polymerase, four nucleoside triphosphates and excess quantities of the oligonucleotide primers are added. The mixture is then cooled sufficiently to allow the primers to anneal to the DNA at either end of the region that is to be amplified. Because they are added in excess, the original parent strands are much more likely to pair with primer than with each other.
3 The DNA is then replicated, producing two double-stranded molecules.
4 The DNA is then heated again to denature it, and the process repeated for 20–30 cycles. After the second cycle, there are four DNA molecules, after the third cycle there are eight, and so on.

The polymerase chain reaction has enabled the DNA of partially-decomposed remains to be recovered in sufficient quantities to be studied. It has even been possible to recover DNA from long extinct insects preserved in amber.

### *Control of transcription*

The mechanisms for switching genes on and off in eukaryotes are different from those in prokaryotes. This does not matter if the bacteria into which a gene has been transplanted are only used to clone the gene. If, on the other hand, the bacteria are used to produce the protein coded for by the gene, the eukaryotic gene must be expressed in the bacterium. To enable this to happen, the gene is usually spliced 'downstream' of the prokaryotic promoter and operator. The gene can then be switched on by addition of the appropriate inducer. If a gene is spliced into the lactose operon of *E. coli*, the gene can be switched on by adding lactose to the medium. In this way the production of the protein can be controlled by the technologist.

## Genetic fingerprinting

Genetic fingerprinting is a technique developed by Alec Jeffries at Leicester University. It makes it possible to determine relationships between individuals by comparing their DNA. The technique depends on the fact that most animals (including humans) have sections of DNA in which a certain base sequence (15 to 100 nucleotides long), is repeated several times forming **tandem repeats**. These occur at various loci on different chromosomes. The number of repeats varies independently at different sites. The overall pattern of repeats is highly characteristic for each individual and acts as a kind of **genetic fingerprint**.

To make a genetic fingerprint, a sample of the DNA of the individual is digested with a restriction enzyme which does not attack any site within a repeated sequence, so the repeats are left intact. The fragments are then separated by gel electrophoresis (see pages 24 and 211). At this stage they are invisible, and are accompanied by many other DNA fragments, which are not wanted.

To reveal the tandemly-repeated DNA fragments, a technique called **Southern blotting** is used (developed by E.M. Southern). The gel is first immersed in sodium hydroxide solution which denatures the DNA, rendering it single-stranded. The gel is overlaid with nitrocellulose or nylon filter and the filter is then covered with a thick wad of paper towelling (see diagram). This acts as a wick, drawing the buffer solution up through the filter leaving the DNA on the filter.

## Genetic fingerprinting *continued*

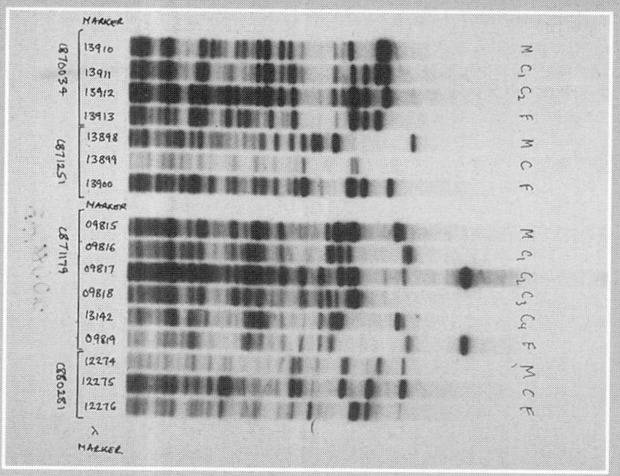

Genetic fingerprinting being used to analyse family relationships

A DNA probe is then used to identify the repeat fragments. This probe consists of single-stranded DNA complementary to the repeat sequence, labelled with radioactive $^{32}P$. The probe binds to the repeat fragments, thus labelling them. The non-specifically bound probe is then washed away, and an X-ray film laid over the filter. After developing, the positions of the repeat sequences are revealed as dark bands on the film. The spacing between the fragments indicates their rates of movement through the gel, and hence their relative sizes. The overall pattern looks like a supermarket bar code, and is different for any individual (except in the case of identical twins).

DNA fingerprinting has been of enormous significance in forensic science (the science of obtaining evidence to be used in court), and also in determining parentage. Whilst comparisons between blood groups of different individuals may make it possible to conclude that Mr X could not have been the father of a particular child, DNA fingerprinting enables one to say that an individual *is* the parent of a child.

### *Post-translational modification*

A number of proteins made by eukaryotes undergo modification after they are produced on the ribosomes. The hormone insulin, for instance, is produced as an inactive precursor which is converted to insulin by a two-step removal of part of the polypeptide chain. Many other proteins are similarly modified after their production – for example by the addition of prosthetic groups. These problems are progressively being solved. Insulin, for example, is now manufactured industrially using bacteria.

## Introducing genes into animal cells

A common method of introducing DNA into an animal cell is by **electroporation**, which involves giving the cells an appropriate electric pulse. This causes minute holes to develop briefly in the plasma membrane, allowing DNA to be taken up from the medium. Another approach is to coat DNA in minute artificial lipid vesicles which are taken up by the plasma membrane.

## Inserting genes into plant cells

Plants have a great practical advantage over animals: their cells are **totipotent**, meaning that, given the appropriate chemical treatment, a single mature cell can grow into an entire plant. Any gene transferred into such a cell can thus be transmitted to the next generation.

The most common method of transferring genes to plant cells is to use *Agrobacterium tumefaciens*. This common soil bacterium causes tumour-like growths in a variety of dicotyledonous plants.

When a bacterium enters the plant (through a wound) it inserts part of a **tumour-inducing (Ti) plasmid** into a host cell. The part of the plasmid that is transferred into the host cell is called **T-DNA**, and this is

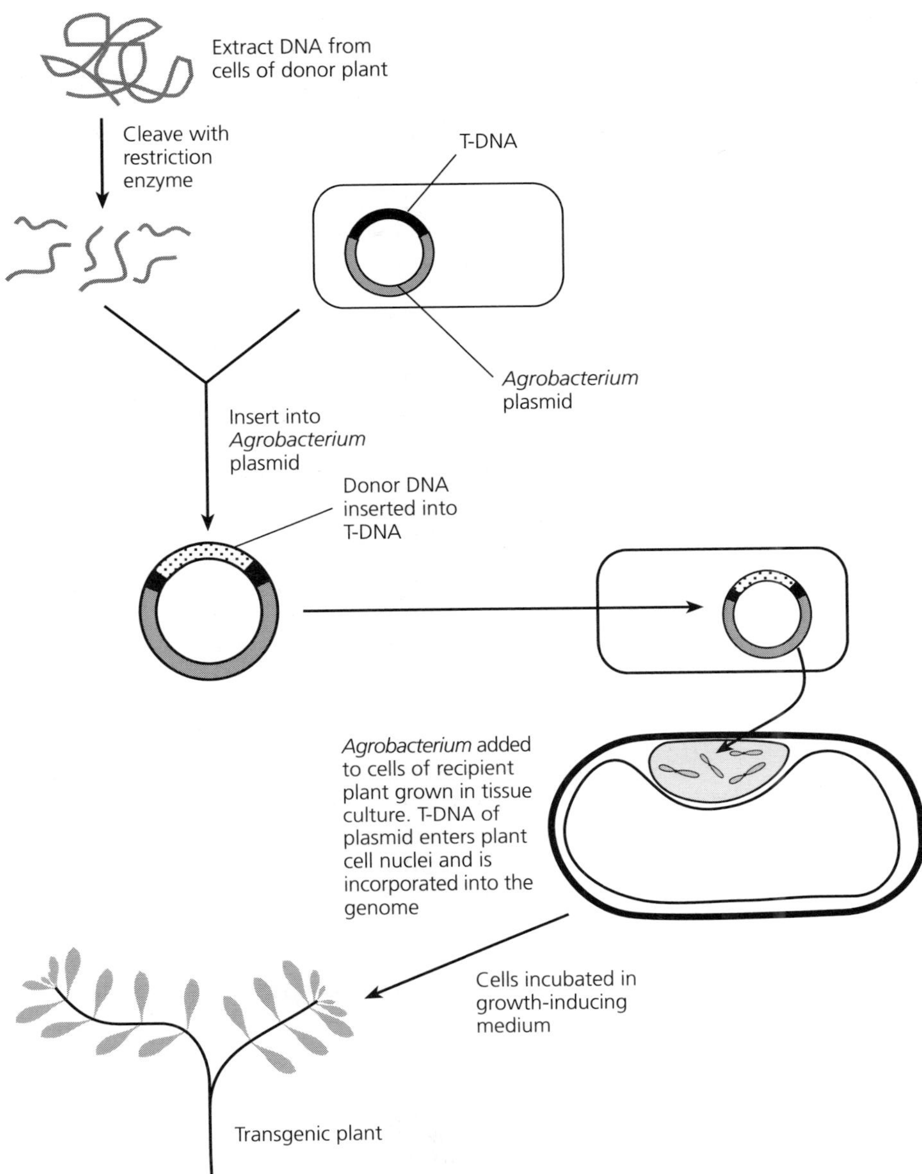
Extract DNA from cells of donor plant
Cleave with restriction enzyme
T-DNA
Agrobacterium plasmid
Insert into Agrobacterium plasmid
Donor DNA inserted into T-DNA
Agrobacterium added to cells of recipient plant grown in tissue culture. T-DNA of plasmid enters plant cell nuclei and is incorporated into the genome
Cells incubated in growth-inducing medium
Transgenic plant

**Figure 13.12**
Use of *Agrobacterium* to insert foreign DNA into a plant cell

then incorporated into a chromosome of the plant (Figure 13.12). Some of the genes in the plasmid are necessary for the infection of a host cell, whilst others induce host cells to multiply to produce a gall.

The bacteria used by the genetic engineer are genetically modified in two ways. First, they have had the gene necessary for gall-formation deleted, whilst retaining the genes needed to infect host cells. Second, they have had genes for resistance to certain antibiotics added.

To use *Agrobacterium* as a vector for introducing a gene into a plant, the desired gene is first isolated, spliced into a plasmid of *Agrobacterium* and cloned in the manner described earlier. Cells of the host plant are grown in culture and are treated with the bacterium and also with the antibiotic for which the plasmid carries a resistance gene. This serves to identify the cells into which the plasmid has been successfully inserted, since only these cells can grow in the presence of the antibiotic. From the cultured cells, entire plants are then grown.

*Agrobacterium* has proved highly successful in producing transgenic plants such as tomato, tobacco, cotton, apple and a variety of others. Unfortunately, it does not infect monocotyledons, which include some of the most important crops such as the cereals. Alternative methods are being developed for these plants. One, the 'shotgun' or **biolistic method**, involves firing minute tungsten or gold beads coated with DNA at intact plant cells. Not only does the DNA enter the cells, but a small proportion of them incorporate it into their chromosomes! Electroporation is another method, but the cell walls have to be digested away first to produce naked cells called **protoplasts**.

Using these methods, plants have been modified in diverse ways. For example:

- crops have been developed that are very resistant to herbicides, so that weeds can be eliminated without damage to the crop.
- *Bacillus thuringiensis* is a bacterium that produces a protein that is highly toxic to caterpillars if it is ingested. The gene for this protein has been transplanted into crops such as tomato, soybean, corn, cotton and tobacco. The result is that any caterpillar eating the plant dies. This also has the great advantage that the toxin is harmless to other animals.
- in response to tissue damage, some plants produce protease inhibitors, which prevent digestion of proteins by insects that eat the crop. The genes for production of these enzymes have been experimentally transplanted into farm crops.
- the antifreeze gene from Arctic fish has been transplanted into soybeans, rendering them highly resistant to frost.

One of the most important goals of the plant genetic engineer is to develop cereals that can act as hosts to nitrogen-fixing bacteria, in the same way that leguminous plants do. This is likely to prove difficult, since about 20 genes are involved. Some are involved in fixation itself, others are needed for infection of the plant by the bacterium.

## Gene therapy

One possibility that is close to being realised is gene therapy, in which a faulty gene is 'repaired' in the test tube and inserted back into human cells. Depending on the kind of cells into which the gene is inserted, there are two quite different approaches: **somatic gene therapy** and **germ-line therapy**.

### *Somatic gene therapy*

In somatic gene therapy the repaired gene is inserted into those body cells in which the gene would normally be expressed. For example, the hereditary disorder **cystic fibrosis** affects cells lining the breathing passages and gut, and results in congestion by large amounts of mucus. The aim of somatic gene therapy in this case would be to insert the normal gene into the affected cells. The means of doing this would probably be a virus that attacks these cells, which had been genetically engineered to render it incapable of causing disease. The normal gene would be spliced into the viral DNA in the hope that it would be incorporated into the host cell DNA.

### *Germ-line therapy*

Though somatic gene therapy can, in principle, cure an individual, the gametes produced by that person would still carry the gene so the children could still be affected. The aim of germ-line therapy is to restore the gamete-producing cells to normal so the person can have normal children. This has not yet been achieved, and has ethical implications.

## Designer genes: directed mutation

In nature, mutation is quite random. If a gene mutates to produce an allele that is beneficial, that is just good luck. More often than not a mutation is harmful.

Genetic engineers are changing all that. It is now possible to produce a specific base change in a particular gene. In outline, the technique is as follows. First, a DNA oligonucleotide containing the desired base sequence is synthesised artificially. This is then allowed to pair with single-stranded DNA from the gene to be changed. Pairing will be imperfect at the point where the oligonucleotide and the natural gene differ. The oligonucleotide strand is then used as primer to synthesise the rest of the gene using DNA polymerase.

## Dangers of recombinant DNA technology

Almost every beneficial scientific discovery has brought with it possibilities for misuse, and genetic engineering is no exception. When genetic engineering first became feasible in the 1970s, it was suggested that new strains of micro-organisms might 'escape' from laboratories and cause outbreaks of uncontrollable disease or other problems. To prevent this, regulations were put into place to ensure that any organisms used in gene technology were so modified that they could not live outside the laboratory. Since then, it has become clear that this 'Frankenstein factor' is far less of a threat than some had feared.

Other concerns remain. Many people consider that producing transgenic animals and plants is an interference with nature (though it could be said that farming and medicine are just that). The **patenting** of transgenic organisms is another problem to which our legal system is currently attempting to adjust.

Perhaps the greatest long-term concern relates to the interference with human genes. Few would argue that the repair of the gene for cystic fibrosis would not be of great benefit. However, some fear that the power to repair defective human genes might be the thin end of a very long wedge. Can we, for example, draw a clear line between genes that are clearly defective and bring great suffering to its possessor, and genes which produce excessive freckling, or some other trifling 'imperfection'?

## QUESTIONS

**1** What kind of enzyme would you use to perform each of the following operations?

- **a)** cutting DNA into fragments
- **b)** joining DNA fragments together
- **c)** making DNA from RNA
- **d)** synthesising DNA in vitro

**2** Suppose you wanted to insert a section of human DNA into a plasmid. Why should you use the same enzyme to cut out the human DNA and to open the plasmid?

**3** Apart from being radioactive, how does a DNA probe differ from normal DNA?

**4** If an animal or plant gene were to be transplanted unaltered into a bacterium, it could not be expressed. Why?

**5** Consider two restriction enzymes: one which recognises a sequence of four bases, and one which recognises a sequence of eight bases.

- **a)** Which would you expect to cut a given length of DNA into (on average) the greater number of fragments?
- **b)** What would be the average ratio of the fragment lengths generated by these two enzymes?

CHAPTER 14

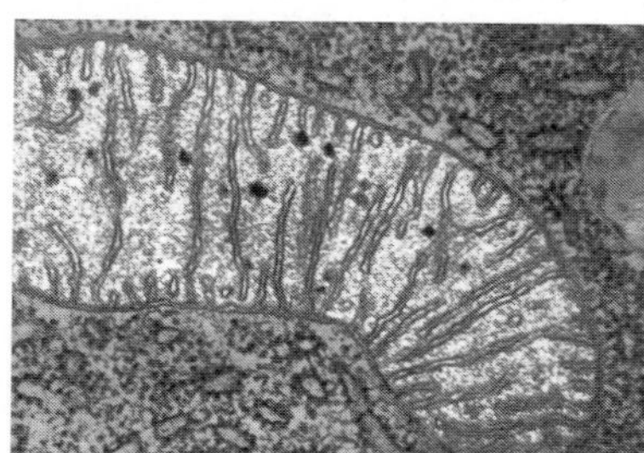

# GENES in POPULATIONS: EVOLUTION

## LEARNING OBJECTIVES

By the time you have completed your study of this chapter you should be able to:

- explain what is meant by the terms gene pool and allele frequency.
- calculate the allele frequency in a population, given the frequencies of all three genotypes for a particular locus.
- state the factors that can affect allele frequencies in a population.
- explain what is meant by Hardy–Weinberg equilibrium, and state the conditions under which it applies.
- calculate genotypic frequencies using the Hardy–Weinberg equation.
- know when to use the Hardy–Weinberg equation and when its use is inappropriate.
- distinguish between stabilising selection, directional selection and disruptive selection, and describe an example of each.
- explain what is meant by a cline, and describe an example.
- explain what is meant by genetic drift, the founder effect and a genetic bottleneck, and give an example of each.
- define what is meant by a species, and give some examples of both pre- and postzygotic isolating mechanisms.
- distinguish between allopatric, parapatric and sympatric speciation, and explain how each is thought to occur.

## The gene pool

Evolution can be defined as a genetic change in a population over a sufficient number of generations to be recognisable as a trend. Ecologists define a population as all the individuals of a given species living in a particular area. Thus, we can speak of the population of trout in a river, or the population of buttercups in a meadow. To a population geneticist, however, members of a population must not only live in the same area – they must also be able to mix their genes. In other words, they must reproduce **sexually**.

Whilst the genetic makeup of an individual is its **genotype**, the genetic makeup of a sexually-reproducing population is the **gene pool**. The gene pool is defined as the sum total of all the alleles of all the gene loci in a population. In most instances, however, we are only interested in the alleles at one particular gene locus.

There are two important differences between the gene pool of a population and the genotype of an individual:

1 an individual's genotype is fixed throughout its life, but a gene pool is ever-changing. Alleles are added to the gene pool every time an

individual reproduces, or when an immigrant arrives. Alleles are lost from the gene pool every time an individual dies or emigrates.

2 a diploid individual cannot have more than two alleles for any given locus, but a gene pool can contain many. Thus each individual only carries some of the alleles in the gene pool.

A gene pool may, to a greater or lesser degree, be isolated from other such pools. In the case of the carp in two neighbouring lakes, isolation is total, except when the fish are artificially transferred by humans. More often than not, however, there is some exchange of genes with other gene pools. This is because many of the inhabitants of ponds and lakes have stages in the life cycle that can travel from pond to pond. Many aquatic insects (such as water boatmen and dragonflies) are terrestrial as adults. The eggs of fresh water crustaceans, such as *Daphnia*, can be dispersed from one pond to another with the mud on bird's feet. Pollen grains may serve as excellent vehicles for the transfer of genes from one population to another.

Because of dispersal, each local population or **deme** is really part of a larger population. Within each deme, there is free exchange of genes. Between demes, gene exchange is less free.

## Allele frequencies

The genetics of wild populations is much more complicated than in breeding experiments. In a breeding experiment, the male and female parents usually have different genotypes, as selected by the experimenter. In a wild population, mating can be between similar or different genotypes, and each genotype can be present in any proportion.

For any particular gene locus there is usually more than one allele. Therefore in the population, each allele represents a certain proportion of all the alleles at that locus. This proportion is the **frequency** of the allele in the population. The term 'allele frequency', incidentally, is preferable to 'gene frequency' because it refers to a specific locus.

As a very simple example, imagine two gardens containing antirrhinums. One garden contains equal numbers of red-flowered (*RR*) and white-flowered (*rr*) antirrhinums. The other contains all pink-flowered (*Rr*) antirrhinums. Although the genotypic composition of the two populations is quite different, the frequencies of the *R* allele and the *r* allele are the same in both 'populations'.

In the above example the frequencies of *R* and *r* are both 0.5. Though allele frequencies can be expressed as percentages, population geneticists prefer to use decimals because it is easier to multiply them in that form.

Now let's imagine a garden containing all three genotypes in the following numbers:

| | | |
|---|---|---|
| red | *RR* | 150 |
| pink | *Rr* | 100 |
| white | *rr* | 250 |

Each of the 500 plants has two alleles for flower colour, so there are 1000 alleles in the population as a whole. The frequencies of *R* and *r* are worked out as follows:

Number of *R* alleles in the *RR* plants $= 2 \times 150 = 300$
Number of *R* alleles in *Rr* plants $= 100$
Hence the total number of *R* alleles in the population $= 300 + 100 = 400$

Number of *r* alleles in the *Rr* plants $= 100$
Number of *r* alleles in the *rr* plants $= 2 \times 250 = 500$

Hence the total number of $r$ alleles in the population = 100 + 500 = 600

The frequency of the $R$ allele is, therefore, 400/1000 = 0.4, and the frequency of the $r$ allele is 600/1000 = 0.6

## Factors that affect allele frequencies

In nature, allele frequencies often change for a variety of reasons:

### *Small populations*

In small populations chance can have a considerable effect on allele frequencies. A change in allele frequency due to the accumulated effects of chance is called **genetic drift** and is dealt with later in this chapter.

### *Immigration*

When organisms (which include spores and seeds) move from one area to another they take their genes with them. This process is called **gene flow**. Immigration will affect allele frequencies if the immigrants come from areas where allele frequencies are different. Emigration, by the way, can only affect allele frequencies if:

- one genotype has a greater tendency to emigrate than the others, or
- the population is very small.

### *Selection*

Successful reproduction depends on two things: the organism must reach reproductive maturity, and, having reached maturity, it must be able to produce offspring.

In each of these respects, not all genotypes are equally successful. Some genotypes have a higher survival rate than others, and, having survived, some individuals are better at reproduction than others. For example one stag may mate with an entire harem of females, leaving others as 'bachelors'. A female leopard may rear three cubs to independence or she may fail entirely.

Whether selection operates on survival or on reproduction, the result is the same: some alleles increase in frequency relative to others.

### *Mutation*

When one allele mutates to another, it obviously affects the frequency of both. The effect is extremely small, since mutation rates are very low. Moreover, if there were no other effects (such as selection) complicating the situation, the rates of mutation of $A \rightarrow a$ and $a \rightarrow A$ would come into equilibrium with each other and there would be *no net change*.

## The Hardy–Weinberg equation

When we know the frequencies of all three genotypes (i.e. when there is no dominance) allele frequencies can be calculated directly in the way shown in the antirrhinum example.

When there are only two phenotypes (when one allele is dominant) we only know the frequency of one genotype (the homozygous recessive). In this situation, an individual showing the dominant phenotype could be either homozygous or heterozygous. Nevertheless, *provided that certain conditions are satisfied*, it is still possible to estimate the frequencies of the two alleles. We will discuss these conditions later, but for the following example we will assume that they are satisfied.

In calculations of allele frequencies, the frequencies of dominant and recessive alleles are conventionally represented by $p$ and $q$, respectively.

The inability to roll the tongue is due to a recessive allele, $r$. People who cannot roll their tongues have the genotype $rr$, and tongue-rollers can be $RR$ or $Rr$. Since $RR$ and $Rr$ genotypes cannot be distinguished, we don't know the actual value of $p$. However, we *do* know the frequency of $rr$, and from this we can estimate the frequencies of the other two genotypes.

In this population each gamete carries either $R$ or $r$. The frequency of the $r$ allele is $q$, so the probability that a given gamete carries $r$ is $q$. The probability that in a given fertilisation both egg *and* sperm carry $r$ is therefore $q \times q = q^2$. This is the frequency of the $rr$ genotype, which we do know. We can thus work backwards from $q^2$ to $q$, since $q = \sqrt{q^2}$

Suppose that 16% of a population cannot roll their tongues. The frequency of the $rr$ genotype ($q^2$) is therefore 0.16, so $q = \sqrt{0.16} = 0.4$.

Now we know $q$, we obtain $p$ by subtraction (since $p + q = 1$, $p = 1 - q$).

The frequency of $R$ is thus $1 - 0.4 = 0.6$.

Knowing the frequencies of both alleles, the frequency of the other genotypes can be worked out. Since the probability of a given gamete carrying the $R$ allele is $p$, the probability that, in a given fertilisation, *both* gametes carry $R$ will be $p^2$. The frequency of the $RR$ genotype is thus $0.6^2 = 0.36$.

The frequency of the $Rr$ genotype $= 2pq$ (it is multiplied by 2 because there are two ways in which this can happen: $R$ egg $+$ $r$ sperm and $r$ egg $+$ $R$ sperm).

The frequency of $Rr$ genotypes is thus $2 \times 0.6 \times 0.4 = 0.48$

The four possible kinds of fertilisation are shown in Figure 14.1. This is just like an ordinary Punnett square except for one thing – the length of the sides is divided up in proportion to the frequencies of the $R$ and $r$ gametes (i.e. $p$ and $q$).

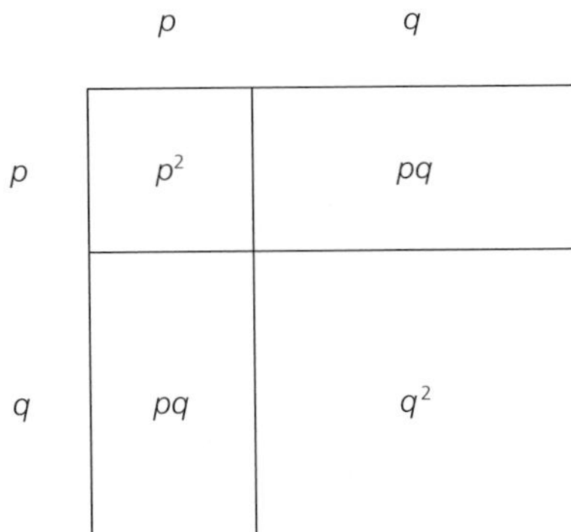


**Figure 14.1**
Punnett square showing how genotypic frequencies can be calculated from allele frequencies

The frequencies of the genotypes are given by the relative areas of the four compartments. The three genotypic frequencies must, of course, add up to 1, so:

$$p^2 + 2pq + q^2 = 1$$

This is the **Hardy–Weinberg equation**, after the British mathematician G.H. Hardy and the German physician W. Weinberg who independently pointed it out in 1908. Mathematically it is simply the algebraic expansion of $(p + q)^2$, shown geometrically in Figure 14.1.

A population in which the frequencies of $RR$, $Rr$ and $rr$ (to use the symbols in the above examples) are equal to $p^2$, $2pq$ and $q^2$, respectively, is said to be in **genetic equilibrium** or **Hardy–Weinberg equilibrium**.

Until 1908 many people thought that a dominant allele should eventually eliminate a recessive one because of its greater 'strength'. Hardy showed that this is not so. He pointed out that the tendency for an allele to be expressed (whether it is dominant or not) has nothing to do with its frequency in the population, which is determined by how useful its effects are. This idea is expressed as the **Hardy–Weinberg Law**, which states that:

> "*In a large, randomly mating population and in the absence of immigration, selection, and mutation, the frequency of alleles and genotypes remains constant from one generation to the next.*"

### *Example:*

Albinism is a condition in which the affected individual cannot make the pigment melanin. The allele responsible ($a$) is autosomal and recessive to the allele for normal pigmentation (*A*). In Western Europe about 1 in 20 000 people is albino. What proportion of the population is heterozygous for this condition?

The albino genotype is *aa*, and its frequency ($q^2$) is 1/20 000 or 0.00005.
Hence $q = \sqrt{0.00005} = 0.0071$
Since $p + q = 1$, $p = 1 - 0.0071 = 0.9929$
Frequency of heterozygotes is $2pq = 2 \times 0.9929 \times 0.0071 = 0.0141$, or 1.41%

### *Sex-linked alleles and the Hardy–Weinberg equation*

In sex-linked recessive traits such as red-green colour-blindness, the situation is slightly more complicated. In females (who have two X-linked alleles) the frequency of recessive homozygotes is $q^2$, as for autosomal genes. For males (who have only one) $q$ is the same as the phenotypic frequency.

## Conditions for genetic equilibrium

Though the Hardy–Weinberg equation is used to calculate genotypic frequencies, the result obtained is actually an estimate rather than a true value. The accuracy of this estimate depends on how fully the following five conditions are met:

1 the population must be *large*. In small populations chance plays an important part in determining which alleles are passed on to the next generation.

2 there must be *no selection*. In other words, all genotypes must be equally likely to reproduce.

3 there must be *no immigration* from populations with different allele frequencies.

4 there must be *no mutation*. More precisely, the rate of mutation of $A \rightarrow a$ must equal the rate of mutation of $a \rightarrow A$, so there is no *net* change due to mutation.

5 mating must be *random*. This means that with respect to the alleles under consideration, each individual must have no preference for mating with any particular genotype. As will be shown in a moment, a preference for mating with similar genotypes leads to the frequency of heterozygotes being lower than $2pq$. Preference for a different genotype, on the other hand, results in a higher than expected proportion of heterozygotes.

## Mate choice

An organism can show preferential mating in two different ways depending on whether it prefers a similar or a different genotype to its own. The former tends to lead to **inbreeding**, the latter to **outbreeding**.

### *Inbreeding*

Inbreeding is mating between relatives which, therefore, share more of their genes in common than unrelated individuals. Mating with a similar genotype tends to increase the proportion of homozygotes. We can see this if we consider the three genotypes *AA*, *Aa* and *aa* in Figure 14.2. Half the offspring of $Aa \times Aa$ matings are homozygous, but *all* the offspring of homozygote $\times$ homozygote matings are homozygous. If every genotype

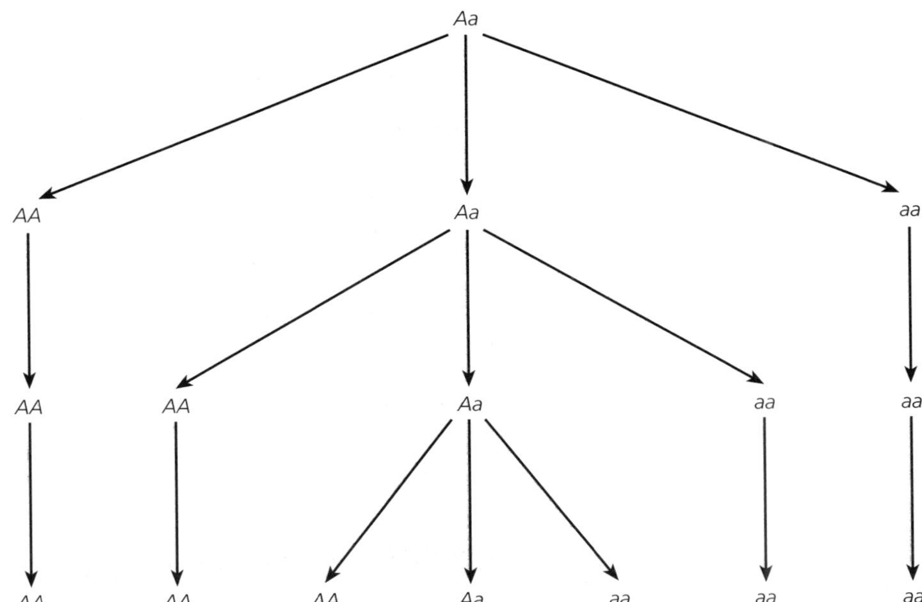


**Figure 14.2**
How inbreeding decreases the proportion of heterozygotes

mated with its own kind, the proportion of heterozygotes would halve each generation.

### *Outbreeding*

A preference for mating with unrelated individuals leads to outbreeding. As a result there is a higher than expected frequency of heterozygotes. Many flowering plants have mechanisms that promote outbreeding, and these are discussed in Chapter 30.

You should now be able to see that non-random breeding and non-random mating are quite different things. Non-random breeding is another term for selection, with some genotypes having more offspring than others. Selection affects both allele frequencies and genotypic frequencies.

Non-random mating, on the other hand, refers to which *kind* of partner each individual is likely to mate with. It does not necessarily imply that individuals have differing reproductive success (though of course they may have). If all individuals are equally successful in getting their gametes fertilised, but mating partners are not chosen at random, there will be no effect on allele frequencies. There will, however, be an effect on genotypic frequencies (as in the case of inbreeding, which produces an excess of homozygotes).

## When not to use the Hardy–Weinberg equation

Where there is complete dominance the frequency of only one genotype, the recessive homozygote, is known. In these circumstances the only way to estimate $p$ and $q$ is to use the Hardy–Weinberg equation. The frequency of the recessive allele is only equal to the square root of the homozygous recessive genotype if certain conditions (listed above) are satisfied – that is, if the population is in Hardy–Weinberg equilibrium. In using the Hardy–Weinberg equation, it has to be assumed that these conditions hold (unless you are told that they do). If they are not satisfied, then estimates of $p$ and $q$ will be inaccurate.

If there is no dominance, the frequencies of all three genotypes are known, so $p$ and $q$ can be calculated directly without resorting to the Hardy–Weinberg equation and its attendant assumptions.

## What use is the Hardy–Weinberg equation?

Although there are many cases in which these conditions are satisfied in nature, there are almost certainly just as many (if not more) in which they are not. After all, gene flow, selection and non-random mating are all common biological processes!

What, then, is the practical value of the Hardy–Weinberg equation? The answer is that if a population is not in genetic equilibrium then one or more of the above biological phenomena must be at work. Moreover, the extent to which a population departs from genetic equilibrium is an indicator of the intensity of these effects.

For example, if there is an unexpectedly low proportion of heterozygotes (i.e. lower than $2pq$), we might suspect that there is a tendency for mating between relatives. On the other hand, if one of the two homozygotes is at a lower than expected frequency, it could be due to selection against that allele.

### Genetic load

In a haploid species any allele that is even mildly harmful in the present environment will be eliminated by selection. In diploid species the situation is quite different because of **dominance.** Because they are not expressed in heterozygotes, recessive alleles are protected from selection. Every diploid species therefore carries a **genetic load** of harmful recessive alleles. If the allele is uncommon, homozygotes are only likely to be produced by mating between heterozygotes. If, say, 1 in 100 individuals is heterozygous for a given locus, only in 1 in 10 000 matings ($\frac{1}{100} \times \frac{1}{100}$) will *both* parents be heterozygous, and even then only a quarter of the offspring will be subject to selection.

However, diploidy and a reservoir of recessive alleles can have its advantages. An allele that is mildly harmful in one environment may be beneficial in another. If there is an environmental change that renders such alleles beneficial, they are already present in the population. Consequently the population does not have to 'wait' for mutation to bring such alleles into existence. Diploid species are thus in a better position to respond to changing conditions than are haploid species.

### 'Improving' the human gene pool

From time to time it has been suggested that the human gene pool should be 'improved' by preventing people with harmful alleles from reproducing. The Hardy–Weinberg equation shows how fruitless this would be (quite apart from any moral implications).

An important consequence of the Hardy–Weinberg equation is that recessive alleles causing uncommon conditions, such as albinism, are actually much more common than the phenotypic frequencies would suggest. Since $p$ and $q$ are both less than 1, $p^2$ is always less than $p$ and $q^2$ is always less than $q$. When $q$ is small, $q^2$ is *very* small. This means that almost all rare recessive alleles exist in the heterozygous state. In the example given earlier, only 1 in 20 000 people is albino but 1 in 71 is a carrier. In other words, for every albino there are about 281 carriers.

Any plan to 'improve' the gene pool by preventing albinos from reproducing would tackle only $\frac{1}{281}$ of the problem, and would thus be doomed to failure. In the case of very rare conditions such as alkaptonuria, in which only one in about a million people has the condition, one in 500 people carries the allele. There are, therefore, 2000 times as many carriers as people who actually have the condition. (Use the Hardy–Weinberg equation to check this, remembering that $q^2 = 10^{-6}$ and that the frequency of heterozygotes is $2pq$.)

# How gene pools change: selection

A key feature of Darwin's Theory of Natural Selection was the proposition that some individuals are more likely to reproduce than others. In modern terms we say that some genotypes make a greater contribution to the gene pool than others.

It is important to realise that, although it is the genes that are passed on to the next generation, it is always the *phenotype* (i.e. the body they help to build) that is subject to selection. It is not the genes themselves that are tested by selection, but their effects. The best genes are those that help to build bodies which produce the most surviving offspring – which, remember, contain copies of those genes.

The raw material upon which selection acts is genetically-based variation, the sources of which have been discussed in Chapters 8 and 12. The simplest cases of selection involve **discontinuously** varying characters which, you may remember, are usually controlled by single gene loci. In selection for discontinuously varying characters, selection affects the relative frequencies of two or three sharply distinct phenotypes.

## Types of selection

Selection can be of three types: **stabilising**, **directional** and **disruptive** (Figure 14.3).

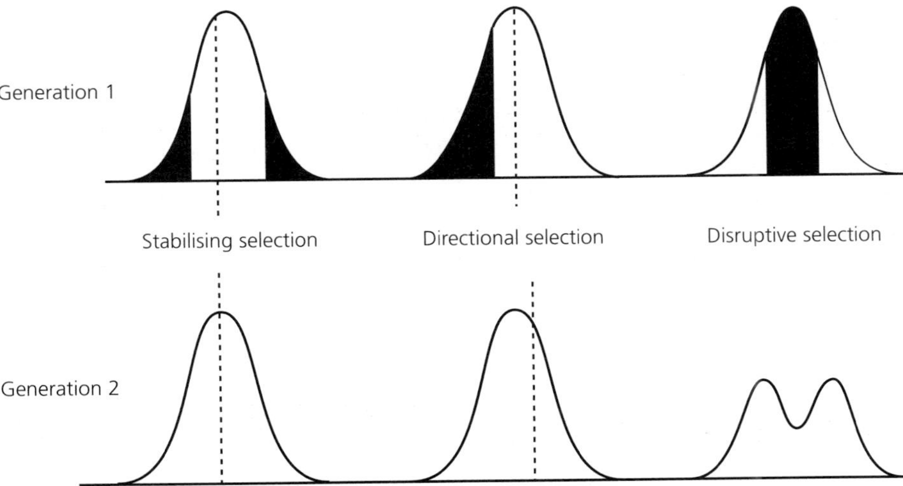


**Figure 14.3**
Different kinds of selection. Each shape represents a frequency distribution for a particular character. Shaded portions represent the phenotypes selected against

### *Stabilising selection*

Stabilising selection is that which favours the 'middle' types, or, in the case of discontinuously varying characters, heterozygotes. Because it favours the average, stabilising selection acts to prevent change. Though we tend to associate selection with evolution, most cases of selection are of the stabilising type.

We should not be surprised by this. As Richard Dawkins (author of *The Selfish Gene*) puts it, every living thing is a product of a long line of ancestors. Every one of those ancestors was successful – otherwise it would not have reproduced and become an ancestor! An organism that succeeds in reproducing has, by definition, a 'good' phenotype. Most random genetic changes result in phenotypes on either side of the 'optimum'.

The fact that stabilising selection acts to prevent change does not mean that it is associated with a constant environment – all environments change, from minute to minute, day to day, season to season and year to year. Stabilising selection occurs when there is no change in the *range* of environments experienced by successive generations. In other words, environmental change can only be an agent of evolutionary change when it occurs over a time scale that is greater than one generation.

## Sickle cell haemoglobin

A classic example of selection is the case of sickle cell haemoglobin. As detailed in Chapter 2, in sickle cell haemoglobin the β chain is abnormal. In homozygotes the haemoglobin tends to crystallise at lower oxygen concentrations causing the red cells to become sickle-shaped as the blood flows through the systemic (non-lung) capillaries. Sickle cells have a much shorter life span than normal red cells, causing severe anaemia. The normal allele ($Hb^A$) is incompletely dominant to the sickle cell allele ($Hb^S$), and the red cells of heterozygotes show only a mild sickling tendency.

Individuals homozygous for the sickle cell allele have a short life expectancy, and few, if any, survive to reproduce. As a result, in families in which both parents are heterozygous, about one in four of the children die. This represents a loss of about half the $Hb^S$ alleles possessed by the parents – the other half are in heterozygotes.

One would expect that because of this selection against the $Hb^S$ allele, it would disappear – yet in parts of Africa and Asia its frequency may be as high as 0.1. The explanation is that compared with $Hb^AHb^A$, heterozygotes are resistant to the most dangerous form of malaria (caused by *Plasmodium falciparum*). As a result the distribution of the sickle cell allele closely follows the distribution of *falciparum* malaria (Figure 14.4). In heterozygotes, then, the $Hb^S$ allele is a 'good' allele, but in homozygotes it is 'bad'. Sickle cell anaemia illustrates the fact that the 'quality' of a gene depends not only on the external environment of the body but also on other genes – what could be called the 'genetic environment'.

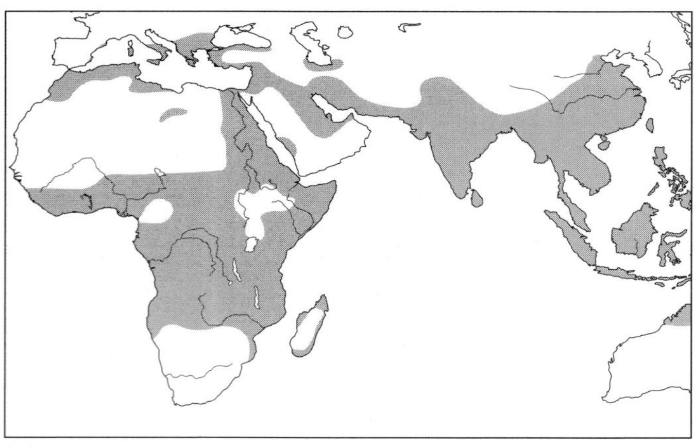

(A)

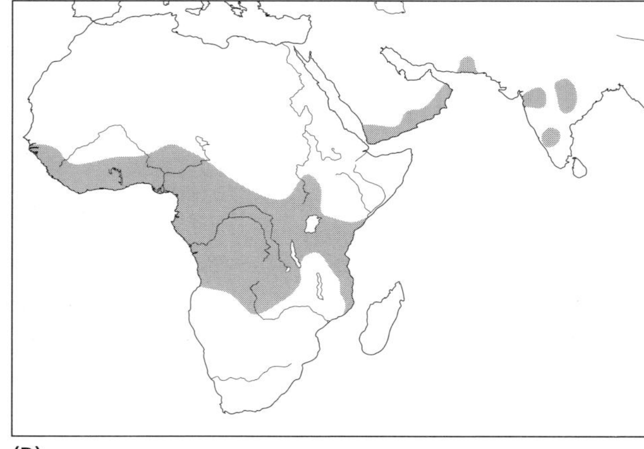

(B)

**Figure 14.4**
Distribution of malaria (A) and the sickle cell allele (B)

Selection begins soon after birth when synthesis of adult haemoglobin begins. Babies are born in Hardy–Weinberg proportions, but soon afterwards the frequency of heterozygotes begins to rise as sickle cell homozygotes succumb to anaemia and normal homozygotes catch malaria.

### *Selection in sparrows*

A particularly interesting example of selection was investigated by Bumpus in the United States at the end of the last century (Figure 14.5). He found 136 sparrows lying on the ground, severely battered by a storm,

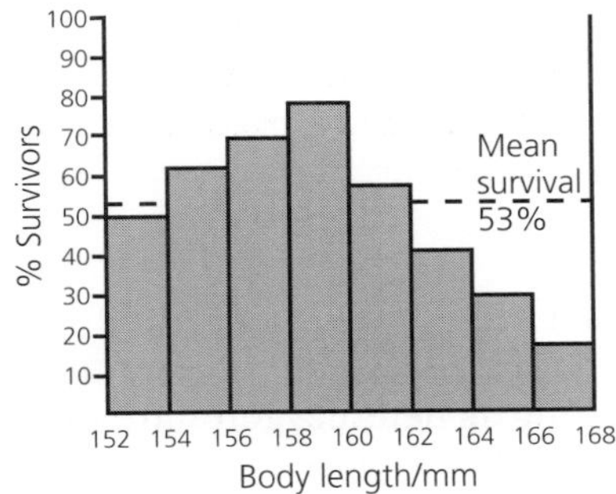


**Figure 14.5▲**
Histogram showing stabilising selection in sparrows after a storm (after Lewis and John)

and measured them. Sixty-two died, corresponding to an overall mortality of about 47%. If death had been random, then roughly the same percentage of each size group would have died. In the event the death rate was much higher amongst the largest and smallest individuals. Those of near average size had a better chance of survival.

Of course, storms are only one of many environmental hazards. Resistance to extreme cold would favour large size (a smaller surface area to volume ratio would reduce heat loss), whilst other factors would favour smaller size. The 'optimum' phenotype is, therefore, a compromise between many different factors.

### Selection for birth size in humans

A well-documented example of stabilising selection is human birth mass (Figure 14.6). The data were collected at a London hospital in the 1940s, before it became possible to save extremely premature babies. In western countries, modern medicine has changed the situation radically since then.

Babies with the best survival chances were those of average size. Thus there was selection against both extremes. The reason why the average is the 'best' is not hard to see. Large babies have greater difficulties being born, while small babies are less likely to survive afterwards. The 'best' birth size is thus a compromise between the hazards of being too small and too large.

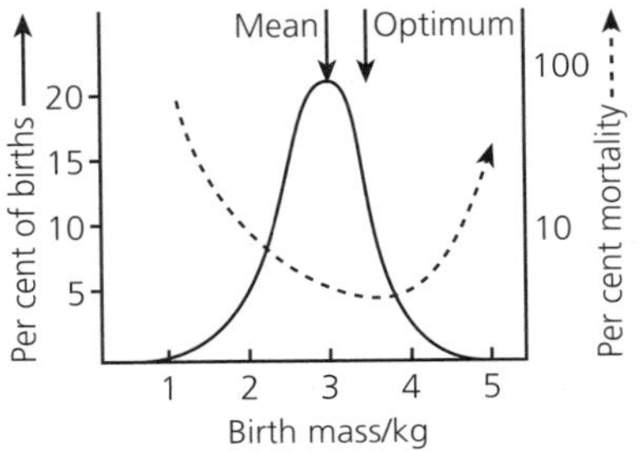


**Figure 14.6▲**
Selection for birth size in human babies in a London hospital 1935–45. (Note that mortality scale is logarithmic)

## Directional selection

When an environmental change is sustained over several generations, directional selection may occur. In the case of discontinuously varying characters it results in an increase in the frequency of one phenotype relative to another. With continuously varying characters, selection results in an increase or decrease in the mean measurement for that character.

Before taking a look at some actual cases it would be useful if we first consider two hypothetical situations: selection for a dominant allele and selection for a recessive allele.

### Selection for a dominant allele

Figure 14.7A shows how the frequency of an advantageous dominant allele $A$ would be expected to change with selection. From the moment it first appears in the gene pool as a result of mutation its frequency rises. The steepness of the gradient will depend on how favourable the allele is – in other words, the intensity of the selection pressure. Eventually the frequency of $A$ approaches 1 (i.e. 100%). However, it never quite replaces

**Figure 14.7▼**
Selection on a newly-produced, advantageous allele if it is (A) dominant and (B) recessive

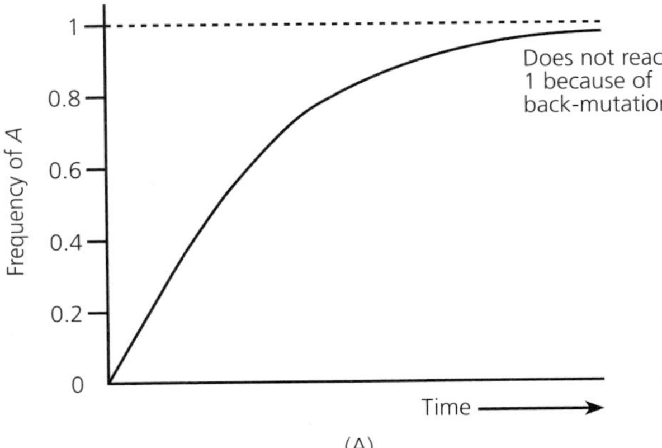


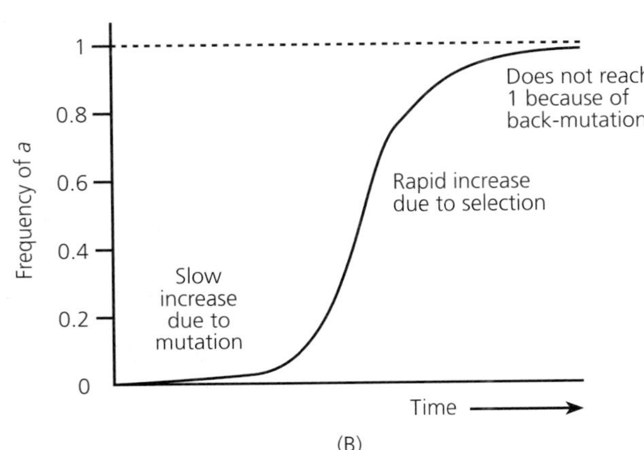

the recessive allele *a* because at low frequencies of *a* it is only rarely present in homozygotes and is thus protected from selection. Moreover, some *a* alleles are continually entering the gene pool due to back-mutation from *A*.

### Selection for a recessive allele

When an advantageous recessive allele enters a gene pool as a result of mutation, selection cannot act on it initially because it is only present in heterozygotes (Figure 14.7B). Moreover, as long as heterozygotes are rare, the probability of two heterozygotes mating is also low. The frequency of *a* continues to rise slowly because of recurrent mutation, until heterozygotes become common enough for mating to occur between them. The *aa* offspring are favoured by selection and the frequency of *a* rises, the rate depending on how advantageous it is. Finally, an equilibrium is established between removal of *A* from the gene pool by selection and its addition by back-mutation.

### Industrial melanism

One of the most fully investigated examples of natural selection occurred in the industrial areas of Europe. During the last 150 years or so, dark forms have appeared in a number of species of moth. The dark colour is due to the pigment melanin, so the dark forms are said to be **melanic**.

The first example of industrial melanism was the peppered moth (*Biston betularia*). The normal form has speckled wings which render it almost invisible against a background of lichen-covered tree bark. Studies of moth collections showed that in the 19th Century a black form appeared in urban areas and within a few decades had replaced the normal speckled type. In rural areas the speckled form remained the normal one.

The greatest environmental changes in urban areas during the 19th Century were undoubtedly a result of the industrial revolution. In country areas lichens normally cover the bark of trees but, since they are extremely sensitive to sulphur dioxide, they virtually disappeared from urban areas. Moreover, tree trunks became covered with soot.

It was suggested that the replacement of the speckled or typical form by the melanic form in urban areas lay in the inability of the speckled moths to hide from bird predators by blending against the background. The speckled pattern (which had previously been so effective as camouflage) had become highly disadvantageous. The melanic or *carbonaria* form, on the other hand, was highly adapted to this new environment (Figure 14.8).

**Figure 14.8**
Cryptic colouration in the peppered moth (*Biston betularia*)

### Testing selection by experiment

Plausible though this idea was, the explanation for industrial melanism remained a hypothesis until it had been tested experimentally. In the 1950s, H.B.D. Kettlewell carried out a series of investigations that gave powerful evidence that predatory birds were involved. He marked and released both kinds of moth in urban Birmingham and in rural Dorset, and then recaptured as many as possible by trapping on subsequent nights. Some of his results are shown in Table 14.1.

Differential predation by birds was confirmed by extensive observation through binoculars. In Dorset, birds were seen to capture nearly six times as many melanic moths as typical ones, whilst in Birmingham about three times as many typical moths as melanic ones were captured. If any further evidence were needed, it has been the effect of clean air legislation which has resulted in the return of the speckled form to many urban areas in Britain, as lichens have recolonised these areas.

**Table 14.1**
Results of Kettlewell's experiments on *Biston betularia*

| | | *typica* | *carbonaria* | Total |
|---|---|---|---|---|
| Dorset | Released<br>Recaptured<br>% Recaptured | 496<br>62<br>12.5 | 473<br>30<br>6.3 | 969<br>92 |
| Birmingham | Released<br>Recaptured<br>% Recaptured | 64<br>16<br>25.0 | 154<br>82<br>53.2 | 218<br>98 |

Breeding experiments have shown that the *carbonaria* allele is dominant to the normal one. Further work has shown that both alleles have pleiotropic effects (effects in addition to the primary one) extending to behaviour. Melanic forms show a strong preference for settling on dark backgrounds and the speckled form shows a strong preference for mottled backgrounds. This is a neat illustration of how form and behaviour are tightly interlinked.

It seems unlikely that the *carbonaria* allele had these pleiotropic effects from the outset. It is more probable that the first melanic moths behaved no differently from the typical form. Those melanics that tended to prefer to settle on dark backgrounds would however be favoured by selection.

### Other examples of directional selection

A number of other cases are known where species have adapted to human-induced changes. Most demonstrate resistance to chemicals released into the environment by human activity, for example:

- resistance to warfarin poison by rats and mice.
- resistance to antibiotics by bacteria.
- after the introduction of myxomatosis into Britain and Australia, rabbits developed resistance and the myxoma virus became less virulent (produced less severe symptoms).

## Disruptive selection

Disruptive selection is the opposite of stabilising selection. In the case of continuously varying characters it is the extremes that are favoured (Figure 14.3). In discontinuous variation heterozygotes are at a disadvantage.

One example of disruptive selection is seen in the coho salmon which breeds in streams on the East coast of North America. Males are genetically of two kinds: large 'hooknoses' which mature at 3 years of age, and 'jacks' which mature at 2 years and are much smaller. During spawning hooknoses compete to get closest to the female, the largest hooknoses usually winning and fertilising most of the eggs. Jacks on the other hand are small enough to sneak through gaps and fertilise some of the eggs, the smallest jacks being most successful. Thus there is selection against the largest jacks and the smallest hooknoses.

In the above example disruptive selection only applies to one sex. If both sexes were to occur in two kinds and there was preferential mating between them, disruptive selection could possibly lead to speciation occurring within the same habitat (see sympatric speciation). This could happen if heterozygotes were less 'fit' than homozygotes because selection would favour a mating between individuals of similar genotype. If each type became adapted to a different niche (way of life) they could eventually evolve into different species.

**Figure 14.9**
Some of the forms of the banded snail, *Cepaea hortensis*

### *Polymorphism*

Industrial melanism and sickle cell anaemia are examples of **genetic polymorphism**. This is the occurrence within a species of two or more distinct forms, the rarest being more common than could be accounted for by mutation alone. Other examples of polymorphism are the human blood groups and the colour and banding patterns of species of the banded snails *Cepaea nemoralis* and *C. hortensis* (Figure 14.9).

In all these examples, the various forms or **morphs** differ from each other by alleles at a very small number of gene loci. In some cases (such as melanic moths and sickle cell haemoglobin) the adaptive significance of the polymorphism is obvious. But what could possibly be the advantage of having blood of Group B compared with Group A or O?

Unlikely as it may appear, research has shown that one's susceptibility to certain pathological conditions depends on one's blood group. For example, people of Group O are significantly more likely to develop gastric and duodenal ulcers than people of other blood groups. On the other hand, those with blood Group A are more susceptible to stomach cancer and pernicious anaemia, and people with blood Group B are more resistant to bubonic plague.

Two kinds of polymorphism are recognised. Sickle cell and normal haemoglobin provide examples of a **stable** or **balanced polymorphism**. In this type, the frequencies of the various morphs do not change. Industrial melanism is a case of **transient polymorphism** in which the frequencies of the various morphs change over time as one replaces another.

## Selection along a gradient – clines

It often happens that the distribution of alleles in a population follows an environmental gradient – usually from north to south, or with altitude. One example is the ability of white clover (and some other members of the legume family) to produce hydrogen cyanide (HCN) when injured. The HCN is produced by breakdown of a glucoside – a sugar containing a cyanide (CN) group.

$$\text{glucoside} + H_2O \xrightarrow{\textit{enzyme}} \text{glucose} + HCN$$

The enzyme which catalyses the reaction is normally safely sealed up in lysosomes and so cannot reach the glucoside. When leaf tissues are injured, however, cell membranes are ruptured, the enzyme is released from the lysosomes and cyanide is produced. Plants which can produce cyanide are said to be **cyanogenic**, while those that cannot are **non-cyanogenic** or **acyanogenic**. Of course, to be cyanogenic, a plant must be able to make the glucoside in the first place, and this requires another enzyme.

$$\text{precursor} \xrightarrow{\textit{enzyme}} \text{glucoside}$$

Cyanogenesis thus requires the action of two genes, one (*P*) controlling the production of the enzyme that produces the glucoside, and the other (*H*) controlling the production of the enzyme that hydrolyses the glucoside to HCN. The two genes are thus complementary – when plants with the genotype *PPHH* are mated the offspring consist of $\frac{9}{16}$ cyanogenic plants:$\frac{7}{16}$ acyanogenic plants.

An interesting feature of cyanogenesis in clover is that cyanogenic plants become less and less common as environmental temperatures decrease from south to north, or with increasing altitude. For every 1°C fall in mean January temperature there is approximately, a 7% decrease in cyanogenic plants (Figure 14.10).

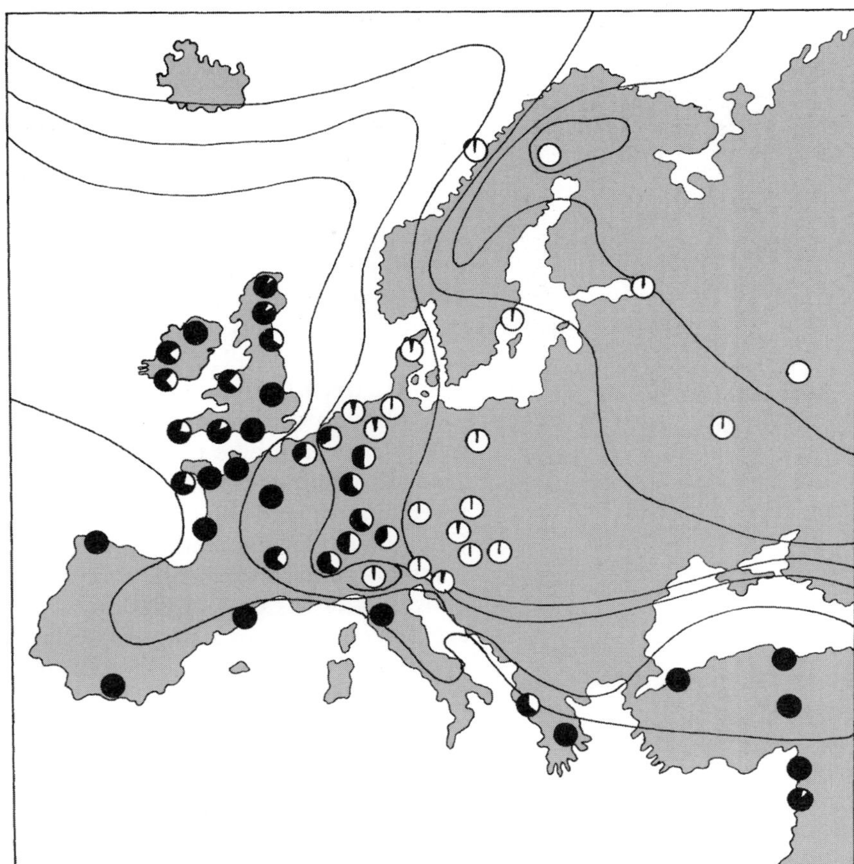

**Figure 14.10**
Geographical variation in the frequency of cyanogenic clover plants (frequency is proportional to the amount of black in each circle)

Why should these plants produce cyanide, and why should its advantages seemingly decrease in colder conditions? HCN is toxic and might be expected to poison slugs and snails feeding on cyanogenic plants. Experiments have shown that in bird's foot trefoil (a relative of clover) slugs and snails do indeed prefer acyanogenic plants. But what could be the explanation for the temperature effect? One hypothesis is that severe frost produces ice crystals in leaf cells which break cell membranes, producing cyanide and inhibiting the metabolism of the leaves (though of course this would also happen during browsing by slugs).

## The effect of chance on allele frequencies

Whatever the size of the population, chance plays a large part in determining the fate of individuals. There is very little that an individual krill can do if it happens to be caught in a whale's mouth – this is just the bad luck of being caught in the wrong place at the wrong time. Countless deaths or other reproductive failures are due to similar chance events. The effect of chance on a **gene pool**, however, depends very much on the size of the population. In large populations chance events tend to cancel each other out, but with small numbers of individuals, the effects can be considerable.

Chance can affect the composition of a gene pool in three ways: **genetic drift**, **the founder effect**, and by a **genetic bottleneck**.

### *Genetic drift*

Suppose that last year 100 oak trees in Britain were struck by lightning before they produced acorns. Even if they were of the same genotype

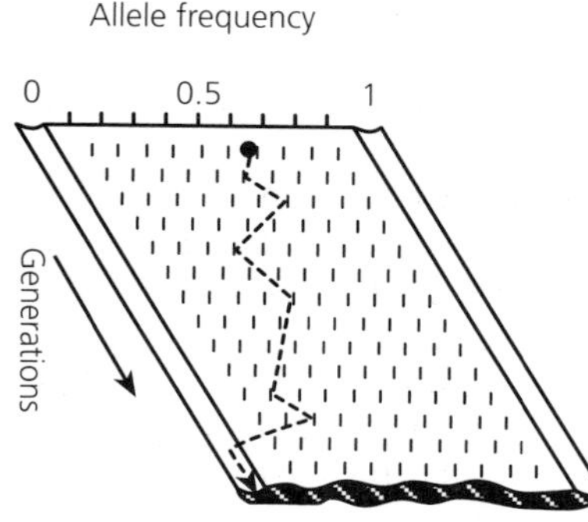


**Figure 14.11**
'Bagatelle' model of genetic drift (after Ehrlich and Holm). After release, the ball ricochets randomly from one pin to another, and sooner or later drops off the table into one of the grooves along the side (representing the loss of the allele from the population)

the effect on the composition of the gene pool would be extremely small.

It is quite a different matter with small populations. Imagine a population of eight individuals, two with the genotype *AA*, four with the genotype *Aa* and two with the genotype *aa*. The frequency of *A* and *a* is the same (0.5). Suppose that one *aa* individual dies as a result of some chance event unrelated to its genotype. That *aa* individual carried one quarter of the *a* alleles in the population and, as a result of its failure to reproduce, the frequency of *a* has declined from $\frac{8}{16} = 0.5$ to $\frac{6}{14} = 0.43$. In very small populations, a single death can have a large effect on allele frequencies. Over a number of generations, the accumulated effects of chance events on allele frequencies can be considerable and are called **genetic drift** (Figure 14.11).

Although the imaginary population above is an extreme case it is possible for numbers to be reduced even below this. In 1979 the New Zealand Chatham Island robin population was reduced to two males and three females and the robin was, at that time, the rarest bird in the world. One female disappeared and the other was an unsuccessful breeder, leaving a single successfully-breeding female. After intensive efforts by conservationists the population has risen to about 150 birds, all of which are descended from a single breeding pair. Nevertheless even in populations of several hundred individuals, considerable changes in allele frequency can result from drift.

## *The founder effect*

It sometimes happens that a population becomes subdivided into two or more smaller ones. Although the 'daughter' populations will between them contain the same alleles as the 'parent' gene pool did, the 'daughter' gene pools are unlikely to contain the same allelic mix. By pure chance the allelic frequencies are likely to be different.

Just how different they are will depend on the size of the daughter populations. If one of the new populations is very small – as, for example, when a small number emigrate from the parent population – they are most unlikely to carry the full genetic variety of the parent population.

Something very like this almost certainly happened when the first humans crossed from Asia to North America during the ice ages when sea levels were considerably lower than they are today. Amongst American Indians the frequency of blood Group O is extremely high, and in South American Indians the alleles for Groups A and B are almost absent. The first colonists may have consisted of only a few families who were quite possibly related. It is possible therefore that they were all Group O, in which case their descendants would all be Group O. By the time the population had built up to substantial numbers, any Group A genes brought in by later arrivals would only have had a small effect.

The composition of the new gene pool would thus be determined by the alleles that the first arrivals just happened to carry. This **founder effect**, as it is called, is really a result of sampling error. Had a different 'sample' left the parent population, the gene pool of the new population would have been different.

## *Genetic bottlenecks*

A population is said to go through a **genetic bottleneck** when it is temporarily reduced to a very small number. This could be due to some catastrophe such as an epidemic (Figure 14.12).

The few survivors contain only a small sample of the genetic variety of the original population. When favourable conditions return the number of

**Figure 14.12**
A cheetah: product of a genetic bottleneck

individuals may increase rapidly, but the gene pool still contains the same alleles as the original survivors. Genetic variety is regained slowly by mutation, but it could take thousands of years to restore the original genetic diversity.

An interesting example of a species that appears to have gone through a genetic bottleneck is the cheetah. Cheetahs are so alike genetically that they can accept skin grafts from each other (something that is normally only possible between identical twins). At some stage, then, the population must have undergone a catastrophic decline, probably to a few individuals. Although numbers have increased since then the species remains highly vulnerable because it lacks the genetic diversity needed to cope with change.

# The origin of species

We have seen how gene pools can change as a result of selection and, in small populations, chance effects such as genetic drift. Over a long period of time a species may undergo considerable change. The greatest evolutionary change, however, occurs when one species gives rise to two or more – in other words, at the branching points of the evolutionary tree.

The process by which one species gives rise to two or more species is called **speciation**. Before delving any deeper into this we must first be clear about what we mean by a species.

For most purposes, a species can be defined as a group of organisms that are sufficiently similar to each other to interbreed and produce fertile offspring. Genes can therefore be exchanged between populations that belong to the same species, but cannot be exchanged between populations belonging to different species. Speciation is, therefore, a process in which a gene pool gives rise to more than one gene pool. The gradual change over long periods *within* a single species is not normally considered to be speciation.

There is one obvious problem with the above definition of a species – a surprising number of organisms do not reproduce sexually – for example *Amoeba*, *Euglena*, dandelions, brambles and a great many others. Such organisms have to be named purely on the basis of anatomical similarities and differences, since the test of the ability to interbreed and produce fertile offspring cannot be applied. For these organisms the only source of genetic variation is mutation.

## What keeps species distinct?

As we have seen, different species cannot normally interbreed. When species occupy geographically separate areas they are said to be **allopatric**, and the barrier between gene pools is obviously a physical one. This may take the form of a mountain range, a desert or a stretch of water. When species live in the same area they are said to be **sympatric**. Here, the barriers to gene flow are biological and are called **isolating mechanisms**. Except in the case of speciation by polyploidy, speciation is essentially the evolution of isolating mechanisms.

Theodosius Dobzhanski, a pioneer in evolutionary biology, distinguished between two kinds of isolating mechanism depending on whether they act before or after fertilisation.

### *Prezygotic mechanisms*

- Different species breed at different times of the year.

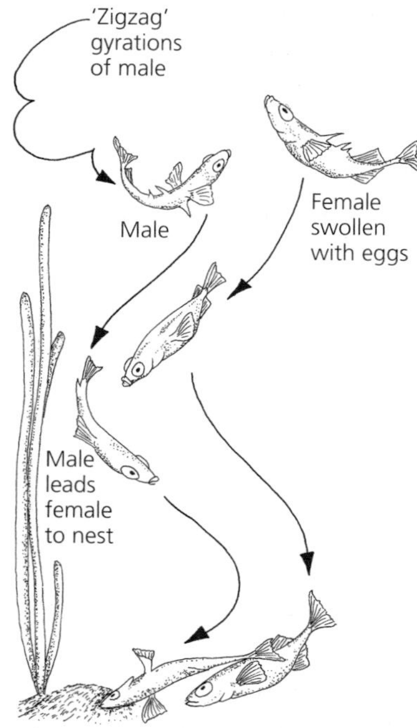


**Figure 14.13**
Zigzag dance of the male three-spined stickleback

- Courtship and mating behaviours differ. For example, the songs of the males of different species of grasshopper sharing the same habitat are different, and the chemical sex attractants of moths and many mammals are also species-specific. In many animals courtship behaviour is quite elaborate and serves to enable species recognition (Figure 14.13).
- The genitalia (external sexual organs) of many insects are shaped in a species-specific way, and it has been suggested that they may act as a kind of 'lock and key' mechanism in which only male and female parts of the same species are compatible.
- Different species of flowering plant are often adapted for pollination by particular types of animal. Red clover, for example, is mainly pollinated by bumble bees, and honeysuckle by moths. Because pollinators tend to specialise on flowers whose nectar they are best adapted to reach, pollen transfer tends to be between flowers of the same type (Chapter 37).
- Many flowers open and secrete nectar and scent at particular times of the day. It is said that the taxonomist Linnaeus could tell the time from the kinds of flower that were open. As a result of this many insects tend to concentrate on visiting one species of flower at any given time.
- Sperms may not be able to fertilise eggs of other species, either because the egg membrane does not have the appropriate chemical makeup or because the sperms cannot survive in the sexual ducts of the female. An analogous situation exists in many flowering plants in which the pollen tube is unable to grow down the styles of other species.

### *Postzygotic mechanisms*

- Hybrid inviability. If fertilisation does occur between species, development usually does not go to completion.
- Hybrid infertility. A hybrid may reach maturity, but will be infertile, as in the case of the mule (the offspring of a male donkey and a female horse). The two sets of chromosomes are unable to pair at meiosis so no viable gametes are formed.
- Hybrid breakdown. The hybrids are fully fertile but $F_2$ and subsequent generations are partly infertile, as in the case of cotton species. The explanation may be that only in the $F_2$ and later generations are genes from the two parent species combined in the same chromosome strand (as a result of crossing over in the $F_1$ hybrids), creating disharmonious gene complexes.

There are cases in which species will interbreed in captivity but not in the wild (e.g. lions and tigers). It seems probable that although one isolating mechanism may be individually insufficient to maintain isolation, in nature several mechanisms operate simultaneously.

It is important to realise that isolating mechanisms are *biological* phenomena and are the result of genetic differences between species. A mountain range is therefore *not* an isolating mechanism. As we shall see, a physical barrier may create the preconditions for isolating mechanisms to evolve.

## How are new species formed?

New species can form in two quite different ways: gradually, by the accumulation of small genetic differences, and suddenly, by changes in chromosome number. Since gradual speciation has not been observed directly, mechanisms are inevitably based partly on hypothesis.

Three possible types of gradual speciation are recognised:

- **allopatric speciation**, in which the population first becomes divided into two separate – and thus allopatric – populations.
- **parapatric speciation**, in which diverging populations occupy habitats that are contiguous (in contact with each other).
- **sympatric speciation**, in which the new species share the same habitat from the beginning.

## Allopatric speciation

Allopatric speciation is thought to be the most common type of speciation. It involves the following stages:

- a period of geographical isolation which prevents gene flow.
- a period of allopatry, during which the populations diverge genetically and postzygotic mechanisms evolve.
- the populations may subsequently merge again, provoking a further period of genetic change involving evolution of prezygotic mechanisms.

Each of these will now be discussed in more detail.

### *Geographical isolation*

Geographical barriers can be of several kinds. A mountain range (which can be pushed up by movement of the Earth's crust in only a few million years) forms a very effective barrier to most animals (some birds excepted) and plants. Moreover, because of the 'rain shadow' on the leeward side, climatic conditions on either side may be very different, leading to different selection pressures.

Mountains can also separate populations in a more subtle way. Temperatures decrease and rainfall increases with altitude. The upper part of a mountain is thus an ecological 'island' separated from neighbouring mountain tops by lowland vegetation and warmer temperatures. As a result plants and invertebrates may become genetically isolated (Figure 14.14).

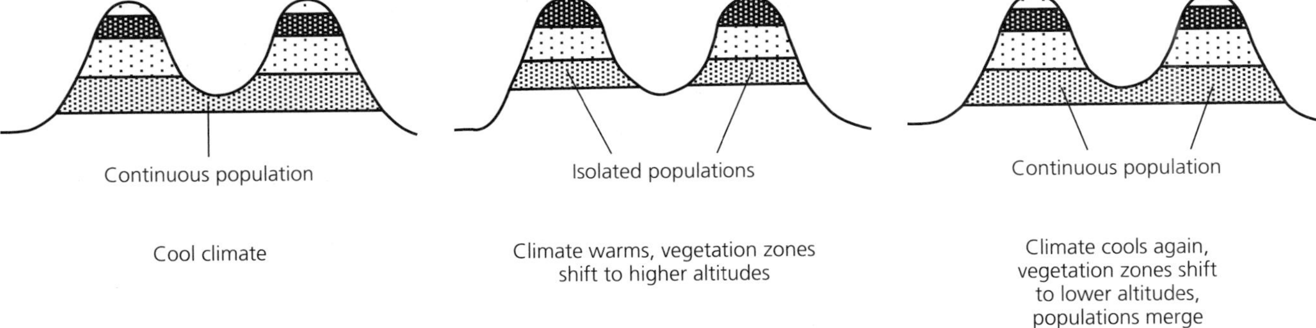
Continuous population
Cool climate
Isolated populations
Climate warms, vegetation zones shift to higher altitudes
Continuous population
Climate cools again, vegetation zones shift to lower altitudes, populations merge

**Figure 14.14**
How mountains can isolate populations

Ecological 'islands' can also be created in another way. A change to a drier climate can create forested areas of higher rainfall separated by arid grassland. This is believed to have occurred several times in the East African highlands and also in some mountain areas in Brazil.

Another barrier can result from climatic warming. After an ice age the ice melts releasing huge amounts of water and sea levels rise. During the most intense glaciations, sea level was about 100 metres lower than today

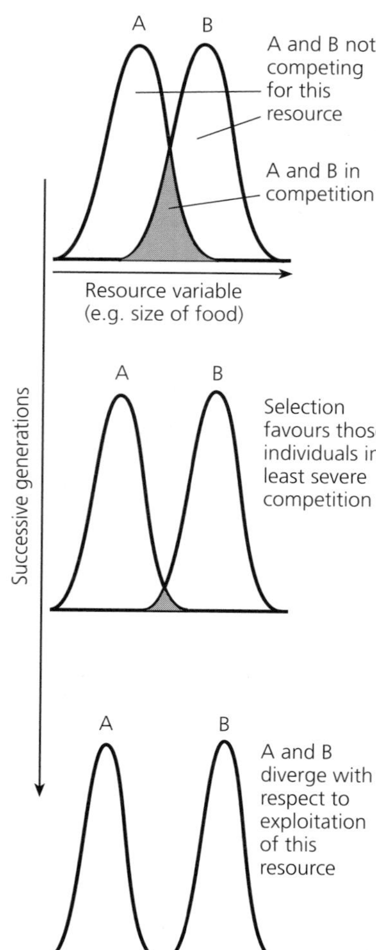


**Figure 14.15**
How competition can favour divergence

and Great Britain and continental Europe were a single land mass. With the melting of the ice caps 12 000 years ago sea levels rose, separating Britain from the rest of Europe.

A rise in sea level can separate fresh water organisms. During the last ice age river mouths were much lower than they are today – the Thames for example flowed into the Rhine. Inhabitants of these two rivers would thus have been able to intermingle, whereas today they are isolated by salt water.

### Period of allopatry

Whatever the nature of the barrier to gene flow, it is highly unlikely that environmental conditions (and therefore selection pressures) on either side of it would be identical. The composition of the two gene pools would thus tend to diverge. The more the conditions differ, the more rapidly the populations will diverge. Depending on how great the differences are between them, the two populations may be considered to be races, subspecies or full species.

Among the genetic differences that evolve during this period will be postzygotic mechanisms. The existence of such mechanisms only becomes apparent, however, if and when the populations become sympatric again and have the opportunity to interbreed. As long as the populations are allopatric this test cannot be applied, and it may never be known whether or not they are separate species.

### Period of sympatry

If the geographical barrier breaks down and the two populations merge, one of three things may happen:

1 there is some interbreeding but the hybrids are less fertile than non-hybrid offspring. Hybridisation would thus be disadvantageous since it would lead to fewer surviving offspring. Selection would favour those individuals which showed a preference for mating with their own kind. Thus individuals whose reproductive biology (breeding season, mating signals, etc.) differed most from that of the other type would be less likely to hybridise. *The evolution of postzygotic isolating mechanisms therefore provides the main impetus for the evolution of prezygotic isolating mechanisms.*

   Besides diverging in their reproductive biology it is likely that diet and other requirements would overlap to some extent, so there would be competition between them. As a result there would be strong selection favouring divergence in diets. Thus contact between two incipient species ('almost species') can cause them to diverge still further and become full species (Figure 14.15).

2 the two populations do not hybridise at all because during the period of allopatry, prezygotic mechanisms also evolve (prezygotic mechanisms can evolve without sympatry). Nevertheless, the two species may be closely similar in their niches, giving rise to rapid divergence as in (1) until competition between the species is minimised.

3 they hybridise freely. In some cases the two populations may reunite, reforming the original single species. In this case there has been insufficient genetic divergence during the period of geographical isolation. In some cases a stable hybrid zone may be established. One example is that of the carrion crow and hooded crow which are regarded as subspecies (Figure 14.16). The hybrid zone is believed to have persisted since the end of the last ice age 12 000 years ago.

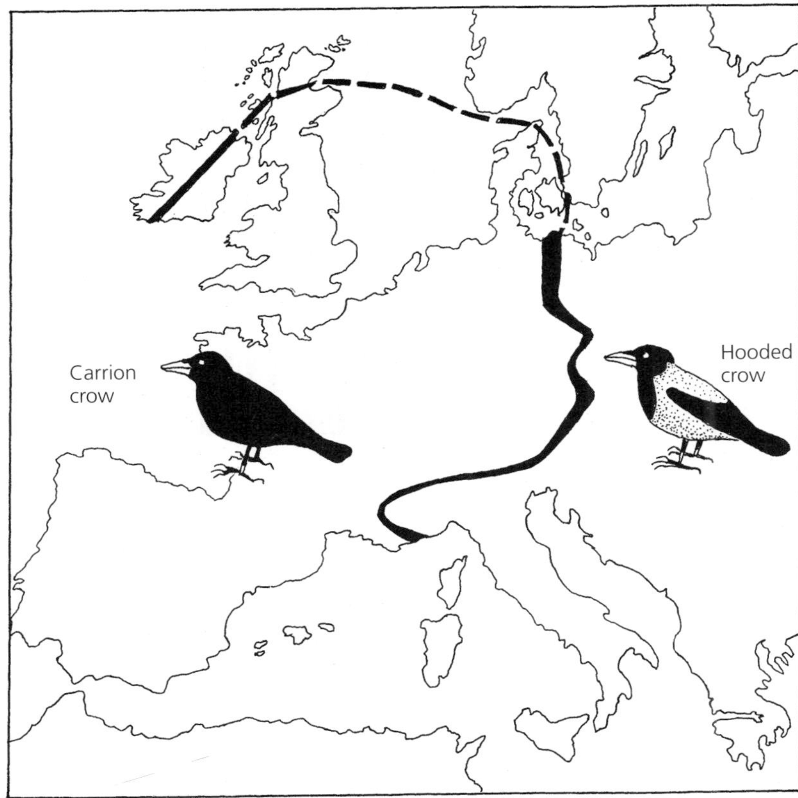


**Figure 14.16**
Zone of hybridisation between carrion and hooded crows

## Parapatric speciation

It is thought that parapatric speciation may occur when populations occupy immediately adjacent habitats separated by sharp environmental boundaries. Because selection pressures change over very short distances, mating between the two populations yields offspring that are ill-adapted to either habitat.

A possible example of the beginnings of parapatric speciation is that of grass populations adapted to heavy metal pollution adjacent to mines. In such areas the soil may contain up to 1% of copper, zinc or lead – over a thousand times the concentration sufficient to poison normal plants, which are quite unable to grow in such conditions. On the other hand, although resistant plants can survive in normal soil, they grow less well than normal plants and are eliminated by competition (Figure 14.17).

An important feature of these environments is that the boundaries between polluted and normal soil may be relatively sharp, with resistant and susceptible plants growing as close as a few metres apart. Since grasses are wind-pollinated, a resistant population is constantly subject to 'immigration' by pollen grains carrying 'non-resistance' genes from unpolluted areas. The reverse is also true – normal populations growing in unpolluted soil receive pollen from plants growing in polluted soil.

As might be expected, the offspring of matings between resistant and susceptible plants have intermediate resistance and are not well adapted to either polluted or normal soil. In both kinds of soil selection acts against partially-resistant types. On the upwind side of a spoil heap the hybrid zone may be as narrow as a metre; on the downwind side it may be 150 metres wide.

Interestingly, it has been shown that in some cases barriers to interbreeding appear to be developing. For example, the common bent grass *Agrostis capillaris* cannot normally fertilise itself, but resistant plants

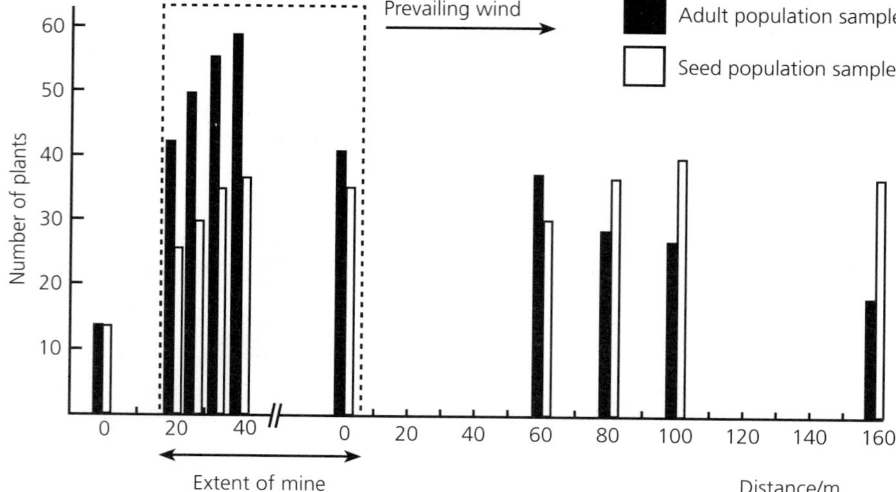


**Figure 14.17**
Distribution of copper-tolerance in *Agrostis capillaris* at the edge of a copper mine

on spoil heaps are partly self-fertile. Moreover, it has been shown that resistant plants flower about a week earlier than susceptible ones. Both these features would help to reduce the likelihood of interbreeding.

*Point to note:* Heavy metal tolerance is often cited as an example of disruptive selection, since the types of intermediate resistance are being selected against. However, resistant and non-resistant plants occupy different habitats and can thus be considered to be different populations. Within each population selection is directional – for resistance in polluted soil and against resistance in unpolluted soil.

## Sympatric speciation

Sympatric speciation is the splitting of one species into two or more species within a single population whose individuals are, to begin with, able to mate freely with each other.

Sudden sympatric speciation is known to have occurred by polyploidy in many plants by hybridisation followed by chromosome doubling (Chapter 12). Gradual sympatric speciation, however, remains a controversial topic, and there are no cases in which it is known with certainty to have occurred.

The problem is this: how can barriers to mating evolve between members of a population whose members can freely interbreed? One possible example is the case of *Rhagoletis pomonella*, a fly whose larvae eat apples and hawthorn. The flies are host-specific – adults of both sexes seek out fruits of the species on which they grew up. Here they mate and the females lay eggs. Because of this host preference, the species is effectively divided into two genetically isolated populations.

Studies of their enzymes have shown that the two populations have many genetic differences. These differences have evolved within recent historical times because apples are not native to North America, and the insect was first found on apples only about 130 years ago.

Even though the two populations are genetically isolated in nature, the host preference is learned rather than genetically determined, depending as it does entirely on the environment of the larva. When put together in the laboratory, the two forms show no mating preference. Given more time host preferences might be expected to become genetically determined. It seems then that the two populations represent the very beginnings of sympatric speciation.

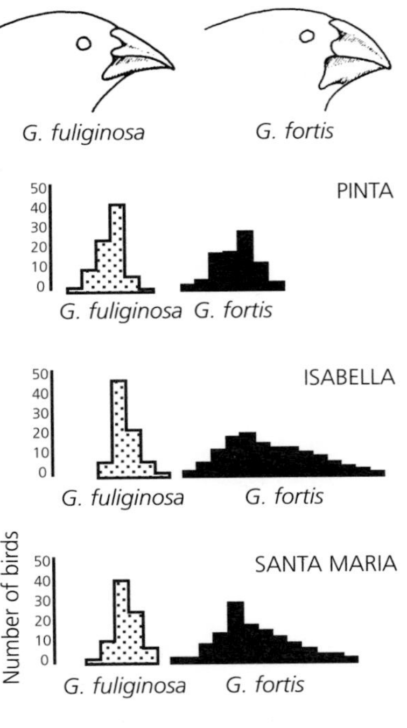


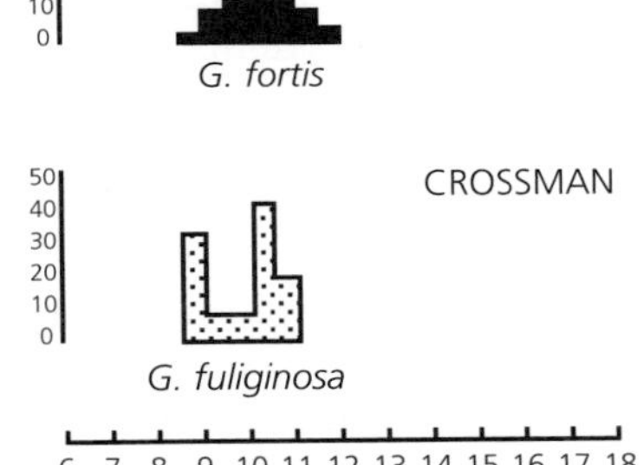


**Figure 14.18▲**
Variation in beak depth in two *Geospiza* species with and without competition (after Lack)

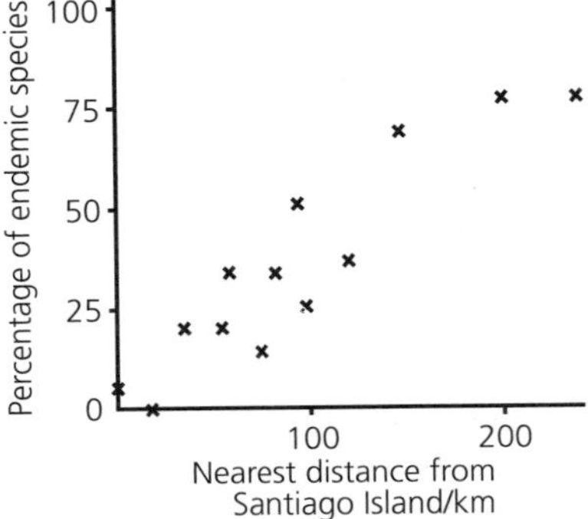


**Figure 14.19▲**
Scattergram showing the relationship between degree of isolation and percentage of endemic species of Darwin's finches

## A case study: Darwin's finches

Though Darwin did not fully appreciate it at the time of his visit in the *Beagle*, the finches of the Galápagos islands provide some fascinating evolutionary insights (Figure 1.8). Some of the 14 species are vegetarians and feed on the ground (*Geospiza* spp), whilst others are insectivorous and feed in trees (*Camarhynchus* spp). *Geospiza* species breed in the arid uplands whilst *Camarhynchus* species breed in the humid forest, although outside the breeding season there is no habitat separation.

Amongst the seed-eating ground finches there is a strong correlation between beak depth and diet. *G. magnirostris*, for example, has a very stout beak and can eat large fruits and seeds, whilst the more slender beaked *G. fuliginosa* takes much smaller seeds.

In an extensive study of Darwin's finches David Lack found evidence that competition for food was a selective force. Figure 14.18 shows the variation in beak depth of three *Geospiza* species – *G. magnirostris*, *G. fortis* and *G. fulignosa* – on a number of islands. On those islands in which only one of these species is present, there is very little difference in mean beak depth, but on islands in which both *G. fortis* and *G. fuliginosa* are present there is little overlap in beak depth.

If beak depth is an indication of diet, the interpretation seems reasonably clear. *G. fortis* and *G. fuliginosa* have similar diets where the species occur on different islands, but on islands where they share the same habitat, their diets are different. This phenomenon in which two species differ more where they are sympatric than when they are allopatric is called **character displacement**.

When we consider the distribution of species among the different islands, two patterns emerge. First, the islands nearest the centre of the archipelago tend to have the largest number of species. Second, the proportion of **endemic** species – those found on no other islands – tends to increase with distance from the central islands (Figure 14.19).

These facts can be explained as follows. The great similarity between the 14 species of finch suggests that they are derived from a common ancestor. A small flock of the ancestral species must have been blown from the mainland by unusually strong and persistent winds. Once established on one of the islands the colonists would have spread to other islands. If these latter invasions were rare, the secondary colonists would have been genetically isolated and would thus have differentiated into new species. Over time, birds would have spread back to the original islands, but by this time reproductive isolation would have been achieved. Repeated recolonisations a sufficiently long time apart could lead to the number of species increasing.

The essential requirement for speciation would therefore be that islands must be sufficiently isolated from each other to make transfer of birds between them a rare event. Only if gene flow between islands is prevented could genetic divergence occur. If the islands were too close together they would effectively act as a continuous habitat. This seems to have occurred on the central islands, three of which have no endemic species. On the more remote islands colonisation is likely to be rare, so there would be fewer species. For the same reason, though, those species are likely to be endemic because of a lack of gene flow from other islands during the period of isolation.

The Galápagos finches provide an example of **adaptive radiation** (Figure 14.20). This is the relatively sudden (in geological terms) proliferation of an ancestral form into a number of types which become adapted to a variety of ways of life occupying a variety of niches (Chapter 37).

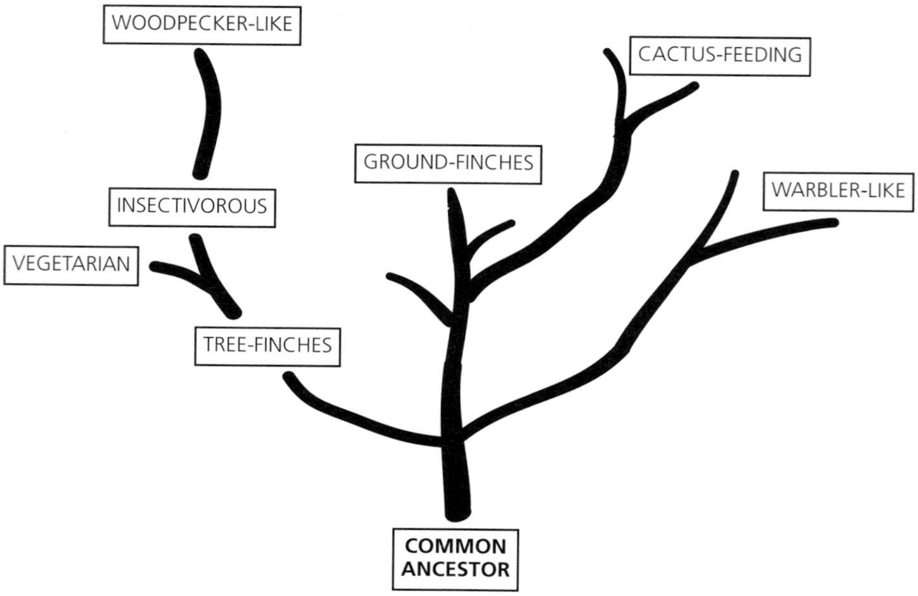

WOODPECKER-LIKE
CACTUS-FEEDING
GROUND-FINCHES
WARBLER-LIKE
INSECTIVOROUS
VEGETARIAN
TREE-FINCHES
COMMON ANCESTOR

**Figure 14.20**
Hypothesised adaptive radiation in the Galápagos finches

Adaptive radiation is a recurrent theme in evolution. The origin of all the major taxonomic groups was marked by their diversification into a wide variety of types. The placental mammals provide an excellent example. They originated about 70 million years ago, and within 15 million years or so they had diversified into bats, whales, rodents and all the other orders of placental mammals.

How do we explain these sudden explosions of diversity? The answer seems to be that each radiation occurs in response to a sudden relaxation of competition. In the case of the Galápagos finches it was the arrival of their ancestors on isolated islands. In the case of major taxonomic groups it was probably the evolution of some fundamentally new organisational feature. The development of the arthropod exoskeleton, for example, would have enabled rapid movement by means of jointed limbs working as levers. This would have opened up enormous possibilities for new ways of life and, in particular, new diets.

## Some problems with species

Though the definition of a species (given earlier in the chapter) works reasonably well for most organisms, there are some problem cases, as the following example shows.

### *Ring species*

A ring species is a special case of a cline which forms a loop, with the two ends joining up with each other. The fact that all known examples are birds is due to their great mobility, making it difficult to achieve geographical isolation. The classic example is the case of the herring gull (*Larus argentatus*) and the lesser black-backed gull (*L. fuscus*). In Western Europe they rarely interbreed and thus behave as distinct species. However, these two forms are simply the ends of a horseshoe-shaped cline extending around the North Pole (Figure 14.21). They are connected by a series of intermediate and interbreeding populations around the pole, with the ends of the cline meeting in Britain and neighbouring areas of Western Europe. Genes can, therefore, be exchanged between the two species, but only over many generations between neighbouring populations in Northern Asia and America.

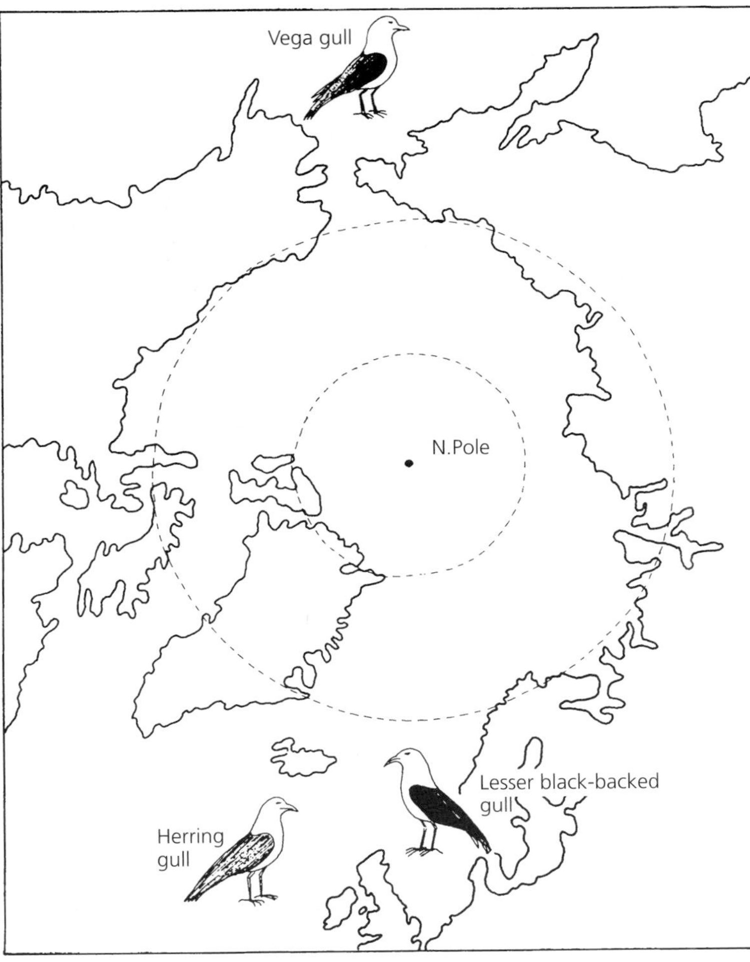


**Figure 14.21**
Distribution of the herring gull and lesser black-backed gull

## QUESTIONS

**1** In each of the three populations **X**, **Y** and **Z** the alleles *A* and *a* each have a frequency of 0.5, however the frequencies of the genotypes *AA*, *Aa* and *aa* differ.

| | X | Y | Z |
|---|---|---|---|
| *AA* | 0.2 | 0.3 | 0.25 |
| *Aa* | 0.6 | 0.4 | 0.5 |
| *aa* | 0.2 | 0.3 | 0.25 |

**a)** Which of these populations is in genetic equilibrium?
**b)** In which population could there be a tendency for each genotype to mate with an organism of the same genotype?
**c)** In which population could there be a tendency for each genotype to mate with organisms with a genotype different from its own?

**2** For reasons explained in Chapter 27, pregnancies in which a mother with Rh-negative blood is expecting a child with Rh-positive blood can be hazardous. 85% of the European population are Rh-positive, and Rh-positive is dominant to Rh-negative.

**a)** What is the frequency of the Rh-negative allele in the population?
**b)** What proportion of men would you expect to be heterozygous for the Rh allele?
**c)** What proportion of total pregnancies would you expect to be at risk?

**3** The life expectancy of children of first cousin marriages is less than that of children of unrelated parents. Suggest an explanation.

## QUESTIONS

**4** The graphs illustrate various factors relating to the reproductive success of great tits in a wood in Oxfordshire.

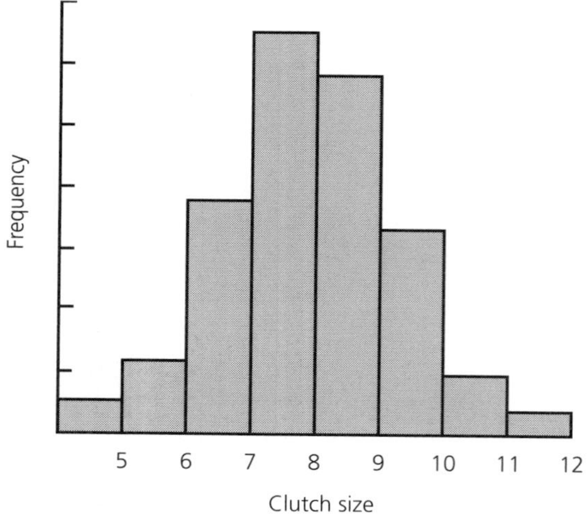


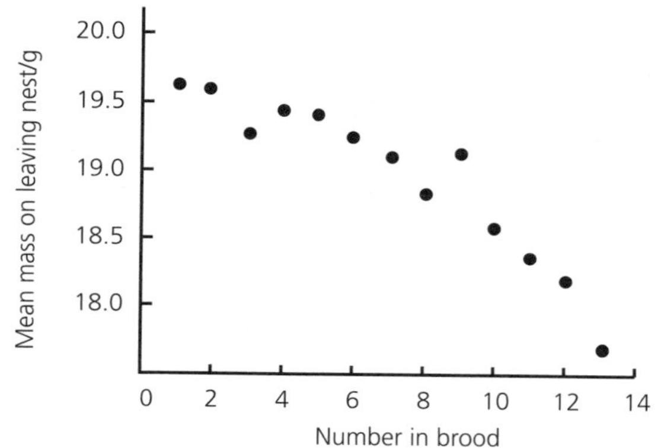


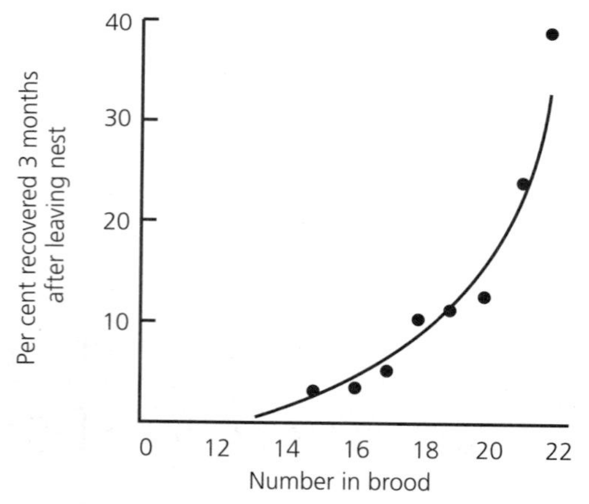


**a)** Why do nestlings in larger clutches tend to be smaller?

**b)** Why is the life expectancy of fledglings (young birds at time of leaving the nest) affected by clutch size? Explain your reasoning.

**c)** Explain why there is an optimum clutch size.

**d)** Of what kind of selection is this an example?

**5** The diagram below shows the distribution of two species of tree frog, *Hyla ewingi* and *H. verreauxi*, in South-East Australia. The area where both species occur is shown in black. Oscillographs (sound traces) of the male mating calls are also shown.

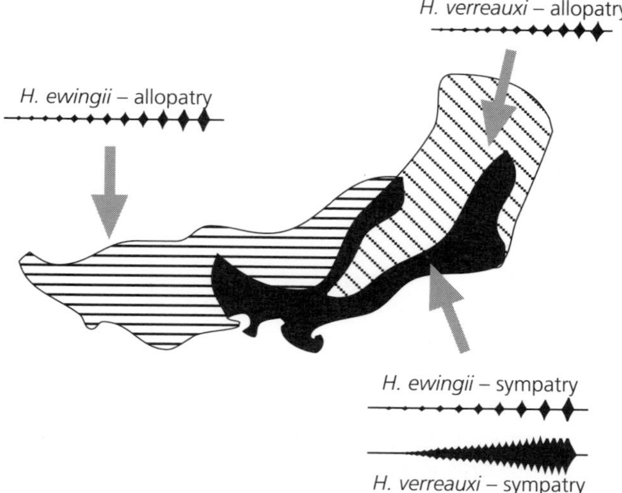


**a)** Explain what is meant by the terms 'allopatric' and 'sympatric'.

**b)** Suggest an evolutionary explanation for the fact that the mating calls of the two species are similar in areas where only one of the two species is present, but markedly different in areas where both species live.

## EXAM QUESTIONS

**1** Chromatography and electrophoresis are techniques which can be used to separate molecules.

**a)** Briefly describe how electrophoresis separates molecules. *(2 marks)*

**b)** The diagram shows the distribution of peptides produced by digesting normal haemoglobin (HbA) and sickle cell haemoglobin (HbS) with an enzyme. The peptides were separated in one direction by chromatography and in the other by electrophoresis.

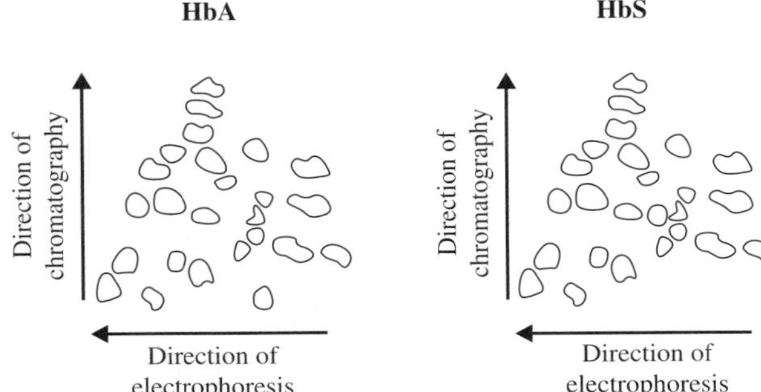


The difference between HbA and HbS is the result of a substitution of a single base pair in the haemoglobin gene. Explain how the diagram supports this statement. *(2 marks)*

**AEB**

**2 a)** Active transport, facilitated diffusion and osmosis are three of the ways in which substances move into cells. Complete the table with a tick (✓) if the statement is true for the particular method of transport.

| Statement | Method of transport into cell: Active transport | Facilitated diffusion | Osmosis |
|---|---|---|---|
| Requires energy from ATP | | | |
| Involves protein carrier molecules | | | |
| Involves movement of molecules from where they are in a high concentration to where they are in a lower concentration | | | |

*(3 marks)*

**b)** A turgid plant cell was placed in a solution of sucrose. The diagram shows the appearance of the cell after one hour.

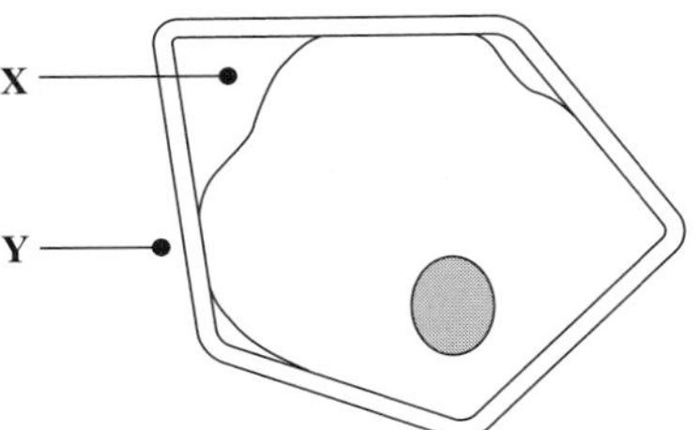


**i)** From the diagram, what is the evidence which shows that the water potential of the cell sap must be higher than that of the sucrose solution? *(1 mark)*

**ii)** Explain why the water potential at point **X** is equal to that at point **Y**. *(1 mark)*

**AEB**

**3** The polymerase chain reaction is a technique used by biologists to make large amounts of DNA from very small samples. The process is explained in the diagram.

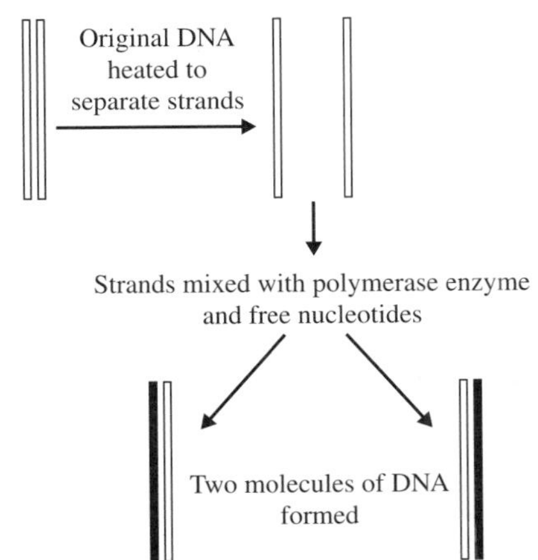

## EXAM QUESTIONS

a) Explain why the DNA produced in this reaction is exactly the same as the original DNA. *(2 marks)*

b) At the end of the first cycle of this reaction, there will be 2 molecules of DNA. How many molecules of DNA will there be at the end of 5 cycles? *(1 mark)*

c) Give *two* ways in which this process differs from transcription. *(2 marks)*

**AEB**

**4** The diagram represents the flow of information in a cell.

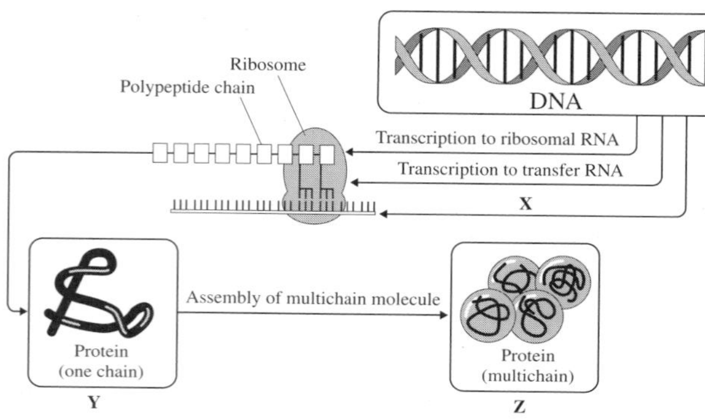


a) What is produced by the process labelled X? *(1 mark)*

b) i) Describe how a polypeptide chain is assembled from free amino acids. *(4 marks)*

ii) What is the highest level of organisation of a protein molecule represented in the diagram by Box Y; Box Z? *(2 marks)*

**AEB**

**5** The diagram represents a tRNA molecule.

a) Explain why, in this molecule, the complementary base to the cytosine base must be:

i) a purine; *(1 mark)*

ii) guanine. *(1 mark)*

b) Briefly describe the part played by the tRNA molecule in protein synthesis. *(3 marks)*

**AEB**

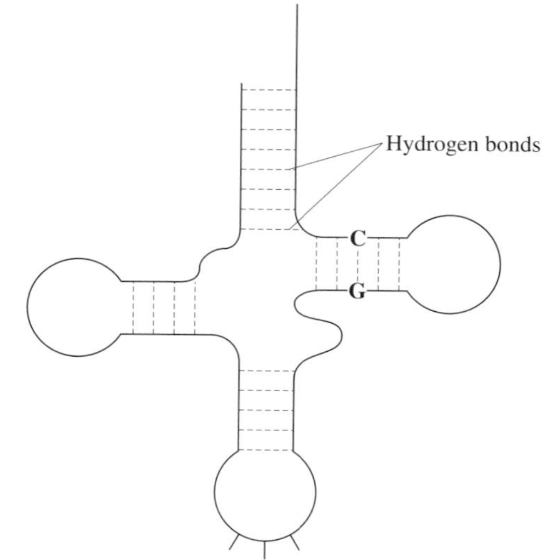


**6** The diagram shows a simple respirometer. It is being used to measure the rate of oxygen uptake by a suspension of yeast cells in tube **A**. There are no yeast cells in tube **B**. A muslin bag containing soda lime is suspended in each tube. The syringe in tube **B** can be used to level the fluid in the manometer.

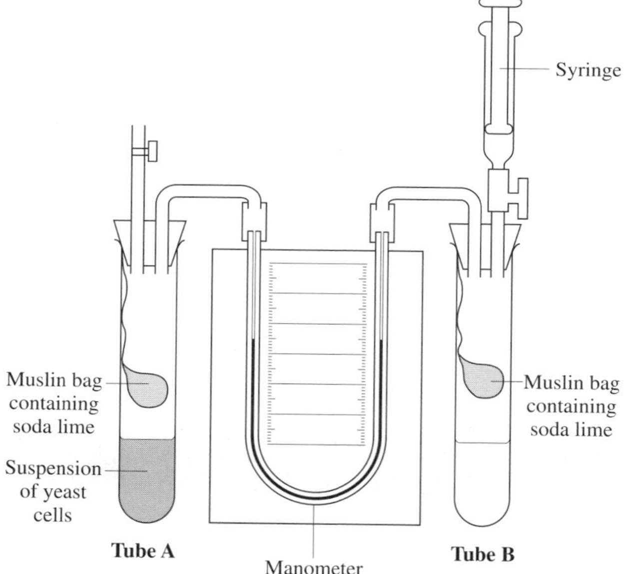


a) i) What is the purpose of the soda lime in tube **A**? *(1 mark)*

ii) Explain how tube **B** acts as an experimental control. *(2 marks)*

## EXAM QUESTIONS

**b)** The table shows the readings that were obtained over a period of 50 minutes.

| Time/minutes | Syringe reading/mm$^3$ |
|---|---|
| 0 | 0 |
| 10 | 3.1 |
| 20 | 5.9 |
| 30 | 9.2 |
| 40 | 12.1 |
| 50 | 14.9 |

Use these readings to calculate the respiratory rate of the yeast suspension. Show your workings. *(2 marks)*

**AEB**

**7 a)** The resistance of houseflies to an insecticide is controlled by a single gene with two alleles. The allele *R* for resistance is dominant to the allele *r* for susceptibility (no resistance). In a survey, 1300 houseflies were trapped and 630 of these were found to be resistant to the insecticide. Calculate:

**i)** the frequency of flies susceptible to the insecticide; *(1 mark)*

**ii)** the frequency of the *r* allele. *(1 mark)*

**b)** Explain why a dominant allele of one gene would become established in a population faster than a recessive allele of another gene even though they might both give the same advantage. *(3 marks)*

**AEB**

**8** The table below refers to components of the cell surface membrane (plasma membrane) and to their roles in transporting substances across the membrane.

Complete the table by inserting an appropriate word or words in the empty boxes.

**9** The diagram below shows the structure of a bacterial cell as seen using an electron microscope.

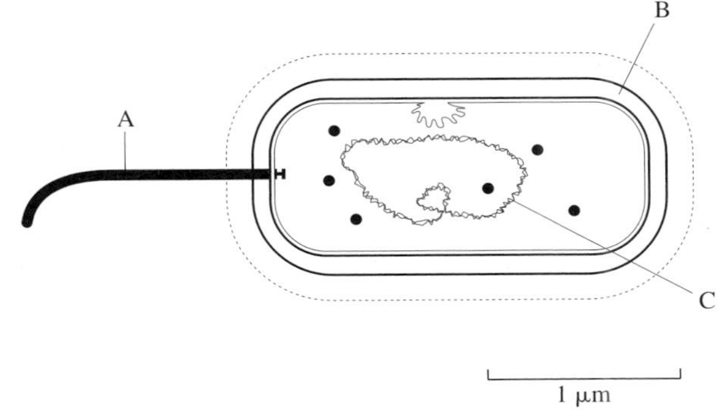


**a)** Name the parts labelled A, B and C. *(3 marks)*

**b)** Give *two* reasons why viruses are an exception to the cell theory. *(2 marks)*

**ULEAC**

**10** Explain what is meant by each of the following terms.

**a)** Point mutation *(2 marks)*

**b)** Reverse transcription *(2 marks)*

**11** The monosaccharides glucose and fructose are *reducing sugars*. Sucrose is a disaccharide which is not a reducing sugar.

The Benedict's test is used to detect reducing sugars. When reducing sugars are boiled with Benedict's solution a red precipitate is produced. This precipitate can be filtered from the solution, dried and weighed. If excess Benedict's solution is used, the mass of precipitate produced is proportional to the concentration of reducing sugar in the solution. The enzyme sucrase is a hydrolase and does not react with Benedict's solution.

| Component | Subunits | Chemical bond between subunits | Role in transport |
|---|---|---|---|
| Phospholipid | Fatty acids, glycerol and phosphate | | |
| Carbohydrate side chain | | | Receptor |
| Protein | | Peptide | |

*(6 marks)*

**ULEAC**

## EXAM QUESTIONS

**a)** In an experiment, sucrase was added to a solution of sucrose and incubated for five minutes. The Benedict's test was then carried out on the resulting solution and a red precipitate was produced.

Suggest an explanation for this result. *(2 marks)*

**b)** A further experiment was carried out to investigate the effect of silver nitrate on the activity of sucrase. The procedure described above was repeated, but different concentrations of silver nitrate were added to the sucrase. The solutions were kept at the same pH for the same time. The mass of precipitate produced by the Benedict's test at each concentration was measured. The results are shown in the table below.

| Concentration of silver nitrate/mol $dm^{-3}$ | Mass of precipitate/ mg |
|---|---|
| 0 (control) | 50 |
| $10^{-6}$ | 37 |
| $10^{-5}$ | 27 |
| $10^{-4}$ | 10 |

**i)** Calculate the percentage decrease in the mass of precipitate produced in the solution containing $10^{-5}$ mol $dm^{-3}$ silver nitrate compared with the control test. Show your working. *(2 marks)*

**ii)** Suggest an explanation for the effect of the silver nitrate solution on the activity of the enzyme sucrase. *(2 marks)*

**c) i)** Explain why it is important to maintain constant pH when investigating enzyme activity. *(2 marks)*

**ii)** State *three* precautions, other than maintaining constant pH, which should be taken to produce reliable results in the above investigation. *(3 marks)*

**ULEAC**

**12 a)** Distinguish between the terms *gene* and *allele*. *(3 marks)*

**b)** The diagram below shows a family tree in which the blood group phenotypes are shown for some individuals.

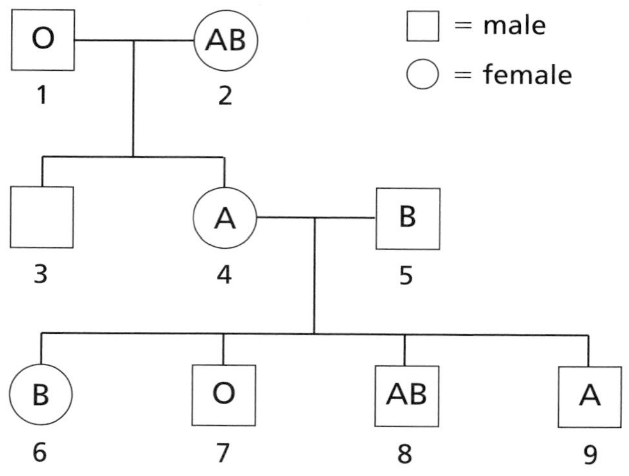


**i)** Using the symbols $I^A$, $I^B$ and $I^O$ to represent the alleles, indicate the genotypes of people 1–6. *(5 marks)*

**ii)** State the possible blood groups of person 3. Explain your answer. *(3 marks)*

**ULEAC**

**13** The diagram below shows the structure of the virus T2.

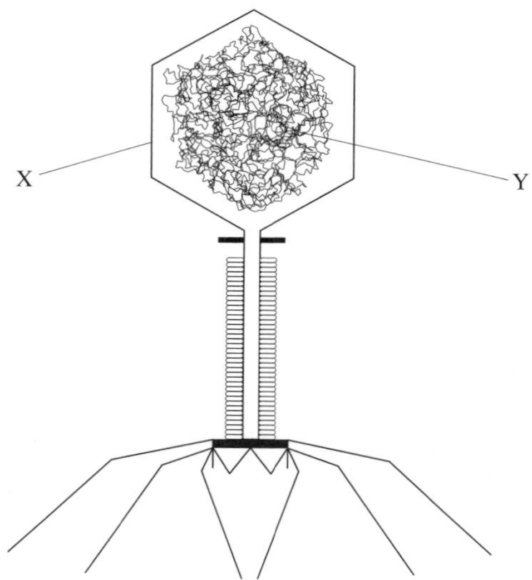


**a)** State the name and chemical composition of the structures labelled **X** and **Y**. *(2 marks)*

T2 cannot reproduce by itself and relies upon a host cell for reproduction.

EXAM QUESTIONS

EXAM QUESTIONS

**b)** State *three* reasons why all viruses rely on host cells for their reproduction. *(3 marks)*

T2 uses a bacterium as its host. The diagram below shows the results of an experiment in which T2 was added to a culture of bacteria. Samples of the culture were then taken at intervals to determine the number of free T2 viruses present.

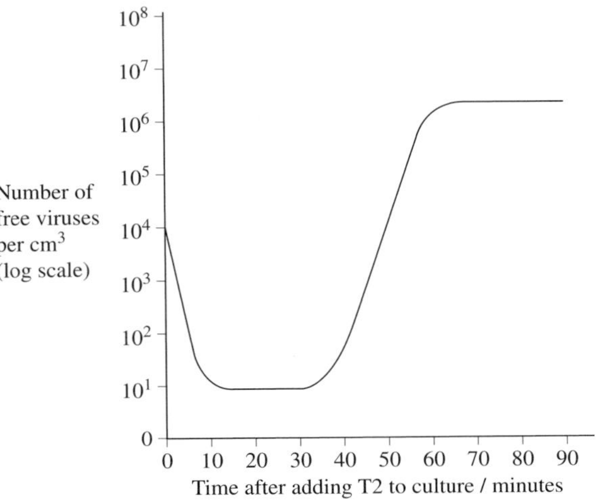


**c)** With reference to the diagram above, describe and explain the changes in number of free T2 viruses

**i)** in the first 10 minutes; *(2 marks)*

**ii)** between 30 and 60 minutes. *(3 marks)*

**UCLES**

**14** The diagram shows a model representing the structure of a phospholipid molecule.

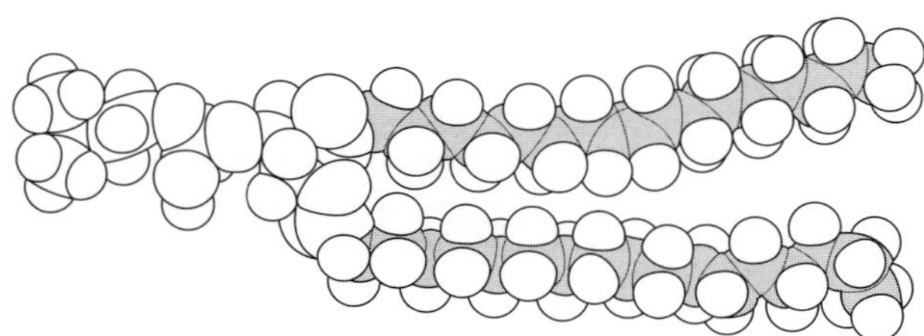

**a)** Where in a chloroplast would phospholipid molecules be found? *(1 mark)*

**b)** On the diagram, label the hydrophilic **and** hydrophobic parts of the phospholipid molecule. *(1 mark)*

**c)** Give *two* ways in which the structure of a phospholipid molecule differs from that of a triglyceride molecule. *(2 marks)*

**NEAB**

**15** The diagram shows some of the processes that occur within a mitochondrion and some of the substances that enter and leave it.

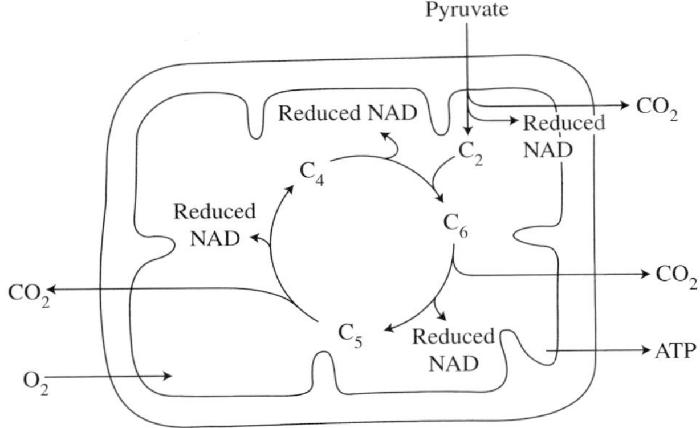


**a)** Name the process that

**i)** produces pyruvate in the cytoplasm;

**ii)** produces reduced NAD and carbon dioxide in the matrix of the mitochondrion. *(2 marks)*

**b)** Describe how ATP is produced from reduced NAD within the mitochondrion. *(3 marks)*

**NEAB**

**16** The diagram shows the structure of a tRNA molecule.

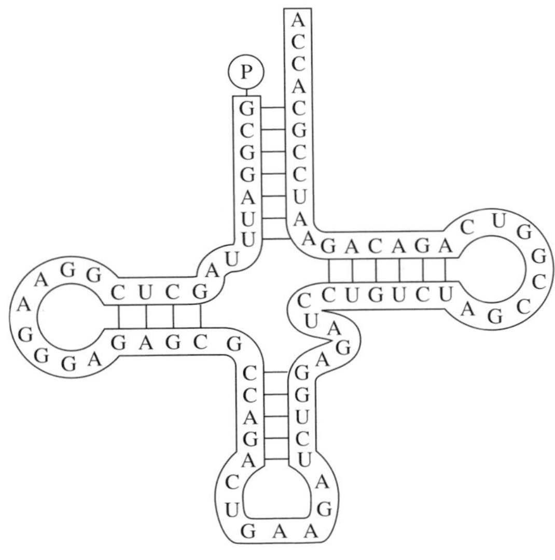

EXAM QUESTIONS

## EXAM QUESTIONS

a) Give *two* ways in which the structure of a tRNA molecule differs from that of a DNA molecule. *(2 marks)*

b) Explain how the specific shape of the tRNA molecule shown in the diagram is determined by the pattern of bonding. *(2 marks)*

c) i) Give the base sequence of the anticodon of this tRNA molecule.

ii) Which mRNA codon would correspond to this anticodon? *(2 marks)*

**NEAB**

**17** Red-green colour blindness is an inherited condition which causes affected individuals to confuse the colours red and green.

The table at the bottom of the page shows the approximate percentage of red-green colour blindness in different populations.

a) i) State *two* general conclusions which can be drawn from the data. Suggest an explanation for each. *(6 marks)*

ii) Explain why the percentages shown in the table may be unreliable. *(1 mark)*

iii) Suggest *one* possible selective disadvantage of red-green colour blindness in the *evolution* of Man. *(1 mark)*

b) i) State the name given to the way in which colour blindness is transmitted. *(1 mark)*

ii) The table suggests that about 0.4% of women in Britain are red-green colour blind. Use appropriate symbols to complete the genetic diagram to show the *most likely* way in which this defect arises in women. *(7 marks)*

| | Male | Female | | |
|---|---|---|---|---|
| Parental phenotype: | .......... | .......... | | |
| Parental genotype: | .......... | .......... | | |
| Gametes: | ○ ○ | ○ ○ | | |
| Offspring genotypes: | .......... | .......... | .......... | .......... |
| Offspring phenotypes: | .......... | .......... | .......... | .......... |
| | .......... | .......... | .......... | .......... |

**WJEC**

| Population | Sample size | Percentage colour blind<br>Male | <br>Female |
|---|---|---|---|
| Arabs | 337 | 10.0 | 1.00 |
| Swiss | 2000 | 8.0 | 0.64 |
| British | 16 180 | 6.6 | 0.36 |
| Japanese | 259 000 | 4.0 | 0.16 |
| Eskimos | 297 | 2.5 | 0.06 |
| Fiji Islanders | 608 | 0.8 | 0.006 |

*Ref: Adapted from Human Biology by Harrison, Weiner, Tanner and Barnicot (1982).*

## EXAM QUESTIONS

**18** Only micro-organisms can fix nitrogen into forms which plants can use. A shortage of 'fixed' nitrogen may reduce productivity of food crops.

One possible way to overcome this problem is to modify the harmless bacteria, which normally live on cereal crops, using the technique of genetic engineering. Scientists are now able to introduce the nitrogen-fixing gene complex (*nif* gene) into bacteria and then clone these bacteria.

**a) i)** Name *one* organism which contains the *nif* gene. *(1 mark)*

**ii)** What is meant by the term *clone*? *(1 mark)*

**b)** The diagram shows six stages, (l) to (q), of one possible method which could be used to produce clones of bacteria.

**i)** In stage (m), what was the purpose of adding

1. start and stop signals;
2. sticky ends? *(3 marks)*

**ii)** In stage (n) the plasmid vector has been cut open so that its DNA can hybridise with the *nif* DNA. Name the *type* of enzyme used to cut open the plasmid. *(1 mark)*

**iii)** Stage (o) shows details of the *nif* plasmid. Use the stages (o)–(q) on the diagram to:

1. name the marker gene carried by the plasmid;
2. suggest why this marker is important in the process. *(3 marks)*

**iv)** Suggest how *you* would screen the bacteria shown in stage (q) to identify those able to fix nitrogen. *(1 mark)*

**v)** In experiments, the bacterium used to clone the *nif* gene was *Escherichia coli*, an organism which is widely distributed and capable of exchanging genetic material with other types of bacteria.

Suggest why some scientists think that the production of recombinant DNA in stage (n) could be dangerous. *(1 mark)*

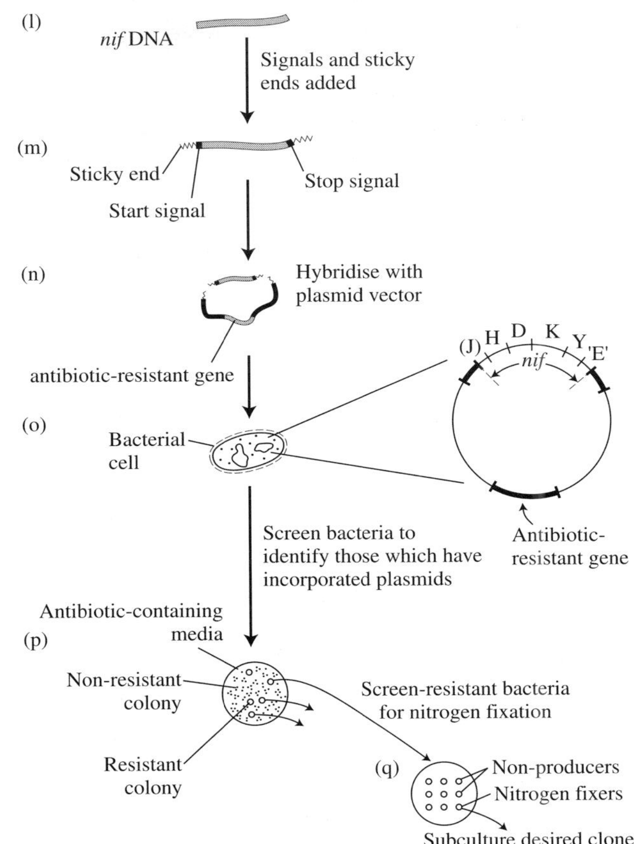


**c)** The procedures described would increase the productivity of cereal crops. Suggest and explain *one* other advantage of this genetically engineered crop. *(2 marks)*

**WJEC**

**19** In an experiment a common zebra was crossed with a donkey to produce a hybrid animal called a zebronkey. The diagram over the page shows the parental phenotypes, the chromosomes contributed by each parent and the offspring phenotype. The animals are not drawn to the same scale.

**a) i)** Describe *two* differences in parental phenotype. *(2 marks)*

**ii)** State the sex of the zebronkey. *(1 mark)*

**iii)** How many chromosomes would you expect to find in a zebronkey skin cell? *(1 mark)*

**b)** Zebronkeys cannot produce offspring. Use the information in the diagram to suggest an explanation for this. *(3 marks)*

## EXAM QUESTIONS

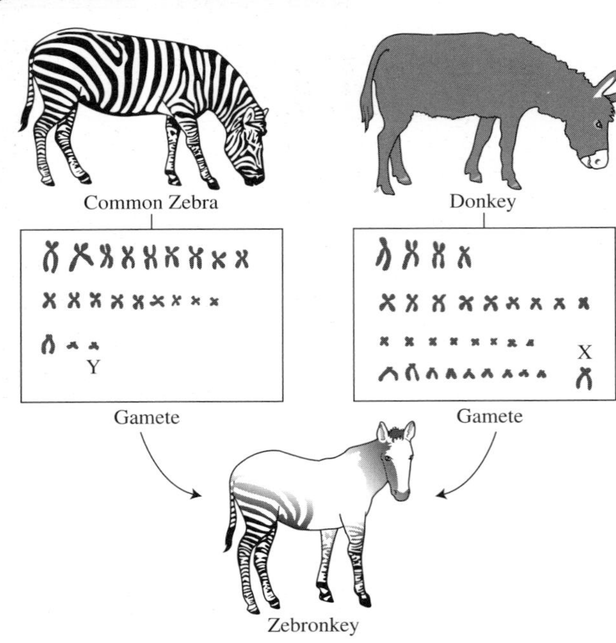


c) Suggest why horse breeders are interested in producing hybrids between different members of the horse family. *(1 mark)*

**WJEC**

**20** The simplified diagram shows the relationship between the main stages of respiration (**X**, **Y** and **Z**) and the regions within the cell where they take place.

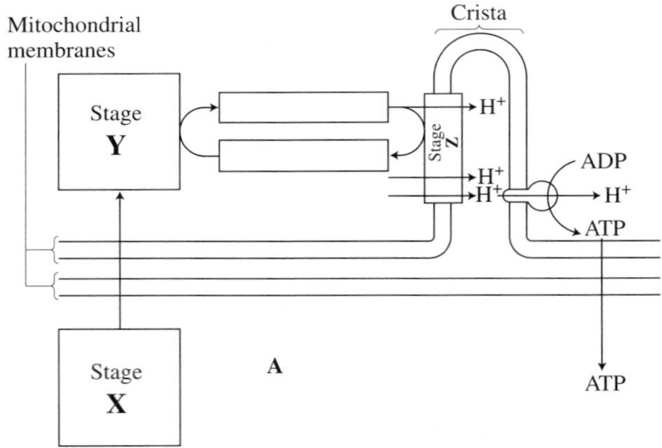


a) Name the
   i) region of the cell labelled **A** in the diagram. *(1 mark)*
   ii) stage of respiration labelled **X**. *(1 mark)*
   iii) chemical compound which passes from stage **X** for use in stage **Y**. *(1 mark)*

b) i) Complete the *two* empty boxes in the diagram. *(2 marks)*
   ii) Name the molecule produced at the end of stage **Z**. *(1 mark)*

c) Protons and electrons are vital to the function of the process labelled **Z** in the diagram.

   State the importance of:
   i) protons
   ii) electrons *(2 marks)*

d) In a pollution incident cyanide, a metabolic poison which binds irreversibly with a large enzyme molecule in stage **Z**, escaped into the River Wye. Suggest why this resulted in the *immediate* death of thousands of fish. *(1 mark)*

**WJEC**

**21** The simplified diagram represents part of the process in the production of a crop plant resistant to the broad-spectrum herbicide, glyphosate. **M** represents the gene for glyphosate resistance.

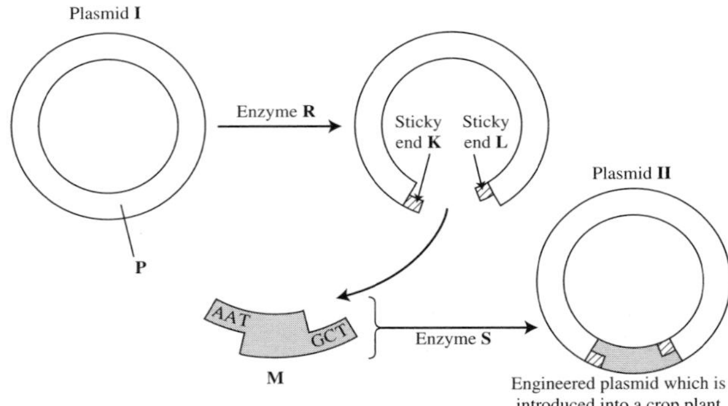


a) i) Name the specific biological molecule **P** which forms the plasmid. *(1 mark)*
   ii) Name the enzyme types represented by **R** and **S**. *(2 marks)*
   iii) The diagram shows the sticky ends on **M** in detail. Enter the base sequence you would expect to find on sticky end **K** and **L** of the plasmid. *(2 marks)*

b) i) Suggest *one* advantage to farmers of growing crops resistant to glyphosate. *(1 mark)*
   ii) Suggest *one* disadvantage of this particular strategy. *(1 mark)*

**WJEC**

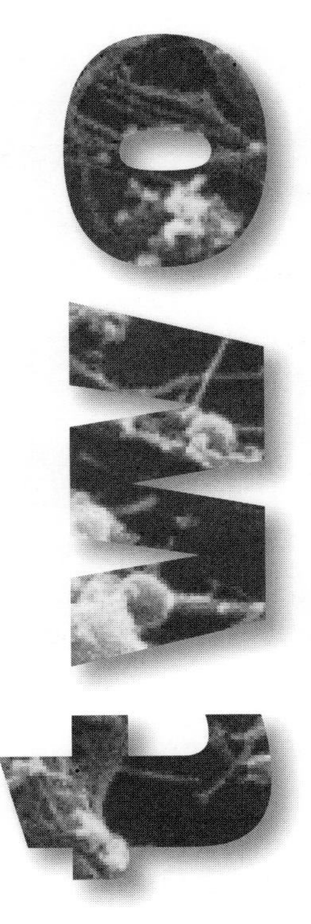

Part two

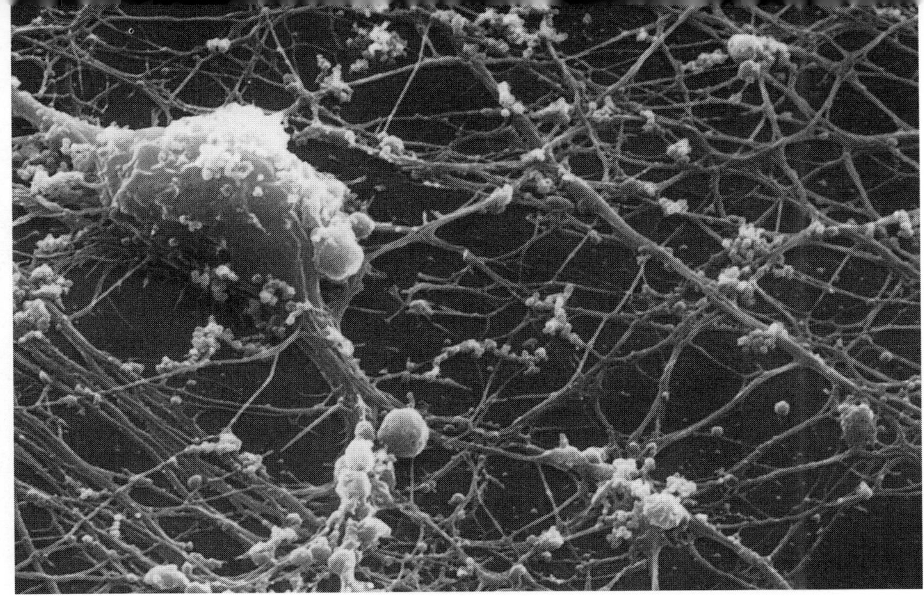

# The FUNCTIONING ANIMAL


**CHAPTER 15**
The VARIETY of LIFE . . . . . . . . . . . . . . . . . . . . . . . . . . . . . . . . . . . . . . 257

**CHAPTER 16**
ANIMAL ORGANISATION – SIZE and COMPLEXITY . . . . . . . . . . . 295

**CHAPTER 17**
The NERVOUS SYSTEM . . . . . . . . . . . . . . . . . . . . . . . . . . . . . . . . . . 303

**CHAPTER 18**
DETECTING CHANGE – RECEPTORS . . . . . . . . . . . . . . . . . . . . . . . . 328

**CHAPTER 19**
MOVEMENT and SUPPORT in ANIMALS . . . . . . . . . . . . . . . . . . . . 348

**CHAPTER 20**
CHEMICAL COMMUNICATION – the ENDOCRINE SYSTEM . . . . . . 372

**CHAPTER 21**
HOMEOSTASIS . . . . . . . . . . . . . . . . . . . . . . . . . . . . . . . . . . . . . . . . 382

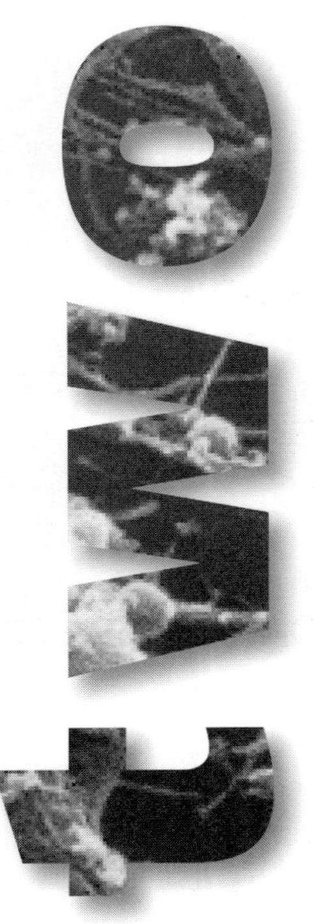

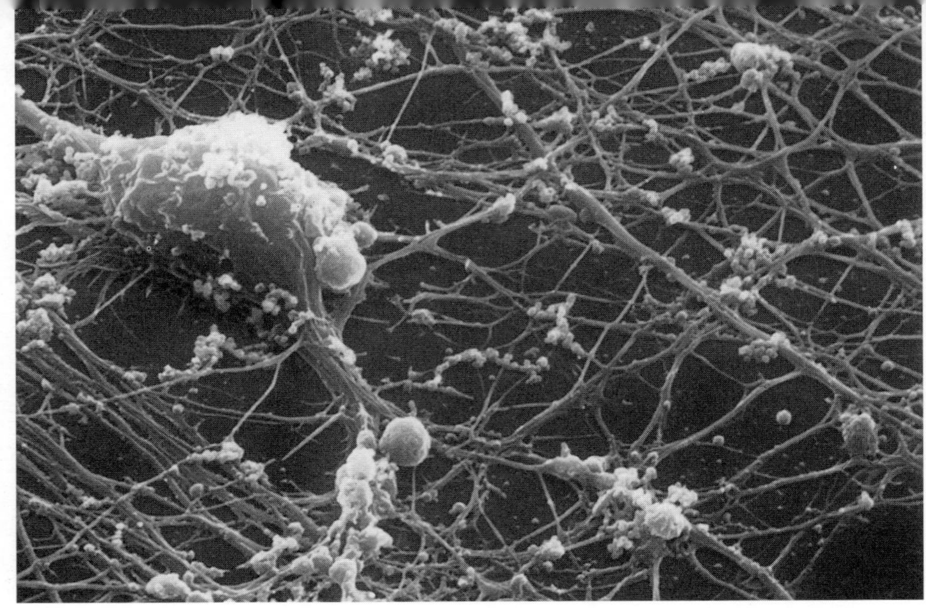

# Part two

# The FUNCTIONING ANIMAL


CHAPTER 22
TRANSPORT in MAMMALS – the BLOOD SYSTEM ............ 388

CHAPTER 23
GAS EXCHANGE and BREATHING ........................ 412

CHAPTER 24
NITROGENOUS EXCRETION and OSMOTIC REGULATION ....... 431

CHAPTER 25
REGULATION of BODY TEMPERATURE ..................... 447

CHAPTER 26
NUTRITION in MAMMALS ............................... 461

CHAPTER 27
DEFENCE AGAINST MICRO-ORGANISMS .................. 486

CHAPTER 28
REPRODUCTION .......................................... 501

CHAPTER 15

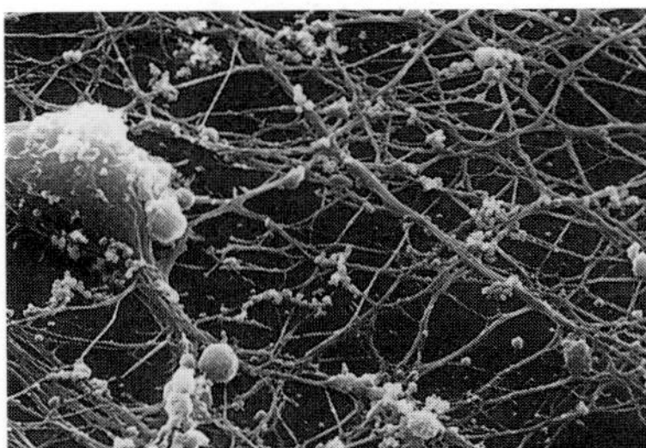

# The VARIETY of LIFE

## LEARNING OBJECTIVES

By the time you have completed your study of this chapter you should be able to:

- explain the practical necessity for scientific names in Biology.
- write the scientific name of an organism.
- explain what is meant by a hierarchical classification, and how the natural classification reflects evolutionary relationships.
- give examples of some of the practical difficulties in establishing taxonomic relationships.
- distinguish between homologous and analogous characteristics, and give examples of each.
- understand the principles underlying a dichotomous key, and be able to use simple keys to identify organisms.
- know the distinguishing features of the Kingdom Prokaryotae.
- distinguish the main shapes of bacteria and give examples of each.
- know the basis for the distinction between Gram-negative and Gram-positive bacteria.
- explain how bacteria reproduce, both sexually and asexually.
- with regard to bacteria, distinguish between the terms autotrophic and heterotrophic, aerobic and anaerobic, and obligate and facultative.
- know the various ways in which bacteria are important both in nature and to humans.
- state the distinguishing characteristics of the Kingdom Fungi, and of its four constituent phyla.
- know the names, chief distinguishing characteristics, and examples of the phyla making up the Kingdom Protoctista.
- give the chief distinguishing characteristics of the Kingdom Plantae and the phyla Bryophyta, Filicinophyta, Lycopodophyta, Coniferophyta and Angiospermophyta.
- give the distinguishing characteristics of the Kingdom Animalia and of the phyla Porifera, Cnidaria, Platyhelminthes, Nematoda, Annelida, Arthropoda, Mollusca, Echinodermata and Chordata.

## The diversity of living things

The diversity of life is enormous. To date, over a million different species of living things have been distinguished, but there are certainly many more. Some estimates put the number of kinds of living thing as high as 100 million. When extinct species are taken into account, the number must run into hundreds, perhaps even thousands, of millions.

### Classifying living things

Classifying things – putting them into groups according to their similarities and differences – is something we all do. Thus, we divide motor vehicles into cars, buses, lorries and so on. Members of the 'car' group are not all the same, but they have features in common that set them apart from members of the 'bus' group and members of the 'lorry' group. Grouping things simply helps us to create order in the world around us.

For our ancestors, the classification of living things was essential to survival. It was vital, for example, to know which plants were edible and which were poisonous, which could be used for medicines and which could not, and so on. This kind of classification had a purely practical use, because it was based on criteria that related to how living things affected people.

Beginning in the 17th Century, naturalists began to classify organisms according to their intrinsic characteristics, particularly their anatomy. The Swedish botanist Carl von Linné (usually known by the Latinised name Carolus (Carl) Linnaeus) divided all living things into two great groups, the plant and animal **kingdoms**. He classified animals by grouping similar kinds, or **species**, into larger groups called **genera** (singular, genus). He placed groups of similar genera into larger groups called **orders**, and similar orders into **classes**. Modern biologists have found it necessary to add further categories, with a group of similar genera being placed in a **family**, and a group of similar classes being placed in a **phylum**.

Each of these groups is called a **taxon** (plural, taxa). These seven taxa are called the **obligate** taxa because by international agreement, they are the minimum that must be used in the classification of every organism. The classification of two common organisms is illustrated in Table 15.1.

**Table 15.1**
The classification of the common frog and the annual sunflower

| | Common frog | Annual sunflower |
|---|---|---|
| Kingdom | Animalia | Plantae |
| Phylum | Chordata | Angiospermophyta |
| Class | Amphibia | Dicotyledonae |
| Order | Anura | Asterales |
| Family | Ranidae | Asteraceae |
| Genus | *Rana* | *Helianthus* |
| Species | *temporaria* | *annuus* |

In practice, the obligate taxa are often insufficient, and so others are added. Thus, phyla may be lumped into **superphyla** and subdivided into **subphyla**. Even these taxa are, in some cases, not enough, and some new groups are used such as a **tribe** (a group of genera) and a **cohort** (a group of orders).

As the lowest taxon in the hierarchy, a species contains organisms that are more alike than members of higher taxa. A species is usually defined as a group of organisms that are sufficiently similar to each other to interbreed and produce fertile offspring, but which cannot interbreed with other such groups. At least in principle, the ability to interbreed is a straightforward way of deciding whether two organisms belong to the same species. However, as we saw in Chapter 14, there are problems with this simple definition.

At the top of the taxonomic hierarchy is the kingdom. Of all the taxa, this includes the greatest diversity, so its members have the fewest features in common. The animal kingdom, for example, includes creatures as diverse as sponges, spiders and elephants.

You might think that because a phylum contains more species than any of its component classes, all phyla contain more species than all classes.

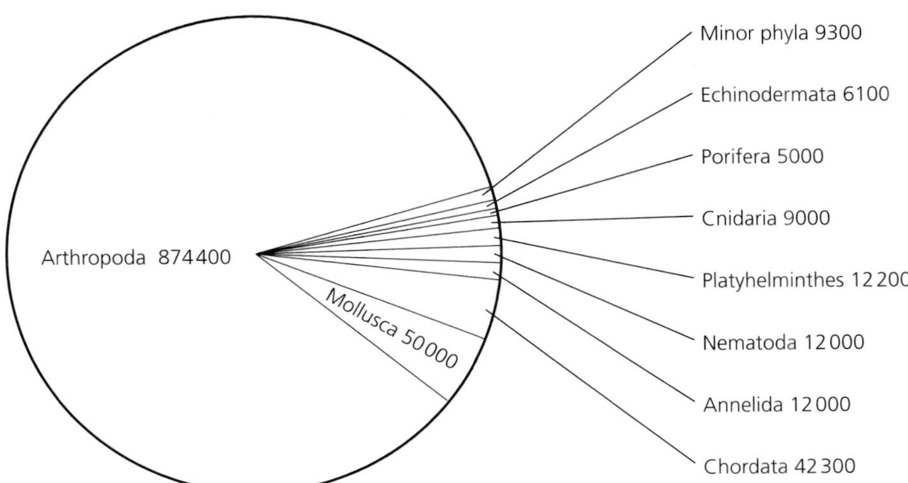


**Figure 15.1**
Relative sizes of different animal phyla

This is not always so, because phyla vary greatly in size (Figure 15.1). The Phylum Arthropoda, for example, accounts for about 80% of the animal kingdom. One of its component classes, the insects, accounts for about 80% of the arthropods. In fact, one order of insects, the Coleoptera (beetles), contains far more species than most phyla. The same inequality exists right down to species level; some genera contain hundreds of species, while some (e.g. our own genus, *Homo*) contains only one.

## Classification and evolution

There is almost no limit to the number of ways in which living things could be classified. To take just one example, we could divide animals into those with wings and those without, or into those with blood systems and those without. In the first case, we would place sparrows and dragonflies in one group, and fleas and rats in the other. But it makes little sense to base a classification on a single characteristic. If we ignore wings, fleas are far more like dragonflies than sparrows. In any case, the wings of a dragonfly and a sparrow are structurally quite different.

If we take into account all available characters, certain ways of classifying organisms are clearly much more 'natural' than others. Take, for example, the eight animals shown in Figure 15.2. All but the earthworm have a vertebral column or 'backbone' protecting a hollow nerve cord, a heart and a number of other features. The shark, rat, beaver, badger, cheetah, lion and leopard have more in common with each other than any of them does with an earthworm. Accordingly they are placed in the same group, the subphylum Vertebrata.

Of these seven vertebrates, the shark clearly stands apart from the other six, all of which suckle their young on milk and have a four-chambered heart and a diaphragm. Because of these and other shared features, the rat, beaver, badger, cheetah, lion and leopard are all placed in a different class from the shark – the Mammalia.

When we compare the six mammals, the rat and beaver stand apart from the others because they have gnawing incisors. For this and other reasons, they are placed in the Order Rodentia. The remaining four share other features – for example, the same three wrist bones are fused. Accordingly, they are all placed in a different group from the rat and beaver – the Order Carnivora.

Of these four, the cheetah, lion and leopard are even more similar to each other and are placed in a different family from the badger – the Felidae (cats). Finally, the lion and leopard – unlike the cheetah – both

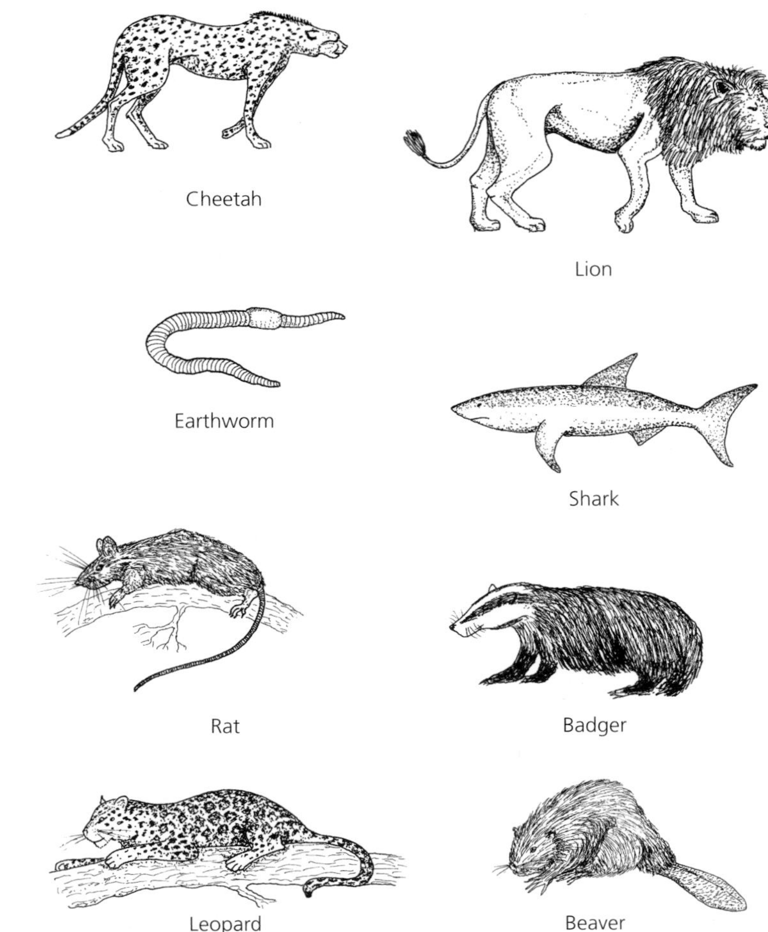


**Figure 15.2**
The eight animals classified in Figure 15.3

have retractile claws. Accordingly, the cheetah is placed in the genus *Acynonyx* and the lion and leopard in the genus *Panthera*.

The result of this classification is a system of groups within groups, represented in Figure 15.3 by a series of nested boxes. Most taxonomists would agree that this way of classifying these animals is more logical than any other. Why should this be so? As briefly discussed in Chapter 1, it occurred to Darwin that shared features were probably the result of shared ancestry or **phylogeny**.

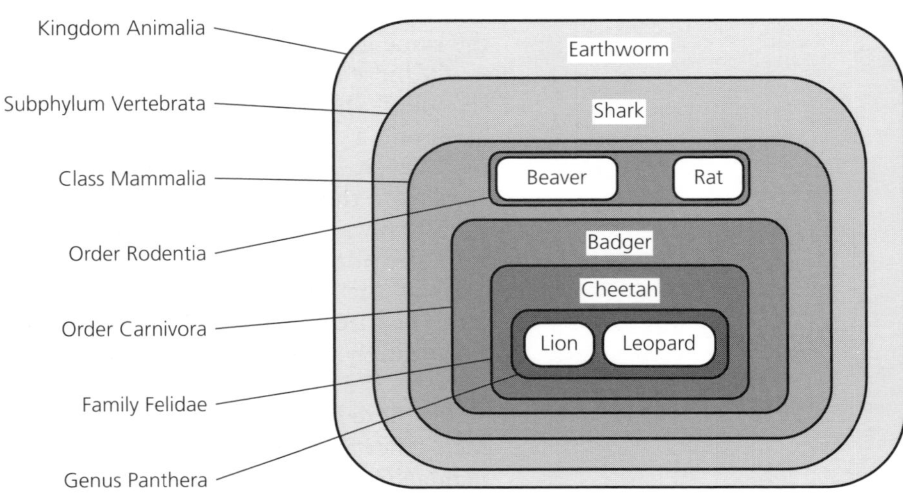


**Figure 15.3**
How the number of shared features increases up the taxonomic hierarchy. All animals within a box share certain features with each other which are not shared with animals outside the box

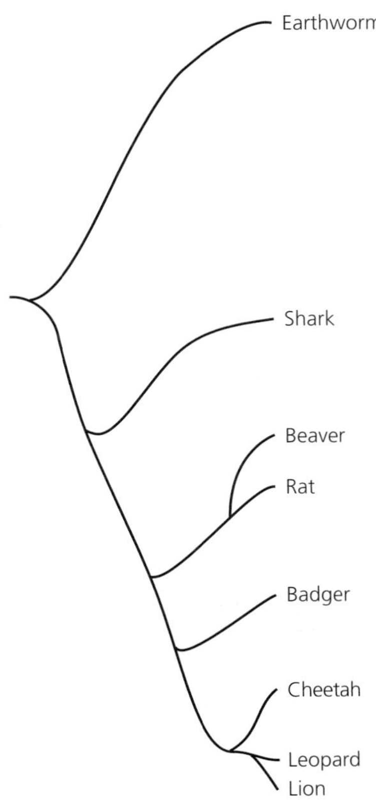


**Figure 15.4**
An alternative way of representing the relationships shown in Figure 15.3

We can see this more clearly if the similarities between the eight animals in Figure 15.2 are shown as a tree (Figure 15.4). Each branching point represents a point of divergence between two lines of evolutionary descent. When one group splits into two, each 'daughter' line retains features of the common ancestor, but begins to evolve features unique to that line.

It therefore follows that the higher the taxon, the further back in time we would have to go to find the common ancestor. Thus, members of the genus *Panthera* have a more recent common ancestor than members of the family Felidae. The common ancestor of the Carnivora lived even further back in time – about 60 million years ago – and the common ancestor of the modern mammals lived about 70 million years ago. To find the common ancestor of the vertebrates we need to go back about 500 million years (Figure 15.4).

Since there has only been one evolutionary history of life there is only one truly natural, or **phylogenetic**, classification. 'Phylogeny' means 'evolutionary history', so a phylogenetic classification is one based on evolution. So why do biologists disagree on matters of classification? The reason is that many details of the evolutionary tree are still unknown. As long as this is so, taxonomists will probably continue to disagree amongst themselves.

Knowing an organism's evolutionary relationships gives a great deal of information about it. Thus, the single word 'bird' implies possession of feathers, a beak, a four-chambered heart, scaly legs, large yolky eggs with a limy shell and so on. The natural classification is thus very useful in that it enables a great deal of information to be conveyed in a very few words.

## Problems for the taxonomist

Whilst biologists agree on the general outline of how organisms should be classified, they often disagree about the details, and accepted classifications sometimes change. In fact **taxonomy**, or the study of the methods of classification, is a discipline in its own right.

One reason for this is that there is no objective criterion for deciding how much difference should separate two groups into, say, phyla rather than subphyla. Some taxonomists, called 'splitters', prefer to have a larger number of smaller groups. 'Lumpers', on the other hand, prefer to combine smaller groups into larger ones. Thus the group of flowering plants that includes peas, clover, lupin and gorse may be given the status of a family (Papilionaceae) by some taxonomists, whilst others prefer to consider them a subfamily – the Faboidea.

A different problem arises from **convergent evolution**, where a particular feature has evolved independently in two or more groups. One example is the resemblance between the wolf and its marsupial counterpart, the thylacine or Tasmanian 'wolf' (Figure 15.5). The resemblance is, however, superficial; except for teeth and jaws, the thylacine resembles the marsupial mole and pouched mouse much more closely than it does the placental wolf.

Another example of convergent evolution is the compound eye, which appears to have evolved independently in both insects and some crustaceans (e.g. crabs and shrimps). A number of other characteristics of the arthropods such as tracheae and Malpighian tubules, may have evolved independently in more than one group. Even the exoskeleton itself – the most characteristic arthropod feature – may have evolved more than once. If this is true the arthropods would be **polyphyletic**, meaning

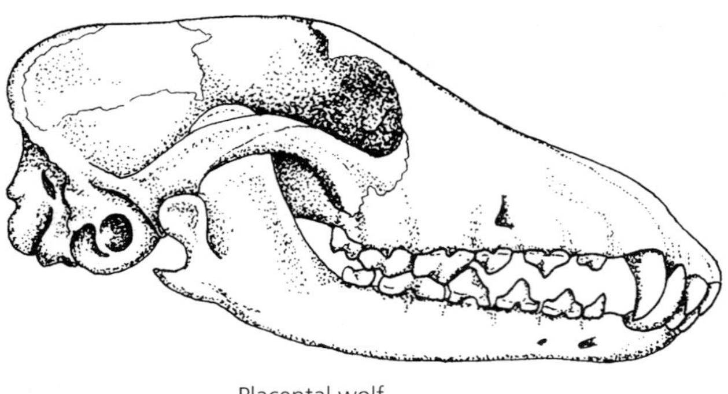


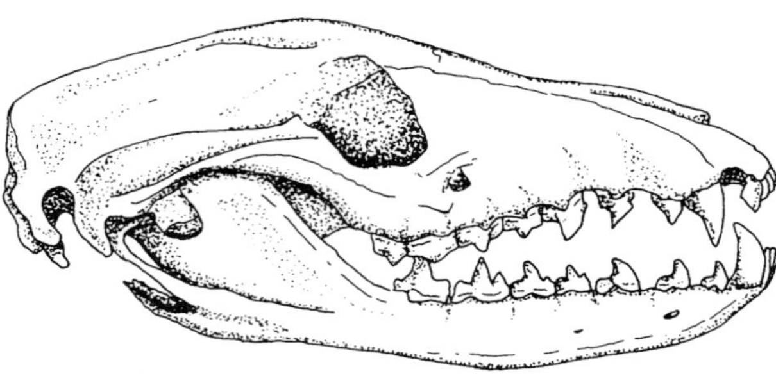


that they have more than one evolutionary origin. The arthropods would thus embrace several groups that have independently reached a similar **grade** of organisation.

## Homologous and analogous structures

At first glance, the limbs of a seal, mole, bat, human and horse look very different. Yet close examination of the skeletons reveals an underlying similarity of structure and embryonic development, despite their very different functions (Figure 15.6). Structures that have underlying similarity, even when carrying out different functions, are said to be **homologous**.

**Figure 15.5▲**
Skulls of placental wolf and thylacine (marsupial 'wolf') along with a photograph of each (placental wolf, above, thylacine wolf, below)

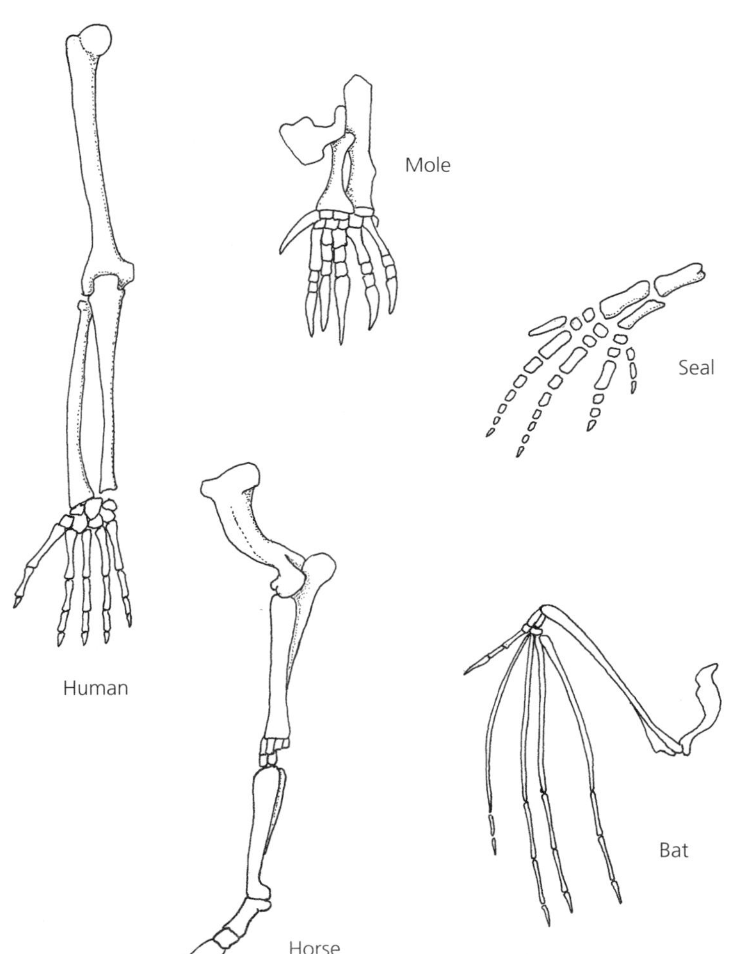


**Figure 15.6►**
Forelimb skeletons of five mammals with very different lifestyles

Homologous structures are in marked contrast to **analogous structures**, which have similar functions but fundamentally different structures. The wings of insects and birds are an obvious example. Analogous structures are of no taxonomic use since resemblances are so superficial.

## Molecular taxonomy

Until recent decades taxonomists relied heavily on anatomy to classify organisms. Despite the obvious convenience of using visible features, taxonomists may disagree on the relative importance of different features. Species A may resemble species B in one respect and species C in another. Should A be classified with B or with C? It all depends on which characteristics are considered to be the more important – and this is often a matter of opinion.

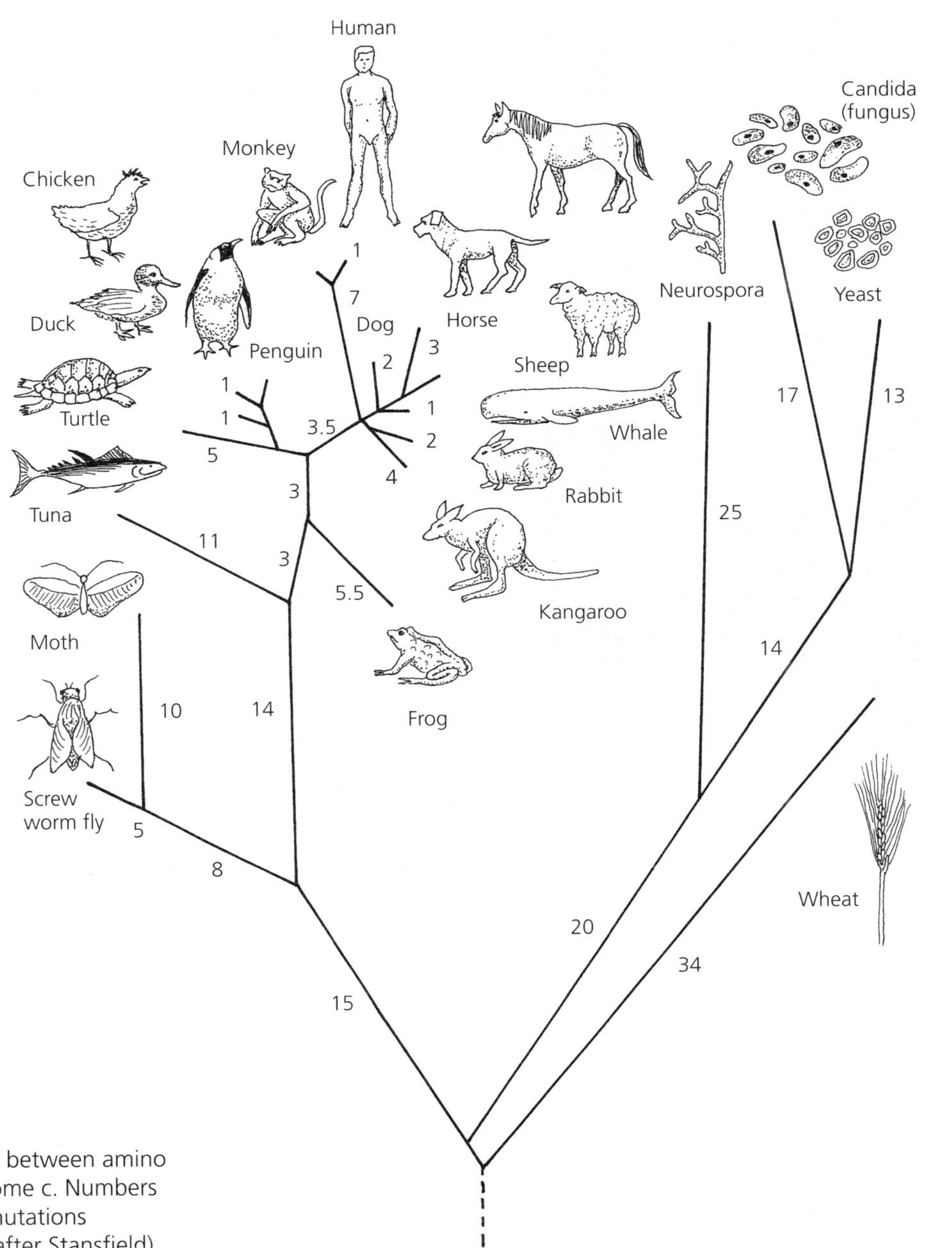


**Figure 15.7**
Evolutionary tree based on comparison between amino acid differences in the protein cytochrome c. Numbers represent the most likely numbers of mutations between successive branching points (after Stansfield)

Problems like these are greatly reduced when comparisons are made between molecules such as DNA, RNA and proteins. All three have large molecules consisting of many smaller units strung together in specific sequences.

In the case of proteins, the building units are **amino acids**. The sequence of amino acids in a protein can now be determined quite quickly. Two proteins can be compared in terms of the number of amino acid differences. Unlike comparisons based on anatomy, numerical comparisons are not a matter of opinion.

Comparisons between proteins, DNA and RNA are becoming increasingly important in classification and have spawned a new discipline – **molecular taxonomy**. Although the findings of molecular taxonomists have usually supported the orthodox classification, there have been some surprises and some conflicts (Figure 15.7).

Our own ancestry is a case in point. When molecular taxonomists began to compare the blood proteins of humans with those of the chimpanzee, gorilla and orangutan, the results suggested that chimpanzees, gorillas and humans are about equally closely related, with the orangutan being more distantly related. This was quite contrary to the traditional view that the chimpanzee, gorilla and orangutan should be placed in the same family, with humans in a family of their own (Figure 15.8).

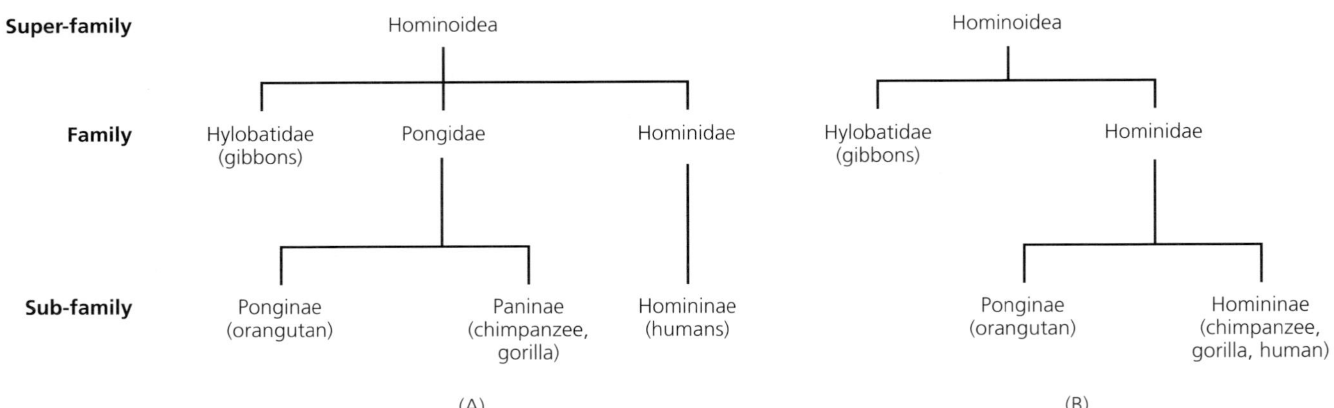


**Figure 15.8**
Traditional (A) and modern (B) classification of humans

## Grades and clades

Problems such as those outlined above highlight a major difficulty in classification, and one that has divided taxonomists into two contrasting groups.

On the one hand there are the cladists, who maintain that classification must reflect solely the results of evolution. Defined in this way, a taxon represents a cluster of distant relatives descended from a common ancestor. That ancestor, together with all its descendants, is called a **clade**. All members of a clade share a more recent common ancestor than any individual does with a member of any other clade.

The alternative, more traditional, view is that since we can never know all the details of a group's evolutionary history, taxa should be defined in terms of similarities and differences between present day organisms, regardless of whether such similarities have evolved independently. Thus, even though the exoskeleton may have evolved independently in different arthropod groups, they should all be placed in the same phylum because they have all attained a similar level or **grade** of organisation.

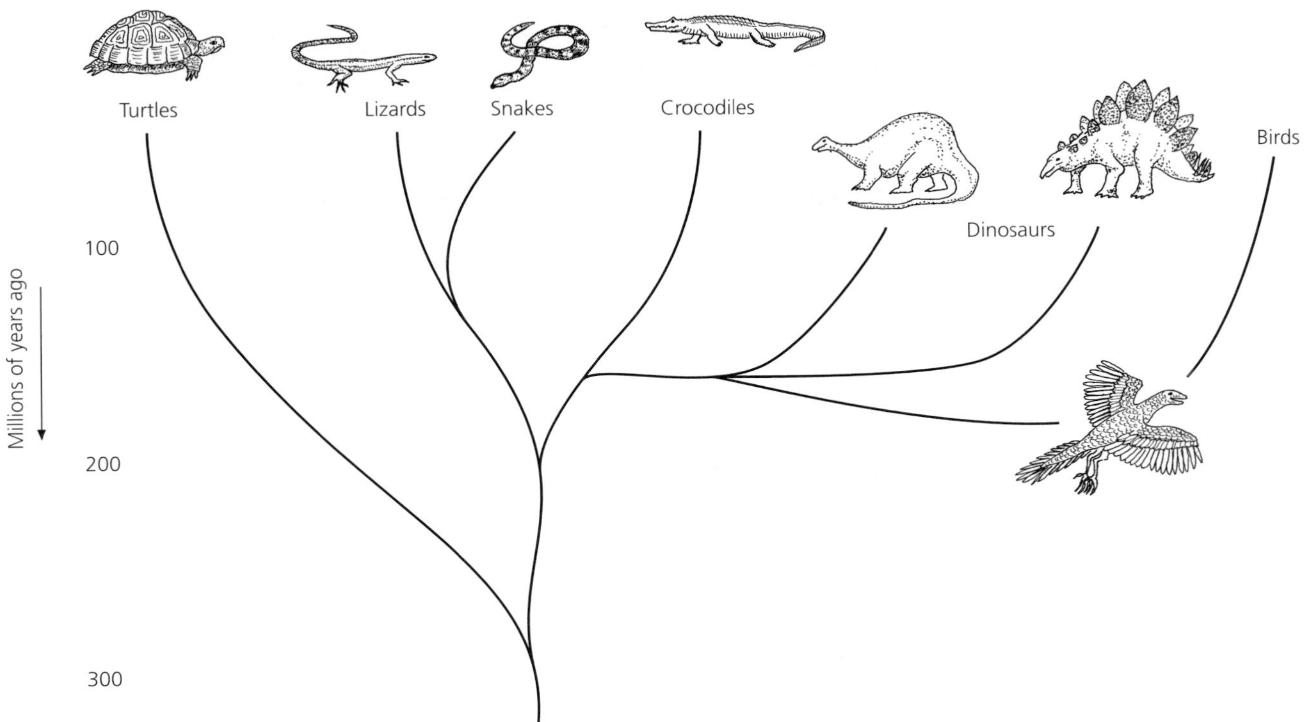


**Figure 15.9▲**
Evolutionary history of the reptiles

These problems can be illustrated by the fossil history of the reptiles (Figure 15.9). Since birds originated as an offshoot of the dinosaurs, and crocodiles arose as an earlier branch of the same stock, a cladist would classify crocodiles, dinosaurs and birds as a single large clade. Lizards and snakes on the other hand would represent a second clade, and turtles a third. Mammals, and the reptiles from which they evolved, would constitute a fourth clade. Together these four clades would constitute an even larger clade.

## Naming living things

Scientists cannot study living things unless they have names for them. Unfortunately common names are unreliable because they vary from one society to another. Thus, in Europe and America a 'glow worm' is a beetle, but in New Zealand, it is the larva of a cave-dwelling gnat (Figure 15.10). Neither animal is a worm, and in any case the term 'worm' has no precise meaning – tapeworms and roundworms, for instance, have little more in common than their shape.

To avoid such confusion, biologists give living things names that are the same in all countries. The scientific or **systematic** way of naming was developed by the Swedish botanist Carl Linnaeus (1707–1778), and has proved so useful that it is still the standard system in use today. It is called the **binominal system** of nomenclature because each organism has two names (*nomen* is Latin for name). The first name denotes the **genus** to which the organism belongs, and the second indicates the **species**. The small white butterfly, for example, is called *Pieris rapae*.

There are certain rules that should be followed when writing systematic names. The genus begins with a capital letter and the species with a small letter. Genus and species are both printed or typed in italics, or in normal type if the text as a whole is in italics. When hand-written, genus and species should be underlined with separate lines – for example, Pieris brassicae, the large white butterfly. After the first mention of an

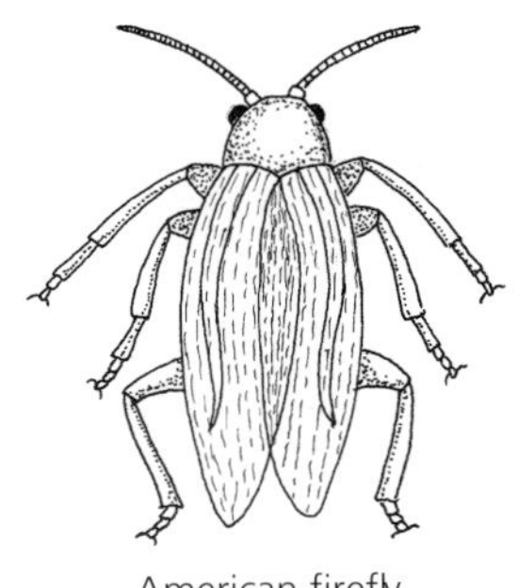

American firefly

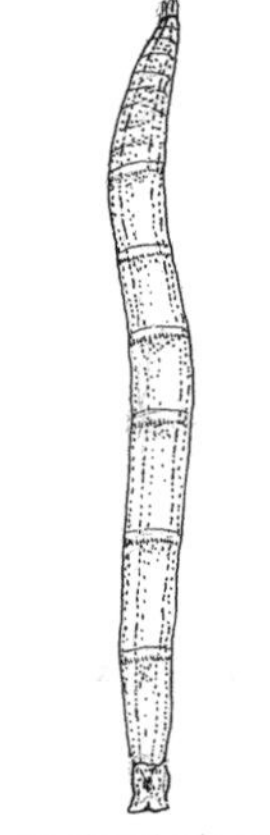

New Zealand glow worm

**Figure 15.10▲**
Two very different kinds of 'glow worm'

organism's systematic name in a text, the genus is usually abbreviated to the first letter. Thus *Pieris brassicae* may be abbreviated to *P. brassicae*. 'Species' may be abbreviated to 'sp' if it is singular and 'spp' if plural. Thus '*Ranunculus* spp' means several species of the genus *Ranunculus*.

# Artificial classifications

Whereas the natural classification existed long before scientists began to study it, an **artificial classification** is invented for the purely practical purpose of identifying an organism quickly. In such a classification, organisms are placed in groups according to a very limited number of characteristics. Consequently members of a group may have very little in common.

## Keys to identity

Suppose we want to design a key to enable anyone to distinguish between animals commonly found in houses, such as those shown in Figure 15.11.

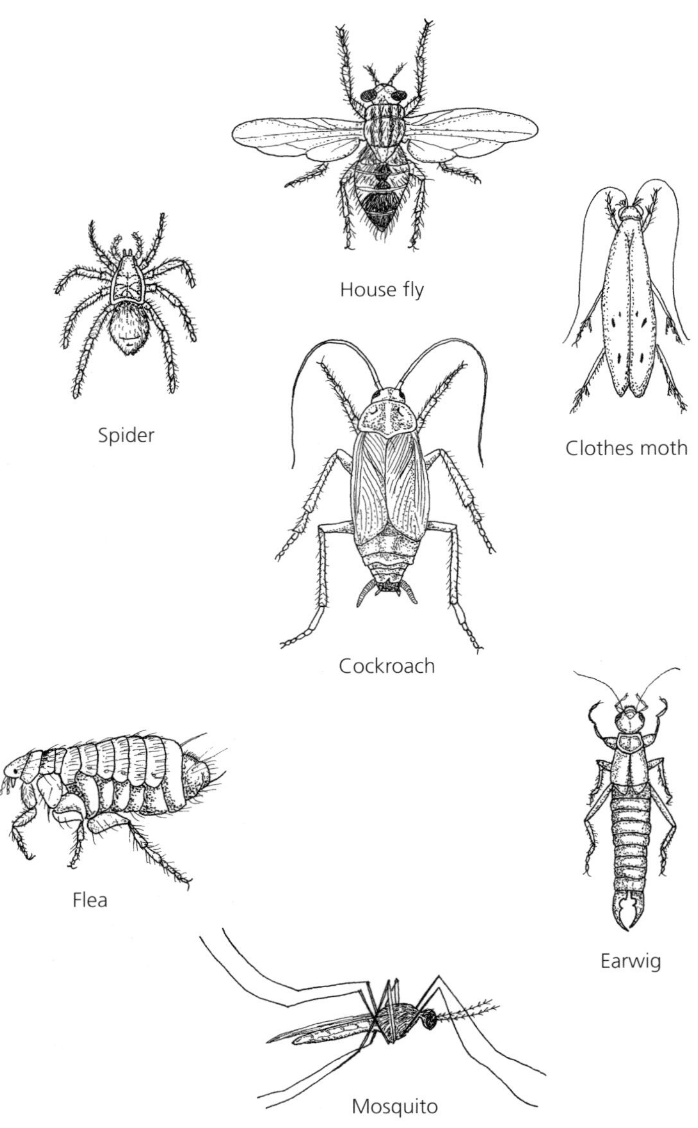


**Figure 15.11**
Some common animals found in houses

The first step is to divide them up into two groups according to some clearly visible character, such as presence or absence of wings. Each of these groups is again subdivided according to some other character, and so on. Figure 15.12 is just one of many ways these creatures could be classified.

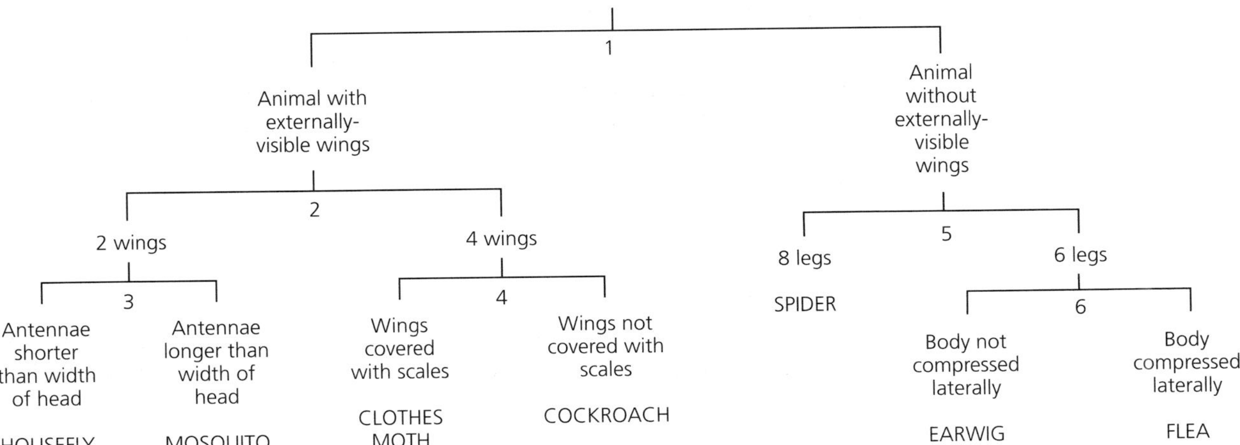


**Figure 15.12**
An artificial classification of the animals shown in Figure 15.11

The classification shown in Figure 15.12 is deliberately short, but most artificial classifications are very long and cannot be accommodated on a single page. It is therefore more practical to rewrite it in the form of a **key** which can span several pages. A simple example is shown below. Since most of the branching points are two-way, such a key is said to be **dichotomous**.

1. Animal with externally visible wings...........2
   Animal without externally visible wings......5
2. Two wings................................................3
   Four wings ...............................................4
3. Antennae shorter than width of head..........HOUSEFLY
   Antennae longer than width of head...........MOSQUITO
4. Wings covered with minute scales..............CLOTHES MOTH
   Wings not covered with scales....................COCKROACH
5. Eight legs..................................................SPIDER
   Six legs.....................................................6
6. Body not compressed laterally ...................EARWIG
   Body compressed laterally .........................FLEA

Note the following points:

- the first classification bears little resemblance to biological relationships. Thus, the earwig is placed in the same group as the spider because it does not have externally visible wings.
- a given classification can only be used to distinguish members of a defined list of organisms. Mice are also found in houses but are not included in the list. The classification would have to be redesigned to identify a mouse.
- this classification would only work for adult stages – young cockroaches, for example, are wingless (though they have small wing buds).

- the most suitable characters to be used are of the 'present or absent' type. When organisms are distinguished on the basis of size or proportions, these must be quantified. Thus 'antennae longer than head' is much less ambiguous than 'antennae long'.
- behavioural characters, such as 'animal nocturnal', should be avoided, since the specimen may be dead.

## Taxonomic turmoil: how many kingdoms?

For nearly two centuries after the death of Linnaeus, biologists continued to classify living things as either animals or plants. Included with the plants were the bacteria, fungi and seaweeds, and many (though not all) single-celled organisms.

Although fungi lack chlorophyll and feed heterotrophically, they were included in the plant kingdom because of their filamentous, branching growth, and the presence of a cell wall and large sap-filled vacuoles. Bacteria were also considered to be plants because they are enclosed by a cell wall.

This simple, two-kingdom view of life has since undergone a complete upheaval. The most glaring anomaly was the position of bacteria, which, in a number of ways, have a simpler level of organisation than other living things. For example, they have no clearly defined nucleus and their chromosomes form closed loops.

For these reasons bacteria are referred to as **prokaryotes**, (*pro* = before, *karyon* = nucleus) and comprise the Kingdom **Prokaryotae** (also known as the Monera). All other living things have open-ended chromosomes enclosed by a nuclear envelope. Organisms other than bacteria are said to have a **eukaryotic** level of organisation (meaning 'having a well formed nucleus').

### More than five kingdoms?

One of the difficulties with classifying microscopic organisms is that there is less structural complexity to work with. Increasingly, taxonomists are using chemical characteristics such as proteins and nucleic acids. Particularly revealing have been comparisons of the base sequences of one of the subunits of ribosomal RNA. Ribosomal RNA has the great advantage that it performs such a fundamental and exacting function, that it evolves only very slowly over immense periods of time, and can thus show up branchings near the base of the evolutionary tree.

These studies have shown that there are really three main branches, or **domains**, to the tree of life; two of which include all the prokaryotes and the third all the eukaryotes. What is really surprising is that within the eukaryotic domain the plants, animals and fungi are but three of ten branches, and each of the two other domains is likewise many-branched. If each branch is accorded kingdom status, then there may be as many as 26 kingdoms!

The fungi have also been removed from the plant kingdom. Unlike plants, their cytoplasm is surrounded by a wall of chitin, and their storage carbohydrate is glycogen rather than starch. For these and other reasons, fungi have also been placed in their own kingdom.

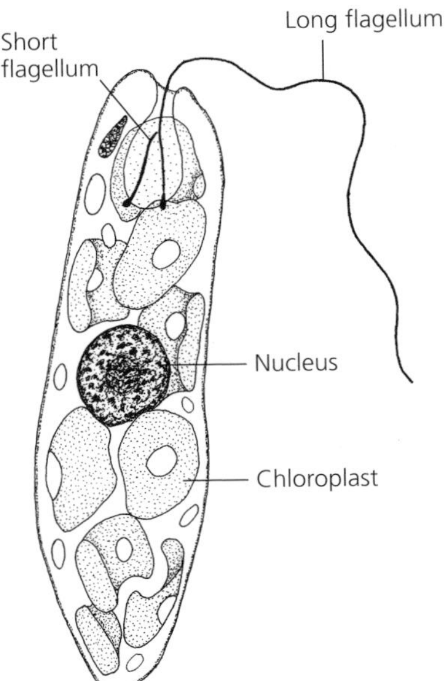


**Figure 15.13**
*Euglena*, a protoctist with both plant-like and animal-like characteristics

There remained two groups to be sorted out – the simple photosynthetic organisms called algae, and the single-celled animal-like organisms called protozoa. The algae included forms ranging from unicells to large seaweeds, and were included in the plant kingdom because of their photosynthetic nutrition. The terms 'algae' and 'protozoa' have no taxonomic status nowadays but, like 'tree', they are still useful terms when used informally, and will be used from time to time in this book.

Both groups presented considerable problems to taxonomists because there seemed no clear-cut dividing line between them. Some unicellular organisms, for example *Euglena*, are animal-like in having no cellulose wall, but plant-like in being able to photosynthesise (Figure 15.13).

A scheme which attempted to deal with these problems was devised by Lynne Margulis and Karlene Schwarz. They proposed that all unicellular eukaryotes should be united with multicellular 'algae' in the Kingdom **Protoctista**. Though this solves one problem, some of its members, such as the single-celled *Chlamydomonas*, have less in common with the other protoctists than they do with members of the plant kingdom. Indeed, there is some truth in the statement that the Protoctista are a ragbag of groups whose true affinities (evolutionary relationships) are as yet uncertain. In the absence of something better, most biologists are prepared to stay with the present five kingdom classification.

# Kingdom Prokaryotae

The prokaryotes include bacteria and the Cyanophyta. The latter used to be called 'blue-green algae' but they are now known to be photosynthetic bacteria and are thus more properly called blue-green bacteria. Bacteria are structurally much simpler than eukaryotic cells, and are also much smaller, averaging about 1 μm across and only just visible under the light microscope.

The difference in size is more impressive if we consider volume. An average bacterium has a diameter less than a twentieth that of a liver cell, so its volume is less than $(\frac{1}{20})^3$ or $\frac{1}{8000}$ that of the liver cell. As explained in Chapter 16, this means that a bacterium has a surface/volume ratio of more than 20 times that of a liver cell. So each unit mass of bacterium has 20 times more cell surface through which to absorb raw materials for growth. This may be one reason why bacteria can divide so much more quickly than eukaryotic cells.

## Bacterial shapes

One of the few distinguishing features of bacteria that is visible under the light microscope is shape, with four distinguishable forms (Figure 15.14). Rod-shaped bacteria are called **bacilli** (singular, bacillus), e.g. *Salmonella*, a cause of food poisoning. Spheres are called **cocci** (singular, coccus), e.g. *Staphylococcus*, the cause of boils. Less common are spiral-shaped bacteria called **spirilla**, e.g. *Treponema*, the cause of syphilis, and a bent rod called a **vibrio**, for example the cholera bacterium.

## Cell walls and staining

The shapes of bacteria are less important in their classification than their chemistry. A particularly useful distinguishing feature is their reaction to the **Gram stain** (named after Christian Gram, a Danish physician who introduced it in 1883). The cells are first stained with the active ingredient

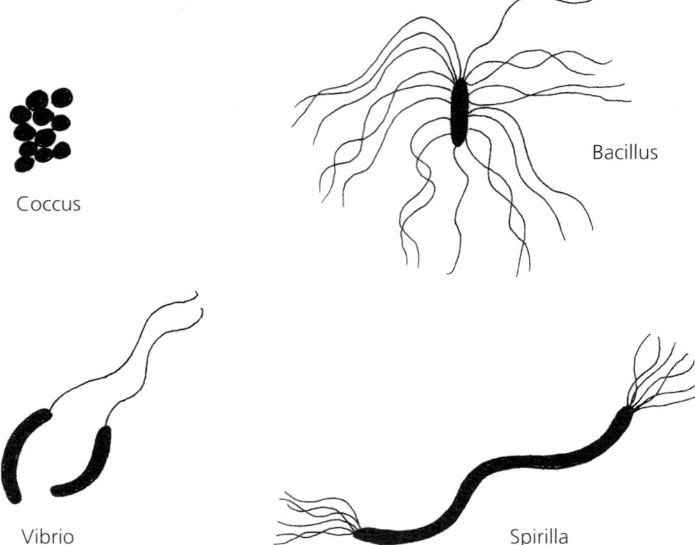


**Figure 15.14**
Some common bacterial shapes

crystal violet, after which they are treated with iodine and then with alcohol or acetone. In **Gram-negative bacteria** the alcohol washes out the stain, while in **Gram-positive bacteria** it is retained. The preparation is then counter-stained with a second stain, such as saffranin. This leaves Gram-negative bacteria pink and Gram-positive bacteria dark purple. The basis for the difference lies in the cell wall, which in Gram-negative bacteria has an additional outer membrane not present in Gram-positive types.

## Movement

Many bacteria can move by means of one or more threadlike **flagella** that extend outward from the cell surface (Figure 15.15). Bacterial flagella are quite different from those of eukaryotes, being much thinner (about 20 nm) and lacking the highly characteristic structure of eukaryotic flagella. Whereas a eukaryotic flagellum undergoes active bending movements, a bacterial flagellum is like a stiff corkscrew and moves by rotating on its axis at its base. Although it is not understood quite how the rotatory force is developed, it is known that it is not powered by ATP but by a proton gradient across the plasma membrane.

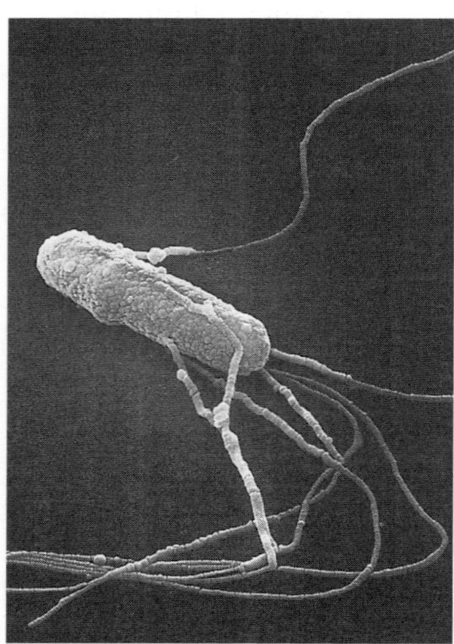

**Figure 15.15**
Electron micrograph showing flagella on the bacterium, *Salmonella typhimurium*

## Fimbriae and sex pili

Many Gram-negative bacteria have projecting threads called **fimbriae**, which are even finer than flagella (Figure 15.16). They consist of protein and seem to play a part in anchoring the cell to solid surfaces. **Sex pili** are another kind of projection, and are required for sexual conjugation.

## Metabolism

Bacteria show extraordinary diversity in their metabolism and inhabit virtually every conceivable environment, from the near-boiling water emerging from hot springs to the guts of animals. They may be present in vast numbers – a gram of soil may contain over a hundred million.

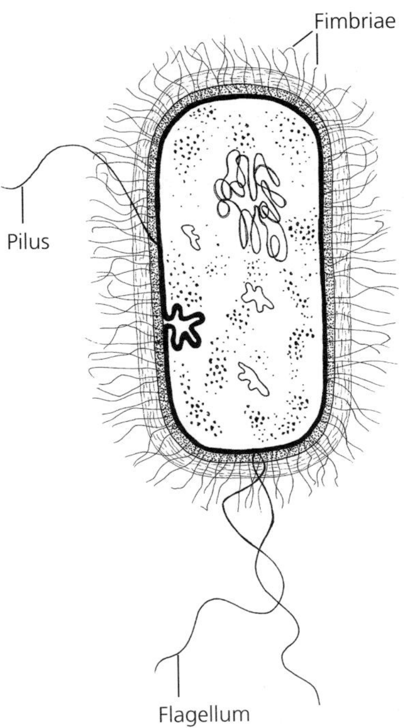


**Figure 15.16**
Fimbriae and pili of a bacterium

Bacteria are carried on air currents and may even be found at altitudes of up to 10 km.

Nutritionally, bacteria fall into two major groups:

- **autotrophs**, which manufacture their own organic matter from inorganic raw materials. Those that use light as an energy source are **photoautotrophs**, for example the purple sulphur bacteria and the blue-green bacteria. Others, called **chemoautotrophs**, use the energy released by the oxidation of inorganic compounds such as ammonia, hydrogen sulphide or ferrous iron ($Fe^{2+}$).
- **heterotrophs**, which feed on ready-made organic matter. Those feeding on dead matter are called **saprotrophs** or **saprobionts**, and help the return of nutrients to the soil by decay.[1] They secrete enzymes into the surrounding organic matter, and absorb the products of digestion over their body surfaces. Some bacteria are **parasites**, feeding on living organisms. Of these, a relatively small number cause disease and are said to be **pathogens**. Examples are the bacteria causing typhoid, cholera and plague.

Bacteria also vary with regard to their need for oxygen. Some are **obligate aerobes**, meaning that they cannot live without oxygen, while others are said to be **obligate anaerobes** because they cannot live in the presence of oxygen. All the photosynthetic bacteria and a number of others are obligate anaerobes. **Facultative anaerobes**, on the other hand, can survive in the presence or absence of oxygen, and include the majority of bacteria.

## Growth and reproduction

Bacteria can reproduce by **binary fission**, splitting into two (Figure 15.17). This kind of reproduction is **asexual** because it does not involve the 'reshuffling' of genetic material that occurs in sexual reproduction. The time taken for each division is the **generation time**, and varies greatly with the species and the environmental conditions. One, *Beneckia natriegens*, is said to have a generation time of just less than 10 minutes under optimal conditions. A bacterium dividing every 20 minutes could increase its numbers four times every 40 minutes and eight-fold every hour. To put this in perspective, bacteria can double their mass in less than an hour, but a young calf takes about 2 months to double its mass.

Of course, such explosive increases in numbers can only occur for short periods of time before food becomes exhausted or their numbers are reduced by predators or other factors.

Bacteria are also able to undergo a sexual process called **conjugation**, in which two cells become attached and a section of replicated chromosome moves across from one cell into another. As a result, DNA from the 'donor' bacterium becomes spliced into that of the recipient producing new combinations of genes. Also passing across are small 'mini-chromosomes' called **plasmids** which carry the genes for resistance to antibiotics. It is by means of conjugation that resistance can spread from one bacterium to another.

### Deadly anaerobes

The ability of many bacteria to live without oxygen can be important in terms of health. Normally, food is sterilised before it is sealed from the air by canning. If sterilisation fails and anaerobic bacteria get into the food they may multiply. *Clostridium botulinum* is an anaerobic soil bacterium that can live on some kinds of food and can cause a potentially lethal form of food poisoning.

[1]The older term 'saprophyte' was used when bacteria and fungi were considered to be plants (*phyton* means plant).

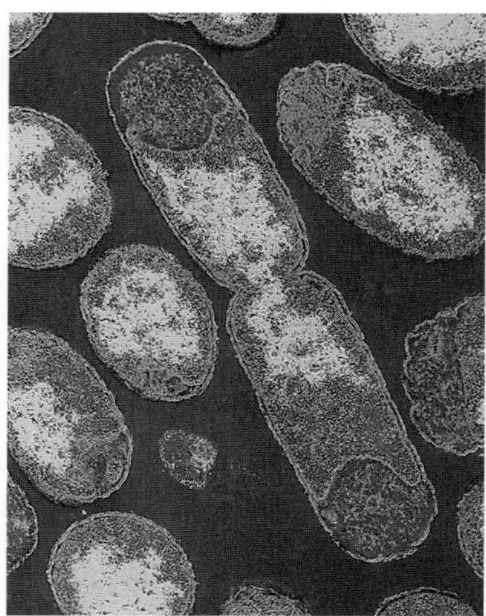
**Figure 15.17▲**
Binary fission in the bacterium, *Escherichia coli*

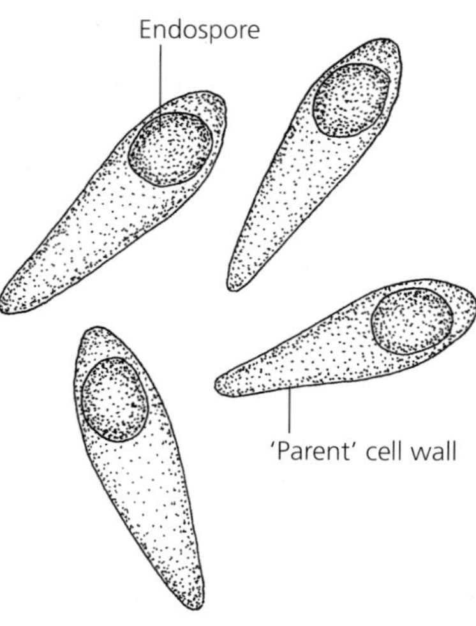

**Figure 15.18▲**
Bacterial endospores

In adverse conditions, some *Bacillus* and *Clostridium* species can form **endospores** within the parent cell wall (Figure 15.18). These are extremely resistant to drying and heat and can survive for long periods in a dormant state. This is why medical and microbiological equipment has to be sterilised by heating in an autoclave (a kind of pressure cooker), which raises the boiling point of water to about 110°C. At this temperature all spores are killed within 15 minutes by the superheated steam.

### The importance of bacteria

Bacteria are of enormous economic importance in a number of ways:

- saprotrophic bacteria play a major role in the decay of organic compounds, releasing simple inorganic substances into the soil that become available to green plants. In the breakdown of sewage, decay is actively encouraged and put to practical use.
- many bacteria enter into relationships with other organisms in which both partners benefit. Some of these **mutualistic** relationships are important in agriculture. Examples are the nitrogen-fixing bacteria in the root nodules of plants belonging to the family Papilionaceae (e.g. clover, peas and beans), and the cellulose-digesting bacteria in the guts of herbivores (Chapter 37).
- a relatively small number of bacteria cause disease in animals.
- increasingly, bacteria are being used in industry (see information box on page 273).

## Kingdom Fungi

There are around 100 000 species of fungi, therefore **mycology** (the study of fungi) is a large discipline. Fungi include the mushrooms, bracket fungi, puffballs, moulds and yeasts. Like most plants, fungi typically have branching, filamentous bodies. This finely-divided shape enables the absorption of nutrients over a large surface area. However, while plants absorb inorganic nutrients such as carbon dioxide and minerals with which to make their own organic compounds, fungi secrete digestive enzymes into their environment and absorb the organic products. This is the most fundamental difference between plants and fungi, for although some fungi may be brightly coloured – and some are even a green colour – none has chlorophyll. Fungi are therefore **heterotrophic**, feeding on organic matter produced by other living things.

The cell wall consists not of cellulose, as in plants, but of **chitin** – the same nitrogenous carbohydrate that forms the principal component of the cuticle of arthropods. The storage carbohydrate is glycogen rather than starch, as is the case in plants. Cilia and flagella are entirely absent, even from the male gametes. The downy mildews which used to be included in the fungi, do have flagella and are now included with the Protoctista.

### *Rhizopus*, a common saprotrophic fungus

*Rhizopus* is a common mould found frequently on rotting fruit and old bread. Its close relative *Mucor* often grows on rabbit (and other herbivore) dung. In both fungi, structure and reproduction have adapted to exploit a food supply that is widely scattered and strongly competed for by other fungi.

## Bacteria in industry

The ability of bacteria to ferment food has been exploited by humans for thousands of years in cheese making. More recently, bacteria have begun to play an increasing part in industry, for example:

- a number of enzymes used in industry are produced commercially by bacteria. Especially important are various digestive enzymes such as proteases, amylases and lipases, some of which are used in washing powders. Glucose isomerase, used in the conversion of glucose to fructose sweetener, is also produced by bacteria.
- a number of vitamins (e.g. Vitamin $B_{12}$ and Vitamin C) and amino acids are produced commercially by bacteria.
- some bacteria produce medically-important antibiotics. For example streptomycin, erythromycin and tetracycline are made by species of *Streptomyces*.
- *Bacillus thuringiensis* is used in a spray as a biological 'insecticide'; it produces spores that contain an extremely toxic protein which kills caterpillars that ingest it.
- *Thiobacillus ferro-oxidans* is used in the extraction of copper from low grade ores.
- some bacteria, such as species of *Pseudomonas*, break down petroleum, and this has been exploited in the clearing up of oil spillages.
- manufacture of single-cell protein (SCP). Agriculture and the food industry produce significant quantities of waste organic matter such as straw, whey and animal faeces. These would normally go to waste but can be used to feed bacteria. These bacteria convert it to their own protein which can then be fed to farm animals. Moreover, bacteria can convert organic matter fed to them far more efficiently and rapidly than animals can.

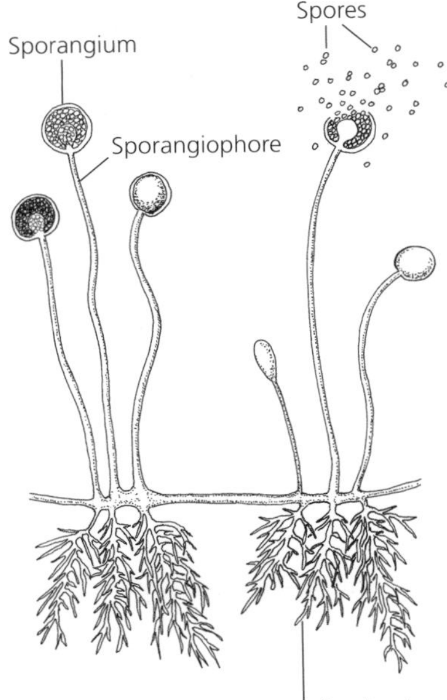


**Figure 15.19**
*Rhizopus*, a common mould

Except for the yeasts which are single-celled, the body of most fungi consists of a meshwork of branching threads called **hyphae**, collectively making up the **mycelium** (Figure 15.19). The feeding hyphae are produced in branching clusters from larger hyphae or **stolons** which extend over the substrate but do not grow into it. The vegetative hyphae are **coenocytic**, many nuclei sharing a common envelope of cytoplasm. Hyphae that have no cross walls are said to be **non-septate**. Since hyphae grow from their tips, the fungus can penetrate and extend within the food supply more rapidly than competing bacteria (but not, of course, other moulds, which also have apical growth).

The finest hyphae secrete starch-digesting and protein-digesting enzymes and absorb the resulting glucose and amino acids. Some of the products are used to build proteins and other complex molecules, but a high proportion are converted to $CO_2$ in respiration.

The speed with which the mycelium grows means that within a short time most of the available food in the substrate is exhausted. Dead fruit and similar foods are scattered, and, since many other fungi are competing for them, do not last long. As in many other organisms living fixed to one spot, reproduction is linked to dispersal, and after a few days of vegetative growth, reproduction begins.

Asexual reproduction involves the production of **spores**, each of which has several nuclei. Spores are produced in spore sacs or **sporangia** at the tips of vertically-growing hyphae called **sporangiophores**. In *Rhizopus*, the sporangium wall breaks open allowing the spores to be dispersed by wind. In *Mucor*, the wall disintegrates leaving the spores adhering in a mass; these are not dispersed by wind but are picked up by passing insects. Each spore is surrounded by a resistant wall that enables it to survive in dry conditions for several years. Their small size (about 3 micrometres, or 3 μm) has two advantages: they can be produced in vast numbers and, in *Rhizopus*, are readily carried on air currents.

Like many other fungi, *Mucor* and *Rhizopus* also reproduce sexually. The details of the process are beyond the scope of this book, but it involves two complementary processes: **fertilisation**, in which nuclei join together, doubling the number of chromosomes, and **meiosis**, in which the number of chromosomes is halved again. Between them, these processes result in new combinations of genes. Both fertilisation and meiosis are described in more detail in Chapter 7.

The ecological and economic importance of fungi rivals that of bacteria:

- saprotrophic fungi feed on dead organisms and, in so doing, return simple inorganic nutrients to the soil, making them available to plants.
- some decay-causing fungi can be harmful, for example those that cause rotting of wood and spoilage of food.
- some fungi enter into mutualistic relationships with other organisms. One of the most important is called a **mycorrhiza**, a relationship in which a fungus lives on the roots of plants (Chapter 37). A lichen is another mutualistic partnership consisting of fungal hyphae intertwined with a single-celled alga.
- many fungi are parasites, some causing diseases of crop plants – for example the rust diseases.
- some fungi are important in the food industry. Yeast is used in bread making, brewing and wine making, and some fungi are used in the commercial production of citric acid. The vitamin riboflavin is produced commercially by the fungus *Ashbya gossypii*.
- yeast is used in the commercial production of ethanol ('alcohol').
- some fungi produce medically-important antibiotics – for example the green mould *Penicillium* makes penicillin.
- some fungi are research tools in genetics – for example *Neurospora crassa*, a pink mould commonly found on bread.

The fungi are divided into three groups:

## Phylum Zygomycota

Examples are *Mucor* and *Rhizopus*, described previously.

## Phylum Ascomycota

Otherwise known as sac fungi, the ascomycetes include some important parasites of plants, such as the powdery mildews, and *Ceratocystis ulmi*, the cause of Dutch elm disease. The yeasts and the edible truffles, parasitic on the roots of certain trees, are also ascomycetes.

Ascomycetes are so-called because the sexual spores are produced in a sac-like **ascus**, the spores being called **ascospores**. Except for the single-celled yeasts, ascomycetes consist of hyphae subdivided by cross walls. These are perforated by fine pores which allow the cytoplasm of adjacent compartments to communicate, so these fungi are actually coenocytic. Asexual spores are produced by the budding off of specialised hyphae called **conidia**, and are called **conidiospores** (Figure 15.20).

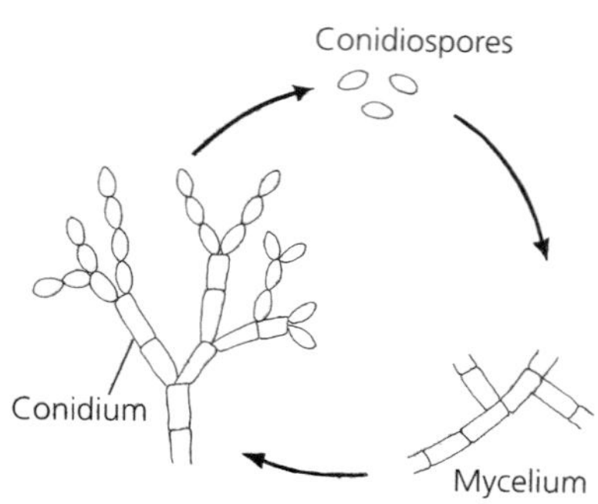


**Figure 15.20**
Asexual reproduction in *Neurospora*

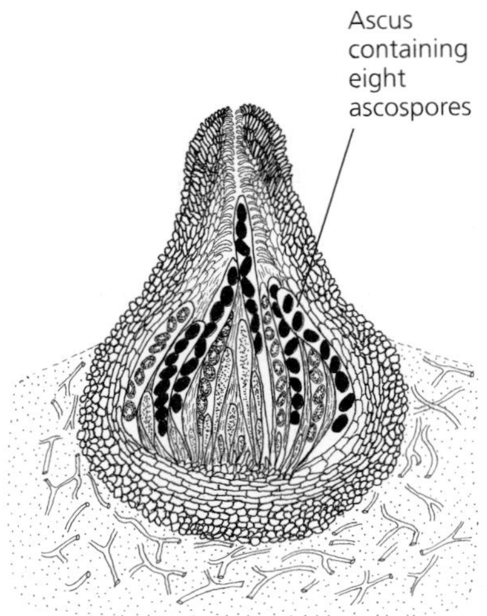


**Figure 15.21**
Perithecium of *Neurospora* containing asci and ascospores

The life cycle of *Neurospora crassa*, a fungus commonly used in genetics illustrates the basic features of the ascomycete life cycle. When mycelia of opposite mating types come into contact, sex organs are produced. These then fuse but the nuclei remain separate. The resulting binucleate cell or **dikaryon** divides repeatedly. Its nuclei divide also, thus maintaining the dikaryotic condition. Mononucleate hyphae and dikaryotic hyphae become interwoven to form a conspicuous fruiting body or **perithecium** (Figure 15.21). Inside the perithecium the nuclei of dikaryotic hyphae fuse to form zygotes. A zygote has two sets of chromosomes and is **diploid**. Each zygote undergoes meiosis followed by mitosis, forming a row of eight haploid ascospores contained within an elongated ascus. The spores are then discharged into the air and dispersed.

## Phylum Basidiomycota

The basidiomycetes include the puffballs, rust fungi, bracket fungi and, best known of all, the 'toadstools' (more properly called mushrooms). Their chief distinguishing characteristic is the production of sexual spores on the tips of club-shaped hyphae called **basidia** (singular, basidium).

The life cycle of the field mushroom is shown in Figure 15.22. It begins with the germination of a spore to form a non-septate **primary mycelium** which forms a fine branching network throughout the soil. If a primary mycelium comes into contact with another of opposite mating type, fusion occurs between hyphae.

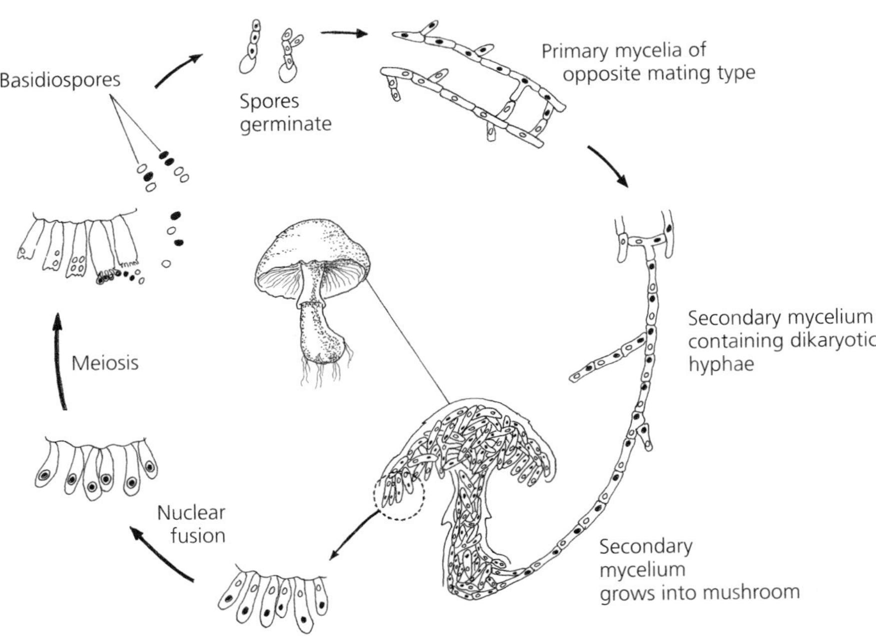


**Figure 15.22▶**
Life cycle of the field mushroom, *Agaricus bisporus*

The result is the formation of a **secondary mycelium** consisting of septate, dikaryotic hyphae. Each hypha contains two nuclei of opposite mating types, one originating from each parental primary mycelium. Each time a nucleus divides, a peculiar 'clamp connection' ensures that the dikaryotic condition is maintained, each daughter cell receiving a nucleus of each mating type.

All these events occur underground. Eventually, hyphae of the secondary mycelium aggregate to form button-like structures on the surface which grow into the familiar mushroom with 'gills' on the underside. On the gills, millions of basidia develop within each of which

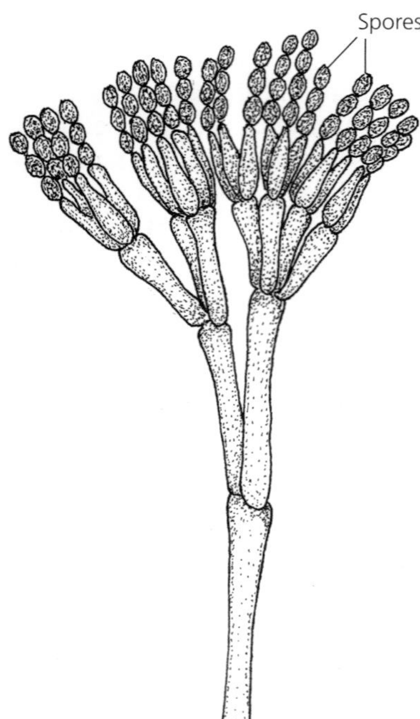


**Figure 15.23**
Reproductive hyphae of *Penicillium*, the source of the first commercially-produced antibiotic

the two nuclei fuse to form a zygote. This then undergoes meiosis and each of the four daughter nuclei becomes the nucleus of a haploid **basidiospore**.

A large mushroom can produce hundreds of millions of wind-dispersed spores. Mushrooms are most characteristically produced in autumn when the surface of the ground is damp and temperatures are still warm enough to support growth.

## Phylum Deuteromycota

These used to be called *Fungi imperfecti*, named because the sexual or 'perfect' stage of the life cycle has not been observed. They include such common saprotrophic moulds as *Penicillium* and *Aspergillus* (Figure 15.23).

# Kingdom Protoctista

The Kingdom Protoctista includes the single-celled eukaryotes, the multicellular seaweeds and certain fungus-like organisms such as the slime moulds and water moulds. These organisms were originally classified as members of the plant, animal and fungal kingdoms, but their membership had never been considered to be natural.

The creation of the Kingdom Protoctista has caused new difficulties. For example, some of the single-celled green algae are clearly related to plants. It seems likely that as molecular taxonomy develops, more changes are on the way. At the moment, the Kingdom Protoctista includes 27 phyla, only a few of which will be dealt with here. The first four of those mentioned below used to be placed in the animal kingdom in the phylum protozoa ('first animals'). They are unicellular, lack cellulose walls and are heterotrophic. The next three phyla are autotrophic and have cellulose walls. They used to be placed in the plant kingdom in the 'Algae'. Though this term now has no scientific meaning, it is still frequently used informally to embrace these simple plant-like organisms.

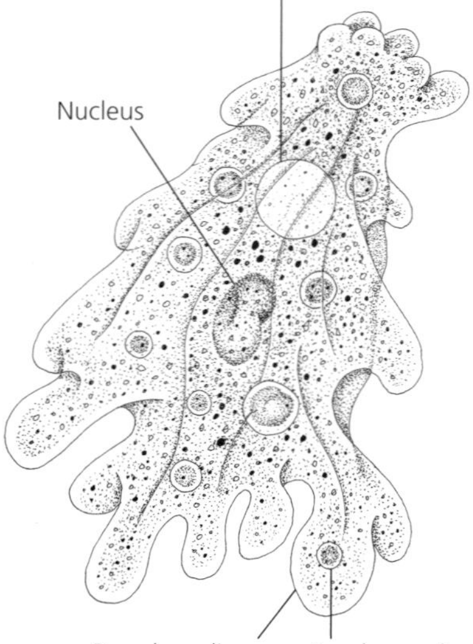

Contractile vacuole

Nucleus

Pseudopodium

Food vacuole

**Figure 15.24**
*Amoeba proteus*, a fresh water protoctist

## Phylum Rhizopoda

This group includes the well-known *Amoeba proteus* and similar forms (Figure 15.24). One member of this group, *Entamoeba histolytica*, is a gut parasite, which causes amoebic dysentery. Members of this group are only known to reproduce asexually by binary fission.

### *Movement in Amoeba*

Amoebae move and engulf food particles by means of slowly flowing cytoplasmic extensions called **pseudopodia** ('false feet'). The mechanism of this **amoeboid movement** is not fully understood, but seems to involve alternations in the state of the cytoplasm between the more solid, outer **ectoplasm** and the inner, more fluid **endoplasm**. The ectoplasm contains many free-moving granules, and, because of its fluid consistency, is called **plasmasol**. The ectoplasm is clear and consists of semisolid **plasmagel**.

A pseudopodium slowly extends by endoplasm flowing into it. At the tip of the pseudopodium, plasmasol appears to change to plasmagel. In other regions of the body the reverse change takes place, plasmagel changing to plasmasol.

These changes involve alterations in the state of *actin*, a protein that also plays a major part in muscle contraction (Chapters 2 and 19). In the

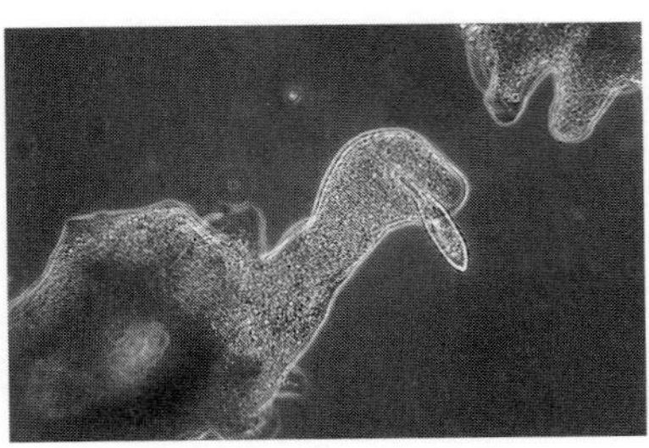

**Figure 15.25**
A single-celled *Amoeba* (bottom left) engulfing a *Paramecium* by phagocytosis. Note that the *Amoeba* has extended a pseudopodium (false foot) towards its prey

change from plasmasol to plasmagel, actin molecules become extensively cross-linked, and when these cross links break down plasmagel is converted to plasmasol. Just how the formation and destruction of cross links is controlled is not fully understood, but it is known to involve changes in the concentrations of $Ca^{2+}$ and $H^{+}$ ions.

For these changes to bring about propulsion they must involve the generation of forces which must in some way be transmitted to the outside world via the plasma membrane. How this is accomplished is not yet fully understood.

### *Nutrition in* Amoeba

*Amoeba* feeds on small, single-celled organisms such as diatoms, engulfing them by **phagocytosis** (Chapter 6 and Figure 15.25). In response to contact with the food, part of the plasma membrane invaginates and encloses the food particle in a *food vacuole*. Lysosomes then fuse with the vacuole, releasing hydrolytic enzymes into it and digesting the food. Eventually undigested remains are egested by exocytosis.

### *Osmoregulation in* Amoeba

In *Amoeba* and other single-celled fresh water organisms, water is continuously passing from the higher water potential outside the body to the lower water potential inside by osmosis. The cytoplasm would very quickly become diluted and the animal would burst, were it not for the ability of the animal to pump water out as fast as it comes in by means of a **contractile vacuole**. The vacuole slowly fills and its contents are periodically expelled by fusion of the vacuole with the plasma membrane. Water enters the vacuole via many tiny 'feeder' vesicles which surround the main vacuole.

Though the precise mechanism of osmoregulation is uncertain, some facts are beyond dispute:

- in dilute salt solutions the rate of pulsation of the vacuole is reduced.
- it involves active transport, since its activity ceases under anaerobic conditions or in the presence of respiratory poisons such as cyanide. This is consistent with electron micrographs showing mitochondria clustered round the vacuolar membrane.

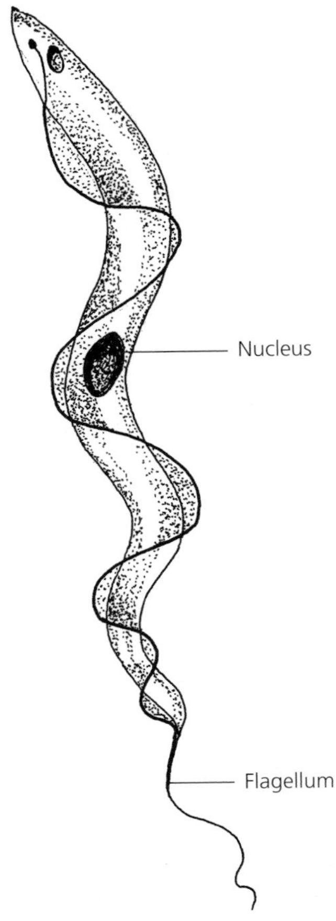


**Figure 15.26**
*Trypanosoma*, the cause of sleeping sickness

## Phylum Zoomastigina

This group includes a variety of forms, all of which swim by means of flagella – which can number several thousand in some species. All are heterotrophic. Some are important parasites of humans, such as species of *Trypanosoma*, the cause of sleeping sickness (Figure 15.26). Others live in the guts of insects such as termites, digesting cellulose.

## Phylum Ciliphora (ciliates)

The ciliates are characteristically covered with cilia by which they create feeding currents and (except in attached species) move through the water. Most have two nuclei – a large **meganucleus** and a small **micronucleus**. Despite consisting of a single cell, many have a highly complex structure.

A common freshwater ciliate is *Paramecium* (Figure 1.2). Each cilium beats slightly out of phase with the one in front and the one behind, producing an effect similar to wind blowing over a field of long grass. This **metachronal rhythm** implies some form of communication between adjacent cilia. The direction of beat is not quite along the axis of the body, causing the organism to swim in a corkscrew movement.

*Paramecium* is a filter feeder, with certain large cilia on one side serving to propel bacteria and other finely-suspended particles into an **oral groove** along one side of the body, and thence into the **cytopharynx**. Here, particles are enclosed in food vacuoles in which they are digested by the action of lysosomal enzymes. Undigested remains are voided by emptying of a food vacuole at the body surface, a process that involves fusion of the vacuolar membrane with the plasma membrane.

Osmoregulation involves the cyclical action of two contractile vacuoles each with feeder vesicles radiating out from the main vacuole.

## Phylum Apicomplexa

This group consists entirely of parasitic forms and includes the malaria parasites (Chapter 37).

## Phylum Phaeophyta (brown algae)

This group includes the largest members of the kingdom Protoctista. All are multicellular and most are marine, living attached to rocks. Like plants, they contain chlorophyll *a*, but instead of chlorophyll *b* they have chlorophyll *c*. The brown colour is due to **fucoxanthin**, a yellow pigment related to the carotene of carrots. They all reproduce sexually and many have a life cycle in which a diploid generation (with two sets of chromosomes) alternates with a haploid generation (with a single set of chromosomes).

Figure 15.27 shows *Fucus vesiculosus*, the bladder wrack, which is common on rocky shores.

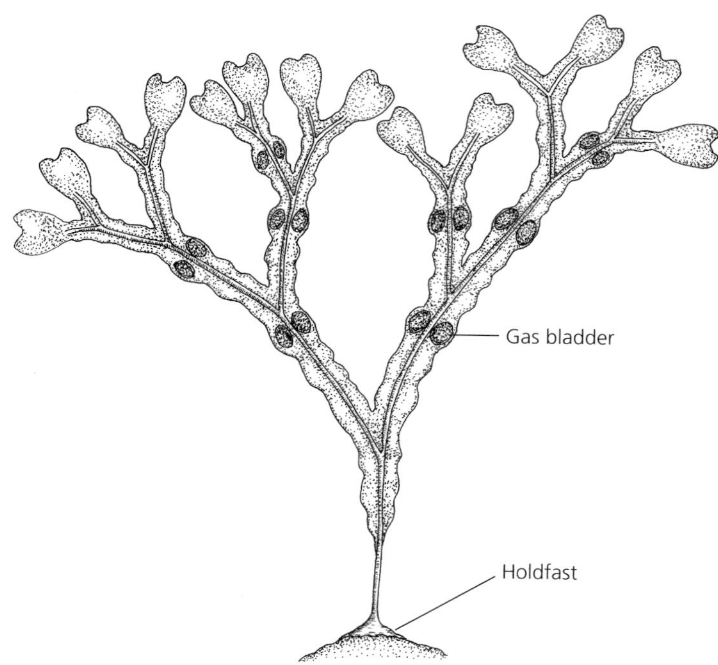


**Figure 15.27**
*Fucus vesiculosus*, a common brown seaweed

## Phylum Rhodophyta (red algae)

The Rhodophyta are the red seaweeds. They are all multicellular and are mainly marine. Only one chlorophyll is present – chlorophyll *a*. Two other photosynthetic pigments are present – the red **phycoerythrin** and the blue **phycobilin**. All members of the group reproduce sexually but flagella are never developed – not even by the male gametes.

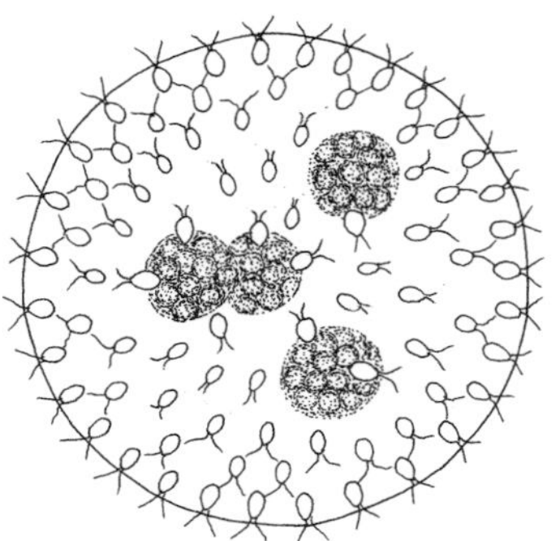

*Volvox*, a colonial green alga

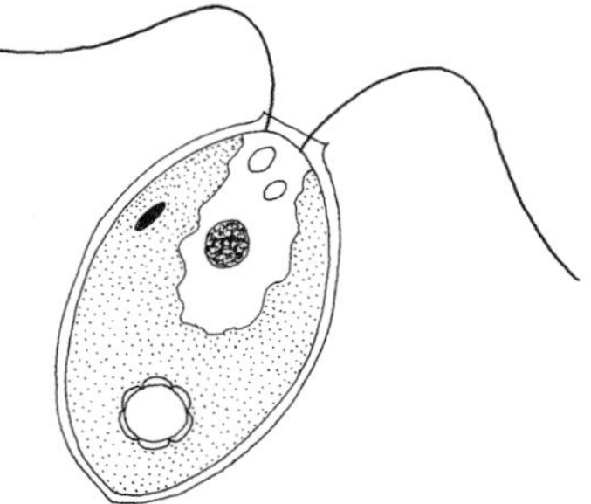

*Chlamydomonas*

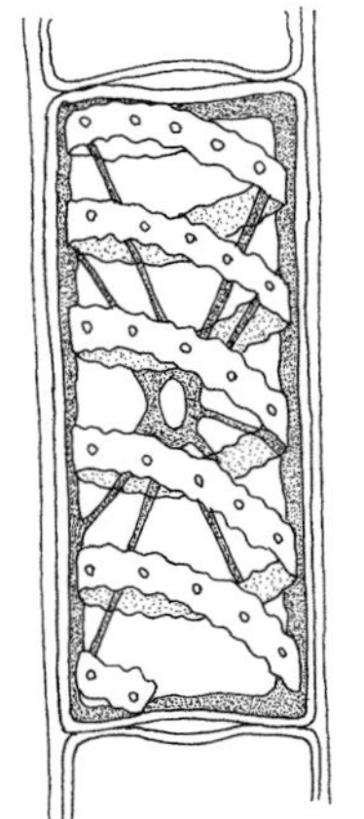

*Spirogyra* – part of a filament

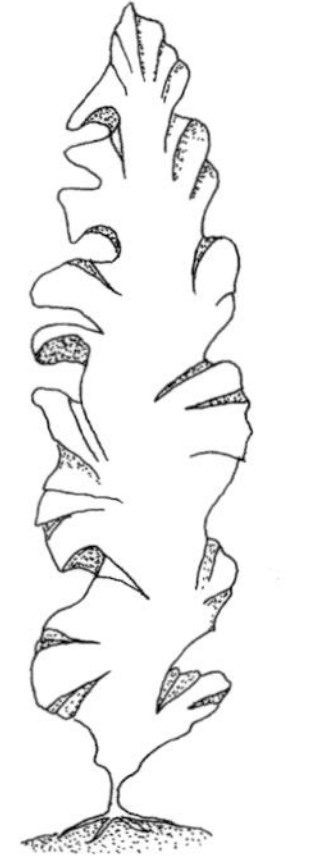

*Ulva*, sea lettuce

**Figure 15.28**
Some common Chlorophyta

## Phylum Chlorophyta (green algae)

Like the plant kingdom, the Chlorophyta have chlorophylls *a* and *b*, and store starch. Unicellular types include *Pleurococcus*, a terrestrial form living on the shaded side of tree trunks, and *Chlamydomonas*, a fresh water form that swims by means of two flagella. Some forms are filamentous, such as *Spirogyra*, whilst others, such as *Ulva*, the sea lettuce, form sheets (Figure 15.28).

## Phylum Euglenophyta

Members of this group have a mixture of plant-like and animal-like features, for example *Euglena* (Figure 15.13). Most are photosynthetic, but some lack chlorophyll and are, consequently, heterotrophic. Instead of a stiff cellulose wall, euglenoids have a flexible **pellicle** which enables them to undergo wriggling movements. Euglenoids swim by lashing movements of a flagellum. There is a second, shorter flagellum which seems to have a sensory function. As far as we know, they only reproduce asexually.

## Phylum Oomycota – fungus-like protoctists

The Oomycota are coenocytic protoctists which used to be included in the fungi. They differ from true fungi, however, in having cell walls of cellulose and flagellated asexual spores. They include the downy mildews, which are parasitic on plants, and the water moulds which are parasitic on fish. A notorious example is *Phytophthora infestans*, the cause of potato blight (Chapter 37).

# Kingdom Plantae

The plants include mosses, ferns, conifers and the flowering plants. In a number of respects members of the plant kingdom resemble the green algae, to which they are believed to be related. They store starch and their chloroplast pigments are chlorophylls *a* and *b*, xanthophyll, and β carotene. Also, like some algae, they have a life cycle consisting of a haploid plant alternating with a diploid plant, a phenomenon called **alternation of generations**. The diploid plant produces spores by meiosis and is called the **sporophyte**. The haploid plant produces gametes by mitosis and is called the **gametophyte**.

All these features are shared with at least some algae. What distinguishes plants is the development of the zygote into a multicellular **embryo**, which begins its development on the parent plant.

## Phylum Bryophyta

The bryophytes (Figure 15.29) are the liverworts (Class Hepaticae) and mosses (Class Musci). Some liverworts hug the ground as flat, leaf-like structures, but others (the leafy liverworts) are superficially moss-like.

Of the two generations, only the gametophyte is a fully-independent plant. Bryophytes are poorly adapted for maintaining their water content in dry conditions, for the following reasons:

- over most of their area, the leaves are only one cell thick, and cannot regulate their water loss. This is because they lack stomatal pores and, with few exceptions, have no water-resistant cuticle.

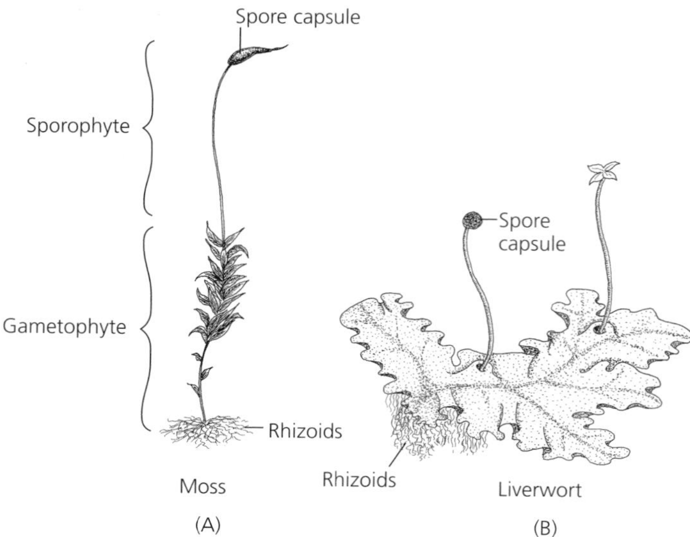


**Figure 15.29**
A moss (A) and a liverwort (B)

- the majority of bryophytes lack specialised water-conducting tissue.
- they have no roots for absorbing water, but have hair-like **rhizoids**, which only penetrate a few centimetres.
- they have no *lignin* – a substance that stiffens cellulose walls of other plants and gives support against gravity and wind.
- fertilisation depends on the presence of an external film of water in which the male gametes can swim (Chapter 30).

Despite these limitations a substantial proportion of mosses grow in habitats that are only intermittently wet, such as sunny rock faces and walls. Instead of *avoiding* desiccation, these plants *endure* it by tolerating dehydration. For long periods in summer the tissues are dry and seemingly dead, but within minutes of rain the cells become active and growth resumes. The fact that water is needed for external fertilisation is thus no barrier to the colonisation of dry places, provided that water is available for long enough for gamete formation and fertilisation to occur.

## Phylum Filicinophyta

The ferns, together with the remaining plant phyla, are often placed in the subkingdom **Tracheophyta**, or vascular plants, because they have specialised vascular (conducting) tissues. The water-transport tissue is called **xylem**, which enables water lost from the leaves to be rapidly replaced from the roots, so ferns can 'afford' to have their leaves higher up where water loss tends to be greater. They also have a tissue specialised for transport of organic materials, the **phloem**. Water loss is greatly reduced by specialisation of the outer layer of the leaf (epidermis). It has a waxy, water-resistant **cuticle**, penetrated by pores called **stomata** (singular, stoma) which close at night and during times of water shortage. The well-developed **roots** can tap deeper, more reliable supplies of water than the surface water upon which mosses and liverworts rely. They also have the ability to stiffen their cell walls by **lignin**, enabling many to grow into quite tall trees (Figure 15.30).

Like the bryophytes, they need a thin film of external water for fertilisation. Unlike the bryophytes, the gametophyte is short-lived and the sporophyte becomes fully independent, persisting from year to year (Chapter 30).

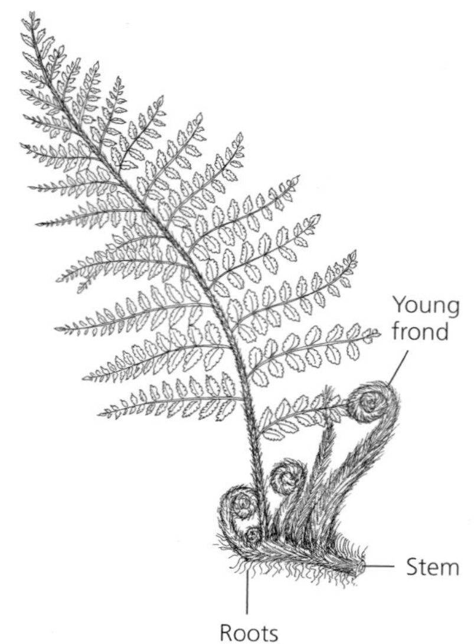


**Figure 15.30**
A fern

## Phylum Lycopodophyta

The clubmosses (Figure 15.31) are an important group because some of them provide clues to the evolution of the life cycle of seed-bearing plants (Chapter 30). Some of the lycopods are **heterosporous**, producing spores of two kinds. In these plants, **microspores** (homologous with pollen grains) grow into male gametophytes which produce sperms. **Megaspores** grow into egg cell-producing female gametophytes.

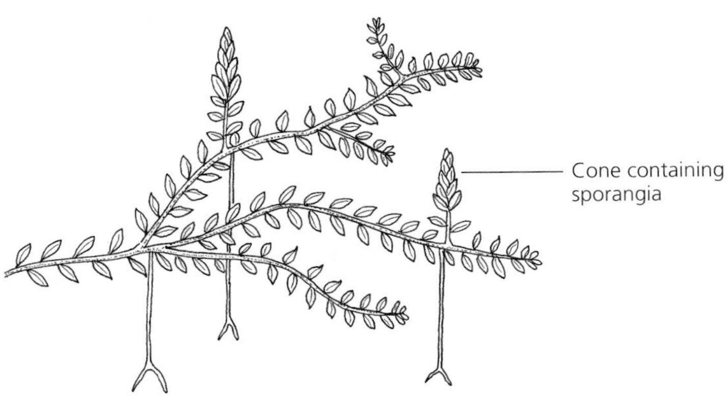


**Figure 15.31▶**
*Selaginella*, a clubmoss

## Phylum Coniferophyta

The conifers, or cone-bearing plants (Figure 15.32), reproduce by **seeds**. In seed-bearing plants the megaspores are retained and develop on the parent sporophyte, so the female gametophytes gain protection and nourishment. External water is not necessary for fertilisation because the sperms are brought from the pollen grains to the egg cells by means of a **pollen tube**, eliminating the need for external water. After fertilisation, the embryo plant is retained for a while on the parent sporophyte, protected and nourished by it. Eventually the embryo is released from the parent, together with protective parental tissue, as a seed (Chapter 30).

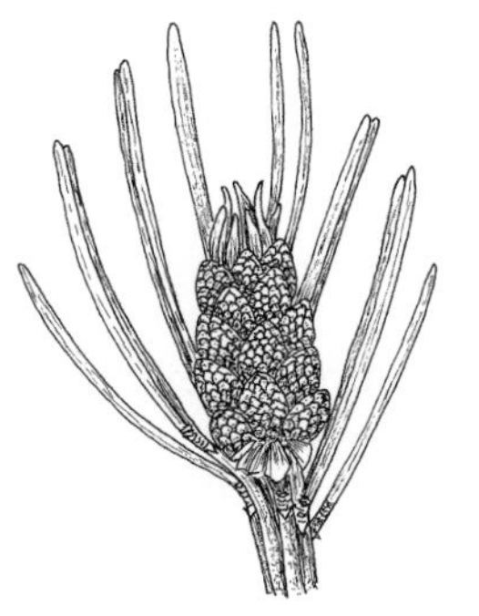


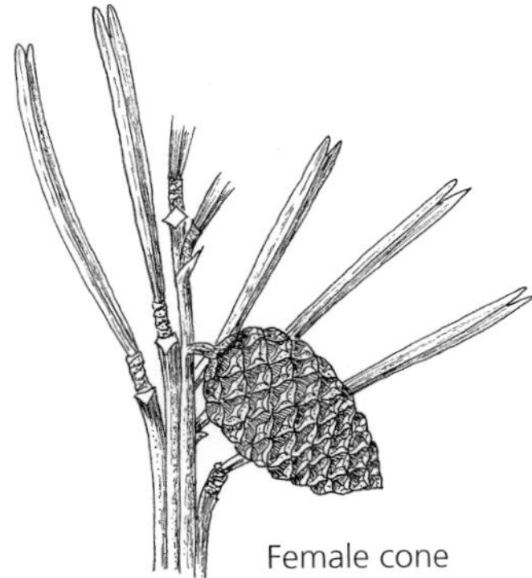


**Figure 15.32▲**
Cones of *Pinus*

## Phylum Angiospermophyta

The angiosperms are the flowering plants and are by far the largest phylum of plants. The male and female organs are borne on highly modified shoots called **flowers** (Figure 15.33). These are often adapted to make the plant more conspicuous to animals, which are needed to convey pollen grains from male to female organs. The seeds develop within the protection of an **ovary**, which develops into a **fruit**.

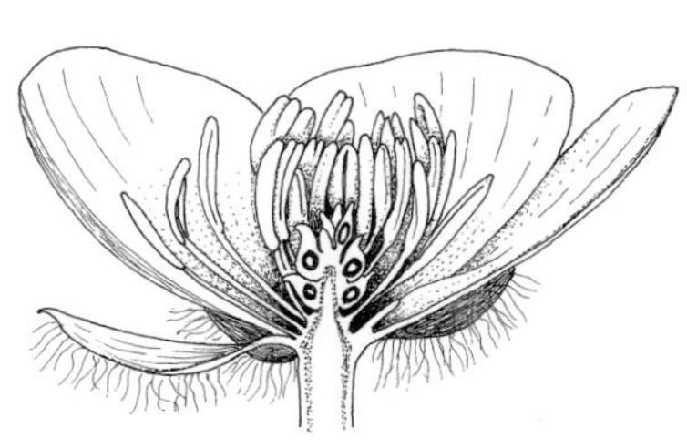


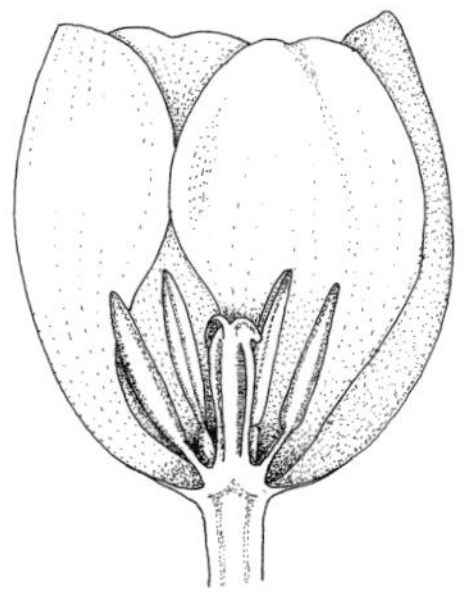


**Figure 15.33▶**
Dicotyledon and monocotyledon flowers

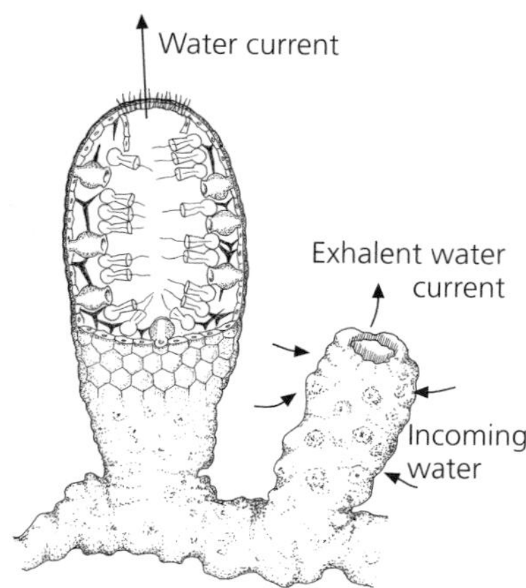


**Figure 15.34**
External and internal anatomy of a sponge

The angiosperms are divided into two classes. The **Monocotyledoneae** have one embryonic leaf, and flower parts are usually in threes or multiples of three. They include the grasses, rushes, orchids, lilies and many other plants bearing narrow leaves with parallel veins. With few exceptions, the **Dicotyledoneae** have two embryonic leaves and their flower parts are usually in fours or fives. They include most plants whose leaves are broad with a network of veins.

# Kingdom Animalia

Members of the animal kingdom are diploid, multicellular heterotrophs, typically feeding on other organisms (or their remains) and digesting them in an internal cavity. Most animals move relatively rapidly from place to place in search of food. All animals develop from an embryonic stage consisting of a hollow ball of cells called a **blastula**.

The animal kingdom is divided into just under 30 phyla, most of which are small (one contains a single species). Those that follow are the most important.

## Phylum Porifera

These are the **sponges** (Figure 15.34). They are all filter-feeding (and hence aquatic) animals. Most are marine and all are sessile, living permanently attached to rock. Lacking both nervous system and blood system, there is very little capacity for integrating activities, and although the cells show some specialisation for different functions, they are not organised into tissues.

## Phylum Cnidaria (Coelenterata)

The **cnidarians**, or **coelenterates**, include the sea anemones, jellyfish, corals and the fresh water *Hydra*. The body is **radially symmetrical** (meaning that it can be cut in many planes to give two equal halves). This feature is related to the fact that cnidarians are either sedentary (stationary) or drifting, so the environment is similar on all sides.

### Body symmetry

All animals mentioned after the cnidarians (except adult echinoderms) have **bilateral symmetry**, meaning that the body has left and right sides, and **dorsal** (upper) and **ventral** (lower) surfaces. There are also **anterior** (front) and **posterior** (rear) ends. A bilaterally symmetrical animal can only be cut down one plane to give two equal halves – there is only *one* plane of symmetry.

Bilateral symmetry is associated with movement in one direction in which important events such as the presence of food, predators or mates, are likely to be detected by the same end of the body.

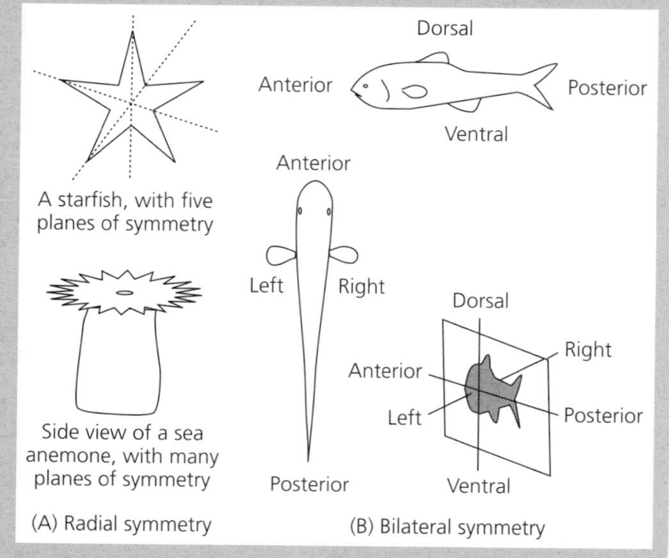


Types of body symmetry

Cnidarians are **diploblastic**, meaning that they have a body wall built of two layers of cells. There is an outer **ectoderm** and an inner **endoderm**, with a jelly-like, non-cellular **mesogloea** between (Figure 15.35). The cells are organised into tissues (see information box below). The digestive cavity or **enteron** is quite different from the tube-guts of most other animals because it has only one entrance. Though it is called the mouth, it also serves as the exit for undigested remains. The mouth is fringed with tentacles.

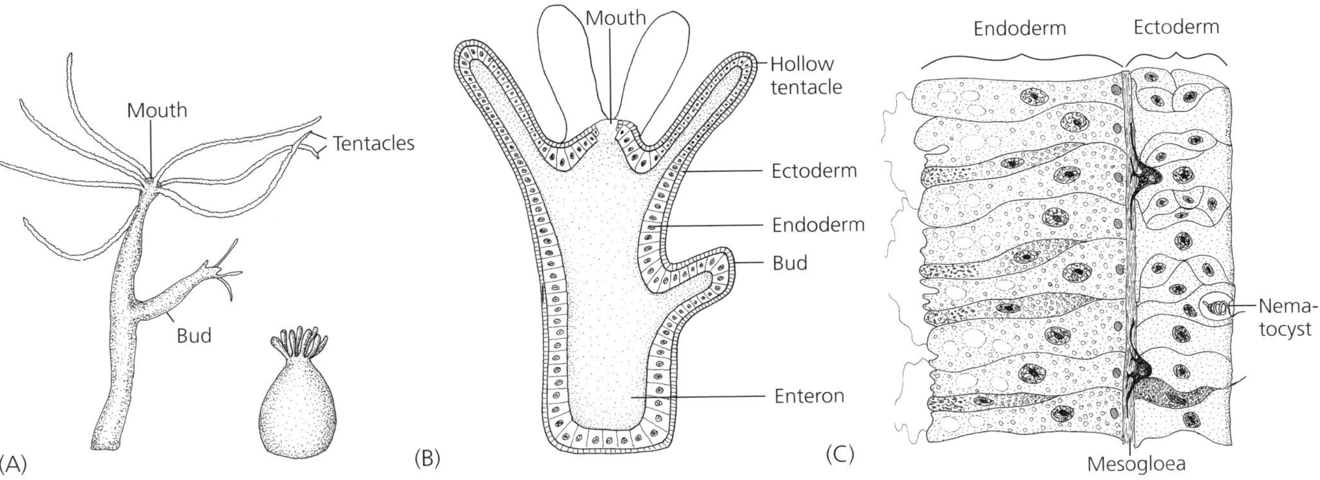


**Figure 15.35**
*Hydra*, a fresh water cnidarian. (A) External features. Whole animal extended (left) and contracted (right). (B) Longitudinal sagittal half. (C) Section through body wall

There is clear **division of labour**, with cells being specialised for carrying out different functions such as contraction, secretion and digestion.

Co-ordination of activities is achieved by a simple nervous system taking the form of a **nerve net**. The ectoderm is studded with highly specialised cells (nematoblasts) equipped with stinging threads called **nematocysts** (Figure 15.36). Transport between cells is slow since there is no blood system.

## Tissues and organs

To a certain extent, all multicellular animals show **division of labour**, different kinds of cell being specialised for carrying out particular functions. Except in sponges, cells of a given type are organised into **tissues.** A tissue is a group of cells of similar type specialised for carrying out a particular function. For example, many of the cells in the ectoderm have contractile extensions called 'muscle tails', all running longitudinally. Since they run in the same direction they can be said to *co-operate*. The cells in the endoderm have muscle tails running round the circumference.

A higher level of organisation is an **organ.** An organ is a group of tissues co-operating together to carry out a function that none of the individual tissues can. In the cnidarians, a tentacle can be regarded as a simple organ. The activity of ectodermal and endodermal muscle tails is controlled by another tissue, the nerve net, enabling the tentacle to move in a co-ordinated manner.

A common feature (except in *Hydra* and sea anemones) is the occurrence of two alternating forms in the life cycle, the sessile **polyp** and the drifting **medusa** (Figure 15.37).

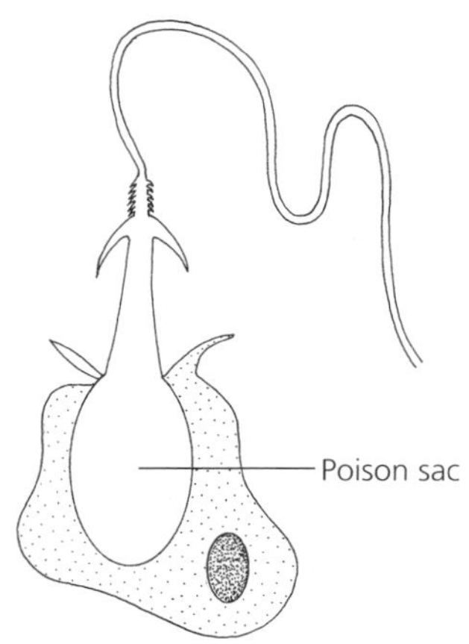


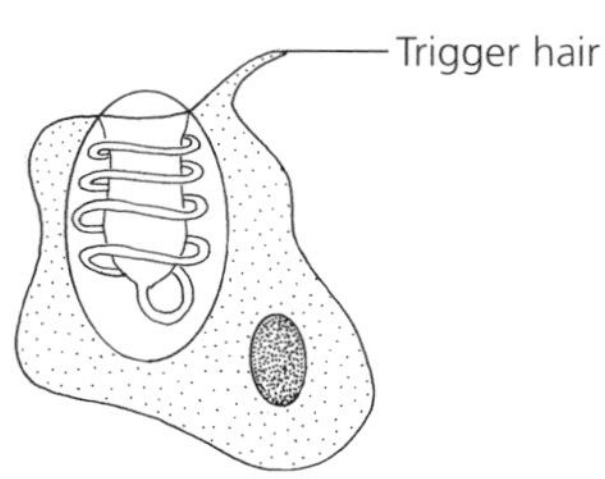


**Figure 15.36▲**
Undischarged (below) and discharged (above) nematocyst

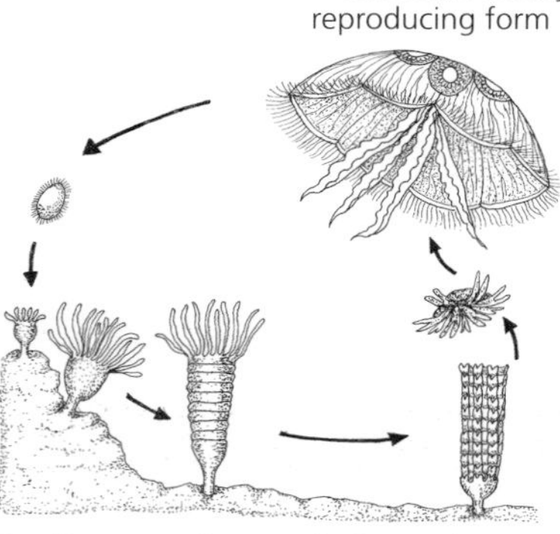


**Figure 15.37▲**
Life cycle of a jellyfish

There are three classes of cnidarians (Figure 15.38):

1 in the Hydrozoa the polyp form reproduces asexually and persists from year to year. The medusoid phase is reduced and has a limited life. In some species, e.g. the freshwater *Hydra*, the medusa is absent altogether. The Hydrozoa include the Portuguese man o' war which, although resembling a jellyfish, is actually a colony of polyps specialised for different functions.

2 in the Scyphozoa (jellyfish) the medusa is large and reproduces sexually, while the small polyp reproduces asexually, for example *Aurelia*, the common jellyfish.

3 the Actinozoa include the sea anemones and corals and have no medusoid stage.

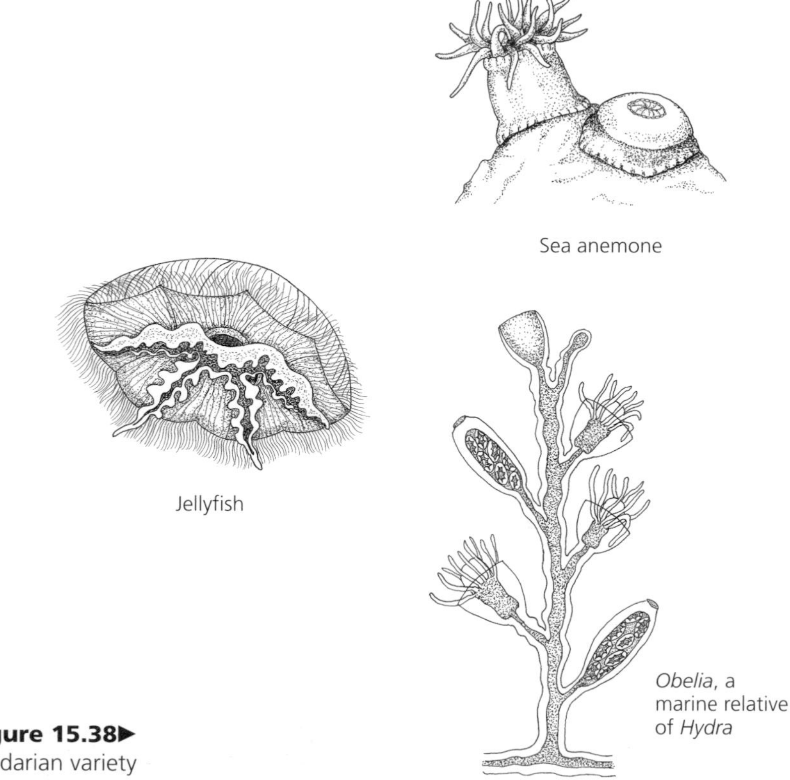


**Figure 15.38►**
Cnidarian variety

## Phylum Platyhelminthes (flatworms)

The flatworms include the flukes and tapeworms which are all parasitic, and also some inconspicuous free-living forms (Figure 15.39).

The platyhelminthes, and all phyla mentioned after them, have a feature not shown by the cnidarians – they are **triploblastic**, meaning that the body develops from three body layers in the embryo. The ectoderm develops into the skin and the endoderm develops into the lining of the gut. A middle layer, the **mesoderm**, develops into the muscles and, in vertebrates, the skeleton.

The digestive cavity has only one entrance (in tapeworms a gut is absent). There is a relatively simple central nervous system and there is no blood system. The lack of a transport system is compensated for in two ways. The flat body shape increases the flatworm's surface to volume ratio, and also minimises distances over which oxygen and $CO_2$ have to

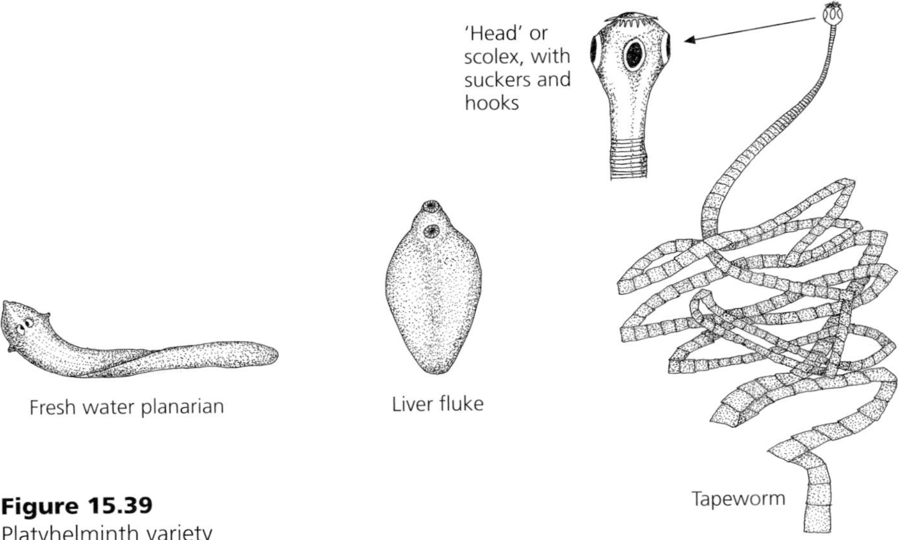


**Figure 15.39**
Platyhelminth variety

diffuse. Transport distances for digested food are reduced by the highly divided gut, excretory and reproductive systems which are interwoven with each other (Figure 15.40).

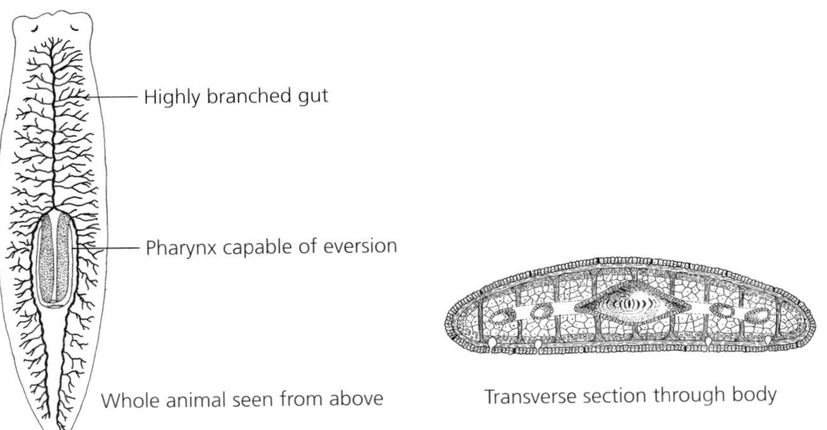


**Figure 15.40**
Organisation of *Planaria*, a free-living flatworm

## Two openings to the gut

**All phyla mentioned after the flatworms have a tube-like gut with two openings. This allows one-way movement of food along the gut and hence, division of labour into parts with different functions such as storage, digestion and absorption. It also allows a meal to be eaten before the previous one has been fully digested.**

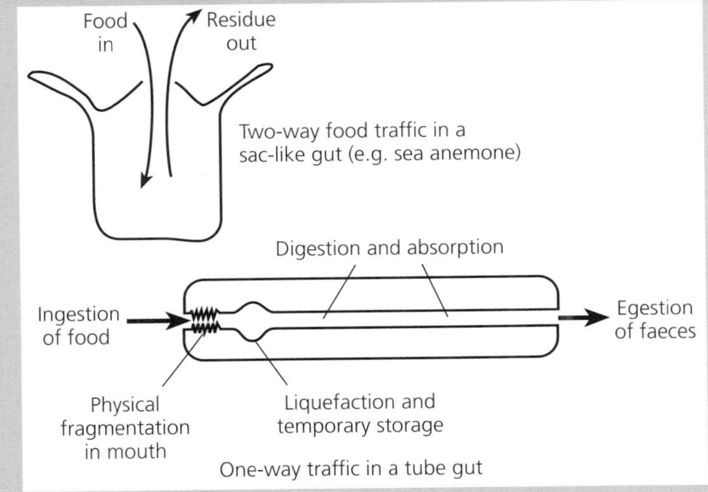


Sacs and tube guts

There are three classes:

1 Turbellaria – small, free-living creatures found creeping on the bottom of ponds and streams.

2 Trematoda, or flukes, all of which are parasitic as both young and adult stages. They all have suckers by which they hold on to the host, for example *Fasciola hepatica*, a liver fluke which lives inside the bile ducts of sheep and cattle.

3 Cestoda, or tapeworms. Like the flukes, they have complex life cycles involving more than one kind of host. The adults are parasitic in the gut of vertebrates (Chapter 37).

## Phylum Nematoda (roundworms)

Though not conspicuous, the roundworms are of great economic importance. They include a number of agricultural pests (such as the potato root eelworm) and species of *Ascaris*, gut parasites of pigs and other domestic animals.

## Phylum Annelida (segmented worms)

These are the earthworms, leeches and bristle-worms (Figure 15.41). The body is divided up into many repeating **segments**. Each segment typically bears a number of bristles called **chaetae**. The skin is thin and, in terrestrial species, moist. There is a **closed blood system** in which blood is confined entirely within the walls of blood vessels (unlike the open blood system of arthropods). As in nematodes there are two openings to the gut. Annelids are said to be **coelomate** animals, having a body cavity called the **coelom** formed within the mesoderm (see information box on page 287). Except in leeches, the coelom is large and serves as a hydrostatic skeleton (Chapter 19).

**Figure 15.41**
Annelid variety

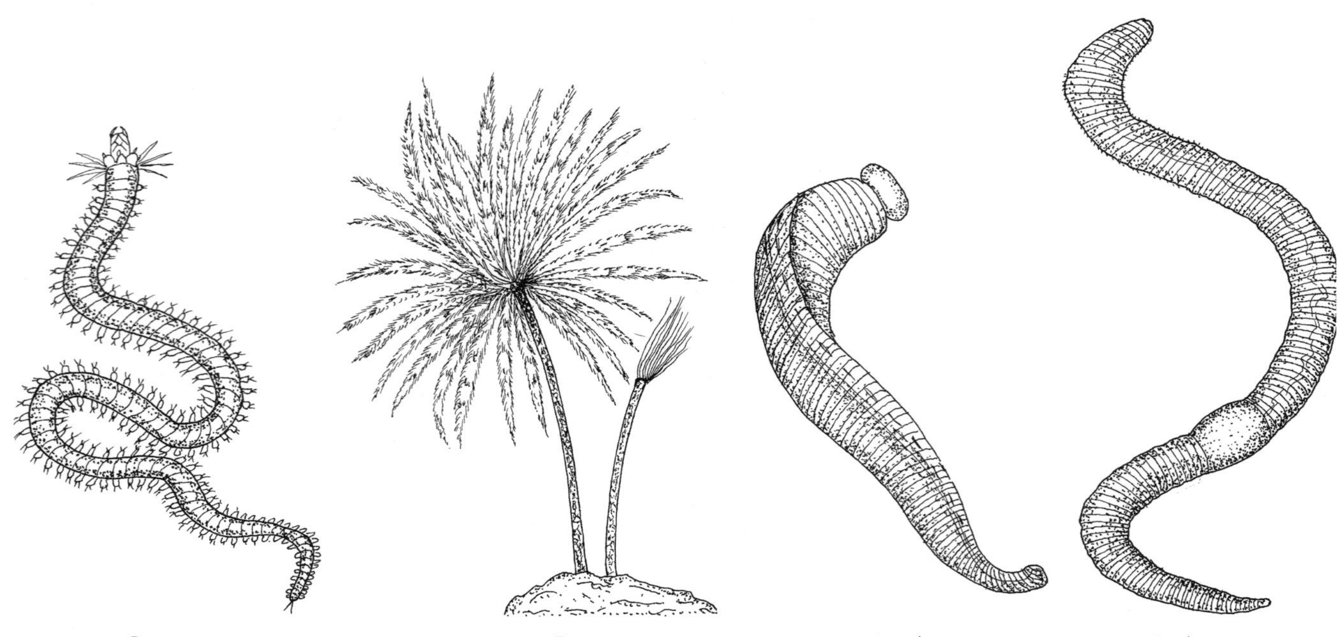

## Blood system and coelom

The annelids, and all animal phyla mentioned thereafter, have a blood system and a body cavity called a **coelom.** A blood system enables substances to be transported rapidly over long distances and is associated with increased body size (Chapter 22). Rapid transport allows increased interdependence between different parts of the body, and hence greater division of labour.

A coelom is a fluid-filled cavity that divides the mesoderm into an outer and an inner layer. In the annelids and the vertebrates it forms a cavity between the gut and the body wall. This enables muscles of the gut and body wall to operate without interfering with each other. In some phyla, the coelom is greatly reduced (as in the molluscs) or vestigial (as in the arthropods).

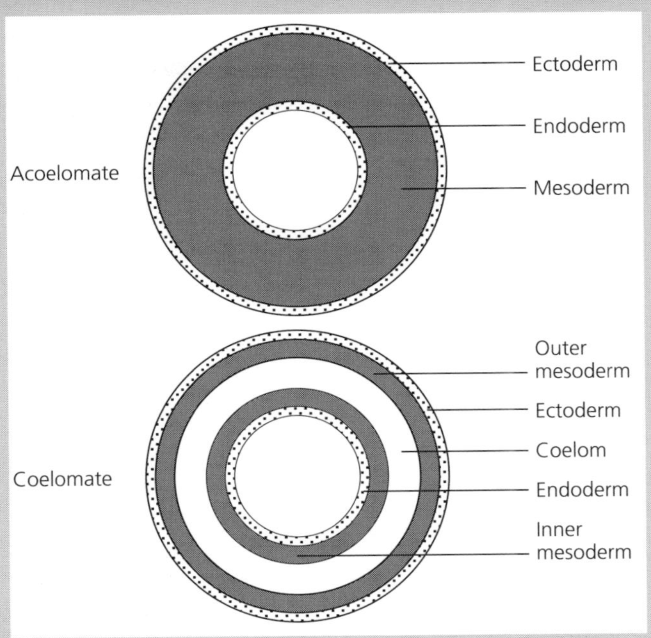


Coelomate and acoelomate organisation as seen in cross section

The three classes of annelids are:

1 Polychaeta, or bristle-worms, so-called because each segment has many chaetae. Most have a fairly well developed head. There is a planktonic **trochophore** larva very similar to that of many marine molluscs. Examples include *Arenicola* (lugworm) and non-burrowing polychaetes such as *Nereis* (ragworm).

2 Oligochaeta, or earthworms, so-called because each segment has a small number of chaetae (*oligos* means 'few').

3 Hirudinea, or leeches. These are all external parasites, sucking blood from their hosts. They have no chaetae and the coelom is greatly reduced.

## Phylum Arthropoda

The arthropods make up by far the largest phylum, and include more than three-quarters of the animal kingdom. The epidermis secretes a cuticle of **chitin**, hardened in parts by the addition of other materials to form a jointed **exoskeleton**. Since it cannot grow, the cuticle has to be periodically shed, a process called **ecdysis**. The body is segmented (though in some groups this is not apparent because of fusion of segments). Each segment primitively bears a pair of jointed limbs, though in most species some segments have become limbless. The body is differentiated into regions called **tagmata**, each tagma consisting of a group of segments specialised for a particular function or group of functions. There is an **open blood system,** in which the organs lie in a system of blood-filled spaces called the **haemocoel**. Collectively these blood spaces form the body cavity, the coelom being vestigial.

There are five major classes:

1 Crustacea, for example water 'fleas' (e.g. *Daphnia*), barnacles, sand hoppers, crabs, shrimps, woodlice (Figure 15.42). The head bears two pairs of antennae and the anterior part of the body is protected by a dorsal shield or **carapace**. In the larger crustaceans the cuticle is hardened with lime.

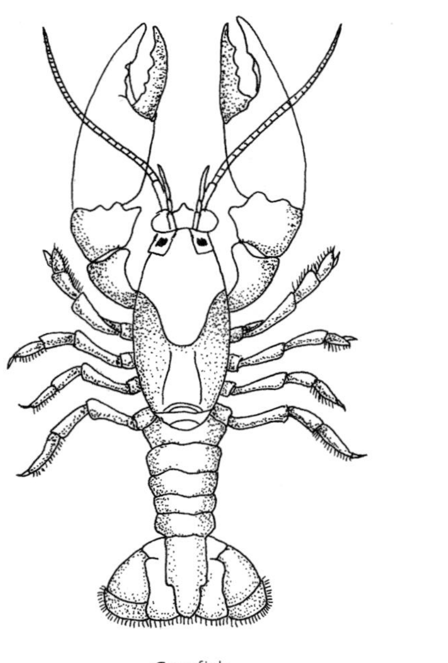


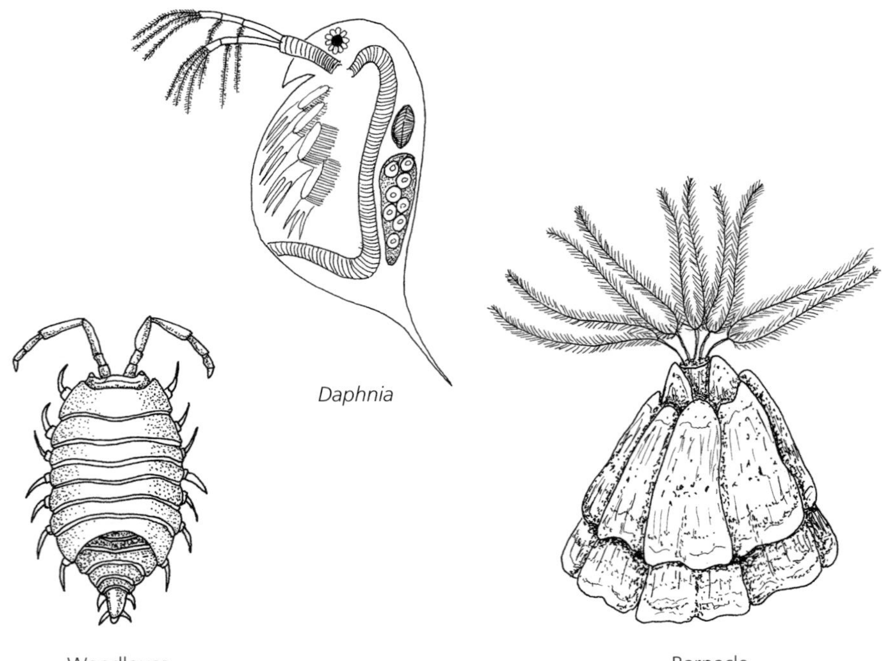


**Figure 15.42**
Crustacean variety

2 Insecta. There are three tagmata: head, thorax and abdomen. The head consists of six fused segments and is concerned with feeding and sensation. It bears one pair of antennae and, typically, a pair of compound eyes and three pairs of mouthparts. The thorax is made of three segments and is specialised for locomotion, typically bearing three pairs of legs and two pairs of wings (though in some insects these have been lost in evolution). The abdomen consists of eleven segments and is concerned with digestion, reproduction and excretion. The blood is colourless, transport of gases occurring through a tree-like system of air-filled tubes called **tracheae** which open on the sides of the body at **spiracles**.

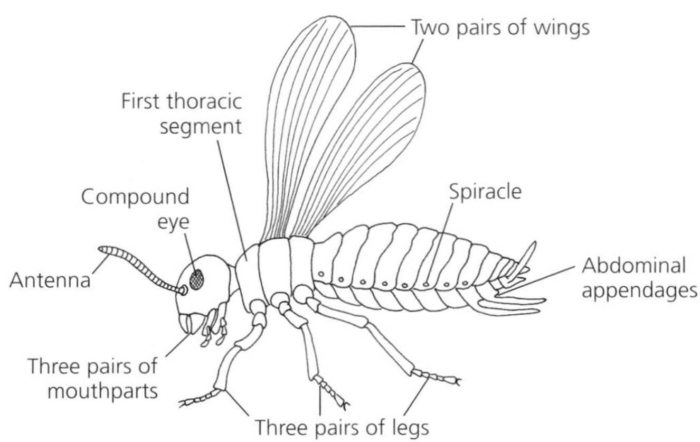


**Figure 15.43**
A generalised insect

Insects are divided into two groups:

(i) primitively wingless insects (whose ancestors never had wings), e.g. silverfish and springtails.

(ii) winged insects (or wingless insects with winged ancestors). These include the great majority of insects and are divided into two groups:
   - Insects in which the young stage, or **nymph**, has a similar diet to the adult. The wings develop externally and are thus visible in the nymph as wing buds. The life cycle of these insects is said to show **incomplete metamorphosis** because the young are (except for the lack of functional wings and their sexual immaturity) like miniature adults. Examples include cockroaches, grasshoppers, locusts, bugs and lice.
   - Insects in which the young stage, or **larva**, has a totally different diet from the adult and is structurally quite dissimilar. The wings develop internally and the adult form develops via an intermediate stage called the **pupa**. Because the young are completely different from the adults, the life cycle shows **complete metamorphosis**. Examples include butterflies and moths, flies, fleas, beetles, ants, bees and wasps.

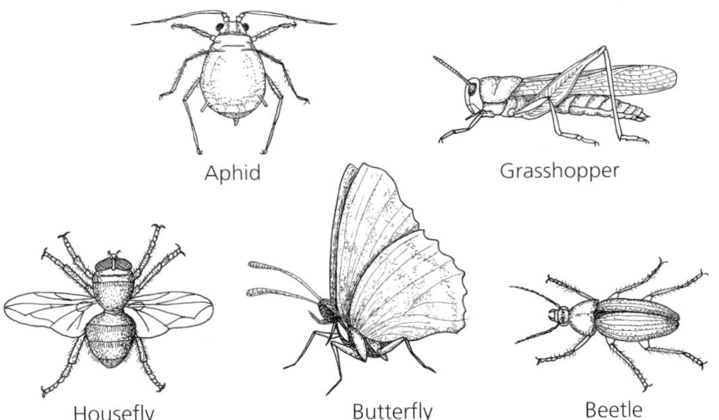


**Figure 15.44**
Insect variety

3 Chilopoda, or centipedes. Carnivorous, usually fast-running animals with many pairs of legs (not 100!). They have poison jaws and a single pair of antennae. Gas transport is via tracheal tubes (Figure 15.45).

**Figure 15.45**
A centipede

4 Diplopoda, or millipedes (Figure 15.46). Herbivorous, slow-walking animals with a single pair of antennae. Gas transport is via tracheal tubes. The body is cylindrical and successive pairs of segments are fused so that each 'segment' has two pairs of legs.

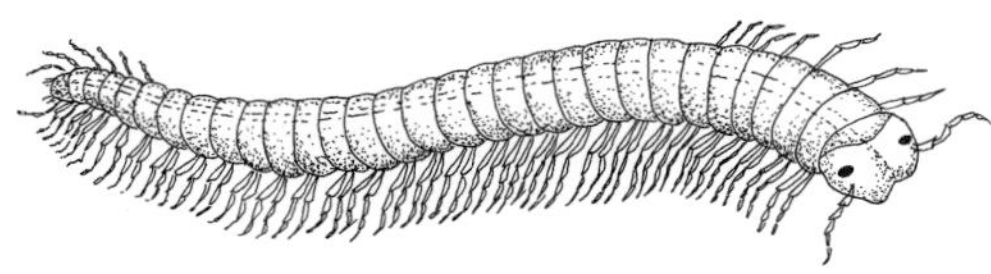

**Figure 15.46**
A millipede

5 Arachnida, for example the scorpions, spiders, ticks and mites (Figure 15.47). There are no antennae. There are typically two tagmata: an anterior **cephalothorax** bearing four pairs of legs and a pair of jaw-like appendages called **chelicerae**, and a posterior abdomen.

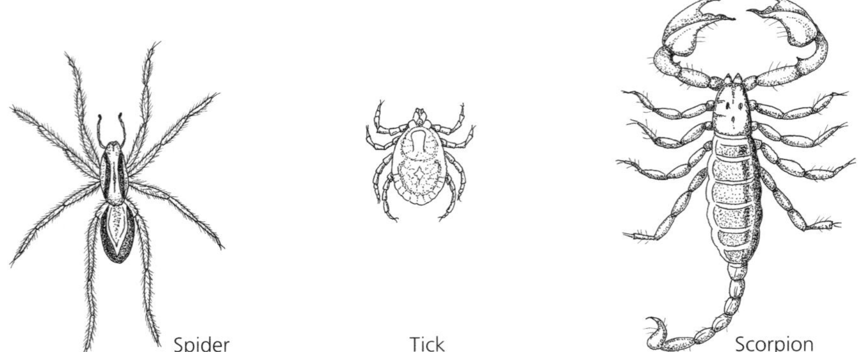


**Figure 15.47**
Arachnid variety

## Phylum Mollusca (molluscs)

The molluscs are the second largest phylum. They are unsegmented animals, with a body divided into a **head**, a muscular **foot**, which is typically propulsive, and a **visceral mass** containing digestive and other organs (Figure 15.48).

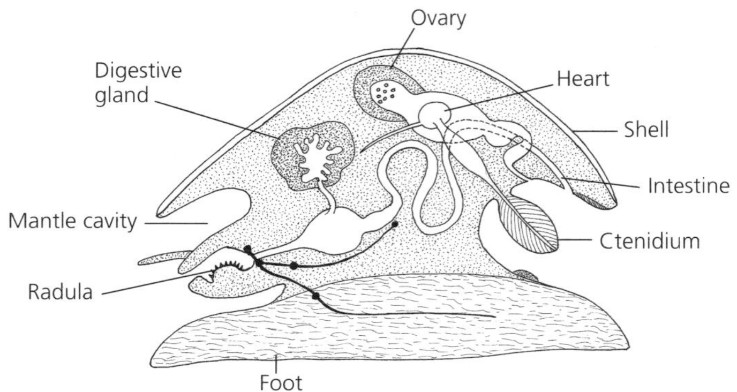


**Figure 15.48**
Basic molluscan body plan

Most molluscs have a shell secreted by a **mantle** enclosing a **mantle cavity**. Except in terrestrial snails and slugs, the mantle cavity contains a pair of gills or **ctenidia**. All molluscs except the bivalves have a rasping tongue-like structure called a **radula**. In most molluscs there is an **open blood system**, the organs being bathed in an extensive blood-filled cavity, the **haemocoel**. Many marine molluscs have a trochophore larva similar to that of polychaete worms. The coelom is greatly reduced. There are three main classes (Figure 15.49).

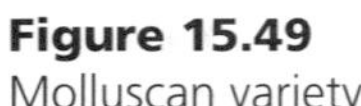

**Figure 15.49**
Molluscan variety

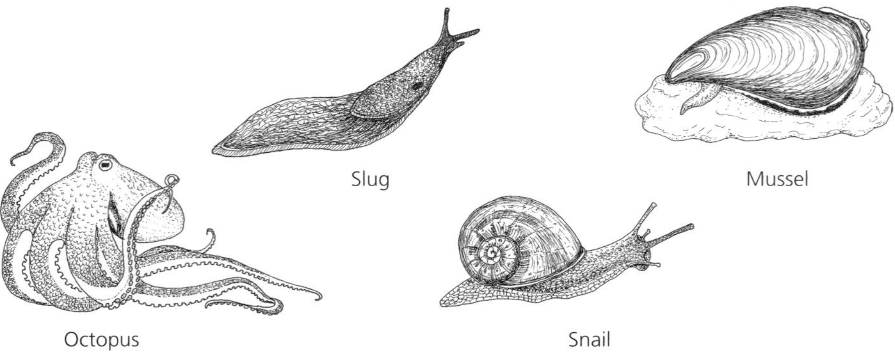

1 Gastropoda – the snails and slugs. There is a well-developed head bearing tentacles. The foot forms a flat creeping organ and the shell is typically coiled into a spiral, though in some it is reduced or lost. Early in development the visceral mass becomes twisted so the anus comes to face anteriorly – a process called **torsion**. In the terrestrial snails the mantle cavity forms a lung.

2 Pelycopoda – the bivalves. Most burrow in mud or sand using the muscular foot, though some live attached to rock (e.g. mussels, oysters). The shell consists of two halves or **valves** and the ctenidia are greatly enlarged, functioning as filter-feeding structures. There is no recognisable head and, consequently, no radula.

3 Cephalopoda – the octopus, cuttlefish and squid. These animals are rapid swimmers, propelling themselves by squirting water from the mantle cavity. As would be expected with such active creatures the eyes and brain are highly developed and the blood system is closed, enabling blood to be delivered to the tissues at high pressure. The shell is greatly reduced or absent and the foot is modified to give a series of eight or 10 tentacles bearing suckers. Besides the radula there is a horny beak with which the animal kills its prey.

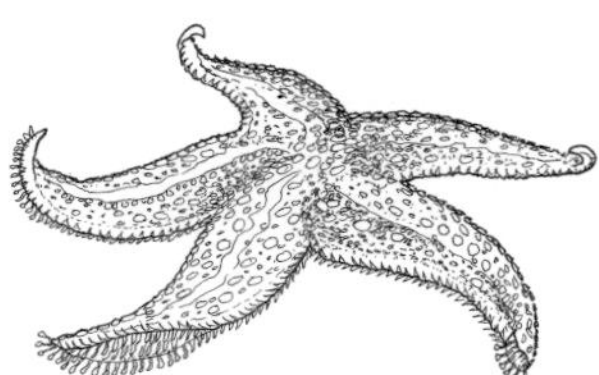
Starfish

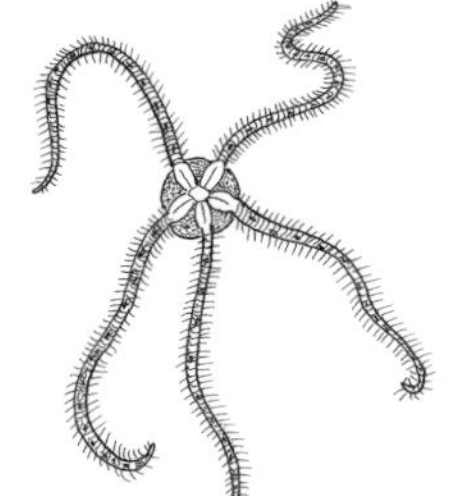
Brittlestar

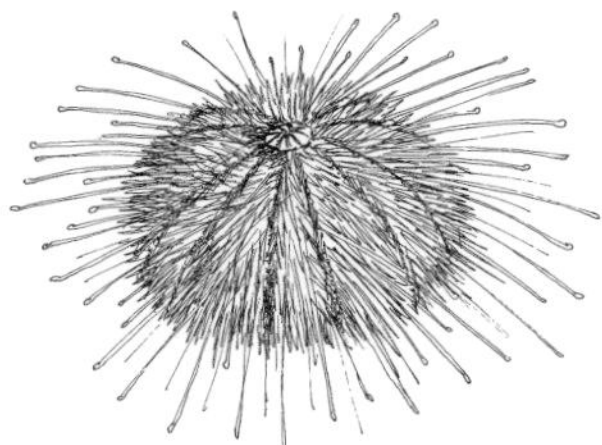
Sea urchin

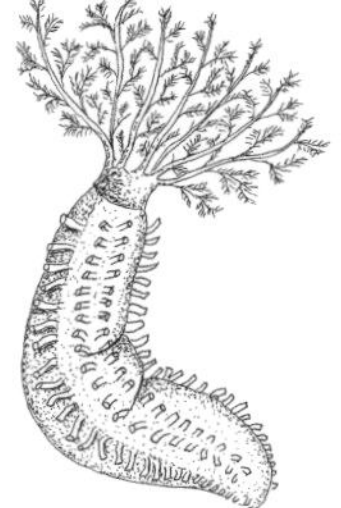
Sea cucumber

**Figure 15.50▲**
Echinoderm variety

## Phylum Echinodermata

The echinoderms include the starfish, sea urchins, sea cucumbers and brittle stars (Figure 15.50). The body has pentaradial symmetry (the body can be cut into two equal halves along five planes) though there is typically a bilaterally-symmetrical planktonic larva. There is an endoskeleton of limy plates secreted by a very thin layer of living tissue just beneath the skin (*echinoderm* = 'spiny skin').

There is a **water vascular system** which contains sea water because it communicates with the exterior. Extensions of the water vascular system form the **tube feet** by which many echinoderms move (Figure 15.51). All echinoderms are marine.

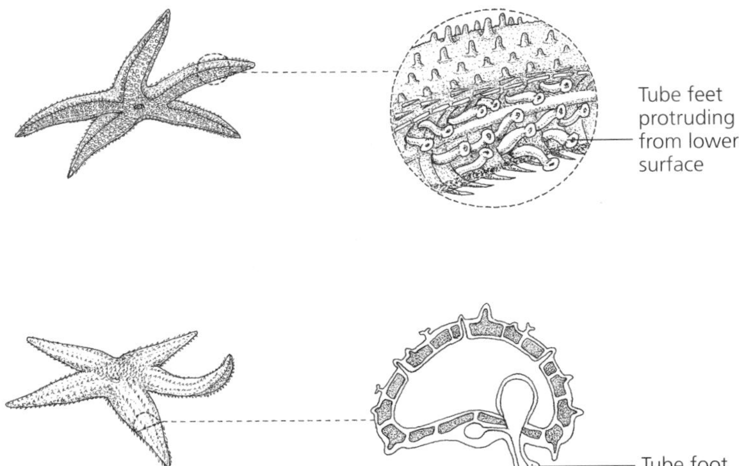


**Figure 15.51►**
Tube feet of starfish. Above, lower surface of body showing tube feet. Below, section through an arm showing a single tube foot

## Phylum Chordata

The chordates include the vertebrates, or backboned animals, and a few much less familiar animals such as sea squirts. At least in the early stages of development, all chordates have the following characteristics (Figure 15.52):

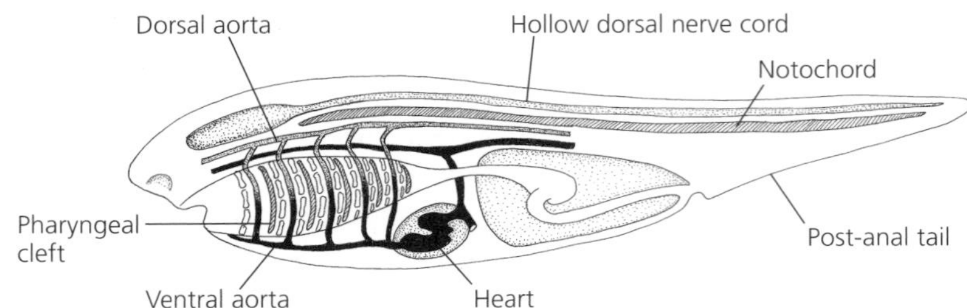


**Figure 15.52**
Chordate characters

- a rod-like **notochord** running under the dorsal surface of the body.
- a hollow dorsal **nerve cord** just above the notochord.
- a segmented body.
- a series of **pharyngeal pouches** which, in the adults of some chordates, perforate the wall of the pharynx (anterior part of the gut) forming gill slits.
- a closed blood system in which there is a ventral heart pumping blood forward on the ventral side and backward on the dorsal side.
- a post-anal **tail**.

The major subphylum consists of the **vertebrates**, in which the notochord is replaced by a vertebral column and there is a brain surrounded by a skull. There are two superclasses: the jawless vertebrates (lampreys and hagfish) which have no paired limbs, and the jawed vertebrates. The latter are divided into six classes (Figure 15.53):

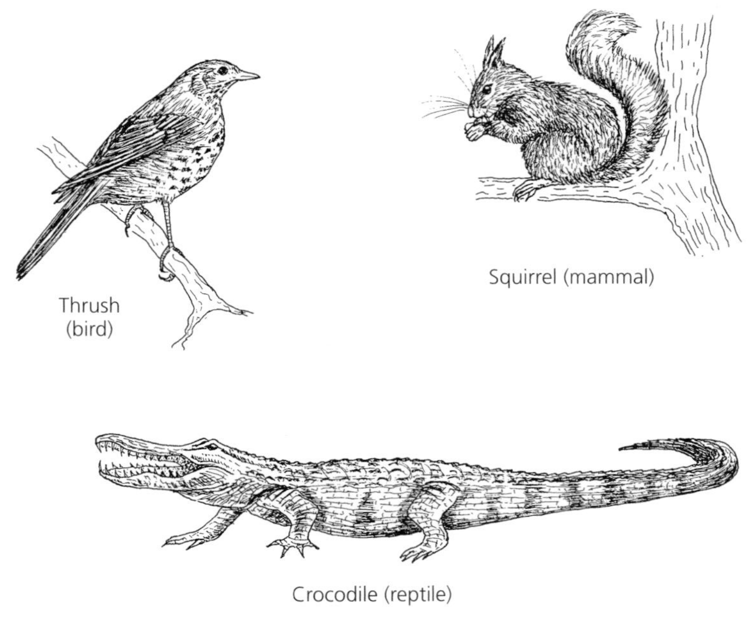

**Figure 15.53**
Vertebrate variety

1 Chondrichthyes, or cartilaginous fish – the sharks, skates and rays. The skeleton is composed of cartilage (though it may be hardened by calcification). The scales are tooth-like. The mouth is ventral and the gill slits are visible externally. The fins are fleshy and the tail is asymmetrical.

2 Osteichthyes, or bony fish. The skeleton is composed of bone and the scales are thin bony plates. The mouth is terminal and the gill slits are concealed by an **operculum** or gill cover. The fins are thin and are supported by bony fin rays and the tail is typically symmetrical. A gas-filled **swim bladder** enables the animal to hang in the water without having to expend muscular energy.

3 Amphibia. The amphibians are typically four-footed vertebrates with a slimy skin lacking scales. The eggs are usually laid in water and hatch into a fish-like tadpole. This later undergoes a metamorphosis into a four-legged adult form.

4 Reptilia. The reptiles are four-footed vertebrates which, except for aquatic species, are much less dependent on external water than amphibians. They have scaly skins which, except in aquatic types, are dry. Fertilisation is internal and the young develop inside large yolky eggs protected by a leathery shell. There is no tadpole stage. In some species the eggs hatch inside the female, and are born as well-developed young.

5 Aves, or birds. The most distinctive features of birds are associated with flight. The high body temperature is regulated by balancing heat production against heat loss. The high metabolic rate is made possible by a double circulation in which the heart is divided into left and right sides, the right pumping blood through the lungs and the left pumping it round the rest of the body. The skin is covered with heat-insulating feathers which also play a crucial role in flight. The sternum (or breastbone) has a keel-like extension to which are anchored the large flight muscles. The larger limb bones are lightened by air sacs which are extensions of the lungs. There is a horny beak and, in modern birds, no teeth. The large yolky eggs are protected by a limy shell.

6 Mammalia, or mammals. The young are suckled on milk secreted by mammary glands. As in birds, the high metabolic rate is sustained by a highly efficient system for obtaining oxygen and distributing it round the body. The thorax and abdomen are separated by the muscular **diaphragm**. There is a double circulation with a four-chambered heart. There are different types of teeth specialised for different functions. The body is typically covered with an insulating layer of hair. There are three bones in the middle ear (other four-footed vertebrates have one). There are three subclasses:

- Monotremata. The duckbill platypus and spiny anteaters of Australasia, both of which lay eggs.
- Metatheria, or pouched mammals. The young are born in a very immature state and complete their development attached to the nipple inside a pouch or marsupium.
- Eutheria, or placental mammals. The young are nourished inside the mother by a placenta and are born at a much later stage of development.

## QUESTIONS

1. Which of the following taxa would include the greatest variety amongst its members?

   phylum, species, kingdom, family, order, genus, class

2. In which of the following would you expect its members to have the most recent common ancestor?

   Order Rodentia, Phylum Chordata, Family Muridae, Genus *Rattus*

3. Which *two* of the following are homologous, and which are analogous?

   an elephant's tusk, a cow's horn, a deer's antler, a tiger's incisor, a spine of a hedgehog, a spine of a sea urchin.

4. The forelimbs of the European and Australian mole are much more similar to each other than are the other parts of the body. Why?

5. What are the advantages of molecular taxonomy over more traditional methods of classification?

6. A student wrote the scientific name for the lesser celandine as ranunculus ficaria. Re-write it correctly.

7. Why are breeding habits not a good character to use when designing an artificial key?

8. Using external features alone, how would you distinguish between members of each of the following pairs:

   a) a moss and a fern
   b) a conifer and a flowering plant
   c) a bony fish and a cartilaginous fish
   d) a reptile and an amphibian
   e) a cnidarian and an echinoderm
   f) an annelid and a fly larva, such as maggot

9. Name the phylum to which animals with each of the following combinations of characters would belong:

   a) an open blood system and a segmented body
   b) an open blood system and an unsegmented body
   c) a water vascular system
   d) a body wall with two layers of cells
   e) one opening to the gut and a body wall with three layers of cells

CHAPTER

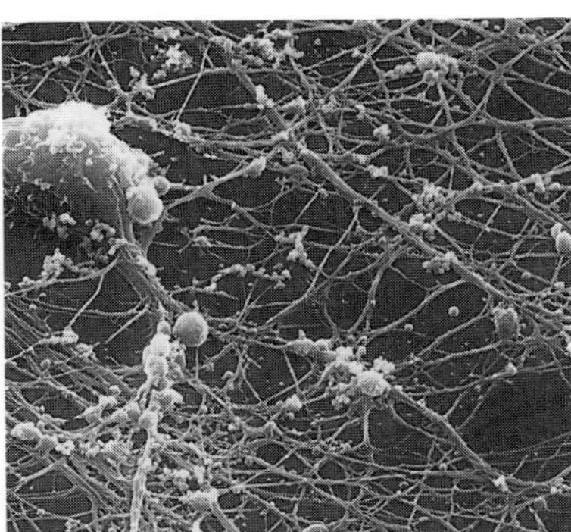

# 16 ANIMAL ORGANISATION – SIZE and COMPLEXITY

**LEARNING OBJECTIVES**

By the time you have completed your study of this chapter you should be able to:

- explain how the organisation of animals differs from that of plants.
- explain how the size of objects affects their surface area relationships, and the significance this had for the evolution of multicellularity.
- explain what is meant by cell differentiation, and describe the organisations of cells into tissues and organs.
- name the four categories of mammalian tissue and describe the functions and distinguishing features of each.

## The animal way of life

Animals and plants make their living in very different ways and, in relation to this, have very different organisations. For convenience these differences are discussed under separate headings, but it is important to appreciate that they are interrelated.

- Nutrition. Animals are **heterotrophic**, obtaining their carbon in the form of organic compounds from the bodies of other organisms. Plants, on the other hand, are **autotrophic**, obtaining their raw materials in the form of simple, inorganic substances – carbon dioxide, water and mineral ions. These are reduced to organic compounds using light energy.
- Locomotion. Animals feed on other organisms which, in most cases, must be obtained by active movement, usually from place to place. In contrast, plants are surrounded by their raw materials and have no need to move from place to place.
- Cellular organisation. Unlike plant cells, animal cells have no cellulose walls. This makes it possible for muscular movement involving rapid changes of shape of specialised muscle cells.
- Body shape. To absorb raw materials, organisms have to expose a relatively large surface area to their surroundings, but animals and plants do this in very different ways. Animals typically feed on concentrated 'packets' of food in the form of other organisms. Moving

around to obtain food requires a compact body shape – yet a large surface area is needed to digest and absorb it. An animal achieves this by processing its food inside a deep intucking into the body called the **gut**, inside which digestive enzymes and the products of digestion can be concentrated (Figure 16.1).

The situation in plants is the opposite of that in animals. The carbon dioxide and mineral ions needed by plants are thinly dispersed in the environment. The best shape with which to absorb such dilute nutrients is a finely dissected body which protrudes into the environment. Thus, whereas animals surround their food, plants are surrounded by theirs.

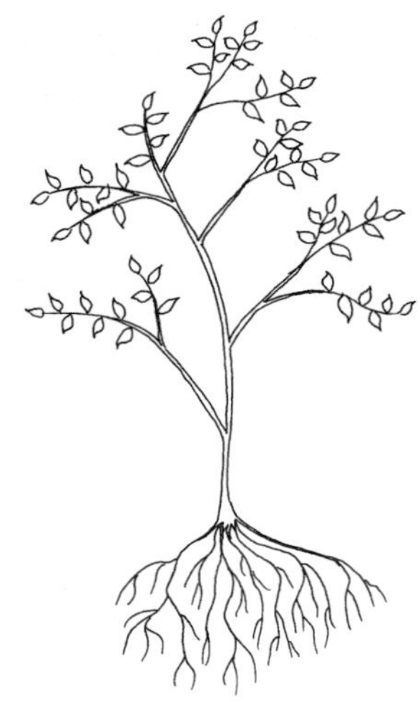

Plant – dissected shape, large external surface

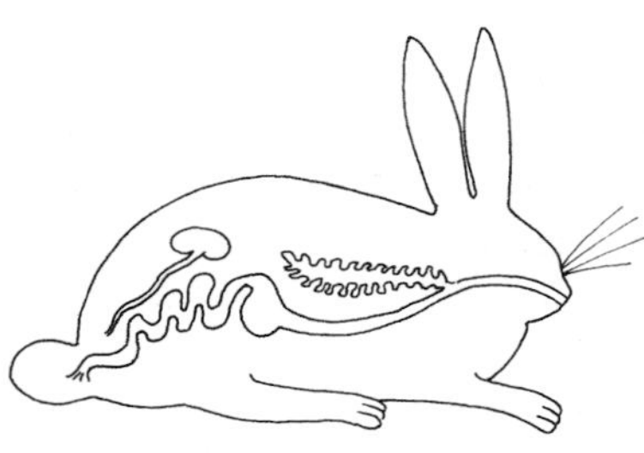

Animal – compact shape, large internal surface area

**Figure 16.1**
Body shape in animals and plants

There are other important differences between animal and plant organisation – in particular, growth (Chapter 29).

# Why be multicellular?

All animals are multicellular. It seems likely that the first eukaryotes were unicellular, and that more complex, multicellular bodies evolved later. Since most multicellular organisms are larger than most unicellular types, it seems highly probable that the answer has something to do with size.

The explanation depends on the fact that when anything increases in size without changing its shape, its length, area and volume increase at different rates. This **scaling effect** has such important consequences for organisation in living things that we must devote a little space to it.

Look at Figure 16.2, and Table 16.1, which show the effect of increasing the size of a cube. Suppose its side is initially 1 cm. Its surface area is $1 \times 1 \times 6 = 6$ cm$^2$. Its volume is $1 \times 1 \times 1 = 1$ cm$^3$. Now suppose we double the length of each side, thus keeping the shape the same. Its surface area has increased by four times, from 6 cm$^2$ to 24 cm$^2$. Its volume has increased eight times from 1 cm$^3$ to 8 cm$^3$.

The reason why volume increases faster than area is that volume has three dimensions whilst area has only two. This is true for any shape – even awkward ones like rabbits and frogs. If, for instance, a frog increases its length three times, its surface area goes up $3 \times 3 = 3^2 =$ nine times, but its volume increases $3 \times 3 \times 3 = 3^3 = 27$ times.

To take an actual example, a newly-hatched caterpillar of a peacock butterfly is about 1.6 mm long. During the next few weeks it grows to about 40 mm long, an increase of 25 times. If its shape doesn't change (and it doesn't change much), its surface area increases $25^2 = 625$ times and its volume increases $25^3 = 15\,625$ times! The volume, therefore, increases 25 times more than the surface area. This means that the surface:volume ratio (surface/volume) of the full-grown caterpillar is only $\frac{1}{25}$ that of the newly-hatched one.

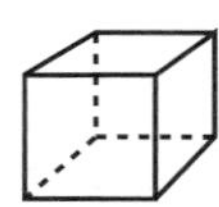



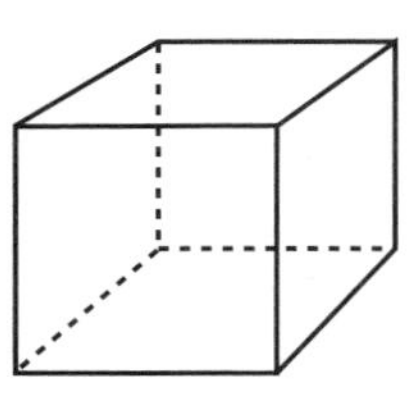



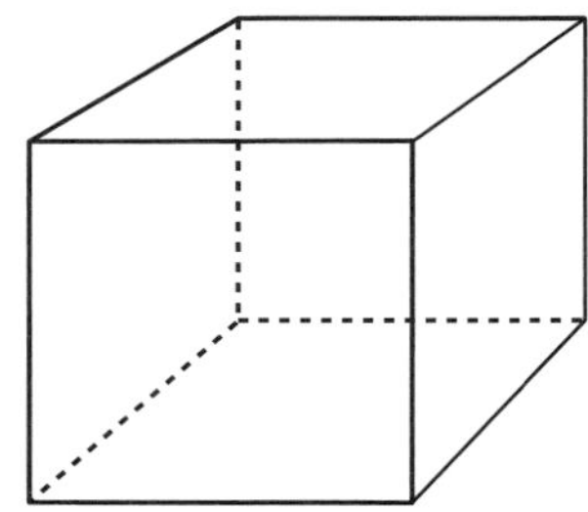



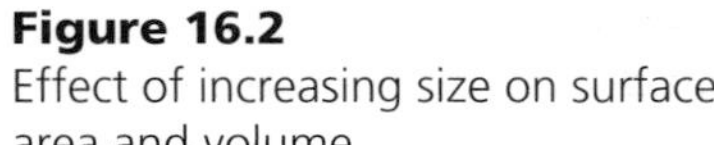

**Figure 16.2**
Effect of increasing size on surface area and volume

**Table 16.1**
The effect of size increase in a cube

| Length (cm) | Surface ($cm^2$) | Volume ($cm^3$) | Surface/ volume ratio |
|---|---|---|---|
| 1 | 6 | 1 | 6 |
| 2 | 24 | 8 | 3 |
| 3 | 54 | 27 | 2 |
| 4 | 96 | 64 | 1.5 |
| 5 | 150 | 125 | 1.2 |
| 6 | 216 | 216 | 1 |
| 8 | 384 | 512 | 0.75 |
| 10 | 600 | 1000 | 0.6 |

**CLEAR THINKING**

**It is sometimes said that one reason why larger organisms are multicellular is to maintain a large enough surface/volume ratio for gas exchange at the cell surface. If you look at Figure 15.36 you will see that in a simple multicellular animal such as *Hydra*, most of the cell surfaces form boundaries *with other cells*, and are thus not available for gas exchange.**

These numbers may be impressive, but what practical importance do they have for living things? The answer is, a great deal. In fact, surface and volume relationships affect most processes in organisms, from gas exchange to support. For the moment, however, we are concerned with the question of why large organisms are multicellular.

The nucleus of a cell can only communicate with the cytoplasm via the pores in the nuclear envelope. The number of pores depends on the *area* of the nuclear envelope. The amount of cytoplasm under the control of the nucleus depends on its *volume*. Thus, if a cell increases in length five times without changing its shape, the nucleus would have $5^3 = 125$ times as much cytoplasm to control. But the area of the nuclear envelope would only have increased $5^2 = 25$ times. The only way the nucleus can maintain adequate control over the cytoplasm is for the cell to divide into two.

## Multicellularity and complexity

A sexually reproducing multicellular organism begins life as a single cell – a fertilised egg. This divides mitotically into two, then into four, and so on to produce a multicellular body.

The earlier stages of growth are concerned with building up cell numbers. Later on the cells begin to specialise, taking on different functions. Experiments have shown that all the cells of the body are genetically identical and carry the entire genome. As a result, the process of cell specialisation – or **differentiation** as it is called – occurs without any genetic change.

How, then, do cells become different if they all carry the same instructions? The answer is that different cells 'read' different parts of the message, ignoring the rest. Thus a skin cell, specialised for producing keratin, ignores all those genes which are not required by a skin cell, such as the genes for making the hormone insulin and the blood pigment haemoglobin.

## Tissues and organs

A mammal the size of a human consists of about six million million cells of about 200 different kinds. Each is specialised for performing a particular function or – in some cases – a number of functions. A group of cells carrying out a particular function is called a **tissue**. Some tissues contain only one kind of cell, others contain several.

In many tissues the cells are not arranged randomly, but are organised in such a way that they can perform a function that none of the individual cells can. This co-operative action of cells can be seen particularly clearly in epithelia, muscle and nervous tissues (see below).

In all the higher animals, tissues are organised into more complex structures called **organs**. An organ can perform a function that none of its constituent tissues can perform by itself. The stomach, for example, contains **glandular tissue** that produces a digestive juice, and **muscle tissue** which mixes and propels the food. It also contains a network of **nervous tissue** which helps to co-ordinate the activity of the muscle. Helping to bind these tissues together is a network of **connective tissue**. Together, these four kinds of tissue can carry out a range of functions that none of the individual tissues can by itself.

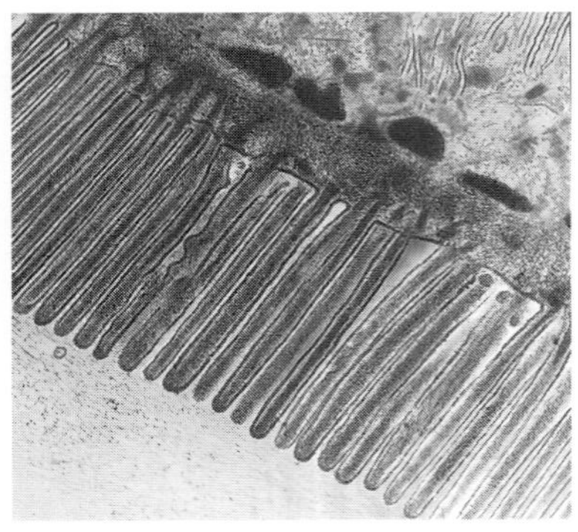

**Figure 16.3▲**
TEM of epithelial cells from the small intestine showing microvilli on their surface

**Figure 16.4▲**
Development of a gland from an epithelium

## A brief survey of mammalian tissues

The study of tissues is called **histology**. Mammalian tissues can be grouped into four major categories: **epithelial** tissues, **connective** tissues, **muscular** tissues and **nervous** tissues. Tissues such as bone, nerve tissue, and certain types of muscle are major topics in themselves, and are dealt with in later chapters.

### Epithelial tissues

Epithelial tissues, or simply **epithelia** (Figure 16.3), form coverings of surfaces and linings of internal cavities such as the breathing passages, gut and the various reproductive tubes. Epithelia that line internal cavities are called **endothelia**.

An important feature of epithelial cells is their ability to adhere to each other to form sheets. This is due to the presence of many **desmosomes** in the plasma membranes. Some epithelia, such as the epidermis and lining of the oesophagus, are further strengthened by bundles of **intermediate filaments** running from cell to cell (Chapter 3).

All epithelia lie on and adhere to an extremely thin, non-cellular layer called the **basement membrane**. This is produced by the epithelial cells themselves and consists of a jelly-like liquid penetrated by a fine network of collagen.

One of the most important functions of epithelia is **protection**. Epithelia may protect against abrasion, or they may act as barriers to water loss or the entry of micro-organisms, as in the epidermis of the skin. Some endothelia act as selective barriers, preventing the passage of some substances whilst promoting the passage of others.

Epithelia may also form the **secretory** tissue of glands. A gland develops as an outgrowth of the epithelia lining an internal cavity such as the gut (Figure 16.4). Some of the most important glandular epithelia occur in the alimentary canal. Some epithelia have a **sensory** function, such as those concerned with the sense of smell.

In some parts of the body, epithelia combine several roles. For example, in the breathing passages mucus-secreting cells are scattered amongst the ciliated cells that propel the mucus (p. 418).

Structurally epithelia are of two general types. **Simple epithelia** consist of a single layer of cells which can be of three general shapes – flat, cubical or columnar (Figure 16.5). The cells of some epithelia have **microvilli** on one side, which greatly increase the surface area for active transport of solutes. Examples are the columnar epithelia lining the small intestine (p. 476) and the cuboidal epithelia forming the walls of kidney tubules (p. 433).

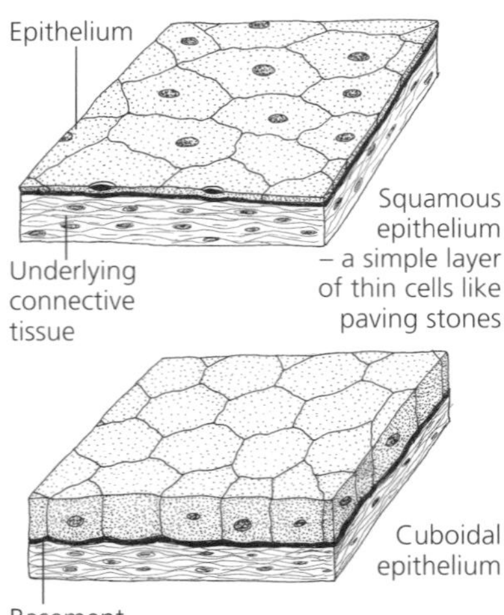


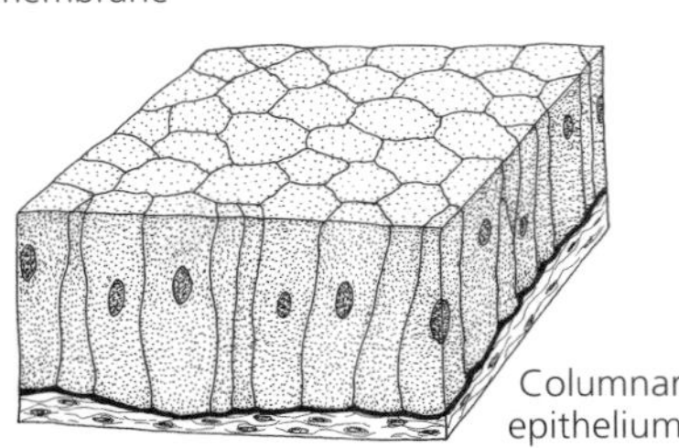


**Figure 16.5▲**
Various kinds of simple epithelia

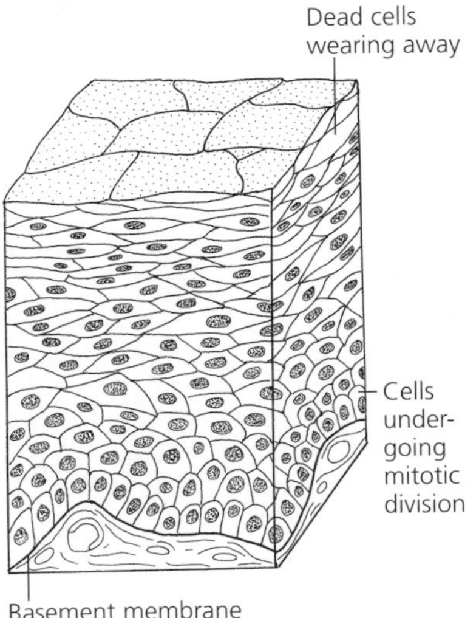


**Figure 16.6**
Compound epithelium

**Compound** or **stratified epithelia** are more complex, consisting of several layers of cells (Figure 16.6). They occur in places prone to mechanical wear and tear, for example the lining of the upper alimentary canal and the epidermis. Cells worn away at the surface are replaced by new ones produced by cell division in the **generative layer** at the base.

## Connective tissues

Connective tissues are the most widely-distributed tissues in the body and include tendon and ligament. Also included in this category are the skeletal tissues, bone and cartilage, which provide support and protection.

Unlike other tissues, a considerable proportion of connective tissue is made up of an extracellular **matrix** secreted by the connective tissue cells. The matrix is made up of a semiliquid **ground substance** containing glycoproteins (p. 29), permeated by a network of **protein fibres**. In bone the matrix also contains large amounts of the mineral **hydroxyapatite**, a form of calcium phosphate.

All connective tissues have a common embryological origin. Though functionally quite different, blood is, for this reason, usually considered to be a connective tissue. It is also the only liquid tissue.

Whereas epithelia never contain blood vessels, all connective tissues, except cartilage, have a rich blood supply and a considerable capacity for repair.

### *Areolar tissue*

Of all connective tissue, **areolar tissue** is the most widespread, forming the internal supporting framework of every organ (Figure 16.7). Besides this purely mechanical function, areolar tissue is important in inflammation, playing a key part in defence against infection.

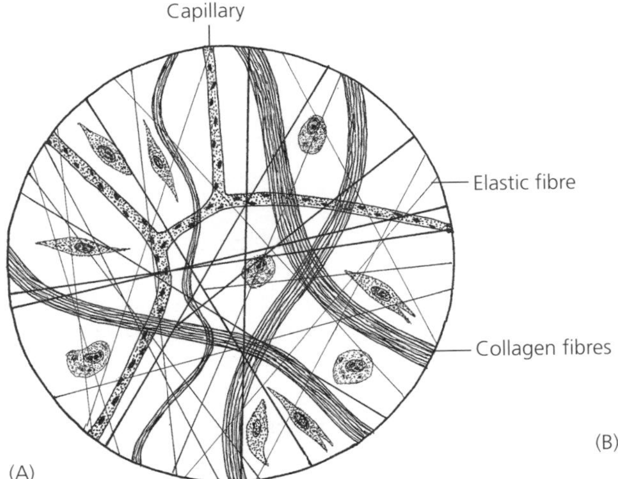


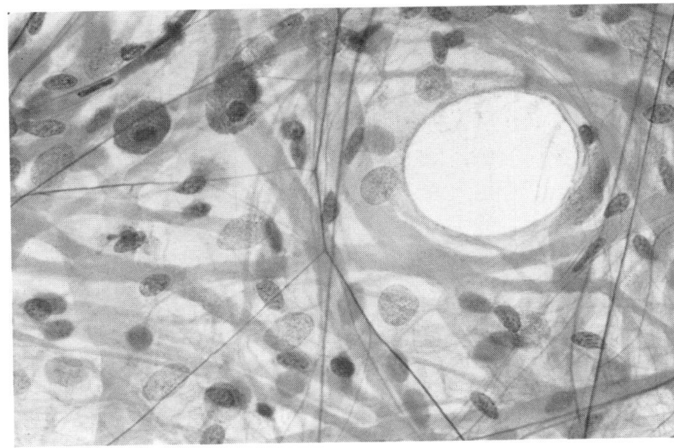


**Figure 16.7**
Areolar tissue; a diagrammatic view (A), under the light microscope (B)

The supporting role of areolar tissue is performed by two kinds of protein fibre. **Collagenous fibres** consist of bundles of the fibrous protein **collagen** and have an extremely high tensile strength. They are relatively inelastic and are the major constituent of tendons and ligaments. **Elastic fibres** are branched and consist of the rubbery protein **elastin** which, as the name suggests, is highly elastic. They can, therefore, stretch when an external force is applied and return to their original length when it is removed.

The cells of areolar tissue include **fibroblasts**, which secrete the collagen and elastin fibres, and a number of types of cell concerned with defence, such as **mast cells** and **macrophages** (p. 489).

## Adipose tissue

Usually referred to as 'fat', adipose tissue occurs beneath the skin and around some internal organs. It consists largely of fat cells, each of which contains a large globule of fat (Figure 16.8). Differences in body fat are due to differences in the size of fat cells rather than number.

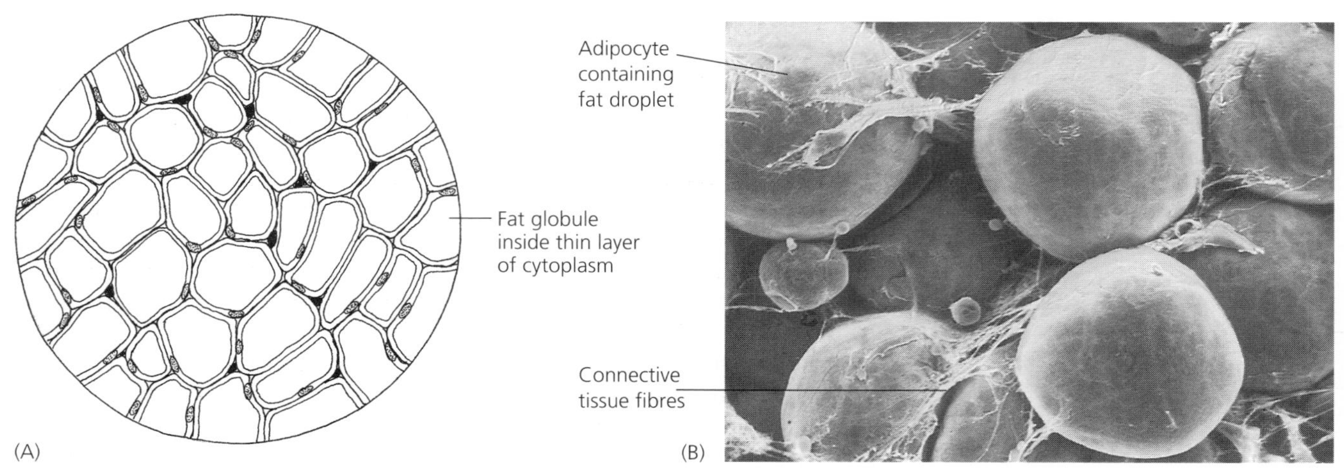


**Figure 16.8**
Adipose tissue; a diagrammatic view (A), under the electron microscope (B)

## Cartilage

Together with bone, cartilage is a **skeletal tissue** (Figure 16.9). Since bone is closer in function to skeletal muscle than cartilage, bone is dealt with later (Chapter 19).

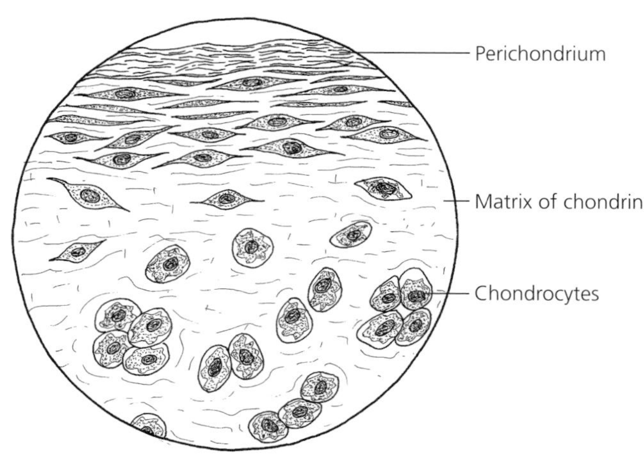


**Figure 16.9**
Hyaline cartilage; a diagrammatic view

The most common kind of cartilage is **hyaline** cartilage, so-called because of its glassy appearance (*hyalos* is Greek for glass). It supports the walls of the trachea and bronchi in the lungs, and covers the ends of long bones. Finally, it forms the embryonic skeleton, though most of it is later replaced by bone.

Like other connective tissues, the bulk of cartilage consists of an extracellular matrix (*extra-* is a prefix meaning 'outside'). The matrix consists of jelly-like **chondrin** penetrated by collagen fibres. The cells that secrete the matrix are called **chondroblasts** while they are actively secreting the matrix as the cartilage grows, and **chondrocytes** when the cartilage is mature.

Cartilage is surrounded by a tough, fibrous **perichondrium** containing densely-interwoven collagen fibres, and a network of capillaries. During growth, cells in the perichondrium divide to produce new cells, which eventually become surrounded by the matrix they secrete.

Together, the perichondrium and matrix act as a **hydrostatic skeleton** (Chapter 19). Cartilage resists change in shape because, when distorted, the pressure on the semifluid matrix causes tension to develop in the perichondrium – just as a balloon resists distortion because the rubber wall resists stretch.

During growth of the cartilage the cells continue to divide slowly, forming clusters. Each cluster represents the descendants of an original parent cell. Unlike other connective tissues, cartilage has no internal network of capillaries, so the chondroblasts inside have to rely on the diffusion of nutrients from the capillaries in the perichondrium. Probably for this reason cartilage cannot repair itself (as many athletes know to their cost).

**Fibrocartilage** occurs in the intervertebral discs, which have to bear the body weight. In this tissue, the collagen fibres are particularly highly developed. The body weight compresses the semiliquid centre of the disc, placing the densely woven fibres in the wall under tension. If a disc is loaded unevenly (for example when lifting a weight by bending the back instead of the knees and hips), its fibrous wall may partially rupture, causing it to bulge out against a spinal nerve – a so-called 'slipped disc'.

**Elastic cartilage** occurs in the pinna (ear flap) and in the end of the nose. It differs from fibrocartilage in that the matrix fibres are made of elastin.

## Muscle tissues

Muscle is specialised for converting the chemical energy in ATP to mechanical energy, resulting in increased tension and – usually – shortening. Muscle consists of large numbers of **muscle fibres**. As befits their high activity, these contain many mitochondria and are richly supplied with blood vessels. There are three main categories: smooth muscle, skeletal muscle and cardiac muscle.

### *Smooth muscle*

Smooth muscle forms a component of the walls of many internal cavities such as the gut, blood vessels, the bladder and the uterus. It also forms the iris of the eye and the muscles that raise the hairs in the skin. It consists of elongated, spindle-shaped cells (Figure 16.10) without any of the cross bands of skeletal and cardiac muscle (see page 302).

**Figure 16.10**
Smooth muscle; a diagrammatic view (A), under the light microscope (B)

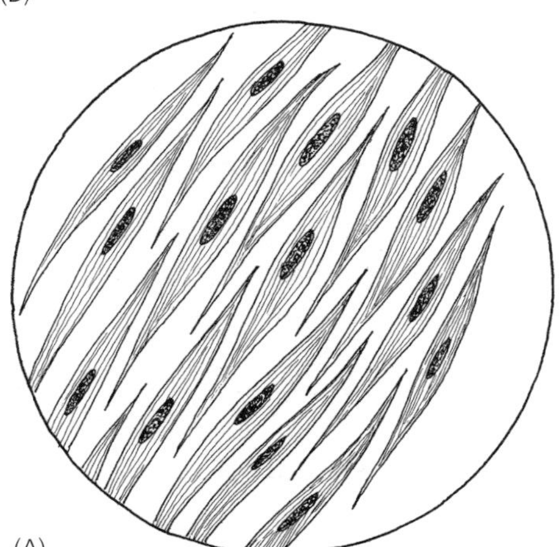

(A)

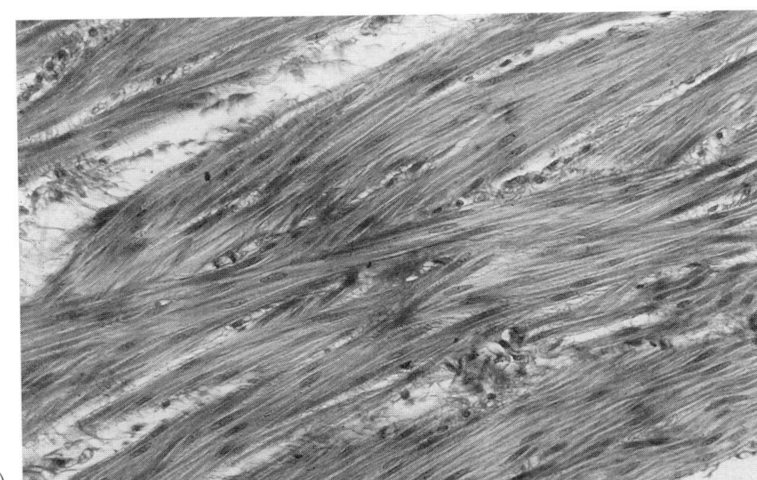

(B)

Smooth muscle can develop sustained contractions and is immune to fatigue. Some smooth muscle, such as that of the gut, has a dual nerve supply, with one set of nerves stimulating contraction and the other inhibiting it. Gut muscle can also contract without any nervous stimulation at all.

### Skeletal muscle

Skeletal muscle is so-called because it exerts 'pull' on the bones, and is dealt with in detail in Chapter 19. It is also known as **striated** or **striped muscle** because of the pattern of striations or bands running across the fibres. It is also called **voluntary** muscle because it is under control of the will. However, since we cannot know whether other animals have 'will', the term is best restricted to humans. In contrast to cardiac muscle and some kinds of smooth muscle, skeletal muscle only contracts when stimulated via nerves.

### Cardiac muscle

Cardiac muscle makes up the bulk of the heart, and is dealt with in detail in Chapter 22. It shares some of the properties of both smooth and skeletal muscle in that it can produce powerful, rapid contractions, but is immune to fatigue.

## QUESTIONS

**1** A young mouse had a body 4 cm long and a mass of 8 grams. After 4 weeks it had grown to 8 cm long.

- **a)** What would you expect it to weigh after this period of growth?
- **b)** By how many times would you expect its surface area to have risen?
- **c)** On what assumption do your answers to **(a)** and **(b)** depend?

**2** Which of the following statements applies to the daughter cells immediately after a cell has divided into two daughter cells of similar shape to the parent cell?

- **A** The combined surface area of the daughter cells is twice that of the parent cell.
- **B** The surface area to volume ratio of each daughter cell is twice that of the parent cell.
- **C** The combined length of the daughter cells is equal to that of the parent cell.
- **D** The combined surface area of the daughter cells is greater than that of the parent cell.
- **E** The surface area to volume ratio of each daughter cell is half that of the parent cell.

CHAPTER

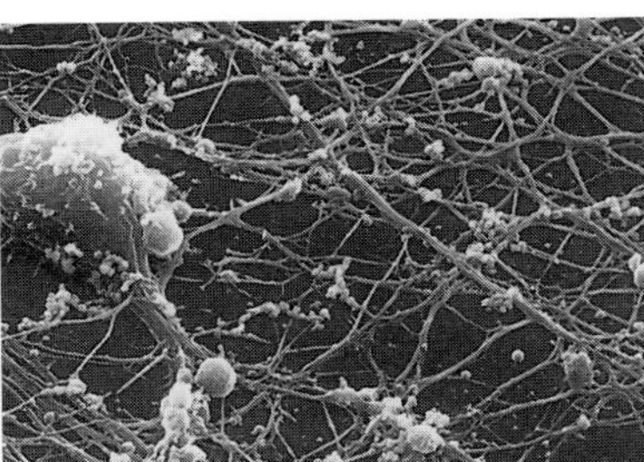

# 17

# The NERVOUS SYSTEM

## LEARNING OBJECTIVES

By the time you have completed your study of this chapter you should be able to:

- state the functions of the nervous system and describe its gross anatomy.
- distinguish between the functions of motor, sensory and intermediate neurons.
- describe the structure of a motor neuron and a sensory neuron.
- explain what is meant by a reflex action, and distinguish between a simple reflex and a conditioned reflex.
- describe a reflex arc and explain what happens in a simple reflex.
- describe how nerves are stimulated artificially and how impulses are recorded.
- explain what is meant by resting potential, threshold, all-or-nothing and refractory period.
- explain the resting potential and how an action potential is produced and propagated along an axon.
- explain why impulses travel faster in thicker axons and in myelinated axons.
- describe the role of synapses in integrating excitatory and inhibitory impulses.
- describe the fine structure of a synapse, and the essential events in synaptic transmission.
- give examples of the ways in which drugs can interfere with the functioning of synapses.
- give a brief account of the structure and functioning of the autonomic nervous system.
- name the principal areas of the brain, together with their locations and functions.

## What do nervous systems do?

All living things show **irritability** or **sensitivity**, meaning that they respond to changes, both in their surroundings and in the interior of the body. A change to which an organism can respond is called a **stimulus** (plural, **stimuli**). Responses to stimuli are nearly always **adaptive**, meaning that they increase the likelihood of survival and reproduction. Sensitivity to change enables an animal to avoid dangerous situations and to take advantage of favourable ones. An earthworm, for example, moves away from light towards the dark where it is less likely to be eaten by a predator. On the other hand, the worm may respond to contact with a dead leaf by eating it.

A response to a stimulus involves a sequence of three events:

1 the stimulus is detected by one or more **receptors**. These may either be nerve endings or entire receptor cells.

2 a receptor then sends signals via a **communication system** to one or more **effectors** situated in some other part of the body.

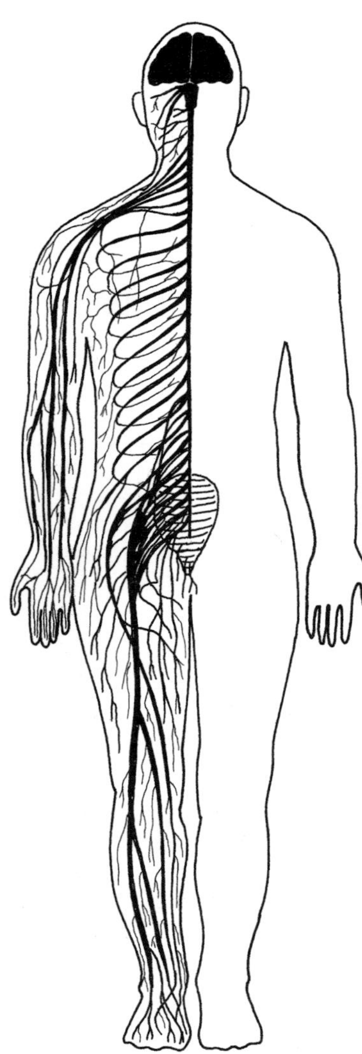

**Figure 17.1**
Gross structure of the nervous system

3 the effectors carry out the response to the stimulus. The most important effectors are **muscles** which produce movement, and **glands** which secrete chemicals.

Communication systems in animals are of two types:

- the **nervous system**, which sends rapid, brief signals along fixed paths.
- the **endocrine system**, which sends slow, long-lasting signals through the blood.

As we shall see in Chapter 20, each has its own advantages and disadvantages. This chapter is concerned with the nervous system, especially that of mammals.

Nervous systems have three functions: communication, integration and storage of information. All three can be illustrated by the sudden withdrawal of one's hand when one accidentally touches a hot object.

The stimulus is detected by receptors in the skin of the fingers; the response is the contraction of muscles in the arm. Since receptors and effectors occupy different parts of the body, there must be some kind of link or communication between them.

Though the withdrawal of the hand may seem simple, it actually involves the contraction of a number of specific muscles. For these muscles to be effective, other muscles must be relaxed. Thus muscles work together in a co-ordinated or **integrated** way.

Finally, touching hot objects is dangerous, and the experience is remembered. As a result, similar hot objects tend to be avoided in the future.

## Gross anatomy of the nervous system

In a dissection seen with the naked eye, the nervous system of a vertebrate can be divided into two anatomically and functionally distinct parts (Figure 17.1):

- the **central nervous system**, or **CNS**, consisting of the **brain** and **spinal cord**. Its functions are the integration and storage of information. It receives input from receptors in various parts of the body and, in the light of information received, sends out the appropriate signals to effectors. It also stores information, enabling the animal to modify its behaviour in the light of past events.
- the **peripheral nervous system**, consisting of a series of paired **cranial nerves** extending from the brain, and paired **spinal nerves** which emerge from the spinal cord (Figure 17.2). The function of the peripheral nervous system is mainly that of transmitting information from receptors to the central nervous system, and from the central nervous system to effectors.

# The cells of the nervous system

The cells making up the nervous system are amongst the most specialised and peculiar to be found anywhere in the Animal Kingdom. They are divided into two major categories: **neurons**, which transmit and process information, and **glial cells**. Glial cells, or **neuroglia** as they are also called, are not directly involved in nervous activity. Unlike neurons, glial cells can divide (it is from these cells that brain tumours originate). In the

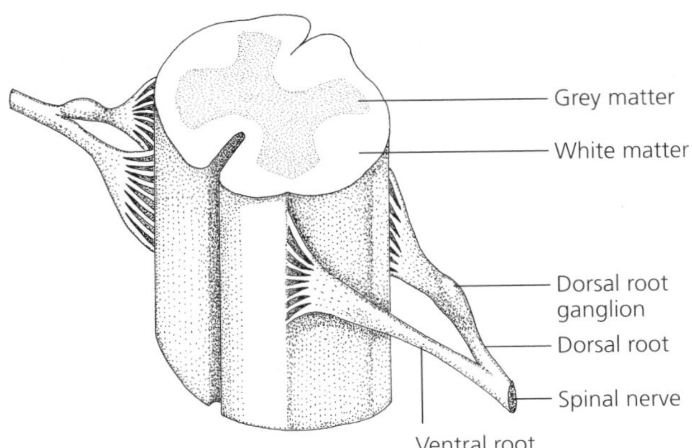


**Figure 17.2**
A spinal nerve and part of the spinal cord

central nervous system, glial cells outnumber neurons by ten to one. In the peripheral nervous system, the most important glial cells are the **Schwann cells** that form sheaths round nerve fibres.

## Neurons

The most distinctive feature of neurons is their extraordinary shape. Each neuron is differentiated into a **cell body** containing the nucleus, and a number of slender **nerve fibres** which extend from it. These are of two kinds: a single long **axon** and many, twig-like **dendrites** which are usually less than a mm long. The fundamental distinction between axons and dendrites lies in the different kinds of signal they carry (see information box on page 308). Incidentally, a nerve fibre should not be confused with a 'nerve' seen in a dissection and which consists of a bundle of axons bound together by connective tissue (Figure 17.3). The collagenous fibres of the connective tissue provide protection against mechanical damage.

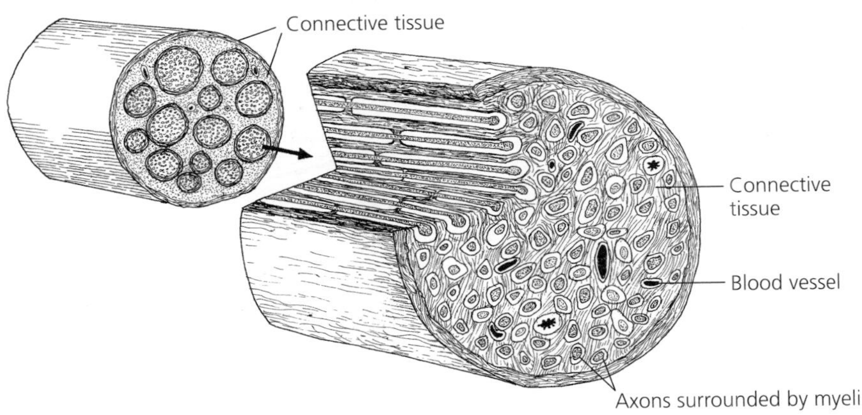


**Figure 17.3**
Structure of a nerve trunk (after Crouch)

A **motor neuron** from the spinal cord has most of the basic features of a neuron (Figure 17.4). Because they carry impulses *away* from the central nervous system, motor neurons are also called **efferent** neurons (*ex*, out; *ferre*, to carry). The nucleus, Golgi bodies and endoplasmic reticulum are all restricted to the **cell body**. In motor neurons supplying skeletal muscle, the cell body is always located in the central nervous system.

The **axon** carries nerve impulses from the cell body to an effector. Since this can be anything from a few centimetres to a metre away, axons can be very long. In mammals and other vertebrates, wider axons are insulated by a fatty **myelin sheath** and are said to be **myelinated** (Figure

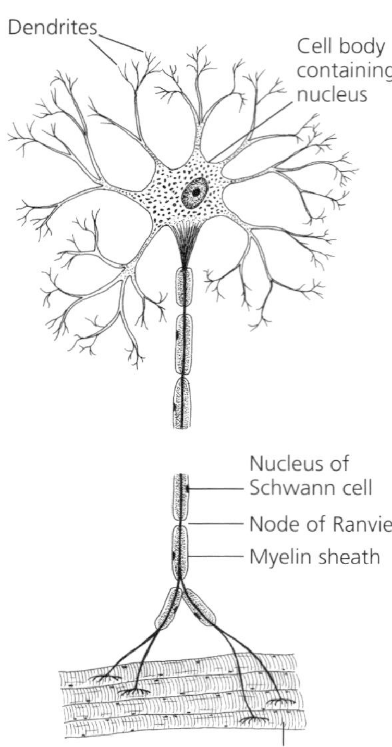


**Figure 17.4▲**
Structure of a motor neuron

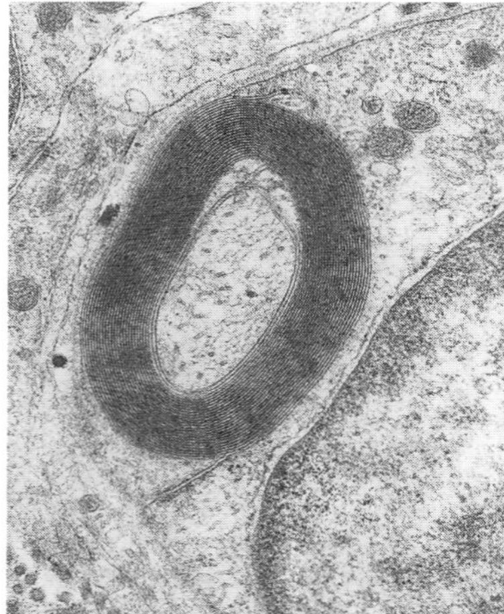

**Figure 17.5▲**
Electron micrograph of a myelinated axon

17.5). The myelin sheath is formed by **Schwann cells**. During embryonic development the latter grow spirally round the axon like a Swiss roll. The result is that the axon becomes enveloped in up to 100 or more layers of plasma membrane (Figure 17.6). Adjacent Schwann cells are separated by **nodes of Ranvier**, up to 1.5 mm apart (Figure 17.7). These are points where the axon membrane comes into direct contact with the intercellular fluid, and they play a vital part in transmission of impulses.

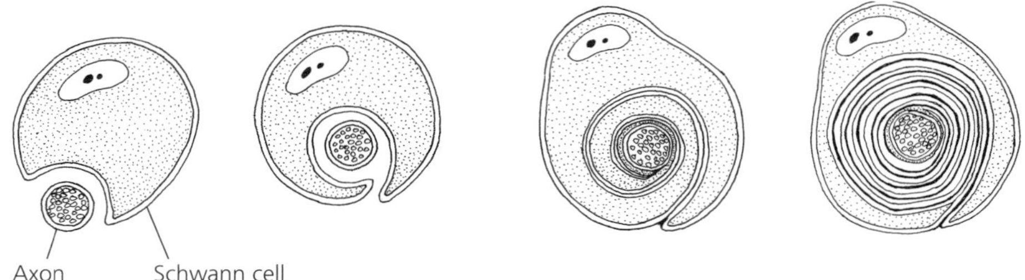


**Figure 17.6▲**
Formation of the myelin sheath

Vertebrate axons narrower than about 1 μm are unmyelinated, with each Schwann cell enveloping a number of axons (Figure 17.8).

While the great length of an axon is ideal for transmitting electrical signals over long distances, it poses problems for chemical communication between different parts of the cell. For example, proteins in an axon terminal may be a metre or more away from the cell body where they are produced. They are transported by microtubules running along the axon cytoplasm, or **axoplasm**.

Whereas a neuron only has one axon, it typically has many tree-like **dendrites** (*dendron* = tree). These act as 'antennae', collecting input from other neurons and may account for 80% of the surface area of the neuron. The dendrites and cell body of a single neuron in the brain may receive input from over 100 000 axon branches of other neurons.

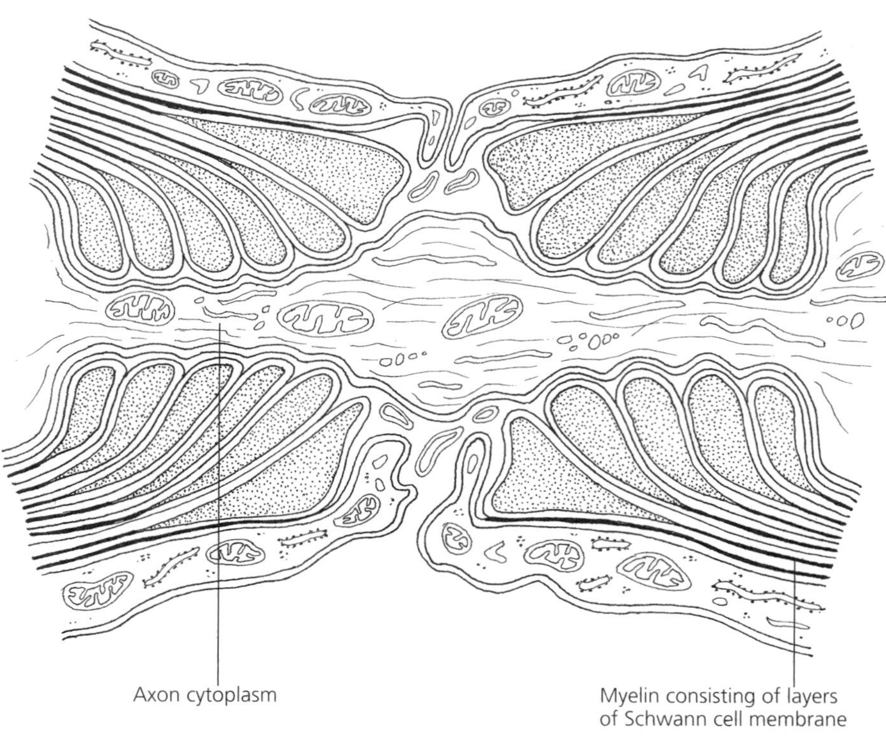


**Figure 17.7►**
Fine structure of a node of Ranvier seen in longitudinal section

Axons

Schwann cell nucleus

**Figure 17.8▲**
Non-myelinated axons sheathed by a Schwann cell

## Other kinds of neuron

If we include those in the brain, there are many other kinds of neuron, some of which are shown in Figure 17.9.

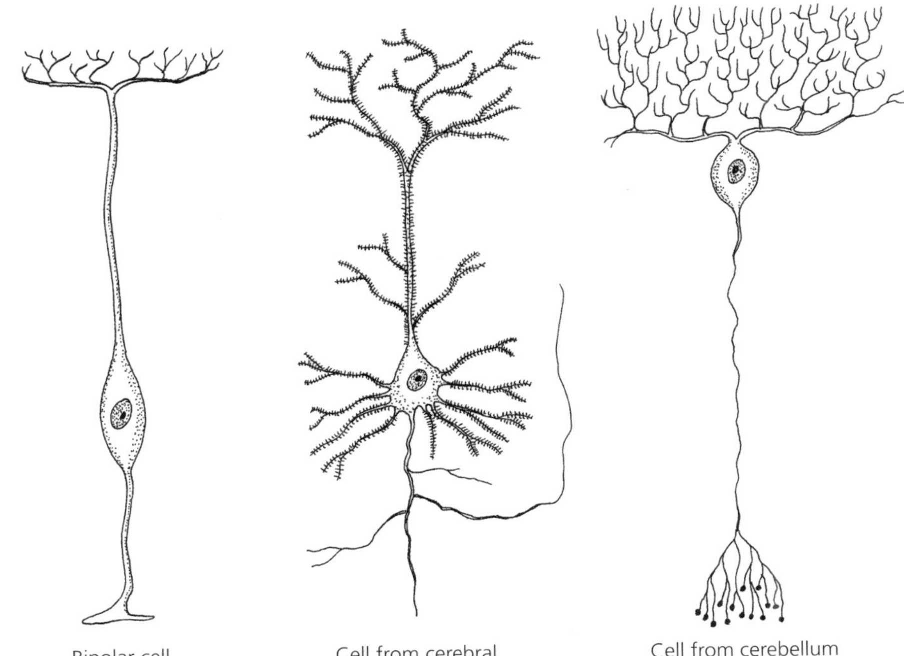


**Figure 17.9▶**
A variety of types of neuron

Outside the brain two kinds of neuron should be noted:

- **sensory neurons** carry information from receptors to the central nervous system (Figure 17.10). Because the impulses go *towards* the central nervous system they are also called **afferent** neurons (*affere* = to bring). Structurally a sensory neuron differs from a motor neuron in that the cell body lies at the end of a short 'cul-de-sac' and receives no dendrites.

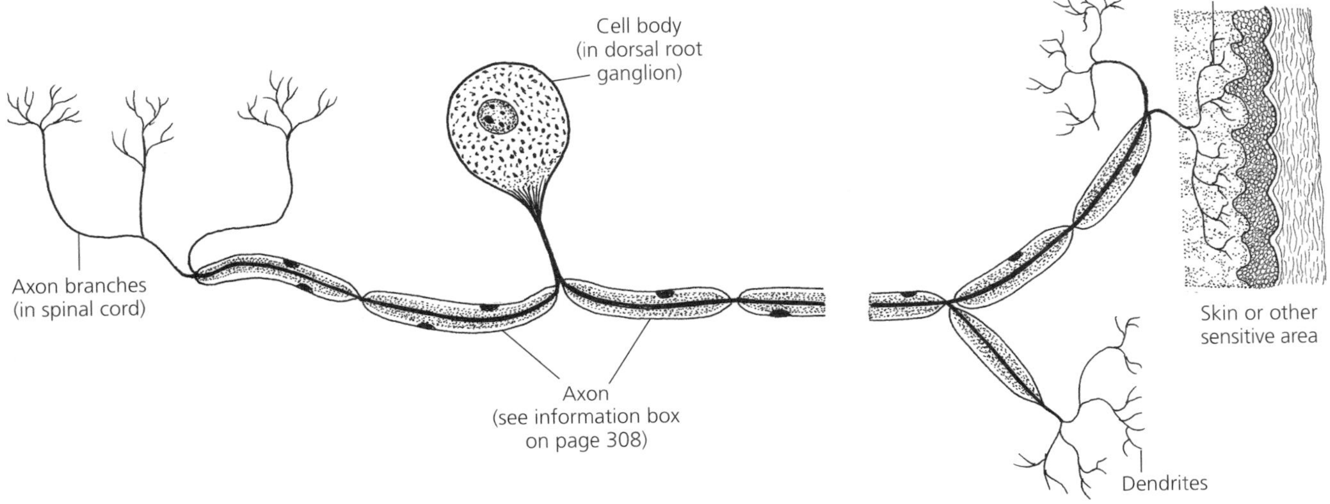


**Figure 17.10**
A sensory neuron

- **intermediate** or **relay neurons** connect sensory neurons with motor neurons and are important in integration.

## Two kinds of fibre; two kinds of signal

The terms 'dendrite' and 'axon' were first applied by anatomists, before it was understood how they worked. The distinction was based on the direction in which they carried signals – axons transmitting them away from the cell body and dendrites carrying them towards it.

There is however a much more fundamental distinction based on the kind of signals they transmit.

- Axons carry all-or-nothing **impulses**, while dendrites carry graded signals whose magnitude depends on the strength of the stimulus.
- The signals carried by dendrites get weaker with increasing distance from the source. An impulse travels along an axon **non-decrementally**, meaning that it does not die away with distance. This is because it is constantly being reamplified as it goes.
- Whilst action potentials travel at speeds of up to 120 m $s^{-1}$, the currents carried by dendrites travel much faster.

## How neurons are organised – reflex arcs

The simplest response that a group of neurons can mediate is called a **reflex action**. Reflexes may be **simple** if they are unlearned, or **conditioned** if they have to be learned.

### *Simple reflexes*

A simple reflex involves an arrangement of neurons called a **reflex arc**. The simplest reflex arcs are called **spinal reflexes** because they involve the spinal cord but not the brain directly. A reflex arc is shown in transverse section in Figure 17.11, together with the neurons involved in the reflex withdrawal of a limb from a hot object.

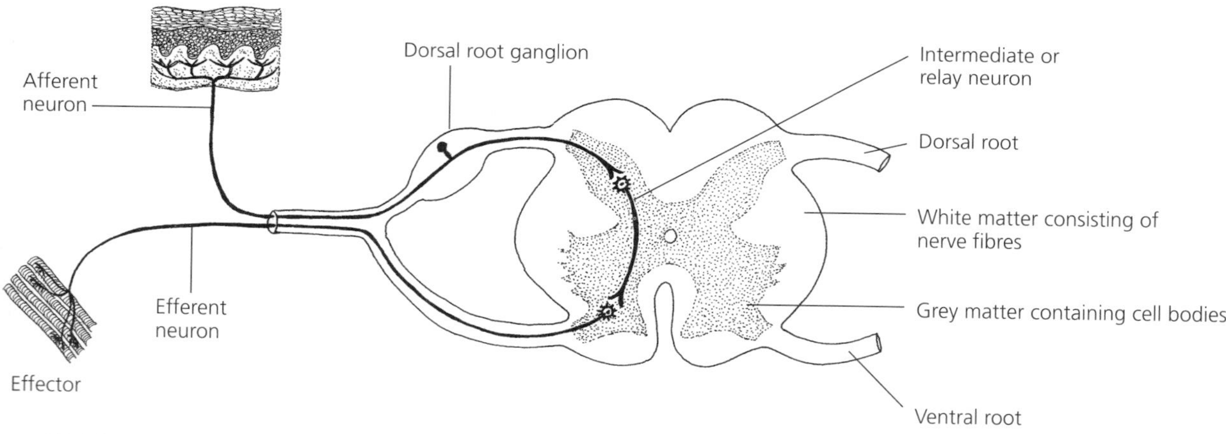


**Figure 17.11**
Transverse section through the spinal cord showing a reflex arc

The outer part of the cord is the **white matter**, consisting mainly of bundles of nerve fibres carrying impulses to the brain (ascending tracts) and from it (descending tracts). It is the myelin sheaths of the fibres that give the white matter its characteristic colour. Inside the white matter is an H-shaped area of **grey matter**, consisting mainly of nerve cell bodies. In the centre of the spinal cord is the **spinal canal**.

Just before connecting with the spinal cord, each spinal nerve divides into a **dorsal root** carrying only sensory fibres, and a **ventral root** carrying only motor fibres. The cell bodies of the afferent neurons are all located in the dorsal root and collectively form a small swelling or **ganglion**.

In the withdrawal of the hand from a hot object, sensory nerve endings in the skin send impulses along afferent fibres to the spinal cord. In the dorsal horn of the grey matter, sensory fibres connect with **intermediate** or **relay** neurons. These in turn connect with the cell bodies of motor neurons whose axons send impulses to the muscles that cause the arm to be pulled away.

The connections between neurons are called **synapses**. A synapse is not a point of actual contact between two neurons since there is always a very narrow gap between them. The way in which signals cross this gap is described later.

The majority of spinal reflexes involve three neurons and two synapses, as in Figure 17.11. The stretch reflex is an exception – there is no intermediate neuron and it is thus a **monosynaptic** reflex. This reflex is provoked by stretching muscle spindles (stretch receptors in skeletal muscles) and results in the contraction of the muscle that had been stretched. The reflex is most usually seen when the quadriceps tendon is tapped below the knee. In normal circumstances this helps to maintain posture since any sagging of the knee results in the stretching of the quadriceps and its reflex contraction.

Even a spinal reflex is more complicated than that described so far, for several reasons:

- almost invariably more than one afferent neuron and more than one efferent neuron are involved.
- besides stimulating the muscles that bring about the response, motor neurons that control **antagonistic** muscles are inhibited. Thus the withdrawal of the hand involves not only the contraction of the muscle that bends the elbow (the biceps), but also the relaxation of the muscle that extends it (the triceps).
- although the brain is not directly involved in the withdrawal reflex, the fact that pain is felt when a hot object is touched shows that impulses are also sent to the part of the brain concerned with consciousness. The value of this is obvious: pain is a powerful way of learning to avoid danger.
- some reflexes can be over-ridden by the brain. For example, the desire to cough or sneeze can, to some extent, be suppressed consciously by the brain when it is important to do so.

Although many simple reflexes do not directly involve the brain, there are many others that do, for example the first three of those shown in Table 17.1. These **cranial reflexes**, however, involve only those brain regions concerned with automatic functions; consciousness and memory play no part.

**Table 17.1**
Some examples of simple reflexes

| Reflex | Stimulus | Response |
|---|---|---|
| Coughing | Irritation of lining of lung passages | Contraction of muscles of expiration |
| Salivation | Presence of food in mouth | Secretion of saliva by salivary glands |
| Narrowing of pupil | Increase in light intensity on retina | Contraction of circular muscles of iris |
| Knee jerk | Stretching of quadriceps tendon | Contraction of quadriceps muscle |

### Conditioned reflexes

Unlike simple reflexes, conditioned reflexes have to be *learned*, but are just as automatic as simple reflexes. All conditioned reflexes, therefore, involve the brain. One example is the increased output of saliva in response to stimuli that have become associated with food, such as the clank of kitchen cutlery.

Conditioned reflexes can be extremely useful tools for investigating the range of stimuli to which an animal is sensitive. For example, we know that dogs can hear sounds of higher pitch than humans, because after being trained to associate high pitched sounds with food, they will salivate in response to high notes even in the absence of food.

## Properties of nerve impulses

Although a nerve impulse is an electrical phenomenon, it differs from a simple electrical current in a number of important ways:

- it travels much more slowly. The fastest nerve impulses travel at speeds of about 120 m $s^{-1}$, whereas an electric current travels along a wire with the speed of light (300 000 km $s^{-1}$).
- when an axon is stimulated artificially, impulses travel in opposite directions away from the point of stimulation (though in the body, impulses travel in one direction because they arise at one end of the axon, and also because the synapses act in a one-way manner). A wire, on the other hand, can only carry current in one direction at a time.
- a wire carries electric current in a purely passive way, since it does not supply the energy for the current. The energy required for an impulse to travel down an axon is supplied by metabolism of the axon itself. If part of an axon is mildly poisoned, for example, the strength of the impulse declines as it moves through the drugged region, but regains its full size as it emerges into the untreated part of the fibre.
- whereas the strength of a current decreases as it passes along a conductor, a nerve impulse is constantly being amplified to retain its 'strength' as it moves along the axon. Nerve impulses are thus said to travel *without decrement*.
- an electric current in a wire involves movement of electrons only; nerve impulses also involve movement of $Na^+$ ions.

In view of these special features, it is more appropriate to speak of an impulse being **propagated** along an axon, rather than 'conducted'.

## How nerve impulses are studied

The idea that nerves carry some form of electrical signal dates from a discovery in 1786 by Luigi Galvani. He observed that a frog's leg muscle twitched when an electric current was applied to the nerve trunk supplying it. Finding out just what happens in individual nerve fibres took a long time, partly because they are so thin. Then, in the 1930s it was discovered that some invertebrates, such as squids, have unusually wide axons as well as normal-sized ones. These **giant axons** supply the muscles responsible for sudden *escape reactions*, in which many muscles must contract simultaneously (Figure 17.12). The largest giant fibres are almost 1 mm in diameter – wide enough for tiny electrodes to be inserted into the interior of the axon.

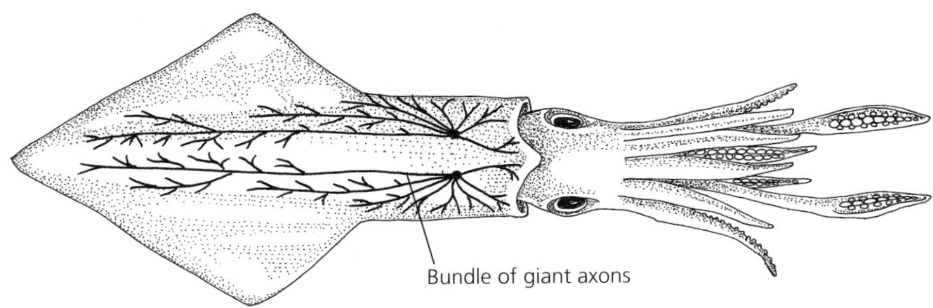


**Figure 17.12**
Giant axons of a squid

To experiment on a giant axon, it is removed from the animal and immersed in sea water, which is similar in composition to the intercellular fluid that normally bathes it.

## *Stimulating an axon*

With the exception of sensory fibres, impulses are normally generated in response to signals received from other neurons. Physiologists must, however, be able to stimulate neurons to produce impulses under laboratory conditions, in many cases after removal from the animal.

An axon can be stimulated to produce an impulse by a variety of artificial stimuli, such as a mechanical shock (as happens when the 'funny bone' – actually the ulnar nerve – in the elbow is tapped). The most convenient stimulus is an electrical one because its strength, frequency and duration can be measured very precisely. For the stimulating and recording electrodes, very fine glass tubes containing concentrated potassium chloride solution are used (Figure 17.13). The tip of the electrode is only about 0.1 μm wide, narrow enough to penetrate an axon without doing significant damage.

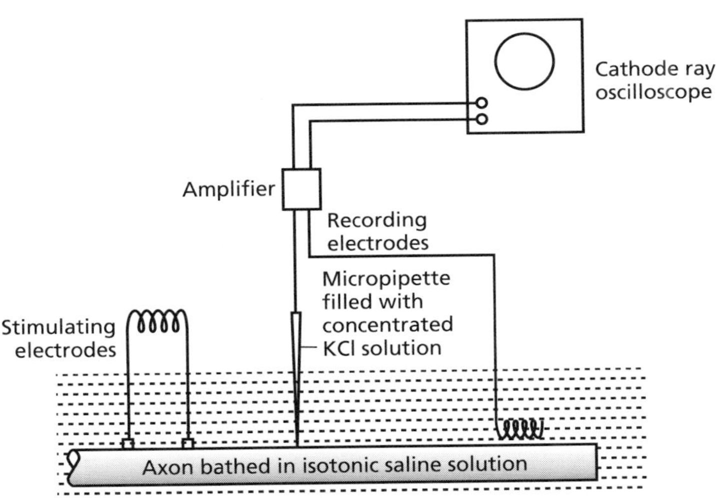


**Figure 17.13**
Apparatus for stimulating and recording electrical events in an axon

## *Recording the response*

Nerve physiologists record the electrical changes in a nerve using a **cathode ray oscilloscope** (CRO). Instead of observing the deflection of a needle, a CRO uses the deflection of an electron beam, recorded as a trace on a cathode ray tube. Since an electron beam has virtually no inertia, it can record changes instantaneously.

## *Resting potential*

When electrodes are placed on either side of an axon membrane, a potential difference of about 70 mV is recorded, the outside being positive with respect to the inside. By convention, membrane potentials refer to the *inside* of the membrane. Thus, a resting potential of −70 mV means that the inside is 70 mV *less positive* than the outside.

A membrane potential is characteristic of all cells, but only in nerve and muscle cells does it have an obvious functional role. In these *excitable cells* the membrane potential is called the **resting potential**.

### Action potential

When an impulse travels past the recording electrodes there is a brief reversal of the membrane potential followed by a rapid recovery. The whole sequence is called an **action potential** (Figure 17.14). At the peak of the action potential, the membrane potential is about +40 mV. The duration of an action potential varies with the kind of fibre and with temperature; in the widest vertebrate fibres it takes about 0.5 ms, but in very small ones it takes several milliseconds.

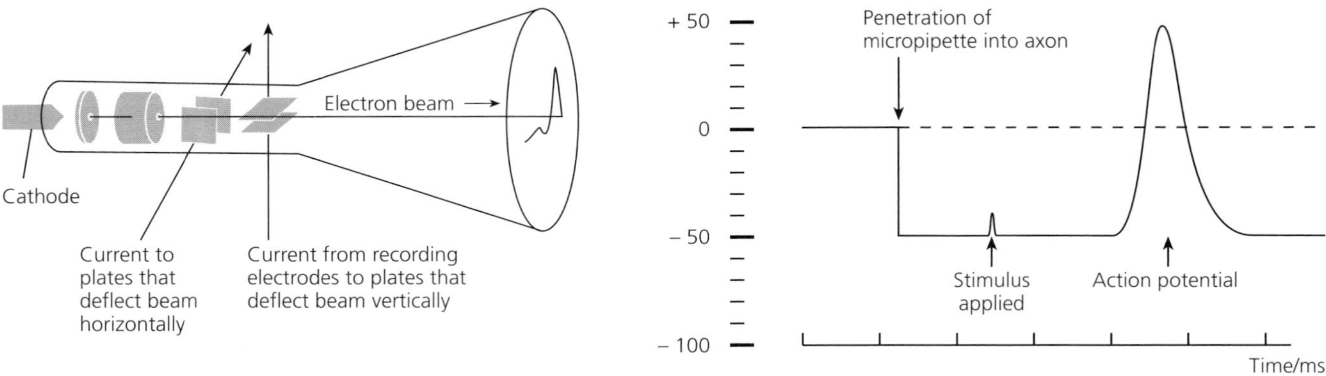


**Figure 17.14**
Cathode ray oscilloscope and recording of an action potential

Under any given conditions, all action potentials in a particular axon are the same; only their *frequency* varies. Impulse frequency is usually a measure of how strongly the neuron has been stimulated.

### Threshold

If a nerve fibre is given shocks of progressively increasing strengths, no impulse is generated until the stimulus reaches a certain **threshold** value, above which a normal impulse is produced. A further increase in stimulus strength makes no difference to the size of the action potential. Nerve impulses are therefore said to be **'all or nothing' events** – they either occur at full strength, or not at all. This is consistent with the idea that an action potential involves the release of energy stored up within the axon itself, with the stimulus simply acting as a 'detonator'. A line of dominoes behaves in a similar way when the first one is pushed. So long as the first domino is pushed hard enough (the 'threshold'), the effect on the rest of the dominoes is the same.

### Refractory period

Immediately after an above-threshold stimulus, an axon cannot respond to further stimuli, no matter how strong. The duration of this **absolute refractory period** corresponds roughly to that of the action potential. In the largest myelinated fibres, it lasts about 0.4 ms; in small fibres it may last 4 ms. The absolute refractory period is followed by a **relative refractory period** lasting several milliseconds. During this period, an action potential can be produced, but it requires a stronger than normal stimulus and is of smaller than normal amplitude.

The absolute refractory period sets an upper limit to the frequency with which a neuron can carry impulses. An absolute refractory period of 0.5 ms would, in theory, allow 2000 impulses per second, but in practice the highest frequencies recorded are about 1000 per second.

## Ionic explanation for the resting potential

The resting membrane potential can largely be explained in terms of two kinds of charged particle in the axoplasm – potassium ($K^+$) ions and negatively-charged organic ions such as proteins. $K^+$ ions are about 30 times more concentrated inside the axon than outside it.

Outside the axon is the extracellular **tissue fluid**. The main solutes in the tissue fluid are sodium ($Na^+$) and chloride ($Cl^-$) ions. As we shall see, neither plays a direct part in the resting potential.

Separating the axoplasm and extracellular fluid is the axon membrane. A resting axon is moderately leaky to $K^+$ ions, so one might expect them to diffuse out. So they would, but for the fact that the axon membrane is completely impermeable to the negatively-charged proteins. Any $K^+$ ions that diffuse out cannot, therefore, be accompanied by negative ions. The outward diffusion of $K^+$ ions thus results in the outside becoming positively charged with respect to the inside. The more $K^+$ ions that diffuse out, the more they tend to be attracted back by the negative proteins left behind. An equilibrium is thus set up between the tendency for $K^+$ ions to diffuse out and the electrical force attracting them back (Figure 17.15).

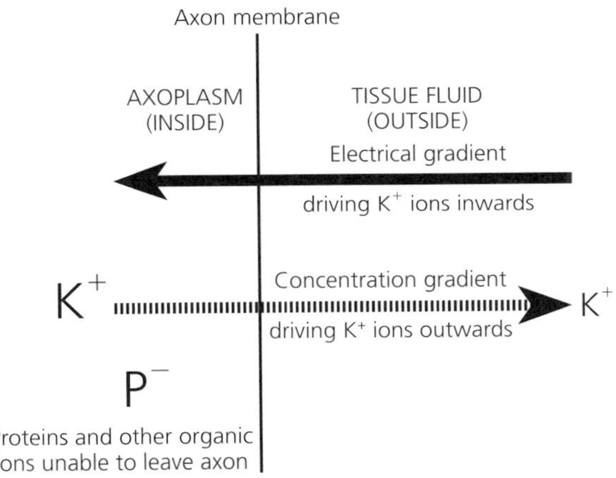


**Figure 17.15**
Ionic explanation for resting potential

If the axon membrane were completely impermeable to all the ions outside it, the resting membrane potential would persist indefinitely without any expenditure of energy. However, the axon is very slightly permeable to $Na^+$ ions (though 100 times less than to $K^+$). $Na^+$ ions are about 10 times more concentrated outside the axon than inside, and thus very slowly diffuse into the axon. The effect of this is to lower the membrane potential and thus to reduce the force attracting $K^+$ ions back into the axon. Every $Na^+$ ion entering the axon allows a $K^+$ ion to leave.

The slow inward leakage of $Na^+$ ions and outward leakage of $K^+$ ions would eventually result in the eventual disappearance of the membrane potential, were it not for the **sodium pump**, which actively pumps $Na^+$ ions out of the axon in exchange for $K^+$ ions.

What about the $Cl^-$ ions in the extracellular fluid? Although they are about six times more concentrated outside the neuron than inside, their inward diffusion is opposed by the negative charge inside the cell.

## Ion movements during an action potential

While the resting potential depends on $K^+$ ions, an action potential depends on the facilitated diffusion of both $Na^+$ and $K^+$ ions. The movement of ions into and out of an axon has been studied using the radioactive isotopes of sodium and potassium.

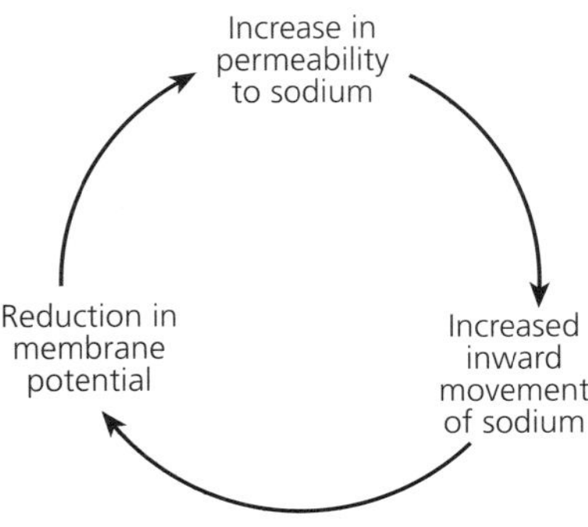


**Figure 17.16**
Positive feedback in an action potential

Lying in the axon membrane are two kinds of protein channel or 'gate', so-called because they can be opened and closed. One kind allows the passage of $Na^+$ ions, while the other is specific for $K^+$ ions.

An action potential results from the ability of the axon membrane to rapidly open and close its $K^+$ and $Na^+$ gates in response to changes in potential difference across the membrane. The effect of a stimulus is to cause the membrane to become more permeable to $Na^+$ ions. This allows some $Na^+$ ions to diffuse into the axon, reducing the membrane potential. This, in turn, reduces the force attracting $K^+$ ions back into the axon, so more $K^+$ ions can diffuse out.

What happens next depends on the strength of the stimulus. If the stimulus is below threshold, the permeability to $Na^+$ ions does not reach the permeability to $K^+$. The inward diffusion of $Na^+$ ions is therefore less than the outward diffusion of $K^+$ ions, so the resting potential is restored.

An above-threshold stimulus raises the permeability to $Na^+$ above that to $K^+$ ions, so $Na^+$ ions diffuse in faster than $K^+$ ions diffuse out. The outward movement of $K^+$ ions is insufficient to counter the effect of the inward movement of $Na^+$ ions. The membrane potential is, therefore, reduced still further, increasing the permeability to $Na^+$ still more, and so on, until it is several thousand times above normal. This 'vicious circle', in which two processes accelerate each other, is called **positive feedback** (Figure 17.16).

*Point to note:* A neuron reaches its threshold when the permeability to sodium exceeds the permeability to potassium.

The entry of $Na^+$ ions swamps the outward diffusion of $K^+$ ions, and the membrane potential momentarily reverses. At the peak of the action potential, the $Na^+$ gates suddenly close again, allowing the exit of $K^+$ ions to restore the resting potential. This recovery is speeded up by a 30-fold increase in permeability to $K^+$ ions (Figure 17.17). This makes it harder to raise the $Na^+$ permeability to the threshold value, giving rise to the relative refractory period.

In the early part of the action potential (when the outside of the axon is positive with respect to the inside), the inward movement of $Na^+$ ions is due to differences in both concentration *and* electrical potential. Once the membrane potential has been reversed, any more $Na^+$ ions that enter are increasingly held back by the negativity outside the axon – just as $K^+$ ions are held back in a resting axon.

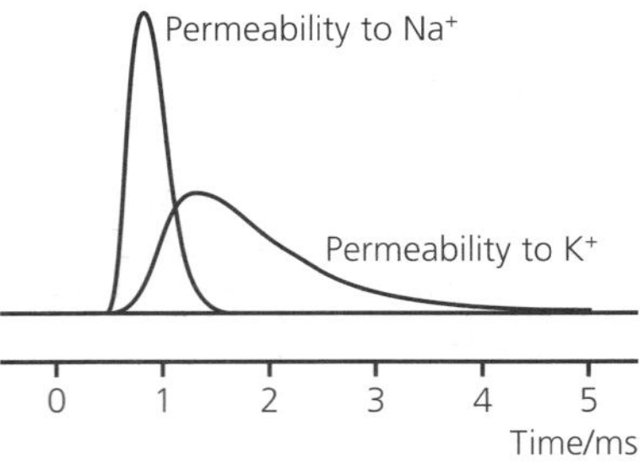


**Figure 17.17**
Changes in permeability of axon membrane to $Na^+$ and $K^+$ during an action potential

As a result of an action potential the axoplasm has gained some $Na^+$ ions and lost some $K^+$ ions. Though the actual quantities involved are only about $10^{-5}$ of the total, the concentration differences would eventually run down if the leakage of ions was not opposed by the sodium pump mechanism. This consumes ATP, produced by mitochondria in the axon. An axon poisoned by cyanide can conduct many thousands of impulses, but the energy stored in the $Na^+$ and $K^+$ gradients gradually runs down without the production of ATP, just as a battery does if it is not continually recharged. In such a poisoned nerve, the size of the resting and action potentials gradually diminishes until no more impulses can be generated.

## How do impulses travel?

At the peak of the action potential the inside of the axon is positively charged with respect to the adjacent resting region of the membrane. There is thus a potential difference *along* the axon membrane as well as across it. This lengthways potential difference causes **local currents** to flow along the axon (Figure 17.18). The flow of local current reduces the potential difference across the membrane in the region ahead of the

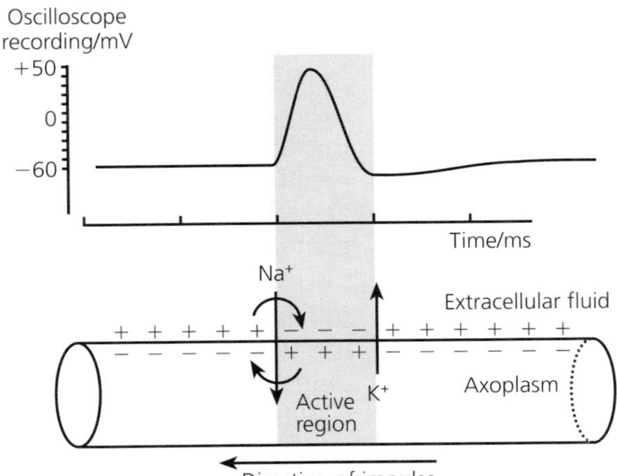


**Figure 17.18**
Role of local currents in propagation of an action potential in an unmyelinated axon

impulse, triggering an action potential. Each region of the axon stimulates the next region to undergo an action potential, and so on, with the result that an impulse sweeps along the axon.

Although local currents occur behind the action potential as well as ahead of it, they cannot trigger an action potential because this region of the axon is in its refractory phase and unreceptive.

The transmission of an impulse thus involves two kinds of current:

- local currents, which occur *along* the axon on either side of the axon membrane. They travel much faster than the impulse, but, at any given instant, they rapidly die away with distance because the axon is electrically 'leaky'.
- ionic currents, which occur *across* the membrane, at right angles to the direction of movement of the impulse. Because they involve the diffusion of ions, ionic currents are relatively slow.

### *The effect of temperature*

The opening and closing of the channels through which ions move involves changes in the shape of the proteins forming the walls of the channels. This is a chemical process and, as such, is faster at warmer temperatures. Impulses thus travel faster in endothermic than in ectothermic vertebrates (such as frogs and fish) in cool environments (Chapter 25). Cooling an axon, on the other hand, increases the duration of an action potential, reducing the frequency with which it can propagate impulses (this is why cooling an injured body part often reduces pain).

## The speed of impulses

Compared with an electric wire, axons are both poorly insulated and have a relatively high resistance. The strength of local currents thus rapidly decreases ahead of an impulse, and so they can only reach threshold a short distance ahead of the action potential. The speed of an impulse is thus limited by how far the local currents can spread ahead of the impulse. There are two ways in which this can be increased: by sheathing it in myelin, or by increasing the diameter of the axon.

### *Saltatory propagation*

In mammals and other vertebrates, wider axons are surrounded by a **myelin sheath** which reduces the leakage of local currents. In myelinated fibres, action potentials occur at the nodes of Ranvier, which are the only

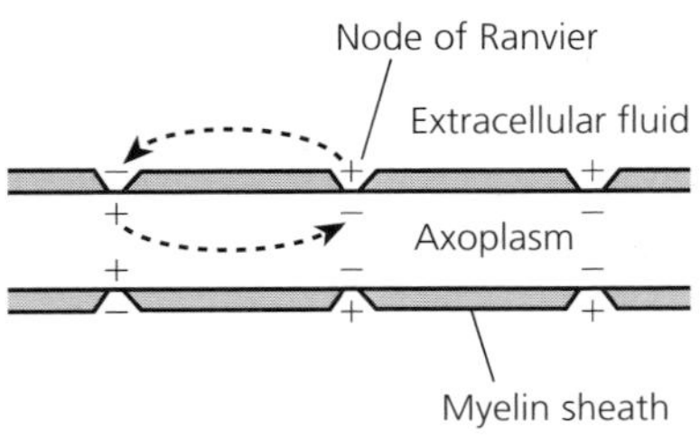


**Figure 17.19**
Saltatory impulse propagation in a myelinated fibre

places where the axon membrane is exposed to the extracellular fluid. The local currents associated with each action potential 'jump' from one node to the next almost instantaneously (Figure 17.19). At each node there is a delay of about 0.1 ms while the action potential occurs. Propagation in myelinated axons is thus said to be **saltatory** (*saltere* = to jump). As a result, the widest myelinated fibres (20 μm in diameter) carry impulses at about 120 m $s^{-1}$, compared with a speed of 25 m $s^{-1}$ for a 500 μm-wide giant squid axon.

### *Wider is not always better*

All other things being equal, the wider an axon the faster impulses travel. One reason for this is that wider axons have lower longitudinal resistance, enabling local currents to trigger an action potential further ahead. In myelinated fibres, this enables wider axons to have longer internodes, so the signal can 'jump' a greater distance from node to node.

In unmyelinated fibres, increasing the fibre diameter yields diminishing returns. This is because impulse velocity is proportional to the *square root* of the axon diameter. To double impulse speed, an unmyelinated fibre has to be four times wider, and increasing the axon diameter 100 times only speeds up impulses 10 times.

Increasing speed must therefore be set against the cost of increasing space. The more space nerve fibres occupy, the fewer the axons that can be accommodated in a given cross-sectional area, and the less information that can be carried.

In mammals and other vertebrates, axons narrower than about 1 μm carry impulses faster if they are non-myelinated. Non-myelinated axons propagate impulses at speeds of about 0.5 m $s^{-1}$, and are usually concerned with functions in which speed of conduction is not so important.

Another advantage of myelination is that, since leakage of ions is confined to the nodes of Ranvier, the energy costs of active transport of $Na^+$ and $K^+$ ions are greatly reduced.

### *Frequency coding*

Nerve impulses are sometimes described as 'messages'. In fact, an individual impulse carries no more information than an individual dot in a message in Morse code. In a given axon under normal conditions, all impulses are identical. The only thing that can vary is the *frequency* of impulses. Impulse frequency is a measure of stimulus intensity. The only kind of information, then, that an individual nerve cell can signal is how strongly it has been stimulated.

How, then, can the nervous system signal complex information such as the shape and texture of an object held in the hand? The answer, as we shall see in Chapter 18, depends on the fact that *collectively*, axons can give a great deal of information, just as a newspaper photograph can be built up from many, individually-meaningless dots.

## Communication between neurons – synapses

There are about $10^{11}$ neurons in the human nervous system. This is a very large number, but the number of possible pathways through the nervous system is far greater because each neuron communicates with more than one other neuron. This can happen in two ways (Figure 17.20). Pathways may *diverge*, when a neuron sends impulses to more than one other neuron, or they may *converge*, when a neuron receives input from more than one other neuron. Because of divergence and convergence, every

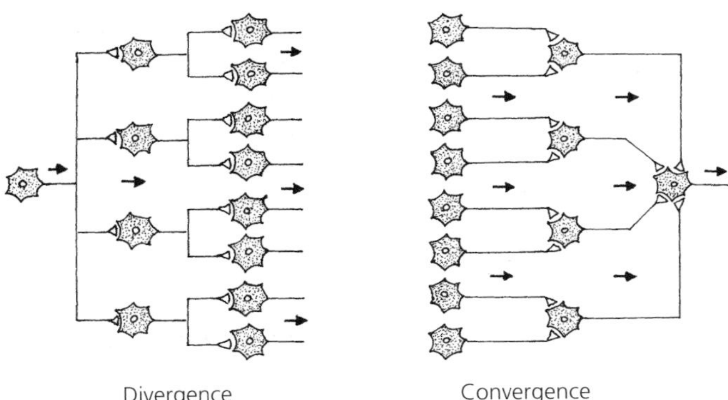


**Figure 17.20**
Divergence and convergence in the nervous system

receptor could, in theory, communicate with every effector, and every effector could receive signals from every receptor.

One can imagine the chaos that would result if this did happen – of course, it normally doesn't. When a receptor is stimulated, only certain effectors – the appropriate ones – actually respond. There must, therefore, be some way of ensuring that signals travel along the right pathways. The vast majority of potential pathways must somehow be blocked.

Once an impulse has started its way along a neuron, under normal circumstances nothing stops it travelling to all other parts of the axon. Selection of pathways must, therefore, involve points *between* neurons, where they communicate with each other. These points are the **synapses**. Most neurons form synapses with hundreds, or even thousands, of other neurons (Figures 17.21 and 17.22).

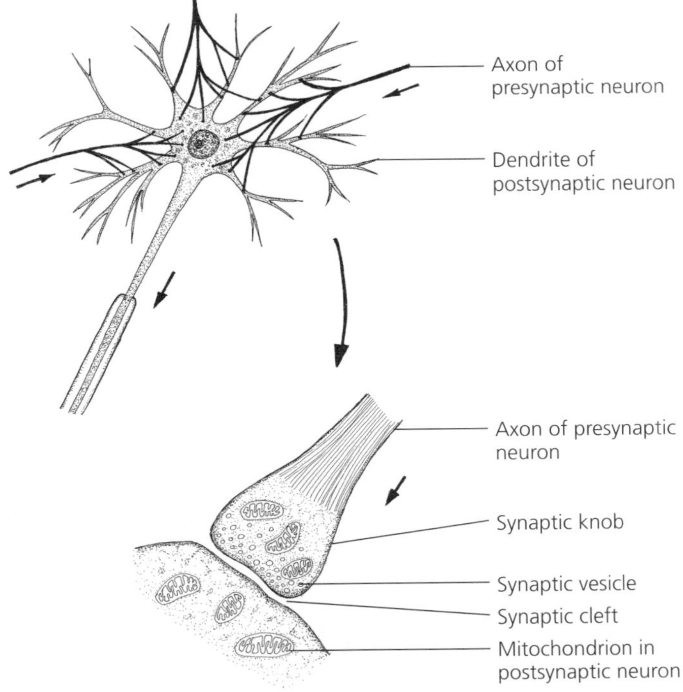


**Figure 17.21►**
Synaptic contacts between neurons, with fine structure inset

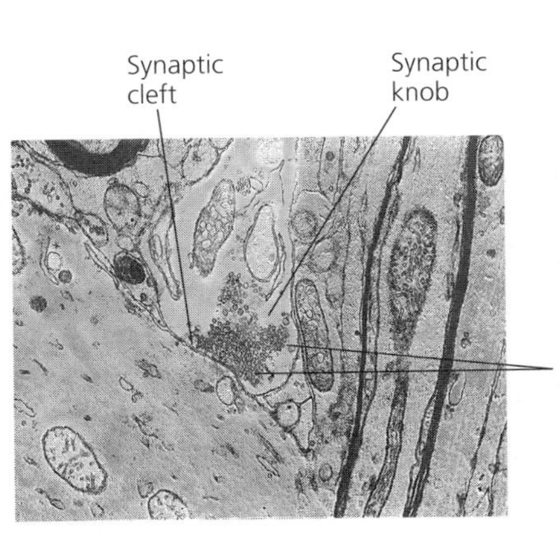


**Figure 17.22▲**
SEM of a synaptic knob. A number of small circular synaptic vesicles are seen within the synaptic knob; these contain the neurotransmitter substance

At a synapse, information travels from a **presynaptic** neuron to a **postsynaptic** neuron. The fine terminals of the presynaptic neuron end as **synaptic knobs** on the dendrites and cell body of the postsynaptic neuron. While they come extremely close, the plasma membranes of the pre- and postsynaptic neurons do not actually touch, but are separated by an extremely narrow **synaptic cleft**. Though only about 20 nm wide, impulses cannot cross it. When we speak of neurons making 'synaptic contact' we are thus referring to a physiological relationship rather than an actual contact.

Communication across synapses is thus indirect, and involves a chemical **neurotransmitter**. The first neurotransmitter to be discovered was **acetylcholine**, but numerous others are now known (see below). Prior to release, the transmitter is stored in **synaptic vesicles** in the synaptic knobs. The latter also contain numerous mitochondria which supply ATP energy for the synthesis and 'packaging' of the transmitter.

Just as the presynaptic terminal is specialised for the release of the transmitter, the postsynaptic membrane is specialised for receiving the chemical and converting it into an electrical signal. Lying in the lipid bilayer of the postsynaptic membrane are proteins that act as channels through which ions can pass. The channels are normally closed, but open when a transmitter chemical combines with them. These channels are said to be **chemically gated**, in contrast to those in axons which respond to electrical changes and are **voltage gated**.

In a motor neuron in the spinal cord, synaptic transmission involves the following events:

- the arrival of an impulse at the synaptic knob results in the opening of calcium 'gates' in the presynaptic membrane, causing calcium ions to diffuse into the synaptic knob from the extracellular fluid in the synaptic cleft.
- the rise in intracellular calcium causes synaptic vesicles to fuse with the presynaptic membrane and to empty their load of acetylcholine into the synaptic cleft – an example of exocytosis.
- the acetylcholine diffuses across the synaptic cleft and combines with receptor proteins in the postsynaptic membrane. This causes a rise in the permeability of the postsynaptic membrane to $Na^+$ ions.
- $Na^+$ ions diffuse into the postsynaptic membrane, producing a small reduction in membrane potential called an **excitatory postsynaptic potential**, or **EPSP**.
- while still bound to the postsynaptic membrane, acetylcholine is hydrolysed by the enzyme **acetylcholinesterase**, as a result of which the sodium gates close.
- acetylcholine is resynthesised and 'repackaged' in new synaptic vesicles.

Only the presynaptic membrane can release the transmitter and only the postsynaptic membrane contains the transmitter receptors that respond to it. A synapse thus acts like a one-way valve, so in the living animal signals travel in the same direction (though as mentioned earlier, in isolated axons they can travel in either direction).

### *Inhibitory synapses*

In synapses of the kind just described, incoming impulses lead to a reduction in the membrane potential of the postsynaptic cell making it more excitable. Synapses like this are said to be **excitatory**. At an **inhibitory** synapse, incoming impulses lead to an increase in membrane potential of the postsynaptic cell (hyperpolarisation) reducing its excitability. One example of the action of inhibitory synapses is the stretch reflex. Here, motor neurons controlling muscles antagonistic to the one being stimulated are inhibited (Figure 17.23).

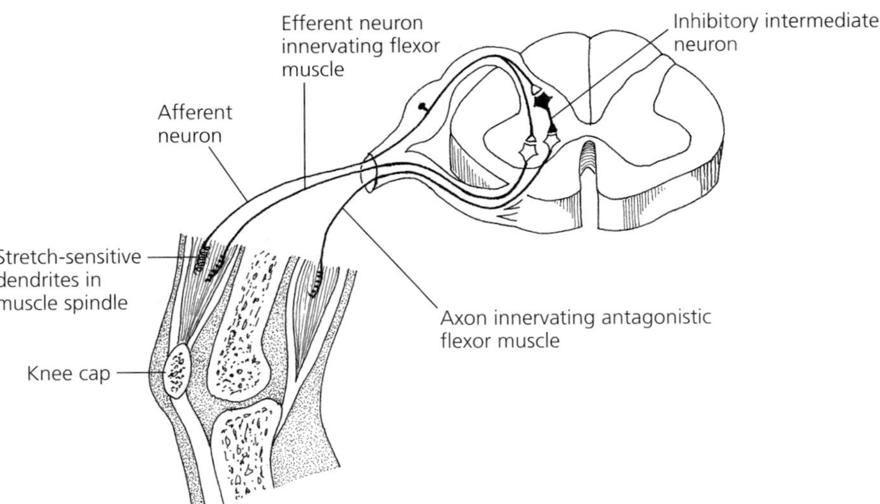


**Figure 17.23**
Inhibitory synapses in the stretch reflex

The effect of the transmitter at an inhibitory synapse is to open 'gates' that are specific to (in most cases) $Cl^-$ ions, causing them to diffuse in. The resultant rise in the membrane potential to a more negative value is called an **inhibitory postsynaptic potential** or **IPSP**, because its effect is to render the postsynaptic cell less excitable.

Though chemical synapses impose a delay in transmission, this is actually very short. The time for the transmitter to be released, diffuse across the synaptic cleft and produce its effect on the postsynaptic membrane is only about 0.5 ms. Diffusion only accounts for a small fraction of this because the distance involved is extremely short (20 nm).

We can see just how important this is when we remember that the time for a substance to diffuse from one point to another is inversely proportional to the *square* of their distance apart. Thus, if a synaptic cleft were 100 times wider (2 μm, or about twice the width of a mitochondrion) the time taken to diffuse across the gap would be $100^2 = 10\,000$ times greater.

## Drugs and the nervous system

A considerable number of chemicals interfere with the functioning of neurons. Some, such as procaine and other local anaesthetics, inhibit the propagation of action potentials, but most act on synapses.

There are various ways in which a chemical can interfere with synaptic transmission.

- It may prevent the release of the transmitter, for example botulinus toxin prevents the release of acetylcholine.
- It may bind to the transmitter receptors in the postsynaptic membrane; barbiturates (used in sleeping tablets) and the tranquillisers Valium and Librium, for instance, bind to GABA receptors. Curare blocks acetylcholine receptors, and strychnine binds to glycine receptors.
- It may compete with the transmitter for binding to the receptor. Nicotine, for example, competes with acetylcholine for the receptor in postsynaptic membranes.
- It may interfere with the enzymic destruction of the transmitter, for example the nerve gas DIPFP inhibits the action of cholinesterase.
- Some inhibitors may antagonise each other. Neostigmine, for example, is an anticholinesterase and can, therefore, be used as an antidote to curare.

## Integration

When a single impulse arrives at an excitatory synapse, the result is a small, **subthreshold** depolarisation of the postsynaptic membrane. It is therefore insufficient to cause the postsynaptic neuron to 'fire' an impulse. Although smaller than action potentials, these **postsynaptic potentials** last much longer (about 15 ms) because of the time taken for the destruction of the transmitter. This enables individual postsynaptic potentials to **summate**, or add together, leading the postsynaptic neuron to produce an action potential. Summation may be **spatial** (in space) if the incoming impulses arrive at different synapses, or **temporal** (in time) if they come via the same synapse. Spatial and temporal summation are not, of course, mutually exclusive; normally a neuron 'fires' because of some combination of the two.

The situation is actually more complicated than this because each neuron receives input via both excitatory and inhibitory synapses. At any given moment, the individual EPSPs and IPSPs are being added together in the cell body of the postsynaptic neuron. Thus the arrival of three EPSPs and one IPSP (two more excitatory potentials than inhibitory ones) would be more likely to result in an action potential than would five EPSPs and four IPSPs (one more excitatory impulse than inhibitory ones).

In a postsynaptic neuron, action potentials are produced at a region at the base of the axon called the **axon hillock**. A neuron will discharge an impulse when the *net* effect of the excitatory and inhibitory input at the axon hillock reaches the threshold. By adding input from several sources simultaneously, a neuron is **integrating** input from those sources.

Superficially similar to temporal summation is **facilitation**. However, whereas summation involves the postsynaptic membrane, the cause of facilitation lies in the presynaptic neuron. When a volley of impulse arrives at a synaptic knob, each impulse liberates more transmitter than the previous one and produces a larger effect on the postsynaptic membrane. The reason is that after the arrival of an impulse, it takes time for the calcium ions to be removed. Calcium ions released by a second impulse are added to those not yet removed after the first impulse, resulting in the release of more transmitter.

When volleys of impulses arriving at the presynaptic membrane are sustained, the supplies of transmitter can run down, causing the synapse to 'fatigue'.

### Neuromuscular junctions

A junction between a nerve and a muscle fibre is called a **neuromuscular junction**. The area of near-contact between axon and muscle membranes is increased by very extensive folding. As you can see from Figure 17.24, the axon terminal contains many vesicles containing transmitter and is separated from the muscle membrane by a narrow gap.

Although a neuromuscular junction is structurally very similar to a chemical synapse, there are important differences. First, most neurons receive input from many other neurons, some excitatory and some inhibitory. A skeletal muscle fibre, on the other hand, is only supplied by one nerve fibre, which is always stimulatory. Moreover, whilst the arrival of a single impulse is never sufficient to produce an impulse in a postsynaptic neuron, it is *always* enough to reach threshold in a muscle fibre.

### Neurotransmitters

Acetylcholine was the first neurotransmitter to be discovered, and neurons that use acetylcholine as a transmitter are said to be **cholinergic**. Besides being the means by which motor neurons in the spinal cord receive their

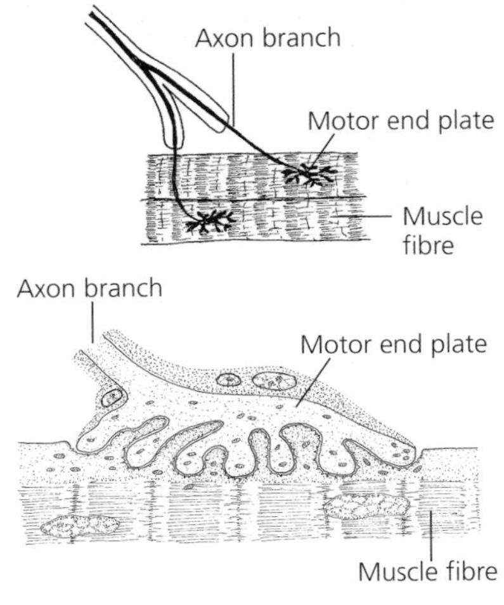


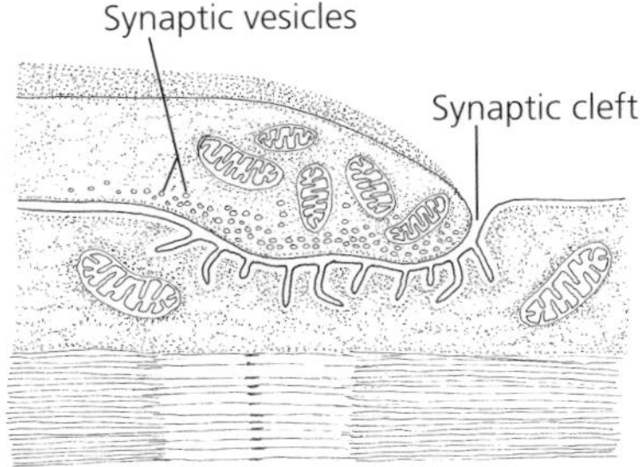


**Figure 17.24▲**
A neuromuscular junction at successively higher magnifications

excitation, it is also the transmitter between motor neurons and skeletal muscle.

Many other neurotransmitters are now known. For example **noradrenaline** is the transmitter released at the endings of most sympathetic neurons, which are therefore said to be **adrenergic**. Some transmitters are amino acids, for example glycine and $\gamma$ **aminobutyric acid (GABA)**, and are produced at many inhibitory synapses in the central nervous system. Some transmitters are peptides which, unlike other transmitters, require mRNA for their synthesis and are made in the cell body.

Although most neurons receive more than one transmitter, any given neuron only produces one kind. Whilst some transmitters seem to be entirely excitatory or inhibitory, this is not always so. For example acetylcholine excites skeletal muscle but inhibits cardiac muscle.

## The autonomic nervous system

Many effectors, such as the heart and glands in the gut, are beyond our conscious control. These involuntary effectors are controlled by the **autonomic nervous system**. The autonomic nervous system is also called the **visceral motor system**, in contrast to the **somatic motor system** which controls the skeletal muscles.

The autonomic nervous system consists of two parts, which are distinct both functionally and anatomically. In general, **sympathetic** nerves trigger responses to various forms of stress, such as exercise, severe blood loss or exposure to cold. **Parasympathetic** nerves are concerned with digestion and other activities that occur during rest and recuperation.

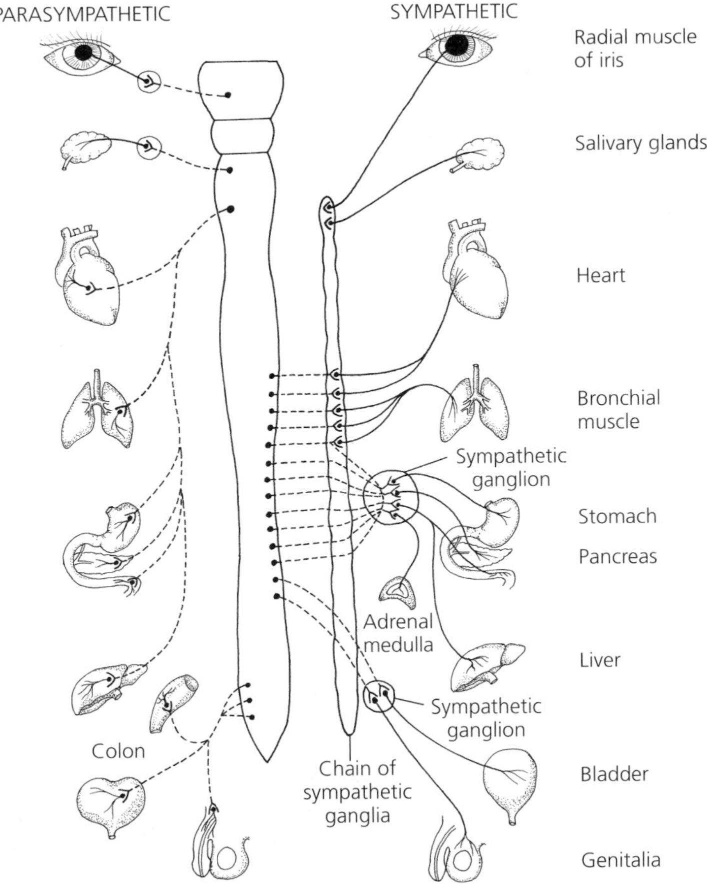


**Figure 17.25▶**
The autonomic nervous system

Besides their different influences on effector tissue, sympathetic and parasympathetic nerves differ anatomically (Figure 17.25). Sympathetic nerves leave the central nervous system via spinal nerves in the thoracic and lumbar regions. Parasympathetic fibres leave via the cranial nerves – particularly the vagus nerve – and also via the spinal nerves of the sacral region.

Whereas a skeletal muscle fibre is supplied by excitatory nerves only, the heart, gut and numerous other internal organs have **dual innervation**, meaning that they are supplied by both sympathetic and parasympathetic fibres. In most cases, the two sets of neurons have antagonistic effects, one being stimulatory and the other inhibitory. Thus the parasympathetic system slows the heart and the sympathetic system stimulates it. The smooth muscle of the gut, on the other hand, is stimulated by parasympathetic nerves and inhibited by sympathetic fibres. At any particular moment, the level of activity of such organs is determined by the net effect of these two competing influences.

**Table 17.2**
Summary of major effects of sympathetic and parasympathetic nerves

| Effector | Sympathetic effect | Parasympathetic effect |
|---|---|---|
| Heart | Stimulation | Inhibition |
| Arterioles | Constriction | No innervation in most |
| Liver | Glucose production | Glycogen synthesis |
| Bronchial muscle | Relaxation | Contraction |
| Gut motility | Inhibition | Stimulation |
| Gut sphincters | Stimulation | Inhibition |
| Pupil | Dilation | Constriction |
| Ciliary muscle | Relaxation | Contraction |
| Salivary glands | Amylase secretion | Potassium and water secretion |
| Tear glands | No innervation | Stimulation |
| Hair erector muscles | Contraction | No innervation |
| Sweat glands | Stimulation | No innervation |

The autonomic and somatic motor systems also differ in their 'wiring'. In the somatic motor system, signals travel all the way from the central nervous system to an effector via a single axon. In the autonomic system signals reach effectors via a pathway consisting of two neurons, a myelinated **preganglionic neuron** and a non-myelinated **postganglionic neuron**. The cell body of the preganglionic neuron is in the central nervous system, whilst that of the postganglionic neuron is in an **autonomic ganglion** (Figure 17.26). The axon of a preganglionic neuron is called a **preganglionic fibre** and each makes synaptic contact with a number of **postganglionic fibres** which terminate on the surface of effector cells.

The sympathetic and parasympathetic systems communicate with their target effectors via different transmitters. Parasympathetic, postganglionic neurons are cholinergic, whilst most sympathetic endings are adrenergic.

In the sympathetic system the preganglionic fibres are relatively short, as a result of which the sympathetic ganglia are close to the central nervous system. Most of the sympathetic ganglia form two chains, one on either side of the spinal cord. Each of these ganglia has two connections with a spinal nerve; a **white ramus** carrying myelinated preganglionic fibres, and a **grey ramus** carrying non-myelinated postganglionic fibres

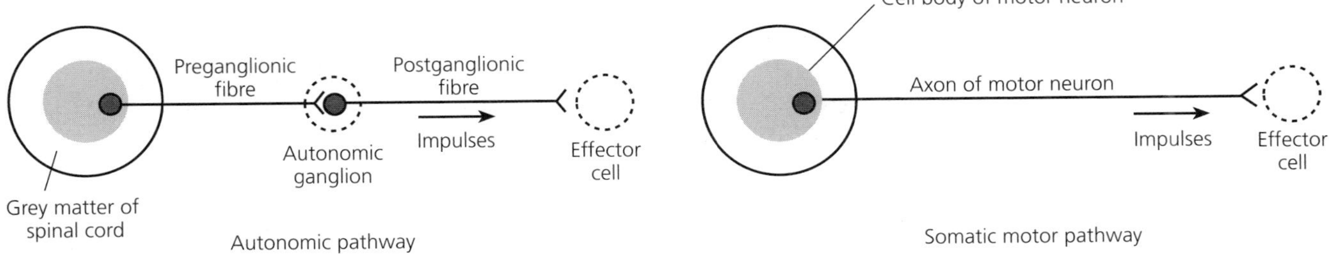


**Figure 17.26**
Layout of autonomic neurons (above left) compared with that of somatic motor neurons (above right)

(Figure 17.27). Other sympathetic ganglia are unpaired, and lie scattered among the abdominal organs.

A peculiar part of the sympathetic system is the **adrenal medulla**, or inner part of the adrenal gland. It is really an enormous ganglion, consisting of a mass of postganglionic neurons without axons. Each postganglionic neuron receives synaptic connections from preganglionic fibres. When stimulated, each postganglionic neuron releases **adrenaline**, a close chemical relative of noradrenaline. Adrenaline is released into the blood which carries it to all parts of the body. As a result, the effects of the sympathetic system tend to be more widespread than those of the parasympathetic system.

In the parasympathetic system, the relative lengths of pre- and postganglionic fibres are the reverse of those in the sympathetic system. The preganglionic fibres are so long that the ganglia are in, or very close to, the target organ, so the postganglionic fibres are very short.

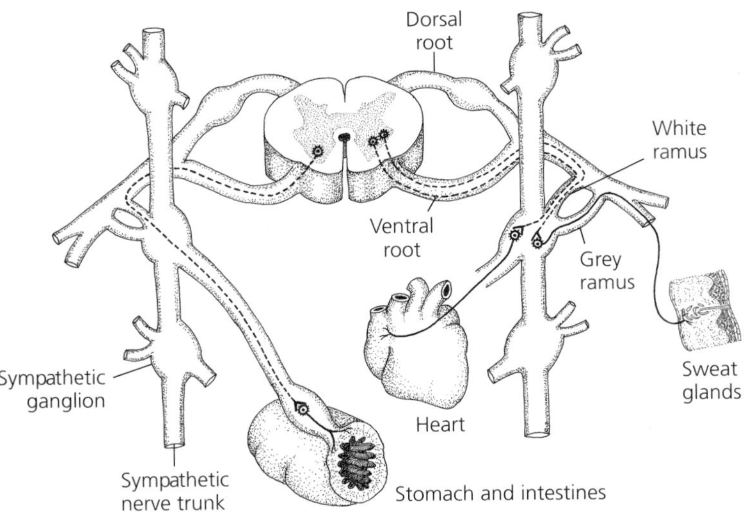


**Figure 17.27**
Relationship between a sympathetic ganglion and a spinal nerve (after Crouch)

## The brain

As bilaterally symmetrical animals, mammals have a clearly defined anterior end, at which the sense organs are concentrated. The more effective processing of this sensory information during the early evolution of the vertebrates involved a concentration of central nervous tissue, the **brain** (Figure 17.28).

In the earliest vertebrates it is likely that the brain was little more than an expansion of the spinal cord, dealing mainly with feeding and the detection of enemies. In later vertebrates the more posterior parts of the body would have been under the direct control of the spinal cord. The brain would have exerted an indirect influence, regulating the overall level of activity of parts rather than their detailed co-ordination. In modern

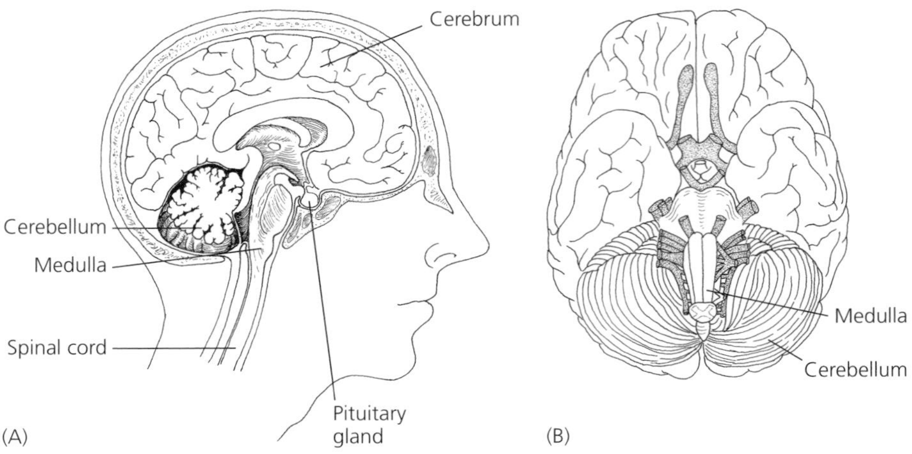


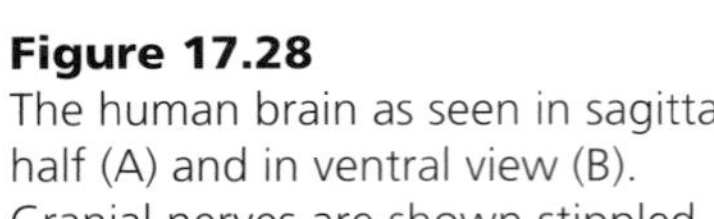

**Figure 17.28**
The human brain as seen in sagittal half (A) and in ventral view (B). Cranial nerves are shown stippled

vertebrates there has been a strong tendency for the brain to expand and to exert an increasingly direct control over the more posterior parts of the body.

The potential for more complex activities can be seen even in the spinal cord, where intermediate neurons, with their many connections, increase the range of possible pathways. The presence of many neurons between sensory input and motor output greatly increases the number of stimuli that can activate a given effector and also increases the number of responses that can result from the stimulation of a given receptor. The brain can be thought of as a great extension of this principle.

It goes without saying that the brain is an organ of great complexity, and exactly how it works is only partially understood. Indeed, it is possible that it will never be completely understood.

Nevertheless the brain is far from being a mysterious 'black box'. Damage to particular regions is often associated with loss of particular faculties, so it is possible to relate a particular function to a particular region. Furthermore, by probing the living brain with electrodes it is possible to produce particular kinds of response.

In the brain, neurons concerned with dealing with particular kinds of information are clustered into functional centres called **nuclei** (quite different from cell nuclei). These range in size from microscopic clusters of neurons to massive regions such as the cerebral cortex, which reaches its greatest size in humans.

Some brain centres operate in a kind of hierarchy, as in the relationship between the cerebellum and the cerebral cortex. The initiation of an action, such as reaching out to pick up an object, originates in the cerebral cortex. However, the business of co-ordinating all the muscles is carried out by the cerebellum. The roles of the cerebral cortex and cerebellum in voluntary movement can thus be likened to different ranks in an army. The High Command (equivalent to the cerebral cortex) issues the overall orders, but the lower ranks in the hierarchy (equivalent to the cerebellum) co-ordinate the fine details.

## Why is the central nervous system hollow?

In the embryo the central nervous system develops from a flat sheet of cells which rolls up to form the hollow **neural tube.** The posterior part of the neural tube retains its simple, tubular form and becomes the spinal cord, whilst the anterior part enlarges to form the brain. Both brain and spinal cord retain their hollow interior, the spinal canal of the cord becoming continuous with the cavities, or **ventricles**, of the brain.

## Parts of the brain

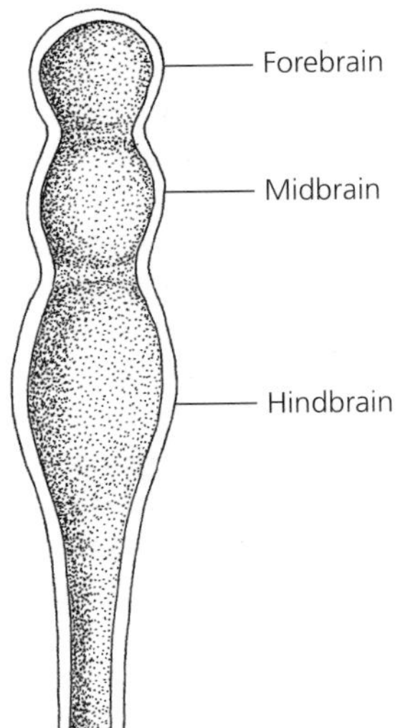


**Figure 17.29**
The three primary regions of the embryonic brain

Early in development the brain becomes differentiated into three regions: the **forebrain, midbrain** and **hindbrain** (Figure 17.29). Each of these three primary divisions develops a dorsal expansion associated with one of the major senses – smell, vision and hearing. In monkeys and apes, and most especially in humans, the cerebellum and cerebrum are so large that they hide other parts of the brain.

### *Hindbrain*

The posterior part of the brain stem develops into the **medulla oblongata**. It is the least modified part of the brain and resembles an anterior enlargement of the spinal cord. Via autonomic nerves its nuclei control blood pressure, heart and breathing rates, digestion and a number of other functions. All cranial nerves except the first two leave the medulla.

Just anterior to the medulla is the **pons** (meaning 'bridge'). This is a wide band of fibres connecting the medulla to more anterior regions of the brain.

The roof of the hindbrain expands into a large dorsal outgrowth, the **cerebellum**. This plays a central role in co-ordinating muscular activities and in the regulation of posture. In the latter connection, the cerebellum receives information from the balance organs of the inner ear.

### *Midbrain*

In mammals the midbrain is relatively small, with many of its functions being taken over by the cerebrum. It contains the **corpora quadrigemina**, which control the eye muscles.

### *Forebrain*

The floor of the forebrain forms the **hypothalamus**, whose importance is out of all proportion to its small size. Unlike other parts of the brain, some of its neurons also secrete hormones which communicate with the **pituitary gland** immediately below it. Via the pituitary, the hypothalamus regulates the activity of the kidney, thyroid gland, gonads and adrenal cortex (Chapter 20).

The hypothalamus also controls much of the autonomic nervous system. It contains nuclei controlling hunger, thirst, body temperature and internal (endogenous) rhythms.

The lateral walls of the midbrain form the **thalamus**, a centre which processes and relays sensory information to the cerebral cortex.

In the roof of the forebrain is the small, stalk-like **pineal body**. This endocrine gland secretes **melatonin**, a hormone involved in maintenance of the sleep–wake cycle. A primitive feature of the pineal body is its sensitivity to light. In birds, the skull is thin enough for the pineal to detect the changes in day length and thus trigger seasonal changes in behaviour such as breeding and migration.

The most anterior part of the forebrain consists of the **olfactory lobes** which are concerned with the analysis of information from the organs of smell. In mammals, by far the most prominent part of the forebrain is the **cerebrum** which is formed of two **cerebral hemispheres**. In humans, these are so large that they hide most of the brain.

### *The cerebral hemispheres*

Like the cerebellum, the cerebrum is differentiated into the outer **cortex** containing nerve cell bodies and an inner area containing nerve fibres (the opposite arrangement to the grey and white matter in the spinal cord). It is in the cortex that the synaptic connections between neurons occur.

In monkeys, apes and especially humans, the increase in the size of the cerebrum has been accompanied by extensive folding of the cortex. The reason is that with an increase in size, surface area increases more slowly than volume unless there is a change of shape. Folding has enabled the evolutionary increase in the surface layer of the cerebrum (the cortex) to keep pace with the expansion of the volume of the cerebrum.

To some extent the cerebral cortex is functionally organised into a number of areas. It has been known for a long time that injury to a particular part of the cerebral cortex may result in loss of functions associated with that area (Figure 17.30). For example, injury to Broca's area on the left side of the brain results in loss of the ability to put words together into meaningful sequences, whilst damage to the visual area results in impaired vision.

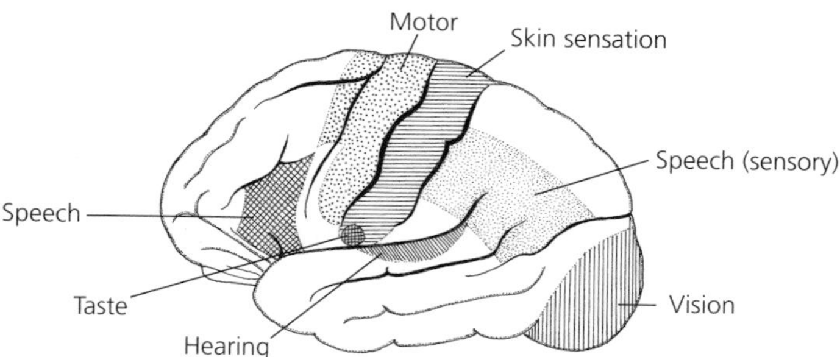


**Figure 17.30**
Regional specialisation in the cerebral cortex

A characteristic of the cerebral cortex is that motor areas (those concerned with initiating voluntary movements) are the opposite way round with respect to the areas they control. For example, the left hand is controlled by the right side of the cortex, and so on.

A feature of the brain peculiar to humans is its **asymmetry**. The two hemispheres are not only slightly different in shape, but there are functional differences between the two sides. For example, the right side is concerned more with spatial and musical skills, whereas the left hemisphere deals more with mathematics and language.

### *The corpus callosum*

Although some brain functions (such as language) can operate independently in the two hemispheres to some degree, others require that each hemisphere knows what the other is doing. Communication between the two hemispheres is the role of a broad band of nerve fibres called the **corpus callosum**. People who have had the corpus callosum surgically cut cannot integrate two processes involving different halves of the brain.

## Protection of the central nervous system

The central nervous system is in special need of protection, not only because of its vital role, but also because, unlike many other parts of the body, it cannot repair itself. The outermost protective layers are the skull and backbone. Within these, the brain and spinal cord float in a layer of **cerebrospinal fluid**, which gives protection against mechanical shock.

The cerebrospinal fluid is enclosed between two membranes, the **meninges** (Figure 17.31). The outer meninx is the tough, fibrous **dura mater**, which lines the skull and vertebral canal. The inner one is the delicate **pia mater** and covers the brain and spinal cord. Crossing the cerebrospinal fluid are the delicate strands of the **arachnoid layer**,

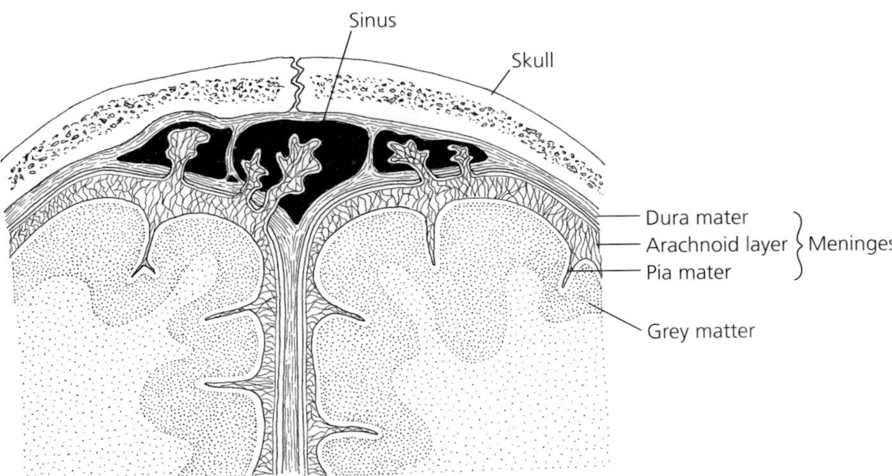


**Figure 17.31**
The meninges as seen in a section through the front part of the head (after Crouch)

so-called because it resembles the cobweb of a spider. Infection of the meninges results in the potentially lethal **meningitis**.

Cerebrospinal fluid also occupies the spinal canal and ventricles of the brain. It is continuously formed by seepage through the walls of networks of blood vessels called **choroid plexuses** in the roof of the ventricles. Since it is also reabsorbed into veins between the meninges, it slowly circulates.

Besides being protected from blows to the head, the brain is also protected from harmful chemicals by a blood–brain barrier (Chapter 22).

## QUESTIONS

**1** What would be the effect on the magnitude of the resting potential of addition of each of the following substances to the extracellular fluid? If immediate and longer term effects differ, say so, and explain the reason for the difference.
**a)** KCl solution.
**b)** NaCl solution.
**c)** cyanide.

**2** What effects would each of **(a)**, **(b)** and **(c)** in Question 1 have on the magnitude of the action potential?

**3** Given suitable apparatus for stimulating an axon and recording action potentials, what two measurements would you need to take to find the velocity of propagation of an impulse?

**4** The main nerve trunk supplying the hind limb of a frog was given electrical stimuli of two different intensities. Recording electrodes were placed 13 cm away from the stimulating electrodes. The recorded action potentials are shown in the diagram. State *two* conclusions that can be drawn from the traces.

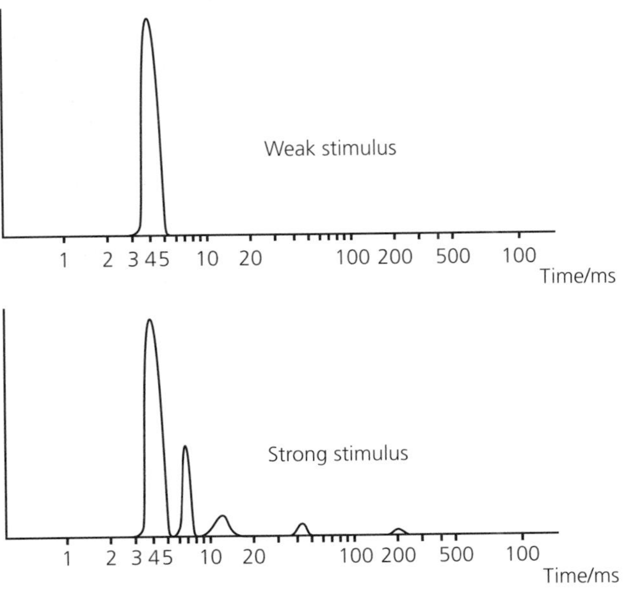


**5** In a myelinated axon carrying impulses at 100 m $s^{-1}$, the action potential lasts 0.5 ms. If the nodes of Ranvier in this axon are 1 mm apart, how many internodes are involved in the action potential at a given instant?

CHAPTER

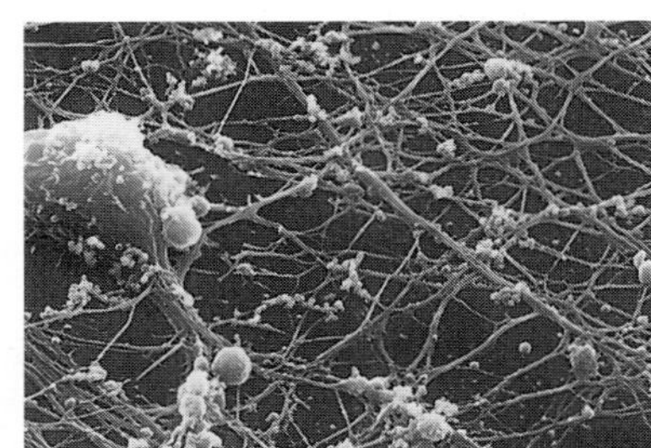

# 18

# DETECTING CHANGE – RECEPTORS

## LEARNING OBJECTIVES

By the time you have completed your study of this chapter you should be able to:

- explain what is meant by sensitivity and explain its significance to living things.
- distinguish between mechanoreceptors, chemoreceptors, thermoreceptors, photoreceptors and nociceptors.
- describe the events by which the detection of a stimulus leads to the production of action potentials in an afferent nerve fibre.
- explain how receptor potentials differ from action potentials.
- explain how the brain distinguishes between different kinds and different intensities of stimuli.
- explain what is meant by adaptation, and its significance in different kinds of receptor.
- explain how the sense of touch is more highly developed in some parts of the body than in others.
- describe how the lens system of the eye forms an image on the retina.
- describe the events that occur when the eye adjusts for different light intensities and when it accommodates for near and distant vision.
- distinguish between the major defects of vision and explain how they are remedied.
- describe the structure of the retina and explain why it is said to be 'inverted'.
- describe the events that occur in photoreception in rods.
- explain the basis for visual sensitivity and visual acuity.
- explain the role of the brain in colour vision and stereoscopic vision.

## The need for sensitivity

Sensitivity – the ability to respond to change – is a fundamental characteristic of life. A change to which an organism is sensitive is called a **stimulus** (plural, stimuli). Responding to a stimulus does not simply mean being affected by change; a dead leaf is affected by wind, but it is clearly not responding to it.

Biological responses to change have two important features; they are **active**, involving the expenditure of energy, and they are **adaptive**, meaning that they increase the chances of the organism surviving and reproducing.

Responses to stimuli may be adaptive in two different ways. First, they may help the organism to avoid hazards. Second, they may enable it to take advantage of potential benefits such as food or a mating partner.

Although we tend to think of stimuli in connection with the external environment – as these are the ones we are most aware of – some of the most important changes occur inside the body and without our awareness.

**Stimulus and sensation**

Stimuli are often confused with the sensations they evoke, but they are actually quite different. A stimulus is a change that stimulates a receptor and gives rise to nerve impulses. A sensation is the feeling we experience when those nerve impulses arrive in the brain. A sugar solution on the tongue is a stimulus; a sweet taste is the sensation it produces.

For example, every time we stand up a slight fall in blood pressure in the arteries supplying the brain is detected by pressure receptors and corrected by an increase in the strength of the heart beat.

## Receptors

Animals detect stimuli using **receptors**. A receptor may be a nerve ending or it may be a specialised receptor cell which makes synaptic contact with a sensory neuron. A **sense organ** contains a cluster of many receptors, together with certain accessory structures which help the receptors to detect the stimulus.

### *Types of receptor*

Although most receptors can be triggered by a variety of stimuli, each is usually particularly sensitive to one type or **modality**. On this basis we can distinguish various types of receptor.

#### *Mechanoreceptors*

Mechanoreceptors detect mechanical changes. They include a variety of types such as the touch and pressure receptors in the skin, and receptors that detect vibration, for example those serving the sense of hearing. Other important mechanoreceptors are the **proprioceptors** which give information about body position and orientation. The muscle spindles (which are stretch receptors in muscles) and the balance organs of the inner ear are examples of proprioceptors.

#### *Chemoreceptors*

Chemoreceptors are sensitive to chemicals. Although the best known chemoreceptors serve the senses of smell and taste, the most important chemoreceptors detect internal changes in body chemistry. These include the receptors that monitor the oxygen and carbon dioxide concentration of the blood.

#### *Thermoreceptors*

Thermoreceptors detect temperature changes, for example those in the skin. Though we are not consciously aware of their input, the most important thermoreceptors are those in the hypothalamus, which is the body's thermostat.

#### *Photoreceptors*

Photoreceptors are sensitive to light. The most important vertebrate photoreceptors are in the retina of the eye. In birds, the pineal organ in the brain is also sensitive to light and plays an important part in seasonal behaviour by providing information on photoperiod (daylength).

#### *Nociceptors*

Nociceptors are sensitive to injurious stimuli. They are sometimes called 'pain receptors', but since pain is a **sensation** rather than a stimulus, 'pain receptor' is an unsatisfactory term.

### *Internal and external receptors*

The stimuli that make the greatest impact on our consciousness are in the external surroundings and are detected by **exteroceptors**. However, many of the most important changes take place inside the body, and are detected by **enteroceptors**. Changes in the pressure or glucose concentration of the blood, for instance, are detected (and usually corrected) without our realising a change has occurred.

**Table 18.1**
Types of receptor based on stimulus sensitivity

| Receptor | Stimulus | Sensation |
|---|---|---|
| Mechanoreceptors | Mechanical distortion | Touch, pressure, hearing, balance |
| Photoreceptors | Light | Vision |
| Chemoreceptors | Chemicals | Smell and taste |
| Thermoreceptors | Change in temperature | Warmth, cold |

### Dendrites as receptors

Structurally, receptors are of two kinds. One type consists of the endings of afferent neurons whose cell bodies lie in the dorsal root ganglia in the spinal cord. Alternatively they may be receptor cells which make synaptic contact with afferent neurons (Figure 18.1).

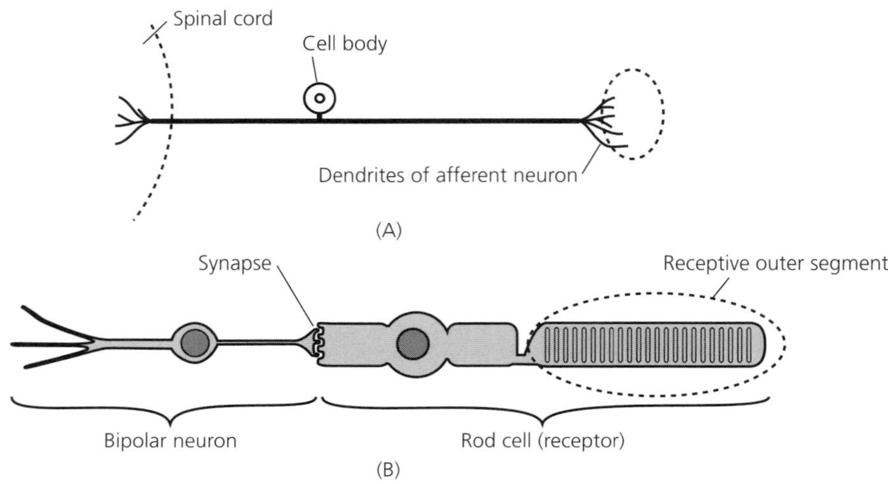


**Figure 18.1**
Two types of receptor shown enclosed in dotted lines: a dendrite (A) and a specialised cell (B), in this case, a retinal rod cell

An example of a dendritic receptor is the **Pacinian corpuscle**, a pressure receptor found in the dermis of the skin and in certain other parts of the body such as the mesenteries. Its relatively large size (almost 2 mm) makes it a convenient structure for study. The receptive dendrite is non-myelinated and is surrounded by an onion-like capsule consisting of concentric layers of collagenous fibres (Figure 18.2). When steady pressure is applied, the capsule slowly deforms, relieving the stress and allowing the dendrite to slowly regain its original shape.

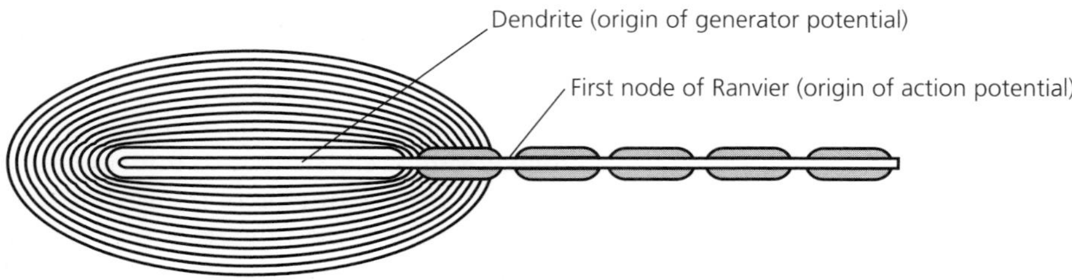


**Figure 18.2**
A Pacinian corpuscle

### Receptor cells

The senses of taste, smell, sight, hearing and balance are all served by receptor cells. Each receptor cell makes synaptic contact with a dendrite of a sensory neuron. When stimulated, the receptor cell undergoes a change in membrane potential, which gives rise to one or more action potentials in the adjacent afferent neuron.

## Receptors as transducers

Receptors act as **energy transducers** – that is, they convert some form of energy from the environment into an electrical change in the receptor membrane called a **receptor potential**.

Unlike an action potential, a receptor potential is of **graded intensity**, meaning that its magnitude increases with the strength of the stimulus; there is no threshold. The receptor potential then gives rise to one or more action potentials in an afferent neuron, but the steps by which this occurs depend on whether the receptor is a dendrite or a specialised receptor cell.

In a receptor cell the process occurs in two steps. The receptor potential causes the release of a transmitter which then induces the production of a **generator potential** in the dendrite of the afferent neuron. It is called a generator potential because, if it is large enough, it generates one or more action potentials in the axon.

In a dendritic receptor such as a Pacinian corpuscle, stimulation of the dendrite induces a receptor potential which spreads along the dendrite. If it is large enough it induces an action potential at the first node of Ranvier (Figure 18.3). In this one-step process, receptor potential and generator potential can be considered to be the same thing.

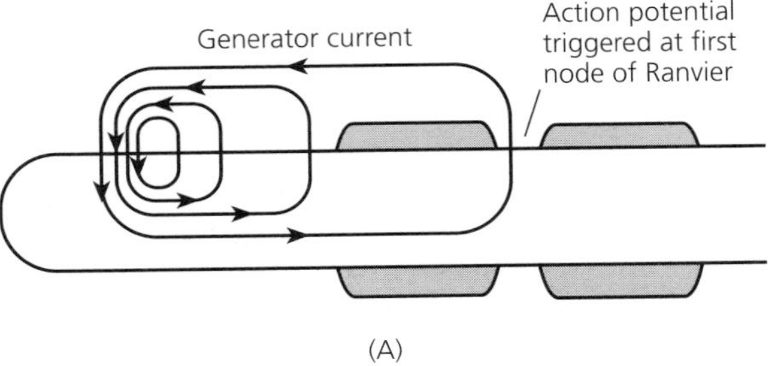


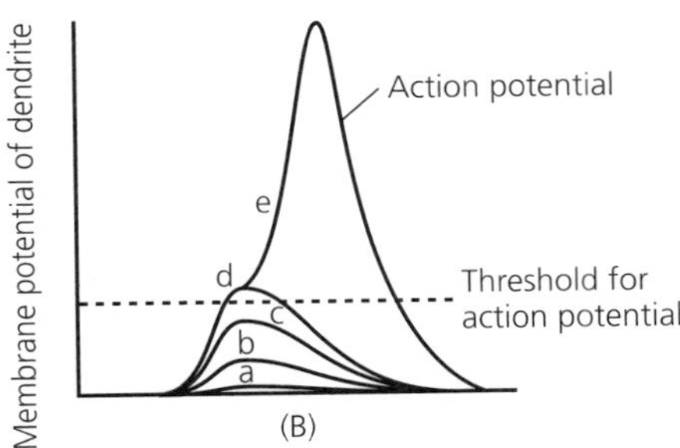


**Figure 18.3**
Effect of stimulation of a Pacinian corpuscle. (A) Production of generator potential in a dendrite. Notice that the intensity of the generator potential decreases with distance from the point of stimulus. (B) Traces a, b, c and d show responses to pressure stimuli of intensities x, 2x, 3x and 4x, respectively. The stimulus of intensity 4x was followed by an action potential (e)

A receptor potential differs from an action potential in two important ways:

1 like a postsynaptic potential, its magnitude decreases rapidly with distance; action potentials travel without decreasing in intensity.

2 whereas an action potential lasts only a millisecond or so, a receptor potential persists for several milliseconds after removal of the stimulus. As a result, two subthreshold stimuli in quick succession can add together to provoke an action potential.

## What information can a receptor provide?

An individual nerve impulse can carry no more information than a single dot in a message in Morse code. In an individual nerve fibre, only the *frequency* of impulses can vary. Yet the fact is that our receptors provide us with detailed information about the world, such as the shape, colour, surface texture and weight, of an object. In this chapter we shall see how it is possible for individually-identical impulses to provide complex information.

### What kind of stimulus?

If impulses from the ear are the same as impulses from the eye, how does the brain distinguish between 'sound' impulses and 'visual' impulses? The answer depends on the fact that each kind of receptor is preferentially sensitive to one stimulus modality. Impulses coming from the retina of the

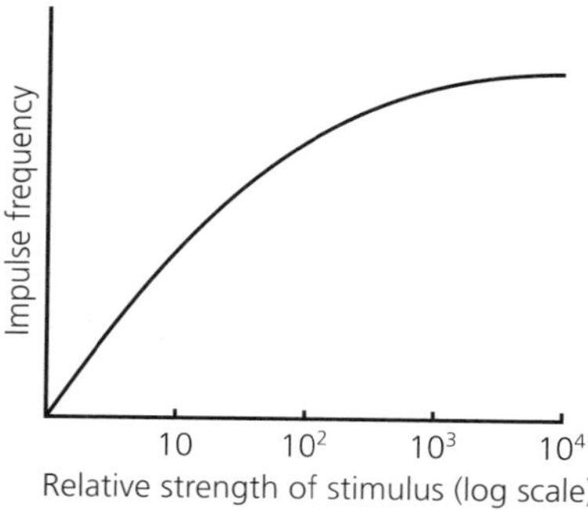


**Figure 18.4**
Effect of stimulus intensity on impulse frequency

eye are normally produced by light, and impulses from chemoreceptors in the nose are normally produced by chemicals. The brain, then, knows what kind of stimulus gave rise to the impulses from the route through which they arrive.

The preferential sensitivity to a particular type of stimulus is not absolute. If, for example, you close your eye and *gently* press against the eyelid on the side nearest the nose, you may 'see' a patch of light. What is happening is that the photoreceptors in the retina are responding to pressure, but since the impulses generated are the same as if light were the stimulus, the sensation is a visual one.

An interesting 'thought experiment' illustrates this point. Imagine that you could cut the optic nerve from the eye and the olfactory nerve from the nose, and stitch the ends 'wrong way round'. If all the axons could be joined up, then shining a torch into the eye would produce the sensation of odour, and holding a tasty meal under the nose would give rise to a visual sensation!

## Stimulus intensity

When a very weak stimulus is applied to a receptor, the generator potential is too weak to induce an action potential. With strong stimuli, the generator potential increases and eventually reaches the threshold for the production of an action potential. As long as the generator potential remains above threshold, action potentials continue to be produced. The stronger the stimulus, the bigger the generator potential and the higher the frequency of the action potentials it induces (Figure 18.4).

Above the threshold, stronger stimuli evoke higher impulse frequencies. However, the relationship is not a linear one – with stronger stimuli the receptor becomes less sensitive to further increases in stimulus strength. The advantage of this is that it enables a receptor to respond to a much wider range of stimulus intensity than if impulse frequency were in direct proportion to stimulus strength. Some receptors can respond to a 100 000-fold range of stimulus intensity.

Besides the frequency of incoming impulses, the brain has another way of knowing how strong a stimulus is because receptors may differ in their thresholds, and thus in sensitivity. Amongst a population of mechanoreceptors in the skin, some begin firing at the slightest touch and others only respond to more intense stimuli.

## Adaptation

In many receptors the frequency of impulses declines even if the stimulus is maintained at a steady level. We all know, for example, how we may notice an odd smell on entering a room, but after a while it may not be detectable. This phenomenon of **adaptation** is developed to different degrees in different kinds of receptor (Figure 18.5).

At one end of the scale are the rapidly-adapting, or **phasic**, receptors, such as the touch receptors in the skin. These adapt so quickly that a steady stimulus may evoke only a few impulses. Phasic receptors are not concerned with the absolute intensity of a stimulus, but with its *rate of change*. These receptors instantly inform us of 'new' and possibly more important events, such as an insect crawling up the arm, but fail to register the less important stimulus of steady contact with clothing.

At the other extreme are **tonic** receptors, which adapt slowly or not at all to steady stimuli. Examples are the balance organs in the inner ear, which continue to discharge action potentials even when the head remains motionless. Nociceptors, which give rise to the sensation of pain, also adapt slowly, if at all.

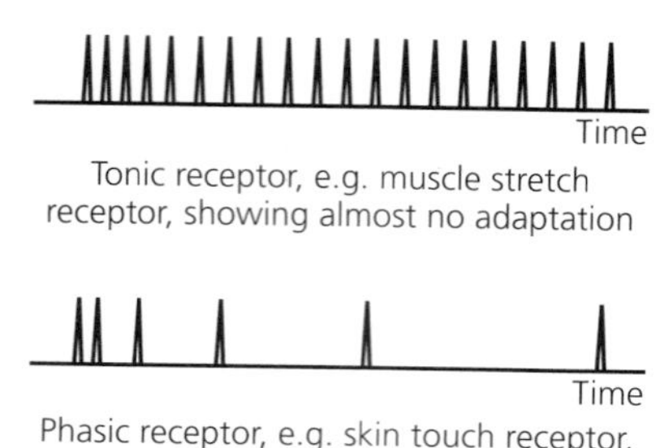


**Figure 18.5**
Rapidly- and slowly-adapting receptors

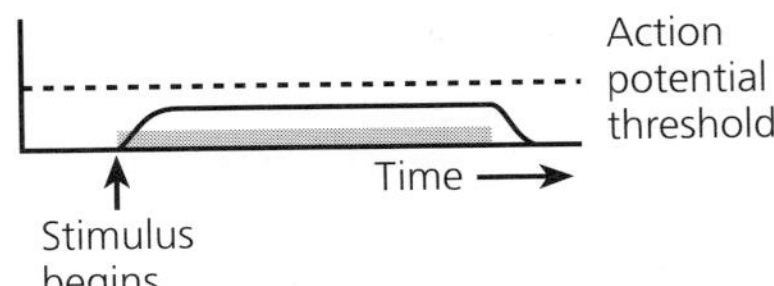


Weak stimulus produces a small generator potential but no action potential

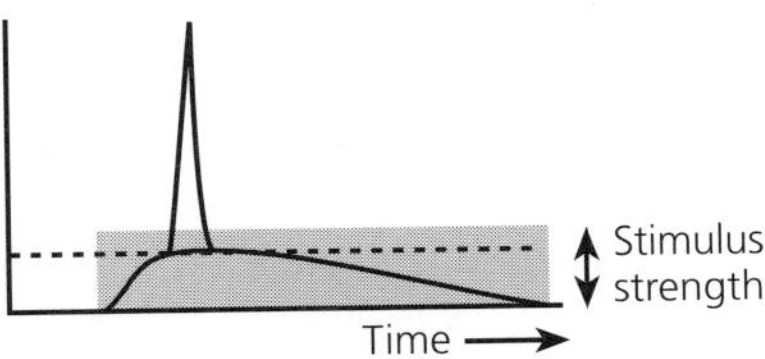


An above-threshold stimulus produces an action potential but the generator potential decays (due to adaptation) to below threshold before any more action potentials are produced

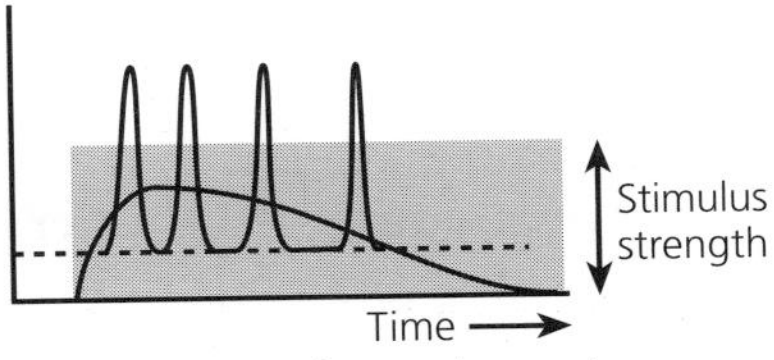


A stronger stimulus produces a larger generator potential and a succession of action potentials which continue until the generator potential decays to below threshold.

**Figure 18.6▲**
Effect of adaptation on the generator potential. The more the generator potential exceeds threshold, the higher the impulse frequency. With a steady stimulus, the generator potential decays, and action potentials cease when it falls below threshold

The cause of adaptation is the decline in the generator potential rather than any decrease in the ability of the generator potential to produce action potentials (Figure 18.6).

## Discrimination of detail

An individual receptor can only inform the brain of the intensity of the stimulus. How, then, does the brain get complex information that enables us to tell the difference between a pebble and a block of wood?

This is where *groups* of receptors come in. The way the brain does this can be illustrated by touch receptors in the skin, though the underlying principles also apply to the discrimination of detail by the eye.

You may have noticed that some parts of the skin can give more detail than others. For example, even when blindfolded, you can easily distinguish between the feel of velvet and silk using your fingertips, but it is harder to do so using your calf or elbow. When we touch an object, it makes many separate contacts with the skin, but to illustrate the essential principle, we only need to consider two (Figure 18.7).

For the brain to distinguish two or more points as distinct, they must stimulate the endings of two or more sensory neurons *separated by at least one unstimulated one.*

Two features of the nerve supply of the skin favour fine discrimination of detail. First, there must be many separate 'lines' to a given area of skin. Second, the area innervated by each neuron should overlap as little as possible with areas innervated by neighbouring neurons.

The part of the skin most sensitive to touch is the ridged part of the fingertips. When the fingertip is passed over a rough surface, each ridge vibrates up and down. The intensity of the stimulus is therefore constantly changing, so the phasic skin receptors do not fall silent as they would with a steady stimulus.

# The eye

Most organisms are sensitive to light, but only arthropods, molluscs and vertebrates can use light to provide detailed information about their environment by forming an image. The enormous importance of vision in humans (and most other vertebrates) is shown by the fact that while the auditory nerve from the ear contains over 30 000 axons, the optic nerve from the eye contains about a million – more than all the sensory fibres entering the spinal cord. Moreover, about a quarter of the cerebral cortex is devoted to processing visual information.

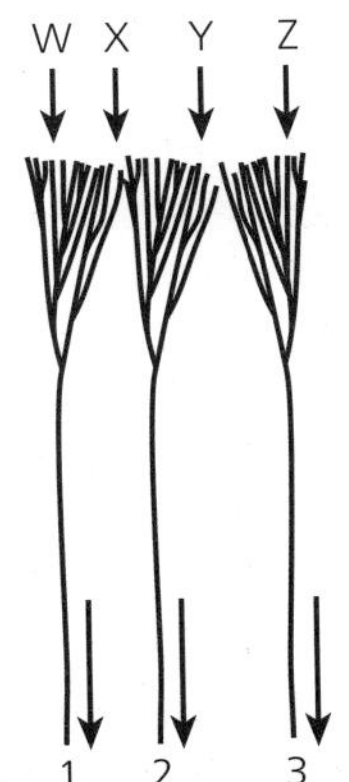


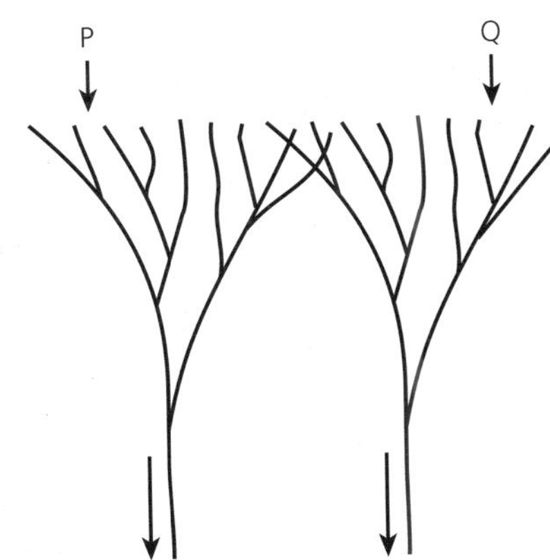


**Figure 18.7►**
Coarse-grained and fine-grained distribution of sensory nerve endings in different areas of skin. In (A), two separate stimuli at W and X or W and Y cannot be distinguished from a single stimulus at X, since both give rise to impulses along neurons 1 and 2. Stimulation at W and Z are perceived as distinct since the two active neurons 1 and 3 are separated by a 'silent' one (2). In the coarse-grained field (B), stimuli at P and Q cannot be distinguished even though they are further apart than W and Z in the fine-grained field

Like X-rays and radio waves, light is a form of **electromagnetic radiation** (Figure 18.8). Light is defined as that part of the electromagnetic spectrum that is visible to humans. We should remember, however, that not all animals share the same sensitivity spectrum as humans (bees, for instance, can see ultraviolet but not red).

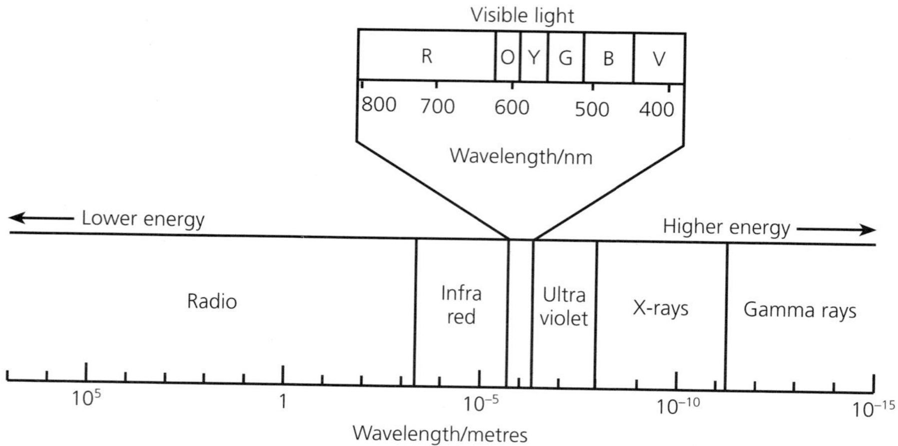


**Figure 18.8**
The electromagnetic spectrum, of which light is but a very small part

Electromagnetic radiation can also behave like particles, called **photons**. The energy of a photon depends on its **wavelength**; the shorter the wavelength the higher the energy. The visible spectrum extends from violet (380 nm) to red (760 nm). To be detected a photon must be absorbed by a pigment in a photoreceptor, producing a chemical change in the pigment. Having been absorbed, the pigment must be able to revert to its original form so that it can absorb another photon: the change must, therefore, be reversible.

This explains why wavelengths shorter than 200 nm or longer than about 700 nm cannot be used in vision. Very short wavelength radiation (X-rays and $\gamma$-rays) cannot be used as sensory information; these photons have so much energy that they break covalent bonds in molecules absorbing them. Low energy photons (microwaves and radio waves) have insufficient energy to produce any chemical change in a pigment. Visible light has just enough energy to excite pigment molecules but insufficient energy to do any damage.

## Structure of the human eye

The eye is protected in a bony socket in the skull called the orbit. The wall of the eyeball consists of three layers: an outer **fibrous layer**, a middle **vascular layer**, and an inner, nervous layer – the **retina** (Figure 18.9). The retina is dealt with later in this chapter.

The outer layer of the eye is tough and fibrous, held in a near-spherical shape by the slight internal pressure in the eye. The transparent front of the eye is the **cornea**, whose forward bulge enables it to act as a convex lens. The rest of the outer layer is the white **sclera**, into which are inserted three pairs of **extrinsic eye muscles** which can rotate the eyeball (intrinsic eye muscles are *inside* the eyeball). The anterior part of the sclera is covered by a thin membrane called the **conjunctiva** which also lines the inner surface of the eyelids. The conjunctiva is continuously being washed by tears secreted by **lachrimal glands**.

The middle layer of the eye consists of the choroid, the ciliary body (or ciliary muscle) and the iris. The **choroid** is rich in blood vessels which nourish the outer layer of the retina. It contains the pigment **melanin** which helps prevent internal reflection.

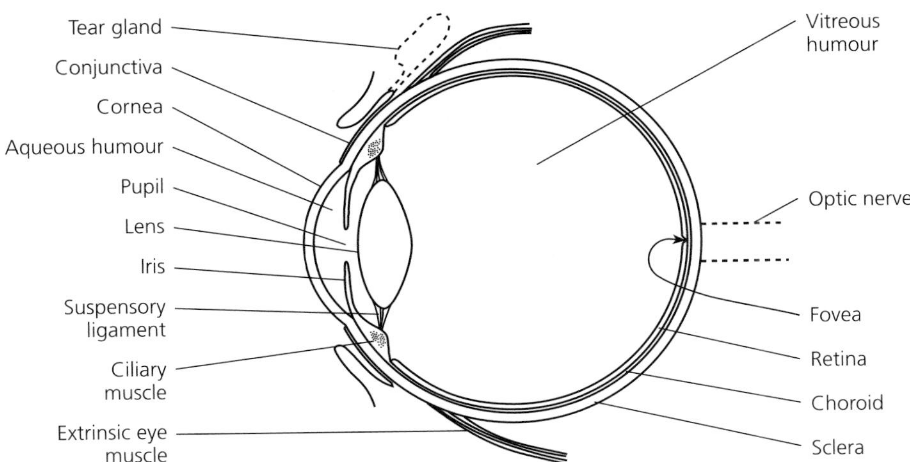


**Figure 18.9**
Vertical section through the eye. Note that the optic nerve is on the same horizontal level as the fovea, but is to one side, so it is shown as a dotted line

The ciliary body and iris contain the intrinsic eye muscles, enabling it to adjust for vision in different conditions. The **ciliary body** is a ring of smooth muscle and adjusts the refracting power of the eye for viewing objects at different distances. The **iris** is a ring of muscle surrounding the **pupil**. By varying the size of the pupil, the iris regulates the intensity of light reaching the retina and also increases the depth of focus when viewing near objects.

The elastic **lens** is anchored to the ciliary body by the fibres of the **suspensory ligament**. The shape of the lens can be changed, enabling the refracting power of the eye to be adjusted (see below). Like the cornea, it consists of living cells, and has no blood supply; capillaries would seriously interfere with light transmission.

The lens and cornea are supplied with nutrients by the slowly-circulating **aqueous humour**, whilst the cornea obtains oxygen directly from the atmosphere. The aqueous humour is a watery liquid containing salts, glucose and amino acids. It is continuously being secreted by part of the ciliary body and drains into a small vein in front of the iris. If drainage is impeded, the pressure in the eye rises, compressing the axons in the optic nerve and also the retinal arteries. If this is prolonged, a form of blindness called **glaucoma** may result.

The **vitreous humour** behind the lens has no nutritive function since it is jelly-like and does not circulate. Its incompressibility, together with that of the aqueous humour, helps maintain the shape of the eyeball.

## How the eye forms an image

Though the sensation of vision is experienced in the brain, the 'picture' we see is produced on the retina. To form an image of an object, rays of light reflected from it must be brought to a focus on the retina.

Light rays reflected from any point in the field of view are always diverging when they enter the eye, though rays from distant objects are almost parallel. To bring them into focus they must be made to converge on a single point on the retina. The result of producing a sharp image is that each photoreceptor in the retina receives light from a single point in the visual field.

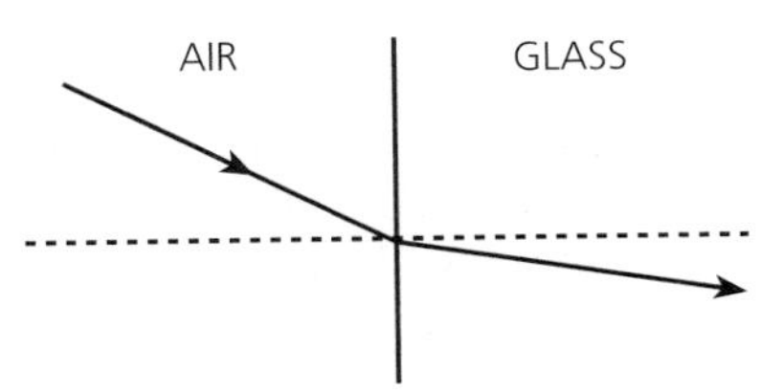


**Figure 18.10**
Refraction of light when it crosses a boundary between two transparent media, such as air and glass

Light normally travels in straight lines, but if it crosses from one transparent material to another it changes direction by **refraction**. The amount of refraction depends on two things:

- the **angle of incidence** (Figure 18.10). The greater the angle of incidence, the greater the change in direction of the light rays. When parallel rays strike the curved surface of a lens, the peripheral rays are refracted more than the central ones, with the result that all rays converge on the same point at the **focus** (Figure 18.11).

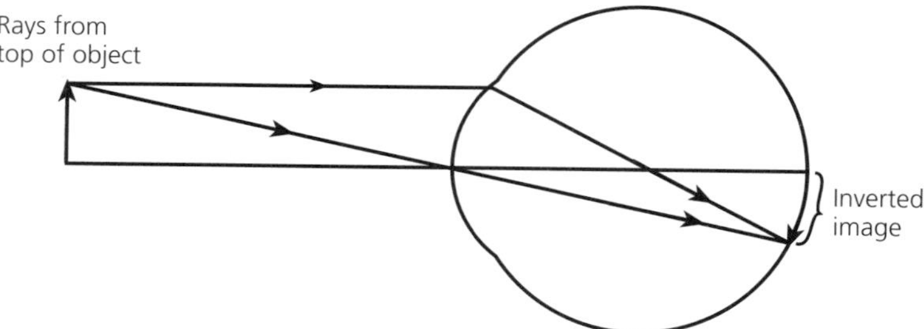


**Figure 18.11**
How an image is formed on the retina

- the difference in the **refractive index** of the two materials. The greater this difference, the more the light is bent.

The eye contains two 'lenses': the cornea and the lens itself. Since the greatest difference in refractive index is at the boundary of the air and cornea, most (about two-thirds) of the refraction occurs at the cornea. You can see the effectiveness of the cornea in focusing by opening your eyes under water. Objects appear blurred because water is much denser than air and has a refractive index closer to that of the cornea.

An interesting point to note is that although the image on the retina is **inverted** (upside-down and back-to-front), we don't see an inverted picture because the brain somehow 'turns' it the right way up.

To maintain a sharp image on the retina it is important that the shape of the eyeball is constant. This is ensured by the slight internal pressure in the eye.

### *The pupil reflex*

The pupil has two functions:

- it regulates the intensity of the light reaching the retina, helping the eye to adjust to different brightnesses.
- it increases the depth of focus when viewing near objects (see below).

The size of the pupil is controlled by two sets of antagonistic smooth muscle fibres: circular fibres, which make it smaller, and radial fibres, which make it larger (Figure 18.12). Both sets of fibres are under autonomic control.

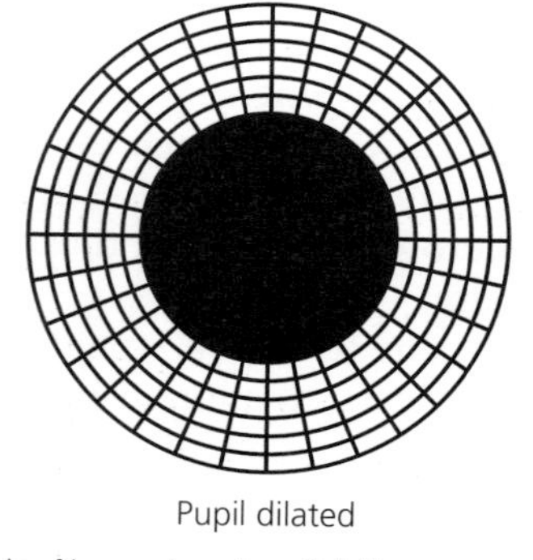

Pupil dilated
Circular fibres relaxed, radial fibres contracted

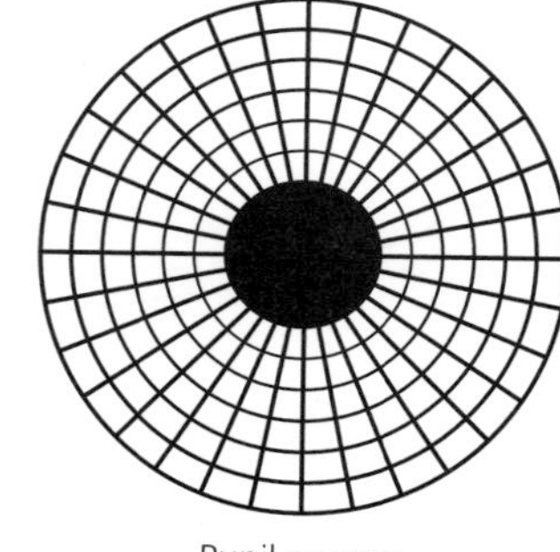

Pupil narrow
Circular fibres contracted, radial fibres relaxed

**Figure 18.12**
How the iris alters the size of the pupil

### Accommodation

Light rays from near objects are diverging more as they enter the eye than rays from distant objects. To be brought to a focus, rays from near objects must therefore be refracted through a greater angle than rays from more distant objects (Figure 18.13).

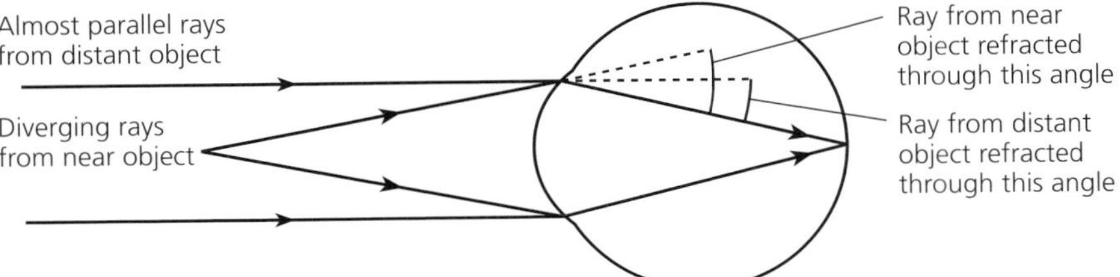


**Figure 18.13**
Why light rays are bent more when viewing near objects

To see a near object, the refracting power of the eye must therefore be increased. Of the two refractive surfaces, only the lens can change its curvature and hence its refracting power. This process is called **accommodation** and involves the intrinsic eye muscles (Figure 18.14).

When we view a distant object the lens is held under tension by the suspensory ligament. If this tension is relieved the lens becomes more convex, increasing the refracting power of the eye. This is what happens when you look at a near object; the circular muscle fibres in the ciliary body contract, decreasing its circumference and slackening the suspensory ligament. The contraction of the ciliary muscle during accommodation also stretches the choroid slightly, creating a tension that is important in returning the lens to a less convex shape.

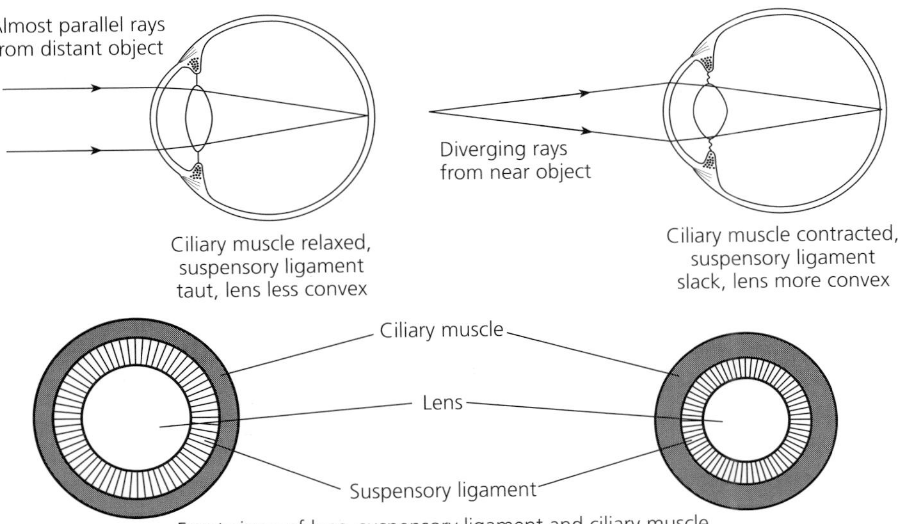


**Figure 18.14**
Mechanism of accommodation (with changes in the shape of the lens greatly exaggerated)

To change from near to distant vision the lens has to be made less convex. The ciliary muscle relaxes and the lens is pulled back into a less convex shape by the tension in the choroid (relaxation of the ciliary muscle cannot do this because muscle fibres cannot actively elongate).

A problem with near vision is that the depth of focus is reduced. This means that two objects only need to be at slightly differing distances from the eye for one of them to be slightly out of focus. This problem is reduced by the narrowing of the pupil when looking at near objects. This has the effect of increasing the depth of focus, making the eye more like a pinhole camera.

Table 18.2 summarises these changes.

**Table 18.2**
Changes that occur during accommodation

| | Distant objects | Near objects |
|---|---|---|
| Ciliary muscle | Relaxed | Contracted |
| Lens | Less convex | More convex |
| Suspensory ligament | Taut | Slack |

## Defects of vision

In a normal eye, rays from distant objects come to a focus on the retina with the ciliary muscles *relaxed*. There are various reasons why the eye may not focus the light properly. The commonest are described below.

### Short sight (myopia)

In short sight, the image is formed in front of the retina because the focusing power of the eye is too great (Figure 18.15). This can be caused by the eyeball being too long, or the curvature of the cornea and/or lens being too great. Distant objects cannot be seen clearly unless concave glasses are worn. Another effect of short sight is that the near point is less than normal – i.e. objects closer than normal can be seen clearly.

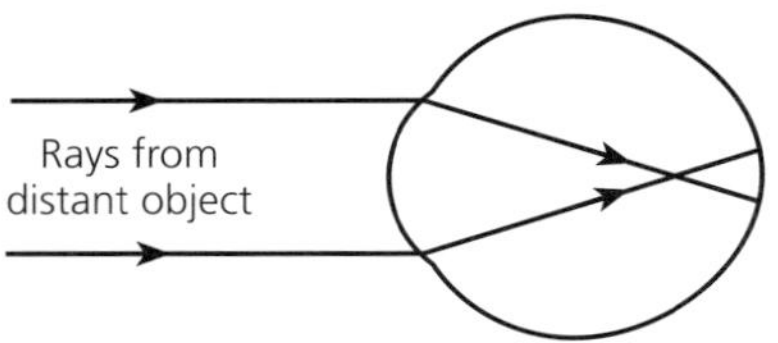


In myopia (short sight), distant objects cannot be seen clearly but near objects can be seen clearly closer to the eye than in people with normal vision.

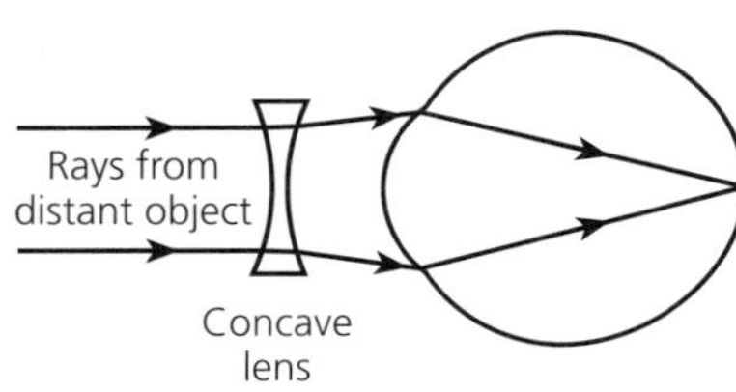


The remedy is to wear concave spectacles, which increase the divergence of the rays entering the eye.

**Figure 18.15▲**
Short sight and its correction

### Long sight (hypermetropia)

In long sight, the focusing power of the eye is insufficient (Figure 18.16). Either the eyeball is too short, or refractive surfaces are insufficiently curved. Even to see distant objects the converging power of the eye has to be boosted by contraction of the ciliary muscles, so the eye is never at rest unless convex lenses are worn.

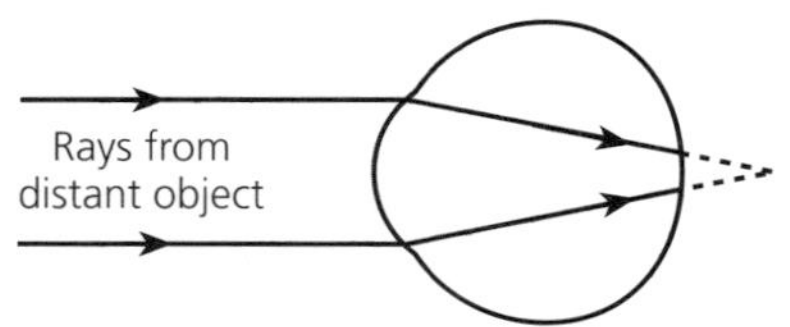


In long sight (hypermetropia) the eyeball is too short for its refracting power, so rays converge towards a point behind the retina. Distant objects can be seen clearly but only by using the ciliary muscles.

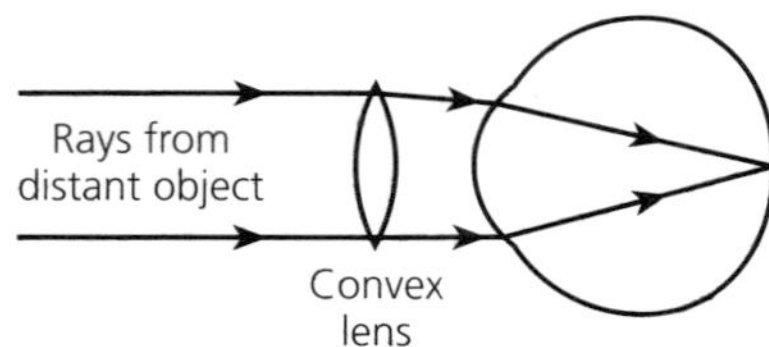


The remedy is to wear concave spectacles which increase the refracting power of the eye.

**Figure 18.16▲**
Long sight and its correction

### Loss of accommodating power

In middle age the lens becomes less elastic and is less able to become fatter when viewing near objects. The remedy is to wear convex lenses for reading. A person who is short-sighted and has lost some accommodating power may use **bifocal** lenses, the upper part concave for distant vision, and the lower part convex for near vision.

### Astigmatism

Astigmatism is caused by either the cornea or the lens (or both) being unequally curved in different directions. Instead of being curved like the surface of a soccer ball, it is more like part of a rugby ball. The remedy is to use a lens that is distorted in the opposite way.

## The retina

The retina is the photosensitive layer of the eye (Figure 18.17). Its outermost layer consists of a single layer of cells containing the pigment **melanin** which helps to prevent internal reflection. This layer extends forward to form the rear of the iris, contributing to its colour. Depending on the amount of pigment, the colour of the iris may appear various shades of brown, hazel or blue. Albinos have no melanin behind the iris (which thus appears pink) and the internal reflection reduces the ability to see detail.

Just inside the pigment layer is a layer of photoreceptors. These synapse with **bipolar neurons**, which in turn synapse with **ganglion cells** whose axons converge to form the optic nerve. At the extreme rear of the retina is a tiny depression called the **fovea,** caused by a thinning of

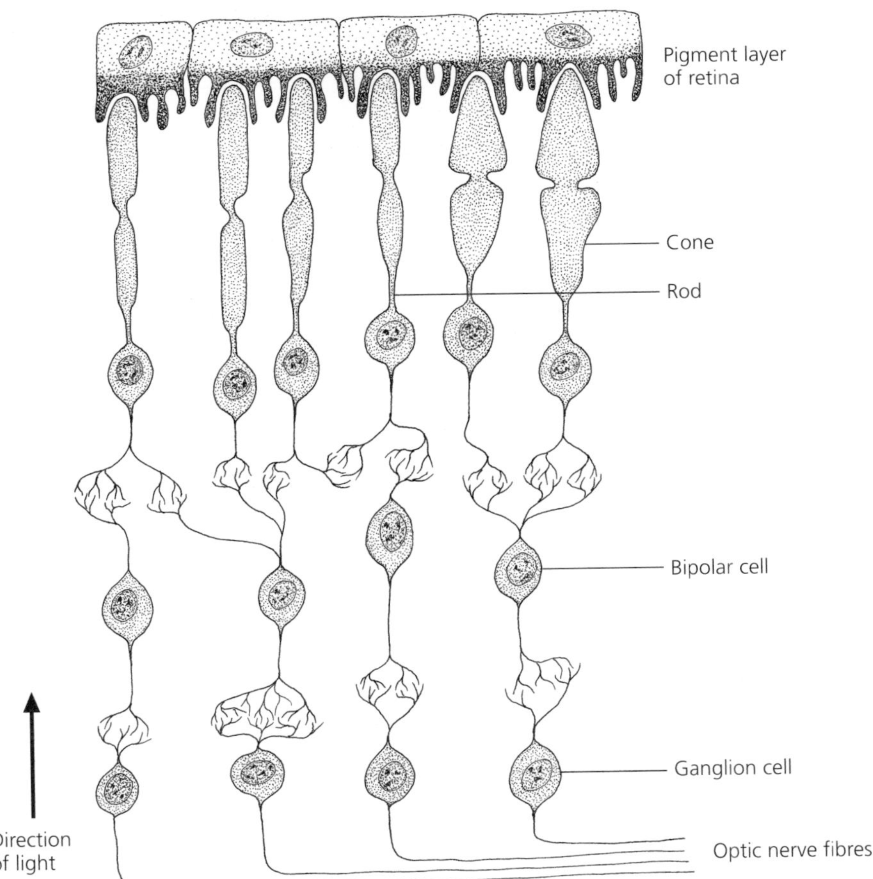


**Figure 18.17**
Structure of the retina

the neuronal component of the retina. For reasons to be discussed later, the fovea is very sensitive to detail. Just to one side of the fovea is the **optic disc** where the retinal blood vessels and nerve fibres leave the eye. Since there are no photoreceptors at this point, it marks the position of the **blind spot** (see below).

### *The inverted retina*

A peculiar feature of the vertebrate retina is that it is **inverted**, meaning that the photoreceptors are on the side furthest from the light.

The reason for this peculiar state of affairs lies in the way in which the vertebrate central nervous system develops, by the rolling up of a plate of cells to form a tube (Figure 18.18). The photoreceptors develop from cells that were originally on the outside of the plate, but which finish up on the *inside* of the tube.

The retina develops from a cup-shaped outgrowth of this **neural tube**. The outer layer of the cup develops into the pigmented layer of the retina and the inner layer develops into the nervous layer. As Figure 18.18 shows, the photoreceptors retain their original position, so the incoming light has to pass through nerve cells before it reaches them.

### *The blind spot*

An interesting consequence of the inverted retina is that to leave the eye, the nerve fibres must pass through the layer of photoreceptors, creating an area devoid of photoreceptors, the blind spot. You can demonstrate its existence by closing the left eye and looking directly at the cross in Figure 18.19. When the page is about 20 cm from your eye the spot seems to disappear.

**Figure 18.18**
Why vertebrates have an inverted retina

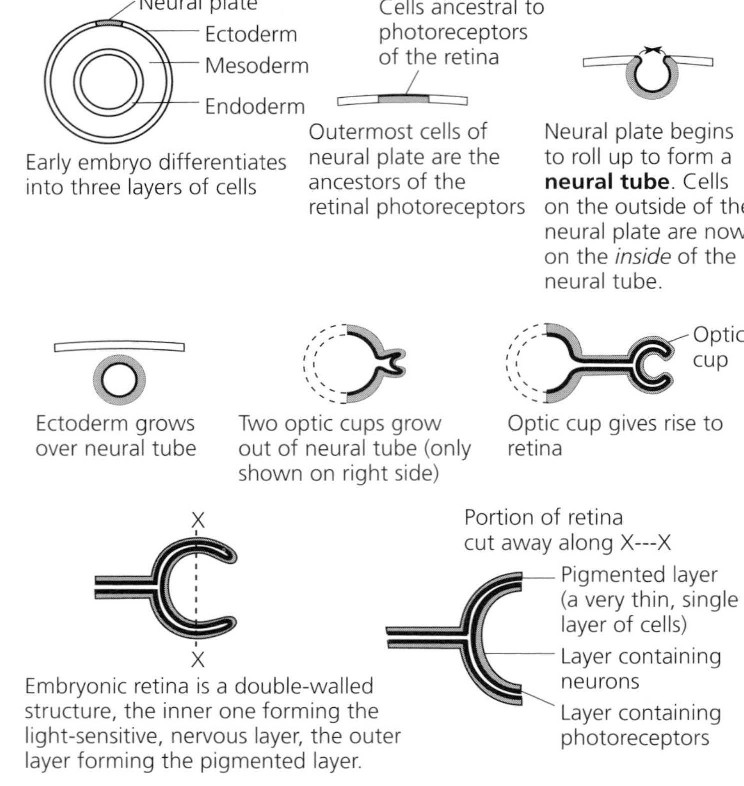


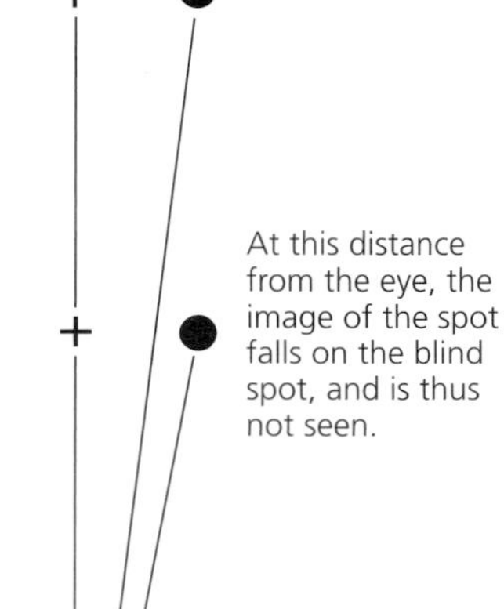


Neural plate
Ectoderm
Mesoderm
Endoderm
Early embryo differentiates into three layers of cells
Cells ancestral to photoreceptors of the retina
Outermost cells of neural plate are the ancestors of the retinal photoreceptors
Neural plate begins to roll up to form a **neural tube**. Cells on the outside of the neural plate are now on the *inside* of the neural tube.
Ectoderm grows over neural tube
Two optic cups grow out of neural tube (only shown on right side)
Optic cup
Optic cup gives rise to retina

Embryonic retina is a double-walled structure, the inner one forming the light-sensitive, nervous layer, the outer layer forming the pigmented layer.
Portion of retina cut away along X---X
Pigmented layer (a very thin, single layer of cells)
Layer containing neurons
Layer containing photoreceptors

At this distance from the eye, the image of the spot falls on the blind spot, and is thus not seen.

**Figure 18.19**
Demonstration of the blind spot

Close the left eye and look at the cross with the page at varying distances from the eye. In one position the spot seems to disappear from view.

## The photoreceptors

The inner layer of the retina contains both photoreceptors and neurons. In humans and other apes and in monkeys the photoreceptors are of two kinds – called **rods** and **cones** because of the shapes of their photoreceptive, outer segments (Figure 18.20).

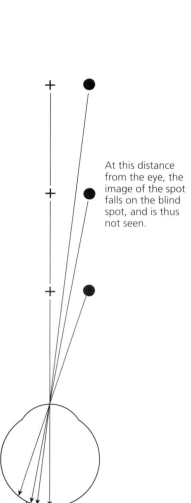

The two kinds of receptor serve different functions. Rods are involved in dim light vision in which we cannot see colour or detail, while cones operate in bright light and enable us to see colour and fine detail.

### *Rods*

The photosensitive region of a rod is the **outer segment**. This contains large numbers of free-floating flattened sacs formed by intuckings of the plasma membrane. The membranes of these sacs are packed with the pigment **rhodopsin**, or **visual purple**. This consists of a protein, **opsin**, joined to the pigment ***cis*-retinal**, which is synthesised from Vitamin A.

The location of the pigment in flat membrane stacks has an interesting, adaptive consequence. It results in all pigment molecules being oriented in the same plane, maximising their absorption of light.[1]

[1]In the eyes of bees and some other insects, pigmented molecules are oriented not only in the same plane, but in the same *direction*. This enables the insect to detect the plane of polarisation of the light – an important navigational aid when the sun is obscured.

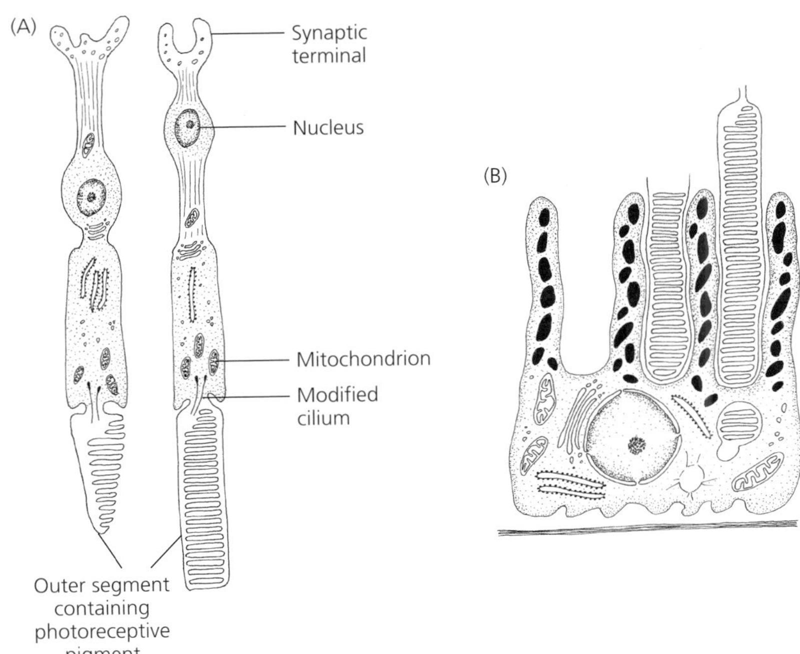


**Figure 18.20**
A rod and a cone (A). Outer segments of rods embedded in pigment cells, preventing reflection (B)

**CLEAR THINKING**

**Diagrams often show the eye in vertical section with the optic nerve below the fovea, yet the dot and cross test for the blind spot shows that the fovea and the blind spot are in the same *horizontal* plane. It follows that the optic nerve (which leaves the retina at the blind spot) and the fovea can only both be shown in a horizontal section of the eye. If the optic nerve were in the same vertical plane as the fovea, the dot would have to be vertically *above* the cross.**

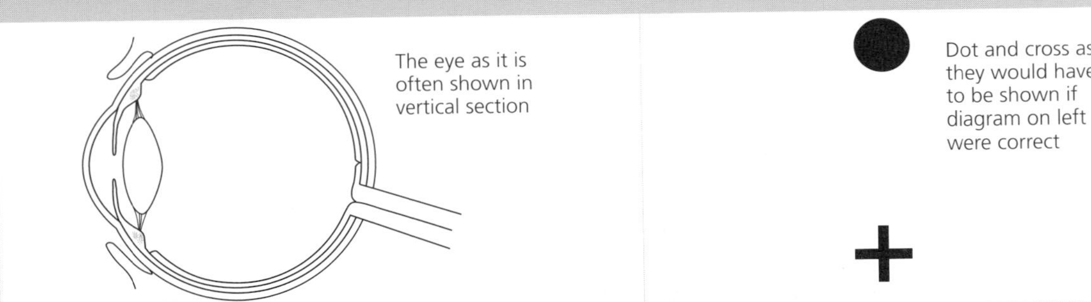


The outer segment is connected to the rest of the cell by a 'waist' consisting of a highly modified cilium. Though the retina is an odd place to find cilia, the evidence is convincingly provided by the presence of nine paired microtubules (although the central pair is absent).

In the inner segment are numerous **mitochondria** which supply energy for the regeneration of rhodopsin. At the end of the inner segment each rod makes a synaptic connection with a bipolar neuron. Many rods synapse with each bipolar cell, the importance of which will become apparent shortly.

### *From photon to receptor potential*

When a photon of light is absorbed by *cis*-retinal it is bleached via a number of intermediates to ***trans*-retinal**. This then separates from the opsin and triggers a complex series of events resulting in the development of a receptor potential (see information box on page 342). The *trans*-

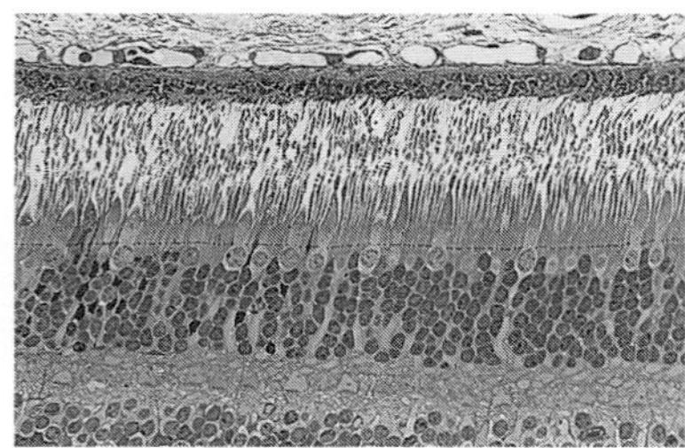

**Figure 18.21▲**
Light micrograph of the retina of the human eye. At the top is a layer of pigmented cells below which lie the cones and beneath these, the rods

retinal is then reconverted to *cis*-retinal using energy from ATP supplied by the mitochondria (Figure 18.22). The resynthesis of rhodopsin is one of the processes involved in the increase in sensitivity that occurs during adaptation to very dim light. Complete dark-adaptation may take half an hour.

Unusually amongst receptors, stimulation of a photoreceptor leads to a *decrease* in the permeability of the cell membrane to $Na^+$ ions, causing a **hyperpolarisation** (increase in membrane potential). The result is a *decrease* in the amount of transmitter released at synapses with adjacent bipolar neurons (Figure 18.23). The bipolar neurons are also unusual: they do not generate action potentials but, like the photoreceptors, undergo graded changes in their membrane potentials. Only in the ganglion cells is information encoded in the form of all-or-nothing action potentials which are then conveyed to the brain.

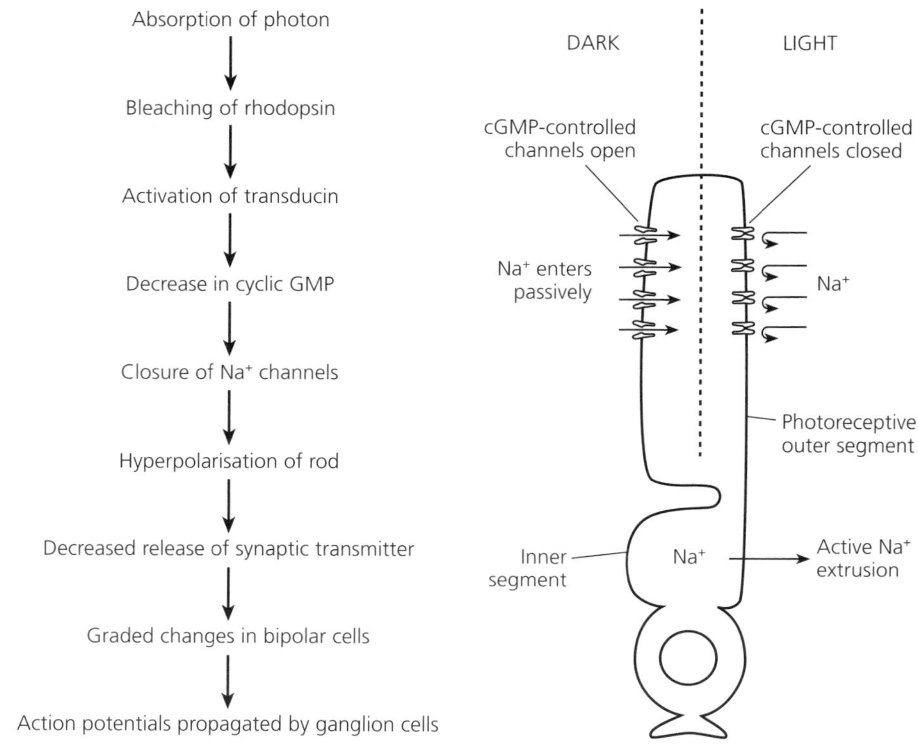


**Figure 18.22►**
Summary of chemical events after a rod absorbs a photon (left)

**Figure 18.23►**
Effect of light on rod cell membrane (right)

## Amplifying the signal

Rods are extremely sensitive to dim light – in fact it has been shown that a single photon causes a rod to respond. One reason for this extreme sensitivity is the ability of rods (and to a lesser extent, cones) to convert a weak incoming stimulus to a much stronger signal.

The amplification of the signal occurs in the following way. The bleaching of each rhodopsin molecule activates several hundred molecules of a protein called **transducin**. Each transducin molecule activates several hundred molecules of an enzyme that destroys **cyclic guanosine monophosphate**, or **cGMP**. Each enzyme molecule removes many cGMP molecules. Each cGMP molecule normally keeps a sodium channel open, so the destruction of cGMP closes the channels, hyperpolarising the membrane. The absorption of a single photon leads to the destruction of tens of thousands of cGMP molecules and the closing of a correspondingly large number of $Na^+$ channels in the rod membrane.

This kind of 'cascade' of events, in which the effect is multiplied at each step, is also a common theme in hormonal communication between cells.

## *Rods and twilight vision*

Rods are superbly adapted for night vision, in a number of ways:

- rods contain more photosensitive pigment than cones, increasing the probability that a photon will be absorbed.
- each rhodopsin molecule absorbs photons more readily than an iodopsin molecule (cone pigment).
- the effect of a photon is amplified by a cascade effect, which is greater in rods than in cones (see information box on page 342).
- receptor potentials in rods last much longer than in cones. Receptor potentials produced by successive photons can thus add together to produce a greater effect on the bipolar cell – an example of **temporal summation** (Chapter 17).

Another factor promoting sensitivity in dim light is the way rods are connected to the brain. Since there are about 120 million rods but only about 1 million axons in the optic nerve, many rods must share each 'line' to the brain. This **retinal convergence** enables the input of many rods to be added together (Figure 18.24). Many rods synapse with each bipolar neuron and many bipolar cells connect with each ganglion cell. Although one rod may only release a small amount of transmitter, the effects of a group of them can add together. This is an example of **spatial summation**, described in Chapter 17.

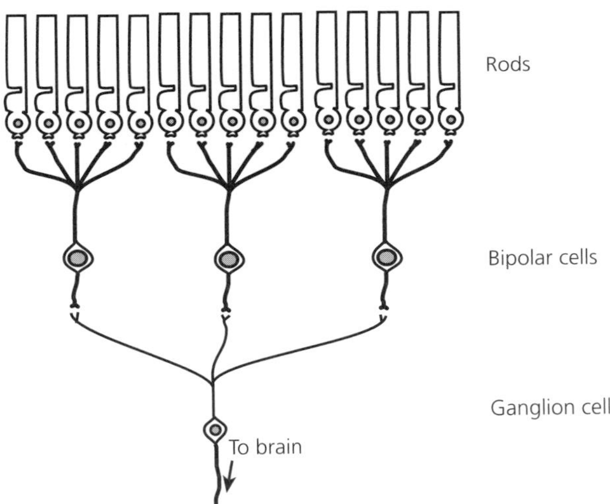


**Figure 18.24**
How retinal convergence promotes sensitivity to dim light

There is a cost of retinal convergence. The brain cannot tell which receptor of a group sharing the same optic nerve fibre is being stimulated. The result is that rods give poor **acuity**, or perception of detail.

## *Cones and the perception of detail*

Cones are outnumbered by rods by nearly 20 to 1. They are involved in daylight (photopic) vision and in the perception of colour and detail. Though present in many other vertebrates, cones are absent from most mammals, except for monkeys, humans and the other apes.

The photosensitive pigment is located in infoldings of the plasma membrane of the outer segment.

Cones have a number of features adapting them to vision in bright light:

- cone pigments absorb photons less readily than rhodopsin, so even in full daylight there is plenty of unbleached pigment.

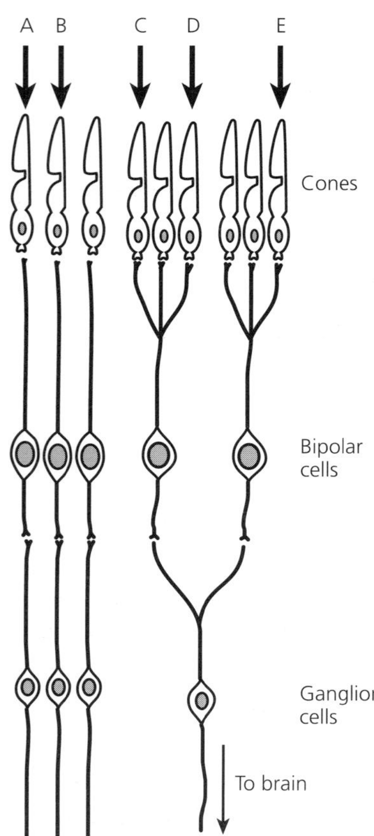


**Figure 18.25**
How reduced retinal convergence in cones increases visual acuity. Points of light at A and B can be distinguished at the fovea because the cones have separate lines to the brain and the stimulated cones are separated by an unstimulated one. C cannot be distinguished from D or E because, even though the stimulated cones are separated by unstimulated ones, they all share the same ganglion cell axon to the brain

- receptor potentials decay much more quickly in cones than in rods (about 20 ms compared with 200 ms), so temporal summation can only occur if photons follow in quick succession.
- their narrowness means that many can be packed into a small area. Note that this feature is only important when each photoreceptor has its own 'line' to the brain, as is the case in the cones at the fovea (see below). Rods are also very narrow, but since many share the same optic nerve axon, this has no significance.

In addition to these features of individual cones, fewer cones share each optic nerve fibre – in other words there is far less retinal convergence in cones than in rods. Cones thus provide a much more detailed picture. The principle is essentially the same as the two-point discrimination in the skin described earlier in this chapter. Figure 18.25 illustrates the point.

## Colour vision

A given photoreceptor can signal only one thing to the brain – stimulus intensity – it cannot signal wavelength. Colour vision (or the ability to discriminate between different wavelengths) requires more than one type of photoreceptor. Figure 18.26A shows the response curve for rods. Although green light stimulates the rods more strongly than blue light, strong blue light could give the same effect as weaker green light.

(B) shows the responses for two imaginary types of receptor, maximally sensitive to green and orange, respectively. Although the green-sensitive receptors are equally stimulated by 440 nm and 560 nm light, the orange-sensitive receptors are affected very differently. The brain thus perceives colour by comparing the frequencies of impulses from different kinds of photoreceptor. According to the now generally accepted **trichromatic theory**, colour vision in humans depends on sensory input from three kinds of cone. These are most sensitive to blue, green and red, respectively. By mixing light of these three **primary colours** and independently varying their intensities, any colour can be produced. The brain perceives colour by comparing the input from the three cone populations.

All three pigments contain retinal, but its light-absorbing properties are modified by the protein joined to it, which is different in each type of cone. If any of the three cone pigments is defective, a form of **colour-blindness** results. The deficiency lies in the protein part of the pigment (opsin) and is hereditary. The most common form of colour-blindness is the inability to distinguish red from green. There are actually two forms of red-green colour-blindness due to defective green-sensitive and red-sensitive cones, respectively. (Can you think of a way in which the two types could be distinguished?) Inability to see blue is a rare third kind.

## Regional differences in the retina

Everyday experience suggests that different parts of the retina are specialised for different functions, for example:

- you can only see detail in the centre of your field of view.
- to see a very faint star you need to look slightly to one side of it.
- colour discrimination decreases sharply in the periphery of the visual field.

These observations can be explained in terms of variations in the organisation of photoreceptors and their nervous connections.

When you look directly at a small object its image falls on the **fovea**. This is adapted for high acuity vision in a number of ways:

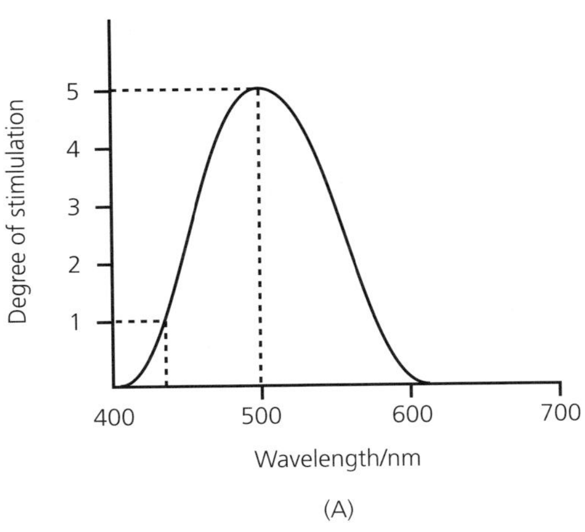


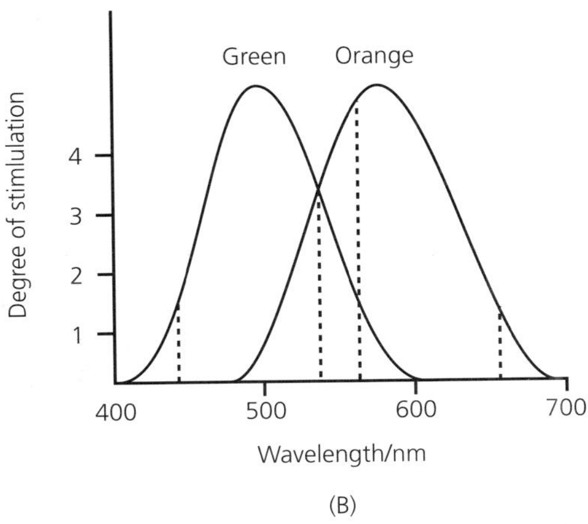


**Figure 18.26**
Why more than one kind of photoreceptor is needed for colour vision. (A) shows the spectral sensitivity curve for rods. 500 nm light would stimulate rods as strongly as 435 nm light of five times the intensity. (B) shows the response curves for two imaginary kinds of cone. With light of any wavelength but 540 nm they would be stimulated to different extents, and so could be distinguished. Light of 540 nm stimulates the two kinds of cone equally, and so could be duplicated by a mixture of 440 nm and 660 nm light

- it contains cones only, and these are more slender (allowing more cones per unit area) than in other parts of the retina.
- in the centre of the fovea each cone has its own 'private' line to the brain – there is no convergence. The receptive field in this part of the retina is thus very 'fine-grained'.
- retinal blood vessels go round the fovea, so the blood does not interfere with light transmission.
- the bipolar and ganglion cells radiate away from the centre of the fovea, creating the characteristic foveal depression (Figure 18.27). Because of the thinning of the retina here, light reaches the cones almost unimpeded.

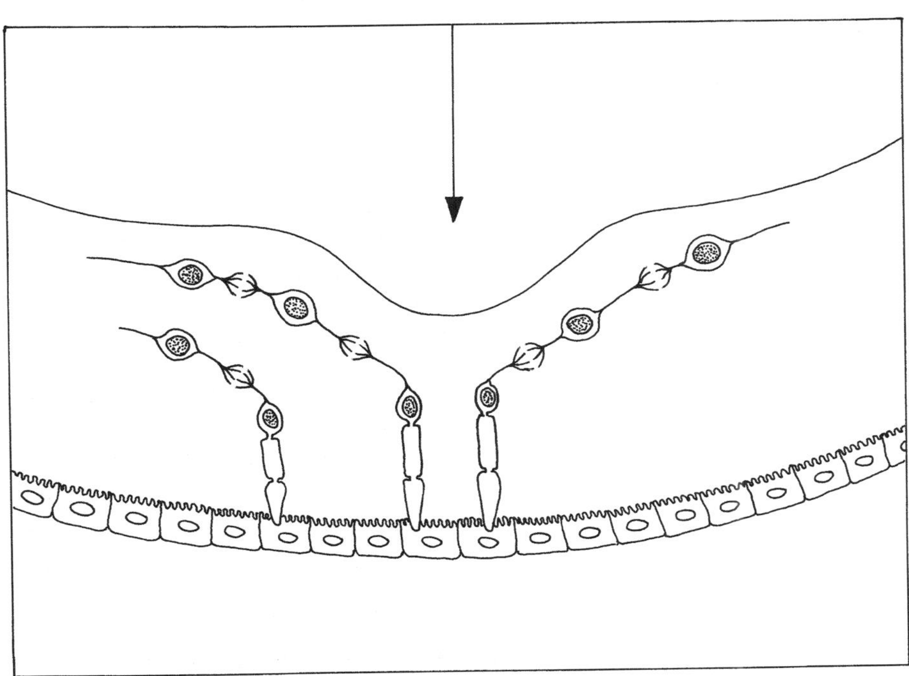

**Figure 18.27**
How light reaches the foveal cones unimpeded by neurons

With no rods, and not even spatial summation amongst the cones, the fovea needs bright light to function. This explains why a very faint star cannot be seen by looking directly at it. When you look slightly to one side of it, its image falls outside the fovea where there are rods.

Just outside the fovea, rods begin to make their appearance, and beyond 10° from the fovea cones become much less common. Towards the periphery of the retina cones are almost absent, which is why we cannot perceive the colour of objects at the edge of the visual field.

### *Information processing by the retina*

If you look at Figure 18.17 you will notice that there are lateral connections between photoreceptors and neurons. This is related to the fact that the retina is not merely a light-receiving and relay station – it is also responsible for preliminary processing of the visual information. This is accomplished by neurons such as bipolar and ganglion cells.

An example of this processing is the enhancement of contrast where there is a boundary between regions of different brightness or colour. A strongly stimulated neuron may *inhibit* its lateral neighbours, thus sharpening the boundaries between different parts of the visual field.

## Role of the brain

Though an image is formed on the retina, it does not send an actual picture to the brain. The situation can be likened to a newsprint photograph which is made up of thousands of tiny dots. The size of each dot is an indication of the light coming from the corresponding point in the object of the photograph – the larger the dot, the lower the intensity of light coming from that point of the object. Individually each dot means nothing, but collectively they form a picture.

In a similar way the frequency of impulses in an optic nerve fibre is an indication of the intensity of light stimulating the photoreceptors 'feeding' it. The brain then somehow assembles the information into a mental picture. To do this, each part of the visual cortex in the cerebrum must correspond to a particular part of the retina.

### *Stereoscopic vision*

Whilst the perception of the colour, shape and size of an object only needs one eye, to accurately judge its distance you need two (there are no top one-eyed tennis players or cricketers).

In a carnivorous animal such as a hawk or cat, and in monkeys and apes, the ability to judge the distance of an object is clearly important. It enables a carnivore to judge the distance to the prey, and a monkey to judge the distance to another branch.

Distance perception, or **stereoscopic vision**, depends on the fact that eyes facing the front have overlapping fields of view, so each eye sees a slightly different picture (Figure 18.28). By comparing the two pictures the brain can judge the distance of the object – the nearer it is, the greater the difference between the two views. Of course we don't see two pictures; in some way the brain merges them into one.

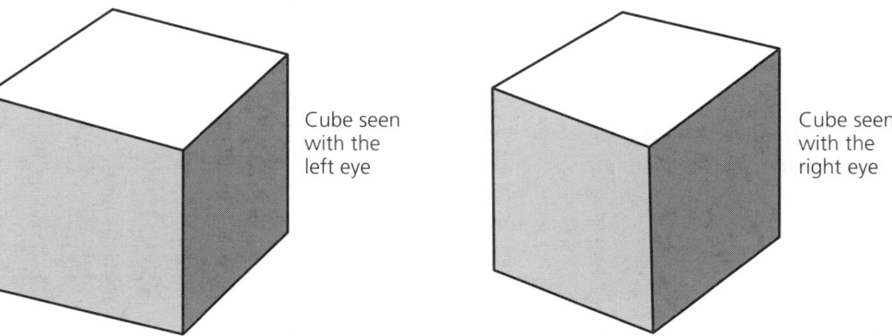


**Figure 18.28**
How an object appears slightly different to each eye

The brains of primates (monkeys, humans and other apes) compare the input from the two eyes in a different way from other vertebrates (Figure 18.29). In non-primates the left side of the brain receives fibres from the right retina, and vice versa, because the optic nerve fibres cross over at the optic chiasma. To compare the input from the two eyes, the two halves of the brain must therefore communicate with each other.

In primates all the fibres from the left side of both retinas go to the left side of the brain, and vice versa. The result is that each half of the brain receives information from both eyes.

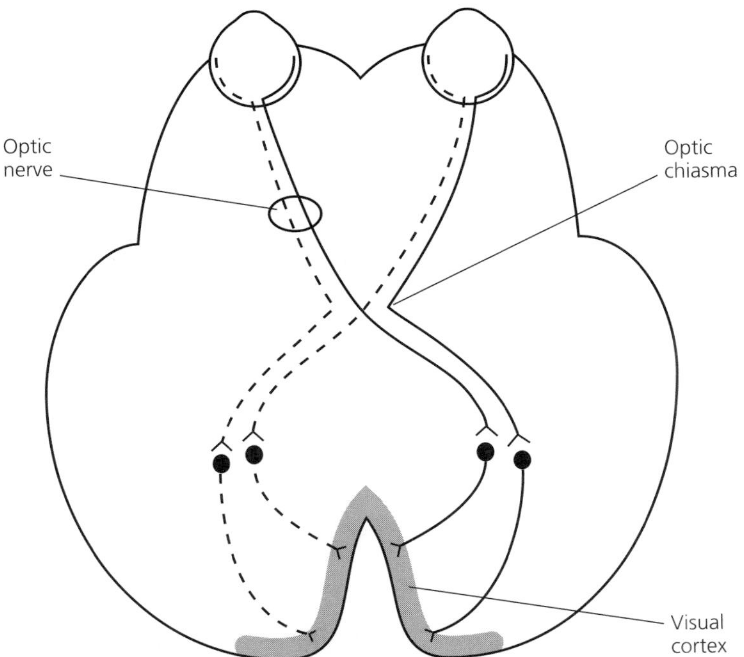


**Figure 18.29**
The course of the optic nerve fibres in a primate, as seen from below

## QUESTIONS

1. What is wrong with the phrase 'taste receptor'?
2. What is wrong with the statement that when changing from near to distant vision, 'the relaxation of the ciliary muscle pulls the lens into a flatter shape'?
3. Who would be able to see more clearly under water without goggles: a short-sighted person, a long-sighted person or a person with normal vision? Explain your answer.
4. It takes about 30 minutes for the retina to adapt fully to very dim light. During the Second World War, night-fighter pilots relaxed in bright red light while they waited for action. Use Figure 18.26A to explain how this would greatly improve their ability to see in the dim light of the night sky.
5. The diagram shows how the frequency of impulses from the retina varies with the intensity of the incident light. Use the information in the graph to suggest an explanation for the fact that when using a light microscope, contrast may be increased by closing the iris diaphragm slightly.

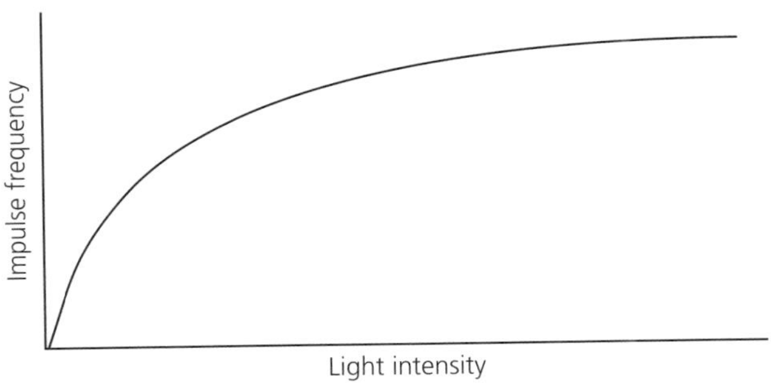

CHAPTER 19

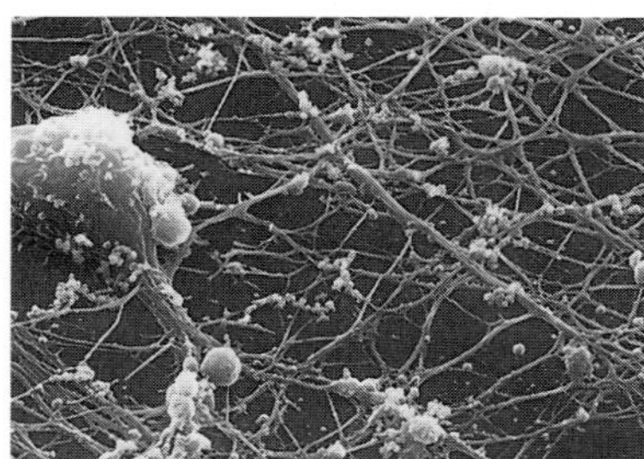

# MOVEMENT and SUPPORT in ANIMALS

## LEARNING OBJECTIVES

By the time you have completed your study of this chapter you should be able to:

- describe the structure of skeletal muscle and explain the pattern of banding in terms of its constituent proteins.
- explain the sliding filament hypothesis of muscle contraction.
- describe the sequence of events between the arrival of a nerve impulse and the end of a muscle contraction.
- describe how the contraction of an isolated muscle can be studied, and explain the effects of varying the intensity and frequency of stimuli.
- show how voluntary muscle contraction can be explained in terms of the action of its constituent muscle fibres.
- explain the roles of creatine phosphate and the formation of lactate in enabling skeletal muscles to work extremely hard for short periods.
- distinguish between jointed and hydrostatic skeletons and explain their functions.
- explain how the microscopic structure of compact bone adapts it to its function.
- describe how the macroscopic structure of a long bone and a lumbar vertebra adapt them to their functions.
- describe the structure of a synovial joint and explain how it is moved by muscles.
- describe how levers can modify muscular forces, giving examples.
- describe the essential features of the arthropod skeleton and compare and contrast it with the vertebrate endoskeleton.
- with reference to an earthworm, describe the role of a hydrostatic skeleton in movement, and explain its limitations compared with a jointed skeleton.

Of all the features that people most readily identify with animal life, movement is the one that first comes to mind. Movement in animals is intimately linked to their mode of nutrition; the need to find a meal while at the same time avoid becoming someone else's meal.

Animals have two kinds of propulsive system: **cilia** and **muscles**. Cilia act by lashing against the surrounding water, but they are ineffective in animals larger than flatworms for reasons explained in the information box on page 350. In all larger animals, propulsion is by muscles. Unlike cilia, which are situated on the body surface and push directly against the surroundings, muscles are on the inside and act by *pulling*. A muscle can thus only move the body if there is some means of converting its pull into a *push* against the outside world. The conversion of pulls to pushes is one of the fundamental functions of a skeleton. Muscles and skeletons thus work as an integrated whole, and, for this reason, are dealt with in the same chapter.

# Structure of skeletal muscle

As its name suggests, skeletal muscle tissue is anchored to bone. It is also called **striated** muscle because of the bands or striations running across its fibres. In humans, it is called **voluntary** muscle because it is under the control of the will, though this term is inappropriate for other species.

Skeletal muscle contracts quickly but soon tires. Although most skeletal muscles produce movement, they can also act to *prevent* it. We would, for example, be unable to maintain upright posture were it not for the sustained contraction of abdominal and back muscles. In some cases a muscle contracts while being *lengthened*, for example when you lower a weight, the muscle acts as a brake rather than an engine.

Skeletal muscle, or 'meat', consists of bundles of **muscle fibres** bound together by connective tissue continuous with the tendon which anchors the muscle to the bone. Branching within the connective tissue are blood vessels and nerve fibres (Figure 19.1).

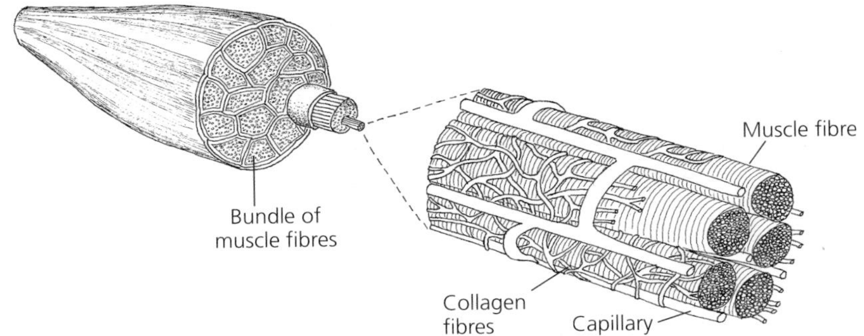


**Figure 19.1**
Structure of skeletal muscle at the level of resolution of the light microscope

A skeletal muscle fibre develops by fusion of many cells in the embryo. The result is a **syncytium**, where many nuclei share the same cytoplasm. When you consider that muscle fibres may be up to 100 μm wide and in some muscles may be as long as 30 cm, it is clear that there is far too much cytoplasm for one nucleus to control. The nuclei are only just beneath the surface of the fibre where they do not interfere with the action of the contractile mechanism below. Mitochondria are also a prominent feature of skeletal muscle fibres.

Just as we all have about the same number of fat cells we are born with, our skeletal muscle fibres do not multiply either – training simply increases their diameter.

The plasma membrane is modified to form the **sarcolemma** and extends as a continuous sheath along the entire length of the fibre. Like an axon membrane, the sarcolemma can propagate action potentials which activate the fibre.

The most characteristic microscopic feature of skeletal muscle fibres is the repeated pattern of light and dark bands crossing each fibre. Under the light microscope the bands seem to cross the entire width of the fibre, but the electron microscope has shown this to be an illusion.

Running the length of each fibre are many **myofibrils**, 1–2 μm wide (Figure 19.2). Each consists of alternating lighter **I bands** and darker **A bands**. Across the middle of each I band is a **Z disc**. In an extended muscle fibre, each A band contains a paler **H zone** with a narrow **M zone** in the middle. The region between adjacent Z discs is called a **sarcomere**, and when fully extended is about 2.5 μm long. Since the bands in adjacent sarcomeres are more or less level with each other, they *appear* to extend all the way across the fibre.

**Figure 19.2►**
Structure of a skeletal muscle myofibril in side view (A) and in transverse sections at three levels (B)

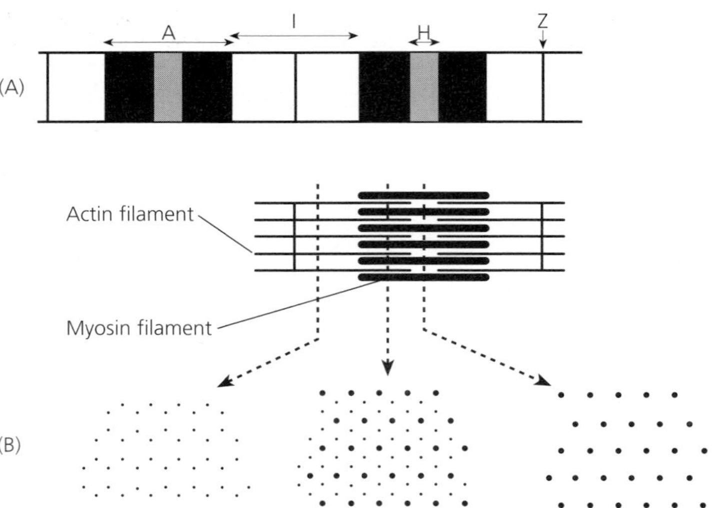

**Figure 19.3▲**
Electron micrograph showing longitudinal section through several adjacent myofibrils

The explanation for the bands lies in the organisation of the protein filaments making up each sarcomere. A longitudinal section shows many filaments of two kinds. The thick filaments extend the length of the A band whilst the thin ones extend from the Z disc to the edge of the H zone.

Transverse sections at different levels in a sarcomere confirm this. Sections through an I band show thin filaments, sections through the H zone show thick filaments only, whilst sections through an A band just to one side of the H zone show both thick and thin filaments.

The discovery of the two kinds of filament was made in the early 1950s. A couple of decades earlier it had been shown that a large proportion of the total protein of muscle is accounted for by two proteins – **actin** and **myosin**. Each protein can be selectively extracted using appropriate solvents. When myosin-free fibres were examined under the electron microscope it was found that the A bands had disappeared. If, on the other hand, actin was removed, a large part of the matter in the I bands had disappeared. The conclusion was clear – the thin filaments contain actin and the thick filaments contain myosin.

## Muscles vs cilia

Cilia and flagella (undulipodia) are very common locomotory organelles in unicellular organisms. Cilia also propel some animals, such as the planktonic larvae of many annelids and molluscs and the adults of free-living flatworms, but all larger animals are propelled by muscles.

The chief limitation of cilia stems from the fact that they are *surface* structures. The number of cilia depends on the surface area of the body, but the bulk to be moved depends on the volume of the body. With increase in size, bulk increases faster than surface area, so the number of cilia fails to keep pace with the load to be moved. Larger animals could not accelerate or change direction quickly enough if they were powered by cilia.

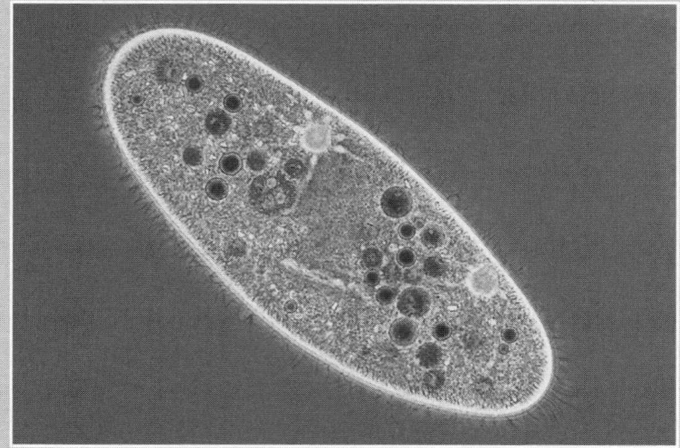

Light micrograph of the ciliate protozoan *Paramecium sp.*

**Muscles vs cilia *continued***

Another limitation of cilia is that although an increase in body size increases the number of cilia and thus the total force the cilia develop, their speed of beat is unaffected. A muscle fibre, on the other hand, consists of sarcomeres that are *in series* with each other, so their individual speeds of shortening add together. A long muscle can thus shorten more quickly than a short one, since it has more contractile units. If an animal doubles in size but its proportions remain unchanged, the cross-sectional area of its muscles – and thus the number of muscle fibres – increases four times. Its muscles could thus pull four times harder. Since the fibres would be twice as long, they could shorten twice as quickly.

The power developed by a muscle is given by its speed of shortening multiplied by the force it develops. If an animal doubles in length and keeps the same shape, the power output of its muscles thus increases $4 \times 2 = 8$ times. Muscle power thus increases in direct proportion to the cube of the increase in body length, but ciliary power increases in proportion to its square.

## The sliding filament hypothesis

So much for the arrangement and chemical identity of the filaments. What is their role in contraction? There are two alternative possibilities: either the filaments shorten, or they remain the same length and slide between each other.

The question was resolved by Hugh Huxley and Jean Hanson of King's College London, and also by Andrew Huxley and R. Niedergerke of Cambridge University. In 1954, working independently, the two groups proposed the **sliding filament hypothesis**.

The evidence was based on measurements of the lengths of the various bands after a muscle fibre had contracted or been extended to different degrees. Two key facts emerged. First, during contraction the A bands remain the same length, as does the distance between the Z disc and edge of the H zone. This shows that neither the thick filaments nor the thin filaments change in length. Second, the H zone and the I band shorten to the same extent, suggesting that during shortening the thin filaments on each side of the H zone slide towards each other.

The big question however still remained: how is the force of contraction developed? Two important clues were already available. First, in 1939 it had been discovered that myosin had **ATPase** activity – that is, it can catalyse the hydrolysis of ATP. Second, in the 1940s Albert Szent-Györgyi had shown that when solutions of actin and myosin are mixed they spontaneously combine to form an **actomyosin complex**. Moreover, when ATP is added the complex breaks down to actin and myosin again. It thus seemed likely that contraction of muscle involves some form of chemical reaction between actin and myosin involving the breakdown of ATP by myosin.

A further clue was provided by electron micrographs of muscle that had been depleted of ATP. These showed the presence of cross-bridges extending from myosin filaments to actin filaments about every 6 nm. If the chemical bonds between actin and myosin were alternately broken and reformed further along, each myosin filament would, in effect, 'walk' along an actin filament. To see how this happens we must first take a more detailed look at the structure of the two kinds of filament.

### *The thick filaments*

Each thick filament is 1600 nm long and 15 nm thick, and consists of many individual myosin molecules. A myosin molecule is like two

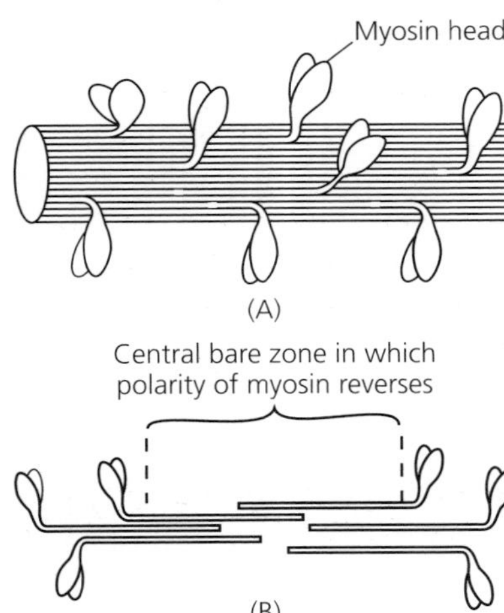


**Figure 19.4**
(A) A small part of a thick myofilament; (B) the reversal of polarity that occurs in the middle of the thick filament

tadpoles with their tails wrapped round each other (Figure 19.4). All the myosin molecules have their heads facing the end of the filament, so there is a **bare zone** in the centre where there are no heads. This corresponds to the M zone in the centre of the H zone.

The head is the part that combines with actin and contains the site of ATPase activity. The function of the tails is to anchor the molecule to the other myosin molecules.

### The thin filaments

The thin filaments are about 6 nm thick and about 1000 nm long. They are more complicated than the thick ones because they consist of significant amounts of two other proteins besides actin. Actin takes the form of two chains of globular proteins wrapped round each other (Figure 19.5). Individual globular protein units are called **G-actin** (G for 'globular'), each with a myosin binding site. The assembled helical filament is called **F-actin** (F for 'fibrous').

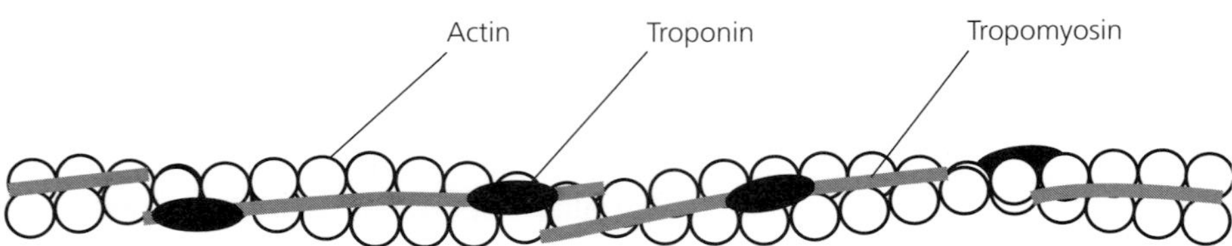


**Figure 19.5**
Fine structure of a thin myofilament

Other proteins include **tropomyosin** and **troponin**, which are involved in the switching on and off of contraction rather than contraction itself. Tropomyosin molecules are rod-like and fit at intervals into a groove between the two chains of actin. In resting muscle they prevent the myosin from combining with the actin by covering the part of an actin molecule which combines with myosin. In active muscle the tropomyosin is displaced to one side, enabling myosin to combine with actin.

Each tropomyosin is linked to a molecule of troponin. In a way to be described shortly, the arrival of a nerve impulse causes troponin and its associated tropomyosin to be displaced slightly, allowing myosin to interact with the actin.

### Myosin 'walking'

For the Z discs at the ends of a sarcomere to be pulled towards each other, the two halves of each myosin filament must 'walk' in opposite directions. The two halves of each myosin filament must therefore have opposite polarity or 'endedness', as must the thin filaments in each half of a sarcomere.

The myosin molecules are aligned head to tail, so each myosin molecule faces the direction in which it 'walks'. A similar polarity exists in the actin filaments which point in opposite ways in the two halves of a sarcomere.

How do myosin molecules 'walk' along actin? Some of the details are not yet understood, but there is general agreement that it involves repeated cycles of the following events (Figure 19.6):

- each myosin head briefly combines with an actin molecule.
- the orientation of the head then changes, causing it to pull the actin past it.
- ATP then combines with myosin, displacing the actin.
- the ATP is hydrolysed to ADP and phosphate.

- myosin recombines with actin, beginning another cycle.

It is worth noting that ATP is necessary to *break* the cross-bridges between actin and myosin. Inadequate supplies of ATP affect a muscle's ability to relax. This can be seen when a muscle is worked to exhaustion; its relaxation is slowed as well as its contraction. After death, the ATP in the muscles slowly breaks down, with the result that myosin and actin remain permanently linked by cross-bridges. This is the cause of **rigor mortis** – the stiffening of muscles after death.

## More evidence for the sliding filament hypothesis

**If the sliding filament hypothesis is true, then the maximum tension that a muscle fibre can develop should be proportional to the number of cross-bridges formed, and thus on the length of overlap between thick and thin filaments. The graph shows how the maximum tension developed by a muscle fibre varies with sarcomere length. Each part of the graph can be related to one of the accompanying diagrams. At the maximum length, overlap is minimal, as is tension. At progressively shorter sarcomere lengths, the tension increases until the actin filaments meet the bare zone of the thick filaments. Here there are no myosin heads and no more cross-bridges can be formed. At shorter lengths still, actin filaments begin to overlap with myosin in the other half of an A band, where the myosin heads are the 'wrong' way round.**

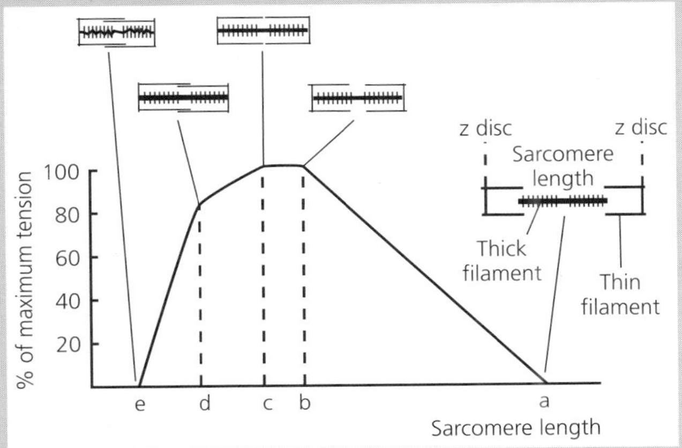


How maximum muscle tension varies with sarcomere length. Each change in gradient corresponds to one of the figures. Tension rises steadily between (a) and (b) as the overlap between actin and myosin increases. Between (b) and (c) there is no further increase since no more cross links can be formed. Between (c) and (d) tension declines because of interference between overlapping actin. Beyond (d) myosin filaments enter other half of sarcomere, in which polarity of thin filaments is reversed and cross links cannot be formed

## Controlling contraction

A muscle can only be useful if the contractile mechanism can be switched on and off. How does the arrival of a nerve impulse at the neuromuscular junction (Chapter 17) lead to the combination between actin and myosin, and how is the process switched off?

The 'decision' that leads to contraction is not taken in the muscle itself, but in a motor neuron in the central nervous system. Each neuron innervates a number of muscle fibres scattered throughout the muscle. Since these act together they form a **motor unit** (Figure 19.7). A muscle contains many motor units, so it is under the control of many motor neurons.

In muscles carrying out finely-graded contractions, each motor unit usually has only a small number of fibres. In the muscles of the larynx, for example, each motor unit contains two or three muscle fibres. At the other extreme are the large limb muscles, in which each motor unit may contain over 1000 muscle fibres.

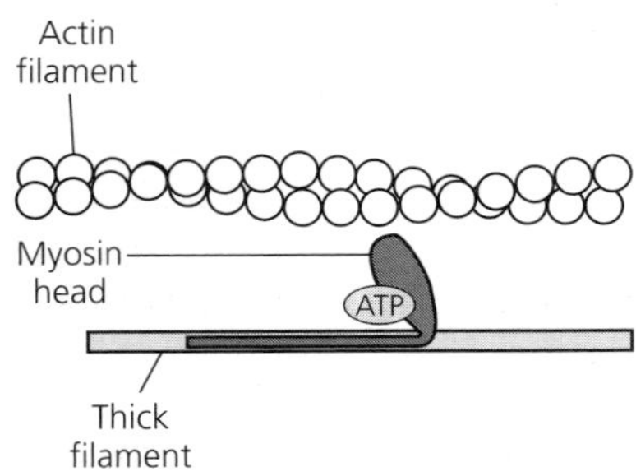


1. In relaxed muscle, ATP is bound to myosin, preventing it combining with actin.

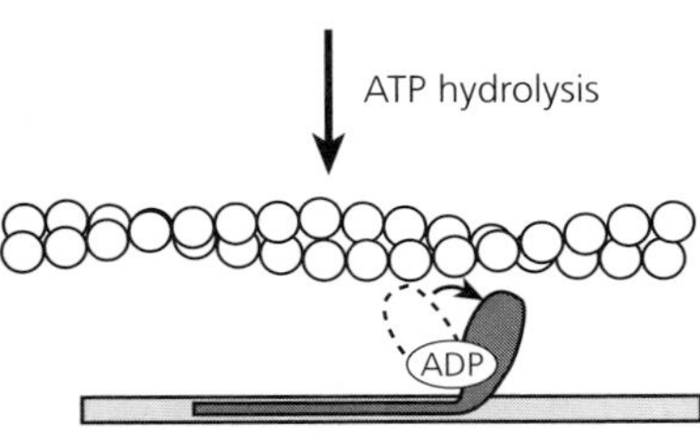


2. ATP is hydrolysed and tip of myosin head swings to new position.

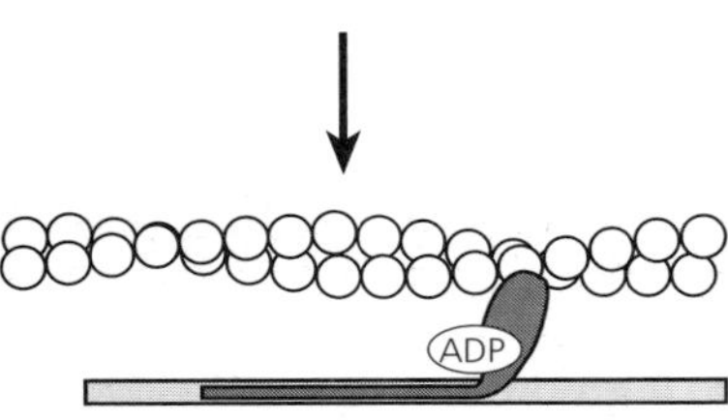


3. Myosin head forms new link with actin.

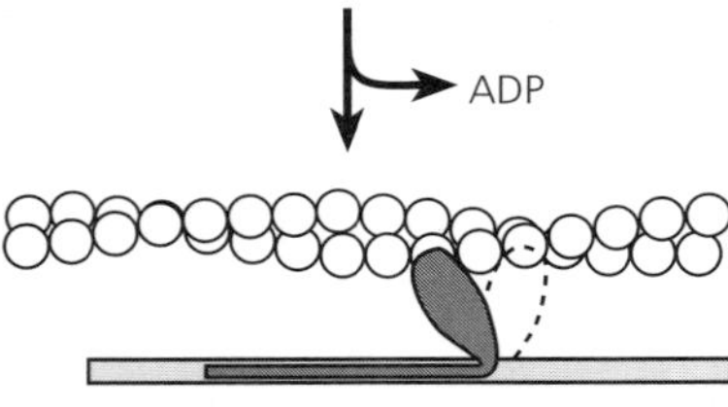


4. Maintaining its attachment to actin, myosin head swings to new position.

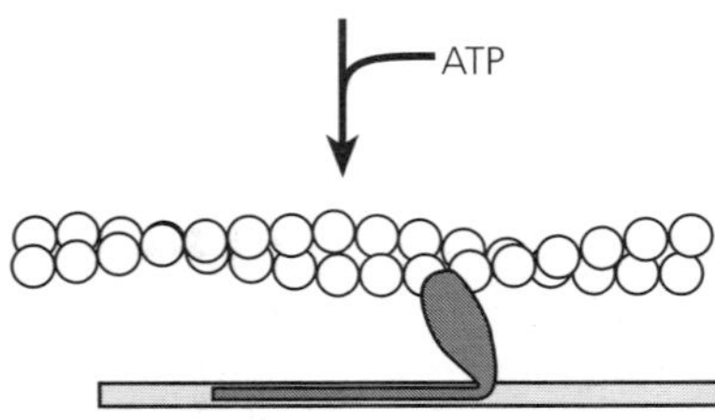


5. Myosin binds to actin and new cycle begins.

**Figure 19.6▲**
Cycle of events during myosin 'walking'. For simplicity, the myosin heads are shown as single structures

The arrival of a nerve impulse at a neuromuscular junction causes the release of acetylcholine. Unlike synapses between two neurons, a single impulse is always sufficient to trigger an action potential in the sarcolemma. The impulse sweeps in both directions away from the neuromuscular junction at about 5 m $s^{-1}$. Though slower than most nerve impulses, this is fast enough to ensure that all parts of the fibre receive the signal almost simultaneously.

So much for how the stimulus spreads along the fibre; how does it reach the contractile machinery within? This is the role of two closely-related sets of tubules: the **transverse tubules**, or **T-system**, which extend across the fibre, and the tubules of the **sarcoplasmic reticulum** which extend lengthways (Figure 19.8).

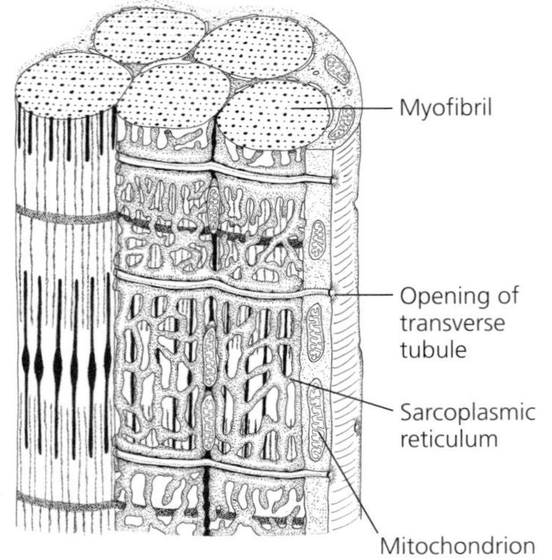


**Figure 19.7◀**
The T-system and the sarcoplasmic reticulum (after Crouch)

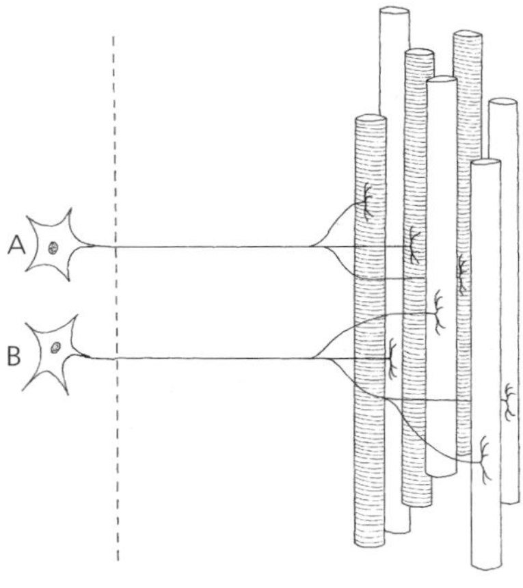


**Figure 19.8▶**
Motor units. All the muscle fibres innervated by neuron A are shaded and form a single functional unit. Non-shaded fibres constitute another motor unit

The T-system consists of deep intuckings of the sarcolemma extending into the interior of the fibre. In mammals, the T-tubules open at points level with the ends of the A bands. Inside the fibre the T-tubules come into close contact with tubules of the sarcoplasmic reticulum, which is the endoplasmic reticulum of the muscle fibre. The sarcoplasmic tubules form a longitudinal network round each myofibril, so no myofibril is further than about 1 μm from the nearest sarcoplasmic tubule. In a resting muscle fibre the tubules of the sarcoplasmic reticulum accumulate $Ca^{2+}$ ions, so their concentration in the sarcoplasm is low – too low to trigger contraction.

As an impulse sweeps along the sarcolemma, it is conveyed into the fibre along the T-tubules. This causes the tubules of the sarcoplasmic reticulum to release $Ca^{2+}$ ions into the sarcoplasm. These combine with troponin, causing it and its associated tropomyosin to be displaced, enabling myosin to interact with actin.

## The behaviour of whole muscle

Muscle can be kept alive for several hours in a solution of suitable ionic composition (e.g. Ringer's solution). Figure 19.9 shows the apparatus used to study the behaviour of an isolated muscle. A favourite muscle used by physiologists is the gastrocnemius, or calf muscle, of a frog. Since it is small, it has a reasonably large surface area through which oxygen can be absorbed. Furthermore, muscles of ectothermic ('cold-blooded') animals survive longer outside the body than those of mammals or birds.

One of the muscle tendons is anchored with a pin, whilst the other is attached to a lever and pen that makes a trace on a revolving drum, or **kymograph** (Figure 19.9). The muscle is stimulated electrically, either directly by placing the electrodes on the muscle, or indirectly via the nerve supply.

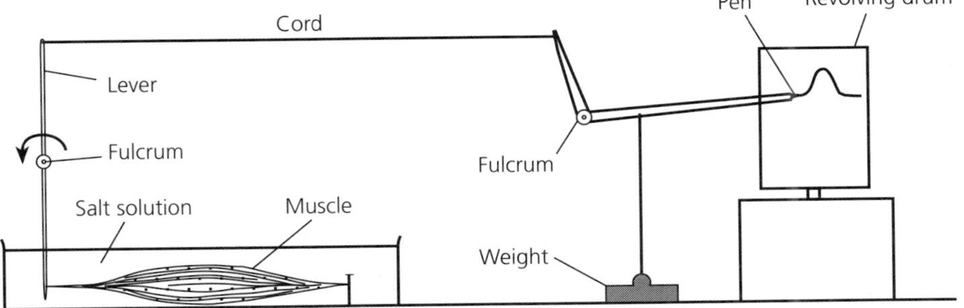


**Figure 19.9**
Diagrammatic representation of the apparatus for stimulating and recording muscle contraction

### Recording muscle contraction

**When a muscle fibre contracts, two things happen – the fibres develop tension, and they shorten. If there is no external load, the increase in tension is very small – just enough to overcome the viscosity and inertia of the muscle itself. If the load is very large, the muscle fibres shorten only to the extent that they can stretch the tendons. These two situations are the basis for two ways in which muscle contractions can be recorded:**

- **isometrically, by recording changes in tension with the ends of the muscle anchored to prevent it shortening (*isometric* = equal length).**
- **isotonically, by recording changes in length with no changes in tension (*isotonic* = equal tension).**

**In practice, every muscle contraction involves some change in both length and tension. Graph A shows the changes in tension in a muscle lifting a light and a heavy weight. Graph B shows the changes in length of the same muscle as it lifts the two weights. The first part of the response is isometric because the tension is insufficient to lift the weight. As soon as the tension exceeds the weight the muscle begins to shorten and the tension stops rising, contraction becoming isotonic. With the heavy weight, the tension has to rise higher before shortening can begin.**

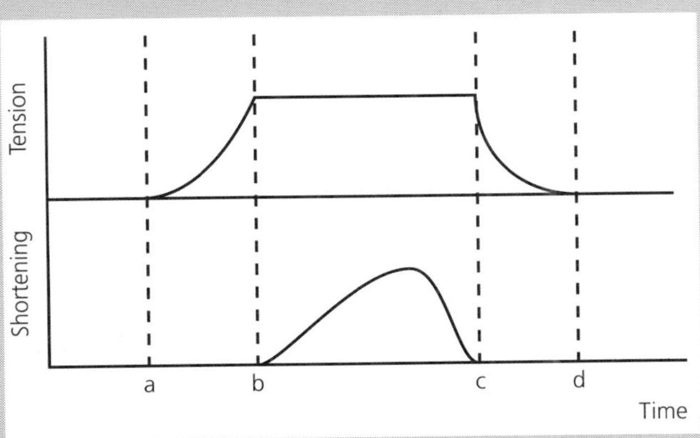


Changes in tension and in length during a muscle contraction. Between a and b tension is rising, but the muscle does not begin to shorten until the tension equals the load at b. After the muscle begins to relax, the tension decreases, the weight is lowered and the muscle extends. When the weight reaches the ground muscle, change in length ceases, and tension falls to zero at d

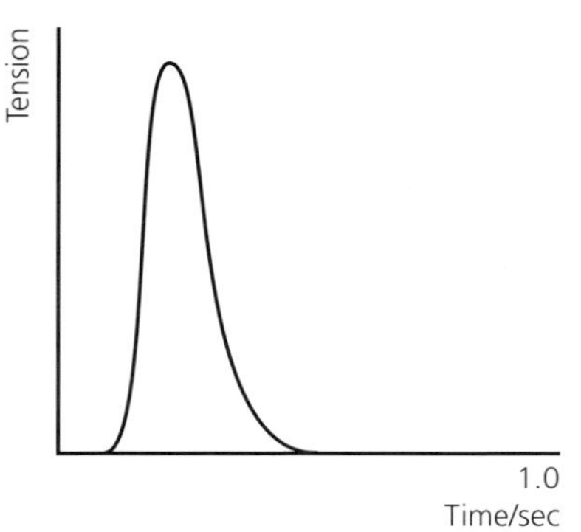


**Figure 19.10▲**
Recording of a muscle twitch

## *The effect of a single stimulus – the twitch*

When a muscle is given a single stimulus of adequate strength, it responds by a short contraction called a **twitch** (Figure 19.10). The time taken for a twitch to be completed varies with the type of muscle and the temperature, but 200 ms (0.2 seconds) is an average interval.

The very short delay (about 20 ms) between the stimulus and the beginning of the response is called the **latent period**. It is due to such factors as the time for the action potential to reach the sarcoplasmic reticulum and the time taken to initiate contraction (longer with indirect stimulation). Another factor is the elasticity and inertia of the recording pen.

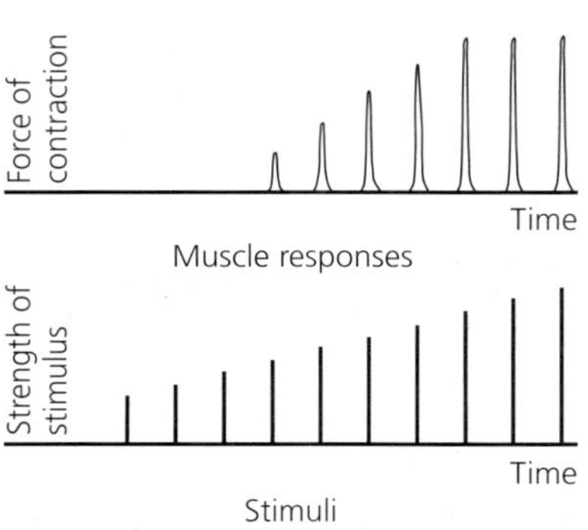


**Figure 19.11▲**
The effect of varying the strength of a single stimulus on muscle response. Though stimuli continue to increase, the last three responses are of the same size

## *Varying the strength of a single stimulus*

When a muscle is given single stimuli of progressively increasing strength, it is found that there is a certain minimum **threshold** below which no response occurs. This is a reflection of the behaviour of the individual nerve fibres in the nerve supply, or, if stimulation is direct, of the sarcolemmas of individual muscle fibres.

Above the threshold the size of the twitch increases (Figure 19.11) up to a maximum, beyond which there is no further increase in the response. Is this a breach of the 'all-or-nothing rule'? Actually, no. With indirect stimulation it is simply the effect of different motor axons having different thresholds; stronger stimuli cause more motor units to respond, each one in an all-or-nothing manner. With direct stimulation stronger shocks trigger responses in more muscle fibres.

## *Varying the stimulus frequency – summation*

If a number of stimuli are given increasingly close together, the muscle behaves as shown in Figure 19.12. To eliminate the effect of varying numbers of muscle fibres responding, stimuli must be strong enough to ensure that all fibres respond.

As the twitches get closer they begin to fuse and the tension increases. With high frequency stimuli, fusion of twitches becomes complete and there is a smooth, steady tension called **tetanus**.

Though it may seem as if the all-or-nothing rule is being broken, it is not. The all-or-nothing rule applies to action potentials, not to the contractile system itself. A trace of the electrical changes in the muscle shows that discrete action potentials continue to travel along the muscle fibres, even though tension is steady.

The explanation is that the muscle fibres are in series with the connective tissue fibres of the tendon (Figure 19.13). Together they behave like a spring (albeit a very strong one). The more the muscle shortens, the more the tendon is stretched and the greater is the tension developed. With a single stimulus the muscle does not shorten enough to develop full tension. Without this elastic component, a single twitch would generate as much tension as a series of fused twitches.

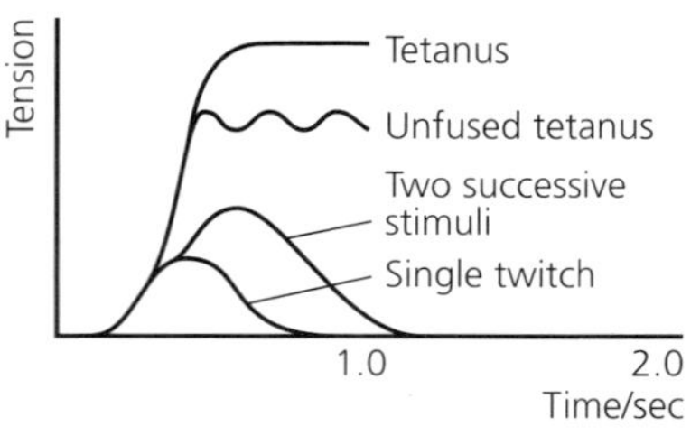


**Figure 19.12▲**
The effect of increasing stimulus frequency on muscle tension

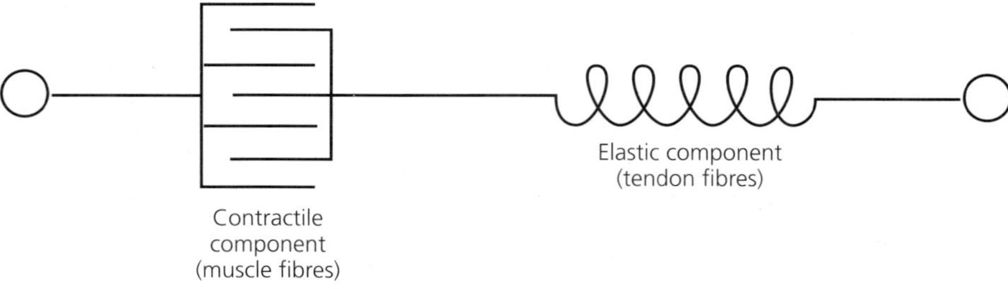


**Figure 19.13►**
The effect of elastic tissue on the force of tetanic muscle contraction. The more the muscle fibres shorten, the more the tendon fibres are stretched, and the greater the tension developed

### *Voluntary muscle contraction*

It may have struck you that muscle responses recorded on a kymograph are most unlike the contractions in voluntary movements in the living body. Normal contractions are smooth and graded in speed and force to suit demand. How can we explain this in terms of muscle twitches?

In experiments on isolated muscles it is normal to vary only one factor at a time. For example, when investigating the effect of impulse frequency, stimuli are strong enough to ensure that all motor units respond. Conversely when stimulus strength is investigated, single stimuli are used to prevent summation.

In a normal muscular contraction two things can vary:

1 the number of active motor neurons, and hence the number of motor units contracting. Muscles in which motor units consist of only a few fibres are capable of more finely-graded contractions than muscles in which there are many motor units.

2 the frequency of impulses sent by each motor neuron, and hence the tension developed by each motor unit.

## Types of skeletal muscle fibre

In reading the description of skeletal muscle tissue given earlier, you may have thought that all skeletal muscle tissue is similar. In fact there are different types of muscle fibre adapted for working in different ways. Three main categories are recognised:

- **fast anaerobic fibres** adapted for generating large forces over very short time periods such as those needed for a 100 m sprint or high jump. They are also called **white fibres** because they have little of the myoglobin that characterises red fibres (see below). ATP generation is largely anaerobic, using glycogen. Mitochondria are fewer than in other types of fibre, and the capillary network is less well developed. The sarcoplasmic reticulum is very well developed, allowing the rapid switching on and off of the contractile mechanism. As a result, twitch times are very short (as short as 10 ms).
- **fast red fibres** are the kind used in long-distance (and therefore aerobic) running. Their colour is due to **myoglobin** in the sarcoplasm. This oxygen-carrying protein is similar to haemoglobin except that it consists of a single polypeptide. Fast red fibres split ATP rapidly and have fast twitch times. They store little glycogen, most of their ATP being generated via the Krebs cycle. Linked to this is the presence of many mitochondria and a dense supply of capillaries. Rapid uptake of oxygen is also promoted by the relatively narrow diameter of the fibres (giving a larger surface area to volume ratio).
- **slow red fibres** are adapted to maintaining the prolonged steady contractions needed for the maintenance of posture, and are highly resistant to fatigue. They split ATP slowly and so shorten slowly and have a correspondingly long twitch time (about 100 ms). They generate ATP mainly aerobically, are correspondingly rich in mitochondria and myoglobin, and are supplied by a dense capillary network.

Most muscles contain more than one type of fibre, though all fibres of one motor unit are of the same kind. The proportions of the three types vary with the muscle. For example, the neck muscles that hold the head up contain a high proportion of slow red fibres, whilst the quadriceps (thigh) muscle contains more white fibres.

The proportions of fibre types are responsive to training, the leg muscles of sprinters generally having a higher proportion of white fibres and those of long-distance runners having more fast red fibres.

## The energetics of muscular exercise

During intense exercise energy expenditure far outstrips the capacity of the heart to deliver oxygen and fuel, even when the latter is pumping as hard as it can. A very active muscle is therefore using energy faster than it is being delivered to it in the blood. A muscle can do this because it carries its own chemical energy in the form of two energy storage compounds: **creatine phosphate** and **glycogen**.

### Creatine phosphate (phosphocreatine)

Like ATP, creatine phosphate can donate a phosphate group to another molecule, but the tendency for it to do so is much stronger than in ATP. As a result creatine phosphate can give up its phosphate to ADP to form ATP:

In active muscle ATP is regenerated from creatine phosphate as fast as it is used up:

$$\text{creatine phosphate} + \text{ADP} \rightarrow \text{creatine} + \text{ATP}$$

Because this reaction proceeds so rapidly, the supply of ATP is maintained even in very active muscle. Creatine phosphate thus represents a store of energy that is almost as readily available as ATP.

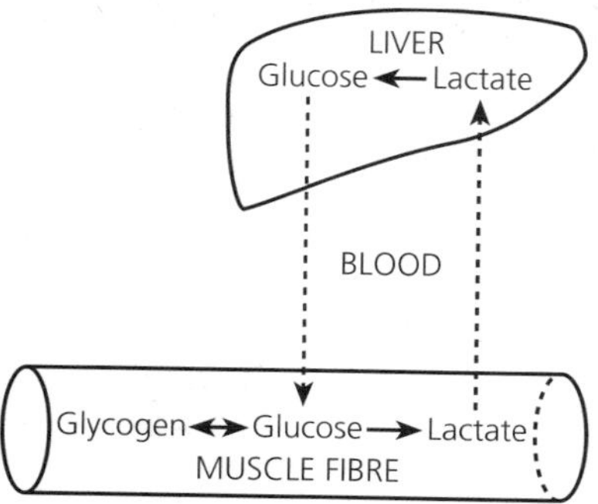


**Figure 19.14**
Recycling of lactate to glucose by the liver

### Glycogen

The supply of creatine phosphate is only enough to last about the first 4 seconds of a 100 metre sprint. The supply of ATP to power the remaining 6–7 seconds is maintained by fermentation, in which glycogen stored in the muscle is converted to lactate. Since the muscle stores glycogen, for short periods it is independent of the blood with regard to supplies of sugar, enabling glycolysis to occur extremely rapidly.

This works well enough for sprints, but the build-up of lactate in the muscles soon begins to interfere with their working (which is why the upper limit for 'pure' anaerobic running is about 200 metres). After a sprint, the lactate in the blood is oxidised by the liver back to pyruvic acid. Some of the pyruvic acid is further oxidised via the Krebs cycle. This produces enough ATP to convert the remaining pyruvic acid back into glucose (Figure 19.14).

During the recovery period, the oxygen that would have been used had the exercise been aerobic is being used *after* the event, and the athlete is said to be repaying an **oxygen debt**. The debt is actually repaid with 'interest', because it costs more energy to convert pyruvic acid to glucose than is released when glucose is converted to pyruvic acid.

Since lactate contains considerable energy, only a small fraction of the energy in the glucose is released. In this sense, lactate fermentation is a very inefficient process. For every glucose unit broken down, anaerobic breakdown of glycogen yields only about a tenth as much ATP as aerobic breakdown.[1] To produce ATP anaerobically at the same rate as aerobically, glycogen would have to be used up 10 times as fast. But maximum anaerobic ATP production is twice as fast as maximum aerobic ATP production. A muscle working at maximum rate anaerobically would therefore deplete its glycogen stores 20 times faster than it would if it were working aerobically.

Even when glycogen is being used aerobically, the muscle stores are insufficient to last the duration of a Marathon (about 40 km), and after

[1] Anaerobic breakdown of glycogen yields a net gain of three ATPs – one more than if glucose itself is being used. This is because conversion of glycogen to glucose phosphate uses only one ATP, whereas glucose to glucose phosphate uses two ATPs.

about 30 km, the muscles make a partial switch to fatty acid as fuel. Though fatty acids yield more ATP than carbohydrate, their breakdown is slower.

## Muscle spindles

Muscles contain mechanoreceptors called **muscle spindles.** Each is about 4–10 mm long and consists of a connective tissue capsule containing modified muscle fibres. These are called **intrafusal fibres** ('within the spindle') to distinguish them from the **extrafusal fibres** of the main body of the muscle.

Each intrafusal fibre has two parts. The central part is non-contractile and is surrounded by nerve endings sensitive to stretch. On either side the muscle fibres are contractile and are supplied by motor axons.

Muscle spindles provide the sensory input in the stretch reflex. Stretching a muscle stimulates the sensory nerve endings. These send impulses to the spinal cord, causing the reflex contraction of the extrafusal muscle fibres.

Muscle spindles are particularly common in postural muscles, for example limb extensor muscles and neck muscles. Any tendency of the body to sag stretches these muscles, bringing about their reflex contraction.

So much for the sensory part of the muscle spindles, what about the motor part? When a muscle shortens, the tension on the sensory part of the muscle spindle would be reduced were it not for the simultaneous contraction of the intrafusal fibres which 'take up the slack'.

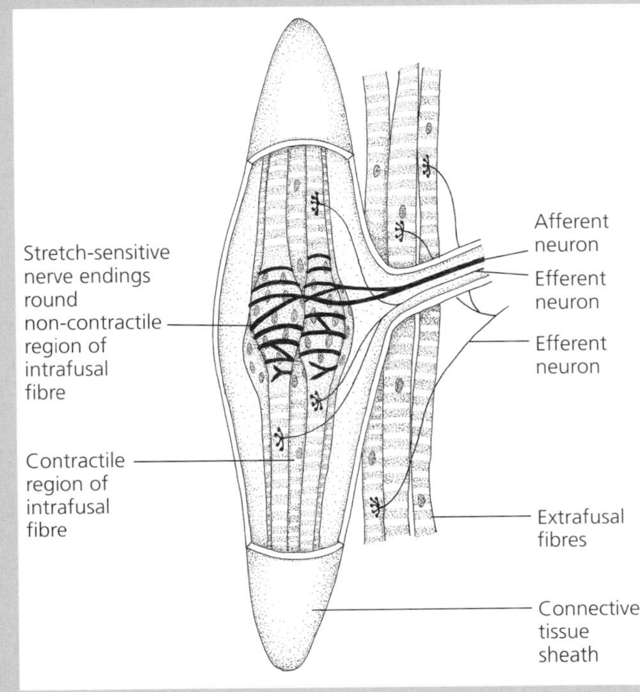


A muscle spindle

# The skeleton

Some form of skeleton is a feature of most animals. Skeletons may perform a variety of functions, which are discussed below. Their role in movement is so intertwined with the action of muscles, that muscles and skeletons can be regarded as a single **skeletomuscular system**.

## Types of skeleton

Skeletons are of two fundamentally different types:

- **hydrostatic skeletons**, characteristic of annelids, molluscs and many other soft-bodied animals.
- **jointed skeletons**, which are themselves of two kinds:
  - **exoskeletons** (external skeletons), characteristic of arthropods.
  - **endoskeletons** (internal skeletons), typical of vertebrates.

Exoskeletons and hydrostatic skeletons are dealt with towards the end of this chapter.

## Functions of skeletons

Skeletons may have diverse functions, not all of which are mechanical. The first three are perhaps the most fundamental since they are characteristic of all skeletons.

- Skeletons help to maintain *body shape*. Tissues and organs can only function if their shape is not distorted too much. This is particularly important on land where there is no buoyancy support by water.
- They transmit muscular forces to the environment. Muscles develop forces *inside* the body, but much animal movement requires forces to act against the *outside* of the body. If we liken a muscle to a car engine, the skeleton can be compared to the drive shaft, back axle and wheels, which transmit forces from the engine to the road.
- Skeletons enable pulls to be converted into pushes. For instance, when you walk uphill, pulls in the thigh muscle are converted into a backward push against the ground.
- The skeleton transmits forces between antagonistic muscles (see below).
- Though muscles can develop very strong forces (up to 20 N $cm^{-2}$), they can only shorten through about a third of their extended length, and then relatively slowly. By acting as lever systems, skeletons enable these slow, strong forces to be converted into weaker but faster forces. The feet of a sprinting cheetah move at 60 mph relative to its body, but its muscles are shortening at about an eighth of this speed.
- Protection against impact, as in the protection of the brain by the skull. In arthropods, much of the cuticle protects internal organs.
- In vertebrates, the jaw bone anchors the teeth.
- Certain bones in the middle ear of land vertebrates are important in transmitting sound.
- In vertebrates, blood corpuscles are made in the bone marrow.
- Bones act as a store of calcium. In times of shortage, calcium is removed from the bones, thus keeping up its concentration in the blood.

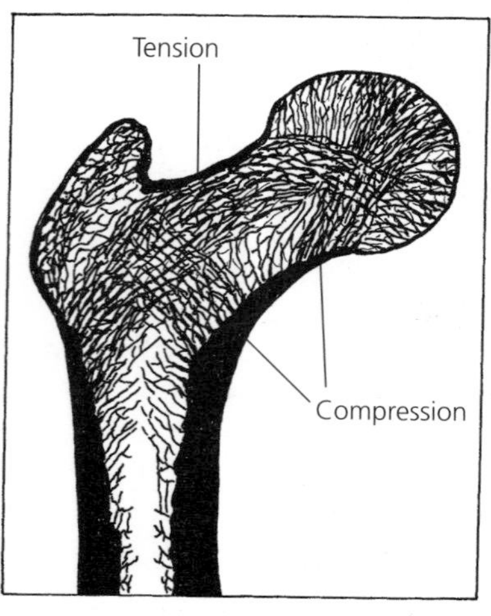


**Figure 19.15**
Longitudinal section through the head of a human femur showing trabeculae oriented along lines of stress

## Bone

Bone is the major vertebrate skeletal tissue. Because it resists both tension and compression, it is not easily bent. Unlike cartilage, bone has a rich internal blood supply which enables it to be repaired after injury. Bones are adapted to withstand deformation in both their macroscopic and microscopic structure.

### *Macroscopic structure of a long bone*

A long bone, such as the femur (thigh bone), has the form shown in Figure 19.15. It consists of two kinds of bone. The ends consist of **spongy bone**, in which an extensive system of narrow blood spaces is criss-crossed by a fine network of bony struts or **trabeculae**. The shaft consists of **compact bone**, which resists forces acting along the axis of the bone.

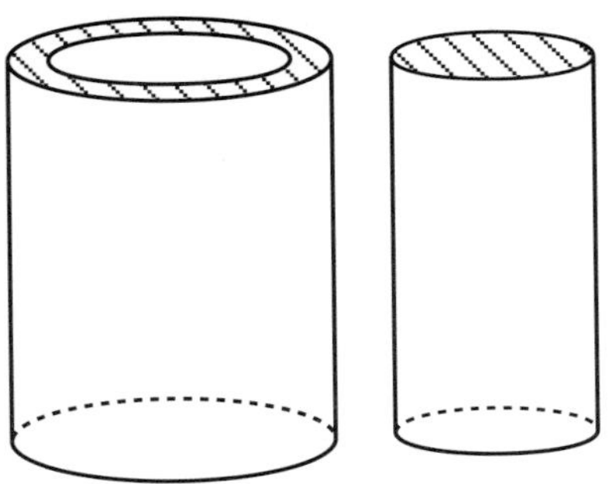

**Figure 19.16▲**
Hollow cylinders are stiffer than solid rods of the same cross-sectional area. To put it another way, for a given stiffness, a hollow cylinder is lighter than a solid rod

Three features of its macroscopic architecture adapt it to bearing weights:

1 the shaft or diaphysis is *hollow*. A hollow shape is stiffer than a solid one of the same cross section of material, so hollow bones *save weight* (Figure 19.16). A similar engineering principle is employed in scaffolding.

2 the ends are expanded, giving a greater surface for articulation with other bones. By spreading forces over a wider area, the pressure on the ends is reduced, enabling greater loads to be carried.

3 as can be seen in Figure 19.15, the trabeculae tend to grow along the lines of stress.

## *The fine structure of bone*

Bone consists of bone cells or **osteocytes**, surrounded by the **matrix** which they secrete. About 65–75% of the matrix is inorganic, consisting of tiny needle-like crystals of **hydroxyapatite**, a form of calcium phosphate. The crystals are separated from each other by a dense weave of **collagen** fibres which form the organic part of the matrix. If the inorganic part is removed by prolonged soaking in acid, bones become very flexible (Figure 19.17).

The inorganic and organic components of the matrix have complementary properties. Calcium phosphate is very resistant to compression but by itself is very brittle. Collagen, on the other hand, is very resistant to tension, but like string it cannot withstand compression. Bone is thus a composite material, strongly resisting both tension and compression.

The bone is organised into many cylindrical **Haversian systems**, each comprising a number of concentric bony cylinders or **lamellae**. At the centre of each Haversian system is a **Haversian canal** containing a capillary and usually a nerve fibre (Figures 19.18 and 19.19). Each

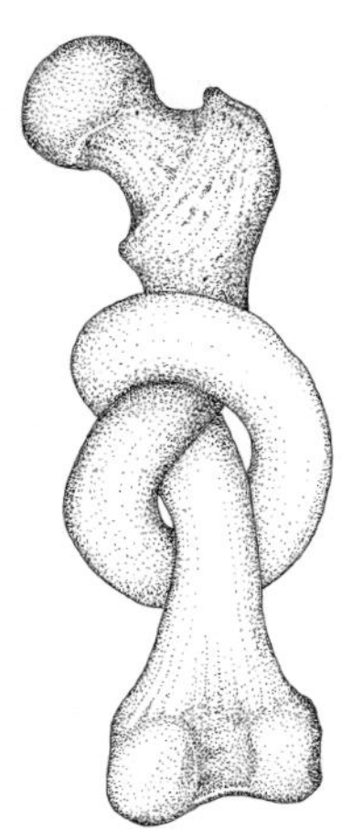

**Figure 19.17▲**
A demineralised femur

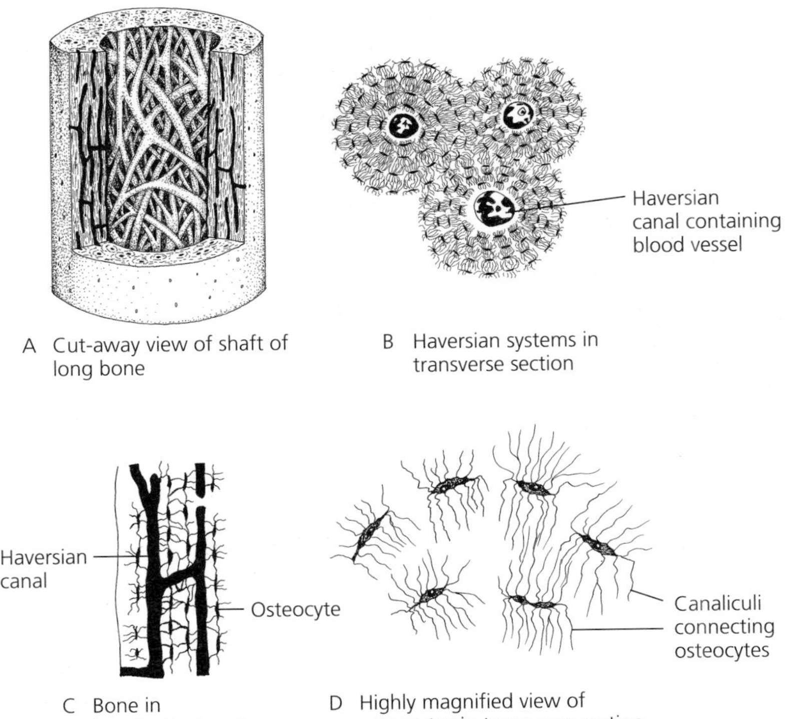


**Figure 19.18►**
Fine structure of compact bone (after Crouch)

**Figure 19.19▲**
Photomicrograph of a transverse section of compact bone showing a Haversian canal, the dark central area, containing blood vessels

osteocyte lies entombed in a cavity or **lacuna** in the matrix it helped to secrete. The osteocytes are interconnected by narrow **canaliculi** containing threadlike outgrowths. The canaliculi are channels through which materials diffuse between the osteocytes and the blood.

Haversian systems average about 200 μm in diameter, so no osteocyte is more than about 100 μm from the nutrients supplied by its nearest capillary. The Haversian canals are interconnected forming a network, and are connected to the marrow cavity and to blood vessels in the **periosteum** – a tough, fibrous membrane covering the bone.

Bone is ever-changing and, even in adults, new Haversian systems are being produced and old ones broken down. Experiments with radioactive phosphorus have shown that about two-thirds of the calcium phosphate in bone is replaced every three weeks.

This dynamic state enables bone to adapt to new stresses. When bones are not stressed by muscle action they become decalcified and weak, and it is now considered sensible for elderly people to keep their bones strong by moderate weight training.

## Spaces make bone stronger

**The fine structure of compact bone gives it great strength. You might think that the honeycomb system of microscopic spaces would make bone weaker than if it were solid, but the cavities actually make it stronger by preventing cracks from running.**

**When a crack starts to run through a material, the stress at the tip of the crack is far greater than elsewhere (this is why paper is much easier to tear once a tear has started). When a crack encounters a space (such as Haversian canal or a lacuna and its entombed osteocyte) the stress is dispersed over a wider area, and in many cases the crack stops. The same principle operates at the submicroscopic level. A crack cannot easily spread from one crystal of hydroxyapatite to its neighbour because they are separated by a dense weave of collagen fibres.**

**It's also worth noting that although the Haversian canals can be regarded as tiny cracks, in the shaft of a long bone they run predominantly lengthways – at right angles to the direction in which cracks would tend to form when a bone is bent. Compact bone thus has a 'grain' like wood. Both are strong when bent across the grain but very weak when bent along it.**

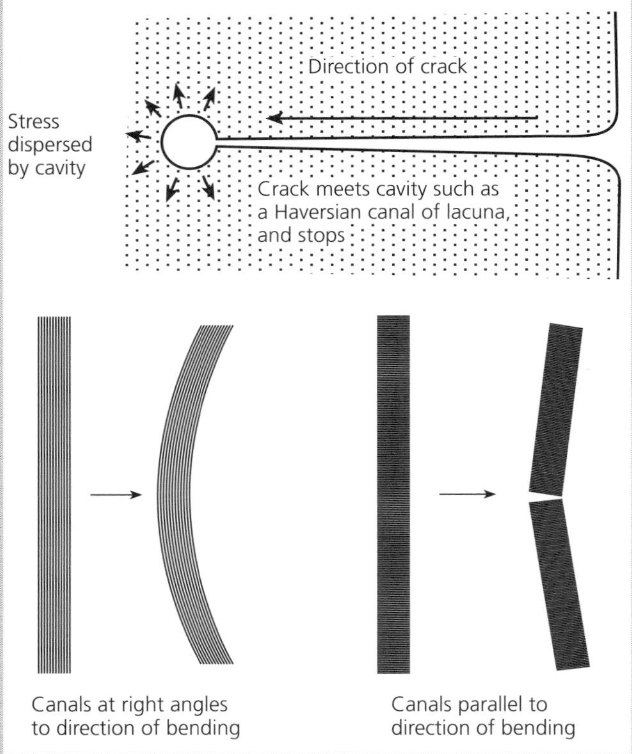


Above: How microscopic cavities in bone help prevent cracks from spreading. Below: A bone is less likely to break if the Haversian canals run at right angles to the direction of bending

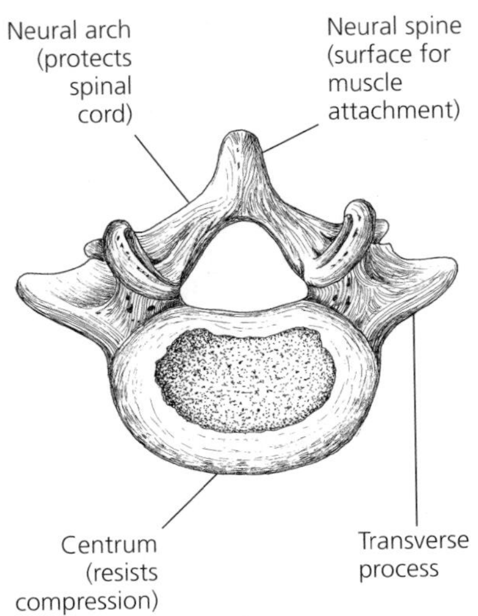


**Figure 19.20**
A human lumbar vertebra seen from above

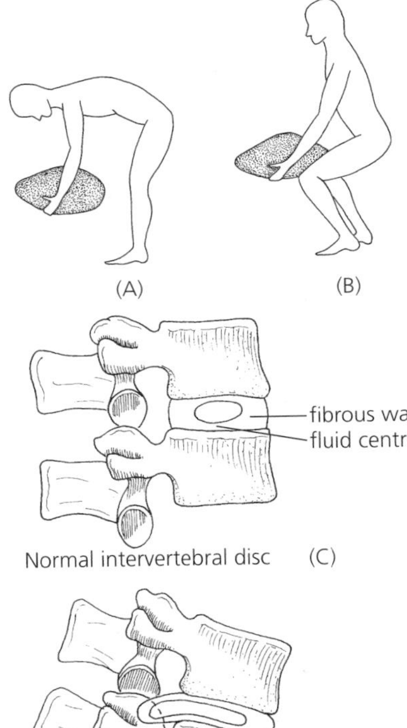


**Figure 19.21**
(A) How lifting a heavy weight the wrong way can damage your intervertebral discs. (B) The correct way to lift a weight. (C) What may happen if you lift the wrong way – the top diagram shows a normal intervertebral disc, while the bottom diagram shows a prolapsed disc, with outer wall pushed out

# Parts of the human skeleton

The skeleton is usually divided into two main parts:

- the **axial skeleton**, consisting of the **skull**, **backbone**, **ribs** and **breastbone**.
- the **appendicular skeleton**, comprising the **limbs** and **limb girdles**.

## *The skull*

The skull houses and protects the brain and certain sense organs such as the eyes and ears. It also anchors the jaw muscles and teeth. It consists of a number of bones that meet at immovable joints called **sutures**. Because of their jigsaw-type interlocking, these joints are very strong yet allow the bones to grow during youth.

## *The vertebral column or backbone*

The vertebral column usually consists of 33 individual bones or **vertebrae**. A typical vertebra (from the lumbar region) is shown in Figure 19.20.

The solid, block-like **centrum** resists compression due to the body weight, and the **neural arch** protects the spinal cord. The **neural spine** and two **transverse processes** project from the neural arch and increase the surface area for the attachment of ligaments and muscles. Also projecting from the neural spine are small processes which restrict the movement of vertebrae relative to each other. In humans these are called superior and inferior articular processes. In other (four-footed) mammals they are called pre- and postzygapophyses.

The backbone is divided into five regions:

- the seven **cervical** (neck) vertebrae carry least weight and are the smallest. The first is called the **atlas** and allows the skull to nod backwards and forwards. The second is the **axis**, which permits the skull and atlas to rotate, as when shaking the head.
- the twelve **thoracic** (chest) vertebrae are those that articulate (form a joint) with the ribs.
- the five **lumbar** (lower back) vertebrae have the largest centra since they carry the greatest weight. Most of the bending of the back occurs here.
- the **sacrum** consists of five vertebrae fused into a solid mass. The first articulates with the hip girdle and thus transmits the body weight to it.
- the **coccyx** consists of four fused bones, which are all that is left of the tail.

## *Intervertebral discs*

Between the vertebrae are pads which act as shock absorbers. Each is like an extremely tough, thick-walled balloon, with a fluid centre surrounded by a thick wall of fibrocartilage. Fibrocartilage is like hyaline cartilage except that the matrix is much more densely laced with collagen fibres. When bearing the body weight the fluid is compressed and the fibrous wall is under tension, just as the wall of a balloon is.

Provided that the pressure on the discs is evenly distributed, they can withstand considerable compression. This is why you should not bend your back when lifting a heavy weight, but should bend at the knee and hip joints, keeping the back straight and the discs evenly loaded (Figure 19.21).

## *The human forelimb skeleton*

The basic structure of the forelimb skeleton follows the **pentadactyl** plan which, in variously modified form, is characteristic of all mammals, birds, reptiles and amphibians (Figure 19.22).

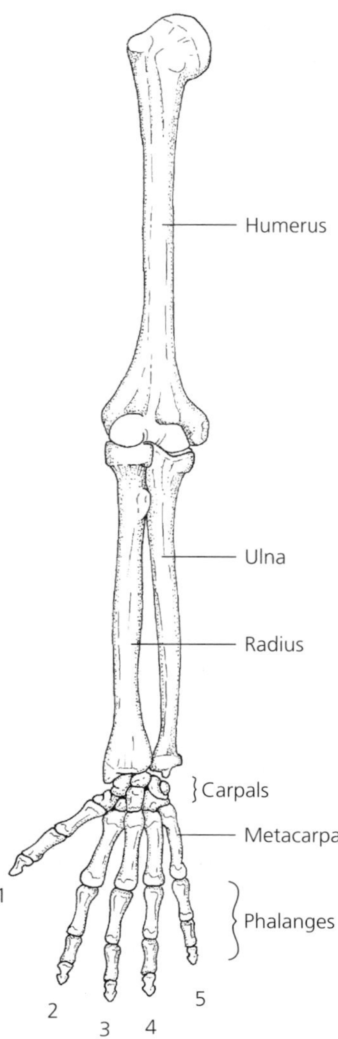


**Figure 19.22**
The human forelimb skeleton

Unlike the hip girdle which articulates with the backbone, the **scapula** (or shoulder blade) is mobile, anchored only by muscle. The scapula forms a large platform for the anchorage of muscles that move the **humerus** or upper arm bone. This articulates with two lower arm bones, the **radius** and **ulna**. The forearm articulates with the wrist, made up of a number of **carpal** bones, and this in turn articulates with the **metacarpals**, or hand bones. Each metacarpal articulates with a **digit** (the term for finger or toe) made up of smaller bones called **phalanges** (singular, phalanx).

## Joints

A joint is formed where two or more bones meet. Joints can be divided into three classes according to the kind of tissue separating the bones. In **fibrous joints**, the bones are held together by fibrous tissue, for example the bones of the skull. In this type, little or no movement is possible. In **cartilage joints**, the bones are held together by cartilage, for example the joints between the ribs and breastbone, and between the vertebrae. **Synovial joints** allow free movement between the ends of bones (Figure 19.23). The joint surfaces are covered by a smooth layer of **articular cartilage**. Movement is restricted by **ligaments**, which consist of dense bundles of collagen fibres. Ligaments are immensely resistant to tension, and sometimes it is a bone that breaks rather than the ligaments that tie them together. Ligaments have a poor blood supply and so appear white. This is why they heal more slowly than bone. An injury to a ligament is a **sprain**.

The ligaments form a capsule containing a lubricating **synovial fluid**, secreted by the **synovial membrane** which forms the lining of the joint capsule. Normally the amount of synovial fluid is very small (about a teaspoonful in the knee joint), but if the synovial membrane becomes inflamed it produces more fluid and the joint becomes swollen.

Synovial joints can be subdivided according to the kind of movement that occurs. A common type is a **hinge joint**, in which movement is possible in one plane only. Examples are the joints of the fingers, elbow and knee. A second type is a **ball and socket joint**, for example the hip and shoulder joints, which allow movement in more than one plane.

### *Transmitting forces to bones – tendons*

Muscles are not anchored directly to bones, but are connected to them by dense bundles of collagen fibres called **tendons**. In a limb muscle, the tendon nearest the trunk (chest and abdomen) is considered to be fixed and is called the **origin**, whilst that nearest the end of the limb is moveable and is called the **insertion**.

Unlike muscles (which actively generate tension), tendons are like cables, transmitting pulls passively. Tendon fibres do not end at the surface of the muscle – they enter and branch inside it (forming the 'gristle' of meat). A muscle and its tendon thus grip each other like two ropes spliced together, their fibres coming into contact over an immense area. Where tendon meets bone there is a similarly strong connection, the tendon fibres being continuous with the fibres of the periosteum.

Tendons enable muscles to be situated a considerable distance from the load they move. This can have two advantages:

- it enables the fingers to carry out fine movements. Most of the muscles that move our fingers are located in the forearm; if they were in the fingers, the muscles would make the fingers too bulky to manipulate objects.

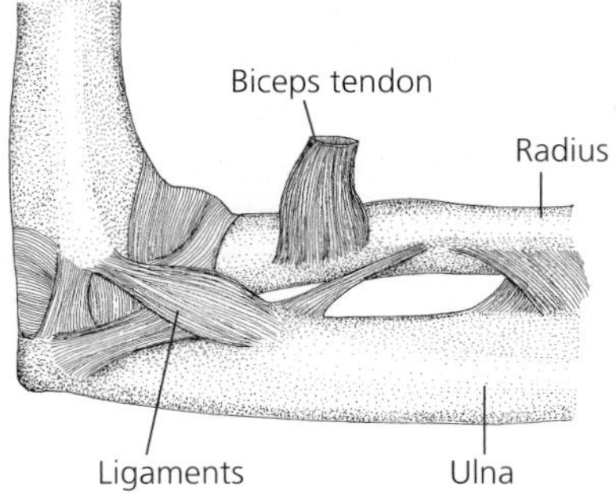


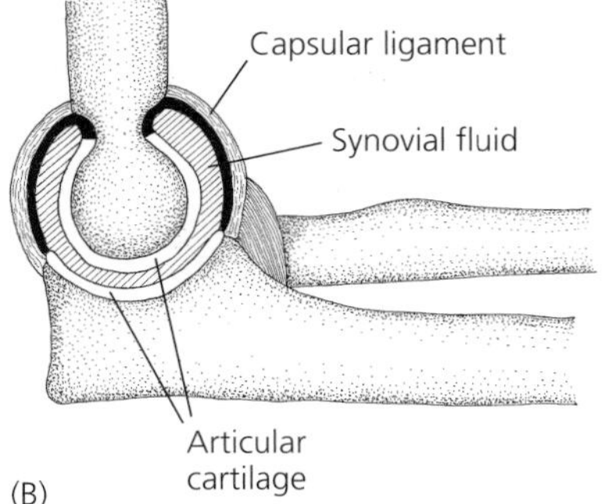


**Figure 19.23**
(A) External view of human elbow joint; (B) human elbow joint in section

- it enables the end of a limb to be almost free of muscle, thus lightening it and making it easier to be moved backward and forward very rapidly with minimal energy expenditure during running.

### *Muscle antagonism*

Since a muscle cannot actively lengthen, it has to be returned to its original length by some other agency, which is usually another muscle. For example, the **biceps** muscle on the inside of the elbow joint flexes (bends) the joint and the triceps muscle on the other side extends (straightens) it. The biceps is a **flexor** muscle and the triceps is an **extensor**. Since these two muscles work against each other, they are said to be **antagonists**.

## Bones as levers

One of the most important functions of a jointed skeleton is to act as a system of levers. An example is a crowbar, which converts a relatively weak force, developed by a puny human, into a force strong enough to lift a very heavy weight. There is a cost though – the effort has to be moved through a much greater distance than the load. The lever, therefore, does not supply energy – it changes a weak force moving through a large distance into a stronger one moving through a shorter distance.

Most levers in a vertebrate limb work in the opposite way, enabling a slow but powerful force produced by a muscle to be converted into a fast (but weaker) force acting on the load. Levers have another advantage: a load can be moved with very little change in sarcomere length, enabling muscles to operate close to the length at which they develop maximal force.

The point about which the lever rotates is called the **fulcrum** or pivot. Depending on the relationship between the fulcrum, load and effort, three orders of levers can be distinguished (Figure 19.24).

In a **first order lever** the load and effort are on either side of the fulcrum, for example a crowbar. An example in the human body is the support of the head by the trapezius muscle (Figure 19.24A).

**Figure 19.24**
Three kinds of lever in the human body

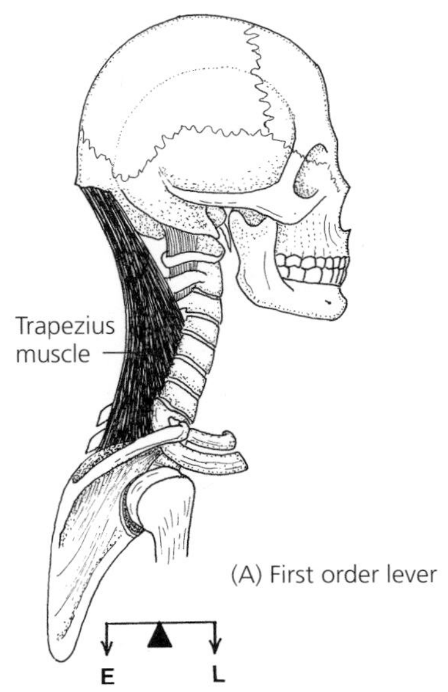


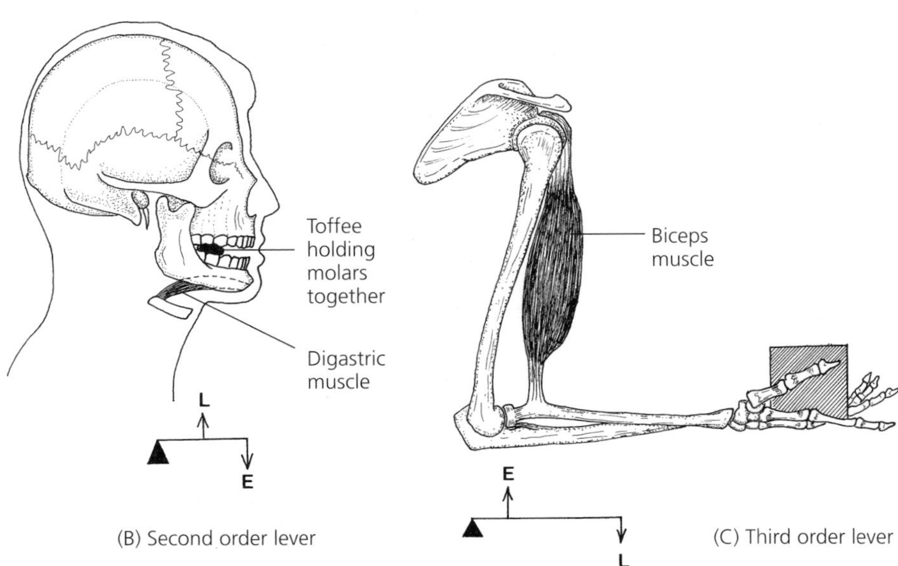

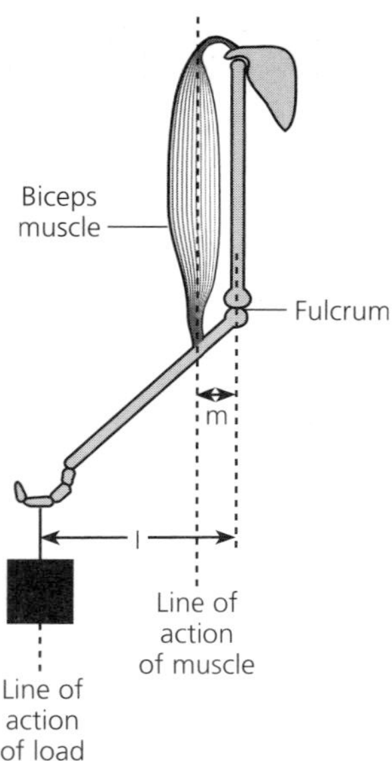


**Figure 19.25**
An example of forces acting at an angle to a lever. In this case the turning effect of the muscle is proportional to the distance m, and that of the load is proportional to the distance l

In a **second order lever** the load is between the effort and the fulcrum, as in a wheelbarrow. Here the force of the effort must be less than that of the load. A human example is when the back teeth are stuck together by, say, toffee, and the muscle that lowers the jaw is used to prise them apart (Figure 19.24B). The ankle joint, which is often compared to a wheelbarrow, is not, as commonly stated, a second order lever (see Clear Thinking box below).

Most limb bones act as **third order levers**, in which the effort acts between the load and the fulcrum. The effort is thus always greater than the load. This arrangement is biologically advantageous since it enables a slow but strong muscular force to be converted into a faster (but weaker) force acting on the load.

One example of a third order lever in the human body is the forearm (Figure 19.24C). The weight shown is about seven times as far from the fulcrum as the attachment of the tendon, so it moves about seven times as fast as the muscle shortens. The cost is that the muscle has to pull seven times as hard as the weight of the load.

In Figure 19.24C both load and effort act at right angles to the lever. In most cases the situation is more complicated than this, with forces acting at an angle to the lever. In such cases it is the distance between the fulcrum and the **line of action** of the force that matters (Figure 19.25).

## CLEAR THINKING

**As a lever, the ankle joint is usually compared to a wheelbarrow, which is a second order lever. The essential feature of a second order lever is that, since the effort is further from the fulcrum than the load, it moves through a greater distance. The force of the effort is therefore less than that of the load.**

**At first sight, the ankle joint appears to resemble a wheelbarrow. There is, however, a crucial difference – the muscles that raise the body are part of the body, and so *move with it*. Not only does the insertion of the calf muscle rise – so does the origin. The distance moved by the effort is the distance through which the calf muscle shortens, which is considerably less than the distance through which the ankle rises.**

**It is easier to see this if you lie on your back with your leg pointing upward. When you point your toes by contracting the calf muscle, the foot is acting as a first order lever. In the normal position, with the sole of the foot on the ground, the load is the reaction of the ground to the body weight.**

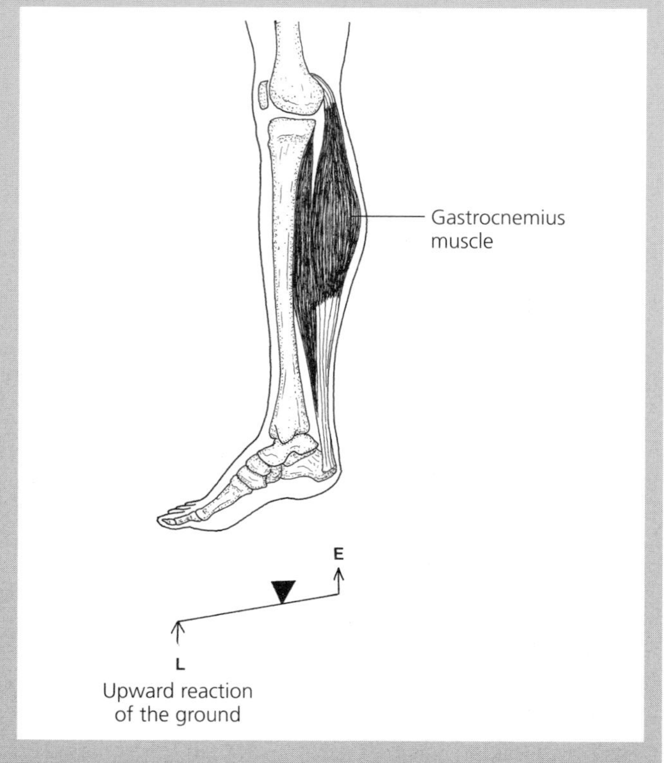


A misunderstood joint

## Levers for sprinting and levers for digging

Levers change the characteristics of forces. Where speed is a priority (as in escaping from enemies or catching prey) muscles work close to the fulcrum so that the end of the limb moves quickly. In the cheetah, the teres major muscle that moves the forelimb backwards is inserted about a tenth of the way along the limb. As a result, the foot moves backwards 10 times faster than the muscle shortens (though with a 10-fold decrease in force).

In the badger, a squat animal that can excavate a burrow, or 'sett', the emphasis is more on force and less on speed. The teres major muscle is inserted about a sixth of the way from the shoulder joint, so the force delivered to the load (the resistance of the earth) is reduced to just a sixth rather than a tenth of the force developed by the muscle.

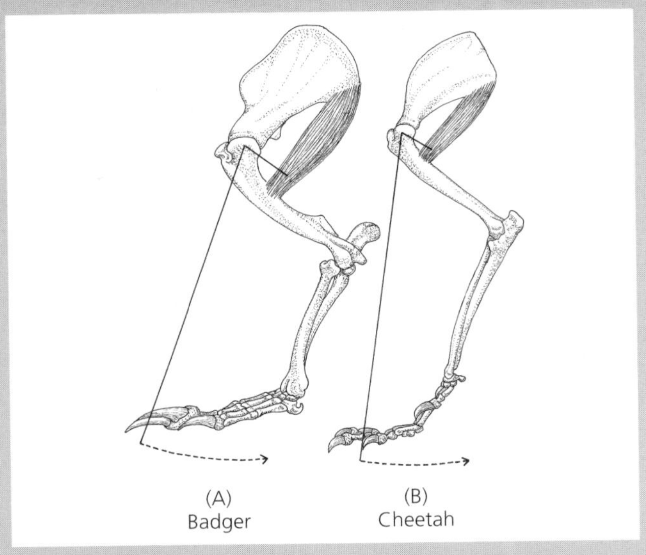


Levers for digging (A) and for sprinting (B)

## How muscles make bones stronger

Bone is much more resistant to compression than to tension. When subjected to pure compression, bones can bear very heavy weights. In many cases bones are subjected to bending, in which one side is stretched and the other is compressed. It is under these conditions that a bone is more likely to break.

One of the most interesting functions of muscles is that they can reduce bending forces on bones, thus effectively making them stronger. A good example is the muscle which runs from the ilium (hip bone) to the outside of the tibia (shin bone). Because the head of the femur is to one side of the shaft, the body weight tends to bend the femur outwards.

If you place your finger on the outside of your leg just above the knee and shift your weight onto that leg, you may feel the tendon of this muscle tighten. The effect is to reduce the tension on the outside of the femur, making it less likely to break.

Another example of this **bracing** is in the forearm. When the biceps contracts in lifting a weight, the radius tends to bend downwards. Or rather it would, but for the simultaneous contraction of the brachioradialis muscle.

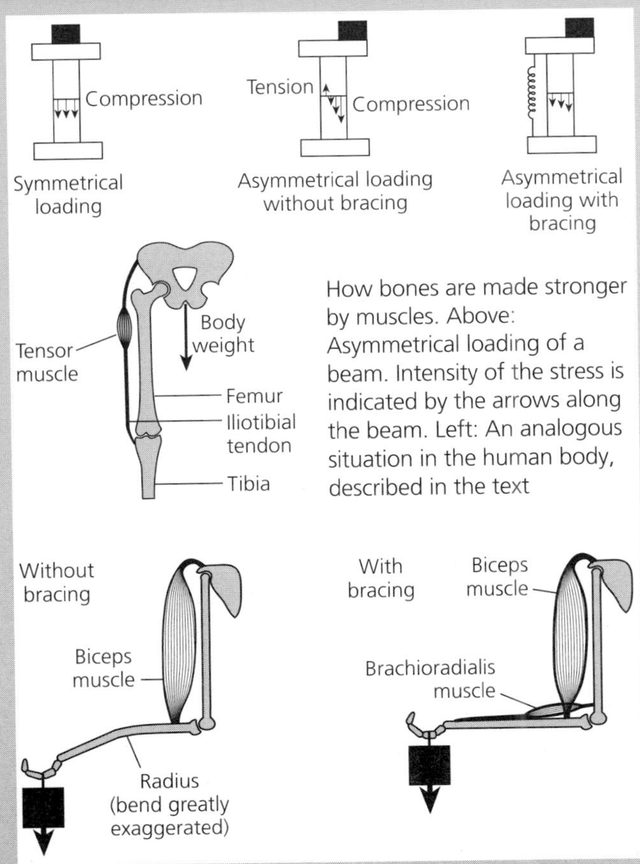


How bones are made stronger by muscles. Above: Asymmetrical loading of a beam. Intensity of the stress is indicated by the arrows along the beam. Left: An analogous situation in the human body, described in the text

How the brachioradialis muscle strengthens the radius when lifting a weight

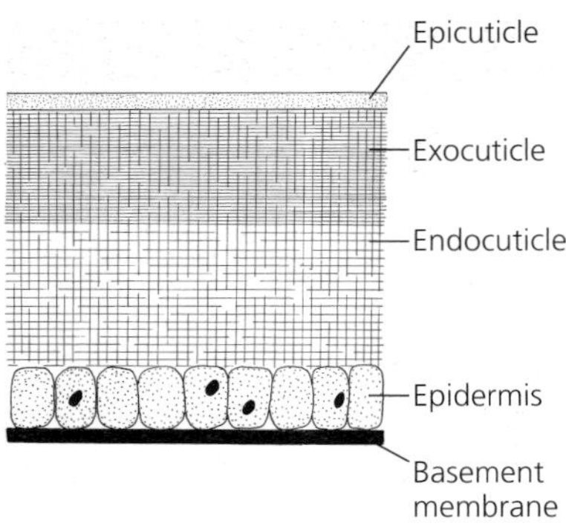


**Figure 19.26**
Section through the arthropod cuticle

## Jointed exoskeletons

Arthropods typically have an **exoskeleton** or external skeleton. Though structurally very different, it shares a number of functions with the endoskeleton of vertebrates. Besides giving support and mechanical protection, it acts as a series of jointed levers allowing rapid movement. In insects and arachnids it also greatly reduces water loss.

The exoskeleton of an arthropod consists of the non-living **cuticle,** which is secreted by the epidermis underneath (Figure 19.26). Most of the cuticle consists of two materials: **chitin,** a nitrogenous polysaccharide, and proteins. In most parts of the body, the outer part of the cuticle – the **exocuticle** – is stiffened by a process called **tanning,** in which the protein molecules become cross-linked to each other. Tanning is most marked where great hardness is required, as in biting mouthparts. The inner part of the cuticle is the **endocuticle** and remains unhardened. At the joints, the exocuticle is absent.

Outside the exocuticle is a very thin **epicuticle** in which chitin and protein are absent. In insects and arachnids the epicuticle consists of wax, which greatly reduces water loss.

Whereas muscles are attached to the outside of an endoskeleton, they are anchored to the inside of an exoskeleton. The arthropod equivalents of tendons are inward projections of the cuticle called **apodemes.** Movement is permitted at the joints, where areas of cuticle remain untanned and soft (Figure 19.27).

The arthropod skeleton must be extremely effective, since 80% of animal species are arthropods. It does however have one significant limitation: since it is non-living it cannot grow or repair itself. Arthropods therefore have to shed their cuticle periodically by a process called **ecdysis** (Chapter 29). The new cuticle rapidly stretches while it is still soft. Only when it has hardened can the new skeleton give the animal the support and protection it needs.

A skeleton that is soft – even if it is only for a short period after ecdysis – imposes a major limitation to the evolution of large body size on land. This is probably the main reason why all land arthropods are small (the heaviest, the goliath beetle, has about the same mass as a mouse).

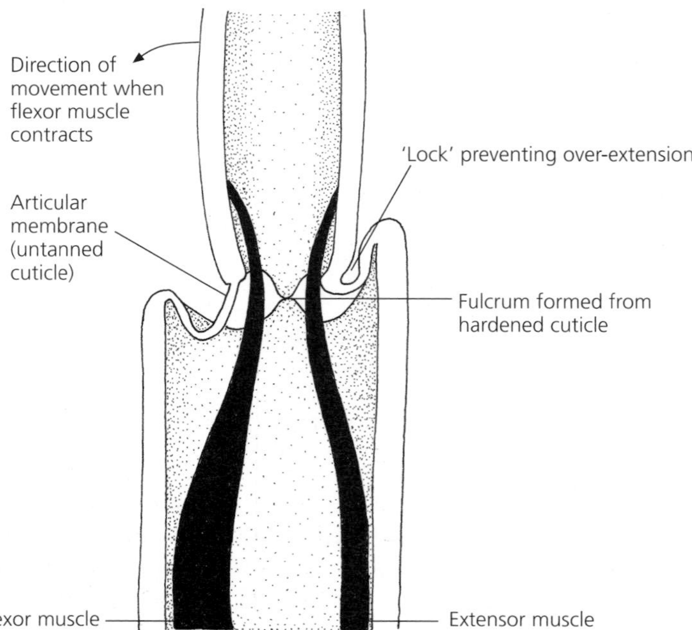


**Figure 19.27**
Section through an arthropod joint (after Russell-Hunter)

## Small is stronger

The photograph shows an ant carrying an object much heavier than itself. What is so special about ant organisation that makes them so strong? Actually, nothing at all. It is simply that in general, small animals are stronger mass for mass than larger ones.

The explanation has to do with a theme that crops up repeatedly in biology – the fact that length, area and volume increase at different rates (Chapter 16). The force that a muscle can exert depends on the number of its fibres and is proportional to its cross-sectional area. On the other hand, the mass of the load being lifted is proportional to its volume.

Now imagine that the ant and the load it is carrying both double in length. If the shapes of the body and load remain the same, then all three dimensions (length, height and thickness) double. Since area has two dimensions, the cross-sectional area of its muscles will increase $2 \times 2 = 4$ times, so its muscles become four times stronger. Since volume has three dimensions, the volume (and hence the weight) of the load increases $2 \times 2 \times 2 = 8$ times. The ant has thus become four times stronger but its load has increased eight times. Doubling the animal's length thus makes it half as strong relative to its mass.

When it comes to comparing ants with people, size is not the only factor at work, because the body plans are quite different. Still, if we make comparisons between animals of similar body plan, we find that the larger ones are relatively weaker than smaller ones.

A similar factor is at work when we look at skeletons. The bones of a mouse account for about 2% of its body mass, but the skeleton of an elephant is about 20% of its body mass. As with muscles, the explanation depends on the fact that, with increasing size, area grows more slowly than volume. If its proportions remain unchanged, doubling the animal's length increases the cross-sectional area (and hence the strength) of its bones four times. Its volume (and hence the load carried by the bones) has increased eight-fold. The only way to keep the bones strong enough to carry the body is for their cross-sectional area to increase in proportion to the volume of the body, which means they must become relatively more bulky.

Ants carrying a large butterfly larva

The largest dinosaurs, such as *Apatosaurus* ('Brontosaurus'), were probably close to the maximum possible size for land animals. In aquatic animals buoyed up by the water, the problem of weight does not arise, and much greater size is possible. A full-sized blue whale has a bulk several times larger that of the largest known land dinosaur.

## Hydrostatic skeletons

A hydrostatic skeleton works very differently from a jointed skeleton. Instead of stiff material, such as bone or hardened chitin, it consists of a fluid enclosed by muscle. Hydrostatic skeletons occur in many soft-bodied animals such as sea anemones, earthworms and many molluscs, and also in maggots and caterpillars. The shape of the vertebrate eyeball and the erect mammalian penis are also due to internal fluid pressure.

### Support and movement in the earthworm

Hydrostatic skeletons are dependent on the fact that although a fluid can take up any shape, it is incompressible and so has a fixed volume. This principle can be seen in the earthworm. An earthworm moves by changes in the shape of its segments in a wave that spreads down its body. Between the body wall and the gut is a fluid-filled space called the **coelom** (Figure 19.28). Each segment is separated from the next by thin sheets of muscle called **septa** (singular, septum). These partially restrict changes in pressure to individual segments.

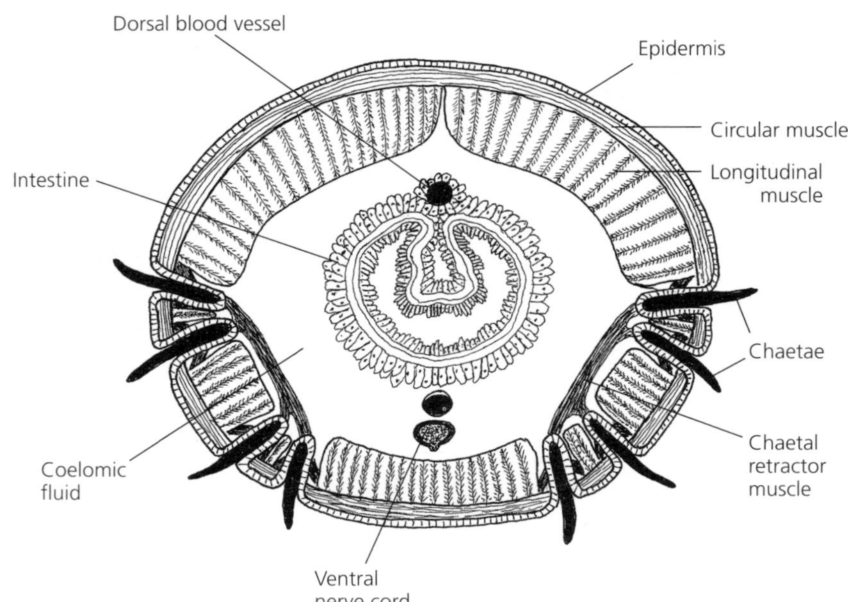


**Figure 19.28**
Transverse section through an earthworm

Each segment changes its shape using two sets of antagonistic muscles. A layer of **circular muscle** runs round the body and a layer of **longitudinal muscle** runs along it. When the circulars contract and the longitudinals relax, each segment not only becomes thinner, it also elongates since fluid is incompressible. Similarly, when the longitudinals contract and the circulars relax, a segment must become wider as well as shorter.

If you watch an earthworm crawling over damp paper you will see that each segment alternates between two shapes. When a segment is moving forward it is long and thin. When it is stationary it is short and fat, and grips the ground by **protracting** (pushing out) short bristles called **chaetae** on the ventral and lateral surfaces of the body. These are **retracted** (withdrawn) when a segment is moving forward. The movements of the segments and chaetae are co-ordinated by the ventral nerve cord.

Besides their use in crawling, the longitudinal muscles are used to rapidly draw the worm back into the burrow and to anchor the posterior segments while doing so. The anterior circular muscles are also used to force the first few segments forward during burrowing.

Compared with jointed skeletons, hydrostatic skeletons have several limitations. For one thing they depend on an adequate water supply – an earthworm deprived of water is less able to move than a fully hydrated one. And since forces cannot be speeded up using the lever principle, movement is slow.

## QUESTIONS

**1** What is the functional importance of each of the following structural features of skeletal muscle?

**a)** The sarcolemma is closely associated with a fine network of collagen fibres.

**b)** The motor end plate or neuromuscular junction lies roughly half way along a skeletal muscle fibre.

**c)** The nuclei lie just beneath the sarcolemma rather than deep within the cytoplasm, and the mitochondria lie in longitudinal files between the myofibrils.

**2** Which of the traces **A–D** below could show how varying the load on a muscle would affect:

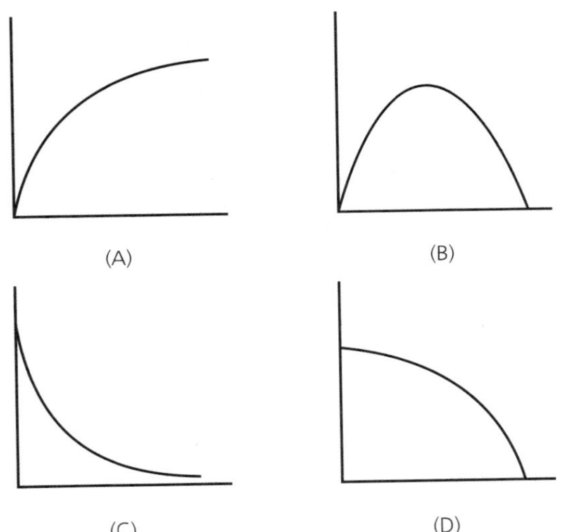


**a)** the maximum speed at which the muscle can shorten;

**b)** the power produced by the muscle (power = speed × load)?

In each case the horizontal axis represents the mass of the load.

**3** Red muscle fibres are narrower than white fibres. Read the information box on page 357 and suggest why this may help red muscle fibres to function more effectively.

**4** From the alternatives **A–D**, choose the one which would limit:

**i)** the velocity with which the muscle can move a very light load.

**ii)** the tension the whole muscle can develop.

**iii)** the maximum power it can develop.

**A** the total mass of muscle fibres.

**B** the number of fibres in a cross section of the muscle.

**C** the length of the fibres.

**D** the length of the A bands.

**5** The data below give the approximate numbers of muscle fibres in the quadriceps (the main knee extensor) and a laryngeal ('voice box') muscle, together with the number of axons in the nerve supplying each muscle.

| | Muscle fibres | Motor axons |
|---|---|---|
| Laryngeal muscle | $2 \times 10^3$ | 850 |
| Quadriceps muscle | $4 \times 10^6$ | 8000 |

Explain how the differences in nerve supply relate to the different roles of the muscles.

**6** The diagram below shows part of a muscle fibre with a mitochondrion. The letters **A–I** indicate materials that enter and leave the muscle fibre or the mitochondrion. Chemical changes are indicated by solid arrows, and movement of materials is indicated by dotted arrows.

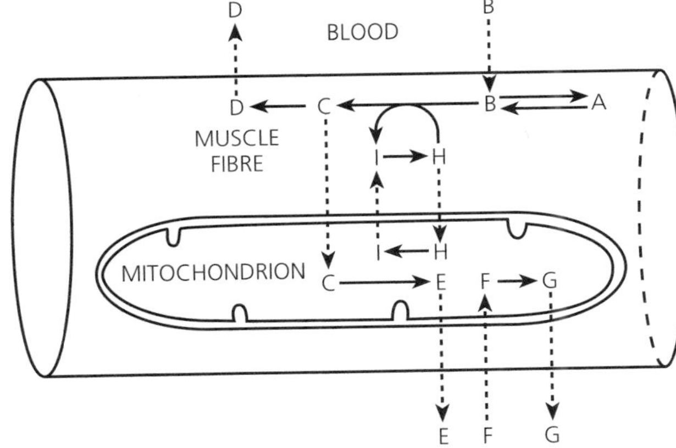


Identify each of the substances **A–I** from the following list:

glucose, glycogen, pyruvate, lactate, carbon dioxide, oxygen, water, ADP, ATP

CHAPTER 20

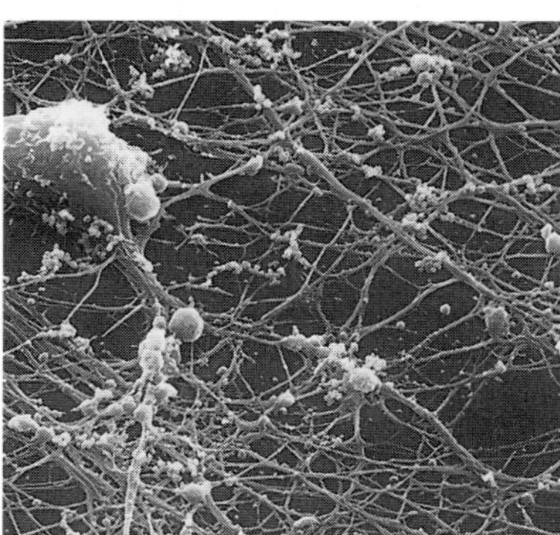

# CHEMICAL COMMUNICATION – the ENDOCRINE SYSTEM

**LEARNING OBJECTIVES**

**By the time you have completed your study of this chapter you should be able to:**

- **define a hormone and explain how an endocrine gland differs from an exocrine gland.**
- **compare the characteristics of hormonal and nervous communication.**
- **explain the concept of a second messenger, and why some hormones can influence their target cells without one.**
- **explain how a cascade effect can enable minute amounts of hormone to have large effects.**
- **explain the importance of hormone destruction by the body.**
- **describe the unique role of the hypothalamus in linking the nervous and endocrine systems.**
- **describe the structure of the pituitary gland and how the activities of its two lobes are controlled by the hypothalamus.**
- **name the eight most important hormones secreted by the pituitary gland and state their functions.**
- **describe the structure of the thyroid gland, the effects of its hormones, and how its activity is controlled.**
- **name the most important hormones secreted by each part of the adrenal gland, and state their functions.**

## The nature of the endocrine system

Whereas the nervous system transmits electrical signals rapidly along fixed pathways, the endocrine system uses chemical signals that travel in the blood. These chemical messengers are called **hormones** and are secreted by **ductless glands**. A hormone is a chemical that is secreted into the extracellular fluid by a cell or group of cells, and is carried in the blood to other parts of the body where it exerts specific *regulatory* effects.

Most glands (like sweat glands and salivary glands) are called **exocrine** glands because they secrete their products into an external tube or **duct**. Ductless glands do not secrete their products into ducts, but do so directly into the blood. Because they secrete their product internally, ductless glands are also called **endocrine** glands (*endon* is Greek for 'within'). The main endocrine glands are the hypothalamus, pituitary, thyroid, parathyroids, Islets of Langerhans in the pancreas, adrenals, gonads and the placenta (Figure 20.1). This chapter deals with the general principles of endocrine control, together with some of the endocrine glands. Other endocrine glands are dealt with in later chapters.

Some endocrine tissues form parts of organs that also have non-endocrine functions. For example, the gonads produce gametes as well as

Pituitary
Thyroid
Parathyroid
Adrenal
Islets of Langerhans in the pancreas
Ovary
Testis

**Figure 20.1**
Locations of the main endocrine glands in the human body

hormones, and the pancreas is also an exocrine digestive gland. To complicate things further, some endocrine glands actually consist of two glands in one, such as the pituitary, adrenal and thyroid glands.

A number of other organs not shown in Figure 20.1 have endocrine activity – for example, the kidney and the lining of the stomach and small intestine.

Hormones are involved in the control of a wide variety of processes. They play a central role in the maintenance of stable conditions inside the body and the control of growth and development. Such processes tend to have time scales from minutes upwards (the changes in puberty take several years) and often involve tissues that are widely dispersed round the body.

## Differences between endocrine and nervous signals

Hormonal signals differ in a number of important ways from nerve impulses (Figure 20.2):

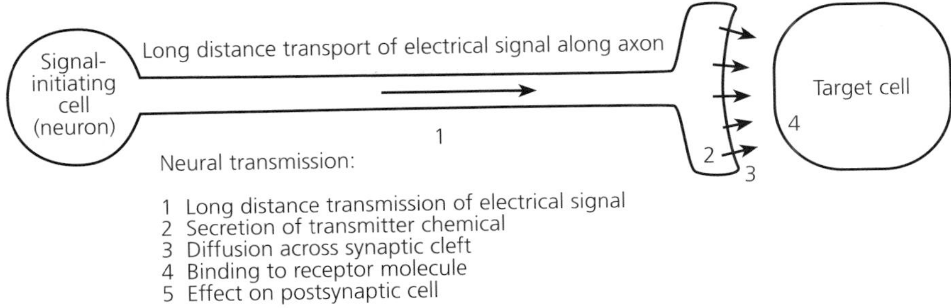


Signal-initiating cell
Target cell
1
2
3
4
BLOOD
Hormonal transmission:
1 Secretion of hormone
2 Long distance transport in blood
3 Diffusion to target cell
4 Binding to receptor molecule
5 Effect on postsynaptic cell

**Figure 20.2**
Comparison between nervous and hormonal pathways

- hormones take seconds rather than milliseconds to reach their destinations, which means that they travel about a thousand times more slowly.
- nerve impulses travel along fixed pathways, but hormones reach all parts of the body.
- whereas nervous signals vary in *frequency*, the strength of a hormonal signal is indicated by its *concentration*.
- a nerve impulse lasts only a millisecond or so; endocrine signals last for as long as the hormone is circulating, which may be from minutes to years.

In view of these differences, nerve impulses are clearly suited to controlling rapid, short-lived responses in particular parts of the body (such as muscles). Hormones are more suitable for producing slow, widespread and long-lived effects. For example insulin affects not only the liver, but most other parts of the body. The sex hormones influence growth in many parts of the body and over long periods of time.

Whilst the nervous system is structurally far more complex than the endocrine system, the latter has complications of its own. Whereas no effector cell is under the influence of more than two kinds of nerve ending (and many receive only one), some cells are affected by numerous hormones. Moreover, hormones can interact with each other so that the effect of one may depend on the level of another. One hormone may antagonise another, or it may only be able to exert its effect if the target cell has previously been exposed to another hormone. Another complication is that endocrine glands may influence each other, and examples of these **feedback effects** will be described in this and later chapters.

## Similarities between endocrine and nervous communication

Despite the differences between nervous and endocrine signals, there are some fundamental similarities. In fact the more that is learned about hormones, the less distinct the two systems appear.

Like hormones, nerve impulses exert their effect chemically. When a cell receives a nerve impulse it is not the impulse itself that causes the response, but the transmitter chemical released by the presynaptic membrane. This binds to a receptor protein in the postsynaptic membrane, triggering changes in the postsynaptic cell.

Although a hormone reaches all cells of the body it can only act on cells that are sensitive to it – the **target cells**. Having reached its target cell, a hormone binds to a specific receptor molecule. Only those cells with the appropriate receptors are sensitive to a particular hormonal signal. A hormone is thus like a neurotransmitter, except that it is released much further away from the target cell.

### *Hormone 'families'*

Chemically speaking, hormones fall into three chemical families:

- **proteins and polypeptides**, which include insulin and the various pituitary hormones. When given medicinally, these have to be injected rather than taken by mouth to prevent them being digested.
- **steroids**, which include the sex hormones and the hormones of the adrenal cortex.
- **catecholamines**, for example adrenaline.

These three categories of hormone differ in their modes of secretion, and also in their modes of action on their target cells. Polypeptides, proteins and amines are secreted by exocytosis of secretory vesicles and act indirectly on their target cells via a second messenger (see below). Thyroid and steroid hormones are exported by simple transfer through the plasma membrane (both are lipid-soluble), and actually penetrate their target cells.

### *How do hormones act?*

All hormones modify the activity of their target cells, but they do so in two quite different ways. On the one hand they may act directly on the activity of existing proteins, such as enzymes and membrane transport proteins. Peptide and protein hormones have predominantly this mode of action. Alternatively, they may act more directly on the genes, causing a change in the rate of synthesis of new proteins. Steroid hormones and thyroxine tend to act in this latter way.

## Second messengers

To influence its target cell, a hormone must first bind to a specific protein **receptor molecule**. Depending on whether the hormone can penetrate the target cell, the receptor may be in the plasma membrane or in the interior of the target cell. Steroid and thyroid hormones are lipid-soluble and readily penetrate the plasma membrane. Adrenaline and the various peptide and protein hormones, on the other hand, cannot enter the target cell because they are not readily soluble in lipid. Receptors for these hormones are situated in the plasma membrane and, since they span its thickness, can communicate with the interior of the cell.

When adrenaline or a peptide hormone binds to its receptor, it triggers the formation of a **second messenger** inside the cell (the hormone being the first messenger). The best known second messenger is **cyclic adenosine monophosphate** (**cAMP**). The binding of the hormone with the receptor activates an enzyme called **adenylate cyclase** located on the inner side of the membrane. This catalyses the conversion of ATP to **cAMP**, which then initiates changes in the chemical activity of the target cell (Figure 20.3).

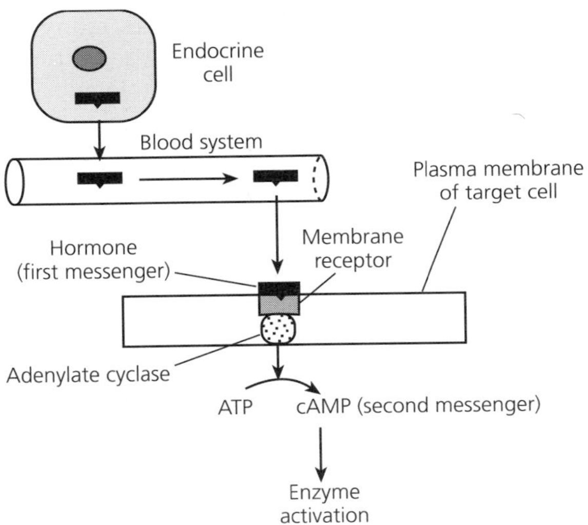


**Figure 20.3**
Role of a second messenger in hormonal communication

In carrying the signal between endocrine and target cells, the hormone is acting as an intercellular messenger. The second messenger is an intracellular signal, relaying the message *within* the cell. cAMP acts as second messenger for a number of hormones, including thyroid stimulating hormone, glucagon, follicle stimulating hormone, luteinising hormone and parathyroid hormone, but other second messengers are also known.

### *How are hormones effective in such minute concentrations?*

Insulin is present in the blood at concentrations about $10^{-11}$ times lower than the glucose whose concentration it helps to control. How can a hormone molecule affect so many molecules in the target cell?

The answer lies in a **cascade effect**, similar to that which occurs in blood clotting and in the rods in the retina (Chapter 18). The general idea can be illustrated by one of the effects of adrenaline – the stimulation of glycogen breakdown in the liver. The key enzyme involved is glycogen phosphorylase, which converts glycogen to glucose-1-phosphate (Figure 20.4).

**Figure 20.4►**
How a second messenger can greatly amplify a hormonal signal. Amplification consists of a 'cascade' in which enzymes activate other enzymes, producing an estimated 25 million-fold magnification of the incoming signal. Amplification steps are represented by double arrowheads

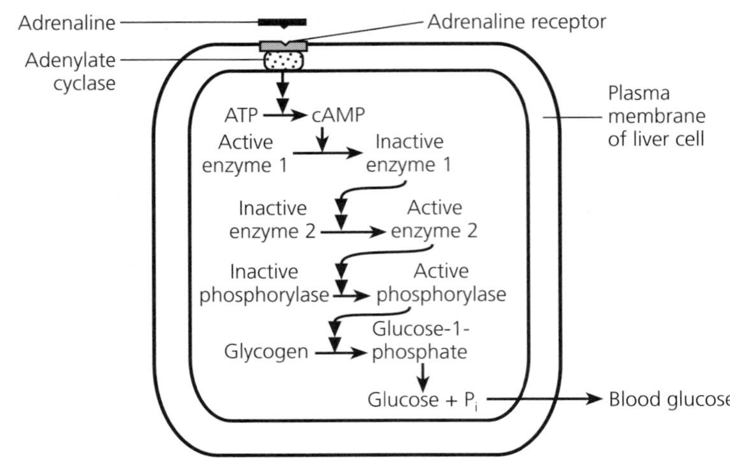


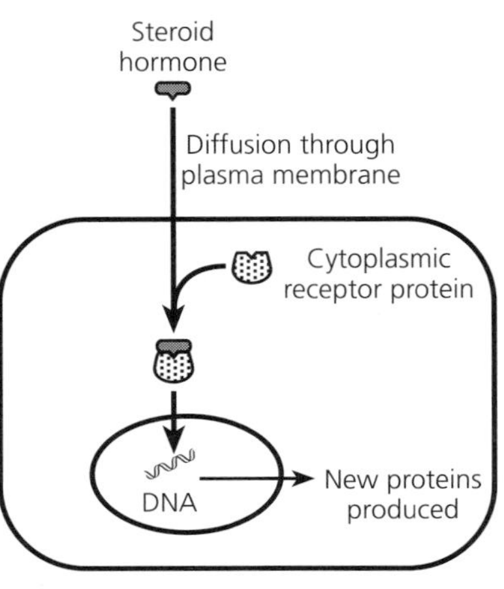


**Figure 20.5▲**
How steroid hormones influence their target cells

The formation of cAMP triggers a chain of catalysis in which enzymes act on enzymes. Each cAMP molecule activates an enzyme molecule, each of which activates many molecules of a second enzyme. Each of these enzyme molecules activates many molecules of a third enzyme, and each of these activates many molecules of glycogen phosphorylase. The result is that for each adrenaline molecule reaching a liver cell, millions of glucose-1-phosphate molecules are produced. Other enzymes then convert glucose-1-phosphate to glucose which is released into the blood. As a result of this catalytic cascade, the effect of adrenaline is relatively rapid – clearly important in the response to emergencies.

### *Intracellular receptors*

Because of the cascade effect, hormones with second messengers exert their effects within minutes. Lipid-soluble hormones freely enter the target cell and do not employ a second messenger but bind to receptors in the cell interior (Figure 20.5). These hormones take hours or even days to exert their effects.

## Changing the strength of the signal

At any given moment the concentration of a hormone depends on the balance between the rate of secretion into the blood and its rate of removal. Removal occurs by excretion via the urine, destruction by the liver and use by the target cell. If there is a constant rate of removal, hormone concentration can be changed simply by changing its rate of secretion.

The rate at which hormones are removed from the blood varies greatly, and so therefore do the rates at which hormone concentrations can be changed. A measure of the rate at which the concentration of a hormone can change is its **half-life** – the time taken for half the hormone to be removed from the blood. Hormones with short half-lives (such as adrenaline, with a half life of about 2 minutes) are secreted and destroyed rapidly, so a fall in output is followed rapidly by a fall in its blood concentration. Hormones with long half-lives (e.g. thyroxine, with a half life of nearly a week) are removed slowly and so their concentrations cannot be decreased rapidly.

One reason why hormones such as thyroxine and steroids are removed slowly is that a high percentage of the hormone is carried bound to a protein. Only the free hormone is active and only this form can be excreted or destroyed. The free hormone is in equilibrium with the bound form, so removal of the free hormone results in the slow release from the carrier protein. Peptide hormones, on the other hand, are carried unbound and are rapidly removed from the blood.

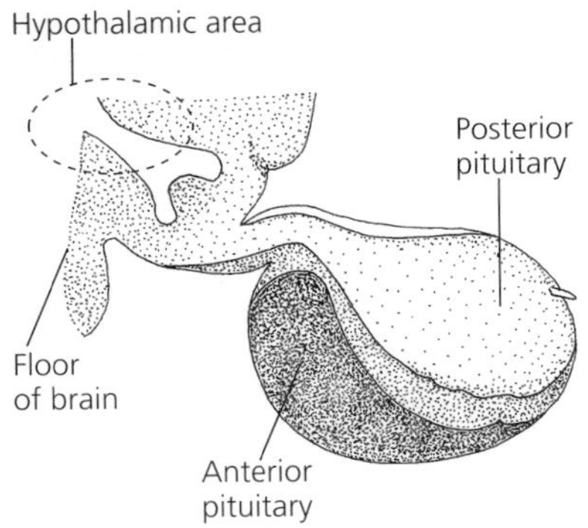


**Figure 20.6**
The pituitary gland and the hypothalamus

## The role of the hypothalamus

Despite the differences between the nervous and endocrine systems, they are by no means independent of each other. For example, the reproductive cycles of many animals are initiated by seasonal cues such as photoperiod (day length). These external stimuli lead to changes in behaviour and in the reproductive organs, involving hormonal changes. The endocrine system can therefore respond to changes in the external environment. Consequently, there must be a link between the neurosensory system and the endocrine system.

This link is the **hypothalamus**, a small area in the floor of the forebrain. The hypothalamus contains **neurosecretory cells**, which are unusual in that they produce both impulses and hormones. Besides being part of the nervous system, therefore, the hypothalamus is also an endocrine gland.

Some of the hormones secreted by the hypothalamus control the activity of another endocrine gland, the **pituitary**, which lies directly below it (Figure 20.6). In response to hormones from the hypothalamus, the pituitary secretes hormones that regulate the activity of the thyroid gland, gonads (ovaries and testes) and the adrenal cortex.

*The hypothalamus is thus the link between the two communication systems of the body.* As part of the nervous system, it receives input from receptors monitoring events in the outside world. Its close connection with the pituitary gland and its hormones also puts it in communication with the internal environment. This unique relationship between the nervous and endocrine systems enables the hypothalamus to play a key role in integrating many aspects of homeostasis (Chapter 21).

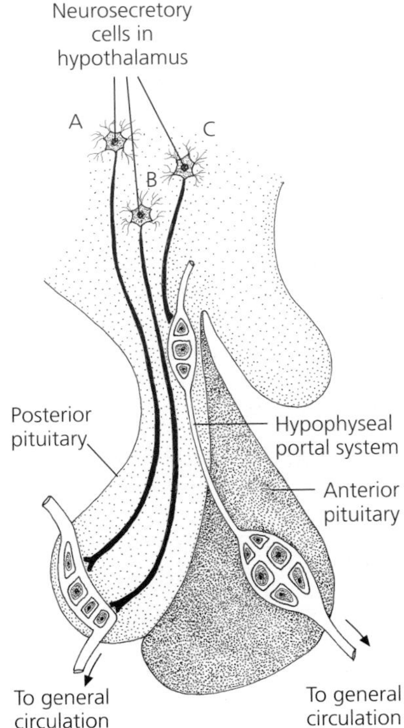


**Figure 20.7**
How the hypothalamus communicates with the two lobes of the pituitary. Cells A and B produce hormones which travel down the axons and are secreted by the axon terminals in the posterior pituitary. Cells of type C secrete releasing hormones into the hypophyseal portal system and stimulate the anterior pituitary to secrete trophic hormones

## The pituitary gland

The pituitary gland (or hypophysis) lies just beneath the hypothalamus, and is about the size of a large pea. It is sometimes described as the 'master gland' of the endocrine system because it controls three other endocrine glands – the thyroid, the gonads and the adrenal cortex.

It consists of two parts which are under different modes of control and have different embryological origins. The **anterior lobe** arises as an upgrowth of the anterior part of the gut and receives hormonal signals from the hypothalamus. The **posterior lobe** developed as a downgrowth of the brain and is controlled by nerve impulses from the hypothalamus.

### The anterior pituitary

The anterior pituitary secretes six important protein and polypeptide hormones. Four are called **trophic** hormones because they stimulate the action of other endocrine glands (*trophic* = 'feeding'). The most important anterior pituitary hormones and their chief effects are summarised in Table 20.1.

The production and secretion of these hormones depends on **releasing hormones** secreted by the hypothalamus. Unlike other hormones, these are not secreted into the general circulation but reach their destination by a tiny system of blood vessels called the **hypophyseal portal system** (Figure 20.7). For each anterior pituitary hormone there is a corresponding releasing hormone. For example, the secretion of thyroid stimulating hormone, or thyrotrophin, is stimulated by thyrotrophin releasing hormone (TRH).

### The posterior pituitary

The posterior lobe of the pituitary secretes two hormones. **Antidiuretic hormone (ADH)** stimulates the reabsorption of water by the kidney

**Table 20.1**
The main hormones produced by the anterior pituitary gland and their effects

| Hormone | Action |
|---|---|
| Thyroid stimulating hormone (TSH) | Stimulates production of thyroxine by the thyroid |
| Adrenocorticotrophic hormone (ACTH) | Stimulates secretion by the adrenal cortex |
| Follicle stimulating hormone (FSH) | Stimulates production of gametes by ovary and testis |
| Luteinising hormone (LH) | Stimulates production of hormones by ovary and testis |
| Growth hormone | Promotes protein synthesis and cell division in a wide variety of tissues, especially bones |
| Prolactin | Stimulates growth of mammary tissue and secretion of milk |

tubules, thus promoting water conservation. **Oxytocin** stimulates the contraction of smooth muscle in the uterus during labour. It also stimulates contraction of smooth muscle in the walls of the ducts of the mammary glands, squeezing out the milk.

Unlike the anterior pituitary hormones, the hormones secreted by the posterior pituitary are actually produced by neurosecretory cells in the hypothalamus, whose axons extend down into the posterior lobe. After synthesis in the cell bodies of the neurons the hormones are transported down the axons. On receipt of impulses at the axon terminals they are released by exocytosis into the capillaries of the posterior lobe of the pituitary.

## The thyroid gland

The thyroid lies just below and on either side of the larynx. It produces two iodine-containing hormones, **thyroxine** and **tri-iodothyronine**. Although tri-iodothyronine is produced in smaller quantities than thyroxine, it is about four times as potent. The thyroid gland also produces **calcitonin**, a peptide hormone which helps to regulate the concentration of calcium in the blood. In this chapter, 'thyroid hormones' refers to thyroxine and tri-iodothyronine.

Thyroid tissue consists of many **follicles** about 150–300 μm in diameter, each consisting of a sac containing a jelly-like **colloid** made of a protein called **thyroglobulin**. Each follicle has a wall of cubical epithelial cells which produce the thyroid hormones. Between adjacent follicles are the many capillaries which give the gland its bright red colour (Figure 20.8).

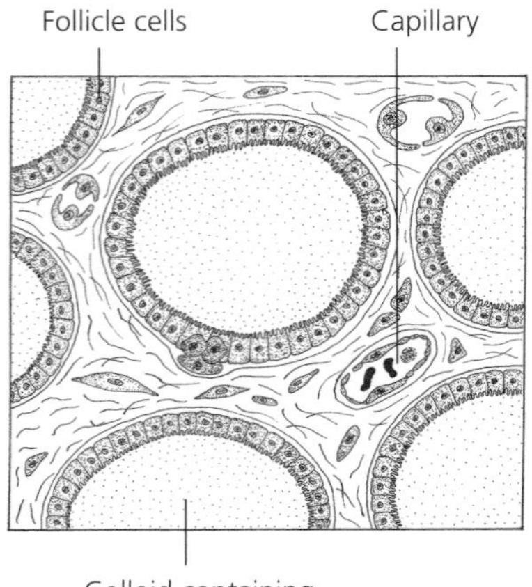


**Figure 20.8**
Fine structure of the thyroid gland

The concentration of iodine in the thyroid cells is normally about 30 times higher than in the blood. This ability to scavenge iodine from the blood and accumulate it is the reason why the radioactive isotope of iodine, $^{131}I$ (present in nuclear fallout) can lead to thyroid cancers.

Like adrenaline, both thyroid hormones are amines derived from the amino acid tyrosine. They are produced in an unusual way (Figure 20.9). First, tyrosine and other amino acids are used by the follicle cells to make thyroglobulin, which is secreted into the cavities of the follicles. Next, iodine atoms are added to tyrosine residues in the thyroglobulin. The iodinated tyrosines in adjacent polypeptide chains are then linked in pairs to form either tri-iodothyronine or thyroxine (depending on the number of iodine atoms added).

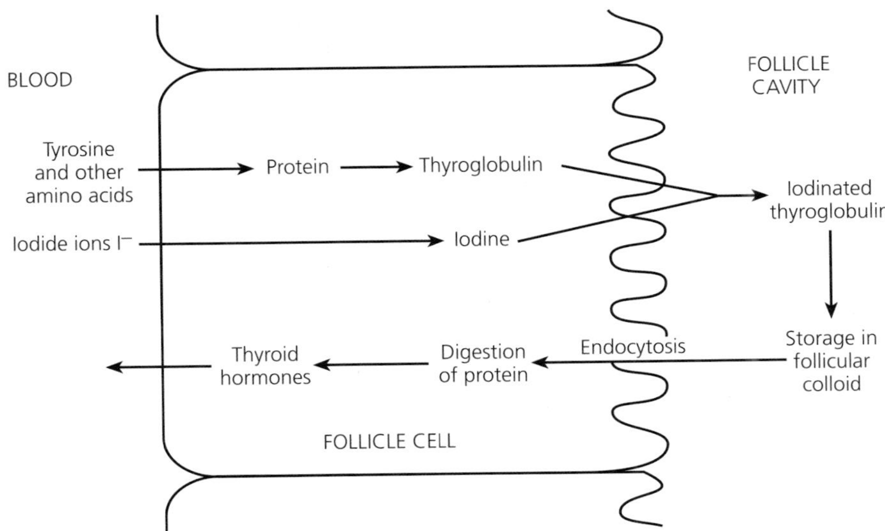


**Figure 20.9▶**
Production of thyroxine

The thyroxine and tri-iodothyronine are still 'locked up' as part of the peptide chain of the thyroglobulin. To produce free thyroxine, the follicle cells absorb the thyroglobulin by endocytosis and digest it, releasing the thyroxine and tri-iodothyronine. After secretion into the blood the hormones bind to plasma proteins, only a tiny proportion remaining as free hormone.

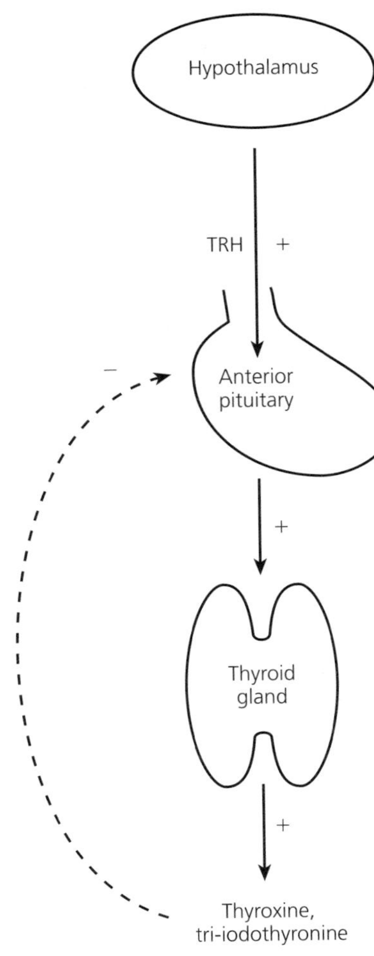


**Figure 20.10▲**
Feedback regulation of thyroid activity

## *Effects of thyroxine*

Thyroxine has two major effects. First, it stimulates respiration in most tissues causing the basal metabolic rate (BMR) to rise. Second, it stimulates protein synthesis, promoting growth in children. An interesting developmental role of thyroxine in amphibians is that it is necessary for metamorphosis – the change from a tadpole to the adult form.

One of the most important roles of thyroxine in mammals and birds is the regulation of body temperature. By adjusting heat production to balance heat loss, body temperature is maintained within narrow limits.

## *Controlling thyroid activity*

The activity of the thyroid gland depends on thyrotrophin or thyroid stimulating hormone (TSH) secreted by the anterior pituitary. The secretion of TSH is in turn *inhibited* by thyroid hormones. The higher the concentration of thyroid hormones in the blood, the more the secretion of TSH is inhibited (Figure 20.10).

As a result, any change in the output of thyroid hormones is automatically countered. Suppose the thyroid secretion increases; the rise in thyroid hormone inhibits the anterior pituitary, which reduces its output of TSH, which in turn causes a reduction in output of thyroid hormones. If thyroid output decreases, the reverse chain of events occurs, causing the level of secretion to rise back to normal.

The upshot of all this is that the more the concentration of thyroid hormone departs from the normal level, or **set point**, the stronger the counteractive effects. This kind of situation, in which the result of a process is to inhibit that process, is called **negative feedback**. Negative feedback is a component of all homeostatic mechanisms, and a number of other examples will be described in Chapter 21.

Negative feedback neatly explains why the concentration of thyroid hormone in the blood does not change very much, but it leaves an important question unanswered. How is the output of thyroid hormone adjusted according to different circumstances?

This is where the hypothalamus comes in. At any given moment the secretion of TSH by the pituitary results from a balance between stimulation by the hypothalamus and feedback inhibition by the thyroid. In response to a slight decrease in body temperature, the hypothalamus steps up its output of TRH. If, on the other hand, body temperature rises, TRH output decreases. The role of the hypothalamus is, therefore, to regulate the set point of thyroid hormone concentration in accordance with the needs of body temperature regulation.

## When the thyroid gland malfunctions

Although the thyroid is not essential to life, it is necessary for normal body functioning. The thyroid gland can malfunction in two different ways: it may be underactive, secreting too little hormone, or overactive, secreting too much.

The effects of **hypothyroidism** (or underactivity of the thyroid) depend on the age at which it first occurs. A very young child with an underactive thyroid is normal at first because it has obtained thyroid hormone from its mother. Soon, however, growth begins to be retarded, especially growth of the brain, causing a form of impaired mental development called **cretinism**. If thyroxine is given soon enough the condition can be cured, but it becomes irreversible if delayed for too long.

In adults, hypothyroidism causes **myxoedema**. In this condition, the metabolic rate is abnormally low, so fuel is stored rather than used in respiration. As a result the person is somewhat overweight, lethargic and sensitive to cold.

Underactivity of the thyroid may result from insufficient iodine in the diet rather than any defect in the thyroid itself. In some areas, the water supply may contain insufficient iodine for normal thyroid function. The reduced level of thyroid hormone causes the level of TSH to be higher than normal, resulting in an enlarged thyroid, or **goitre**. Before iodine was added to salt, such goitres were common in parts of the Pennines, giving rise to the term 'Derbyshire neck'.

Hyperthyroidism, or overactivity of the thyroid, results in an abnormally high metabolic rate, with fuel being respired rather than stored. As a result the person is thin, restless and often nervous.

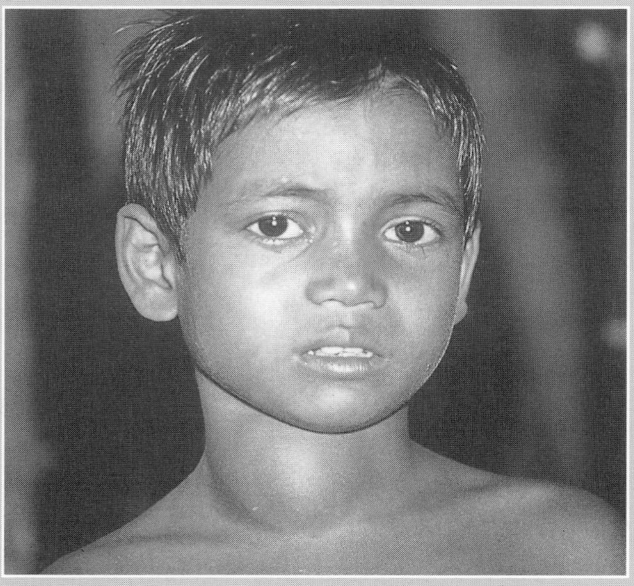

A young boy suffering from goitre due to underactivity of the thyroid

## The adrenal glands

The adrenal glands lie close to the kidneys – immediately on top of them in humans (hence *ad* renal). Each consists of two parts, an inner **medulla** and an outer **cortex**, which have no relationship other than their intimate position (Figure 20.11). Their hormones are chemically quite different, and are controlled in completely different ways. Their embryological and evolutionary origins also differ.

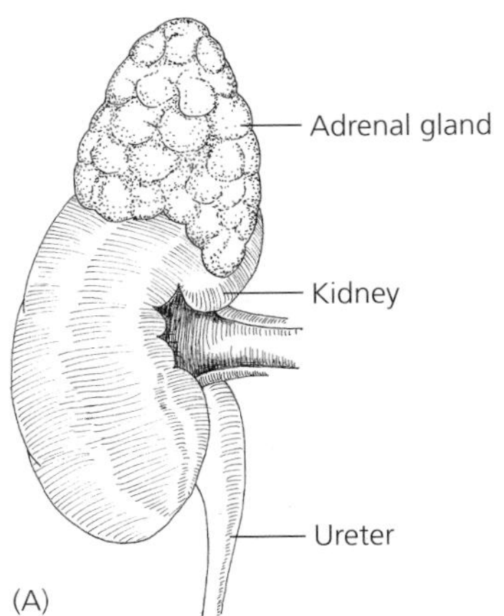


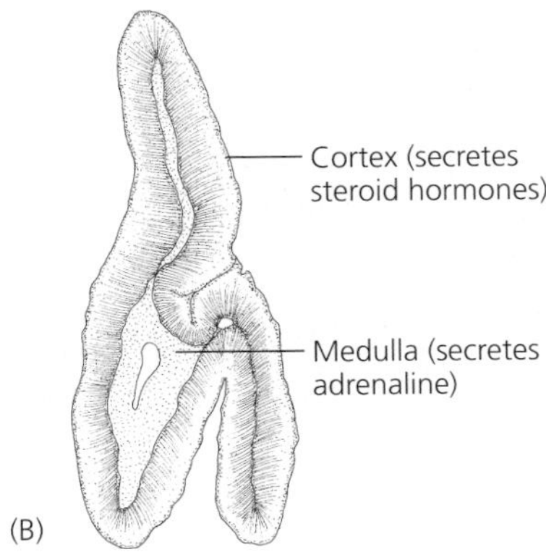


**Figure 20.11**
The adrenal gland; (A) external view, (B) longitudinal section

## The adrenal cortex

Whereas the adrenal medulla can be removed without serious consequences, the cortex is essential to life. The cortical hormones are all steroids, and, on the basis of their functions, can be divided into two main types: mineralocorticoids and glucocorticoids.

- **Glucocorticoids** These are involved in the metabolism of glucose and amino acids, and are secreted by the middle layer of the cortex. The most important is **cortisol** (hydrocortisone), which stimulates the production of glucose from amino acids, raising the blood glucose concentration. It is also essential if the body is to cope with almost any kind of stress such as injury, exposure to cold and infection. The secretion of glucocorticoids is dependent on the supply of **adrenocorticotrophic hormone** (**ACTH**) from the anterior pituitary. For this reason, the most serious effect of removal of the pituitary is on the ability of the body to cope with stress.
- **Mineralocorticoids** These are involved in the regulation of the plasma sodium and potassium levels. The most important mineralocorticoid is **aldosterone**, secreted by the outermost layer of the cortex. Aldosterone stimulates the reabsorption of sodium ions from the kidney tubules in exchange for potassium. Under-secretion of aldosterone leads to **Addison's disease**, in which the plasma sodium level is low, leading to reduced blood pressure. Unlike the glucocorticoids, the secretion of aldosterone does not depend on the pituitary gland, but is regulated in a different way (Chapter 24).

## The adrenal medulla

As explained in Chapter 17, the inner part, or **medulla**, is actually a greatly enlarged sympathetic ganglion in which the postganglionic neurons have lost their axons and secrete their product directly into the blood rather than across a synapse. As such, the medulla is under direct control by the brain. It secretes two catecholamine hormones: **adrenaline** and smaller quantities of **noradrenaline**. Both prepare the body to deal with emergencies, particularly those requiring violent physical activity. The effects of adrenaline are summarised in Table 20.2.

**Table 20.2**
Effects of adrenaline

| Target | Effect |
|---|---|
| Cardiac muscle | Stimulates |
| Blood glucose | Stimulates conversion of glycogen to glucose |
| Metabolic rate | Increases |
| Brain | Increased alertness |

## QUESTIONS

**1** Draw up a table summarising the similarities and differences between hormonal and nervous communication.

**2** Give two examples of neurons that secrete a hormone into the blood.

**3** Name a gland that
**a)** is both an endocrine gland and an exocrine gland.
**b)** is two endocrine glands in one.

**4** Why is it important that hormones are destroyed in the body?

**5** After removal of the thyroid gland it takes several days for thyroxine levels to fall significantly, but after removal of the pancreas the level of insulin falls within a few minutes. Suggest an explanation based on how these hormones are carried in the blood.

CHAPTER 21

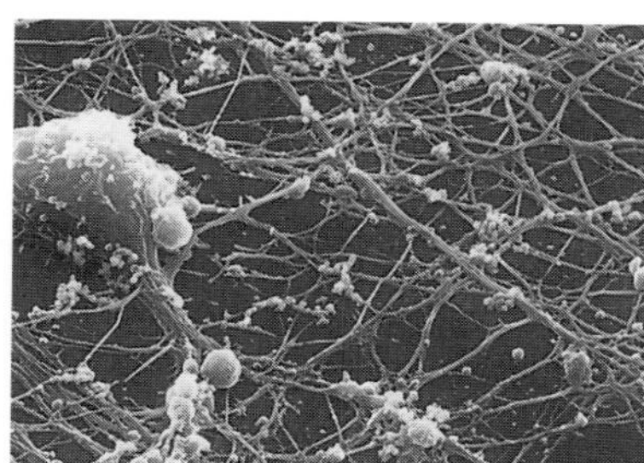

# HOMEOSTASIS

## LEARNING OBJECTIVES

By the time you have completed your study of this chapter you should be able to:

- explain what is meant by homeostasis, with reference to a unicellular organism, a flowering plant and a mammal.
- explain the requirement for energy and sensitivity in homeostasis.
- explain what is meant by the term 'internal environment' in a complex organism such as a mammal.
- describe the role of the pancreas, adrenal glands and liver in the regulation of blood glucose, showing how it illustrates the concept of negative feedback.

## What is homeostasis?

Life processes can only occur under a relatively narrow range of conditions. This is largely because of the chemical conditions which globular proteins require for their activity. Enzymes, for example, are very sensitive to pH, and many require the presence of metal ions such as potassium, magnesium and cobalt.

What this means is that conditions inside and outside an organism always differ from each other, and in environments such as fresh water and dry land they differ widely. For life to continue, organisms must be able to maintain the special internal conditions that make metabolism possible.

You might think that one way of doing this would be for an organism to surround itself with an impermeable outer layer. Some organisms do cover parts of their bodies in this way. Insects and the young shoots of flowering plants, for instance, have an almost impermeable cuticle.

The trouble is that while this would greatly reduce disadvantageous outward leakage of useful materials and the inward leakage of harmful materials, it would also prevent the uptake of raw materials and the getting rid of waste products. Impermeable barriers can, therefore, never completely cover an organism – there must be some parts of the body surface through which nutrients can be absorbed and waste products can be got rid of.

Organisms are thus **open systems**, continuously exchanging materials with their surroundings. Some disadvantageous leakages are inevitable, yet despite this organisms are, to some extent at least, able to regulate the compositions of their insides.

An organism illustrating some of these ideas is *Amoeba*. As in all freshwater organisms, ions and other solutes are much more concentrated

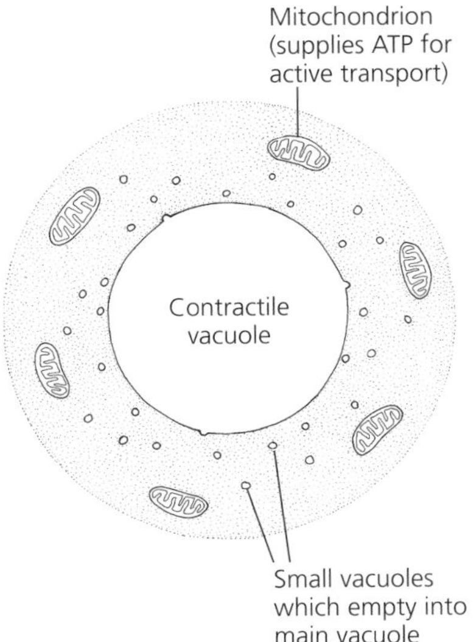


**Figure 21.1**
The contractile vacuole of *Amoeba*

inside the body than outside, so water enters by osmosis. If this were to continue unopposed, the interior of the cell would become diluted and it would swell and eventually burst.

The reason why this does not happen in *Amoeba* is that a **contractile vacuole** pumps water out of the body as fast as it enters (Figure 21.1). The vacuole slowly fills with water and eventually empties its contents by fusing with the plasma membrane.

This ability to maintain stable internal conditions in a hostile external environment is called **homeostasis** (from the Greek *homoios* meaning 'alike' and *stasis* meaning 'standing'). The term was coined in 1929 by Walter Cannon, a Harvard University physiologist. Although Cannon's ideas centred around the human body, it is important to appreciate that to some degree homeostasis is a fundamental property of all living things. One of the most familiar examples is the ability of birds and mammals to maintain their body temperature in an environment that may be cooler or warmer than the body.

### Homeostasis requires energy

To maintain a difference between the inside of the body and the external environment requires energy. To see why, think of *Amoeba* again. The contractile vacuole contains a solution that is more dilute than the cytoplasm. In passing from cytoplasm to vacuole, water is moving from a region of low water potential (the cytoplasm) to a region of higher water potential (the vacuole). Any movement of a substance against a concentration gradient requires the expenditure of energy. Evidence for this comes from the fact that in the presence of cyanide (a respiratory inhibitor) the contractile vacuole ceases activity and the organism swells and bursts, as a result of the accumulation of water.

### The detection of change

Though energy expenditure is a fundamental requirement for homeostasis, in many organisms such energy expenditure is not the only requirement. Even a fresh water *Amoeba*, which lives in an osmotically-stable environment, can adjust to change. In dilute salt solutions the rate of pulsation of the vacuole is slower, and in more concentrated solutions it stops altogether. In some way therefore the rate at which water is pumped out is adjusted to meet the rate at which it enters by osmosis.

In land habitats (and even more so on the sea shore) conditions are highly changeable, so the organism must be able to detect changes and to make the necessary adjustments. In land plants, for example, severe wilting of leaves results in the release of a substance which causes the stomata to close, reducing further water loss.

## The internal environment

In animals with a blood system, the cells are bathed in fluid. In arthropods and other animals with an open blood system, this fluid is the blood itself. In vertebrates and other animals with a closed blood system, the cells do not come into direct contact with the blood, but are bathed by **tissue fluid**, which acts as an intermediate between blood and cells. Most solutes in the tissue fluid (except proteins) are freely exchanged with those in the blood, so the composition of these two are similar and they can be considered together as the **extracellular fluids**.

All animals with a blood system can to some degree regulate the composition of their extracellular fluid. In most marine invertebrates this is similar to that of sea water. Except in estuaries and the sea shore, sea water has a stable composition. Most marine invertebrates consequently have little need – or capacity – to regulate the composition of their extracellular fluid.

Whilst the cells of marine invertebrates are osmotically similar to the extracellular fluid, their ionic compositions are quite different. Just as in unicellular organisms, the individual cells of marine invertebrates must, therefore, be capable of homeostatic regulation.

Fresh water animals and marine vertebrates are another matter. In these creatures the composition of the extracellular fluid is very different from the fluid outside the body. It was Claude Bernard, a 19th Century French physiologist, who likened tissue fluid to an **internal environment** or 'milieu interieur'. It is the regulation of the internal environment, rather than any particular toughness of the cells, that enables an animal to remain active in a hostile external environment.

The capacity to regulate the composition of the internal environment is greatest in the vertebrates, particularly in mammals and birds. The features of the internal environment that can be regulated are numerous and the list below includes only the most important ones.

- Osmotic concentration, or solute potential
- Sodium and potassium concentrations
- Calcium concentration
- Glucose concentration
- Carbon dioxide concentration
- Body temperature

Some aspects of homeostasis – such as osmotic and ionic regulation and the regulation of body temperature – are quite complex and merit chapters to themselves. This chapter is concerned with some of the general principles by which the composition of the blood is maintained, as illustrated by the regulation of the glucose level in the blood.

# The control of blood glucose

Although fatty acids are the most energy-rich fuel, glucose is the major energy-provider in humans; and for the central nervous system, it is the only fuel. Though the need for glucose is continuous, its dietary supply is intermittent. As a result, the concentration of glucose in the blood leaving the small intestine is much higher after a meal than during a fast. Despite this, the concentration of glucose in the general circulation is relatively stable, ranging between 70 and 100 mg 100 $cm^{-3}$ of blood (0.07–0.1%). During prolonged fasting it rarely falls below 50 mg 100 $cm^{-3}$, and even after a very heavy carbohydrate-rich meal it seldom rises above 150 mg 100 $cm^{-3}$. An abnormally low blood glucose level is called **hypoglycaemia**, and an abnormally high level is called **hyperglycaemia**.

Before dealing with the hormones that control blood glucose level, it would be helpful to review the processes that produce and remove glucose (Figure 21.2). Glucose is removed from the blood by the following processes:

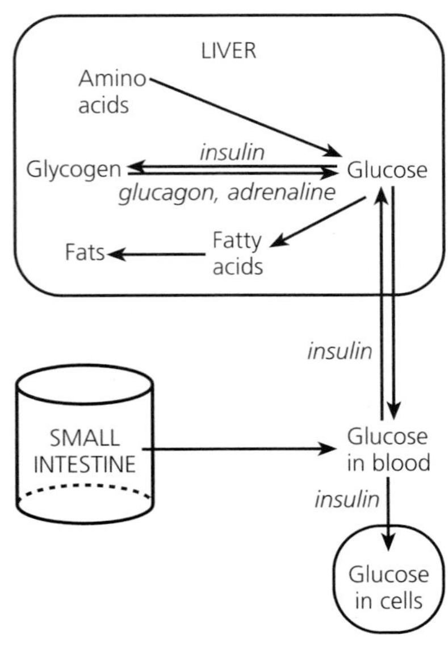


**Figure 21.2**
Processes that increase and decrease blood glucose concentration

- some is used in energy metabolism.
- it may be converted to glycogen in the liver and muscles for temporary storage.
- it may be converted to fat in the liver. The fat is then carried to the adipose tissue beneath the skin where it is stored.

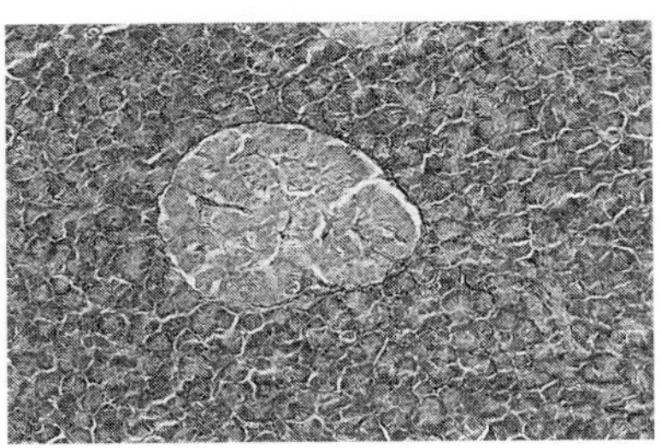

**Figure 21.3**
Light micrograph of a normal human pancreas, showing a section through an Islet of Langerhans

Blood glucose comes from three sources:

- it is absorbed from the gut.
- it is produced from glycogen in the liver.
- it is produced by deamination of amino acids in the liver. This is called **gluconeogenesis** (*neo* = new, *genesis* = origin), because the glucose is produced from a non-carbohydrate source.

To maintain a stable blood glucose level, glucose must enter and leave the blood at equal rates. The balance between these processes is regulated by a number of hormones, the most important being secreted by the pancreas and adrenal glands.

## The islets of Langerhans

The pancreas is both an exocrine and an endocrine gland (Figure 21.3). The endocrine tissue takes the form of about a million tiny clusters of cells called **islets of Langerhans**, named after Paul Langerhans, the German medical student who discovered them in 1869. Although they account for only about 1% of the pancreatic tissue, the islets take 10% of the blood flow, so each gram of endocrine tissue takes about 10 times as much blood as each gram of exocrine tissue. The blood leaving the islets enters the hepatic portal vein, which carries it to the liver. Since the liver is directly 'downstream' of the islets, it can respond more immediately to pancreatic hormones if they are secreted into the hepatic portal vein than if they entered the general circulation.

Each islet contains several distinct cell types. The most important are the central **β cells** which secrete **insulin**, and the outer **α cells** which secrete **glucagon**, which is antagonistic to insulin.

### *Insulin*

An islet is a small island and it is from the Latin *insula*, meaning 'island', that insulin takes its name. Insulin is a polypeptide of 51 amino acids, whose effect is to lower the blood glucose concentration. It does this in a number of ways:

- it stimulates the uptake of glucose by all cells (except those of the central nervous system, which can take up glucose independently of insulin).
- it promotes the synthesis of glycogen from glucose in liver and skeletal muscle.
- it stimulates the synthesis of fatty acids from glucose in the liver.
- it inhibits the production of glucose from glycogen and amino acids.

Though the islets of Langerhans receive autonomic nerve fibres they are independent of the nervous system for their output of insulin, the β cells responding directly to blood glucose level. The initial effect of a rise in glucose is to stimulate the release of insulin stored in secretory vesicles. This is followed by a longer-lasting increase in the rate of synthesis of insulin.

If either the secretion of insulin or the response of its target cells is impaired, **diabetes mellitus** results. *Mellitus* means 'honeyed', referring to the presence of glucose in the urine (in the days before chemical tests, glucose had to be identified by taste!). As explained in Chapter 24, the reason why glucose is excreted is that the blood glucose level rises so high that it cannot all be reabsorbed from the kidney tubules. This in turn makes it harder for the kidney to reabsorb water, so urine production increases. In severe diabetes the patient may produce over 20 litres of urine per day, resulting in intense thirst.

Without insulin the cells cannot use glucose so they switch to fatty acids. In the absence of carbohydrate these are burned inefficiently, producing acidic **ketone bodies** as waste products. Besides giving the breath a peculiar odour, they lower the blood pH, causing **acidosis** and coma and, if there is no treatment, death.

There are two kinds of diabetes mellitus with quite different causes. **Type I** or **early onset diabetes** mainly affects young people under 20. It is associated with the partial or complete loss of β cells in the islets, and so can only be treated by regular injections of insulin (it would be digested if taken orally). Because of this it is also known as **insulin-dependent diabetes**. Before insulin became available for injection, Type I diabetes was invariably fatal. The cause of β cell loss is thought to be either destruction by the body's own antibodies (an autoimmune condition), or an attack by a virus.

**Type II** or **late onset diabetes** affects much older people. In this type of diabetes, the β cells remain intact. Diabetic symptoms arise for one of two reasons. In some individuals, the level of insulin in the blood is normal, but the target cells become less sensitive to the hormone. In other cases, the level of insulin is low because its secretion is impaired. Type II diabetes develops gradually and can usually be managed by a strict diet.

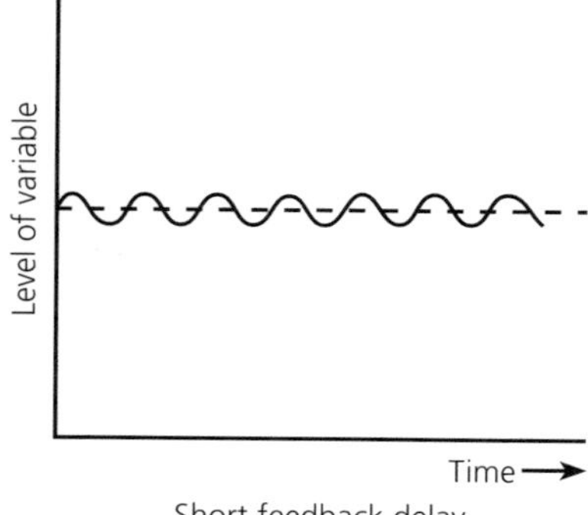


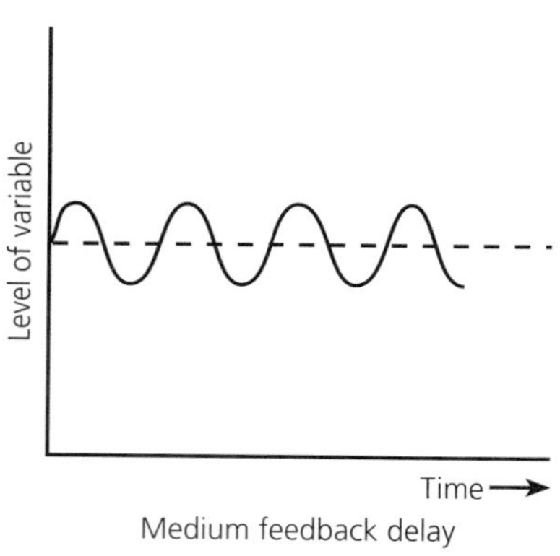


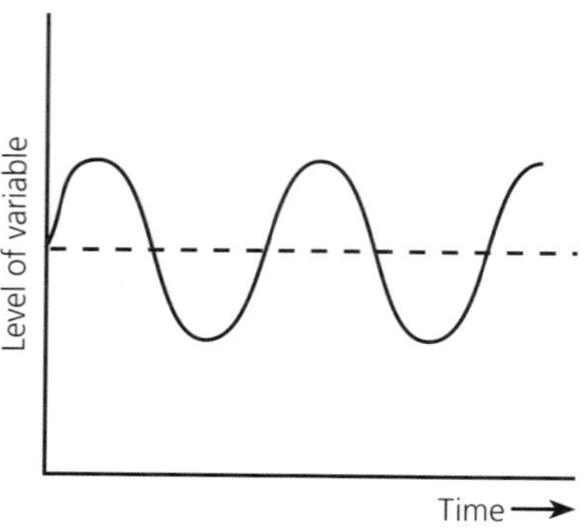


**Figure 21.4**
Overcorrection in feedback control brought about by delayed response

## *Glucagon*

Glucagon is a polypeptide hormone secreted by the α cells of the islets. It raises blood glucose level by mechanisms that are antagonistic to the actions of insulin. Its most powerful effect is on the liver, in which it promotes the production of glucose from glycogen and amino acids.

Like the β cells, the α cells respond directly to the glucose concentration of the blood. When blood glucose level falls, glucagon output rises, and vice versa. The effect of glucose on glucagon output is strongly influenced by insulin; a lowering of blood glucose has a much greater stimulatory effect on glucagon secretion if insulin concentration is low. Conversely, a rise in blood glucose inhibits glucagon secretion much more strongly if insulin concentration is high.

## *Feedback relationships in the control of blood glucose*

The direct sensitivity of the islets of Langerhans to glucose enables the blood glucose level to be finely controlled. When blood glucose rises, insulin output rises and glucagon secretion falls. Both changes bring about a corrective fall in blood glucose concentration. A fall in blood glucose causes the reverse effects. Moreover, the greater the change in blood glucose, the stronger the corrective effect. This self-correcting mechanism is an example of **negative feedback**, in which a change in the level of a factor has effects that counteract the change (Figure 21.5).

Although negative feedback is a self-correcting mechanism, it cannot produce absolute stability because there is inevitably some delay between a change and the correction. As a result the level of glucose – or any other aspect of the internal environment – fluctuates about the normal level or **set point**. The greater the delay between disturbance and response, the greater the overshoot (Figure 21.4).

To minimise delay in feedback, a hormone must both act and be destroyed quickly. The first effects of insulin, for example, occur within seconds, and half the circulating insulin is subsequently destroyed within

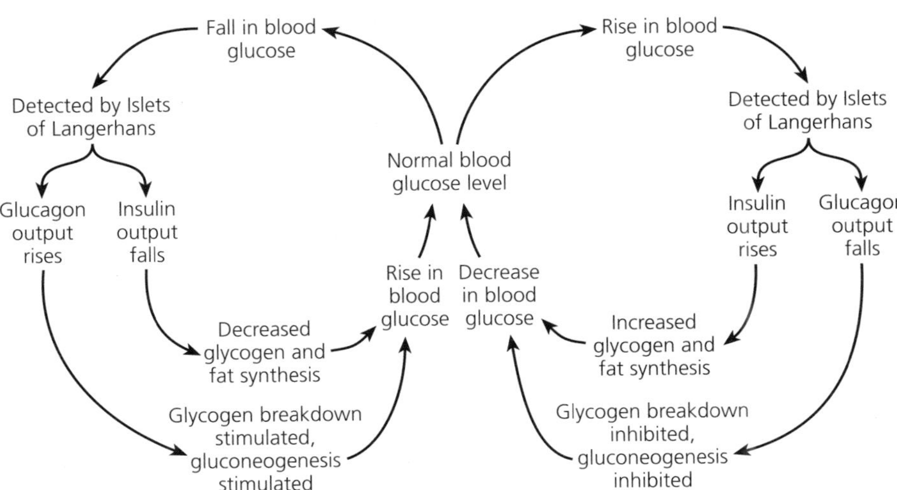


**Figure 21.5**
Feedback control of blood glucose

five minutes. Rapid action and destruction are both helped by the fact that, like other polypeptide hormones, insulin and glucagon circulate as free hormones rather than being bound to protein (Chapter 20).

## Cortisol

Cortisol is a steroid hormone secreted by the adrenal cortex. It is less important in the minute-by-minute fine-tuning of blood glucose than insulin and glucagon for two reasons. First, unlike insulin and glucagon, its secretion is not directly responsive to blood glucose, but is under the control of the hypothalamus via the anterior pituitary (Chapter 20). Second, like other steroid hormones it is slow-acting – its effects take hours or days.

Nevertheless, cortisol strongly influences blood glucose. Its most important effect is to stimulate the breakdown of muscle protein to amino acids and their conversion to glucose. This is particularly important in starvation in which muscle protein forms an important final energy reserve. If cortisol secretion is inadequate a person cannot use body protein as fuel, and death may result from hypoglycaemia.

### *Adrenaline*

Adrenaline is secreted by the adrenal medulla and plays a vital role in helping the body to cope with physical stress. Amongst other effects it stimulates the breakdown of glycogen to glucose.

## QUESTIONS

**1** Which of the following observations does *not* illustrate the concept of homeostasis?

- **A** The body temperature of a worm living at the bottom of the ocean changes little.
- **B** The osmotic concentration of the blood of many estuarine animals changes less than that of the surrounding water.
- **C** Earthworms burrow more deeply in winter.
- **D** The rate and depth of breathing rises during exercise.

**2** Give an example of homeostasis in a unicellular organism and a flowering plant.

**3** Why is it not possible to regulate any factor in the body so precisely that there are no fluctuations?

CHAPTER

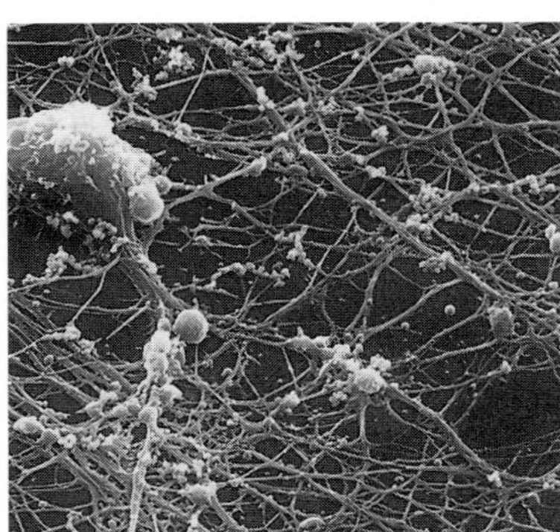

# 22 TRANSPORT in MAMMALS – the BLOOD SYSTEM

**LEARNING OBJECTIVES**

By the time you have completed your study of this chapter you should be able to:

- explain the relationship between body size and the possession of a blood system.
- describe the constituents of mammalian blood and state the functions of each.
- explain the functional significance of the dissociation curves for oxyhaemoglobin and oxymyoglobin, and the effect of carbon dioxide on oxygen transport.
- explain how different forms of haemoglobin can adapt animals to different environments, including that of the foetus.
- describe how carbon dioxide is transported, and explain how it is linked to the transport of oxygen.
- describe the main features of the clotting mechanism.
- use knowledge of the structure of the heart to interpret a graph of the cardiac cycle.
- describe how cardiac muscle differs structurally and physiologically from skeletal muscle, and explain the significance of the differences.
- explain how the contraction of the chambers of the heart is co-ordinated.
- describe the role of the autonomic and endocrine systems in the regulation of cardiac output.
- distinguish between the three types of blood vessel and explain how their structure adapts them to their functions.
- describe the three mechanisms by which substances enter and leave the blood.
- explain how tissue fluid is formed and how it is returned to the blood.
- describe how the blood system adapts to changing posture, to exercise and to bleeding.

## Constituents of the mammalian blood system

Like all vertebrates, mammals have a closed blood system. Unlike the open blood system of arthropods and most molluscs, in a closed blood system the blood is always separated from the body tissue by the walls of the blood vessels (Figure 22.1).

The mammalian blood system has three essential components:

- the transport medium, or **blood**, which carries substances in solution.
- the **heart**, which propels the blood.
- the **blood vessels**. These are of three types: **arteries**, which carry blood from the heart, **capillaries**, in which substances enter and leave the blood and **veins**, which bring blood back to the heart.

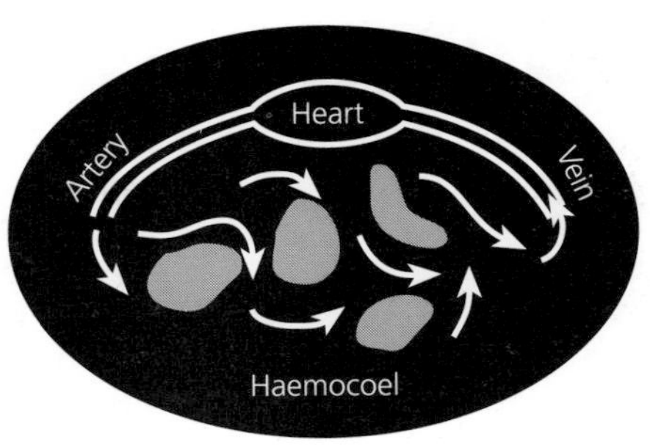


Open blood system, in which organs float in a 'lake' of blood (black) which flows sluggishly between the organs (shown here as grey islands).

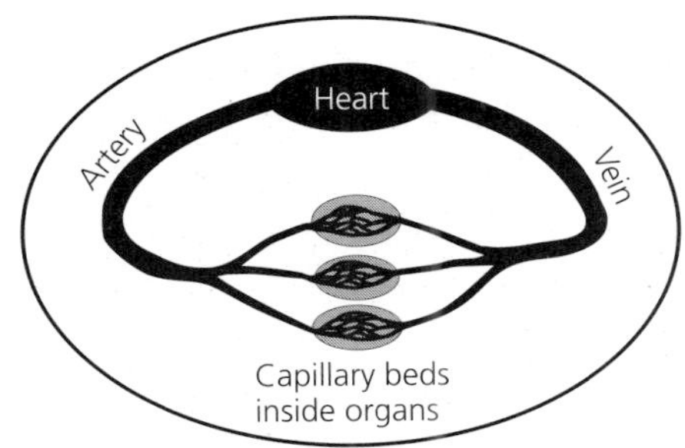


Closed blood system, in which blood is separated from the tissues by the walls of capillaries

**Figure 22.1**
Open and closed blood systems compared

## The need for a transport system

In all living things, substances move from one part of the body to another. Oxygen, for example, moves from the surface of the body to the mitochondria, and carbon dioxide in the reverse direction. In very small organisms these gases are transported by diffusion alone, but in larger animals this would be too slow. There are two reasons for this:

- the rate at which a substance diffuses from one point to another is inversely proportional to the distance between them (Chapter 6). This means that for an animal of given shape, doubling its thickness would halve the rate at which a substance would travel from the surface of the body to its centre.
- since oxygen enters the body through its surface, the *surface area* of the body limits oxygen supply. The need for oxygen is governed by (amongst other things) the *volume* of the body. As explained in Chapter 16, with increase in size, surface area increases more slowly than the volume. Provided its shape does not change, doubling an animal's length increases the volume (and hence oxygen demand) by $2^3 = 8$ times. The surface area (and thus oxygen supply) only increases $2^2 = 4$ times, so its oxygen supply/demand ratio is halved. In other words, each cell would get only half as much oxygen as before.

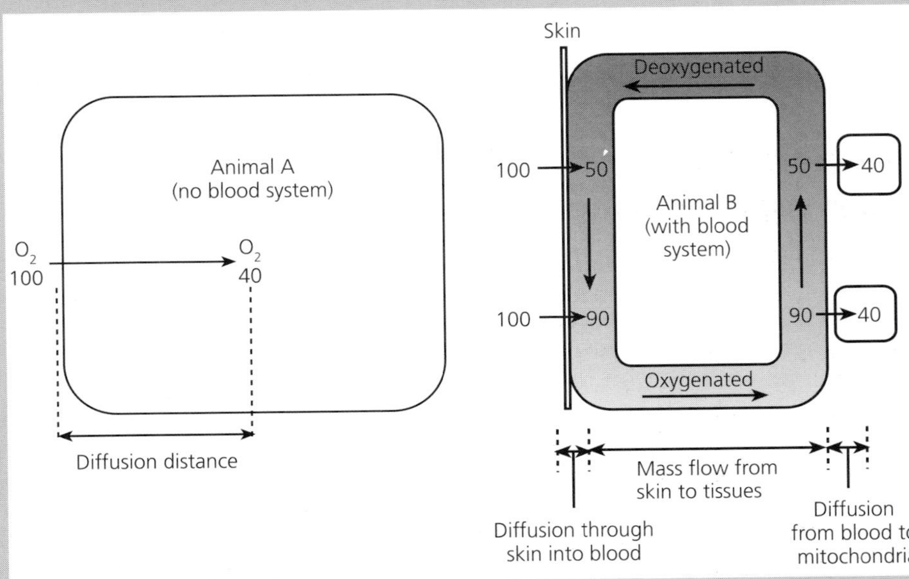


How a blood system speeds up diffusion. In both animals A and B the external oxygen concentration is the same (100 units) and the oxygen concentration in the mitochondria is 40 units. In A, oxygen travels the entire distance by diffusion, but in B, only the first and last stages are by diffusion

## The need for a transport system *continued*

Diffusion is fast enough for an active animal no thicker than a millimetre or so. In most larger animals (except for insects, Chapter 23), transport is by some form of **vascular system**, in which substances are carried in the **blood.**

As the diagram shows, a blood system greatly speeds up transport. Animals A and B are shown in cross section. In animal A, oxygen transport is by diffusion only, along a shallow concentration gradient extending from the body surface to the mitochondria. In animal B it occurs in three stages: diffusion across the body surface, mass flow in the blood and diffusion from blood to tissues. What the blood system does is to speed up diffusion at both ends of the journey by steepening the concentration gradient.

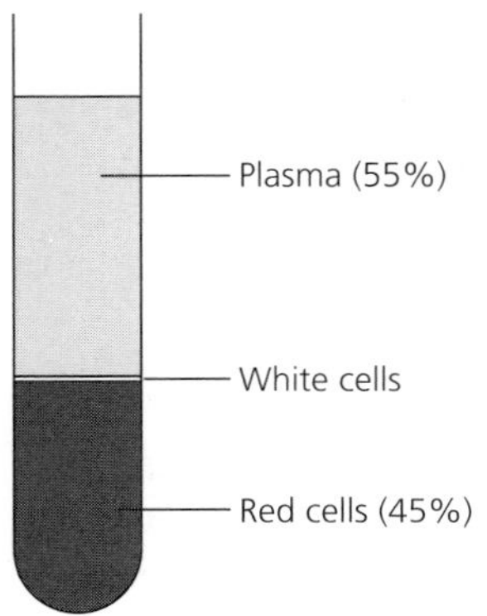


**Figure 22.2**
Centrifuged blood

Neutrophil Eosinophil
Blood platelets
Large lymphocyte
Basophil
Monocyte
Red corpuscle (erythrocyte)
Small lymphocyte

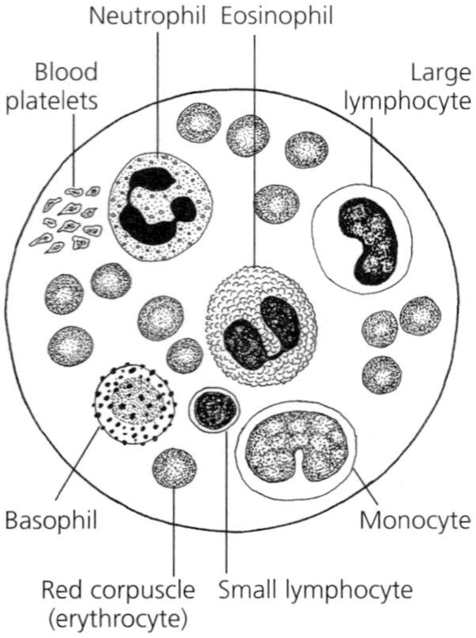

**Figure 22.3**
The main cell types in mammalian blood. In reality, red cells are far more numerous (after Freeman and Bracegirdle)

# Blood

By centrifugation, blood can be shown to be a suspension of cells or **corpuscles** in a pale yellow liquid or **plasma** (Figure 22.2). The corpuscles are of two categories: **red corpuscles**, or **erythrocytes**, and several kinds of **white corpuscles**, or **leucocytes**. There are also small **platelets** (thrombocytes) which play an important part in blood clotting (Figure 22.3). They are not cells but fragments that are budded off from cells in the bone marrow. Leucocytes are involved in defence and are dealt with in Chapter 27.

## Plasma

Plasma is a pale yellow aqueous solution containing a variety of substances, the most important of which are as follows:

- nutrients, such as glucose, amino acids, fatty acids, lipids and vitamins
- urea, a nitrogenous waste produced by the liver
- ions, such as sodium ($Na^+$), potassium ($K^+$), calcium ($Ca^{2+}$), magnesium ($Mg^{2+}$), chloride ($Cl^-$) and hydrogencarbonate ($HCO_3^-$)
- hormones
- antibodies
- fibrinogen and other clotting proteins
- gases, such as oxygen, carbon dioxide and nitrogen.

## Red corpuscles (erythrocytes)

Mature mammalian red corpuscles are biconcave discs, about 7.5 μm in diameter and about 2 μm in thickness (Figure 22.4). Though slightly wider than the narrowest capillaries, their shape can easily be distorted to allow them to squeeze through. They are by far the most numerous type; on average, each $mm^3$ of blood contains about 5 million red cells (men have slightly more than women). Their colour is due to **haemoglobin**, an iron-containing conjugated protein which combines reversibly with oxygen to form **oxyhaemoglobin**. Red cells also contain the enzyme **carbonic anhydrase** which is important in the transport of carbon dioxide.

When blood absorbs or gives up oxygen it changes its colour. **Deoxygenated** blood (blood low in oxygen) is dark red. If you pass oxygen through a flask of deoxygenated blood it changes to the scarlet of

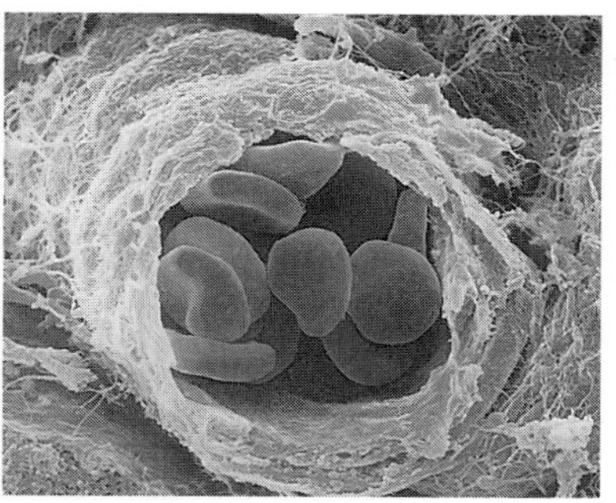

**Figure 22.4**
SEM of mammalian red corpuscles travelling through an arteriole

**oxygenated** blood. If you bubble nitrogen, or better still carbon dioxide, through oxygenated blood, it becomes dark red again.

Unlike those of other vertebrates, mature mammalian red cells lack a nucleus and other organelles typical of eukaryote cells. While this makes room for more haemoglobin, there is a price to pay in that red cells cannot repair themselves and only live about 4 months before being broken down in the liver. The high death rate of red cells is balanced by an equally high rate of production. This occurs by cell division in the red marrow of bones such as the vertebrae, ribs and sternum. Young red cells have a nucleus, but this is pushed out leaving only the haemoglobin-rich cytoplasm.

Despite their lack of mitochondria and other organelles, red cells do have some limited metabolic activity. For example, there is a sodium pump mechanism in the plasma membrane. The ATP needed to drive this is provided by glycolysis.

## Haemoglobin

Haemoglobin is a conjugated protein consisting of four subunits. Each subunit consists of a polypeptide **globin** bound to an iron-containing **haem** group which can combine with an oxygen molecule. A haemoglobin molecule can thus carry up to four oxygen molecules. In human haemoglobin, the four polypeptide chains are of two types: two **$\alpha$ chains** and two **$\beta$ chains**.

In both haemoglobin and oxyhaemoglobin the iron atom is in the iron (II) state (haemoglobin in which the iron is in the oxidised iron (III) state cannot carry oxygen). When haemoglobin combines with oxygen it therefore becomes *oxygenated* rather than *oxidised*.

There are actually many different haemoglobins in the animal kingdom which differ in their oxygen-binding properties. The differences result from varying amino acid sequences in the globins.

## Oxygen transport

A litre of plasma can dissolve only about 3 $cm^3$ of oxygen if shaken with air, but a litre of human blood can carry 200 $cm^3$ of oxygen – almost 70 times as much. The difference is accounted for by the fact that, whereas plasma can only carry oxygen in physical solution, red corpuscles carry it chemically combined with haemoglobin:

$$\text{oxygen} + \text{haemoglobin} \rightleftharpoons \text{oxyhaemoglobin}$$

Since haemoglobin can carry four oxygen molecules, oxygenation takes place in four stages. In the following equation haemoglobin is abbreviated to Hb:

$$\begin{aligned} Hb + O_2 &\rightleftharpoons HbO_2 \\ HbO_2 + O_2 &\rightleftharpoons HbO_4 \\ HbO_4 + O_2 &\rightleftharpoons HbO_6 \\ HbO_6 + O_2 &\rightleftharpoons HbO_8 \end{aligned}$$

In the lungs, where oxygen concentration is high, the formation of oxyhaemoglobin predominates over its dissociation into haemoglobin and free oxygen. In the tissues, where there is less oxygen, the reverse is true.

Haemoglobin is one of a number of **oxygen transport pigments** found in the animal kingdom. They are alternatively known as 'respiratory pigments', a term coined when the distinction between respiration and breathing was confused (and in medical circles, still is). Haemoglobin is also found in earthworms, some crustaceans and some molluscs. Another such pigment is **haemocyanin**, a copper-containing protein found in many crustaceans and molluscs. It is blue when oxygenated and colourless when deoxygenated.

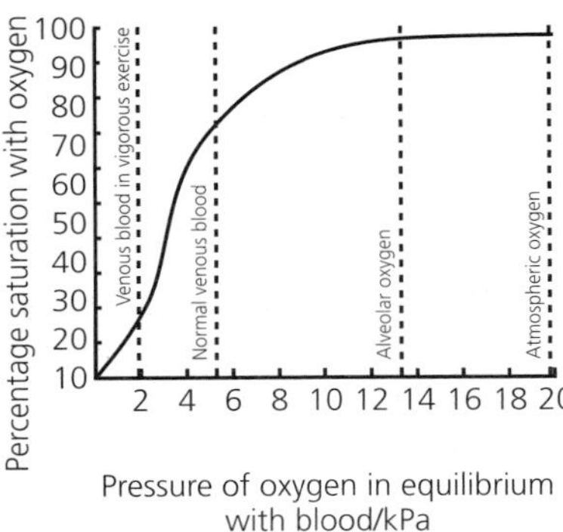


**Figure 22.5**
The oxyhaemoglobin dissociation curve for humans

In most invertebrates oxygen transport pigments are dissolved in the plasma. The disadvantage of this is that dissolved protein increases the viscosity of the blood, placing a limit on how much oxygen-carrying pigment the blood can contain. In vertebrates haemoglobin is carried in concentrated 'packets' in corpuscles. This allows the blood to contain a high concentration of haemoglobin without increasing its viscosity.

The efficiency of blood as an oxygen transporter depends on the *difference* between the oxygen content of the blood leaving the lungs and the blood leaving the tissues. To see how efficiently blood performs read the information box below and then look at Figure 22.5, which shows the **dissociation curve** for human oxyhaemoglobin. This is a graph showing how the percentage saturation of blood with oxygen changes with the partial pressure of oxygen.

## A word about units

The concentration of oxygen in blood is expressed as the pressure of oxygen in a gas mixture that would be in equilibrium with the blood. The pressure of oxygen in a gas mixture is expressed as its **partial pressure**, abbreviated to $pO_2$. This is the proportion of the total pressure that is contributed by the oxygen in the mixture. Normal atmospheric pressure is 101.3 kPa. Since approximately 21% of this pressure is due to oxygen, the partial pressure of oxygen in normal air is 21% of 101.3, or 21.3 kPa. Blood in equilibrium with normal air is said to have a $pO_2$ of 21.3 kPa.

An alternative way of expressing gas pressure (long favoured by physiologists) is in terms of millimetres of mercury (mm Hg), or the height to which the pressure can push a column of mercury. Atmospheric pressure = 760 mm of mercury = 101.3 kPa, and 1 mm Hg = 0.133 kPa.

To obtain an oxyhaemoglobin dissociation curve, samples of blood are left in sealed vessels with gases containing known partial pressures of oxygen. After each blood sample has come into equilibrium with the gas, the percentage saturation of the blood is determined.

The most significant feature of the graph is the fact that it is **sigmoid** or 's'-shaped (*sigma* is the Greek letter 's'). This means that it is steep at lower pressures of oxygen and relatively flat at higher pressures. This has two important consequences:

1 at the higher oxygen pressures in the lungs, a decrease in the partial pressure of oxygen ($pO_2$) has little effect on the oxygen content of the blood. This means there is a **safety factor**; if you hold your breath, even though the $pO_2$ in the lungs falls, there is little immediate effect on the oxygen content of the blood leaving the lungs.

2 at lower oxygen pressures (such as those found in the tissues) the curve becomes steeper. A small fall in $pO_2$ therefore causes a larger decrease in the percentage saturation of the blood, so more oxygen is released. Blood leaving most tissues is still about two-thirds saturated with oxygen, so there is a considerable reserve. In very active tissues such as skeletal muscle, the $pO_2$ may fall to very low levels.

It is also worth noting that for reasons explained in the next chapter, $pO_2$ in the lungs is only about two-thirds that of the atmosphere outside – yet at this $pO_2$ the blood is still 98% saturated with oxygen. Breathing pure oxygen, therefore, has little effect on the oxygen content of the blood leaving the lungs (though in certain medical conditions it is of great benefit).

## Why a sigmoid curve?

**A sigmoid oxyhaemoglobin dissociation curve facilitates the transport of oxygen by the blood. This, however, does not explain its shape; advantages are *consequences*, not causes. The curve is sigmoid because the polypeptide chains interact with each other. Combination with the first oxygen molecule causes a shift in the position of the polypeptide chains relative to each other. This increases the affinity of the other haem groups for oxygen and accounts for the increasing gradient of the curve at lower oxygen concentrations. This co-operative binding of oxygen cannot happen with myoglobin because it consists of a single polypeptide chain only. Consequently the dissociation curve for oxymyoglobin is not sigmoid.**

## Carbon monoxide poisoning

**Carbon monoxide is a colourless, poisonous gas, and the fact that it is also odourless makes it particularly dangerous. It is poisonous because haemoglobin combines with it 250 times more readily than it does with oxygen, forming carboxyhaemoglobin. This is even brighter red than oxyhaemoglobin and cannot carry oxygen. A patient suffering from carbon monoxide poisoning is given pure oxygen, which slowly displaces the carbon monoxide from the blood.**

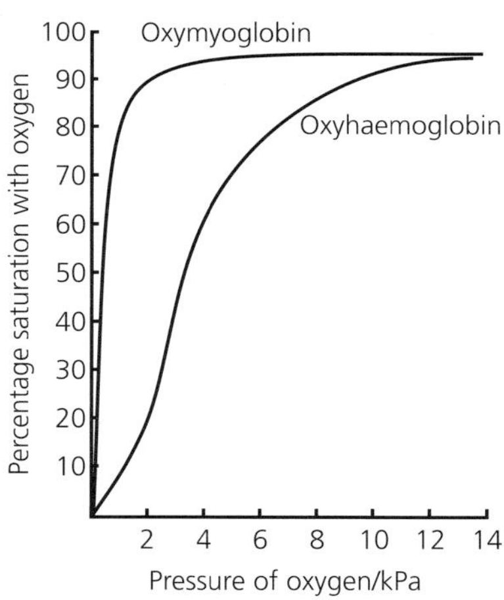


**Figure 22.6**
Dissociation curves for oxymyoglobin and oxyhaemoglobin

### *Myoglobin, or 'muscle haemoglobin'*

Unlike haemoglobin, myoglobin consists of a single polypeptide chain. Strictly speaking it is not a *blood* pigment since it occurs inside the fibres of skeletal and cardiac muscle. Myoglobin is responsible for the characteristic colour of 'red muscles' (Chapter 19), and is particularly abundant in the muscles of diving mammals such as whales and seals, in which it acts as an oxygen reservoir between periods at the surface.

Myoglobin differs from haemoglobin in two important ways:

- it has a much greater affinity for oxygen than haemoglobin. When a muscle contracts it squeezes the blood vessels, reducing the oxygen supply. As the oxygen level in the muscle falls, myoglobin releases its stored oxygen. When the muscle relaxes again, oxygen is transferred from haemoglobin to myoglobin.
- its dissociation curve is not sigmoid (Figure 22.6). The reason for this is that myoglobin has only one polypeptide chain, so there is no co-operative binding as there is in the case of haemoglobin.

## Haemoglobins in other species

**Oxyhaemoglobin dissociation curves for different species show considerable differences. The graph given below shows the dissociation curves for oxyhaemoglobin of a pigeon, a human and a lugworm. Because of structural differences between the lungs of mammals and birds, oxygen is more concentrated at the gas exchange surface in birds than it is in mammals. This is reflected in the properties of the**

## Haemoglobins in other species continued

haemoglobin. Pigeon blood requires a higher concentration of oxygen to saturate it. But because it releases its oxygen at a higher concentration, pigeon tissues can function at a higher oxygen concentration than human tissues.

At the other extreme is the lugworm which lives in burrows in tidal mud flats. When the tide is out there is no flow of sea water to bring fresh oxygen, and oxygen fall to low levels. Lugworm haemoglobin combines avidly with oxygen, only giving it up at very low oxygen concentrations. At low tide, lugworm blood is able to carry oxygen from the low concentration outside the body to the even lower concentration in its tissues.

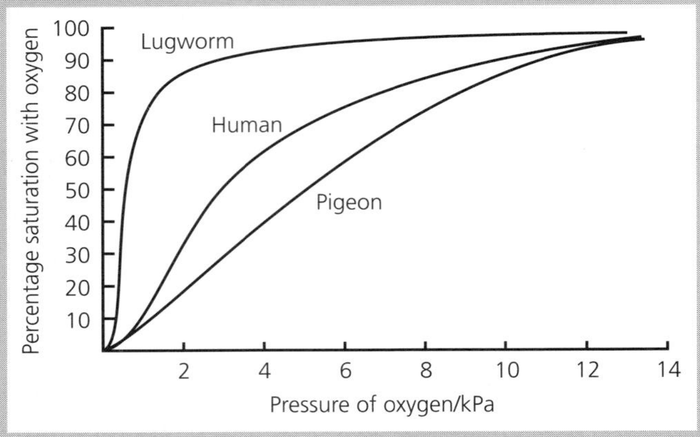


Oxyhaemoglobin dissociation curves for pigeon, human and lugworm

### *The effect of carbon dioxide – the Bohr effect*

To obtain an oxyhaemoglobin dissociation curve, only the partial pressure of oxygen is varied. When the oxyhaemoglobin dissociation curve is obtained for different partial pressures of carbon dioxide ($pCO_2$), the results shown in Figure 22.7 are obtained.

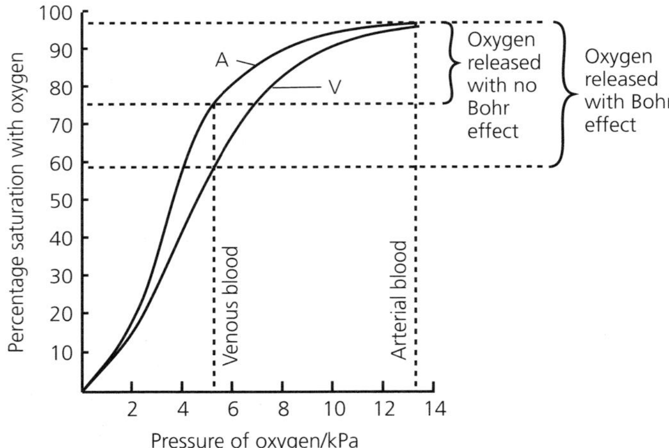


**Figure 22.7**
The effect of carbon dioxide on the dissociation of oxyhaemoglobin and on the amount of oxygen delivered to the tissues. A = Dissociation curve for haemoglobin at $pCO_2$ of arterial blood. V = Dissociation curve for haemoglobin at $pCO_2$ of venous blood

As you can see, raising the $pCO_2$ level shifts the dissociation curve to the right – a phenomenon called the **Bohr effect**. It means that at a given partial pressure of oxygen, a rise in $pCO_2$ level reduces the percentage saturation of the blood with oxygen. In other words, if the carbon dioxide level increases, the blood can carry less oxygen so more is released. An increase in $pCO_2$ is what happens in the tissues, so the blood releases more oxygen than it would do otherwise.

The effect of carbon dioxide is actually due not to carbon dioxide itself, but to the reduced pH as a result of carbonic acid. Lactic acid has the same effect, so in very active muscle the release of oxygen by the blood is further increased.

Another point worth noting is that the Bohr effect is most pronounced at lower partial pressures of oxygen, such as those found in the tissues. At the high oxygen levels in the lungs, a rise in $pCO_2$ has little effect on the ability of the blood to carry oxygen. This is important because if breathing is briefly interrupted, the rise in the carbon dioxide level in the lungs does not significantly affect the ability of the blood to pick up oxygen.

## Unloading pressures

A useful way of comparing the relative strength with which an oxygen transport pigment binds with oxygen is in terms of the **unloading pressure**. This is the partial pressure of oxygen at which the pigment is 50% saturated with oxygen, and is referred to as the $P_{50}$. Rather confusingly, physiologists often refer to the partial pressure of a gas as its 'tension'. This term is best avoided since tension is the precise opposite of pressure.

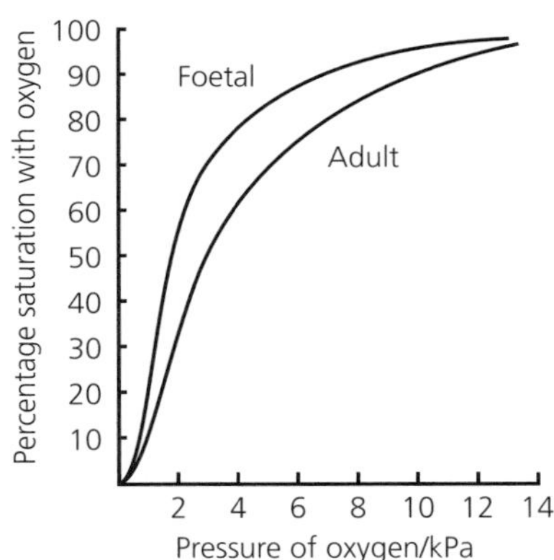


**Figure 22.8**
Dissociation curves for foetal and adult oxyhaemoglobin

## Haemoglobin in the foetus

In its early development a young mammal obtains its oxygen from the mother via the placenta. If foetal and adult haemoglobin were the same, no more than half the oxygen in the maternal blood would diffuse into the foetal blood. In reality the dissociation curve for foetal oxyhaemoglobin is to the left of that for maternal oxyhaemoglobin (Figure 22.8). This is because foetal haemoglobin has a stronger affinity for oxygen than maternal haemoglobin – in other words, it takes it up more readily and gives it up more reluctantly. The result is that more oxygen diffuses from maternal to foetal blood than would otherwise be the case.

The explanation for the difference is that foetal haemoglobin has two γ (gamma) polypeptide chains instead of two β chains (the α chains are the same). After a baby is born, the gene for the γ chain is somehow switched off and the gene for the β chain is switched on.

## Transport of carbon dioxide

Most carbon dioxide is carried in the plasma in the form of hydrogencarbonate ions. In the tissues, carbon dioxide diffuses into the red cells where it reacts with water to form carbonic acid. Normally this reaction is very slow, but the red cells contain an enzyme, **carbonic anhydrase**, which greatly speeds it up. The carbonic acid then dissociates into hydrogencarbonate and hydrogen ions:

$$CO_2 + H_2O \rightleftharpoons H_2CO_3 \rightleftharpoons H^+ + HCO_3^-$$

The hydrogen ions are buffered by the haemoglobin (see Chapter 2), and the hydrogencarbonate ions then diffuse out of the red cell into the plasma. The loss of negative ions encourages an inward diffusion of chloride ions into the red cell, a process called the **chloride shift**. The net result is that the plasma becomes slightly richer in hydrogencarbonate ions and the red cells become richer in chloride ions.

The process just described may seem a rather roundabout way of converting carbon dioxide into hydrogencarbonate ions, but it is actually of great significance. When oxyhaemoglobin absorbs hydrogen ions it promotes the release of oxygen, as shown in Figure 22.9. This is the explanation for the Bohr effect, and results from the fact that the reactions between haemoglobin and hydrogen ions, and between haemoglobin and oxygen, are linked by common intermediates.

Now we can see the significance of carbonic anhydrase. A red corpuscle takes no more than a second to travel the length of a capillary, so it must release its oxygen in this time. Without carbonic anhydrase the $H^+$ ions associated with $CO_2$ production would be produced too slowly for the Bohr effect to come into play before the blood has reached the end of the capillary.

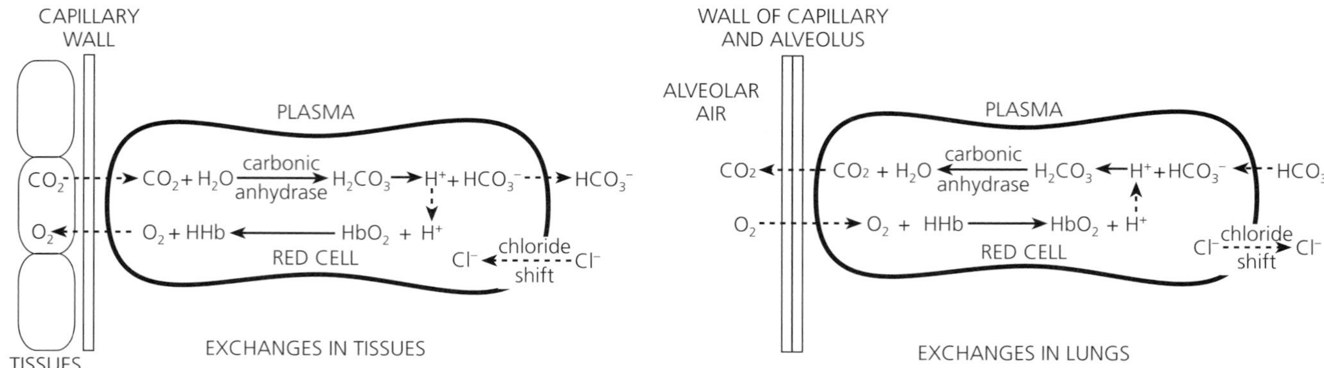


**Figure 22.9**
Interlinking of reactions involved in oxygen and carbon dioxide transport. Note that the reactions are not unidirectional as indicated here, but are equilibria

It is important to realise that all these reactions are equilibria, so each reaction can be driven in the reverse direction by a change in the relative concentrations of oxygen and carbon dioxide. This means that in the lungs, the uptake of oxygen encourages the release of carbon dioxide.

Besides travelling in the plasma as hydrogencarbonate ions, about a third of the carbon dioxide is transported as **carbamino compounds**, formed by combination with amino groups of haemoglobin and other blood proteins. In the following equation X represents the rest of the protein molecule.

$$X{-}NH_2 + CO_2 \rightarrow X{-}NH{-}COOH$$

## Blood clotting

Blood is always under pressure, so injury to a blood vessel causes loss of blood and provides a potential entry point for pathogenic (disease-causing) micro-organisms. These dangers are minimised by the ability of blood to seal the wound by forming a solid plug or **clot**.

The complexity of the clotting mechanism is linked to two requirements. As a protective mechanism it must occur as rapidly as possible. Second, the dangers of forming a clot in intact blood vessels are very great, so there must be ways of ensuring that clotting only occurs after injury.

The primary event in clotting is the conversion of **fibrinogen** (a soluble plasma protein made in the liver) into strands of insoluble **fibrin** (Figures 22.10 and 22.11). This forms a dense meshwork in which red corpuscles become trapped. Later the fibrin slowly contracts, pulling the sides of the wound closer together. The contraction of the clot also squeezes out fluid called **serum**, which is identical to plasma minus fibrinogen.

Obviously it is essential that the conversion of fibrinogen to fibrin only happens at an injured site. The catalyst for this process is the enzyme **thrombin**. This is normally present as an inactive precursor, **prothrombin**, whose synthesis in the liver requires Vitamin K. A key event in clotting is, therefore, the conversion of prothrombin to thrombin. This is brought about by a group of substances collectively called **prothrombin activator**.

Formation of prothrombin activator is very complex and involves a self-accelerating cascade of chemical events. Each conversion involves a chemical factor that is given a Roman numeral. It is lack of one of these, factor VIII, that is the cause of the most common form of **haemophilia**. Two events trigger the cascade:

- substances collectively called **thromboplastin** are released when tissues are damaged.

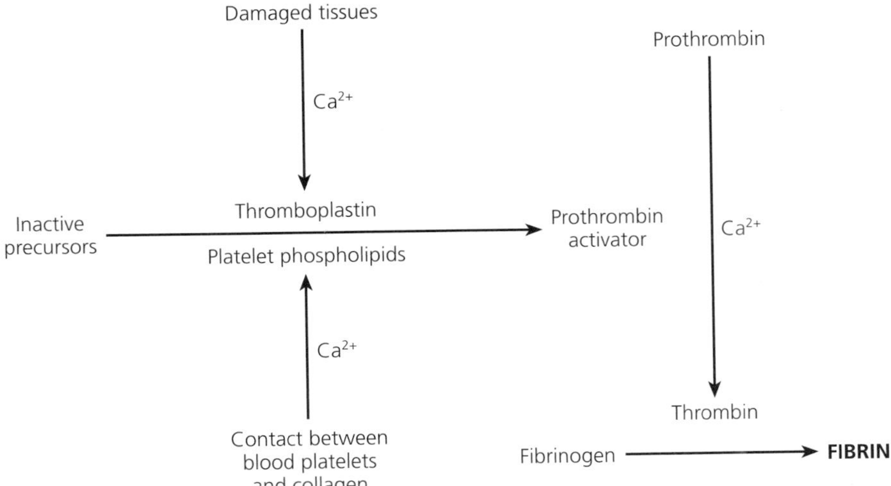


**Figure 22.10►**
Formation of a blood clot (the process has been highly simplified)

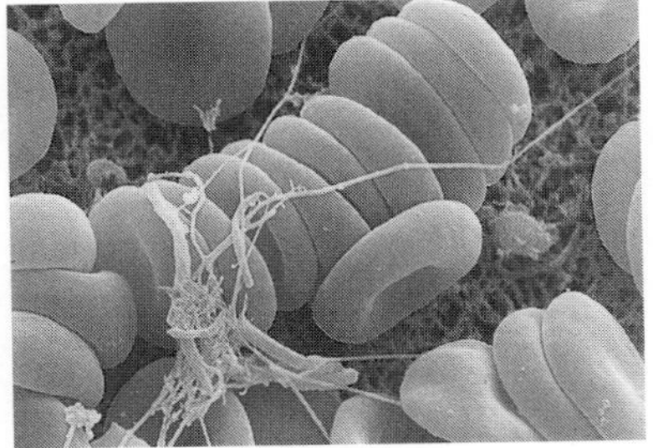

**Figure 22.11▲**
SEM of an early blood clot. The red blood cells form rolls as they become enmeshed in the fibrin fibres

- platelet phospholipids are released when blood **platelets** come into contact with collagen, as occurs when blood vessel walls are ruptured.

Most of the conversions in the cascade require calcium ions. A sample of blood can therefore be prevented from clotting by the addition of sodium citrate, which precipitates calcium ions as calcium citrate.

Besides these mechanisms there are various factors that inhibit clotting. These normally prevent the process from occurring in the absence of injury. Sometimes these mechanisms fail resulting in **thrombosis**, when a clot forms inside an intact blood vessel, usually a vein. If the clot breaks free it is carried along and blocks an artery further along the circulatory system.

## The circulation of the blood

To understand the circulation of a mammal, it is helpful first to briefly consider that of a fish (Figure 22.12A). The fish heart consists of a very thin-walled receiving chamber, the **sinus venosus**, a slightly thicker-walled **atrium** and a much more muscular **ventricle** which pumps blood out of the heart.

Notice two important things:

- the fish has a **single circulation** because the blood passes once through the heart in any complete circuit.
- all blood must pass through *two* capillary networks before returning to the heart.

By the time the blood reaches the second set of capillaries in a fish, its pressure has fallen considerably. A single circulation thus prevents blood being delivered to the tissues at sufficient pressure to sustain a high metabolic rate.

In mammals, the blood leaving the lung capillaries is returned to the heart before being pumped to the rest of the body (Figure 22.12B and C). Mammals have a **double circulation**, the blood passing twice through the heart in each complete circulation.

The mammalian heart is really two pumps in one. The right side receives deoxygenated blood and pumps it along the **pulmonary circuit** through the lungs. The left side receives oxygenated blood from the lungs and pumps it along the **systemic circuit** round the rest of the body.

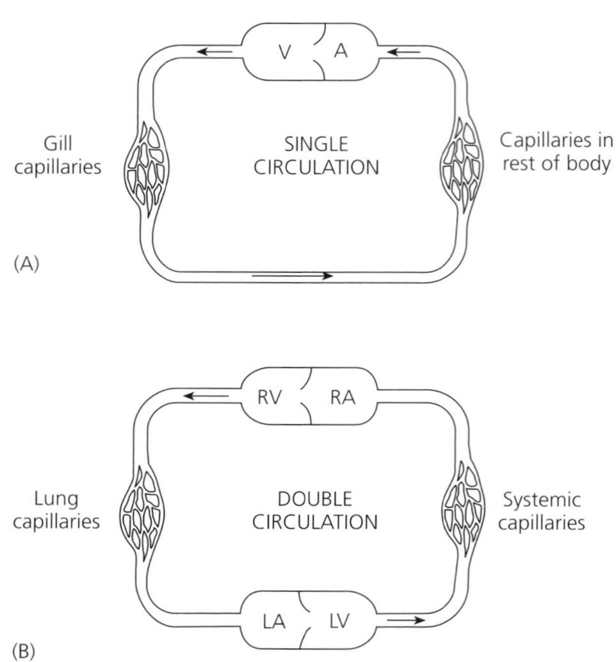


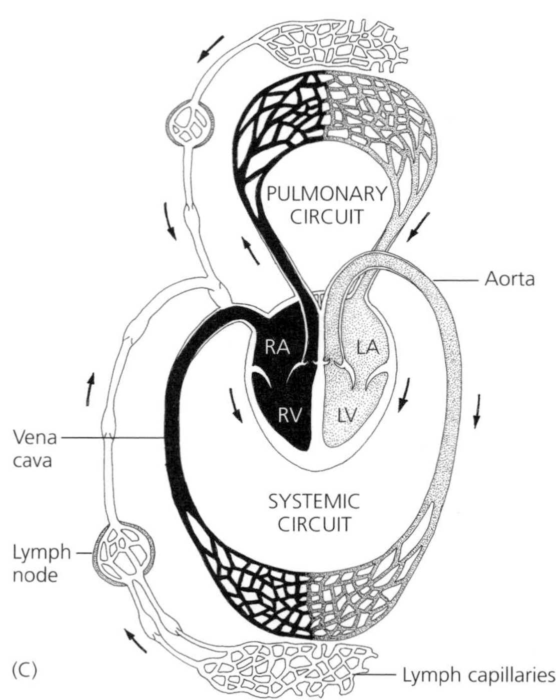


**Figure 22.12**
Single circulation of a fish (A) and double circulation of a mammal (B). A = atrium, V = ventricle, RA = right atrium, RV = right ventricle, LA = left atrium, LV = left ventricle. The double circulation as it actually is, with the left and right sides of the hearts combined in a single muscle mass (C)

## The heart

The heart lies in a membranous sac called the **pericardium** which contains a thin layer of lubricating fluid. The wall of the heart consists of **cardiac muscle**. Each side of the heart consists of a thin-walled **atrium** and a thicker-walled **ventricle** (Figure 22.13). The right atrium receives blood from large veins called **venae cavae,** and the right ventricle pumps the blood to the lungs along the **pulmonary artery**. Blood returns from the lungs along the **pulmonary veins** and enters the left atrium. The left ventricle pumps blood along the **aorta**, the trunk of the arterial tree whose branches distribute blood to all parts of the body except the lungs.

Although the two sides of the heart carry blood with different chemical composition and develop different pressures (see the Clear Thinking box on page 399), they must pump *equal volumes*. This is because they are in *series*; all the blood leaving one side must enter the other side.

Back flow of blood is prevented by four valves of two kinds (Figure 22.14). Between each atrium and ventricle is a large **arterioventricular valve** or **A-V valve,** which prevents blood flowing into the atrium when the ventricle contracts. The flaps of each A-V valve are supported by tendons which act as guy ropes, preventing the flaps from being pushed back into the atria when the ventricle contracts. The tendons are anchored by **papillary muscles**. The right A-V valve is called the **tricuspid valve** because it has three flaps, whilst the left, with two flaps, is called the **bicuspid valve**. Back flow into each ventricle is prevented by a **semilunar valve,** so-called because it consists of three half-moon-shaped flaps (three flaps make up *one* valve).

Though the heart is full of blood, it needs its own capillary network for two reasons:

- the wall of the heart is far too thick for oxygen to reach the muscle fibres fast enough by diffusion
- the right side contains deoxygenated blood.

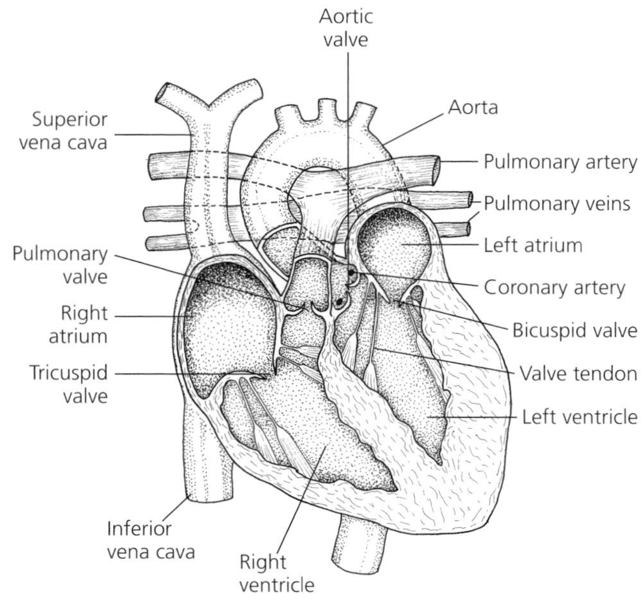


**Figure 22.13**
Vertical half of the human heart

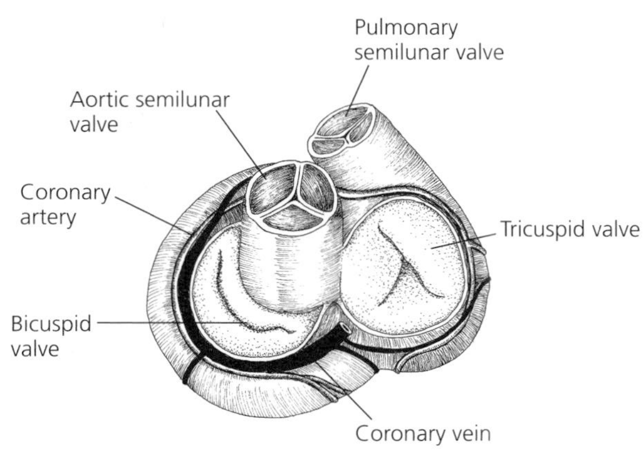


**Figure 22.14**
The heart with the atria cut away to show the valves

## CLEAR THINKING

**The wall of the left ventricle is several times thicker than that of the right, and develops a correspondingly higher pressure. It is often stated that the reason for this is that whereas the right side only pumps blood through the lungs, the left side has to pump it 'all round the body'. Let's think critically about this for a moment.**

**While it is true that the total length of systemic capillaries is much greater than that of the pulmonary capillaries, the additional tubing consists of vessels in *parallel* (which actually decreases resistance to flow). Extra length only increases resistance if it is in *series*.**

**Of course, many parts of the body are much further from the heart than the lungs, so we would expect there to be a greater fall in blood pressure by the time the blood gets there. In fact, most of the journey away from the heart is accomplished with very little drop in pressure. By far the greatest fall in blood pressure occurs in the arterioles, which are only centimetres in length.**

**The reason why the left ventricle develops such a high pressure is because the kidneys require blood at very high pressure (Chapter 24). In fact, if the other organs were to receive blood at the same pressure as the kidney, tissue fluid would leak out faster than it could be drained away by the lymphatic system.**

**In the lung capillaries it is essential that blood pressure be low; if it were too high, tissue fluid would accumulate in the air sacs and the person would be in danger of drowning.**

**It is not that the lungs do not need such a high pressure as the other organs – a high blood pressure would be positively dangerous. If the right ventricle were to develop the same pressure as the left, tissue fluid would leak out of the capillaries so quickly that it would accumulate and interfere with gas exchange. In the event, the mean pressure in the pulmonary capillaries is only a third of that in the glomerular capillaries in the kidney.**

The capillaries inside the heart are supplied by branches of two **coronary arteries** which leave the aorta just above the semilunar valve. The cardiac capillaries are drained by **cardiac veins** which drain into the right atrium.

### Cardiac muscle

The ability of the heart to function with almost total reliability is dependent on the special structural and physiological properties of cardiac muscle (Figure 22.15). Cardiac muscle fibres consist of chains of

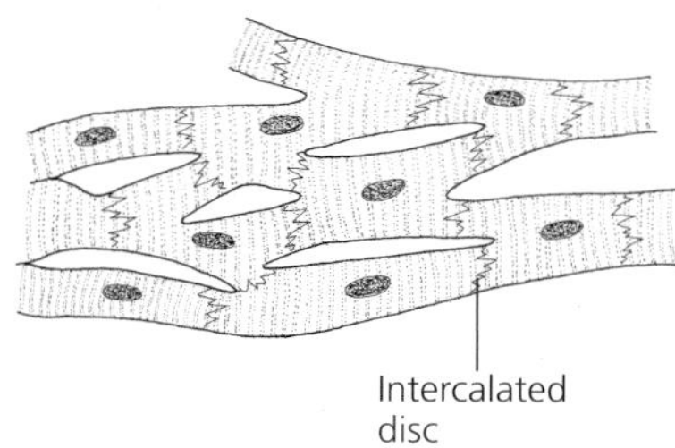


**Figure 22.15**
Fine structure of cardiac muscle

uninucleate cells which are rich in mitochondria, myoglobin and glycogen. They have a pattern of cross striations similar to that of skeletal muscle. The cells branch at their ends, forming cross-connections linking the fibres into a single continuous network.

Longitudinally, junctions between cells are marked by **intercalated discs** which are rich in **desmosomes** and **gap junctions** (Chapter 3). The desmosomes hold the cells together, allowing force to be transmitted from one cell to the next. The gap junctions allow action potentials to spread directly from cell to cell so that the entire heart muscle constitutes a single **electrical syncytium**.

When isolated in nutrient solution, cardiac muscle cells develop action potentials spontaneously, each action potential provoking a contraction. Since contraction occurs without any external stimulus the heart is said to be **myogenic**, meaning that the excitation originates within the muscle itself. The hearts of crustaceans, spiders and some insects require nervous stimuli to beat, and are thus said to be **neurogenic**.

A very important feature of cardiac muscle is that it is immune to fatigue and cramp. This is because it has a refractory period as long as the twitch itself – about 250 ms (Figure 22.16). The heart thus cannot begin another contraction until the previous action potential has subsided, by which time it has relaxed. By forcing the heart to take a rest between contractions, the long refractory period prevents twitches being fused into the steady tetanic contractions that are characteristic of skeletal muscle contraction.

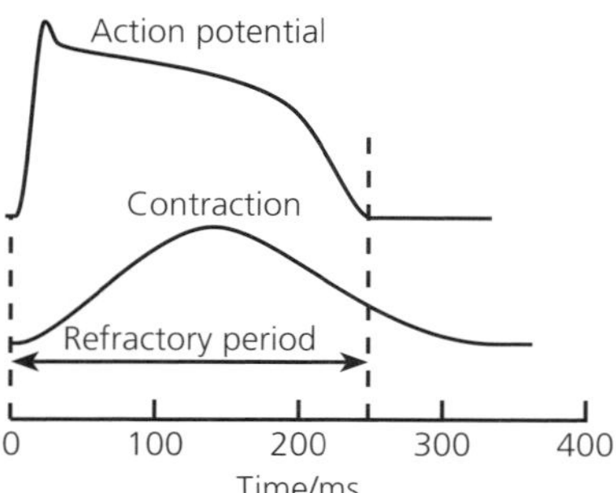


**Figure 22.16**
Mechanical and electrical events in a cardiac muscle twitch. Note that since the refractory period is almost as long as the twitch, tetanus is impossible (unlike skeletal muscle)

### *How are the chambers co-ordinated?*

One of the important features of cardiac muscle is that it continues to contract spontaneously when removed from the body. The rate of this spontaneous contraction depends on the part of the heart from which the muscle is taken. Atrial muscle contracts with a faster rhythm than ventricular muscle, and a small patch of tissue in the wall of the right atrium called the **sinoatrial node** (**S-A node**) has an even faster rhythm.

In the intact heart, of course, all chambers beat to the same time, even when the nerves supplying the heart are cut. It follows that there must be some kind of overall control and that this must be in the heart itself. This is the function of the S-A node, which is called the **pacemaker** because it imposes its own rhythm on the rest of the heart.

The human S-A node is about 15 mm long and 5 mm wide, and is an evolutionary vestige of the sinus venosus of the fish heart. In fish, the atrium and ventricle adopt the rhythm of the sinus venosus. In mammals, the S-A node retains this primitive function of pacemaking.

The pacemaker cells undergo action potentials at their own spontaneous rate, and these spread to the surrounding atrial muscle cells, which thus contract before they would do so spontaneously.

If heart muscle cells are in direct electrical continuity, why don't the ventricles contract at almost the same time as the atria? They would do, were it not for the fact that atria and ventricles are separated by a layer of fibrous tissue encircling each A-V valve.

There is, however, one route by which atrial impulses can reach the ventricles. This is the **atrioventricular node**, or **A-V node**, a small patch of specialised muscle cells in the wall of the right atrium (Figure 22.17).

The A-V node propagates action potentials much more slowly than ordinary cardiac muscle cells. This delay gives the atria time to relax before the excitation has crossed the A-V node.

Leading from the A-V node is a bundle of modified heart muscle cells

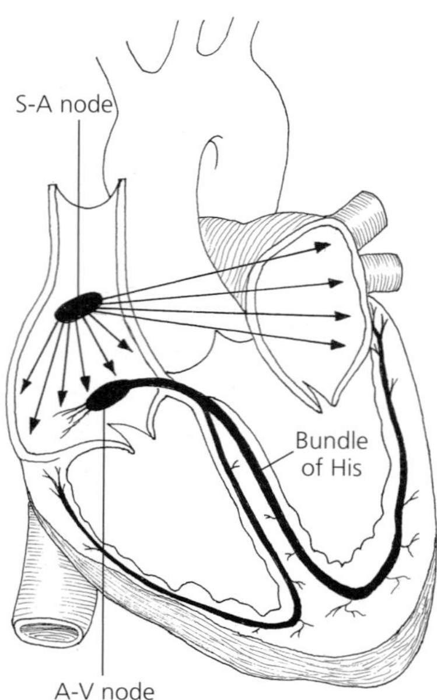


**Figure 22.17**
Conducting system of the heart

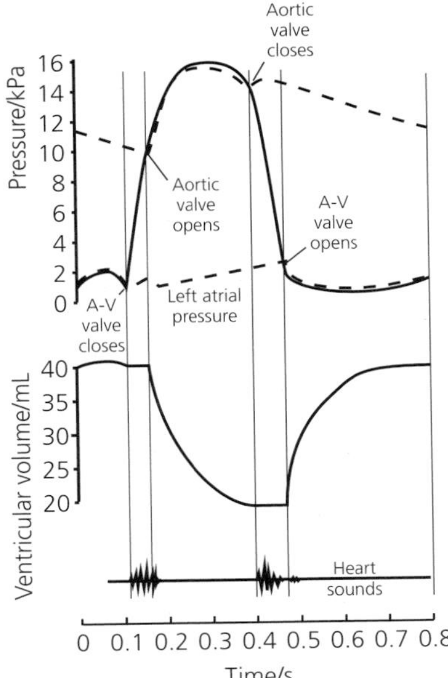


**Figure 22.18**
Events of the cardiac cycle

called the **bundle of His**. These cells are only weakly contractile, and are specialised for propagating action potentials. The bundle runs down the septum between the two ventricles, sending branches to the walls of the ventricles. The twigs of the system consist of **Purkinje tissue** (named after Johannes Purkinje, the Bohemian naturalist who discovered them). The bundle of His and Purkinje tissue propagate impulses much more rapidly than unmodified cardiac muscle cells (3–5 $m\ s^{-1}$ compared with 0.3 $m\ s^{-1}$), so the entire ventricular muscle contracts more or less simultaneously.

## The cardiac cycle

The average resting heart beats around 72 times a minute, males having, on average a slightly slower beat than females. The average figure conceals considerable variation – anything between 50 and 90 is considered to be normal. The four chambers work in co-ordinated sequence, the two ventricles contracting while the atria are relaxing, and vice versa.

The action of the heart follows a sequence of events called the cardiac cycle, shown in Figure 22.18. During a complete cycle each chamber undergoes a contraction, or **systole** (pronounced 'sistolee') and a relaxation, or **diastole**. As with any cycle it has neither a beginning nor an end, so the starting point is arbitrary. The cycle can be divided into two main stages: ventricular systole and ventricular diastole.

### Ventricular systole

As the ventricles begin to contract, blood pressure is initially lower than that in the arteries, so the semilunar valves are closed. Pressure soon rises above that in the atria, forcing the A-V valves to close. For a brief moment all four valves are closed, so blood is neither entering nor leaving the ventricles. At this time, the muscle cells are thus developing tension but volume is not changing. Since the muscle cells are not shortening, contraction is **isometric**. When the pressure has risen above that in the arteries the semilunar valves are forced open. Blood enters the arteries, allowing the muscle cells to begin shortening.

### Ventricular diastole

As the ventricles relax, the blood pressure falls. When it drops below that in the arteries, the semilunar valves close. Pressure continues to fall, but for a moment the A-V valves remain closed because the pressure is still higher than in the atria. During this brief period the ventricular muscle cells are relaxing but are unable to increase in length. This is therefore a period of **isometric** relaxation. When the pressure has dropped below that in the atria the A-V valves open and blood enters the ventricle. For a moment the atria remain relaxed, but because the pressure in the veins is higher than in the ventricles, blood enters the ventricles. When the ventricles are about two-thirds filled the atria contract (atrial systole), pushing more blood into the ventricles. Thus the atria do no more than 'top up' the ventricles, and their contraction is not essential. The atria then relax but the A-V valves remain open until the ventricles begin to contract again.

### The heart sounds

The closing of the heart valves produces two characteristic sounds, described as 'lub' and 'dup'. The first (and softer) sound is produced during ventricular systole by the closing of the A-V valves. The second, sharper sound is produced in ventricular diastole by the clapping shut of the semilunar valves.

### *Controlling the output of the heart*

Cardiac output is the product of two things: the heartbeat rate and the volume of blood pumped by each ventricle in each contraction (the stroke volume):

$$\text{cardiac output} = \text{stroke volume} \times \text{number of beats per minute}$$

In a resting person, the stroke volume averages about 80 $cm^3$, and the heart rate about 70 beats per minute (though there is considerable variation with body size, age and physical fitness). The average resting cardiac output is between 4 and 6 $dm^3$ per minute, but in trained athletes it can rise up to seven times higher. This increase involves a rise in both stroke volume and pulse rate. For example, if the heart rate and the stroke volume each double, the heart output has risen by four times.

#### *Nervous control of the heart*

Although the nerves supplying the heart do not initiate the beat, they do modify its output. The heart is supplied by both sympathetic and parasympathetic nerves which have antagonistic effects (Chapter 17). The sympathetic nerve endings extend throughout the heart muscle but are particularly concentrated at the S-A and A-V nodes. Parasympathetic nerves are densest in the S-A node.

Sympathetic nerves stimulate the heart in three ways: they increase the rate of discharge of the S-A node, the speed of impulse propagation by the conducting system and the force of contraction of the cardiac muscle. Parasympathetic endings have the opposite effects.

#### *Hormonal control of the heart*

The most important hormone affecting the output of the heart is adrenaline, though thyroxine also stimulates it. Adrenaline has similar effects to the noradrenaline released by sympathetic nerve endings.

#### *Autoregulation of heart output*

To some extent the heart can adjust its output independently of nervous and endocrine control. This form of autoregulation depends on the fact that if cardiac muscle fibres are stretched more, the subsequent contraction is more forceful. This happens during exercise: the more blood is returned to the heart, the more the ventricles are distended during diastole and the more strongly they contract during the subsequent systole.

The mechanism of this autoregulation is dependent on the fact that the maximum force a muscle fibre can develop depends on the number of cross links formed between actin and myosin filaments. This in turn depends on the degree of overlap between the filaments, which varies with sarcomere length (see the information box on page 353).

## The blood vessels

Blood vessels are of three kinds:

- distributing vessels, or **arteries**, which carry blood from the heart.
- collecting vessels, or **veins**, which return blood to the heart.
- exchange vessels, or **capillaries**, in which substances enter and leave the blood.

The lining of arteries and veins is formed from a single layer of squamous epithelium, the **tunica intima**. Outside the tunica intima is the **tunica**

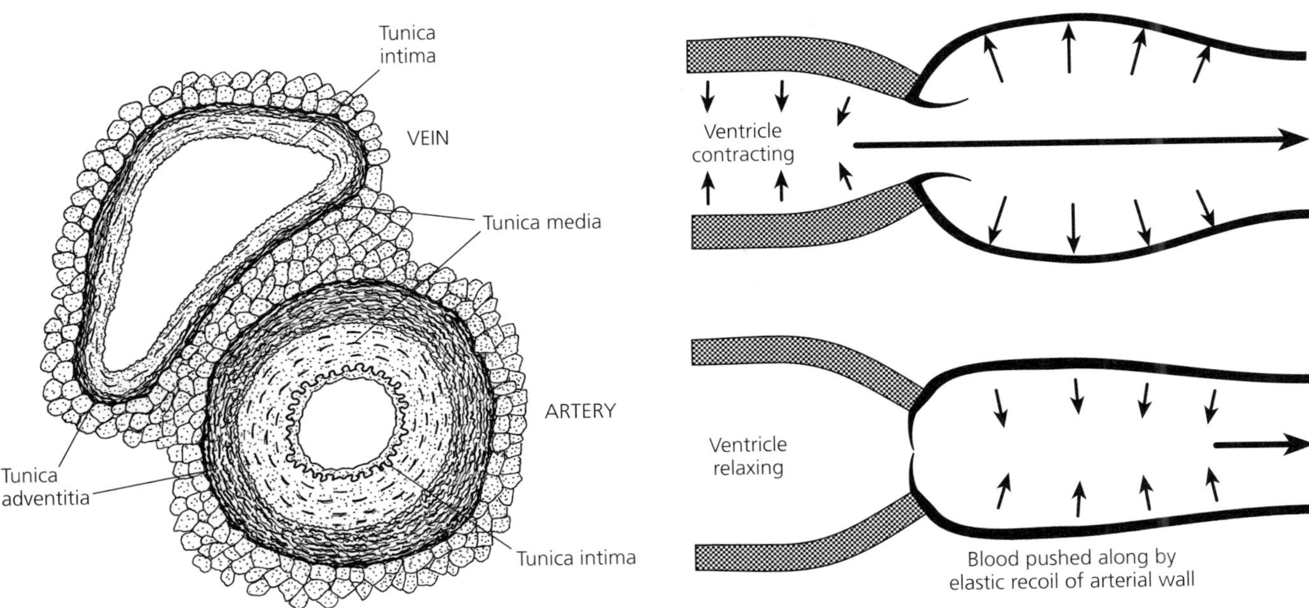


**Figure 22.19**
Structure of an artery and a vein

**Figure 22.20**
How large arteries store and release energy

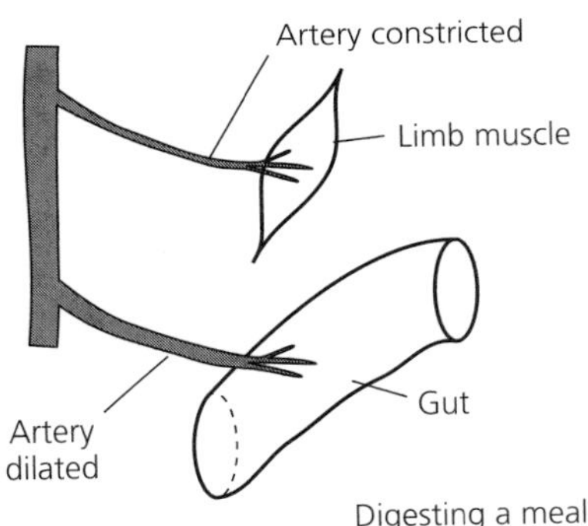


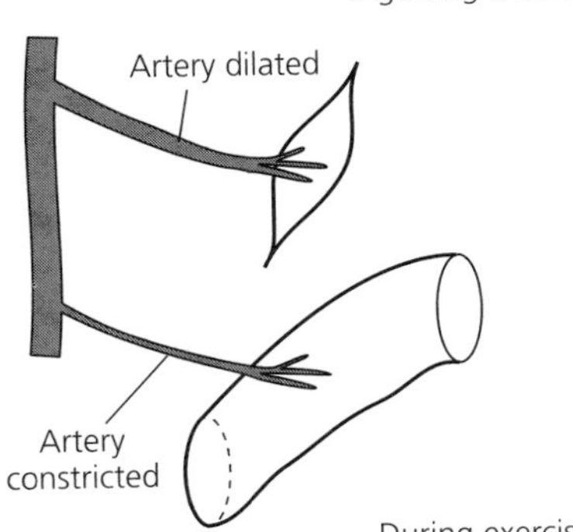


**Figure 22.21**
How arterial smooth muscle controls the distribution of blood. When digesting a meal, blood is diverted to the gut by relaxation of arterial smooth muscle in the mesenteric artery and contraction of smooth muscle in arteries to limb muscles. During exercise, the reverse occurs

**media**, consisting of varying proportions of smooth muscle and elastic tissue (Figure 22.19). The outermost layer is the **tunica adventitia**, consisting mainly of collagenous fibres which set an upper limit to the diameter of the vessel.

## *Distributing vessels: arteries and arterioles*

Arteries carry blood *from* the heart. They are adapted for this function in three ways:

1 the thick layer of collagen fibres in the outer wall enables them to resist high pressure (enough to push blood from a cut artery to a height of over a metre).

2 the walls of larger systemic arteries contain much elastic tissue, enabling the walls to stretch during systole. When the ventricles relax, the stretched arteries contract elastically, pushing the blood onwards (Figure 22.20). The elasticity of the arteries thus enables them to act as energy reservoirs, temporarily storing some of the energy of the heart's contraction and releasing it during diastole. By the time the blood reaches the smallest twigs of the arterial tree, the **arterioles**, the pressure and rate of flow are steady.

3 arteries, particularly the smaller ones and the arterioles, contain smooth muscle which enables the flow of blood to be controlled. Contraction of the muscle causes **vasoconstriction**, or narrowing of the vessel, reducing blood flow. Relaxation results in **vasodilation**, or widening of the vessel, increasing blood flow. By varying the resistance to flow, blood can be diverted to where it is most needed (Figure 22.21).

As the blood flows along the arterial tree it loses speed and pressure, though for very different reasons. Pressure falls because energy is lost in overcoming the resistance to flow. This resistance is not, as is often stated,

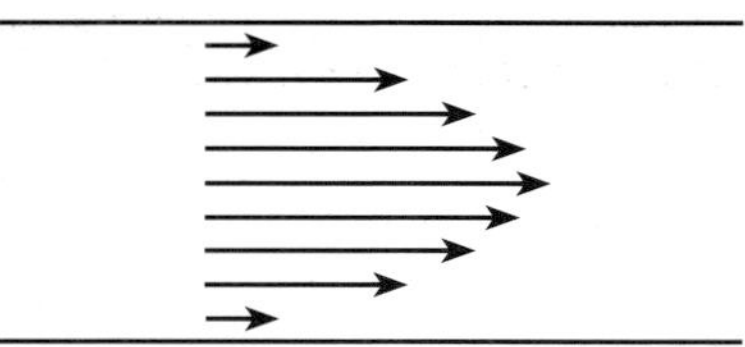

**Figure 22.22**
How speed of flow varies across a blood vessel. Blood velocity is maximal in the centre and zero next to the endothelium

due to friction between the blood and the vessel. The blood touching the endothelium is actually stationary, the maximum rate of flow occurring in the centre (Figure 22.22). It is the movement of blood relative to blood that produces the resistance to flow.

Speed decreases along the arterial tree for the following reason. Every time an artery branches, the sum of the cross-sectional areas of the two branches is greater than that of the trunk. By the time blood reaches the capillaries, their total cross-sectional area is many times greater than that of the aorta.

## Atherosclerosis – a preventable cause of death

One of the greatest causes of premature death in industrialised countries is **atherosclerosis.** This is the narrowing of an artery by the deposition of fatty material in its wall, reducing the flow of blood. It is a progressive condition and usually begins many years before symptoms develop.

The trigger that starts the process of deposition is thought to be damage to the endothelium lining the artery, associated in some way with such factors as carbon monoxide (from cigarette smoke), high blood cholesterol level, high blood pressure and diabetes mellitus. In response to such damage, monocytes begin to adhere to the lining of the artery and penetrate between the endothelial cells. Inside the tunica media they develop into macrophages and begin to accumulate fatty materials such as cholesterol and triglycerides.

In response to growth substances produced by the macrophages, the smooth muscle cells begin to proliferate. The result is the development of an **atherosclerotic plaque** – a mound of fatty material covered by smooth muscle and fibrous tissue. If the fibrous cap is thin it may rupture, and the exposure of the fatty material below triggers the formation of a clot or **thrombosis**. This grows quickly and within minutes may block the artery completely. When this occurs in the heart the affected heart cells die, causing a **cardiac infarction** or 'heart attack'. If it occurs in the brain it causes a 'stroke'.

Plaques that contain more fibrous tissue generally do not rupture but continue to grow slowly, producing a localised chronic starvation of the tissues. In the heart this condition is called **ischaemia.** It may produce pain called **angina** during exertion, which may subside after rest.

Though the exact mechanism of plaque formation is not fully known, studies of the epidemiology (pattern of occurrence) of atherosclerosis show that it is associated with a number of **risk factors**, the most important of which are:

- smoking
- diet
- obesity
- high blood pressure (hypertension)
- diabetes mellitus
- hereditary factors

It seems likely that there are two components to the mechanism: the initial damage to the endothelium, and the subsequent growth of the plaque. Factors which seem to contribute to endothelial damage include carbon monoxide associated with smoking, high blood pressure and diabetes mellitus.

Since cholesterol is a major constituent of atherosclerotic plaque, it is not surprising that atherosclerosis is associated with a high blood cholesterol level. The important question is why do some people have higher cholesterol levels than others. A high cholesterol diet is only a partial explanation because most blood cholesterol does not come from the food but is made in the liver. After export to other parts of the body it is used as a constituent of cell membranes, and in the synthesis of steroid hormones and bile salts.

There seem to be two main reasons why some people with moderate cholesterol intake can have high blood cholesterol levels. One is that cholesterol is produced

## Atherosclerosis – a preventable cause of death *continued*

from saturated fats, so people on high-fat diets tend to convert some surplus dietary fat to cholesterol.

The second is more interesting, since it provides a possible mechanism for the observed fact that high cholesterol levels seem to have a hereditary component. Cholesterol is carried in the blood combined with a **lipoprotein**. There are two main classes of lipoprotein carrier: **low-density lipoproteins** (**LDLs**) and **high-density lipoproteins** (**HDLs**). Surveys show that elevated levels of LDL are linked to an increased risk of coronary artery disease (CAD), but that higher levels of HDL are associated with a reduced risk. The explanation is thought to be that LDLs pick up cholesterol and deposit it in cells (including smooth muscle cells in blood vessel walls) whilst HDLs remove cholesterol from body cells and transport it to the liver.

To take up this lipoprotein–cholesterol complex from the blood, a cell needs specific receptor molecules in its plasma membrane. It is thought that some people have too few of these receptors, and so the removal of cholesterol from the blood is reduced. Since the production of receptor proteins is under genetic control this would explain – at least in part – the tendency for CAD to be hereditary.

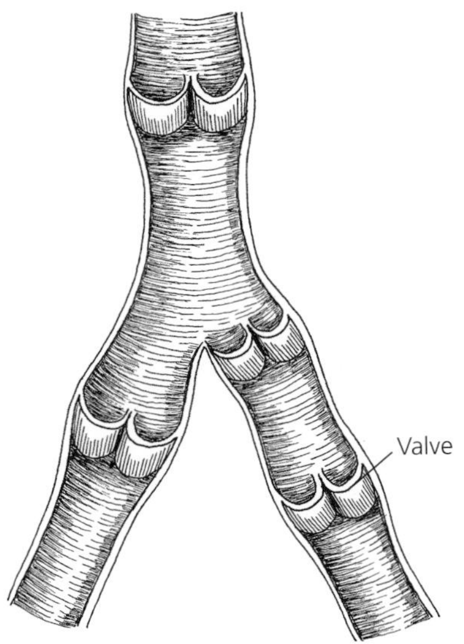


**Figure 22.23▲**
A vein cut open to show the valves

### Collecting vessels: veins and venules

Veins carry blood *towards* the heart, so the blood is under low, steady pressure. Their structure reflects this function in a number of ways (Figure 22.19):

1 they have much thinner walls than arteries, reflecting the low pressure they have to withstand.

2 they have a wider cross-sectional area, reducing resistance to flow. Resistance is further decreased by the fact that there are more veins than arteries (in your forearm there are numerous veins but only two arteries).

3 most veins have valves which prevent the blood intermittently flowing backwards during body movement (Figure 22.23).

Valves are particularly important in the return of blood during exercise. Active limb movements massage the veins, pumping the blood back to the heart (Figure 22.24). The changes in pressure round the venae cavae in the chest and abdomen during breathing have a similar effect.

For the same reason as blood slows down as it travels along the arterial tree, it speeds up along the veins; wherever veins meet, the total cross-sectional area of tributaries is greater than that of the vein they combine to form.

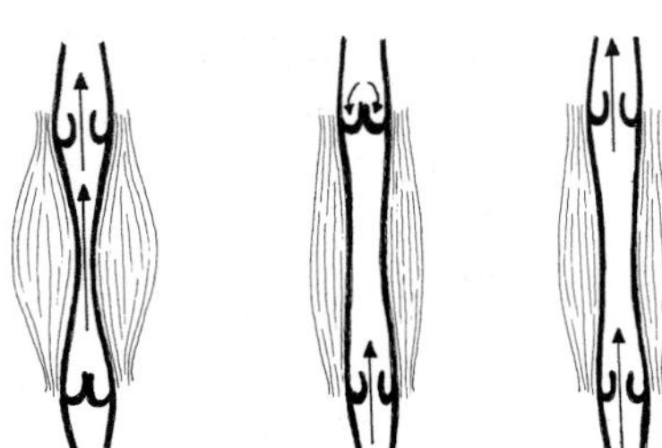

**Figure 22.24▲**
Action of the valves in the return of blood to the heart

### Controlling the arteries and veins

The smooth muscle in blood vessels is under nervous, endocrine and local control. Sympathetic pathways originate in the **vasomotor centre** in the medulla of the brain, and bring about vasoconstriction. Most blood vessels receive no parasympathetic fibres, so control is by variations in frequency of sympathetic impulses. Arterial smooth muscle is also stimulated to contract by adrenaline, released by the adrenal medulla.

Arterioles are also under a considerable degree of local control. When an organ is particularly active, the accumulation of waste products and other substances causes the relaxation of smooth muscle in the arterioles. During exercise this local effect overrides the effect of sympathetic impulses, with the result that blood is diverted to the skeletal muscles.

## Blood pressure

One of the things a doctor measures in a medical examination is one's blood pressure, using an inflatable cuff placed round the arm. Arterial blood pressure rises and falls with the pulse. Pressure peaks correspond to ejection of blood from the ventricles and are the **systolic pressure.** The troughs between the surges are the **diastolic pressure.**

Although blood pressure is most properly measured in kiloPascals, medical people still prefer to use 'millimetres of mercury' (see the information box on page 392). In a healthy young adult systolic and diastolic pressures are 120 and 80 mm Hg, respectively, and blood pressure is referred to as '120/80'.

Both systolic and diastolic pressures increase steadily with age, but in people with **hypertension**, or high blood pressure, this is abnormally so. Hypertension results from excessive and sustained contraction of the arterial muscle, though what causes this is not yet known. Hypertension may contribute to the rupture of a blood vessel. If this occurs in the brain, it results in loss of blood supply to the affected part leading to a 'stroke'.

### *Exchange vessels: the capillaries*

Capillaries are microscopic vessels from which substances enter and leave the blood (Figure 22.25). They permeate every tissue except the lens, cornea and fovea of the eye, and the epidermis. Although they contain only about 5% of the blood volume, it is only in these vessels that substances can enter and leave the blood.

Pinocytic vesicles

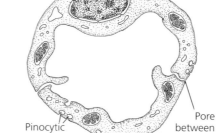

Pore between adjacent endothelial cells

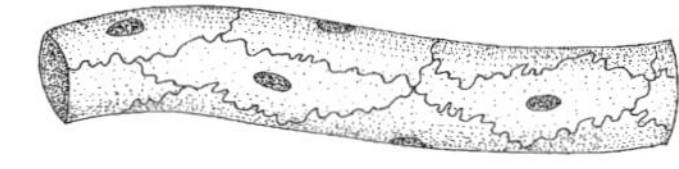

**Figure 22.25◄▼**
Structure of a capillary in transverse section (left) and in side view (below). The walls are actually proportionately thinner than shown

**Figure 22.26▼**
Red cells squeezing through a capillary. Due to their distorted shape, most of the haemoglobin is displaced to the periphery of the cells, reducing the distance oxygen has to diffuse to the surrounding cells

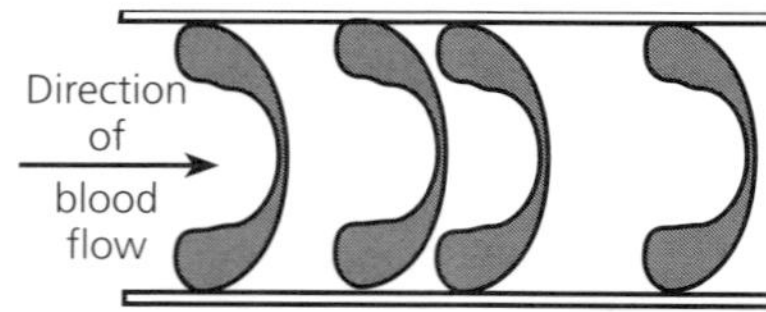


Capillaries average about 0.5–1 mm long and are between 6 μm and 8 μm in diameter, so the corpuscles have to travel in single file. To squeeze along the narrowest capillaries, red cells become bell-shaped, increasing their surface area still further (Figure 22.26). Their numbers run into billions, and, since few cells are more than 100 μm away from the nearest capillary, diffusion distances are short. Some estimates put the total capillary surface area at over 6000 $m^2$.

## Small vessels are stronger

As blood flows from capillaries to veins the pressure drops, yet the vessel walls get thicker. Moreover, capillary walls are about $\frac{1}{4000}$ the thickness of the wall of the aorta, yet they withstand pressures about a quarter that of the aorta. Clearly there is more to the thickness of blood vessel walls than the pressure they have to withstand.

The explanation is that the narrower the vessel, the more curved are its walls. A greater proportion of the tension in the walls is therefore directed inward to oppose the blood pressure. Wide blood vessels need relatively thicker walls than narrow ones in order to withstand the same blood pressure.

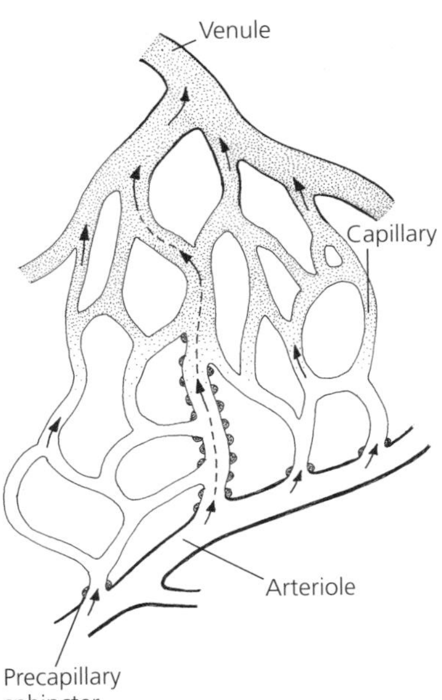


**Figure 22.27**
The capillary bed

Capillary walls consist of a single layer of squamous epithelium on a basement membrane. Adjacent cells are in close contact except at minute, slit-like pores. The presence of these pores means that the basement membrane is the only continuous layer separating blood from the surrounding tissues. The pores are about 4 nm wide, big enough to allow passage of all plasma solutes except proteins. The absence of these pores in brain capillaries explains the inability of many introduced drugs to reach the brain (the 'blood–brain barrier').

Some capillaries, such as those in the glomeruli in the kidney, also have large holes in the endothelial cells which further increase their permeability (Chapter 24).

Another factor helping the exchange of substances between blood and tissues is the slow flow of blood due to their large total cross-sectional area. It takes about a second to travel the length of an average capillary, giving plenty of time for substances to enter and leave the blood.

### Controlling the capillaries

Capillaries have no muscle, so they cannot control the blood flowing through them. Blood flow is thus entirely dependent on the pressure of blood delivered to them. There are two ways in which the distribution of blood can be varied. One is by the degree of contraction of the smooth muscle in the arterioles. The other is the **precapillary sphincter** muscle which lies at the beginning of each capillary. In inactive tissues, each precapillary sphincter closes every so often, shutting the capillary down for a brief period (Figure 22.27).

### How do substances enter and leave the blood?

Substances can move through capillary walls in three different ways:

- by simple diffusion through the plasma membranes of the endothelial cells, for example oxygen and carbon dioxide.
- through the slits between adjacent capillary endothelial cells (see below). Simple substances such as glucose and amino acids pass through in this way.
- by endocytosis across the inner membrane of endothelial cells and exocytosis through the outer membrane (Chapter 6). This is the only way in which protein hormones can pass from the plasma to their target cells.

## Tissue fluid and its formation

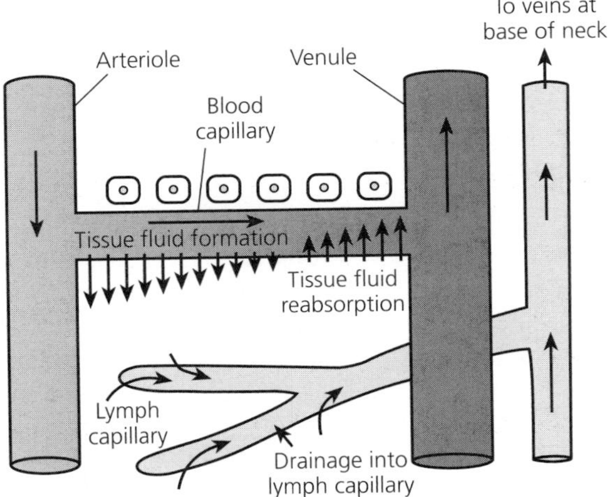


**Figure 22.28**
Formation and drainage of tissue fluid

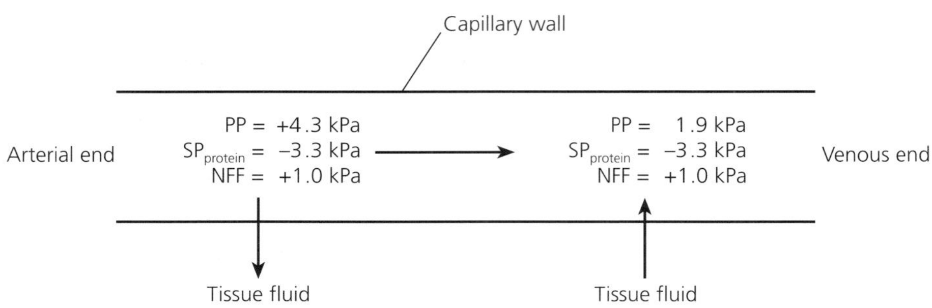


**Figure 22.29**
The various factors affecting formation and reabsorption of tissue fluid (NFF = net filtration force, PP = pressure potential, SP = solute potential)

The pores between adjacent capillary wall cells are wide enough for all plasma solutes except proteins to pass through. At the arterial end of a capillary the blood pressure is high enough to force liquid out through these pores into the surrounding tissues. This **tissue fluid** bathes the cells and is identical to plasma minus its proteins (Figure 22.28). Further down a capillary, tissue fluid drains back into the blood.

It is easy to see why tissue fluid should seep out of capillaries, since the pressure inside the capillary is higher than outside. It is less obvious why it should seep back into the blood further along the capillary, when the pressure inside the capillary is still higher than outside.

The explanation depends on the fact that there are actually *two* forces at work, one tending to drive liquid out of the capillary and the other acting in the opposite direction. At the arterial end of the capillary the outward force exceeds the inward force, whilst at the venous end the reverse is true (Figure 22.29).

The outward force is blood pressure. At the arterial end of a capillary this is about 4.3 kPa, and at the venous end it is about 1.9 kPa. The inward force is an osmotic one and is due to the fact that proteins do not leave the capillaries. Thus although the concentrations of ions, glucose and other simple substances are equal in the plasma and tissue fluid, the protein concentration is higher in the blood. This produces a protein solute potential of about −3.3 kPa. This is less than the blood pressure at the arterial end of the capillary so the net force is outward. At the venous end it is greater than the blood pressure so the net force is inward.

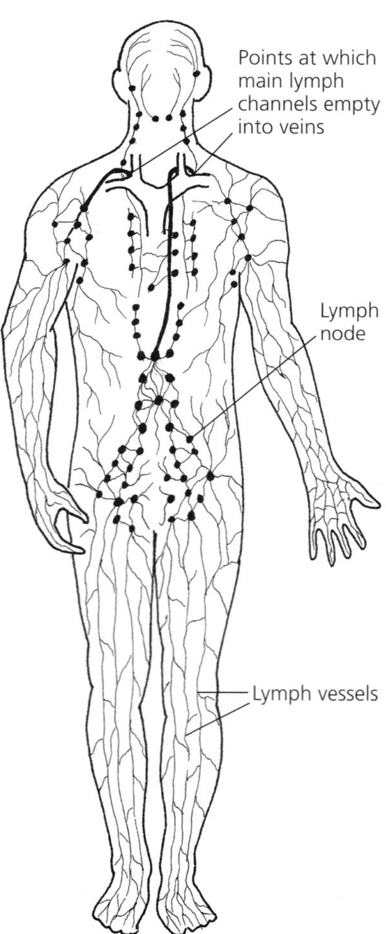


**Figure 22.30**
Return of lymph to the blood stream by the lymphatic system

## *The lymphatic system*

Tissue fluid is formed slightly faster than it re-enters the capillaries. It would therefore accumulate in the tissues were it not drained away by the **lymphatic system**, a tree-like system of tubes extending throughout the body. Once tissue fluid enters a lymphatic vessel it is called **lymph**. All the lymphatic vessels eventually converge into two trunks which empty into the veins at the base of the neck (Figure 22.30). Excess tissue fluid is thus eventually returned to the blood.

The pressure inside the lymph vessels is very low, and lymph flow is helped by valves. Any muscular movement, even the pulse of a neighbouring artery, squeezes the lymph vessels, pushing the lymph along. If lymph flow is impeded the tissues become swollen with surplus tissue fluid, causing **oedema**. This can happen in the ankles when you stand for a long time.

Another function of the lymphatic system is to provide a route by which proteins can enter the blood. Clotting proteins, antibodies, protein hormones and other plasma proteins are not secreted directly into the plasma but are released into tissue fluid. Since proteins cannot enter capillary walls they have to enter the blood indirectly via the lymph system. In the small intestine the lymph capillaries are the route by which fats enter the blood (Chapter 26).

At intervals the lymphatic vessels are interrupted by **lymph nodes**. These are packed with white corpuscles and are very effective at filtering out invading bacteria (Chapter 27).

## Adjusting to change

The demands on the blood system are seldom the same for long, and it must be able to adjust rapidly to change. Changes in posture, exercise and loss of blood are just three examples of changing demands on the heart and blood vessels. Some responses are rapid and are controlled by the autonomic nervous system (Chapter 17) while others are slower and involve hormones (Chapter 20). Since some of the brain centres concerned are not clearly defined, in this account they will be referred to collectively as 'cardiovascular centres'.

Like most internal organs, the heart is innervated by two sets of nerves, one stimulatory and the other inhibitory. The sympathetic, stimulatory fibres leave the central nervous system via the spinal cord. The parasympathetic, inhibitory fibres leave the brain via the vagus nerve. The nervous regulation of the heart involves shifting the balance between these two antagonistic influences.

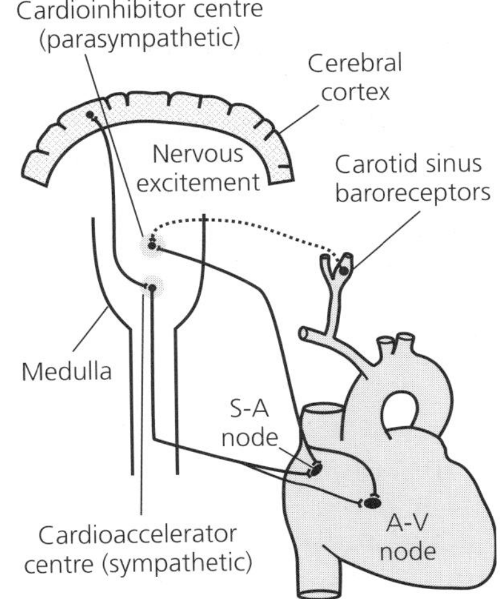


**Figure 22.31**
Neural control of the heart

### *Changing posture*

When you stand up, blood tends to collect in the abdomen and legs causing a brief fall in blood pressure in the arteries supplying the brain. This is almost immediately corrected by a rise in frequency and strength of the heart beat. The reverse happens when you lie down; there is a brief rise in blood pressure followed by a slight drop in heart rate.

These responses are brought about by varying the level of inhibition of the heart. The receptors mediating this reflex are sensitive to blood pressure and are called **baroreceptors**. These are situated in the walls of the **carotid sinus**, a small swelling at the base of each internal carotid artery (Figure 22.31). Afferent neurons from the baroreceptors lead to cardiovascular centres in the medulla of the brain. Parasympathetic fibres leave these centres and reach the S-A node of the heart via the vagus nerve.

When you lie down, the rise in blood pressure in the head stimulates the baroreceptors to increase the frequency of afferent impulses to the cardiovascular centre. This increases the frequency of parasympathetic impulses to the heart causing blood pressure to fall slightly. Thus the higher the blood pressure in the carotid sinuses, the more the heart is inhibited and vice versa – an example of homeostasis by negative feedback.

### *Adjusting to exercise*

Whereas the resting heart rate is controlled mainly by adjustments to the level of its inhibition, exercise involves stimulation of the heart. When you take hard, physical exercise, your energy expenditure rises to several times above normal. Although muscles can work for short periods without increased blood supply, prolonged hard exercise requires an increase in cardiac output and the redistribution of blood flow from the gut to the skeletal muscles.

In important sporting events, increased activity of the sympathetic nervous system (responsible for pre-race 'nerves') initiates a number of adjustments in the circulatory system:

- it causes vasoconstriction throughout most of the blood system, especially the veins, causing a rise in blood pressure. The veins contain about 60% of the blood volume so their constriction is particularly important, increasing the pressure of blood returning to the heart.

- it increases cardiac output. Note that this could not happen without the increased venous return just mentioned, since the heart can only pump out as much blood as returns to it.

Once the exercise is under way, further changes occur:

- the contraction and relaxation of limb muscles massages the veins, increasing the return of blood to the heart, so even without sympathetic stimulation the heart pumps more forcefully.
- the increased production of $CO_2$ and other wastes by the limb muscles inhibits the smooth muscle in the arterioles. The resulting dilatation increases the blood flow to the active muscles. This local effect overrides the effect of sympathetic stimulation.
- sympathetic stimulation increases cardiac output – by up to six times in trained athletes.

It is worth nothing that although the sympathetic nervous system is essential for intense exercise, the needs of moderate exercise (such as jogging) can be met without any additional nervous or hormonal activity. The output of the heart can be increased two- or threefold as the direct result of the increased venous return caused by squeezing of veins by active limb muscles. The dilatation of arterioles in the muscles in response to generation of wastes is another factor. Third, even without any increase in blood supply to the muscles, the lower oxygen concentration in active muscle causes more oxygen to be released by the blood.

### *Giving blood*

A blood donor gives about 600 $cm^3$, or about 10% of total blood volume. A person can lose this quantity without any ill-effects because of a number of adaptive changes in the blood system. Some involve the sympathetic nervous system and occur within minutes, whilst others involve hormones and take longer.

The immediate effect of loss of blood is a fall in blood pressure, which is detected by the baroreceptors in the walls of the atria and pulmonary vessels. The decrease in the frequency of impulses from the baroreceptors results in an increase in frequency of impulses in sympathetic nerves. This stimulates the heart and brings about vasoconstriction of all blood vessels, except those in the brain and heart. Constriction of the veins helps to maintain the pressure of blood returning to the heart, while constriction of the arteries reduces the fall in arterial blood pressure.

These responses take a little time to come fully into effect, which is why blood donors are asked to sit quietly for a few minutes after giving blood. The loss of blood volume takes longer to correct, since this involves additional fluid intake (helped by a cup of tea). This is brought about by the kidneys, which increase their fluid retention in response to increased output of the hormones ADH and aldosterone (Chapter 24). These effects take several hours. The recovery of the number of red cells takes a month or more, and is brought about by another hormone, erythropoietin, secreted by the kidney.

## QUESTIONS

**1** A certain volume of whole blood from which all oxygen had been removed was shaken with an equal volume of normal atmospheric air in a sealed vessel. The process was then repeated in exactly the same way except that plasma was used instead of whole blood. In which case would the plasma contain the higher concentration of dissolved oxygen? Give a reason.

**2** A slight rise in temperature promotes the dissociation of oxyhaemoglobin.

**a)** How would the dissociation curve at 40°C in active muscle compare with that at 37°C, all other conditions being the same?
**b)** Under what conditions in the body would this be advantageous?

**3** Draw graphs showing how each of the following changes with distance along the systemic circulatory system:

**a)** blood pressure
**b)** blood velocity
**c)** oxygen content.

**4** Large arteries have much elastic tissue and less muscle; in small arteries the situation is reversed. Explain the functional significance of this.

**5** Which of the following could describe the flow of blood through a certain artery?

A oxygenated blood under high, steady pressure
B oxygenated blood under low, steady pressure
C deoxygenated blood under low, steady pressure
D deoxygenated blood under moderate surging pressure
E deoxygenated blood under high, surging pressure

**6** In what ways does cardiac muscle **(a)** resemble **(b)** differ from, skeletal muscle?

**7** Which muscle antagonises arterial smooth muscle?

**8** If, during exercise, the blood flow through the lungs is compared with the total flow through the limb muscle, it would be found that:

A the flow through the limb muscles would exceed that through the lungs.
B the flow through the lungs would exceed that through the limb muscles.
C the flow would be equal.
D the relative rates of flow would depend on the intensity of the exercise.

**9** When all four heart valves are closed, the pressure in the chambers is, in descending order:

A ventricles → arteries → atria
B arteries → ventricles → atria
C atria → arteries → ventricles
D arteries → atria → ventricles

**10** Read the information box on page 406 and then decide which *two* of the following statements are correct. As blood flows along the venous system:

A its speed increases and the tension in the vein walls decreases.
B its speed of flow and the tension in the vein walls both increase.
C its oxygen content doesn't change and the tension in the walls increases.
D its oxygen content decreases and the tension in the walls increases.
E its speed of flow decreases and the tension in the walls increases.

**11** The thickness of capillary walls (about 0.5 μm) is a small fraction of the total diffusion path (5–100 μm). What, then, is the advantage of them having such thin walls (hint: think of the mechanism of tissue fluid formation).

CHAPTER

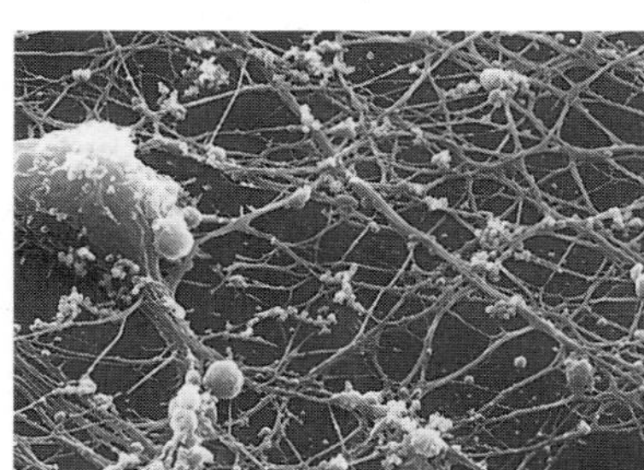

# 23 GAS EXCHANGE and BREATHING

## LEARNING OBJECTIVES

By the time you have completed your study of this chapter you should be able to:

- distinguish between respiration, gas exchange and breathing.
- explain why the presence of gas exchange organs is related to body size.
- explain the essential similarity and difference between a gill and a lung.
- explain how the structures of the gills of a bony fish are adapted for gas exchange.
- explain how a continuous flow of water is maintained over the gills, and how the direction of blood flow maximises the efficiency of gas exchange.
- describe how the fine structure of the mammalian lung adapts it for gas exchange.
- explain how air is pumped into and out of the lungs during breathing.
- describe the nervous control of breathing.
- explain the role of negative feedback in regulating the concentration of carbon dioxide and oxygen in the blood.
- distinguish between the various subdivisions of the lung capacity, and calculate minute volume and rate of oxygen uptake from a spirometer trace.
- describe how the lungs are protected against pathogenic bacteria.
- describe how the structure of the tracheal system of insects adapts it to the transport of oxygen and carbon dioxide.

## Gas exchange and body size

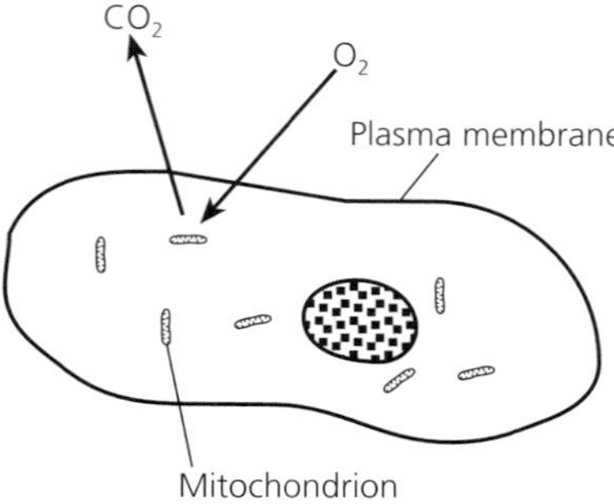


**Figure 23.1**
Gas exchange in unicellular organisms. The diffusion path is between the environment and the mitochondria, only a few of which are shown

With the exception of a few animals that obtain their energy by fermenting sugar to lactic acid, animals obtain their energy by **respiration** (see Chapter 5 for the distinction between fermentation and anaerobic respiration). Respiration occurs in the mitochondria and, besides generating ATP, uses oxygen and produces carbon dioxide. As a result, oxygen is depleted inside the cells and diffuses in, whilst carbon dioxide accumulates in the cells and diffuses out. The diffusion of oxygen and carbon dioxide in opposite directions across the surface of a cell or organism is called **gas exchange**.

In small organisms oxygen and carbon dioxide travel the entire distance between mitochondria and external environment by diffusion alone (Figure 23.1). This is also true of some quite large animals, such as jellyfish and sea anemones. These animals obtain enough oxygen by diffusion, either because they are relatively inactive, as in sea anemones, or because a high proportion of the body consists of metabolically-inactive mesogloea, as in jellyfish (see Chapter 15).

In larger and more active animals, gas exchange over the general body surface cannot satisfy the animal's needs for the following reason. All other factors being equal, the rate at which oxygen can enter the body is limited by the *area* of the surface through which it enters. Oxygen demand and carbon dioxide production, on the other hand, are governed by the mass of the body, which is proportional to its *volume*. As explained in Chapter 16, with increasing body size volume increases faster than surface area. If its shape remains constant, doubling the length of the body increases its surface area (and hence its oxygen supply) four times, but its volume (and therefore its oxygen requirement) goes up eight times. Doubling body length would thus halve the ratio of oxygen supply to oxygen demand, so each cell could only obtain oxygen at half the speed.

There are two ways in which larger animals can obtain sufficient oxygen:

- by having a body shape which provides a large surface area relative to volume. Flatworms, as their name suggests, have a large surface by virtue of their leaf-like bodies (Figure 15.42). This steepens the concentration gradient of oxygen by reducing the distance between the interior and exterior of the animal.
- by means of a blood system, as in annelids, arthropods, molluscs and vertebrates. By removing oxygen from the gas exchange surface, a blood system increases the rate of diffusion per unit area of body surface (see the information box on page 389).

With the exception of insects, most animals use a combination of both methods. Oxygen and carbon dioxide are carried in the blood to and from a large expansion of the body surface called a **gas exchange organ** or 'respiratory organ'. The former term is preferable to the long-established alternative, since its function is not respiration (which is a chemical process occurring inside cells), but gas exchange.

## *Gills and lungs*

Gas exchange organs are of two types: **gills**, which are filamentous outpushings of the body and **lungs**, which are intuckings of the body (Figure 23.2). Gills work well in aqueous environments because the filaments are supported and held apart from each other by the buoyancy of the water. In air, however, they collapse and stick together, greatly reducing their surface area. A lung does not suffer this disadvantage, but for different reasons is ineffective in water.

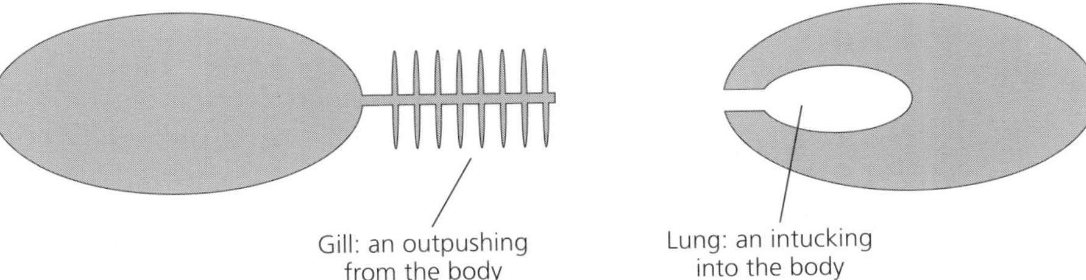


**Figure 23.2**
Difference between a gill and a lung

Gills and lungs have a number of features which promote gas exchange:

- they have a *large surface* compared with their volume.
- with the exception of insects, they are *richly supplied with blood.*
- their boundary layers are very *thin*, thereby shortening the diffusion distance between the blood and the external environment.

- even in terrestrial animals, their surfaces are *moist*. This is important because diffusion depends on free movement of molecules and so cannot readily occur in a solid. Before entering the body, then, oxygen must first dissolve in a thin layer of liquid.

### *Ventilation*

As a result of gas exchange, the concentration of oxygen immediately outside the gas exchange surface tends to fall and carbon dioxide tends to accumulate. In most animals this is counteracted by the **ventilation** of the gas exchange surface by pumping fresh supplies of water or air over it. The pumping movements that bring about ventilation are called **breathing**.

There are two kinds of ventilation (Figure 23.3). In most aquatic animals water flow is pumped **unidirectionally**.

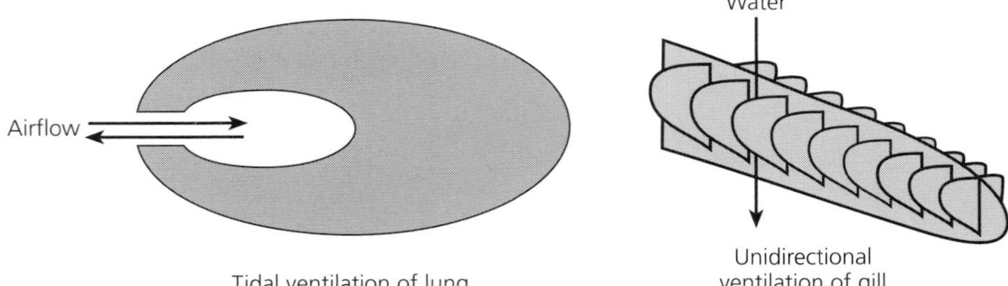


**Figure 23.3**
Unidirectional and tidal ventilation

In land animals ventilation is **tidal**, air entering and leaving the lung by the same route. A lung is never emptied completely – some 'stale' air always remains. The air inside a lung therefore contains less oxygen and more carbon dioxide than the air outside. This has an important implication: the gas transport pigment must become saturated at an oxygen concentration that may be considerably lower than that of the atmosphere.

## Gas exchange in a bony fish

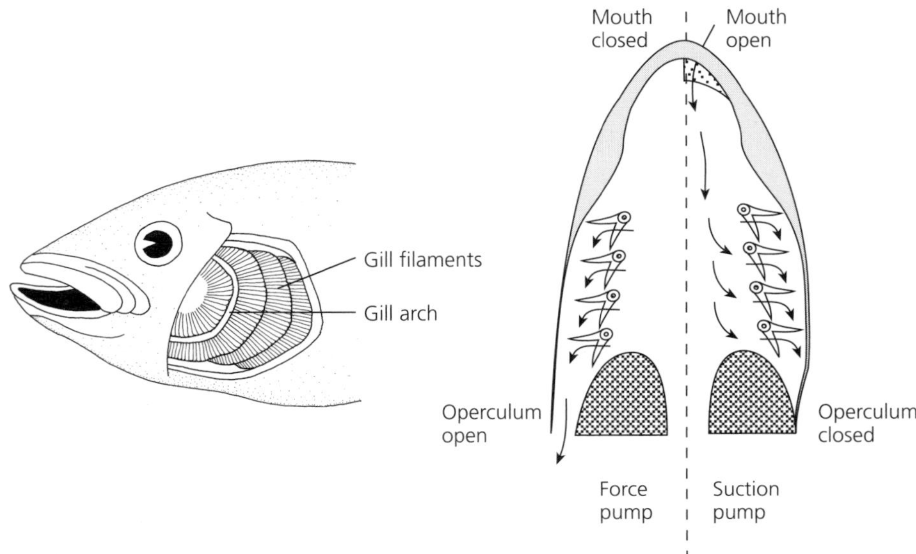


**Figure 23.4**
Gills of a bony fish seen (A) with operculum removed and (B) in a horizontal section through the pharynx

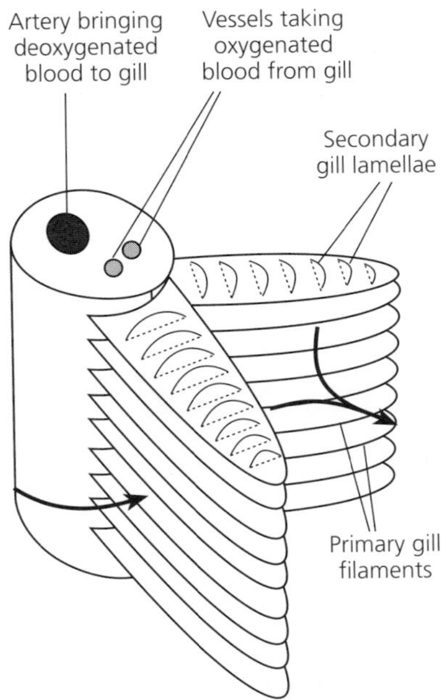


**Figure 23.5**
Part of a gill of a bony fish.

In fish, the anterior part of the gut is concerned with both breathing and feeding and is called the **pharynx**. Water enters through the mouth and leaves via five pairs of **gill slits** in the side walls of the pharynx (Figure 23.4). In bony fish these are invisible externally because they are covered by a flap-like **operculum**, or gill cover.

The partition between two adjacent gill slits is a **gill**, and bears two rows of plate-like **primary filaments** (Figure 23.5). In cross section the primary filaments look like the two halves of a letter 'V'. The surface area of each primary filament is further increased by many even thinner **secondary lamellae**. Each primary filament is supported by a **gill ray** which extends from a **gill arch**.

### Ventilation and gas exchange

Despite the disadvantages of water as a gas exchange medium (see the information box on page 416), bony fish manage to extract nearly 80% of the dissolved oxygen from the water. Two features of the gas exchange system enable them to do this.

1 The flow in the secondary lamellae is in the opposite direction to that of the water. As a result of this **counterflow** a concentration gradient exists along the entire length of each capillary, with the most oxygenated water adjacent to the most oxygenated blood and the least oxygenated blood adjacent to the least oxygenated water (Figure 23.6).

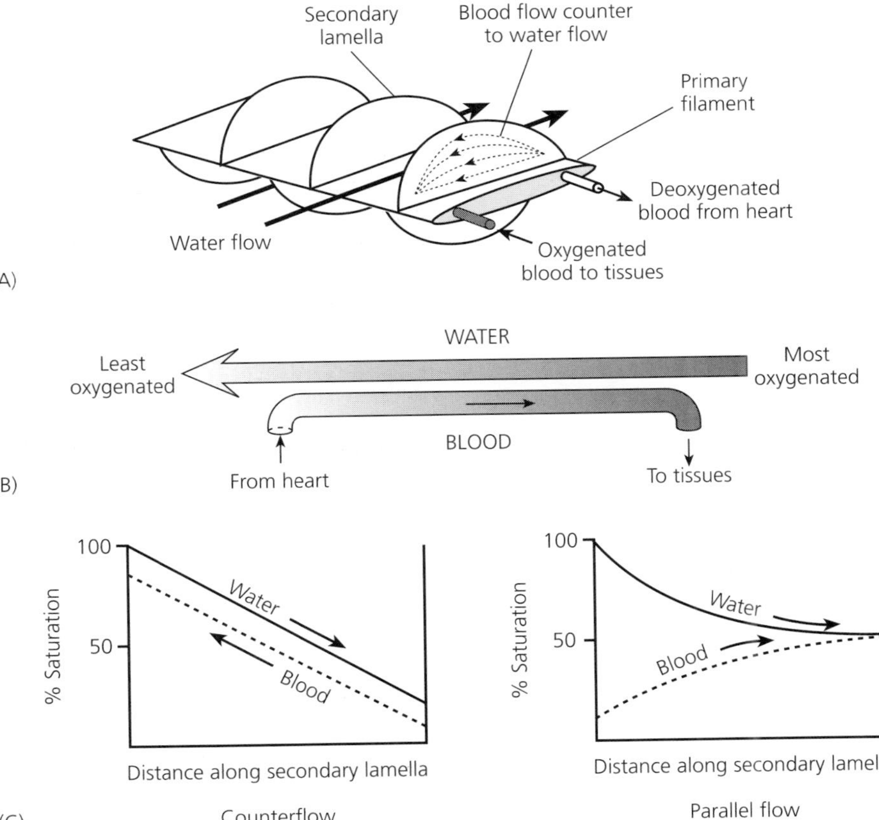


**Figure 23.6►**
Counter-flow in a gill of a bony fish. (A) Gill lamellae showing flow of blood and water (after Hughes); (B) diagram showing how the most oxygenated water is opposite the most oxygenated blood; (C) theoretical graphs showing how blood can pick up more oxygen with counterflow. Though the % changes in oxygenation shown assume that blood and water travel at equal speeds and have equal oxygen carrying capacities (actually not the case), the general principle holds

2 A nearly continuous water flow is maintained over the gills. Oxygen is sparingly soluble in water, so any cessation of the current would quickly lead to the depletion of oxygen immediately next to the gill surface. Continuous water flow depends on the alternate action of two pharyngeal pumps – a suction pump which operates during inspiration, and a force pump which acts during expiration.

Expiration begins with the closing of the mouth. Pressure in the pharynx increases as the floor of the pharynx rises and opercular flaps move inwards. Water is forced over the gills and out past the opercular flaps. Inspiration begins with the opening of the mouth and expansion of the operculum. The reduction in pressure causes water to enter the mouth and, because of the simultaneous expansion of the operculum, water movement over the gills is maintained by suction.

## Air- and water-breathing

Why doesn't a gill work well on land or a lung work well in the water? The reasons lie in a number of key properties of water and air, summarised in the table below.

| | Water | Air |
|---|---|---|
| Density | high | low |
| Viscosity | high | low |
| Oxygen content | low | high |
| Diffusion rate | low | high |

- Oxygen is only sparingly soluble in water. Air contains about 27 times as much oxygen as an equal volume of water saturated with air at 20°C.
- Water is about 770 times denser than air and about 100 times as viscous. It therefore costs a lot more energy to move a current of fresh water over a gas exchange surface than to move the same volume of air.
- Diffusion of gases in solution is many thousands of times slower than diffusion in air. This is important because no matter how well a gill is ventilated, there is always a very thin layer of stationary water adjacent to the gill surface through which oxygen and carbon dioxide must move by diffusion. For this reason ventilation of a gill needs to be more or less continuous.

When we consider these factors together, the picture for fish looks even worse. A kilogramme of air contains 209 000 $cm^3$ of oxygen – over 30 000 times as much as the same mass of water saturated with air at 20°C. A fish thus needs to move a large mass of water to obtain a small quantity of oxygen. Taking into account the greater density and viscosity of water, breathing is much more energy-expensive for a fish than for a mammal. A trout uses about 20% of its energy in breathing (and some fish use more), whereas a mammal uses less than 5%.

A further disadvantage of water-breathing is the fact that oxygen becomes less soluble with a rise in temperature, being only half as soluble at 35°C as it is at 0°C. Oxygen availability is thus more variable in water than it is on land.

Though terrestrial gas exchange seems to have all the advantages, things are not quite as one-sided as they might appear. Buoyed-up by the high density of water, gills can protrude into it without sticking together as they would on land. As an *outpushing* from the body, a gill can be ventilated by a unidirectional flow of water. As we have seen, this makes it possible for water to flow in the opposite direction to that of the blood, which allows a higher percentage of oxygen to be extracted from the water.

Another important point is the fact that even in terrestrial animals, gas exchange surfaces are always moist. The concentration of oxygen in this film of moisture is thus limited by the solubility of oxygen in water. The oxygen concentration next to the gill capillaries in a trout is actually higher than the oxygen concentration in the film of moisture lining the air sacs in a mammal. The difference is that if a fish stops breathing, the oxygen concentration in the water just outside the gills falls quickly because there is so much less to replace it, and what little there is, diffuses into the body more slowly. In a mammal, there is continuous replenishment from the large reservoir of gaseous oxygen immediately adjacent to the gas exchange surface.

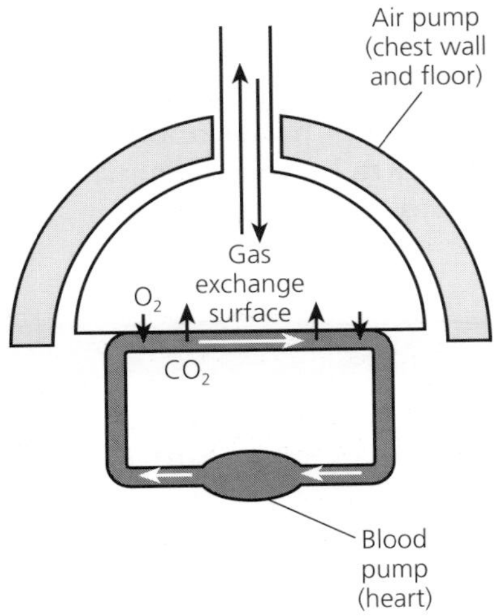


**Figure 23.7**
Essential features of the human gas exchange system

# The human gas exchange system

From a functional point of view, the human gas exchange system can be considered to consist of five parts (Figure 23.7):

1 the gas exchange surface itself,

2 the bronchial tree – a branching system of tubes that deliver air to and from the gas exchange surface,

3 the walls and floor of the chest, which pump air to and from the gas exchange surface,

4 the heart, which pumps blood through the lungs,

5 the pulmonary arteries and veins, which carry blood to and from the gas exchange surface.

## Structure of the lungs

The lungs are two large, pink, spongy organs situated on either side of the heart (Figure 23.8). Each lung lies in a separate **pleural cavity** containing a thin layer of fluid. Each cavity is bounded by a **pleural membrane**, which is folded back on itself to form inner and outer layers. The outer pleural membrane lines the chest cavity and the inner pleural membrane forms the covering of the lung. The pleural cavities are extremely narrow and contain just enough fluid to allow the pleural membranes to slip past each other during breathing. The lungs are elastic and tend to pull away from the chest wall, causing the pressure in the pleural cavities to be slightly lower than atmospheric pressure, except in forced exhalation.

Air is delivered to and from the lungs by the **trachea**, or windpipe, at the top of which is the **larynx**, or 'voice box'. The trachea is held open by incomplete rings of cartilage. The spaces between allow it to extend slightly during breathing movements. The trachea divides into two main **bronchi**; each bronchus branches repeatedly in the lung tissue, forming a **bronchial tree** (Figure 23.9). The bronchi are held open by irregular

**Figure 23.8▼**
The human chest showing the breathing organs. The last two ribs do not extend far enough forward to be visible

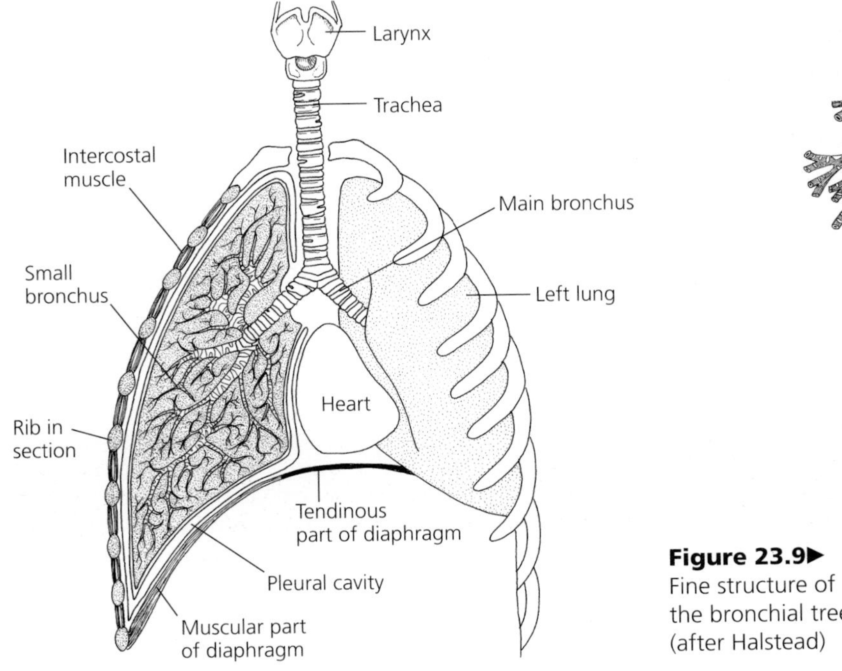


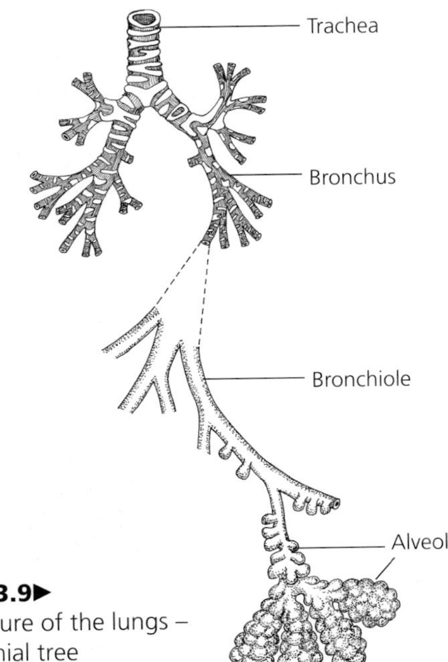


**Figure 23.9►**
Fine structure of the lungs – the bronchial tree (after Halstead)

plates of cartilage and, like the trachea, are lined by ciliated epithelium which provides protection against micro-organisms (Chapter 27). The smallest bronchi give rise to **bronchioles**, which have no cartilage and are held open by elasticity of the surrounding tissue. They have smooth muscle in their walls which enables their diameter to be controlled. It is the excessive contraction of this muscle that gives rise to breathing difficulties in **asthma**.

## Cleaning the lungs

**The epithelia lining the trachea and bronchi contain many goblet cells which secrete mucus in which bacteria and dust particles become trapped. In the manner of an escalator, the mucus sheet slowly carries particles towards the pharynx where it is periodically swallowed. The bacteria are subsequently killed by acidic gastric juice.**

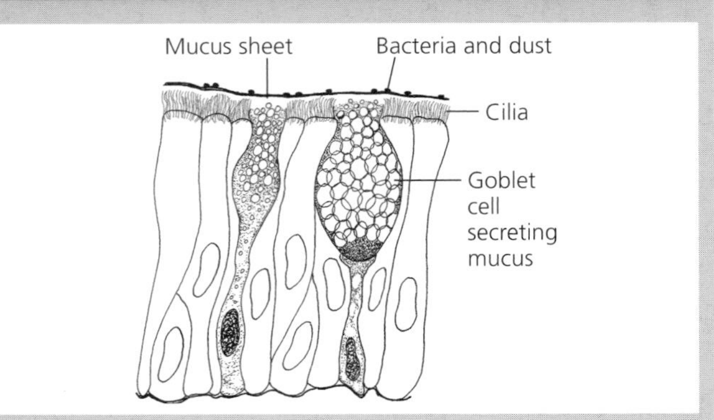


Ciliated epithelium lining the trachea and bronchi

The finest bronchioles open into tiny, thin-walled air sacs or **alveoli** (Figure 3.6). These average about 100 μm in diameter, and an average human has about 350 million in each lung. The total area of the alveoli has recently been estimated to be just over 140 $m^2$ – about half the area of the oft-quoted tennis court (266 $m^2$), and about 90 times that of the skin.

Alveoli are richly supplied with capillaries and since each capillary lies between adjacent alveoli, it is almost surrounded by air. So dense are the capillaries that they form an almost continuous 'lake' of blood, with an area recently estimated to be about 125 $m^2$, or about 87% of the alveolar surface.

The alveolar wall consists of squamous epithelium, but other kinds of cell are also present (Figure 23.10). Amongst these are **macrophages** – white corpuscles which help defend the body against pathogens. Other cells, called **type II** cells, produce a detergent-like substance that plays a vital role in breathing (see the information box on page 420).

### *Gas exchange in the lungs*

Table 23.1 shows that, as we might expect, blood in the pulmonary artery is richer in carbon dioxide and poorer in oxygen than alveolar air. Under these circumstances oxygen and $CO_2$ diffuse down their respective concentration gradients – oxygen from the alveoli into the blood, and $CO_2$ in the reverse direction.

**Table 23.1**
Comparison between partial pressures (kPa) of gases in the alveoli with those in the blood entering and leaving the lungs

| | Air in alveoli | Blood entering lungs | Blood leaving lungs |
|---|---|---|---|
| $O_2$ | 13.6 | 5.3 | 13.6 |
| $CO_2$ | 5.3 | 6.1 | 5.3 |
| $N_2$ | 76.1 | 76.1 | 76.1 |

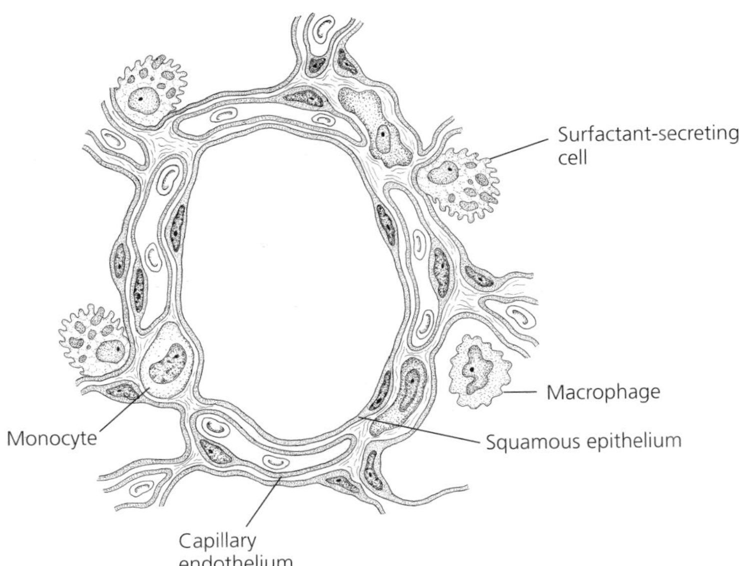


**Figure 23.10**
Diagrammatic section through an alveolus. For clarity, the thickness of the walls has been drawn disproportionately thick

In Chapter 6 we learned that Fick's First Law of diffusion states that the rate of diffusion of a gas is directly proportional to the area of the diffusion path and to the concentration difference, and inversely proportional to the distance over which it occurs. In the lungs, two of these factors are highly favourable to rapid diffusion of gases between blood and air for the following reasons:

- the alveolar and capillary walls are extremely *thin*, so the diffusion path is very short – in fact the distance between air and the red corpuscles averages only about 0.5 μm (Figure 23.11).

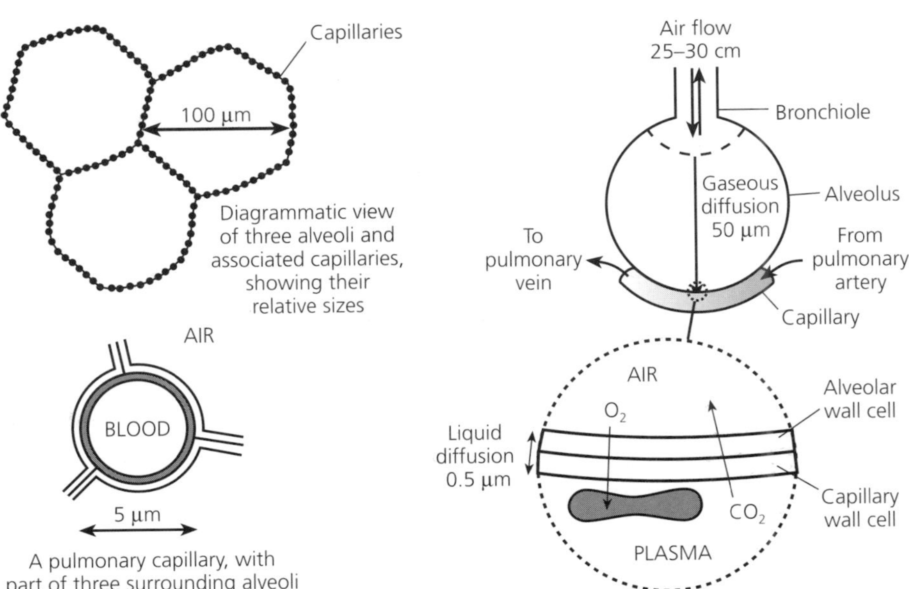


**Figure 23.11**
The gas exchange surface

- the *area* of near-contact between blood and air is very large.

The third factor – the concentration difference – is not as great as theoretically possible for reasons to be discussed shortly.

Table 23.1 also indicates that the gases in the blood leaving the lungs have the same partial pressures as in the alveoli, showing that by the time the blood leaves the alveoli it has taken up as much oxygen and given up as much carbon dioxide as it can. In fact, even in intense exercise when blood flows more quickly through the lungs, it is fully equilibrated with the air in the alveoli well before it leaves the capillaries.

## Emphysema, a very slow death

One of the many harmful effects of smoking is **emphysema**, a condition in which there is a breakdown of elastic tissue in the lungs. The alveolar walls break down causing them to coalesce into a smaller number of larger spaces. This greatly reduces the gas exchange surface, resulting in a chronic shortage of oxygen and the accumulation of carbon dioxide, which renders the person permanently short of breath. In destroying alveolar capillaries, it increases the resistance to pulmonary blood flow, causing the right side of the heart to enlarge. With less elastic tissue in the lung to hold them open, the bronchioles tend to become narrower making it harder for the patient to breathe. The disease, which may last 20 years or more, is usually fatal.

Alveoli from normal lung (above) and from a person with emphysema below)

## Surface tension in the lungs

As explained in Chapter 2, wherever water meets air the water behaves as if it has an invisible elastic 'skin'. This **surface tension** is due to the attraction between water molecules and tends to reduce the surface to a minimum. In a curved surface, such as in a soap bubble, the tension is directed partly inwards. If the bubble is connected with an outlet, surface tension tends to deflate the bubble. The smaller the bubble, the greater the proportion of the tension that is directed inwards, so the greater its tendency to deflate.

The alveoli are like tiny balloons, lined with an extremely thin layer of fluid. Since they are not sealed they tend to deflate, and the smaller they get the greater their tendency to shrink further. Inhaling is, therefore, like inflating millions of microscopic balloons.

Surface tension has another potentially harmful effect. By exerting an inward pull, it sets up a slight negative pressure round the capillaries, tending to draw tissue fluid from them.

In reality, neither of these effects is significant because certain cells in the alveolar wall produce **surfactant**, a phospholipid with detergent-like properties, which reduces surface tension. However, premature babies, whose lungs are not mature enough to produce surfactant, suffer from **breathing distress syndrome.** Not only is it hard for them to inflate the lungs, but the lungs may become congested with tissue fluid.

### *Evidence that gas exchange occurs in the alveoli*

The fine structure of the lungs is superbly adapted for gas exchange. However, lung anatomy only provides circumstantial evidence that the alveoli are the site of gas exchange. Direct evidence is provided in Table 23.2, which gives the percentage composition of inhaled air in comparison with that of bronchial air and alveolar air.

**Table 23.2**
Comparison between percentage atmospheric, alveolar and bronchial air

| | Inspired air | Bronchial air | Alveolar air | Exhaled air |
|---|---|---|---|---|
| $O_2$ | 20.92 | 20.92 | 14.5 | 16.89 |
| $CO_2$ | 0.04 | 0.04 | 5.55 | 3.5 |
| $N_2$ | 79.04 | 79.04 | 79.95 | 79.61 |

The first part of an exhaled breath is called **bronchial air**, and consists of air which had occupied the trachea and bronchi at the end of the previous exhalation. The last part consists of **alveolar air**, or air that had been in the alveoli. An entire exhaled breath is a mixture of alveolar and bronchial air. Table 23.2 shows that only the composition of the alveolar air is changed, so gas exchange must occur in the alveoli.

You may notice that the data in Table 23.2 appear to suggest that the body is producing nitrogen, since it is more concentrated in exhaled air than in inhaled air. It isn't, of course; the increase in nitrogen concentration results from the fact that, since the body normally burns some fat as well as carbohydrate, slightly more oxygen is used than carbon dioxide is produced, giving an RQ of slightly less than 1 (Chapter 5). The concentration (but not the *amount*) of nitrogen is thus slightly increased.

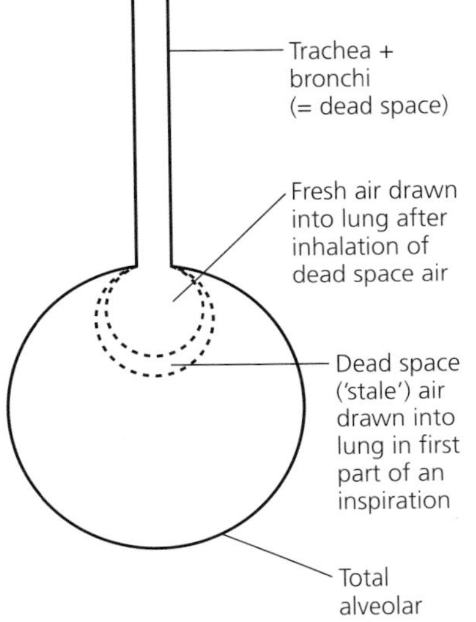


**Figure 23.12**
The dead space

## *The dead space*

Since they play no part in gas exchange, the trachea, bronchi and bronchioles collectively form the **dead space** (Figure 23.12). At the end of an expiration, the dead space contains alveolar air which re-enters the alveoli at the beginning of the next inspiration. Since the first air to enter the alveoli is 'stale', only a part of each breath is actually exchanged with the environment. The volume of the dead space is about 150 $cm^3$. An average breath in a resting person is about 500 $cm^3$, so only about 70% of each breath at rest is useful. When the depth of breathing is greater, the proportion of dead space air is less.

Although the trachea and bronchi play no part in gas exchange, they do have two very important functions. First, they allow the air to be humidified and warmed before it reaches the alveoli. Were it not for this, the cells lining the alveoli and the smallest bronchioles would die. Second, the mucous lining cleans the air of bacteria and other potentially harmful particles (see the information box on page 418).

### The alveolar atmosphere and the cost of saving water

At the end of each expiration a considerable volume of air remains in the lungs. In a quiet inspiration, the fresh air reaching the alveoli is only about $\frac{1}{7}$ the volume of air already present. Each inhaled breath merely 'tops up' the air already in the lungs so it has little effect on its composition, which changes little during the breathing cycle. An important result is that, as Table 23.2 shows, alveolar air is very different from atmospheric air, having over 100 times as much carbon dioxide and about two-thirds as much oxygen.

You might think it would be better if all the air in the lungs could be changed with each breath. Actually it would not. Quite apart from the fact that it is anatomically impossible to empty the lungs, it would have a serious disadvantage. Whilst it would certainly raise the $pO_2$ in the alveoli, it would greatly increase water loss. Even in quiet breathing the amount of water vapour lost is nearly half of the total unavoidable water loss. The price of saving water is thus a lower $pO_2$ at the gas exchange surface.

## Matching air flow and blood flow

Suppose an athlete and an unfit person of similar body mass run up the same number of steps in the same time. They do a similar rate of work and use a similar volume of oxygen – yet the unfit person is much more out of breath than the athlete. Why is this?

The answer is in the heart, which in an unfit person cannot pump as much blood through the lungs as the heart of an athlete can. If the blood flow is insufficient to get rid of the extra carbon dioxide generated, carbon dioxide builds up and generates the feeling of breathlessness. For maximum efficiency, therefore, ventilation and blood flow must be matched. Much of the increased breathing of the untrained person is wasted because the blood pump cannot keep pace with the air pump.

The problem of matching air flow and blood flow also occurs in different parts of the lung. The pressure in the pulmonary artery is so low that when standing, the upper part of the lung gets less blood than the lower part. When lying down, all parts of the lung receive similar amounts. This is where the smooth muscle in the bronchioles comes in. If one part of the lung receives less blood than another part, the bronchioles supplying that part constrict. In reducing the air supply to that part of the lung, more air is diverted to those parts with a better blood flow. In this way, changes in the distribution of blood are matched by changes in the distribution of air.

## The breathing mechanism

Breathing involves the alternate increase and decrease of air pressure in the lungs relative to that outside. A fall in air pressure in the lungs causes **inspiration**, or breathing in; a rise in pressure causes **expiration**, or breathing out.

Pressure changes result from changes in volume – expansion of the lungs decreases the air pressures inside them, and vice versa (Boyle's Law). The lungs are unable to contract or expand by themselves, but do so passively in response to forces generated by muscles in the chest wall. These forces are transmitted to the lungs by the fluid in the pleural cavities. Fluid has a fixed volume, so any expansion of the chest wall causes the lungs to expand by the same amount.

### *Inspiration*

Two sets of muscles increase the chest volume: the **diaphragm** and the **external intercostal muscles**. The diaphragm is the more important, and is a dome-shaped structure forming the floor of the chest. The central, non-muscular part forms a white, tendinous sheet. Radiating from it are muscle fibres anchored to the vertebral column, sternum and lowest ribs. When the diaphragm contracts it pulls the lungs downwards. At the same time it pushes the abdominal viscera against the muscles of the abdominal wall, which thereby become stretched (Figure 23.13).

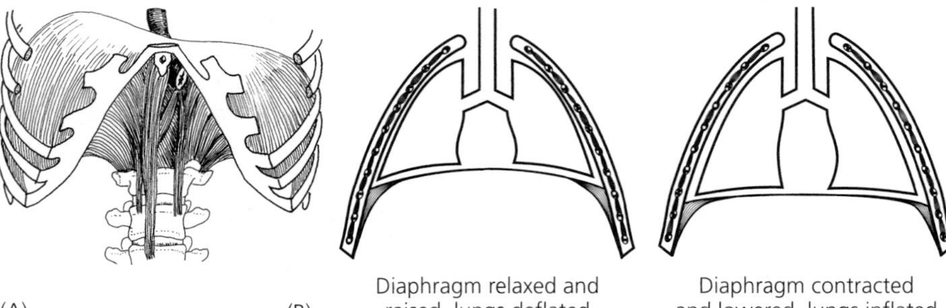


**Figure 23.13**
The diaphragm seen from the front with ribs cut away (A) and a diagrammatic representation of its movement (B)

The external intercostal muscles run downwards and forwards between adjacent ribs. Each rib forms a synovial joint with a thoracic vertebra and at the other end is linked to the sternum (breastbone) by hyaline cartilage. These cartilaginous joints give springiness to the chest.

Figure 23.14 shows how contraction of the intercostal muscles raises the ribs, though their movement is actually more complex than indicated since they are curved. To see how the ribs are raised, notice that shortening of the muscle fibres decreases the angle between the rib and the region of the vertebral column above it.

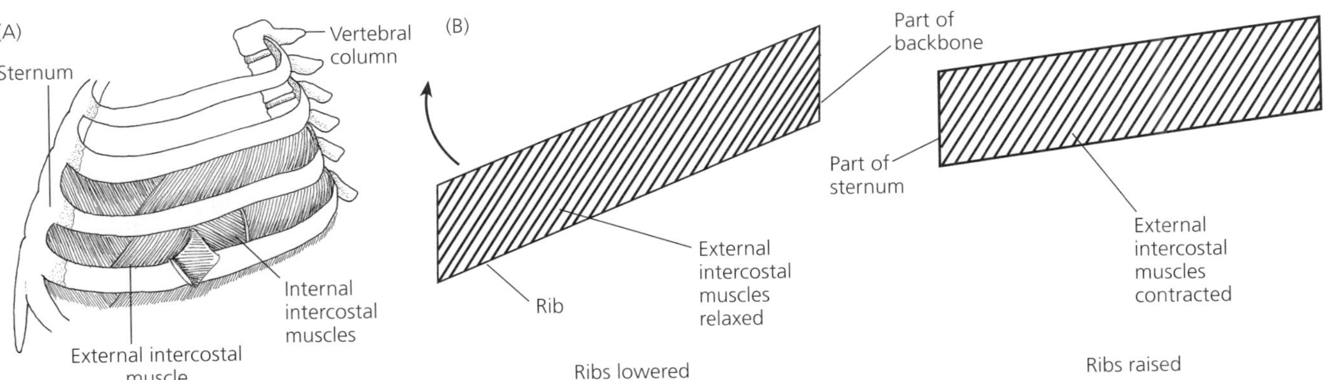


**Figure 23.14▲**
The intercostal muscles, with a slip of external intercostal muscle peeled back (A) and how they move the ribs (B)

## *Expiration*

Inspiration stretches the walls of the alveoli and elastic fibres in the lungs, and also the muscles of the abdominal wall. When the inspiratory muscles relax, the lungs shrink under their own elasticity and the abdominal wall muscles push the organs up against the diaphragm, raising it. The energy for expiration thus comes from energy stored as elastic tension generated during the previous inspiration.

In forced breathing (or when coughing) elastic forces cannot deflate the lungs quickly enough and expiration is aided by muscle contraction. Ribs are forced downwards by **internal intercostal muscles**. At the same time the muscles of the abdominal wall contract, forcing the abdominal organs up against the diaphragm. You can feel the abdominal muscles contract when you try forcibly to exhale with the glottis (entrance to the larynx) closed.

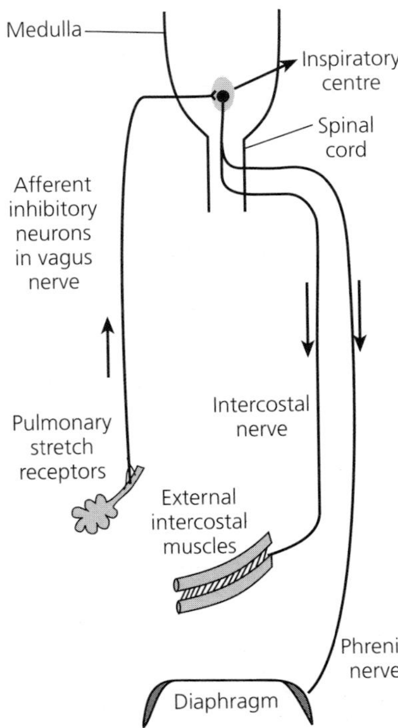


**Figure 23.15**
Control of breathing movements in a resting person

## *Co-ordination of breathing*

Though we can inhale or exhale voluntarily, breathing is largely automatic. The neurons responsible occupy the **breathing centre**, which lies in the medulla of the brain stem. It is differentiated into two regions: a dorsal **inspiratory centre** and a ventral **expiratory centre** (Figure 23.15).

The expiratory centre is normally only active in forced breathing, since quiet expiration is passive. The inspiratory centre communicates with the diaphragm by the left and right **phrenic nerves**, and with the external intercostal muscles by the **intercostal nerves**. The expiratory centre communicates with the expiratory muscles. At regular intervals, the neurons in the inspiratory centre send volleys of impulses, causing inspiration. At the end of each volley the inspiratory muscles relax again and expiration occurs.

This rhythmicity seems to be intrinsic to the breathing centre, since it continues even when all afferent connections have been severed. However, the activity of the inspiratory centre is influenced by sources external to it.

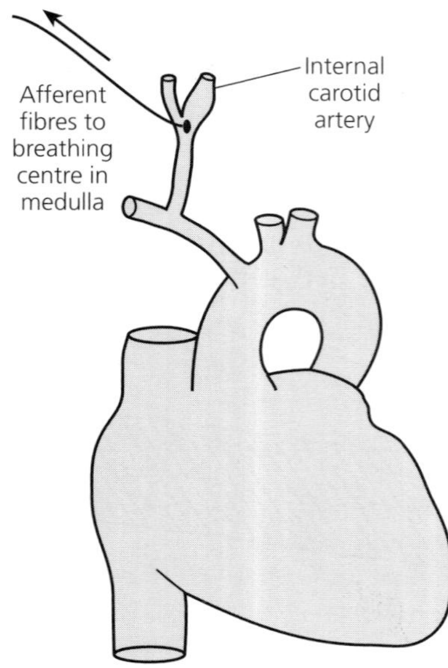


**Figure 23.16**
The carotid bodies

One such source consists of stretch receptors in the walls of the bronchial tree. These are relatively inactive in quiet breathing, but during deep inspiration they send inhibitory impulses via the vagus nerve to the inspiratory centre. At the same time they discharge impulses to the expiratory centre causing contraction of the expiratory muscles. With the pulmonary stretch receptors no longer stimulated, the inspiratory centre is freed from inhibition and fires another succession of impulses to the inspiratory muscles, beginning another cycle.

Breathing can of course be deliberate or, for a limited period, prevented. In this case, impulses from the cerebral cortex bypass the breathing centre and travel directly to the spinal motor neurons controlling the breathing muscles.

## *Regulation of breathing*

We are all aware that the rate and depth of breathing increase during exercise, and that when we hold our breath the desire to breathe increases. Both these observations are evidence of some kind of control system by which the ventilation rate is matched to the needs of the body. In some way the breathing centre must be kept informed of changes in the composition of the blood. This raises two obvious questions: what aspect of blood chemistry is being monitored, and where are the receptors located?

There are two obvious possible stimuli: an increase in the carbon dioxide level, and/or a decrease in the oxygen level. Experiments performed early this century established that carbon dioxide is the main stimulus. When air containing 15% oxygen but normal carbon dioxide level is inhaled, there is no detectable effect on breathing. Only when the oxygen level falls to 12% is breathing stimulated. In contrast, even a small increase in the proportion of carbon dioxide in inhaled air strongly stimulates breathing.

Where are the receptors? The most important site for carbon dioxide detection is in the medulla of the brain, very close to the breathing centre itself. A small group of cells near the surface of the medulla monitors the carbon dioxide and pH level of the cerebrospinal fluid (CSF). (The pH of CSF changes more readily than tissue fluid since it contains much less protein buffer.)

In response to a rise in $pCO_2$ or a fall in pH, the activity of the inspiratory centre is increased in two ways. First, the frequency of impulses in each volley increases, provoking greater contraction of the inspiratory muscles. Second, the time between successive volleys decreases, reducing the time between breaths. Because of the effect of the pulmonary stretch receptors, the expiratory centre is also stimulated.

Another group of chemoreceptors consists of the **carotid bodies** – small clusters of receptors in the wall at the base of the internal carotid artery. They communicate with the breathing centre by the IXth cranial nerve (Figure 23.16). Like the breathing centre itself, the carotid bodies are stimulated by a rise in $pCO_2$ or a fall in pH, but they are also stimulated by a fall in $pO_2$. Similar chemoreceptors, the **aortic bodies**, are located in the arch of the aorta, but they are thought to be less important and are insensitive to pH.

Under normal circumstances the carotid and aortic bodies are probably subordinate to the breathing centre. In their sensitivity to oxygen shortage, however, they have no substitute, and play an essential role in regulation of breathing at high altitudes (see the information box on page 425).

The regulation of breathing is an example of **homeostasis** (Chapter 21). The higher the $pCO_2$, the higher the rate and depth of breathing. If

$pCO_2$ falls below normal, breathing is reduced and may temporarily stop. This is an example of **negative feedback** – the greater the deviation from normal, the stronger the tendency to correct it (Figure 23.17).

The precision with which ventilation rate is matched to carbon dioxide production is shown by the fact that in vigorous exercise, the composition of exhaled air hardly changes (see Question 3 at the end of this chapter).

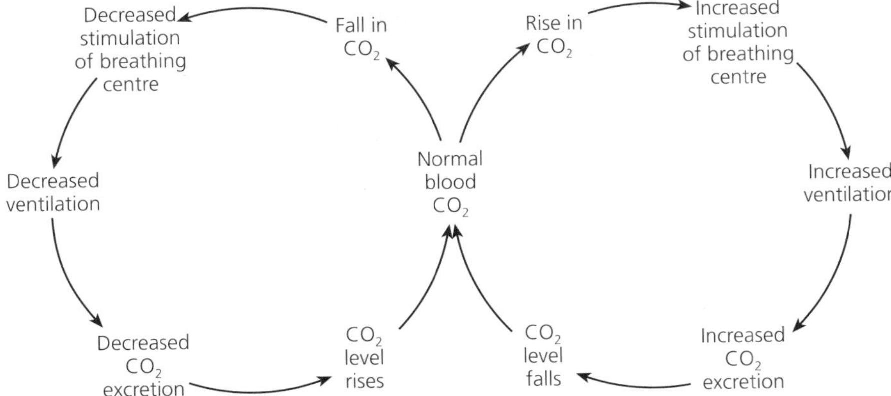


**Figure 23.17**
Homeostatic control of the carbon dioxide content of the blood

## High altitude breathing

**Under normal conditions a rise in oxygen uptake is accompanied by an increase in carbon dioxide production, and both needs are met by increased ventilation. At high altitudes the situation becomes more complicated. In order to obtain a given amount of oxygen, breathing has to increase. The trouble is that this lowers the $pCO_2$ in the blood, reducing the stimulus to breathe. After a few days the breathing centre adjusts to this situation by becoming more sensitive to carbon dioxide, so the stimulus to breathe is maintained at lower carbon dioxide levels.**

**Other adaptations to high altitude involve two changes in the blood itself. Within hours the concentration of 2,3-diphosphoglycerate in the red cells increases, which raises the affinity of haemoglobin for oxygen. By shifting the oxyhaemoglobin dissociation curve to the left, the blood leaving the lungs carries more oxygen than it would normally.**

**A much slower response involves an increase in red cell production. The slight fall in plasma $pO_2$ leads to an increase in the secretion of the hormone erythropoietin by the kidneys and liver, resulting in increased red cell output by the bone marrow. People adapted to high altitudes may have up to 8000 red cells per $mm^3$, in contrast to the average figure of 5000 per $mm^3$ for non-adapted people. High red cell counts persist for some weeks after descending to normal altitude, and many distance runners increase their performance by training at high altitude before an important event.**

### *Lung volumes and capacities: spirometry*

Breathing movements and oxygen uptake can be recorded using a spirometer (Figure 23.18). The apparatus is normally filled with medical grade oxygen. The lid of the apparatus is pivoted at one end and sealed from the environment by water. When the subject breathes into the apparatus, the lid rises up and makes a recording of the breathing movements on a revolving drum or **kymograph**. It is important that the breathing effort is normal, so the lid is counterweighted so that it does not exert pressure on the air beneath.

To ensure that the subject does not rebreathe the same air, the mouthpiece contains a one-way valve. One of the two hoses connected to

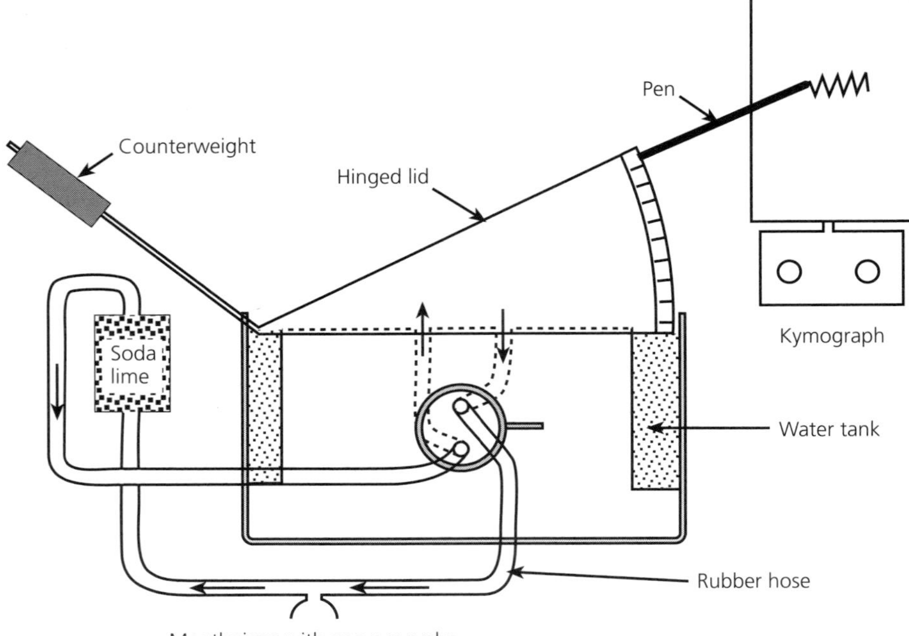


**Figure 23.18**
Simplified sectional view of a spirometer

the mouthpiece contains soda lime which absorbs carbon dioxide. Although the amount of oxygen in the apparatus falls, its *percentage* remains the same. The subject wears a nose clip to ensure that air cannot enter or leave the body except via the apparatus.

A tap at the end of the hose enables the subject to be isolated from, or connected to, the apparatus. After the tank has been filled with medical grade oxygen, the subject is allowed to breathe quietly whilst isolated from the apparatus. At the end of an expiration, the tap is opened so that the next inspiration draws gas from the spirometer.

*Point to note:* It is quite safe to use air instead of oxygen, but ***only if the soda lime is first taken out of the circuit***. If air is used *with* soda lime, the percentage of oxygen falls but there is no rise in carbon dioxide. There is, therefore, no desire to breathe until the $pO_2$ has fallen low enough to stimulate the carotid bodies, by which time the person may have lost consciousness.

Figure 23.19 is a kymograph trace showing the various lung capacities. The vertical axis represents changes in volume and the horizontal axis represents time. The time scale is worked out from the speed of rotation of the kymograph drum. The vertical scale must first be calibrated by blowing known volumes of gas into the spirometer. Notice that changes in volume are *inverted*, since a decrease in lung volume pushes the recording pen upwards.

The trace shows the following lung volumes:

- the **vital capacity** – this is the largest single breath the subject can take, and is obtained by taking a maximum exhalation after a maximum inhalation (or the other way round). In large, young athletes, vital capacities may be 6000 $cm^3$ or more.
- the **tidal volume** – this is the volume breathed in or out when the subject is not consciously trying to breathe. At rest, tidal volume averages about 500 $cm^3$, but it can rise to 2000 $cm^3$ during exercise.
- the **inspiratory reserve volume** – this is the additional air that can be inhaled after a normal inspiration.

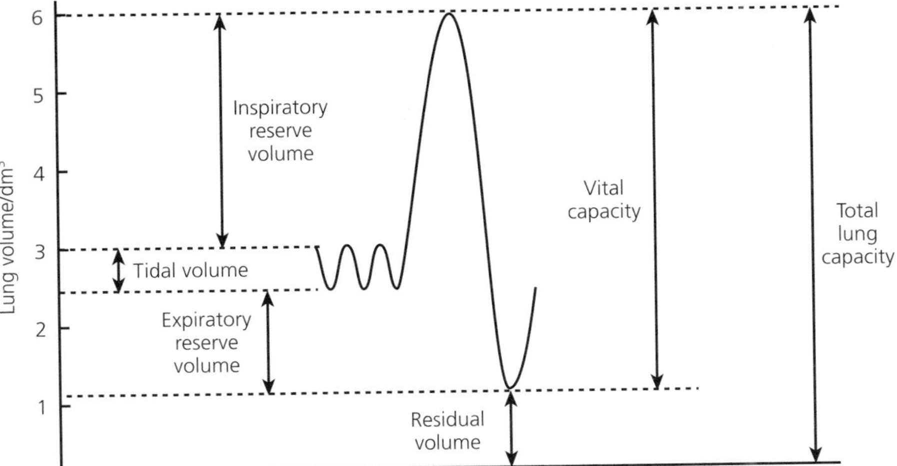


**Figure 23.19**
A spirometer trace showing lung volumes and capacities of an average adult. The trace is recording volume of air in the spirometer rather than in the lungs, so a rise in the trace indicates a fall in lung volume. The trace has therefore been inverted

- the **expiratory reserve volume** – this is the additional air that can be exhaled after a normal expiration.
- the **residual volume** – this is the amount of air remaining in the lungs after a maximum exhalation, and is about 1500 cm$^3$ in an average adult.

Though shown in Figure 23.19, residual volume cannot be measured directly, although it can be estimated in the following way. The person first exhales as much as possible, and then inhales a known volume of an inert gas such as helium. The air is then exhaled and the concentration of the helium in the exhaled gas is then measured. If, say, 2000 cm$^3$ of helium is inhaled, and its concentration in the exhaled gas is 50%, then the residual volume would be equal to the volume of the helium inspired.

A spirometer can also be used to measure two other things:

- the **minute volume**, or the amount of air inhaled (or exhaled) in one minute. This is equal to the tidal volume × the number of breaths per minute.
- the rate of **oxygen uptake**, indicated by the gradient of the trace (Figure 23.20). To obtain the gradient, the line of best fit is drawn along the trace, and then a right-angled triangle is drawn with the gradient as the hypotenuse. The gradient is given by:

$$\frac{\text{change in volume}}{\text{time taken}}$$

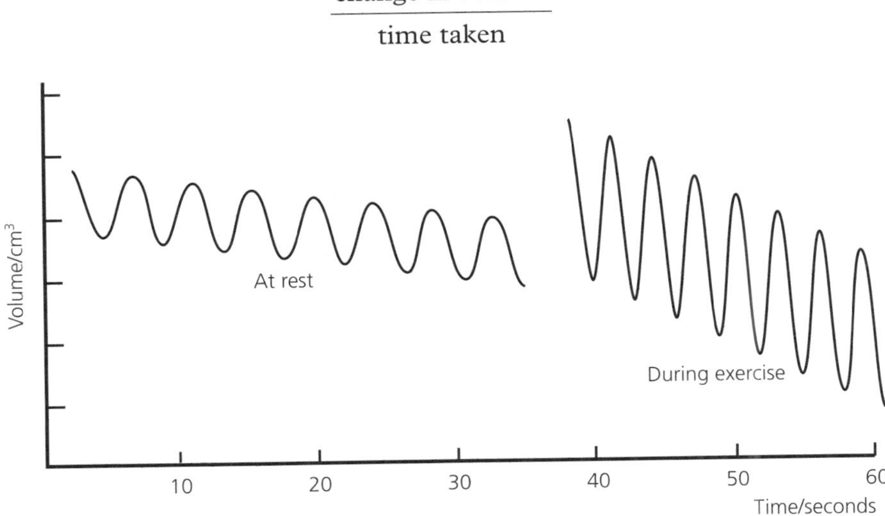


**Figure 23.20**
Spirometer trace showing oxygen uptake. Each division on the vertical scale represents 100 cm$^3$

## The dangers of breath holding

After taking a single large breath, few people can hold their breath for more than three minutes. However, by **hyperventilating**, or taking a dozen or so rapid, deep breaths beforehand, breath-holding time can be increased. This is because hyperventilation lowers the $pCO_2$ of the blood, so it takes longer to reach the level at which the desire to breathe becomes pressing.

*Point to note:* Breath-holding without hyperventilation is safe because the desire to breath becomes irresistible well before the $pO_2$ of the blood falls to dangerous levels. *Hyperventilating before endurance breath-holding under water, however, is extremely dangerous, and should under no circumstances be attempted.* This is because the person may run short of oxygen before $pCO_2$ has risen enough to force the person to the surface. The result may be blackout and **drowning.** Although oxygen shortage is detected by the carotid bodies, the signal is weaker than that generated by the build up of carbon dioxide, so the person may not respond in time.

Hyperventilation may lead to slight dizziness and so it should not be attempted unless sitting in an armchair. The reason is that a lowering of the $pCO_2$ reduces the frequency of sympathetic impulses to the arterial muscle, which consequently relaxes. This decreases the blood pressure, and thus the blood supply to the brain.

## Gas exchange in insects: the tracheal system

Oxygen transport in insects is quite different from that in most other animals in that the blood plays no part. The breathing system consists of a series of air-filled, tree-like tubes called **tracheae**. Each trachea is really a deep intucking of the epidermis and is lined by cuticle, thickened in the form of a spiral of chitin. The main tracheae are connected by longitudinal trunks and open along the sides of the body at paired holes called **spiracles** (Figure 23.21). In most insects there are ten pairs of spiracles, one in each of the second and third thoracic segments and one in each of the first eight abdominal segments. The spiracular openings are protected from entry of dust particles by tiny, densely-packed bristles.

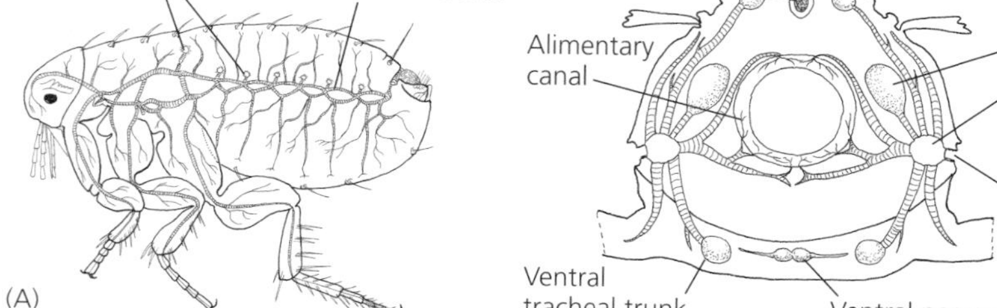


**Figure 23.21**
Tracheal system of a flea (after Wigglesworth) (A) and transverse section of the thorax of an insect to show tracheal tubes (after Essig) (B)

The smallest tracheae give rise to even finer **tracheoles**, the ends of which are filled with fluid. Because of their great numbers and extremely thin walls, tracheoles are the main site of gas exchange. Tracheoles are about 0.2–1 μm in diameter, and vary between 200 μm and 350 μm in length. Figure 23.22 shows that, unlike tracheae, they are **intracellular**, each tracheole being a tunnel through an individual cell! Electron microscope studies have shown that even the tracheoles are lined by a thin spiral of chitin.

The tracheal system is rather like a highly branched lung that penetrates every part of the body, no cell being more than a few micrometres from the nearest tracheole. In the tissues, the partial pressure

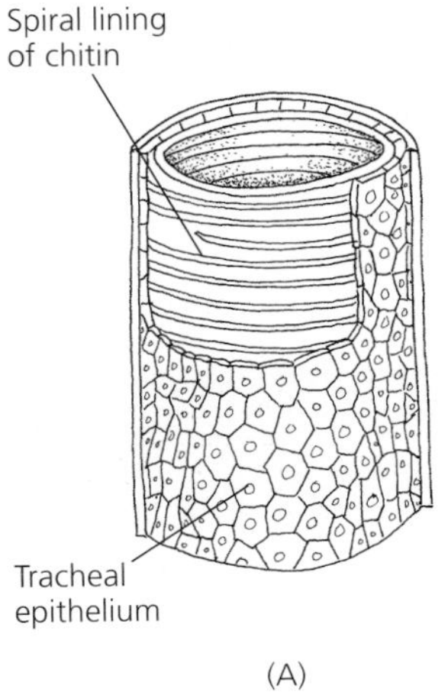


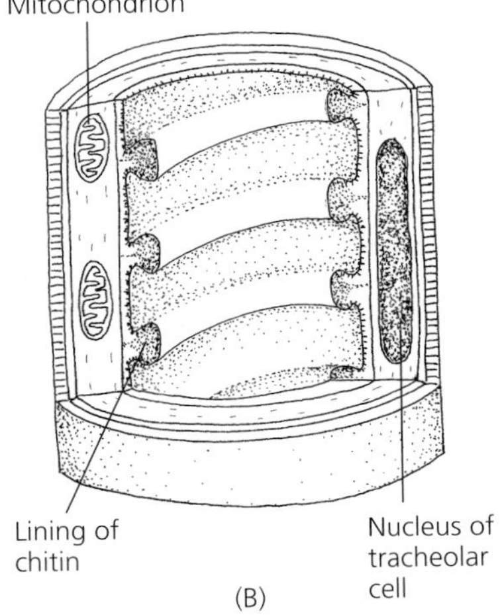


**Figure 23.22▲**
Part of a trachea (A) and a tracheole (B). The trachea is an intercellular tube; the tracheole is intracellular

of oxygen is lower than that in the atmosphere outside, so it diffuses along the tracheae to the tissues. Since tracheae are gas-filled, diffusion is rapid over distances of several millimetres. Whilst diffusion from tracheoles to mitochondria is in solution (and therefore much slower) the length of this final stage is very short.

Most $CO_2$ leaves the insect body via the spiracles, but a small proportion is thought to diffuse through the tissues and out through the cuticle. This is because carbon dioxide is much more soluble in water and diffuses more rapidly through cells than oxygen does.

In small insects, in which the diffusion path is no more than a millimetre or two, gas transport is entirely by diffusion. In larger insects such as bees and locusts, and especially during periods of high activity, the tracheal system is actively ventilated by pumping movements of the abdomen. In many larger and more active insects the volume of each 'breath' is increased by **air sacs**, which are expansions of the main tracheal trunks.

Gas exchange in very active tissues, such as flight muscles, is assisted by the withdrawal of fluid from the ends of the tracheoles. The accumulation of lactic acid in the muscle fibres lowers the solute potential, causing water to move by osmosis from the tracheoles into the surrounding muscle fibres (Figure 23.23).

As in all terrestrial organisms, gas exchange is achieved at the price of water loss. However, whilst the diffusion of water vapour through the spiracles cannot be prevented, it is minimised by regulating the spiracular aperture by moveable flaps. At rest, spiracles are kept nearly closed, or are opened intermittently. In active insects the increased $pCO_2$ in the tissues stimulates chemoreceptors, bringing about the reflex opening of the spiracles. The higher the $pCO_2$, the wider the spiracles and the more rapidly $CO_2$ leaves the body. Low $pCO_2$ levels tend to close the spiracles, so that at any given time the spiracles are no wider than is necessary for gas exchange. By means of this feedback mechanism, a balance is maintained between the opposing needs of obtaining oxygen and conserving water.

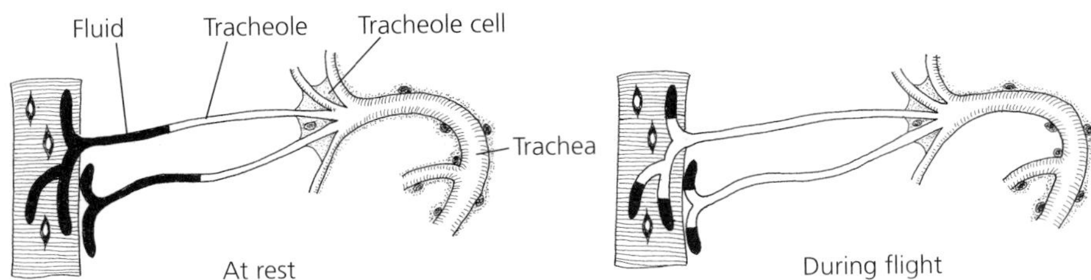


**Figure 23.23►**
Osmotic withdrawal of fluid from tracheoles during flight (after Wigglesworth)

## CLEAR THINKING

**It is often stated that the tracheal system limits the size to which insects can grow, because diffusion along the tracheae is slow. There are two reasons why this is unlikely to be so.**

**First, tracheal gas exchange in larger insects does not by any means depend entirely on diffusion, since the tracheae of large, active insects are often ventilated by abdominal pumping movements. Second, and more important, oxygen diffuses 300 000 times faster through air than it does in solution. This means that for a given concentration difference, oxygen would diffuse through 1 μm of liquid at the same rate as it would diffuse through 300 000 μm, or 30 cm, of gas! It is, therefore, the final stage of diffusion from tracheoles to cells that is likely to be the rate-limiting stage. A much more probable factor limiting insect size is the weakness of the new cuticle immediately after the old one has been shed.**

## QUESTIONS

**1** The table shows the tidal volume, breathing rate (number of breaths per minute) and the oxygen and carbon dioxide content of air exhaled by an athlete at rest and during exercise.

| % Composition of exhaled air | | | | Tidal volume ($dm^3$) | | Breathing rate | |
|---|---|---|---|---|---|---|---|
| Rest | | Exercise | | | | | |
| Oxygen | $CO_2$ | Oxygen | $CO_2$ | Rest | Exercise | Rest | Exercise |
| 16.5 | 4.5 | 16.5 | 4.5 | 0.5 | 2.5 | 15 | 30 |

a) What is the minute volume **(i)** at rest, **(ii)** during exercise?
b) What can you conclude from the fact that exercise does not affect the composition of exhaled air?
c) By how many times did the oxygen consumption increase as a result of the exercise?

**2** Pure oxygen is administered to patients suffering from a variety of conditions, but there are some conditions in which it is of little use. In each of the following, predict whether oxygen would help the patient or not, giving a reason for your answer.
a) Cyanide poisoning
b) Severe iron-deficiency anaemia
c) Pulmonary oedema (accumulation of fluid in the lungs)
d) Emphysema

CHAPTER 24

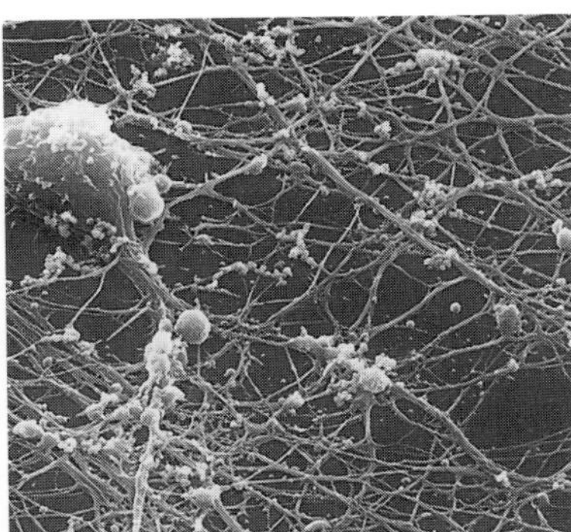

# NITROGENOUS EXCRETION and OSMOTIC REGULATION

## LEARNING OBJECTIVES

By the time you have completed your study of this chapter you should be able to:

- distinguish between excretion and egestion.
- name the main excretory products and the processes responsible for their production.
- state the functions of the kidney.
- describe the gross anatomy of the human urinary system and the fine structure of the kidney.
- distinguish between the two principal stages of urine production, and explain how the structure of the nephron is adapted to carry them out.
- explain the mechanism by which urine is concentrated or diluted according to circumstances.
- explain how the composition and output of urine is controlled by the hormones ADH and aldosterone.

## Excretory products

All living organisms produce wastes in their metabolism. In very small animals, waste products may simply leave the body by diffusion. In larger animals, most metabolic wastes are actively got rid of – a process called **excretion**.

It is important to appreciate that excretion is quite different from **egestion**, the getting rid of indigestible material as faeces. Excretory materials have been produced inside cells; faeces have not.

In animals there are two processes that generate significant amounts of waste:

- **respiration** produces carbon dioxide, which is excreted via the gas exchange surface.
- **deamination** of amino acids in the liver, described in more detail in Chapter 26. The initial product is ammonia, but this is extremely toxic and, in mammals and many other animals, it is immediately converted into less toxic compounds (see the information box on page 434). The getting rid of these substances is called **nitrogenous excretion**. In mammals, the main nitrogenous waste is **urea**. Humans also excrete small quantities of **uric acid**, produced in the breakdown of the purines of nucleic acids.

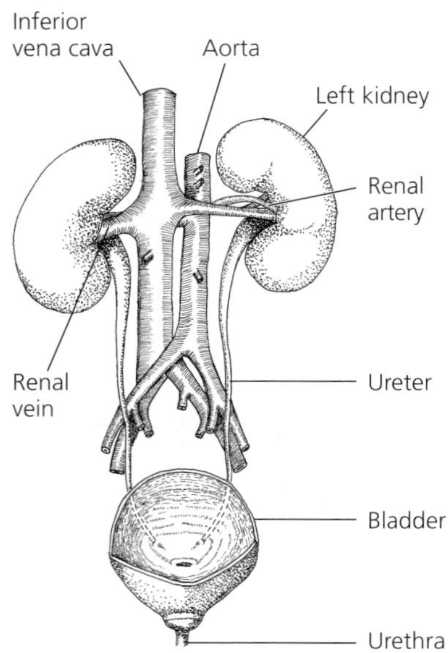


**Figure 24.1**
The human urinary system

## Functions of the kidney

In mammals and other vertebrates the organs of nitrogenous excretion are the **kidneys**. These have a number of other homeostatic functions besides nitrogenous excretion, namely:

- osmoregulation – the regulation of the water potential of the blood.
- ionic regulation – maintaining the proportions of the various ions in the blood, particularly $Na^+$ and $K^+$.
- the long-term regulation of blood pressure.
- the regulation of blood pH.
- the production of certain hormones.

Some of these functions are inter-related and may have conflicting consequences. For example, even in times of water shortage some water loss is unavoidable since urea is excreted in solution.

An indication of how important the kidneys are in helping to maintain the composition of the blood is the fact that although they account for only 0.5% of the body mass, they take approximately 25% of the resting cardiac output. Each gram of kidney therefore takes 25/0.5 = 50 times as much blood as the average for the rest of the body!

# Structure of the human excretory system

The kidneys are bean-shaped organs, each about two-thirds the size of one's clenched fist (Figure 24.1). Each is supplied with blood by a **renal artery** and is drained by a **renal vein**. The waste liquid produced by the kidneys is called **urine**, and is carried away by muscular tubes, the **ureters**, to the **bladder**. Here it is temporarily stored until it is carried to the exterior by the **urethra**.

A ureter is not a passive drainpipe, but actively propels the urine along by waves of muscular contraction called **peristalsis**. The opening to the bladder is surrounded by two ring-like **sphincter** muscles, the outer being under voluntary control. Distension of the bladder wall stimulates stretch receptors, and brings about the reflex relaxation of the inner sphincter muscle and the desire to urinate. Urination is brought about by the voluntary relaxation of the outer sphincter muscle, helped by the contraction of the smooth muscle in the bladder wall.

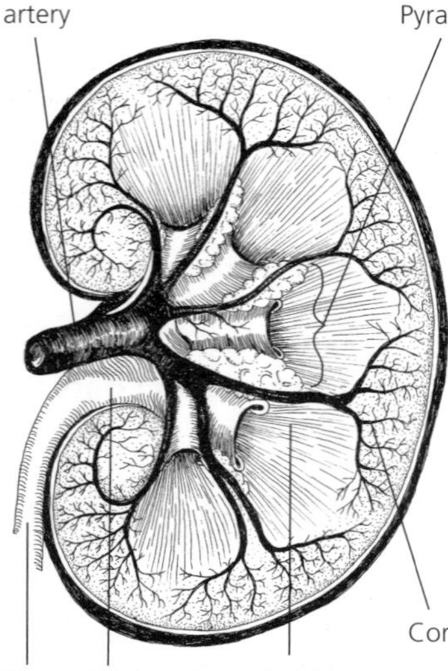

Renal artery
Pyramid
Cortex
Ureter Pelvis
Medulla

**Figure 24.2**
A kidney with its arterial blood supply seen in longitudinal section (after Cormack)

## Structure of the kidney

When cut longitudinally, a kidney is seen to consist of two layers – an outer red-brown **cortex** and a pink, inner **medulla** – surrounding a central cavity, the **pelvis** (Figure 24.2). The medulla is subdivided into a number of **pyramids** whose apices protrude into the pelvis.

The kidney consists of about 1–1.5 million blind-ending tubes called **nephrons**, together with their blood supply (Figure 24.3). Since a single nephron and its blood supply can produce urine, the nephron can be considered to be the functional unit of the kidney. However, as we shall see, an isolated nephron cannot concentrate the urine without the co-operative action of other tissues.

Functionally a nephron consists of two parts: a blind-ending, cup-shaped **Bowman's capsule** and a long **renal tubule**. Bowman's capsule

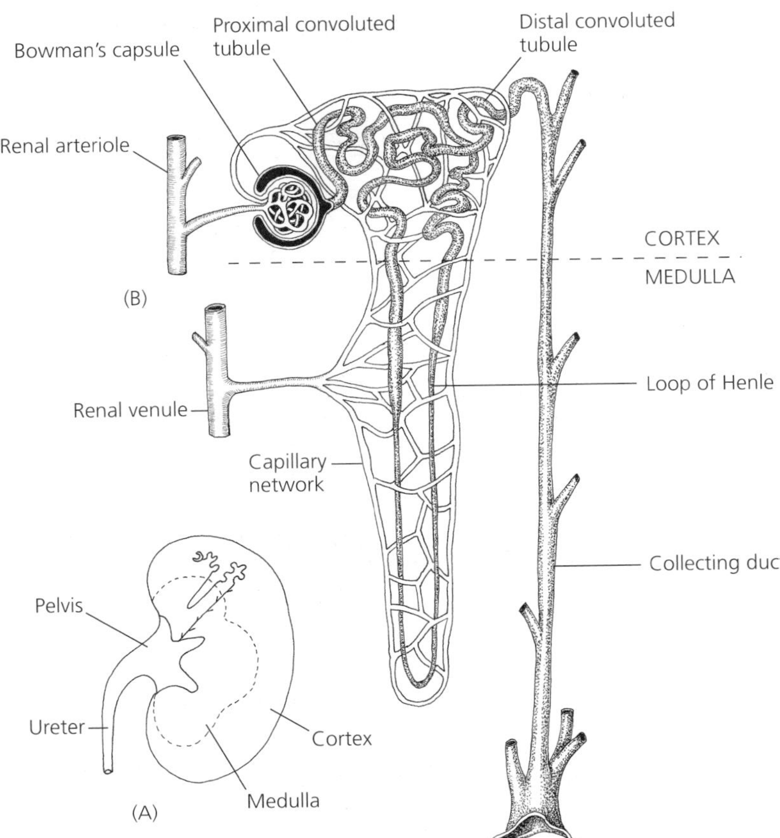


**Figure 24.3**
Main regions of a kidney (A) and its functional unit, the nephron (B)

is concerned with filtering the blood, whilst the renal tubule adds and removes substances from the filtrate converting it into urine.

The renal tubule is further divisible into three functionally-distinct parts. Leading from Bowman's capsule is the **proximal convoluted tubule** ('proximal' means 'near', 'convoluted' means 'twisted'). This leads into a **loop of Henle** (named after F.J. Henle, a German pathologist). Henle's loop dips deeply into and out of the medulla and is continuous with the **distal convoluted tubule**, which marks the end of the nephron ('distal' refers to the fact that it is furthest from the beginning of the nephron). The distal tubules of many nephrons join to form **collecting ducts**, which open into the pelvis near the apices of the pyramids.

The blood supply to the nephron is unusual in that it consists of two capillary networks in series. Dipping into Bowman's capsule is a cluster of capillaries called a **glomerulus**, supplied by an **afferent arteriole** and drained by an **efferent arteriole**. Together, Bowman's capsule and its glomerulus form a functional unit called a **renal corpuscle** (also known as a **Malpighian corpuscle**). The efferent arteriole divides again into a second capillary network round the rest of the nephron, which leads into branches of the renal vein.

Although all nephrons follow the same structural plan, there are actually two kinds. About 85% are **cortical nephrons**, with Bowman's capsules in the outer cortex and short loops of Henle which only extend to the outer region of the medulla. The second kind, **juxtamedullary nephrons**, play an essential role in producing the most concentrated urine and are so-called because their capsules are close to the medulla. They have very long loops that penetrate almost to the tips of the pyramids (see the information box on page 444).

## Urine formation

Urine is produced in two stages involving quite different processes:

- **filtration**, in which plasma in the glomerulus is filtered into Bowman's capsule. Since filtration discriminates solely on the basis of particle size, the filtrate contains both useful and harmful substances.
- **active transport**, by which the composition of the filtrate is further modified. Although most active transport consists of *reabsorption* of useful substances back into the blood, there is also some *secretion* of substances from the blood into the tubule.

Energy for the two stages comes from different sources. Filtration depends on blood pressure and is thus powered by the heart, whilst the energy for active transport is supplied by respiration in kidney tubule cells.

### Nitrogenous excretion and water conservation

When amino acids are used as fuel in respiration, they must first be deaminated by the removal of the amino group as ammonia. Ammonia is extremely toxic and as such can only be excreted in a very dilute solution. For fresh water animals which have to contend with water surplus this is no disadvantage. Animals which excrete ammonia are said to be **ammonotelic**, and include fresh water fish, crustaceans and the tadpole stage of amphibians.

For terrestrial animals the main factor in osmoregulation is the conservation of water, and so ammonotely is out of the question. Mammals and terrestrial amphibians convert ammonia into urea, which is much less toxic than ammonia and can thus be excreted in more concentrated solution. Animals that excrete urea are **ureotelic**.

Even if urea were completely non-toxic, there are limits to the concentration of urine an animal can produce. Many terrestrial snails, insects, reptiles and birds, excrete **uric acid**. Besides being almost completely non-toxic, it is highly insoluble and so can be excreted as a paste of crystals, needing only water to carry the crystals along the excretory ducts. Animals whose main nitrogenous waste is uric acid are said to be **uricotelic**. Humans produce some uric acid, but this is the product of the breakdown of purines in the nucleic acids in the diet.

It is interesting to note that uricotelic animals all lay large, yolky eggs protected by a shell. In such a **cleidoic egg**, the developing animal has to make do with the small amount of water it receives from its mother. The accumulation of a soluble waste, even if non-toxic, would produce osmotically disadvantageous effects. For a developing animal in this situation, uric acid is the ideal nitrogenous waste since it is virtually insoluble. It may be that uricotely evolved in response to selection pressures on the embryo, but has been retained in the adult.

### Filtration

To find out what is going on in a renal corpuscle necessitates the analysis of samples of the liquid inside Bowman's capsule. This was first achieved by J.T. Wearn and A.N. Richards in 1921 using amphibian kidneys, which have large and easily accessible capsules. Richards showed that the liquid is like tissue fluid, containing all the plasma constituents but with negligible protein. The glomerulus and Bowman's capsule thus act like a filter, allowing ions, glucose, amino acids and other simple substances to pass through, but retaining all but the smallest plasma proteins. The liquid in Bowman's capsule is accordingly called the **glomerular filtrate**.

Whereas conventional filtration separates suspended solids from solutes, filtration through capillary walls separates substances according to the sizes of their molecules. This can only be achieved by forcing the filtrate through under pressure, a process called **ultrafiltration**. We have already met ultrafiltration in the formation of tissue fluid (Chapter 22), but it occurs much faster in the kidney. There are two reasons for this: the blood pressure is higher and the filter is more permeable.

An effective filter needs three characteristics: holes small enough to be **selectively permeable**, an **extensive area** and an adequate **pressure**.

### *The driving force*

The blood pressure in the glomeruli is about 6 kPa, compared with 4.3 kPa at the arterial end of other systemic capillaries. There are two reasons for this high pressure:

- the renal artery divides up more rapidly than other arteries, so the arterioles are shorter and offer less resistance.
- the efferent arterioles are narrower than the afferent arterioles, creating a 'bottleneck'. This effect can be increased by contraction of a sphincter muscle at the glomerular end of the efferent arteriole.

Not all of the glomerular blood pressure is available for filtration. Two other forces act in the opposite direction:

- the solute potential due to the unfiltered plasma proteins, which is about −3.3 kPa.
- the hydrostatic 'back-pressure' in Bowman's capsule, due to the resistance to flow along the nephron, which is about 1.33 kPa.

The net filtration force is the difference in water potential (Ψ) either side of the filter, i.e. water potential of plasma minus water potential of filtrate. It is calculated as follows:

$$\Psi_{\text{plasma}} = \underset{\text{pressure}}{\text{hydrostatic}} + \underset{\text{potential}}{\text{protein solute}} = 6 + (-3.3) = 2.7 \text{ kPa}$$

$$\Psi_{\text{filtrate}} = \underset{\text{pressure}}{\text{hydrostatic}} + \underset{\text{potential}}{\text{protein solute}} = 1.3 + 0 = 1.3 \text{ kPa}$$

$$\text{Therefore the net filtration force} = \Psi_{\text{plasma}} - \Psi_{\text{filtrate}} = 1.4 \text{ kPa}$$

By the time the plasma has reached the end of the glomerular capillaries, the filtration force has been reduced to zero. This is because the plasma proteins are not filtered, causing the solute potential of the plasma to become more negative until there is no net driving force.

### *The filter*

Each renal corpuscle is a microscopic filter, able to separate large molecules from small ones. Some idea of the size of the pores has been gained by injecting harmless substances of different molecular size into the blood stream and comparing their concentration in the plasma and in the filtrate. The filtrate/plasma (F/P) ratio is given by

$$\frac{\text{concentration in filtrate}}{\text{concentration in plasma}}$$

and is a measure of how readily a substance is filtered. For a freely filtered substance, the plasma and filtrate concentrations are equal and the filtrate/plasma ratio is 1. If it is not filtered at all, the ratio is zero.

Table 24.1 shows the F/P ratios for substances of various molecular size. Serum albumin (the smallest of the plasma proteins) is virtually absent from glomerular fluid in healthy people. Evidently the pores are about the size of a haemoglobin molecule, which is about 6 nm in diameter.

**Table 24.1**
Filtrate/plasma concentrations for substances of various molecular size

| Substance | Molecular weight | F/P ratio |
|---|---|---|
| Glucose | 180 | 1 |
| Inulin | 5000 | 1 |
| Myoglobin | 16 500 | 0.75 |
| Egg albumin | 43 000 | 0.22 |
| Haemoglobin | 68 000 | 0.03 |
| Serum albumin | 69 000 | 0.002 |

Where are the holes? There are only three layers separating plasma from filtrate:

- the capillary endothelium.
- the basement membrane on which the capillary cells lie.
- the inner layer of Bowman's capsule.

Studies with the electron microscope have shown that the filter must be the basement membrane. The capillary endothelial cells are unusual in that they are **fenestrated**, having 'windows' in them. These are about 70 nm in diameter, which is large enough to retain cells but much too large to hold back any plasma proteins.

Whilst the outer layer of Bowman's capsule consists of ordinary squamous epithelium, the inner layer – or **visceral layer** – is quite extraordinary. Instead of forming a continuous layer like the lining of a hollow cup, the cells completely surround the glomerular capillaries (Figure 24.4). Each cell of the visceral layer is called a **podocyte** (meaning 'foot cell'). It bears a number of **major foot processes**, each with **minor foot processes** resting on the basement membrane of the capillary (Figure 24.5). A podocyte can be visualised if it is likened to a starfish, the major foot processes representing the 'arms' and the minor foot processes representing the 'tube feet'.

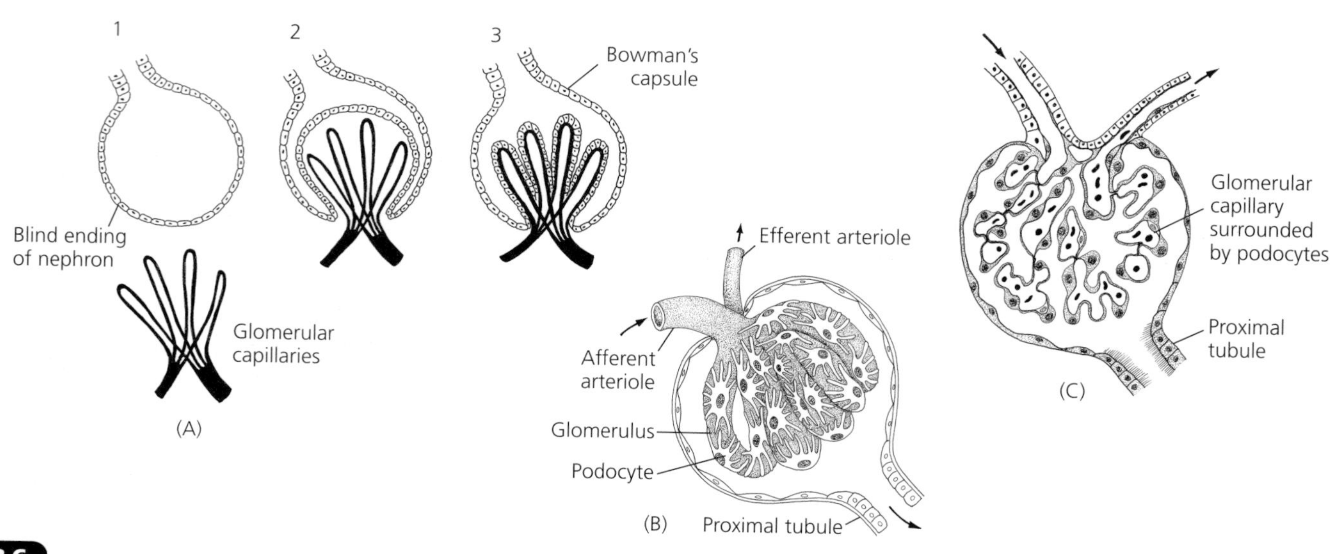


**Figure 24.4**
A renal corpuscle. (A) How it is formed by intucking of a tuft of glomerular capillaries into the blind end of a nephron. In reality, the glomerular capillaries are completely surrounded by the cells of the inner layer of Bowman's capsule. These latter cells are not of simple shape as shown, but are starfish-shaped as shown in (B). (B) Cutaway view through outer layer of capsule showing podocytes. (C) In longitudinal section through centre of glomerulus

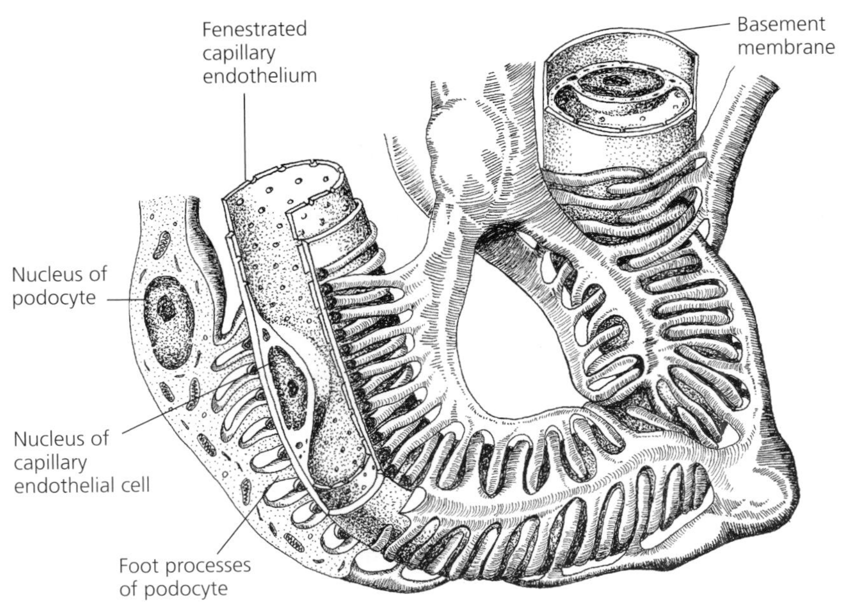


**Figure 24.5**
Podocytes surrounding a glomerular capillary (after Ham)

The gaps between the minor foot processes are called **slit pores** and are about 25 nm wide – big enough to let even large protein molecules through.

A third important factor is the total area of the filter. There are about 50 capillaries in each glomerulus. With a total of about a million glomeruli in each kidney, the total filtration area is over a square metre.

## Measuring filtration rate and renal blood flow

Physiologists have measured the rate of filtration using the fact that certain substances are freely filtered but are neither reabsorbed into the blood nor secreted into the nephron. **Inulin**, a polysaccharide (molecular weight 5000), is such a substance. After inulin is injected into a person, it begins to be filtered and travels down the nephron. Since it only enters the nephron by filtration and only leaves it via the urine, it is filtered at the same rate as it is excreted. It follows that its rate of movement in mg per minute remains constant. Thus, if its *velocity* of movement down the nephron decreases by half, its concentration must double.

This is the basis for the calculation of filtration rate. The rate of excretion of inulin is equal to the volume of urine produced per minute ($V_{urine}$) × concentration in urine ($C_{urine}$).

Similarly, the rate of delivery of inulin to the nephrons is equal to the volume of filtrate produced per minute ($V_{filtrate}$) × inulin concentration in the filtrate ($C_{filtrate}$).

Since the rate of delivery of inulin equals its rate of excretion, it follows that

$$V_{urine} \times C_{urine} = V_{filtrate} \times C_{filtrate}$$

Since inulin is known to be freely filtered, its concentration in the filtrate is the same as that in the plasma, which can easily be measured. This leaves the rate of filtration as the only unknown, so it can easily be calculated.

The total blood flow to the kidney can also be determined by a similar method using para-aminohippuric acid (PAH). The method depends on two things. First, unlike inulin, PAH is excreted by secretion into the tubules as well as by filtration. Second, all the blood in the renal artery goes to the capillaries supplying the renal tubules. The kidney, therefore, excretes all the PAH delivered to it (provided this is below a certain maximum). Renal blood flow is calculated using the same principle as with inulin:

$$\text{plasma concentration} \times \text{renal blood flow} = \text{urine concentration} \times \text{urine output}$$

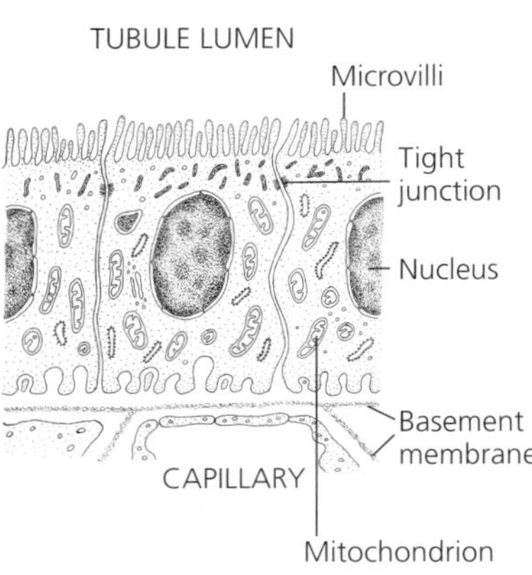


**Figure 24.6**
A proximal tubule cell

## Reabsorption in the proximal tubule

Approximately 7.5 $dm^3$ of filtrate are produced each hour (see the information box on page 437 for how the filtration rate is calculated). At this rate the entire plasma volume would be lost in about an hour. Besides water, the filtrate contains many other essential materials such as glucose, amino acids, various ions such as $Na^+$, $K^+$, $Ca^{2+}$ and $PO_4^{3-}$, and vitamins. Most of these are reabsorbed into the blood in the dense capillary network surrounding the proximal tubules.

Directly or indirectly, most reabsorption from the proximal tubule depends on active transport. This is reflected in the structure of the cubical epithelial cells forming the walls of the proximal tubule (Figure 24.6). On the side facing into the tubule there are many **microvilli**, which are absent in other parts of the nephron. These increase the total area of plasma membrane to about 50 $m^2$. Energy for transport is provided by the abundant mitochondria.

### Glucose and amino acids

Under normal circumstances all glucose is reabsorbed before it reaches the end of the proximal tubule. As described in detail in Chapter 6, the mechanism for glucose reabsorption is linked to sodium reabsorption. It depends on three types of protein carrier, each restricted to a particular part of the plasma membrane. Amino acids are absorbed by a mechanism that is similarly coupled to the uptake of sodium.

Since there is a finite number of carriers in the proximal tubule cells, there is a limit to the rate at which glucose can be absorbed. This is called the **tubular transport maximum ($T_m$)**. Under normal circumstances the rate at which glucose is delivered to the nephron is well below $T_m$, so no glucose appears in the urine. In diabetes mellitus, however, the blood glucose level may exceed $T_m$, the excess glucose 'spilling over' into the urine. In severe diabetes this results in the reabsorption of less water (see below) and the production of large amounts of urine.

### Sodium and chloride

In addition to the sodium that is cotransported with glucose, sodium ions are also reabsorbed by other carriers, all of which require energy. The removal of sodium ions causes chloride ions to follow passively. Altogether about 75% of the filtered sodium and chloride are reabsorbed in the proximal tubule.

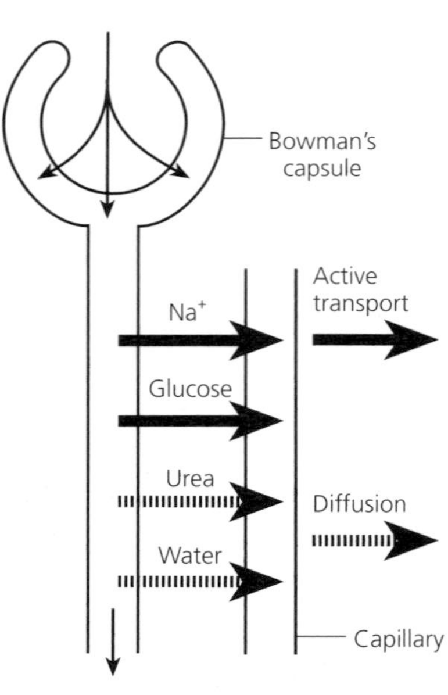


**Figure 24.7**
How water reabsorption is driven by solute reabsorption

### Protein

Although the concentration of plasma protein in the filtrate is negligible, the amount of filtrate formed is large enough to result in a potential daily loss of about 30 g protein. This does not happen because protein is recovered by pinocytosis at bases of the microvilli in the proximal tubule. After reabsorption, proteins are hydrolysed and the amino acids recycled.

The scavenging of filtered protein is not totally effective. Human chorionic gonadotrophin – a protein hormone with a molecular mass of 39 000 – is excreted in sufficient quantities to form the basis for the pregnancy test.

### Water

Since the walls of the proximal tubule are highly permeable to water, water moves across the tubule cells by osmosis in response to any difference in solute potential (Figure 24.7). Although osmosis is a passive process, the osmotic difference causing it is set up by active processes.

These are of two kinds. The most important is the reabsorption of solutes from the proximal tubule. Since about 75% of the total dissolved solutes are reabsorbed here, water follows by osmosis in the same proportion.

A second factor creating a difference in water potential is the concentration of plasma proteins which, you may remember, are not filtered to any great extent. Since about 25% of the plasma entering the glomerulus is filtered, the protein concentration rises, so the solute potential falls proportionately. The solute potential in the tubular capillaries is thus lower (more negative) than in the tubules, so water moves by osmosis back into the blood.

The situation is comparable to the formation and reabsorption of tissue fluid in other parts of the body, but with an interesting difference. In the kidney there are *two* capillary beds in series – the glomerulus which operates at high pressure and the tubular capillaries which operate at low pressure. Filtration occurs in the first, and reabsorption in the second. In other parts of the body, tissue fluid is formed and reabsorbed at the arterial and venous ends of the same capillary (Figure 24.8).

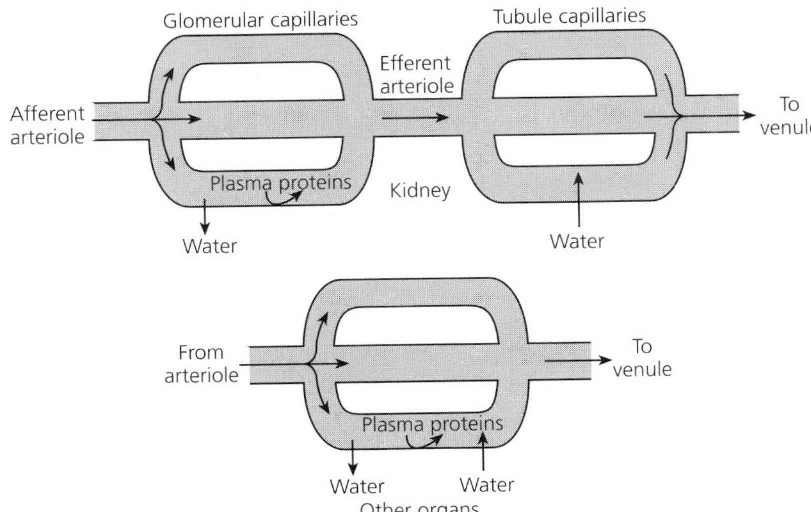


**Figure 24.8**
Filtration and reabsorption in renal capillaries (above) and in non-renal capillaries (below). In the kidney, tissue fluid leaves and re-enters the blood in different capillary beds; in other organs tissue fluid is formed and reabsorbed in the same capillary bed

In a normal person the same amount of water is reabsorbed from the proximal tubules, regardless of the state of hydration of the body. This water reabsorption (accounting for about 75% of the filtrate) is therefore **obligatory**. Variation in urine output is brought about by the activities of the distal tubules and the collecting ducts.

### *Urea*

As a result of the reabsorption of water from the proximal tubules, the concentration of urea in the tubular fluid rises well above that in the blood. The proximal tubules are sufficiently permeable to urea to allow about half of it to diffuse back into the blood.

### *Secretion into the nephron*

Though reabsorption is the main activity of the tubule, a number of substances are secreted into the fluid. $K^+$ and $H^+$ are secreted in exchange for $Na^+$. Other secreted substances include uric acid, creatinine and some 'foreign' materials such as penicillin.

## Osmoregulation by the kidney

The kidney helps regulate the solute potential of the blood and tissue fluid by varying the output and solute concentration of the urine. When the body is overhydrated, larger volumes of more dilute urine are produced, and when the body is dehydrated, less urine is produced and its solute concentration increases. Depending on the circumstances, human urine can have a solute potential of between a quarter and four times that of the plasma, and rates of production vary between 0.240 $dm^3$ and over 24 $dm^3$ per day.

In a healthy individual, about 75% of the water in the filtrate is reabsorbed from the proximal tubules. This proportion is constant, irrespective of the osmotic conditions of the body, and is thus obligatory. It is the distal tubules and the collecting ducts that are responsible for varying the amount and concentration of the urine.

### *Diluting the urine – the distal tubule*

The distal convoluted tubule is about 5 mm long and, when the body is overhydrated, brings about the final dilution of the urine. It does this by reabsorbing more salt from the distal tubule. Since it is almost impermeable to water, water cannot follow, making the urine very dilute.

Cortex

Intercellular fluid isotonic with plasma in the outer medulla

Pelvis

Intercellular fluid up to 4× the concentration of the plasma in the inner medulla

**Figure 24.9**
Salt gradient in the renal medulla

## Concentrating the urine

All vertebrates can produce a urine more dilute than the blood, but only birds and mammals can produce urine more concentrated than the blood. Some desert rodents can produce urine more than 25 times as concentrated as the plasma, though humans can only produce urine four times that of the plasma concentration.

The ability to concentrate the urine depends on Henle's loop, which only birds and mammals have. Its role is to generate a gradient of salt concentration in the medulla (Figure 24.9). The deeper down the medulla, the more concentrated the fluid. The mechanism by which this gradient is built up will be explained shortly, but for the moment let's see how the salt gradient enables the urine to be concentrated.

When the body is short of water the liquid leaving the distal convoluted tubule is isotonic with the blood. As it enters the collecting ducts and passes down into the medulla, it enters an environment with a high salt concentration. The walls of the collecting ducts become more permeable to water, allowing water to pass by osmosis from the collecting ducts into the peritubular fluid, raising the concentration of the urine (Figure 24.10). Water reabsorption is also aided by the fact that, unlike the cortical collecting ducts, the medullary collecting ducts are permeable to urea. The urea diffuses into the peritubular fluid, raising its solute concentration and increasing the reabsorption of water. Up to 99% of the water in the original filtrate can be recovered.

What determines the permeability of the walls of the collecting ducts? This is under the control of the hypothalamus in the brain, but before dealing with this we need to understand how the medullary concentration gradient is built up and maintained.

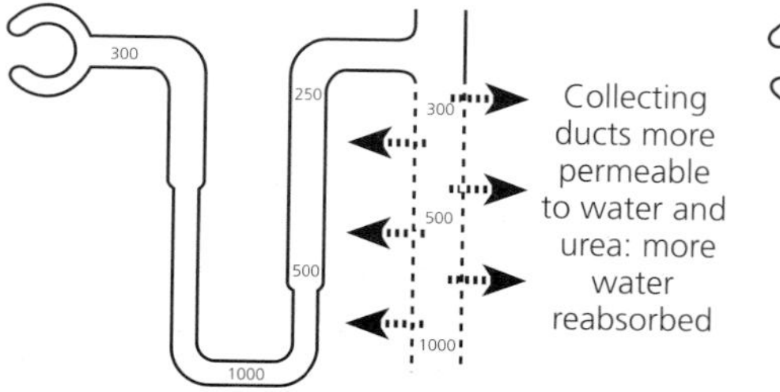


Body dehydrated, ADH concentration maximal

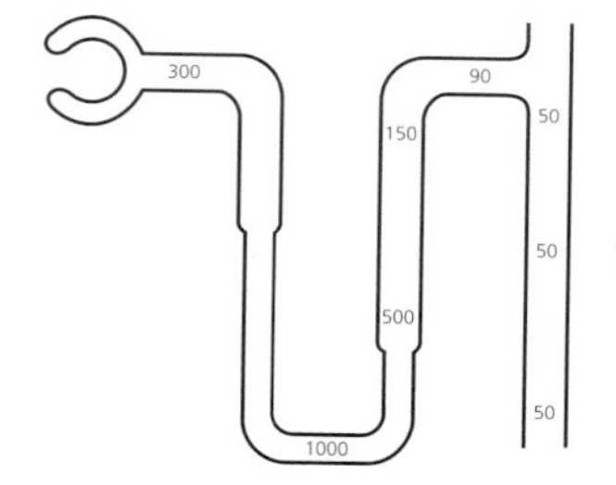


Collecting ducts less permeable to water and urea: less water reabsorbed

Body overhydrated, ADH absent

**Figure 24.10**
Reabsorption of water from the collecting ducts. Under normal circumstances the body is somewhere between the two extremes illustrated above. Concentrations are in milliosmoles. ADH = antidiuretic hormone

### *How is salt concentrated in the medulla?*

Although the juxtamedullary nephrons only account for some 15% of the nephrons, they are the most important in the production of a concentrated urine. The cortical nephrons illustrate the basic principle, and involve fewer complications.

The loop of a cortical nephron consists of two parts:

- the **descending limb** receives fluid from the proximal tubule and carries it down into the outer medulla. Its cells do little transport work and have few mitochondria. The walls are permeable to water but relatively impermeable to salt.
- the **ascending limb** has much thicker walls, and its function is to transport salt from the tubule into the peritubular fluid. The mechanism is complex, involving cotransport of $Na^+$ with $K^+$ and $Cl^-$ ions. Energy for this transport is supplied by the abundant mitochondria.

Now let's see how these two parts work together (Figure 24.11). The ascending limb pumps salt from the tubule into the peritubular fluid. This raises the salt concentration round the descending limb causing water to leave it by osmosis, raising *its* salt concentration. The liquid delivered to the ascending limb therefore has a higher salt concentration than the liquid leaving the proximal tubule. This enables the ascending limb to raise the salt concentration outside the tubule to a higher value than would otherwise be possible, since there is a limit to the concentration difference that tubule cells can establish.

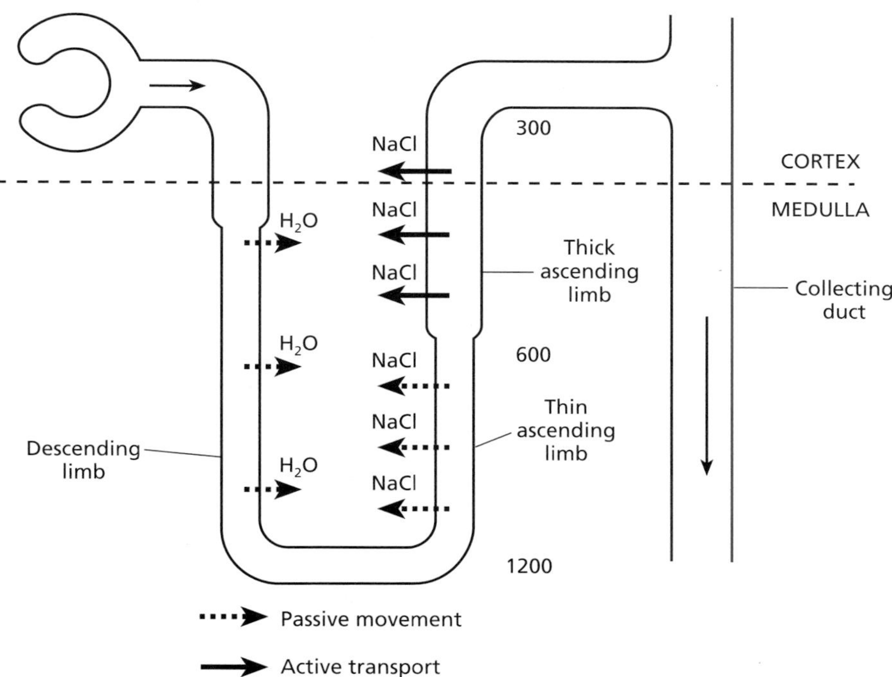


**Figure 24.11**
The loop of Henle as a countercurrent multiplier. Concentrations are in milliosmoles

The higher the salt concentration in the medullary peritubular fluid, the more water leaves the descending limb, and the higher the concentration of salt delivered to the ascending limb and so on. Thus the salt concentration built up at one level in the loop is added to the concentration built up at the next level down.

This arrangement acts as a **countercurrent multiplier**. 'Countercurrent' refers to the fact that liquid flows in opposite directions in the two limbs of the loop. 'Multiplier' refers to the fact that the small

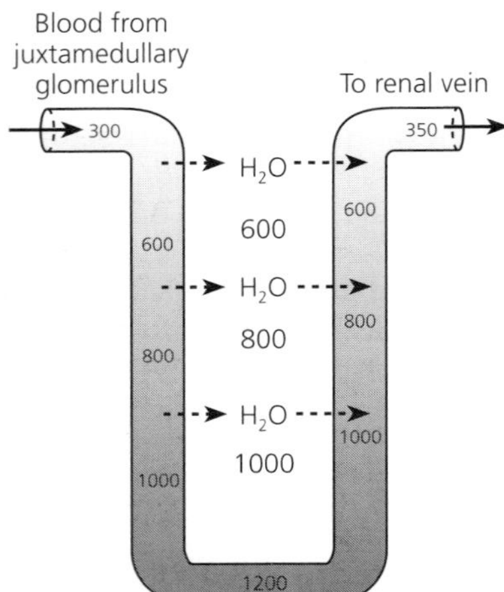


**Figure 24.12**
The vasa recta as countercurrent exchangers. Concentrations are in milliosmoles. Normal blood plasma has a concentration of 300 milliosmoles

concentration differences developed at each level are accumulated down the loop. At any given level, the difference in the solute potential inside and outside the ascending limb is much smaller than the difference between the two ends of the loop.

In the juxtamedullary nephrons, the multiplier effect is greater – partly because of their extra length, but also because of the presence of an additional **thin segment** in the ascending limb. This region (which is absent in cortical nephrons) is highly permeable to salt but impermeable to water. This enables salt to diffuse out from the high concentration in the tubule into the peritubular fluid, enabling much of the salt to leave the ascending limb without expenditure of energy.

### *Maintaining the medullary gradient: the vasa recta*

One problem that has been glossed over until now is this: why isn't the high salt concentration built up by the loop of Henle flushed out by the flow of blood through the medulla? The answer lies in the peculiar blood supply to the medulla.

In cortical nephrons, the capillaries supplied by efferent arterioles are more or less confined to the cortex, weaving around the proximal and distal convoluted tubules. In the juxtamedullary nephrons (those nearest to the medulla), the efferent arterioles supply capillaries that dip deeply into the medulla, forming long, straight loops called **vasa recta** (from the Latin, meaning 'straight vessels'). Blood flow in them is very sluggish, accounting for only about 2% of the renal blood flow.

The vasa recta (singular, vas rectum) are freely permeable to ions and to water. As shown in Figure 24.12, they act as **countercurrent exchangers**. As the blood descends into the medulla it loses water and gains salt, but as it passes up the ascending limb it gains water and loses salt. The blood therefore changes little in composition as it flows through the medulla.

Notice that whereas a countercurrent multiplier actively builds up a salt gradient, a countercurrent exchanger acts purely passively to reduce the rate at which the gradient is dissipated. You can now see why it is so important for the blood flow in the vasa recta to be sluggish; if it were too rapid there would be insufficient time for the exchanges to occur and salt would be flushed out of the medulla.

## *Controlling the collecting ducts: the role of the hypothalamus*

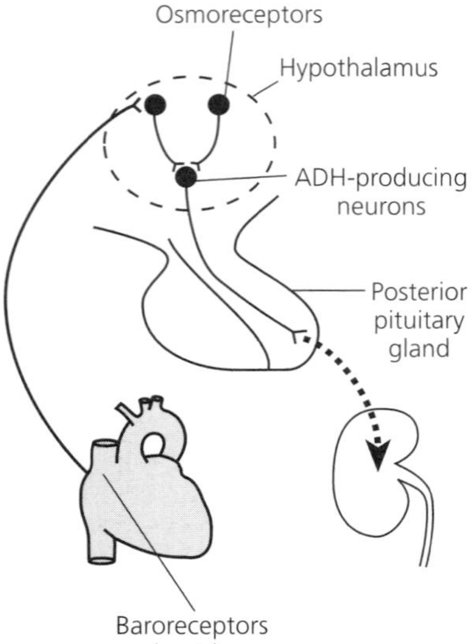

Though the kidney is able to respond to osmotic changes, it cannot detect them. It acts on information sent to it by the posterior lobe of the **pituitary gland**. This contains the axon terminals of neurons whose cell bodies lie in the **hypothalamus** immediately above. These neurons are called **osmoreceptors** since they are sensitive to changes in the solute potential of the blood. The osmoreceptors also produce the polypeptide hormone **ADH** (**antidiuretic hormone**) which slowly travels down the axons to the posterior pituitary, where it is temporarily stored in secretory vesicles.



**Figure 24.13**
Hypothalamic control of the kidney

When the body is becoming dehydrated, the solute potential of the blood falls (becomes more negative). This is detected by the osmoreceptors, which respond by sending impulses along their axons to the posterior pituitary and triggering the release of ADH into the blood (Figure 24.13).

Acting via cAMP as second messenger (Chapter 20), ADH raises the permeability of the kidney collecting ducts to water. This allows water to pass by osmosis into the more concentrated medullary interstitial fluid, so more water is reabsorbed. When the body is overhydrated, the rise in solute potential of the blood inhibits the release of ADH. As the existing ADH is destroyed by the liver and kidney, its concentration in the blood falls. This decreases the permeability of the collecting ducts to water, so less is reabsorbed. The secretion of ADH is also inhibited by alcohol, which therefore stimulates urine production.

It sometimes happens that the axons from the hypothalamus to the posterior pituitary are damaged, for example by a head injury. Since ADH cannot reach the kidney, very large amounts of dilute urine are produced even if the person is dehydrated. This condition is called **diabetes insipidus**, and is quite different from diabetes mellitus because there is no glucose in the urine.

Under normal circumstances, it is not a case of ADH being either 'present' or 'absent' as is often implied. ADH is only secreted at maximum rate when urine production is minimal, and likewise its secretion only ceases completely when urine output is running at its maximum rate of over 1 $dm^3$ per hour. Most of the time ADH output is between these two extremes.

Besides being sensitive to solute potential of the blood, the release of ADH is also stimulated by a fall in blood pressure, detected by baroreceptors in the walls of the atria. The adaptive value of this is clear: by promoting the retention of water, it helps to restore blood pressure.

### *Thirst*

Osmoregulation involves balancing water gain and water loss. The kidney and the osmoreceptors controlling it are only concerned with the loss side of the water 'budget'. Other osmoreceptors in the hypothalamus control water intake by stimulating the desire to drink. There is inevitably a delay between drinking and restoration of normal solute potential because it takes time for water to be absorbed, and also because it takes time for existing ADH in the bloodstream to be destroyed.

Like other aspects of homeostasis, osmoregulation involves constant adjustments brought about by negative feedback. The main processes involved are shown in Figure 24.14.

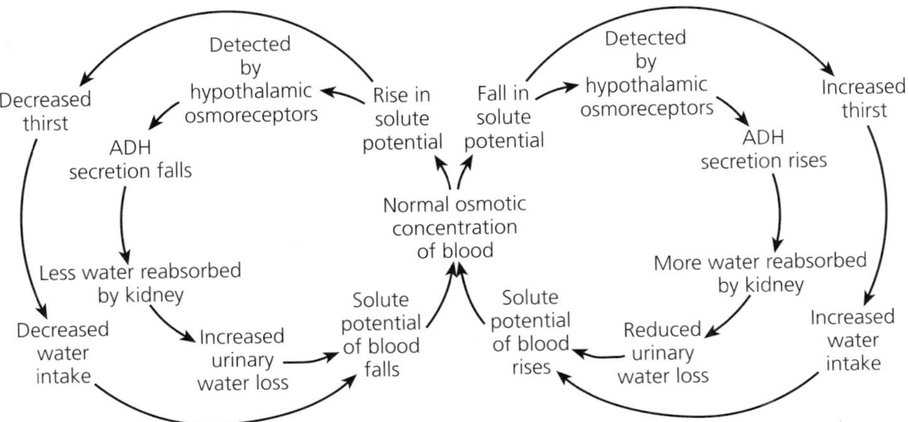


**Figure 24.14**
Negative feedback control of body water content

### *Other factors in water balance*

Besides drinking and urinary loss, there are other routes by which water is lost from the body:

- evaporation from the skin. This is highly variable, ranging between 50 $cm^3$ per hour when there is no sweating, to 1.5 $dm^3$ per hour when sweating is maximal.

- evaporative loss in the breath. This is also highly variable, being higher during exercise and when atmospheric humidity is low.
- faeces, accounting for about 100 $cm^3$ per day on average. In land animals, faecal water loss is considerably reduced by reabsorption of water from the colon.

Although most water is taken in through the mouth, a small amount of **metabolic water** is produced in respiration. This is greater when fat is the fuel, but since this requires more oxygen to burn, it involves greater ventilation and hence more evaporative loss via the lungs. Metabolic water is of little significance in most mammals, but in some desert rodents (such as the kangaroo rat) it is a major source of water. This animal can survive without ever drinking, and relies on water in the food (such as seeds) and metabolic water.

## The role of urea in concentrating the urine

Although the cortical nephrons illustrate the general principle of the countercurrent multiplier, maximum urine concentrations depend on the juxtamedullary nephrons. They have much longer loops and can build up medullary salt concentrations four times higher than that of the plasma. Although only 15% of the nephrons are of this type, the concentrated peritubular fluid they produce promotes water absorption from fluid leaving the cortical nephrons.

One of the factors in achieving this is the accumulation of urea in the medulla. It occurs in the following way. During water shortage, water is reabsorbed as the urine travels down the collecting ducts under the influence of ADH, so the concentration of urea in the urine rises. In addition to its effect on water permeability of the collecting ducts, ADH also raises the permeability to urea in those parts of the collecting ducts deepest in the medulla. Urea consequently diffuses into the medulla, where it raises the solute concentration of the medullary interstitial fluid. This causes water to leave the descending limbs of the juxtamedullary nephrons, raising their salt concentration. The salt concentration delivered to the ascending limb is thus raised, causing salt to diffuse out into the medullary interstitial fluid. In this indirect way, therefore, urea assists the build up of salt in the medulla. When the liquid reaches the thick segment of the ascending limb, more salt is actively extruded in the same way as in the cortical nephrons.

## Sodium regulation and blood pressure

Osmoregulation is the control of the amount of water in relation to all solutes added together, the *proportions* of the various ions being unimportant. *Ionic* regulation is concerned with the maintenance of the relative proportions of the various ions such as sodium, potassium, calcium, phosphate and magnesium. Sodium is the major plasma cation, and under normal circumstances its intake and loss are balanced. Since the intake in the food and loss via sweat both vary, the regulation of its urinary loss is essential.

In one respect sodium regulation is more complicated than that of other ions. Because it is the major plasma cation, its regulation interacts with that of water content and blood pressure. To make this clear, imagine that a person were to be injected with 1 $dm^3$ of salt solution of exactly the same ionic composition as the plasma. Though this would have no effect on the plasma composition, its volume would increase, and so therefore would the blood pressure. Short-term dilatation of the blood vessels reduces the increase, but the long-term solution is to get rid of more water *and* salt.

## Sodium regulation and blood pressure *continued*

This is exactly what the body does, in the following way. First, less sodium is reabsorbed by the nephron, so sodium loss increases. The fall in plasma salt concentration dilutes the plasma slightly, inhibiting the release of ADH and thus increasing the loss of water in the urine.

The opposite situation occurs in severe bleeding or when a person donates blood. The fall in blood pressure sets in train a number of responses which help to reduce the fall in blood pressure.

Besides the rapid neural responses mentioned in Chapter 22, there are slower, hormonally-mediated effects. Apart from an increase in the secretion of ADH, the most important responses involve the kidney itself.

Following a fall in blood pressure, the smooth muscle cells in the afferent arteriole release into the blood an enzyme called **renin**. This converts an inactive plasma protein into **angiotensin**, which has two principal effects:

- it stimulates arterial smooth muscle, causing a generalised vasoconstriction which helps to reduce the fall in blood pressure.
- it stimulates the secretion of aldosterone by the adrenal cortex. This steroid hormone promotes the reabsorption of sodium ions from the distal tubules and the collecting ducts in exchange for potassium ions. The reduced loss of sodium ions helps to raise the salt concentration in the plasma, which in turn stimulates ADH production and the retention of water.

The part of the nephron responsible is the **juxtaglomerular (JG) apparatus**, formed where the distal tubule comes into contact with the afferent arteriole supplying the *same* nephron. The JG apparatus thus consists of two parts: the afferent arteriole, and a small part of the distal tubule, the **macula densa**. The afferent arteriole contains stretch receptors which monitor blood pressure. The macula densa is sensitive to the concentration of $Na^+$ ions in the tubular fluid.

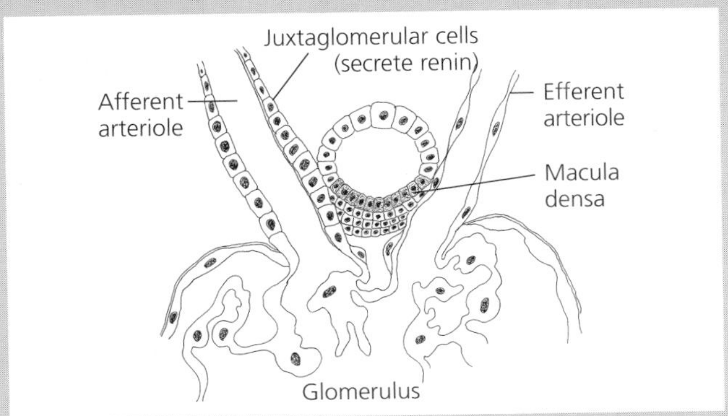


The juxtaglomerular apparatus

The working of the JG complex is as follows. If blood pressure falls, so does the filtration rate. Less salt is filtered so less reaches the macula densa. In some way this is communicated to the walls of the adjacent afferent arteriole which release renin, triggering the changes described above.

Besides being responsive to changes in blood pressure detected by the kidney, the adrenal cortex is directly sensitive to the level of plasma sodium. A rise in sodium causes a decrease in aldosterone output and an increased excretion of sodium.

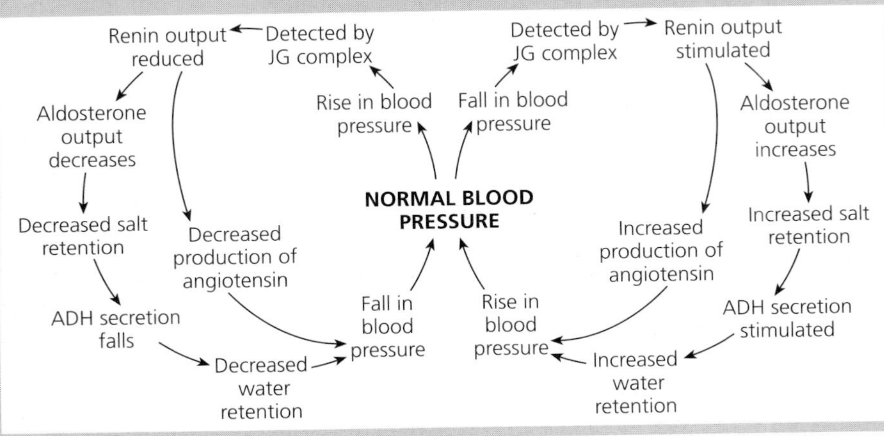


Role of the kidney in the control of blood pressure

## The kidney and pH regulation

One of the most important functions of the kidney is to maintain the pH of the body fluids at a pH value of about 7.2. The body is continuously generating acidic substances. Of these, carbon dioxide is eliminated via the lungs, and lactic acid is metabolised to carbon dioxide and water. Some acids cannot be got rid of in this way, such as phosphoric acid produced in the breakdown of nucleic acids. The kidney prevents a fall in pH by the secretion of hydrogen ions by the tubule cells into the urine. This is why the pH of urine is normally slightly acid (about pH 6). If, on the other hand, the body fluids become alkaline, the kidney responds by excreting $HCO_3^-$ ions, making the urine slightly alkaline.

## QUESTIONS

**1** Each of the columns **A–D** in the table below gives the average percentage of protein, glucose, urea, sodium and water taken from one of the following locations: renal artery, Bowman's capsule, end of the proximal tubule and bladder. Identify each of the locations **A–D**.

| | A | B | C | D |
|---|---|---|---|---|
| Protein | 0 | 7 | 0 | 0 |
| Glucose | 0 | 0.1 | 0.1 | 0 |
| Urea | 2 | 0.03 | 0.03 | 0.15 |
| Sodium | 0.3 | 0.3 | 0.3 | 0.3 |
| Water | 96 | 92 | 99 | 99 |

**2** What would be the effect of drinking a large volume of water on the composition of the liquid in **(a)** Bowman's capsule and **(b)** the liquid arriving at the beginning of Henle's loop?

**3** What would be the effect on the composition of the urine of eating a large meal rich in **(a)** protein and **(b)** carbohydrate?

**4** When a kidney is supplied with blood cooled to 15°C, there is a marked increase in the output of urine, and glucose is present in the urine. Suggest an explanation for these two facts.

**5** The table below shows the concentrations in grams per $dm^3$ of various substances in the plasma and the urine of a normal person on an average diet after injection of inulin, which is not a normal urinary constituent. Inulin is freely filtered but is neither secreted into nor reabsorbed from the nephron, which means that its rate of filtration is the same as its rate of excretion. The rate of urine production in this person was 1 $cm^3$ per minute.

| | Concentration in grams per $dm^3$ | | |
|---|---|---|---|
| Substance | Urine (U) | Plasma (P) | U/P |
| $Na^+$ | 34.0 | 34.0 | 1 |
| Glucose | 0 | 1.0 | 0 |
| Urea | 9.0 | 0.15 | 60 |
| Creatinine | 0.15 | 0.01 | 150 |
| Inulin | 120 | 1.0 | 120 |

a) What is the rate of filtration of inulin in this person?
b) Which of these substances *must* be secreted into the nephron? Explain how the data support your answer.
c) What proportion of the $Na^+$ ions are reabsorbed from the filtrate?

**6** Mannitol is a soluble, non-toxic carbohydrate. It is filtered by the kidneys but is neither reabsorbed nor secreted into the renal tubules. After an injection of a suitable quantity of mannitol, a person's urine output rises considerably. Suggest why.

**7** A blood donor gives 600 $cm^3$ of blood. Explain the part played by the hypothalamus, pituitary gland, adrenal cortex and kidney in restoring blood pressure to normal.

CHAPTER

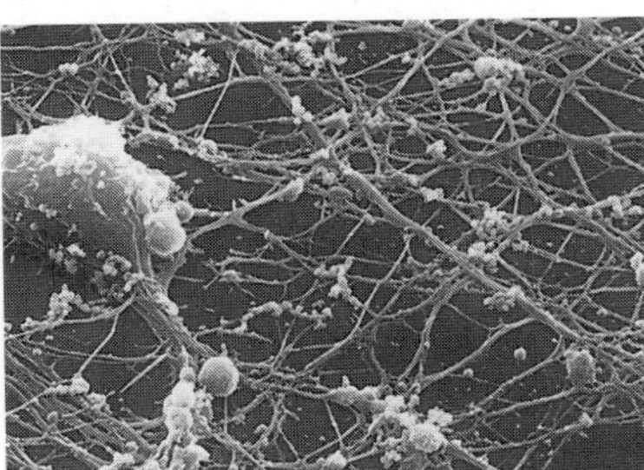

# 25 REGULATION of BODY TEMPERATURE

## LEARNING OBJECTIVES

By the time you have completed your study of this chapter you should be able to:

- explain why life is only possible at moderate temperatures.
- distinguish between poikilotherms and homeotherms, and between ectotherms and endotherms.
- discuss the costs and benefits of homeothermy.
- describe the various ways in which heat can be lost from a terrestrial mammal.
- state the functions of the skin and explain how it enables the rate of heat loss to be varied.
- explain what is meant by basal metabolic rate, how it can be measured and how it is controlled.
- describe the roles of the autonomic nervous system and endocrine system in the regulation of body temperature.
- describe anatomical adaptations to cold and warm climates.

## Life and temperature

With the exception of some bacteria, life can only exist between temperatures of −2°C and about 50°C. There are several reasons for this:

- enzymes that catalyse metabolic reactions are very temperature-sensitive (Chapter 4). If too cold, they are stable but do not work fast enough. If too warm, they are unstable and are rapidly denatured. Between these two extremes is an optimum at which enzymes are stable and work quickly. Most species have enzymes with optimum temperatures somewhere between 0°C and 45°C.
- many membrane proteins can only function properly if they are floating freely in the lipid bilayer, which is only possible if the latter is liquid; at low temperatures the bilayer freezes.
- at around 85°C, DNA 'melts', meaning that the double helix separates into two single strands.

### *Temperature homeostasis*

To a varying degree all organisms can actively maintain a relatively constant internal environment, a phenomenon called **homeostasis**. In addition to regulating the composition of their interiors, some animals are able to maintain a body temperature that is different from the surrounding, or **ambient**, temperature.

The ability to regulate body temperature is called **homeothermy**. Homeotherms include birds and mammals and are said to be 'warm

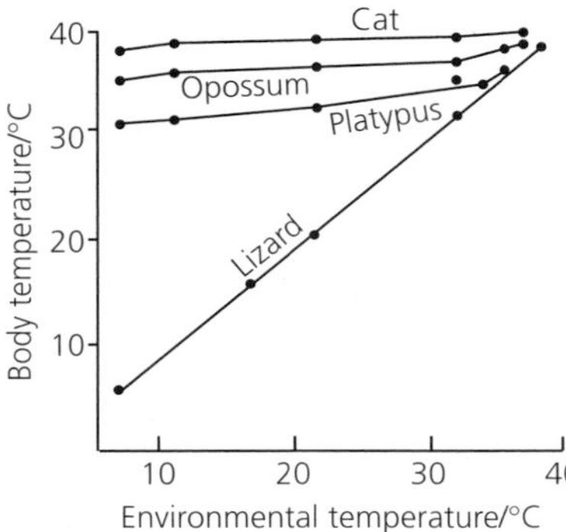


**Figure 25.1**
Effect of external temperature on body temperature in a variety of animals

blooded', in contrast to other animals which are 'cold blooded'. These terms are not very helpful – the blood of a Nile crocodile at midday, for example, is likely to be warmer than that of a mammal. It is more useful to distinguish between **poikilotherms**, whose body temperatures vary with that of the environment, and **homeotherms**, which regulate their body temperatures (*poikilos* is Greek for 'varied', and *homoios* is Greek for 'similar'). Figure 25.1 shows the effect of changes in external temperatures on the temperature of the body in different species.

Even the term 'homeothermic', however, is not so easy to define, because many land animals can partially regulate their temperatures by their behaviour. These **behavioural homeotherms** can adjust the amount of heat they absorb from the sun by varying their posture, or, in very cold or very hot weather, by retreating into crevices (see the information box below).

## Behavioural homeotherms

**Many terrestrial animals can regulate their body temperature by their behaviour. On hot summer days honey bees keep the hive temperature several degrees cooler than the outside by fanning the hive. In winter they crowd together, keeping themselves up to 10°C warmer than the temperature outside the hive.**

**The desert locust (*Schistocerca gregaria*) regulates its temperature by adjusting its position. Between about 17°C and 38°C it maximises absorption of solar heat by orientating itself at right angles to the Sun, but above about 40°C it reduces the absorption of radiant heat by aligning itself parallel to the rays. If it gets very hot the animal raises itself off the ground by extending its legs or, in extreme cases, climbing vegetation. These last two responses utilise the fact that although the ground may be hot, air temperature decreases very rapidly with height.**

Larvae of the desert locust, *Schistocerca gregaria*, displaying behavioural homeothermy

The most useful distinction is based on *how* temperature is regulated. **Ectotherms** regulate body temperature by their behaviour; they can only maintain their body temperatures above the ambient temperature by absorbing radiant heat from *outside* (*ecto* = outside). **Endotherms** include mammals and birds, which keep warm using heat generated *inside* the body (*endo* = inside), and regulate their heat loss by physiological mechanisms in the skin.

## Body temperature in endotherms

Most mammals have body temperatures between 37°C and 39°C, while birds have slightly higher temperatures between 40°C and 42°C. Though we often say that mammals have a 'constant body temperature', this is actually not strictly so for several reasons:

## Endotherms that are not birds or mammals

Not all endotherms are birds or mammals. Bumblebees are also endotherms, using their insulating coat of fur to retain heat generated by the flight muscles. When active, the temperature inside a bumble bee's thorax is maintained at about 32–36°C, in air that may be 15°C cooler. This is why bumblebees can gather nectar in weather that is too cold for honey bees.

Some large, fast-swimming fish are a partial exception to the rule that endothermy is impossible in fish. Tuna and some large sharks have muscles that may be as much as 14°C higher than that of the surrounding water. They manage this seemingly impossible feat by cooling the blood leaving the muscles through a system of countercurrent heat exchange vessels, so that blood is warmed on its way to the muscles and cooled as it leaves.

- temperature shows rhythmic variations in the same individual, even under constant environmental conditions. In healthy humans the average is about 35.8°C in the early morning and about 37.3°C in the evening. For precise comparisons, then, body temperatures should be taken at the same time of the day.
- temperature varies with level of activity, and may rise to 40°C during vigorous exercise.
- women in the second phase of their menstrual cycle have temperatures about 0.3°C higher than in the first phase of their cycle.
- even when comparisons are made between people for whom the above factors are the same, body temperature is found to vary, with a standard deviation of about 0.2°C suggesting that there may be genetic factors at work.
- many endotherms hibernate in winter, during which time body temperature falls to a degree or two above the ambient temperature. Humming birds and some bats become poikilothermic at night, entering a state of **torpor**.

Temperature also varies from one part of the body to another. Though the 'core' temperature in the deeper parts of the body does not fluctuate much, in the outer 'shell' (especially in the limbs) it varies considerably (Figure 25.2). Even in different organs there are slight differences, reflecting variation in heat production.

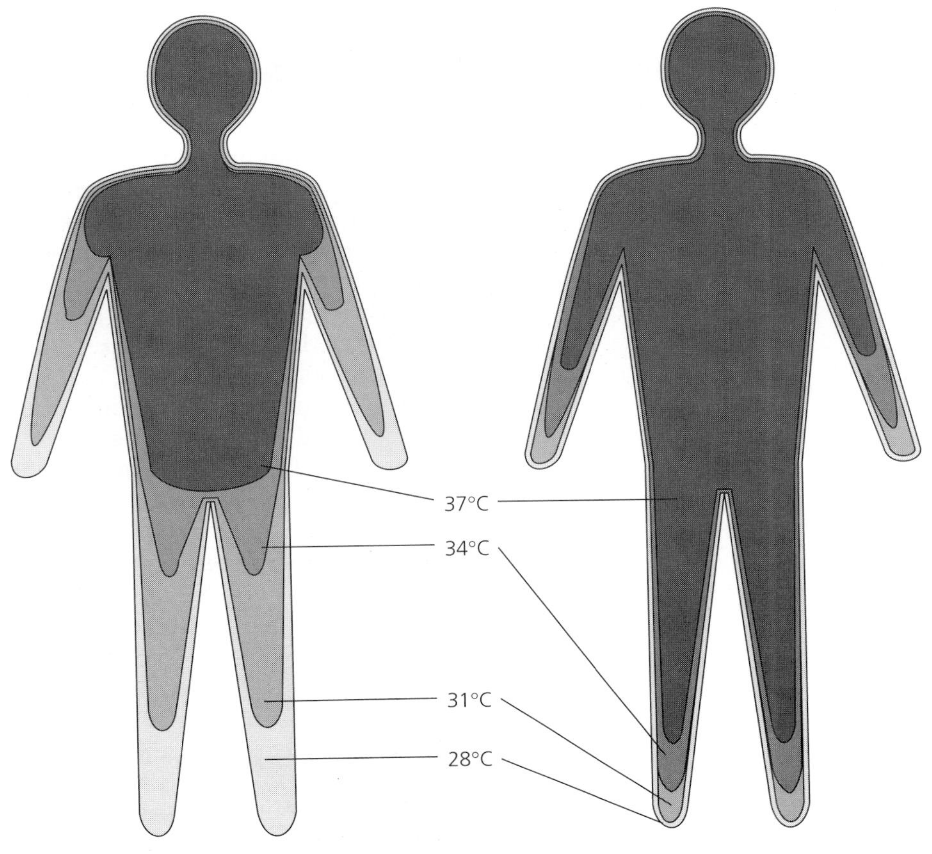


**Figure 25.2**
Temperature distribution in different parts of the body of a naked human in air at 20°C (left) and 35°C (right)

## Thermoregulation in water and on land

With the exception of secondarily aquatic mammals (e.g. whales and seals) and birds (e.g. penguins) whose ancestors were terrestrial, there are no fully aquatic animals that are completely endothermic. There are two reasons for this.

First, the advantages of endothermy are much less in the water than on land. Temperature changes in large bodies of water are considerably less than those on land because water has a high heat capacity, so the lakes and oceans warm up slowly in summer and cool down slowly in winter. Aquatic animals are thus to a large extent insulated against extreme temperature changes. In large continental land masses such as central Asia, summer and winter temperatures may differ by as much as 80°C, and in some deserts the *daily* temperature range may be 35°C.

A second reason is that although temperatures are much harsher on land, the possibilities for its regulation are more favourable than they are in water. Only on land is it possible for an animal to remain cooler than its environment by evaporation. Keeping warm is also easier, since air has a much smaller heat capacity than water and is an even poorer conductor of heat. A terrestrial mammal or bird, therefore, loses body heat more slowly than an aquatic one. In fact it would be quite impossible for a fish to be fully endothermic, because the blood would cool down to the ambient temperature as it passed through the gill capillaries.

Another important point is that temperatures on land vary over short distances, so animals can find shelter from extreme cold or heat by hiding in crevices or underground. Water, on the other hand, tends to be thermally uniform, so there are no 'pockets' to which animals can retreat. Behavioural homeothermy is therefore impossible in aquatic environments.

## Balancing the heat budget

As in other cases of homeostasis, thermoregulation involves balancing incomings and outgoings – in this case heat gain and heat loss. The immediate result of a decrease in environmental temperature is an increase in heat loss and, initially, a slight fall in body temperature. This leads to a corrective decrease in heat loss, and, in cases of prolonged cold, an increase in heat production.

To control heat loss and heat gain the body does not actually need to measure either. Instead it measures the body temperature. If this rises, then heat gain exceeds heat loss, and vice versa. A fall in body temperature is countered by an increase in heat loss and, if prolonged, also by decreased heat production.

The control of body temperature requires some form of **thermostat**. This is located in the **hypothalamus**, just above the pituitary gland. Any departure from the normal body temperature, or **set point**, results in the hypothalamus sending out signals to correct the situation.

To do this, the hypothalamus must be informed of changes in body temperature, detected by **thermoreceptors**. The most important thermoreceptors are in the hypothalamus itself, but there are also temperature-sensitive nerve endings in the skin.

Having been informed of a change in body temperature, the hypothalamus must send out the appropriate instructions to **effectors**, which bring about corrective responses in the form of changes in heat loss and heat production. Heat loss is controlled by the autonomic nervous system, and involves fairly rapid responses by effectors in the skin. Long-term changes in heat production, on the other hand, are under endocrine control and involve widespread and prolonged responses.

## Endothermy, size and shape

Endothermy is an expensive way of life, and is only possible in animals in which heat is lost at a rate that can be replaced by metabolism. There are obviously limits to how rapidly an animal can obtain, digest and absorb food, so there are limits to the rate of heat loss an animal can sustain.

This is why size and shape are so important. All other things being equal, the rate at which an animal loses heat is governed by its surface area, but the amount of heat-producing tissue is governed by its volume. If an animal's shape does not change, doubling its length increases the area for heat loss four times, but the volume of the body goes up eight times. Thus eight times as many cells have to produce only four times as much heat, so each cell can be less active. Looking at things the other way, making an animal smaller increases the rate at which each cell has to produce heat for the body to stay at the same temperature. The smallest mammal, the pygmy shrew, has to eat nearly its body weight of food every 24 hours, and most of its time is spent in the search for food. The smallest humming birds are even smaller, but they become poikilothermic at night, and their body temperature drops to that of the surroundings.

Body shape is another factor affecting heat loss. The external ears of a mammal are potentially a major source of heat loss, but, since they give early warning of predators they cannot be dispensed with. This conflict between thermoregulation and avoiding predation is particularly intense in the Arctic hare, which has short ears. In the Mexican jack rabbit with very long ears, keeping warm is no problem.

Ear length in rabbits from different geographical zones

# Heat loss and the mammalian skin

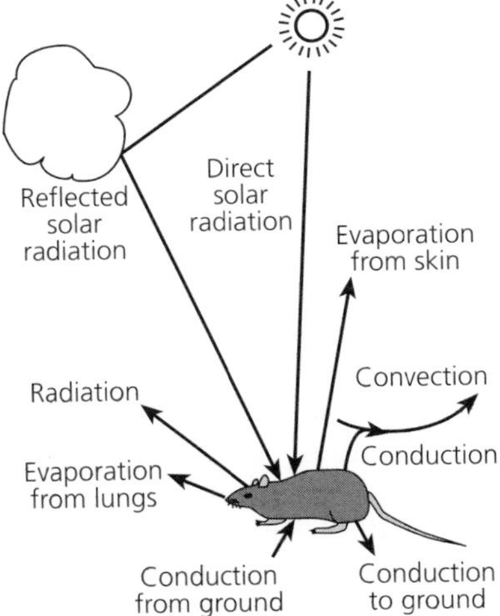


**Figure 25.3**
Various ways in which heat can enter and leave the body of a mammal

Heat can leave an animal's body in three ways: by radiation, by conduction and by evaporation (Figure 25.3). The rate of the last two processes is greatly influenced by a fourth process, **convection**, which conveys heat away from just outside the body (see below).

### *Radiation*

A warm body emits heat in the form of infrared radiation. At 20°C, about half the heat lost by a naked human body in air is due to radiation. Heat can also be gained by radiation from hotter objects in the environment such as the Sun. Under most circumstances there is a net loss of heat by radiation. Notice that radiation heat loss or gain has nothing to do with air temperature – you can still absorb heat in strong sunshine even if the air is very cold.

### *Conduction*

When you sit on a cold rock you lose heat by conduction. If the air is cooler than the body, heat is conducted from the body through the air. Heat is also lost by conduction through the gut wall when an animal eats food or drinks water that is cooler than the body. It is also lost in warming inhaled air. Like diffusion, conduction depends on random collision between molecules; a 'hot', fast-moving molecule can transfer some of its energy to a 'cooler', slower-moving molecule.

The rate at which heat is lost by conduction is affected by two factors. First, it depends on the thermal conductivity of the material next to the skin. A naked human can remain comfortable in air at 20°C, but quickly feels cold in water of the same temperature. This is mainly because water conducts heat about 25 times faster than air does. People with more fat lose heat more slowly than lean people because fat conducts heat at only about a third the rate of lean tissue.

Second, it depends on the **temperature gradient**, or how quickly the temperature changes with distance (Figure 25.4). For a given temperature difference between two points, the closer they are together the steeper the thermal gradient.

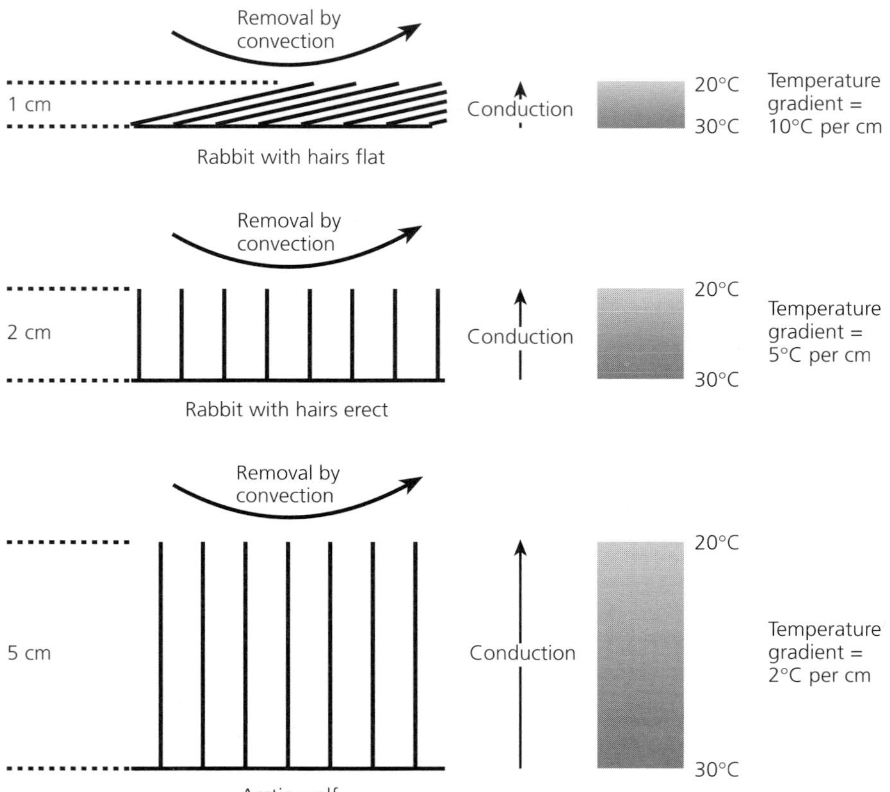


**Figure 25.4**
How a temperature gradient affects the rate of heat conduction through the fur of three mammals (assuming that the fur differs only in its thickness)

## *Evaporation*

Except in air that is saturated with water vapour, all animals on land lose water by evaporation. The change from liquid to gas absorbs **latent heat**, which produces a cooling effect (you can feel this when you hold a wet finger in a strong wind). For each gram of water evaporated at 20°C, 2441 Joules of energy are absorbed. Evaporation is the *only* way by which heat can be transferred from a lower to a higher temperature. Therefore to keep cooler than its surroundings, an animal has to lose water.

At cool temperatures, evaporation occurs from the lining of the breathing passages and from the general surface of the skin. This evaporation is uncontrolled, and is said to be **insensible**. In an average sized human it is about 50 $cm^3$ per hour. At higher temperatures, most mammals produce **sweat**, which cools the skin when it evaporates. Evaporation is slower in humid air and sweat may accumulate as liquid. This is why you may feel hotter in warm, humid weather than in hot, dry conditions. In very hot conditions a person may produce as much as 1.5 $dm^3$ of sweat per hour.

### *Convection*

Convection is the bulk flow of gas or liquid. When air next to the body is warmed, its density decreases and the air rises, carrying heat with it. In practice, convection currents due to heating are usually swamped by wind or water currents, which therefore have a much greater effect on heat loss. Physiologists usually distinguish between two kinds of convection:

- **free convection**, in which gas or liquid moves as a result of local heating from a warmer to a cooler area.
- **forced convection**, in which air or liquid moves from a higher to a lower **pressure**. Forced convection includes wind, breathing and also blood flow.

Whether free or forced, convection greatly influences the rate of heat loss. Convection acts indirectly by its effect on conduction and evaporation. Imagine that there are no air currents next to the skin. As heat is conducted to the outside it warms up, reducing the temperature difference and hence the rate of conduction. Air currents prevent this, by maintaining a temperature gradient between the skin and the environment and thus the rate of conduction through the layer of air immediately outside the body. By preventing convection of air next to the skin, a coat of fur (or in humans, clothing) traps a layer of still air, reducing heat loss by conduction. The thin layer of water between the skin and a diver's wetsuit acts in a similar way.

Forced convection can increase heat loss in another way. By preventing humid air from accumulating, it encourages evaporative cooling.

## The skin

Though dealt with here because of its role in homeothermy, the skin has several other diverse functions:

- it provides *protection* against hazards, such as invasion by pathogens, mechanical injury, ultraviolet light and water loss.
- it is an important *sense organ*, containing receptors involved in the sense of touch and pressure.
- in many (perhaps all) terrestrial mammals it produces *pheromones*, chemicals concerned with communication with other members of the species.
- in the presence of ultraviolet light, it is the site of Vitamin D production.

### *Structure of the skin*

The skin is the largest organ in the mammalian body. It consists of two distinct layers – an outer **epidermis** and an inner **dermis** (Figure 25.25).

#### *The epidermis*

As the outer layer of the skin, the epidermis has to withstand wear and tear. The innermost cells are in constant mitotic division and form the **Malpighian layer**, or **germinative layer**. The cells of this layer contain the black pigment **melanin**, which absorbs and protects against ultraviolet light. Melanin is produced in cells called **melanocytes** which lie in the deepest layer of the epidermis. The melanocytes export their

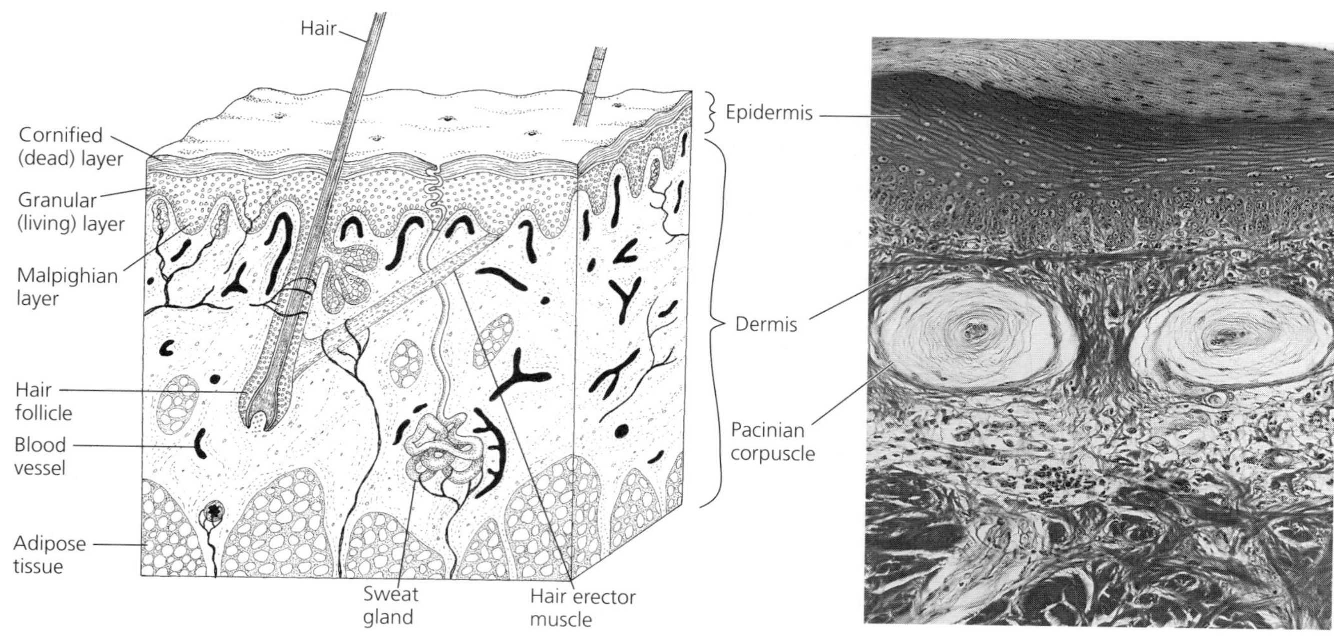


**Figure 25.5**
Vertical section of the human skin.

melanin via outgrowths which penetrate the surrounding cells. The outgrowths periodically pinch off small melanin-containing sacs. All people (except albinos, who cannot make melanin) have about the same number of melanocytes, but in dark-skinned populations the melanin is more spread out, making the skin appear darker.

As a result of cell division, cells are slowly pushed up towards the surface. As they do so they produce large amounts of the tough, fibrous protein, **keratin**, and become strongly bonded to each other by **desmosomes** (Chapter 3). Cells nearer the surface are older and make up the **cornified layer**. Unless broken, this tough, dry layer is difficult for bacteria to penetrate, though many live on its surface. The thickness of the cornified layer varies with the amount of wear to which the skin is exposed, and may be very thick, as on the soles of the feet and palms of the hands. It takes about four weeks for cells produced in the Malpighian layer to reach the surface of the skin, by which time they are dead and flake-like and are carried away by air currents (accounting for most of the dust on the tops of wardrobes).

### *The dermis*

The dermis is much thicker than the epidermis and contains a network of two kinds of fibrous proteins. **Collagen** which gives considerable strength, whilst **elastin**, as the name suggests, gives elasticity. Unlike the epidermis, the dermis contains a network of blood vessels. These are much more extensive than is needed to nourish the skin – their additional function is to bring heat to the body surface. The smooth muscle in the arterioles is supplied by sympathetic nerves; stimulation of this muscle tissue results in constriction of the arterioles, and hence a reduction in the rate of delivery of heat to the skin. The dermis also contains sensory nerve endings of various kinds, some of which are sensitive to touch and pressure, and others to changes in temperature.

**Hair follicles** are pits from which hairs grow, and are deep intuckings of the epidermis into the dermis. The hair root consists of a specialised cluster of cells which, instead of producing a sheet-like layer of epidermis, produce a hair. Under the microscope the epidermal scales of the hair are visible. The angle of the hair can be adjusted by contraction of the small

**erector pili muscle** attached to each follicle. In most mammals other than humans, this enables the thickness of the fur to be increased. This thermoregulatory role has been lost in humans, though the muscles still contract when the body is cold, producing 'goose pimples'. The erector muscles are supplied by sympathetic nerve fibres, and are also stimulated by fear as occurs, for example, when a dog's 'hackles' are raised.

Though hairs are dead, they can play an important role in the sense of touch by acting as *levers*. Even a minute force, such as would be produced by a tiny insect crawling up the skin, is magnified at the base of the hair and is detected by nerve endings round it.

The skin also contains vast numbers of **sweat glands**. Like hair follicles, these are deep epidermal infoldings into the dermis and are supplied by sympathetic nerve fibres. They are of two kinds – **eccrine** and **apocrine** glands. Eccrine glands are the most abundant kind in humans and open to the outer surface of the skin. In other mammals, the most common ones are called **apocrine** glands and open into hair follicles. Their secretion is believed to contain **pheromones**, chemicals used in communication with other members of the same species. In humans these are confined mainly to the armpits and pubic region.

Also in the skin are the **sebaceous glands**. These are outgrowths of the hair follicles and secrete a yellow, oily substance called **sebum** which helps to keep the coat water-resistant.

Beneath the dermis is a layer of **adipose** (fat-storage) tissue. The cells owe their large size to the enormous fat droplet each contains. Besides acting as an energy reserve and providing mechanical protection by padding, adipose tissue acts as a thermal insulator. This is not only because it conducts heat less efficiently than other tissues, but because it has a poor blood supply, so little heat is convected through it. Consequently, in cold conditions (when the blood supply to the skin is restricted) not much heat crosses it by convection.

## Heat production

With the exception of the urine and faeces, most of the chemical energy taken into the body is eventually lost as heat. Some is lost as heat produced in respiration. More is lost as ATP energy which is 'spent' in driving cellular processes. Even the energy that is used to make proteins and other polymers is eventually converted to heat because all polymers (except DNA) are continually being broken down as fast as they are being built up. In ectotherms this heat production is insufficient to raise the body temperature much above that of the surroundings, but in endotherms it is the major source of body heat.

Heat is not produced at equal rates in all organs (Table 25.1). The heart, liver, kidney and brain are all very active, but skin is much less so. Heat production may also vary a good deal in a given organ, such as skeletal muscle. Skin, for example, has a low metabolic rate, yet the blood flow through it may be very high because of its role in getting rid of surplus heat.

Metabolic rate varies somewhat with circumstances. The minimum level of activity required to sustain life-essential functions is the **basal metabolic rate** (**BMR**). BMR is measured when the person has been motionless for some hours, and has not had a meal for at least 12 hours, since digestion and absorption require energy. Even under these standard conditions and in a constant environment, BMR varies rhythmically with the time of day. This is but one of many examples of **endogenous rhythms**, driven by an internal 'clock'.

**Table 25.1**
Heat production by different parts of a resting human body

| Organ | Heat production (Joules $kg^{-1}$ $min^{-1}$) |
|---|---|
| Liver | 400 |
| Kidney | 1200 |
| Heart | 1940 |
| Brain | 660 |
| Skeletal muscle | 40 |
| Skin | 60 |
| Whole body | 80 |

### Measuring metabolic rate

All other things being equal, larger animals obviously have higher energy turnover than small ones. To have any meaning, BMR values must therefore take account of differing body sizes. BMR correlates more closely with body surface than with body mass, so values are usually expressed in terms of heat production per square metre of body surface.

The simplest way to measure an animal's heat production is to determine it indirectly through measurement of its oxygen consumption, measured in a spirometer. The energy value of oxygen varies slightly with the kind of food being respired, but for a person on an average diet, each $dm^3$ of oxygen used releases 20.2 kJ of energy. If greater accuracy is required, the proportions of carbohydrate, fat and protein in the diet must be taken into account.

## Responses to cold

Adaptations by endotherms to heat and cold operate over very different time scales and involve quite different mechanisms:

- at one end of the scale are responses to rapid changes, which involve muscles and are mediated by the nervous system.
- slower responses involve metabolic changes and are under endocrine control.
- longer-term responses to seasonal (and thus predictable) changes involve changes in growth, such as the production of a thicker winter coat.
- operating on the evolutionary time scale are permanent anatomical features of animals that spend their entire lives in cold or hot climates. Elephants, for example, are able to lose heat more quickly because of the large surface area of their ears, while penguins reduce heat loss by a thick layer of fat under the skin. These permanent adaptations have a genetic base and develop in response to natural selection over many generations. They include anatomical features such as those described in the information box below.

### Conserving heat by counterflow

In an endotherm, heat is lost most rapidly from the limbs, which have relatively large surfaces. Many endotherms living in temperate or cold climates reduce heat loss from their extremities by exchanging heat between arteries and veins, which normally run close together. The venous blood has lost some heat to the exterior and is therefore cooler than the arterial blood. The vein is so close to the artery that some heat is conducted from artery to vein. This **counterflow** of blood allows heat to short-circuit the extremity of the limb, and is thus conserved. In the flippers of whales and seals and also in the legs of many birds this principle is further elaborated, the artery being surrounded by a network of veins.

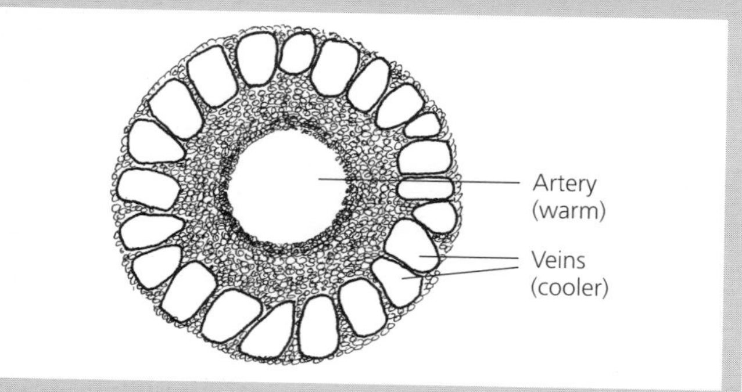


Blood vessels in a whale flipper as seen in transverse section. Heat is transferred from the warm arterial blood to the cooler venous blood. By the time it has reached the capillaries, it has lost much of its heat to the venous blood, so less is lost to the environment

## Neural responses

The response to cooling is to decrease heat loss and increase heat production. Heat loss is reduced by lowering the rate at which heat is delivered to the skin by the blood, and the rate at which it is conducted away from the skin through the air. Thermoreceptors in the hypothalamus detect a fall in core body temperature, and corrective signals are sent out via sympathetic nerves to effectors in the skin. If chilling is sudden (as for example when falling into cold water) the hypothalamus receives advance information from thermoreceptors in the skin, enabling it to anticipate its own cooling. There is an increase in frequency of impulses to the smooth muscle in the arterioles, causing vasoconstriction and hence reduced flow of heat to the skin. As a result, there is a marked decrease in temperature of the outer regions of the body (Figure 25.6). The piloerector muscles also contract, increasing the thickness of the still air next to the skin, and there is a decreased outflow of impulses to the sweat glands.

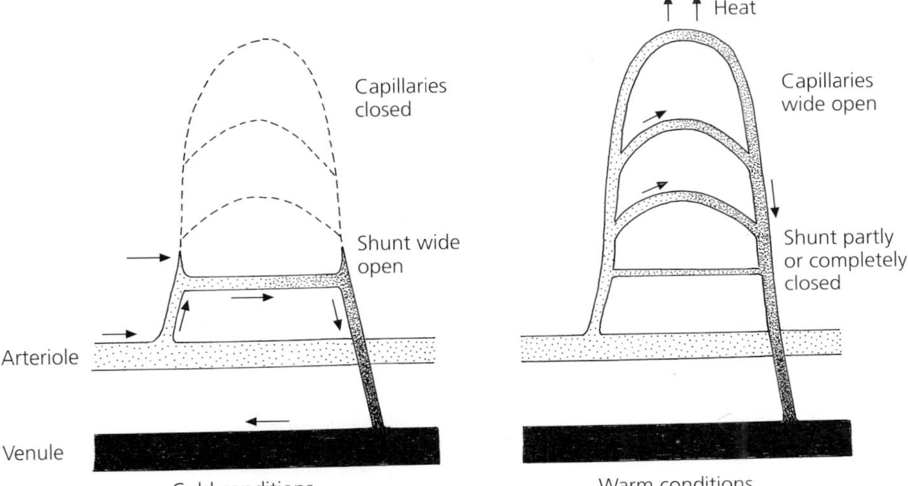


**Figure 25.6**
Changes in blood flow in the skin in different external conditions

The second response to chilling is an increase in heat production brought about in two ways.

- Shivering – the repeated involuntary contraction in the muscles – which, like any muscle contraction, releases heat.
- A widespread increase in metabolic rate brought about by noradrenaline, released by sympathetic nerve fibres.

In human babies, sympathetic stimulation also raises heat production in **brown fat**, a fatty tissue concentrated between the shoulders. Unlike adipose tissue, brown fat cells have abundant mitochondria and are capable of a high metabolic rate, made possible by a rich blood supply. Unlike other tissues, most of the energy released in brown fat is in the form of heat.

## Hormonal responses

If cold conditions extend over hours or longer, further responses occur. There is increased emphasis on heat production under the influence of hormones. The most important are **adrenaline** and **thyroxine** (Figure 25.7). Adrenaline is more concerned with short-term changes in temperature and its effects are almost immediate.

Thyroxine is more concerned with seasonal changes in temperature, as its effects are delayed and persistent. In response to cooling of the blood,

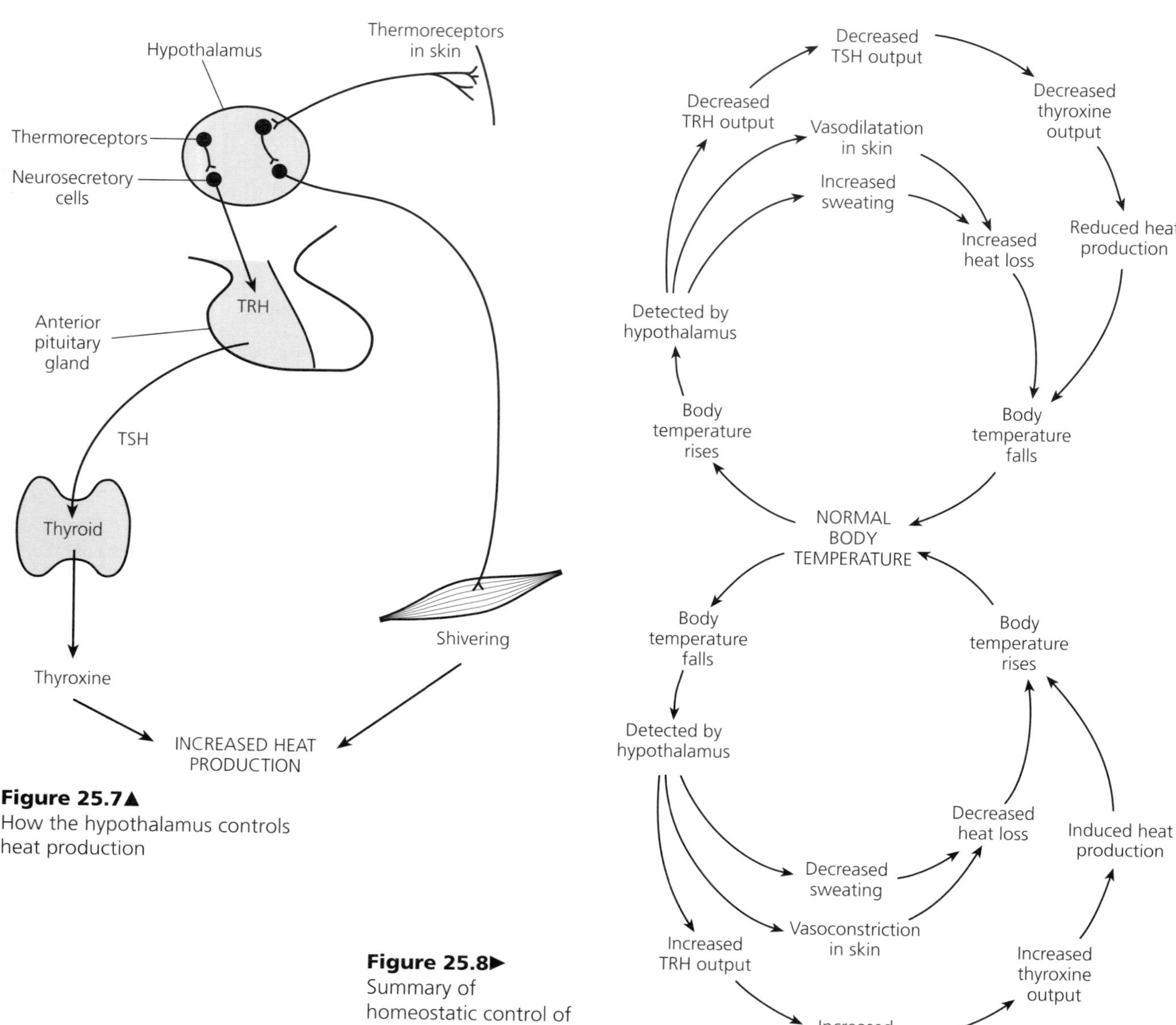


**Figure 25.7▲**
How the hypothalamus controls heat production

**Figure 25.8▶**
Summary of homeostatic control of body temperature

the hypothalamus increases its output of thyrotropin releasing hormone (TRH). This stimulates the anterior pituitary to increase its output of thyrotropin, or thyroid stimulating hormone (TSH). Thyroxine secretion rises, bringing about a sustained increase in the basal metabolic rate and consequent heat production.

Figure 25.9 shows the effect of external temperature on the metabolic rate of a human and a weasel. Over a certain temperature range – the **thermoneutral zone** – the body can regulate its temperature by varying its rate of heat loss, with heat production remaining constant. At temperatures below the **lower critical temperature** (Lc) it can only thermoregulate by raising its heat production.

The graph shows two important things:

- extrapolating the graph gives a line that cuts the x-axis at body temperature. This means that *if* all heat production was needed to maintain body temperature, at an air temperature of 37°C there would be no need for any heat production. In reality, some waste heat is produced in non-thermoregulatory processes so the graph does not cut the x-axis.

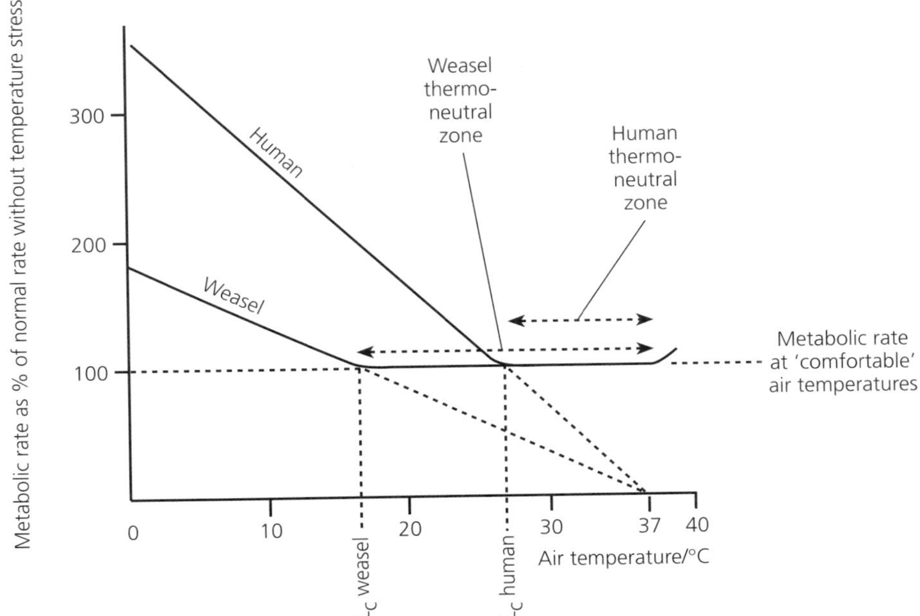


**Figure 25.9**
Changes in metabolic rate with environmental temperature for an unclothed human and a weasel (adapted from Scholander)

- the weasel has a lower critical temperature than the human, a reflection of its better thermal insulation.

## Response to heat

The responses to heating are essentially the reverse of those to cooling. Heat production falls because of a decrease in metabolic rate. There is increased delivery of heat to the skin surface due to dilatation of dermal arterioles. This extra heat is used to evaporate the increased output of sweat. At the same time the hair erector muscles relax, increasing the steepness of the temperature gradient next to the skin.

Some animals (notably birds and dogs) increase evaporative loss by panting. This consists of rapid, *shallow* breaths of not much more than the volume of the dead space. Since most of each breath is dead space air, loss of carbon dioxide is not excessive and the blood pH does not rise significantly.

## Negative feedback in thermoregulation

We have seen how both heat loss and heat production are controlled by nervous and endocrine signals originating in the hypothalamus. The more the temperature drops below the set point (or normal value) the greater the tendency to correct it. This is **negative feedback** and is an essential component of any homeostatic system (Chapter 21). Figure 25.8 illustrates the main features of body temperature control.

Body temperature can only be controlled within certain limits. If it falls beyond a certain point, the various control mechanisms become inadequate and **hypothermia** occurs. The metabolic rate falls reducing heat production, causing a further fall in temperature, and so on. Unless gentle heat is supplied by rescuers, the downward spiral continues until death occurs. A similar **positive feedback** cycle occurs at high temperatures: increased temperature raises metabolic rate, which increases heat production, and so on, leading to **hyperthermia**, or **heat stroke**.

## The costs and benefits of endothermy

The advantage of having enzymes working at or near the optimum temperature is obvious. You only have to recall how much more difficult it is to swat a fly or catch a butterfly on a hot day than in cool weather to appreciate this. Endothermy enables an animal to be active in a wide range of environmental temperatures so that it can remain active at night, and also (provided there is adequate food) in winter.

Despite the advantages there is a heavy cost. Endothermy uses a lot of energy, requiring the rapid intake and distribution of food and oxygen round the body. Indirectly, a high metabolic heat production also costs water if the diet is rich in protein, since the high urea production requires more water to excrete.

Keeping cool also has a price tag in the form of loss of water, and, in the case of prolonged sweating, salt. Reptiles, birds and mammals all lose heat by evaporation from the lining of the lungs. Most land mammals also keep cool by the evaporation of sweat. Birds and some mammals, such as dogs, keep cool by panting, losing latent heat in the evaporation of water from the breathing passages.

## QUESTIONS

**1** Give *two* reasons why a human loses heat much faster in water than in air at the same temperature.

**2** Name three effector tissues in the skin that are involved in the regulation of body temperature.

**3** In an experiment by Benzinger, a resting human subject was maintained in warm air at 45°C, and the skin temperature was continuously measured. The subject then consumed enough ice to have a slight cooling effect on the blood reaching the brain. A few minutes later the skin temperature rose from 36.9°C to 37.6°C, before returning to the previous value.

**a)** Suggest an explanation for this observation.
**b)** Predict what would have happened had the subject been maintained in air cool enough to produce slight 'goose pimples'.
**c)** What hypothesis do you think Benzinger was testing in this experiment?

**4** The blood in the right side of the heart in a resting person is about 0.2°C warmer than that in the left side.

**a)** Suggest an explanation.
**b)** Predict what would happen to this temperature difference if a person were to undertake vigorous exercise, and give an explanation.

CHAPTER

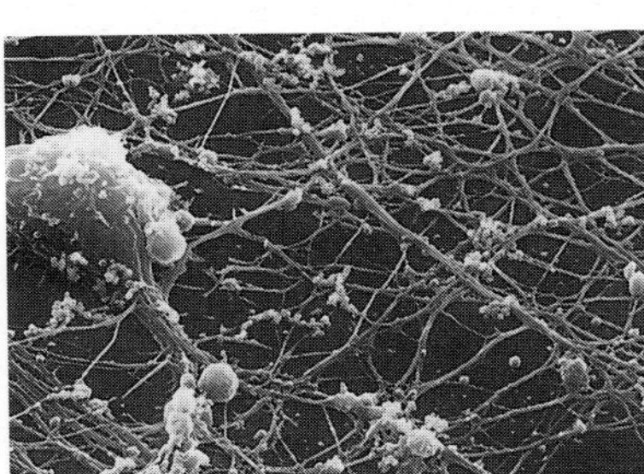

# 26 NUTRITION in MAMMALS

## LEARNING OBJECTIVES

By the time you have completed your study of this chapter you should be able to:

- distinguish between holozoic, parasitic and saprotrophic nutrition.
- distinguish between the general roles of the different classes of nutrients in the human diet.
- explain why the requirements for energy vary between individuals.
- carry out simple calculations of the energy value of foods based on the results of calorimetry.
- explain how a diet may contain a large enough quantity of protein but may be qualitatively inadequate in terms of essential amino acids, and how the intake of carbohydrate may influence protein requirement.
- discuss the importance of fibre in the human diet.
- explain the roles of calcium, iron, phosphate, Vitamins A and D and nicotinamide in the mammalian diet.
- explain the role of hydrolytic enzymes in the digestion of food, and explain why so many enzymes are necessary for digestion.
- explain how the muscle of the gut differs from and resembles that of the heart.
- explain the role of the nervous and endocrine systems in the co-ordination of the stomach, pancreas, liver and gall bladder.
- describe how starch, protein and fat are digested, and how nutrients and water are absorbed into the blood.
- describe the structure of the large intestine and explain its role in osmoregulation.
- describe how glucose, fats and amino acids are utilised.
- describe the structure of the liver and explain its role in regulating blood glucose.

## Types of nutrition

Like all fungi and most bacteria, animals are **heterotrophic**. Literally, this means 'other feeding', because heterotrophs are dependent on other organisms for their organic materials. Ultimately all heterotrophs depend on **autotrophs** for their energy. Autotrophs synthesise organic materials using external energy sources such as light, as in **photoautotrophs**, or oxidation of inorganic chemicals in the environment, as in **chemoautotrophs**. This dependence of heterotrophs on other organisms can be expressed as a **food chain**, in which the first organism is always an autotroph (Chapter 39).

Most heterotrophs feed on organic matter that is too complex to be absorbed into the body directly, and is first broken down by enzymes into simpler substances. Heterotrophs are conventionally thought of as belonging to three major nutritional types:

- **parasitic nutrition**, in which one organism, the **parasite**, feeds at the expense of another, the **host**, harming it to some degree but not usually killing it.
- **holozoic nutrition**, in which animals search for or capture food and digest it in a cavity inside the body. Holozoic nutrition is characteristic of non-parasitic (free-living) animals. Animals may be further subdivided into **carnivores** (which feed on other animals), **herbivores** (which feed on plants) and **omnivores** (which feed on both animals and plants).
- **saprotrophic nutrition**, in which the organism feeds on dead matter, digesting it outside the body bringing about its **decay**. Many fungi and most bacteria are saprotrophic.

Like many other classifications this is far from perfect, and some of the distinctions break down when studied closely. The broad distinctions do, however, have some use.

# Nutritional needs of humans and other mammals

Though animals vary widely in their diets, chemically speaking all food consists of varying proportions of six categories of nutrients, or **foodstuffs**. These are carbohydrates, lipids, proteins, vitamins, minerals and water. The chemistry and role of carbohydrates, lipids and proteins is dealt with in Chapter 2. Though nucleic acids are constituents of the diet, animals can synthesise their constituent nucleotides from amino acids.

Broadly speaking, nutrients may fulfil one or more of three fundamentally different roles in animals:

- they may be broken down in respiration to yield **energy**, for example carbohydrates, lipids and proteins. These are required in large amounts (measured in tens to hundreds of grams per day), and are thus called **macronutrients**.
- they may be converted into **structural components** of cells or extracellular materials, for example proteins and phospholipids of cell membranes.
- some play a role in **catalysis**, either as enzymes or their cofactors. Most, if not all, vitamins and many minerals act in this way. Vitamins and minerals are needed in amounts measured in micrograms to milligrams per day, and are called **micronutrients**.

## The need for energy

All cells need energy for activities such as active transport, synthesis of complex molecules and movement. In addition mammals and birds need energy for heat generation to maintain body temperature. Energy is provided by the oxidation of organic molecules in respiration. The most important respiratory fuels are carbohydrates and fats, though a considerable amount of protein is also used as fuel (see the Clear Thinking box on page 463).

## CLEAR THINKING

It is often stated that proteins are only used as respiratory fuel in times of severe shortage of energy. Everyday experience shows that this cannot be so.

Consider the following facts. At the height of the adolescent growth spurt the average male teenager is growing at a rate of about 10 kg per year. Most of this new flesh is muscle, of which about 20%, or 2 kg, is protein. Hence body protein is increasing at a rate of about 2 kg per year, or about 5 grams per day. The recommended daily protein intake for such a teenager is 75 grams – 15 times the actual increase in body protein. What happens to the extra 70 grams?

Some protein is lost via the faeces (as, for example, sloughed off cells), and some is lost as dead skin, but by far the greatest proportion of dietary amino acids are eventually deaminated to form keto acids and urea. The keto acids are used for energy and the urea is excreted in the urine.

What happens is as follows. A high proportion of dietary amino acids may, in the first instance, be used to make proteins, *but* body proteins are being broken down and used for energy almost as fast as they are being synthesised. In an adult, protein breakdown and protein synthesis are approximately equal, so there is no *net* protein breakdown and the person is said to be in nitrogen balance. During growth, the rate of protein synthesis slightly exceeds the rate of breakdown, and the person is said to be in positive nitrogen balance. During protein shortage, breakdown of tissue proteins exceeds synthesis, so there is a state of negative nitrogen balance.

Overall, then, most of the carbon in dietary protein is lost from the body as $CO_2$, and most of the nitrogen finishes up as urea.

As indicated in Tables 26.1 and 26.2, human energy needs vary considerably. In particular they vary with the following factors:

- physical activity.
- age. Since growth requires energy, young people require more energy than adults of similar body size and level of physical activity.
- body size. All other things being equal, larger people need more energy than smaller people. Resting energy requirements are approximately proportional to surface area, and are expressed in kJ per $m^2$ of body surface rather than per kg of body mass.
- environmental temperature. Since heat loss is faster in cold conditions, more energy is needed to maintain body temperature.
- gender. Females tend to have more fat than males, so maintenance of body temperature requires less heat production.
- pregnancy. ‘Eating for two’ obviously requires an increased intake of all nutrients.

**Table 26.1**
Typical energy needs of people of various ages

| | Body mass (kg) | Resting energy requirement (kJ $m^{-2}$ $day^{-1}$) |
|---|---|---|
| 1 year old baby | 10 | 5350 |
| 8 year old child | 25 | 4560 |
| Adult woman | 55 | 3820 |
| Adult man | 65 | 3480 |

The energy value of foods is measured in a **calorimeter**, so-called because energy used to be measured in calories (Figure 26.1). A known mass of food is burned in oxygen inside a chamber surrounded by a

**Table 26.2**
Energy expenditure in various activities by a 65 kg man

| Activity | Energy expenditure ($kJ\ min^{-1}$) |
|---|---|
| Sitting | 6 |
| Standing | 7 |
| Dressing | 15 |
| Quick walking | 21 |
| Climbing stairs | 38 |

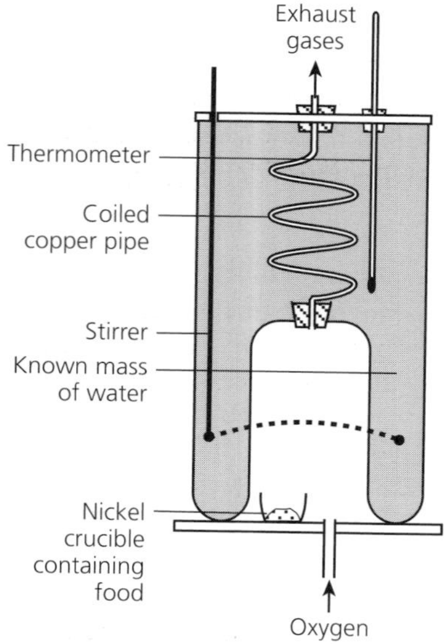


**Figure 26.1**
A calorimeter

known mass of water, and the rise in temperature is measured. Since it takes 4.18 Joules to heat a gram of water through 1°C, the energy content of the food can be calculated thus:

$$\text{Energy content (kJ) per gram of food} = \frac{\text{rise in temperature of water} \times \text{mass of water in kg}}{\text{mass of food in grams}}$$

Complete combustion is ensured by using oxygen rather than air. The food is ignited using an electrical heating coil. Although the heat it gives off is a potential source of error, it can be calculated and a correction made by subtracting it from the heat produced in the apparatus.

The energy values of foods burnt in a calorimeter are not always the same as the energy yield in the body. For example, some of the energy in proteins is lost to the body in the form of urea. After excretion this energy is released, as the urea is oxidised to nitrates by various types of nitrifying bacteria. Similarly fats may not yield their full energy value when metabolised; if insufficient carbohydrate is present, fatty acids are oxidised incompletely, producing organic **ketone bodies** (such as propanone). These may be excreted in the breath giving it a characteristic odour.

### *The main energy-yielding fuels: carbohydrates and fats*

Carbohydrates are **energy-yielding foods**, yielding 17 kJ per gram on oxidation. Most dietary carbohydrate is in the form of polysaccharides such as starch. This is abundant in tubers, such as potatoes, and in cereals and in other seeds, such as peas and beans. In herbivorous mammals, cellulose is a major dietary carbohydrate, though this is not digested by the animal but by micro-organisms in the gut. Animals cannot convert fats into sugar, and carnivores (which eat little carbohydrate but much protein) obtain most of their sugar from amino acids as explained later in this chapter.

Fats yield 39 kJ per gram – more than twice as much energy as carbohydrates. However, although most mammalian tissues can use fatty acids as fuel, cells must be supplied with some glucose for two reasons:

- the brain has an absolute requirement for glucose.
- the respiratory breakdown of fatty acids involves the consumption of small amounts of carbohydrate.

Despite the absolute requirement of cells for glucose, this does not have to be derived from the diet – some is normally produced in the liver from amino acids and some is produced from the glycerol component of dietary fat.

Though carbohydrate can substitute for fat as an energy source, some dietary fat is indispensable for two reasons:

- three unsaturated fatty acids, linoleic, linolenic and arachidonic acids, are constituents of both membrane phospholipids and **prostaglandins**

– substances which act like local hormones. Because these **essential fatty acids** cannot be made from carbohydrate, they must be obtained in the diet from plant and fish oils.

- the fat-soluble vitamins A, D, E and K are normally only present in fatty foods.

## Amino acids

Though animals eat proteins, it is the constituent amino acids that are actually needed by the cells. Although many proteins do not contain all 20 types of amino acid, cells need the full range to produce all their proteins. Eleven amino acids are not essential in the diet because they can be produced in the liver from other amino acids by **transamination**. In this process, an amino group is transferred from an amino acid to a keto acid. For example, alanine can transfer its amino group to α-ketoglutaric acid to form pyruvic acid and glutamic acid:

$$\text{alanine} + \alpha\text{-ketoglutaric acid} \rightarrow \text{pyruvic acid} + \text{glutamic acid}$$

Nine **essential amino acids** cannot be produced by transamination and must be present in the diet (Table 26.3). It is worth noting that the term 'non-essential' refers only to the diet; in the biochemical sense *all* amino acids are needed for making proteins.

**Table 26.3**
Amino acids essential in the human diet. *Though some arginine and histidine can be produced in the body, they are essential for normal growth in infants because they cannot make enough for themselves.

| Amino acid |
|---|
| Arginine* |
| Histidine* |
| Isoleucine |
| Leucine |
| Lysine |
| Methionine |
| Phenylalanine |
| Threonine |
| Tryptophan |
| Valine |

In general, plant proteins contain fewer essential amino acids than animal proteins, but by eating different kinds of plant food, deficiencies can be avoided. For example, cereals are low in lysine but contain adequate methionine, whilst legumes (such as peas and beans) are low in methionine but have sufficient lysine. By eating legumes with cereals, deficiencies in amino acids can thus be avoided. They must, incidentally, be eaten at the same meal, since amino acids cannot be stored.

The relative deficiency of some amino acids in plant food means that herbivores would be deficient in some amino acids were it not for micro-organisms in their gut. These not only digest cellulose, but also synthesise the full range of amino acids from those in the diet. Although the bacteria use the amino acids to build up their own proteins, the amino acids become available to the animal when the bacteria are digested (Chapter 37).

The amount of protein needed in the diet depends on four factors:

- body size – larger people need more than smaller people.
- age – young, growing people need more than adults of similar size.
- protein quality – strict vegetarians may have to eat more protein to ensure that they get enough of the essential amino acid in shortest supply relative to its requirements.
- amount of carbohydrate and fat – if these are in short supply, more energy is obtained from amino acids so less are available for making proteins. Carbohydrates and fats are thus 'protein sparers'; a person may become deficient in protein if the diet is deficient in fat or carbohydrate.

## Fibre

In recent decades it has become clear that indigestible plant matter, called **roughage** or **fibre**, is an important part of the human diet. Fibre consists of the indigestible substances making up plant cell walls, such as cellulose, hemicelluloses, lignin and pectin, and is important *because* it cannot be digested. By adding 'bulk' to the food, fibre speeds up its transit through the gut (see the information box on page 466).

## The importance of fibre

Diseases such as cancer of the colon, coronary heart disease, gallstones, diverticulitis and conditions such as obesity and constipation, are almost unknown in societies which eat unrefined vegetable foods. Epidemiological studies (studies of the correlation between the incidence of diseases and the environment) show that the risk of such diseases is greatly increased if the diet is low in fibre.

All plant foods contain fibre, but some are better sources than others. Particularly good sources include bran, coconut, peas and beans, broccoli, fruit, wholemeal bread and brown rice.

How does fibre have these beneficial effects? The time for indigestible food to travel from mouth to anus is called the **transit time.** On a low fibre diet, transit times may average more than 48 hours, but with adequate fibre it is reduced to 30 hours or less. It is thought that by decreasing transit time, fibre reduces the exposure of the colon to carcinogens produced by colon bacteria. It is also known to reduce the absorption of cholesterol.

Too much fibre can also be harmful. Fibre can bind certain minerals such as iron, copper, and zinc, reducing their absorption.

## Inorganic ions or 'minerals'

Most minerals have a catalytic role as enzyme activators, which explains why most are needed in very small amounts. A total of 20 inorganic ions are needed, including sodium, potassium, calcium, iron, magnesium, phosphate, chloride, manganese, cobalt, zinc and copper.

### *Calcium*

About 99% of the body's calcium is present as calcium phosphate in the bones, and as such is essential for the growth and mechanical strength of the skeleton and teeth. More fundamental than this are its varied chemical roles. For example, it triggers muscle contraction and the release of neurotransmitters, and is essential in blood clotting.

Dietary requirements for calcium are higher during growth and pregnancy and also during lactation, since milk production represents a loss of calcium from the body. In men and in women who are not 'eating for two', calcium deficiency is rare because it occurs in so many foods. Good sources include cereals, vegetables and especially milk and cheese.

The plasma concentration of calcium is maintained by regulating its uptake from the gut and the balance between its removal from and deposition within the bones. A fall in plasma calcium results in an increase in its absorption from the gut, and release from the bones. A rise in plasma calcium causes decreased absorption from the gut and thus increased excretion, and its deposition in the bones.

Calcium homeostasis is under the control of two hormones, **parathyroid hormone** and **calcitonin**, and also **Vitamin D**. Parathyroid hormone is secreted by the **parathyroid glands** (four small patches of tissue embedded in the thyroid) and stimulates the release of calcium ions from the bones into the plasma. Calcitonin has the opposite effect and is produced by cells in the thyroid gland. The role of Vitamin D is described later in this chapter.

The mineral content of bone begins to decrease in middle age. In women this is associated with the menopause, and oestrogen is known to play a part in maintaining bone calcium.

### *Iron*

By far the greatest amount of iron is present in the body as haemoglobin, but it is also a component of the cytochromes in mitochondria. Shortage results in the most common form of **anaemia**. Though it is present in most diets, only a small proportion of iron is absorbed from food, mostly from cereals. What constitutes a good source, therefore, depends on how readily the iron is absorbed as well as its amount. Liver, kidney and red meat are the best sources. Milk contains very little, and babies have to rely on iron stored in the liver to tide them over until they are on solid food. Women need more iron than men because they lose blood each month during menstruation.

### *Phosphate*

Besides being a mineral component of bone, phosphate is a key constituent of nucleotides. These are the units from which DNA and RNA are built, and are also constituents of ATP and cofactors such as NAD and FAD. As a constituent of phospholipids, phosphate is also an essential constituent of cell membranes. Because of its importance in living matter it is present in a wide variety of foods.

### *Sodium*

Sodium is the principal cation in the blood and tissue fluid of vertebrates and other animals with a blood system. Ionically, it is balanced chiefly by chloride and hydrogencarbonate, and together these ions enable the inside of the cells to be in osmotic equilibrium with their surroundings. Sodium plays a key part in the development of action potentials in excitable cells such as nerve and muscle. Its concentration in extracellular fluids is maintained by the kidney (Chapter 24).

Sodium is so widespread in foods that its deficiency only occurs during prolonged sweating without replacement of salt. Most people actually take in far more salt than they need, partly due to the addition of salt to many processed foods.

## Vitamins

Until early this century it was believed by most nutritionists that the only essential constituents of the human diet were proteins, carbohydrates, fats, minerals and water. There were some, however, who believed that other materials may be necessary. One of the first to put this idea to the test was Frederick Gowland Hopkins, whose classic experiment is the subject of Question 2 on page 484. Hopkins showed that certain organic substances are also necessary in the diet, though in minute quantities.

At first these substances were thought to be amines, and their name was shortened from 'vital amines' to **vitamins**. It was later shown that there were several vitamins, some water-soluble and some fat-soluble. Before their chemical formulae were established they were given letters of the alphabet (e.g. A, $B_1$, $B_2$, C, D, E and K). Their chemical formulae are now known as are, in most cases, their biochemical roles. Most (possibly all) vitamins act as enzyme cofactors or their precursors, a fact that is in keeping with their minute dietary requirements.

Lack of a particular vitamin interferes with metabolism and results in characteristic **deficiency symptoms**.

Vitamins (or the cofactors made from them) are recycled. Even so, they are slowly lost from the body and need to be replaced in the diet. Deficiency symptoms may take some time to appear because some vitamins, such as A, D, and $B_{12}$, are stored in the liver.

The daily requirements vary with the vitamin, but are measured in milligrams, micrograms or even nanograms. Some vitamins can be made in the body so their dietary requirements are not easy to define. Vitamin K for instance is produced by bacteria in the colon, and Vitamin D is produced in the skin when it is illuminated with ultraviolet light. Some, such as Vitamins A and D, are toxic in high doses.

Certain vitamins are unstable to cooking and other treatments. Ascorbic acid is easily oxidised when heated, especially in alkaline solution. Riboflavin is also unstable in alkaline conditions and is destroyed by ultraviolet light.

Some substances are vitamins for one kind of organism but not for others. Ascorbic acid, for instance, is a vitamin for monkeys and apes, but not for most other mammals, which synthesise it for themselves. Most vitamins are produced by plants, and pass along the food chain with other nutrients. Vitamin $B_{12}$, required for the production of deoxyribonucleotides and hence for cell division, is absent from plants but is produced by micro-organisms. Herbivores obtain their supplies from mutualistic bacteria in their gut, and carnivores obtain it from their prey.

### *Vitamin A (retinol)*

Vitamin A is fat-soluble and is required for the synthesis of the photosensitive pigments in the rods and cones in the retina. The rods are more severely affected by deficiency which causes **night blindness** – the inability to see in dim light. The vitamin is also necessary for the healthy functioning of epithelia, because deficiency causes hardening and drying of mucous membranes and a reduction in resistance to infection. Its exact biochemical role in epithelia is not known.

Whilst animal foods contain the vitamin itself, in green leaves it is available as the provitamin β-carotene, which is converted in the intestinal mucosa into two molecules of Vitamin A.

Good sources of the vitamin (or carotene) include fish liver oil, liver, eggs, carrots and green leaf vegetables.

### *Thiamine (Vitamin $B_1$)*

Thiamine is the precursor of a cofactor in an enzyme complex involved in the entry of pyruvate into the Krebs cycle. As such it occupies a central role in metabolism. Deficiency leads to **beri beri**, characterised by muscular weakness and nerve paralysis. Good sources include yeast, liver and meat.

### *Riboflavin (Vitamin $B_2$)*

Riboflavin is a precursor of the respiratory cofactor **flavine adenine dinucleotide** (**FAD**). It is thus of fundamental importance in oxidative phosphorylation in respiration. It is found in foods derived from metabolically-active tissues, such as liver, heart, kidney, green leaf vegetables, yeast and egg yolk (food for a metabolically-active embryo).

### *Nicotinic acid (Vitamin $B_3$)*

Nicotinic acid, or 'niacin', is a water-soluble vitamin used to produce **nicotinamide adenine dinucleotide** (**NAD**). This is a hydrogen acceptor in respiration (Chapter 5), and also a hydrogen donor in a number of reduction reactions. It can be produced from dietary tryptophan (an essential amino acid) so provided there is an adequate intake of tryptophan, the vitamin may not be needed in the diet.

Nicotinic acid occurs widely, but rich sources include yeast, wheat germ, nuts, fish, lean meat, peas and beans. Deficiency results in **pellagra**

(Italian for 'rough skin'), characterised by malfunctioning of the skin and gut lining, and also psychological disturbances. Pellagra is especially common in societies in which the staple crop is maize, as maize is low in tryptophan so this tends to be diverted into production of proteins rather than nicotinic acid.

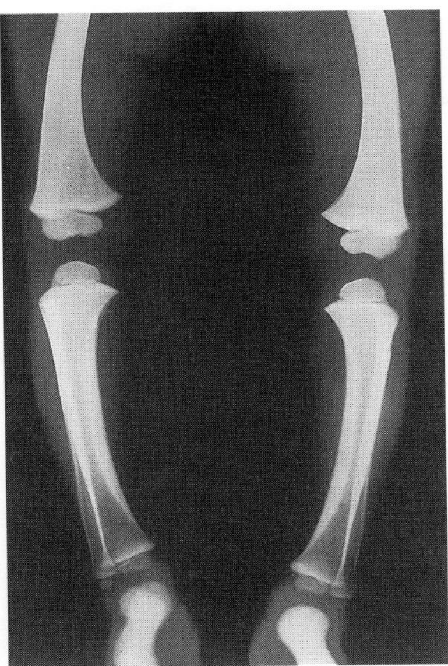

**Figure 26.2**
An X-ray showing the weakened bones and bowed legs of a child suffering from rickets

## *Ascorbic acid (Vitamin C)*

Ascorbate is not a vitamin for most animals since they can produce it for themselves. Early in the evolution of monkeys and apes (including humans) however, this ability was lost, and in these animals it is essential in the diet. Good sources are fresh green vegetables and fruit. Potatoes are less rich in ascorbate than many fruits, but they are an important source for many people because they form a major part of the diet.

The precise biochemical role of ascorbate is uncertain, but it is known to be essential for the formation of collagen and deficiency leads to **scurvy**, in which the connective tissues do not function properly, and there is leakage of blood from the capillaries.

## *Calciferol (Vitamin D)*

Vitamin D is required for the absorption of calcium from the gut and its deposition in the bones. In children, deficiency results in **rickets**, in which bones grow in abnormal shapes (Figure 26.2). Adults develop **osteomalacia**, in which the bones are of normal shape but are weak.

Vitamin D is actually one of a group of related steroids, the metabolically-active form being produced by two sequential steps in the liver and kidney. The vitamin itself (Vitamin $D_3$) is **cholecalciferol**, a steroid. This is produced by ultraviolet irradiation of **7-dehydrocholesterol** in the skin, a plant steroid present in most diets. Cholecalciferol is then converted by two steps into **1,25-dihydroxycholecalciferol**, which promotes the uptake of calcium from the gut and thus a rise in plasma calcium.

These conversions are subject to several feedback controls. Parathyroid hormone stimulates the production of 1,25-dihydroxycholecalciferol in the kidney, which results in a rise in plasma calcium. This in turn inhibits the secretion of parathyroid hormone (Figure 26.3).

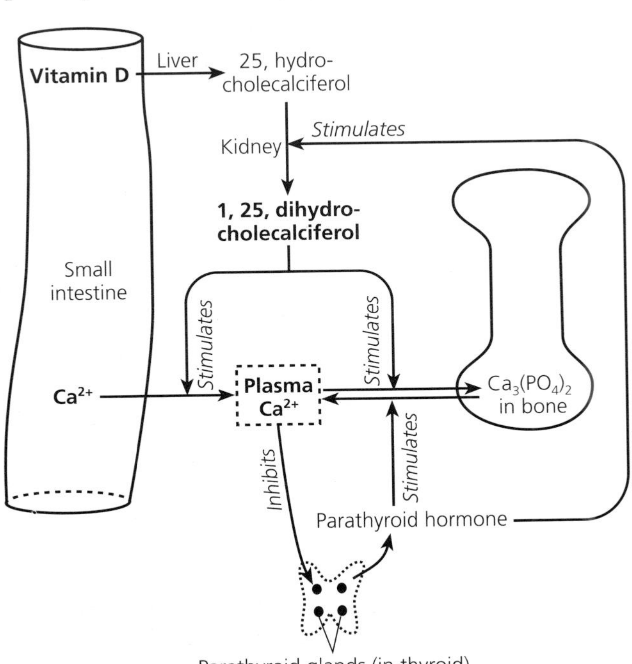


**Figure 26.3**
Control of calcium uptake from the gut

If the calcium level falls, parathyroid hormone output rises, stimulating the production of 1,25-dihydroxycholecalciferol and increasing the uptake of calcium from the gut. Increased plasma calcium has the reverse effect.

### *Water*

Though water is both produced and used in metabolism, it is generally regarded as a nutrient. In ways explained in Chapter 2, it plays an essential part in maintaining the structure of membranes and the tertiary and quaternary structure of globular proteins. It also performs other roles such as that of a transport medium in the blood and a raw material for the various hydrolytic reactions of digestion.

## The balanced human diet

Though all people need all nutrients, the amounts required vary with a person's circumstances. For example a pregnant or nursing mother needs extra amounts of most nutrients, particularly protein and calcium.

A balanced diet contains all the necessary nutrients in amounts and proportions appropriate to the specific needs of the individual. A diet that is unbalanced is a source of **malnutrition**. In poor countries malnutrition most often results from a combination of inadequate energy and protein, while in affluent countries it generally results from too much energy.

In a balanced diet the proportions of the various nutrients are just as important as the actual amounts. The situation is complicated by the fact that the required amount of one nutrient may depend on the intake of another. For example, dietary proteins supply the amino acids for synthesis of cell proteins, but they can also be used as sources of energy. If there is insufficient carbohydrate and fat in the diet, a higher proportion of the amino acids are used as fuel rather than in protein synthesis. A shortage of carbohydrate and fat may thus lead indirectly to the symptoms of protein deficiency.

The greatest source of imbalance in Western society is too much energy intake in relation to expenditure. The additional energy is stored as fat, causing overweight or **obesity**. Obesity is strongly linked to cardiovascular disease and thus life expectancy.

Just as important as the amount of energy in the diet is the form in which it is taken in. Much research into diet and disease suggests that saturated fatty acids present in animal fats, and cholesterol, both lead to increased risk of atherosclerosis, in which arteries become partly blocked by fatty deposits in their walls (see Chapter 22).

# The mammalian alimentary canal

Like most animals, mammals feed on other living things or parts of them. Chemically the food consists of proteins, fats, polysaccharides and other materials too complex to enter the body directly. Food must first undergo a process of **digestion**, in which complex foodstuffs are broken down by enzymes into simpler ones which are then absorbed into the body. This takes place in a tube called the **alimentary canal**, or **gut**, which extends from the mouth to the anus (Figure 26.4).

Before it can truly become part of the body the food has to undergo a number of processes in sequence.

- **Ingestion**, the taking in of food into the mouth.

- **Chewing**, by the teeth into smaller pieces. This makes the food easier to swallow and greatly increases the surface area over which digestive enzymes can subsequently act. Though sometimes misleadingly called 'physical digestion', chewing is fundamentally different from digestion, which is a *chemical* process.
- **Digestion**, the hydrolysis of large molecules into smaller, soluble ones.
- **Absorption**, the transfer of soluble food items from the gut cavity across the gut lining. The gut cavity is a deep tubular intucking of the outside world, and, since this is technically outside the body, food does not actually enter the body until it has crossed this boundary layer.
- **Assimilation**, the incorporation of the digested food into the cells.

Undigested food is **egested** as faeces via the anus. Though egestion is often confused with excretion (as explained in Chapter 24) they are quite different, since faeces consists mainly of matter that has not been produced in metabolism.

## Digestive enzymes

Digestion is a series of hydrolytic reactions in which larger molecules are broken down into smaller ones. Although molecules in the food fall into four classes – carbohydrates, triglycerides (fats), proteins and nucleic acids, over 30 different hydrolases are required to digest a normal meal. This is because enzymes are specific for breaking certain kinds of linkage.

Take carbohydrases, for example. The most abundant carbohydrate in the human diet is starch. This contains $\alpha$ 1,4 and $\alpha$ 1,6 links, which require different enzymes to split. There are also several disaccharides to be digested, each of which requires a different enzyme.

Proteins need even more enzymes for digestion. **Endopeptidases** (for example pepsin) attack internal linkages and yield polypeptides. There are several different endopeptidases, each specific for the kind of amino acid adjacent to the bond it breaks. **Exopeptidases** break terminal peptide bonds to split off single amino acids. Those that split off the amino acid at the carboxyl end of the chain are **carboxypeptidases**, whilst those that attack the amino end of the chain are **aminopeptidases**.

There is a danger that these enzymes might digest the organ that secretes them, so proteases are all produced in an inactive form called a proenzyme, or **zymogen**. When it reaches the cavity of the intestine, this is activated by enzymatic removal of a short section of the polypeptide chain.

## General structure of the gut wall

Though different parts of the gut are specialised for different functions, all parts have the same basic layers, shown in Figure 26.5:

- the **mucosa** – the lining of the gut. This consists of an epithelium which in parts is extended into deep, glandular intuckings. Like the skin, the gut lining is constantly renewing itself by cell division.
- the **muscularis mucosae** – a thin layer of smooth muscle at the base of the mucosa.
- the **submucosa** – a layer of connective tissue and glands which are actually intuckings of the mucosa. It also contains a network of neurons which regulate the activity of the muscle and glands.

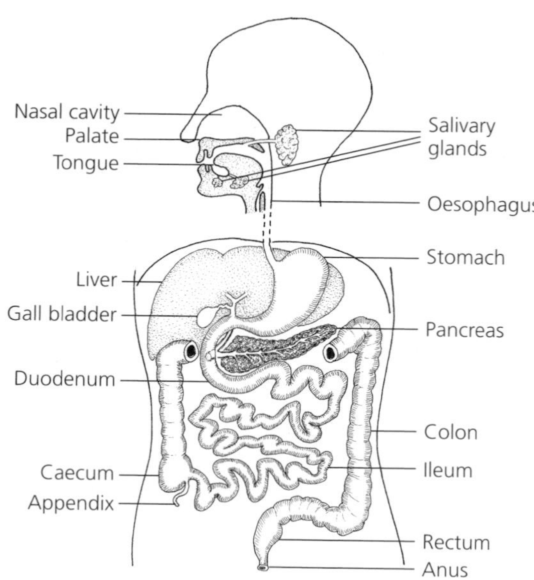


**Figure 26.4▲**
The human alimentary canal

- a layer of **circular muscle**.
- a layer of **longitudinal muscle**.

### The muscle of the gut

Food is propelled along the alimentary canal by the co-ordinated, antagonistic activity of the longitudinal and circular muscle. Gut muscle resembles cardiac muscle in three important ways:

1 contraction is **myogenic**, meaning that the muscle contracts spontaneously without the need for nervous stimulation.

2 it has a *dual innervation*, with one set of nerves stimulating it and the other inhibiting it. The role of the nerve supply is thus to speed up or damp down *endogenous* (internally originating) activity.

3 the individual muscle cells are linked to each other by numerous **gap junctions** which allow action potentials to pass directly from cell to cell, so that the entire gut musculature is an **electrical syncytium**.

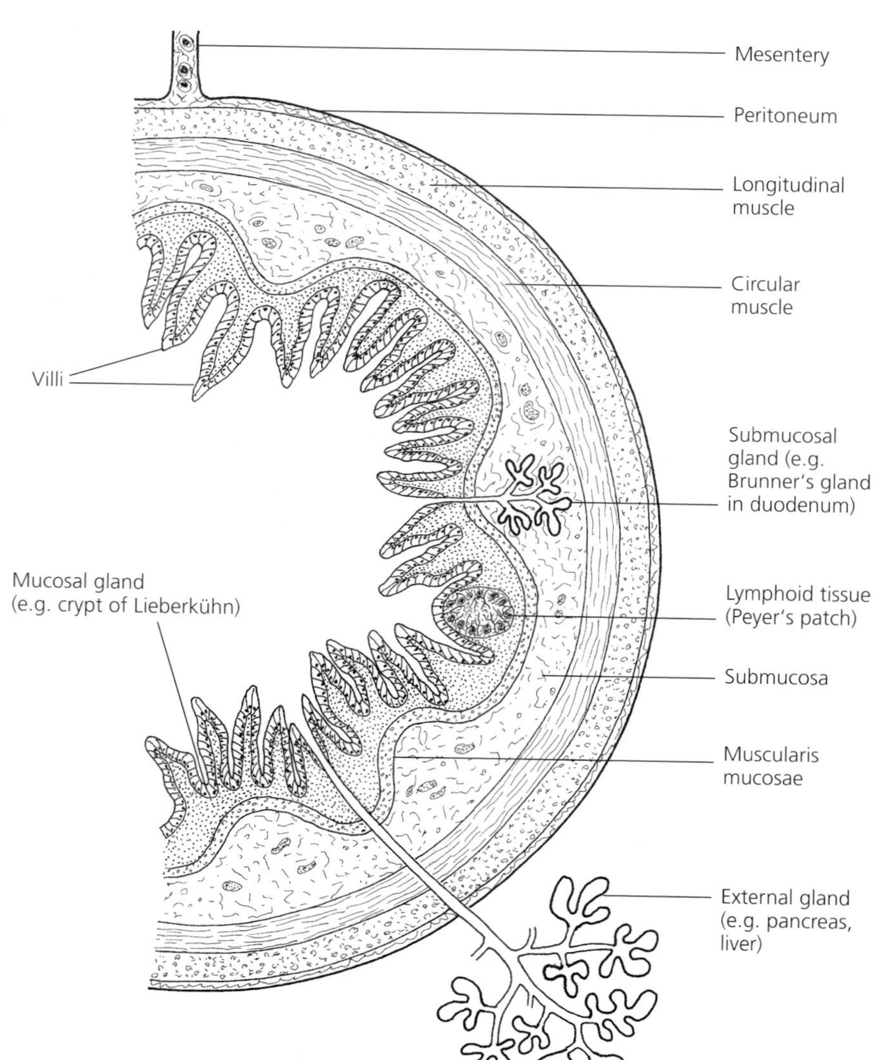


**Figure 26.5▶**
General tissue layout of the intestine

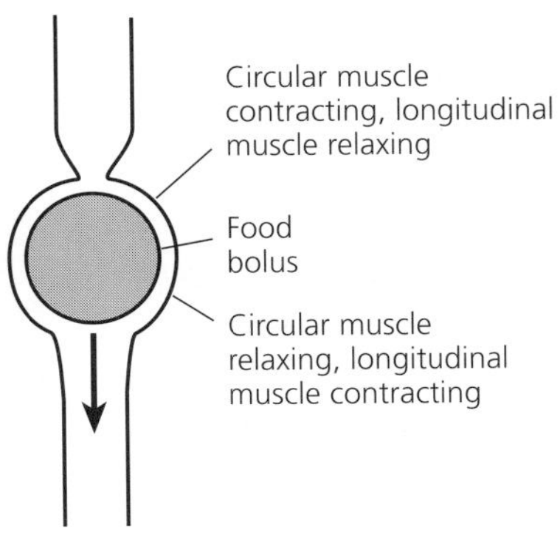


(A) Peristalsis in the oesophagus

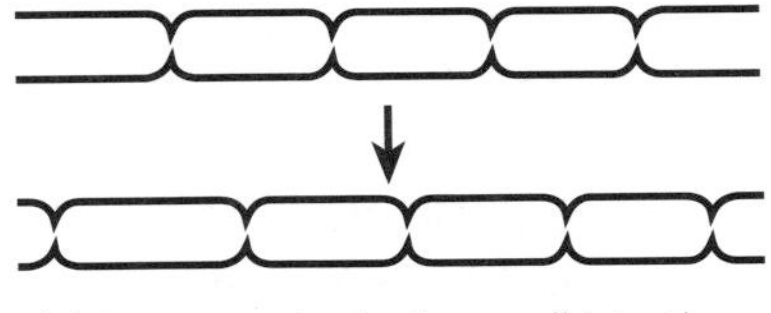

(B) Segmentation in the small intestine

**Figure 26.6**
Movements of the gut

The muscle of the gut performs three basic functions: propulsion, mixing and retention. Food is propelled by **peristalsis,** in which waves of contraction of the circular muscle are preceded by contraction of the longitudinal muscle (Figure 26.6A). In the small intestine food is also mixed by movements called **segmentation,** in which the intestinal contents are 'chopped' into sections by contraction of bands of circular muscle a short distance apart (Figure 26.6B). In certain parts of the gut the muscle is organised into strong, circular bands called **sphincters,** which temporarily *prevent* movement of contents to the next region by sustained contraction.

### The mesentery

Like the heart and lungs, the stomach and intestines are suspended in a narrow, fluid-filled cavity, which allows them to move without interference from neighbouring organs and skeletal muscle. The abdominal cavity is enclosed by a membrane which is folded back on itself, so that the lining of the abdominal wall is continuous with the covering of the gut. This membrane is called the **peritoneum,** and where it is doubled back on itself, it forms a double layer called the **mesentery,** in which are found blood vessels, lymphatics and nerves.

### The digestive secretions

Altogether the food is subjected to the action of five juices, whose total output is about 7 $dm^3$ per day. This large volume is necessary because the soluble products of digestion lower the solute potential of the gut contents, so without this dilution water would enter the gut from the blood by osmosis.

Secretion of digestive juices is under both autonomic and endocrine control. The duration of the various stages increases along the gut, requiring more sustained stimulation and, accordingly, the emphasis on hormonal control increases. Thus digestion in the mouth, which is of relatively short duration, is entirely under nervous control, whilst that in the small intestine, which lasts several hours, is predominantly under the control of hormones.

## Food processing in the mouth

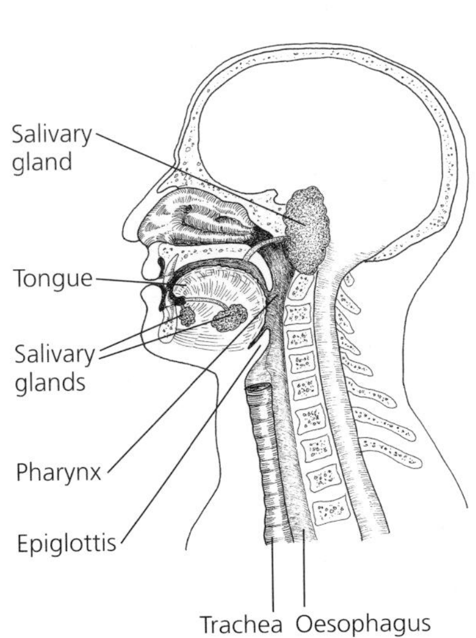


**Figure 26.7**
Vertical section of the head showing the location of the salivary glands

In the mouth, food is broken up by chewing. This not only facilitates swallowing – it also increases the surface area for the subsequent action of enzymes in the various digestive juices. The first of these is **saliva,** a neutral liquid secreted by three pairs of **salivary glands** (Figure 26.7).

About 1.5 $dm^3$ of saliva is secreted per day. It contains **mucin,** a slimy mucopolysaccharide which helps lubricate the food in preparation for swallowing. Though the saliva of most mammals contains no enzymes, human saliva contains the enzyme **salivary amylase,** which hydrolyses alternate $\alpha$ 1,4 links in starch. Amylose is converted to maltose and amylopectin is converted to **dextrins** by removal of successive maltose molecules as far as the 1,6 links at the branching points. Though food is normally only chewed for a short time, the action of salivary amylase continues for a while in the stomach, until the acidic gastric juice penetrates the food and inhibits its action.

Saliva acts as a solvent, enabling food to be tasted, and by keeping the mouth moist it helps speech. It buffers acids produced by bacterial action on sugars. In this way it protects the enamel from erosion, which is the prelude to decay. It also helps the tongue to work the food into a soft lump, or **bolus,** and lubricates its passage down the oesophagus during swallowing.

Secretion of saliva continues at a low rate even when fasting, but its output greatly increases during eating. The salivary glands are controlled by both sympathetic and parasympathetic branches of the autonomic nervous system. Secretion is stimulated by the taste or smell of food or, when hungry, even by the thought of it.

### *Swallowing*

Swallowing is the propulsion of food from mouth to stomach. After working the food into a bolus, the tongue pushes it to the back of the mouth and into the **pharynx**, where the trachea and food channels meet. There would be a risk of food going down the 'wrong way' into the trachea, were it not for certain protective reflexes. As the pharyngeal muscles push the bolus downwards, breathing is inhibited. At the same time the larynx rises up, covering the glottis (entrance to the larynx) with the flap-like **epiglottis**. The bolus is pushed into the **oesophagus**, a muscular tube leading to the stomach.

Though the oesophagus is normally closed, its highly folded walls permit considerable expansion during swallowing. It is lined with stratified epithelium which, like the epidermis, allows some resistance to mechanical wear and tear. The bolus is propelled by peristaltic contractions and its passage is lubricated by mucus secreted by glands and individual goblet cells in its walls. After about half a minute the bolus reaches the stomach.

## Digestion in the stomach

The stomach is a muscular, 'J'-shaped bag. It is highly distensible, enabling food to be eaten less frequently but in larger quantities. The stomach liquefies the food, which is then 'fed' to the small intestine at a steady rate over a relatively long period. A person can survive after removal of the stomach, but food has to be taken in much smaller amounts and more frequently.

The gastric mucosa contains millions of **gastric pits**, into which **gastric glands** open (Figure 26.8). These secrete **gastric juice**, which has a number of important functions. First, it liquefies the food by partially digesting it. This is important because the small intestine can only handle liquid food. Second, it sterilises the food. The thin lining of the gut would be an ideal invasion route for pathogenic organisms were it not for the gastric juice, which has a pH of about 2 – low enough to kill most bacteria. Some pathogens, such as the bacteria which cause typhoid, cholera and other diseases spread by food and drinking water, are able to survive the acid.

About 2.5 $dm^3$ of gastric juice is secreted per day. It contains two major constituents: pepsin and hydrochloric acid.

- **Pepsin**. This endopeptidase is produced by the **peptic**, or **chief**, **cells** of the gastric glands. There are three very similar forms of this enzyme. All convert longer polypeptide chains into shorter ones by breaking peptide links adjacent to aromatic amino acids such as tyrosine. Pepsin has an optimum pH of about 2–3, provided by hydrochloric acid. Pepsin is particularly important in disintegrating animal food due to its action on collagen, which forms an internal supporting framework of animal tissues.

  Pepsin is secreted as an inactive precursor, **pepsinogen**. Under the influence of hydrochloric acid, this is converted into pepsin by cleavage of a short section of polypeptide which 'unmasks' the active site of the enzyme. Pepsin can itself activate more pepsinogen, so the process is self-accelerating or **autocatalytic**.

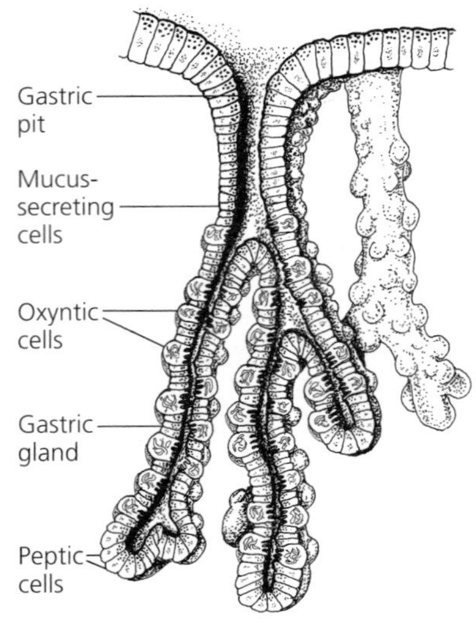


**Figure 26.8**
A gastric pit and gastric glands

**Why doesn't the stomach digest itself?**

Gastric juice readily digests tripe, which is the wall of a cow's stomach, so why doesn't it digest the living stomach that secretes it? There are several protective mechanisms. First, the stomach lining secretes a layer of thick alkaline mucus which helps to separate it from the gastric juice. Second, the cells of the stomach lining are continuously dividing, so that the entire lining renews itself every 3–4 days. The third mechanism depends on the fact that pepsinogen and its activator, hydrochloric acid, are produced by different cells, so pepsin is probably not formed in significant amounts until the juice reaches the stomach cavity. Even without this safeguard, pepsin is inactive at the pH of about 7.2 that exists inside cells.

- **Hydrochloric acid**. This is produced by **oxyntic** or **parietal cells** of the gastric glands. Besides killing bacteria, hydrochloric acid activates pepsin (see below) and provides the low pH necessary for its action. It also catalyses the hydrolysis of disaccharides such as maltose and sucrose (albeit rather slowly at body temperature).

The secretion of hydrochloric acid involves the enzyme **carbonic anhydrase**. Under its influence, carbon dioxide produced in the oxyntic cells combines with water to produce carbonic acid, which dissociates into $H^+$ and $HCO_3^-$ ions. The $H^+$ ions are extruded from the side of the cell nearest the stomach cavity, with $Cl^-$ ions following passively. On the other side of the cell, $HCO_3^-$ ions are pumped out into the blood accompanied by $Na^+$ ions. The $NaHCO_3$ secreted into the blood causes it to become more alkaline.

In calves and some other unweaned mammals (but probably not in human babies), gastric juice contains **rennin**, an enzyme that converts the soluble milk protein **caseinogen** into insoluble **casein**, allowing it to be retained in the stomach long enough for it to be digested by pepsin.

Gastric juice also contains **intrinsic factor**, a protein essential for the absorption of Vitamin $B_{12}$ from the small intestine. Complete removal of the stomach thus necessitates injections of the vitamin.

During its stay in the stomach the food is subjected to peristaltic action which mixes it and gradually converts it into a creamy liquid called **chyme**. The exit from the stomach, the **pylorus**, is guarded by a sphincter muscle which is always in a state of at least partial contraction. The sphincter allows liquids to leave the stomach but ensures that large lumps of solids are retained. With the arrival of each peristaltic wave, a small squirt of chyme enters the duodenum. The emptying of the stomach is controlled by reflexes which ensure that chyme leaves the stomach at a rate at which it can be accepted by the duodenum.

The length of time it takes the food to pass through the stomach depends on the kind of food. Liquids tend to go straight through, while food rich in fat (which inhibits gastric emptying) may be retained for several hours.

### *Control of gastric secretion*

The gastric glands are under both nervous and hormonal control. The brain communicates with the stomach via parasympathetic fibres of the vagus nerve. Following the taste, smell and even the thought of food, the vagus nerve endings stimulate the gastric glands to secrete, so some gastric juice is already present in the stomach before food arrives. Since this first phase of gastric secretion is mediated entirely by the brain it is called the **cephalic phase**.

The second phase, or **gastric phase**, of secretion is mediated by both nervous and endocrine mechanisms. It begins with the chemical and mechanical stimuli associated with the arrival of food in the stomach. Stimulation is conveyed to the gastric glands by three routes (Figure 26.9):

- impulses are carried directly to the gastric glands by the nerve network in the submucosa. In these **local reflexes**, excitation is entirely confined to the stomach wall.
- excitation is also carried to the brain by afferent neurons, and back to the gastric mucosa by fibres of the vagus nerve.
- activation of the local nerve network also leads to secretion by cells in the mucosa of a polypeptide hormone **gastrin**. After circulation in the blood, this brings about a sustained secretion of gastric juice.

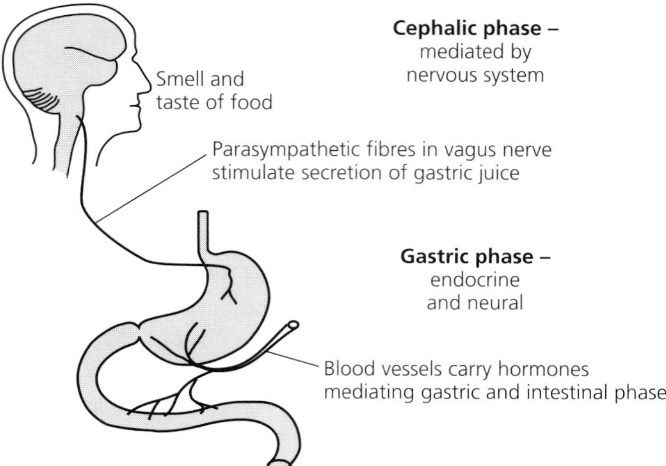


**Figure 26.9**
Control of gastric secretion

## Digestion in the small intestine

The small intestine averages about 3 metres long in life (though after death, relaxation of the longitudinal muscle causes it to extend to about 7 metres). Three juices enter the small intestine: bile, pancreatic juice and intestinal juice. Digestion in the small intestine occurs under slightly alkaline conditions, and all three juices secreted into the small intestine are made alkaline by sodium hydrogencarbonate. The sodium ions are later reabsorbed with the products of carbohydrate and protein digestion.

The first part of the small intestine is the **duodenum**, so-called because it is supposed to have a length approximately the width of 12 fingers (Latin: *duodecim* = twelve). The duodenum differs from the rest of the small intestine by the presence of coiled, tubular **Brunner's glands** in the submucosa (Figure 26.10). These secrete an alkaline mucus which helps protect the mucosa against the acidity of the chyme.

Opening into the duodenum is a duct formed by the joining of the bile duct and the main pancreatic duct (Figure 26.11).

### *Bile*

Bile differs from other digestive juices in that it contains no enzymes, yet it is essential for the digestion of fat and absorption of the products. It is produced continuously in the **liver** and temporarily stored (and also concentrated) in the **gall bladder**. The contraction of the gall bladder is stimulated by the hormone CCK (see the information box on page 478).

About 0.5 $dm^3$ of bile is secreted per day. It contains the following constituents:

- **sodium hydrogencarbonate**, which makes it alkaline and therefore helps in the neutralisation of the acidic chyme.
- **bile salts**. These are derivatives of steroids and act like detergents, converting the large globules of fats into minute droplets about 1 μm in diameter called **micelles** (Figure 26.12). In the resulting **emulsion** the fat has a greatly increased area of contact with the lipase in the surrounding aqueous solution. Each time the diameter of fat droplets is halved, the total surface area is doubled, so the conversion of a 1 cm diameter globule to droplets 1 μm diameter increases the surface area 10 000 times.
- **bile pigments**, which give bile its golden-yellow colour. These are excretory products derived from the breakdown of haemoglobin in worn out red corpuscles.

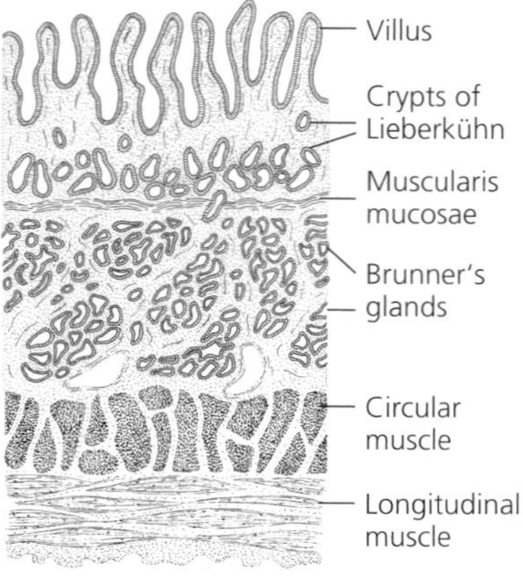


**Figure 26.10**
Fine structure of the duodenum

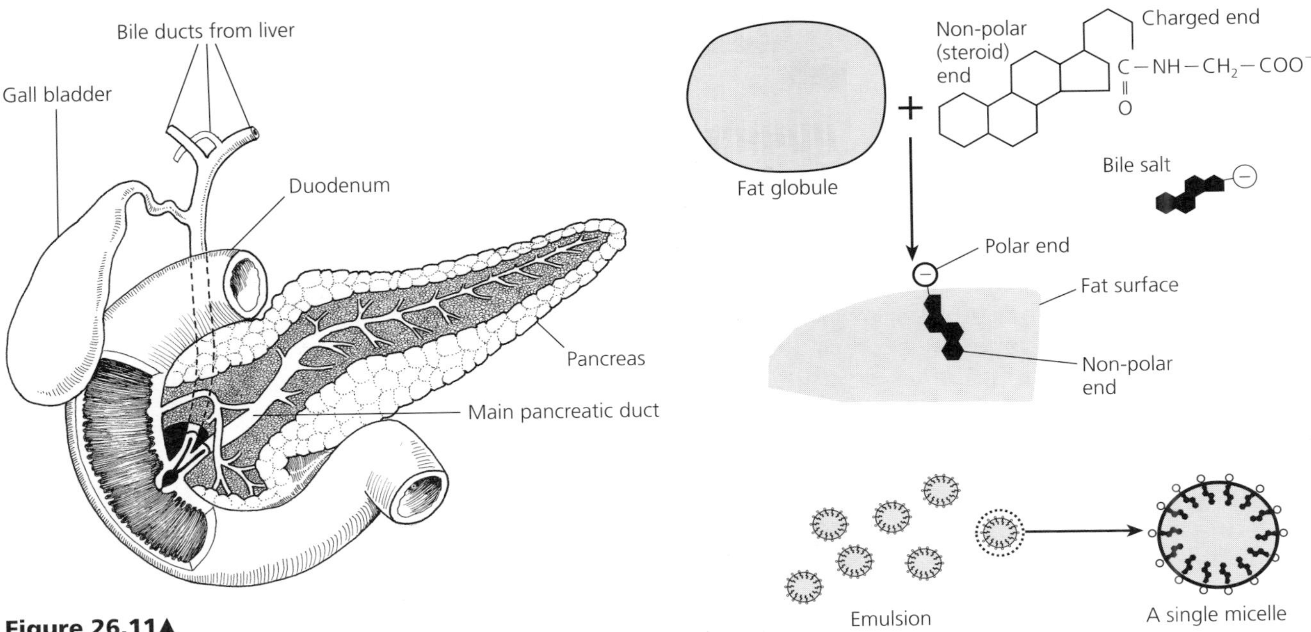


**Figure 26.11▲**
The duodenum, pancreas and associated ducts (after Bell, Davidson and Scarborough)

**Figure 26.12▲**
Emulsification of fats by bile salts

- **cholesterol**. This is fairly insoluble in bile, and on occasions is precipitated in the gall bladder as **gallstones**.

## *Pancreatic juice*

The pancreas is the size of a banana, and daily produces about 1.5 $dm^3$ of an alkaline juice rich in enzymes. These include a number of proteases, all of which (like pepsin) are secreted as inactive precursors. One, **trypsin**, is secreted as **trypsinogen**, and plays a key role in activating the rest. When it reaches the small intestine it is activated by **enteropeptidase** (enterokinase), an enzyme produced by the intestinal mucosa. Once formed, trypsin can convert more trypsinogen into trypsin so, like pepsin activation, the process is autocatalytic. Trypsin also activates all the other pancreatic proteases: **chymotrypsinogen** is converted into **chymotrypsin**, **proelastase** to **elastase** and **procarboxypeptidase** to **carboxypeptidase**.

Though all proteases attack peptide bonds, they show some degree of specificity. Trypsin, chymotrypsin and elastase, for example, are **endopeptidases** since they cleave internal peptide bonds. Each has some specificity – trypsin attacks links adjacent to basic amino acids such as arginine and lysine for instance. Carboxypeptidase, on the other hand, is an exopeptidase since it splits off individual amino acids at the carboxyl end of the polypeptide chain.

A number of other enzymes are secreted by the pancreas. **Lipase** hydrolyses fats into monoglycerides and fatty acids. **Amylase** hydrolyses $\alpha$ 1,4 links in starch, converting amylose to maltose and amylopectin to dextrins (the $\alpha$ 1,6 links in amylopectin prevent its complete conversion to maltose). **Deoxyribonuclease** hydrolyses DNA to deoxyribonucleotides, and **ribonuclease** hydrolyses RNA to ribonucleotides.

In descending the rest of the small intestine, there is a gradual change in microscopic structure, and, on this basis, the upper two-fifths and lower three-fifths are sometimes distinguished as the **jejunum** and the **ileum**, respectively. There is, however, no sharp boundary between them.

## Integration of stomach, pancreas and liver

On entering the duodenum, chyme is mixed with pancreatic juice, bile and intestinal juice. How is the emptying of the stomach integrated with the secretion of these juices?

Both nervous and endocrine mechanisms are involved. Two peptide hormones play an important role. The presence of acid stimulates the duodenal mucosa to produce **secretin**, which stimulates the secretion of the alkaline component of pancreatic juice and bile. The more acid that enters the duodenum, therefore, the more alkali enters it as a result – an example of **negative feedback**. Another feedback effect of secretin is the inhibition of peristalsis in the stomach, slowing the delivery of chyme to the duodenum.

A second peptide hormone is **cholecystokinin-pancreozymin**, or **CCK-PZ**, often shortened to CCK. It is secreted by the duodenal mucosa in response to the presence of fatty acids and monoglycerides. Its double-barrelled name stems from the fact that it was once believed that cholecystokinin (CCK) and pancreozymin (PZ) were different hormones, but they are now known to be one and the same. CCK stimulates the secretion of the enzymic component of pancreatic juice, and also stimulates the contraction of the gall bladder.

The entry of chyme into the duodenum also triggers a number of nervous reflexes involving the vagus nerve. One of the most important is the inhibition of gastric emptying. As a result, the stomach empties at a rate at which the chyme can be processed; any tendency for the stomach to be emptied too quickly automatically inhibits further emptying – an example of negative feedback. Another reflex response to the arrival of chyme in the duodenum is the contraction of the gall bladder.

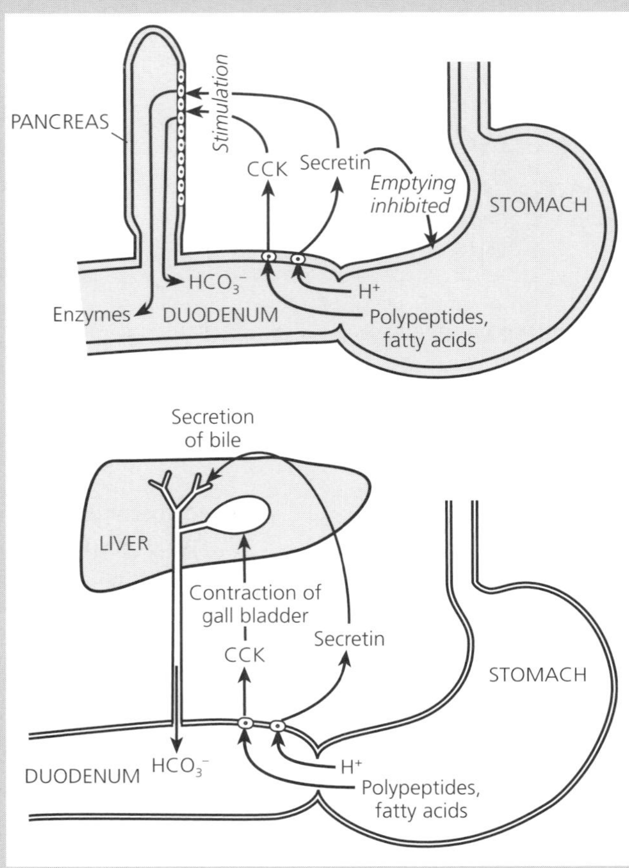


Hormonal feedback between stomach and pancreas (above) and between stomach and liver (below)

The small intestine is the main site of absorption of digested food. The mucosa is thrown into many transverse folds which increase its area about threefold. Its surface is increased another 10-fold by (in humans about 5 million) finger-like or leaf-like **villi**. With a density of about 30–40 $mm^2$, the villi make the intestinal lining resemble a tubular carpet (Figure 26.13). The area of the columnar epithelium covering the villi is increased another 20 times by thousands of **microvilli**, forming a **brush border**. The total area of the lining of the small intestine is about 250 $m^2$, or about 300 times greater than its external area.

Each villus is about 1 mm long and about 0.1 mm wide and contains blood capillaries, and lymphatics called **lacteals**. It also contains a thin strand of smooth muscle which enables it to be waved around and allows it to undergo contractions that help propel lymph out of the lacteals.

Between the bases of the villi are the openings of deep **intestinal glands**, or **Crypts of Lieberkühn** (Figure 26.14). These secrete about

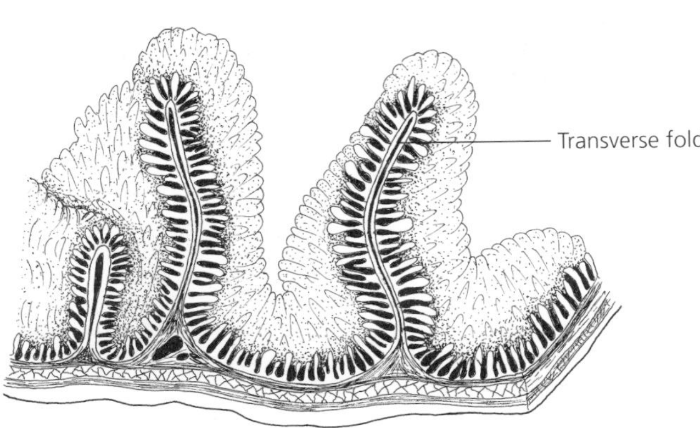

(A) Cutaway view of intestinal lining

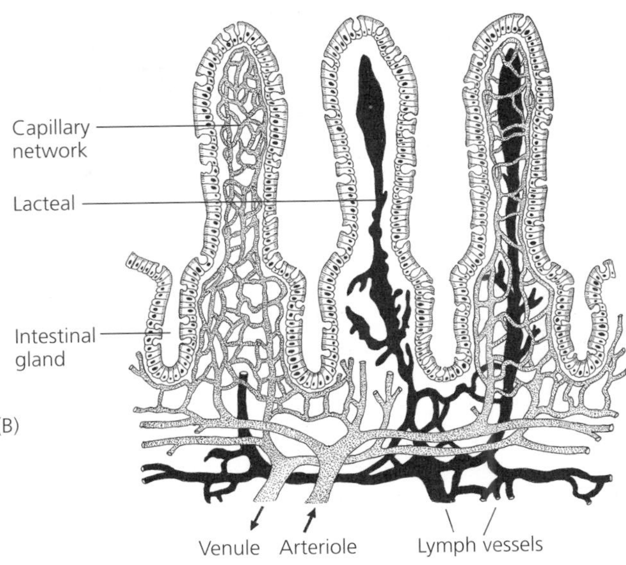


**Figure 26.13**
Two views of the lining of the small intestine to show how its area is increased (A) by folds and (B) by villi (in one villus the lacteal has been omitted; in another, the capillaries) (after Crouch)

1 $dm^3$ per day of an enzyme-free juice containing mucus and salts. The cells lining the bottom of the crypts are in a continuous state of division, so that the new cells are continuously being pushed up by the division of cells deeper down. After a life of about 5 days, the cells are sloughed off when they reach the tips of the villi.

### *The intestinal enzymes*

At one time it was believed that the intestinal glands produce a digestive juice ('succus entericus') containing enzymes in solution. Actually, enzymes produced by the intestinal mucosa are not secreted, but remain bound to the microvilli of the cells that produce them. Amongst these enzymes are the following:

- **enteropeptidase**, which triggers the activation of the pancreatic proteases, starting with trypsin.
- **maltase**, which hydrolyses maltose to glucose.
- **lactase**, which hydrolyses lactose to glucose and galactose.
- **sucrase**, which hydrolyses sucrose to glucose and fructose.
- **α-dextrinase**, which removes the branches of dextrins by breaking the α 1,6 links.
- **aminopeptidases**, which split amino acids away from the amino end of polypeptide chains.
- **dipeptidases**, which hydrolyse dipeptides into amino acids.

The final stages of digestion thus occur on the surface of the same cells that are responsible for absorption, so the products are absorbed almost instantaneously (Figure 26.15).

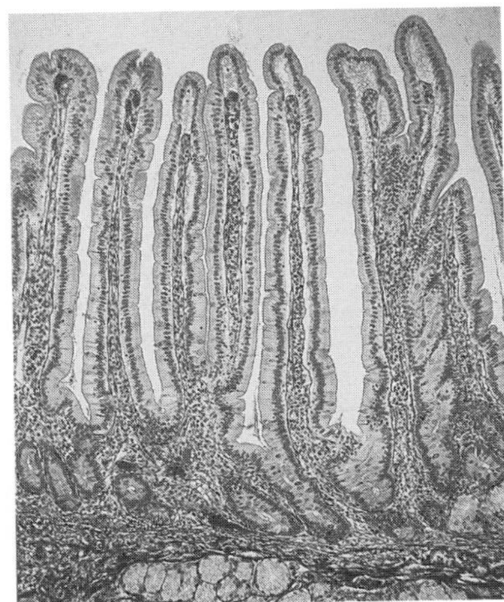

**Figure 26.14**
Section through the wall of the human small intestine showing finger-like villi and crypts of Lieberkühn – the circular structures at their bases

## Absorption of the digested food

As a result of digestion, carbohydrate is converted to glucose with small amounts of fructose and galactose, proteins are broken down into tri- and dipeptides and some free amino acids and fats are broken down into monoglycerides and fatty acids.

Glucose and galactose are actively absorbed by a pump involving cotransport with sodium ions, whilst fructose is absorbed by facilitated

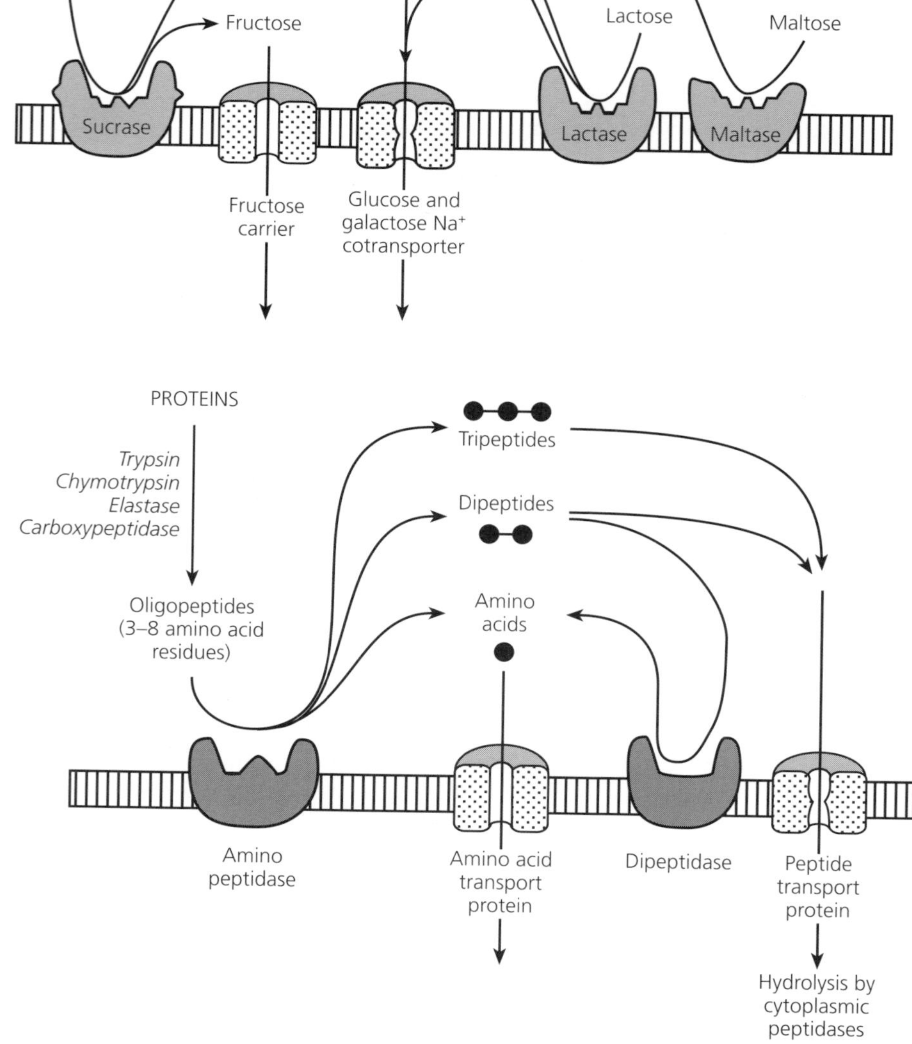


**Figure 26.15**
Digestion and absorption of disaccharides (above) and peptides (below)

diffusion. Di- and tripeptides and amino acids are cotransported with sodium ions into the epithelial cells where the peptides are hydrolysed to amino acids.

Once absorbed, glucose and amino acids enter the capillaries in the villi, after which the blood carries them to the liver via the **hepatic portal vein**.

Fatty acids and monoglycerides are absorbed rather differently. In the intestine they combine with bile salts to form minute complexes called **micelles**, about the size of an average globular protein molecule. Each bile salt molecule acts like a detergent, with its polar end facing outward into the water and its non-polar end associating with the fatty acids and monoglyceride. At the surface of the microvilli, the fatty acids and monoglycerides leave the micelles and enter the epithelial cells by diffusion. Short chain fatty acids, which are more water-soluble, enter the blood directly. Longer chain fatty acids combine with monoglycerides to form triglycerides. These become coated with lipoprotein and enter the lacteals as tiny droplets called **chylomicrons**. In the lower ileum most of the bile salts are reabsorbed into the blood and recycled to the liver.

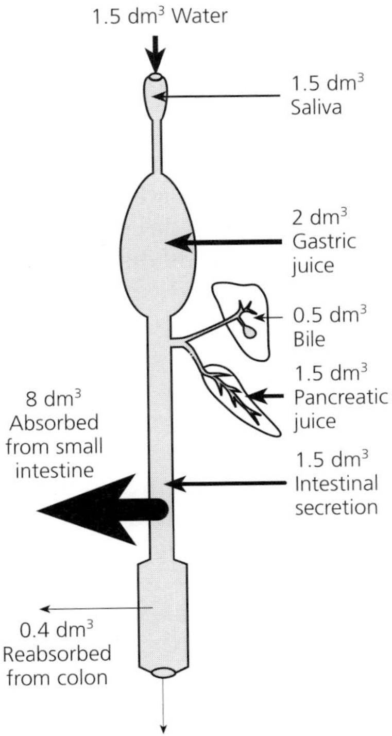


**Figure 26.16▲**
Daily water budget for the human alimentary canal. Of the daily total of 8.5 $dm^3$ entering the gut (1.5 $dm^3$ in food and drink and 7 $dm^3$ as digestive juice), about 95% is absorbed from the small intestine, most of the remainder being absorbed from the colon. (Actual figures vary depending on the circumstances)

### *The absorption of water and ions*

Though they are not the products of digestion, ions are essential nutrients and must be absorbed. Each has its own specific active transport mechanism.

The volume of digestive juice secreted into the gut in each 24 hour period is about 7 $dm^3$ – too much to be easily replaced by drinking. About 95% of the water is reabsorbed from the small intestine, the remainder from the colon (Figure 26.16). Water reabsorption is passive, with the water moving down the osmotic gradient created by the absorption of solutes such as sugars and ions.

## The large intestine

The large intestine (Figure 26.17) is about 1 metre long (though it extends to about 1.5 metres after death due to loss of muscle tone). In humans the **caecum** and **appendix** are small with little function.

Though it has no digestive role, the colon has the important function of absorbing most of the remaining water. Its mucosa contains many mucous glands but has no villi – most of the digested food has already been absorbed (Figure 26.18).

The colon contains huge numbers of bacteria. The minority that are facultatively aerobic (use oxygen if present) such as *Escherichia coli*, use up all the oxygen. This creates ideal conditions for the huge numbers of obligate anaerobes such as *Bacteroides* and *Clostridium*. Some gut bacteria are beneficial in that they produce Vitamin K. Most people are partially dependent on this source because they do not obtain enough in the diet. The bacteria also produce gases such as carbon dioxide and methane.

By the time it reaches the rectum most of the remaining water has been reabsorbed, leaving the rest of the material to be egested as **faeces**. Faeces consist largely of bacteria and sloughed off epithelial cells, together with undigested cellulose. The brown colour of faeces is due to bacterial action on bile pigment, and the unpleasant smell is due to chemicals such as skatole, also produced by bacteria.

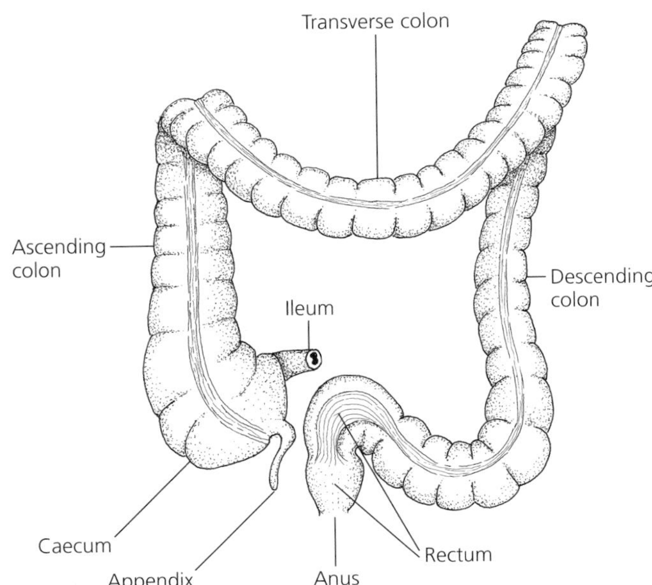


**Figure 26.17**
The large intestine

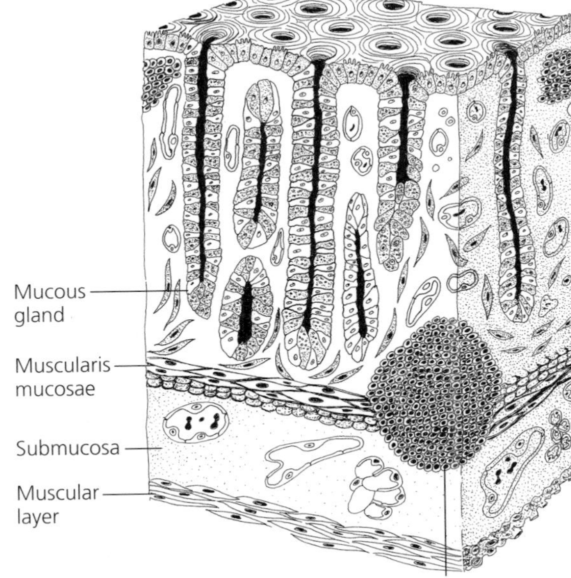


**Figure 26.18**
Fine structure of the colon

**Functions of the liver**

The liver carries out a number of diverse functions, most of which are essential for life:

- storage of glycogen.
- production of glucose from amino acids with the formation of urea.
- conversion of glucose to fatty acids.
- breakdown of most hormones.
- inactivation of drugs such as penicillin, and toxins, such as alcohol.
- synthesis of non-essential amino acids by transamination.
- production of bile.
- production of all plasma proteins except antibodies.
- breakdown of worn out red corpuscles.
- storage of certain vitamins, A, $B_{12}$, D and K, and minerals, such as iron and copper.
- production of cholesterol.

## The assimilation of the digested food: the liver

After passing through the epithelium of the intestinal mucosa, the products of protein and carbohydrate digestion are carried to the liver via the **hepatic portal vein**. The liver thus receives blood from two sources – the heart and the gut (Figure 26.19). This unusual arrangement is linked to the fact that the liver is important in the homeostatic regulation of the concentration of glucose in the blood leaving the gut. Being directly 'downstream' of the intestine, the liver can exert a more immediate influence on the blood than it could if the blood went to the liver *after* the heart.

### Structure of the liver

Except for the skin, the liver is the largest organ in the body and takes about 25% of the resting cardiac output. It consists of many functional units called **lobules**, each about a millimetre in diameter (Figure 26.20). At the centre of each lobule is a branch of the hepatic vein called the **central vein**. At the edges of the lobule are branches of the hepatic artery, hepatic portal vein, bile duct and a lymphatic vessel. Radiating from the central vein are irregular plates of liver cells called **hepatocytes**. Between the hepatocytes are fine **bile canaliculi**, which converge to form branches of the bile duct. Running between the plates of hepatocytes are blood spaces called **sinusoids** containing phagocytic **Kupffer cells**. Blood enters the sinusoids from branches of the hepatic portal vein and hepatic artery and flows into the central vein.

### The fate of glucose

On reaching the liver, glucose undergoes one of two possible fates – it may be used as fuel in respiration, either in the liver or in other tissues, or it may be built up into glycogen or fat. Glycogen synthesis occurs mainly in the liver and in skeletal muscle and is called **glycogenesis**. If the blood glucose level falls, glucose is released into the blood by the liver by the breakdown of glycogen, a process called **glycogenolysis**.

The liver cannot store more than about 100 g of glycogen, which is enough to last about 24 hours of resting metabolism. Any additional glucose is converted into fat as indicated in Figure 21.2. Glycerol is produced from triose phosphate, an intermediate on the glycolytic pathway. Fatty acids are produced from acetyl coA, which is produced from pyruvate.

The rates of the various biochemical processes by which glucose, amino acids and fats are added to and removed from the blood are controlled by hormones and are dealt with in Chapter 20.

### The fate of fat

Most fat is used as fuel in respiration although it may, in the meantime, be temporarily stored in adipose tissue. Some fat is used in the synthesis of the lipid components of membranes or to make cholesterol. The glycerol component of fat can be converted into glucose via triose phosphate.

### The fate of amino acids

Amino acids in the blood can be thought of as a 'pool' which is in a continuous state of **turnover**, in which amino acids are simultaneously being added to and removed from the blood (Figure 26.21). Amino acids are added to the pool by two processes: protein breakdown, which occurs continuously in all tissues, and dietary intake from the intestine. Amino acids are removed from the pool by two processes: protein synthesis,

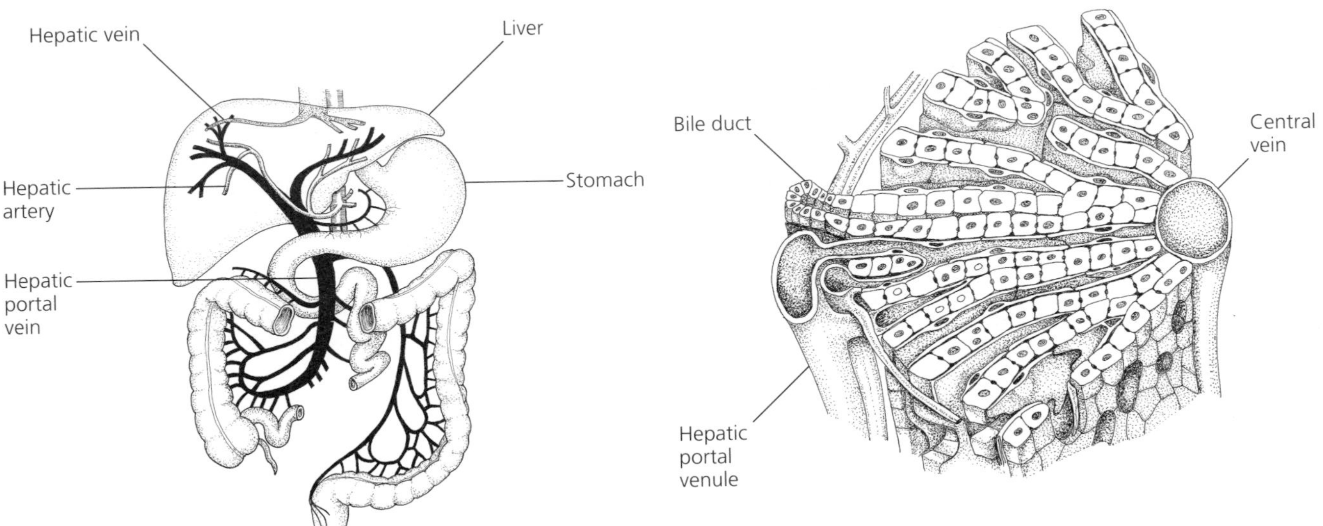


**Figure 26.19**
The liver and its blood supply (after Kappas and Alvares)

**Figure 26.20**
Part of a liver lobule

which occurs continuously in all tissues, and the breakdown of amino acids in the liver through deamination.

### *Amino acids as a source of energy: deamination and urea formation*

Amino acids cannot be used directly in respiration – they must first undergo a process called **deamination**, in which the amino groups are removed. First, the amino acid undergoes a transamination in which its amino group is transferred to α-ketoglutaric acid, forming **glutamic acid** and a keto acid (Figure 26.22). Most amino acids yield keto acids that can be converted into glucose. Others yield keto acids that can be converted into fatty acids. In either case the keto acids can be used for energy.

Amino acids in gut
Amino acids in blood
Tissue proteins
Urea
Keto acids
LIVER
Glucose

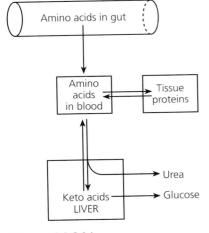

**Figure 26.21▲**
The amino acid pool

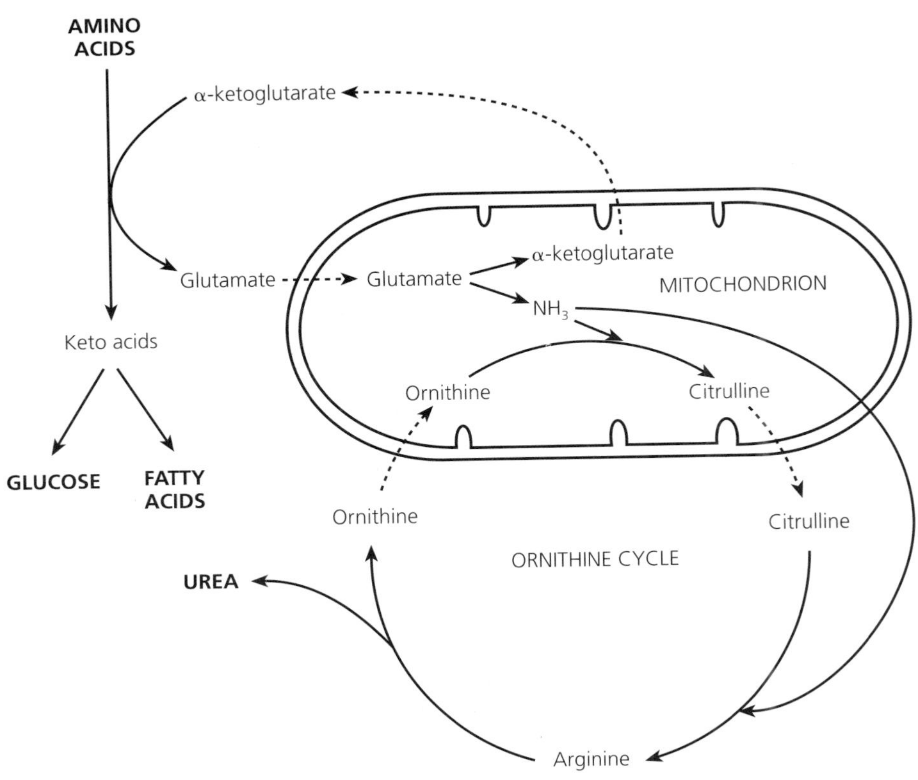


**Figure 26.22►**
How amino acids are used as fuel

The production of glucose from non-carbohydrate sources, such as amino acids and glycerol, is called **gluconeogenesis**. Some amino acids are used as fuel, even when energy intake is adequate. During starvation, most glucose enters the blood in this way. Then glycerol from stored fat and amino acids from tissue protein become the main source of blood glucose. During such times, gluconeogenesis is stimulated by **cortisol**, a hormone secreted by the adrenal cortex. In carnivores, whose diet contains little carbohydrate, gluconeogenesis is normally the main source of glucose.

So much for the carbon part of the amino acids. What about the amino part? As previously mentioned, this combines with α-ketoglutaric acid to form glutamic acid. The latter is then deaminated, losing its amino group as ammonia. Ammonia is extremely poisonous and is immediately converted to the much less toxic **urea** in a series of reactions called the **ornithine cycle**.

The overall process can be summarised thus:

$$\text{amino acid} + H_2O + NAD^+ \rightarrow \text{keto acid} + NH_3 + NADH + H^+$$

$$2NH_3 + CO_2 \rightarrow CO(NH_2)_2 + H_2O$$

The keto acids are then either used in the Krebs cycle or are converted into glucose.

## QUESTIONS

**1** Each of the descriptions **A–G** applies to one (and in some cases more than one) of the components of the human diet. For each of the nutrients **(i)–(vii)** choose the letter of an appropriate description. Each description can only be used once.

| | |
|---|---|
| **i)** fats | **A** give energy |
| **ii)** carbohydrates | **B** synthesised by other living things; do not supply energy |
| **iii)** minerals | **C** all insoluble in water |
| **iv)** fibre | **D** organic; does not provide energy; needed in large amounts |
| **v)** vitamins | **E** inorganic; needed in small amounts |
| **vi)** water | **F** some are soluble; contain carbon, hydrogen and oxygen only |
| **vii)** proteins | **G** inorganic; needed in large amounts |

**2** In an investigation into the nutritional requirements of rats, two groups of eight newly-weaned rats were fed on different diets. Group A was fed on purified casein (milk protein), lard, starch, sucrose, mineral salts and water. Group B was given the same diet but with a daily supplement of 3 cm³ of milk. After 18 days the diets were reversed. The growth of the rats is shown in the graph.

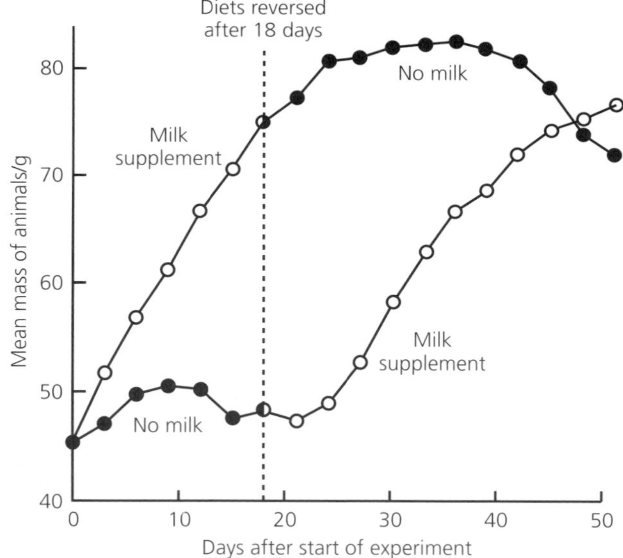


**a)** What hypothesis was Hopkins testing?
**b)** What can you conclude from the graphs?
**c)** Suggest why the effect of the change of diet was delayed in Group B.
**d)** Give *three* features of the experimental design that increased the reliability of the results.

## QUESTIONS

**3** Three newly-weaned rats A, B and C were fed adequate amounts of starch and fat, and were given vitamin and mineral supplements. In addition, Group A was given 1 g of purified maize protein; Group B was given 1 g pea protein and Group C was given 0.5 g maize protein + 0.5 g pea protein. After 3 weeks the average change in mass for each group was as follows:

Group A = −5.6 g
Group B = −4.3 g
Group C = +14.3 g

**a)** Explain these results as fully as you can.
**b)** Which group of rats would be expected to excrete the greatest quantity of urea? Give a reason for your answer.

**4** In an investigation into the action of bile on fat, six tubes were set up with contents as indicated in the table. After shaking, they were left at 37°C for an hour and after this time the contents of each tube were tested for the presence of fat and fatty acid. In the table, '+' indicates presence of a substance and '−' indicates absence.

Give the letter(s) of the tube(s) which indicate that

**a)** fatty acids must have been produced from fat;
**b)** digestion of fat requires the presence of pancreatic juice;
**c)** the active constituent of the pancreatic juice is an enzyme;
**d)** the action of pancreatic juice is enhanced by the presence of bile;
**e)** the active constituent of bile is a non-protein substance.

| | | BEFORE INCUBATION | | | | AFTER INCUBATION | |
|---|---|---|---|---|---|---|---|
| | | Pancreatic juice | | Bile | | | |
| Tube | Fat | Fresh | Boiled | Fresh | Boiled | Fat | Fatty acid |
| A | + | + | − | + | − | − | + |
| B | − | + | − | + | − | − | − |
| C | + | + | − | − | − | + | trace |
| D | + | − | − | + | − | + | − |
| E | + | − | + | + | − | + | − |
| F | + | + | − | − | + | − | + |

**5** In an investigation into the absorption of carbohydrate by the small intestine, standard areas of intestine were immersed in sugar solutions of known molar concentrations, and the initial rates of uptake were measured. Two sugars were used: xylose (a pentose) and galactose (a hexose). The results, in which concentrations and rates are in arbitrary units, are given in the table.

| | Rate of uptake (arbitrary units) | |
|---|---|---|
| Relative concentration | Xylose | Galactose |
| 1 | 0.22 | 2.12 |
| 2 | 0.43 | 3.91 |
| 5 | 1.13 | 4.04 |
| 10 | 2.39 | 4.03 |

**a)** Give *two* ways in which the uptake of these two sugars differs.
**b)** Which of these two sugars appears to be absorbed by simple diffusion? Give a reason for your answer.
**c)** Suppose a similar experiment were done with fructose, which is absorbed by facilitated diffusion. Would you expect the effect of concentration to resemble galactose or xylose? Give a reason for your answer.
**d)** How would you expect the uptake of fructose to be affected by:
**i)** a reduction in temperature from 37°C to 30°C;
**ii)** treatment with cyanide?
Give a reason for each answer.

CHAPTER

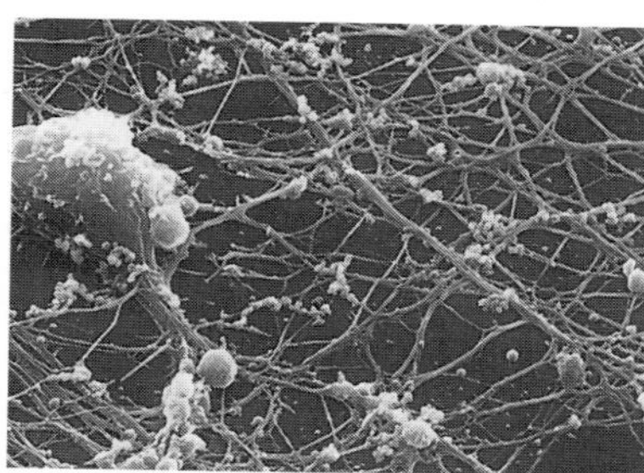

# 27

# DEFENCE AGAINST MICRO-ORGANISMS

## LEARNING OBJECTIVES

By the time you have completed your study of this chapter you should be able to:

- distinguish between innate and acquired resistance.
- list the external barriers to infection and explain how each works.
- describe the destruction of bacteria by phagocytosis.
- explain what is meant by the terms antigen and antibody.
- distinguish between cell-mediated and humoral-mediated immunity, and explain the different roles of T cells and B cells.
- describe the structure of an antibody molecule, and explain the functions of its variable and constant regions.
- list the various ways in which antibodies can lead to the destruction of pathogens, and explain how they interact with phagocytosis and complement.
- explain the essential differences between the clonal selection hypothesis and the instruction hypothesis.
- describe how the immune system learns to distinguish between 'self' and 'non-self'.
- explain what monoclonal antibodies are, and how they can be used to detect specific proteins.

## What is immunity?

One of the most pressing needs for an animal is to avoid being eaten, not only from the outside by predators, but also from the inside by micro-organisms such as bacteria, fungi and various kinds of worms. These organisms are **parasites**, feeding at the expense of another organism, called the **host**. Although many parasites do little harm and live in reasonable harmony with the host, some cause disease and are called **pathogens**.

Hosts have a variety of ways by which they can resist pathogens. These mechanisms may be innate or acquired.

- **Innate resistance** operates without any need for previous contact with a pathogen. It involves mechanisms that are **non-specific**, meaning that they do not distinguish between different kinds of pathogen. Mechanisms of this kind include various barriers that prevent micro-organisms from entering the body, and some chemical mechanisms that act against micro-organisms in general.
- **Acquired** or **adaptive resistance** develops following contact with a particular type of pathogen, and only works for that kind. Acquired resistance or **immunity** in the strict sense, is thus **specific**.

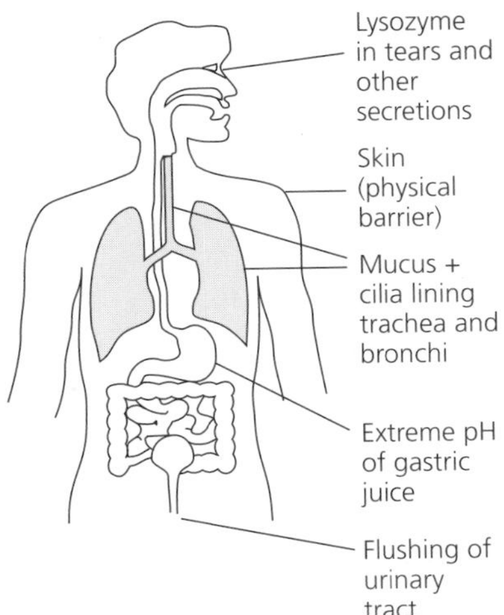


**Figure 27.1**
The main barriers to micro-organisms

Incidentally, immunity is quite different from **non-susceptibility**. Whereas immunity depends on the *presence* of certain defence chemicals, non-susceptibility depends on the *absence* of one or more key requirements of the parasite. To enter a host cell, a virus has to attach itself to a certain chemical marker in the plasma membrane. This specificity of viruses even extends to particular kinds of cell. The human immunodeficiency virus (HIV), for example, can only enter certain kinds of cell because only these have the specific chemical marker that the virus 'recognises'. This is probably why humans do not get myxomatosis, a viral disease that affects rabbits.

## The outer barriers

The body's first line of defence against invaders are its boundary layers, such as the skin and the linings of the breathing passages and gut (Figure 27.1).

### The skin

The outer tissue of the skin, the **epidermis**, is a highly effective barrier to micro-organisms (Figure 25.5). The cells are packed with keratin, a protein that micro-organisms cannot easily digest. The cells in its deeper layers are bound together by many **desmosomes**, making it difficult for micro-organisms to squeeze between them. The outermost cells are continually flaking off, carrying with them any bacteria on them.

The glands of the skin also render it chemically hostile to bacteria. Cells lining the **sebaceous glands** break away into the follicles and break down to produce a yellow, oily **sebum**. This contains unsaturated fatty acids which lower the pH of the skin and inhibit the growth of some bacteria and fungi. **Sweat**, secreted by the **sweat glands**, contains **lysozyme**, an enzyme which dissolves the cell walls of some bacteria. It is also present in tears, saliva, nasal mucus and some internal tissue fluids. Most areas of skin are also unfavourable to bacterial growth due to periodic drying. In certain parts of the body, such as the armpits, ears and genitourinary areas, the skin is moist enough to support a resident population of bacteria. Most are harmless, and some are beneficial because they keep potentially pathogenic micro-organisms in check.

### The alimentary canal

The intestinal lining is thin, warm, wet and richly provided with blood vessels. It would be an ideal invasion route for micro-organisms were it not for the hydrochloric acid in gastric juice which kills most bacteria before they reach the intestine. Yet despite the extremely hostile pH in the stomach, a few pathogens manage to survive. These include the bacteria which cause food- and water-borne diseases like typhoid and cholera, and the infective stages of various parasitic worms.

### The breathing passages

The lining of the lungs is thin, moist and very well supplied with blood vessels – an ideal entry route for pathogenic bacteria. Fortunately, the breathing passages are well protected by a cleaning mechanism, described in Chapter 23.

## Non-specific internal defences

Once inside the body a number of mechanisms help to deal with invaders. Though they work without any previous exposure to a micro-organism, they are closely integrated with the specific immune responses.

### *Phagocytosis*

Micro-organisms that evade the body's outer defences are 'eaten' by **phagocytes**. These constitute one of two main groups of **leucocytes**, or white blood corpuscles. They engulf any small particles, whether living or not. In the lung, for example, they 'eat' inhaled soot particles, leading in extreme cases to 'black lung'.

When a phagocyte comes into contact with a bacterium, it may engulf it by **phagocytosis**. The phagocyte develops outgrowths called **pseudopodia**, which engulf the microbe in a **phagosome** (Figures 6.18 and 27.7). In the phagocyte are lysosomes containing digestive enzymes and a variety of bactericidal substances such as hydrogen peroxide. One or more lysosomes then fuse with the phagocytic vesicle to form a larger digestive vesicle, in which most bacteria are killed and digested. Not all pathogens are killed by this treatment. Some, such as tuberculosis bacteria, may survive and multiply, eventually destroying the phagocyte.

Under certain conditions, phagocytes can respond to pathogens that have previously been identified as foreign and 'labelled' by the immune system. In this connection, phagocytes play an essential role in acquired immunity and are described in more detail later in this chapter.

### *Complement*

Complement is the collective name for a group of about 20 defensive plasma proteins. Certain polysaccharides in bacterial cell walls activate complement proteins, causing them to undergo a 'catalytic cascade'. This leads to the production of **perforin**, a protein that makes holes in the outer walls of many bacteria, causing them to lyse (lose contents) and die.

Like phagocytosis, complement can operate as just described in a non-specific manner, but it also participates in specific defence responses (see below).

### *Inflammation*

Immune cells are widely dispersed around the body, but when there is local infection they converge on the area. The processes by which this occurs are collectively called **inflammation** and are caused by **histamine**, released by **mast cells**. Local blood flow increases due to vasodilatation, causing the area to become red. The capillaries become more permeable, causing an increase in tissue fluid formation and swelling. Nerve endings become more sensitive, so the area is painful to the touch. A most important aspect of inflammation is **chemotaxis**, in which phagocytes move towards chemicals called **chemokines** that are released by cells in the inflamed area.

## Acquired resistance – immunity

Acquired resistance is dependent on the fact that the surface of every cell is chemically unique. This provides the basis for the ability of the body to distinguish between its own cells and those of an invader. Acquired resistance differs in three important respects from innate resistance:

1 it only develops *after* exposure to a pathogen.

2 resistance developed against one kind of pathogen is ineffective against another. Immunity to chicken pox, for example, does not give resistance to measles.

3 a second exposure to a pathogen produces a more powerful, rapid response than the initial one. This **immunological memory** is the basis for vaccination against disease.

## Parts of the immune system

Unlike the blood system, the immune system has no clearly defined boundaries. Despite its diffuse nature, its total mass is comparable to that of a large organ such as the liver. Functionally it is considered to consist of two parts. In the **primary lymphoid tissues** – the **bone marrow** and **thymus gland** – the immune cells complete their development. The mature immune cells then migrate to the **secondary lymphoid tissues** – the **spleen**, **lymph nodes**, **tonsils** and other patches of tissue in the lining of the gut (Figure 27.2).

### *Cells of the immune system*

The immune system consists of various kinds of leucocyte. Though they are called white blood corpuscles, they are only functionally active *outside* the blood, which simply serves to transport them from one tissue location to another. They are divided into three main groups:

- **granulocytes** (polymorphonuclear leucocytes or 'polymorphs'). These have lobed nuclei and granular cytoplasm. The different kinds are named according to whether they stain with neutral, acidic or basic dyes. **Neutrophils** have characteristic multilobed nuclei, and are by far the most abundant, accounting for over 90% of the circulating granulocytes. **Eosinophils** stain with eosin, an acidic dye, and account for 2–5% of the white cells, whilst **basophils** form about 0.2% of white cells.
- **monocytes** and **macrophages**. Monocytes are large cells with kidney-shaped nuclei. They are non-phagocytic, but, after leaving the blood and entering sites of infection, they enlarge and develop into actively phagocytic macrophages. Macrophages are the most effective phagocytes – they are long-lived and capable of eating many more bacteria than neutrophils.
- **lymphocytes** are spherical cells with very little cytoplasm. They look insignificant, and for many years their key role in immunity was unrecognised. It is the lymphocytes that provide the specificity of the immune system. Each lymphocyte has in its plasma membrane many copies of a protein which acts as a receptor. This has a surface with a complex shape and, as with a lock and key, only one other kind of molecule can fit it.

**Figure 27.2**
Parts of the immune system (A) with an enlarged view of a lymph node (B)

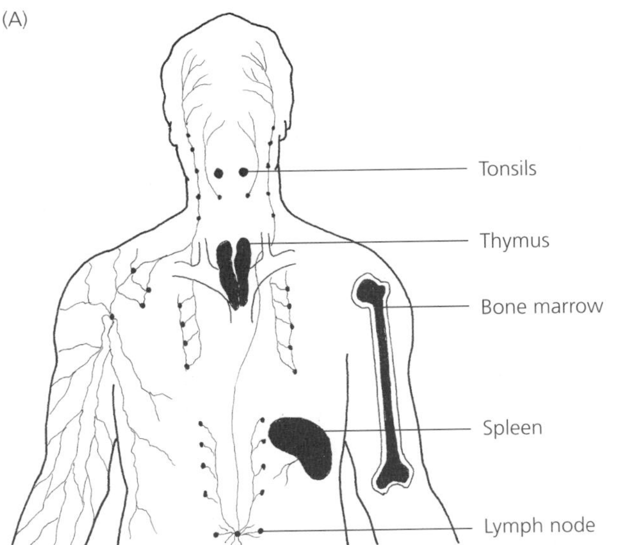


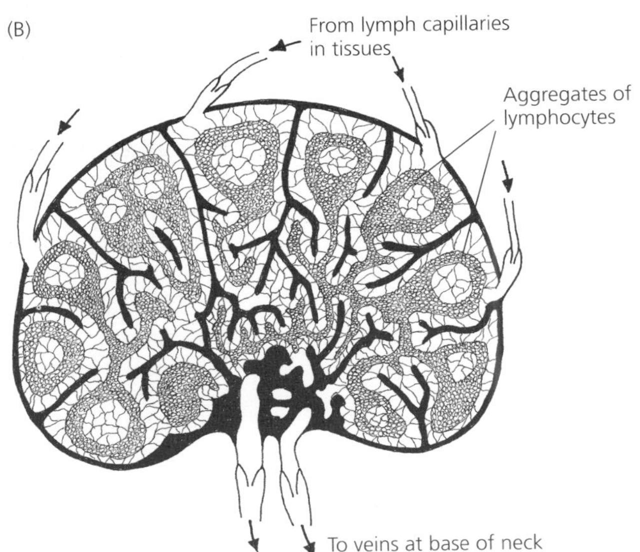

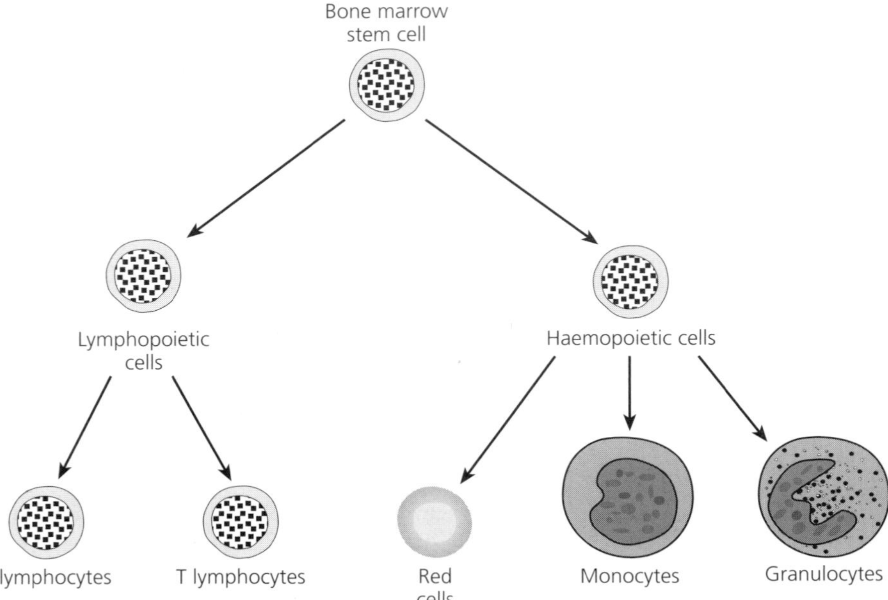


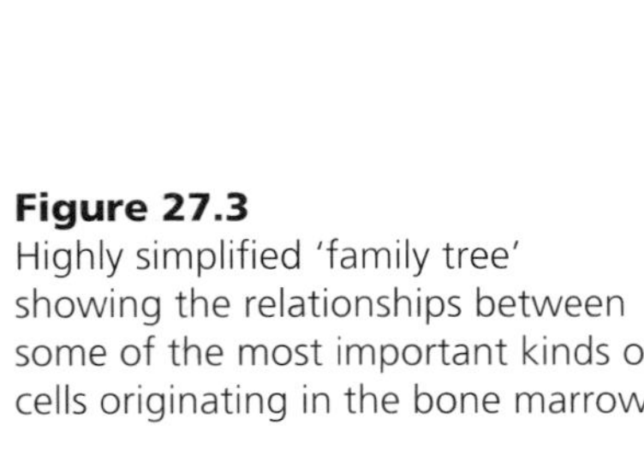

**Figure 27.3**
Highly simplified 'family tree' showing the relationships between some of the most important kinds of cells originating in the bone marrow

All immune cells originate by division of **stem cells** in the bone marrow. Descendants of the stem cells differentiate along three main pathways (Figure 27.3). One line eventually differentiates into red blood corpuscles. Another leads to the various kinds of phagocyte, and the third gives rise to lymphocytes. Because stem cells have the capacity to differentiate into any of several kinds of cell, they are said to be **pluripotent**.

## Antibodies and antigens

A central feature of the immune system is its ability to distinguish the body's own chemicals ('self') from foreign chemicals ('non-self'). It depends on the fact that the surface of every cell and every virus is chemically unique, and therefore distinguishable from the body's own cells. Any substance that the body can recognise as foreign is called an **antigen**. The body responds to the presence of an antigen by producing an equally specific substance called an **antibody**.

The ability of the immune system to distinguish 'self' from 'non-self' is not infallible. It sometimes happens that substances on the surface of the body's own cells trigger an **autoimmune** response. In view of this, it is better to define an antigen as *any substance that can trigger an immune response.*

### *Antigens*

An antigen is a substance that can stimulate the production of an antibody. To do this, it must have enough specificity to distinguish it from other antigens. This is why only fairly large molecules, such as proteins and some polysaccharides, have antigenic properties.

Most antigens are found on the surface of micro-organisms. In a virus, for example, it is the protein coat that is antigenic. Some antigens are bacterial products, for example the toxins produced by tetanus and diphtheria bacteria.

An antibody only binds to a small part of an antigen molecule, so it is possible to chemically alter a non-binding part of an antigen without it losing its antigenic properties. As explained later, this is the basis for some kinds of immunisation in which a bacterial toxin is made harmless without affecting the body's ability to mount an antibody defence.

## Antibodies

Whereas a variety of substances can be antigenic, antibodies are *always* proteins. Antibodies are produced in the lymph nodes by **plasma cells,** which develop from certain kinds of lymphocyte. Antibodies travel in the lymph to the blood, which distributes them round the body (Chapter 22).

Antibodies form the **globulin** fraction of the plasma proteins, and are called **immunoglobulins**. There are several classes of antibody, but all are built on a common plan. The most abundant kind (accounting for 20% of plasma protein) is **immunoglobulin G**, or **IgG**. It consists of two identical halves. Each half consists of two polypeptide chains – a **heavy chain** and a **light chain** – held together by disulphur bonds (Figure 27.4). Both heavy and light chains consist of a **constant portion** that is the same (or nearly so) for all antibodies of a given category, and a **variable portion** that is unique to each antibody.

Functionally an antibody molecule can be divided into two parts. The two variable portions of each limb of the 'Y' provide the specificity for binding to an antigen. Because it has two identical 'recognition' sites, antibody and antigen can form a lattice or network, precipitating antigens or **agglutinating** cells bearing them (Figure 27.5). This is helped by the flexible 'hinge' at the base of each arm of the antibody, which allows the molecule to adopt slightly varying shapes.

Binding to an antigen is only the first step in the functioning of an antibody. Subsequent events, such as phagocytosis and complement activation, are set in motion by the constant part of the molecule.

There are several other classes of antibody with slightly different roles. IgM is the first immunoglobulin to be produced in response to a pathogen. IgA is secreted in saliva, tears and the mucus of the breathing passages and gut, enabling pathogens to be dealt with before they enter the body.

When an antibody binds to an antigen it can have one of several effects:

- it may act as an **opsonin**, rendering bacteria more susceptible to attack by other agents. Phagocytes respond far more vigorously to bacteria that have been coated with antibody (Figure 27.6). Another effect of opsonisation is to render bacteria susceptible to lysis by complement.
- it may bind to the protein coat of virus particles, preventing the viruses from entering new host cells.
- bacterial flagella contain proteins that are antigenic, and are paralysed after binding with antibody.

**Figure 27.4▼**
Structure of immunoglobulin G. The molecule is symmetrical, with an antigen binding site on each arm

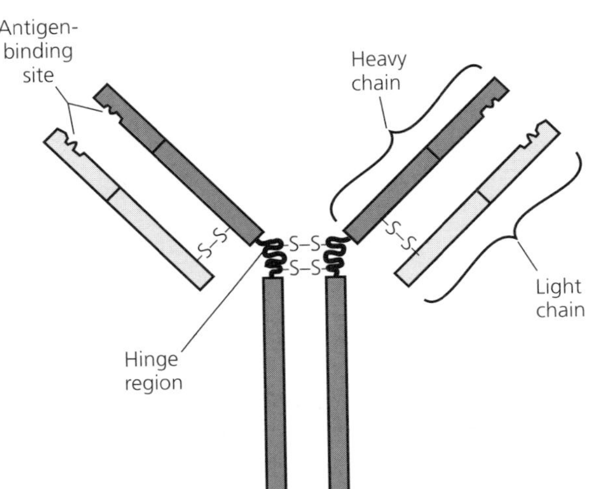


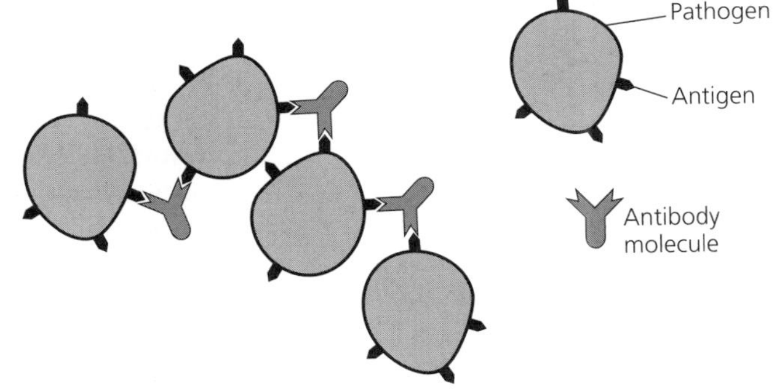


**Figure 27.5▲**
Agglutination of bacteria by antibody–antigen binding. Because each antibody molecule has two antigen binding sites, it can form links with antigen molecules on different invading cells, binding them together in a lattice which, in a test tube, may form a precipitate

- because each antibody molecule has more than one binding site, it can bind to more than one antigen molecule. If the antigens are on different pathogens, bacteria or viruses may be linked together into clumps. This process of **agglutination** restricts the ability of pathogens to spread through the tissues.
- bacterial toxins that are sufficiently complex to act as antigens may be neutralised by antibody. Antibodies acting in this way are called **antitoxins**.

## Two arms to the immune system

There are two quite different (but closely integrated) arms to the immune system, mediated by different groups of structurally indistinguishable lymphocytes, respectively.

**T-lymphocytes**, or **T cells**, interact closely with their target cells and are involved in **cell-mediated immunity**. They are called T cells because they complete their development in the thymus gland ('T' for thymus). In a way to be described later, T-lymphocytes attack cancer cells, cells infected with viruses and also transplanted tissues.

In mammals, **B-lymphocytes**, or **B cells**, complete their development in the bone marrow. They act against pathogens at long range, only coming into contact with them by accident. When provoked, B-lymphocytes develop into cells which produce antibodies. Because antibodies are carried to their targets by the blood and lymph, the B-lymphocytes are involved in **humoral immunity** (*humor* is Latin for 'liquid').

## Trial and error or made to measure?

How does the immune system 'tailor' an antibody to fit a specific antigen? In the first half of this century, there were two competing hypotheses. According to one view, each antigen is directly involved in the production of the antibody. This **instructional hypothesis** stated that the presence of antigen somehow enables the shape of the antibody to be moulded to fit it. In some way, then, the antigen 'teaches' the immune system how to produce the appropriate antibody. An antibody, according to this hypothesis, is made *after* exposure to an antigen.

A completely different idea, the **clonal selection hypothesis**, was put forward independently by Sir McFarlane Burnet and David Talmage in 1957. They proposed that a wide variety of antibodies are produced *before* exposure to an antigen. In the presence of an antigen, any cell that just happens *by chance* to be producing the correct antibody is stimulated to multiply, producing a line of genetically-identical descendants called a **clone** (Figure 27.7). As the number of cells in the clone increases, the production of the antibody rises.

Experimental evidence has shown that both humoral and cell-mediated arms of the immune system operate by clonal selection rather than by instruction. A lymphocyte that just happens to produce an antibody fitting the antigen is stimulated to divide and become the founder of a clone of descendants.

### *Antibody diversity*

The immune system is believed to be capable of producing millions of different antibodies. Even when presented with an artificially-made antigen that no organism has encountered before, it produces a complementary antibody.

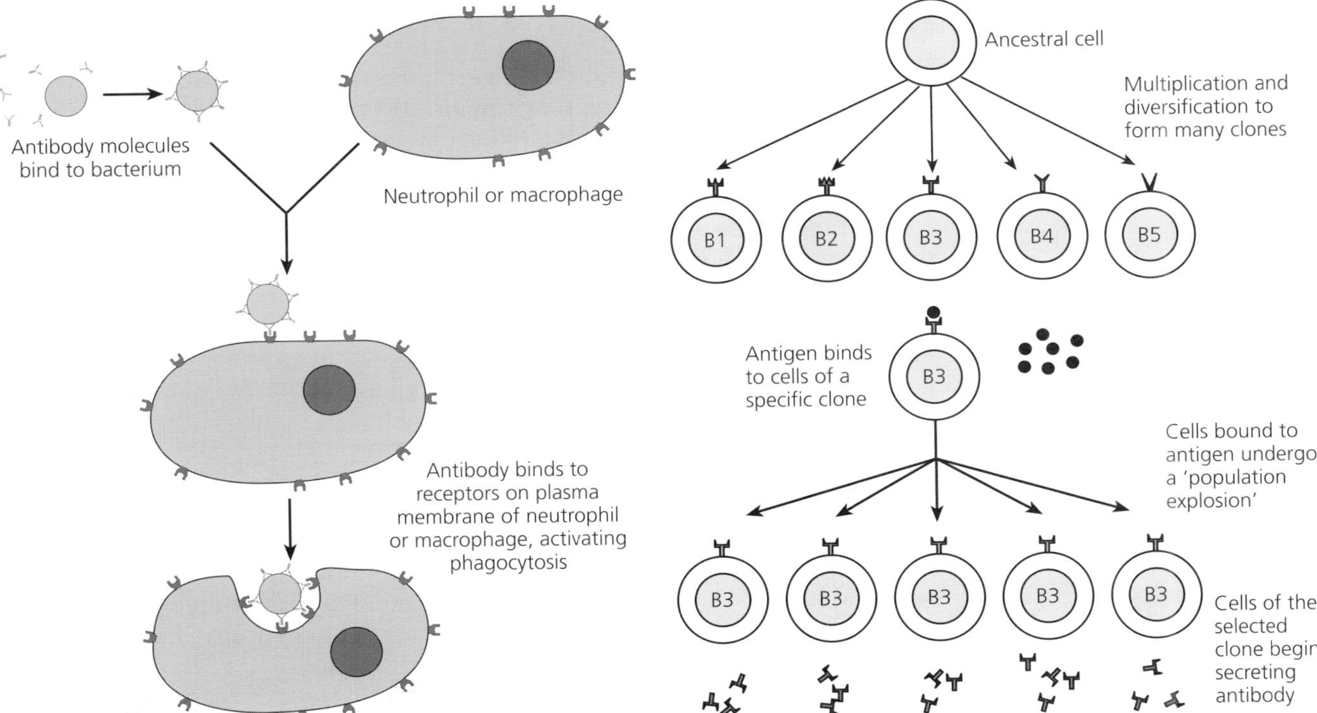


**Figure 27.6▲**
Phagocytosis of bacterium labelled with antibody. Phagocytosis can occur without presence of antibody on an invading cell, but it occurs much less actively (above)

**Figure 27.7▲▶**
The clonal selection hypothesis (modified from Alberts, *et al.*) (above right)

Since antibodies are proteins, and proteins are specified by genes, this poses an obvious problem: how can the human genome, with only about 100 000 genes, specify the structure of millions of different antibodies?

The answer lies in the ability of young lymphocytes to undergo rapid gene splicing. With each cell division, new genetic combinations are produced, eventually generating millions of new types.

## Learning to distinguish 'self' from 'non-self' – lymphocyte preprocessing

In the latter part of foetal development and in the few months after birth, the immune system has to learn the distinction between 'foreign' and 'indigenous' substances. This is what happens when the young lymphocytes undergo a period of **preprocessing** (Figure 27.8).

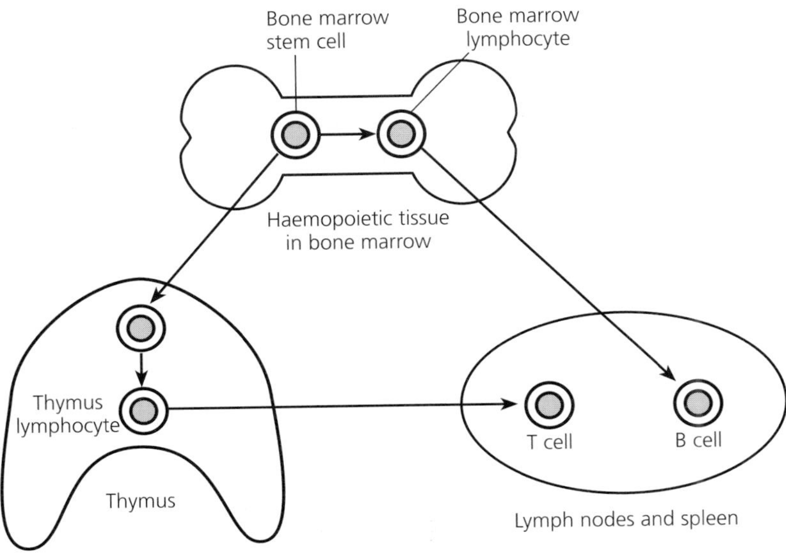


**Figure 27.8**
Origin of B- and T-lymphocytes

To begin with, there is a single population of lymphocytes in the bone marrow. From here, some young lymphocytes travel to the thymus gland where they multiply rapidly. With each division the DNA concerned with producing the T-cell receptor proteins undergoes rapid and random recombination. The result is millions of genetically different T cells, each displaying a receptor potentially complementary to one – and only one – antigen.

During their stay in the thymus, each T cell comes into contact with many of the body's own chemicals. A high proportion of T cells have receptors that 'recognise' and bind to these 'self' substances. These T cells are then destroyed, leaving those that recognise 'non-self' antigens as survivors. These are then carried by the blood to lymphoid tissue in various parts of the body where they await activation by 'non-self' antigens. Unactivated or **virgin** T cells can thus be likened to experimental prototypes, only a small proportion eventually going into mass production.

B cells are preprocessed in the liver during mid-foetal life, and in the bone marrow in late foetal development and after birth. Each B cell receptor is identical to the antibody it is programmed to produce, except that it is anchored in the plasma membrane. Preprocessing of B cells is similar to that of T cells, producing millions of genetically-different cells each with a different receptor. Also, as in the case of T cells, only cells with receptors that do not recognise 'self' substances survive. These migrate to lymphoid tissue throughout the body to await activation.

## The AIDS virus

Unlike other diseases, AIDS is caused by an agent that attacks the immune system itself. It is caused by the human immunodeficiency virus, or HIV. Like other retroviruses, it uses its RNA to make DNA which is inserted into the host cell DNA and is subsequently used to make more viral RNA. It is transmitted either by semen during sexual contact, or by blood as a result of transfusion or intravenous drug injection.

Despite an enormous amount of research, neither a cure nor a vaccine has yet been found. One reason is that the virus is ever-changing because it mutates much faster even than the 'flu virus. Finding a vaccine for such a 'moving target' has so far proved impossible.

The host cells of HIV are helper T cells which normally stimulate other cells of the immune system. After a long incubation period during which no symptoms appear, there is a steep fall in helper T cells and other immune cells. Diseases which the patient would normally fight off easily, such as certain forms of cancer, begin to take hold. Eventually the person dies from these, rather than from the direct effects of the virus itself.

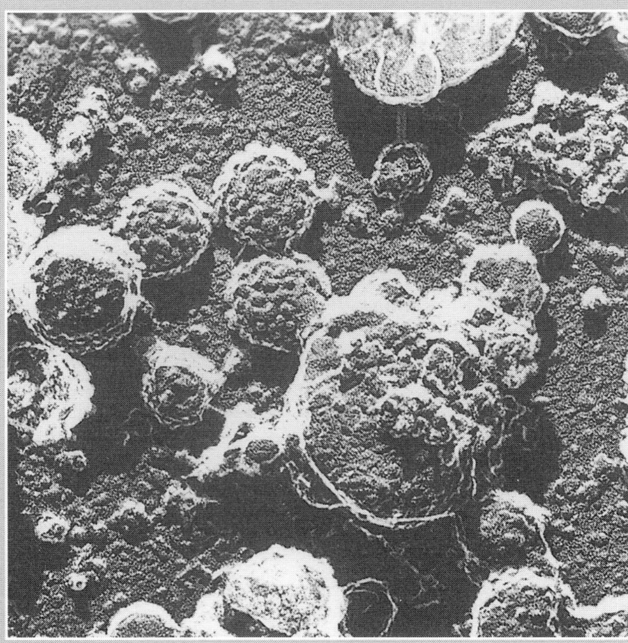

The AIDS virus

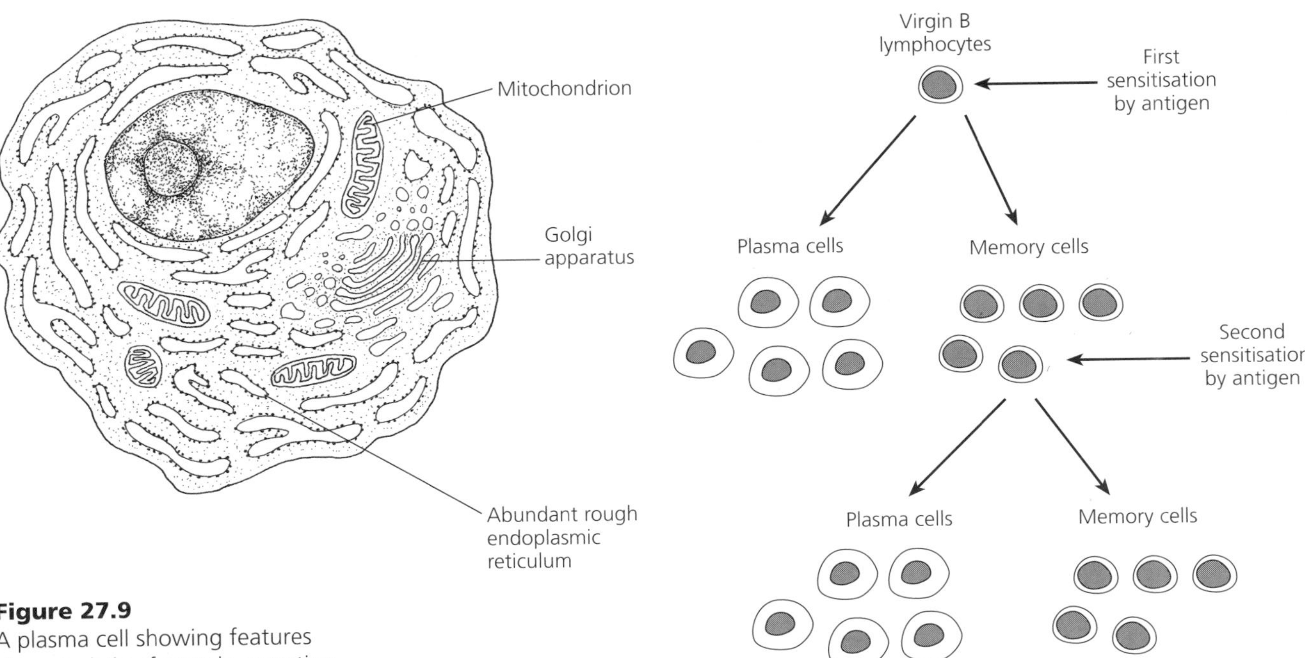


**Figure 27.9**
A plasma cell showing features characteristic of protein-secreting cells

**Figure 27.10**
Formation of B memory cells. Comparable changes occur after sensitisation of T cells

## *Activation of lymphocytes*

To be activated, a lymphocyte must come into contact with an antigen that its receptor can recognise. This process is explained in the information box on page 496. If the antigen binds to the lymphocyte's receptor, the lymphocyte is stimulated to divide actively, forming a clone. After activation, clones of T and B cells behave differently.

B cells divide and form two types of clone, both remaining in the lymph nodes. In one type, the cells develop into **plasma cells** by greatly enlarging the endoplasmic reticulum (Figure 27.9). Instead of producing a membrane-bound receptor they produce an antibody. This is essentially the same as the receptor protein, except that it is soluble and is secreted. Each B cell produces large quantities of one kind of antibody and dies after a few days. The remaining B cells develop into **memory cells** (Figure 27.10). These remain inactive for many years but retain the capacity to be reactivated and can develop into plasma cells if activated by a later contact with antigen.

Each T cell clone differentiates into several kinds of T cell. Among these are **cytotoxic T cells**, or **killer cells**, which leave the lymph nodes. When a killer cell comes into contact with a cell bearing the target antigen, it releases perforin into the plasma membrane of the target cell causing it to develop a hole and killing it. Cytotoxic T cells are responsible for killing cancer cells, cells infected by viruses, and also cells of transplanted organs. Some T cells differentiate into memory cells, while others develop into **helper T cells** which are the host cells of HIV, the virus that causes AIDS (see the information box on page 494). Helper T cells secrete a variety of locally active chemicals called **lymphokines** and **interleukins**, which promote the activity of other immune cells.

## Recognising antigens

Virgin T cells cannot be activated by antigens that are 'unaccompanied' – the antigens must be 'presented' to them by macrophages. When a macrophage encounters an antigen-bearing cell it engulfs it. The antigen is only partially digested, and is transported to the plasma membrane of the macrophage where it is 'displayed' for recognition by virgin lymphocytes.

On the surface of the macrophage, the antigen becomes closely associated with one of a group of 'self' proteins called **human lymphocyte antigens (HLA)**. The HLA, in effect, acts as a label, enabling the T cell to recognise the antigen. HLA proteins are specified by genes of the **major histocompatibility complex**, or **MHC**. These genes were first identified because of their importance in matching organ donors and recipients, before their role in defence against pathogens was understood. Though HLA proteins are 'self', in organ transplants they act as antigens because they are 'foreign' to the recipient.

Having been presented with antigen by a macrophage, a sensitised T cell then begins to divide repeatedly to form a clone as described earlier.

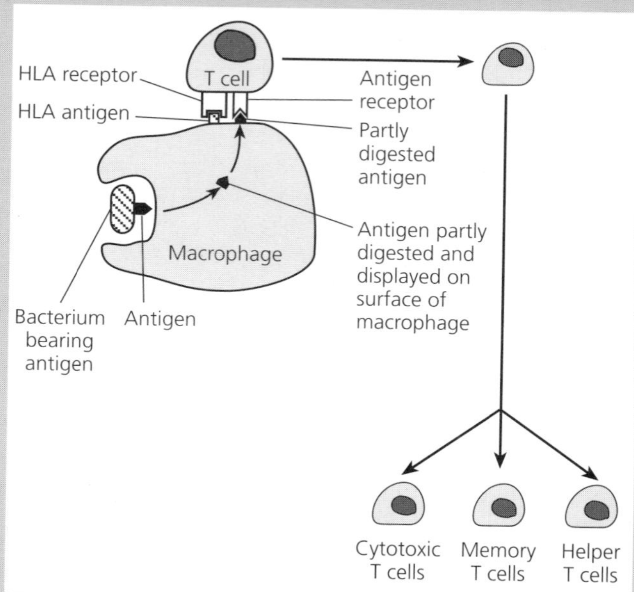


Antigen presentation

## Immunological memory

The ability of the immune system to remember previous experience is illustrated by Figure 27.11, which shows the responses to a first inoculation with antigen and a second dose. The response to a second exposure differs in four important ways from the primary response. The delay is shorter, it rises more steeply, reaches a higher maximum and lasts much longer. The shorter delay and steeper rise are due to the memory cells, which are already there, 'waiting' for a second dose of antigen.

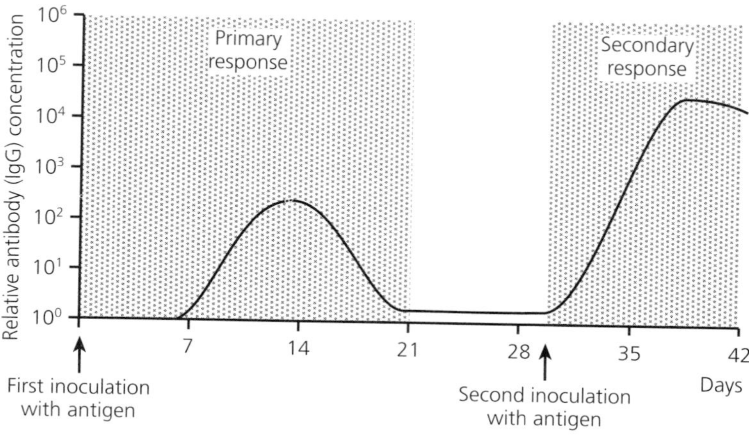


**Figure 27.11**
Primary and secondary responses to an antigen. Note that the secondary response is much more than twice the primary response since the scale is logarithmic

## Monoclonal antibodies

An antibody is so specific for a given antigen that it can be used to 'label' it and distinguish it in a population of thousands of other molecules. To do this, one needs pure antibody. The trouble is that even though an antigen stimulates the secretion of antibodies specific for it, the antibody molecules are not identical. Different virgin B cells, for example, may be activated by different parts of the antigen molecule. The result is that a single antigen may activate many clones of plasma cells. The antibodies produced in response to a given antigen are therefore **polyclonal.**

It is now possible to isolate a single clone of B cells, and thus to produce identical **monoclonal antibodies.** The method of production combines the ability of a B cell to produce a specific antibody with the potential immortality of a cancer cell. First, a mouse is immunised with the antigen of interest, causing the mouse to produce many different clones of B cells. B cells from the spleen are then fused with **myeloma cells** (myeloma is a kind of cancer). The resulting **hybridoma cells** continue to make large quantities of the same monoclonal antibody as the original B cell. Since they also have the potential immortality of other cancer cells, they are easy to culture.

Monoclonal antibodies are now an extremely useful research tool. For example, they can be used to identify and separate cells bearing a particular cell surface antigen. The antibody is covalently bonded to a fluorescent dye (which does not affect its specificity), and then added to a mixed population of the cells under investigation. When illuminated with the appropriate light the labelled cells fluoresce, and can be isolated using an automatic cell-sorting device. Fluorescent antibodies are also routinely used to detect the distribution of an otherwise invisible protein inside a cell.

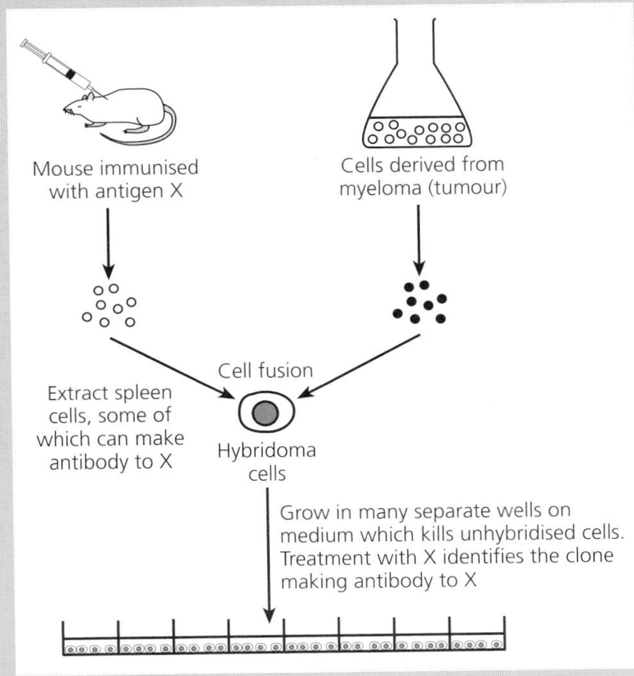


Making monoclonal antibodies

## Defence against viruses and cancer

One might think that, once inside the host cell, a virus would be beyond the reach of the immune system. Actually, even though viral proteins are foreign, the host cells have a way of ensuring that any viral proteins they are producing are made available for inspection by immune cells.

The mechanism involves the proteins produced by a group of MHC genes which are different from those mentioned in the information box on page 496. Some of the proteins made in the endoplasmic reticulum – including viral proteins – are partly digested and transported with MHC proteins to the cell surface. Facing outward from the plasma membrane they can come into contact with T-cell receptors. This triggers the production of cytotoxic killer cells which destroy other cells displaying the same viral 'label'.

In a similar way cancer cells are also subject to immune surveillance. Even though a cancer cell only makes 'self' proteins, some of these are only normally produced early in

## Defence against viruses and cancer *continued*

development, before the immune system learns to distinguish 'self' from 'non-self'. When these proteins are presented with MHC proteins for inspection by cytotoxic T cells, they may be identified as foreign and destroyed.

The immune system is clearly not effective against all cancer cells. Yet the fact that one of the symptoms of AIDS is the development of certain forms of cancer is evidence that some cancers at least are dealt with by the body's own defences.

MHC proteins are also important in organ transplants – in fact it was their importance in rejection that led to their discovery. These proteins are extremely variable – so much so that, except for identical twins, no two people have identical MHC proteins. When an organ is transplanted, the recipient's T-cell receptors react to the MHC proteins on the donor cells as they would to foreign antigens. Unless the recipient's immune response has been suppressed by drugs, the T cells are activated and destroy the donor cells.

## Active and passive immunity

Immunity can arise in one of two ways. **Passive immunity** arises when antibodies that have been made by another organism are introduced into the body. Since 'imported' antibodies are soon destroyed, passive immunity only lasts a few weeks.

Passive immunity can develop naturally or artificially. Young mammals receive antibodies from their mothers via the placenta and also in the first milk (colostrum). In humans, antibodies in the milk are not absorbed and act only within the gut, but in calves they are absorbed into the blood by pinocytosis. For a few weeks the infant is immune to every disease its mother has had, after which time its own immune system has to take over.

Antibodies made by another animal, such as a horse, can also be introduced by injection. This is frequently done after a person is suspected of being in danger of developing tetanus. The injection should not be repeated, because the horse antibodies can themselves act as antigens, causing immunity to develop in the patient.

Introduction of an antigen into the body leads to **active immunity**, in which the body makes its own antibodies. Active immunity is prolonged, since antibodies continue to be produced. Active immunity can be developed artificially by **vaccination**, in which a person is usually given a dose of pathogen that has been weakened or **attenuated**. The pathogen

## When the immune system goes wrong

Most of the time the immune system gives excellent protection against infectious disease, but sometimes it can malfunction. In **autoimmunity** the immune system fails to discriminate between self and non-self, and attacks some of its own proteins. An example of an autoimmune disease is **myasthenia gravis**, in which the body produces antibodies to the acetylcholine receptors at motor end plates. The result is progressive muscular weakness and eventually paralysis. Multiple sclerosis is also thought to be an autoimmune disease.

Another kind of malfunction is **hypersensitivity**, in which the immune system over-reacts to an antigen. One kind of hypersensitivity leads to **allergies**, in which the humoral branch of the immune system over-reacts to an antigen. In severe cases the person has difficulty in breathing and suffers a severe drop in blood pressure, which can be fatal. Antigens associated with allergies are called **allergens** and lead to the production of IgE antibodies.

cannot produce disease, but retains antigenic properties and so provokes the production of antibodies. Alternatively the person may be given a **toxoid**, which is a toxin that has been rendered harmless chemically but is still able to elicit an antibody response. Diphtheria toxin, for instance, can be made harmless by treatment with formalin whilst retaining its antigenic specificity.

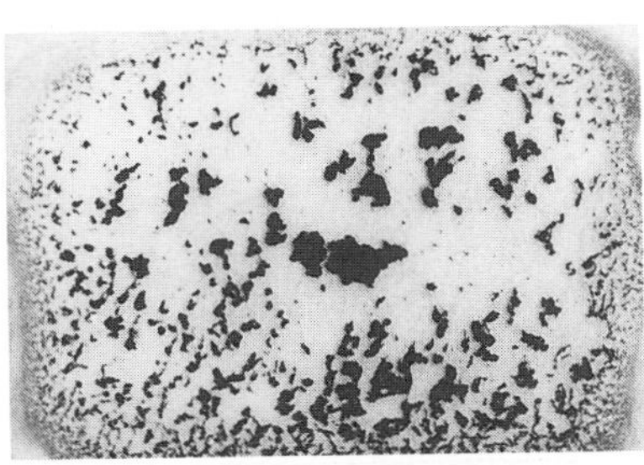

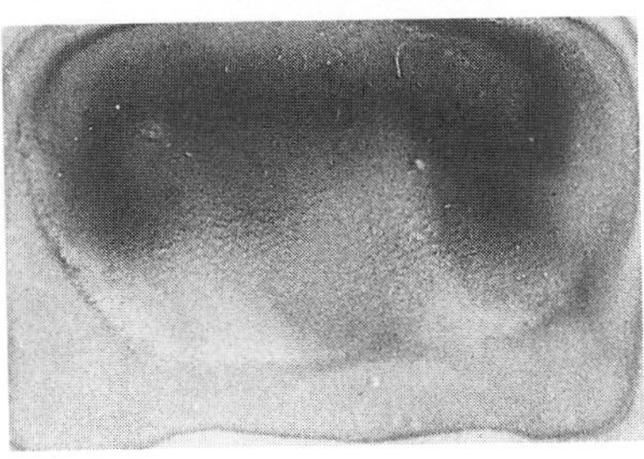

**Figure 27.12**
Agglutination of red corpuscles (top); non-agglutination (below)

## Blood groups

Whilst organ transplants are still fairly uncommon, blood transfusions have been routine since 1900 when Carl Landsteiner discovered the rules governing successful transfusions. Before that time transfusions frequently resulted in the **agglutination**, or clumping together, of the red corpuscles (Figure 27.12). This can have serious consequences, such as kidney failure.

Agglutination results from an interaction between red cells and plasma, and is quite different from **clotting**, which can only occur in plasma. Landsteiner discovered that every person belongs to one of four blood groups, depending on the type of antigen on the surface of the red cells. Group A blood has antigen A, Group B has antigen B, Group AB has both antigens and Group O has neither.

The four blood groups also differ with regard to the antibodies to these antigens (blood group antibodies are also called **agglutinogens**, since they can cause agglutination). Group A plasma contains the antibody to B antigen (anti-B); Group B plasma contains antibody to A antigen (anti-A); Group AB plasma contains neither a nor b antibody, and Group O plasma contains both anti-A and anti-B antibodies.

**Table 27.1**
Safe and unsafe blood transfusions

| RECIPIENT | DONOR | | | |
|---|---|---|---|---|
| | A | B | O | AB |
| A | ✓ | × | ✓ | × |
| B | × | ✓ | ✓ | × |
| O | × | × | ✓ | × |
| AB | ✓ | ✓ | ✓ | ✓ |

Table 27.1 shows which transfusions are safe and which are unsafe. Notice that the safety of some transfusions depends on which way they are performed. A Group O person can safely donate blood to a Group A recipient, but the reverse is dangerous. This is because the donor plasma is greatly diluted by the much greater volume of the recipient's plasma. Although donor antibodies still bind to the recipient's red cells, the red cells have too few antibody molecules on them to bind them together, so they cannot agglutinate.

A peculiar feature of the ABO blood group system is the fact that anti-A and anti-B antibodies are normally present in people who have never had a blood transfusion. The reason is that certain gut bacteria have A and B antigens on their surfaces, and it is in response to these that the person develops antibodies.

Since Landsteiner's discovery, many other blood group systems have been discovered. They are clinically far less important than the ABO system because antibodies are only produced *after* exposure to an incompatible blood type. First transfusions are thus normally safe.

The most important of these is the **Rhesus system**, so-called because it was first discovered in rhesus monkeys. Individuals having a rhesus antigen are said to be **Rh-positive**, whilst those without are **Rh-negative**. There are several antigens, but antigen D is by far the most important, and a person is said to be Rh-positive if he or she has the D antigen. Since the allele for the D antigen is dominant, a rhesus-positive person may be heterozygous or homozygous.

Problems with the Rh antigen occur when a Rh-negative mother and an Rh-positive father have a child. If the father is heterozygous there is a 50% chance that he will pass the Rh antigen on to the child. At around the time of birth, the placenta loosens its hold on the uterus and some of the baby's red cells may cross into the mother's blood stream causing her to produce antibodies to the Rh antigen. Although these have no effect on the first child, they may put subsequent babies at risk. Some maternal antibodies may cross the placenta causing the baby's red cells to agglutinate. The condition is called **haemolytic disease of the newborn** and may be fatal.

## QUESTIONS

**1** Match each of the terms **(1)–(20)** with the letter of the appropriate descriptive phrase.

**1)** Antibody
**2)** Antigen
**3)** T-lymphocyte
**4)** B-lymphocyte
**5)** IgG
**6)** Macrophage
**7)** Cytotoxic T cell
**8)** Thymus
**9)** Bone marrow
**10)** Perforin
**11)** Lysozyme
**12)** Agglutination
**13)** helper T cell
**14)** Toxoid
**15)** Phagocytosis
**16)** Complement
**17)** Monocyte
**18)** Memory cell
**19)** Plasma cell
**20)** Opsonin

**A** Chief phagocytic cells
**B** Responsible for humoral mediated immunity
**C** Site of maturation of T cells
**D** Protein that forms holes in target cell
**E** Clumping of cells by antibody
**F** Protein produced in response to a foreign substance
**G** Cells attacked by HIV
**H** Site of maturation of B cells
**I** Responsible for cell-mediated immunity
**J** Vaccine consisting of bacterial product made harmless
**K** Substance provoking production of an antibody
**L** Precursor of macrophage
**M** Secretes antibodies
**N** Bactericidal enzyme in tears
**O** Lymphocyte that kills virus-infected cells
**P** Enables more rapid response to second attack by pathogen
**Q** Makes bacteria more palatable to phagocytes
**R** The most abundant immunoglobulin
**S** Plasma proteins that co-operate with antibodies
**T** Eating of pathogen by defence cell

**2** What is wrong with the phrase 'disease-causing antigens'?

**3** In highly-inbred strains of rats (or other mammals), it is possible to transplant organs without rejection. Suggest a reason.

**4** How do B cells assist phagocytes?

**5** The instructional and clonal selection hypotheses have strong parallels with the conflict between the evolutionary ideas of Darwin and Lamarck. Which of the two immunological hypotheses is essentially Darwinian and which is Lamarckian? Explain your answer.

CHAPTER

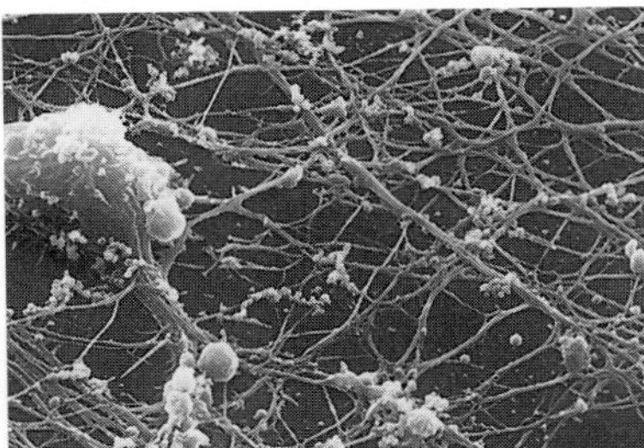

# 28

# REPRODUCTION

## LEARNING OBJECTIVES

By the time you have completed your study of this chapter you should be able to:

- distinguish between sexual and asexual reproduction.
- explain how the reproduction of mammals differs from that of other vertebrates.
- describe the anatomy of the male and female reproductive systems in a mammal.
- describe the process of gametogenesis in the testis and ovary, and comment on the similarities and differences between them.
- describe the changes that occur in puberty, and explain the role of hormones in these changes.
- describe the changes that occur in the uterus and ovary during the menstrual cycle, and explain the role of gonadal, pituitary and hypothalamic hormones in mediating these changes.
- explain the basis for the hormonal contraceptives.
- describe the events immediately prior to fertilisation, and events leading to implantation.
- state the functions of the placenta.
- explain how the structure of the placenta facilitates the transfer of materials between foetus and mother.
- describe the changes that occur in the mother during pregnancy, and explain the role of hormones in these changes.
- describe the process of labour.
- explain how the secretion of milk is initiated and maintained, and comment on the nutrient composition of milk.

## Sexual and asexual reproduction

Living things essentially reproduce in two different ways. The simplest, quickest and the cheapest energetically is **asexual reproduction**, in which genetically-identical offspring are produced by **mitosis** (Chapter 7). Asexual reproduction is widespread in both animals and plants, though it does not occur in birds or mammals.

A more complicated way of producing offspring is by **sexual reproduction**. This involves the reshuffling of genes by two complementary processes:

- **fertilisation**, in which the chromosome number is doubled. This involves the fusion of haploid sex cells, or **gametes**, to form a diploid **zygote**.
- **meiosis**, in which each diploid cell divides to produce four haploid cells.

Mitosis and meiosis are described in detail in Chapter 7. For present purposes the essential result of sexual reproduction is that the offspring are genetically different from each other. In conditions that do not change from one generation to the next this is a disadvantage, since many of the offspring would not be as well adapted as the parents were. But in most environments, conditions do change, so genes that are best for one generation may not be best for another. Sex ensures that at least some of the offspring are likely to be better adapted to new conditions.

In all organisms sexual reproduction is more complicated than asexual reproduction. In mammals it involves the following processes:

- **gametogenesis**, or gamete production. This occurs in the sex organs, or **gonads**. As in all other animals and in all plants the gametes are of two kinds – tiny, motile male gametes, or **sperms**, and large, stationary female gametes, or **eggs**.
- **fertilisation**, or the joining of egg and sperm to form a zygote. In mammals as in birds, reptiles and some other vertebrates, this occurs inside the body of the female. The introduction of the sperms into the female is called **copulation** (or **coitus** in humans). It involves a complex sequence of behavioural and physiological events integrated by the nervous and endocrine systems.
- after fertilisation the zygote divides repeatedly, forming a ball of cells, or **blastocyst**.
- the blastocyst becomes **implanted** in the lining of the **uterus** of the mother, and forms a tissue connection called the **placenta**. Here nutrients, oxygen and wastes are transferred between the blood of the **foetus** and that of the mother.
- when fully developed the foetus is expelled from the mother. After birth there is a period of **parental care**, in which the mother continues to feed the young on **milk** secreted by **mammary glands**.

## The male reproductive system

The human male reproductive system is shown in Figure 28.1. Sperms are produced in the male gonads, or **testes** (called testicles in humans), which also produce male sex hormone. In most mammals the testes lie in an extension of the abdominal cavity, the **scrotal sacs**, enclosed in a bag called the **scrotum**. Their semi-external position enables them to be maintained at a temperature about 5°C lower than the core body temperature which, for reasons unknown, is essential for sperm production. Another feature which reduces their temperature is a heat exchange system in the arteries and veins supplying the testes, similar to that in the limbs of many birds and mammals (Chapter 25). The temperature of the testes is fine-tuned by the **dartos muscle** in the scrotum. It contracts in cold weather, pulling the testes up against the abdomen, and relaxes in warm weather.

The testis is subdivided internally into about 200 compartments, each containing about three **seminiferous tubules** in which sperms are produced (Figure 28.2). Each tubule is about 80 cm long, so their combined length is about 500 metres! The seminiferous tubules unite to form a network which leads to a 6 metre-long, highly convoluted tube, the **epididymis**, where sperms complete their development and are stored. The epididymis leads into a muscular tube, the **vas deferens**. The two vasa deferentia join near the base of the bladder to

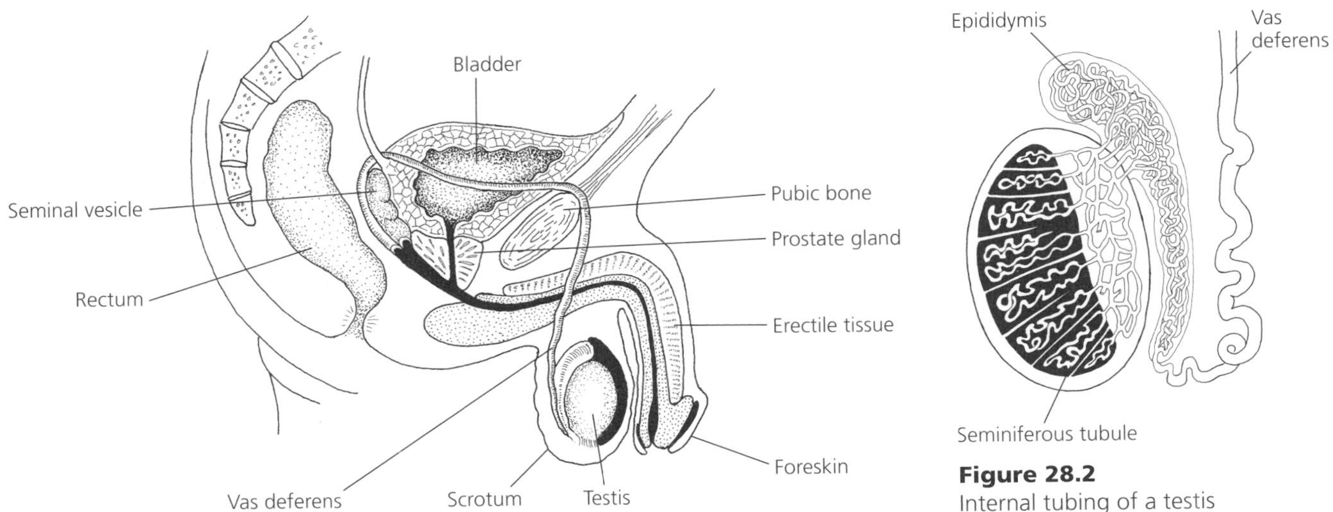


**Figure 28.1**
The human male reproductive organs as seen from the side

**Figure 28.2**
Internal tubing of a testis



form the **urethra**, which is shared by both urinary and reproductive systems.

Sperms are carried to the exterior in **seminal fluid**, or **semen**. This contains a variety of substances, none of which appears to be essential for sperm function since sperms removed from the epididymis are capable of fertilisation. Semen is produced by three kinds of gland:

- two **seminal vesicles**, one opening into the base of each vas deferens. Secretion from seminal vesicles accounts for over half the volume of the semen and contains fructose (which probably acts as an energy source for sperm) and prostaglandins (locally acting, hormone-like substances derived from essential fatty acids).
- the **prostate gland**, which surrounds the base of the urethra, and discharges its secretion via many tiny openings.
- **Bulbourethral** or **Cowper's glands** open into the urethra below the prostate.

Semen is introduced into the female via the **penis**. This contains three tracts of spongy blood spaces called **erectile tissue**. During sexual excitement, the pressure in the blood spaces rises because the arteries supplying them dilate and the veins draining them constrict. As a result, the penis becomes stiff and erect, enabling it to enter the vagina during intercourse. The end of the penis, the **glans**, is covered by the **foreskin**. When this is too tight for easy urination, it may be removed by a simple operation called **circumcision**.

**Undescended testes**

A peculiar feature of most male mammals is that, towards the end of embryonic development, the testes descend from their original position in the abdomen into the scrotal sacs. This migration leaves evidence in the form of the looping of the vasa deferentia over the bases of the ureters. In humans this is normally complete by birth, but in some cases the testes need a helping hand in the form of a simple operation. If the testes remain in the abdomen they fail to achieve the lower temperature needed for spermatogenesis. This condition is called **cryptorchidism** ('hidden testis') and, if not treated, results in greatly increased chances of testicular cancer.

## The male gametes, or sperms

The smallest mammalian cells are spermatozoa, or sperms (Figure 28.3). Each consists of three parts: **head**, **middle piece** and **tail**. The flattened head consists mainly of the haploid nucleus, but also has at its tip a sac called the **acrosome**. This is actually a greatly enlarged lysosome and contains digestive enzymes used to enter the egg. The middle piece contains the mitochondria which produce the ATP needed for movement of the tail. The tail is simply a large flagellum and contains the nine pairs of peripheral filaments and the two central filaments typical of cilia and eukaryotic flagella.

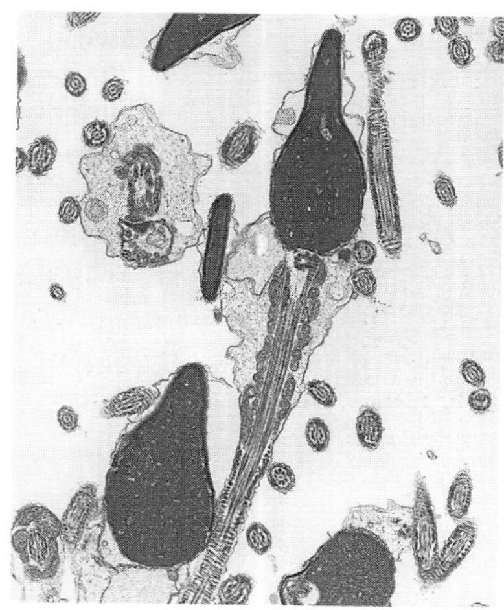

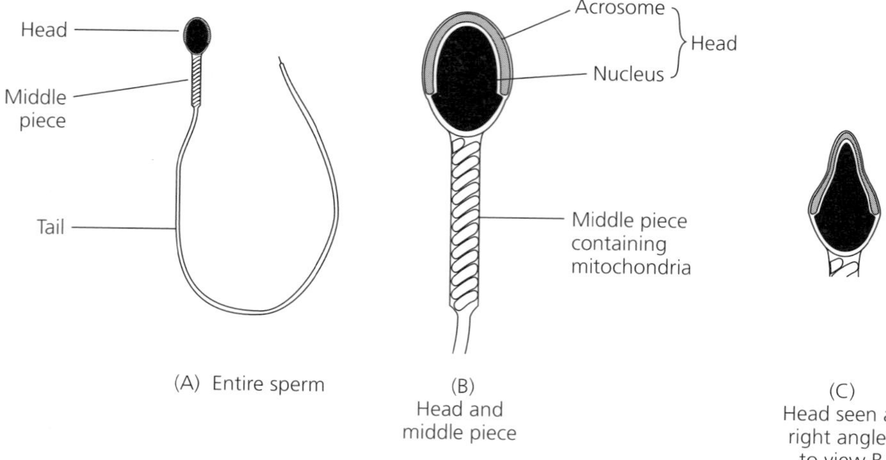


**Figure 28.3**
Structure of a sperm. Entire sperm (A); head and middle piece (B); head at right angles to B (C); and TEM of human sperm showing mitochondria in middle piece (above)

## Spermatogenesis, or sperm production

**Spermatogenesis** takes place in the **seminiferous epithelium** which forms the walls of the seminiferous tubules (Figure 28.4). Early in development the ancestors of the gametes, the **primordial germ cells**, migrate to and colonise the embryonic testis. Eventually they come to lie near to the basement membrane of the seminiferous epithelium. At puberty they become **spermatogonia**, which continue to multiply mitotically. Of the two products of each division, one remains as a source of new cells. The other may divide again, but eventually enters a phase of growth, becoming a **primary spermatocyte**.

Each primary spermatocyte then undergoes the first division of meiosis, producing two **secondary spermatocytes** (Figure 28.5). Each secondary spermatocyte then undergoes a second meiotic division to form two haploid **spermatids**. The spermatids then develop into sperms by shedding most of their cytoplasm and reorganising the rest into the middle piece and tail. The entire process of spermatogenesis takes an average of about 64 days. From puberty onwards, sperms are produced at a rate of about 300 million per day.

**Figure 28.4▼**
Cross section of two seminiferous tubules (below)

**Figure 28.5▼▶**
The stages of spermatogenesis (below right)

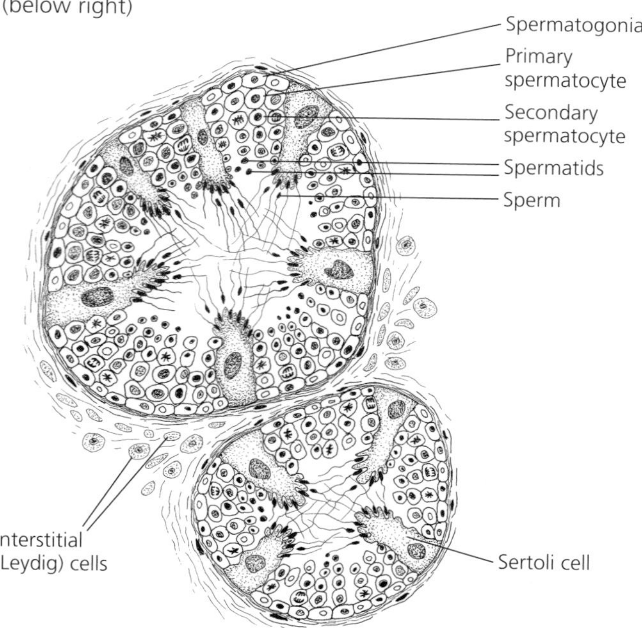


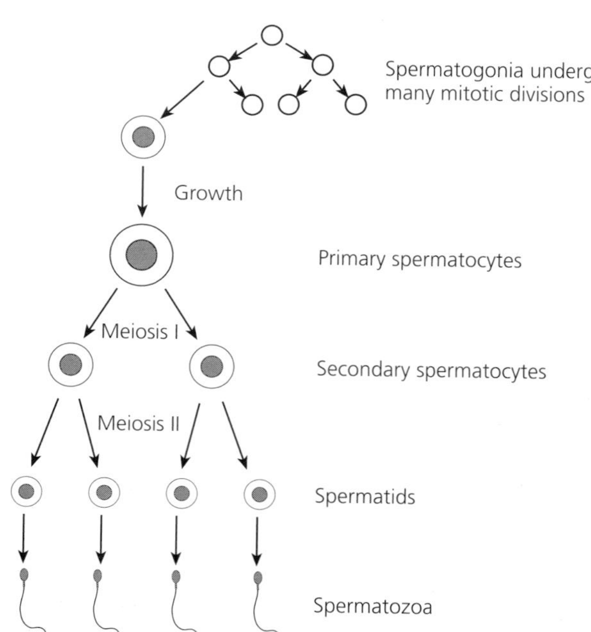

Since the sperms are genetically different from body cells, they would be in danger from the immune system were it not for **Sertoli cells** which act as 'chaperones'. The Sertoli cells extend across the thickness of the seminiferous epithelium and are held together by tight junctions, except where spermatocytes and spermatids are squeezed between them (Figure 28.6). Nutrients and other substances can thus only reach the developing sperms by passing through the Sertoli cells. Spermatocytes and spermatids slowly move along the spermatogenic 'production line' towards the lumen of the tubule. As they move between adjacent Sertoli cells, existing tight junctions between these cells are broken on the side nearest the tubule lumen and new ones are made on the other side.

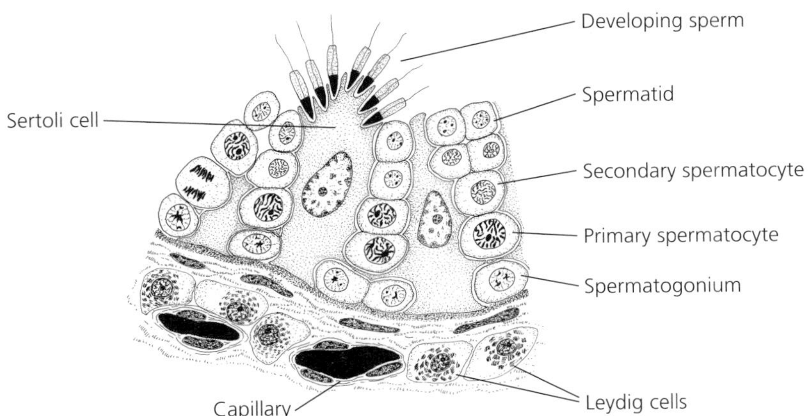


**Figure 28.6**
The role of Sertoli cells as a barrier between blood and sperm (after Junqueira, Carneiro and Kelly)

When sperms are shed into the seminiferous tubules they are not yet fully motile, and are carried passively into the epididymis where they acquire the ability to swim. Even then, a sperm cannot immediately fertilise an egg until it has spent some time in the female reproductive tract.

### Vasectomy

One of the simplest and most reliable methods of contraception is **vasectomy**, in which the vasa deferentia are cut. The testis continues to produce sperms and testosterone, so there is no effect on the production of seminal fluid (except that it contains no sperms) and sexual desire remains unaffected. Its disadvantage is that it is often difficult to reverse by rejoining the tubes, so men who undergo the operation are told that they should regard it as permanent.

## The female reproductive system

Anatomically the female reproductive system is simpler than that of the male (Figure 28.7).

The ovaries lie in the abdominal cavity and each is about the size and shape of an almond. Like the testes they have a dual function, producing both gametes and sex hormones. The ovary consists of a dense **stroma** of connective tissue, in which are embedded many **follicles** at varying stages of development (Figure 28.8). Each follicle consists of an immature egg surrounded by a glycoprotein envelope, the **zona pellucida**. The egg and zona pellucida are surrounded by **granulosa cells**, embedded in a capsule called the **theca**. The theca is differentiated into an inner, vascular **theca interna** and a more fibrous **theca externa**.

Close to each ovary is the funnel-shaped opening of an **oviduct**, or **Fallopian tube**, in which fertilisation occurs. Its walls are lined with ciliated epithelium and contain smooth muscle. The opening of the oviduct consists of tentacle-like **fimbriae**. At ovulation the ciliated surfaces of the fimbriae move over the surface of the ovary, and the cilia sweep the egg into the oviduct.

A combination of ciliary action and peristaltic muscle contractions serves to transport eggs down to the **uterus**, or **womb**. The walls of the uterus have two distinct layers, reflecting its two functions. The bulk of

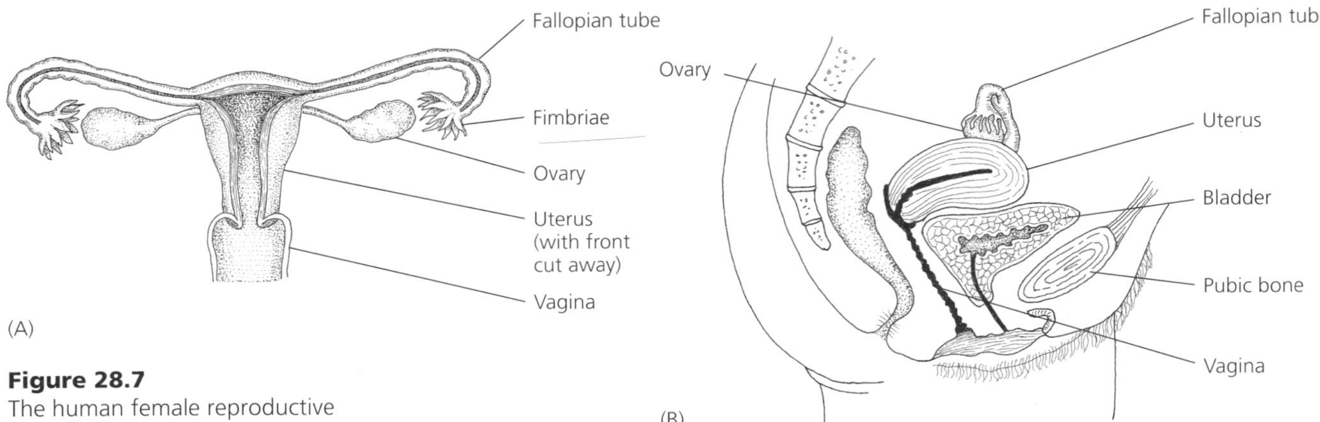


**Figure 28.7**
The human female reproductive system as seen from the front (A) and the side (B)

the uterine wall is the **myometrium** which consists of smooth muscle, and only functions to expel the infant at birth. The inner lining, or **endometrium**, is glandular and is concerned with the anchorage and nutrition of the embryo. The neck, or **cervix**, of the uterus opens into the **vagina**. Close to the opening of the vagina is the **clitoris**, a small structure containing erectile tissue which is homologous with the penis. The opening to the vagina is covered by folds of tissue forming the **vulva**.

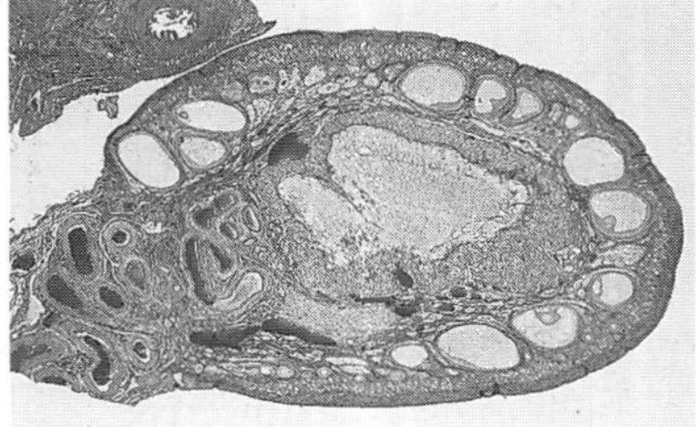

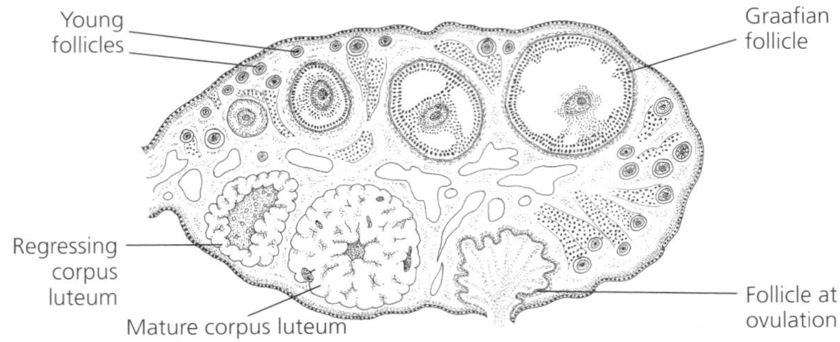


**Figure 28.8**
Internal structure of the ovary showing stages in follicle development (right) and a mammalian ovary as it appears under the microscope (above)

## The ovary and oogenesis

As in spermatogenesis, egg production, or **oogenesis**, involves three phases in which the germ cells go through multiplication, growth, and maturation (Figure 28.9). There are however two important differences. First, oogenesis begins before birth but is not completed until fertilisation, which could be as much as 50 years later. Second, the cytoplasmic divisions accompanying meiosis are very unequal. The result is that only one functional ovum is produced, the other products being tiny, functionless **polar bodies** (two or three, depending on whether the first polar body undergoes meiosis II).

Early in embryonic development the primordial germ cells migrate to the ovary from the yolk sac. Now called **oogonia**, they enter the phase of multiplication. After repeated mitotic divisions they enter a phase of growth, enlarging to become **primary oocytes**. On completion of its growth, the primary oocyte enters prophase of the first division of meiosis, but development is then suspended until after puberty. Meanwhile, cells round the oocytes have differentiated into little clusters called **primordial follicles**. At the time of birth there are about a million primordial follicles in each ovary. Like the primary oocytes within them, the primordial follicles remain in 'suspended development' until puberty, by which time

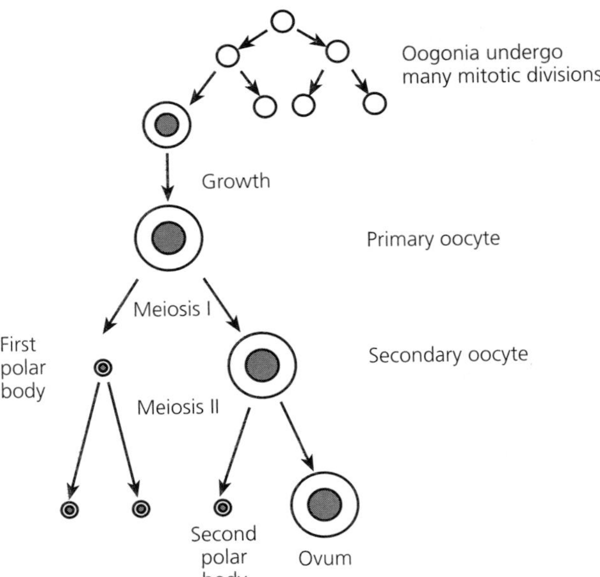


**Figure 28.9**
Stages in oogenesis. Note that oogenesis overlaps with fertilisation, since the entry of the sperm triggers meiosis II

all but about 400 000 have degenerated, of which only about 400 will be released into the oviduct.

At monthly intervals about 20–25 follicles begin to develop further. The follicle cells multiply to become granulosa cells and begin to secrete oestrogens. Eventually one follicle outstrips the others and becomes dominant, the others degenerating to form **atretic follicles**. The follicle cells multiply and become several layers thick. Fluid begins to accumulate in a cavity inside the expanding follicle, which slowly moves towards the surface of the ovary. It eventually reaches a diameter of about 15 mm and is called a **Graafian follicle**. Shortly before the follicle reaches its maximum size the primary oocyte completes the first meiotic division, becoming a **secondary oocyte**. The other product of this division receives very little cytoplasm and becomes the tiny **first polar body**.

### *Ovulation and formation of the corpus luteum*

By the time the follicle has matured into a Graafian follicle it is ready for release from the ovary, a process called **ovulation**. The thin layer of tissue separating the follicle from the abdominal cavity is digested by enzymes secreted by the theca, releasing the oocyte still surrounded by a layer of follicle cells. Stimulated by hormones, the fimbriae move over the surface of the ovary and normally 'catch' the oocyte, which is swept by ciliary action down into the Fallopian tube.

Meanwhile back at the ovary, the majority of the follicle cells that were left behind develop into a **corpus luteum**, or 'yellow body'. It is so-called because of the yellowish, lipid-rich material that develops within it. The corpus luteum secretes hormones which, in the event of the egg being fertilised, are vital for its further development. If fertilisation occurs, the corpus luteum persists and plays an important part in early pregnancy. If there is no fertilisation, it dies.

## The sex hormones and puberty

Besides producing gametes, the ovaries and testes secrete the steroid **sex hormones**. These are responsible for the changes that occur at puberty, and in the female they also play a central part in pregnancy. These hormones are **testosterone** in the male, and **oestrogens** and **progesterone** in the female.

## Testicular feminisation

To act as a signal, a hormone must have a specific receptor molecule in the target cell. If the receptor is not functional the cell cannot respond. It is probably the lack of such a specific receptor that is the cause of testicular feminisation, in which a person has the male (XY) genotype and produces testosterone, but the target tissues are unable to respond. The person is born with a vagina and later develops breasts because of the small amounts of oestrogens that males produce. The testes are present but do not descend (and so should be removed to prevent testicular cancer).

### *Testosterone*

Whereas sperms are produced in the seminiferous tubules, **testosterone** is produced by the **Leydig cells** between the tubules. Testosterone begins to exert two of its effects before birth. It stimulates the embryonic sex organs (which are indistinguishable at first) to differentiate into the male pattern. In the absence of testosterone the sex organs develop into the female pattern. Another early effect is that it causes the foetal brain to differentiate into the male pattern.

The most obvious effects of testosterone begin in adolescence producing the changes associated with puberty, for example:

- enlargement of the reproductive organs;
- increased skeletomuscular growth;
- enlargement of the larynx with associated deepening of the voice;
- thickening of the hairs on chest, legs and face;
- a tendency towards more aggressive and assertive behaviour;
- increased interest in the opposite sex.

Some effects of testosterone take much longer, for example balding in men (though genetic factors are also important).

Besides testosterone, the testes also secrete the hormone **inhibin** which, as described later, plays an important part in regulating the feedback control of the testis by the hypothalamus and pituitary.

### *Oestrogens*

The female counterpart of testosterone is **oestrogen**. This is actually not one hormone but three, with similar activity, the most important being **oestradiol**. Oestrogens are secreted by the granulosa cells of follicles, and later by the corpus luteum. In pregnancy they are also secreted by the placenta.

In humans the following effects of oestrogens occur at puberty:

- enlargement of the uterus;
- development of the pelvis into the female type;
- growth of the duct tissue of the mammary glands;
- deposition of fat round the mammary tissue of the breasts;
- growth of the endometrium of the uterus during the early phase of the menstrual cycle (see below);
- increased interest in the opposite sex.

### *Progesterone*

The ovary also secretes another steroid hormone, **progesterone**, which has the following effects:

- it stimulates the growth of the glandular tissue of the breast.
- it inhibits the contraction of uterine muscle.
- it prepares the endometrium for the possible arrival of a fertilised egg.

### *The role of the pituitary and hypothalamus*

How do the testes and ovaries ‘know’ that the time has come for puberty to begin? The answer lies not in the gonads, but in the hypothalamus of the brain. As explained in Chapter 20, the hypothalamus secretes **releasing hormones** which stimulate the secretion of hormones by the

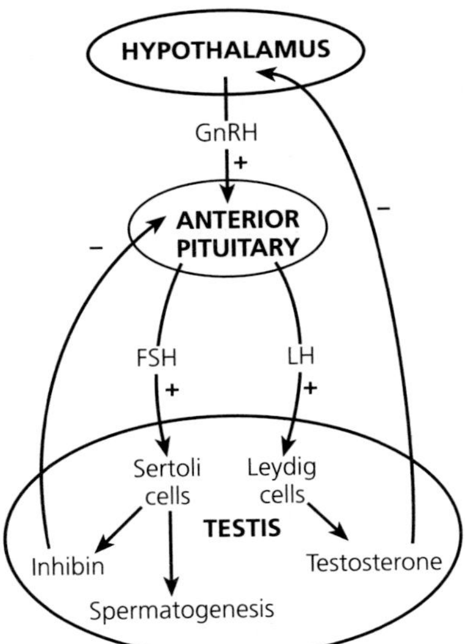


**Figure 28.10**
Feedback control of the testis. '+' indicates stimulation, '–' indicates inhibition

anterior pituitary. These anterior pituitary hormones in turn stimulate the activity of the thyroid, adrenal cortex and the gonads.

In some way (as yet not understood) the hypothalamus 'knows' that it has reached the age of puberty, and begins to secrete **gonadotrophin releasing hormone**, or **GnRH**. This hormone stimulates the anterior pituitary to secrete its two **gonadotrophic hormones, follicle stimulating hormone** (**FSH**) and **luteinising hormone** (**LH**). Both are glycoproteins, and although they were named according to their roles in the female, they stimulate the activity of both ovaries and testes.

In the testis, FSH stimulates the Sertoli cells to secrete substances that induce spermatogenesis. In the ovary, it stimulates the division of granulosa cells in follicles and their secretion of oestrogen.

In the testis, LH stimulates the Leydig cells to secrete testosterone. In the ovary, its role is more complicated. It stimulates young follicles to secrete oestrogen and stimulates the final stages of follicle maturation, ovulation, and the development and maintenance of the corpus luteum.

### *Feedback between testis and brain*

The interaction between the brain and the gonads is not a one-way affair – the ovaries and testes also exert feedback effects on the hypothalamus and the anterior pituitary gland.

In the male, testosterone has an inhibitory effect on the secretion of GnRH by the hypothalamus and, therefore, on the secretion of LH and FSH. There is thus a negative feedback control of the testes – a rise in testosterone output results in a fall in LH and FSH, which in turn causes testosterone secretion to fall (Figure 28.10).

Another negative feedback effect involves **inhibin**. This glycoprotein hormone is secreted by the Sertoli cells in response to FSH and inhibits the secretion of FSH. If FSH secretion rises above normal, the rise in inhibin output automatically checks this by negative feedback.

Feedback mechanisms between the gonads and the brain also occur in the female and are described in the section on the menstrual cycle.

## Sexual cycles

In many animals, young are produced in spring and early summer, when food is most abundant. Like other kinds of seasonal behaviour, the onset of the breeding season is usually indicated by changing daylength, or **photoperiod**, which is monitored by the hypothalamus. As explained above, the hypothalamus communicates with the gonads via the anterior pituitary. As a result of the increased output of FSH and LH, the gonads become active and begin producing gametes.

Gametogenesis is of no use in itself unless it is accompanied by the various anatomical, physiological and behavioural changes necessary to achieve fertilisation and, in many animals, for care of the offspring. These are brought about by the other product of the gonads – the sex hormones.

Most mammals are **polygamous**, where one male mates with many females – unsuccessful males remaining as 'bachelors'. After mating, a male's role is over, but this is not so with the female. She must support the young within her and also after birth. Reproduction for a female mammal is thus a longer-term business, requiring preparatory growth changes in the uterus and mammary glands.

Because of these demands, female mammals are unable to sustain their bodies in a permanent state of sexual readiness. Instead, females have one or more cycles of sexual activity, called **oestrous cycles**. The most obvious part of the cycle is when the female shows interest in the male. She advertises her fertility by producing chemical signals, or

**pheromones**, which (as anyone who has owned an unspayed bitch knows) have a powerful stimulatory effect on males. This period is called **oestrus** or 'heat' and in most mammals coincides with ovulation. (Note that 'oestrus' is a noun, and 'oestrous' is an adjective).

Some mammals such as deer and seals have only one oestrous cycle per breeding season. Others such as dogs normally have two (though their wild relatives, the wolves, have one). Others such as mice reproduce continually as long as conditions are favourable.

A special kind of female cycle is the **menstrual cycle**, which occurs in humans and other apes and in some monkeys. It is so-called because its most obvious event is the breakdown of the lining of the uterus, which comes away as blood and debris during **menstruation**.

In humans (unlike other mammals) there is no oestrus. Women do not advertise the fact that they are ovulating and, in fact, they are usually unaware of it. Nevertheless the menstrual cycle *is* a cycle of fertility, with menstruation and ovulation usually occurring at the opposite 'ends' of the cycle. The menstrual cycle in humans averages 28 days, though with considerable variation.

## The menstrual cycle

Although the menstrual cycle takes its name from its most evident signs, it involves integrated cycles in the ovary and in the anterior pituitary (Figure 28.11).

The cycle can be conveniently divided into three stages:

1 the **menstrual phase**, or the 'period', which arbitrarily is regarded as Day 1 of the cycle. It is precipitated by the sustained vasoconstriction of the long, coiled **spiral arterioles** which supply all but the deepest layer of the endometrium. The tissue supplied by these arterioles dies and comes away as blood and debris, the **menses**. The resulting menstrual flow usually lasts 3–5 days and involves the loss of about 60–80 $cm^3$ of blood. The deepest layer of the endometrium does not die because it is supplied by arterioles (the **straight arterioles**) which do not constrict (Figure 28.12).

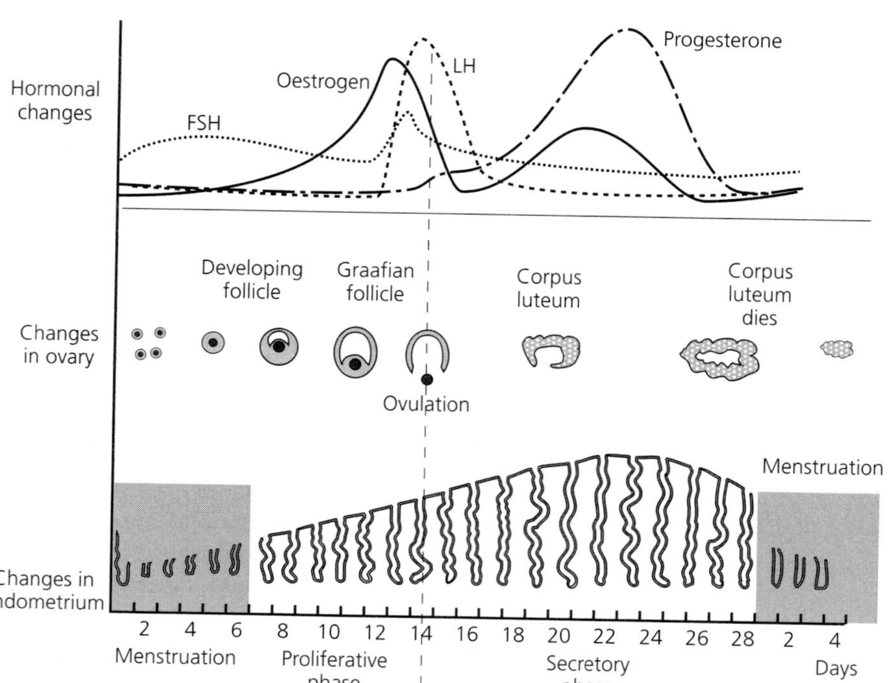


**Figure 28.11**
Changes in the endometrium and ovary during the menstrual cycle

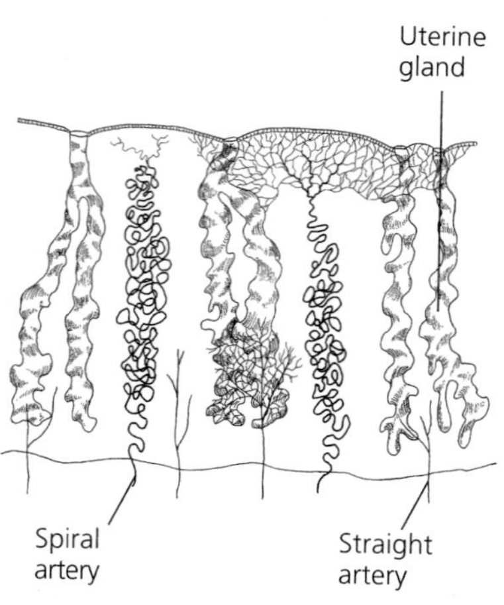


**Figure 28.12**
Blood vessels in the endometrium (after Daron)

2 the **proliferative**, or **follicular**, **phase**. This is so-called because of the rapid cell division and growth of the endometrium. Under the influence of oestrogens secreted by developing ovarian follicles, the surviving endometrial tissue regenerates, reaching a thickness of about 3 mm.

3 the **secretory**, or **luteal**, **phase**. After ovulation, the corpus luteum begins to secrete large amounts of progesterone, hence the alternative name for this part of the cycle. In response to this, the endometrial glands become coiled and begin to actively secrete liquid. If fertilisation occurs, this may provide some temporary nourishment for the embryo until it has implanted in the endometrium.

The secretory activity of the endometrium depends on progesterone. When (as is usually the case) the corpus luteum dies and stops secreting progesterone, the spiral arteries again go into spasm precipitating another 'period'.

### Controlling the menstrual cycle

The changes in the endometrium are caused by cyclical changes in a hormonal interplay between the anterior pituitary and the ovary (Figure 28.13). The beginning of the cycle is marked by a slow rise in the output of FSH and LH by the anterior pituitary. FSH stimulates the growth of a number of follicles, and also the secretion of oestrogens by the granulosa cells. The rising output of oestrogens begins to inhibit the secretion of FSH by negative feedback, and FSH secretion begins to decline. Then, for reasons not fully understood, the self-restraining negative feedback between oestrogen and the pituitary changes to positive feedback. In this situation, oestrogen stimulates LH output, and LH stimulates oestrogen output. This self-accelerating cycle leads to an explosive burst in the output of LH (and to a lesser extent FSH), triggering ovulation.

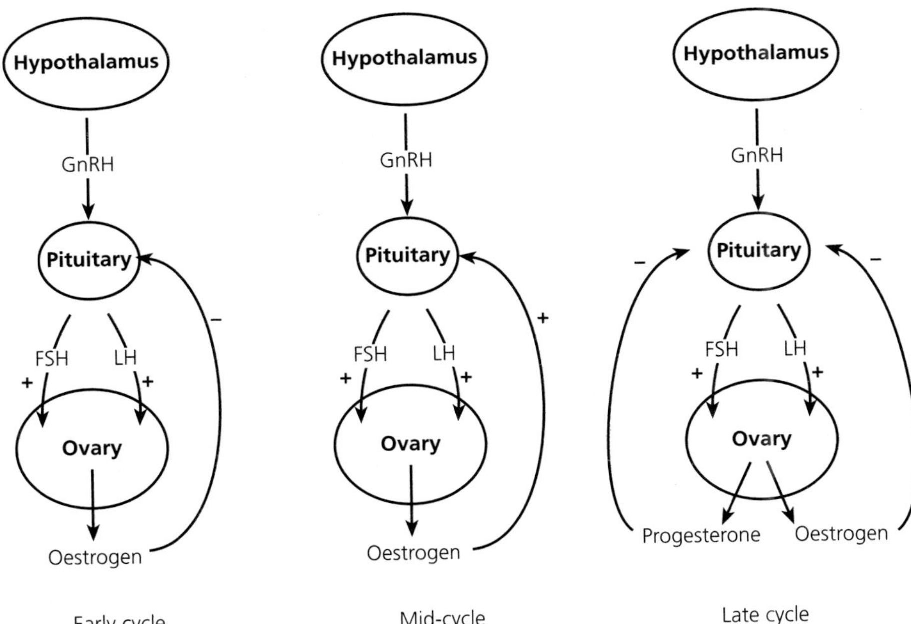


**Figure 28.13**
Role of the anterior pituitary in the menstrual cycle. '+' indicates stimulation, '−' indicates inhibition

Though the discharged oocyte is surrounded by granulosa cells, most of the granulosa cells are left behind in the ovary. Under the influence of LH these multiply and grow into a **corpus luteum**. The corpus luteum manufactures increasing amounts of progesterone and oestrogens.

### Hormonal contraception

**During pregnancy, ovulation is inhibited by high levels of oestrogen and progesterone. This is the 'pill' method of female contraception. It consists of synthetic oestrogens and progesterone in proportions that depend on the kind of pill. In effect, the ovary is 'deceived' into 'thinking' that the woman is pregnant. The natural hormones cannot be used since they are rapidly degraded in the liver. The synthetic hormones have similar biological activity, suppressing the secretion of FSH and LH, and hence the development and maturation of ovarian follicles. Some women can suffer side effects such as an increased tendency for thromboses, but otherwise it is very safe.**

Progesterone stimulates the growth and activity of the endometrial glands, inhibits the contractility of the uterine muscle, and stimulates the growth of the secretory tissue of the breast. It also influences the thermoregulatory centre in the hypothalamus, raising the temperature set point by about 0.5°C. As a result, ovulation is followed by a small rise in body temperature.

Progesterone and oestrogens also have another important effect – they inhibit the secretion of LH by the pituitary. The corpus luteum needs LH to survive, so the decline in LH leads to the death of the corpus luteum. The resulting drop in progesterone output causes the endometrium to die, precipitating another 'period'. Then, freed from inhibition by oestrogen and progesterone, the pituitary begins to secrete FSH and LH, beginning another cycle.

### *Menopause*

Whereas spermatogenesis continues throughout life in men, women start their reproductive lives with a finite number of follicles, many of which die every month. By the time she reaches her late forties or early fifties, a woman has few follicles left. Her menstrual cycles become irregular and then cease altogether, as does the output of sex hormones. This transition is called the **menopause**.

One of the effects of the reduction in oestrogen after menopause is that the bones undergo some demineralisation. This can be prevented by taking oestrogen supplements (hormone replacement therapy).

## Courtship and copulation

In many animals, fertilisation is preceded by some form of **mating behaviour** in which signals are exchanged between male and female. Mating signals may be visual (e.g. most birds), vocal (e.g. grasshoppers, many birds), chemical (e.g. many moths), or more usually, some combination of these. Complex mating behaviour is called **courtship**. Courtship may have a number of functions:

- it enables the female to identify the male as a possible mating partner rather than a meal (as in spiders) or a predator. Copulation is not possible unless the flight or aggressive responses of the female are first suppressed.
- it may enable the male and female systems to be brought into a state of readiness at the same time.
- since mating signals are usually species-specific, it helps to prevent interspecific mating which is wasteful of resources, since any resulting zygotes are usually less viable.
- when there is competition for females, it may also help females to select the 'best' males.

Though copulation in humans is essentially similar to that in other animals, courtship is frequently extremely complex. In addition to physical attraction, it usually involves a complex weave of physical and psychological factors. In males, sexual excitation leads to engorgement of the spongy tissue in the penis causing erection. In females, it leads to the erection of the clitoris and breasts, and the secretion of mucus by the vaginal wall, facilitating the insertion of the penis into the vagina. Rhythmic movements of the penis eventually lead to a peak of sexual excitement called **orgasm**, characterised by ecstatic pleasure and psychological release. In males, orgasm involves the contraction of the smooth muscle in the vasa deferentia and walls of the seminal vesicles, causing the **ejaculation** of semen into the vagina. In females, orgasm is not a prerequisite for fertilisation and involves rhythmic contractions of the walls of the vagina and uterus.

## Male and female strategies

From a genetic point of view, males and females make equal genetic contributions to the next generation, but in many other respects they are very different. Sperms are minute and can be produced in huge numbers with little cost in energy. Eggs are thousands of times larger and cost correspondingly more energy to produce.

A female thus invests more in each egg than a male invests in each of his sperms. It is therefore to the female's advantage to ensure that her eggs are fertilised by the highest quality sperms, so she is more choosy about whom she mates with. The male, with his superabundance of sperms, is under no such constraints. It is in his genetic interests to fertilise as many eggs as possible, regardless of their quality, since the cost of wastage is so small.

The differing 'aims' of males and females are reflected in their mating behaviour, with males competing for females. For the winner the genetic advantage may be very large. A successful male gorilla, for example, has exclusive reproductive 'rights' to a harem of a dozen or more females, with unsuccessful males living solitary lives as bachelors. When the competition is settled by physical force (as it is in many mammals) there may be strong selection in males for large size; an adult male gorilla, for instance, may be twice the size of an adult female, and male elephant seals may be three times as large as females.

Differences between males and females that are in addition to those of the sexual organs are called **sexual dimorphism**. In many birds, sex differences are in plumage, colourful males often displaying before drab females.

Photo of male and female elephant seals showing difference in body size as a result of differing sexual strategy

## Fertilisation

On entering the female genital tract, the semen coagulates due to an enzyme in the prostatic fluid, but within a few minutes it liquefies again, enabling the sperms to swim freely. The acidic pH in the vagina is hostile to sperms, but is partially neutralised by the alkaline seminal fluid.

In the 15–18 hours in which the egg remains viable after ovulation, it can get no further than the upper part of the oviduct, so fertilisation normally occurs here. The 200 million or so sperms in a typical ejaculate are deposited just outside the cervix of the uterus, about 10–15 cm away from the egg.

One mechanism which helps the sperms is the cervical mucus. Whereas normal cervical mucus is thick, the mucus produced around the time of ovulation is thin and extends as long strands which serve as guides by which the sperms can swim up the cervix.

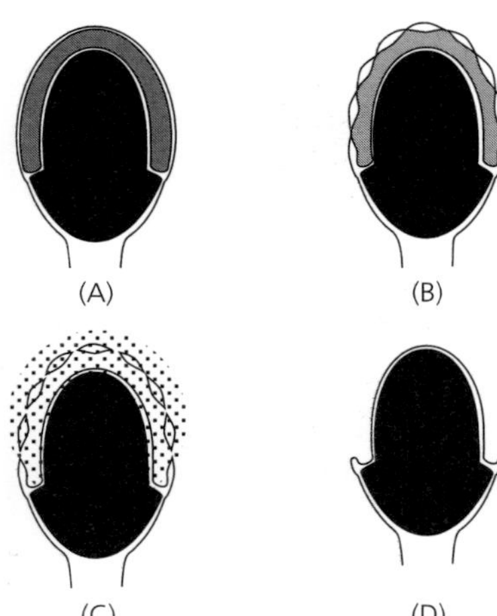


**Figure 28.14**
The acrosome reaction. The acrosomal membrane fuses at several points with the plasma membrane (B), releasing acrosomal enzymes (C)

Though sperms can survive in the female reproductive tract for up to 48 hours, the average time to reach the upper oviduct is 4–6 hours. It is known that sperms are able to reach the upper oviduct within 5 minutes of coitus, so they must be helped in some more active way. This is now thought to be the active muscular contractions of the uterus during orgasm, possibly stimulated by prostaglandins in the seminal fluid.

During the journey through the female reproductive tract the sperms undergo a process called **capacitation**, in which they acquire the ability to fertilise the egg. Capacitation takes about 5–7 hours, and is thought to involve the removal of certain proteins from the membrane overlying the acrosome.

Usually no more than a few hundred sperms reach the vicinity of the egg, and, for reasons to be explained, only one normally fertilises it. Contact between egg and sperm seems to be a matter of chance – there is no evidence that human sperms can orientate their movement with respect to the egg.

Before it can effect fertilisation a sperm must undergo the **acrosome reaction** (Figure 28.14). When the sperm head comes into contact with the granulosa cells, it triggers the entry of calcium ions into the sperm plasma membrane. This causes the outer membrane of the acrosome to fuse with the plasma membrane of the sperm, releasing the enzymes from the acrosome. These include **hyaluronidase** and **acrosin**, and together they digest a pathway through the granulosa cells and the zona pellucida.

Next, the sperm head contacts and fuses with the plasma membrane of the oocyte and the sperm enters (Figures 28.15 and 28.16).

The fusion with the egg cell membrane triggers a depolarisation of the plasma membrane of the oocyte, temporarily blocking entry of other

**Figure 28.15►**
The process of fertilisation

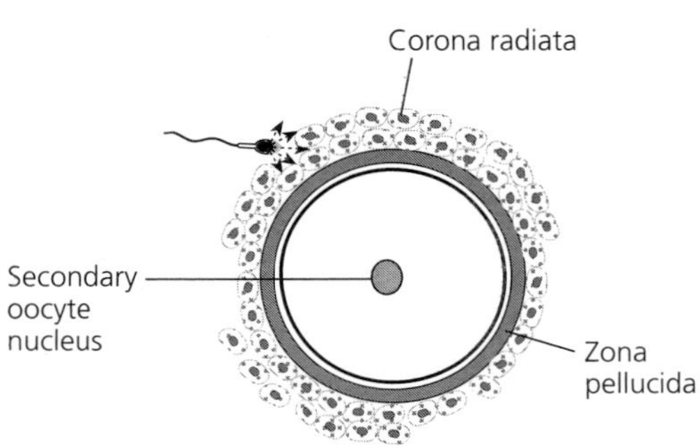


1. Acrosomal reaction enabling penetration of sperm through corona radiata and zona pellucida

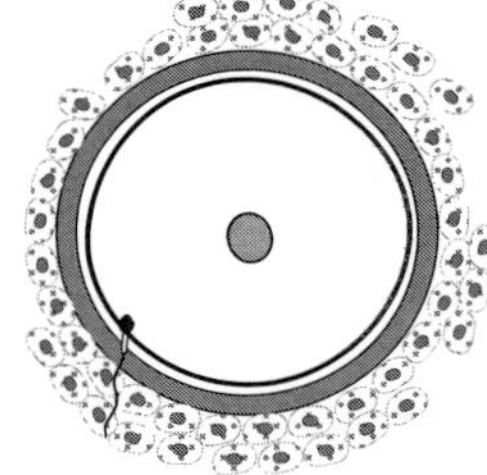

2. Fusion of sperm and oocyte plasma membranes

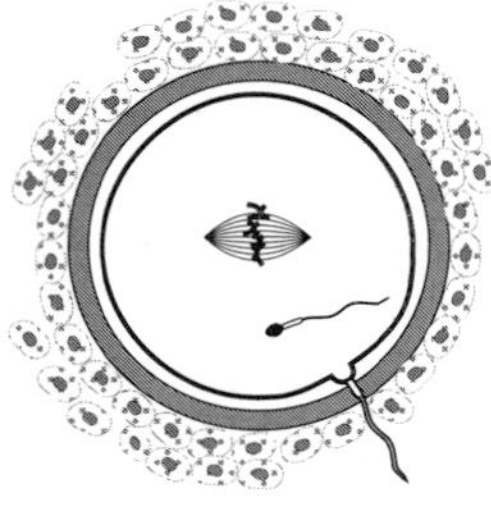

3. Sperm enters egg, leaving its plasma membrane outside. Second meiotic division of oocyte

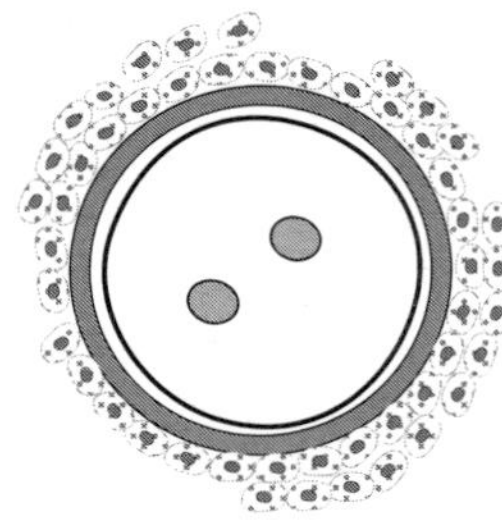

4. Sperm nucleus becomes male pronucleus

5. Spindle forms for first division of zygote

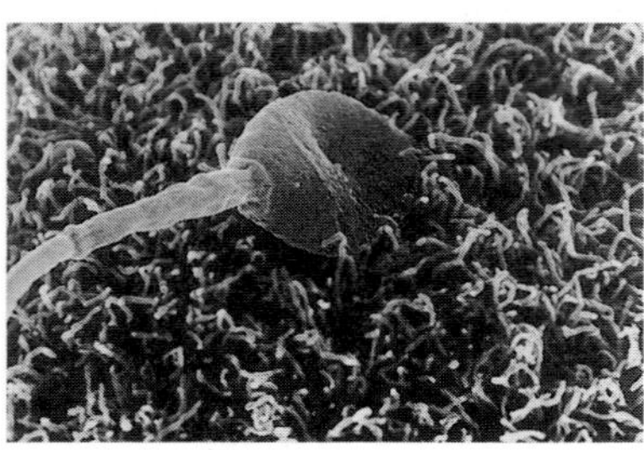

**Figure 28.16▲**
Electron micrograph of a human egg being fertilised by a sperm

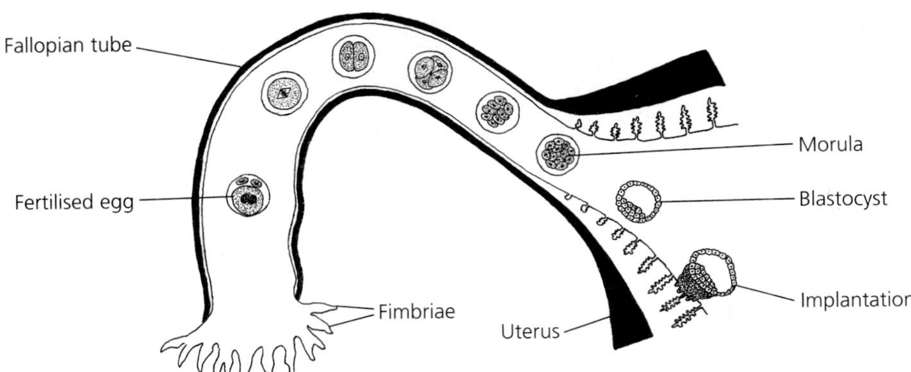


**Figure 28.17▶**
Cleavage in a mammal

sperms. This is followed by a more permanent block called the **cortical reaction**. Certain vesicles in the outer egg cytoplasm, or **cortex**, are exocytosed, rendering the zona pellucida impenetrable to other sperm.

The zona pellucida is also a barrier to interspecific fertilisation – if the zona is digested away from hamster eggs, they can be fertilised by human sperms. Only when a sperm touches the zona of an egg of the same species can it enter the egg.

The entry of the sperm also stimulates the second meiotic division which produces the mature ovum and the second polar body. There are now two nuclei in the egg – the male and female **pronuclei**. In mammals these do not fuse, but remain separate while their DNA replicates. The centrioles of the sperm also duplicate. Several hours after the sperm has penetrated the egg, the nuclear envelopes of the pronuclei break down and the first mitotic spindle forms. After the sperm head has entered the oocyte, the tail degenerates.

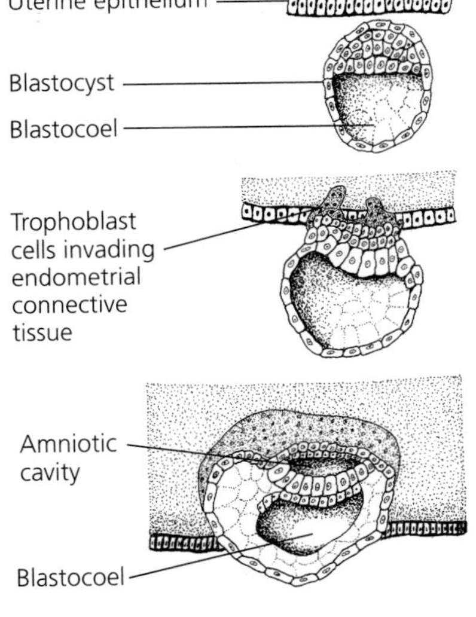


### *Early development*

The zygote (still surrounded by the zona pellucida) is slowly carried down the oviduct by ciliary action, possibly aided by muscular action of the walls of the oviduct. Its first developmental step is the replication of the chromosomes. Next it begins the process of **cleavage**, in which it divides mitotically into two, then four, and so on, forming a solid ball of cells called a **morula** (Figure 28.17). During this time it does not grow, so the cells get progressively smaller.

By the time it has reached the uterus (about 4 days after fertilisation) the little ball of cells has developed a cavity called the **blastocoel** and is now called a **blastocyst**. This is differentiated into an **inner cell mass** from which the embryo eventually develops, and an outer layer of cells, the **trophoblast**, which is concerned with the future nourishment of the embryo. All these cells are derived from the zygote, though not all will develop into the embryo.

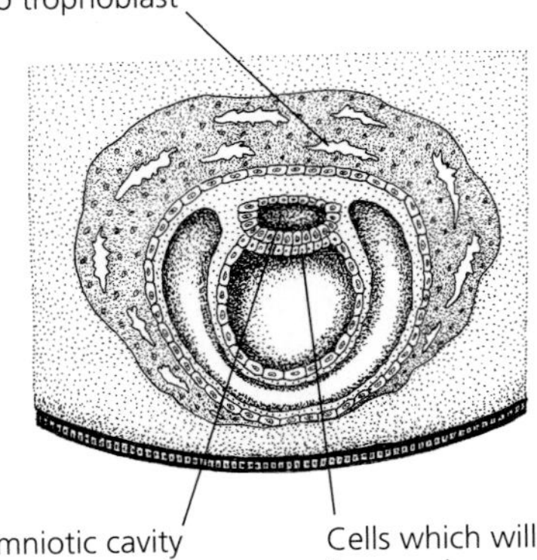


**Figure 28.18**
Implantation of a human blastocyst (after Tuchmann-Duplessis, David and Haegel)

### *Implantation*

By the time the blastocyst arrives in the uterus it consists of about 100 cells. It remains free in the uterine cavity for 2–4 days, nourished by secretions of the uterus. During this time the zona pellucida slowly disintegrates, allowing the blastocyst to come into direct contact with the endometrium. Over a period of about a week the blastocyst then becomes buried within the endometrium, a process called **implantation** (Figure 28.18). The trophoblast begins to develop branched extensions called **trophoblastic villi**. The outer layer of the trophoblast secretes enzymes that digest the superficial tissues of the endometrium releasing nutrients that are used by the embryo. For a week after implantation the products of digestion are the only source of food for the embryo.

As the digestion extends to the walls of the maternal blood vessels, the villi become surrounded by maternal blood spaces (Figure 28.19). Since the villi are finely divided, they provide a large surface area over which exchange of gases and nutrients can occur. Later the trophoblastic villi are replaced by a far larger and more extensive connection between mother and infant – the **placenta**.

The trophoblast has a second, equally important function – it secretes the hormone **chorionic gonadotrophin** (**CG**), which has a similar action to LH. By sustaining the corpus luteum, it maintains the supply of progesterone and oestrogen. Without these hormones the endometrium would break down. It is quite likely that many miscarriages are caused by the failure of these hormones to maintain early pregnancy. By about 2–3 months into pregnancy the endocrine role of the corpus luteum has been largely taken over by the placenta; until then, maintenance of the endometrium depends on the ovary.

The production of CG by the blastocyst can be detected a day or so before implantation and forms the basis for a pregnancy test (Figure 28.20).

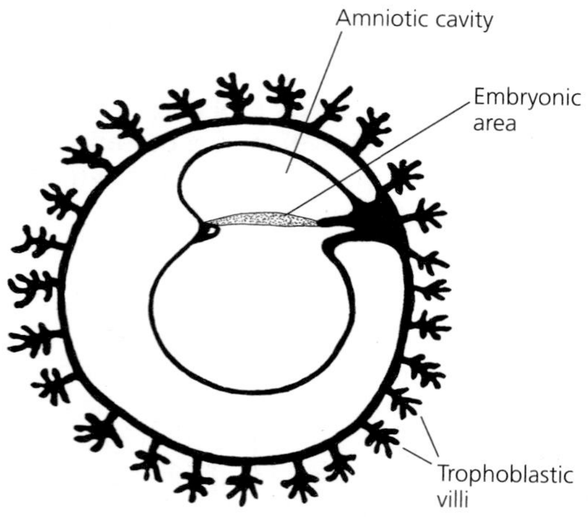


**Figure 28.19**
Three week old human embryo showing trophoblastic villi

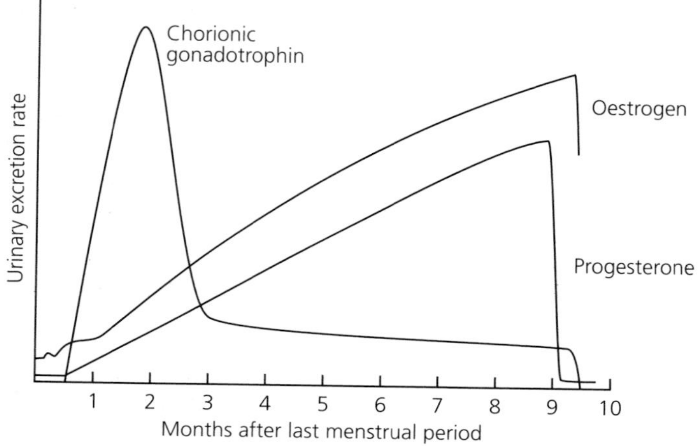


**Figure 28.20**
Urinary concentrations of chorionic gonadotrophin, oestrogen and progesterone during pregnancy

## *Embryonic development*

The period of development is called **gestation**. During this time the new individual increases in size by over a thousand million times, a relative growth rate that is far in excess of growth after birth. Development also involves a great increase in complexity – the relatively uniform cells of the inner cell mass eventually differentiate into over 200 kinds of cells, organised into tissues and organs.

Early development involves not only cell division but also cell movements, in which cells migrate from one location to another. The most critical cell movements are those that occur in **gastrulation** during the third week. In this process the main **germ layers**, the **ectoderm**, **endoderm** and **mesoderm**, are differentiated. The ectoderm gives rise to the epidermis and also the central nervous system, the endoderm gives rise

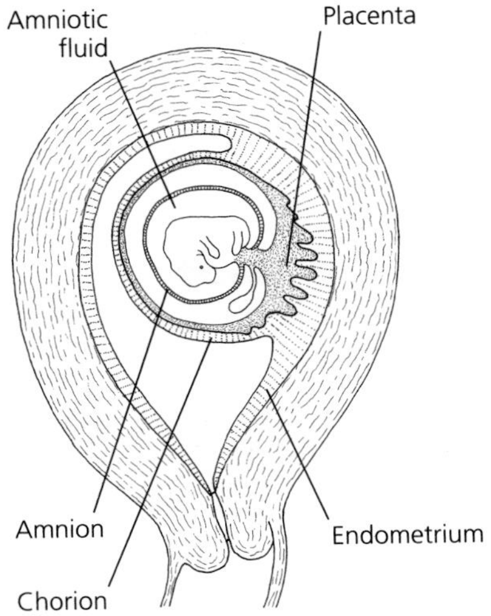


**Figure 28.21**
Two month old human embryo showing the foetal membranes

to the lining of the gut and its associated glands, and the mesoderm gives rise to muscle and connective tissue. Another critically important movement is the rolling up of a plate of cells to form the tubular central nervous system.

By the time it is a month old the embryo has the beginnings of a gut, kidneys, brain and a heart which is already beating. By 2 months the main organ systems have been laid down and the embryo is called a **foetus**.

Paralleling the changes in the embryo proper is the development of a series of **foetal membranes** (Figure 28.21). Although they are continuous with the foetus they are actually outside it, so they are also called **extraembryonic membranes**. The outermost of these is the **chorion** which overlays an inner **amnion**. Inside the amnion is a fluid-filled **amniotic cavity** in which the foetus floats. The fluid supports the foetus by buoyancy, and also helps to protect against physical blows by spreading any applied force over the whole body. Another membrane is the **allantois**, which grows out from the embryonic hind gut and later fuses with the chorion to give rise to the placenta.

## The placenta

The placenta is the place where maternal and foetal tissues come into contact. It has three functions:

- it allows substances to be exchanged between the blood of the foetus and the mother. The placenta therefore acts as the lungs, kidneys and gut of the foetus.
- it is an important endocrine gland, producing hormones necessary for the maintenance of pregnancy – chorionic gonadotrophin, oestrogen and progesterone – and also relaxin, which is important in the preparation for childbirth.
- it allows antibodies to be transferred from the mother to the foetus, so that for some weeks after birth the infant has immunity to the same diseases as its mother.

The placenta is formed from foetal and maternal tissues and, when fully developed, is about 20 cm across and 3 cm thick. It is connected to the foetus by the **umbilical cord** which consists of two **umbilical arteries** carrying blood from the foetus to the placenta, and an **umbilical vein** carrying blood from the placenta back to the foetus.

In the placenta, the blood of the mother and the foetus come very close together but *do not actually mix*. It is vital that they do not do so for two reasons. First, disease-causing organisms would be able to pass from mother to foetus (though HIV, rubella and the bacterium causing syphilis may still do so). Second, the mother and the baby are often of different blood groups. The latter situation can cause problems when the foetus is Rh-positive and the mother is Rh-negative (see Chapter 27).

As the placenta develops, many tree-like **chorionic villi** grow into the endometrium digesting the walls of the maternal blood vessels. As a result villi are surrounded by a very thin, continuous 'lake' of maternal blood, supplied by the **uterine arteries** and drained by the **uterine veins** (Figure 28.22). Each villus contains a network of foetal capillaries.

Two structural features of the placenta promote the transfer of materials between mother and foetus:

- the thousands of finely-divided villi have a large *surface area* (about 10 $m^2$ when fully developed).

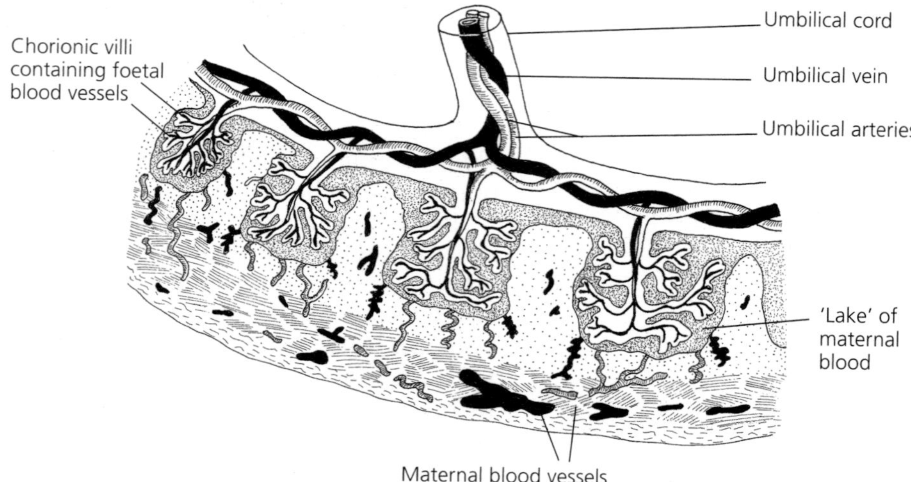


**Figure 28.22**
Structure of the human placenta

- in the fully-developed placenta, only three foetal layers separate foetal and maternal blood: the capillary endothelium, a thin layer of connective tissue and the epithelium covering the villi. The barrier separating foetal and maternal blood is therefore extremely thin, so the diffusion path between foetal and maternal blood is correspondingly short (about 3.5 μm).

Oxygen and foods such as glucose, amino acids, vitamins and minerals, pass from the mother's blood to the foetal blood. Wastes, such as carbon dioxide and urea, pass in the opposite direction (Figure 28.23). Simple substances with small molecules such as oxygen, $CO_2$ and urea, cross between foetal and maternal blood by simple *diffusion*. Amino acids cross the placenta by *active transport* and glucose is transported by *facilitated diffusion*. Maternal antibodies are taken up into the villi by pinocytosis.

Transport of oxygen is helped by the fact that foetal haemoglobin has a greater affinity for oxygen than does adult haemoglobin, and also by the Bohr effect (see Chapter 22).

**Figure 28.23**
Transport across the placenta

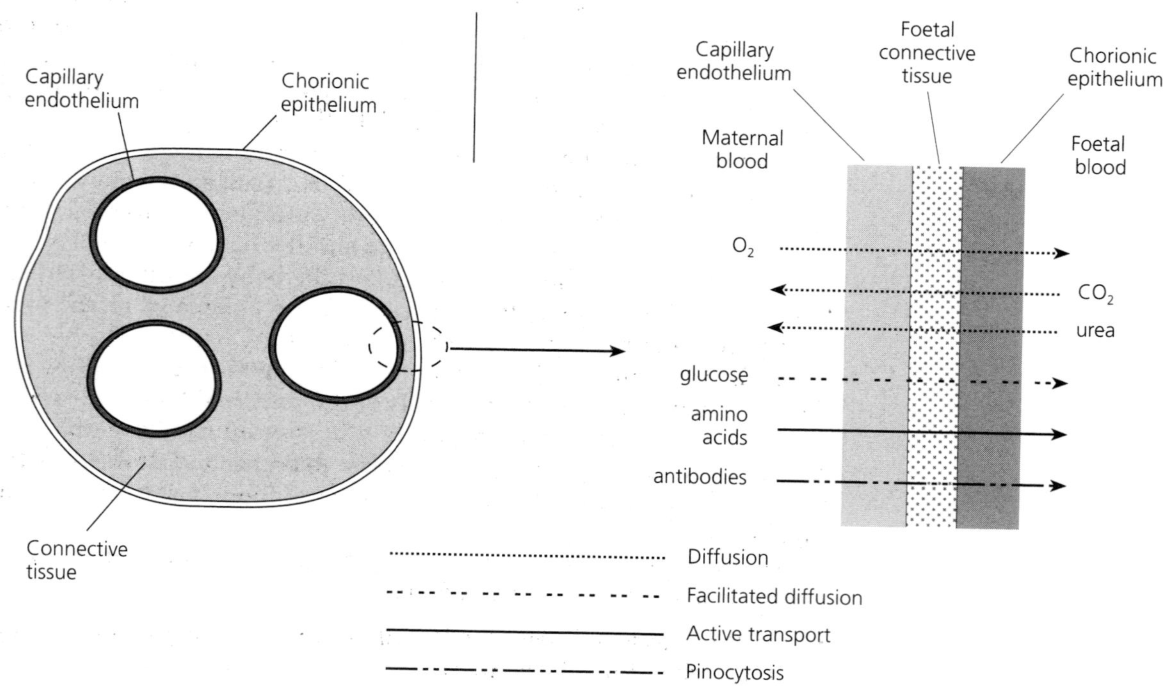

## Twins

Although it is normal for a human to have one baby, two or even more can be produced in one pregnancy. There are two kinds of twins:

- **dizygotic**, or **dissimilar**, **twins** are produced when two eggs are released from the ovary (probably one from each). Since each is fertilised by a different sperm, they are no more alike than other brothers and sisters.
- **monozygotic**, or **similar**, **twins** are produced when the embryo splits into two, each growing into a foetus. Because they develop by mitotic division of the same zygote they are genetically identical, readily accepting tissue grafts from each other. Although they are called **identical** twins, they are never quite identical because no two people grow up in identical *environments*.

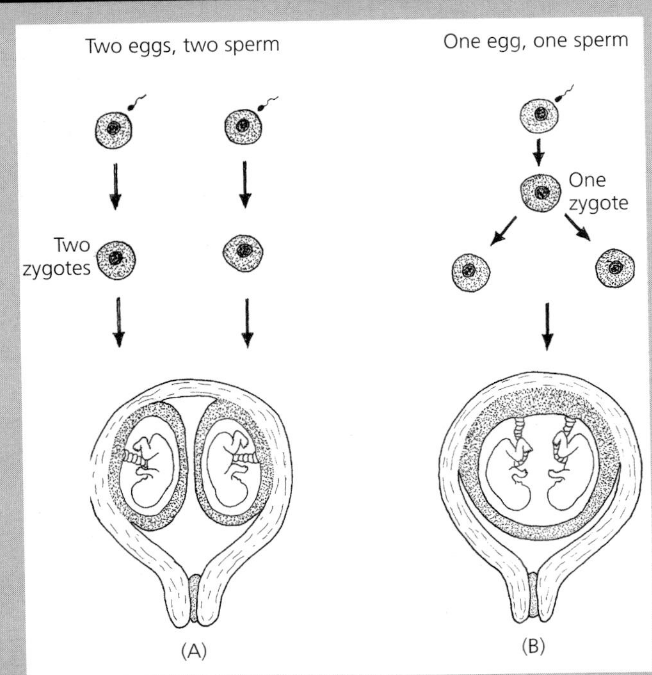


Dizygotic (A) and monozygotic (B) twins

### Maternal changes in pregnancy

As the foetus develops, changes are also occurring in the mother. The uterine smooth muscle fibres enlarge greatly, and the uterus grows from its normal mass of about 50 g to over a kilogram by the time the baby is born. In response to oestrogen, the ducts of the breast tissue enlarge, whilst progesterone stimulates the growth of the secretory tissue. In the last two months of pregnancy, the ligaments holding the two sides of the pelvic girdle at the front – and also at the joint between the sacrum and hip bone – begin to soften. This is brought about by the hormone **relaxin** secreted by the placenta, and makes the process of birth easier. Relaxin also helps to soften the cervix in preparation for its dilatation during childbirth.

### Birth, or parturition

A normal pregnancy lasts approximately 38 weeks from implantation (40 weeks from the last period). During the last few weeks the foetus usually comes to lie with its head down against the **cervix**. The process of childbirth, or **labour**, involves three stages (Figure 28.24).

The first stage is preceded by the rupture of the amnion and the release of the amniotic fluid and is by far the longest, lasting from a few hours to as much as 24 hours. The function of the first stage is to stretch the cervix, a process that usually involves considerable pain. The uterus begins to undergo spontaneous, weak contractions beginning about every 30 minutes, and increasing in strength and frequency until they are coming about every 3 minutes. These contractions are aided by the hormone **oxytocin** secreted by the posterior pituitary gland.

When the baby's head has fully dilated the cervix, the second stage begins. It lasts only a few minutes and results in the expulsion of the baby,

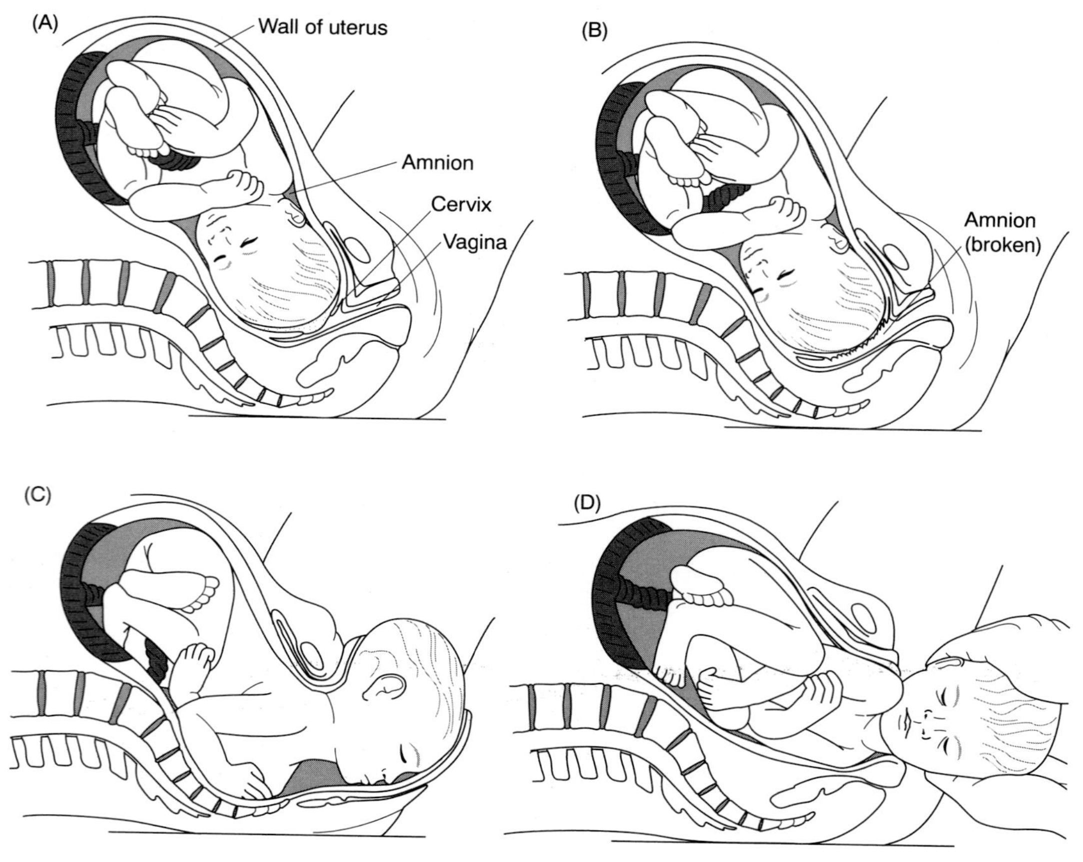


**Figure 28.24**
The stages of labour

aided by voluntary contractions of the mother's abdominal muscles. The umbilical cord is cut, isolating the baby from the mother. The resulting rise in the carbon dioxide content of the blood stimulates the baby to take its first breath.

The third stage consists of the expulsion of the placenta and occurs about half an hour after the baby is born.

Though the mechanics of labour are well known, the exact signal for its onset in humans is not understood. One thing is known though – it is the baby that 'decides', and not the mother. Whatever its exact nature, it involves the release of local hormones called **prostaglandins** in the foetal side of the placenta. These stimulate the uterine muscle to begin contractions.

### *Changes in the circulation at birth*

There are a number of important differences between the circuitry of the foetal and adult blood systems. These are related to the fact that before birth the most oxygenated blood enters the right side of the heart, but the developing brain (which has the greatest need for oxygen) is supplied by the aorta which leaves the left side of the heart (Figure 28.25).

The foetal system is characterised by three shunts or short cuts which close off after birth:

1 the **foramen ovale**, which connects the right and left atria. This is guarded by a flap valve which allows blood to flow from right side to left, but not in the reverse direction.

2 the **ductus arteriosus**, through which blood flows from the pulmonary artery to the aorta.

3 the **ductus venosus**, through which most of the blood in the umbilical vein bypasses the liver and enters the posterior vena cava (inferior vena cava in humans).

Blood enters the right atrium in two streams: a deoxygenated stream from the head and arms, and a partially oxygenated stream from the placenta, trunk and legs. However, mixing of these two streams is only partial. The most oxygenated blood passes through the foramen ovale into the left side of the heart, which pumps it to the brain and forelimbs. The less oxygenated flow passes into the right ventricle and the pulmonary artery. Only a small proportion of this stream goes to the lungs which, in their unexpanded state, offer a high resistance to flow. Most of the blood crosses to the aorta via the ductus arteriosus, and then travels to the legs, trunk and placenta. Because of these two shunts, the left and right sides of the heart work in parallel rather than in series (as they do after birth) and so do not have to pump equal volumes of blood.

After birth these shunts close, and the left and right sides of the heart become separated. First, the smooth muscle in the umbilical arteries contracts, closing them. The $CO_2$ level in the baby's blood begins to rise, stimulating the breathing centre in the medulla. As the baby takes its first breath, the expansion of the lungs lowers the resistance of the pulmonary vessels, increasing the flow of blood through them and raising the pressure of blood in the left atrium. This reverses the pressure gradient across the foramen ovale, pushing the flap valve closed. Over the next few months, closure becomes consolidated by tissue fusion. In some babies closure is incomplete, leading to impaired separation of oxygenated and deoxygenated blood.

The ductus arteriosus also closes soon after birth, though for a different reason. Before birth, the placenta takes 40% of the cardiac output. The loss of placental circulation increases the resistance to blood

**Figure 28.25▼**
The foetal circulation

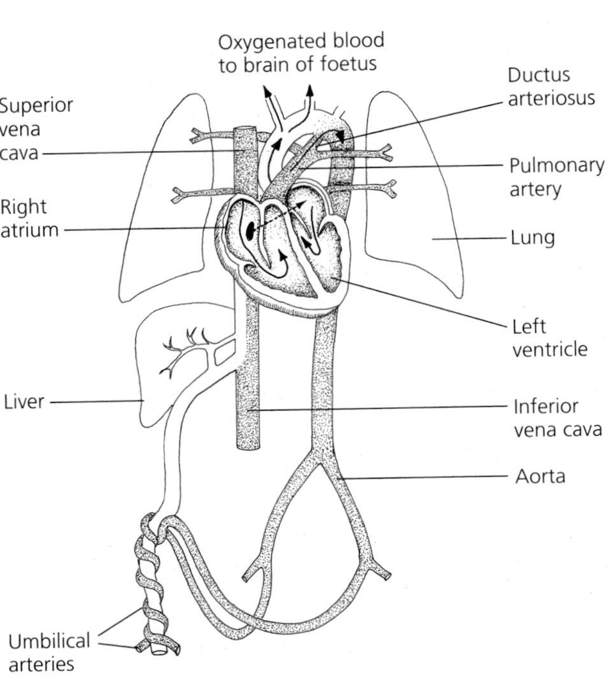


**Figure 28.26▼**
How suckling promotes the secretion and release of milk. Note that suckling promotes milk secretion by inhibiting secretion of prolactin inhibiting hormone

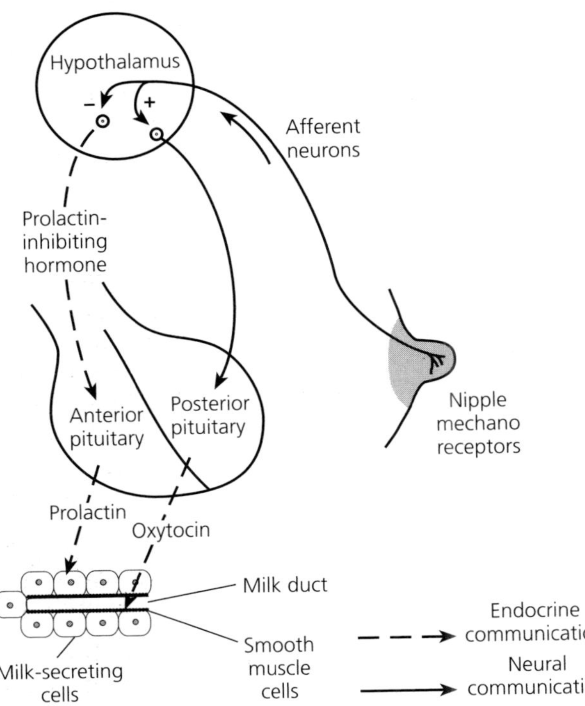

flow in the systemic circuit, so the pressure of blood in the aorta rises. This, coupled with the fall in pressure in the pulmonary artery, reverses the blood flow through the ductus arteriosus. Oxygenated blood from the aorta now flows through the ductus arteriosus. This rise in oxygenation stimulates the smooth muscle in the ductus arteriosus to contract, closing it off.

Besides these structural changes, birth also triggers a switch from the production of foetal to adult haemoglobin (though it takes 3–4 months for all the foetal red cells to be replaced). Since adult haemoglobin binds oxygen less strongly, it releases it more readily in the tissues, allowing them to function at higher oxygen concentrations.

## Lactation and parental care

After the baby has been born, the milk-producing tissue of the breasts begins to secrete milk – a process called **lactation**. When the baby suckles, nerve impulses travel to the mother's brain causing her pituitary gland to produce oxytocin. This stimulates the muscle in the walls of the milk ducts to contract, squeezing milk into the baby's mouth (Figure 28.26). Another hormone, **prolactin** (secreted by the anterior pituitary) stimulates the secretion of milk.

One of the effects of suckling is that it helps to prevent secretion of FSH and LH and hence ovulation. A mother is, therefore, less likely to conceive as long as she continues to feed her baby on her own milk. Breast milk is also better because it is free of bacteria and contains the essential nutrients in the right proportions.

Milk contains all that a baby needs for three or four months and is especially rich in calcium to help its bones grow. Soon, however, the baby will need solid food because milk contains no fibre and practically no iron (a new born baby has enough iron stored in its liver to last several months).

## QUESTIONS

**1** Vasectomy is a form of contraception in which the vasa deferentia (tubes from the testis) are cut. Explain why:

**a)** the man becomes sterile.
**b)** his sexual desires and ability to have intercourse are unaffected by the operation.
**c)** his semen cannot be DNA tested.

**2** Highly inbred strains of rats are so similar genetically that they can accept tissue grafts from each other. What would you expect to be the effect of transplanting a testis of a newly-weaned rat into a mature rat?

**3** Most mammals are polygamous (one male mating with many females) but most birds are monogamous (each male mating with one female). What feature of reproduction in birds makes it advantageous to the male to remain with one mate?

**4** If the ovary is removed a month after implantation the foetus is aborted, but if it is removed when the foetus is 4 months old normal development continues. Explain why.

**5** Consider the following statements:

**a)** In the placenta, the concentration of glucose is higher in the maternal blood than in the foetal blood.
**b)** D-glucose, the naturally occurring form, crosses the placenta considerably faster than L-glucose, which does not occur naturally.
**c)** The concentration of amino acids is higher in foetal blood than in maternal blood.

Use this evidence to suggest the most probable mechanisms for the transport of
**i)** glucose;
**ii)** amino acids, across the placenta.

## EXAM QUESTIONS

**1** Trout are carnivorous freshwater fish which may be reared intensively on fish farms. They are fed on pellets of artificial food.

**a)** Briefly describe the test you would carry out and the results that you would expect which would show that these trout pellets contained protein. *(2 marks)*

**b)** The table shows the amount of food recommended for 18 cm trout when kept at different temperatures.

| Temperature/°C | Amount of food as percentage body mass per day |
|---|---|
| 4 | 0.8 |
| 6 | 1.2 |
| 8 | 1.5 |
| 10 | 1.7 |
| 12 | 1.9 |

Give an explanation for the difference in food required as the temperature increases. *(2 marks)*

**c)** Trout convert a higher proportion of food into body mass than domestic mammals such as cattle and sheep. Suggest *two* explanations for this. *(2 marks)*

**AEB**

**2** The diagram summarises the way in which adrenaline can control a chemical reaction in a liver cell.

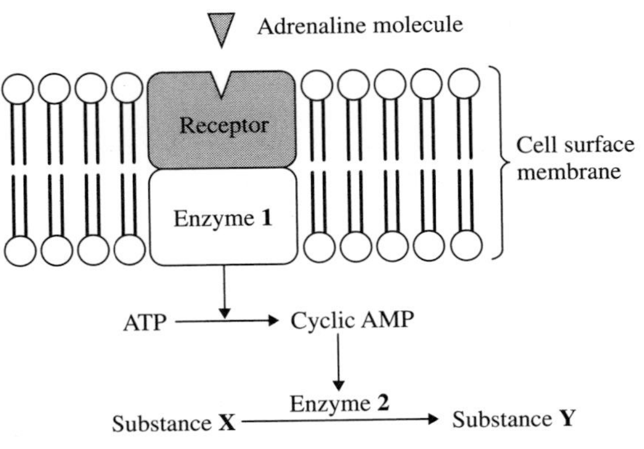


**a)** Describe the function of cyclic AMP in this process. *(1 mark)*

**b)** Give *one* example of a chemical reaction in a liver cell which is controlled by adrenaline by naming:

**i)** substance **X**; *(1 mark)*

**ii)** substance **Y**. *(1 mark)*

**c)** Use the diagram to explain:

**i)** why adrenaline may affect some cells and not others; *(1 mark)*

**ii)** how a single molecule of adrenaline may cause this cell to produce a large amount of substance **Y**. *(2 marks)*

**AEB**

**3** The graph shows changes in the potential difference across the cell surface membrane during an action potential.

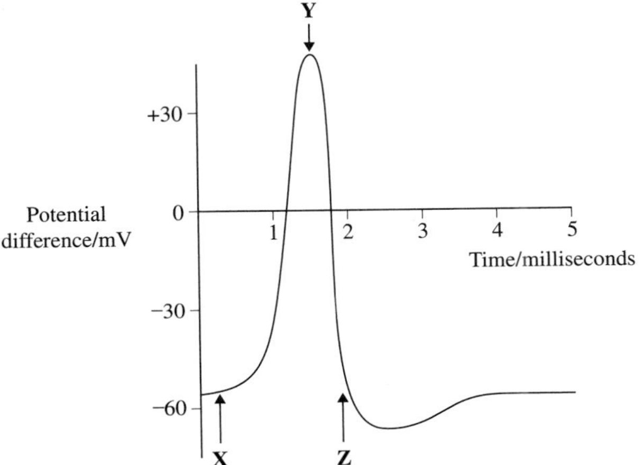


**a)** Describe the events which lead to the change in potential difference between:

**i)** times **X** and **Y**; *(2 marks)*

**ii)** times **Y** and **Z**. *(2 marks)*

**b)** The table shows the rate of conduction of nerve impulses along three different axons.

| Axon | Diameter/μm | Rate of conduction/m $s^{-1}$ |
|---|---|---|
| A | 7 | 1.2 |
| B | 15 | 90 |
| C | 500 | 33 |

Suggest an explanation for the rate of conduction along axon **B**. *(2 marks)*

**AEB**

**4** The diagram over the page shows the way in which temperature is regulated in the body of a mammal.

**a)** Which part of the brain is represented by box **X**? *(1 mark)*

## EXAM QUESTIONS

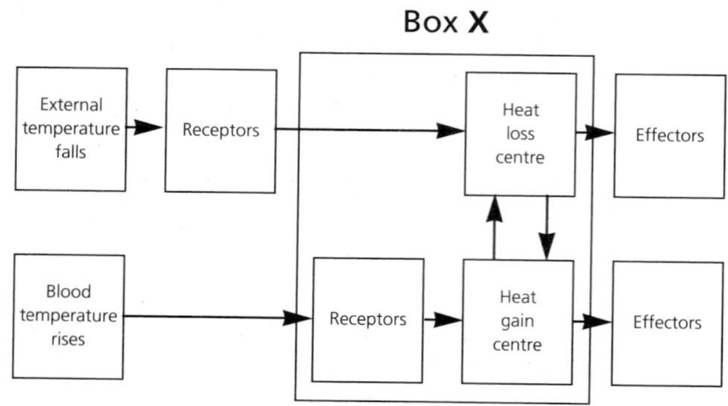


b) i) How does the heat loss centre control the effectors which lower the body temperature? *(1 mark)*

ii) Explain how blood vessels can act as effectors and lower the body temperature. *(3 marks)*

**AEB**

**5** The diagram shows one way in which twins can be formed.

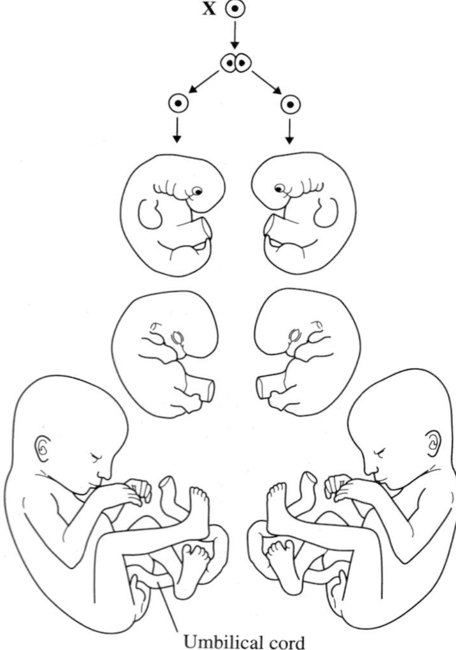


a) i) Give the name of the cell labelled **X**. *(1 mark)*

ii) Why, in this case, will the individuals which develop from the embryos be identical twins? *(2 marks)*

b) The umbilical cord contains two arteries and one vein. Give *one* function of:

i) the umbilical arteries; *(1 mark)*

ii) the umbilical vein. *(1 mark)*

**AEB**

**6** a) The diagrams show how an entire muscle is made up of muscle fibres containing muscle fibrils.

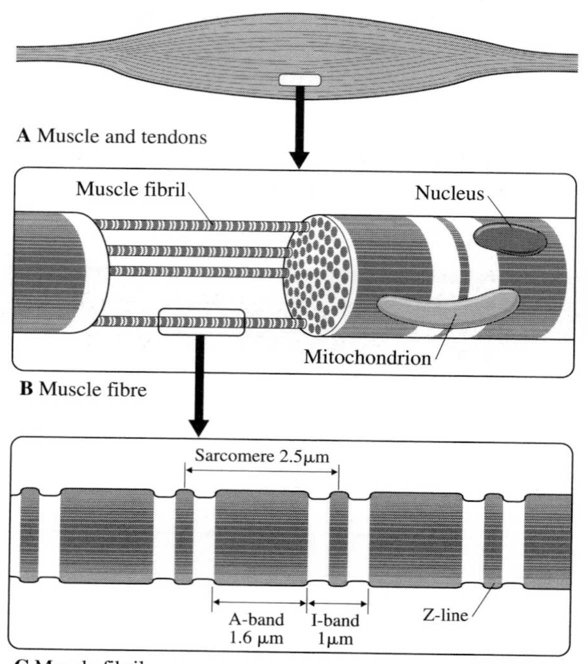


Use the diagrams to explain how:

i) a muscle fibre contracts; *(3 marks)*

ii) the entire muscle is able to move a bone. *(1 mark)*

b) The diagram shows some of the stages in the process of glycolysis and the reaction which links this to the Krebs cycle in a muscle cell.

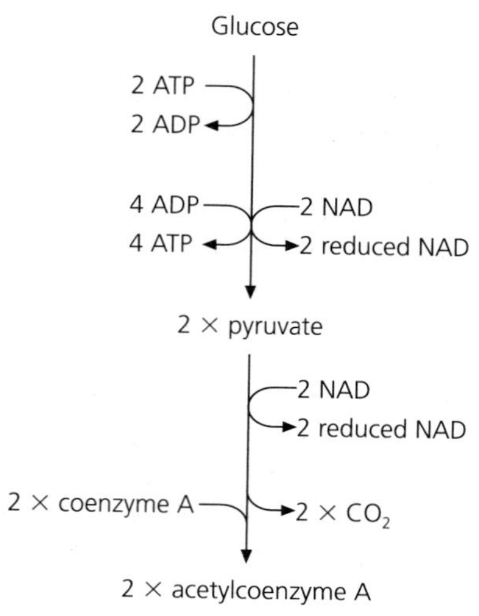

## EXAM QUESTIONS

i) Calculate the net gain of ATP produced, per molecule of glucose, in the process of glycolysis. *(1 mark)*

ii) Describe what would happen in the reactions shown in the diagram if there were a shortage of oxygen. *(2 marks)*

**AEB**

**7 a)** The graph shows changes in the depth of breathing of an adult.

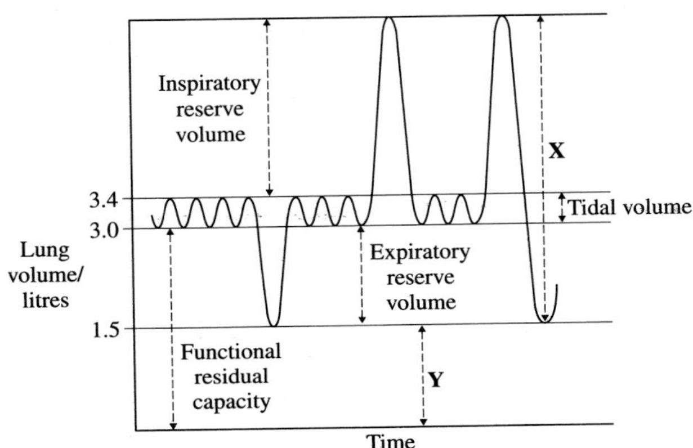


i) Give the names of the lung capacities labelled **X** and **Y**. *(2 marks)*

ii) Give *two* factors which could alter the tidal volume. *(1 mark)*

**b)** Describe how carbon dioxide released from an actively respiring cell would be transported from the cell and eventually pass into the lumen of an alveolus. *(4 marks)*

**AEB**

**8 a)** Give *one* difference between the composition of glomerular filtrate and blood plasma. *(1 mark)*

**b)** In the renal capsule, the effective filtration pressure depends on the hydrostatic pressure in the capillaries of the glomerulus which tends to force fluids out and the osmotic pressure of the blood proteins which tends to draw fluids back into the blood.

i) If the hydrostatic pressure in the capillaries is 5.5 kPa and the osmotic pressure of the blood proteins is 4.2 kPa, calculate the effective filtration pressure. *(1 mark)*

ii) Describe *one* structural feature of the blood vessels associated with the glomerulus and explain how it helps to maintain a high effective filtration pressure. *(2 marks)*

iii) Suggest an explanation for the fact that the urine of patients suffering from hypertension may contain protein. *(2 marks)*

**c)** The diagram summarises the mechanism by which sodium is reabsorbed from the first convoluted tubule.

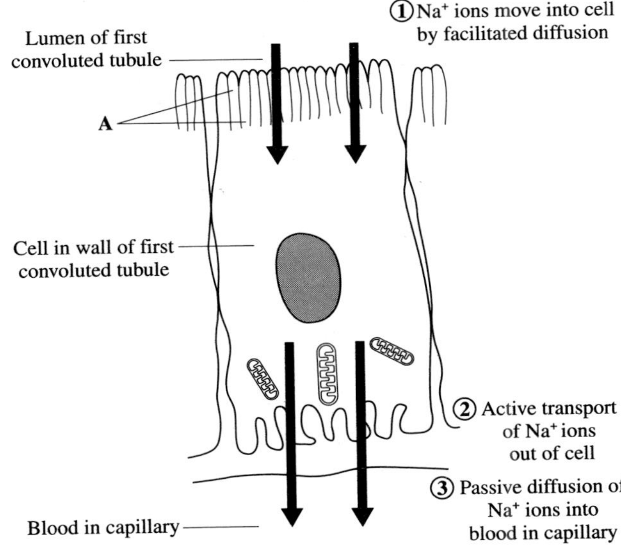


i) What is the function of the structures labelled **A**? *(2 marks)*

ii) Sodium ions move into the tubule cell by facilitated diffusion and out into the blood by active transport. Give *two* differences between facilitated diffusion and active transport. *(2 marks)*

iii) Explain the importance of the active transport mechanism in enabling diffusion of sodium ions into this cell. *(1 mark)*

**d)** Explain why, although large amounts of glucose and sodium ions are reabsorbed from the first convoluted tubule, its overall solute concentration remains constant. *(2 marks)*

**e)** The graph over the page shows how the rates of filtration of glucose by the glomerulus and its reabsorption from the first convoluted tubule vary with plasma glucose concentration.

EXAM QUESTIONS

## EXAM QUESTIONS

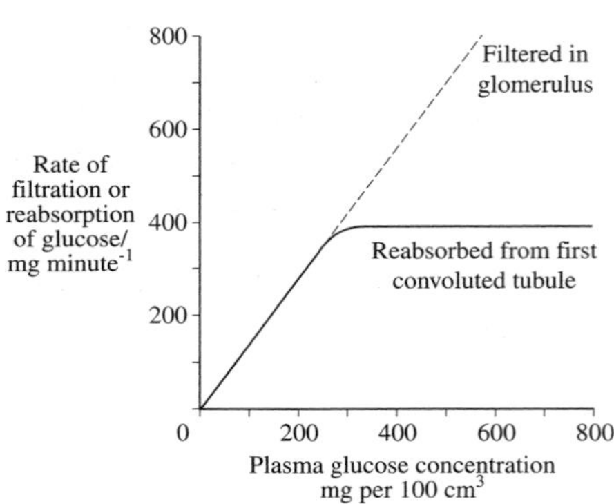


i) Describe and explain the difference in the shape of these two curves at plasma glucose concentrations greater than 300 mg per 100 cm$^3$. *(3 marks)*

ii) Copy the graph and sketch a curve on it to show how the amount of glucose in the urine would vary with plasma glucose concentration. *(2 marks)*

f) Explain why, in a non-diabetic individual, glucose does not normally appear in the urine. *(2 marks)*

**AEB**

**9** a) The diagram shows a posterior view of the human heart.

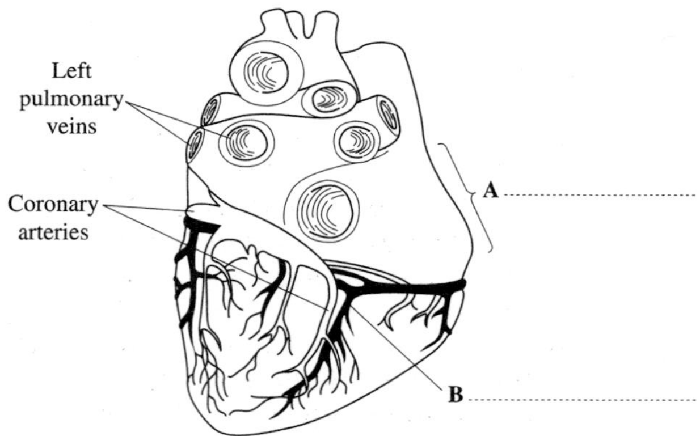


i) On the diagram, name the structures labelled **A** and **B**. *(2 marks)*

ii) Give the name of the tissue which makes up the bulk of the wall of the left ventricle. *(1 mark)*

b) The graphs show the changes in pressure in the left ventricle and aorta and the changes in volume of the left ventricle during the cardiac cycle.

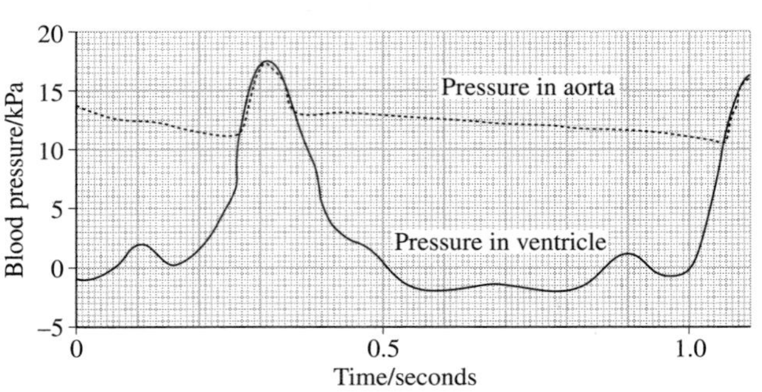


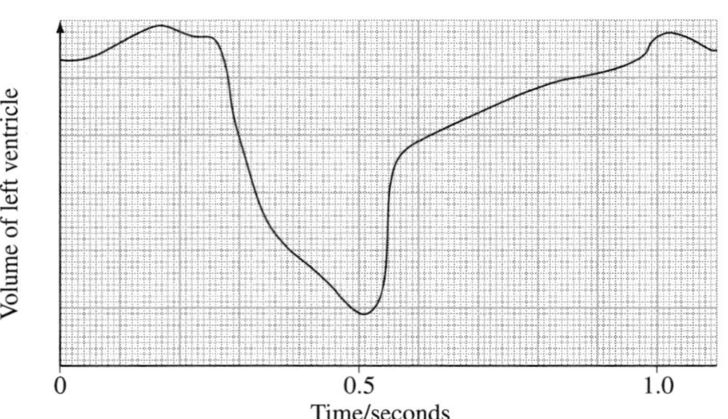


Explain how the changes in blood pressure in the left ventricle and in the aorta between 0.2 and 0.5 seconds are brought about. *(4 marks)*

**AEB**

**10** When foreign antigens enter the body, they often initiate an immune response which leads to the production of antibodies and an immunity. Vaccinations have been used for many years as a means of inducing an immune response and immunity without contracting the disease. The vaccines used in the United Kingdom against influenza are based on inactivated viruses which must be injected. In other countries, the vaccines used contain 'live', attenuated viruses which can be delivered by a nasal spray.

a) i) What is an antigen? *(1 mark)*

ii) What is an antibody? *(2 marks)*

b) Suggest a biological advantage in being able to deliver an influenza vaccine by means of a nasal spray. *(1 mark)*

c) Suggest a possible danger in using 'live', attenuated viruses in vaccines. *(2 marks)*

**AEB**

## EXAM QUESTIONS

**11** The graph shows the changes in pH caused by the action of the enzyme lipase on milk fat in the presence and in the absence of bile salts.

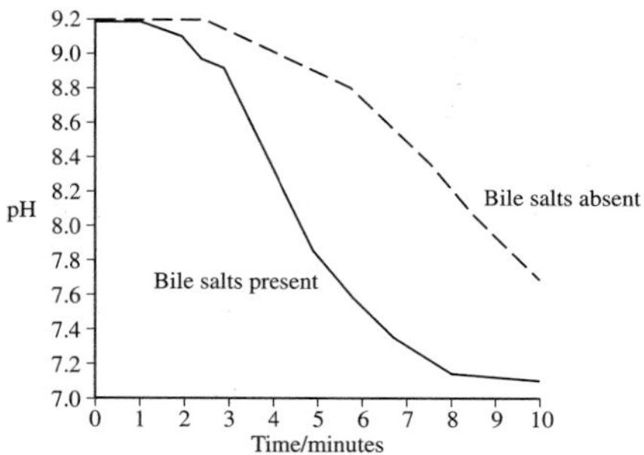


**a)** Explain what causes the fall in pH during fat digestion. *(1 mark)*

**b)** Explain the difference between the two curves. *(3 marks)*

**c)** The rate of action of lipase in the presence of bile salts is seen to slow down after eight minutes. Suggest *two* reasons for this. *(2 marks)*

**AEB**

**12** The relative sensitivity of the human eye to different wavelengths of light varies under different conditions of illumination. The graph shows the relative sensitivity to different wavelengths when exposed to bright light and to dim light.

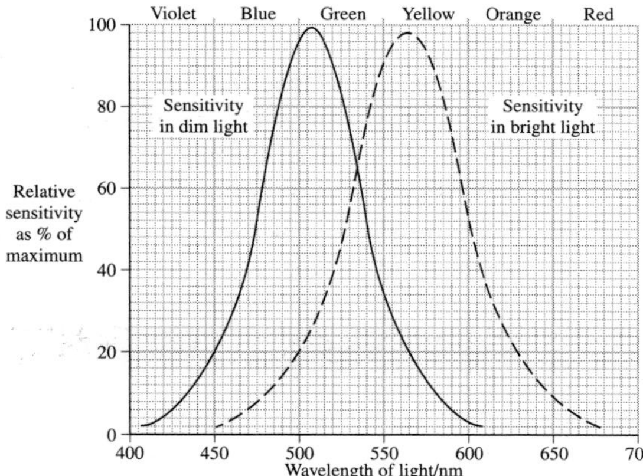


**a)** To which wavelength of light is the eye most sensitive:

**i)** in dim light; *(1 mark)*

**ii)** in bright light? *(1 mark)*

**b)** Suggest an explanation for this difference in sensitivity based on your understanding of the structure of the human retina. *(4 marks)*

**AEB**

**13** During exercise, the body adapts to meet the new demands placed upon it. Diagram **A** shows how the distribution of blood to various organs of the body alters as more and more vigorous exercise is undertaken. Diagram **B** shows some of the monitoring and control mechanisms which bring about these changes.

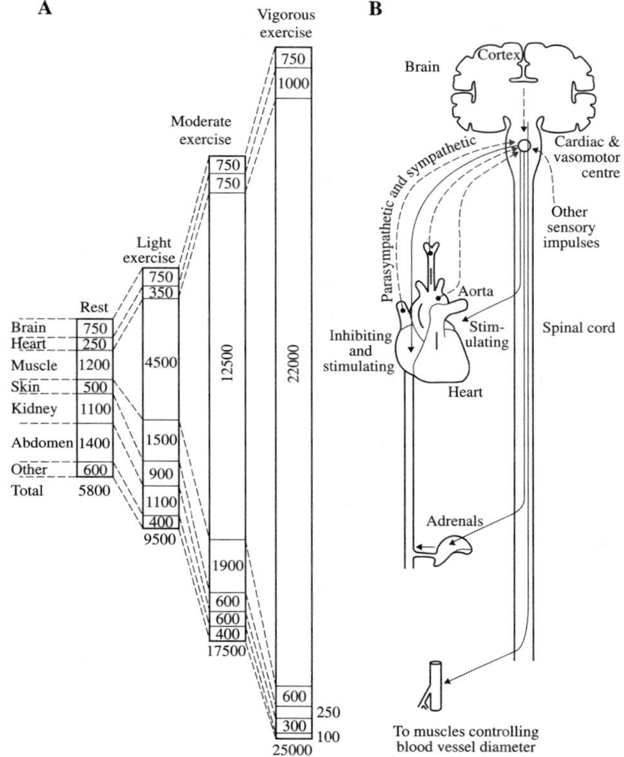


**a)** What percentage of the total cardiac output passes to the muscles:

**i)** at rest;

**ii)** during vigorous exercise? *(2 marks)*

**b)** Why is it an advantage to have more blood flowing to the muscles during vigorous exercise than at rest? *(4 marks)*

**c)** Use Diagram **B** and your own knowledge to explain:

**i)** how the increase in cardiac output is achieved; *(4 marks)*

**ii)** the mechanisms which alter the distribution of blood to the various organs. *(4 marks)*

All exercise involves the contraction of muscles. The diagram over the page represents part of a myofibril of striated muscle.

## EXAM QUESTIONS

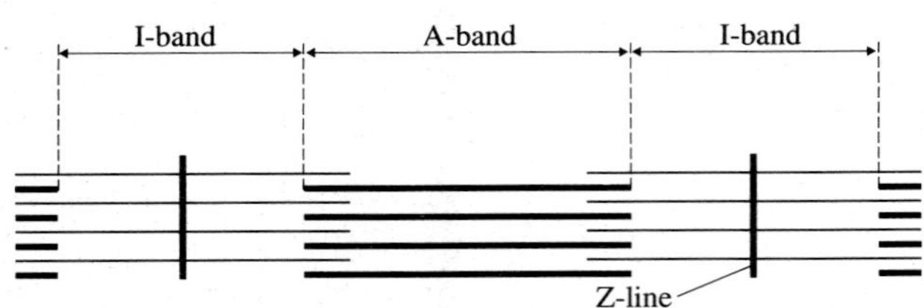


d) Give the name of the main protein which is found in the thin filaments. *(1 mark)*

e) When the myofibril contracts, what will happen to the size of:
   i) the **A** band;
   ii) the **I** band? *(2 marks)*

f) Describe how, within a myofibril, energy is constantly made available for contraction. *(3 marks)*

**AEB**

**14** The diagrams show transverse sections through a chordate, a platyhelminth and an annelid to show their relative size and shape.

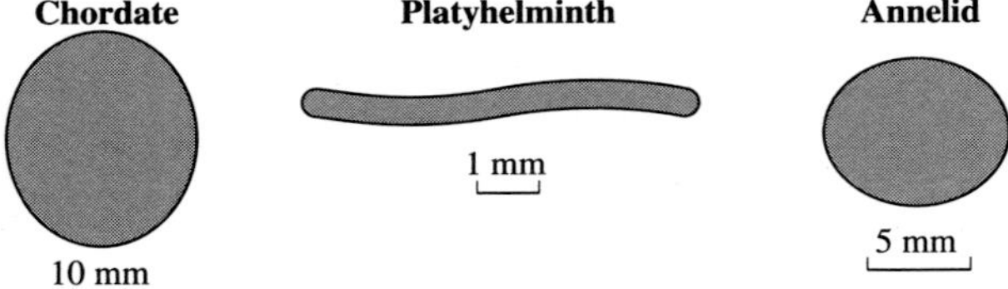


a) Complete the table to show the presence (✓) or absence (✗) of each of the systems given in these three organisms. *(2 marks)*

| System | Chordate | Platyhelminth | Annelid |
|---|---|---|---|
| Specialised gas-exchange system | | | |
| Blood system | | | |

b) Explain the relationship between body size and the presence of:
   i) a specialised gas-exchange system; *(2 marks)*
   ii) a blood system. *(2 marks)*

**AEB**

**15** Small desert mammals can survive without drinking. They meet their water requirements mainly from metabolic water.

a) What is metabolic water? *(1 mark)*

b) Describe *one* other way in which these animals are able to gain water. *(1 mark)*

c) The camel has a characteristic hump of fat on its back. One suggestion is that it functions as a means of providing the camel with water.

The table gives some figures relating to metabolism of fat in a camel over a period of time.

| | |
|---|---|
| Mass of fat respired/kg | 1.06 |
| Mass of water formed on oxidation/kg | 1.13 |
| Volume of oxygen required/$dm^3$ | 2130 |
| Mass of water evaporated from lungs/kg | 1.80 |

Use the figures in this table to explain why:

i) one gram of fat can produce more than one gram of water on oxidation; *(1 mark)*

ii) the suggestion that the fat in the camel's hump provides the animal with water cannot be correct. *(2 marks)*

**AEB**

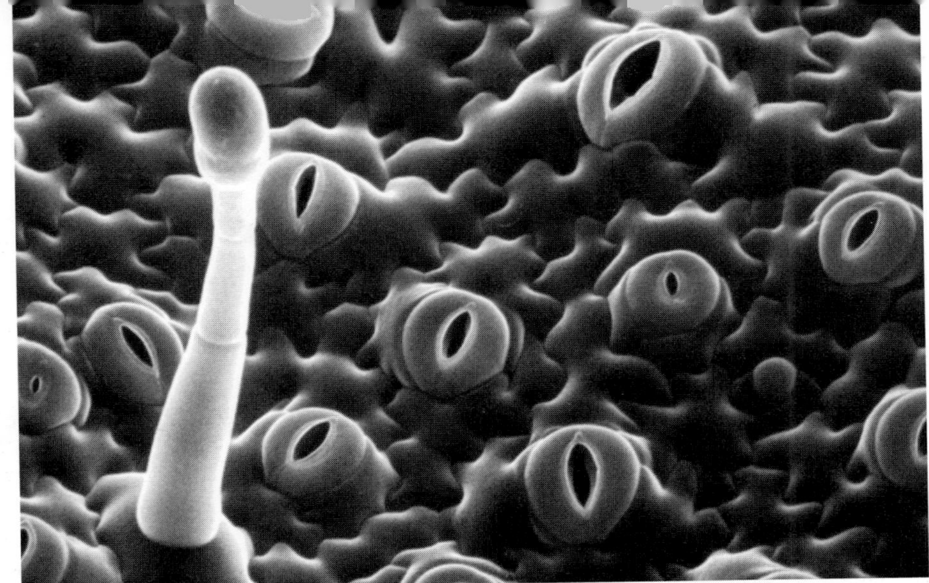

# The FUNCTIONING PLANT

CHAPTER 29
PLANT ORGANISATION . . . . . . . . . . . . . . . . . . . . . . . . . . . . . . . . . . 530

CHAPTER 30
PLANT LIFE CYCLES . . . . . . . . . . . . . . . . . . . . . . . . . . . . . . . . . . 537

CHAPTER 31
GROWTH in PLANTS . . . . . . . . . . . . . . . . . . . . . . . . . . . . . . . . . . 562

CHAPTER 32
The CONTROL of PLANT GROWTH . . . . . . . . . . . . . . . . . . . . . . . . 585

CHAPTER 33
GATHERING ENERGY and RAW MATERIALS . . . . . . . . . . . . . . . . . 606

CHAPTER 34
TRANSPORT in PLANTS . . . . . . . . . . . . . . . . . . . . . . . . . . . . . . . . 641

CHAPTER

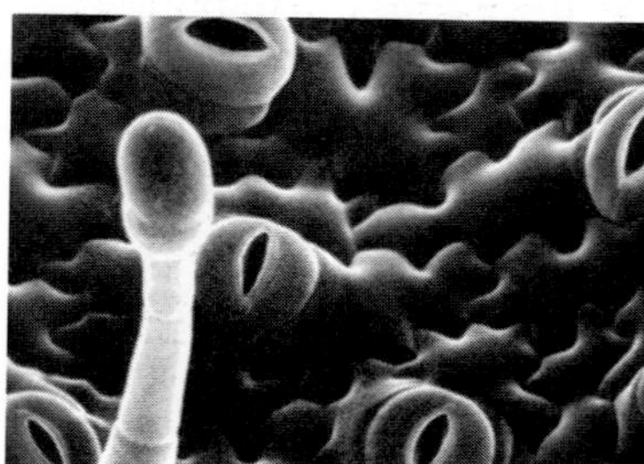

# 29

# PLANT ORGANISATION

**LEARNING OBJECTIVES**

By the time you have completed your study of this chapter you should be able to:

- describe how the nutrition of plants differs from that of animals.
- explain how the external features of a flowering plant relate to its mode of nutrition.
- state how plant cells differ from animal cells when observed under the light microscope, and explain how their structural features are related to the plant way of life.
- explain what is meant by a tissue, and describe the locations and distinguishing features of parenchyma, sclerenchyma, collenchyma, xylem and phloem.

## The plant way of life

Animals are **heterotrophic**, feeding on organic compounds obtained from the bodies of other organisms. In contrast plants are typically **autotrophic**, or 'self-feeding', building up their own organic compounds by the process of **photosynthesis**. This difference in nutrition can be related to a number of other differences.

- Whereas animals generally have fairly compact shapes, plants have branching, finely-divided bodies. Only by presenting a large external surface to the environment can thinly dispersed nutrients like $CO_2$ and mineral ions be absorbed.
- Plants are fixed to one spot – since their nutrients are delivered by diffusion and air or water currents, they have no need to move from place to place.
- In animals increase in size is typically confined to a juvenile period, but plants grow until they die. This is essential to avoid shading by competitors and also to regenerate parts eaten by animals. For the roots, continued growth is necessary to tap new areas of soil for minerals.
- Plants grow from the tips of their shoots and roots. This **apical growth** means that different parts are of different ages, those nearest the tips being the youngest. An oak twig may be only a few months old, but the centre of the trunk may be over a hundred years old.
- Cell division in plants is restricted to certain regions called **meristems**. In animals cell division is more widespread, and in the embryo occurs throughout the entire body. Meristems at the tips of roots and shoots are called **apical meristems** and are responsible for growth in length.

**Lateral meristems**, for example the **vascular cambium**, give rise to an increase in thickness. In trees and shrubs the vascular cambium is located just beneath the bark.

- Whereas animals typically have very complex bodies with many organs, plants have a very small number of different kinds of parts, such as leaves, stems and roots, which are repeated a large but indefinite number of times.

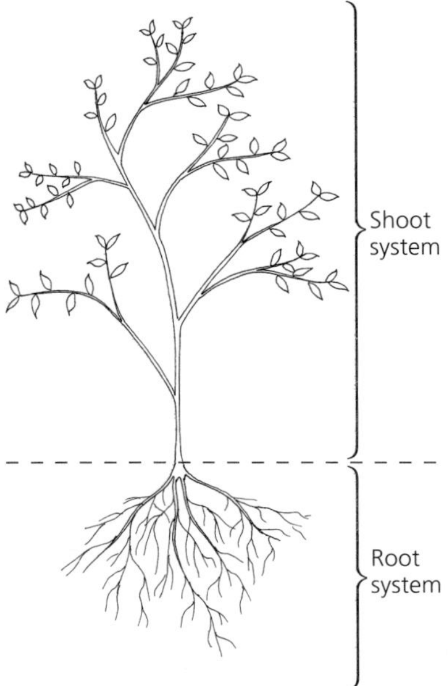


**Figure 29.1▲**
Basic form of a flowering plant

## The land plant

In a sense, a land plant lives in two environments at once: above ground there is light but no reliable water supply, whilst below ground are water and mineral ions but no light.

In adapting to this two-part world, land plants have evolved a body consisting of two interdependent parts: an above-ground **shoot system** specialised for photosynthesis and a below-ground **root system** which absorbs water and minerals, and provides anchorage (Figure 29.1).

### *The shoot system*

A shoot consists of a columnar **stem** bearing the **leaves** and the organs of sexual reproduction. The leaves are specialised for photosynthesis, while the stem supports the leaves and transports materials between leaves and roots. Leaves arise from the stem at the **nodes**, and the region between two nodes is an **internode**. The angle between the stem and a leaf is the leaf **axil**, in which there is an **axillary bud**. This is simply a tiny shoot with unexpanded leaves and very short internodes. Though they can be very small and may never develop any further, axillary buds have the potential to grow out into lateral branches. At the tip of the stem the youngest leaves are clustered into a **terminal bud**.

Leaves may be **simple**, or undivided, or they may be **compound**, in which the blade or lamina is subdivided into **leaflets**. Superficially, a compound leaf of ash or rose looks like an entire shoot, but the absence of any buds reveals its true identity (Figure 29.2).

Although all flowering plants have the same basic parts, they can be so highly modified as to make them difficult to recognise at first glance. Yet plants grow according to a common set of rules that usually make it possible to identify the true nature of their parts. For example, a potato tuber is actually a stem because when young, it has tiny scale-like leaves with axillary buds. The spiny green 'leaves' of a prickly pear cactus are actually stems because they continue to grow and produce more branches (Figure 29.3).

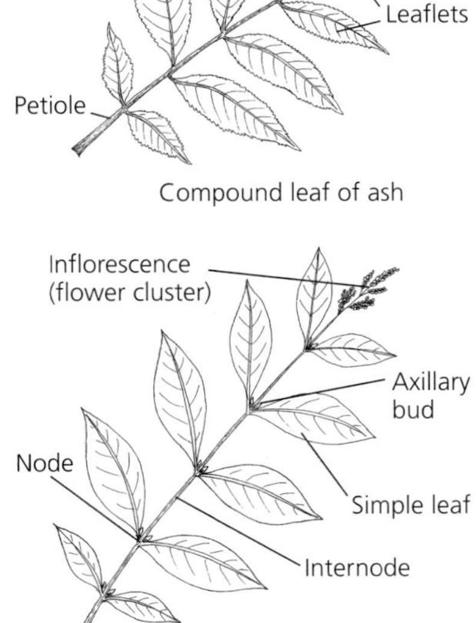


**Figure 29.2▲**
A compound leaf of ash (*Fraxinus excelsior*) and a shoot of osier (*Salix viminalis*)

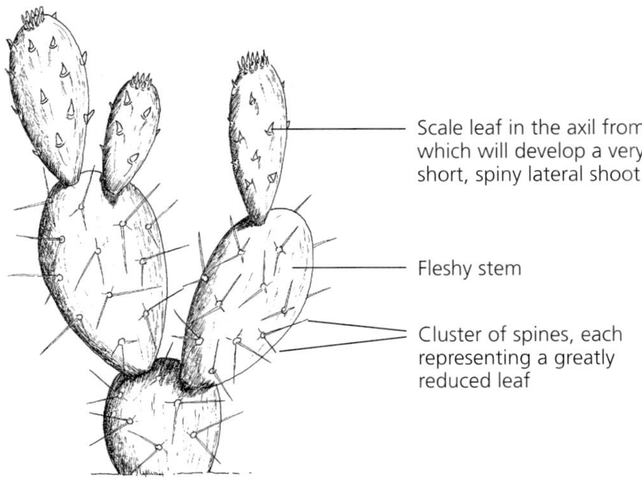


**Figure 29.3►**
Modified stems of prickly pear cactus

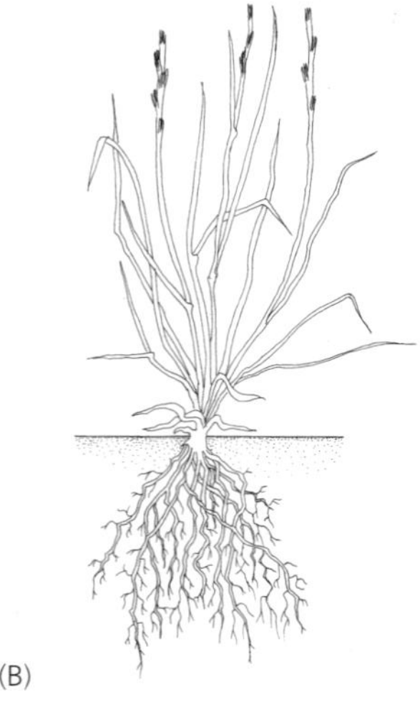


**Figure 29.4**
Tap root system of dandelion (A) and fibrous root system of annual meadow grass (B)

### *The root system*

Roots typically live in darkness and lack chlorophyll. They have two principal functions: the absorption of water and ions, and anchorage. Some roots also store energy for growth the following spring. Root systems are generally of two kinds (Figure 29.4):

- in a **tap root system** one root is conspicuously larger than its side branches, as in dock and dandelion.
- in a **fibrous root system** no root is noticeably larger than the others. In many plants the fibrous roots arise directly from the nodes of creeping or underground stems. These are called **adventitious roots** and are produced in many monocotyledons, such as grasses.

From the point of view of absorption, the most important feature of a root is the large surface area presented by thousands of tiny **root hairs** a few millimetres behind the tip (discussed further in Chapter 33).

## Plant cells

Plant cells have a number of characteristics that distinguish them from those of animals (Figure 29.5).

A plant cell consists of two parts: a metabolically-inactive **cell wall**, surrounding a living **protoplast**. The cell wall consists largely of inextensible cellulose **microfibrils**, together with hemicelluloses and proteins (Chapter 4). By resisting tension, the wall enables the cell to build up high internal pressures. In many plant cells the wall is the only functional part, persisting as an empty shell after the death of the protoplast.

Whereas unmodified cellulose walls have a high tensile strength (strong resistance to pulls), they buckle when compressed because the microfibrils are able to slide over each other. In many cells this is prevented by the deposition of **lignin**, a complex non-carbohydrate polymer which forms cross links between adjacent microfibrils. Lignified cellulose is an example of a **composite material**: it is stronger than either of its two individual constituents.

By filling the spaces between microfibrils, lignin also greatly reduces the permeability of cell walls to water, oxygen and $CO_2$. Yet lignification is not always a death sentence for a cell – many of the parenchyma cells in timber are lignified yet are alive.

Another important substance that may be deposited within cell walls is **cutin**. This waxy material greatly reduces the permeability to water, and is an important component of the **cuticle** that covers the epidermis of leaves

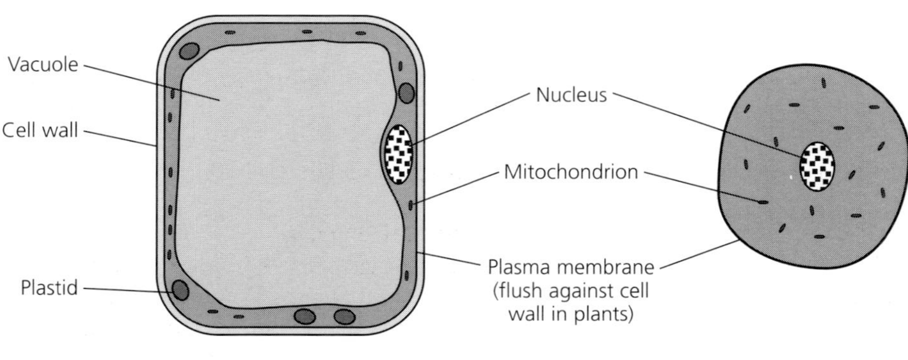


**Figure 29.5**
Plant and animal cells as seen under the light microscope. Structures present in both animal and plant cells are labelled in the centre; features peculiar to plant cells are labelled on the left. Centrioles, which occur in non-seed-bearing plants, are not shown

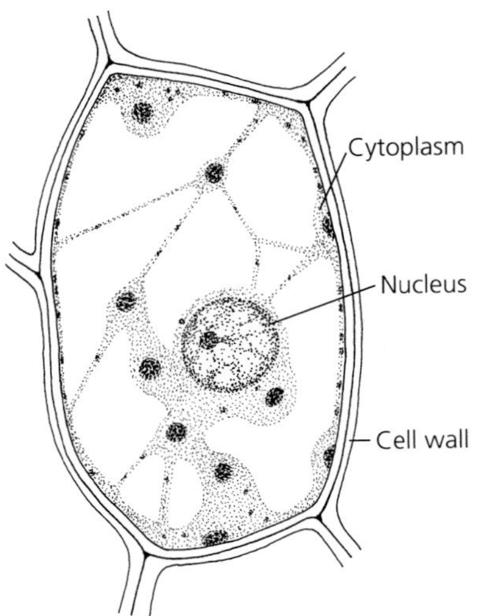


**Figure 29.6**
A parenchyma cell

and stems. A similar substance, **suberin**, impregnates the walls of cork cells.

Though the cellulose wall may appear to isolate a cell from its neighbours, plant cells are actually interconnected by strands of cytoplasm called **plasmodesmata**. These extend in bundles through thin areas in the walls of adjacent cells, called **pits**, which remain permeable to water even when the rest of the wall is heavily lignified.

All living plant cells have organelles called **plastids**. These differentiate from small self-replicating **proplastids** which, like mitochondria, are believed to be descendants of mutualistic bacteria. There are various kinds, specialised for different functions. **Chloroplasts** carry out photosynthesis, whilst **amyloplasts** store starch, and **chromoplasts** are responsible for the orange and yellow pigments of many flowers and fruits.

Mature living plant cells usually have a large space called a **vacuole,** containing **cell sap**. Besides being a storehouse of various solutes, the vacuole has the beneficial effect of displacing the cytoplasm to the periphery of the cell, thus decreasing the diffusion distances between the exterior of the cell and the mitochondria and chloroplasts.

# Plant tissues

As in animals, plant cells occur in organised groups called **tissues**, in which the cells are specialised for a particular function or group of functions. Some tissues contain more than one kind of cell, but are still considered to belong to the same tissue because they develop from the same embryonic population of cells.

Tissues are not just random assemblages of cells – in most cases the constituent cells are organised in such a way that they co-operate to carry out a function that they cannot do as individual cells. The most important tissues are concerned with basic functions such as photosynthesis, transport of water, transport of organic solutes, support and regulation of water loss. In each case the most important thing to appreciate is how the structure of the tissue adapts it to its function. For this reason, only certain tissues are described in detail in this chapter; others are more appropriately dealt with in a functional context in later chapters.

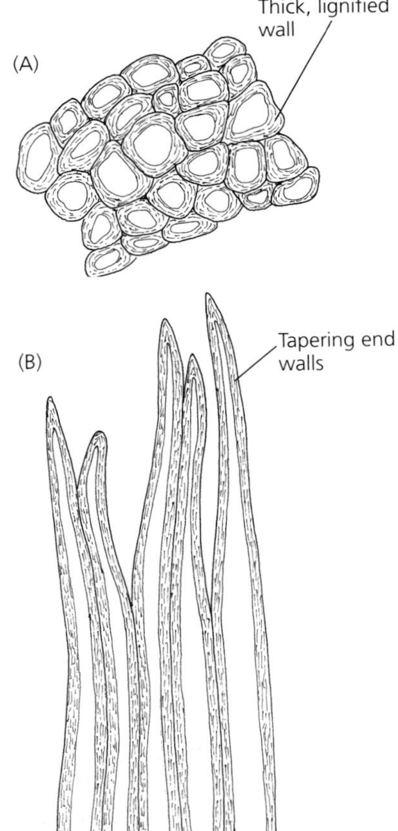


**Figure 29.7**
Fibres in transverse section (A) and in longitudinal section showing tapering ends (B)

## *Parenchyma*

The most widespread tissue in plants is **parenchyma**. It consists of living, usually fairly thin-walled cells that are not particularly elongated (Figure 29.6). It is a kind of 'general purpose' tissue, carrying out a range of functions such as storage and photosynthesis. Because they are usually living, parenchyma cells can develop turgor pressure, and, by forming a firm packing between other tissues, they provide support for soft parts such as leaves and very young stems.

Some parenchymatous tissues are more specialised and are given names in their own right. **Chlorenchyma**, for example, is parenchyma that is specialised for photosynthesis. Another example of specialised parenchyma is the epidermis (or skin) of shoots and roots, dealt with in later chapters.

## *Sclerenchyma*

Sclerenchyma takes its name from the Greek *scleros* meaning 'hard'. The commonest form of sclerenchyma consists of **fibres**. These long, thin cells usually occur in bundles and are generally dead when mature (Figure 29.7). Their walls are thick and usually lignified, making them very resistant to both tension and compression. Adjacent fibres adhere very

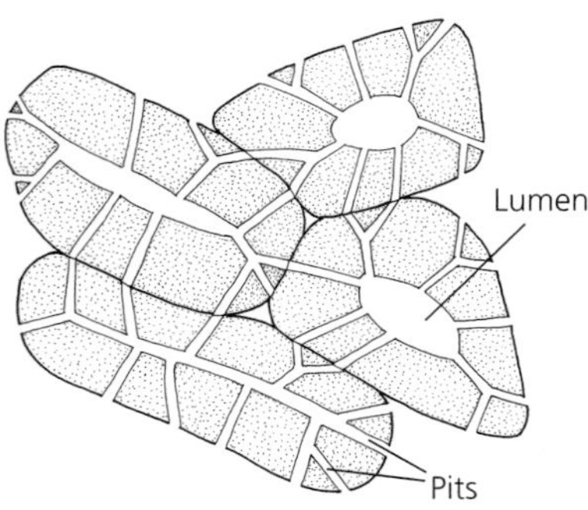


**Figure 29.8**
Sclereids. As the wall is thickened, pits merge, producing the branched appearance shown

strongly because their tapering ends are tightly interlocking, with intercellular air spaces almost absent.

Fibres have one limitation – since they are dead, they cannot grow, and even if they were alive their lignified walls would not be able to extend. Fibres thus only develop in older regions where the surrounding tissues have stopped elongating.

Another kind of sclerenchyma consists of **sclereids**. These are short, squat cells with even thicker walls than fibres. Your teeth may have encountered clusters of them as gritty particles when eating a pear. The 'stones' of fruits such as plums and cherries consist of a solid mass of interlocked sclereids (Figure 29.8).

## Collenchyma

Although fibres are highly effective supporting cells, they are no use in parts that are still elongating, such as young stems. Temporary support in these tissues is provided by **collenchyma**.

Collenchyma cells are elongated and living and, although the walls are thick, are *never lignified*. This enables them to extend and keep pace with surrounding tissues that are still elongating. Like fibres, collenchyma cells adhere strongly because they have chisel-shaped, interlocking end walls. A peculiar characteristic is that the walls are usually thickened in the corners (Figure 29.9). By preventing the walls from bulging into intercellular air spaces, this may help the tissue to resist compression.

Collenchyma forms ribs in the outer parts of stems, and is particularly important in young leaves which must be able to support themselves whilst still expanding.

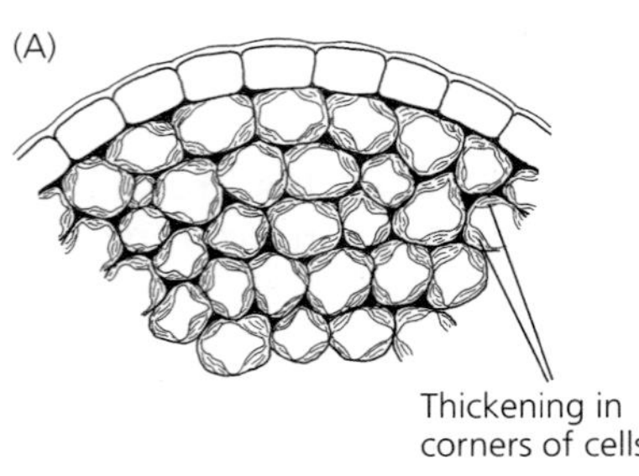


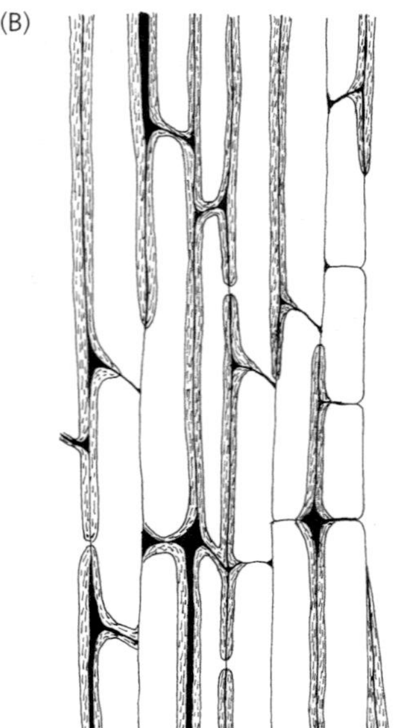


**Figure 29.9**
Collenchyma cells in transverse section through the ridge of a herbaceous stem (A) and in longitudinal section (B)

## Xylem

In bulk, **xylem** (pronounced 'zylem') is familiar as wood (*xylon* is Greek for wood), and is the most abundant tissue on earth. Besides transport of water and minerals it also provides support, and in woody plants some of its cells store starch. Unlike **simple tissues** such as parenchyma and collenchyma, xylem is a **mixed tissue** because it consists of several kinds of cell. In flowering plants these include tracheids, vessel members, fibres and parenchyma cells.

### Tracheids

Tracheids are the sole water-conducting cells in all but a few ferns and conifers, and are also present in most flowering plants. They are long cells (usually 3–5 mm), with tapering walls. As they mature, the nucleus and cytoplasm die leaving a hollow shell (Figure 29.10). The walls are strongly lignified, enabling them to resist caving in under the high tensions that may develop in xylem. Though lignification greatly reduces the permeability of the walls, water is still able to flow through the thin primary walls via the pits.

### Vessels

Besides tracheids, most flowering plants have water-conducting channels called **vessels**. Like tracheids, vessels have lignified and extensively pitted walls, but they offer less resistance to water flow in two ways:

- a vessel is not an individual cell but is formed by end-to-end joining of many individual cells called **vessel members** (Figure 29.11). Early in their development, the end walls are broken down, creating a continuous tube (Figure 29.12).
- vessels are usually wider than tracheids – in some plants they may be

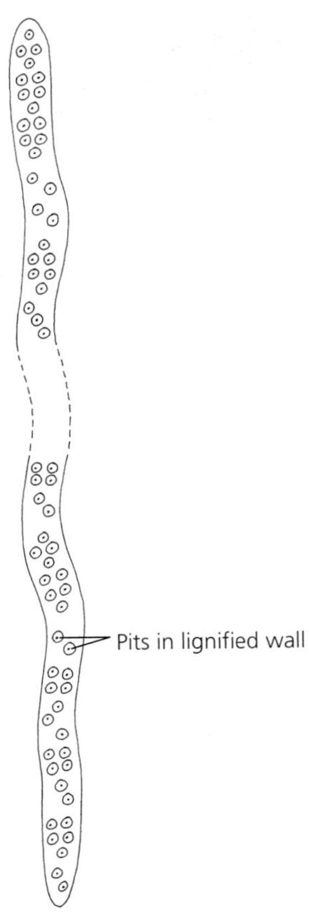


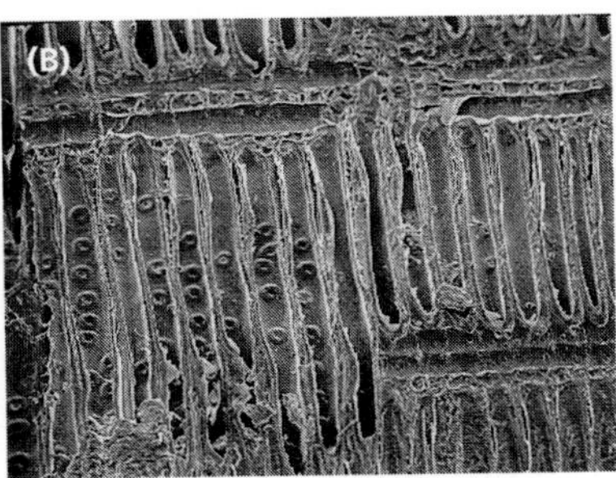


**Figure 29.10▲**
A tracheid in side view (A). The dotted line indicates that the cell is proportionately longer than shown. Tracheids as seen under SEM showing their pitted surface (B)

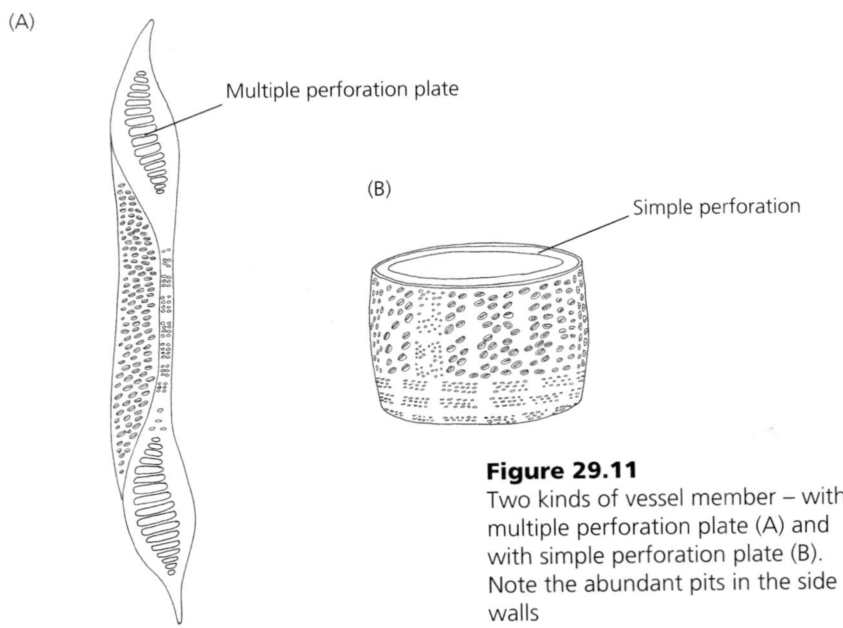


**Figure 29.11**
Two kinds of vessel member – with multiple perforation plate (A) and with simple perforation plate (B). Note the abundant pits in the side walls

400 μm across and are just visible to the naked eye in transverse section.

Although individual vessels may be several centimetres or even metres in length, they do not extend the whole way from roots to leaves. When the sap reaches the end of a vessel, it passes laterally to an adjacent one through the pits.

Besides the water-transporting cells, xylem also contains starch-storing parenchyma cells and, in many flowering plants, fibres.

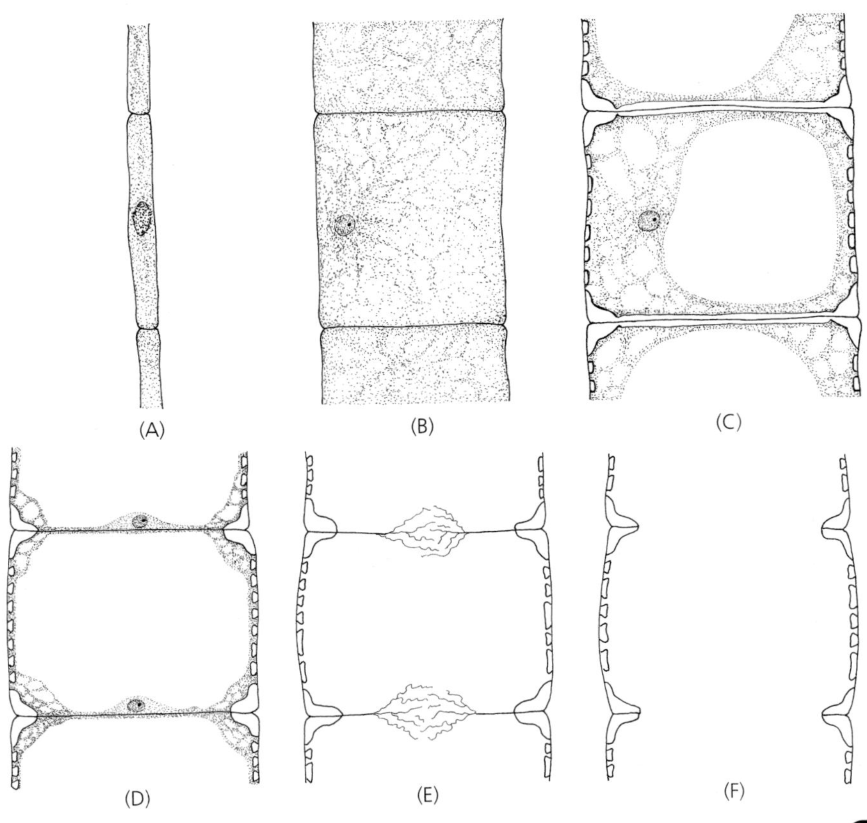


**Figure 29.12►**
Formation of a xylem vessel. A–B, elongated cell increases in width; C–D, thickening of cell wall with formation of pits; E–F, breakdown of end walls with degeneration of nucleus and cytoplasm (after Eames and MacDaniels)

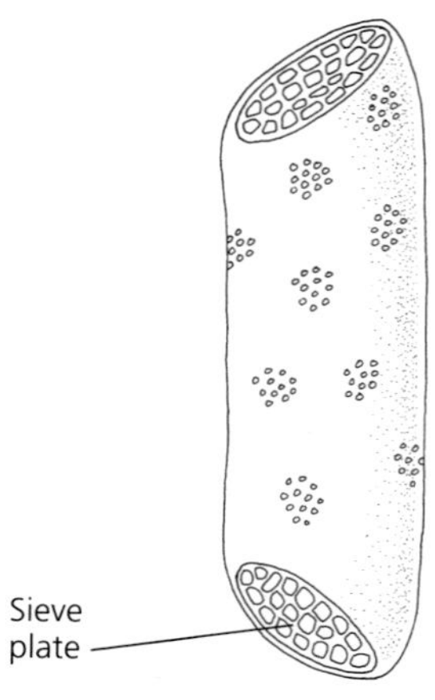


**Figure 29.13**
A phloem sieve tube element

### *Phloem*

Another mixed tissue is **phloem**, which conducts organic materials such as sugar. The conducting cells of phloem are **sieve tube elements** (Figure 29.13). Also present are **companion cells**, **parenchyma cells** and often **fibres**. The detailed structure of phloem is more appropriately dealt with in the context of its function, in Chapter 34.

## Plant organs

An **organ** is a group of tissues which co-operate to carry out a function that none of the individual tissues can perform by itself. Stems, leaves and roots are examples of organs and are dealt with in later chapters in connection with their respective functions.

### QUESTIONS

**1** Explain how the finely-divided shape of a plant relates to its style of nutrition.

**2** Give *three* reasons why colonisation of the land by plants has been associated with an increase in structural complexity.

**3** In the key below, each of the following kinds of cell is represented by one of the letters A, B, C, D, E:

parenchyma cell, collenchyma cell, fibre, vessel member, sieve tube element

Use the key to identify the letter representing each kind of cell.

1. Cell alive when functional ............go to 2
   Cell dead when functional ...........go to 4
2. Nucleus present when functional go to 3
   No nucleus present when functional .....A
3. Cells elongated, walls thickened in corners ....................................................B
   Cells not elongated, walls not thickened in corners...............................................C
4. Entire cell wall present when functional...............................................D
   No end walls when functional ...............E

CHAPTER

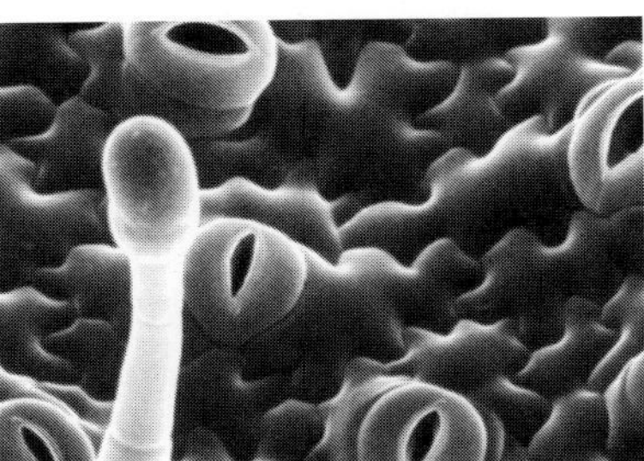

# 30

# PLANT LIFE CYCLES

## LEARNING OBJECTIVES

By the time you have completed your study of this chapter you should be able to:

- state the essential events of sexual reproduction.
- explain the essential features of alternation of generations.
- compare the life cycles of a moss and of a fern, and explain how the fern life cycle is better adapted to life on land.
- know the basic parts of a flower and their functions.
- understand the need for pollination and describe the different ways in which it is brought about.
- distinguish between self- and cross-pollination, and discuss their respective advantages and disadvantages.
- know the distinctive features of wind-pollinated flowers.
- describe the events that occur between pollination and fertilisation, and those that occur after fertilisation.
- describe the structure of a dicotyledonous seed and a monocotyledonous seed.
- explain the advantages of seed dispersal.
- describe the ways in which seeds and fruits are adapted for different modes of dispersal.
- distinguish between ephemerals, annuals, biennials and perennials.

## The sexual cycle

As in all sexually reproducing eukaryotes, sexual reproduction in plants involves two essential events: **meiosis**, in which there is a reduction from the diploid to the haploid number of chromosomes, and **fertilisation**, in which the diploid chromosome number is restored. Between them these two processes bring about a recombination of genes.

Though animal and plant life cycles share these fundamentals, the emphasis on the diploid and haploid phases differs markedly. Amongst eukaryotes there are four kinds of sexual life cycle:

1 the **diplontic cycle**, which occurs in animals and in some protoctists. The entire life cycle is spent in the diploid phase except for the gametes (Figure 30.1).

2 the **haplontic cycle**, which occurs in many unicellular algae such as *Chlamydomonas* and *Spirogyra*. The organism spends almost its entire life cycle in the haploid state. After fertilisation the zygote undergoes meiosis, the haploid products resuming active life. Since there is no dominance, all alleles are subjected to selection and any that are even slightly disadvantageous tend to be eliminated. Haplontic species are

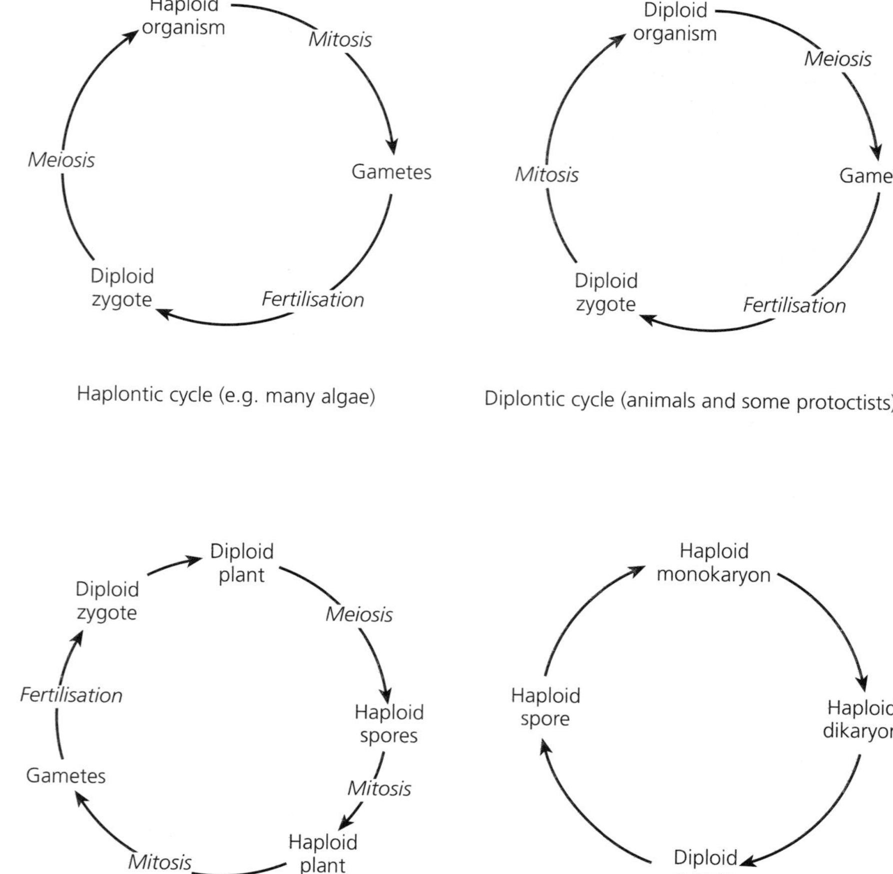


**Figure 30.1**
Four kinds of eukaryote life cycle

therefore less genetically variable than diplontic species, in which recessive alleles are protected from selection by dominant alleles.

3 the **haplodiplontic cycle**, characteristic of plants and of some plant-like protoctists such as *Ulva*, the sea lettuce (Figure 30.2). There are two multicellular stages: a diploid **sporophyte** reproducing by spores, and a haploid **gametophyte** which reproduces by gametes.

These two stages in the life cycle alternate – a state of affairs called **alternation of generations**. In *Ulva*, the sporophyte and gametophyte generations are structurally indistinguishable (except by counting their chromosomes).

In plants the two generations are both structurally and genetically different. Though the life cycles of all plants are fundamentally similar, the colonisation of the land involved considerable reduction in emphasis on the gametophyte, which depends on external water for its reproduction. In flowering plants the gametophyte is vestigial and is scarcely recognisable.

4 the life cycle of Basidiomycete and many Ascomycete fungi resembles the haplontic cycle in that for the major part of the life cycle the nuclei are haploid, but part of this is spent as a **dikaryotic** phase in which cells contain two haploid and genetically-different nuclei (Chapter 16).

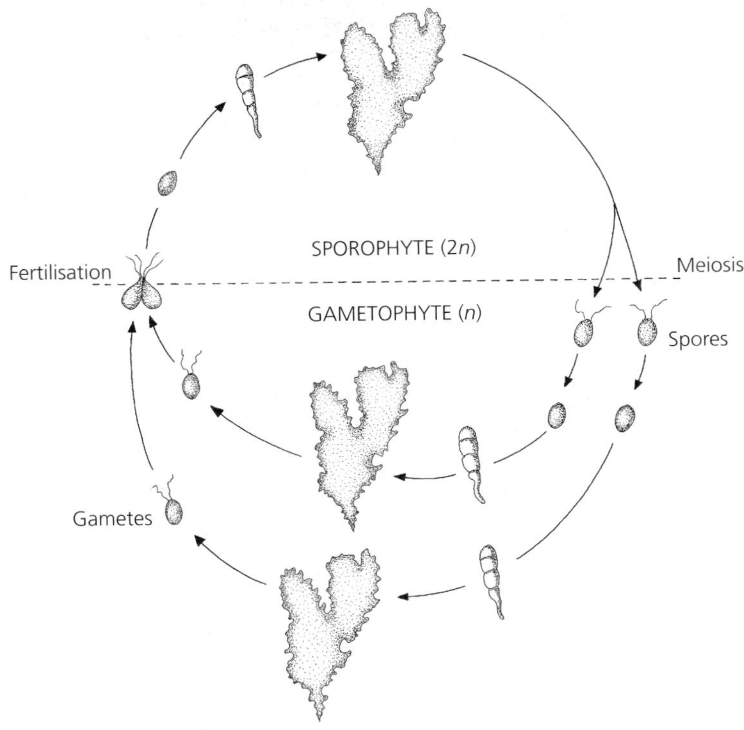


**Figure 30.2**
Life cycle of *Ulva*, the sea lettuce

## The life cycle of mosses (Musci)

Mosses and liverworts belong to the bryophytes. The life cycle is broadly similar and will be illustrated by that of a moss. The familiar moss plant is the haploid gametophyte. Since it is already haploid it produces gametes by **mitosis**, so all the gametes produced by a given plant are genetically identical. The gamete-producing organs are at the tips of the shoots (Figure 30.3). The bag-like male organs, or **antheridia**, produce sperms.

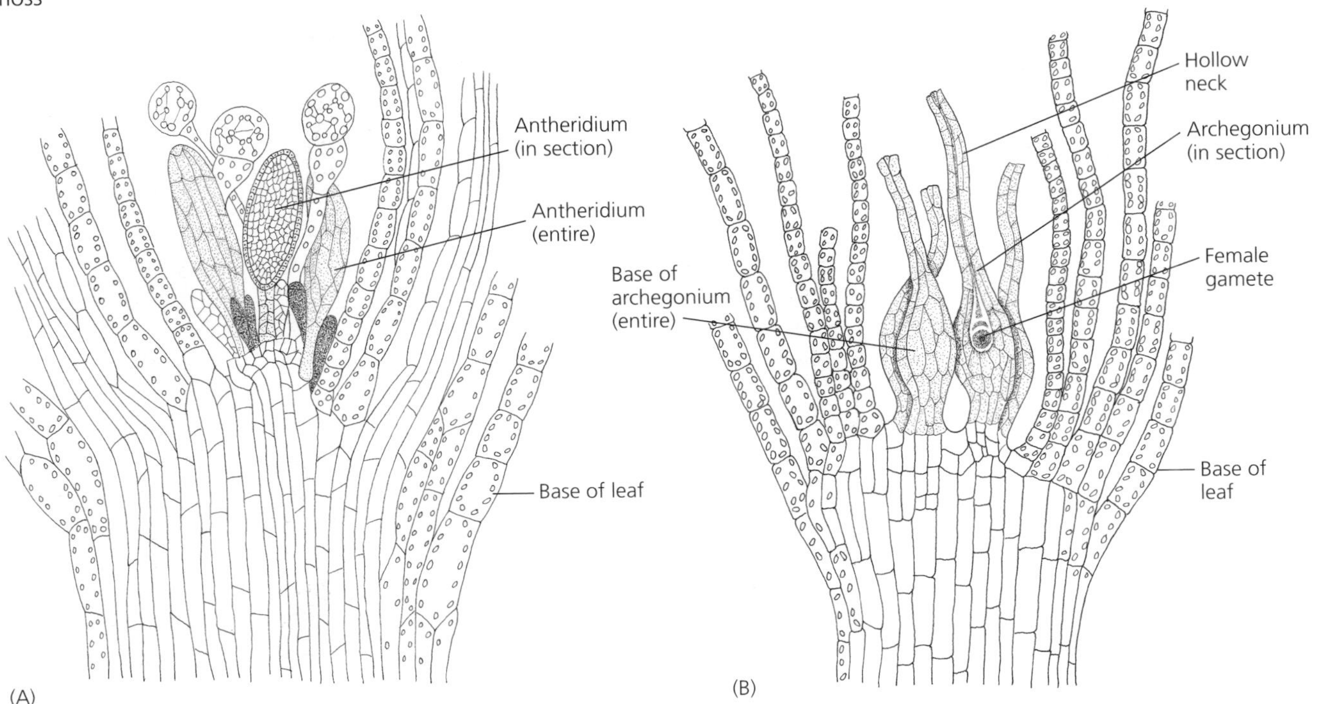


**Figure 30.3**
Male (A) and female (B) organs of a moss

In wet conditions each antheridium ruptures, releasing the sperms into the thin film of surrounding water. The sperms swim using flagella, and some may reach the bottle-shaped female organs, or **archegonia**. The base of each archegonium contains a female gamete or egg which releases a chemical that attracts sperms. The first sperm to swim down the neck of the archegonium fertilises the egg to produce a diploid zygote.

By mitotic divisions, the zygote grows into a sporophyte. This looks quite different from the parent gametophyte, consisting of a **capsule** on a long stalk which remains attached to the parent gametophyte (Figures 30.4 and 30.5). For a while the capsule may be covered by a 'hat' formed from the remains of the archegonium. Though the unripe capsule is green and can thus photosynthesise, it depends on the gametophyte for water and minerals.

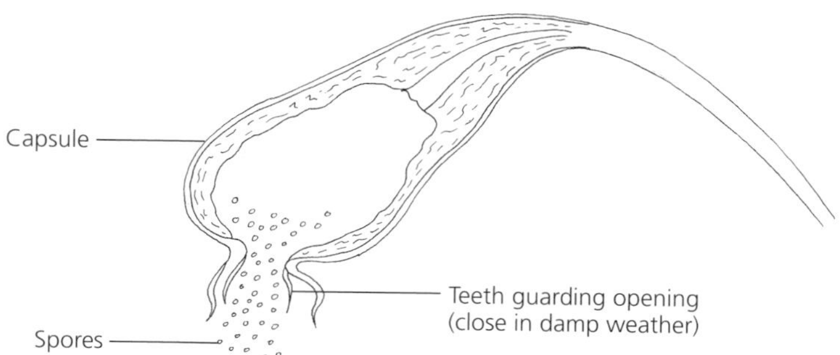


**Figure 30.4►**
Section through a ripe moss capsule

The new plant is called the sporophyte because inside the capsule, haploid **spores** are produced by meiosis and are thus genetically different. In dry weather the teeth guarding the opening to the capsule bend outwards, releasing the spores to the wind. The sporophyte then dies. If a spore lands in a suitably damp place it germinates, growing into a new gametophyte.

The life cycle is summarised in Figure 30.6. Though both generations are essential parts of the life cycle, the sporophyte is never an independent plant and dies after reproducing. The life cycle as a whole still needs wet conditions for long enough for the gametophyte to reproduce.

**Figure 30.5▲**
Moss showing sporophyte growing as a capsule attached to the parent gametophyte

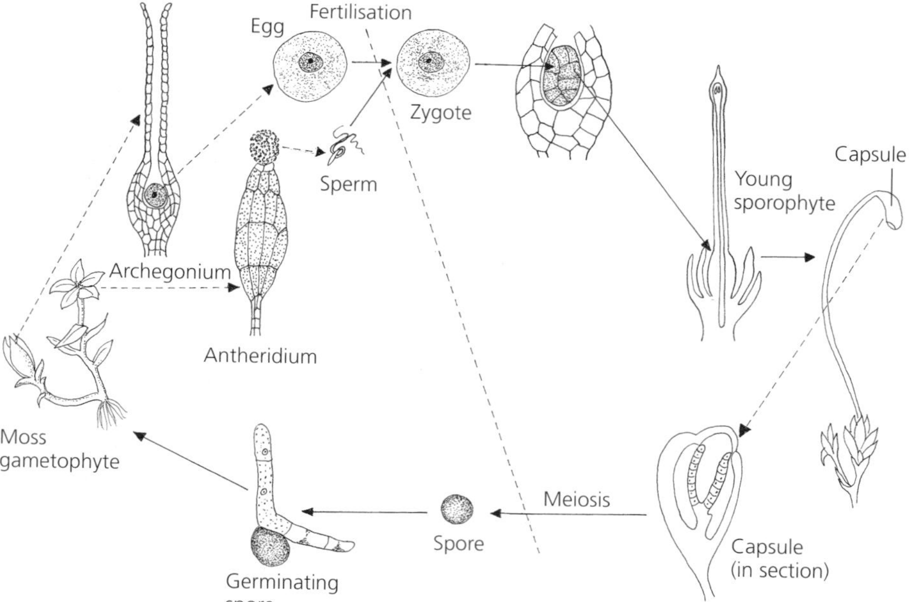


**Figure 30.6►**
Summary of the moss life cycle (after Holman and Robbins)

**CLEAR THINKING**

It is often stated that reproduction by moss and fern sporophytes is asexual, presumably because it does not involve fertilisation. This, however, is to misunderstand the nature of sex. The essence of sex is that it produces new gene combinations by fertilisation and meiosis. To reshuffle a diploid set of chromosomes into a new combination requires *both* processes, meiosis halving the chromosome number and fertilisation doubling it. Since the two processes are complementary, both are integral parts of the sexual process, and neither can be said to be more or less 'sexual' than the other.

In mosses and ferns, meiosis and fertilisation occur in plants of different generations, but this does not mean that spore production is an asexual process. The hallmark of asexual reproduction is the production of genetically-identical copies by mitosis. Rather than being confined to the gametophyte, sex in plants spans *two* generations, meiosis marking the transition from sporophyte to gametophyte, and fertilisation the transition from gametophyte to sporophyte.

## The life cycle of ferns (Filicinophyta)

Together with the bryophytes, the ferns used to be called 'cryptogams' because for many years it was not understood how they reproduced sexually (the word means 'hidden marriage'). Though fundamentally similar to that of mosses and liverworts, the life cycle of a fern shows a major shift in emphasis from the gametophyte to the sporophyte, which is fully independent and lives from year to year (Figure 30.7). The gametophyte, on the other hand, is relatively short-lived, and is so small as to be seldom noticed.

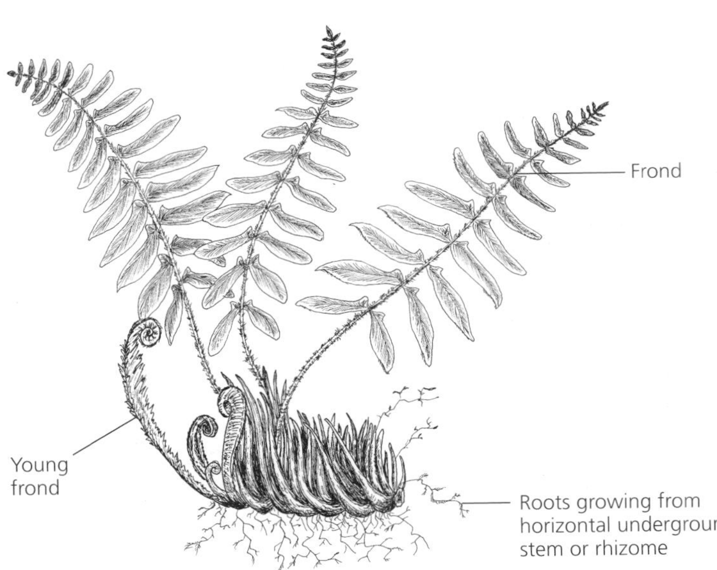


**Figure 30.7**
A fern sporophyte

The sporophyte produces spores in small sacs called **sporangia** on the undersides of the leaves (which in ferns are called **fronds**), in clusters called **sori** (Figure 30.8). The rim of each sporangium is called the **annulus**, and is formed from a line of specialised cells that play an important part in the release of the spores. In most ferns each sorus is protected by a thin layer of tissue called an **indusium**, which shrivels as the sporangia ripen. As the ripe sporangium dries out, tension develops in the cells of the annulus causing it to bend back. There is a sudden release of tension caused by the formation of a water vapour-filled cavity in each annulus cell, jerking the haploid spores out into the wind.

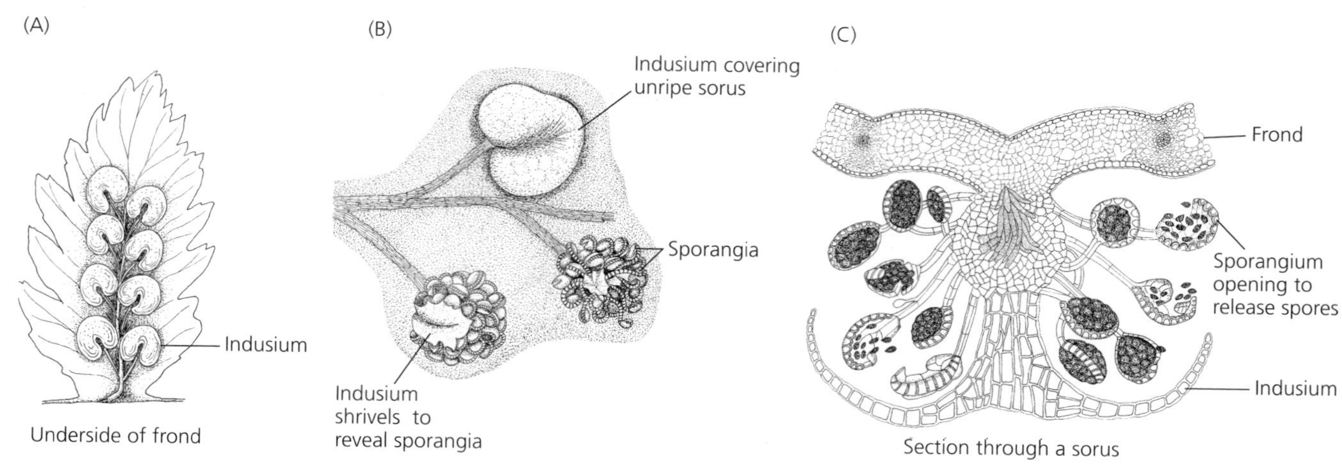


**Figure 30.8▲**
Reproductive structures of a fern sporophyte (after Kny)

**Figure 30.9▼**
Germination of a fern spore to form a young prothallus (A) and mature prothallus from below (B) (after Holman and Robbins)

If a spore lands in a damp place with enough light, it germinates, developing into a small, thin, heart-shaped gametophyte called a **prothallus** (Figure 30.9). It has neither cuticle nor stomata to restrict water loss, and is anchored by rhizoids rather than roots. Moreover it cannot survive drying, so it is less well adapted to terrestrial conditions than the gametophyte of some mosses.

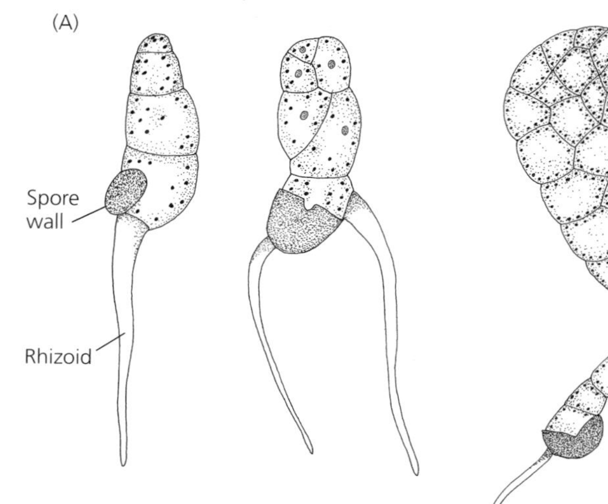


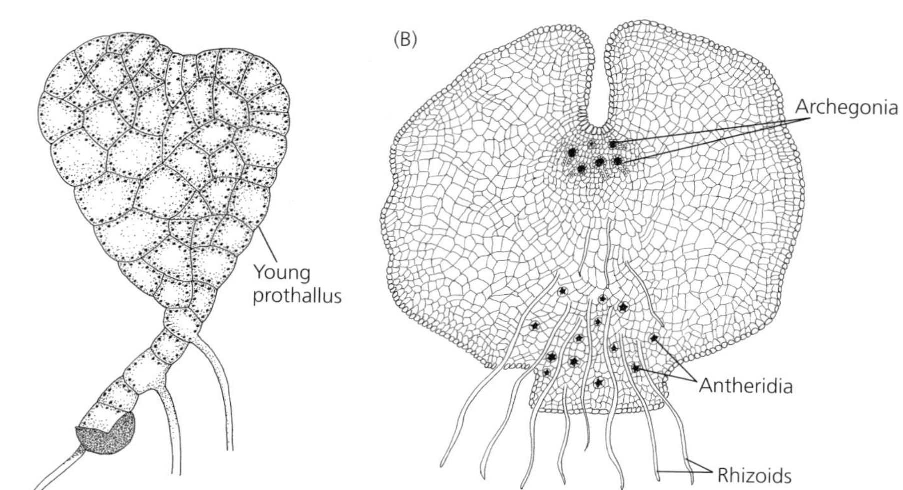


**Figure 30.10▼**
Antheridium (A) and (B), a sperm (C) and an archegonium (D) of a fern

The antheridia and archegonia are produced on the underside of the gametophyte, where it is wettest (Figure 30.10).

After being released from the antheridia, the sperms swim in the surface film of water to an archegonium where the egg is fertilised. In

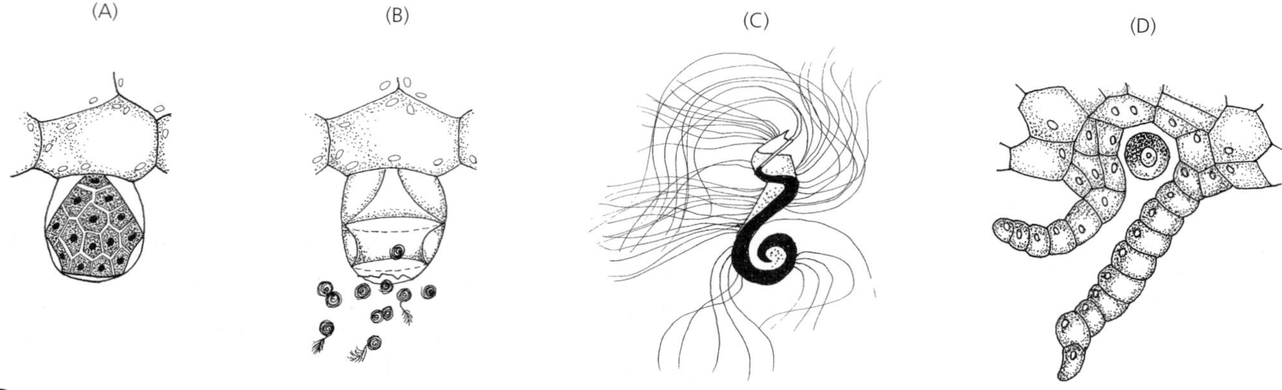

most species, the archegonia and antheridia ripen at different times so, more often than not, cross-fertilisation occurs producing new combinations of genes. The zygote then undergoes repeated divisions, developing into a new sporophyte (Figure 30.11).

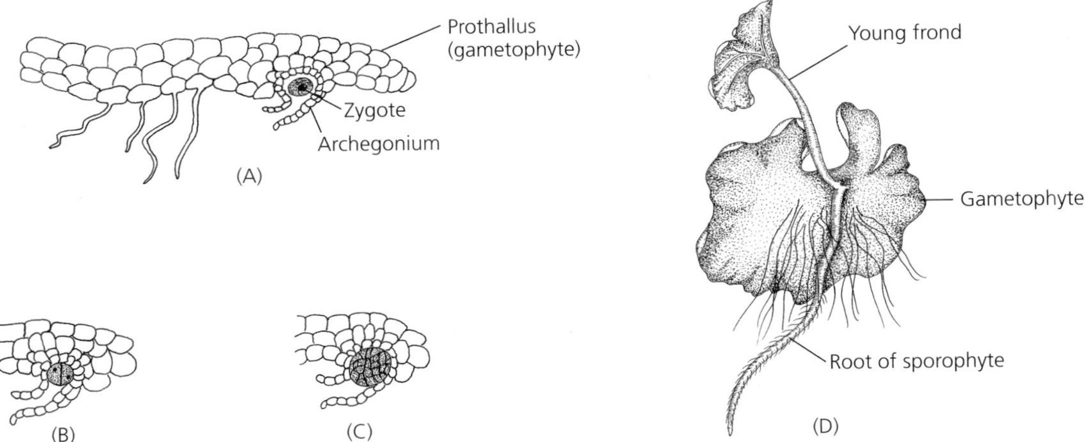


**Figure 30.11▲**
Development of the young fern sporophyte

Though it is nutritionally dependent on the gametophyte at first, the young sporophyte develops its own roots and leaves and becomes independent. The gametophyte then dies. The life cycle is summarised in Figure 30.12.

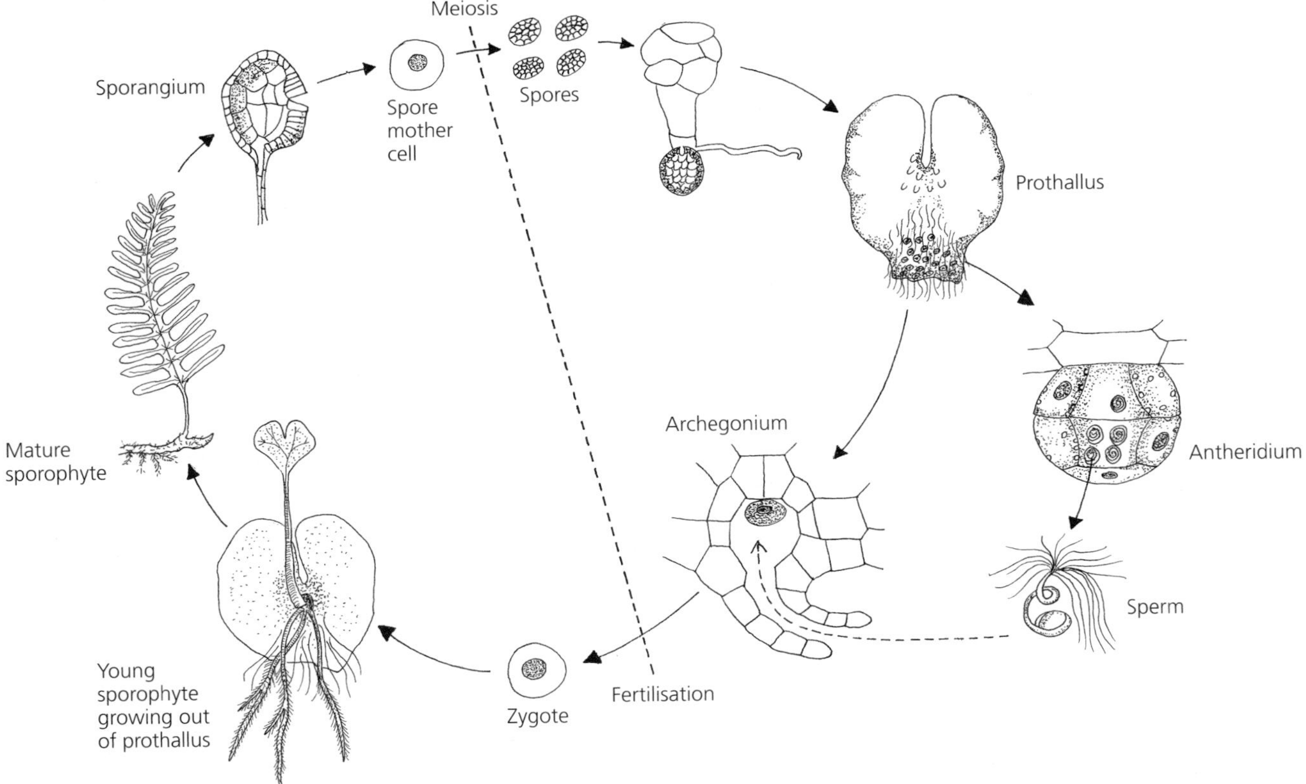


**Figure 30.12▲**
Summary of fern life cycle (after Holman and Robbins)

Though ferns still need external water for fertilisation, the sporophyte is better adapted to land life than is the gametophyte of mosses and liverworts. The success of ferns is due to their superiority in absorbing, transporting and retaining water, which enables them to grow taller and thus intercept more light than the bryophytes.

# Sexual reproduction in the Angiospermophyta, or flowering plants

In flowering plants and in most conifers, the sperms do not swim to the eggs as they do in ferns. The male gametes are produced inside **pollen grains** which are actually spores. Pollen grains have no power of movement, but are transported from male to female parts by wind, animals (usually insects) or, in some cases, by water. This process is called **pollination**, and precedes fertilisation. In insect-pollinated flowers, the plant invests considerable energy in advertising the flowers to pollinators and 'rewards' them with food.

A flower differs from a pine cone in that the seeds develop within a protective **ovary**, formed from one or more reproductive leaves called **carpels**. A carpel is actually a **megasporophyll**, which is a highly modified reproductive leaf (see the information box below). When the ovary ripens it forms a **fruit**, which not only protects the developing seeds but in many cases is adapted for their dispersal.

## From spores to seeds

**Although many fern sporophytes are quite well adapted to life on land, they have not succeeded in colonising dry places. Not only does fertilisation require external water, but the gametophyte is structurally ill-adapted to terrestrial conditions. The evolution of the seed freed the gametophyte from this dependence on external water.**

**We will probably never know all the details of how this happened, but important clues are provided by some peculiar living relatives of the ferns called lycopods, or 'clubmosses'. Among the most interesting in this respect are the various species of *Selaginella*.**

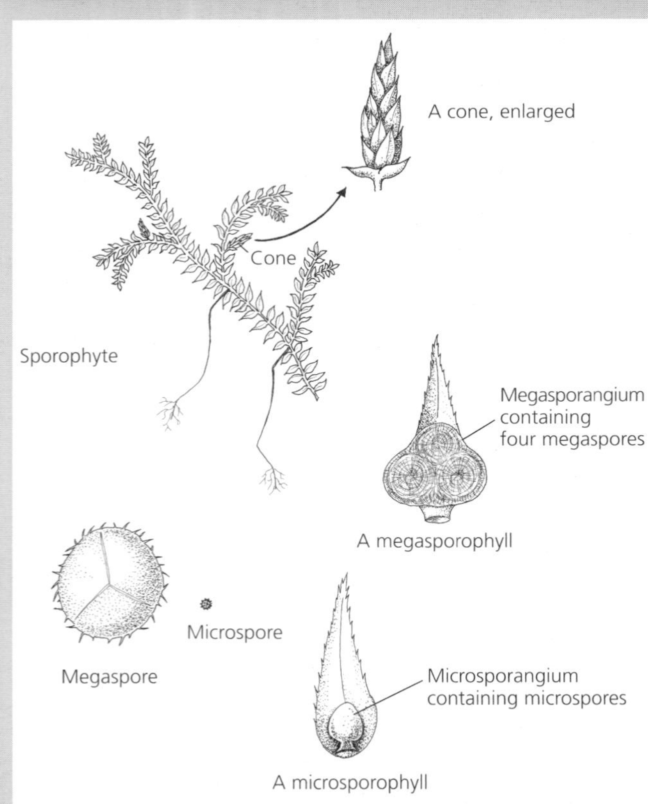


Sporophyte of *Selaginella*, a 'living fossil' (after Holman and Robbins)

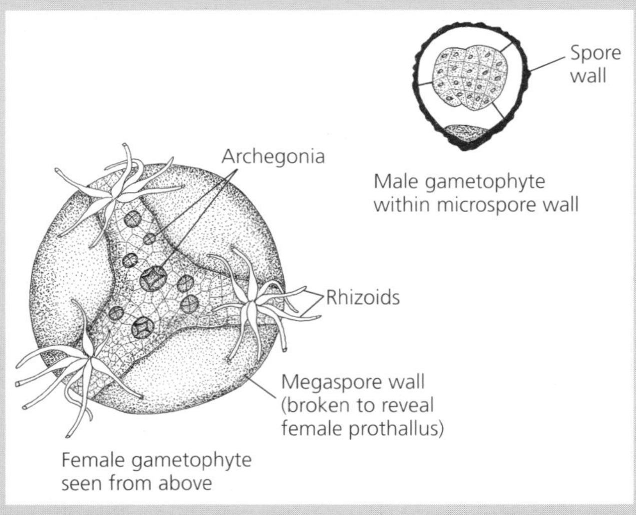


Female and male gametophyte of *Selaginella* (after Holman and Robbins)

## From spores to seeds *continued*

**Selaginella** is a kind of 'living fossil', because its nearest relatives became extinct in the Palaeozoic era over 300 million years ago. One species is native to Britain and is found in upland habitats in the North and West. It is probably a descendant of a group of plants related to those that gave rise to the seed plants. It differs in four important ways from ferns.

1 The sporangia are borne on special reproductive leaves, or **sporophylls**, aggregated into reproductive shoots, called **cones.**

2 Whereas almost all ferns are **homosporous** producing only one kind of spore, *Selaginella* is **heterosporous** producing two kinds of spore. Many small **microspores** are produced in **microsporangia**, borne on **microsporophylls.** Large **megaspores** are produced in **megasporangia** on **megasporophylls.** Microspores develop into **male prothalli** and megaspores develop into **female prothalli**. In most species of *Selaginella*, only one cell in each megasporangium undergoes meiosis, so only four megaspores are produced.

3 Neither male nor female prothallus is a full-independent plant. Both produce gametes whilst inside the spore wall using energy reserves received from the parent sporophyte.

4 Although external water is still needed for fertilisation itself, the female prothallus is less dependent on prolonged damp conditions for development. A fern prothallus has to expose a large surface to the air for photosynthesis, but, for reasons already explained, it can only do so in damp environments. Since a megaspore of *Selaginella* has sufficient reserves for growth, the female gametophyte only needs to expose a small area to the environment, with a resultant reduction in water loss.

Both megaspores and microspores fall to the ground. If there is sufficient water the microspore wall ruptures to release the sperms. These swim to the archegonia on the surface of a nearby female gametophyte, exposed by partial rupture of the megaspore wall. Usually only one zygote develops, using the energy reserves stored in the gametophyte. These reserves were received from the parent of the female gametophyte, which is the *grandparent* of the young sporophyte! The gametophyte thus has no nutritional role in *Selaginella.*

In flowering plants these trends are carried even further:

- of the four potential megaspores produced in each megasporangium, only one develops (this happens in some species of *Selaginella*).
- the megaspore develops into the female prothallus whilst remaining inside the megasporangium.
- the megasporangium does not open, and in flowering plants is called the **nucellus.**
- the female gametophyte receives nourishment directly from the sporophyte, and in flowering plants is represented by the **embryo sac.**
- instead of opening to release the sperms, a microspore landing near the megasporangium develops a **pollen tube.** This grows through the megasporangium wall and carries the sperms directly to the female gamete inside the female gametophyte. There is thus no need for external water for fertilisation.
- collars of tissue grow round the megasporangium forming the **integuments.** The entire structure – integuments, megasporangium and embryo sac forms an **ovule.** After fertilisation this develops into a **seed** which, because it carries considerably more energy reserve than a spore, can germinate in drier habitats (see the information box on page 557).
- the microsporophylls lose all resemblance to leaves and in flowering plants are called **stamens**, the microspores being called the **pollen grains.**
- the megasporophylls enclosing the ovules lose their leaf-like character and are called **carpels.**

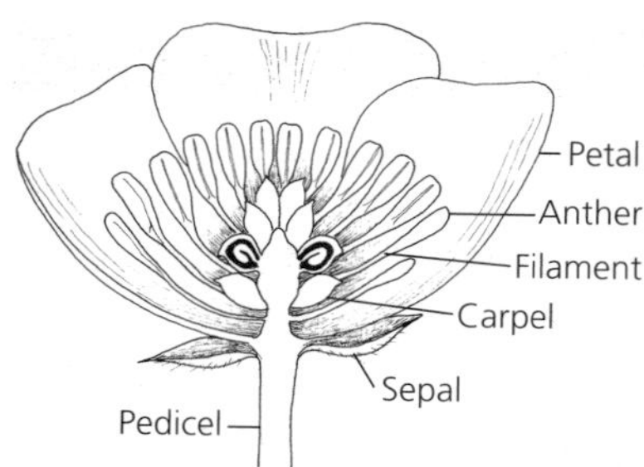


**Figure 30.13**
A buttercup flower cut vertically down the middle

## *A simple flower – the creeping buttercup*

An example of a flower with a fairly uncomplicated structure is the creeping buttercup, *Ranunculus repens* (Figure 30.13). The flower stalk is the **pedicel** and its apex forms the **receptacle,** to which the floral parts are attached. The latter are modified leaves and are of four kinds. Outermost are the **sepals,** which protect the inner parts in the bud stage, and collectively form the **calyx**. When the flower opens, these are bent back to reveal the **corolla,** made up of five large **petals** containing carotenoid pigment. To humans they look a uniform yellow, but the centre part reflects ultraviolet which most insects see as a colour. At the base of each petal is a small flap-like **nectary** – a patch of glandular tissue that secretes **nectar,** a sugary solution. The nectary is easily accessible to insects with a short proboscis (plural, proboscides), such as small beetles and flies.

Inside the corolla is the male part of the flower, the **androecium,** which consists of many **stamens**. Each stamen is actually a **microsporophyll** (see the information box on page 544) and consists of an **anther** held up by a long stalk or **filament** (Figure 30.14). The anther contains four **pollen sacs** in which pollen grains are produced by meiosis. When the anther is ripe it splits down the middle to release the pollen, some of which sticks to the bodies of insects feeding on the nectar.

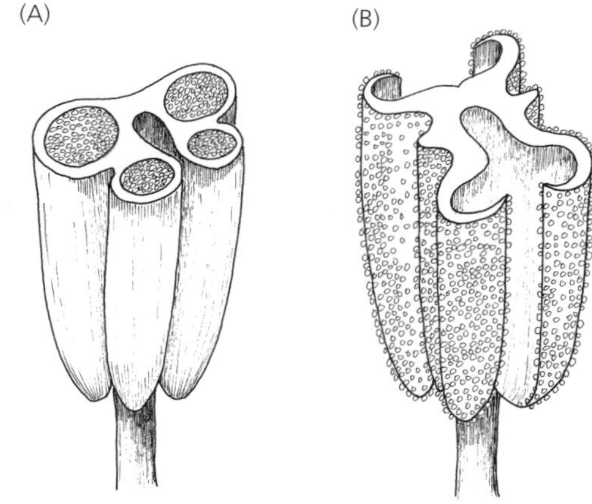


**Figure 30.14▶**
An anther before (A), and after (B) release of pollen (right)

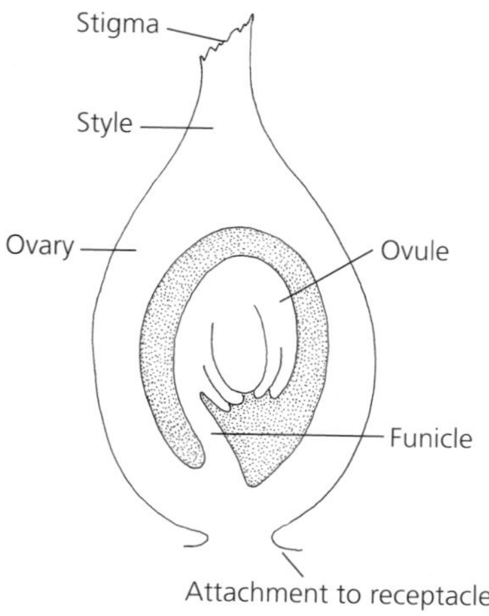


**Figure 30.15▲**
Longitudinal section of a buttercup carpel

The centre of the flower forms the **gynoecium,** or female part of the flower. It consists of many **carpels,** which are the female equivalents of stamens. Each consists of the three parts: stigma, style and ovary (Figure 30.15). The **stigma** not only receives pollen adhering to the body of insects, but also secretes sucrose which stimulates pollen grains to germinate. The **style** is a short stalk holding the stigma up. Each **ovary** houses a single **ovule** which develops into a seed.

### The disgusting behaviour of plants

In the late 17th Century it was generally believed that flowers had been created by God to enrich human life with their beauty. Though a few botanists suspected that the pollen had something to do with producing seeds, a proper understanding of plant reproduction could only be gained by experiment.

The crucial tests were done by Rudolph Camerarius, Director of the Botanical Gardens at Tübingen in Germany. In 1694 he reported the results of experiments in which he deliberately prevented pollen from being transferred from stamens to stigmas. In one experiment he removed the 'tassels' (the long protruding stigmas) of young

## The disgusting behaviour of plants *continued*

female maize flowers, and found that the flowers produced no seeds. In another experiment he removed the stamens of young male castor oil flowers, with the result that the female flowers failed to produce seeds. Though an understanding of fertilisation in plants had to wait for refinements in the microscope, it was clear that (in many plants at least) making seeds involved sex.

The discovery of sex in plants made little impact until the Swedish botanist Carl Linnaeus revolutionised plant classification. Previously botanists had classified plants on the basis of rather arbitrary characters, such as whether they were poisonous or edible, or whether they were trees or herbs. Linnaeus came up with the revolutionary idea of using the structure and arrangement of their sexual parts. He described hermaphrodite (bisexual) flowers as those in which 'husbands and wives share the same marriage bed', and unisexual flowers as those in which 'husbands and wives enjoy separate beds'.

Despite this delicate way of putting it, the Church was outraged at the suggestion that plants have genital organs. To clergymen the idea was disgusting, and some tried to have Linnaeus' works banned, accusing him of having a dirty mind. Some botanists even accused Linnaeus of obscenity.

Before Linnaeus, botany was a suitable subject for nice young people to study. With the acceptance that plants have sex lives, the study of flowers became a way of gently introducing sex to tender young minds. Times have certainly changed.

### Floral variation

The buttercup is often said to be a 'typical' flower but in reality there is no such thing, as flowers vary enormously in almost all their parts. In dicotyledons the floral parts are usually in multiples of four or five, whilst in monocotyledons they are usually in multiples of three. Sepals may be fused (e.g. red campion, *Silene dioica*) or separate (e.g. buttercup). Petals are also often fused, sometimes forming a tube with the nectary at the base (Figure 30.16).

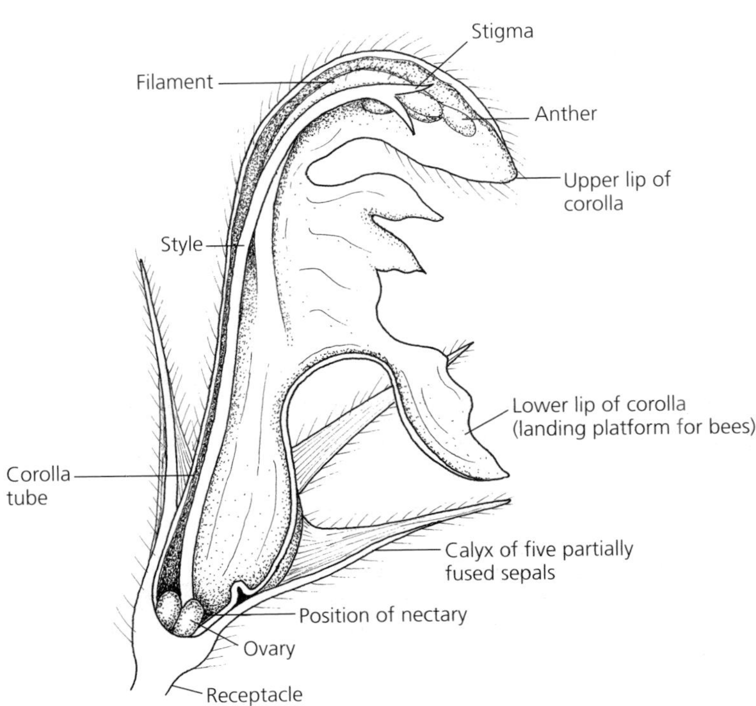


**Figure 30.16**
Half-flower of white deadnettle, *Lamium album*

The outer two layers (whorls) of flower parts are collectively called the **perianth**. In most dicotyledons, the outer whorl is differentiated as the **calyx** and the inner one as the **corolla**. In many monocotyledons, the perianth consists of two similar petaloid whorls of three (Figure 30.17). Only the inner whorl is homologous with the corolla of a buttercup.

The gynoecium may consist of one carpel (as in the pea family) or more than one, in which case they may be joined or separate. When carpels are joined, the composite nature of the ovary may be shown by the presence of internal partitions as in an orange or tomato (Figure 30.18).

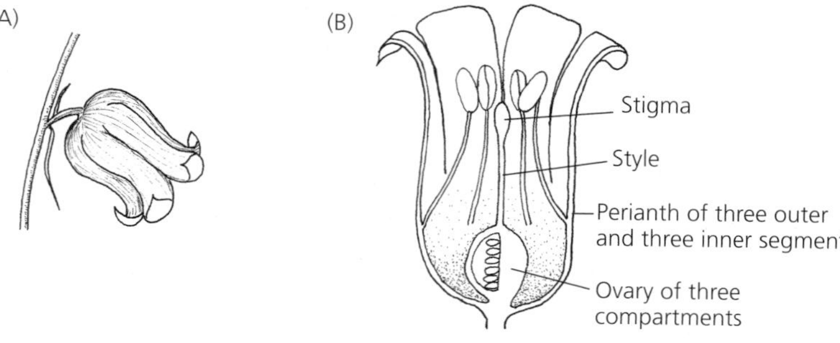


**Figure 30.17►**
A monocotyledon (bluebell) flower, showing parts in threes. Single whole flower (A) and half flower (B)

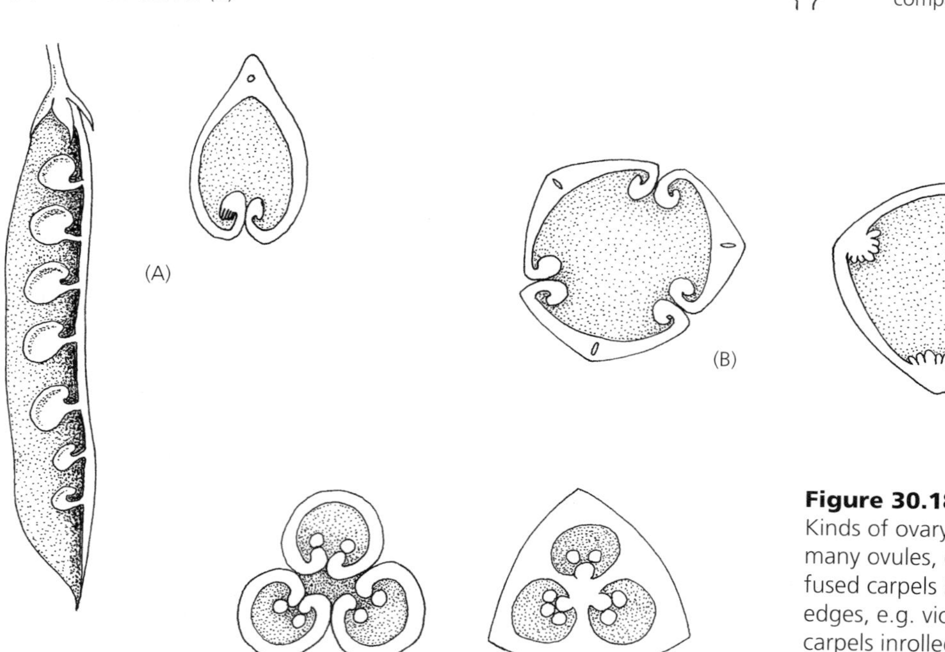


**Figure 30.18▲**
Kinds of ovary. Simple carpel with many ovules, e.g. pea (A), three fused carpels bearing ovules at the edges, e.g. violet (B) and three fused carpels inrolled, e.g. bluebell, tulip (C) (after Coult)

## Flowers, bees and clocks

You will probably have noticed that many flowers close at night and open again the following day. Some flowers are only open for part of the day. For example, the scarlet pimpernel closes its flowers about mid-afternoon, while evening primrose (as its name would suggest) opens in the early evening. It is said that Linnaeus knew the flower timetable so well that he could tell the time by observing which kinds of flower were open.

What Linnaeus did not know was that the clock-like behaviour of flowers extends beyond mere opening and closing. Nectar and scent production also show a daily rhythm, peaking for a few hours each day. Bees take advantage of this because they have a very accurate internal clock, enabling them to visit flowers at the best times. By concentrating on a particular kind of flower at each time of day, they are more likely to transfer pollen to the same species.

## Co-evolution in insects and flowers

Insect pollination involves adaptations by both insect and plant. Some flowers can only be successfully pollinated by certain kinds of insect, which are specialised for visiting those flowers. This is advantageous for both parties. Consider a flower such as honeysuckle, in which the nectar is at the bottom of a corolla tube so long that only certain moths have proboscides long enough to reach it. It is clearly advantageous to the moths to concentrate on flowers like honeysuckle, to which they have a near-monopoly of access. A honeysuckle flower is, therefore, more likely to be visited by an insect that has previously visited another honeysuckle flower.

This relationship between insects and plants is an example of **co-evolution**, in which each species exerts a selective pressure on the other. Insects with longer than average proboscides, for example, will have an advantage in reaching concealed nectaries and will tend to specialise on them. Flowers with nectaries restricting the range of insect visitors will also be selected for, since they are more likely to receive pollen from the same species.

Many plants have flowers clustered into an **inflorescence**, which to an insect probably looks like a single large flower. In the daisy family the inflorescence is so tight that it appears at a casual glance like a single flower (Figure 30.19). Each individual 'petal' is actually the corolla of a tiny flower or **floret**, whose five petals are fused together. The 'sepals' protecting the inflorescence in bud are not sepals at all but **bracts**. In many members of this family the outer florets are sterile, with the corolla serving purely for advertisement.

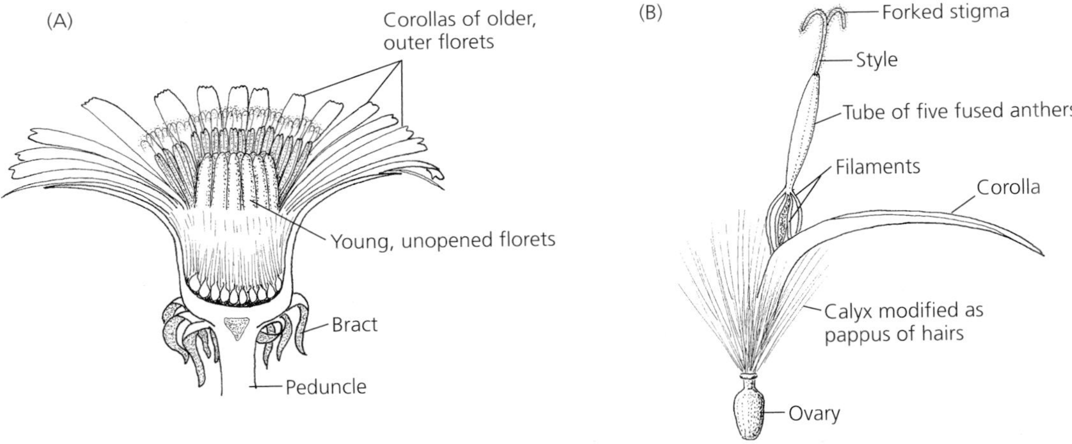


**Figure 30.19**
A dandelion inflorescence (A) and floret (B)

## Pollination

Pollination is the transfer of pollen from an anther to a stigma of a flower of the same species (Figure 30.20). In **cross-pollination** the pollen is transferred between different plants and leads to **outbreeding**. In **self-pollination**, pollen transfer is to a stigma on the same plant (though often to a different flower) and leads to **inbreeding**. Each kind of pollination has costs and benefits, and the balance of advantage depends on ecological circumstances (see below).

In Britain, most flowers are pollinated by insects (**entomophily**) or by wind (**anemophily**). Insect-pollinated flowers produce sticky pollen which adheres in clumps to the body of the animal. When searching for nectar or pollen, the animal brushes against the anthers picking up pollen

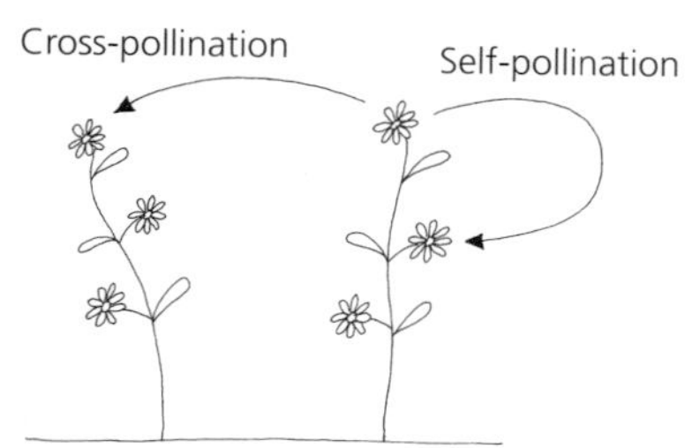


**Figure 30.20**
Cross- and self-pollination

on its body. When visiting another flower, some of the pollen comes off and sticks to the stigma. As far as the animal is concerned, pollination is thus accidental.

By far the most effective pollinators are bees, because they have to collect food for their larvae as well as themselves. Nectar contains only sugar, but larvae need protein to grow. This is supplied by protein-rich pollen. Some bee-pollinated flowers make no nectar, relying entirely on pollen as a lure – for example broom, gorse, poppies and tulips.

### *Insect pollination (entomophily)*

The sweet pea, *Lathyrus odoratus*, has a bilaterally symmetrical, or **zygomorphic**, flower (Figure 30.21). This means that it has only one axis of symmetry (i.e. it can be cut into two equal halves along only one plane). In this respect, it differs from regular, or **actinomorphic**, flowers which have several axes of symmetry (five in the buttercup). The importance of zygomorphy is that the insect can only take up one position on the flower, so the anthers and stigma can be positioned where they are most likely to touch the insect's body. This maximises the chances of pollination and reduces pollen wastage.

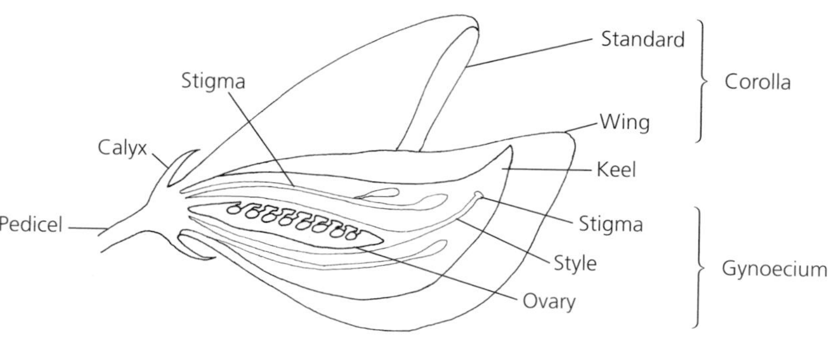


**Figure 30.21►**
Half-flower of a sweet pea

Sweet peas are cultivated for their brightly coloured corollas, which are due to water-soluble pigments called **flavonoids** dissolved in the cell sap. They also produce scent, which makes it easier for an insect to find the nectar, and in so doing to pollinate the flower.

Of the five petals, one is enlarged and forms the **standard**. Two lateral **wing petals** form a landing platform for bees, whilst the inner two form a boat-like **keel** enclosing the androecium and gynoecium. The bases of the keel and wing petals interlock so that when the wings are depressed, so is the keel.

The androecium consists of ten stamens, nine of the filaments being fused to form a trough into which nectar is secreted. Inside the androecium is the gynoecium, consisting of stigma, style and an ovary containing several ovules. When the anthers dehisce (open), the pollen is shed into the top of the keel and adheres to the hairs on the style.

The sweet pea flower is adapted for pollination by bumble bees (Figure 30.22), which are heavy enough to open the flower and have proboscides long enough to reach the nectar. When a bee lands on the wings, its weight depresses the keel and the hairy style sweeps pollen onto the insect's body. At the same time, the underside of the bee picks up pollen from the anthers and deposits it on the stigma.

Though the sweet pea is clearly well developed for cross pollination by bees, the flowers normally pollinate themselves. It seems likely that its wild ancestors were cross-pollinated, but many years of cultivation have led to the plant becoming self-fertile (how do you think self-pollination could develop as a result of domestication?).

**Figure 30.22**
Bumble bee pollinating a flower

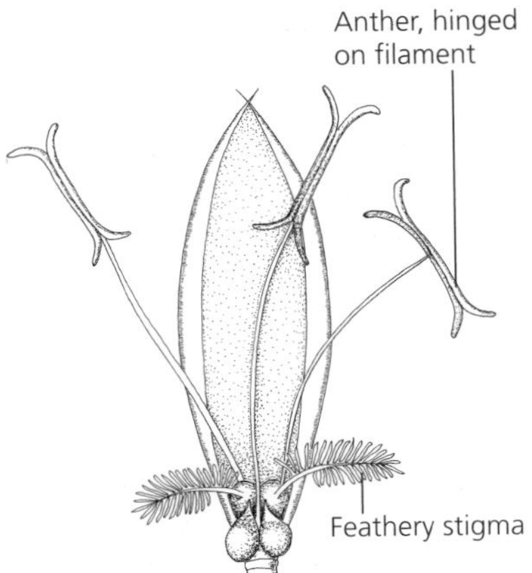


**Figure 30.23**
A flower of meadow fescue (*Festuca pratensis*) (after Rendle)

## *Wind pollination (anemophily)*

All conifers and many flowering plants are wind-pollinated, for example grasses, plantains and nettles. In contrast to the showy animal-pollinated flowers, wind-pollinated flowers are usually small, green and inconspicuous, and never produce nectar or scent.

The energy savings made possible by not having to produce nectar, scent and large petals are offset by the fact that wind carries pollen randomly. Only a tiny proportion of the pollen grains find their target, and the plant has to produce large quantities. Since the pollen grains are very small, they have a high surface to mass ratio, so they fall slowly and are easily caught by the wind. They are also non-sticky, so they do not adhere to leaves, branches and other obstacles. Many wind-pollinated trees produce their flowers before the leaves open, thus helping free air movement.

Meadow fescue (Figure 30.23) illustrates many of the characteristics of wind-pollinated flowers. The inflorescences are produced on long stems, well above the leaves, so the wind is able to catch them. Each flowering stalk bears many miniature inflorescences or **spikelets**, each consisting of several tiny flowers. A fescue flower has three stamens, each with its filament attached midway along the anther so that it rocks in the wind, shaking out the pollen. The stigmas are feathery, exposing a large surface to the wind. Even so, the chances of a stigma intercepting more than one pollen grain are not high and, like most wind-pollinated flowers, meadow fescue has only one ovule in each ovary.

The differences between wind and insect-pollinated flowers are summarised in Table 30.1.

**Table 30.1**
Differences between wind-pollinated and insect-pollinated flowers

| Typical insect-pollinated flower | Typical wind-pollinated flower |
|---|---|
| Flowers large and conspicuous | Flowers small and green |
| Usually produce nectar and scent | Never produce nectar or scent |
| Anthers do not pivot on filaments | Anthers usually pivot on filaments |
| Pollen sticky | Pollen non-sticky |
| Pollen grains relatively large | Pollen grains very small |
| Smaller amounts of pollen | Larger amounts of pollen |
| Stigmas not usually feathery | Stigmas feathery or hairy |

## *Preventing inbreeding*

Darwin's oft-quoted statement that

> “*nature abhors self-fertilisation*”

was based on the fact that many plants have adaptations which reduce the likelihood of self-pollination. As explained in Chapter 15, inbreeding reduces genetic variability, producing a high proportion of homozygotes. Recessive alleles (which are normally protected in heterozygotes by dominance) may be eliminated by selection. The reduction in the genetic variability of the population lowers its capacity to adapt to changing conditions. Plants have a number of ways of reducing the chances of self-pollination.

- Flowers may be **dichogamous**, the anthers and stigmas ripening at different times. The majority of flowers are **protandrous**, the anthers shedding their pollen before the stigma is receptive. Some, such as bluebell (*Hyacinthoides non-scriptus*) and plantains (*Plantago* spp), are **protogynous**, the gynoecium ripening first.

Although dichogamy increases the likelihood of cross-pollination, it does not usually ensure it for two reasons. First, dichogamy in most plants is incomplete, with a period of overlap when the flower is functionally hermaphrodite. Second, most plants have flowers of differing ages, enabling pollen to be transferred from flowers that are functionally male to flowers that are functionally female.

- Many plants are **dioecious**, each plant producing flowers of only one sex, for example holly, mistletoe, willow and perennial nettle. Others are **monoecious** with unisexual flowers of both types on the same plant, as in oak, beech, hazel, annual nettle and maize.
- Some plants are **self-incompatible**, pollen growing more slowly (or not at all) on a stigma of the same plant.

### *Incompatibility*

The inability of a pollen grain to grow normally on a stigma of the same plant can have two quite different causes. It may be due to **incompatibility genes** carried by the pollen grain itself. This is called **gametophytic incompatibility** (because the pollen is the male gametophyte, see the information box on page 544). In clover and in poppies, incompatibility is controlled by a single gene locus with many alleles. A pollen grain cannot grow on stigmatic tissue containing the same allele as itself.

An alternative type of incompatibility is **sporophytic incompatibility**, which is due to proteins produced by the parent (sporophyte). These proteins are deposited on the surface of the exine (outer layer of a pollen grain, see below) while the pollen grains are maturing in the anther.

### *Promoting inbreeding*

Inbreeding is not always disadvantageous. Some woodland plants, for example wood sorrel (*Oxalis acetosella*) which cannot rely on visitation by insects, 'hedge their bets' by producing two kinds of flower. Besides the normal flowers which are adapted for cross-pollination, they produce **cleistogamic flowers**. These remain closed and pollen is shed onto the stigma in the flower bud.

## The formation of pollen

Pollen grains (microspores) are produced within the four pollen sacs (**microsporangia**) of each anther (Figure 30.24). Each pollen sac of a young stamen contains many diploid **pollen mother cells**. These undergo meiosis, the resulting pollen grains often remaining together for a while as **tetrads**. The parent cell wall breaks down allowing the individual pollen grains to separate and adopt their final shape. During this time the developing pollen obtains nutrients released by the breakdown of the inner layer of the pollen sac, the **tapetum**.

The wall of a mature pollen grain is two-layered. The inner **intine** consists of a continuous and unmodified cellulose wall. The outer **exine** is waterproof and consists of **sporopollenin**, which is similar to cutin and suberin but is even more resistant to decay. The exine is perforated by a number of pores and is sculpted in a manner that is characteristic of the species (see the information box on page 553).

The wall of the anther contains a fibrous layer in which the cells have unevenly thickened walls. As the anther ripens, tensions are set up as the fibrous layer dries, causing it to split longitudinally along two lines and releasing the pollen (Figure 30.25).

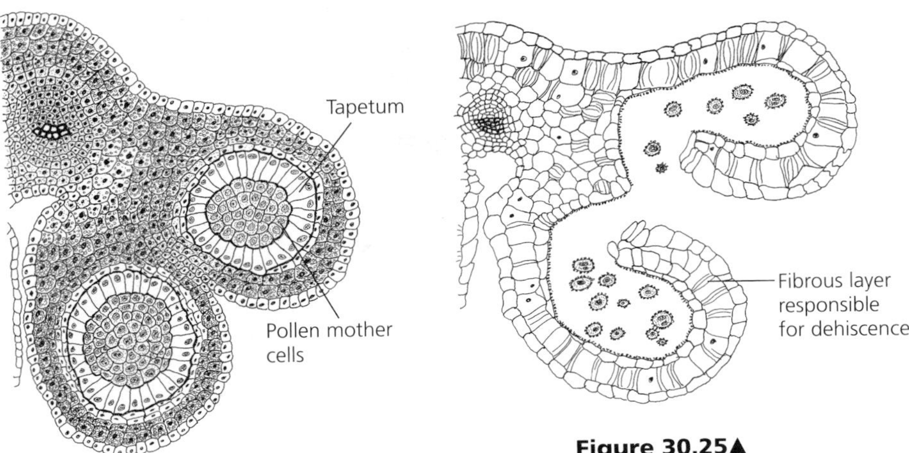


**Figure 30.24►**
Transverse section of half of a ripe anther of lily (after Priestley and Scott)

**Figure 30.25▲**
An anther of lily after dehiscence (after Priestley and Scott)

## Clues from the past from pollen

Pollen grains have two characteristics that are useful to scientists. First, they are extremely resistant to decay, and, under favourable circumstances, may be preserved in sediments for tens or hundreds of thousands of years. Second, their size, shape and surface pattern are so characteristic that an expert can identify the species to which it belongs.

The study of changes in the frequency of different kinds of pollen preserved in deposits such as peat is **palynology**, and it has enabled scientists to learn about changes in the relative abundance of different species of wind-pollinated plant. From changes in flora it has been possible to gain insight into past climates.

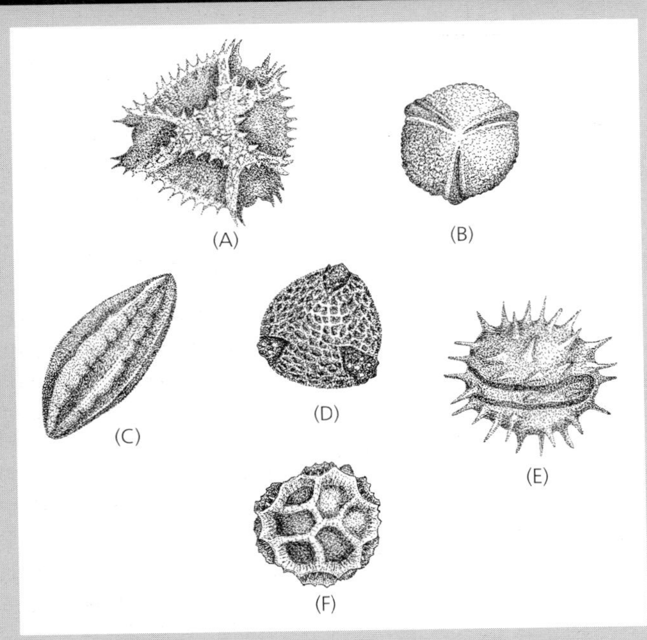


A variety of pollen grains. (A) *Tragopogon pratensis*, (B) *Fagus grandifola*, (C) *Ephedra glauca*, (D) *Salix fragilis*, (E) *Nymphaea advena* and (F) *Stokesi laevis* (after Wodehouse)

As the anther ripens, the nucleus of the pollen grain divides into a **generative nucleus** (so-called because it later gives rise to the male gametes), and a **vegetative nucleus**, which is the vestigial male gametophyte (see the information box on page 544).

## The ovule and egg

An ovule is a potential seed and develops on a part of the ovary wall called the **placenta**. It originates as a swelling called a **nucellus**, and becomes partially surrounded by two (less often one) coverings, or **integuments**.

These develop as collars of tissue that grow up from the base, leaving a pore or **micropyle**. The ovule is attached to the placenta by a stalk, or **funicle**, which carries a strand of xylem and phloem (Figure 30.26).

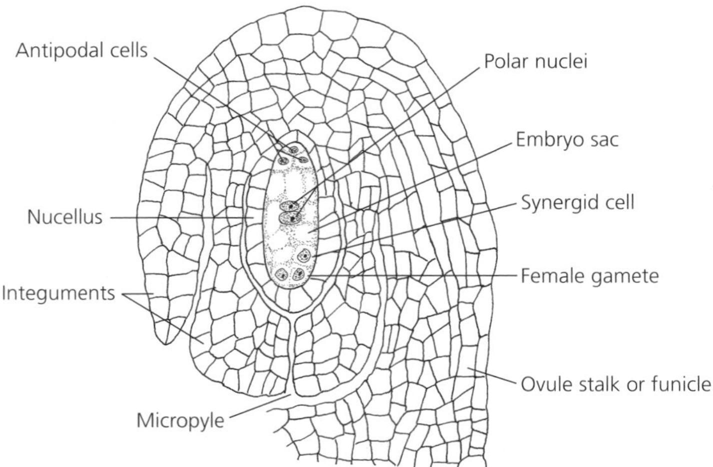


**Figure 30.26**
Longitudinal section through a ripe ovule of *Lilium*

The nucellus is actually a **megasporangium** (see the information box on page 544). Inside it develops a **megaspore mother cell**, which undergoes meiosis to produce four potential **megaspores**. The details of what happens next vary considerably, and the following description applies to one of the more common types.

One of the four haploid products of meiosis becomes a **megaspore**; the other three die. The megaspore than undergoes three consecutive mitotic divisions to produce eight haploid nuclei. These and the surrounding cytoplasm form the **embryo sac**, which is the greatly reduced **female gametophyte**.

The eight nuclei become organised into two groups of four, one group at each end of the embryo sac. Of each group, three nuclei become surrounded by envelopes of cytoplasm and form cells. The remaining two **polar nuclei** move together in the centre of the embryo sac.

Of the three cells nearest the micropyle, one remains naked and is the **egg**, and the other two develop thin cell walls and become **synergid cells**. The three cells at the other end also develop cell walls and are called **antipodal cells**; they appear to have no function.

## Fertilisation

Pollination merely brings the pollen to a stigma (Figure 30.27) while the egg is still some distance away inside an ovule. The next stage is the growth of the **pollen tube**, which brings the sperm to the egg – a process that takes several days.

**Figure 30.27**
SEM of pollen grains on the stigma of a flower

Stimulated by sugar secreted by the stigma, the pollen tube grows out through one of the pores in the outer layer of the pollen grain (Figure 30.28). The tube grows down into the stigma and style, secreting digestive enzymes as it goes. The simple products of digestion (such as sugars and amino acids) are absorbed by the pollen tube and used for growth. The length to which the pollen tube grows depends on the length of the style. In a buttercup, the stigma is little more than a millimetre away from the ovule, but in some species it may be several centimetres.

During its growth down the style, the generative nucleus divides mitotically into two **male gametes**, each of which retains a thin envelope of cytoplasm.

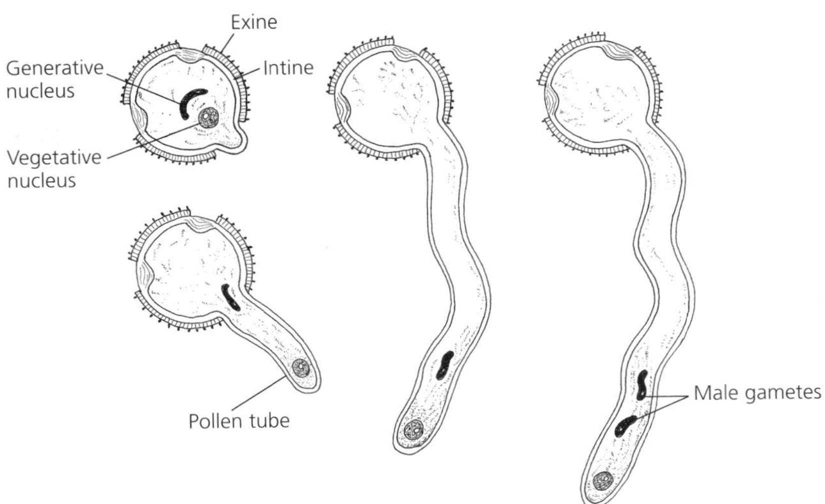


**Figure 30.28**
Germination of a pollen grain

When the pollen tube reaches the ovary it grows towards an ovule, attracted by chemicals produced by the egg. The pollen tube usually grows through the micropyle. On reaching the egg the tip breaks down, releasing the two sperms (Figure 30.29). One joins with the egg to form a **zygote**, and the other fuses with the two polar nuclei to form a triploid **primary endosperm nucleus**. This **double fertilisation** is unique to flowering plants.

In most plants the ovary contains many ovules, so many pollen grains will be needed if they are all to develop into seeds.

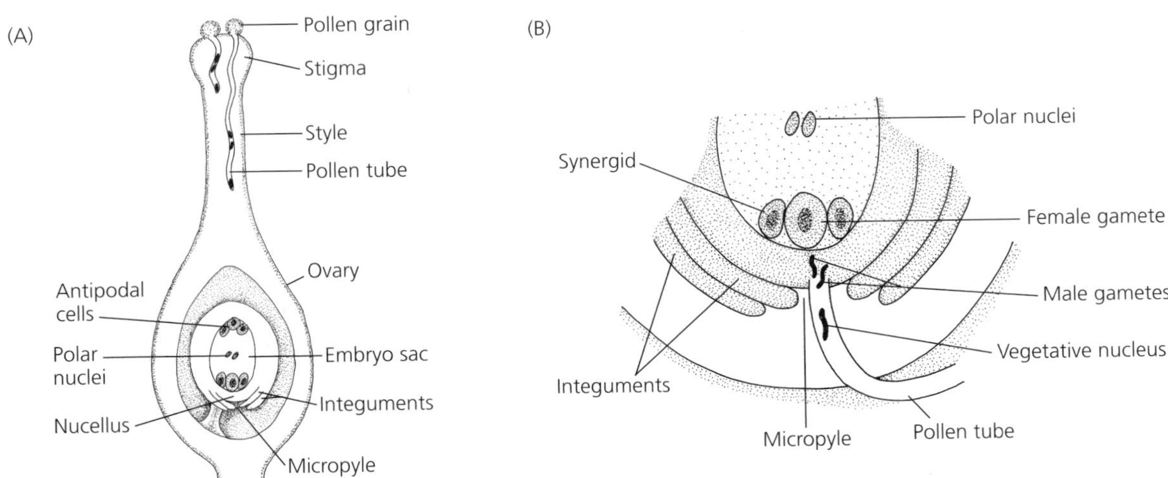


**Figure 30.29**
Fertilisation in a flowering plant (A) and (B)

## *After fertilisation*

After fertilisation the primary endosperm nucleus divides many times to form a spongy storage tissue called the **endosperm**. Though all seeds have an endosperm to begin with, in **nonendospermic seeds** it disappears early in development to be replaced by energy stored within the embryo itself. In **endospermic seeds** such as cereals (e.g. rice, corn), the endosperm continues to develop and forms a large part of the mature seed.

Meanwhile the zygote has been dividing mitotically to produce a chain of cells called the **proembryo**. The details of what happens next vary somewhat, and the following description applies to *Capsella* (shepherd's purse). The pro-embryo differentiates into three parts: a large **basal cell** near the micropyle, an **embryonal cell** at the other end, and the cells connecting them forming the **suspensor** (Figure 30.30).

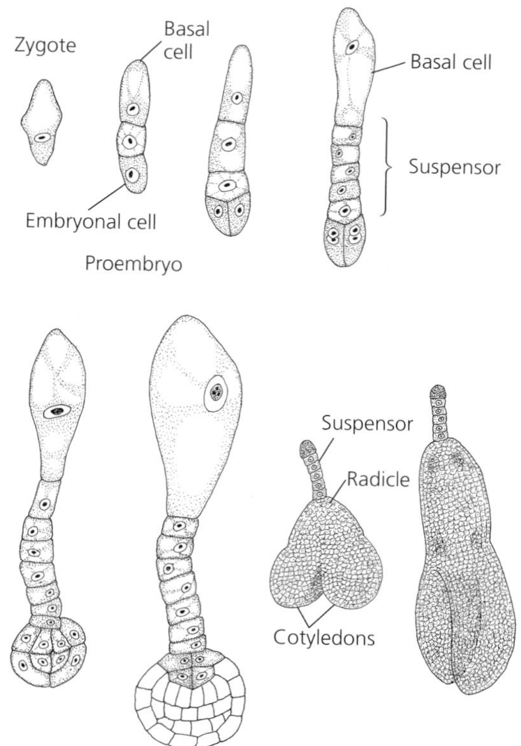


**Figure 30.30**
Embryo formation in shepherd's purse (*Capsella bursa-pastoris*)

The embryonal cell continues to divide, forming a group of eight cells. The four nearest the suspensor give rise to the **radicle**, or embryonic root, the others to the two embryonic leaves, or **cotyledons**, and the embryonic shoot, or **plumule**.

Pollination and fertilisation also trigger a number of other changes.

1 The integuments continue to grow and become grafted together forming a single layer, the seed coat, or **testa**. In some seeds the testa is not merely a protective layer, but plays an active part in maintaining dormancy.

2 Under the influence of **auxin** (Chapter 32) produced by the developing embryo, the ovary grows into the **fruit**, and the ovary wall becomes the fruit wall, or **pericarp**.

3 In most flowers the sepals, petals and stamens fall off.

The developing seed lays down large amounts of nutrient in the form of protein and either fat or starch. In most seeds the final stage of development is marked by a drastic decrease in water content to about 5–20% and the complete cessation of any detectable metabolic activity. These changes are a necessary preparation for dispersal, in which the seed has to travel through conditions hostile to growth.

## *Seeds without fertilisation: apomixis*

Some plants produce seeds without fertilisation by a process called **apomixis**, for example dandelion, hawkweed and most blackberries. The mechanisms by which this occurs are diverse, but the end result is that the offspring are genetically identical to the parent, the long-term benefits of sex being lost. There is however a short-term bonus, in that the plants are not dependent on outside agents like insects or wind for pollen transfer.

If pollination is unnecessary in these plants, why do they expend energy producing brightly-coloured petals and nectar? It seems probable that their ancestors reproduced sexually, but have 'cashed in' on the short-term benefits of doing without sex.

The development of eggs without fertilisation occurs in many animals as well as plants, and is given the more general name of **parthenogenesis**.

## Independence from water

The dominant land plants are the flowering plants. Two evolutionary developments have been especially important in their success. First, the pollen tube freed plants from a reliance on external water for fertilisation. A second development that reduced the reliance on water was the evolution of the seed. A microscopic spore carries very little stored energy and so can only germinate successfully in the light, and, therefore, on the soil surface. But surface water tends to be unreliable, and spores can only germinate and get established where the surface remains damp for reasonably long periods.

As a much larger structure, a seed carries enough energy reserves to support growth in darkness for at least a few days (orchids are an exception). Seeds can, therefore, germinate underground where the water supply is more certain.

## Fruits

A fruit is usually defined as a ripened ovary, and in many plants it plays an important part in the dispersal of the seeds. According to the way the pericarp develops, fruits fall into two categories:

- in **succulent fruits** the pericarp becomes soft and juicy and is eaten by animals.
- in **dry fruits** it becomes quite tough and is not eaten.

Dry fruits containing more than one seed are usually **dehiscent** – that is, they open to release the seeds. In these fruits, the seeds are dispersed individually. **Indehiscent fruits** (which do not open) usually contain only one seed.

### *When is a fruit not a fruit?*

Some fruits are formed from parts of the flower other than the ovary and are called 'false fruits' (Figure 30.31). A strawberry is a greatly swollen receptacle, the actual fruits being the little seed-like structures on the surface. An apple is formed by the ripening of a deeply cup-shaped receptacle, the actual ovary being the 'core'.

Despite the traditional anatomical distinction between 'true' and 'false' fruits, they are alike in so far as they are both parts of a flower that enlarge in response to auxin produced by developing seeds (Chapter 32). If a fruit is defined on this physiological basis, the term 'false fruit' ceases to have any meaning.

### *Dispersal of fruits and seeds*

In most plants the survival chances of seeds are increased by dispersal. This always involves a considerable element of chance, and a large proportion do not reach a suitable habitat. The successful ones do not have to compete with each other or with the parent for resources, and there is always a chance that some seeds may reach a better environment than that of the parent.

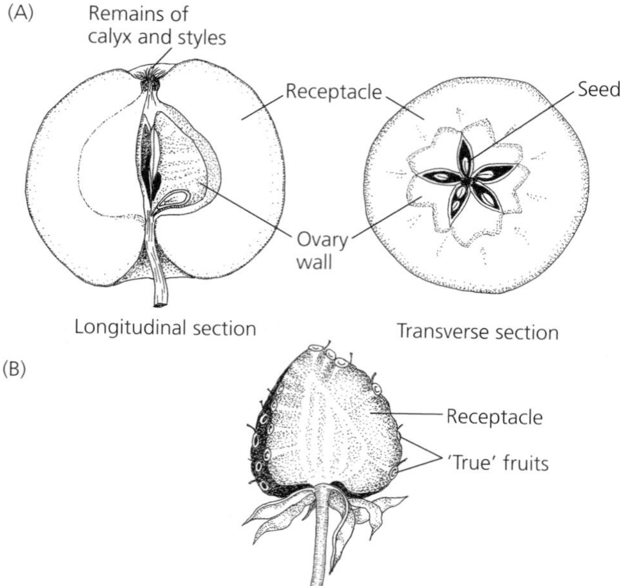


**Figure 30.31**
'False' fruits – apple in section (A) and strawberry in section (B)

The dispersal unit may be a fruit or a seed. A useful term that covers both (and also spores) is a **propagule**. Propagules may be dispersed by wind, by animals, by self-dispersal or by water.

### *Dispersal by wind*

Wind-dispersed propagules have diverse forms but with one essential feature in common: they have a large surface/mass ratio (Figure 30.32). In most fruits and seeds this is achieved by a complex shape in the form of wings (e.g. sycamore) or feathery outgrowths (e.g. dandelion). Although these outgrowths may be superficially similar, in some cases they have evolved by modification of different structures. In willowherb, for example, the parachute is an extension of the testa, and in dandelion it is a highly-modified calyx called a **pappus**.

In orchids a high surface/mass ratio results from the minute size of the seeds, which are blown about like dust. With masses as small as 0.005 mg, orchid seeds carry insufficient stored energy to get established by themselves. Instead they obtain energy provided by a saprobic fungus, with which the orchid lives in a mutualistic relationship.

Wind dispersal of seeds and fruits is obviously wasteful, but this is offset by the fact that because the seeds or fruits have to be light, larger numbers can be produced without great cost.

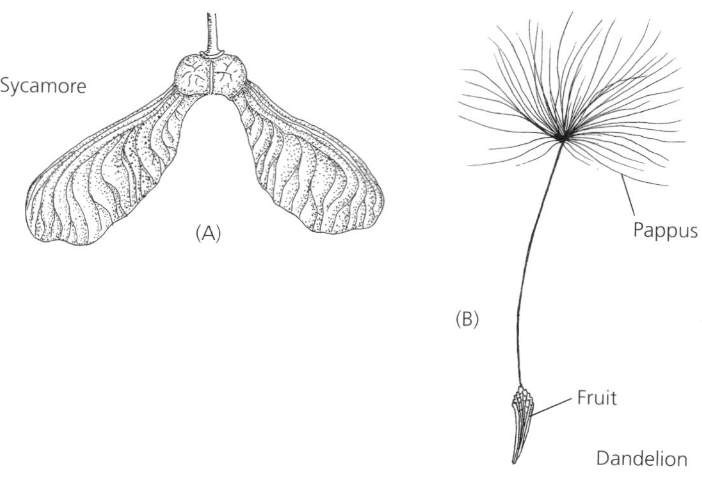


**Figure 30.32**
Wind dispersed fruits. Sycamore (A) and dandelion (B)

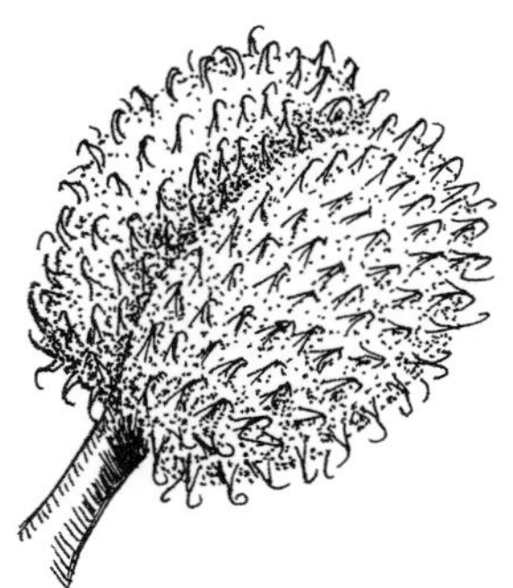

**Figure 30.33▲**
Cleavers, a fruit dispersed by attachment to animals

### *Dispersal by attachment*

Dispersal by attachment to animals is somewhat less wasteful than wind-dispersal because animals tend to frequent areas supporting plant growth. Figure 30.33 shows an example of a fruit that is dispersed in this way. Although the attachment devices are often superficially similar, their biological nature may be quite different. In cleavers (*Galium aparine*), for instance, the hooks are on the pericarp, in *Geum* (avens) they are formed from a modified style, and in burdock they are bracts protecting the ripened inflorescence.

### *Dispersal by being eaten*

Succulent fruits are nutritious and are eaten by animals, especially birds. The seeds pass unharmed through the gut to be deposited when the animal defaecates several hours later (Figure 30.34). The fruits are usually conspicuous, often with bright colours, and the seeds are protected from digestion by a resistant testa (e.g. elder) or by the innermost layer of the pericarp, as in the 'stone' of a cherry.

A completely different dispersal mechanism occurs in the oak, in which the seeds are buried as food by squirrels. Some are forgotten and may germinate.

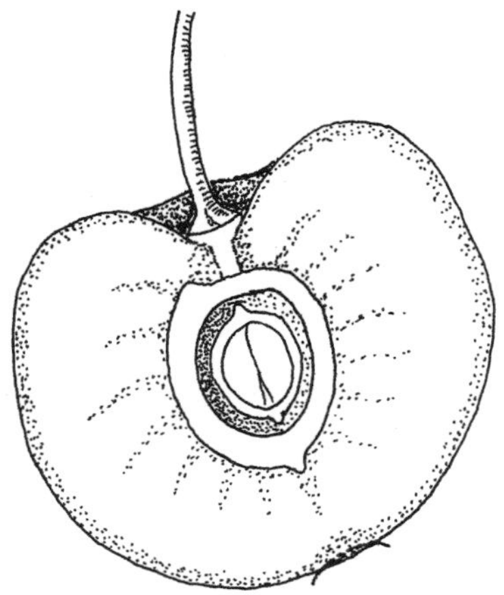

**Figure 30.34▲**
Longitudinal section through a cherry, a fruit with a fleshy pericarp

### *Pepper pot dispersal*

Another form of dispersal is the censer, or 'pepper-pot', mechanism, in which seeds are jerked out of a fruit (such as a poppy) when knocked by a passing animal (Figure 30.35). Wind is probably too gentle to jerk the stems sharply enough.

### *Self-dispersal*

In some dehiscent fruits the pericarp is constructed in such a way that it warps as it dries out. Tensions are set up which are released suddenly with the violent opening of the fruit, jerking out the seeds (Figure 30.36). In Himalayan balsam the tensions are due to turgor.

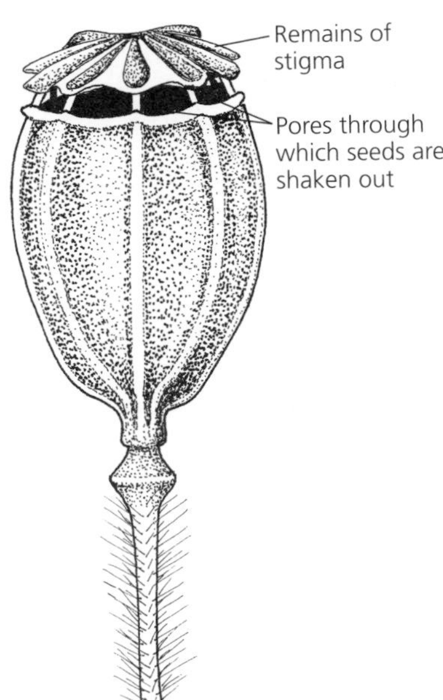


**Figure 30.35▲**
Capsule of poppy

## Different kinds of life cycle

Though all living things have the same 'aim' – to produce as many offspring as possible – they have different life cycle 'strategies' for achieving this. Some concentrate all their resources on rapid growth and early reproduction. Others grow more slowly and delay reproduction until a larger size has been reached.

Based on the length of their life cycles, plants can be divided into four groups:

- **ephemerals** complete their life cycles in a few weeks, and under favourable conditions are capable of several generations a year. Ephemerals are 'opportunists', exploiting habitats that are only available for short periods of time, such as a flower bed or a desert after the heavy rain that may fall every few years. Examples of ephemerals are groundsel, shepherd's purse and many other weeds.
- **annuals** complete one life cycle per year. Like an ephemeral, an annual puts all its resources into a single reproductive effort, after which it dies.
- **biennials**, such as cabbage and carrot, have a life cycle that spans two seasons. In the first year the plant does not flower but concentrates on building up its energy reserves. The following year it uses all its stored energy for flower and seed production, and then it dies.

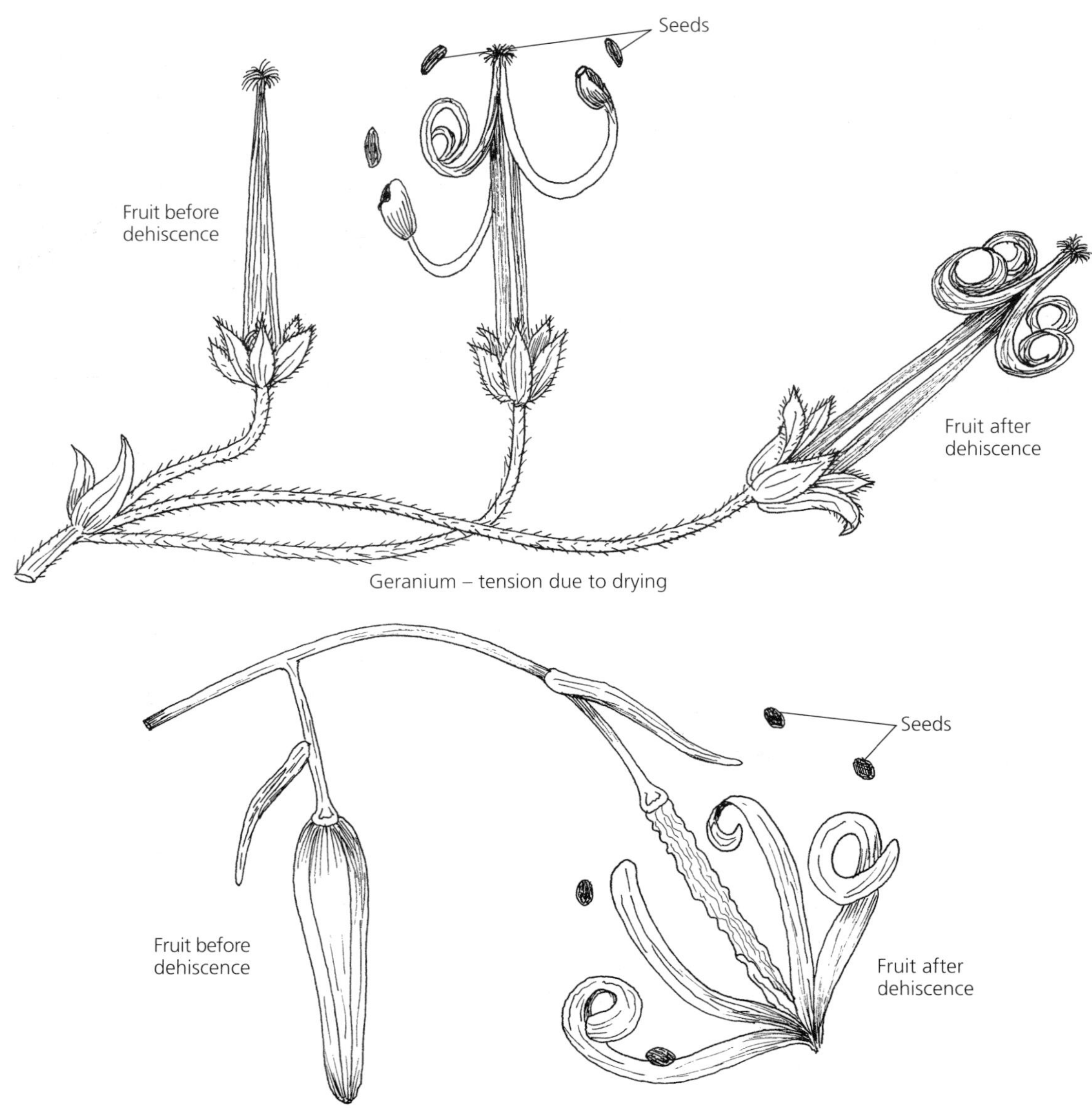


**Figure 30.36**
Self-dispersal mechanisms

- **perennials** survive for many years. The process of resuming growth year after year is called **perennation**. Perennials are of two kinds. **Woody perennials** are trees and shrubs, in which the aerial parts remain alive in winter (though in deciduous plants they lose their leaves). Growth is resumed each spring from buds on the previous year's aerial shoots, which get taller each year. In these plants, energy reserves are stored above ground as well as below. In **herbaceous perennials** the aerial parts die back at the end of each year, growth being resumed in spring from buds near or below ground level. In these plants energy reserves are stored below ground. Note that many plants have stems with 'woody' texture, but which die back at the end of the year and so are strictly speaking herbaceous.

## QUESTIONS

**1** For each of the following, give the name of the part of a flower that has that function:

**a)** makes pollen
**b)** contains one or more ovules
**c)** protects the flower in the bud stage
**d)** anchors the other flower parts
**e)** secretes a liquid that insects seek
**f)** stimulates pollen grains to germinate.

**2** Choose the phrase which best finishes the statement. A flower that produces scent but no nectar is likely to

A have sticky pollen
B be self-pollinated
C be pollinated by butterflies
D produce large amounts of very light pollen grains.

**3** A flower with a stigma that secretes a sugary solution

A could be pollinated by wind, insects or birds
B must be insect-pollinated
C could be wind-pollinated
D could not be wind-pollinated.

**4** 'Red hot poker' is a common garden plant, so-called because of its large, bright orange inflorescences. As in many flowers, the stamens ripen before the ovaries. A foraging bee usually starts at the lower flowers and works its way upward.

**a)** Suggest why this might be an advantage to the plant.
**b)** Suggest how the behaviour of the plant could cause bees to move from lower to upper flowers.

**5** Bees have hairy heads, and it was once suggested that this was an adaptation to aid the transfer of pollen from one flower to another. Why is this unlikely?

**6** If an insect-pollinated flower were to produce larger quantities of nectar, it would be more attractive to insects. Apart from increased energy costs, suggest why it is actually advantageous for flowers to be 'mean' in their nectar output.

**7** Dichogamy does not occur in insect-pollinated flowers that do not produce nectar. Suggest why.

**8** Bilaterally symmetrical flowers tend to have fewer stamens than radially symmetrical ones. Suggest an explanation.

**9** Where precisely in a flowering plant is each of the following located?
**a)** a male gamete, **b)** a female gamete?

**10** Zygote is to embryo as ovule is to .......... and ovary is to ..........

**11** A cell in the developing endosperm of a wheat seed has 63 chromosomes. How many chromosomes are there in a pollen grain of this species?

CHAPTER

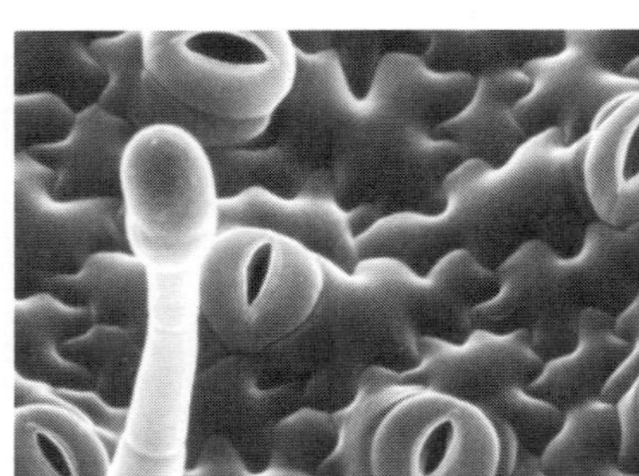

# 31

# GROWTH in PLANTS

**LEARNING OBJECTIVES**

By the time you have completed your study of this chapter you should be able to:

- know the parts of familiar food seeds, such as the garden pea and maize grain.
- describe the structural changes that occur in the germination of a pea, a sunflower and a maize grain.
- explain the physiological changes that occur in germination, and the changes in growth brought about by stimulation by light.
- discuss the respective survival values of small and large seeds.
- explain the different ways of measuring growth, and distinguish between absolute and relative growth rates.
- distinguish between primary and secondary growth.
- explain how a stem grows in length.
- explain how the organisation of tissues in a stem relates to its supporting function.
- describe how a root grows in length, and explain how the differences between the primary structure of roots and stems can be related to their respective functions.
- describe the secondary growth in a dicotyledon, and explain how it can be used to tell the age of a tree.

## Seed structure

A seed is an embryo plant surrounded by a coat of parental tissue, the **testa**. It also carries reserves of protein and energy in the form of either fat or starch. Although the most important crop seeds store starch, many seeds store fat, which has over twice the energy content per gram of carbohydrate. In wind-dispersed seeds this saves weight.

### *The garden pea*

The pea is a non-endospermic dicotyledonous seed (Figure 31.1). On one side is the scar, or **hilum**, marking the site of attachment of the ovule stalk, or **funicle**, to the placenta. In a soaked seed, the position of the micropyle at one end of the hilum can be seen from the drop of water that seeps out when the seed is squeezed.

Inside the testa is the embryo. The two hemispherical **cotyledons**, or embryo leaves, are laden with stored starch and protein. Between the cotyledons are the tiny **plumule**, or embryonic shoot, and the **radicle**, or embryonic root, which is tucked into a pocket in the testa.

### *Sunflower*

The sunflower 'seed' is actually a one-seeded fruit. This is indicated by the presence of *two* scars, left by the attachment to the style and

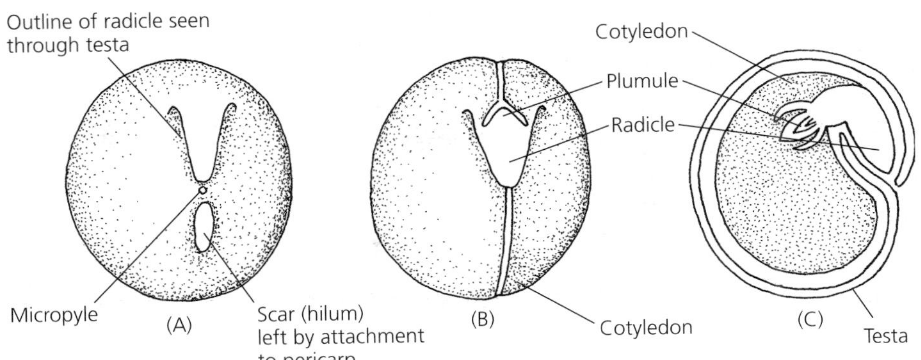


**Figure 31.1**
Structure of the garden pea seed. External features of the whole seed (A), entire embryo after removal of testa (B) and section through entire seed (C)

receptacle, respectively. Inside the pericarp is the seed itself, covered by a very thin testa. The embryo is essentially similar to that of the pea except that the cotyledons store oil instead of starch.

### *Maize grain*

All cereal grains are actually one-seeded fruits in which the pericarp has been grafted on to the testa to form a single composite layer. The bulk of the seed consists of the whitish **endosperm**. The outermost part of the endosperm, the **aleurone layer**, stores protein, while the rest stores starch.

The single cotyledon is called the **scutellum** and lies against the endosperm. The plumule is surrounded by a sheath-like **coleoptile** which protects it during its growth through the soil. Another sheath, the **coleorhiza**, protects the radicle.

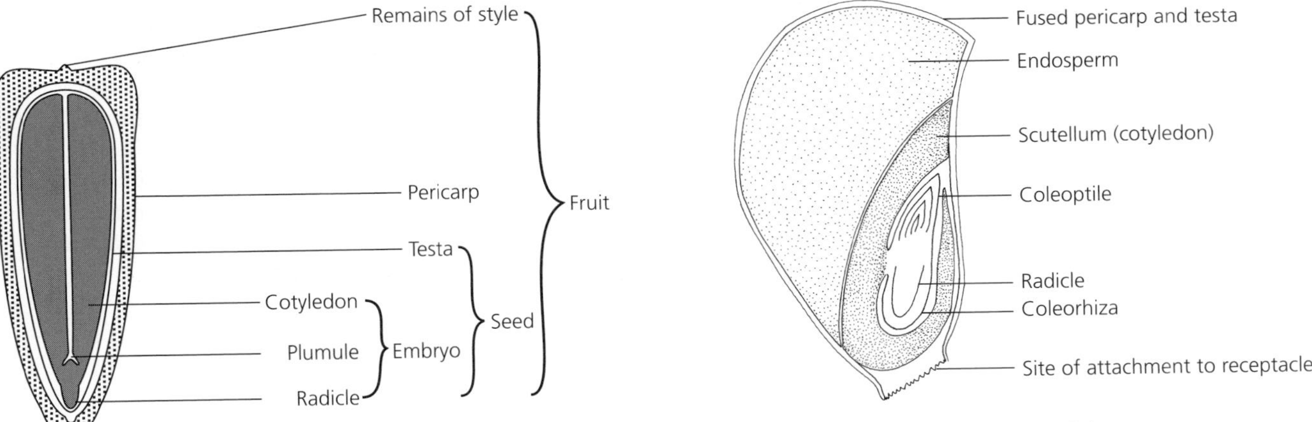


**Figure 31.2▲**
Longitudinal section of a sunflower 'seed'

**Figure 31.3▲**
Longitudinal section of a maize grain

## Germination

Germination is the process by which the young plant emerges from the seed and becomes independent. Like other life processes, germination requires certain environmental conditions.

- There must be adequate water. For reasons explained in Chapter 3, water is a fundamental requirement for all life processes.
- A suitable temperature. Seeds have different ranges of temperature preference, but most germinate between 5°C and 30°C.
- With few exceptions (such as rice, which normally grows in flooded soil), seeds require oxygen to germinate.

Besides these requirements, some seeds require light to germinate; this is discussed in Chapter 32.

### *Germination in the garden pea*

Before germination can begin, the seed has to absorb water. This occurs in two distinct phases. For about the first 12–15 hours this occurs by a process called **imbibition**, in which the proteins and polysaccharides of the seed become hydrated. The seed swells and its wet mass (or live mass) increases. This initial water uptake is reversible – during imbibition seeds can be dehydrated again without ill effects.

Following imbibition water uptake continues at a slower pace, by osmosis. It is associated with the formation of vacuoles and the onset of metabolic activity, and cannot be reversed by dehydration without killing the plant.

As the cells become hydrated, enzymes synthesised before dispersal are activated. New enzymes are synthesised, including hydrolases that digest energy reserves. Starch in the cotyledons is hydrolysed to glucose which is then converted to sucrose, and proteins are hydrolysed to amino acids.

Energy for synthesis is provided by respiration, which increases steeply. Though carbohydrate is the main respiratory substrate in peas, the respiratory quotient may rise above 1 for a while. This is because until the testa has split due to swelling of the seed, oxygen uptake may be impeded, leading to some fermentation (Chapter 6).

The products of digestion are translocated to the radicle and plumule. Here, they are used either in the synthesis of cellulose, proteins and macromolecules, or as fuel in respiration. Because some organic matter is used in respiration, there is an overall decrease in dry mass.

In contrast to dry mass, the wet mass (live mass) of the seed continues to increase because the amount of water absorbed is greater than the amount of organic matter used in respiration.

These physiological events begin before any growth is apparent. The first external change is the emergence of the radicle through the testa (Figure 31.4). Its early emergence is related to the need to secure a reliable water supply and an anchorage for the shoot system. The radicle shows **positive gravitropism** (formerly called geotropism) – no matter which way the seed lands, the radicle grows downwards. As it reaches deeper soil where the water supply is more permanent, lateral branches grow outwards from the main root at an angle to gravity.

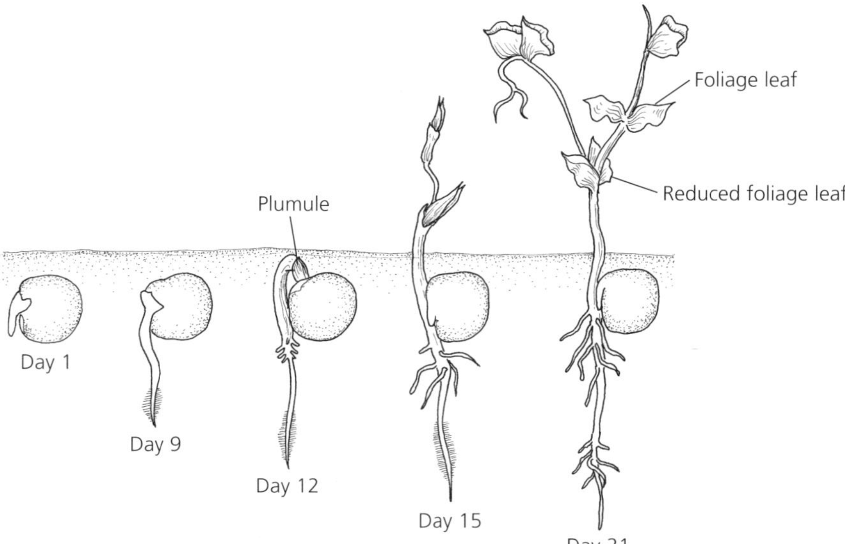


**Figure 31.4**
Germination of the garden pea. The timescale obviously varies with temperature

Since the seed has its own energy store, the need for light is less immediate than the need for water, and the plumule emerges a day or two after the radicle. It elongates by extension of the region immediately above that of the insertion of the cotyledons, the **epicotyl**. Since the cotyledons remain below the ground, this kind of germination is said to be **hypogeal** (*hypo* = below, *geo* = earth).

The upward growth of the plumule is an example of **negative gravitropism**. By growing away from gravity it grows toward the light, which it cannot yet 'see'. While moving through the soil it is pale yellow with very small leaves. It is also hook-shaped, which protects the delicate leaves from abrasion by the soil.

When the plumule emerges above ground, light stimulates a number of changes in its growth:

- the plumular hook straightens out.
- the leaves expand.
- the rate of elongation of the plumule slows down.
- chlorophyll develops and the leaves become green.

All but the last of these changes are mediated by the pigment **phytochrome** (Chapter 32). Once photosynthesis exceeds respiration, the plant is independent. The cotyledons eventually wither and die as all the reserves are removed from them.

When seedlings are grown in darkness, these light-induced changes do not occur. The seedling becomes **etiolated**, in which the leaves are small and pale, and the internodes long.

### *Germination in the sunflower*

Germination in the sunflower differs from that of the garden pea in that it is **epigeal**, that is the cotyledons are carried *above* the ground (*epi* = above). In this case it is the region just below the cotyledons – the **hypocotyl** – that elongates (Figure 31.5). On reaching the light the hypocotyl straightens, and the cotyledons expand and develop chlorophyll.

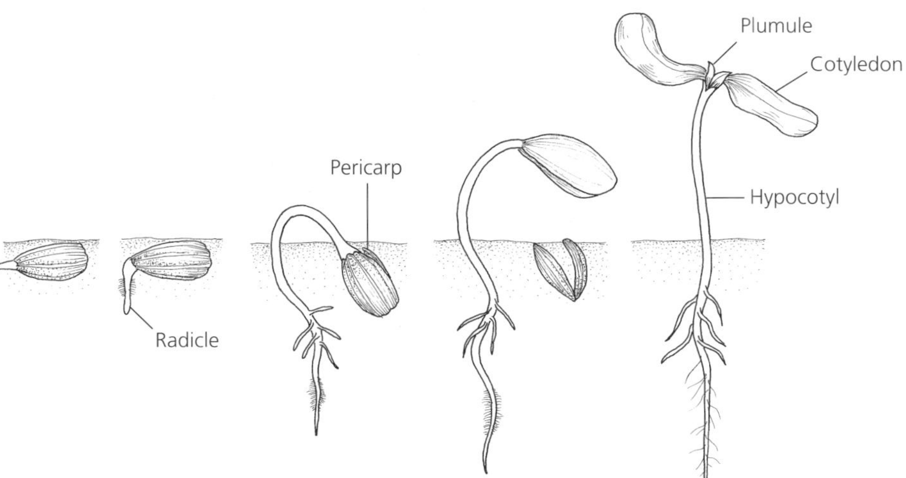


**Figure 31.5**
Germination in the sunflower

Sunflower germination also differs physiologically from that of the pea. The fat stored in the cotyledons is converted into carbohydrate (something which animals cannot do). This involves the addition of oxygen and therefore a gain in mass (1 g of fat yields about 1.25 g of carbohydrate). During the first few days of germination, a fat-storing seed may actually gain more dry mass due to this conversion than it loses in respiration, even though it is in complete darkness!

### *Germination in cereals*

The germination of a maize grain is shown in Figure 31.6. The most distinctive feature of cereals is the sheath-like coleoptile, which protects the first foliage leaf in its passage through the soil. Its extreme sensitivity to light and gravity has made coleoptiles a favourite for investigators into phototropism and gravitropism.

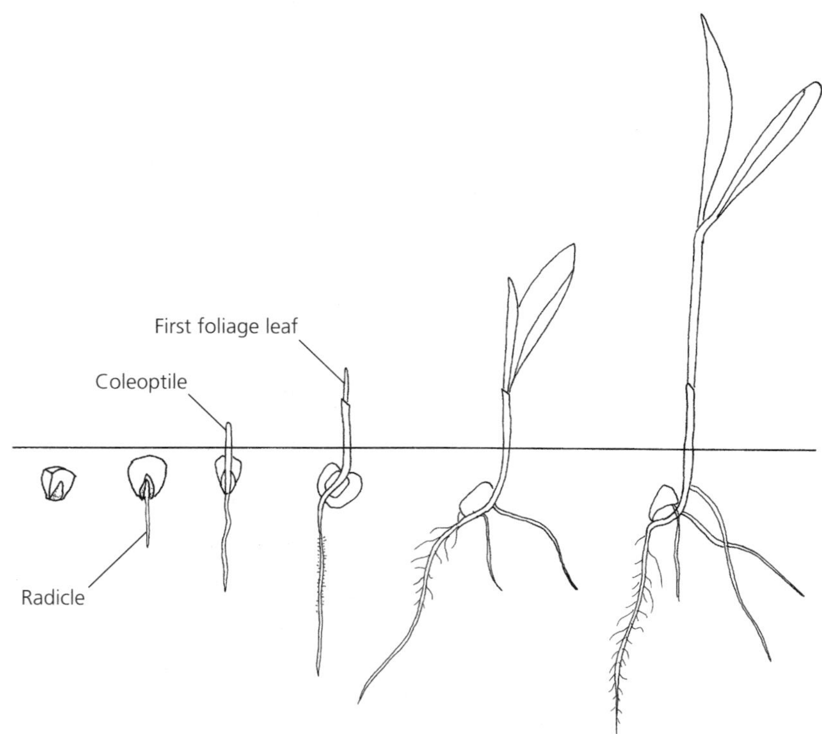


**Figure 31.6►**
Germination in maize

Germination in cereals differs physiologically in an important respect from that of a pea or bean. Unlike the cotyledons of a legume, the cereal endosperm has no vascular supply and, except for the outer aleurone layer, is dead by the time the seed is mature. Starch is digested by enzymes produced in the aleurone layer. The trigger for the production of α amylase and other enzymes is the secretion of gibberellins by the scutellum (Figure 31.7, and Chapter 32). These diffuse across to the aleurone layer, where they stimulate the production of digestive enzymes.



**Figure 31.7▲**
Mobilisation of reserves in a cereal grain

## Measuring growth

Growth can be defined as an irreversible increase in size, involving the production of new organic matter. 'Size' can be measured in various ways, each with its practical advantages and disadvantages.

- Height. This is obviously very easy to measure, but it ignores the root system and only takes into account one of the three dimensions of size.
- Live mass, or mass of living material. This has the disadvantage that live mass includes water, which can vary considerably with time of day. The live mass of a wilting plant at noon may be several per cent lower than at midnight, when at least some of its water deficit has been made up.

- Dry mass. This is by far the best index of growth, since it does not fluctuate with water content. Dry mass is measured by placing material in an oven at 105°C until there is no further change in mass.

## Measuring changes in dry mass

The obvious disadvantage of using dry mass is that you have to kill the tissue. Since measuring a change involves two or more measurements, a change in dry mass cannot actually be measured. Instead it has to be *estimated* by taking repeated samples from a 'population' of seeds.

Suppose, for example, you wanted to measure changes in the dry mass of pea seeds during germination. You would sow a large number of seeds and, at regular intervals, find the mean dry mass of a sample. Differences in dry masses of samples taken at different times will give an indication of changes in dry mass over that time.

The reliability of the results depends on sample size. The larger the samples, the less their mean dry masses will vary at the time of planting. Hence the more confident one can be that differences between successively taken samples are due to changes over time. The cost of larger samples is of course labour and time, so in practice there has to be a compromise.

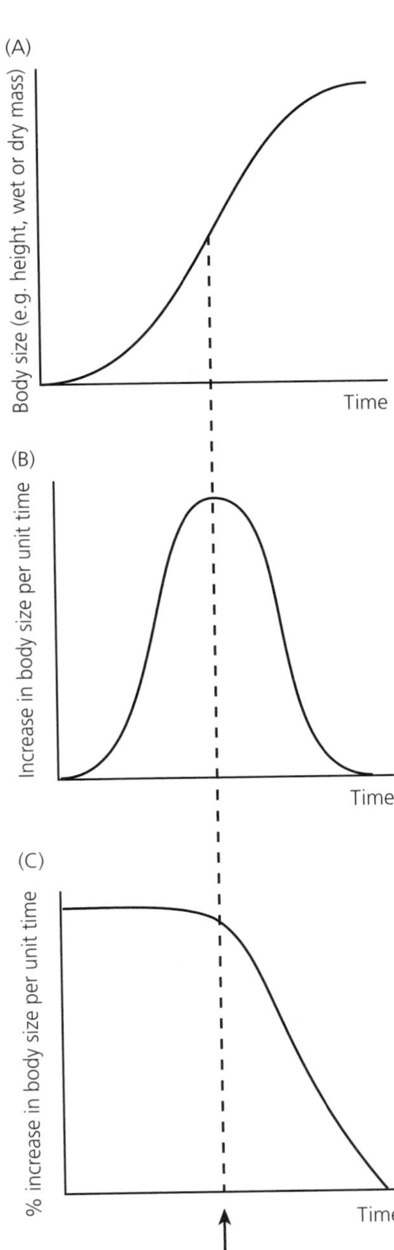


**Figure 31.8**
Three different ways of representing growth. Body size (A); growth rate (B); relative growth rate (C)

## Growth curves

The simplest way of representing growth is a graph of some measure of size (total leaf area, height, live mass, dry mass) against time. Most plants show a **sigmoid**, or 'S'-shaped, curve, in which size increases steeply at first and then levels off (Figure 31.8A).

Whatever kind of measurement is chosen, there are three different ways of representing the data graphically. The increase in size per unit time is the **absolute growth rate** (Figure 31.8B). Notice that in the latter part of the curve, the growth rate is decreasing while size is still increasing.

The **relative growth rate** is the *percentage* increase in size over a given period, divided by the time taken to attain that percentage increase. To illustrate how relative and absolute growth rates differ, consider a meadow buttercup 10 cm high, and an oak tree 10 m high, both increasing in height by 2 cm per week. Both are growing at the same absolute rate, but relative to its size the buttercup is growing much more quickly.

A sigmoid growth curve is very common in both plants and animals. In plants, we expect the curve to get steeper in the early phase of growth for the following reason. A single root may increase in length at a constant rate, but each time it branches the growth of the branches is added to that of the main root, and similarly for shoots. The total absorptive capacity of the root system and photosynthetic capacity of the shoot therefore increase at an increasing rate.

Why doesn't the curve continue to get steeper? There are at least two reasons. First, all species are genetically programmed to attain a certain size before diverting resources to the production of flowers and seeds. The final size represents a compromise between the advantages of rapid growth and early reproduction, and slow growth and delayed reproduction. For each species, the final body size is the optimum.

The second reason relates to the inevitable consequences of increase in size. As a plant grows it produces new cells, only some of which photosynthesise and, therefore, contribute directly to further growth. If the proportion of photosynthetic cells in the plant were to remain the same, growth rate could continue to increase. But increasing size requires an increasing proportion of specialised strengthening tissue, much of which is dead, and therefore non-productive (see the information box on page 369). The proportion of photosynthetic tissue therefore decreases as a plant gets bigger and so, inevitably, does its relative growth rate.

### *Development: increase in complexity*

Whereas growth is a **quantitative** change, **development** is a **qualitative** one, in which there are changes in organisation but not necessarily in size. For example, a germinating seed normally decreases in dry mass (it is respiring but not photosynthesising) but active cell division and differentiation are nevertheless occurring.

## The growth and functional anatomy of stems

A stem is an organ that typically holds leaves, flowers and fruits aloft so that they can carry out their functions. It also contains the transport tissues that service them. Because plants show apical growth, the age of a stem increases with distance from the tip. Growth continues for at least some distance behind the tip, and with this growth the anatomy of the stem also changes. For this reason an understanding of stem anatomy necessitates a brief treatment of how it grows.

Growth in length of a stem or root is called **primary growth**. Using leaves as natural markers, you can see that the zone of elongation is confined to the terminal few centimetres of the stem (Figure 31.9). The source of new cells in primary growth is the **apical meristem**. This is not visible externally since it is normally protected by the youngest leaves.

In most dicotyledons (but very few monocotyledons), primary growth is followed by **secondary growth**, or an increase in thickness. In secondary growth, new cells are produced by a **lateral meristem**. The **vascular cambium** is a lateral meristem; it produces **secondary xylem** and **secondary phloem**. Another lateral meristem is the **cork cambium** which, in trees and shrubs, produces bark.

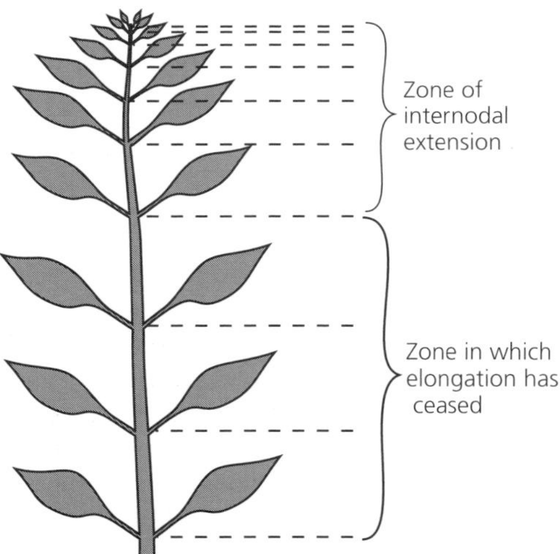


**Figure 31.9**
Elongation of a young shoot using leaves as markers. Dotted lines mark the position of nodes

### *Primary growth of stems*

In primary growth, cells go through three kinds of activity, the peaks of which tend to be located at successively greater distances away from the tip. At the apex the cells are dividing, producing an increase in cell *number*. A little further back, many cells are going through a period of *elongation*. Further still from the tip, cells are going through a phase of *differentiation*. There is, however, considerable overlap between these three activities, some cells beginning to enlarge before others have stopped dividing.

**Figure 31.10**
Longitudinal section of a shoot apex

### *The production of new cells*

The apical meristem of a shoot is a dome-shaped mass of cells, protected by the youngest leaves (Figure 31.10). The cells of an apical meristem differ from mature cells in a number of ways. They are squat in shape and have thin walls with no intercellular air spaces. Since they have no vacuoles, the nucleus accounts for a large proportion of each cell, and chromosomes are visible in many of them.

### *Cell extension*

In stems, the zone of elongation extends for several centimetres behind the tip. Most of the increase in cell size is due to the development of the *vacuole* by the osmotic uptake of water. Although osmosis is a passive process, it can only continue if a gradient of water potential is maintained by active uptake of ions (Figure 31.11).

Whilst the cell is enlarging by vacuolation, the wall is *plastic* – like stretched chewing gum, it does not regain its original shape when the external force is removed. As it stretches, new cellulose is laid down so that the wall thickness is maintained.

The attainment of final cell size marks the completion of the **primary wall**. Any further wall material laid down contributes to an increase in thickness and forms the **secondary wall**. Like rubber, a secondary wall is *elastic*, meaning that when tension is removed it springs back to its original dimensions.

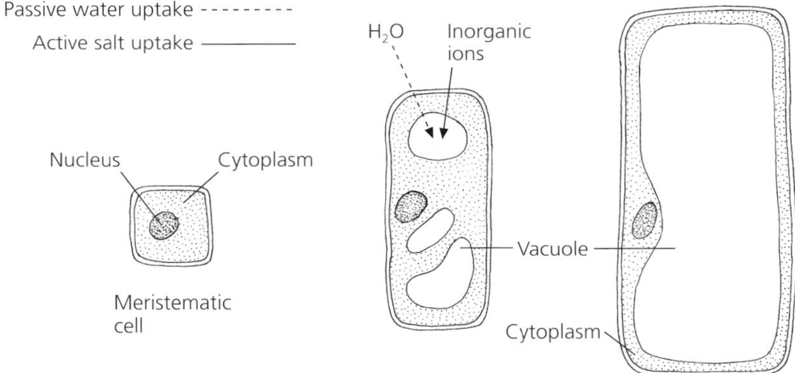


**Figure 31.11**
How a plant cell increases in size by a combination of active and passive mechanisms

## *Differentiation of primary vascular tissues*

Although cell extension continues for several centimetres behind the shoot apex, dividing and enlarging cells must be supplied with raw materials and energy. There is thus an urgent need for some xylem and phloem to differentiate quite close to the shoot apex, whilst surrounding cells are still elongating.

Whereas most cells produced in the apical meristem stop dividing before they enlarge, certain bundles of cells elongate whilst remaining meristematic. These cells make up the **procambium**, and are destined to differentiate into **primary xylem** and **primary phloem**. The procambial cells divide in the longitudinal plane, increasing the thickness of the procambium (Figure 31.12).

At the same time, cells at the inner edge of the procambium begin to differentiate into xylem cells, and those at the outer edge differentiate into phloem cells. These first-formed xylem and phloem cells differentiate only a few millimetres from the apex, and form the **protoxylem** and **protophloem**, respectively (Figure 31.13).

## How cells take their shape

One of the first signs that a cell is beginning to differentiate is the development of a particular *shape*. This is determined by the way the microfibrils are laid down in the cell wall. As the cell enlarges, the wall stretches in the 'easiest' direction. If, for example, the microfibrils are laid down like hoops round a barrel, the cell becomes sausage-shaped.

What determines the orientation of the microfibrils? This depends on the microtubules, which in some way are responsive to the orientation of the cell with respect to neighbouring tissues.

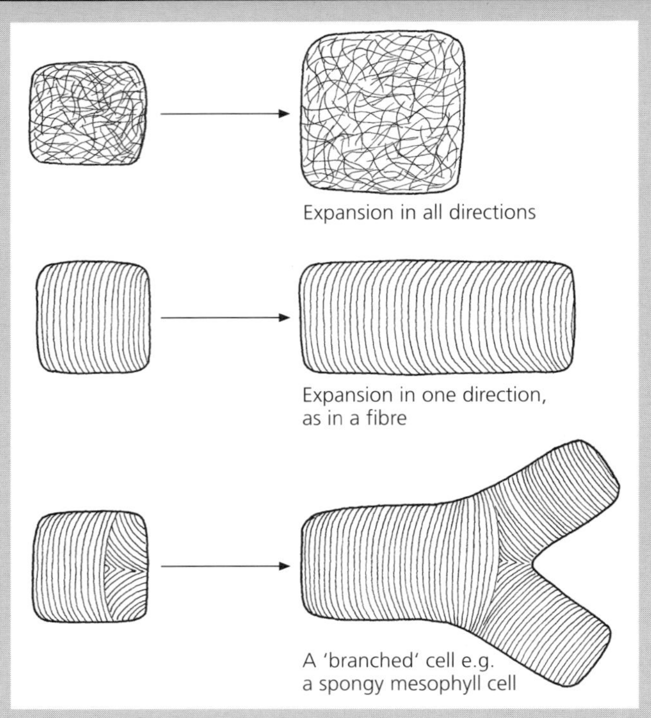


How cells take their shape. As the cell absorbs water, it swells along the direction of least resistance, which is determined by the direction of the cellulose microfibrils

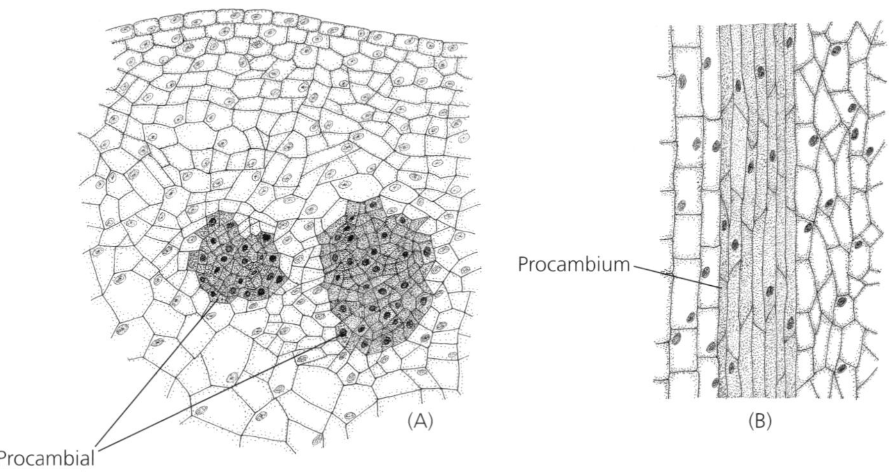


**Figure 31.12**
Procambium as seen in transverse (A) and longitudinal (B) sections (after Eames and MacDaniels)

### *Protoxylem and metaxylem*

Because the surrounding tissues are still elongating, protoxylem and protophloem cells are stretched. Stretching rapidly destroys the protophloem, which can thus only be seen in sections cut close behind the apex. Protoxylem cells, however, can stretch considerably before they finally collapse, because of the way the secondary wall is laid down. It is deposited in the form of spirals or rings of lignified cellulose, so the unlignified cellulose in between is able to stretch (Figure 31.14).

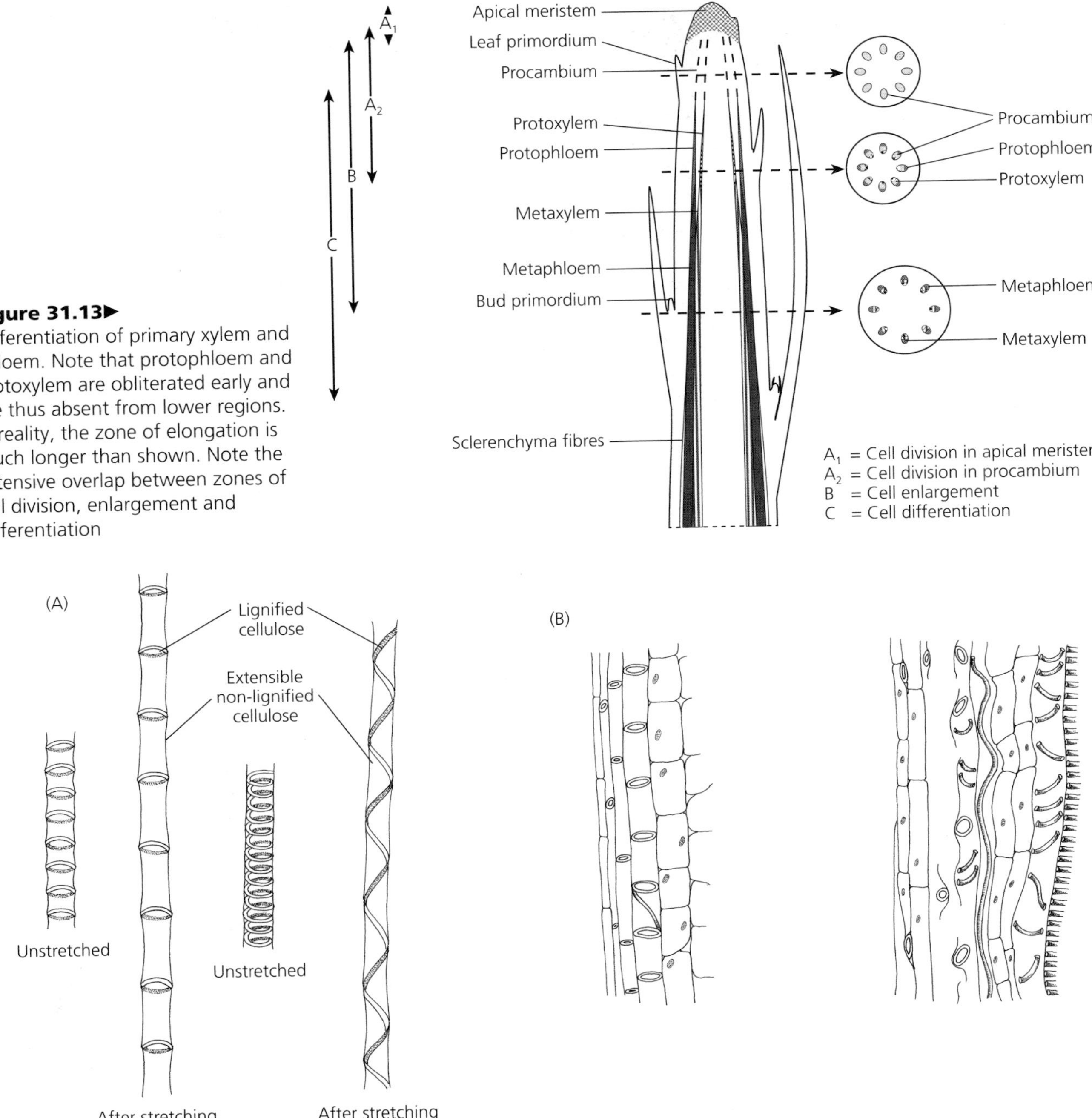


**Figure 31.13►**
Differentiation of primary xylem and phloem. Note that protophloem and protoxylem are obliterated early and are thus absent from lower regions. In reality, the zone of elongation is much longer than shown. Note the extensive overlap between zones of cell division, enlargement and differentiation

**Figure 31.14▲**
Stretching (A) and destruction (B) of protoxylem vessels ((B) after Eames and MacDaniels)

By the time the protoxylem and protophloem have ceased to function they have been replaced by the differentiation of **metaxylem** and **metaphloem**. Metaxylem vessels are more extensively thickened, either with a network of lignified cellulose (**reticulate** thickening), which allows limited elongation, or throughout the entire wall with the exception of the pits (**pitted** thickening) (Figure 31.15).

Eventually all (or almost all) of the procambium differentiates into vascular tissue, forming a series of **vascular bundles** (Figure 31.16).

Meanwhile other tissues have also been differentiating. The outermost layer of cells forms the **epidermis**, which regulates water loss by means of a waxy **cuticle** and adjustable pores called **stomata**. Together with the **cortex** and **pith** the epidermis also plays an important part in hydrostatic support, as described on page 572.

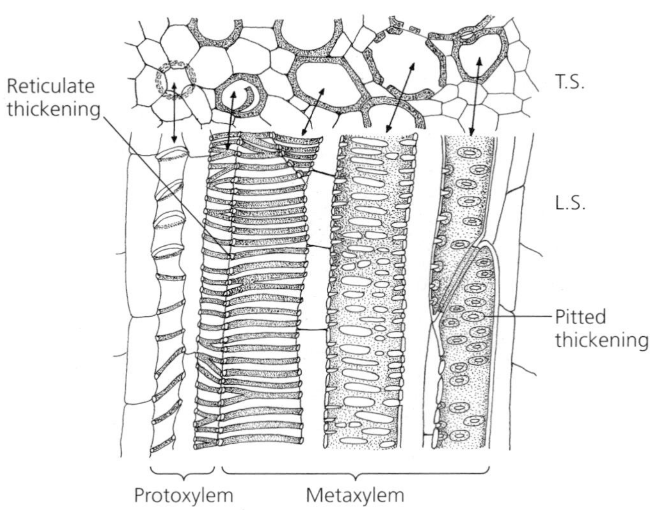


**Figure 31.15**
Protoxylem and metaxylem in TS and LS

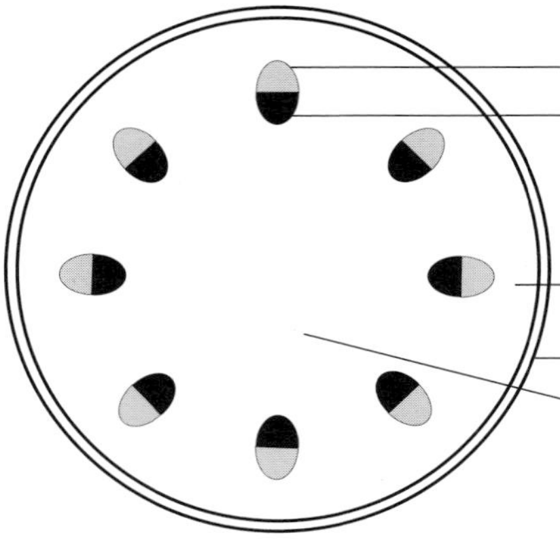

Primary phloem
Primary xylem
Cortex
Epidermis
Pith

**Figure 31.16**
Diagrammatic transverse section of a young dicotyledonous stem showing primary vascular tissue

## The stem as a supporting organ

An organ is a group of tissues organised in such a way that they co-operate to carry out a function that none of the individual tissues can. Like a tissue, an organ is thus more than the sum of its parts. A good illustration of this concept is provided by a herbaceous stem, one of the principal functions of which is to support leaves and flowers against gravity and wind.

Stems are stiffened in two ways: by cell turgor and by specialised strengthening tissues.

### *Support by turgor*

You have probably noticed that young shoots often wilt when short of water, but recover their stiffness an hour or two after the plant is watered. How can water, which is not a stiff material, help in support?

From the point of view of hydrostatic support, a stem can be likened to a sausage balloon. If the balloon is bent, the rubber on the outside of the bend is stretched. Like any elastic material, it obeys Hooke's Law: the greater the increase in length, the more it resists further stretching. By holding the sides of the balloon wide apart, air pressure forces the rubber to undergo more stretching when the balloon is bent than if a limp balloon were bent through the same angle (Figure 31.17). Also, since air pressure stretches the rubber even without bending the balloon, it resists further stretching more strongly than if it were deflated.

Turgor support in plants works in a similar way. Figure 31.18 shows part of a young stem of a dicotyledon. The epidermis and outer cortex are the equivalent of the skin of a balloon. They are adapted to resist tension – the small size of the cells gives them a relatively large surface area over which to adhere together.

The pith, on the other hand, is under compression, consisting of thin-walled parenchyma cells squashed together. Resistance to compression is due to the fluid cell contents. As far as resisting compression is concerned, cell size is immaterial. However, a large cell contains less organic matter and so costs less energy to build than an equal total mass of small cells

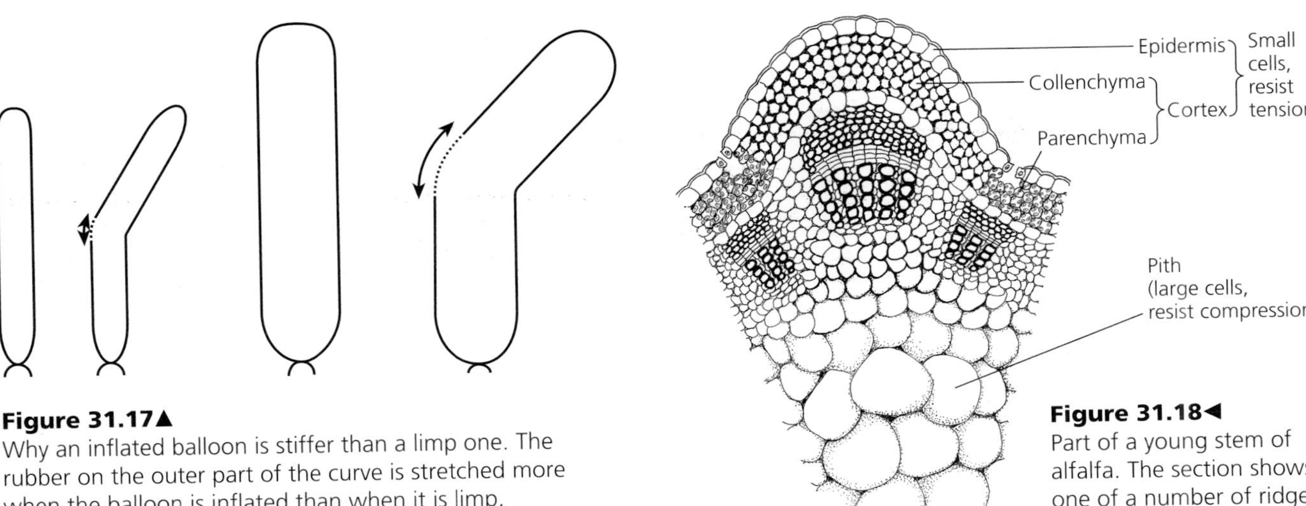


**Figure 31.17▲**
Why an inflated balloon is stiffer than a limp one. The rubber on the outer part of the curve is stretched more when the balloon is inflated than when it is limp, because air pressure forces the inner and outer sides of the balloon further apart

**Figure 31.18◄**
Part of a young stem of alfalfa. The section shows one of a number of ridges occupied by collenchyma

(assuming equal thickness of cell wall and cytoplasm). It is therefore more economical for the pith cells to be large.

Cell turgor is dependent on the osmotic tendency of cells to take up water which, as a physical process, does not depend on the input of energy by the plant. However, turgor cannot be maintained without a steady active uptake of ions to counteract their slow outward leakage, and this in turn requires ATP. This is where the network of intercellular air spaces comes in, by allowing rapid diffusion of oxygen and $CO_2$ between the mitochondria and the atmosphere.

Hydrostatic support has two limitations. First, it is dependent on water supply. Second, although an individual parenchyma cell cannot be compressed, parenchyma *tissue* can be. When parenchyma tissue is compressed, each individual cell remains the same volume but its shape changes by bulging at the corners into the air spaces, whose volume *can* be reduced. This limitation is reduced in collenchyma, in which the walls are thickened in the corners.

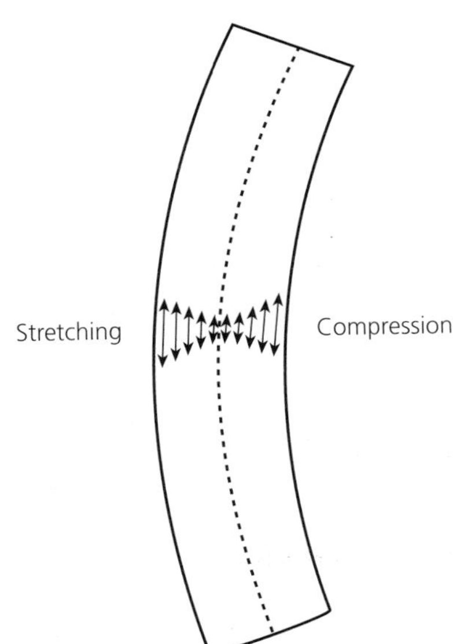


**Figure 31.19**
Changes in length when a stem is bent. The further from the centre, the greater the change in length and the greater the resistance to bending. Tissue down the middle offers no resistance since it does not change in length

### Specialised strengthening tissues

Most stems have specialised supporting tissues whose distribution follows an important engineering principle: for a given cross-sectional area of strengthening material, a hollow cylinder is stiffer than a solid rod. Figure 31.19 shows that when a rod-like structure is bent, the material furthest from the centre undergoes the greatest change in length and, therefore, resists bending most strongly.

It is thus most economical to concentrate collenchyma and sclerenchyma towards the *outside* of a stem (Figure 31.20).

Though the vascular bundles are scattered in monocotyledons, they are still more concentrated towards the outside (Figure 31.21). The same principle operates in limb bones and in scaffolding, both of which are hollow.

## Growth and functional anatomy of roots

Roots are specialised for anchorage and for the absorption of water and mineral ions. Both of these functions are reflected in their anatomy (Figure 31.22).

As in shoots, growth involves the division, expansion and differentiation of cells, but there are differences of detail that are related to function (Figure 31.23).

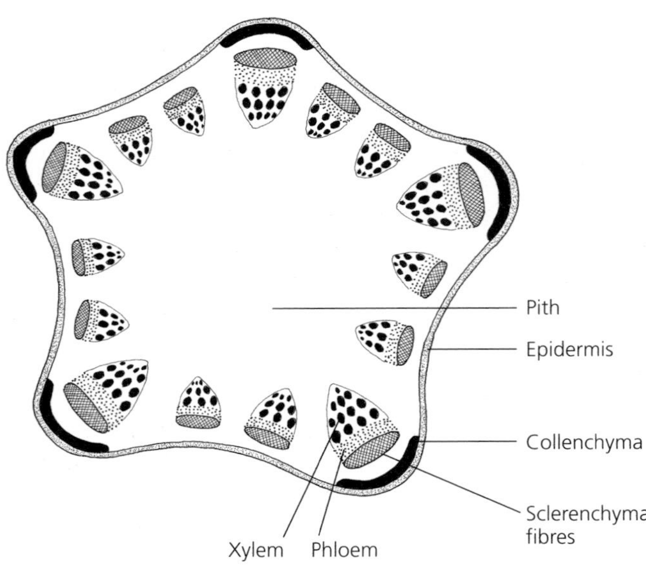


**Figure 31.20**
Transverse section of a young stem of alfalfa

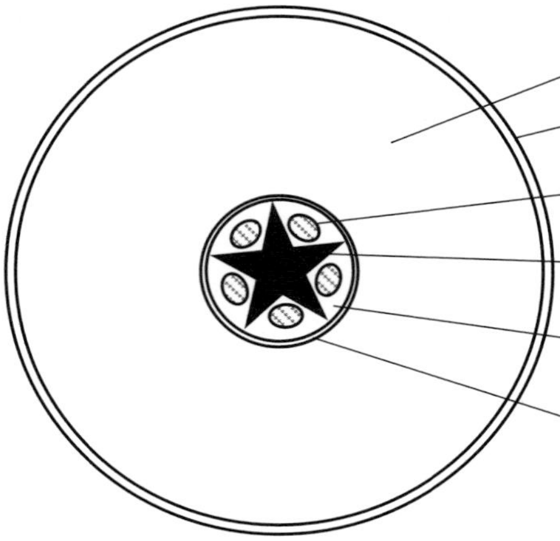

Cortex
Epidermis
Phloem
Xylem
Pericycle
Endodermis

**Figure 31.22**
Transverse section of a young dicotyledon root showing distribution of the strengthening tissue, the xylem

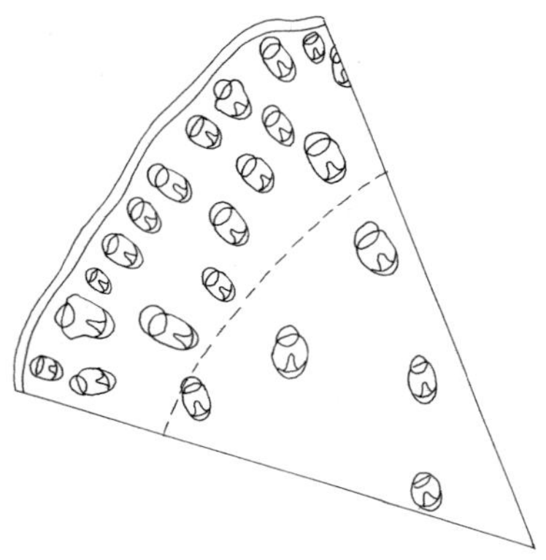

**Figure 31.21**
Transverse section of a segment of a maize stem showing greater concentration of vascular bundles, and thus of supporting xylem, near the outside (dotted line separates equal areas)

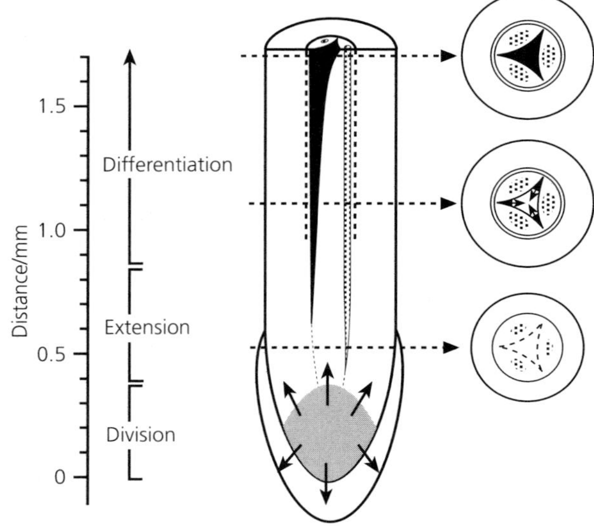


**Figure 31.23**
The main growth zones of a root

The root apex is shown in Figure 31.24. This is simpler in organisation than the apex in the shoot, since it does not give rise to leaves. It does, however, produce a **root cap** – a loose cluster of cells that are continuously breaking free and disintegrating to form **mucilage**. This is a slimy material that lubricates the passage of the root tip through the soil. The root cap also protects the apical meristem from abrasion.

Behind the apex is the zone of elongation, which is only a few millimetres long in roots. A short elongation zone is a necessity in roots to enable them to undergo tropic bending in a resistant medium like soil (Chapter 32).

Behind the zone of elongation, the epidermal cells differentiate by developing **root hairs** which greatly increase the surface area for absorption. Root hairs do not live very long and the epidermis is replaced by the layer beneath, which becomes the **exodermis**.

The vascular tissue develops from a central strand of procambium. When fully differentiated, the primary xylem is star-shaped in transverse

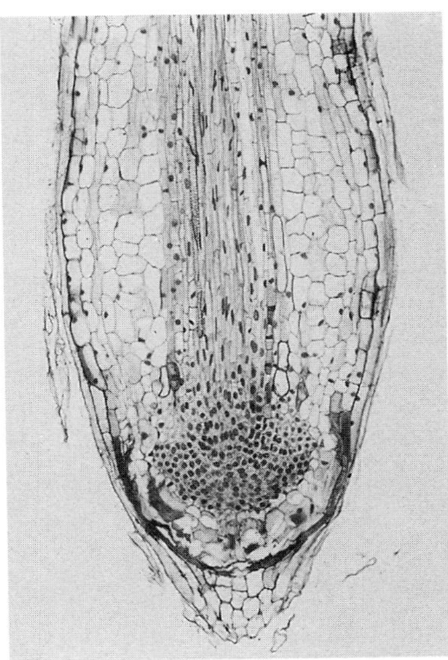

**Figure 31.24**
Longitudinal section of root tip meristem

section, the protoxylem occupying the points of the star and the metaxylem the centre. The phloem lies on alternate radii from the protoxylem, so water can move from root hairs to xylem without passing through the phloem.

Though the metaxylem of a root is essentially similar to that of a stem, spiral and annular thickening are absent. The reason is connected with the fact that whereas the shoot apex has to import water and mineral ions, the root apex absorbs them directly from the soil. There is thus not the same urgent need for early differentiation of xylem as there is in a stem. In fact rapid uptake of water and ions has to await the development of root hairs, which cannot begin until elongation has ceased. Since protoxylem differentiates behind the zone of elongation, the conducting cells do not have to accommodate to stretching.

Phloem differentiation on the other hand cannot wait, since the root tip must import carbohydrate from above. Protophloem is fully functional only a few millimetres behind the tip where elongation is still occurring.

Between the epidermis and the central vascular tissue is the **cortex**. Its parenchymatous cells are more loosely packed than in a stem, and there are no specialised supporting tissues. Both features reflect the absence of compression stresses in roots. The innermost layer of the cortex is the **endodermis**. It plays an essential role in the selective uptake of ions, as described in Chapter 33.

Behind the root hair zone, **lateral roots** develop. Like root hairs, their development has to wait for elongation to cease, and occurs from tissue that has already differentiated. They originate from the **pericycle**, a layer of parenchyma between the xylem and endodermis. The conversion of differentiated cells to meristematic cells is 'differentiation in reverse', or **dedifferentiation**.

Because they originate in deep-lying tissues, lateral roots are said to have an *endogenous development*. This enables the xylem of the lateral roots to be contiguous with that of the main root, which is important for both water transport and support. As a result of their endogenous development, lateral roots have to push their way out through the cortex to reach the soil (Figure 31.25).

## Roots as cables

The root system is an anchor, preventing the plant from being pulled out of the ground by wind or grazing animals. The main strengthening tissue is the xylem, which is concentrated near the centre of the root (Figure 31.22). This enables the root to act as a cable, combining flexibility with resistance to tension. This arrangement is in marked contrast to the distribution of strengthening tissue in stems, whose main mechanical property is stiffness.

## Secondary growth in stems

With continued primary growth, the increasing number of leaves require an increase in both 'plumbing' and support. In most dicotyledons, growth in length is followed by **secondary growth**, or growth in thickness. In this process secondary xylem and secondary phloem are added to the primary tissues already present. It results from the activity of a cylindrical meristem, the **vascular cambium**. In woody plants, secondary growth also involves the production of a new protective layer, the **bark**, which replaces the epidermis (see the information box on page 579).

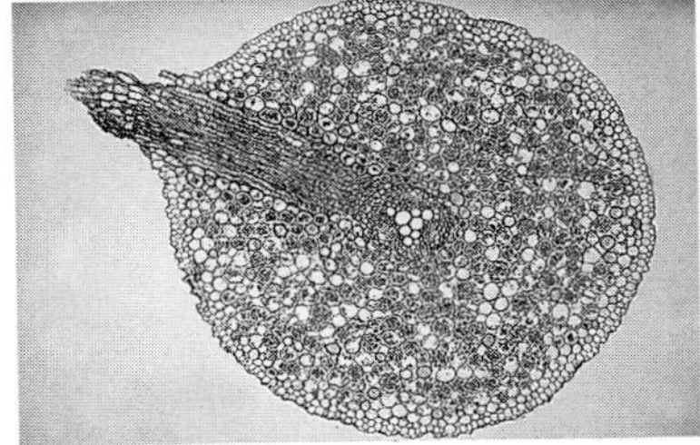

**Figure 31.25**
Origin of a lateral root

The vascular cambium in a stem originates from certain cells in the procambium. Instead of differentiating into xylem or phloem they remain meristematic, dividing repeatedly in the tangential plane (parallel to a

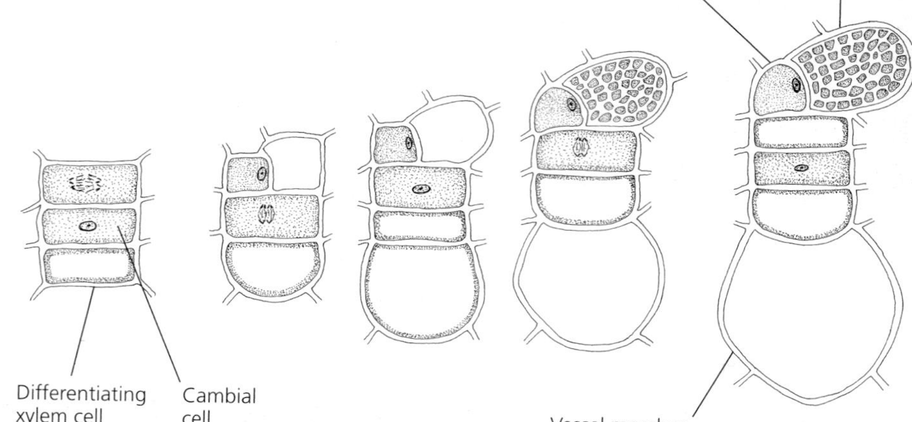


**Figure 31.26**
Production of secondary xylem and phloem by the cambium (after Holman and Robbins)

tangent to the epidermis). Those cells produced towards the inside differentiate into **secondary xylem** cells, while those produced towards the outside become **secondary phloem** (Figure 31.26).

To begin with, the cambium in most herbaceous dicotyledons consists of a broken cylinder of meristematic cells. Soon, however, the parenchyma cells between the vascular bundles revert to a meristematic state and are converted into cambial cells, creating a continuous cylinder of cambium (Figure 31.27).

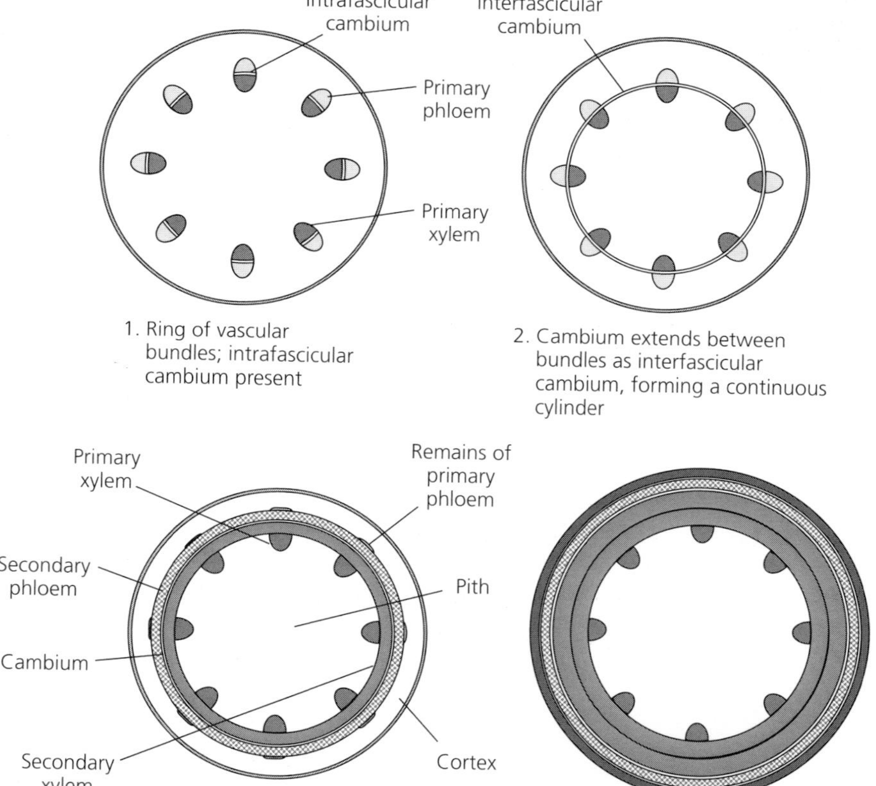


**Figure 31.27**
How the vascular cambium develops in a herbaceous dicotyledon

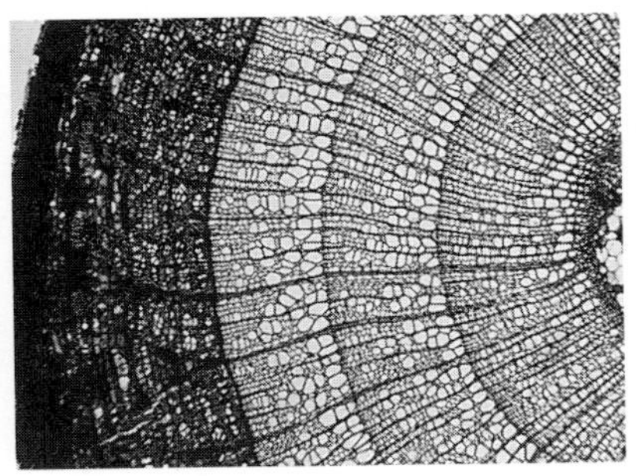

**Figure 31.28**
Transverse section of three-year old lime twig to show annual rings

Although the cambium seems to be several cells thick, in reality only one layer of cells, the **initials**, has the capacity to remain meristematic indefinitely. Cells outside the initials may divide, but are committed to eventually differentiating into phloem, and those inside the initials will form xylem.

In shrubs and trees the cambium produces more xylem and phloem each year. In temperate climates the structure of the secondary xylem changes as the growing season progresses. The vessels of the spring wood are usually wider, while there are often more fibres in autumn wood. As a result, the boundary between autumn wood and spring wood is usually sharp enough for the secondary xylem to consist of distinct **annual rings**, enabling the age of the trunk to be determined (Figure 31.28).

## Tree rings and climate

If you look at the stump of a felled tree you may notice that the growth rings in the wood differ in thickness, 'good' seasons being indicated by wider rings. By taking samples from the trunks of living trees, it is possible to learn about growing conditions during periods before weather records were taken.

Though complicated in practice, the basic idea is simple. Cylinders of wood are bored out of the trunk and later dried and examined in the laboratory. The trees should be of the same species because different species have different requirements. Provided that the samples are taken from trees that are not too near other large trees, the ring sequences from a number of trunks can often be matched.

The use of tree rings to learn about the past is called **dendrochronology** (*dendron* = tree, *chronos* = time). Ring sequences can tell us much more than, say, how warm it was in 1632 in Oxfordshire. In long-dead preserved trees it is sometimes possible to find ring sequences which overlap with those in living trees. In this way, clues to climate can be obtained over periods of time much longer than the life of a single tree. The present age record for tree rings is held by the North American bristlecone pine, for which sequences have been traced back 8000 years – twice the maximum lifespan of an individual tree.

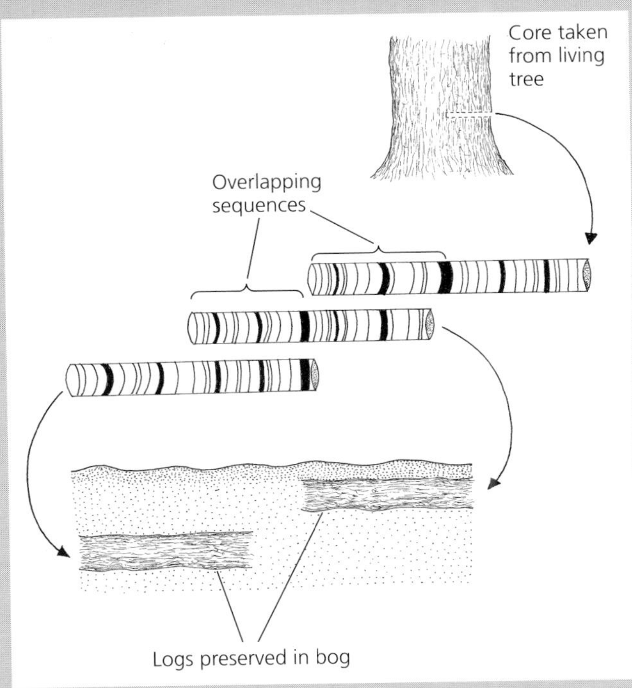


How a long-dead tree can be dated. A tree buried in a bog can be dated if it can be connected with a living tree by overlapping ring sequences

Growth rings do not form in the secondary phloem. Rather than accumulating year after year, secondary phloem is continually being stretched and destroyed by the expanding xylem. As a result most tree trunks contain only one or two year's functional phloem.

In woody plants both secondary xylem and secondary phloem store starch in parenchyma cells. Some of these cells are elongated in the same direction as the vessels and fibres. Others form strands of cells that radiate

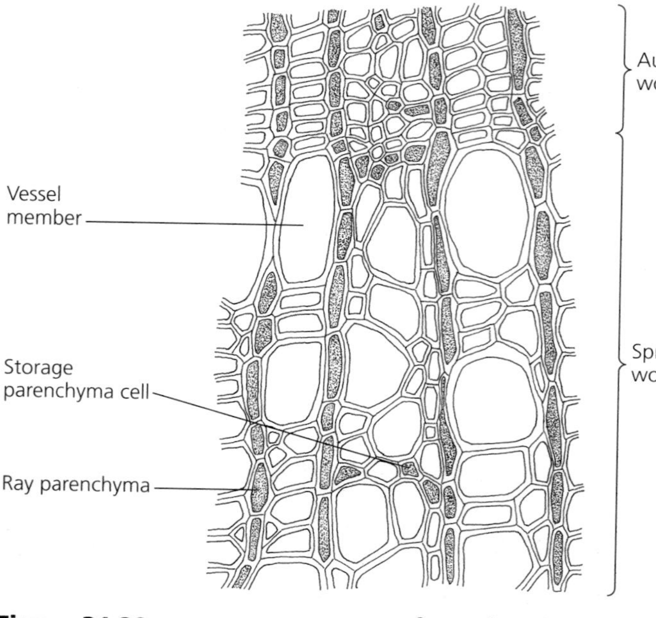


**Figure 31.29**
Transverse section of wood of lime (*Tilia europaea*) showing rays and junction between the annual rings

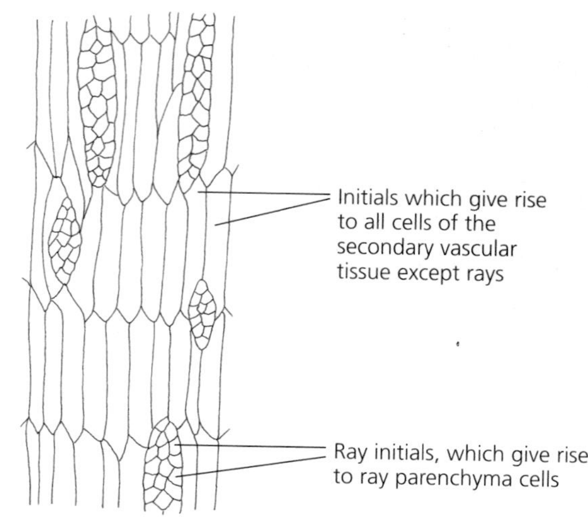


**Figure 31.30**
Face view of the vascular cambium of a woody plant

from the pith like the spokes of a wheel. These radial strands are called **rays** and their function is to transport carbohydrate between the storage parenchyma cells of the secondary vascular tissues and the phloem sieve tubes (Figure 31.29).

Rays are produced by clusters of squat cambial cells (Figure 31.30). The appearance of a ray depends on the way the wood has been cut. In a radial longitudinal section (RLS) rays appear as prominent streaks, and are responsible for the attractive appearance of many timbers cut in this plane. In a longitudinal section cut at right angles to a radius (a tangential longitudinal section, or TLS) they appear like small flecks. Figure 31.31 shows the different appearances of timber cut along three planes, and Figure 31.32 shows the same three faces highly magnified.

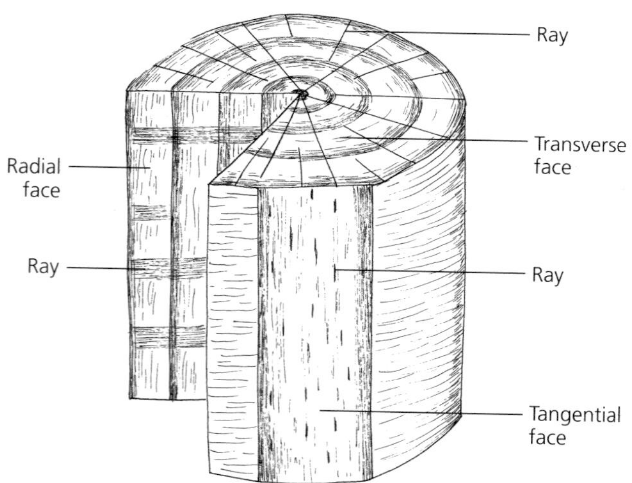


**Figure 31.31**
A piece of timber showing its different faces

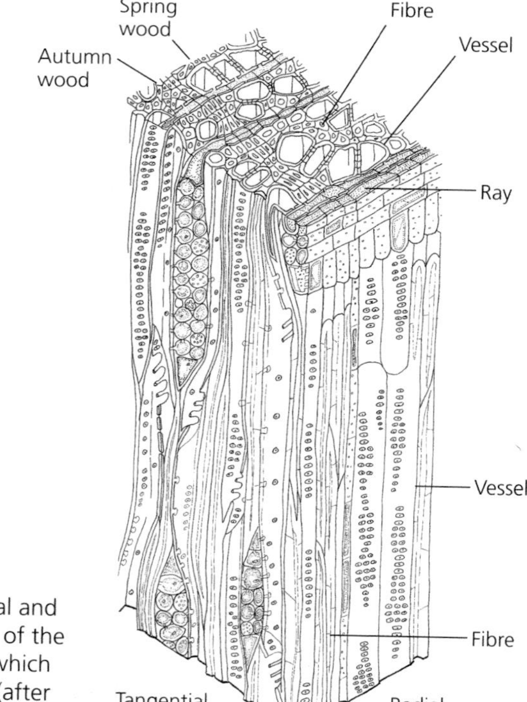


**Figure 31.32**
Transverse, radial longitudinal and tangential longitudinal faces of the timber of a flowering plant which have been highly magnified (after Bailey)

**Figure 31.33**
Logs showing heartwood (dark) and sapwood (light)

To accommodate the continued increase in girth of the secondary xylem, the cambium must increase in circumference. To maintain the tangential distance between rays, the cambium must also give rise to new rays from time to time.

After a number of years, the parenchyma cells of the secondary xylem die. Before this happens, the energy reserves are removed and various toxic chemicals are deposited, making it very resistant to attack by insects and fungi. This is entirely dead and is called **heartwood**. It is non-functional except for the support it provides. It is usually a darker colour than the outer **sapwood**, which contains living parenchyma cells (Figure 31.33). Because it resists decay, heartwood is more valuable as timber than sapwood.

As the tree grows, the conversion of sapwood to heartwood keeps pace with the growth of the tree. In a very big tree the sapwood forms a relatively thin layer, so that by far the greatest part of the tree consists of heartwood.

## Bark

As a differentiated tissue, the epidermis cannot increase very much in circumference and, in woody plants it is eventually ruptured by secondary growth. Before this happens, a new protective tissue, the **cork** or **phellem**, is produced. Cork provides excellent protection against water loss and pathogens for two reasons. Movement of water through the cell walls is prevented by their impregnation with **suberin**, a waxy, water-resistant substance like cutin. Movement between cells is prevented by the absence of intercellular air spaces resulting from their close-fitting, brick-shapes.

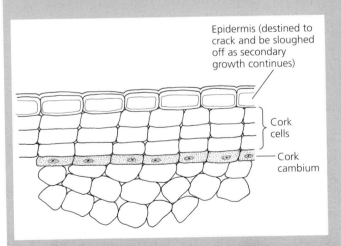

Epidermis (destined to crack and be sloughed off as secondary growth continues)
Cork cells
Cork cambium

An early stage in the production of cork by the cork cambium

**Cork is continually being produced by the activity of a meristem called the cork cambium, or phellogen.** The phellogen arises by division of already-differentiated cells: in stems it arises in the epidermis or the outer cortex, and in roots it originates in more deep-seated tissues. The phellogen cells divide in the tangential plane, the cells produced on the outside giving rise to cork cells, and those produced on the inside differentiating into parenchyma cells.

Mature cork cells are dead, so cork cannot grow to accommodate the continued increase in girth, and eventually cracks. New cork is, however, continually being produced by the phellogen below, which is meristematic and can therefore increase in girth.

Cork is highly impermeable not only to water but also to oxygen and $CO_2$. Respiratory gas exchange is made possible by the presence of **lenticels**. These are small areas in which the suberised cells produced by the cork cambium are rounded, and thus have extensive air spaces between them.

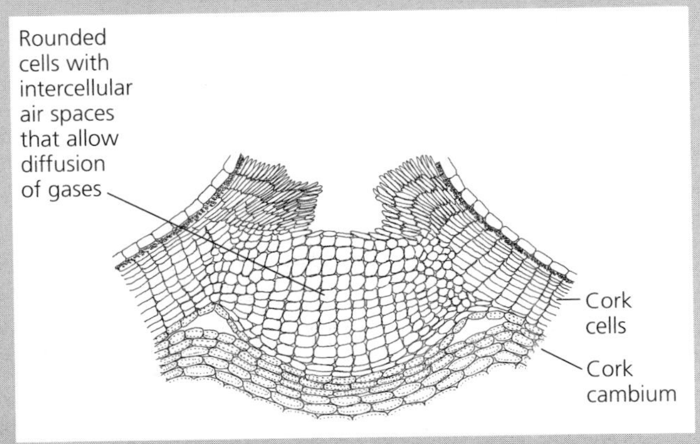


Section through a lenticel

## Making look-alikes: vegetative reproduction

Besides producing flowers, many angiosperms reproduce asexually by **vegetative propagation**, in which shoots become detached from the parent and develop into independent plants. Whereas seed production involves genetic recombination by meiosis and fertilisation, vegetative propagation involves only mitosis, producing genetically-identical progeny. In fact, the offspring of a given parent are really fragments of the same plant which just happen to have become physically separated. All potatoes of a particular variety, for example, are really genetic 'Xerox copies' of a single plant that had originally been produced sexually.

The 'pros and cons' of asexual reproduction and sexual reproduction are complementary – in one kind of situation, the advantages of one are the disadvantages of the other. In an environment in which the offspring experience a different range of conditions from the parents, there are advantages in producing random new genetic combinations. When offspring experience similar conditions to the parents, the 'best' genotypes are probably those of the parents (since they have already proved themselves) and the advantage lies with asexual reproduction.

These considerations apply to all kinds of asexual reproduction, including asexual spore formation in fungi, and binary fission in unicellular protoctists. Vegetative propagation in plants has the additional advantage that mortality is much lower, for two reasons:

- the offspring remain attached to the parent until they are self-supporting, so their mortality is lower. Although they compete with the parent for light, mineral ions and water, this is no different genetically from competition between branches on the same shoot.
- there is no reliance on external agents for dispersal of seeds, and, since there is only one parent, there is no wastage in pollen transfer.

## Perennation

**Perennials** are plants that survive from one year to the next. The process of survival is called **perennation**. Since it involves survival through the winter (or in some tropical areas, through a dry season), perennation always involves the laying down of energy reserves in storage parenchyma, usually as starch. This is not only an adaptation to aid survival over the winter. By enabling rapid spring growth, it helps in the competition with other plants for light.

Perennials are of two general categories – herbaceous and woody.

### *Herbaceous perennials*

In **herbaceous perennials** the aerial parts die back each autumn, with growth being resumed from buds at or below ground level. Some of the main types are described below.

#### *Runners*

After flowering, a strawberry plant produces long side shoots called **runners** which grow rapidly over the surface of the ground (Figure 31.34). The tip of each runner gives rise to a young plant which produces adventitious roots at the nodes and, after the connecting runner has rotted, becomes independent.

#### *Rhizomes*

A rhizome is a horizontally-growing, *underground* stem, which produces adventitious roots at the nodes (Figure 31.35). Some rhizomes (such as

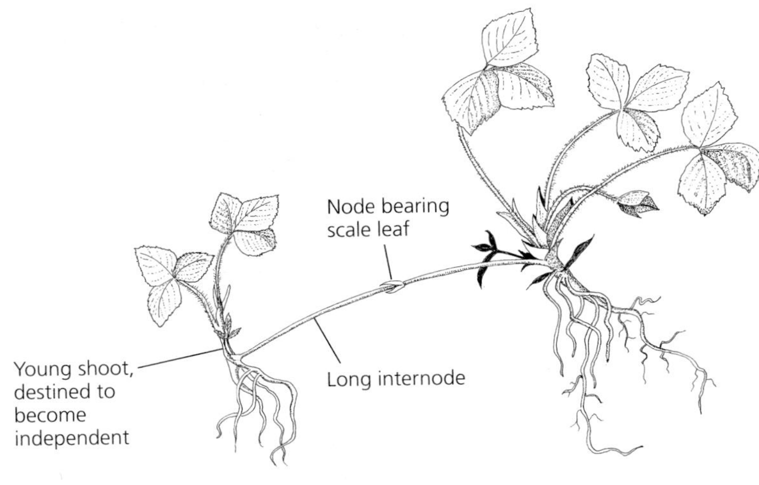


**Figure 31.34▶**
Strawberry plant showing a runner (after Priestley and Scott)

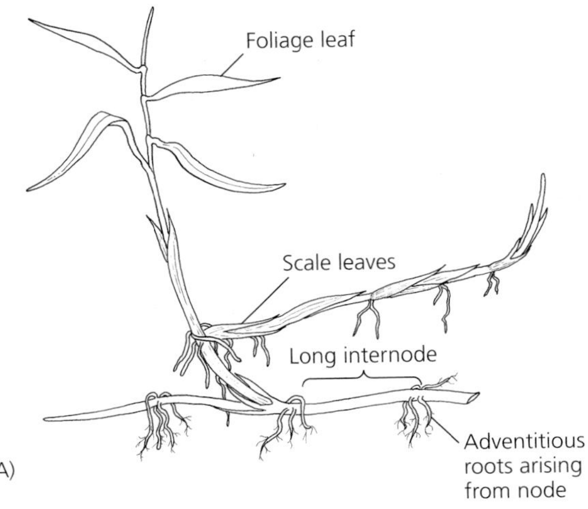


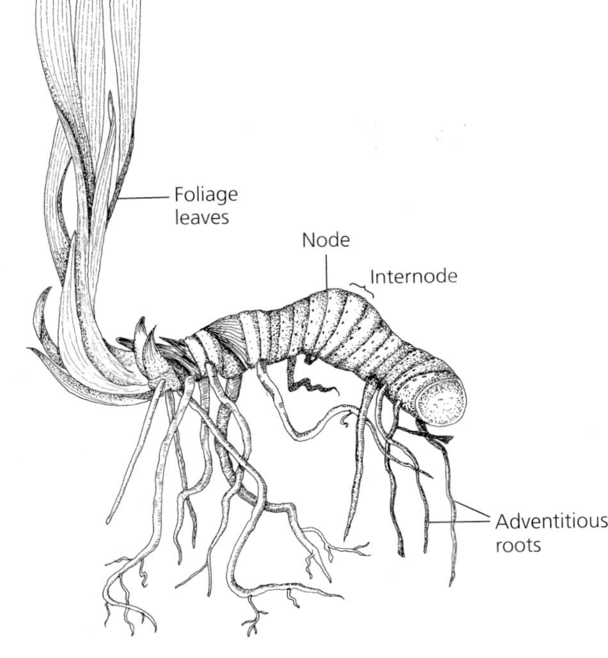


**Figure 31.35▲▶**
Two types of rhizome: couch grass (*Agropyron repens*). Long internodes permit rapid invasion of new areas (A). Iris, with short internodes, resulting in slow invasion of new ground, but dense cover of existing area and strong competitive ability (B)

those of couch grass) are quite thin and grow rapidly so that they quickly spread into new areas. Others, such as those of iris, are squat and grow more slowly, but are strongly competitive in areas already occupied.

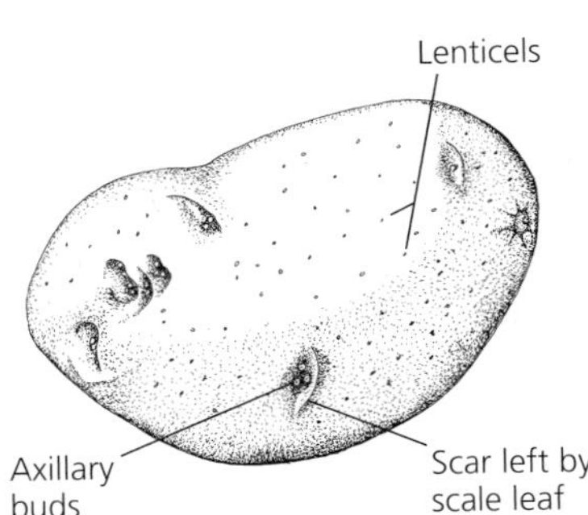


**Figure 31.36**
A potato tuber

### *Stem tubers*

Another kind of underground stem is a **stem tuber**, such as a potato (Figure 31.36). You can tell it is a stem because it has small scale leaves, in the axils of which are buds ('eyes'). In the spring the buds grow into aerial shoots. In late summer these produce side branches which grow downwards, the ends becoming swollen with starch and forming new tubers.

### *Bulbs*

Quite different from rhizomes and tubers are **bulbs**, for example daffodil and onion. These store energy in scales, which are actually the tubular bases of foliage leaves (Figure 31.37). The scales all arise from a very squat stem from which the roots also grow.

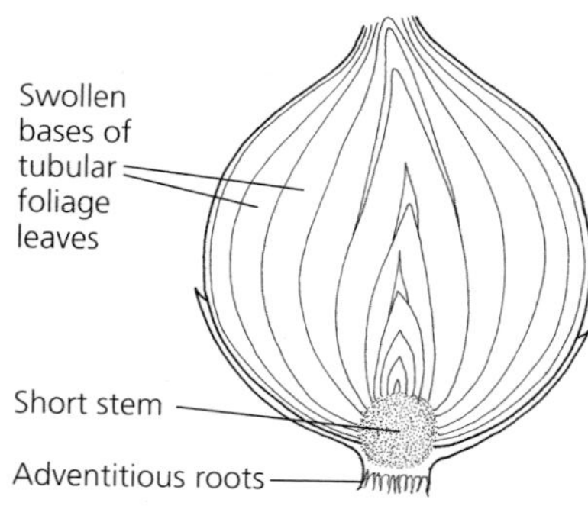


**Figure 31.37▲**
Longitudinal section through an onion bulb

### *Corms*

Superficially a corm resembles a bulb, but it is actually a short, vertically-growing, underground stem, swollen with energy reserves (Figure 31.38).

Though their anatomical details differ, all herbaceous perennials lay down large energy reserves. In the cooler temperatures of winter only a small proportion of these reserves is used in respiration, and the heaviest demands on the plant's 'bank account' actually come in the spring. Starting from buds near ground level, the new shoots must grow as rapidly as possible to compete with neighbouring plants. This surge of growth is largely funded by reserves laid down the previous year.

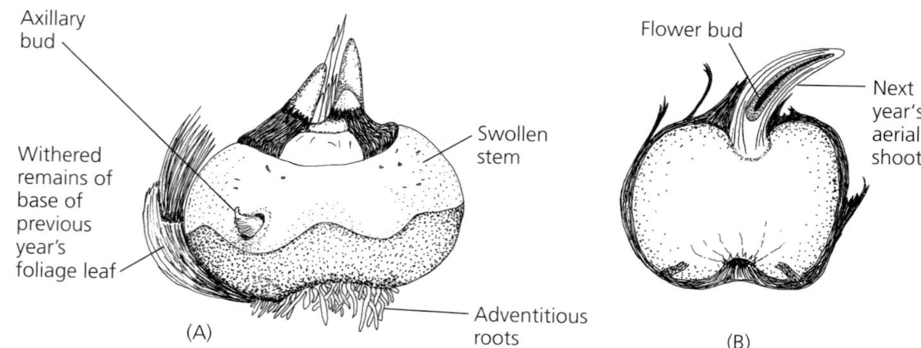


**Figure 31.38►**
Structure of a crocus corm. External features (A); longitudinal section (B) (after Priestley and Scott)

### *Woody perennials*

In **woody perennials**, the aerial parts remain alive during the winter, and energy is stored above and below ground. Each spring growth is resumed from buds above ground level, so both the height and thickness of aerial shoots increase each year.

In woody plants the buds survive in a dormant state over the winter protected by **bud scales**. These modified leaves are produced in the summer and remain small (Figure 31.39). Inside the bud scales next year's foliage leaves are produced, but these remain tiny and are protected by the bud scales during the winter.

In the spring the young foliage leaves expand and the bud scales fall off. The internodes between the foliage leaves elongate greatly – like pulling out a telescope. However the internodes between the lowermost bud scales remain very short, so the scars they leave are close together, collectively leaving a **girdle scar**. Since each girdle scar marks the place previously occupied by a terminal bud, you can often tell the age of a twig by counting the girdle scars. In deciduous trees such as sycamore and ash, you can see the girdle scars even more clearly (Figure 31.40).

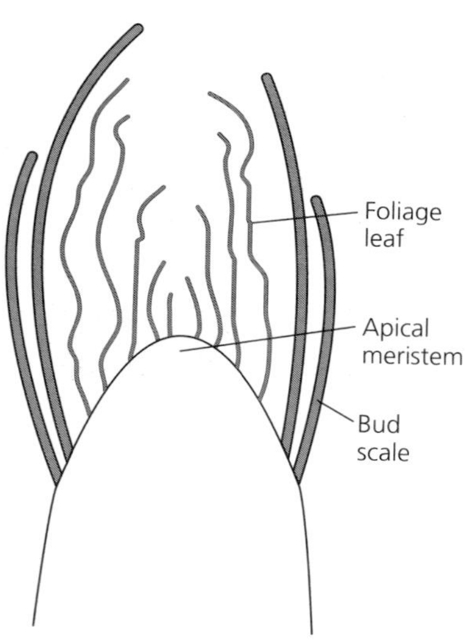


**Figure 31.39▲**
Diagrammatic longitudinal section through a winter bud of a woody plant

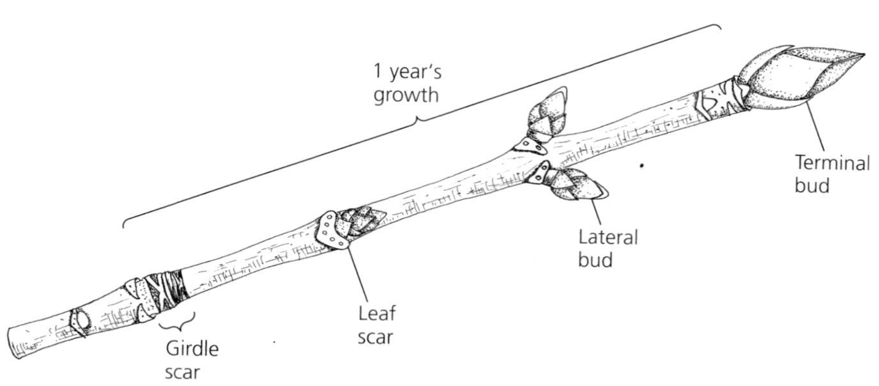


**Figure 31.40►**
A winter twig of sycamore

## QUESTIONS

**1** **a)** What energy reserve do *all* seeds store?
**b)** If a seed is planted too deeply, it dies before it reaches the surface. Why?
**c)** In which part of a germinating pea seed would you expect to find digestive enzymes?
**d)** Most seeds can germinate after months or even many years of dormancy. What does this tell us about the rate of respiration of a dormant seed?

**2** The table shows the results of an investigation into the way peas use energy during the first month of growth. Before sowing in moist sawdust, a sample of 100 peas was picked at random. The cotyledons of each seed were separated from the radicle and plumule, and the mass of each part determined. Their average dry masses were then obtained. After two days' germination, and again at 7 days and 28 days, further random samples of 100 seeds were similarly treated. The results are shown below (masses in grams).

**a)** How would you measure the dry mass of a batch of seedlings?
**b)** Changes in live (wet) mass can be measured directly, but changes in dry mass have to be estimated. Explain why.
**c)** What assumption do you have to make when estimating changes in dry mass, and how can the reliability of such an estimate be increased?
**d)** Which of the following processes accounts for the increase in fresh mass during the first two days?

A transpiration
B respiration
C osmosis
D translocation
E photosynthesis
F germination

**e)** If the first plumule emerged above the soil at 10 days, choose the letter, from the alternatives given in **d)**, which would account for the change in the:
**i)** total dry mass during the first week
**ii)** total dry mass during the next three weeks
**iii)** dry mass of the roots and shoots in the first week.

**f)** What substance would chiefly account for the dry mass of the cotyledons after 28 days?
**g)** Why is it not possible for *all* the organic material in the cotyledons to be used by the root and shoot?

| | | Time after planting | | | |
|---|---|---|---|---|---|
| | | **0 days** | **2 days** | **7 days** | **28 days** |
| **Root + shoot** | **Live mass**<br>**Dry mass** | 0.026<br>0.02 | 0.16<br>0.02 | 0.7<br>0.05 | 3.13<br>0.31 |
| **Cotyledons** | **Live mass**<br>**Dry mass** | 0.24<br>0.21 | 1.2<br>0.2 | 1.1<br>0.04 | 0.8<br>0.01 |
| **Whole plant** | **Live mass**<br>**Dry mass** | 0.266<br>0.23 | 1.36<br>0.22 | 1.8<br>0.09 | 3.93<br>0.42 |

## QUESTIONS

**3** True or false? Germination in darkness is *always* accompanied by a decrease in dry mass.

**4** A leaf begins as a tiny cluster of cells just behind the apex of a stem. As it develops, it undergoes a number of other changes besides increasing in size.

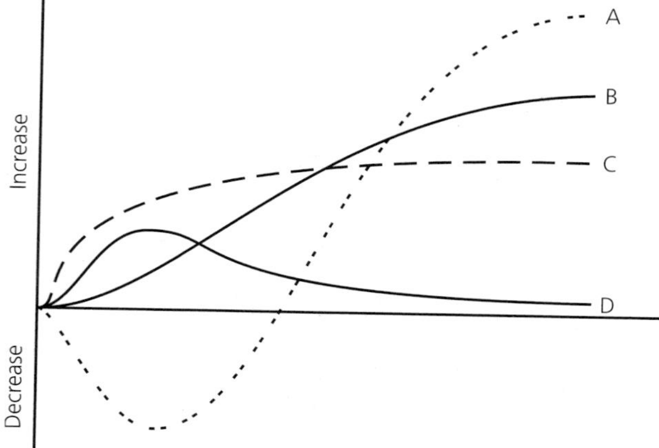


Examine the graph and decide which curve most likely represents each of the following changes in the leaf as it grows:

**i)** area
**ii)** cell number
**iii)** rate of respiration
**iv)** net carbohydrate production per 24 hours.

**5** The hourly increase in height of a tomato plant was measured over 24 hours. For the first 12 hours the plant was kept in strong artificial light, and for the second 12 hours it was kept in darkness. The results are shown in the graph below.

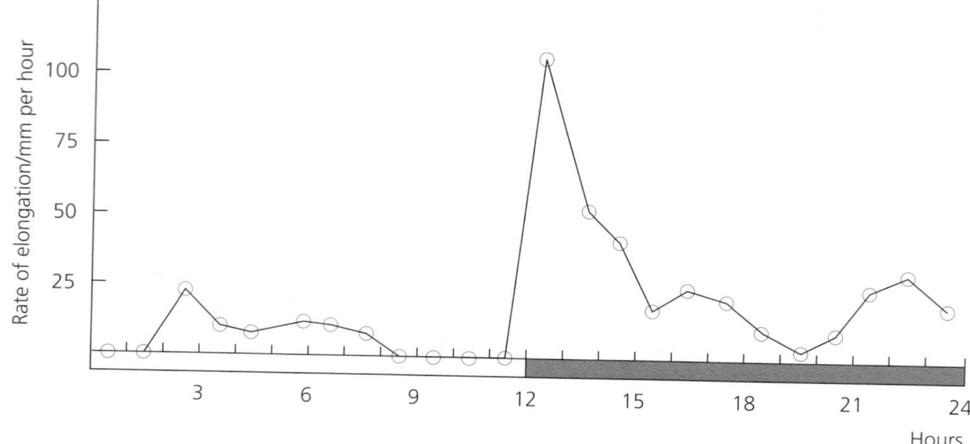


**a)** Using the idea that light influences stomatal aperture, suggest an explanation for the more rapid increase in height in darkness.
**b)** Suggest an explanation for the sudden surge in growth during the first few hours of darkness.
**c)** Why do you think the increase in growth rate was not sustained during the later part of the dark period?

CHAPTER

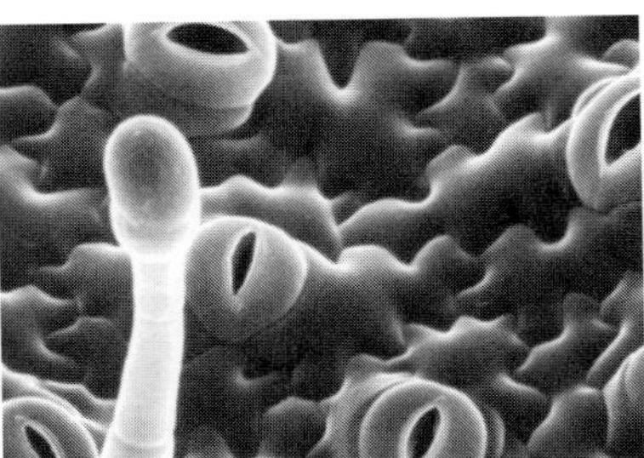

# 32 The CONTROL of PLANT GROWTH

## LEARNING OBJECTIVES

By the time you have completed your study of this chapter you should be able to:

- distinguish between dormancy and quiescence in seeds, and give examples of different kinds of dormancy mechanisms.
- distinguish between nastic and tropic responses.
- explain the basis for naming different kinds of tropism.
- explain the role of auxin in phototropism.
- explain the basis for the bioassay of auxin.
- describe what is known about the transport of auxin and its mode of action in promoting extension.
- discuss the evidence for the identity of the receptive pigment in phototropism.
- describe the evidence for the statolith hypothesis in gravity perception by plants.
- explain how the perception of gravity is believed to lead to a tropic response in a shoot and in a root.
- describe the roles of auxin in apical dominance, fruit development and adventitious root development.
- describe the role of synthetic auxins as herbicides.
- describe the role of gibberellins in stem elongation, seed germination and the breaking of bud and seed dormancy.
- describe the role of cytokinins in cell division and in the initiation of lateral shoots.
- describe the role of abscisic acid in dormancy of seeds and buds and in stomatal closure.
- describe the roles of ethene in fruit ripening, leaf fall and in wound healing.
- distinguish between long-day, short-day and day-neutral plants.
- explain the evidence for the view that plants measure night length rather than day length, and that photoperiodic stimuli are perceived by the leaves.
- describe the evidence for the involvement of phytochrome in flowering.
- explain how vernalisation differs from photoperiodic induction of flowering.
- describe how bud dormancy is initiated and broken, and how internal and external factors influence the timing of leaf fall.

As explained in Chapter 31, plant growth involves the division, enlargement and differentiation of cells. These processes are, of course, sensitive to external factors such as temperature and the supply of raw materials and energy, but they are also affected by environmental factors that operate in more subtle ways, as the following examples show.

- Many seeds will not germinate even when provided with all the conditions normally required for growth.

- Plant organs may change their direction of growth in response to external stimuli.
- When a hedge is trimmed, the removal of shoot apices results in the growth of axillary buds.
- At certain seasons, growth may change from the production of vegetative shoots to the production of flowers.
- In late summer, many plants start to produce dormant buds and store energy reserves, even though conditions remain favourable for growth.
- When a section is removed from a tap root of dandelion or dock, the end that had been nearest the original shoot produces new shoots, and the other end produces new roots. Each end 'knows its place', irrespective of whether the root section is orientated right way up or upside-down.

In all these examples, external stimuli interact with internal growth control systems. Sensitivity and the control of growth are so closely intertwined that it is only logical to deal with them together.

# Hormones in plants?

In animals, different parts of the body are specialised as receptors and effectors, with clear lines of communication between them. Plants do not possess specialised receptor or effector cells, but they do transmit chemical signals from one part of the body to another. These substances are effective in minute concentrations – mostly in the region of few parts per million, but in some instances considerably lower.

These chemicals were called 'plant hormones' in the belief that they were the plant equivalent of animal hormones. However, they differ from animal hormones in a number of important ways. For example, animal hormones are produced by cells specialised for that function but 'plant hormones' are not produced by specialised cells. Also, the effects of 'plant hormones' tend to interact to a greater extent than animal hormones – the effect of one depending on the level of another. For these reasons, many plant biologists have become wary of drawing too close a parallel with animal hormones and prefer to call them **plant growth regulators** (**PGR**), or **plant growth substances**.

Five classes of plant growth regulators have been identified: auxins, gibberellins, abscisic acid, cytokinins and ethene. Each is involved in the control of several aspects of growth, and each growth process may be influenced by more than one PGR.

## Control of seed dormancy

When taken out of a packet and planted in moist, well-aerated soil, the seeds of many domestic plants such as peas or beans normally germinate. Such seeds are **quiescent** – they only require water, oxygen and a suitable temperature for growth to begin.

In contrast, the seeds of many wild plants (and some domestic ones) will not immediately germinate even when provided with conditions favourable to growth. Such seeds are said to be in a state of **dormancy**, in which further conditions must be satisfied before growth can begin.

Dormancy is a means of ensuring that germination only begins when conditions are favourable – not only for early growth, but also for establishment and longer-term survival. Dormancy mechanisms are very diverse, and some of the most important ones are dealt with below.

### *The need for light*

Many small seeds such as foxglove (*Digitalis purpurea*), hairy willowherb (*Epilobium hirsutum*) and some lettuce varieties are **photoblastic** – they will not germinate unless they are first exposed to light for at least a few minutes. Light is only effective on the imbibed (hydrated) seeds; dry seeds are unreceptive.

Seeds with a light requirement are typically small and can only germinate successfully near the soil surface. The effect of light on germination can be seen when large numbers of seedlings appear soon after a flower bed has been dug – digging brings many seeds near to the surface.

Clues to the underlying mechanism were obtained by Flint and McAllister in the 1930s using lettuce. They discovered that if seeds were given a few minutes of red light, 100% germinated, in contrast to the 30% of the control seeds that germinated without any light treatment. Even more interesting, they found that if the seeds were treated with 'far-red' light (light at the very end of the visible spectrum, between 700 and 760 nm), fewer than 5% of the seeds germinated. In some way, far-red light appeared to be *inhibiting* germination.

These findings were followed up by H.A. Borthwick and S.B. Hendricks working for the U.S. Department of Agriculture in the 1950s. They discovered that if red light was immediately followed by far-red, the effect of the red was cancelled out, and that if far-red was followed by red, the effect of the far-red was nullified. In fact, a whole succession of alternating red and far-red could be given, the result depending on the last kind of light given.

The fact that red light is most effective in germination indicated that a blue or blue-green pigment was involved. Though it had not at that time been isolated, Borthwick and Hendricks named it **phytochrome**. The reversibility of red and far-red light suggested that it could exist in two photoreversible states. One (**$P_r$**), absorbs red light and the other (**$P_{fr}$**) far-red. When a dormant seed is given red light, $P_r$ is converted into $P_{fr}$ which is the physiologically active form, and triggers off the events leading to germination.

$$P_r \underset{\textit{far-red}}{\overset{\textit{red}}{\rightleftharpoons}} P_{fr}$$

Although sunlight contains both red and far-red, it is richer in red, so sunlight has the same effect as red light.

The advantage of a light-requirement in small seeds is clear: the seed only has sufficient energy reserves to germinate successfully if it is near the surface. But what could be the significance of the suppression of germination by far-red?

The answer depends on the fact that whilst chlorophyll absorbs red light strongly, it transmits far-red. A seed close to the soil surface under a leafy canopy receives little of the photosynthetically useful blue and red light since these wavelengths have been absorbed by the leaves above. There is plenty of far-red, however, and this inhibits germination. If the canopy above is removed, red light now reaches the seed and germination is triggered.

Phytochrome has since been extracted and purified and, as expected, is a blue ($P_r$) or blue-green ($P_{fr}$) protein. The absorption spectra for $P_r$ and $P_{fr}$ are shown in Figure 32.1.

Phytochrome is believed to be located in the plasma membrane and, since it is not moved around the plant, is not considered to be a plant growth regulator.

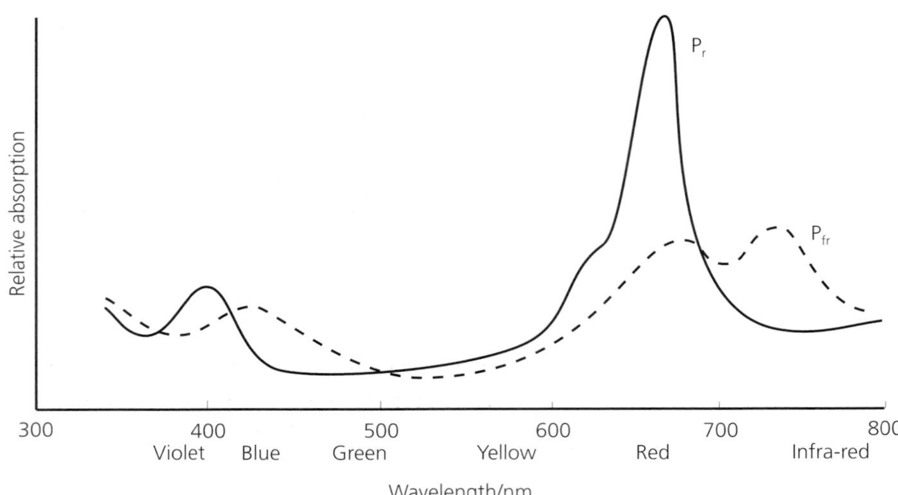


**Figure 32.1**
Absorption spectra for the two forms of phytochrome

### *The need for cold*

Many seeds of trees native to temperate climates, such as sycamore, birch and ash, will not germinate until they have been chilled for several weeks, thus preventing the seeds from germinating in the autumn.

### *Inhibition by chemicals in the seed coat*

In many succulent fruits, germination of the seeds in the fruit is inhibited by chemicals such as abscisic acid (ABA), in the pericarp. As the fruit passes through a bird's gut, the flesh is digested away and, after defaecation, the seeds germinate.

### *Impermeable seed coats*

Some seeds, for example those of lupin, gorse (*Ulex europeus*) and broom (*Sarothamnus scoparius*), have a testa that is impermeable to water. In clover between three and five per cent of the seeds are 'hard', and will only germinate after micro-organisms have broken down the outer impermeable layer of the testa. Dormancy can be artificially broken by scratching the surface of the testa, making it permeable to water.

### *Immature embryos*

Some seeds, such as apple (*Malus*) cannot germinate until they have undergone some further development called **after-ripening**.

## Photomorphogenesis

When a seedling is grown in darkness the shoot is long and thin, the leaves remain very small and pale yellow and the tip (plumule or hypocotyl depending on the kind of germination) remains bent over. A shoot in this condition is said to be **etiolated** (Figure 32.2). A similar effect can be seen in potato tubers that have sprouted in the dark. As soon as the shoot reaches the light, its rate of elongation slows and the leaves expand and develop chlorophyll. These changes are mediated by phytochrome, since they are triggered by red light and reversed by far-red.

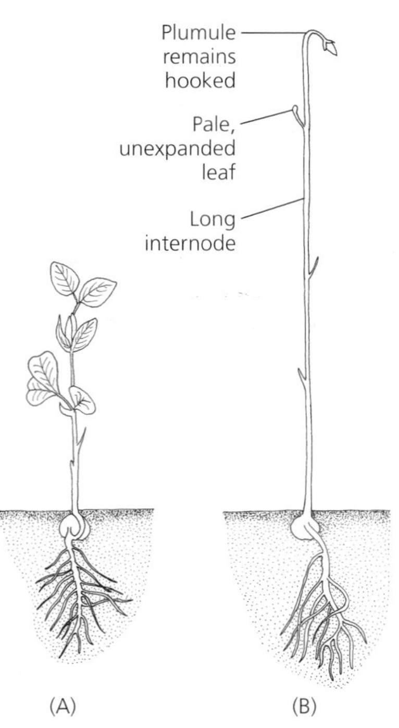


**Figure 32.2**
The effect of light on the germination of a pea seedling: normal (A), grown in darkness (B)

# Plant movements

Plants make two kinds of movement in response to stimuli: nasties and tropisms.

### Nastic movements or nasties

A **nastic movement**, or **nasty**, is a movement in response to a stimulus in which the direction of the movement bears no relation to any direction the stimulus may have. An example of a nastic movement is the opening of many actinomorphic flowers during the day and closing at night. Dandelion inflorescences make similar movements in response to increased light intensity (**photonasty**). Some flowers react similarly to a rise in temperature (**thermonasty**).

Nastic movements may be due to unequal growth, but more often result from turgor changes in specialised cells. An example is the 'sleep' movement of clover leaves. At the base of each leaflet is a swelling called a **pulvinus** (plural, pulvini) containing large parenchyma cells. These undergo cyclical changes in turgor, raising the leaflets vertically at night and lowering them to a horizontal position in the day. 'Sleep' movements occur in numerous other plants (e.g. dwarf bean) and are largely under the control of an endogenous rhythm.

More dramatic examples of nastic movements are the sudden **seismonastic** collapse of the leaflets and leaves of the sensitive plant (*Mimosa pudica*) when touched, and the closing of the leaves of the Venus' fly trap when stimulated by the repeated touch of an insect or other object.

## Tropisms, or tropic movements

In a **tropism**, the direction of the response depends on the direction of the stimulus, for example when shoots bend towards the light. Tropisms are usually the result of unequal growth on two sides of an organ, but some result from turgor changes in pulvini, such as the solar tracking movements of soybean leaves.

Tropisms are named according to the nature of the stimulus and the direction of the response. The most important are as follows:

- **phototropism**, or response to light. Movement towards light is **positive phototropism**, and movement away from it is **negative phototropism**. Orientation at right angles to light is **diaphototropism**.
- **gravitropism** (geotropism), or response to gravity.
- **chemotropism**, or response to chemicals, for example the positive chemotropism shown by the pollen tube as it approaches the ovule. A special case of chemotropism is **hydrotropism**, which is said to be shown by roots growing towards water.
- **thigmotropism**, or response to contact. Tendrils show positive thigmotropism when coiling round a support.

**Figure 32.3**
Demonstration of positive phototropism in cross seedlings. Seedlings with grown in all-round light (right), no light (centre) and light from one side (left)

## Phototropism

Most shoots are **positively phototropic** (Figure 32.3). Roots are generally indifferent to light – though a few, such as the adventitious roots of ivy, are negatively phototropic. Leaves tend to align themselves at right angles to light and are said to be diaphototropic.

Amongst the first to investigate the mechanism of phototropism were Charles Darwin and his son Francis. In 1880 they published a book entitled *The Power of Movement in Plants*, in which they described, amongst others, their investigations into phototropism in canary grass coleoptiles.

By selectively shielding different parts of the coleoptile from the light, the Darwins showed that the tip is by far the most sensitive region (Figure 32.4). Since bending occurs mainly in the lower part, they concluded that some form of 'influence' must move down the coleoptile from the tip.

This experiment showed that phototropism involves three component processes:

- detection of the stimulus;
- transmission of some kind of signal;
- the response itself.

All three processes have proved extremely difficult to investigate. Unlike the situation in animals, which have structures specialised for such functions, none has yet been identified in connection with phototropism.

Early research was directed into the nature of the message. In about 1910 the Dutch plant physiologist Peter Boysen-Jensen began a series of experiments in which he tried various ways of blocking off the route between the tip and the base (Figure 32.5).

He drew two important conclusions. First, the signal must be a water-soluble chemical because it could pass through a block of jelly. Second, it must move down the shaded side. Since this side elongates faster, it must promote extension. Also, since cell division finishes early in the growth of a coleoptile, the chemical must somehow stimulate cell extension.

A year or two later, the Hungarian A. Paàl discovered that even in darkness, if a coleoptile is decapitated and the tip replaced eccentrically, the stump bends towards the side away from the tip. He concluded that light was not necessary for the production of the growth-promoting chemical, but resulted in more passing down the shaded side.

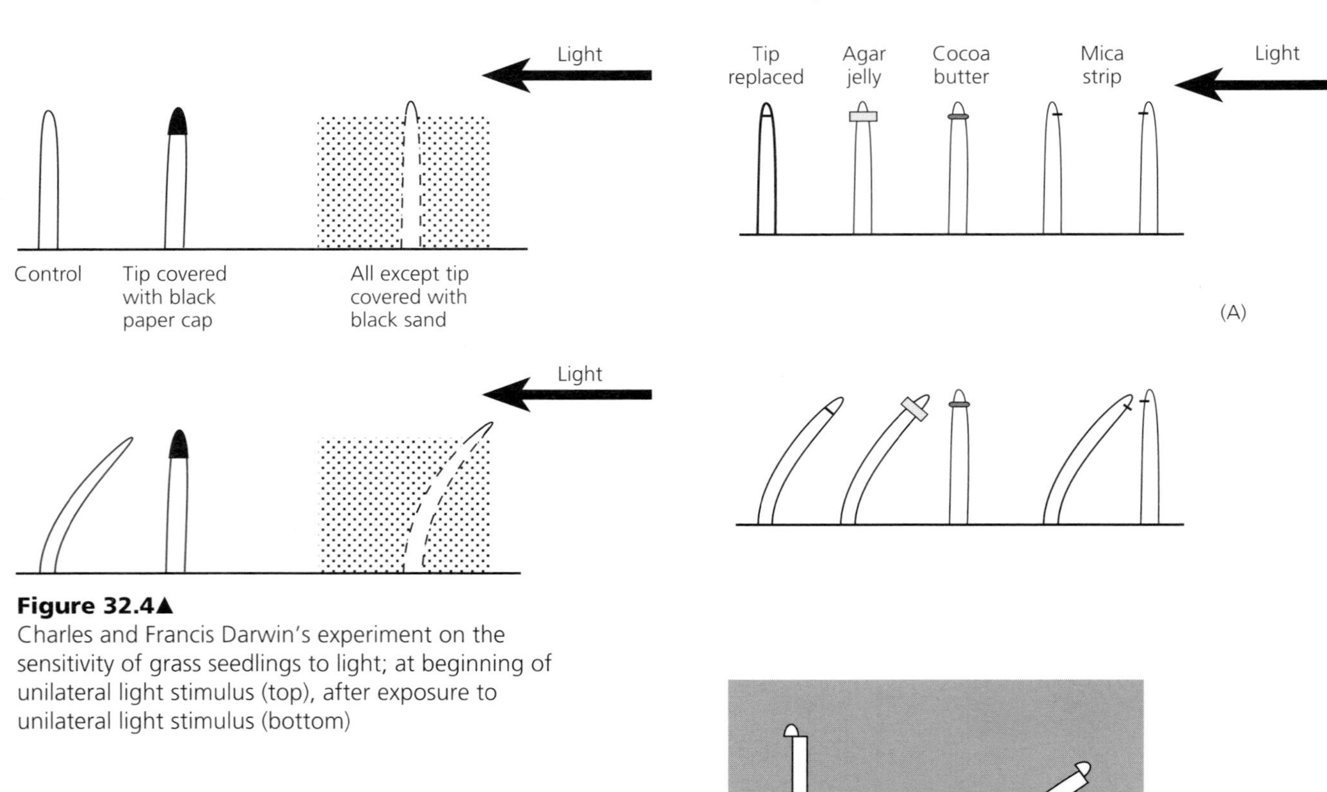


**Figure 32.4▲**
Charles and Francis Darwin's experiment on the sensitivity of grass seedlings to light; at beginning of unilateral light stimulus (top), after exposure to unilateral light stimulus (bottom)

**Figure 32.5►**
Boysen-Jensen's (A) and Paàl's (B) experiments

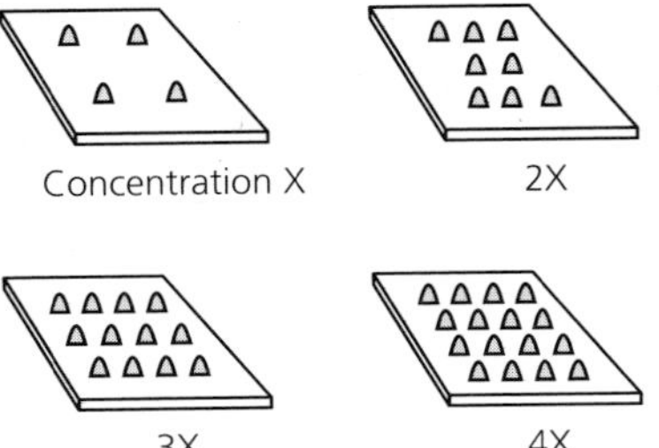


**Figure 32.6**
How Went obtained different concentrations of auxin

## Measuring the messenger

In 1926 the Dutch scientist Frits Went took Paàl's discovery a stage further when he found that a block of agar jelly that had been left in contact with a coleoptile tip had the same effect as the tip itself. This was final proof of the chemical nature of the growth stimulant.

In later experiments, Went found that the curvature induced by the agar was proportional to the number of tips that it had been in contact with, and thus to the concentration of the chemical (Figure 32.6).

This discovery led him to a way of comparing concentrations of the growth-promoting substance, using the angle of curvature induced as a measure of concentration. Went used coleoptiles of *Avena* (oat), and so his method became known as the ***Avena* curvature test**.

The *Avena* curvature test is an example of a **bioassay**, in which the concentration of a substance is expressed in terms of its biological effect, rather than mg per $dm^3$. For such a test to be accurate, stringent precautions had to be taken.

- The coleoptiles were taken from a highly inbred (and therefore genetically-uniform) strain of oat.
- The age of the coleoptile and the level at which it was decapitated were standardised.
- The physical conditions, such as relative humidity (90%) and temperature, were standardised.
- The experiments were conducted in red light (to which plants are phototropically 'blind').
- The period of time over which coleoptile tips were left in contact with the agar was constant, as was the time over which the agar blocks were left in contact with the stumps.
- Large numbers of agar blocks were used, and the average curvature calculated.

A few years after Went devised his *Avena* curvature test, the growth-promoting substance was given the name **auxin** (Greek, *auxein* = to increase), and was isolated and identified as **indole-3-acetic acid (IAA)**. Although a number of other chemicals with similar effects have been isolated from plants, IAA is thought to be the principal auxin.

In recent years, much more sensitive tests have been developed for measuring auxin concentrations.

## Redistribution or destruction?

The *Avena* curvature test enabled Went to test the hypothesis that in phototropism, auxin is redistributed, moving from the illuminated side across to the shaded side, rather than being destroyed on the illuminated side (there are other possibilities). By using a fine partition, he was able to collect the auxin leaving the shaded and illuminated sides of the coleoptiles in separate agar blocks (Figure 32.7).

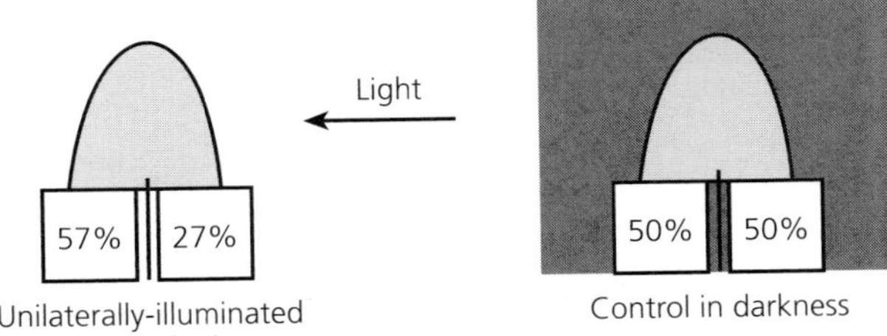


**Figure 32.7**
Went's demonstration of the lateral transport of auxin

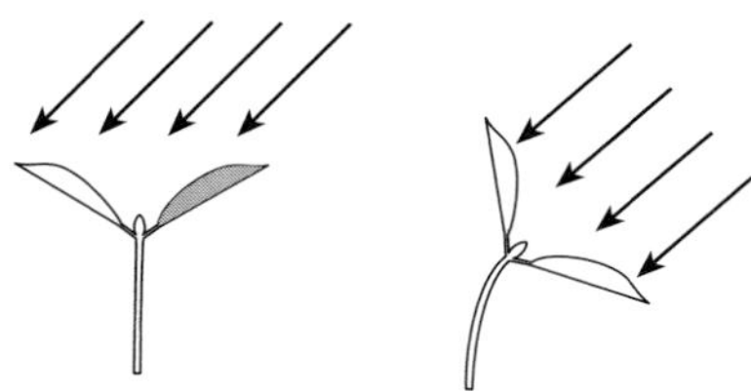

Figure 32.8▲
Phototropism in leafy shoots

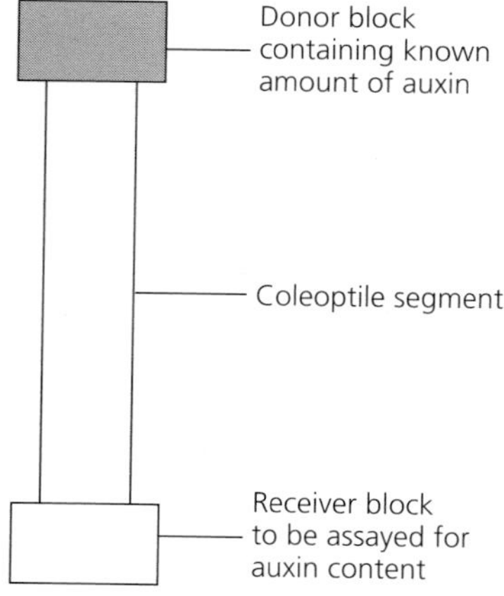


Figure 32.9▲
Method for studying auxin transport

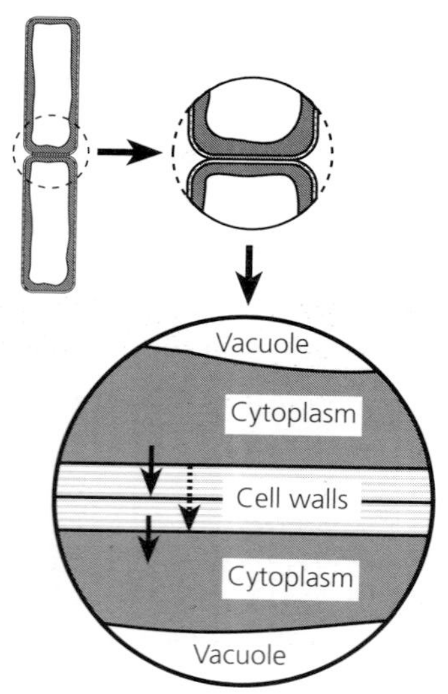


Figure 32.10▲
Cellular basis for polar auxin transport. Adjacent ends of two cells are shown at higher magnifications

Using his test, Went compared the amounts of auxin leaving the two sides of the coleoptile. Controls were set up in darkness; in these, the amounts of auxin extracted were equal on the two sides. The results from the illuminated tips indicated that there was considerable lateral transport of auxin. Though his results suggested that some auxin (16%) is destroyed, later experiments indicated that there is little destruction.

### Phototropism in leafy shoots

Most experiments on phototropism have involved dark-grown coleoptiles. Such material is atypical in two respects. First, etiolated shoots are very much more light-sensitive than non-etiolated plants. Second, coleoptiles have no leaves, which must be taken into account in any general explanation of phototropism.

The 'coleoptile' explanation given above must therefore be modified to account for phototropism in green, leafy shoots. Most of the auxin that stimulates stem elongation comes from the young, expanding leaves near the apex. When a shoot is equally illuminated on both sides, the young leaves export equal amounts of auxin to the two sides of the stem below, which thus remains straight. If one side is more brightly lit than the other, the leaves on this side export more auxin, causing the stem on that side to elongate faster (Figure 32.8).

### Auxin transport

By placing an agar block containing IAA at one end of a length of coleoptile and an 'empty' block at the other, the rate of movement of IAA from 'donor' to 'recipient' block can be studied (Figure 32.9).

Using this technique, certain facts about auxin transport have been established:

- it requires energy, since it is blocked by cyanide or a lack of oxygen.
- it occurs mainly in the parenchyma cells round the vascular bundles (the bundle sheath).
- it is much slower than phloem transport (about 1 cm $hr^{-1}$).
- it is *polar*, since a coleoptile segment can transport auxin from a donor block to a recipient block from the end nearest its original tip to the end nearest its original base, but not in the other direction. In stems, auxin transport is away from the tip, or **basipetal** ('towards the base'), and in roots it is **acropetal** ('towards the tip').

The basis for the polarity of transport seems to be that each cell absorbs auxin at the end nearest the shoot tip and exports it from the other. After secretion from the plasma membrane of one cell, it diffuses across the intervening walls to the next cell, which absorbs it (Figure 32.10).

### The detection of the stimulus

For light to have an effect on a biological process, it must first be absorbed by a pigment. A clue to the identity of the pigment in phototropism can be obtained by comparing the phototropic sensitivity of plants to different wavelengths of light. The result, when plotted on a graph, is an **action spectrum** (Figure 32.11).

The action spectrum for phototropism shows that blue-green light is the most effective, suggesting that a red or orange pigment is involved (a red pigment appears red because it absorbs red light *least*).

Two orange or red pigments are known in shoots: **β carotene** and **riboflavin** (Vitamin $B_2$). Figure 32.12 shows the **absorption spectrum** for each of these pigments. An absorption spectrum is a graph of the

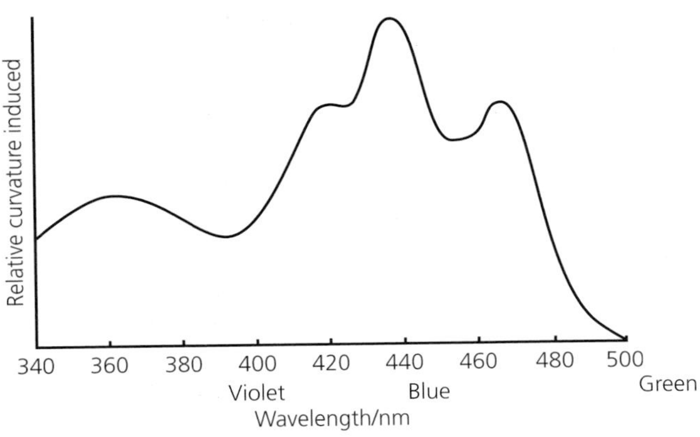


**Figure 32.11**
Action spectrum for phototropism

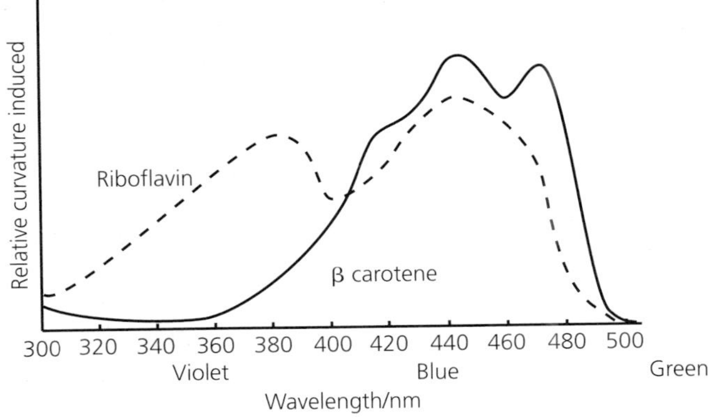


**Figure 32.12**
Absorption spectra for β carotene and riboflavin



proportion of the incident light absorbed at different wavelengths.

Neither curve matches the action spectrum precisely, but the fact that mutant plants lacking carotene exhibit normal phototropic behaviour suggests that riboflavin is more likely to be the pigment involved. Just how the absorption of light leads to auxin redistribution is unknown.

### *How does auxin promote extension?*

Experiments have clearly shown that auxin promotes cell elongation by making cell walls of the epidermal cells more plastic and thus easier to stretch. As explained in Chapter 31, the epidermis of a young stem is normally under tension and the inner tissues are under compression. Elongation of the epidermis would thus allow extension of the inner tissues.

How does auxin make cell walls more plastic? Cellulose microfibrils are not stretchable, but wall extension could result from the breaking of bonds connecting adjacent microfibrils which could then slide past each other.

One hypothesis proposes that auxin stimulates the pumping out of $H^+$ ions from cells. The resultant lowering of the pH in the cell walls increases the activity of enzymes that cut links between neighbouring microfibrils.

**Doubts over the conventional explanation**

**The phototropism explanation given above is the conventional one, and has occupied plant physiology textbooks for many years. In recent decades increasing doubts have been expressed. For one thing, most experiments have been done on coleoptiles, which are highly specialised structures, and the results applied uncritically to stems. Another problem with the conventional 'story' has been the discovery that IAA is synthesised quite rapidly by bacteria from the amino acid tryptophan, its normal precursor in plants. Since none of the classic experiments was done under aseptic conditions, there are doubts as to how much of the IAA was produced by the coleoptiles and how much was produced by bacterial action on tryptophan released by the injured tissue.**

## Gravitropism

Just as heliotropism was renamed phototropism because the stimulus is light rather than the Sun, geotropism is now called gravitropism because the stimulus is gravity rather than the earth.

Gravitropism is particularly marked in germination – plumules and hypocotyls are negatively gravitropic, and radicles are positively gravitropic. The adaptive significance of this is obvious: seeds are randomly orientated, and the only way the plumule can reach light it cannot 'see' is to grow upward. Radicles need to grow downwards to reach a more permanent water supply and to secure better anchorage for the shoot system. With successive branchings, most roots lose much of their gravisensitivity; secondary roots tend to grow roughly horizontally and later branches tend to be insensitive to gravity. Shoots also tend to become less sensitive to gravity with successive branchings. Rhizomes are **diagravitropic**, growing at right angles to gravity.

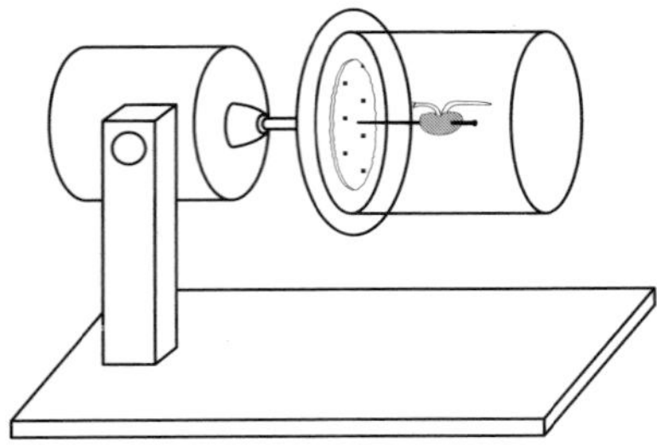

**Figure 32.13**
A klinostat

## *Controlling the gravity stimulus*

Unlike light, gravity cannot be switched on and off. A solution is to place the plant (usually a seedling with radicle just emerged) on a slowly rotating wheel called a **klinostat** (Figure 32.13). If the klinostat rotates every 15 mins or less, the direction of gravity changes faster than the radicle can respond, and it grows straight along the axis of rotation. By stopping the klinostat for varying periods of time, the effect of the duration of a gravity stimulus can be investigated. In this way it has been found that the klinostat must be stopped for a certain minimum **presentation time** (usually several minutes) for any response to occur, an observation that gives clues to the mechanism of detection (see below).

Incidentally, a klinostat is *not*, as is commonly stated, an adequate control for demonstrating gravitropism because it changes the orientation of the plant with respect to every possible unidirectional stimulus; gravity is not selectively eliminated.

## *How is gravity detected?*

In animals, gravity detection involves small grains of mineral material called **statoliths**. Because they are denser than cellular structures, they tend to sink downwards, stimulating receptors. There is good evidence that in plants, the function of statoliths is performed by starch-storing plastids called **amyloplasts**. Since they are denser than the surrounding cytoplasm, they tend to sink onto cytoplasmic membranes, which is thought to trigger the gravitropic response in some way (Figure 32.14).

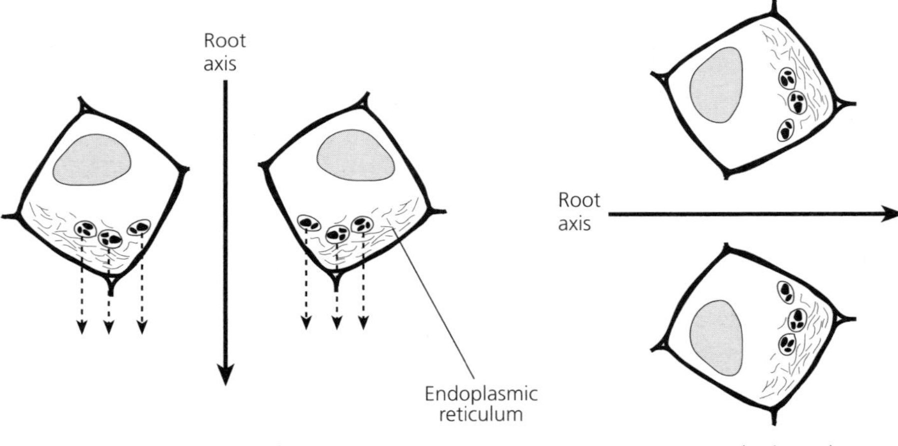


**Figure 32.14**
Effect of orientation on amyloplasts in root cap cells

Evidence supporting the statolith hypothesis includes the following observations:

- amyloplasts occur in all gravity-sensitive organs in plants, being most abundant in those parts that are most sensitive, such as root caps and

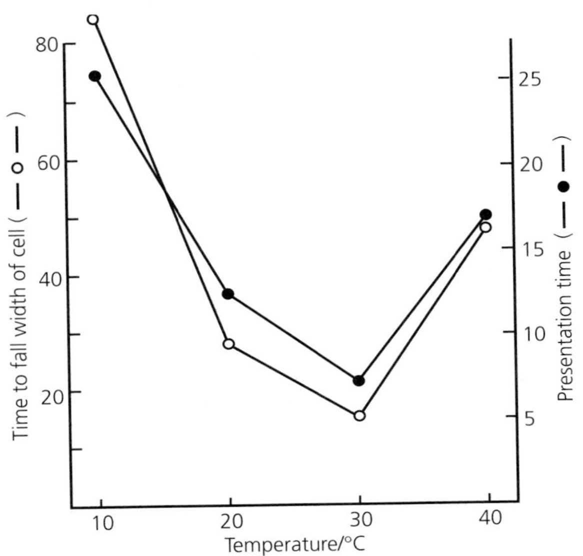


**Figure 32.15▲**
Effect of temperature on presentation time (●) and on rate of amyloplast sinking (○) (after Hawker)

Horizontal partitions – no bending

Vertical partitions – bending results

Coleoptiles

Radicles

**Figure 32.16▲**
Evidence that gravitropism involves the vertical movement of a growth regulator

the nodes ('joints') of aerial stems of grasses. In stems of other plants, they occur in the cells just outside the vascular tissue. Even plants that never produce storage starch (such as onion) produce amyloplasts in gravity sensitive parts.

- when radicles or coleoptiles are treated with gibberellin (a treatment that stimulates the digestion of starch in the amyloplasts) sensitivity to gravity is lost. After removal of gibberellin, starch is resynthesised in the amyloplasts and gravisensitivity is regained.
- presentation times for different plant organs are very similar to the times taken for amyloplasts to settle.
- the effect of temperature on presentation time and the rate of starch sinking are very similar (Figure 32.15).

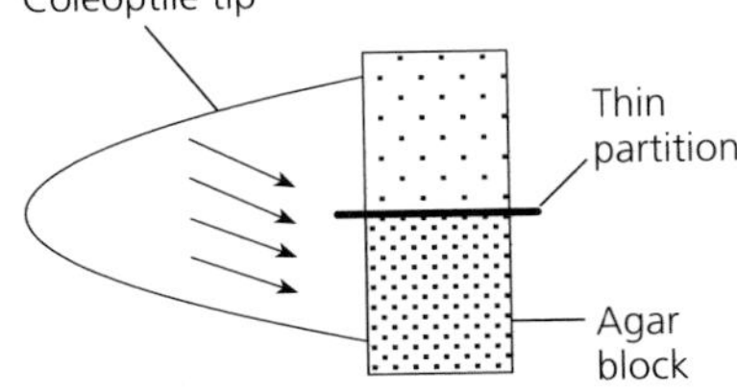


**Figure 32.17▲**
Evidence that auxin is vertically transported in a horizontally placed coleoptile tip. When a coleoptile tip is placed horizontally, more auxin leaves the lower half than the upper

## *Transmission of the signal*

When amyloplasts sink against the lower side of a cell, some kind of signal must be generated, which then passes to the zone of cell elongation. The experiments shown in Figure 32.16 suggest that in both radicles and coleoptiles, gravitropism involves the vertical movement of a growth regulator.

In shoots and in coleoptiles, auxin seems to be involved. When agar blocks are placed in contact with upper and lower halves of a horizontal coleoptile tip, more auxin is collected from the lower block (Figure 32.17).

Experiments suggest that in roots, bending seems to be due to a growth-inhibitor produced by the root cap (Figure 32.18).

Half root cap removed

Insertion of transverse partition with (left), and without (right) root cap

**Figure 32.18►**
Evidence that gravitropism in roots involves a growth inhibitor

- Removal of half the root cap of a vertical radicle causes it to bend towards the intact side.
- Insertion of a transverse piece of metal foil into one side of a vertical radicle just behind the tip causes it to bend away from the foil.

These experiments suggest that a growth-inhibitor passes back from the root tip along the lower side. Went proposed that, slightly modified, a similar mechanism could account for gravitropism in both shoots and roots. The concentrations of auxin which inhibit root growth are similar to those that stimulate growth in shoots, so downward movement of auxin away from the root cap would retard growth on the lower side and cause downward bending (Figure 32.19). A similar hypothesis was proposed independently by N. Cholodny. The **Went–Cholodny hypothesis**, as it became known, was accepted for many years.

One difficulty with auxin as a messenger is that it is transported *towards* the root tip. It had been suggested that abscisic acid (see below) may be the messenger, but more recent opinion is now in favour of $Ca^{2+}$ ions. Experiments have shown that $Ca^{2+}$ ions are redistributed to the lower side of a root cap within 30 minutes of gravistimulation, and treatment with substances that bind $Ca^{2+}$ ions makes radicles insensitive to gravity.

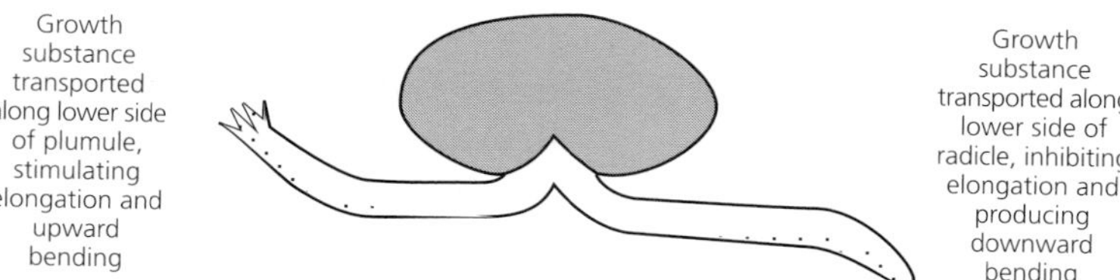
Growth substance transported along lower side of plumule, stimulating elongation and upward bending
Growth substance transported along lower side of radicle, inhibiting elongation and producing downward bending

**Figure 32.19►** The Went–Cholodny hypothesis for gravitropism in a germinating bean seed

## *Other effects of auxin*

Since its discovery, auxin has been shown to have a wide range of biological effects.

- Development of adventitious roots (Figure 32.20). Many stems can produce roots when cut from the parent plant, and gardeners exploit this when taking cuttings. The natural stimulus to produce adventitious roots is auxin. Because its transport is polar, it accumulates near the base of the stem where it induces root formation. Many shoots that do not produce sufficient auxin themselves can be induced to produce roots by treatment with 'rooting powder', which contains auxin-like chemicals such as IBA (indole-3-butyric acid) and NAA (naphthalene-acetic acid).
- Fruit development. Auxin produced by the developing embryo stimulates the growth of the fruit. In seedless fruits (such as bananas) fruit development is stimulated by artificial treatment of the flowers with auxin.
- If the 'seeds' (actually fruits) are removed from a strawberry when it is still very small, the 'fruit' does not develop any further (Figure 32.21). In some way it seems that the developing 'seeds' are necessary for the 'fruit' to develop normally. If, after removing the 'seeds', the young 'fruit' is treated with lanolin paste containing auxin, it develops normally. Controls treated with plain lanolin remain small.
- Auxin stimulates the division of cambial cells and the differentiation of xylem and phloem.
- Apical dominance. Auxin produced by the young leaves near the stem apex inhibits the growth of lateral buds further down.

**Figure 32.20**
Adventitious roots sprouting from a cutting in response to auxin. The cutting on the right was untreated and bears no roots

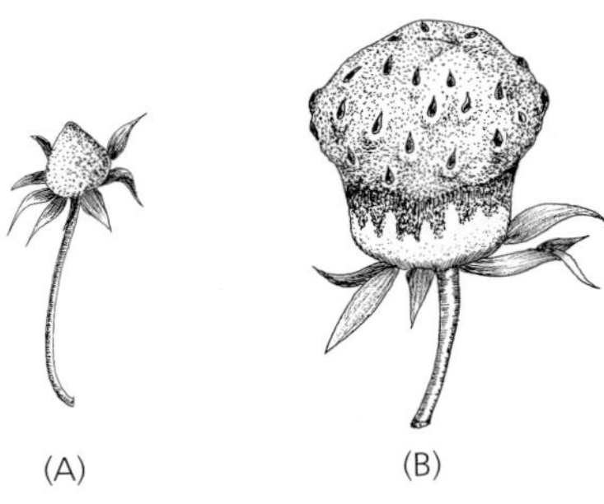


**Figure 32.21▲**
The role of auxin in fruit development in strawberries. (A) Small, very young strawberry; (B) the 'seeds' have been removed from the lower half of the receptacle which does not enlarge fully

- Hormone weedkillers. A number of auxin-like substances have been synthesised in the laboratory. Some disrupt growth so severely that they kill the plant and are used as herbicides, such as **MCPA, 2,4-D** and **2,4,5-T**. Monocotyledons are less sensitive to these substances, so some can be used to selectively kill off dicotyledonous weeds in a lawn.

## Gibberellins

Bakanae is a disease of rice in which the plants grow so tall that they eventually fall over (in Japanese, *bakanae* means 'foolish seedling'). The fungus that was present in the diseased plants is called *Gibberella fujikuroi*. In 1926, Japanese scientists discovered that the fungus was producing a powerful growth-promoting substance. It was called **gibberellin**, or **gibberellic acid** (GA), but since then over 80 substances with gibberellin activity have been discovered, many of which are produced in flowering plants. Most plants produce several, which may have differing activities. To simplify matters, the term 'gibberellin' will be used to cover all of them.

Gibberellins are named according to their order of discovery: gibberellic acid is $GA_3$ and is the best known and most commercially available. All gibberellins have a common molecular framework, shown in Figure 32.22.

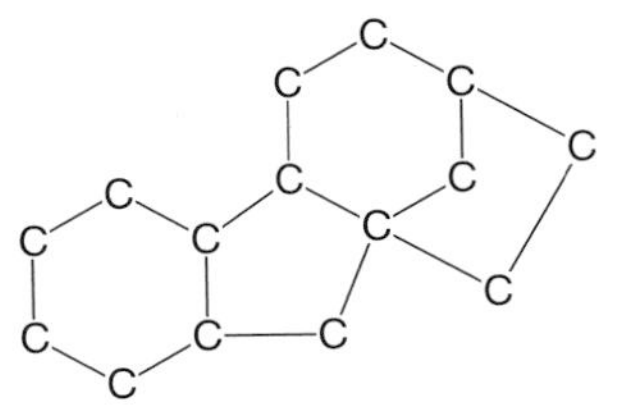

**Figure 32.22▲**
Structure of the gibberellin carbon skeleton

Gibberellins are produced in shoot apices and in developing embryos, and are transported in the xylem and phloem.

Gibberellins have a wide range of effects on plant growth, as shown below.

- They promote internodal extension in dwarf varieties of plants (such as Mendel's dwarf peas). These mutant plants lack one of the genes involved in the synthesis of gibberellin, but grow normally when it is supplied artificially. At least part of the extension results from increased extensibility of the cell walls, but the mechanism is different from that of auxin.
- They cause premature flowering in biennials such as cabbage (Figure 32.23). Normally the stem remains very short in the first year, but 'bolts' in the second year, elongating greatly and producing flowers. Bolting normally occurs after a few weeks of winter cold, but after treatment with gibberellin, the plant flowers in the first year. It seems that gibberellin can substitute for cold and, in some plants, can substitute for long days (see below).
- They help to break the dormancy of buds and seeds. Seeds that require light for germination will germinate in darkness if treated with gibberellin. Similarly, dormant winter buds that normally require several weeks of cold before they grow, can be induced to grow after treatment with gibberellin.
- In many seeds, especially cereals, they promote the synthesis of enzymes that hydrolyse stored reserves.
- Many shrubs and trees will not flower until they have reached a certain stage of development, before which the plant is said to be **juvenile**. Some plants can be induced to enter the adult, reproductive stage by treatment with gibberellins. Plant breeders take advantage of this to obtain progeny more quickly by shortening the life cycle.
- They inhibit the development of adventitious roots.
- They stimulate cell division in the shoot apex.

**Figure 32.23▲**
Bolting of cabbage induced by gibberellin

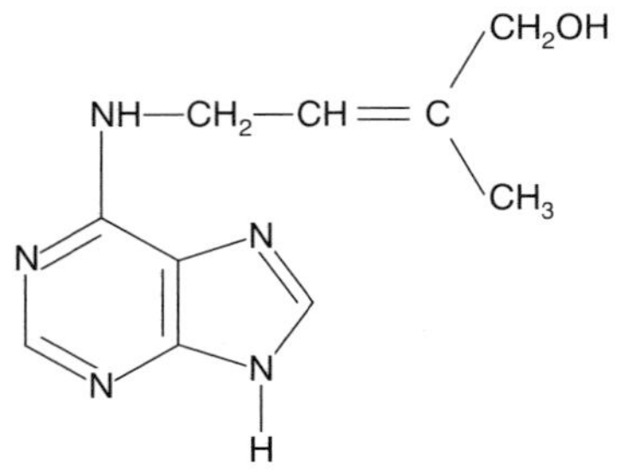

**Figure 32.24**
Structure of zeatin

## Cytokinins

The first naturally-occurring cytokinin was extracted from maize (*Zea mays*) in 1964, and was called **zeatin**. Its structure is shown in Figure 32.24. Others are known, but zeatin is the most active in the majority of plants. Like all cytokinins, it is a derivative of the purine adenine, one of the bases of DNA and RNA.

Cytokinins are synthesised in root tips and are transported to the shoots in the xylem. In the presence of auxin they have the following effects:

- they stimulate cell division.
- they stimulate the production of lateral buds. A high auxin/cytokinin ratio favours the production of lateral roots, and a low ratio favours the production of lateral shoots.
- they delay leaf senescence and leaf fall. When a small area of leaf is treated with cytokinin it remains green longer than the surrounding areas.

## Abscisic acid (ABA)

Abscisic acid was named because it was found to promote the abscission of cotton fruits (abscission is the loss of a part due to the formation of a specialised zone of weakness at its base). It has since been realised, however, that in most plants it plays little part in leaf fall. Its general function seems to be to prepare a plant to withstand stress (due either to cold or to lack of water) by the following mechanisms:

- it is a powerful growth inhibitor, and plays an important part in maintaining dormancy in buds and seeds.
- it brings about stomatal closure (Chapter 34).

ABA is synthesised in plastids from carotenoids and, like gibberellins, is translocated in the vascular tissue, especially the phloem.

## Ethene, a gaseous growth regulator

Since the turn of the century, it has been known that fruit ripens more quickly when stored in a room containing a kerosene stove. The reason (discovered much later) turned out to be that such stoves give off minute amounts of ethene (or ethylene as it used to be called), and it was this gas that induced ripening.

It took a long time before it was realised that plants actually produce ethene. A key observation was that some fruits can be induced to ripen by the presence of a ripe orange or an over-ripe apple in the same storeroom, even if it is not touching the other fruits.

The explanation turned out to be that some fruits produce their own ethene as they ripen, and this induces further ethene production in a self-accelerating cycle. Thus a single ripening fruit not only accelerates its own ripening but triggers the ripening of other fruits.

Besides fruit ripening, ethene is involved in a wide range of plant growth processes. Its effects are complicated by interaction with auxin.

- It accelerates leaf fall. It was first noticed in the mid 19th Century that trees near gas-lit street lamps sometimes lost their leaves early.
- It stimulates cell division in wounded tissue, forming a mass of cells called a **callus**.

- It inhibits the elongation of stems and roots.

Ethene is produced in all parts of the plant and, of course, moves freely through intercellular air spaces.

# Seasonal effects in plant growth

Plant growth varies with the seasons in two quite different ways. First, there are the direct effects of changes in the temperature and energy supply on the rate of photosynthesis. Metabolism is faster at warmer temperatures, and there is more light in summer. For both these reasons, growth is faster in summer than at other times of the year. In this respect, the effect of seasons on plant growth is quantitative, affecting the *rate* of growth.

More subtle are the effects of seasons on the *kind* of growth. Many plants produce flowers at certain times of the year. In late summer, trees produce winter buds, potato plants produce tubers and onion plants produce new bulbs. In the autumn, leaves die and are shed.

These effects are also more subtle in another way – they occur well *before* the onset of the conditions they adapt the plant to meet, before the plant can benefit from them. The production of winter buds occurs while conditions are still favourable for growth. Daffodil bulbs produce flower buds in the autumn, several months before they emerge.

For a plant to adapt its growth in *anticipation* of need, it must have some way of telling the time of year; i.e. some form of *calendar*. The most widespread cue that plants use is daylength, which is more reliable than other environmental factors.

## The control of flowering

The role of photoperiod in the control of flowering was discovered by two American plant breeders, Wightman Garner and Henry Allard, in 1920. A new hybrid tobacco plant, which they called 'Maryland Mammoth' and from which they wanted to obtain seed, had not produced flowers by late autumn. To save the plants from frosts they transplanted them into a greenhouse, and the plants flowered in December.

After experiments in which they artificially manipulated variables such as temperature and light intensity, they showed that it was daylength – or **photoperiod** – that triggered flowering. Evidently, Maryland Mammoth required short days to promote flowering, but winter temperatures prevented it from flowering outdoors.

Garner and Allard called the induction of flowering by daylength, **photoperiodism**. Since then, many other types of seasonal behaviour in both animals and plants have been shown to be under photoperiodic control.

As a result of further work, Garner and Allard concluded that, on the basis of their flowering requirements, plants could be divided into three groups:

- **short-day plants** (**SDP**), which only flower after exposure to photoperiods of *less* than a certain minimum value, called the **critical daylength** (**CDL**). Many short-day plants are native to low latitudes where daylength never exceeds 14 hours in any season, for example sugar cane, rice and maize. In temperate climates, short-day plants flower in spring and autumn.
- **long-day plants** (**LDP**), which only flower when the photoperiod

*exceeds* the critical daylength. Long-day plants flower in late spring or early summer, and are native to higher latitudes where summer photoperiods are long. Examples include radish, lettuce and most temperate grasses.

- **day-neutral plants**, which are insensitive to photoperiod. Examples are dandelion and tomato. Ephemerals, whose life cycle is too short for seasonal factors to be important, are also day-neutral.

The fundamental difference between short-day and long-day plants is not that short-day plants have shorter critical daylengths than long-day plants. Henbane, a long-day plant, has a CDL of about 11 hours, whilst cocklebur, a short-day plant, has a CDL of 15.7 hours. The difference is that cocklebur requires photoperiods of 15.7 hours *or less*, whilst henbane needs photoperiods of 11 hours *or more*. To henbane, an 11.5 hour photoperiod is a 'long day', and to cocklebur, 15 hours is a 'short day'.

If short-day and long-day plants can have overlapping CDLs, why do they flower in different seasons in nature? There are several reasons.

- Long-day plants first experience their CDL when days are *lengthening*, and usually flower in late spring or early summer. Many short-day plants reach their CDL when days are *shortening*, and are induced to flower in autumn. In many spring-flowering perennials such as daffodils, tulips, oak and ash, the flower buds are actually initiated in the autumn and remain dormant over the winter, expanding the following spring.
- Many plants require exposure to cold before they are receptive to photoperiod, and thus flower in the spring. This is called **vernalisation**, and is dealt with later in this chapter.
- Some plants have dual photoperiodic requirements. **Short-long-day plants**, such as white clover, need short days followed by long days and flower in early summer. **Long-short-day plants**, such as *Bryophyllum*, need long days followed by short days and flower in late summer and early autumn.

Another complication is that in some plants a photoperiodic requirement may not be absolute, or **obligate**, but **facultative**. The garden pea, for example, is a long-day plant, but if maintained under short days it will eventually flower.

### Receptiveness to flower

The production of flowers, seeds and fruits requires considerable resources which only a mature plant can devote. Even when all the appropriate environmental conditions have been met, most plants will not flower until they have reached a certain stage of development. Only then does it become receptive to flowering stimuli such as photoperiod. In trees this juvenile period may last 40 years or longer, but in herbaceous plants it may be only a few weeks.

### Night length is what counts

Garner and Allard assumed that it was the length of the day that plants measure. In 1938, Karl Hamner and James Bonner reported some experiments which showed that plants actually measure the length of the night.

In one experiment they interrupted the night with a few minutes of light – SDPs were prevented from flowering and LDPs were induced to flower. On the other hand, interrupting the day with a period of darkness had no effect (Figure 32.25).

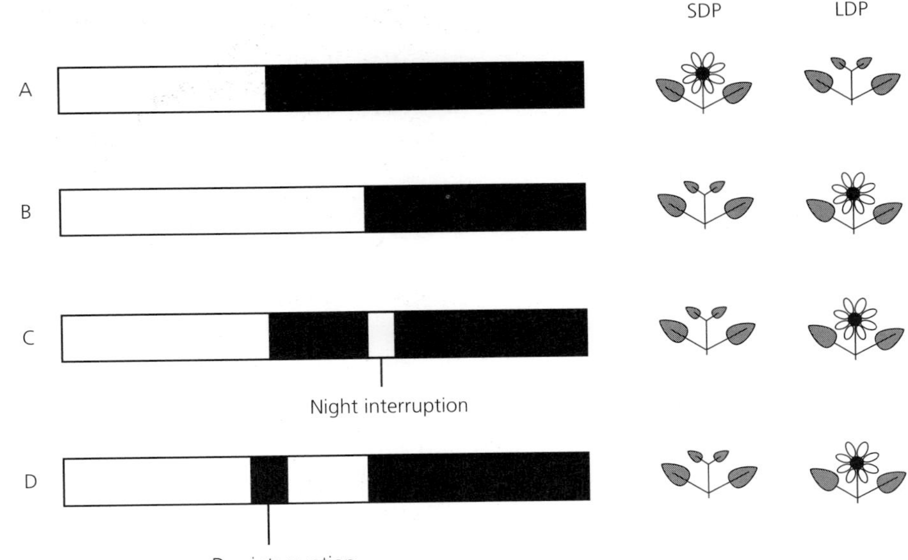


**Figure 32.25▸**
Effect of a night interruption on flowering. Light = unshaded, darkness = shaded. A = short days, B = long days, C = interruption of night with light pulse, D = interruption of day with period of darkness

In a second experiment Hamner and Bonner gave cocklebur (a SDP) non-24 hour cycles consisting of 4 hours light and 8 hours dark. The plants did not flower, even though 4 hours is only about a quarter of the CDL for this species. With cycles of 16 hours light and 32 hours dark the plants did flower, even though the 'day' length was more than the CDL of 15.5 hours (Figure 32.26).

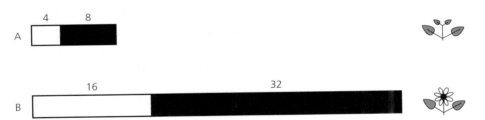


**Figure 32.26▸**
Effect of non-24 hour cycles on flowering. A = 12 hour cycles (4L:8D), B = 48 hour cycles (16L:32D)

These experiments point to the same conclusion: it is the length of the night rather than day length that matters. 'Short day' plants are really 'long night' plants (since the night must be longer than a certain period), and vice versa for 'long day' plants (Figure 32.27). Other work has shown that even this interpretation is not the whole story, and that photoperiodism involves endogenous rhythms (see below).

### *Flowers all the year round*

The ability to artificially manipulate photoperiod has been of great benefit to horticulturalists. At one time, cut chrysanthemums (which are SDPs) could only be bought in the shops in the autumn. Nowadays they can be produced all year round by artificially manipulating the day length.

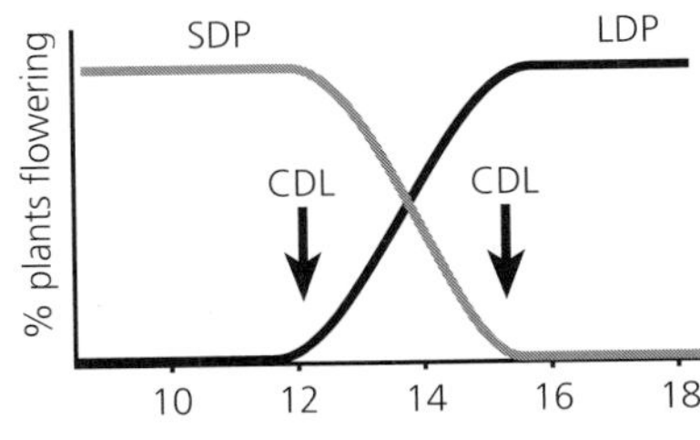


**Figure 32.27**
Critical day lengths in short-day and long-day plants (after Wilkins)

### *Leaves detect the flowering stimulus*

By exposing different parts of both SDP and LDP to differing photoperiods, it was found that photoperiod is detected by the leaves, with the shoot apices (the sites of future flower production) being insensitive (Figure 32.28). Some kind of signal must therefore travel from leaves to stem apices.

In further experiments, closely related SDP and LDP were grafted together. Regardless of the photoperiod, both parts were induced to

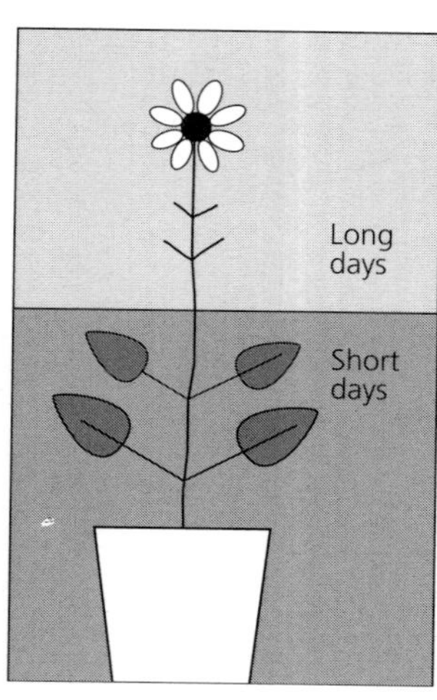


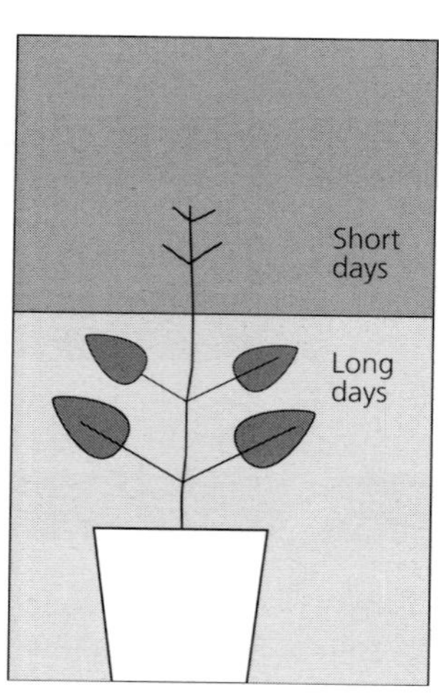


**Figure 32.28**
Evidence that photoperiod is perceived by the leaves. Upper leaves are removed from two *Chrysanthemum* plants, which are SDP. The apex and the leaves are given different photoperiods. Flowering only occurs when the *leaves* are given short days (after Wilkins)

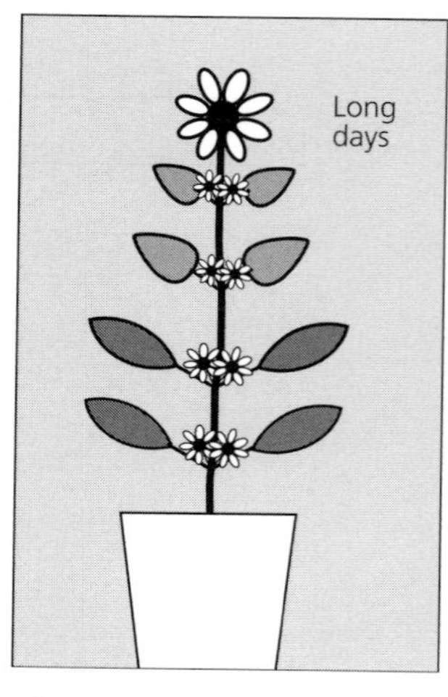


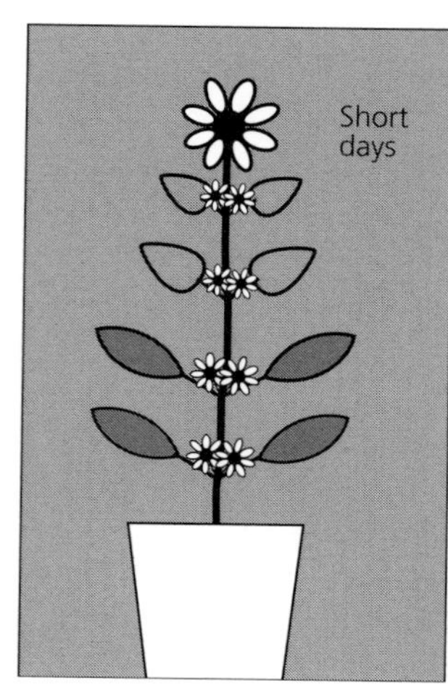


**Figure 32.29**
Grafting experiments show that the flowering message is the same in SDP and LDP (after Wilkins)

flower, showing that the flowering stimulus must be the same for both SDPs and LDPs (Figure 32.29).

On the strength of these results the Russian plant physiologist M.H. Chailakhjan suggested in 1936 that a 'flowering hormone', which he named 'florigen', was involved. Despite intensive efforts no such hormone has been found. All that has been established is that the message is probably transmitted via the phloem.

## *The involvement of phytochrome*

The fact that photoperiodic responses can be reversed by interrupting the night with a brief period of light suggested an obvious experiment. When different wavelengths (but equal intensities) of light were used to interrupt the night, red light (660 nm) was found to be the most effective (Figure 32.30).

This is the wavelength that is most absorbed by **phytochrome**. If phytochrome is involved, the effect of red light should be reversed by exposure to far-red, and this was found to be the case (Figure 32.31).

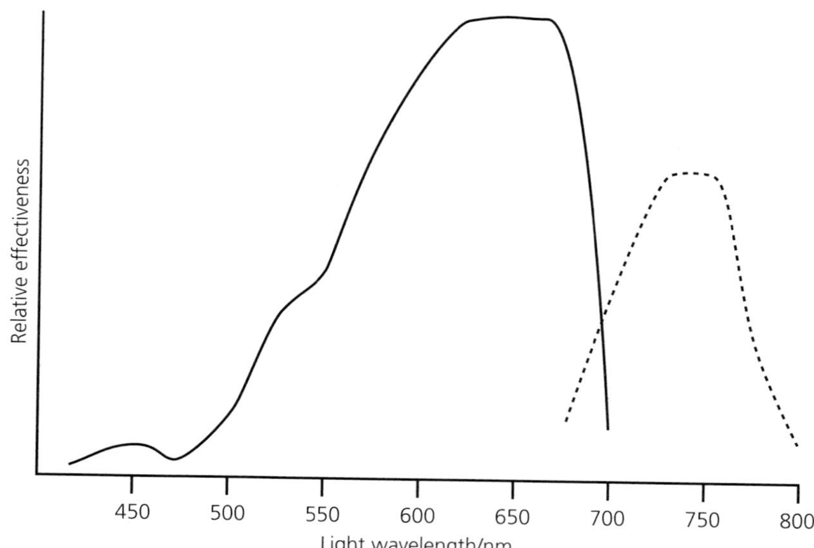


**Figure 32.30**
Solid line: action spectrum for reversal of photoperiodic responses by a night interruption. Dotted line: effectiveness of far-red light in reversing the effect of a previous dose of red light

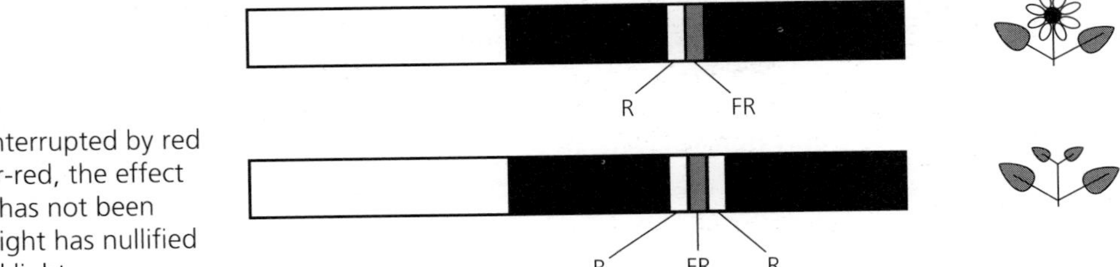


**Figure 32.31**
When the night is interrupted by red light followed by far-red, the effect is as if the red light has not been given – the far-red light has nullified the effect of the red light

## *How do plants measure photoperiod?*

An interesting feature of phytochrome is that $P_{fr}$ slowly changes back to $P_r$ in the dark:

$$P_r \underset{darkness}{\overset{red}{\rightleftharpoons}} P_{fr}$$

At one time it was thought that the slow reversion of $P_{fr}$ to $P_r$ would act as a kind of hourglass, enabling the plant to measure the length of the night. According to this hypothesis, the effect of sunlight would be to 'reinvert the hourglass' and start it going again. This idea has been abandoned for two reasons:

**1** reversion of $P_{fr}$ to $P_r$ is too quick, occurring within 2 hours – much too short for the 6–16 hour nights that occur in temperate regions.

**2** reversion is temperature-sensitive, which the CDL is known not to be.

How plants measure night length is unknown, but there is evidence that the mechanism involves an internal clock.

### *Role of internal clocks in measurement of the photoperiod*

As long ago as 1936, Erwin Bünning suggested that plants measure daylength using the same kind of internal clock that is present in all groups of eukaryotic organisms. Several experiments have given support to this idea.

In one experiment, soybean plants (SDP) were maintained for several 72 hour cycles, each consisting of 8 hours light followed by 64 hours of darkness. The plants were divided into 19 groups and each group was subjected to a 30 minute light break at a particular time during the 'night'. The mean number of flowers produced per plant was then plotted against the time at which the light breaks had been given. The number of flowers produced showed a rhythmic rise and fall, the time between successive peaks being approximately 24 hours (Figure 32.32).

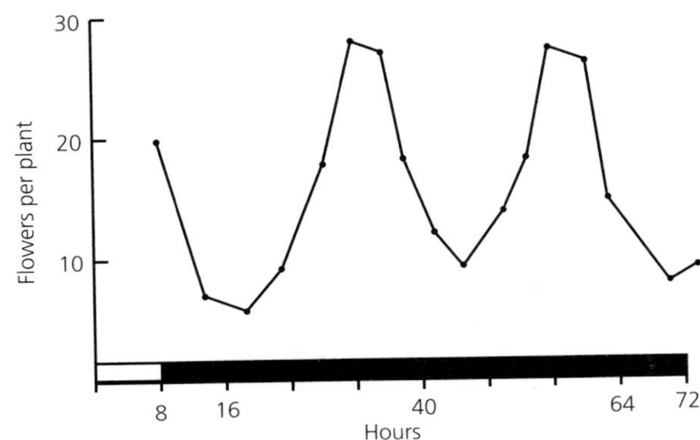


**Figure 32.32**
Evidence that a circadian rhythm is involved in photoperiodic induction of flowering. Effectiveness of night interruption shows peaks at circadian intervals (after Lees)

This and other experiments along similar lines, strongly suggests that although photoperiodic phenomena in plants appear to have an annual periodicity, the underlying mechanism involves a circadian clock (*circa* = around, *dies* = day).

It is believed that the physiologically active form of phytochrome is $P_{fr}$ and that there is an endogenous rhythm in the sensitivity to $P_{fr}$, but how this relates to the induction of flowering is not yet clear.

### *Vernalisation*

Other environmental factors besides photoperiod can influence flowering. Many plants native to temperate climates will not flower until they have been subjected to a prolonged period (several weeks) of low temperatures (between 1°C and 9°C).

Promotion of flowering by chilling is called **vernalisation**, and is particularly common in biennials such as carrots, swedes and cabbages, and also in herbaceous perennials such as Michaelmas daisies, primroses and wallflowers.

Some plants have both a vernalisation and a photoperiodic requirement for flowering. Henbane, for example, can only be induced to flower by long days if it has been previously exposed to cold.

Unlike photoperiodic induction, chilling acts directly on the apical meristem. This has been shown by experiments in which tiny refrigeration coils were inserted into the shoot apex.

## Bud dormancy and leaf fall

In deciduous trees short days trigger the formation of **winter buds**. The outer leaves in a winter bud form the protective bud scales. These protect the inner foliage leaves which remain tiny and dormant during the winter. In the following spring the tiny foliage leaves (which completed all their cell divisions the previous summer) expand by vacuolation, and the internodes between them extend.

Bud dormancy is not a passive absence of growth. It is a state that is actively maintained by a growth inhibitor, **abscisic acid** (**ABA**). Dormancy is normally broken by several weeks of exposure to cold.

Chilling breaks dormancy by increasing the concentration of gibberellin, which stimulates the resumption of growth (coupled with a fall in ABA level). Without chilling (such as can occur in an exceptionally mild winter) a plant may be late in resuming growth next season.

### *Leaf fall*

Even evergreen trees and shrubs lose their leaves, but in deciduous trees and shrubs leaf fall is seasonal. In most woody plants, leaf fall is associated with the ageing of the leaves, and in particular with a decline in auxin content. Leaf fall is also accelerated by short days, which is why some trees retain their leaves longer under street lamps.

## QUESTIONS

**1** Which of the following are tropisms?
- **A** opening of a dandelion inflorescence
- **B** opening of a stoma
- **C** alignment of a leaf at right angles to the incident light
- **D** growth of a pollen tube towards an egg nucleus
- **E** closure of the leaf of a Venus' fly trap when stimulated by an insect.

**2** A radicle was marked with transverse bands of Indian ink at 1 mm intervals. The seedling was then turned so that the radicle was horizontal, and left for 12 hours. It was then turned into its original position and left for another 12 hours. Sketch the appearance of the radicle after
- **a)** the first 12 hours
- **b)** the second 12 hours.

**3** Write the letters of the following events in the order in which they occur in phototropism:
- **A** increasing plasticity of cell walls
- **B** a change in the rate of water uptake by cells
- **C** longitudinal transport of auxin
- **D** lateral transport of a growth-promoting substance

**4** How could apical dominance and gravitropism explain the differences between the shapes of poplar and oak trees?

**5** Cuttings are more likely to produce roots successfully if they have young shoots. Suggest why.

**6** What is wrong with the following statement? 'Plants respond to stimuli by producing hormones which cause growth of the plant in a particular direction, and these growth responses are called tropisms'.

**7** The basis for Went's *Avena* curvature test was that the angle of induced curvature was in direct proportion to the concentration of auxin; double the auxin concentration and the angle of curvature doubled, and so on. Illustrated here is a graph reproduced in many textbooks showing a relationship between auxin concentration and the percentage stimulation of elongation in shoots. Notice that the horizontal scale is *logarithmic*. If Went's *Avena* curvature test was valid, what is wrong with the graph?

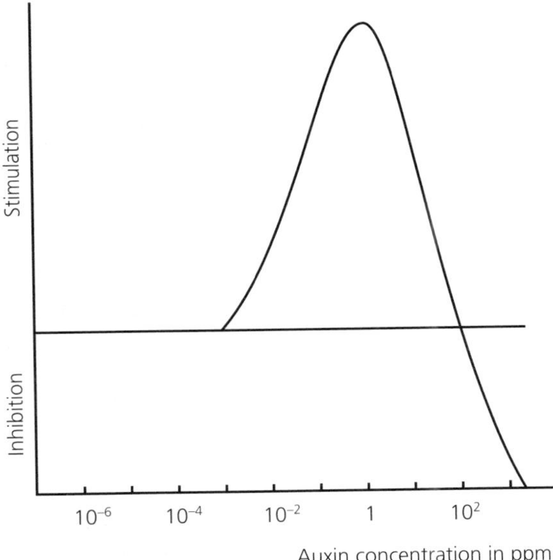

CHAPTER

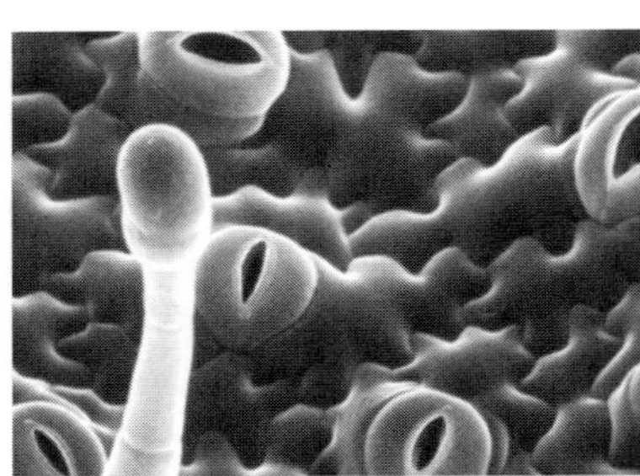

# 33 GATHERING ENERGY and RAW MATERIALS

## LEARNING OBJECTIVES

By the time you have completed your study of this chapter you should be able to:

- know the raw materials and end products of photosynthesis.
- understand the principles underlying investigations into the need for light, carbon dioxide and chlorophyll, and the production of oxygen.
- know the raw materials and end products of the photostage and synthetic stages of photosynthesis, and their sites of occurrence within the chloroplast.
- understand the difference between gross and net photosynthesis.
- understand how the interdependence between the two stages of photosynthesis can explain the effects of light, carbon dioxide and temperature on photosynthetic rate.
- understand the basis for the chromatographic separation of chloroplast pigments.
- distinguish between an absorption spectrum and an action spectrum.
- explain how the structure of a leaf is adapted to carry out photosynthesis and how it represents a compromise with the demands of water conservation, and how leaf structure illustrates the concept of an organ.
- state the names and functions of macronutrients needed by plants, and the general role of micronutrients.
- explain the relationship between the structure of a root and its role as an absorptive organ.
- explain the mechanisms by which mineral ions are absorbed from the soil.

## Light into chemical energy: photosynthesis

Apart from algae and certain bacteria, green plants are the only organisms that do not depend on other organisms for their energy supply. Instead of feeding on organic matter made by other organisms, they make their own by **photosynthesis**. Green plants are therefore **autotrophic** ('self-feeding'). Animals, fungi and most bacteria cannot make their own organic substances from inorganic materials, and are said to be **heterotrophic**.

In photosynthesis, $CO_2$ and water are converted to sugar and oxygen. The energy driving the process is **light**, which is trapped by the green pigment **chlorophyll**. The entire process occurs in the **chloroplasts**, and

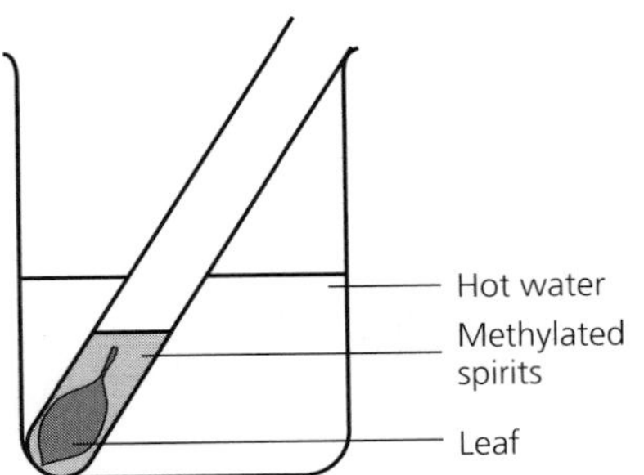


**Figure 33.1**
Removing chlorophyll from a leaf

can be summarised thus:

$$\text{carbon dioxide} + \text{water} + \text{light energy} \xrightarrow{\textit{chlorophyll}} \text{carbohydrate} + \text{oxygen}$$

## Experimental overview

The need for light, carbon dioxide and chlorophyll can easily be investigated using the experiments shown in Figures 33.1–33.4.

### *Testing a leaf for starch*

Although the organic product of photosynthesis is a sugar, it is rapidly converted into starch. This is easy to detect because of the intense blue given by the reaction it has with iodine solution ($I_2$/KI). Since this test involves a colour change, the green chlorophyll must first be removed by boiling the leaf in alcohol. This is easiest with soft leaves that do not have a waxy surface.

1 The leaf is boiled in methylated spirits using an electrically-heated water bath (Figure 33.1). Dipping the leaf in boiling water for about a minute before boiling it in alcohol speeds up the action of the alcohol, though this is not, as frequently stated, 'to kill the leaf' – boiling alcohol does this most effectively.

2 The leaf is placed in a Petri dish and iodine solution is added. If starch is present the leaf turns blue-black.

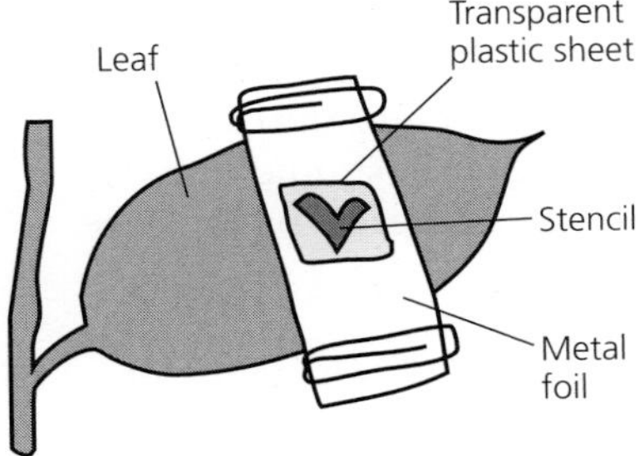


**Figure 33.2**
Apparatus for investigating the need for light in photosynthesis

### *The need for light*

The need for light can be investigated using the apparatus shown in Figure 33.2. For reasons explained in the Clear Thinking box on page 609, the leaf should not be destarched before applying the stencil. If possible, the stencil should be made of metal foil and should have a thin sheet of transparent plastic glued over it, thus ensuring that the part of the leaf under the 'window' of the stencil has the same (and slightly reduced) access to air as the part of the leaf that is in darkness. The stencil is placed on a leaf of a potted plant on a window ledge or other sunny situation. After about two days it is removed, and the leaf tested for starch.

### *The need for carbon dioxide*

In this case, the plant *must* be destarched first by keeping it in the dark for about 48 hours (and in so doing, assuming the need for light). Next, remove two leaves and set up the apparatus shown in Figure 33.3.

**Figure 33.3**
Apparatus for investigating the need for $CO_2$ in photosynthesis

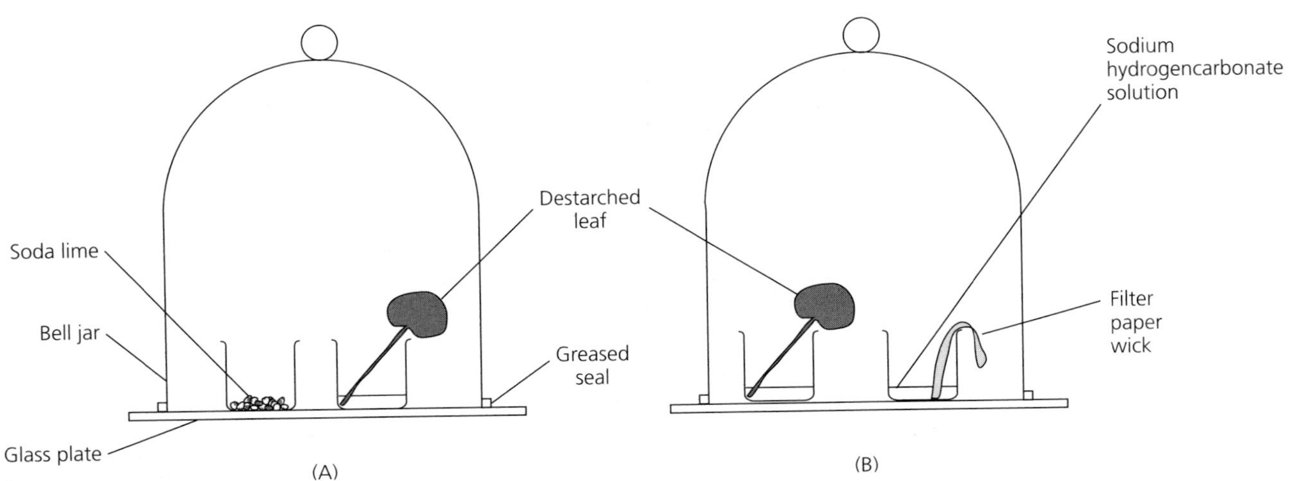

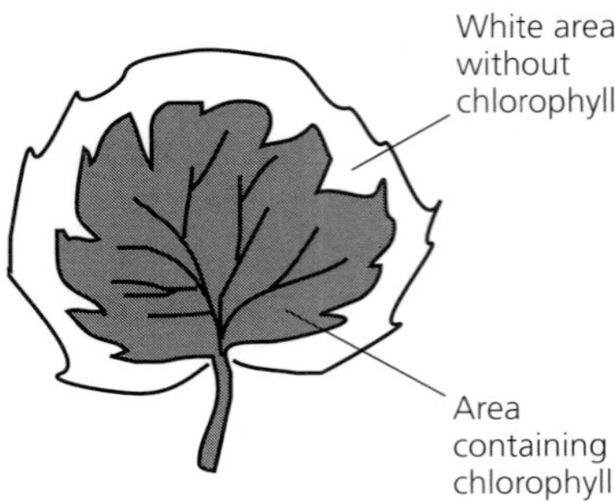


**Figure 33.4**
A variegated geranium leaf

In A, the beaker contains soda lime to remove $CO_2$ from the air. Unfortunately, putting a leaf in a bell jar changes the surroundings of the leaf in other ways – the air inside the jar is more humid and, in bright sunshine, a lot warmer. It is, therefore, necessary to set up a **control** (B), identical in every way except for the factor being investigated. The concentrated sodium hydrogencarbonate solution gives off $CO_2$, replacing any that might be absorbed by the leaf (without this source of $CO_2$, the amount in the flask would be quite inadequate to produce enough starch for a positive result). After leaving both leaves in bright light for several hours, they are tested for starch in the way described above.

### *The need for chlorophyll*

Chlorophyll cannot be removed from a leaf without killing it. The problem is solved by using **variegated** leaves, in which parts of the leaf are naturally lacking chlorophyll (Figure 33.4).

A plant with variegated leaves is left in bright light for a few hours. It is not necessary to destarch it first (see the Clear Thinking box on page 609). Next, a leaf is removed and carefully drawn to show the distribution of chlorophyll, and then tested for starch. The distribution of starch is then compared with the 'map' of the distribution of chlorophyll. If chlorophyll is necessary for photosynthesis, then starch should only be present in the areas that had been green, and this proves to be the case.

### *The need for water*

It is not possible to deny a leaf water without killing it; even at the point of death from dehydration a leaf contains over 50% water. Simple laboratory demonstrations of the need for water as a raw material for photosynthesis are therefore not possible. However, by using a heavy isotope of oxygen it has been shown conclusively that water is used up (see below).

### *The production of oxygen*

It is not easy to detect oxygen production by a land plant because the small quantities given off would be difficult to distinguish from the large amount (21%) already present in the air. With pondweed however it is a simple matter because oxygen is only slightly soluble in water, and during rapid photosynthesis it is given off as bubbles (Figure 33.5). A trace of potassium hydrogencarbonate added to the water acts as a source of $CO_2$.

In most pondweeds there are large air spaces extending throughout the plant, so gas produced in the leaves emerges as bubbles from the cut surface of the stem. Although the gas produced is oxygen, it is rapidly diluted with nitrogen diffusing from the surrounding water. As a result, the gas collected will frequently not relight a glowing splint. The problem is avoided if the nitrogen is first removed by boiling and cooling the water, and then resupplying it with $CO_2$ by bubbling the gas through it.

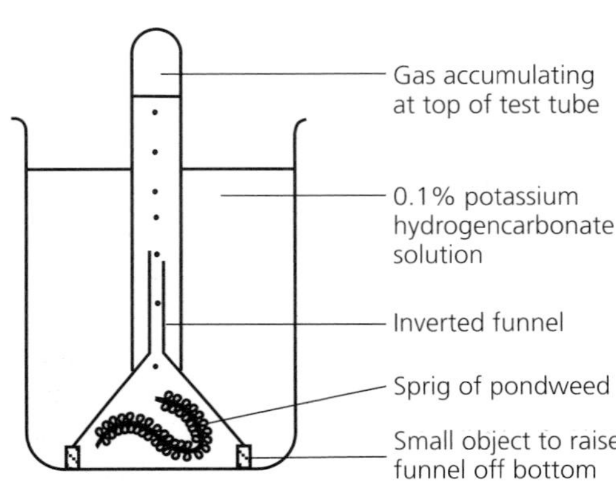


**Figure 33.5**
Apparatus for demonstrating the production of oxygen by an illuminated plant

**CLEAR THINKING**

In experiments to find out whether light is needed for photosynthesis, it is common practice to place the plant in darkness to destarch it before fixing a stencil on a leaf. Destarching occurs because, in darkness, any starch that has accumulated in the chloroplasts during previous exposure to light is converted to sugar. *However, we do not know this before doing the experiment.* To assume what one is trying to prove is bad science. Prior destarching is, however, perfectly legitimate in a *demonstration* that light is needed for photosynthesis.

When investigating the need for $CO_2$, prior destarching is essential (though in doing so one is of course assuming the need for light). This investigation involves artificial manipulation of the $CO_2$ supply of a leaf which, prior to the experiment, had access to $CO_2$. It is thus essential to know that any starch detected has been made *after* the $CO_2$ supply had been altered.

In investigations into the need for chlorophyll using a variegated leaf, the situation is significantly different. There is no artificial manipulation in chlorophyll content; cells without it during the experiment didn't have chlorophyll before the experiment. Destarching is therefore quite pointless.

## Measuring the rate of photosynthesis

Photosynthetic rate can be expressed in three ways:

- the amount of oxygen given off per unit time;
- the amount of $CO_2$ taken in per unit time;
- the increase in dry mass per unit time.

When investigating the effect of different conditions on the rate of photosynthesis by a plant, it is best to use the same individual plant. If photosynthetic rates of different plants are being compared, then the rate must be expressed as the amount of photosynthesis per hour by each gram of plant, or by each $cm^2$ of leaf.

### Gross and net photosynthesis

Photosynthetic rates are not what they seem because respiration occurs at the same time. Suppose a plant makes five units of oxygen but uses one unit in respiration in the same time; only four reach the outside world to be measured. The true rate, or **gross photosynthesis**, is therefore always faster than the apparent, or **net photosynthesis** (in this imaginary example, five units rather than four). To find the true photosynthesis, the rate of respiration is first measured in darkness and then added to the net rate of photosynthesis:

$$\text{true photosynthesis} = \text{net photosynthesis} + \text{respiration}$$

In full sunlight a leaf may be photosynthesising 10–20 times faster than it is respiring (though only perhaps five times as fast as the whole plant is respiring). The small amount of $CO_2$ produced in respiration is used by the chloroplasts before it has time to escape from the leaf. In the daytime, therefore, the leaf does not *seem* to be respiring at all.

If a plant is to grow, its leaves must make more organic matter during the day than the whole of the plant uses over a 24 hour period. Net photosynthesis must, therefore, more than compensate for the organic matter used up in respiration at night.

## The nature of light

When light from the Sun passes through a prism it separates into a mixture of colours called a **spectrum**, ranging from red at one end to violet at the other (Figure 33.6). What we see as light is only a small part of a much wider spectrum of **electromagnetic radiation** which includes X-rays, $\gamma$-rays, infrared and radio radiation.

In the last century it was shown that all electromagnetic radiation travels as waves, the distance between successive peaks being the **wavelength**. Wavelengths of visible radiation extend from 400 nm (violet) to 700 nm (red).

Although the wave theory of light explains much of its behaviour, in some respects light behaves as particles called **photons**. Each photon is a discrete package – or **quantum** – of energy. The shorter the wavelength of the radiation, the larger is each quantum of energy. Thus a photon of violet light (400 nm) has 700/400 times as much energy as a photon of red light (700 nm).

### Chloroplast pigments

Before light can be used it must first be absorbed. This is the function of the four lipid-soluble pigments in the chloroplasts. They belong to two categories – the chlorophylls and the carotenoids.

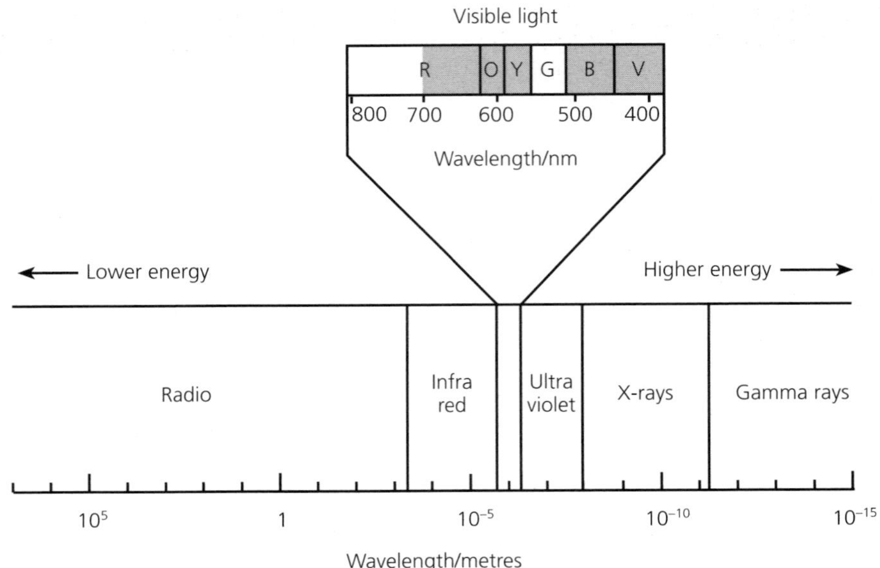

**Figure 33.6►**
The electromagnetic spectrum, of which visible light is a very small part. The photosynthetically useful wavelengths are shaded

**Figure 33.7**
Molecular structure of chlorophyll *a*

1 **Chlorophylls** There are two kinds of chlorophyll in the plant kingdom: **chlorophyll *a*** (bottle green) and **chlorophyll *b*** (olive green). Others occur in protoctists and in photosynthetic bacteria. All chlorophylls have a tadpole-shaped molecule with a 'head' and a long hydrocarbon 'tail' (Figure 33.7). The head consists of a **porphyrin** ring structure containing four smaller nitrogen-containing rings with a magnesium atom in the centre. A similar porphyrin ring structure is found in haemoglobin, but the magnesium is replaced by iron. The head is the light-absorbing part, and in the intact chloroplast is linked to proteins. The tail anchors the molecule in the lipid membranes inside the chloroplast.

2 **Carotenoids** These yellow or orange pigments are chemically quite different from chlorophyll, containing neither nitrogen nor magnesium. In the plant kingdom there are two kinds. **Carotenes**, such as **β-carotene**, are purely hydrocarbon. **Xanthophylls**, such as yellow **lutein**, also contain oxygen. They are normally masked by chlorophyll, but in deciduous trees they are revealed in the autumn as the chlorophyll breaks down, giving the familiar autumn colours.

Besides acting as light-harvesters, carotenoids protect chlorophyll. In the absence of carotenoids, chlorophyll is bleached in the light by oxygen; the carotenoids prevent this. In solution, chlorophyll is not so protected; this is why extraction and separation of chlorophylls should be done in dim light.

## The absorption of light by molecules

When a chlorophyll molecule absorbs a photon, the energy is absorbed by one of its electrons which becomes 'excited'. Almost immediately the excited electron loses its extra energy and regains its normal or 'ground' state.

There are two ways in which an excited chlorophyll molecule can lose its energy. It may hand it on to another molecule, or it may re-emit light as **fluorescence**. The emitted photon always has slightly less energy (longer wavelength) than the photon that had been absorbed, so there is always some heat loss. You can see this fluorescence by strongly illuminating a very dark green solution of chlorophyll and placing yourself on the same side of the chlorophyll as the light source.

## Separating chloroplast pigments by paper chromatography

All the photosynthetic pigments are in the lipid bilayer of the chloroplast membranes and can be removed by grinding up a leaf in ethanol or propanone ('acetone'). The resulting solution is a mixture of four pigments which can easily be separated by paper **chromatography**. The paper with the separated pigments is called a **chromatogram**.

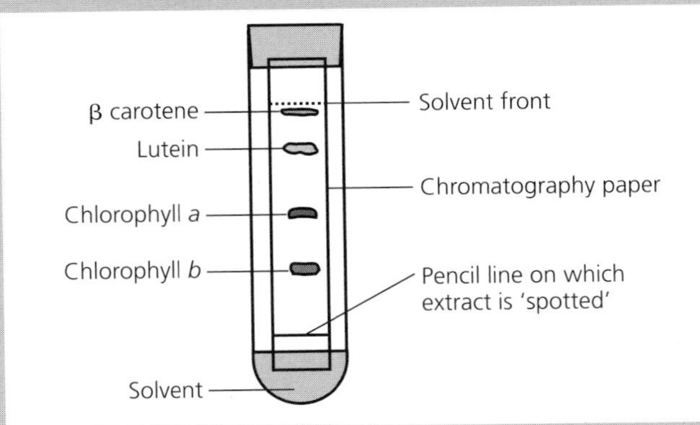


Simple chromatographic separation of chloroplast pigments

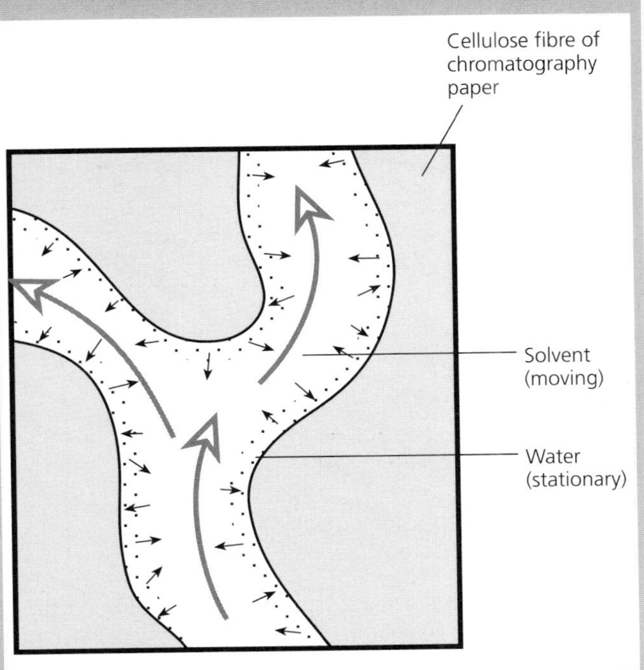


Principle of chromatography (after Harper)

A tiny drop of extract is repeatedly streaked along a prepared pencil line near one end of the paper, allowing the solvent to evaporate each time before adding another drop. After a medium-dark green streak has been produced, the end of the paper is suspended in a suitable solvent (such as 100 parts petrol ether:12 parts propanone). The solvent runs up the paper which acts like a wick, carrying the pigments with it.

Paper chromatography is dependent on the fact that the cellulose fibres of the paper have a thin layer of water molecules clinging to them. Thus when an organic solvent creeps through the paper, it is really moving over an extremely thin, stationary layer of water. When a solution of several substances in an organic solvent is allowed to creep through filter paper, different solutes move at different rates. Each solute molecule spends some of its time attracted to the polar water and some of its time in the moving, non-polar organic solvent. The more polar molecules spend more of their time attracted to the water and move more slowly. The less polar molecules move faster since they spend a higher proportion of their time in the moving organic solvent.

If you look at the formulae for chlorophylls *a* and *b* you will see why chlorophyll *b* moves slightly slower than chlorophyll *a*: it has a polar aldehyde group instead of the non-polar methyl group of chlorophyll *b*. Similarly, the oxygen in lutein makes it slightly polar, so it moves more slowly than β carotene.

The relative speed of movement of substances is expressed as the **$R_f$ value** ($R_f$ stands for retardation factor), and is given by:

$$R_f \text{ value} = \frac{\text{distance moved by solute}}{\text{distance moved by solvent}}$$

In the solvent given above, chlorophyll *a* has an $R_f$ value of 0.64, meaning that it moves at 0.64 of the speed of the solvent.

Each solute has a characteristic $R_f$ value, but only for a given solvent; if this is changed, so is the $R_f$ value. This fact is exploited in **two-way chromatography**, in which a wider separation is obtained by having two runs at right angles to each other in different solvents. Combinations of $R_f$ values for two solutes are useful in identifying substances on a chromatogram.

## Separating chloroplast pigments by paper chromatography *continued*

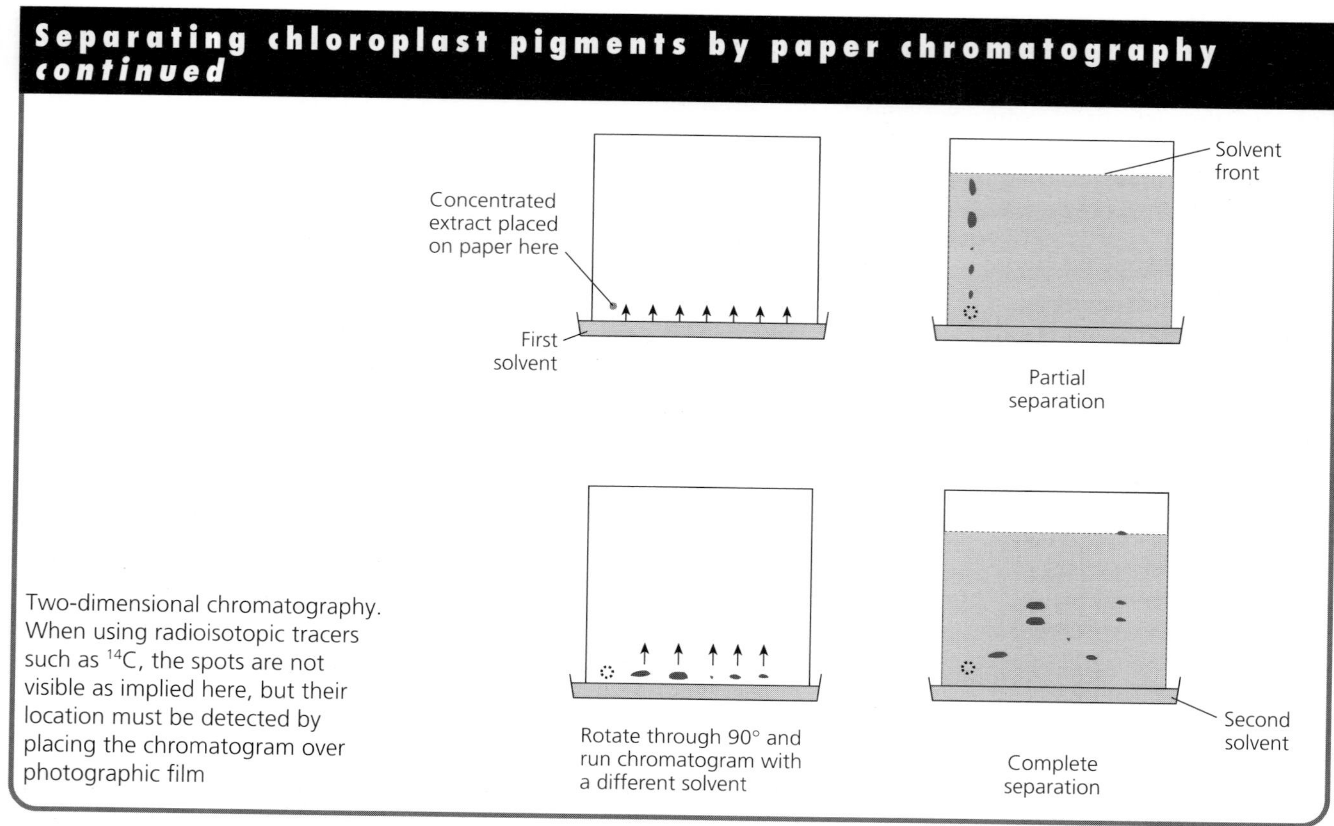


Two-dimensional chromatography. When using radioisotopic tracers such as $^{14}C$, the spots are not visible as implied here, but their location must be detected by placing the chromatogram over photographic film

When a solution of chlorophyll *a* and *b* in an organic solvent is illuminated by light of a wavelength that is absorbed mainly by chlorophyll *b*, the fluorescence is characteristic of chlorophyll *a*. This shows that although the chlorophyll *b* is absorbing the light, it must be handing the energy on to the chlorophyll *a*, which re-emits it.

The transfer of energy from one pigment molecule to another plays an important part in photosynthesis. In the intact chloroplast, energy transfer is much more efficient because pigment molecules are anchored close to each other in membranes, forming 'photosystems' (see below). Because the energy is transferred more efficiently, less is wasted and fluorescence is much harder to detect.

### Absorption spectra

Sunlight consists of a mixture of all the colours of the spectrum, which we see as white light. Many substances only absorb light of certain colours, reflecting or transmitting the others. A solution of chlorophyll appears green when held up to the light because it absorbs blue and red, the green light passing through to the eye. A leaf appears green because, after it has passed through the chloroplasts, the green light is reflected by cell surfaces back to the eye.

After separation by chromatography, the light-absorbing properties of each pigment can be investigated. Light of different wavelengths is passed through a dilute solution of the pigment. For each wavelength, the fraction of the light that is absorbed is measured. When the results are plotted on a graph, an **absorption spectrum** is obtained. Figure 33.8 shows the absorption spectra for chlorophylls *a* and *b* and for β carotene. As you can see, red and blue light are most strongly absorbed.

The two main peaks in the absorption spectrum correspond to the different excited states in which a chlorophyll molecule can exist. The state produced by absorption of a red photon lasts long enough for the

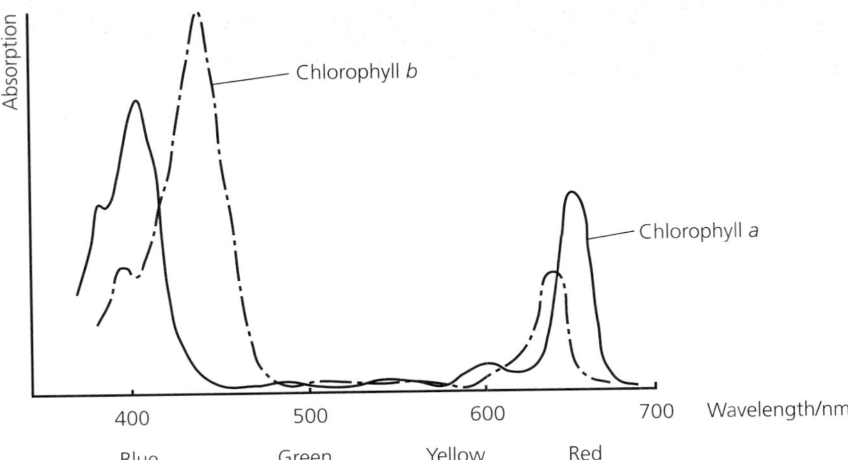


**Figure 33.8**
Absorption spectra of chlorophylls *a* and *b*

energy to be handed on to other molecules. The excited state produced by a blue photon is too short-lived and it quickly drops back to the same energy level as is produced by a red photon, emitting the extra energy as heat. Although blue photons are absorbed as readily as red photons, the additional energy of blue light is wasted.

### Molecular 'pass the parcel'

Besides fluorescence, an excited molecule can also lose energy by passing it to another pigment molecule. When chlorophyll *b* or the carotenoids absorb light, they hand on the energy to chlorophyll *a*, but the reverse transfer cannot occur. This is because chlorophyll *b* absorbs higher energy light than chlorophyll *a* (its red absorption peak is to the left, at a shorter wavelength). Similarly, carotenoids can hand on their energy to chlorophyll *b* and chlorophyll *a*.

### Action spectrum for photosynthesis

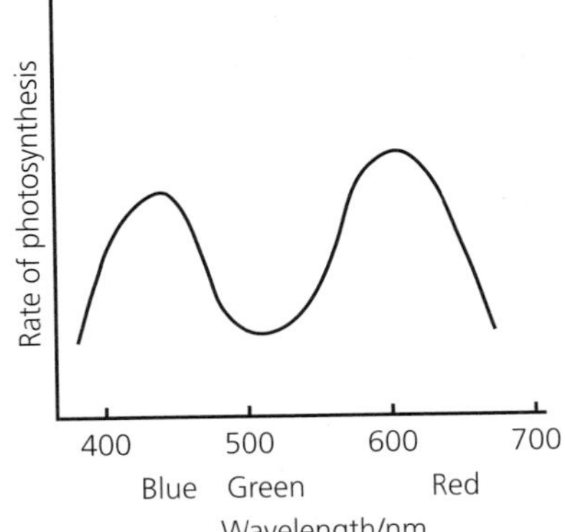


**Figure 33.9**
The action spectrum for photosynthesis

If the rate of photosynthesis is plotted against the wavelength of light given, an **action spectrum** is obtained. If chlorophyll is chiefly responsible for absorbing the energy that drives photosynthesis, one would expect that blue and red light would be the most effective and green the least effective. Figure 33.9 shows that this is so. Although green light is poorly absorbed by chlorophyll, it is about half as effective as blue or red because of the ability of the carotenoids to hand energy on to chlorophyll.

Two important precautions must be taken when obtaining an action spectrum. First, the light should be dim enough to limit the rate of photosynthesis – with bright light, other factors are rate-limiting and wavelength makes little difference (see below). Second, the light given should have the same intensity at each wavelength (expressed as numbers of quanta $cm^{-2}\,s^{-1}$, rather than Watts $cm^{-2}\,s^{-1}$, since energy varies with wavelength).

## The chemistry of photosynthesis

Photosynthesis is the conversion of light energy into chemical energy stored in the two products carbohydrate and oxygen. The process can be represented simply by the following overall equation, where $(CH_2O)$ is a general formula for carbohydrate:

$$\text{light} + CO_2 + H_2O \rightarrow (CH_2O) + O_2$$

## Photosynthetic units

When a dense suspension of unicellular algae is weakly illuminated it begins to produce oxygen almost immediately, even though each chlorophyll molecule has to wait an average of an hour to collect enough photons to produce one molecule of oxygen! This and other observations have led plant physiologists to conclude that chloroplast pigments are organised into **photosynthetic units**, each containing about 300 chlorophyll molecules together with carotenoids. Most act as light harvesting **antenna pigments** – quanta of energy absorbed by these are collected into a common point for use in the next stage.

At the centre of each photosynthetic unit is a molecule of chlorophyll *a* that collects energy quanta from the other pigment molecules. This chlorophyll molecule is linked to neighbouring proteins in a different way from the others, causing it to absorb photons of slightly lower energy than other chlorophyll *a* molecules. This enables it to accept quanta from other chlorophyll molecules, but quanta cannot be transferred in the reverse direction.

This special chlorophyll molecule processes energy from the entire photosynthetic unit and forms the **reaction centre**. In full sunlight each ordinary chlorophyll molecule absorbs only a few photons of light per second. These all converge towards the reaction centre, which is kept busy processing several hundred quanta per second. A photosynthetic unit can thus be likened to a 'funnel', the reaction centre forming the 'spout'.

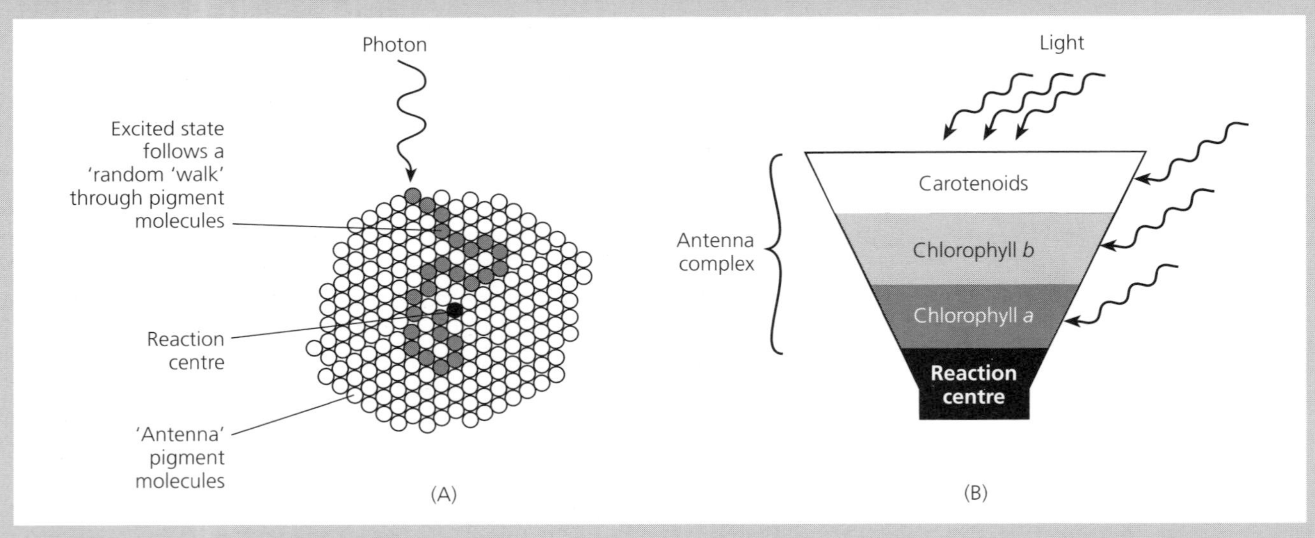


Principle of the photosynthetic unit. (A) Incoming photon is absorbed by 'antenna' pigment molecules (open circles) and the excitation is passed from one pigment to another along a 'random walk' shaded until it reaches the reaction centre (black). (B) The different pigment molecules in a photosynthetic unit are arranged so that excitation is funnelled from carotenoids (which absorb higher energy), converging toward a reaction centre

Of course this equation only indicates what goes into the chloroplasts and what comes out; it says nothing about the many chemical reactions in between the raw materials and end products. These are dealt with in the next section.

## Photosynthesis as a two-stage process

An important clue to the photosynthetic mechanism was obtained by F. Blackman in 1905, who studied the effect of temperature and light intensity on the rate of photosynthesis in pondweeds. In dim light photosynthesis was unaffected by temperature. This is just as would be

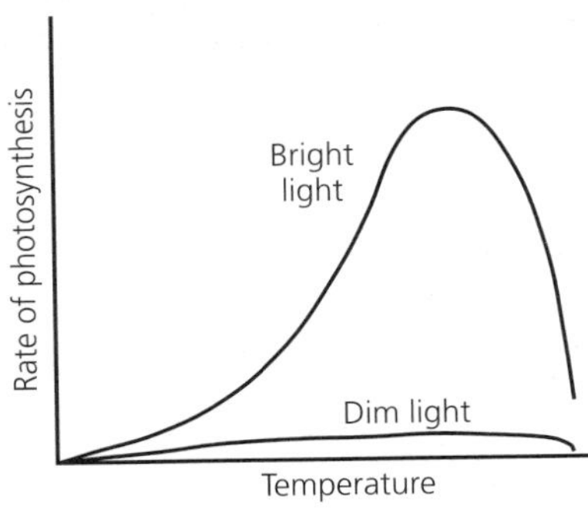


**Figure 33.10**
Effect of temperature on rate of photosynthesis in different light intensities

expected of a photochemical reaction, such as that which occurs when photographic film is exposed to light. In bright light the result was very different, the rate roughly doubling for every 10°C rise in temperature (Figure 33.10). This is the behaviour of 'ordinary' chemical reactions.

These results led Blackman to conclude that photosynthesis consists of two stages:

- a light-dependent, **photochemical stage**, or **photostage**.
- a temperature-sensitive stage involving enzymes, which does not require light.

Blackman called the temperature-sensitive stage the 'dark' stage, since it does not directly require light. It is important to realise that, except in the artificial conditions possible in the laboratory, *the 'dark' stage does not occur in darkness*. For this reason, it is preferable to call it the **light-independent stage**, or, since it results in the synthesis of organic from inorganic carbon compounds, the **synthetic stage**.

Blackman suggested that if the products of the photostage are used by the synthetic stage, the overall process could be represented as follows:

$$A \xrightarrow{\textit{light}} B \xrightarrow{\textit{enzymes}} C$$

This two-stage scheme would neatly explain why photosynthesis is insensitive to temperature in dim light and insensitive to light intensity in cold conditions. In dim light, B would only be formed slowly. Under these conditions, the enzymes catalysing the synthetic stage would be starved of raw materials, so a rise in temperature would have no effect. In bright light there would be adequate photochemical product B, so a rise in temperature would be expected to increase the rate of the synthetic stage.

Blackman summarised his findings by his **principle of limiting factors**. Stated simply it says:

*When a process is dependent on more than one factor, the rate of the overall process is limited by the factor that is in shortest supply relative to the demand for it.*

A few years after Blackman's experiments, further support for his ideas was obtained by Otto Warburg. He compared the amount of photosynthesis in bright, continuous light with the amount in light of the same intensity but given in flashes. He found that the yield per unit amount of light energy given was greater in flashing light than in continuous light. This makes sense if one assumes that during each flash, photochemical product is formed faster than it can be used in the synthetic stage. During the dark periods between flashes, the photochemical product is used up. The periods between flashes thus give the synthetic stage time to 'catch up'.

## The source of oxygen

The overall equation for photosynthesis shows that for every molecule of oxygen produced, one molecule of water is used. Since the water does not contain enough oxygen to account for all the oxygen given off, the equation implies that at least some of the oxygen produced in photosynthesis comes from carbon dioxide.

Until the late 1930s it was thought that the role of light was to split carbon dioxide into carbon and oxygen, the carbon then reacting with water to form carbohydrate. If this were true, oxygen would be an inevitable by-product of the conversion of $CO_2$ to carbohydrate. Logical though this idea appeared, seeds of doubt had already been planted in the late 19th Century, by two discoveries:

- T.W. Engelmann had discovered that certain bacteria (purple sulphur bacteria) photosynthesise without producing oxygen, accumulating sulphur in their cells instead.
- at about the same time, S. Winogradsky had found that some bacteria can make carbohydrate from $CO_2$ *in darkness*! These chemosynthetic bacteria have no chlorophyll, but use energy derived from oxidation of inorganic substances in their environment to reduce $CO_2$ to carbohydrate.

As so often happens in science, these discoveries got in the way of conventional ideas and were ignored until the 1930s. Then, Cornelius van Niel published the results of further studies of the purple sulphur bacteria. By measuring the amounts of hydrogen sulphide used and sulphur produced, he showed that they photosynthesise according to the overall equation:

$$\text{light} + CO_2 + 2H_2S \rightarrow (CH_2O) + 2S + H_2O$$

In an audacious extension to his hypothesis, van Niel then proposed that photosynthesis in green plants and in purple sulphur bacteria were essentially similar processes. He suggested that in both cases light was used to split a hydrogen-containing substance, and that the hydrogen was then used in a synthetic stage to reduce $CO_2$ to carbohydrate:

$$\text{light} + 2H_2X \rightarrow 4H + 2X$$

$$4H + CO_2 \rightarrow (CH_2O) + H_2O$$

The difference is in the source of the hydrogen – hydrogen sulphide in purple sulphur bacteria, and water in plants. The second stage of the process – the reduction of $CO_2$ to carbohydrate using the hydrogen generated in the first stage – is the same in both types of photosynthesis.

The hypothesis of van Niel was finally put to the test in 1941 by Samuel Ruben and Martin Kamen using a heavy isotope of oxygen, $^{18}O$. They illuminated suspensions of the unicellular protoctist, *Chlorella*, in water enriched with $^{18}O$. When the oxygen evolved was collected and analysed, it was found to have the same proportion of $^{18}O$ as the water:

$$2H_2{}^{18}O + C^{16}O_2 \rightarrow (CH_2{}^{16}O) + {}^{18}O_2 + H_2O$$

The two stages of photosynthesis can thus be represented as in Figure 33.11.

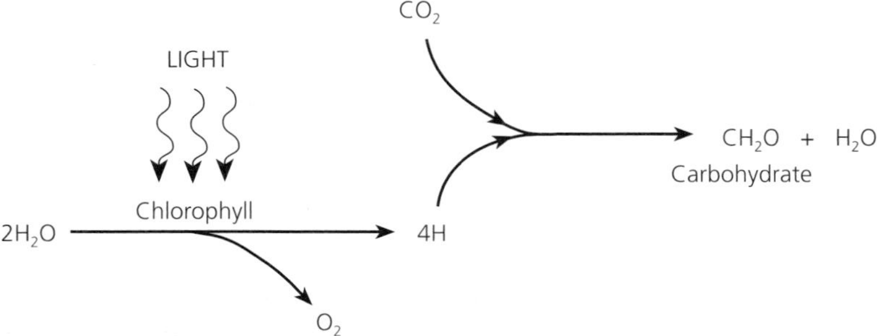


**Figure 33.11**
Summary of photosynthesis

### *The Hill reaction*

According to van Niel, the production of oxygen and the consumption of $CO_2$ occur in different stages of photosynthesis. Powerful support for this idea was obtained in 1937 by Robin Hill. He showed that under the right conditions, an illuminated suspension of chloroplasts could produce oxygen *without using $CO_2$ or producing carbohydrate*. The **Hill reaction** (as it became known) provided the first experimental proof that oxygen production and $CO_2$ consumption in green plants are separable parts of the photosynthetic process.

### The hydrogen acceptor

When water is split by electrolysis it produces oxygen and hydrogen gases. Although photosynthesis produces oxygen, no gaseous hydrogen is formed. An indication of the reason was provided by the fact that Hill's chloroplast suspensions would only produce oxygen if an electron acceptor such as $Fe^{3+}$ ion were added. Each of the two hydrogen atoms in a water molecule consists of an electron and a proton. If the electrons reduce $Fe^{3+}$ ions to $Fe^{2+}$ ions, the hydrogen ions are left in solution:

$$\text{light} + 2H_2O + 4Fe^{3+} \rightarrow 4Fe^{2+} + 4H^+ + O_2$$

$Fe^{3+}$ ions are an artificial electron acceptor. In the early 1950s the natural acceptor was identified as **nicotinamide adenine dinucleotide phosphate**, or **$NADP^+$**. This is chemically very similar to the respiratory hydrogen carrier, $NAD^+$. Like $NAD^+$, each $NADP^+$ molecule can accept two electrons and a hydrogen ion, the other hydrogen ion remaining in solution. The splitting of water by light, or **photolysis**, can be simply represented thus:

$$\text{light} + 2H_2O + 2NADP^+ \rightarrow 2NADPH + 2H^+ + O_2$$

The NADPH is then used in the synthetic stage to reduce $CO_2$ to carbohydrate:

$$2NADPH + 2H^+ + CO_2 \rightarrow (CH_2O) + H_2O + 2NADP^+$$

### The photolysis of water

To understand this part of photosynthesis, it is helpful first to recall the formation of water in respiration (Chapter 5). Here, hydrogen passes from NADH to oxygen forming water. A hydrogen atom consists of a proton and an electron, but so far as energy changes are concerned it is the transfer of the electrons that is important. NADH is a moderately strong reducing agent (readily gives up electrons), and oxygen is a strong oxidising agent (has a powerful tendency to accept electrons). Electrons are thus moving energetically 'downhill', from NADH which 'wants' to give them up, to oxygen which 'wants' to accept them, so there is a release of energy.

In photosynthesis the opposite occurs. Electrons and protons must be removed from water (leaving oxygen) and transferred to $NADP^+$ to form NADPH. Since the oxygen in water doesn't 'want' to let electrons go, and $NADP^+$ doesn't 'want' to accept them, this is an energetically uphill process.

How is this done? The essential event in the photostage – the absorption of light by chlorophyll – leads to the creation of two chemicals:

- an even stronger oxidising agent than oxygen – strong enough to remove electrons (and protons) from water to form oxygen.
- an even stronger reducing agent than NADPH – strong enough to donate electrons (and protons) to $NADP^+$ to form NADPH.

The energy for the formation of these two chemicals comes from light, absorbed by chlorophyll. When a chlorophyll molecule absorbs a photon, one of its electrons becomes 'excited'. In this state the chlorophyll is a powerful reducing agent (represented by chl★ in the equations below), donating an electron to $NADP^+$. By accepting two such electrons and two protons, $NADP^+$ is reduced to NADPH leaving chlorophyll deficient of an electron (represented by $chl^+$):

$$\text{chl} + \text{photon} \rightarrow \text{chl}^\star$$

$$2\ \text{chl}^\star + \text{NADP}^+ + \text{H}^+ \rightarrow 2\ \text{chl}^+ + \text{NADPH}$$

The now electron-deficient chlorophyll is powerful enough as an oxidising agent to oxidise water to oxygen:

$$2\ \text{chl}^+ + \text{H}_2\text{O} \rightarrow 2\ \text{chl} + 2\text{H}^+ + \tfrac{1}{2}\text{O}_2$$

Having regained its normal or 'ground' state, the chlorophyll is now ready to absorb another photon.

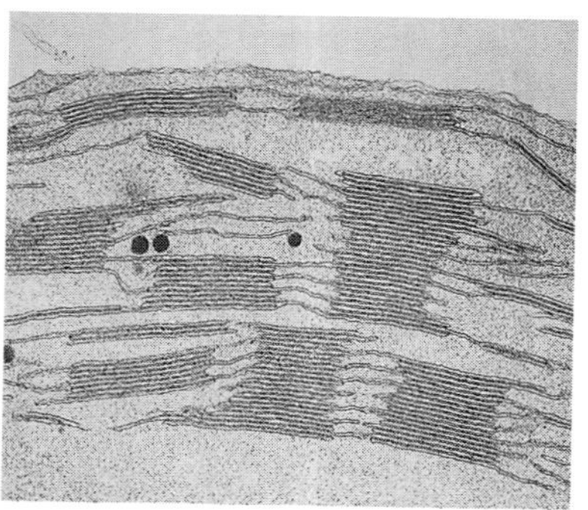

**Figure 33.12**
TEM of the stacks of grana in a chloroplast from a leaf of maize, *Zea mays*

## The fine structure of the chloroplast

As might be expected of an energy-converting organelle, a chloroplast is structurally highly organised. Under the electron microscope it is seen to consists of two outer membranes enclosing a very complex inner one (Figure 33.12). The inner membrane is organised into a large number of flattened sacs called **thylakoids**, arranged in piles called **grana**.

The photosynthetic pigments are entirely located in this complex inner membrane system, and it is here that the photostage occurs. The colourless liquid in between is called the **stroma**, and contains the enzymes catalysing the synthetic stage.

### Two photosystems – the Z-scheme

An interesting feature of the action spectrum for photosynthesis is that with wavelengths longer than about 680 nm, the efficiency of photosynthesis falls markedly, even though chlorophyll still absorbs light up to about 700 nm. In the mid 1950s, Robert Emerson discovered that this sharp drop in efficiency (the 'red drop') could be prevented by giving shorter wavelength light (say 600 nm) at the same time as the longer wavelength light. In fact, a mixture of 600 nm and 700 nm light yields more oxygen than the sum of the yields when the two wavelengths are given separately! Each kind of light must in some way increase the efficiency with which the other is used.

The explanation for this **enhancement effect** was discovered some years later. There are actually *two* photostages in photosynthesis, involving different assemblages of pigment molecules. Both absorb blue light, but their red absorption peaks differ. **Photosystem I (PS I)** absorbs maximally at about 700 nm and is involved in the reduction of $NADP^+$ to NADPH.

**Photosystem II (PS II)** absorbs maximally at about 680 nm and is involved in the photolysis of water. The two systems have a slightly different pigment composition, PS I being richer in chlorophyll *a*.

Enhancement is explained by the fact that the two systems are interdependent, in a manner described as the **'Z scheme'** (shown in the diagram). In PS I, electrons emitted by excited chlorophyll molecules reduce $NADP^+$ to NADPH. The electrons lost from chlorophyll in PS I are replaced by electrons emitted by excited chlorophyll molecules of PS II, which has in turn removed them from water. PS II only weakly absorbs 700 nm light and is too weakly driven to replace electrons emitted by PS I. Conversely, with only shorter wavelength red light, PS I cannot accept electrons from PS II because it is not emitting them itself.

## Two photosystems – the Z-scheme *continued*

As the diagram shows, electrons travel via a number of carriers, each of which is alternately oxidised and reduced as in respiration. A number of the carriers have been omitted for simplicity. The position of each carrier on the vertical scale indicates its ability to donate or accept electrons. The higher up a substance is in the Z scheme, the greater its tendency to donate electrons, and vice versa. Thus NADPH, shown high up, is a moderately strong reducing agent, and oxygen, with a strong tendency to accept electrons, is shown low down.

The 'Z' scheme. In PS I absorption of a photon causes it to lose an electron that is used, via a series of carriers (not shown) to reduce NADP to NADPH. In PS II, absorption of a photon causes P680 to lose an electron, and in doing so it becomes an electron acceptor powerful enough to oxidise water. The electron emitted by P680 is used to replenish the electron lost by P700 in PS II

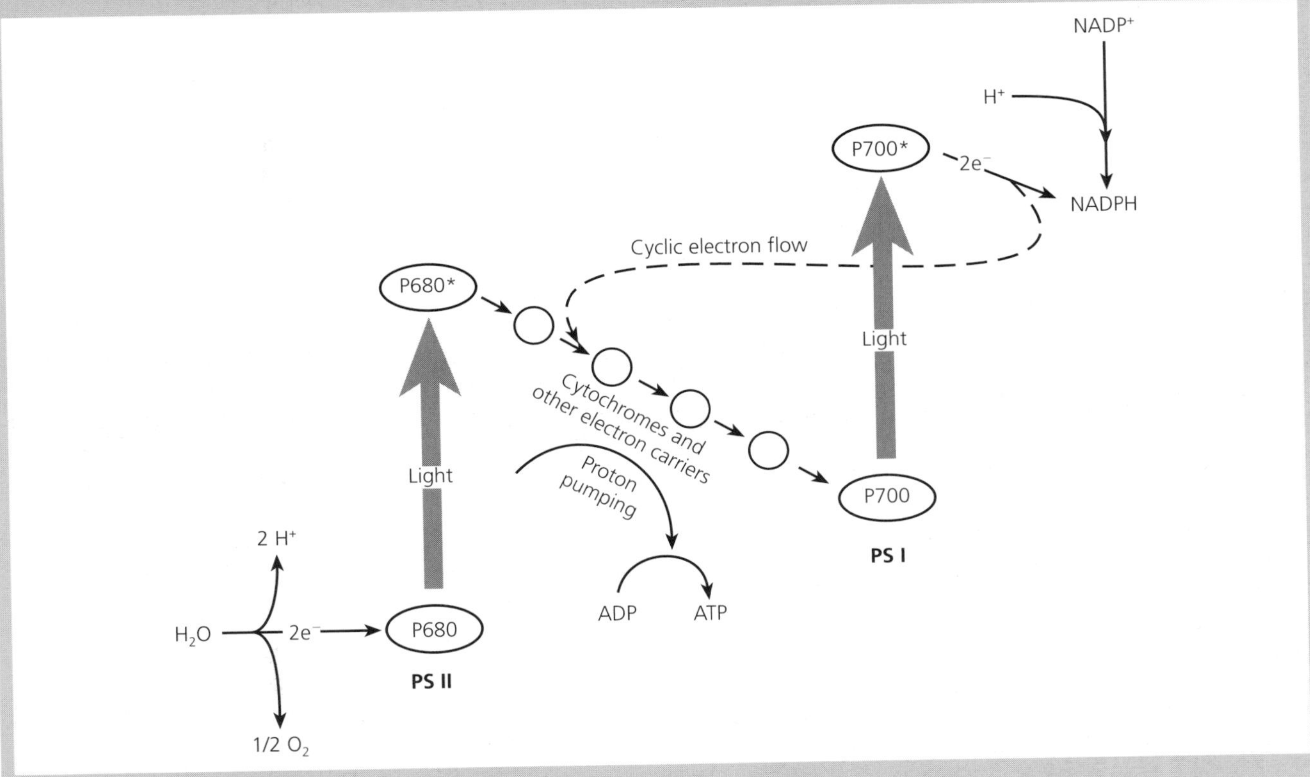


### *Photosynthetic phosphorylation*

NADPH is only one of two energy-rich products of the photostage. The other is ATP, which is made in a process called **photophosphorylation**. This is quite different from the ATP production in respiration which uses energy released as electrons flow from NADH to oxygen. In photophosphorylation, chloroplasts use energy released by electrons that have been excited by light:

$$\text{light} + \text{ADP} + \text{P}_i \rightarrow \text{ATP}$$

The question of how light is used to make ATP is dealt with later in this chapter. For the moment it will be enough to say that the overall result of the photostage is the conversion of light into chemical energy in the form of two energy-rich compounds, NADPH and ATP:

$$\text{light} + 2\text{H}_2\text{O} + 2\text{NADP}^+ + \text{ADP} + \text{P}_i \rightarrow 2\text{NADPH} + 2\text{H}^+ + \text{ATP} + \text{O}_2$$

These are then used in the synthetic reactions to reduce $CO_2$ to carbohydrate.

As Figure 33.13 indicates, the two stages are interlocked and interdependent. For this reason either stage can limit the overall rate of photosynthesis. In dim light the reduction of $CO_2$ is limited by the supply of NADPH. In bright light the supply of NADPH and ATP for the reduction of $CO_2$ are adequate, so an increase in temperature enables the enzymes to work faster. At low temperatures, or if $CO_2$ is in short supply, there is insufficient $NADP^+$ and ADP to use all the available light, so much of the light is wasted.

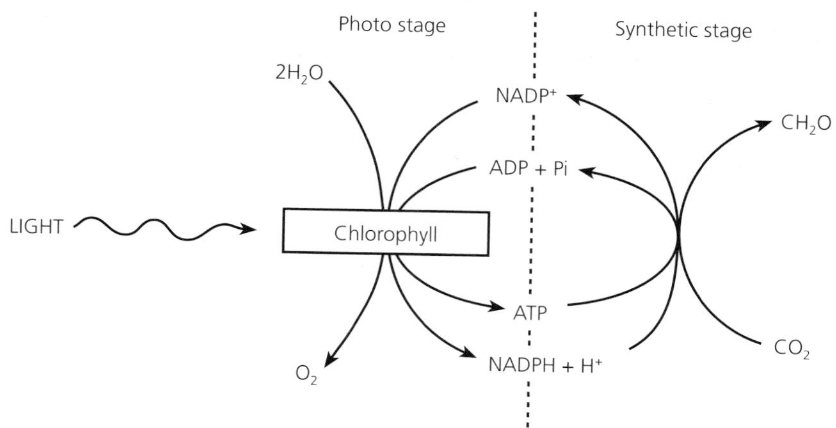


**Figure 33.13**
Interdependence of photostage and synthetic stage of photosynthesis

## How chloroplasts make ATP – photophosphorylation and proton pumping

The reduction of $CO_2$ to carbohydrate requires 'reducing power' in the form of NADPH and ATP. Both are generated in the photoreactions in the grana. ATP is produced by a process called **photophosphorylation**, in which light energy is first used to generate a proton gradient which is then used to generate ATP.

As explained in the information box on page 618, each reaction centre in PS I and PS II acts alternately as an electron donor and an electron acceptor. This happens in two ways:

- in **non-cyclic photophosphorylation**, electrons pass from PS II to PS I via cytochromes and other carriers. PS I donates photoexcited electrons to $NADP^+$, reducing it to NADPH. These are supplied by PS II, which obtains them by removing electrons from water.
- in **cyclic photophosphorylation**, excited chlorophyll molecules in PS I emit electrons that pass through cytochromes and back in a circuitous route to the chlorophyll in PS I.

Whether electrons flow cyclically or non-cyclically, the result is the same. As they flow through the cytochromes, the energy they lose is used to generate ATP.

How does this happen? The mechanism is similar to that by which mitochondria generate ATP using the energy stored in a proton gradient (Chapter 5). There is however an important difference. Whereas in respiration protons are pumped *out* of the mitochondria using energy derived from the oxidation of organic compounds, in chloroplasts protons are pumped *into* the thylakoids using energy derived from light.

As in mitochondria, the mechanism is called **chemiosmosis**. Electrons are passed along a series of carriers, including cytochromes, from a strong reducing agent (electron donor) to a strong oxidising agent (electron acceptor). As explained in the information box on page 618, the strong reducing agent and the strong oxidising agent are both chlorophyll. Chlorophyll that has been excited by light becomes a powerful electron donor, but after it has donated an electron it becomes a strong electron acceptor.

## How chloroplasts make ATP – photophosphorylation and proton pumping *continued*

An essential feature of proton pumping is that one of the carriers **plastoquinone** (PQ in the diagram) transports hydrogen atoms, whilst the others transport only electrons. Plastoquinone accepts two electrons from an electron carrier and two protons from the stroma, forming $PQH_2$. When $PQH_2$ gives up its hydrogen, the electrons are given up to the next electron carrier (a cytochrome) and the protons *are ejected into the thylakoid*. The overall effect is that two protons have been transported from stroma into the thylakoid.

The final stage in the process is the production of ATP. Protons are much more concentrated inside the thylakoids than outside, but the thylakoids are sealed compartments so the protons cannot easily leak out. There are, however, certain channels in the thylakoid membrane which contain an ATP-synthesising enzyme. In a way not yet fully understood, the tendency for protons to leave the thylakoids through these channels is used to make ATP from ADP and inorganic phosphate.

How chloroplasts make ATP: photophosphorylation. Of the several carriers linking PS I and PS II, only plastoquinone (PQ) is shown. Others include cytochromes and other iron-containing proteins. The ATP and NADPH produced are used to drive the synthetic reactions of the Calvin cycle. For simplicity, the relative numbers of electrons, protons and oxygen atoms have not been balanced

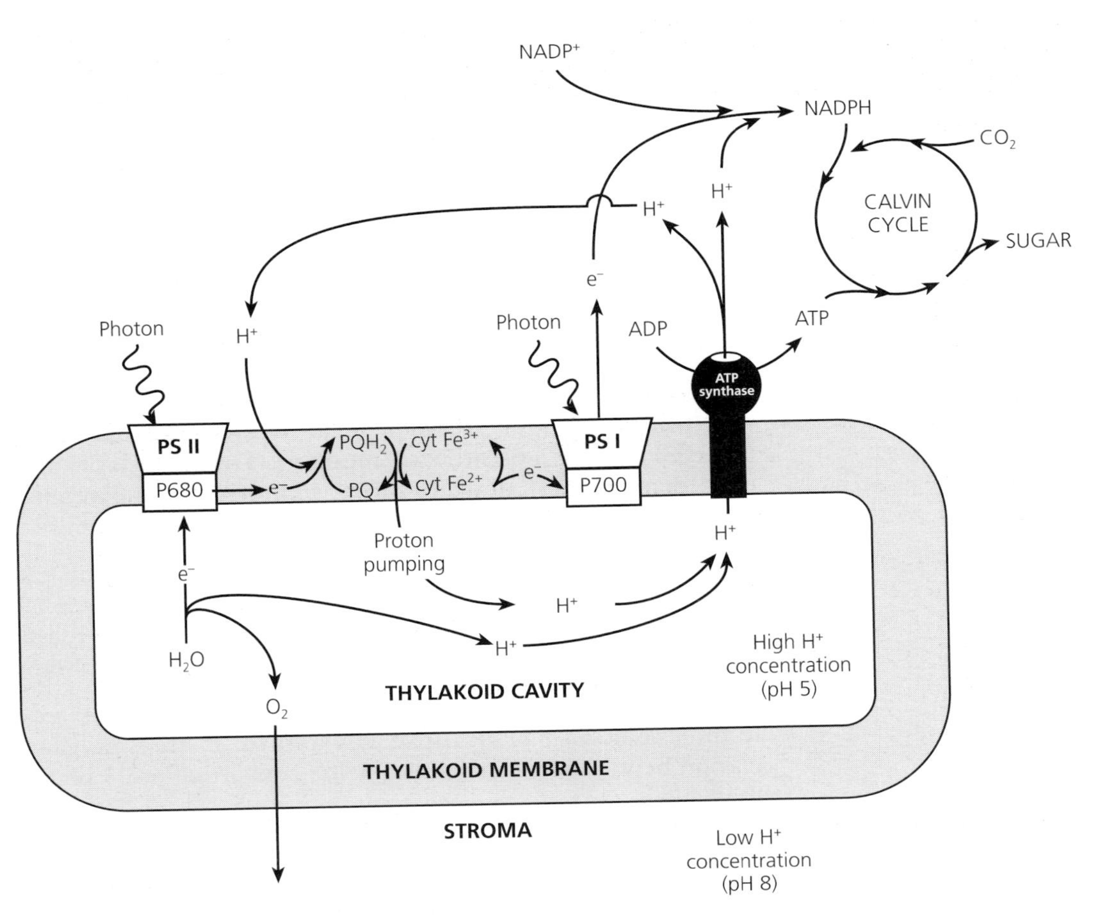

## Herbicides and photosynthesis

About half the commercially used herbicides kill plants by inhibiting particular stages of photosynthesis. **Paraquat**, for instance, interferes with NADPH formation by blocking the flow of electrons from PS I to $NADP^+$, and **DCMU** prevents the flow of electrons through PS II. As with many other pesticides, the benefits have begun to be lost because many weed species have begun to evolve resistance.

## The synthetic stage

In the synthetic stage the energy-rich products of the photostage are used to reduce $CO_2$ to carbohydrate. The chemical pathway by which this occurs is called the **Calvin cycle** after the leader of the team that first elucidated it.

Essentially the cycle consists of the following steps (Figure 33.14):

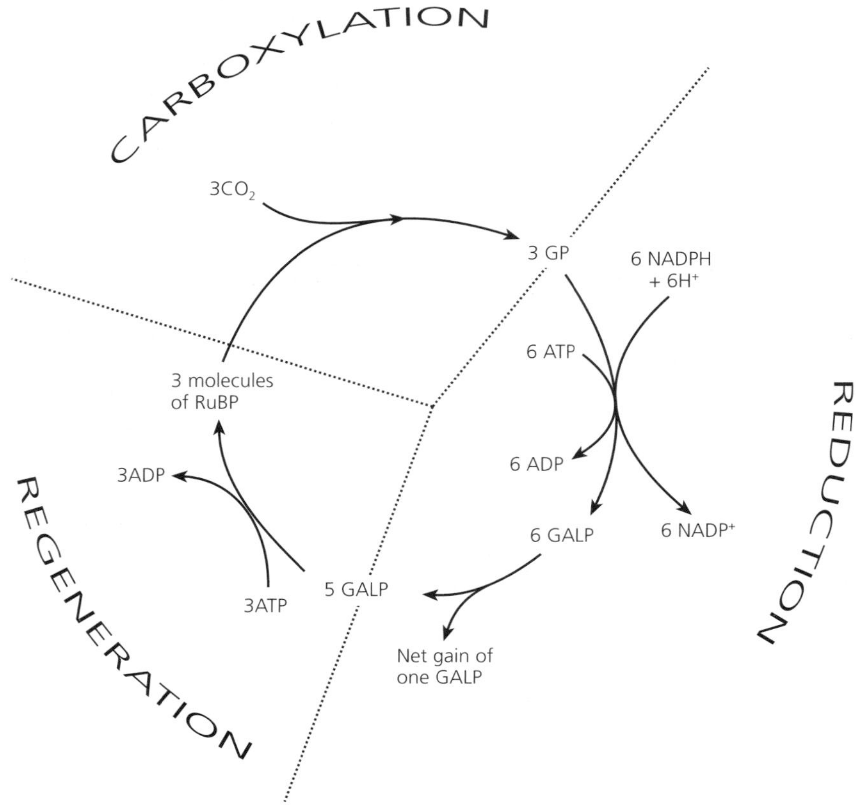


**Figure 33.14**
The Calvin cycle

- $CO_2$ combines with a 5-carbon sugar, **ribulose bisphosphate** (**RuBP**) to form a 6-carbon compound. This is unstable and immediately breaks down to form two molecules of **glycerate-3-phosphate** (**GP**).
- GP is reduced to **glyceraldehyde-3-phosphate** (**GALP**), a triose (3-carbon sugar) phosphate. This is essentially the reverse of the conversion of triose phosphate to GP in glycolysis, and requires NADPH and ATP, both of which are provided by the photostage.
- in a complex series of reactions, five molecules of GALP are converted to three molecules of **ribulose phosphate**. There is no gain nor loss of carbon here, 5 triose molecules being converted into 3 pentose molecules.
- ribulose phosphate is converted to RuBP, using up more ATP. The RuBP is now ready to accept more $CO_2$.

For every three molecules of $CO_2$ that enter the cycle, six molecules of GALP are produced. However since only five of these are used to regenerate the RuBP, there is a net gain of one GALP molecule for every three $CO_2$ molecules that enter the cycle. It is this that is the primary product of photosynthesis, rather than glucose. The cost of producing each triose phosphate is the consumption of six molecules of NADPH and nine molecules of ATP.

Each turn of the cycle also uses up a molecule of water for every three $CO_2$ molecules consumed. However, since the reduction of GP to GALP produces a water molecule, there is no net consumption of water in the synthetic stage.

In bright light, carbohydrate accumulates much faster than it can be exported to other parts of the plant. If it were to accumulate as sugar it would lower the water potential of the cell, causing water to enter the cell by osmosis. This is prevented by temporarily converting the sugar into starch, which is insoluble. At night the starch is converted via glucose into sucrose, which is transported to other parts of the plant. The synthesis of sucrose and starch both require ATP, produced within the chloroplast.

## Tracking the path of carbon

It took Calvin and his colleagues nearly a decade to work out the sequence of intermediate compounds by which $CO_2$ is converted into carbohydrate. There were two major difficulties. The first was to distinguish intermediates on the pathway from the myriad of other compounds already present. The solution came with the discovery of **carbon-14**. This isotope of carbon is chemically identical to the normal isotope, carbon-12, but it can be distinguished by the radiation it emits as it decays to nitrogen-14. It is particularly useful because it has a long half-life (over 5000 years), so its concentration is effectively constant during the course of an experiment.

The second difficulty was that the intermediates are only present in minute quantities – as fast as each is formed, it is converted into the next. The problem of separating and identifying substances in such minute amounts was solved by the use of two-way paper chromatography (see the information box on page 611).

Though radioactive intermediates are not visible on a chromatogram, their positions can be revealed by **autoradiography**. The chromatogram is left between sheets of photographic film for long enough to cause fogging, and the film is then developed. The chemicals can be identified by comparing their $R_f$ values with those of known compounds. The radioactivity of each spot was measured by a Geiger-Müller tube, thus giving a measure of its concentration.

To identify the order of intermediates on the pathway, photosynthesis had to be allowed to occur for very short and precisely-controlled periods, so that only the chemicals directly on the pathway would have had time to become labelled. This meant that Calvin had to be able to start and stop photosynthesis almost instantaneously.

He overcame these problems using rapidly-stirred suspensions of unicellular protoctists, such as *Chlorella*, in the lollipop apparatus shown. At a precise time the light was switched on. The thin shape of the 'lollipop' meant that all cells received similar light intensity, and the minute size of the cells ensured the rapid inward diffusion of $CO_2$. By opening the tap at the bottom after precisely-timed intervals, samples of the suspension could be run off into boiling methanol. This not only instantly killed the cells, it also acted as a solvent for the reacting chemicals which were then separated using chromatography.

By allowing photosynthesis to occur for very short periods (less than 2 seconds), Calvin found that the first intermediate compound is glycerate-3-phosphate (GP). After a few more seconds numerous other compounds became labelled, including glyceraldehyde phosphate (GALP) and fructose bisphosphate. Both were known to be intermediates in glycolysis, and it seemed reasonable to suppose that GALP was produced by reduction of GP – the reverse of the oxidation step in glycolysis.

## Tracking the path of carbon *continued*

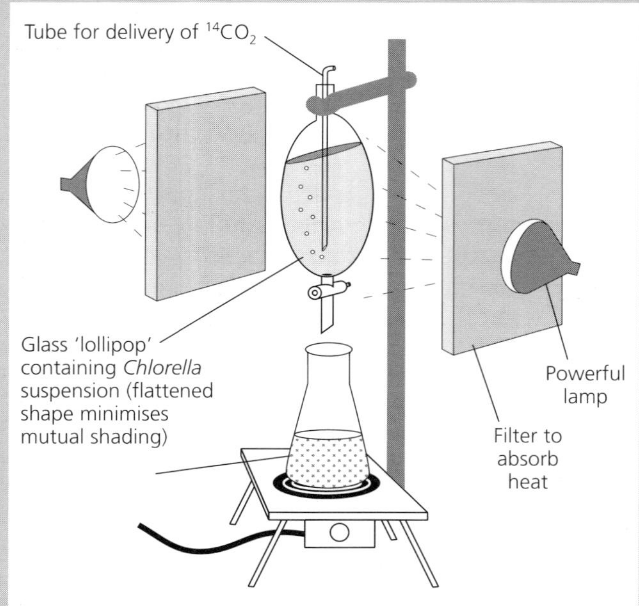


Highly simplified diagram of the 'lollipop' apparatus used by Calvin to track the path of carbon in photosynthesis

The most vexed question was the identity of the $CO_2$ acceptor molecule – the starting point for the process. After less than 2 seconds photosynthesis, only the carboxyl carbon atom of the GP was labelled, suggesting that a 2-carbon acceptor was combining with $CO_2$ to form GP. A five-year search failed to find such a 2-carbon compound.

An important clue was provided by the discovery that after only a few seconds, all three carbon atoms of GP and all six carbon atoms of fructose bisphosphate became radioactive. The inference was that the acceptor was rapidly regenerated from the products of GP reduction.

This suggested two complementary experiments to Calvin's team. If light is required for the reduction of GP, then switching it off should stop the acceptor being regenerated and GP would accumulate until all the acceptor was used up. Conversely, if the supply of $CO_2$ were to be suddenly reduced whilst maintaining illumination, GP should disappear and the acceptor should accumulate.

In both experiments, photosynthesis was first allowed to occur in the presence of labelled $CO_2$ long enough for all intermediates to be fully labelled. The radioactivity of each compound would thus give a measure of its concentration.

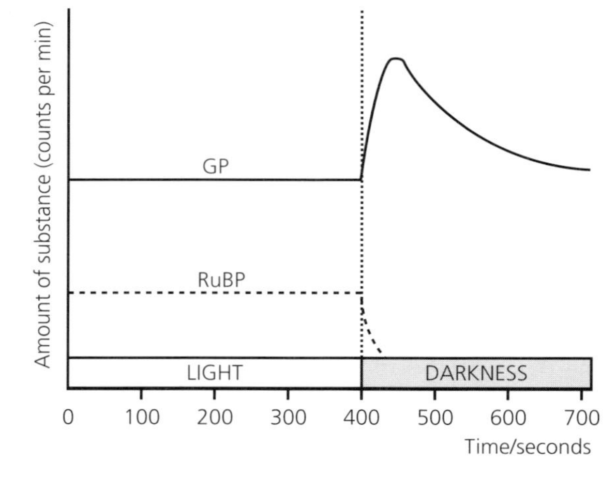


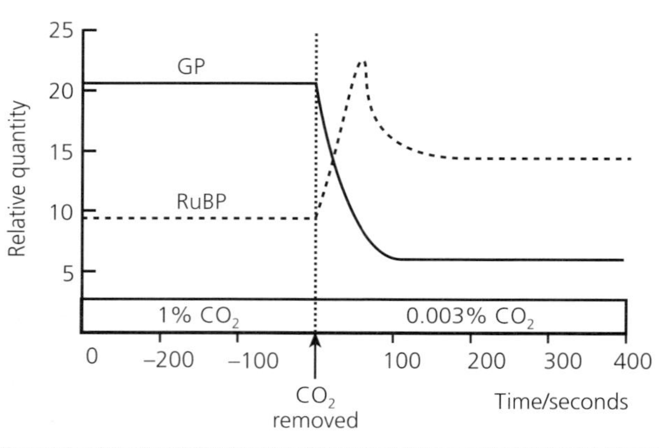


The path of carbon in photosynthesis. Above: The effect of sudden darkness on the concentrations of glycerate-3-phosphate (GP) and ribulose bisphosphate (RuBP). Below: The effect of sudden depletion of $CO_2$ on concentration of GP and RuBP

The results are shown in the diagrams. In the first experiment, switching off the light resulted in a steep rise in GP, showing that light is not needed for its formation but is necessary for its conversion. At the same time the concentration of RuBP fell sharply, suggesting that it was the acceptor molecule. Confirmation of the identity of RuBP as the acceptor was provided by the second experiment, in which sudden depletion of the $CO_2$ supply resulted in a sharp rise in RuBP and a fall in GP.

### Fate of the carbohydrate product

The triose phosphate produced in photosynthesis forms the raw material for the production of glucose, fatty acids, amino acids, nucleotides and a host of other organic compounds. Most photosynthetic product is exported from the leaves as sucrose via the phloem. Triose phosphate is transported from the chloroplasts into the cytosol where it is converted into glucose and fructose phosphates, which are then combined to form sucrose (Figure 33.15).

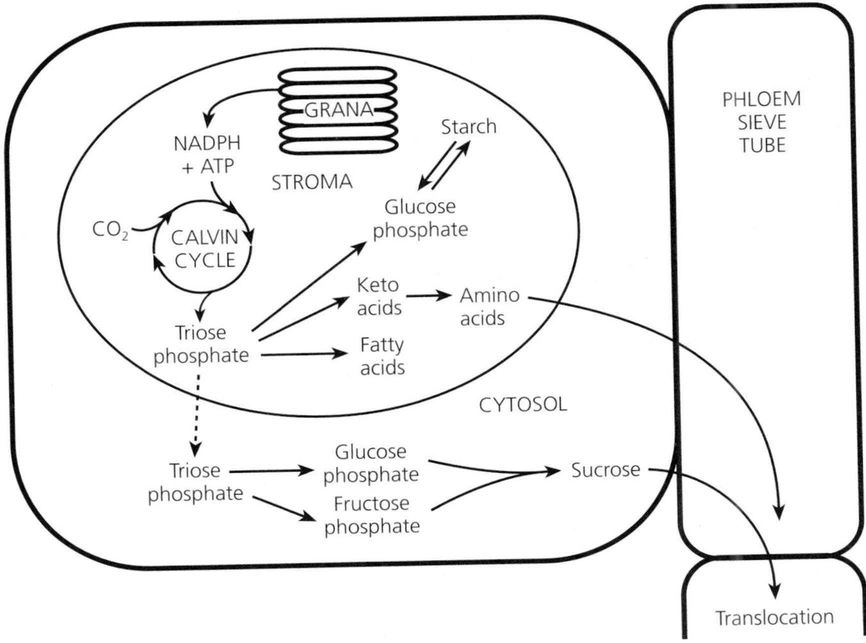


**Figure 33.15**
Fate of the carbohydrate produced in photosynthesis

During rapid photosynthesis, carbohydrate accumulates much faster than it can be exported. Triose phosphate is converted into glucose phosphate and then into starch, which accumulates as grains in the chloroplast stroma. At night, the starch is converted into glucose and fructose phosphates which are then converted into sucrose.

Of the total carbohydrate produced in photosynthesis, about a third is eventually broken down in respiration to supply energy (although much of this respiratory breakdown occurs in other parts of the plant). The remainder (representing the net photosynthetic product) is used to produce all the organic compounds in the plant. Some of these conversions are relatively straightforward, but most require further energy input from ATP formed in respiration. Cellulose and starch are produced by polymerisation of glucose. Amino acids are produced by combining ammonia, produced by reduction of nitrate from the soil, and keto acids produced from carbohydrate. Glycerol is produced from triose phosphate, and fatty acids are produced from oxaloacetate in the Krebs cycle.

### Structure and function in the chloroplast

The reactions of the Calvin cycle occur in the structureless stroma, but the reactions of the photostage are dependent on the highly organised grana. There are two reasons why this molecular organisation is essential to the process:

- since the pigment molecules are anchored in the lipid membrane, they can easily hand on energy from one to another. The same principle holds with the cytochromes and other electron carriers.

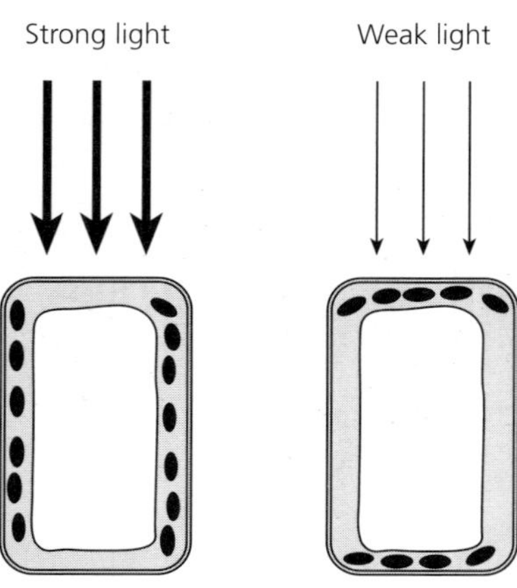


**Figure 33.16**
Changes in chloroplast position in light of different intensities

- as sealed compartments, the grana can maintain the proton gradient needed to make ATP (see the information boxes on pages 618 and 620). The establishment of the proton gradient is also facilitated by the large surface/volume ratio of the thylakoids (resulting from their minute size and flattened shape).

### *Structure and function in the cell*

Adaptation is also apparent at the level of the individual cell in two ways:

- the vacuole displaces the chloroplasts to the periphery of the cell, minimising the distance over which $CO_2$ has to diffuse in solution.
- cells can adjust the position of their chloroplasts to optimise the light intensity. In bright light they are aligned along the direction of the incident light, thus screening each other, whilst in dim light they occupy the sides at right angles to the light (Figure 33.16). The chloroplasts do not move themselves – they are moved along by a mechanism involving microtubules in the surrounding cytoplasm (Chapter 3).

## Structure and function in the leaf

The most obvious adaptation to function is at the level of the whole leaf. Leaves vary somewhat in anatomy, but privet illustrates the features of most dicotyledonous leaves (Figures 33.17 and 33.18). Most photosynthesis occurs in the **mesophyll**, which contains abundant chloroplasts and is honeycombed with air spaces. It is divided into an upper **palisade layer** and lower **spongy layer**. The palisade cells are the main photosynthesisers and each has up to 100 chloroplasts. A dense network of veins, or **vascular bundles**, forms the 'plumbing' or transport system of the leaf. Each vein contains two transport tissues: **xylem** which carries water and minerals to the mesophyll cells, and **phloem** which carries the sugar made in photosynthesis. The fine branches are so extensive that no mesophyll cell is more than a fraction of a millimetre from a vein. The outer layer of the leaf is the **epidermis** which secretes a waxy, water-resistant **cuticle**. The lower epidermis is perforated by large numbers of pores, or **stomata**, each bounded by two **guard cells** that look rather like lips (*stoma* = mouth). By changing their shape the guard cells can adjust the stomatal aperture, thereby regulating water loss.

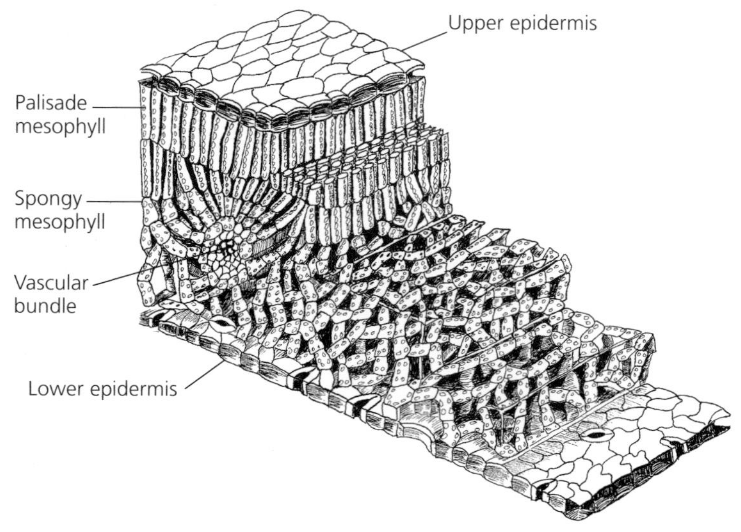


**Figure 33.17**
3D-view of the internal structure of a dicotyledonous leaf (after Eames and MacDaniels)

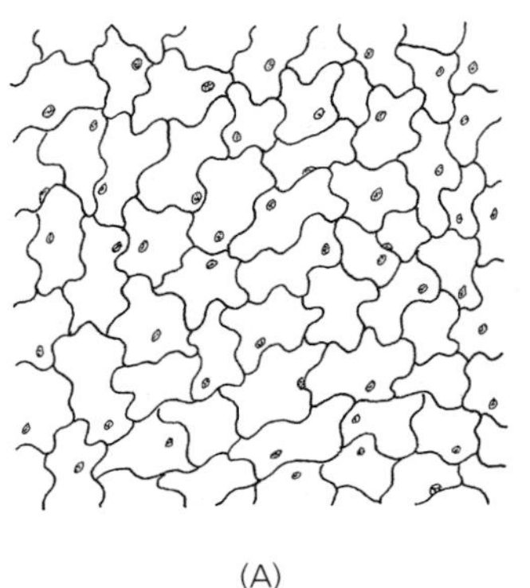

(A)

Interlocking cells enable resistance to tension – important in hydrostatic support

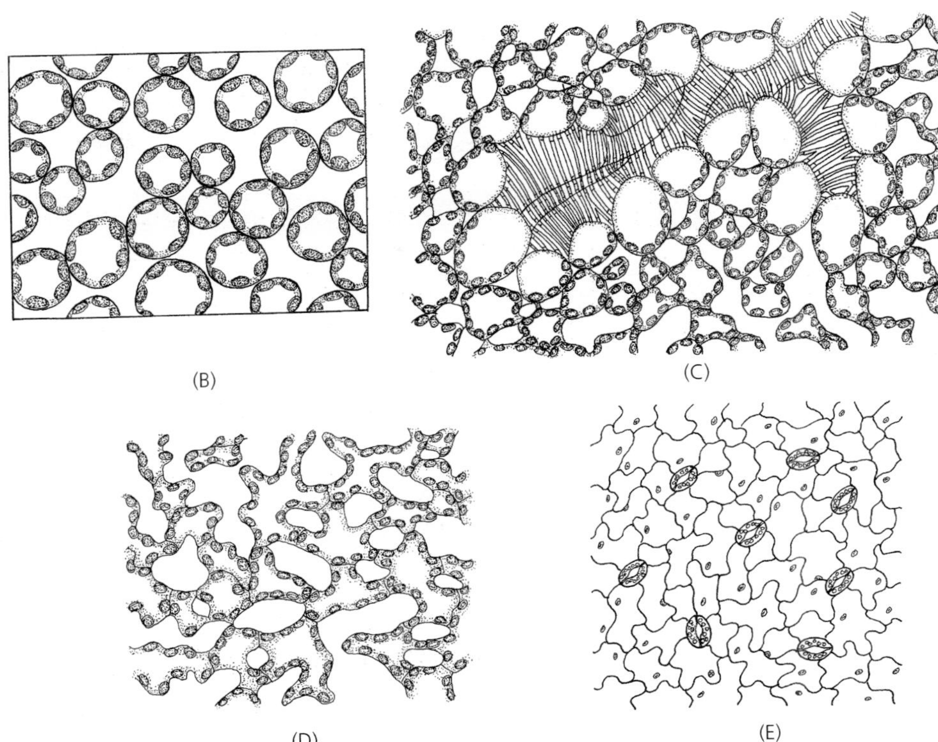

(B)

(C)

(D)

(E)

**Figure 33.18**
Sectional views of a dicotyledonous leaf, cut parallel to the leaf surface at the level of (A) upper epidermis, (B) palisade mesophyll, (C) xylem of a vascular bundle, (D) spongy mesophyll and (E) lower epidermis

Photosynthesis imposes a number of requirements on the leaf. It must be able to:

- absorb light
- absorb carbon dioxide
- import water
- balance water loss and water uptake
- export the photosynthetic products
- support itself and maintain its orientation.

As we shall see, some of these requirements are in conflict, and the structure of the leaf represents a compromise.

## Absorbing carbon dioxide

Normal air contains only about 0.035% (350 parts per million) $CO_2$, or about $\frac{1}{600}$ the concentration of oxygen. Leaves are adapted for absorbing this very dilute gas in a number of ways:

- their thinness gives a large surface area for $CO_2$ uptake compared with the volume of tissue absorbing it. Although the stomata occupy a tiny proportion of the epidermis, $CO_2$ diffuses into a rapidly-photosynthesising leaf about half as fast as it would if the epidermis were completely absent! As explained in the information box on page 628, the substomatal air spaces play an important role in speeding up the diffusion of $CO_2$ through the stomata.

- thinness also results in a short diffusion distance between atmosphere and chloroplasts.

- the network of intercellular air spaces greatly speeds up its diffusion ($CO_2$ diffuses 40 000 times faster through air than in solution). The air spaces even penetrate between the palisade cells, as sections cut parallel to the epidermis show (Figure 33.18C).

- since the chloroplasts are only just beneath the cell wall, $CO_2$ only has to diffuse about 5 μm in solution (Figure 33.18D). Short though this distance is, it offers the same resistance to diffusion as a layer of air 40 000 times as thick, or 40 000 × 0.005 mm = 20 cm!

## Small pores are better

As a thin structure, a leaf has a large surface area compared with its volume. But $CO_2$ enters the leaf through the stomata, which only account for a tiny proportion of the total leaf surface. You might therefore think that the *effective* area for gas exchange is much less than the area of the leaf.

Actually this is not so, for an interesting reason. About the turn of the century, Brown and Escombe discovered that gases diffuse through a large number of small pores in a thin membrane much faster than through a small number of large pores of the same total area. Even though the stomata only occupy about 1 per cent of the leaf surface, gases diffuse through the epidermis *about half as fast as if the epidermis were completely absent*!

The explanation depends on the fact that inside each stoma is a **substomatal air space**, which allows $CO_2$ to diffuse away from the pore towards the edge of an imaginary hemisphere sometimes called a 'diffusion shell'. This *lateral* diffusion away from the pore speeds up the removal of $CO_2$ from just inside the stoma, and thus accelerates diffusion.

The diffusion round the edge of a stoma is proportional to the *circumference* of the pore rather than to its cross-sectional *area*. As a pore gets smaller, its cross-sectional area decreases faster than its circumference (halving its diameter reduces its cross-sectional area to a quarter). A large number of small pores therefore has a greater total circumference than a small number of large pores of the same total cross-sectional area. The fact that a partly-open stoma is slit-shaped gives it an even larger perimeter relative to its area.

Diffusion through small pores. Because of the large air spaces beneath the stomata, $CO_2$ can diffuse sideways towards the edges of an imaginary hemisphere called a diffusion shell. Diffusion shells are imaginary lines of equal concentration and are comparable to the isobars on a weather map or the contours on an ordinary map

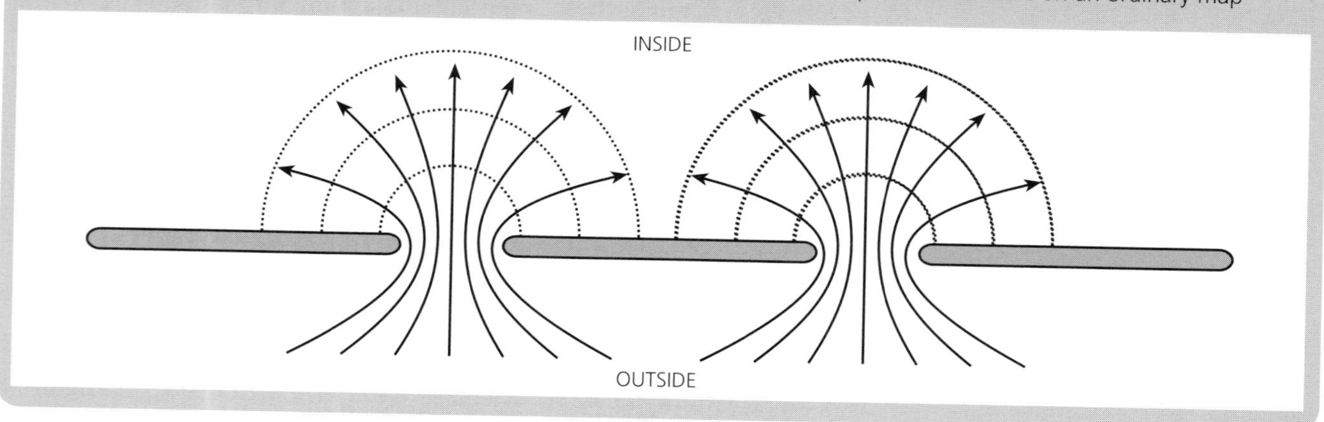


### Trapping light

For maximum light absorption, as little as possible should be scattered by reflection and refraction before it has had time to reach the chloroplasts. Unfortunately $CO_2$ can only be absorbed rapidly if the cells have a large area in contact with air spaces, and these surfaces tend to scatter the light by repeated reflections. In a dicotyledonous leaf (such as privet) this

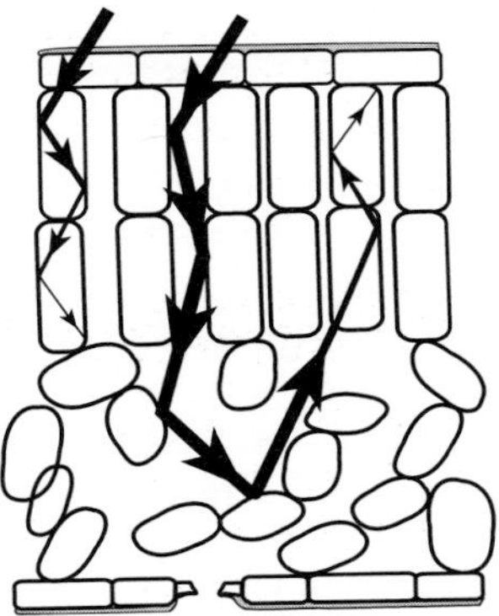

**Figure 33.19**
How mesophyll structure maximises light absorption. Rays passing between the palisade cells are reflected down between them to the spongy mesophyll and back up again, giving them a second chance to be absorbed by the palisade cells. Even light that is not reflected back into the palisade layer undergoes multiple reflections within the spongy mesophyll. By increasing the length of the light path, the probability that a photon will be absorbed is increased

conflict is minimised by the differentiation of the mesophyll into upper palisade and lower spongy layers, in a way to be explained shortly.

Light first strikes the upper epidermis. Some is reflected by the waxy cuticle (especially in glossy leaves) but most passes through the epidermal cells. These lack chloroplasts. Separated from the atmosphere by the cuticle and strongly bonded to each other, they have limited contact with air spaces and so cannot use light effectively.

Light entering the palisade layer can follow two alternative routes. Most enters the palisade cells and is absorbed by the abundant chloroplasts. However, a section through the palisade layer parallel to the epidermis shows that a considerable proportion (about 30%) of the light must pass *between* the palisade cells. The elongated shape and parallel arrangement of the palisade cells ensures that this light reaches the spongy mesophyll by repeated glancing reflections (Figure 33.19). Unlike the palisade cells whose reflecting surfaces run in the same direction, the surfaces of the spongy cells are orientated in every plane, scattering the light in all directions. Some is absorbed by the spongy mesophyll cells, and the rest is either reflected back up into the palisade layer or is scattered downwards out of the leaf. Because some light is scattered upwards by the spongy layer, light passing between the palisade cells gets a 'second chance' to be absorbed.

The organisation of the mesophyll explains why many leaves look paler from below. If the air in a leaf is forcibly replaced by water (which has a refractive index much closer to that of cell walls than air), light-scattering is greatly reduced and the leaf looks equally dark from both sides.

## Sun and shade leaves

**Except for the tallest trees, most woodland plants have to 'make do' with light that has already passed through at least some leaves higher up. This light is not only depleted of blue and red, but, since it has been scattered by the foliage above, it is also diffuse, striking lower leaves at many different angles.**

**Leaves of plants adapted to shade have a number of distinguishing features and are called shade leaves, in contrast to sun leaves that are adapted to high light intensities. Shade leaves differ from sun leaves in a number of ways, both structurally and physiologically.**

**Shade-adapted dicotyledon leaves are generally thinner than sun-adapted leaves, largely because of the poorly-developed palisade tissue. As explained earlier, the shape and orientation of palisade cells increases absorption of the more or less parallel rays of direct sunlight. In diffuse light (in which rays are travelling in all directions) there is no advantage in cells being elongated. Another feature of shade leaves is that stomatal density is usually lower, a fact that can be correlated with the lower rate of uptake of $CO_2$.**

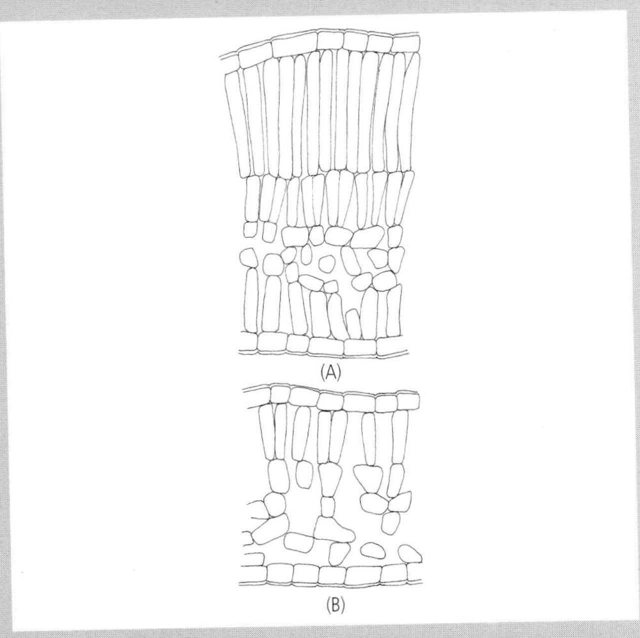


Anatomy of sun and shade leaves of beech. Sun leaf (A) and shade leaf (B)

## Sun and shade leaves *continued*

Physiological differences are more subtle. To be a net exporter of carbohydrate, a leaf must 'earn' more in photosynthesis in the daytime than it 'spends' in respiration over an entire 24-hour period. Shade leaves have low light compensation points, and are able to make a 'profit' on a lower 'income' than sun leaves can.

There are two reasons for this. First, shade plants have lower respiration rates than sun plants. Second, in dim light the rate of photosynthesis rises more steeply with light intensity than it does in sun plants, showing that they use a greater proportion of the light. On the other hand shade leaves cannot use bright light efficiently, becoming light-saturated at much lower light intensities than sun leaves do.

Why can shade leaves utilise a higher proportion of dim light than sun leaves, yet waste much more light in high intensities? The explanation lies in the differing emphasis placed on the production of energy-harvesting and energy-processing molecules. Above a certain light intensity, each reaction centre is accepting quanta of energy as fast as the associated electron carriers can handle them. Shade leaves produce a higher proportion of the energy-harvesting chlorophyll molecules but fewer of the cytochrome and other electron-transporting molecules.

The distinction between sun and shade leaves is not sharp – there is a continuous range of intermediates. Moreover, they can occur in different parts of the same plant. In trees, very young leaves can adapt to a range of light intensities – the leaves at the top of the canopy develop into sun leaves, whilst those near the bottom grow into shade leaves.

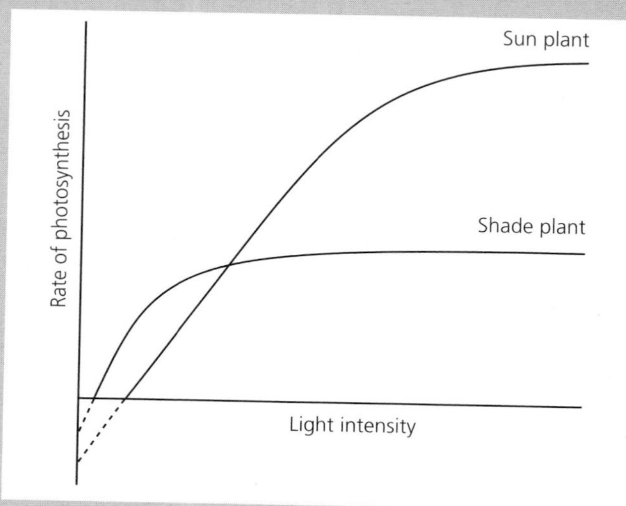


Effect of light intensity on rate of photosynthesis in shade and sun leaves

### *Growth responses in leaves*

By differential growth of the petiole, a young dicotyledonous leaf can maximise the area it exposes to the light by orienting itself at right angles to the prevailing light direction. It can also adjust its position with respect to other leaves so as to minimise mutual shading. These growth responses are examples of tropisms (Chapter 32).

### *How a leaf supports itself*

To be an effective light absorber a leaf has to be stiff enough to hold itself flat, and (as closely as possible) facing the light. Leaves are stiffened by a combination of hydrostatic support and specialised supporting tissues.

As explained in Chapter 6, a plant cell that is well supplied with water is turgid, like a football. The same is true of the leaf as a whole. Figure 33.18E shows a section through the spongy mesophyll of a leaf, cut *parallel* to the epidermis. It forms a network of struts which resist compression, the spaces between them providing ample space for diffusion of $CO_2$.

Counteracting the compression of the spongy mesophyll is the epidermis, the cells of which are strongly bonded together and so resist tension. A leaf can thus be compared to a balloon – the stretched epidermis being the equivalent of the rubber skin, and the spongy mesophyll acting like the compressed air. Just as a balloon sags when the air pressure is reduced, a leaf wilts when the spongy cells lose turgor.

Leaves also obtain support from specialised tissues such as sclerenchyma and collenchyma, which are abundant in the midrib and vascular bundles and form a scaffolding. Leaves of plants living in dry environments usually have larger amounts of these tissues and are less reliant on turgor for support.

### *Balancing water loss and carbon dioxide uptake*

The very features that promote the absorption of $CO_2$ also favour the loss of water by evaporation, a process called **transpiration**. Transpiration is thus the 'price' the plant has to pay for photosynthesis.

Provided that water can be replaced as fast as it is lost, this is no disadvantage. More often the leaf has to compromise between these two requirements, an important topic that is dealt with in Chapter 34.

### *Light absorption vs carbon dioxide uptake*

Besides conflicting with water conservation, the requirements for $CO_2$ uptake are also (to some extent) in conflict with that for light absorption. Rapid $CO_2$ uptake requires an extensive area of contact between the cell surfaces and air; but these surfaces also scatter light by reflection and refraction.

The division of the mesophyll into two layers provides a compromise between these two conflicting requirements. The structure of the palisade layer minimises the amount of light scattered back out of the leaf, whilst the spongy layer maximises scattering, reflecting much of the unabsorbed light back up into the palisade layer.

## The leaf as an organ

The external requirements for photosynthesis are light, $CO_2$, water and the mineral ions needed to manufacture chlorophyll and other components of the photosynthetic system. In many aquatic protoctists these can all be obtained by the same cell. A leaf of a land plant consists of many kinds of cell, only some of which can photosynthesise.

Why do leaves need so many kinds of cell if a protoctist, such as *Chlorella*, can do it in only one cell? In a number of ways photosynthesis is more 'difficult' on land. As explained earlier, to keep the palisade cells turgid and to maintain their supply of $CO_2$ and light requires the co-operative action of other tissues, most of which contribute only indirectly to photosynthesis.

Co-operation between tissues means that a leaf is more than the sum of its parts, because it can carry out a function that none of its individual tissues is able to by itself. As such it is an excellent example of a plant **organ**.

### *Adaptation in the shoot*

At a higher level of organisation than a leaf is the shoot system. This is adapted in two ways to minimise mutual shading of leaves. First, the points on the stem from which the leaves arise – the **nodes** – are usually widely separated. Second, young leaves can make phototropic adjustments to their positions so that there is minimal shading of other leaves. The result is an almost complete canopy, called a **leaf mosaic** (Figure 33.20).

**Figure 33.20**
A leaf mosaic

## Environmental factors affecting photosynthesis

Photosynthesis is affected by a variety of environmental factors, namely light intensity and wavelength, $CO_2$ concentration, temperature, water availability and mineral ions.

### *Light intensity*

Figure 33.21 shows the effect of light intensity on the rate of net photosynthesis. In dim light the graph is a straight line, so the rate is *directly proportional* to light intensity – doubling the light intensity doubles the rate of photosynthesis. At these intensities, light limits the rate of photosynthesis because it is in shorter supply than any other factor. Increasing the temperature or $CO_2$ concentration in dim light therefore makes little difference.

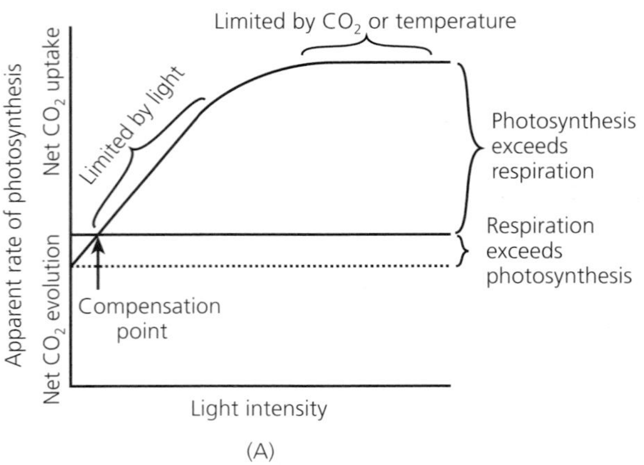


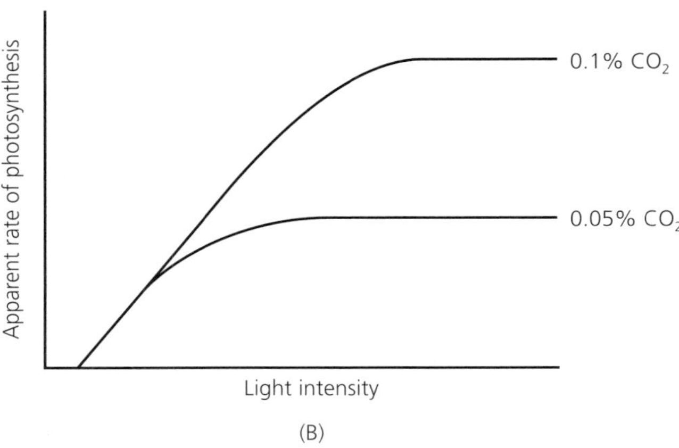


**Figure 33.21**
Effect of light intensity on the rate of photosynthesis, other conditions being held constant (A); effect of light intensity at two different $CO_2$ concentrations (B)

The situation changes in bright light – above a certain light intensity, brighter light makes no difference to photosynthetic rate and the plant is said to be **light-saturated**. This is because some other factor is in short supply, such as $CO_2$, or the temperature is too low. This is shown by the effect of raising the $CO_2$ concentration; light-saturation is reached at a higher intensity.

Since more of the light is wasted in very bright light, photosynthesis is actually less *efficient* under these conditions than in dim light. In most plants adapted to temperate climates, light saturation occurs at about half the intensity of full sunlight.

These effects can be explained in terms of the chemistry of photosynthesis. In dim light, the Calvin cycle is limited by a shortage of NADPH and ATP. Therefore GP, ADP and NADP tend to accumulate. In bright light, the supply of ATP and NADPH are adequate. RuBP is therefore regenerated quickly so there is more than enough to combine with the $CO_2$ available. RuBP molecules are thus 'kept waiting' for $CO_2$ to diffuse into the chloroplast. As a result GP-formation is held up and ATP and NADPH accumulate. This in turn produces a shortage of NADP and ADP, which means that much of the energy absorbed by chlorophyll has 'nowhere to go' and is wasted, being re-emitted as heat and weak fluorescence.

Notice that the line in Figure 33.21 does not pass through the origin – it cuts the horizontal axis at the **light compensation point** (as distinct from the **carbon dioxide compensation point**, which is beyond the scope of this book). This is the intensity at which photosynthesis just balances respiration, so there is no net gas exchange and the plant is 'marking time'. For most plants grown in sunny conditions, the light compensation point is about 2% of full sunlight.

The light compensation point sets a lower limit to the light intensity at which a plant can gain organic matter. However, the plant must be able to produce more organic matter in the daylight hours than it uses during the entire day–night cycle, because of respiratory loss at night.

When comparing photosynthetic and respiratory rates we need to remember that respiration occurs throughout the plant, but photosynthesis occurs mainly in the leaves and, to a much lesser extent, in green stems. Thus although a leaf may be photosynthesising 20 times faster than it is respiring, for the plant as a whole photosynthesis may only be, say, five times faster. Also, respiration continues round the clock, so on a per-24 hour basis, photosynthesis may only be 2–3 times as fast.

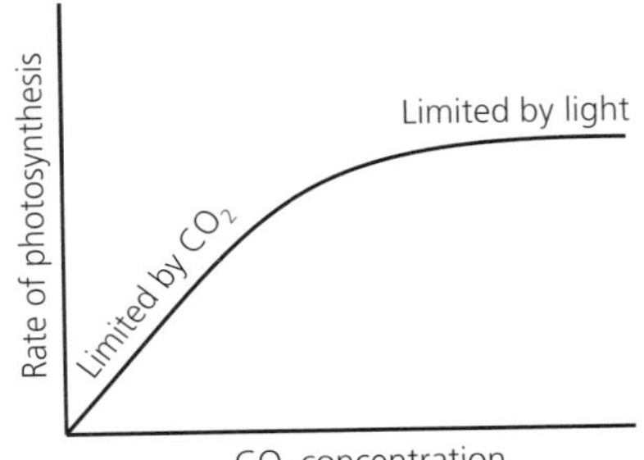


**Figure 33.22**
Effect of $CO_2$ concentration on the rate of photosynthesis

### *Light quality*

For a plant living on the woodland floor, the light is not only dimmer but poorer in quality than full sunlight, as most of the blue and red light has been absorbed by the canopy of leaves above. The effect of wavelength on photosynthesis has already been dealt with (Figure 33.9).

### *Carbon dioxide concentration*

The effect is similar to that of light intensity (Figure 33.22). In the conditions encountered on a warm summer's day, $CO_2$ is rate-limiting, so photosynthetic rate is proportional to $CO_2$ concentration. At higher concentrations some other factor (such as light intensity) becomes rate-limiting.

### *Temperature*

The effect of temperature has already been mentioned (Figure 33.10), but its interaction with light can be more fully explained here. In bright light and with adequate $CO_2$ there is no shortage of RuBP, NADPH and ATP. Under these conditions the enzymes catalysing the Calvin cycle are very responsive to a rise in temperature, though at higher temperatures photosynthesis is slower because enzymes are being denatured. In dim light temperature has little effect on enzyme activity because there is a shortage of ATP and NADPH.

### *Water supply*

In many parts of the world water supply is *the* factor that limits plant growth. The effect however is indirect. Water shortage provokes stomatal closure, which reduces $CO_2$ uptake.

## High efficiency photosynthesis

Most plants living in temperate climates convert about 2% of the incident light energy into chemical energy. At higher temperatures efficiency falls away sharply because of the increasing effects of a light-activated process called **photorespiration.** This resembles mitochondrial respiration only in so far as it uses oxygen and produces $CO_2$ – it generates no ATP, and so is an energy drain to the plant.

Photorespiration results from a peculiarity of **ribulose bisphosphate carboxylase** (RuBP carboxylase) or **'rubisco'** as it is also called. This is the enzyme catalysing the fixation of $CO_2$. It is unusual in that it can also act as an *oxygenase*, catalysing the oxidation of ribulose bisphosphate. One of the end products of photorespiration is $CO_2$, which thus represents a loss of carbon to the plant.

## High efficiency photosynthesis *continued*

These two alternative substrates – oxygen and $CO_2$ – compete for the active site of rubisco. At low temperatures and in the presence of adequate $CO_2$, carboxylase activity predominates and photorespiration is insignificant. But rubisco's oxygenase activity increases more rapidly with temperature than its carboxylase activity, so at higher temperatures oxygen competes more strongly, especially at lower than normal $CO_2$ concentrations. In hot conditions up to 40–50% of the photosynthetic product may be lost in photorespiration.

These limitations of rubisco are overcome in some tropical plants such as maize and sugar cane. They are called **C4 plants** because the first intermediate compound on the carbon pathway is a 4-carbon compound, in contrast to the more common **C3 plants** in which the first product is a 3-carbon compound.

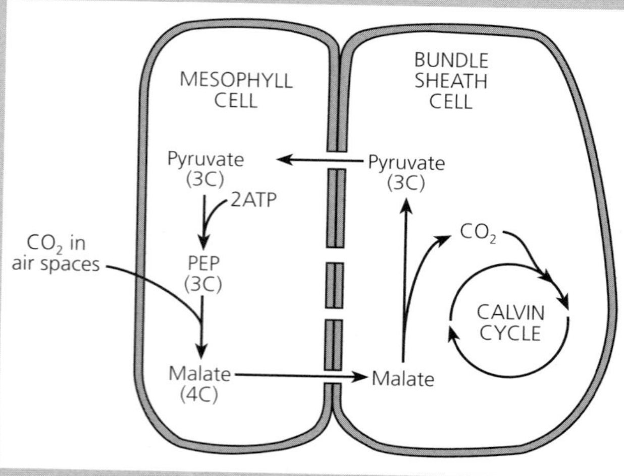


How $CO_2$ is pumped into the bundle sheath cells in C4 plants

In C4 plants, the fixation of $CO_2$ and its subsequent reduction occur in different tissues. Carbon fixation occurs in the mesophyll cells. Here $CO_2$ combines with a substance called **phosphoenolpyruvate (PEP)** to form the 4-carbon compound oxaloacetate, which is then reduced to malate (or in some species, aspartate). Malate is then transported to the cells of the **bundle sheath**. In C4 plants, these cells are large and close-fitting and contain many chloroplasts. In the bundle sheath cells malate is decarboxylated to pyruvate and $CO_2$. The $CO_2$ enters the Calvin cycle as in C3 plants, and the pyruvate is transported back to the mesophyll cells where it is converted back to PEP. The overall result is that $CO_2$ is transported from mesophyll to bundle sheath, where its concentration is raised high enough to competitively exclude oxygen and thus prevent photorespiration.

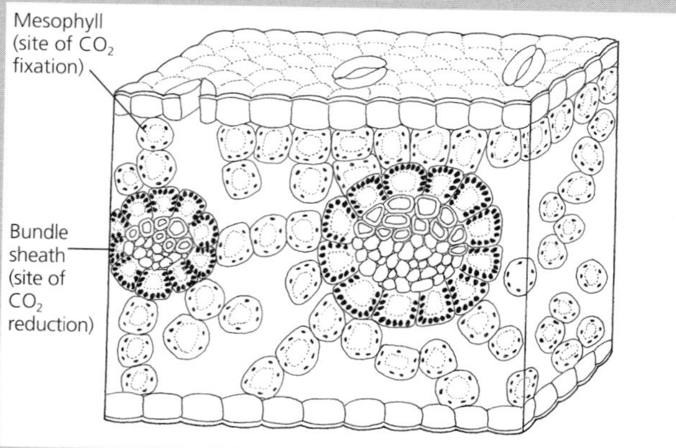


C4 leaf anatomy (after Taiz and Zeiger)

Besides preventing photorespiration, raising the concentration of $CO_2$ in the cells carrying out the Calvin cycle enables higher light intensities to be utilised and, unlike C3 plants, C4 plants do not become light-saturated even in full sun.

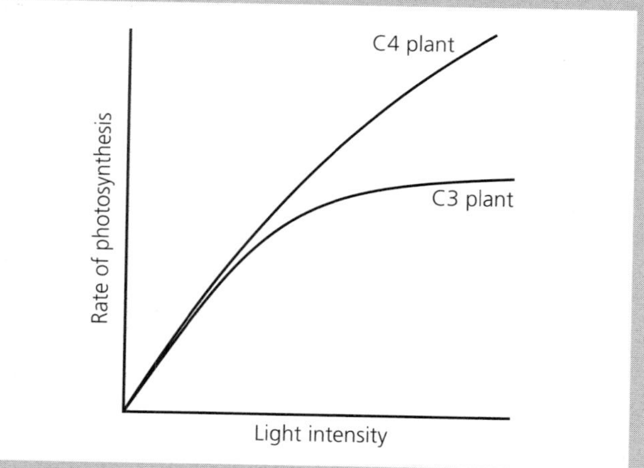


Effect of light intensity on photosynthesis in C3 and C4 plants. C4 plants do not show light-saturation even in full sun

## High efficiency photosynthesis *continued*

The transport of $CO_2$ from mesophyll to bundle sheath is so effective that C4 plants can continue to take up $CO_2$ from the mesophyll even when its concentration falls extremely low. This enables C4 plants to use water more efficiently than C3 plants. In hot, dry conditions plants reduce transpiration by partially closing their stomata, thus reducing the supply of $CO_2$. Below a certain stomatal aperture a C3 plant cannot achieve net photosynthesis because it cannot obtain enough $CO_2$. Because of their ability to scavenge $CO_2$ from the mesophyll, a C4 plant can still take up $CO_2$ when its stomata are almost closed.

If C4 plants are so much more efficient photosynthesisers than C3 plants, why has natural selection not eliminated C3 plants? The answer is that C4 photosynthesis carries a price tag; the regeneration of each PEP molecule uses up 2 ATP molecules. In hot dry conditions, the benefits of the C4 pathway exceed the losses, but in temperate conditions the reverse is true.

In C4 plants the fixation and reduction of $CO_2$ occur simultaneously but in different tissues. In many succulents, this separation occurs in *time*. The plants close their stomata in the daytime and open them at night when water loss is much slower. The $CO_2$ taken in at night is used to convert oxaloacetate into malate. In the daytime malate is decarboxylated, releasing $CO_2$ which is then used in the Calvin cycle.

This kind of photosynthesis was first discovered in plants of the family Crassulaceae (stonecrop family), and was called **crassulacean acid metabolism (CAM)**. Though CAM photosynthesis is now known to be much more widespread, the term 'CAM photosynthesis' is still used.

# Mineral nutrition of plants

Carbon dioxide provides the plant with a source of carbon and oxygen, and water is a source of hydrogen, but many other elements are also required. These are obtained from the soil in the form of dissolved mineral ions.

Long before scientists discovered the reason, farmers knew that if crops were harvested year after year without adding manures – such as dung – to the soil, the soil gradually lost its fertility. With advances in chemistry in the 19th Century, it was realised that in addition to $CO_2$ and water, plants need a number of mineral ions which they obtain from the soil.

Which particular minerals do plants require? Scientists tackled this problem by growing plants in different **culture solutions**, each lacking a different mineral. Without certain minerals, **deficiency symptoms** slowly develop. These are usually sufficiently characteristic for an expert to identify the element that is deficient.

## Macronutrients

Using culture solutions, 14 elements have been shown to be necessary, in addition to carbon, hydrogen and oxygen. The first essential elements to be identified were those needed in the greatest amounts, and are called **macronutrients**.

### *Nitrogen*

Nitrogen is a constituent of amino acids, nucleotides and their respective polymers – proteins, DNA and RNA. It is absorbed as nitrate ($NO_3^-$) and ammonium ($NH_4^+$). Nitrate is reduced to nitrite and then to ammonia, using either NADH produced in mitochondria or NADPH produced in chloroplasts:

$$NO_3^- \xrightarrow{\textit{nitrate reductase}} NO_2^- \xrightarrow{\textit{nitrite reductase}} NH_3$$

Ammonia then combines with the keto acid α-ketoglutaric acid to form the amino acid glutamic acid. This then donates its amino group to various other keto acids in transamination reactions to form the corresponding amino acid. For example, it can donate its amino group to oxaloacetic acid (a keto acid) to form aspartic acid (an amino acid):

glutamic acid + oxaloacetic acid → α-ketoglutaric acid + aspartic acid

### Phosphorus

Phosphorus is absorbed as phosphate ions, $H_2PO_4^-$. As a constituent of $NAD^+$, $NADP^+$ and sugar phosphates it plays a key role in energy metabolism. As a constituent of nucleotides it is an essential component of DNA and RNA.

### Sulphur

Sulphur is absorbed as sulphate ions, $SO_4^{2-}$. As a constituent of two amino acids (cysteine and methionine) it is essential for protein synthesis.

### Potassium

Like all metals it is absorbed as the ion, $K^+$. As an activator for many enzymes, it plays an essential role in metabolism.

### Magnesium

Magnesium is a constituent of chlorophyll, and is a cofactor for some enzymes. Deficiency causes **chlorosis**, in which leaves are abnormally pale.

### Calcium

Calcium is important in regulating membrane permeability, and is a constituent of the middle lamella that binds cells together.

## Micronutrients

As techniques for removing impurities from culture solutions improved, the list of essential elements steadily grew. Some – such as iron, manganese, boron, copper, molybdenum, cobalt, zinc and chlorine – are needed in concentrations measured in parts per *billion*, and are called **micronutrients**. Molybdenum is needed in the smallest amounts – one atom for every 60 million atoms of hydrogen! Such minute concentrations suggest that micronutrients are necessary for the action of enzymes. Molybdenum is a component of nitrate reductase, which catalyses the reduction of nitrate to nitrite. Another that is worth an individual mention is iron, which is a constituent of some of the electron carriers in respiration and photosynthesis.

Some micronutrients are only needed by particular plants. Silicon, for example, is needed by grasses, in which silica is a constituent of cell walls. Sodium is needed for plants with the C4 photosynthetic pathway (see the information box on page 634).

## Roots as absorbing surfaces

In a number of ways, it is more difficult for plants to absorb minerals from the soil than it is to absorb $CO_2$ from the air.

- Some minerals are even more dilute in the soil than $CO_2$ is in the atmosphere – phosphate concentrations, for example, are usually only a few parts per million.
- Minerals move through the soil with extreme slowness. While there are always air currents to bring $CO_2$ to the leaf, the only movement of soil water is by very slow capillary flow brought about by the uptake of water by the roots. Moreover, minerals in solution diffuse far more slowly than $CO_2$ diffuses in air. Not surprisingly, roots soon exhaust the minerals close by and have to grow continuously to tap new soil.

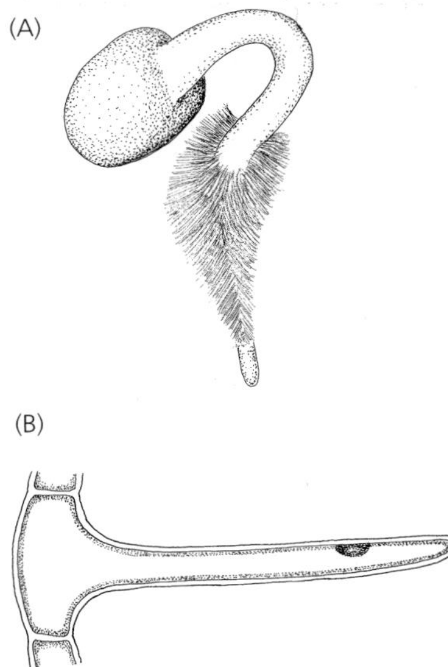


**Figure 33.23**
A sunflower seedling showing root hairs (A); a single root hair (B)

To absorb minerals from very low concentrations in the soil, roots must have a huge surface area – many times that of the shoot system. The root system of a rye plant was estimated to have a total length of 622 km and an area of about 640 $m^2$. Compare this with the 375 $m^2$ through which the average (and much larger) human body absorbs materials (lungs 125 $m^2$ and intestine 250 $m^2$). The root system of a 50 gram rye plant has a surface/mass ratio over 2000 times greater than that of a 70 kg human.

How can roots have such huge surfaces? If you grow seedlings on moist filter paper in a covered Petri dish, you will see the answer. Just behind the tip, each root is covered with thousands of downy **root hairs**, each of which is actually an outgrowth of an epidermal cell. They are so delicate that they are almost always rubbed off when you dig a plant up, so you don't normally see them (Figure 33.23). Even though root hairs occupy only a very small proportion of the length of a root system, they account for about $\frac{2}{3}$ of its total area.

### *How do roots absorb mineral ions?*

Minerals exist in the soil as charged particles, or **ions**, to which the membranes of the root hair cells are almost impermeable. Most of the minerals essential for plant growth are more concentrated in the root cells than in the soil. For both these reasons, uptake of minerals cannot be explained entirely by diffusion.

A number of observations suggest that mineral uptake involves **active transport**, using energy supplied by respiration. For example, it is halted by cyanide or by lack of oxygen. It is also very sensitive to temperature, suggesting a chemical process rather than a physical one such as diffusion (which is only slightly temperature-sensitive).

Ion uptake by roots is complicated by the fact that (as in all cells) the cells of the outer part of the root have a **membrane potential**, the outside being about 100 mV positive to the inside. This means that metal ions, and ammonium ions ($NH_4^+$), are electrically attracted into the cell. The movement of ions into the root is thus subject to two influences:

- a higher concentration inside the root, which tends to cause them to diffuse out;
- the membrane potential, which tends to cause cations to leak into the root and anions to leak out.

For anions, the situation is straightforward. Phosphate, nitrate and sulphate ions have to enter cells with negatively-charged interiors, and thus are electrically repelled. Uptake of anions is therefore against both concentration and electrical gradients and *must* be actively transported.

For cations, the situation is more complicated. Although they are more concentrated inside, they are attracted into the negatively-charged interior of the root cells (Figure 33.24). This electrical force is actually greater than the concentration effect so they diffuse passively into the root, *even though they are more concentrated inside*. Entry of these ions is by facilitated diffusion, either by channel proteins or by carrier proteins (Chapter 6).

Although the entry of cations is strictly passive, it is ultimately dependent on active processes. The generation of the membrane potential results from the pumping out of positively-charged hydrogen ions, the energy for which comes from respiration. Indirectly, then, the uptake of cations depends on respiration just as much as the uptake of anions does.

## *Help from fungi*

Many plants (especially woodland trees) rely on a mutualistic relationship with a fungus to absorb mineral ions, especially phosphate. This relationship is called a **mycorrhiza**, and is dealt with in Chapter 35.

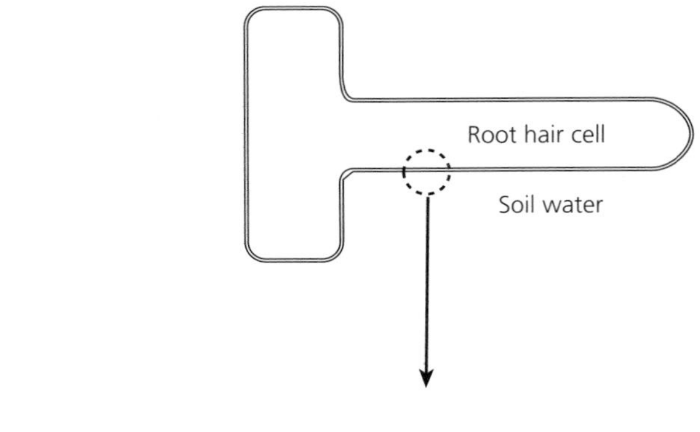


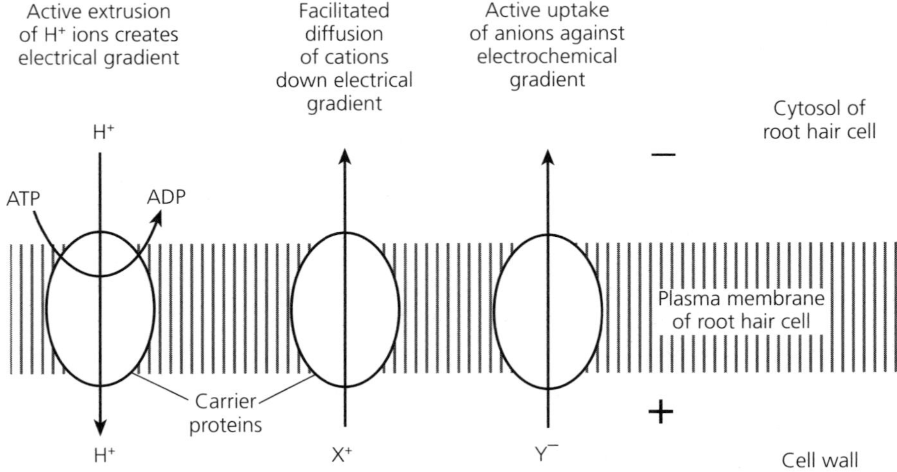


**Figure 33.24**
Interaction of active and passive transport in ion uptake by roots

## QUESTIONS

**1** Complete the following statement:

In photosynthesis ....A.... from the air and ....B.... from the soil are used to make ....C.... which diffuses out of the plant, and ....D.... which is used in respiration, and as building material for growth. The energy which drives the process is in the form of ....E...., which is absorbed by the pigment ....F.... . The entire process takes place within the ....G.... .

**2** Give *three* ways of expressing 'rate of photosynthesis' (take care with units).

**3** Which factor – light, temperature or $CO_2$ – would you expect to limit the rate of photosynthesis under each of the following conditions:

**a)** the woodland floor in summer
**b)** full sun in winter
**c)** a sunny day in summer?

**4** In darkness, a leaf gives off 0.5 mg of $CO_2$ per hour, and in bright light it absorbs 2 mg of $CO_2$ per hour. What is the gross rate of photosynthesis of the leaf?

**5** **a)** What is meant by *light compensation point*?
**b)** Which would have the higher compensation point – a leaf, a shoot or an entire plant? Justify your answer.
**c)** What effect would a rise in temperature have on the compensation point of a plant, and why?

**6** Which of the graphs **A–D** below could show the effect of light intensity on the rate of photosynthesis for red and green light?

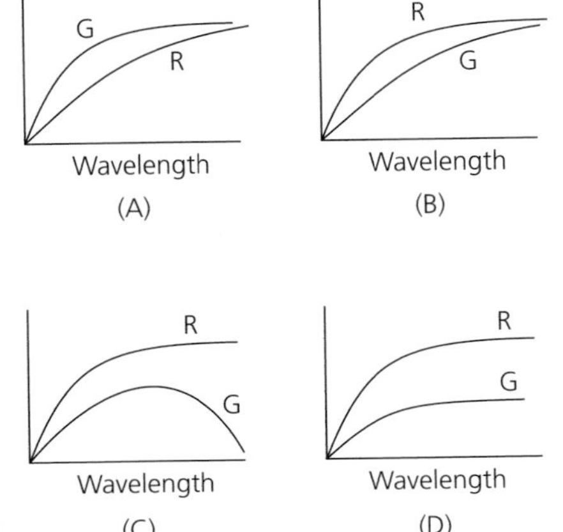


**7** Two leaves (A and B) were placed in water in sealed glass tubes as shown. The two leaves were identical except that B was variegated, chlorophyll being confined to the inner region. Using a probe sealed into the tube (not shown), the concentration of $CO_2$ in each tube was measured at frequent intervals over a three hour period. For the first hour, the light was dim, for the next hour it was bright, and for the third hour it was left in darkness. All other environmental conditions were identical for A and B. The graph shows the changes in $CO_2$ concentration for leaf A.

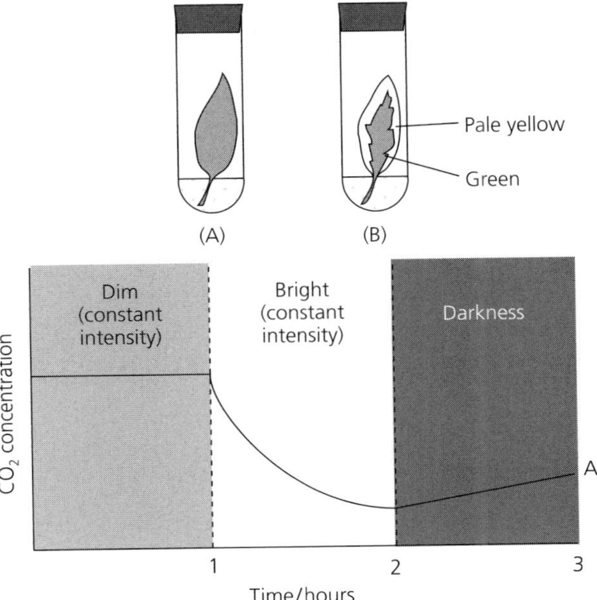


**a)** What two processes affect the $CO_2$ concentration in the air round the leaf?
**b)** What can you say about the rates of these processes during the first hour?
**c)** Why is the graph for the second hour curved rather than straight?
**d)** Copy the graph and on it sketch the curve that you would expect for leaf B, labelling it 'B'.

## QUESTIONS

**8** The diagram represents an outline of a chloroplast, with chemical participants in photosynthesis represented by the letters **A–H**. Different compartments in the chloroplast are represented by **X** (dotted line) and **Y**.

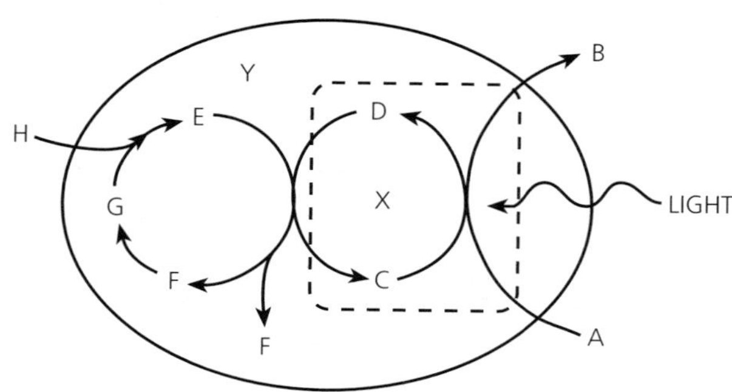


a) Identify the substances **A–H** from the following list:
$O_2$, $CO_2$, $H_2O$, NADPH, $NADP^+$, GP, GALP, RuBP.

b) Identify the compartments **X** and **Y**.

c) Give the letter(s) of those substances that would be in short supply on a cold but sunny winter's day.

**9** The diagram shows a graph drawn by a student.

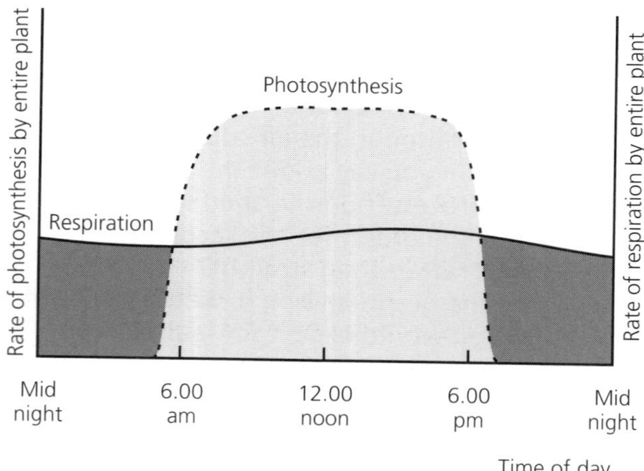


a) What would the area under each curve represent?

b) A plant showing the behaviour indicated in these graphs would be unlikely to survive. Why?

**10** Suppose that only *blue* light were to reach the earth. What would be the consequences for plant growth? Consider phototropism and photomorphogenesis (Chapter 32) as well as photosynthesis.

**11** When soil is flooded, the uptake of ions is reduced. Suggest why.

**12** Is it possible for **(a)** a metal ion and **(b)** a nitrate ion, to diffuse from a low concentration in the soil to a higher concentration in the root? Explain your reasoning in each case.

CHAPTER

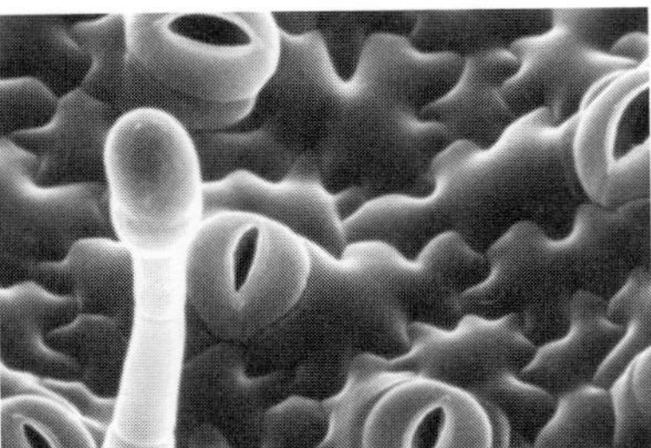

# 34

# TRANSPORT in PLANTS

## LEARNING OBJECTIVES

By the time you have completed your study of this chapter you should be able to:

- explain why, for a land plant, success in the competition for light requires two specialised transport systems.
- define transpiration and explain why it is an inevitable process.
- state the factors that influence transpiration rate and explain how they operate.
- understand the principles underlying the use of a potometer.
- distinguish between apoplastic and symplastic pathways through plant tissues.
- give evidence supporting the cohesion–tension hypothesis.
- explain how transpiration sets up capillary forces in leaf cells, and how these forces are transmitted down to the roots.
- explain how root pressure is developed, and how water enters a root.
- give evidence that organic materials are translocated in the phloem.
- explain the pressure–flow hypothesis and the evidence for it.

### *The need for transport in land plants*

To compete successfully for light a shoot must gain height above the ground, and to obtain a reliable water supply the roots must penetrate deeply into the soil. There is thus a strong tendency for leaves and roots to be some distance apart – too far for materials to be transported between them by diffusion. Ferns, conifers and flowering plants have specialised conducting or vascular tissues through which materials move by bulk flow.

## The transport of water and minerals

Water and mineral ions are transported through tubular channels in the xylem. The propulsive force arises from **transpiration**, the evaporation of water from the shoot. Since transpiration provides the motive power for water transport, we need to consider it in some detail before dealing with the upward movement of water.

## Relative rates of transpiration and photosynthesis

For every molecule of water a plant uses in photosynthesis, about 500 are lost in transpiration (C4 plants use water more efficiently). There are three reasons why water diffuses out so much faster than $CO_2$ diffuses in.

First, the concentration gradient is about 50 times as steep for water as it is for $CO_2$ (the actual value depending on atmospheric humidity). Second, whereas the diffusion path for water is entirely gaseous, for $CO_2$ it includes the liquid in the cell walls and a thin layer of cytoplasm. Though this extra distance is only about 2 μm, $CO_2$ diffuses through it 40 000 times more slowly. A layer of liquid 2 μm thick thus offers the same resistance as a layer of air $2 \times 40\,000$ μm = 8 cm! Third, water molecules are less than half the size of $CO_2$ molecules and so move slightly quicker ($\sqrt{\frac{44}{18}} = 1.56$ times as fast) at a given temperature.

# Transpiration

In general, a land plant absorbs and loses far more water for each kilogram of body mass than an animal does under the same physical conditions. On a hot summer's day a leaf may lose its own mass of water by transpiration in an hour – many times faster than the rate of photosynthesis.

Transpiration is not limited to photosynthetic parts of the plant because all living parts of the plant require oxygen for respiration.

There are three routes through which water can evaporate from a plant:

1 by far the greatest transpiration occurs via the **stomata**. These are tiny, adjustable pores found in the epidermis of both leaves and stems (Figure 34.1).

2 the **cuticle**, which forms a continuous covering of the epidermis except at stomata. The cuticle consists of wax, but its permeability varies with its thickness. In most leaves the rate of cuticular transpiration is about 10% of the total, but in some plants it is much lower than this.

3 **lenticels**, which occur mainly on those parts of a plant in which the epidermis has been replaced by a layer of protective cork. They permit the slow gas exchange necessary for respiration (Chapter 31). Whereas cork cells are close-fitting, in a lenticel they are round with well-developed air spaces between them, allowing diffusion of oxygen and $CO_2$.

### *The photosynthesis–transpiration compromise*

If a leaf is to absorb $CO_2$ there must be a free diffusion path between the mesophyll cells and the atmosphere outside. Living plant cells have moist surfaces, so the air spaces in the mesophyll are very humid. Since the air outside the leaf is usually much drier, water is more concentrated inside the leaf than outside, so it diffuses out.

Photosynthesis thus carries the inevitable price of water loss, but provided it can be replaced at the same rate, this is no disadvantage. When water is in short supply, unrestricted transpiration would rapidly lead to death by dehydration. Plants are able to regulate their water loss in two quite different ways:

- by short-term adjustments to changes in *weather*, carried out by temporary changes in stomatal aperture.
- in the long-term water shortages that occur in dry *climates*, water loss is restricted by permanent anatomical features.

# The role of the stomata

The stomata are the chief means by which plants regulate the balance between the need to obtain $CO_2$ and to conserve water. In most dicotyledons they are more common on the lower surface of the leaf, and in many trees and shrubs they are absent altogether from the upper surface. The situation is reversed in the floating leaves of water lilies, in which stomata are confined to the upper epidermis. In monocotyledons, in which the leaves tend to be held vertically, they are present on both surfaces.

Why do so many plants have stomata mainly on the lower surface? One possible advantage is that the guard cells and substomatal air spaces would tend to scatter incoming light if they were on the upper surface. Also the slightly lower temperature on the lower surface may reduce transpiration.

### The mechanism of stomatal movement

Each stoma is bounded by two guard cells which, unlike other epidermal cells, have chloroplasts. Their most distinctive property is that their shape depends on the degree of turgor. In dicotyledons they are straight when flaccid but become bent when turgid, causing the stoma to open.

An important part of the mechanism is the orientation of the cellulose microfibrils in the cell walls. Most are laid down round the width of the cells like hoops on a barrel (Figure 34.2). Along the inner walls many additional microfibrils run lengthways. As a result, when the guard cells absorb water the outer walls elongate more than the inner walls, causing the guard cells to curve outwards. When they lose water the reverse happens.

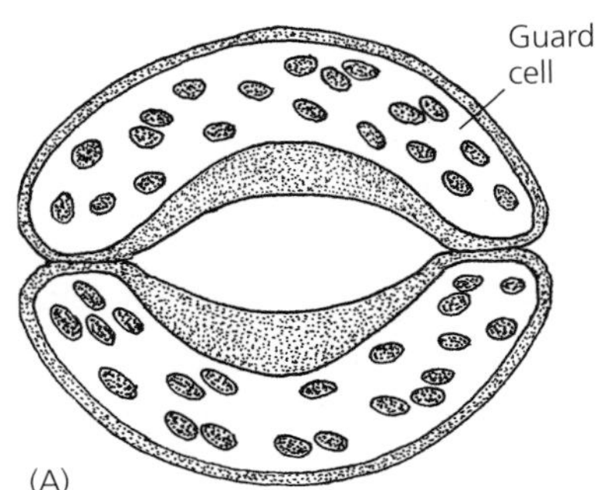


What causes the guard cells to gain or lose water? A number of different stimuli can provoke stomatal movement.

- *Light intensity*. Except for CAM plants (Chapter 30), stomata generally open in the day and close at night.
- *$CO_2$ concentration.* A decrease in $CO_2$ concentration (such as normally occurs in the mesophyll during the day) causes stomata to open.
- *Water stress*. When a plant is short of water it experiences **water stress**, in which the cells lose turgor. Moderate or severe water stress causes stomata to close.

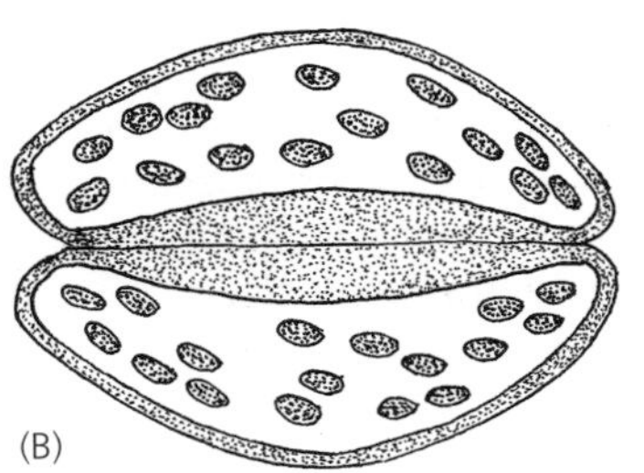


Even under constant environmental conditions, stomata continue to show a regular cycle of opening and closing, with a period of approximately 24 hours. Since they are stimulated by internal factors, these movements are said to be **endogenous**. This is one of many examples of endogenous rhythms shown by animals and plants.

For many years plant physiologists have searched for a common mechanism by which light, $CO_2$ and water stress could act. It has always been clear that changes in water content result from changes in water potential, which in turn result from changes in solute content. Stomata must, therefore, open as a result of an increase in solute content of the guard cells.

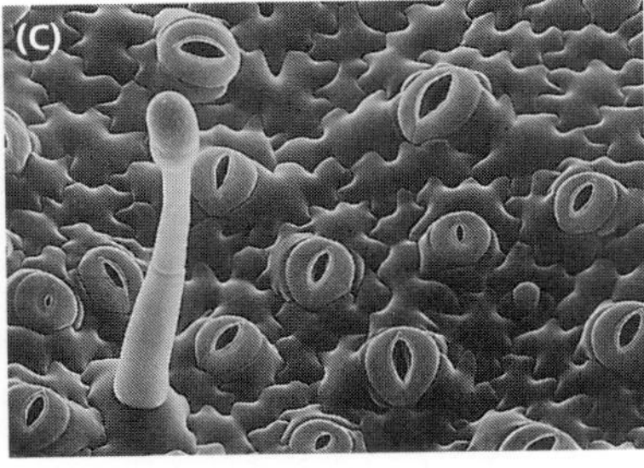


The key question is, what causes changes in solute content of the guard cells? Though ideas have changed considerably over the years, the basic mechanism is now more or less understood. During the daytime (except in times of water stress) $K^+$ ions are actively taken up by the guard cells from neighbouring epidermal cells. In most plants these are balanced electrically by the production of malate ions from starch stored in guard cells. In plants that do not store starch (e.g. onion), $Cl^-$ ions enter alongside the $K^+$ ions.

In plants that are not under water stress, the stimulus for $K^+$ uptake appears to be a decrease in $CO_2$ concentration. Normally this would result from an increase in light intensity acting on photosynthesis.

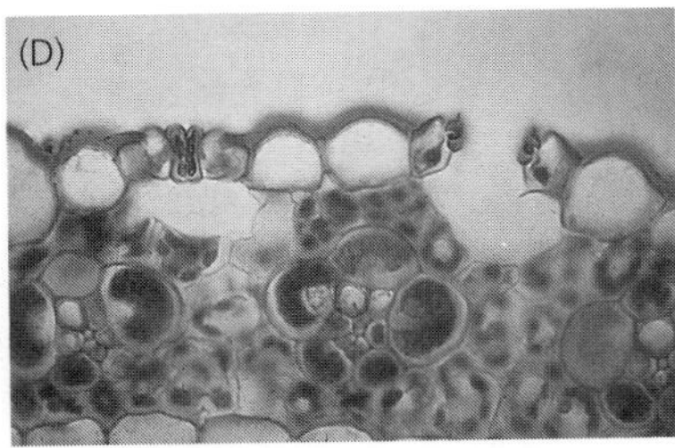


**Figure 34.1**
A stoma, open (A), closed (B) and photomicrographs showing stomata in surface view (C) and in transverse section (D)

What about the closure that occurs in response to water stress? While it is tempting to think that wilting automatically results in loss of guard cell turgor, it turns out to be more complicated. Stomatal closure during water shortage results from the action of **abscisic acid** (**ABA**). During water shortage its concentration in the leaf rises, causing the guard cells to lose $K^+$ ions and hence to close.

Although some of the rise in ABA is due to its increased production in

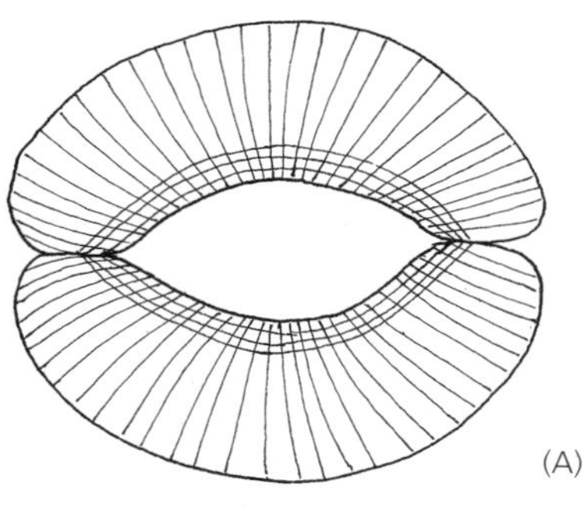


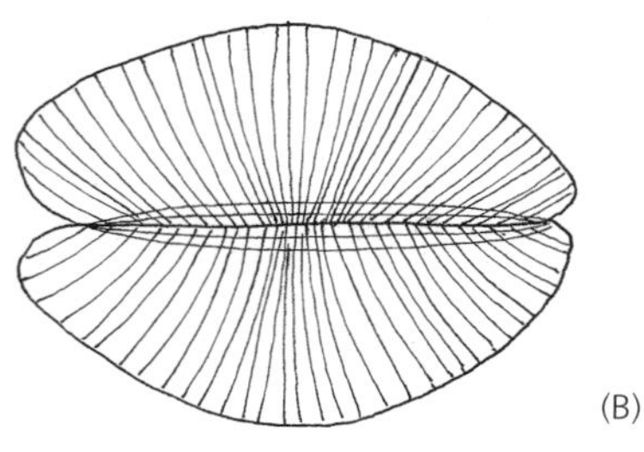


**Figure 34.2**
Orientation of cellulose microfibrils in guard cells when the stoma is open (A) and closed (B)

the leaf mesophyll, there is evidence that some is produced in water-stressed roots and is then carried up in the xylem to the leaves. Since roots are likely to be affected first by drying soils, this would give the leaves an 'early warning' of water shortage.

### *Homeostasis in the leaf*

On a warm sunny day, the stomata are subjected to two conflicting influences: light intensity and water availability. An increase in light intensity tends to increase stomatal aperture, but it also increases water loss. If the soil water supply is not adequate to meet the extra demand, increased ABA production reduces stomatal aperture, bringing water loss back into balance with uptake. In this way, the stomata achieve a balance between satisfying the 'hunger' for $CO_2$ while at the same time minimising the 'thirst' for water.

### *The effects of transpiration*

Since transpiration is inevitable if a leaf is to photosynthesise, it cannot be said to have a *function*. It may however have two *advantages*:

- evaporation uses up *heat*. In hot weather, a rapidly-transpiring leaf may be 5°C cooler than the air round it.
- the upward flow of xylem sap from roots to leaves is caused mainly by transpiration. Since the xylem sap contains dissolved mineral ions, the transport of minerals from roots to leaves is helped by transpiration.

## Factors affecting transpiration

Transpiration rates are influenced by a variety of factors which can act in two quite different ways. First, there are physical factors that fluctuate over the short term, such as humidity, wind, light intensity and other aspects of 'weather'. Second, there are anatomical features of the plant which have evolved in response to long-term, climatic factors.

### Relative humidity

An indispensable term in transpiration is **relative humidity** (R.H.). This is given by

$$\frac{\text{amount of water vapour present}}{\text{amount of water vapour present when the air is saturated at that temperature}}$$

This means that if air has a relative humidity of 50%, it contains half as much water vapour as it could do at that temperature.

Relative humidity is not the same as *absolute humidity*, or *water vapour pressure*, which does not depend on temperature. Imagine a sample of air in equilibrium with water in a sealed vessel, so the air is saturated with water vapour (R.H. = 100%). Suppose that the air is instantaneously warmed. The R.H. *decreases*, since the air becomes capable of holding more water.

At the instant of the temperature rise there is no effect on the water vapour pressure, which is determined solely by the water concentration in the air. But as evaporation speeds up, the water vapour pressure rises until a new equilibrium is established and the air once again has a R.H. of 100%.

## Physical factors

Six physical factors can influence transpiration, but all except altitude act either by changing the steepness of the water vapour gradient, or by stimulating changes in the aperture of the stomata.

### Humidity gradient

The steeper the concentration gradient of water vapour, the faster it diffuses out of the leaf. The steepness of the water concentration gradient depends on two things: the concentration difference and the length of the diffusion path. Three physical factors can affect the concentration gradient.

- *Temperature.* A rise in temperature speeds up transpiration in two distinct ways. First – and by far the more important – it increases the number of water molecules that have enough energy to evaporate from the mesophyll cells into the air spaces. A rise in temperature therefore raises the concentration of water vapour inside the leaf.

  Second, it speeds up the movement of water molecules that are already in the gas phase. This effect is very small because molecular speed is proportional to the *square root* of the *absolute temperature*. An increase from 10°C to 20°C (i.e. from 283K to 293K) speeds up diffusion by $\sqrt{\frac{293}{283}} = 1.0175$ times, or 1.75%.
- *Atmospheric humidity.* Transpiration is faster in drier air. The drier the atmosphere, the lower is the concentration of water vapour *outside* the leaf, and so the greater is the difference in water vapour concentration between the inside and outside of the leaf.
- *Wind.* Wind speeds up transpiration by preventing water vapour accumulating just outside the leaf. Even in windy conditions there is a thin layer of still air immediately next to the leaf. Water vapour crosses this **boundary layer** by diffusion. The stronger the wind, the thinner the boundary layer, and hence the shorter the diffusion path – in strong wind the boundary layer may be a fraction of a millimetre thick and on a still day it may be over a centimetre thick.

### Stomatal aperture

Stomatal aperture is influenced by two environmental factors: light intensity and soil water.

- *Light intensity.* In most plants, stomata open in the light and close in darkness (CAM plants are an exception, as explained in Chapter 30).
- *Soil water.* In drought conditions, stomata close by a mechanism involving ABA (see above).

The effect of stomatal aperture on transpiration is not as simple as might be supposed. Under conditions conducive to transpiration (such as wind and low humidity) stomata are highly effective in regulating transpiration. In still air their effect is much less marked (Figure 34.3). Can you think why?

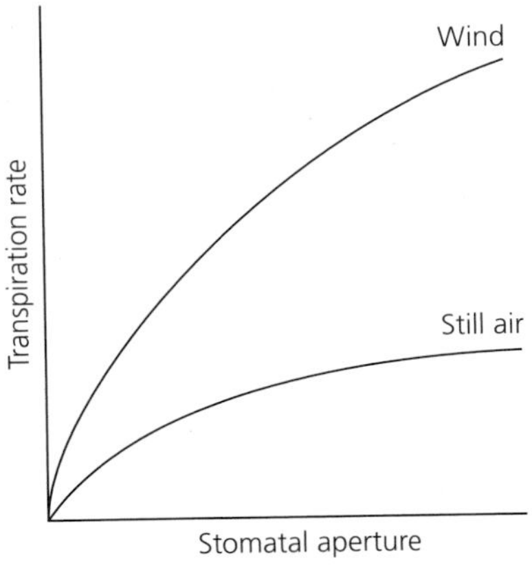

Transpiration rate

Wind

Still air

Stomatal aperture

**Figure 34.3**
The effect of wind on the control of transpiration by stomata

### Altitude

Transpiration is also affected by atmospheric pressure and thus by altitude. At lower atmospheric pressures, molecules collide less frequently. A water molecule moving from water to air is thus less likely to rebound back into the liquid water. Transpiration is therefore faster at higher altitudes, where there may be the additional factor of high winds.

## Leaf structure

Leaves have a variety of anatomical features which reduce transpiration.

- *Low stomatal density.* Plants that are adapted to living in dry soils may have fewer than 150 stomata per $mm^2$ of leaf, while those living in moist climates may have over 500 per $mm^2$.
- *The cuticle.* Since the cuticle is not completely impermeable to water, the rate of cuticular transpiration depends on its thickness. In leaves with a thick cuticle it may account for as little as 1 per cent of total transpiration with stomata open. In leaves with a thin cuticle it may account for up to 40 per cent of total transpiration.
- *Length of the diffusion path.* Plants can increase the distance between the still air in the mesophyll and the moving air outside in various ways (see the information box on page 647).

### Relative humidity and water potential

In Chapter 6, water potential was dealt with purely in the context of aqueous solutions. To understand water movement in plants we need to extend the concept to air.

If an aqueous solution is confined with air in a sealed vessel, water evaporates until an equilibrium is established between the solution and the air above it. In evaporating, water is moving from a higher to a lower water potential, until at equilibrium the water potentials of air and solution are equal. If the liquid is pure water (with a water potential of zero) the air will have a R.H. of 100%.

What is surprising (and important from the point of view of water transport) is just how negative the water potential of air normally is. Imagine a plant cell in equilibrium with air, so their water potentials are the same. If the cell has a water potential of −2.7 MPa, the air in equilibrium with it has a R.H. of 98% at 20°C. Even more important, water potential falls logarithmically with an increase in R.H. At 20°C, air with a R.H. of 50% has a water potential of −215 MPa.

This is important because by far the greatest drop in water potential is between the humid air inside the leaf and the drier air outside. Stomata are thus located precisely at the point where they can exert the greatest influence on the rate of water movement.

## Using a potometer to investigate transpiration

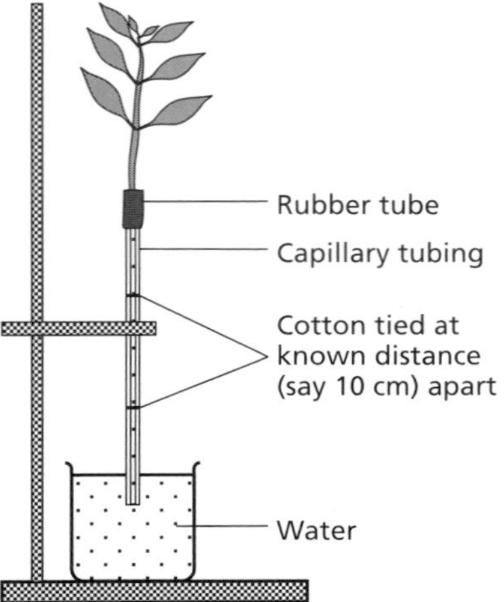


**Figure 34.4**
A simple form of bubble potometer

Under normal conditions the loss and uptake of water are almost exactly equal, so it is possible to study changes in transpiration by measuring changes in uptake using a potometer. There are various types, but all consist of a piece of fine glass capillary tubing containing water, connected to a leafy shoot by an air-tight seal. Figure 34.4 shows a very simple one that is easy to use. Provided that it has been set up properly, the shoot absorbs water at almost exactly the same rate as it loses it.

To set up the apparatus you need a bucket of water and strong, sharp pruning scissors. First, cut a suitable sized leafy shoot from the plant. Although some air is inevitably sucked into the xylem, you can remove the part containing the air by quickly dipping the shoot in water and cutting another 10 cm or so off the end *while it is under water*. Any air entering the shoot would cause an airlock in the xylem vessels, so that the uptake and loss of water would no longer be equal. Connecting the shoot to the potometer must likewise be done under water using a basin. As an extra precaution the shoot can be cut early in the morning while transpiration tensions are relatively low.

The apparatus is then set up as in the diagram, keeping the leaves dry. A small bubble of air is allowed to enter the tube by briefly lifting it out of the water in the beaker and then lowering it again. If two pieces of cotton are tied round the tube, the time for the bubble to travel that distance can be measured. Before the bubble reaches the top of the capillary, it should be pushed out by squeezing the rubber tube. A fresh bubble can then be allowed to enter as before.

To study the effect of external conditions on transpiration you should use the *same* shoot, and all you need to measure is the rate of bubble movement (distance/time) expressed in mm per minute. Remember that before taking a set of measurements you should leave the potometer for five minutes or so to adjust to the new conditions. For each set of conditions, take several measurements and find the average. If possible, change only one set of conditions at a time. Notice that the figures you obtain give the *relative* rates of transpiration. This means you would be able to say that in draughty conditions, transpiration is X times as fast as in calm air, but you would not know the actual rate.

When using different shoots (as when comparing rate of transpiration in different species) you need to take account of their different total leaf areas. Provided that your potometers have capillary tubes of equal cross-sectional area, you can express your results as mm travelled per minute per $cm^2$ of leaf area. If you want to know the actual rate of transpiration, you must know the volume of water taken up per minute, for which you will need the cross-sectional area of the capillary tubing. The rate of transpiration is given by

$$\frac{\text{distance moved by bubble} \times \text{cross-sectional area of tube } (\pi r^2)}{\text{time} \times \text{area of leaf}}$$

## Water shortage and water glut

Plant life can be found in a wide range of degrees of water availability. At one extreme are **xerophytes** that live in dry conditions, and at the other are **hydrophytes**, living in conditions of superabundance of water. Plants living in conditions of moderate water supply are **mesophytes**. These categories are of course not distinct, but merge into a continuous spectrum.

Xerophytes have diverse survival strategies, some structural and some physiological. Features that are anatomical are described as **xeromorphic**. The most common feature is a reduced rate of transpiration. One way of achieving this is to decrease the area over which water can be lost by reducing the leaf area. In gorse, the leaves are reduced to spines. In cacti, the leaves are reduced to spines and photosynthesis is carried out by the fleshy stems. In addition to their role in reducing water loss, spines also provide protection against grazing animals, which are a double hazard in dry environments. First, wounds created by grazing are potentially sites of water loss, and second, since xerophytes grow slowly they cannot regenerate lost tissue as quickly as mesophytes can.

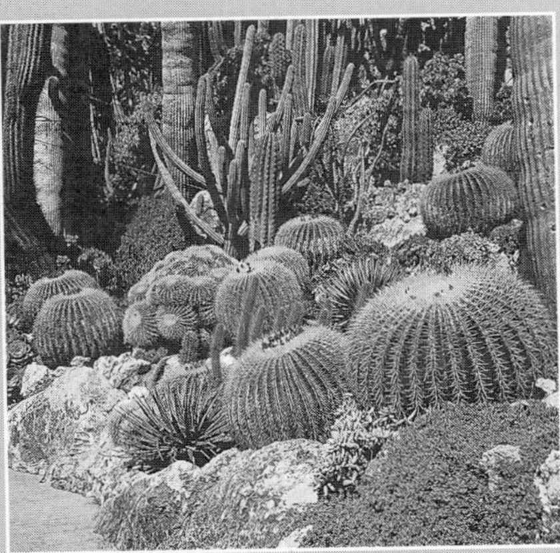

Cactus, showing reduction of leaves to spines, minimising water loss by transpiration

## Water shortage and water glut *continued*

An alternative way of reducing transpiration is to increase the resistance of each unit area of leaf to water loss. The most obvious way is by increasing cuticular thickness. Another way is by decreasing the steepness of the water vapour concentration gradient. *Ammophila arenaria* (marram grass), a plant living in sand dunes, has stomata on the inner surface of the leaves. In dry conditions the leaves roll up, creating a humid microenvironment next to the stomata. This movement is due to changes in turgor of certain large parenchymatous **hinge cells** on the inner surface of the leaf. Some plants have a dense coat of hairs on the lower surface of the leaf. Others have sunken stomata below the leaf surface, as in *Pinus*.



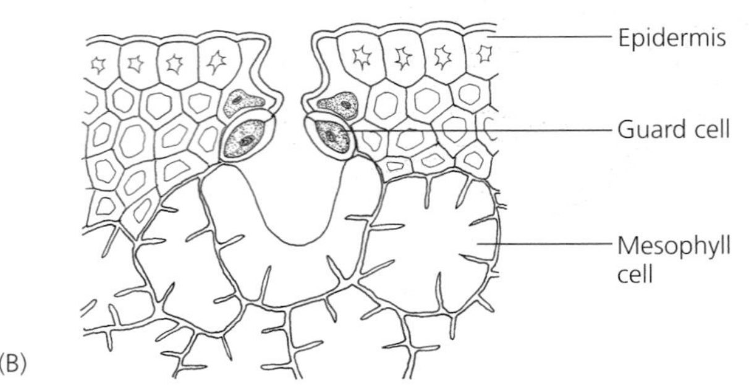



Transverse section through *Ammophila* leaf (A); sunken stomata in a leaf of *Pinus* (B) (after Eames and MacDaniels)

Support by turgor is unreliable in dry environments, and many xerophytes have highly lignified tissues. Living cells with lignified walls are able to resist collapse under severe water stress. Moreover, they can develop *negative* pressure potentials, further increasing their water-absorbing capacity.

Cacti and other succulents can take advantage of infrequent rainfall by rapidly absorbing and storing water in succulent tissue. Many of these have CAM photosynthesis, with $CO_2$ being fixed at night and converted to carbohydrate during the day when the stomata are closed (see the information box on page 633).

Other desert plants, such as the palms that live in oases, have roots that extend 10 or more metres down, enabling them to tap permanent water sources. They transpire at quite normal rates and their tissues do not encounter water stress.

A different strategy is shown by plants that can survive complete drying out of the tissues. Upon rehydration, they recover rapidly and resume normal activity. An example is *Tortula muralis*, a common moss that lives on rocks and walls which may be dry for weeks at a time.

Some plants can survive years of drought as dormant seeds. When the rain does come it is usually heavy enough to support growth for the few weeks it takes to complete their life cycle. Plants with very short life cycles are called **ephemerals.**

At the other extreme are the hydrophytes. In some, such as various species of rush, only the root systems are flooded. Others, such as water lilies, have leaves that float on the water surface. In the most extreme hydrophytes, such as Canadian pondweed (*Anacharis canadensis*), the entire plant is submerged.

Living in water has both costs and benefits. Buoyed up by water and with no need for water transport, floating plants save energy by producing little or no xylem or sclerenchyma. Though roots may be present as anchors, root hairs are absent.

## Water shortage and water glut *continued*

The main disadvantages are in gas exchange. In the still water of a pond, $CO_2$ may be in short supply because it diffuses 40 000 times more slowly in solution than in air. The same applies even more forcefully to respiration, since oxygen is only sparingly soluble in water and diffuses 300 000 times more slowly in solution than in air.

In most hydrophytes these disadvantages have led to the evolution of an extensive internal air space system through which gases can diffuse rapidly. Oxygen produced in the leaves of a pondweed can thus move relatively rapidly down to the stems.

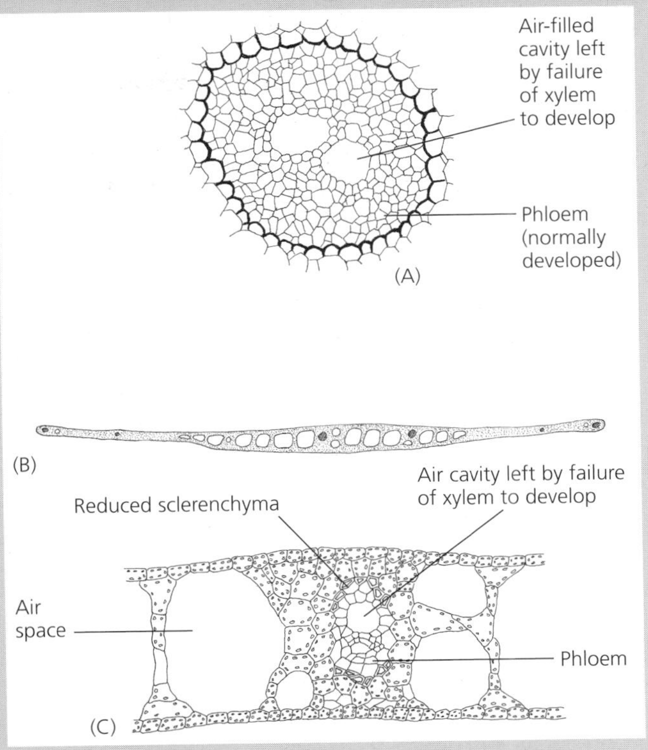


Sections through *Potamogeton* (pondweed). (A) Transverse section of central region of stem showing xylem replaced by air channels. (B) Low magnification view of transverse section through leaf showing air channels. (C) Higher magnification of part of (B)

## The path of water

In moving from roots to leaves, water travels through several kinds of cell, but by far the greatest part of the journey is through the xylem. This can easily be shown by leaving a shoot in eosin or similar dye, and then examining a section under the microscope. Apart from leakage to neighbouring tissue, the stain is seen to be confined to the xylem. The best results are obtained if two cuts are made before putting the shoot in the dye, the second under water (as with setting up a potometer). The rise of the ink can be accelerated by placing the shoot in a draught to speed up transpiration. Some plants naturally have reddish pigments in the stem, so the section needs to be compared with an unstained control.

### *Symplastic and apoplastic pathways*

In passing through plant tissue, water can take three alternative routes (Figure 34.5):

- it may avoid the cytoplasm altogether by travelling through the submicroscopic spaces between the cellulose microfibrils in the cell walls. These spaces form a continuous interconnecting system called the **apoplast**.
- it could travel through the cytoplasm, passing from cell to cell via the plasmodesmata and skirting round the vacuoles. The protoplasts and plasmodesmata form a single, interconnected system called the **symplast**.
- if water moved through the vacuoles it would be travelling along the **transcellular pathway**, though this is just a variation on the symplastic pathway.

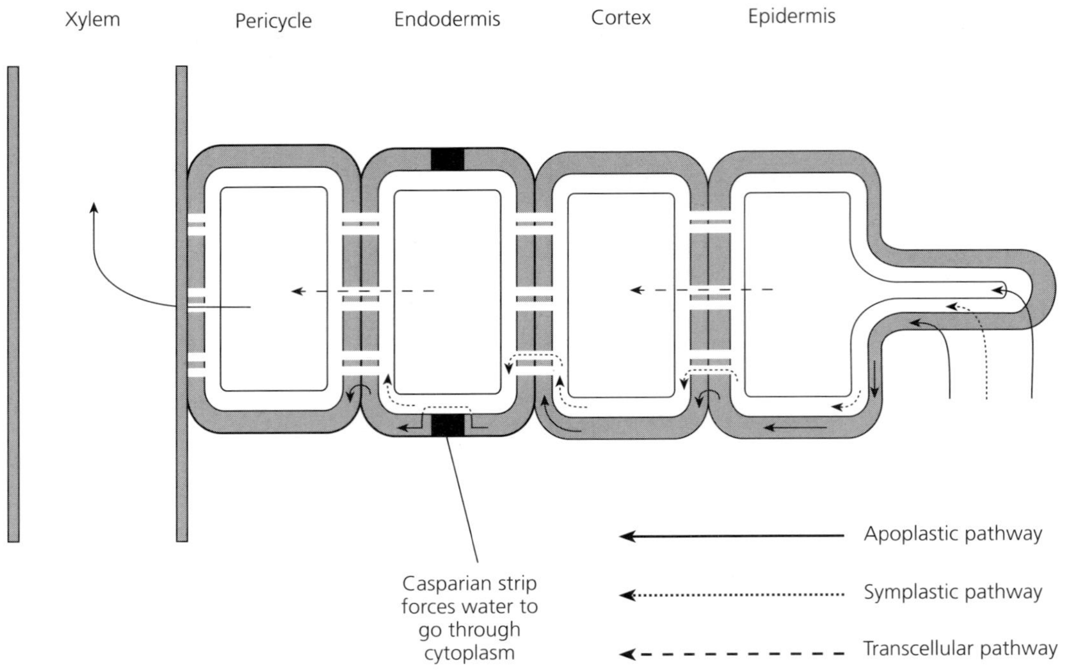


**Figure 34.5**
The alternative routes for water flow across a root. As explained in a later section, the endodermis divides the apoplast into two compartments, inside and outside the endodermis, respectively

Both apoplast and symplast extend throughout the plant. The apoplast offers less resistance and most water appears to travel along this route. The xylem vessels and tracheids are essentially part of the apoplast. Unlike the symplast (which is theoretically a single compartment) the apoplast is divided into two by the endodermis of the root (see below).

### *How fast does the sap move?*

Rates of sap flow in living trees have been measured by applying a brief heat pulse to a branch and recording the time of arrival of the warmer sap a known distance above. Using methods like this it has been shown that in flowering plants, sap may flow at rates of up to about 40 metres per hour compared with about 0.5–1 m $hr^{-1}$ in conifers.

### *Push or pull?*

The tallest trees in the world (the giant redwoods of California and the mountain gum tree of Australia) can be over 100 metres high. To propel water to such a height needs a considerable force.

Wherever this force is produced, it does *not* originate in the xylem itself. Even though xylem contains living parenchyma cells, we know these play no part in movement because sap continues to flow through the xylem even after a twig has been killed with steam or poison. The role of xylem in water transport must therefore be passive, the propulsive force being set up elsewhere.

There are three possibilities – a *push* from below, a *pull* from above or a combination of both. In some plants the xylem sap near ground level can be shown to be under **root pressure** (Figure 34.6). Root pressure is most marked in spring, and can often be seen when sap oozes from the stem of a plant after it has been cut.

Root pressure is also the cause of **guttation**, the seepage of water from the tips of many leaves at night and at other times when relative humidities are high. The 'dew' on a lawn is not due to condensation but guttation.

Is root pressure sufficient to explain the rise of sap? There are several reasons why it cannot be:

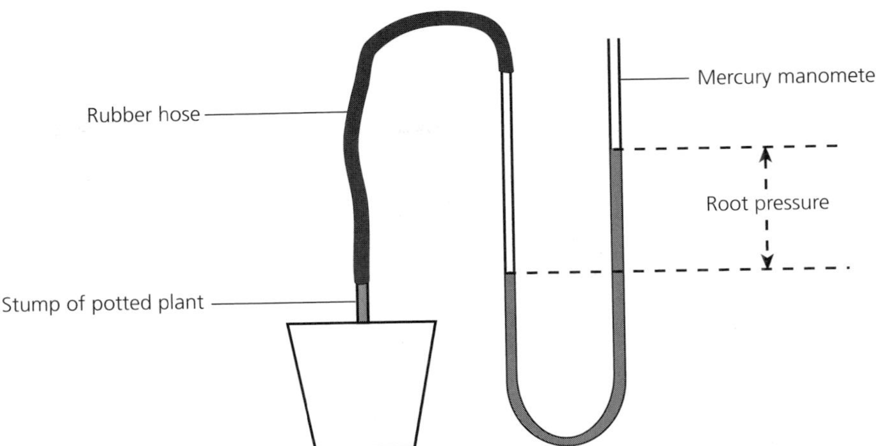


**Figure 34.6►**
Demonstration of root pressure

- the highest recorded root pressures would only be sufficient to push sap up to about 20 metres, and many conifers do not develop root pressure at all.
- the rate of exudation from roots is usually far lower than rates of transpiration, especially in mid-summer when transpiration rates are highest.

### *The cohesion–tension hypothesis*

If root pressure cannot fully account for the rise of sap, there must be a pull from above. According to the **cohesion–tension hypothesis**, transpiration sets up a tension in the leaf cells which is transmitted all the way down to the roots via the xylem vessels, drawing the sap upwards. This tension is called **transpiration pull**.

When Dixon and Joly put forward the cohesion–tension hypothesis in 1894, many doubted that water could withstand a tension. It was well known that lift pumps could not draw water through a vertical distance of more than about 10 metres – the height to which a column of water could be *pushed* by atmospheric pressure. Above this distance, the water column always broke.

What Dixon and Joly were suggesting was that water has a **tensile strength**, so that (like a solid) it can be pulled. To be true, their hypothesis had to satisfy a number of requirements:

1 Cohesive strength of water – only a year after Dixon and Joly put forward their hypothesis, Askenasy showed that water can resist tension. Using the apparatus shown in Figure 34.7, he drew a column of mercury to a height of 130 cm – 54 cm higher than could be pushed by the atmosphere. To do this required a tension of −71 kPa.

The basis for the tensile strength of water is the **cohesion**, or mutual attraction, between water molecules, due to hydrogen bonding (Chapter 3). How great must the tensile strength of water be to be compatible with the cohesion–tension hypothesis? To support the weight of a column of the water, the height of a redwood tree needs a tension of about −1 MPa, but to this must be added the −2 MPa needed to overcome resistance to flow at the observed rates (about 1 metre per hour in redwoods).

Practical measurements have shown that water can withstand tensions as high as −30 MPa, and calculations based on the strength of hydrogen bonds between the molecules give values of over −100 MPa. Clearly, water is easily strong enough to satisfy the requirements of the hypothesis.

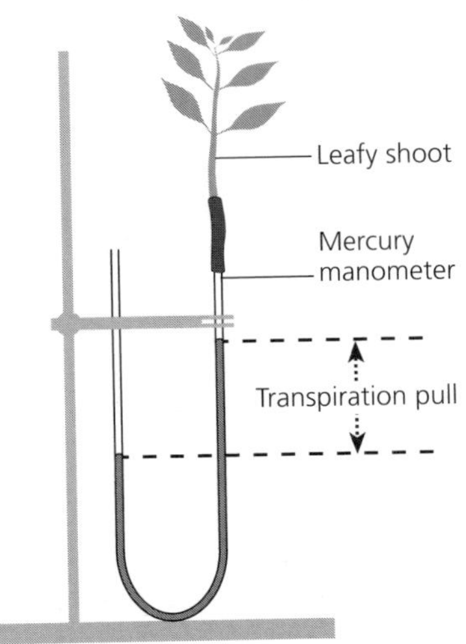


**Figure 34.7**
Demonstration of transpiration pull

## CLEAR THINKING

**It is often said that since xylem vessels are fine tubes with wettable walls, water will tend to rise up the plant by capillarity. Capillarity depends on surface tension, which requires the existence of an open water surface. Since functional xylem vessels are not open to the air, the narrowness of the xylem vessels cannot contribute in any way to the force pulling up the sap.**

**Yet in a different location, capillary forces do haul water up the plant. The diagram shows the surface of a leaf mesophyll cell under extremely high magnification. The spaces between the cellulose microfibrils are filled with water. When water evaporates from the cell, it begins to retreat into these spaces, forming curved menisci. Each meniscus acts like a microscopic 'trampoline', pulling water from deeper in the cell wall.**

**When water evaporates from a plant cell, the water in the cell wall is therefore under tension. Its water potential is thus negative, and equal to that of the cytoplasm and vacuole. Because the mesophyll cells are in contact with each other, tension set up in a leaf cell wall is transmitted through the walls to adjacent cells, and so on to the xylem and all the way down to the roots.**

**The tension developed by a meniscus is inversely proportional to its diameter. The spaces between the microfibrils in plant cell walls are only about 10 nm wide – narrow enough to create menisci that could support a column of water 3 km high!**

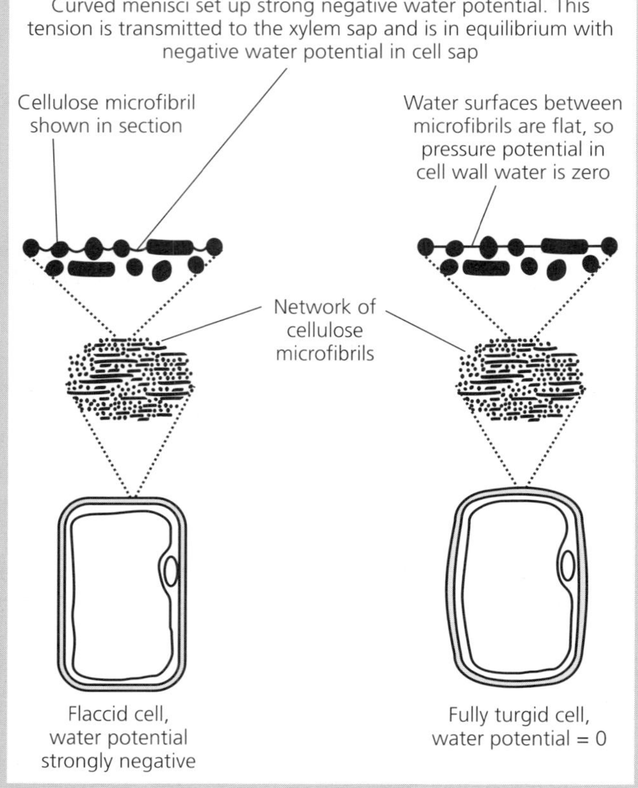
Curved menisci set up strong negative water potential. This tension is transmitted to the xylem sap and is in equilibrium with negative water potential in cell sap
Cellulose microfibril shown in section
Water surfaces between microfibrils are flat, so pressure potential in cell wall water is zero
Network of cellulose microfibrils
Flaccid cell, water potential strongly negative
Fully turgid cell, water potential = 0

The role, and the non-role, of capillarity in the rise of sap

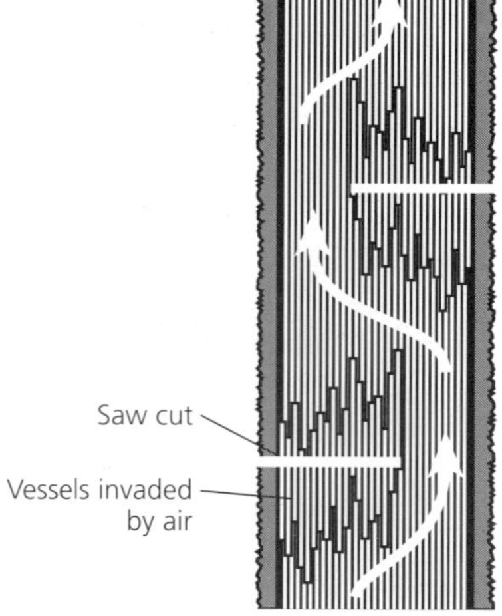
Saw cut
Vessels invaded by air

**Figure 34.8**
Effect of overlapping saw cuts. Although vessels invaded by air become non-functional, water is able to go round the air-filled vessels by passing through the side walls

2 Adhesion to vessel and tracheid walls – no matter how strong a column of water, it will break under tension if the water molecules are not attracted to the vessel walls. Attraction between water and other molecules is **adhesion**. Cellulose meets this requirement because its many free —OH groups readily form hydrogen bonds with water.

3 Continuity of the columns – a crucial requirement of the cohesion–tension hypothesis is that the water columns must be continuous. Making overlapping saw-cuts would appear to be an obvious test (Figure 34.8). If the cuts are made sufficiently far apart the tree is not adversely affected, but if they are close together, the tree wilts and dies.

This can be explained in the following way. Although all the vessels opened by a saw cut instantly become filled with air, the airlock cannot enter adjacent, intact vessels. Provided that the saw-cuts are separated by a vertical distance greater than the length of individual vessels, water is still able to get round the air locks by flowing into adjacent, intact vessels through the submicroscopic spaces in walls. Also the living parenchyma cells in the sapwood may, to some extent, be able to act as a temporary water reservoir during the few days it takes for the cambium to produce another layer of vessels.

### *Why do columns sometimes break?*

Experimental values of around −30 MPa are hard to obtain because of the need to exclude any particles that do not adhere strongly enough to water, such as dust particles and dissolved air (nitrogen and oxygen are only sparingly soluble in water because their molecules have little attraction for water).

While dust cannot enter an intact plant, dissolved gases are a normal component of xylem sap, and it seems that these are responsible for the formation of breaks in the water columns. Though this process of **cavitation** puts the vessel out of action, breaks cannot spread to neighbouring vessels, for reasons explained in the information box below.

Cavitation appears to be a normal phenomenon, and in some trees only the current year's xylem contains sap (which is why most timbers float).

## Vessels vs tracheids

Vessels are generally wider than tracheids, and, correspondingly, sap flows more easily through them. This advantage is actually greater than it might appear, because resistance to flow is inversely proportional to the *fourth* power of the diameter. Thus for a given pressure difference, doubling the diameter of a tube raises the rate of flow by $2^4 = 16$ times, even though its cross-sectional area has only increased four times. The rate of flow through a given cross-sectional area of tubing is thus increased by 16/4, or four times. In some vines, the vessels are 10 times wider than typical tracheids, so the resistance to flow through equal cross-sectional areas would be only $1/10^2 = 0.01$ as great.

Why then have tracheids not been entirely replaced by vessels during evolution? The explanation is probably that vessels have disadvantages. During rapid transpiration the sap in the xylem is under great tension. If the wood of a rapidly transpiring tree is cut, air rushes in and almost instantaneously fills the ruptured vessels. If air could spread from one vessel to another, *all* the vessels would become filled with air and cease working. It would thus only be necessary to break off one twig and the entire water transport system would become functionless.

In fact, an airlock cannot spread from one vessel to another (though individual vessels may extend for a metre or more). This is because the spaces in the walls are incredibly fine. If air enters a vessel, the tension in adjacent vessels tends to pull the sap through the walls, but this is opposed by strong surface tension forces in the menisci set up between the cellulose microfibrils.

One place where the xylem might appear to be particularly vulnerable to damage is in the leaves, which in many plants are continually being eaten by animals. With every munch, a caterpillar opens up dozens of xylem channels which instantly become air-filled. Since however the finest xylem channels are tracheids rather than vessels, airlocks are more localised. In their much greater capacity to localise air locks, tracheids have a significant advantage over vessels.

### *Evidence for the cohesion–tension hypothesis*

- In rapidly transpiring trees, the xylem sap can be shown to be under very high tension. One simple way is to cut a rapidly transpiring shoot under the surface of a dye; the dye entering the shoot penetrates some way below the cut as well as above it.
- Although transpiration and water uptake both increase during the day, in trees the increase in water loss begins *before* the rise in water uptake (Figure 34.9).

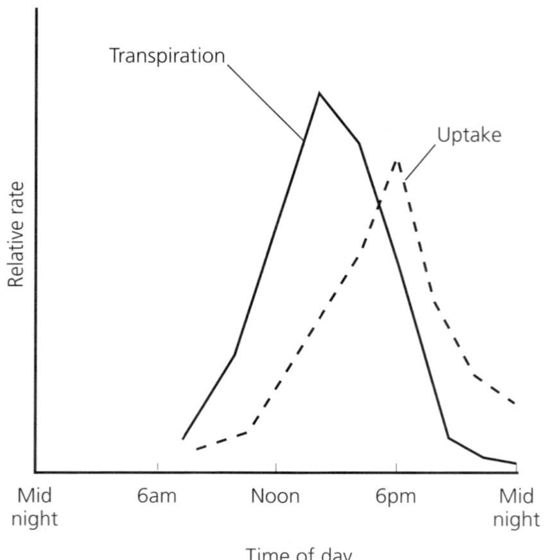


**Figure 34.9**
Changes in rates of transpiration in a tree occur before changes in rates of water uptake

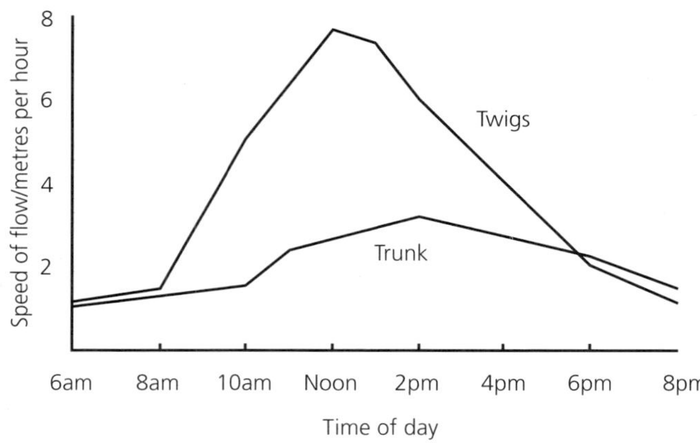


**Figure 34.10**
Changes in the rate of sap flow in the trunk and twigs of a tree during the day

- In the morning the upward movement of sap in trees begins in the smaller branches, with the acceleration of flow occurring last of all in the trunk. This is exactly what would be expected if the driving force were transpiration (Figure 34.10).
- Tree trunks undergo very small daily fluctuations in diameter (less than a mm in a fair sized tree), being minimal around midday when transpiration is fastest, and maximal at night (Figure 34.11). This is consistent with the idea that transpiration generates tension.

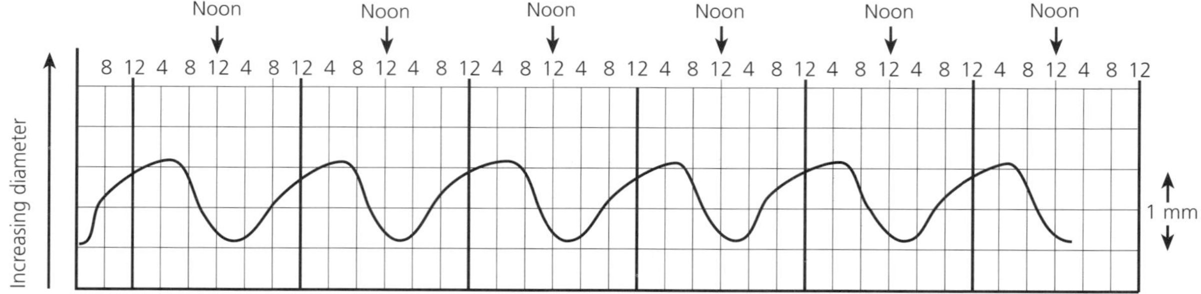


**Figure 34.11**
Daily fluctuations in diameter of a pine tree trunk (based on McDougall)

## *The uptake of water*

In travelling from soil to atmosphere, water follows a simple rule: it moves from a higher to a lower water potential. The first stage is the movement of water from the soil into the root xylem (Figure 34.12).

The root hairs greatly increase the surface area of the root, but water uptake also requires a driving force in the form of a water potential gradient. Provided that the water potential of the sap in the root xylem is lower (more negative) than that of the soil, water will move from soil to xylem.

There are two ways in which the water potential of the xylem sap can be lowered relative to that of the soil. The most important is the transpiration pull developed in the leaves and transmitted via the xylem to the roots.

An alternative mechanism is **root pressure**, resulting from the presence of dissolved mineral ions which reduce the solute potential of the xylem sap. Since the continued movement of water into the xylem

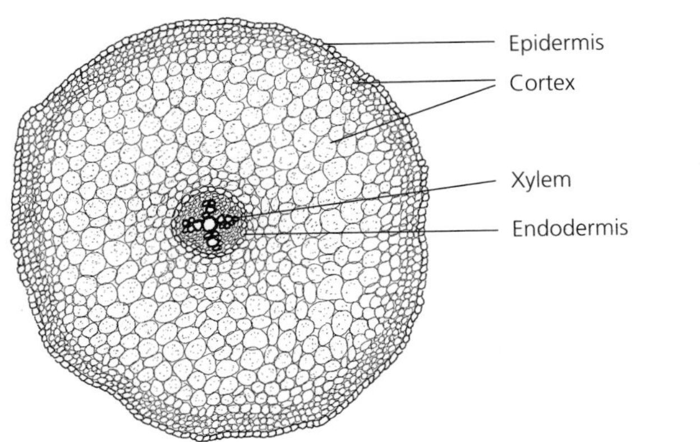


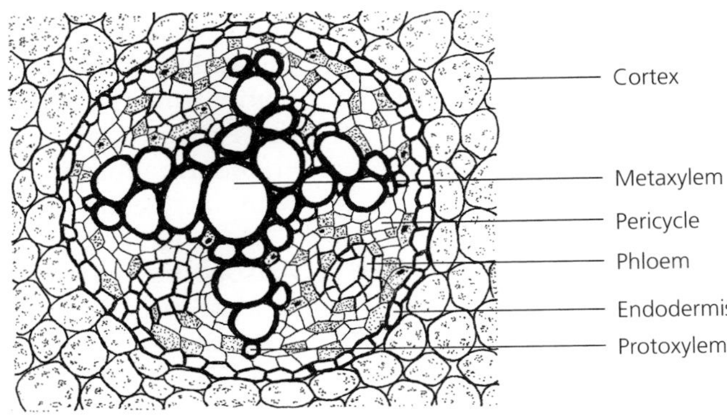


**Figure 34.12**
Transverse section of a buttercup root cut above the root hair zone; entire root (A), central vascular cylinder (B)

must sweep away solutes, a more negative solute potential can only be maintained by continuous entry of ions into the root xylem from the soil. This is known to involve active transport since it is abolished by respiratory inhibitors. Whether ions are actively secreted into the xylem is not known. Since ions are less concentrated in the xylem than in the living root cells, ions actively accumulated by the root cells could diffuse passively into the xylem.

### *The role of the endodermis*

Water can take either of two routes across the cortex. It can travel through the fine spaces in the cell walls (the apoplast), or it may move through the cytoplasm (the symplast).

If water were to travel the entire distance from soil to atmosphere via the apoplast, it would not pass through a single cell membrane. The plant would simply be acting like a wick, and the water reaching the leaves would contain ions in the same proportions as in the soil.

This does not happen because of the inner layer of the cortex, the **endodermis**. In these cells, the walls that are at right angles to the epidermis (the anticlinal walls) are thickened by a band of suberin called the **Casparian strip** (Figure 34.13).

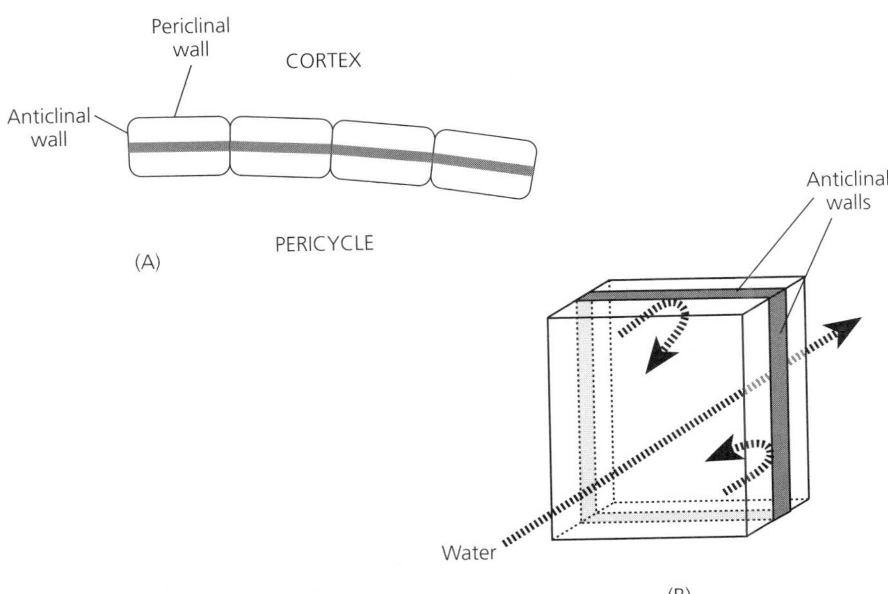


**Figure 34.13**
Endodermal cells showing the Casparian strip; in transverse section through a root (A) and a 3-D view of an endodermal cell showing how the Casparian strip blocks water movement through anticlinal wall (B)

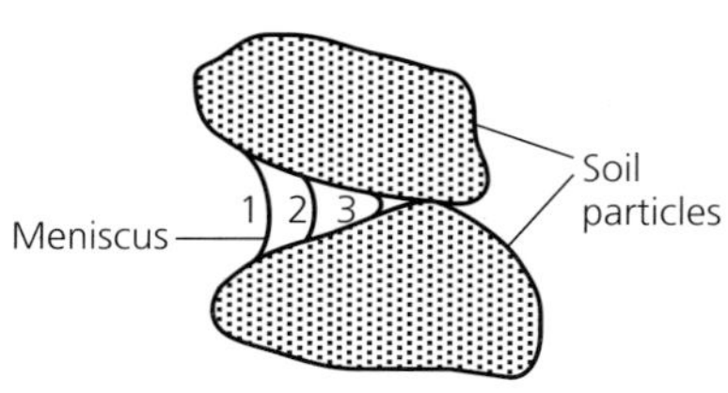


**Figure 34.14**
How surface tension makes it harder for a plant to absorb water as a soil dries out. When there is plenty of water, the water surfaces are curved as in (1). As the water retreats into narrower spaces, the curvature of the water surfaces increases (2) and (3). The more strongly curved the menisci, the greater the tension they exert and the more negative the pressure potential, and hence the water potential, of the soil water

Suberin is a waxy substance like cutin, and is impermeable to water and thus to dissolved ions. By rendering the anticlinal walls impermeable, the endodermis divides the apoplast of the root into outer and inner compartments, separated by at least two cytoplasmic membranes. It is these membranes that exert control over uptake. Once through the endodermal barrier, water can travel all the way to the leaf mesophyll without passing through any membrane barrier.

### *Absorbing water from dry soils*

In a well-watered soil, the only significant factor affecting its water potential is its solute potential, typically of the order of 0.2 MPa. As the soil dries out, the water retreats into finer and finer spaces between the soil particles, increasing the curvature of the menisci (Figure 34.14). The pressure potential becomes increasingly negative, until an equilibrium is set up between the water potential inside the roots and that in the soil, and no more water can be extracted.

### *The gradient of water potential*

In passing through the plant from soil to atmosphere, water moves down a gradient of water potential (Figure 34.15). By far the greatest drop in water potential occurs at the leaf epidermis, so it is here that the greatest force driving the water is located. Since the drop in water potential across the root surface is small, so is the force driving water across each unit area of root. Root systems thus need a large area to absorb water at the same rate as it is being lost through the shoot. Transplantation usually results in considerable loss of root surface, but consequent wilting can be reduced by placing a plastic bag over the shoot system for a few days until new rootlets and root hairs have developed.

Figure 34.16 gives an overview of the route followed by water through the plant.

## *Translocation of ions*

By the early decades of the 20th Century, plant physiologists were convinced that mineral ions were transported from roots to leaves in the xylem. Not only does xylem sap contain small quantities of mineral ions, but the upward movement of ions continues after removal of the bark

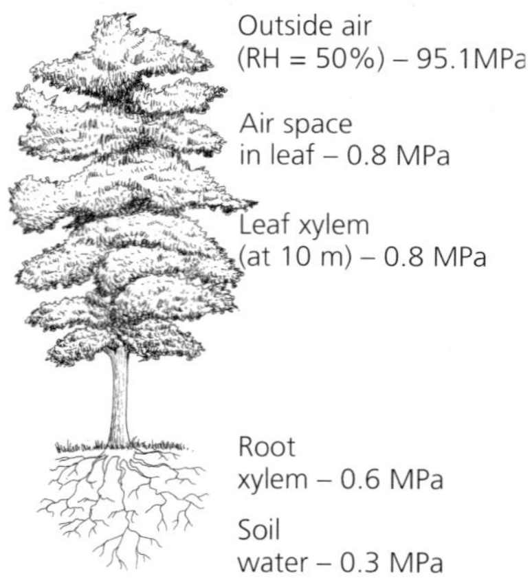


**Figure 34.15▲**
Changes in water potential as water flows through a tree (modified from P. Nobel). Notice that the greatest drop in water potential occurs across the stomata

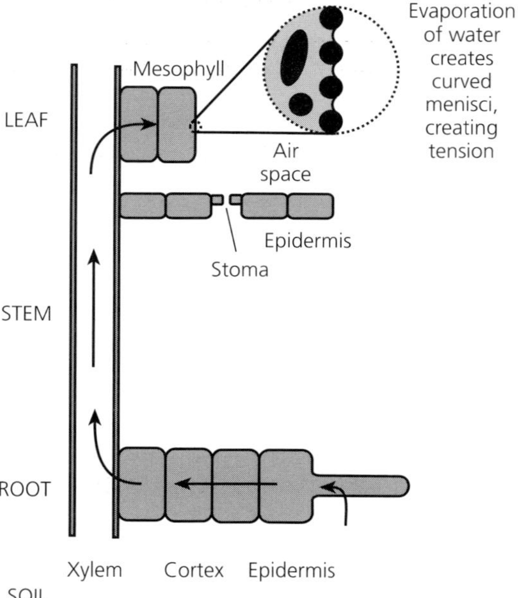


**Figure 34.16◀**
Overview of water flow through a plant

from a section of branch, leaving only the wood (secondary xylem). However, this does not exclude the possibility that some upward movement of minerals occurs in the phloem.

That the phloem is not involved to any significant extent in mineral transport was demonstrated by the experiment shown in Figure 34.17.

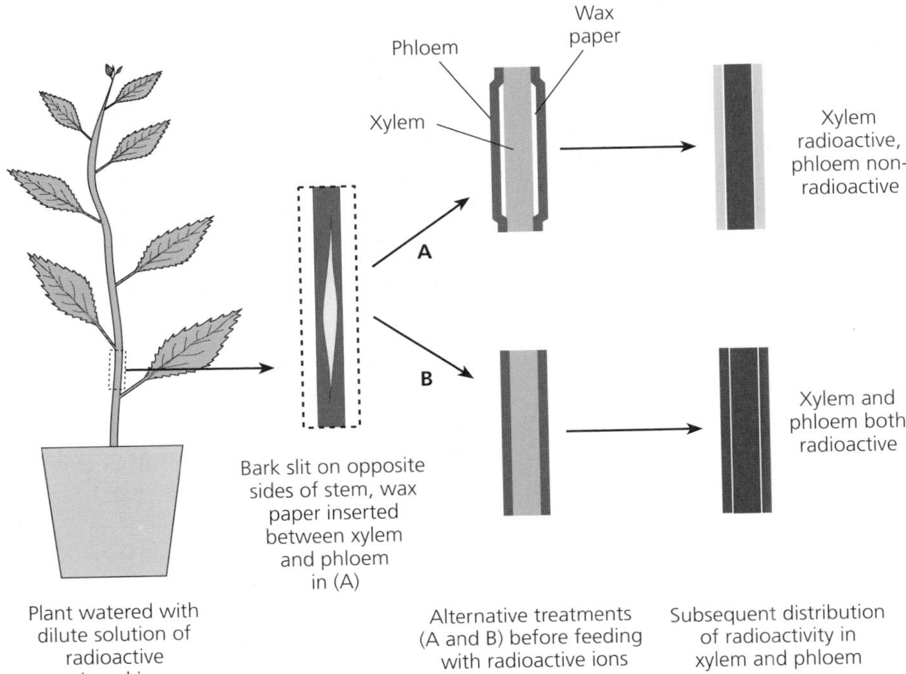


**Figure 34.17**
Experimental demonstration that upward movement of mineral ions occurs in the xylem rather than the phloem

Woody plants, such as willow or cotton, were grown in culture solutions. Two 20 cm-long slits were made in the bark on opposite sides of certain shoots and the bark gently prised away from the wood. In some of these plants a length of waxed paper was inserted between the xylem and phloem. In the control plants the slit bark was left in contact with the xylem. Other plants were left with the bark uninjured. All the plants were then watered with radioactive phosphate or potassium ions, and after a few hours the radioactivity of the phloem and xylem in the treated region were analysed. Only where the phloem had been in contact with the xylem did it contain significant radioactivity.

## The translocation of organic materials

Whereas water always moves from roots to shoots, the pattern of movement of organic materials is more variable and frequently changes over time. Underground parts normally import carbohydrate, but storage tissues frequently export carbohydrate to the young shoots in the spring. Young leaves import sugar and amino acids, but mature leaves export them.

### The structure of phloem

The transport of organic materials by plants is referred to as **translocation**, and occurs in the phloem. Like xylem, phloem is a complex tissue containing several kinds of cell. In flowering plants, the

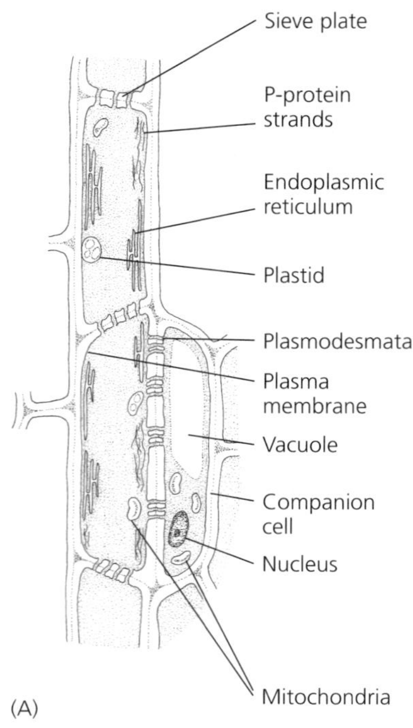


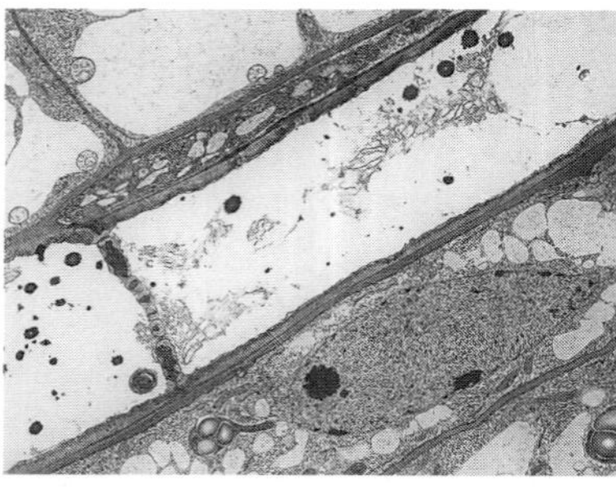


**Figure 34.18▲**
Sieve tube elements and a companion cell (A) (after Taiz and Zeiger) and electron micrograph of a sieve tube element (B)

conducting tubes are called **sieve tubes**, and are formed by end-to-end joining of individual cells called **sieve tube elements** (in non-flowering vascular plants the conducting cells are called **sieve cells** and are not linked into tubes). The end walls of the sieve tube elements form **sieve plates**, perforated by many small holes (Figure 34.18). Sieve tube elements are lined by a thin layer of cytoplasm containing mitochondria and sometimes plastids, but there is no tonoplast. Unique among plants, the nucleus degenerates during early development.

Next to each sieve tube element are one or more **companion cells** – specialised parenchyma cells derived by division of the same mother cell as the sieve tube element. Their precise function is unclear. Also present in phloem are unspecialised parenchyma cells and, frequently, fibres.

## How do we know that organic materials are transported in the phloem?

One of the first experiments implicating phloem in transport was carried out by Marcello Malpighi over 300 years ago. He removed a ring of bark from a tree, an operation which leaves the xylem intact but removes the phloem. During the subsequent months he observed that the bark above the ring became swollen, while below the ring it became withered (Figure 34.19). Malpighi concluded that the ring had prevented nutrients from moving down the trunk.

Two and a half centuries later (1928), T.G. Mason and E.J. Maskell obtained further evidence by analysing changes in the sugar content of the phloem and the leaves of cotton plants. A rise in leaf sugar was soon followed by a rise in sugar content of the phloem (Figure 34.20).

Direct proof that sugar is translocated in the phloem was obtained in the 1940s with the introduction of radioisotopic tracers. Sections were cut through a stem below a leaf that had been 'fed' with $^{14}CO_2$ (Figure 34.21). The sections were covered with a fine photographic film for long enough for the radioactivity to affect the film. After the film was developed, the distribution of radioactivity was found to correspond with that of the sieve tubes.

An ingenious method of directly analysing phloem sap was invented in 1953 by Mittler, an entomologist studying aphids. These insects have stylets so fine that they pierce a single sieve tube. By anaesthetising a feeding aphid with $CO_2$, the animal can be cut away from its mouthparts, after which sap continues to exude from the cut surface for hours (and even days) enabling it to be analysed (Figure 34.22).

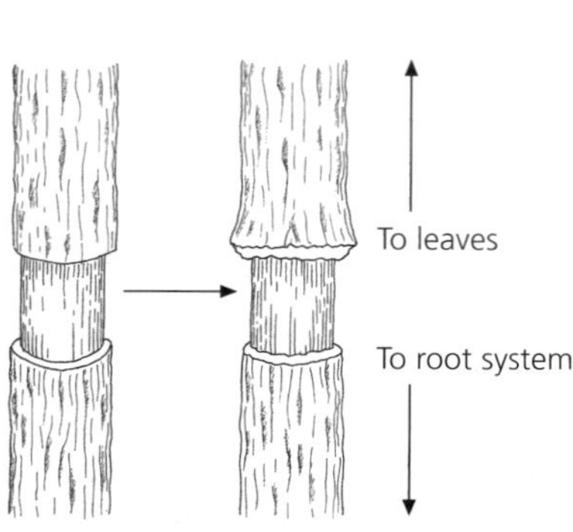


**Figure 34.19▲**
Effect of ringing a branch of a tree

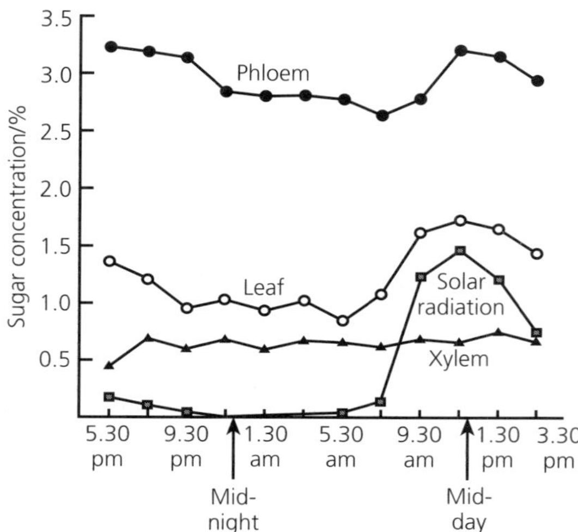


**Figure 34.20◀**
Changes in sugar content of leaves and phloem of cotton plants. Note the near synchrony of the peaks for solar radiation and leaf sugar content, and the slight delay before the peak in phloem sugar (after Mason and Maskell)

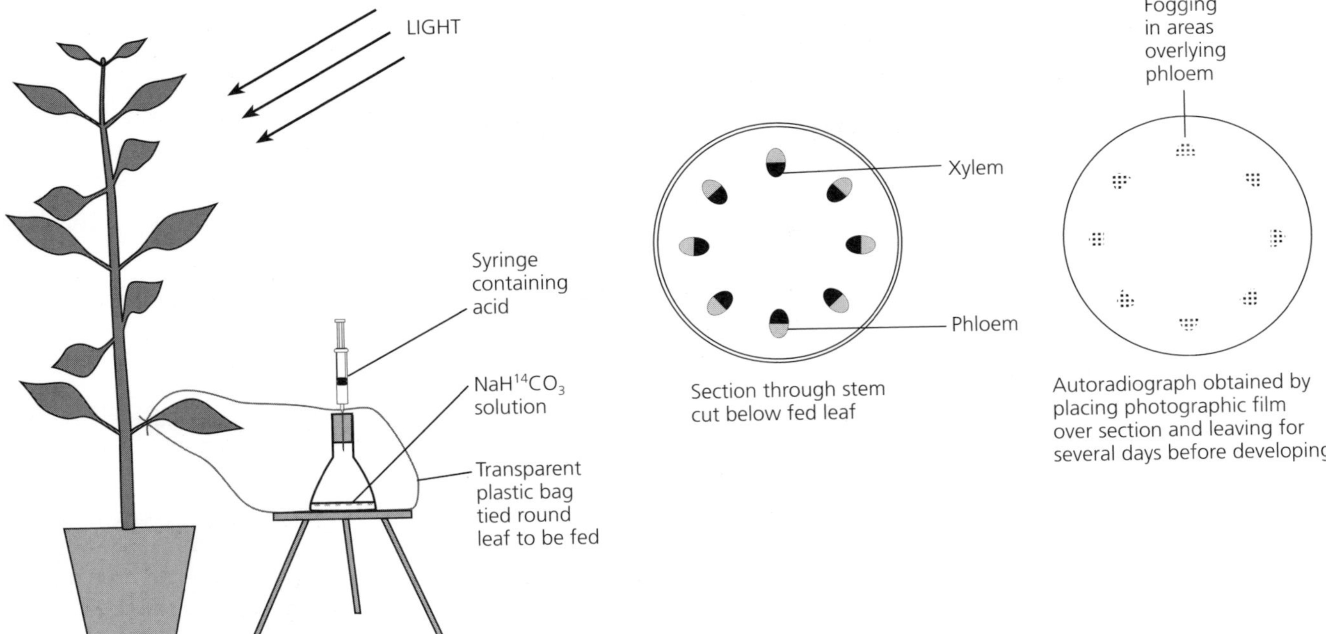


**Figure 34.21▲**
Feeding a leaf with radioactive carbon and tracing its subsequent distribution in the stem

Aphid stylets have become a favourite research tool of phloem physiologists, and studies have shown that in most plants, phloem sap contains about 20–30% sucrose (and in some cases, other non-reducing sugars), with smaller amounts of amino acids and amides.

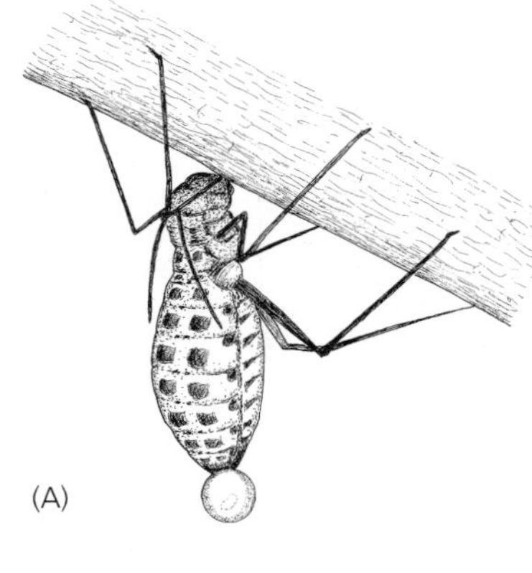


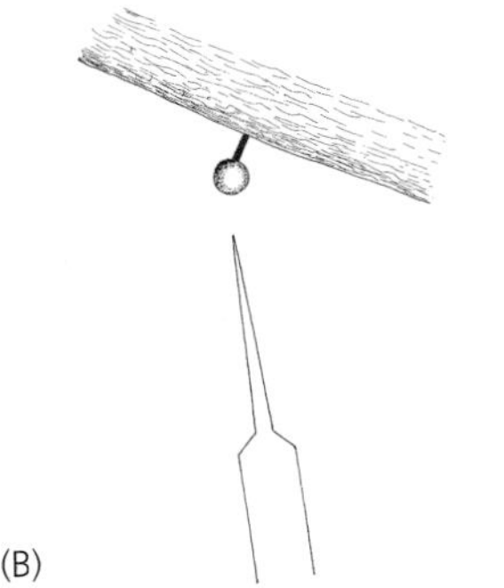


**Figure 34.22▲**
Using aphid stylets to collect phloem sap. Aphid allowed to feed, before anaesthetising and gently cutting away from base of stylets (A); sap collected from stylets using a micropipette (B)

## Characteristics of phloem transport

Phloem transport differs in a number of important ways from xylem transport.

- Whereas vessels and tracheids are dead, sieve tube members are living and carry out metabolic processes. Killing the phloem with steam stops phloem transport immediately and permanently.
- Phloem sap is under high pressure (about 2 MPa, or 150 times human arterial blood pressure).
- Sugar moves more slowly in phloem than water does in xylem – about 50–150 cm $hr^{-1}$.
- Whereas water almost always moves from roots to leaves, the sugar pathway is more variable. Sugar is translocated from **sources**, or sites of sugar production, to **sinks**, or sites of sugar consumption. Sources include rapidly photosynthesising leaves and storage tissues in which sugar is being mobilised, such as the cotyledons and the endosperm in germinating seeds. Sinks include most roots, young leaves and developing seeds and fruits.

## The mechanism of phloem transport

Research into phloem transport has been bedevilled by the extreme fragility of the sieve tubes. While tracheids and vessel members are self-sealing, sieve tubes actively produce their own protection in the form of two substances: P-protein ('P' for phloem) and callose.

In functioning sieve tubes, **P-protein** is present as delicate strands crossing the sieve tube. Immediately after a sieve tube is punctured, the surge of sap following the release of pressure causes the P-protein to pile up against the sieve plates, sealing them.

A longer-term defence substance is **callose**, a polysaccharide. This is absent from functioning sieve tubes but within minutes of injury it accumulates over the sieve plates and blocks them. So sensitive is the trigger for callose production that it was always produced during the preparation of tissues for examination. Only with the development of rapid freeze-drying techniques has it been conclusively shown that sieve plates are normally open.

### *The pressure-flow hypothesis*

The first of several competing hypotheses was the **pressure-flow hypothesis**, first proposed in 1927 by Ernst Münch. According to this view, phloem sap is pushed down a pressure gradient, from a high pressure in the source to a lower pressure in the sink.

A physical model is shown in Figure 34.23. The two compartments, A and B, are each separated from the surrounding water in C by a differentially permeable membrane. A contains a dilute sugar solution and B contains a more concentrated sugar solution. Water passes from C to B by osmosis, raising the pressure inside B. Since they are connected, the pressure also rises in A. The pressure rise in A is sufficient to reverse any tendency for water to enter A from C, so water leaves A.

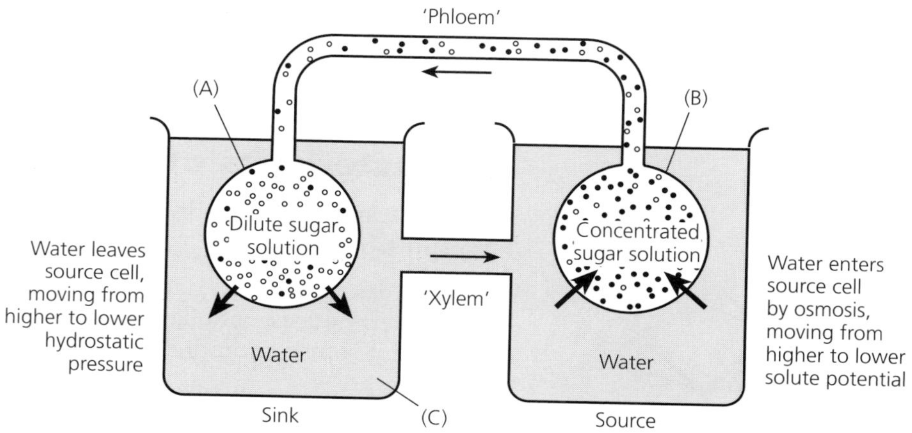

**Figure 34.23**
Laboratory model of the pressure-flow mechanism. Sugar solution flows down a gradient of hydrostatic pressure, generated by the greater tendency for water to enter source cells than sink cells

Notice that the external solution does not have to be pure water; it does not even have to be less concentrated than in A. The only necessary condition is that the sugar is more concentrated in B than in A, so water has a greater tendency to enter B than it has to enter A.

With continued movement of sugar from B to A, the laboratory model will eventually 'run down' as the difference in sugar concentrations decreases. In the living plant the concentration difference is maintained by active transport. In the source (represented by compartment B) sugar is actively transported from the parenchyma cells into the sieve tubes. Water follows by osmosis, raising the pressure in the source. In the sink (represented by A) sugar is removed, either by respiration or in the synthesis of other molecules, and water follows.

Notice that although sugar moves from source to sink, water *circulates*, moving back from sink to source via the xylem, as shown in Figure 34.24.

Like any good hypothesis, the pressure-flow hypothesis made certain specific predictions that could be tested experimentally:

1 the pressure must be lower in the sink sieve tubes than in the source sieve tubes, and the rate of flow should be correlated with the magnitude of the pressure gradient.

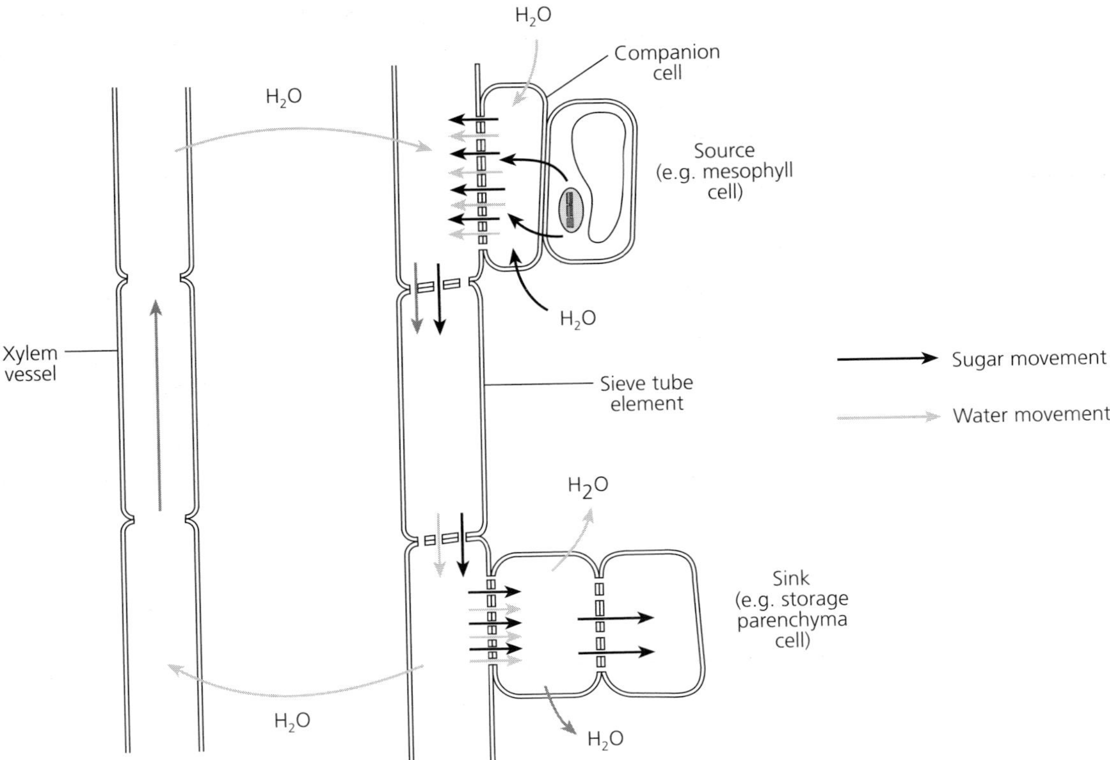


**Figure 34.24**
Role of the xylem in phloem translocation

2 within a moderate range of temperatures, the rate of transport should not be closely correlated with metabolic rate of the sieve tubes.

3 bidirectional transport must not occur in the same sieve tube.

Direct measurements have shown the first prediction to be true, but for many years it was believed that the second provision was not met because cooling a stem frequently resulted in severe slowing of transport. For this reason a number of alternative hypotheses have been suggested, all of which maintain that the sieve tubes themselves provide the energy for transport.

None of the hypotheses based on active movement has received much experimental support, and in the last few years the objections to the pressure-flow hypothesis have been dealt with. The effect of cooling is now known to be temporary – if cooling is maintained, the rate of translocation recovers, even though the metabolic rate of the cooled sieve tubes is only a fraction of normal.

Finally, while neighbouring sieve tubes can transport materials in opposite directions, it has never been convincingly shown that transport can occur in both directions in the same sieve tube at the same time.

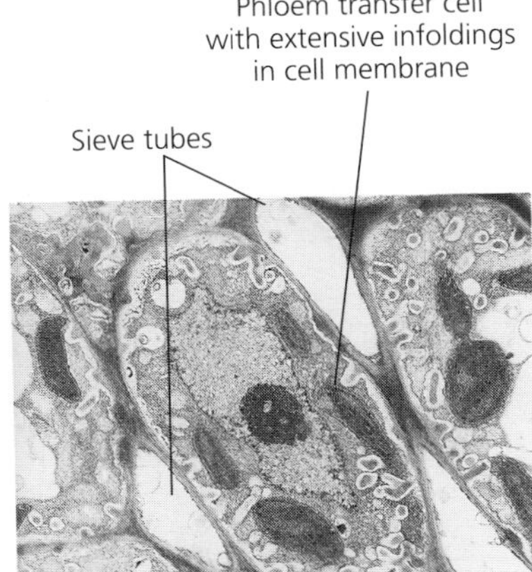


**Figure 34.25**
A phloem transfer cell – a modified companion cell

### *Phloem loading*

Although sieve tubes play a passive part in translocation, the generation of the pressure gradient depends on the active secretion of sugar into the sieve tubes in the source. This is called **phloem loading**, and is thought to involve cotransport with protons. Protons are actively extruded by the sieve tube–companion cell complex, and diffuse back through channel proteins accompanied by sucrose.

In certain flowering plants, some of the phloem companion cells are modified as **transfer cells** (Figure 34.25). They have extensive infoldings of the cell walls, suggesting that they are involved in some active transport role. Unfortunately they are less common in the vein endings (where they would be expected to be the most common) and they are not restricted to phloem.

## The transport of mineral ions

Although mineral ions are predominantly transported in the xylem, some also move in the phloem. For example before leaf fall, nitrogen, phosphorus and potassium are all removed from the leaves via the phloem and are thus retained by the plant. Others, such as boron, iron and calcium, are immobile and are lost at leaf fall.

This difference in mobility shows in the effects of deficiency. Shortage of nitrogen, potassium and phosphorus is shown in the oldest leaves because the young shoots are kept supplied by translocation from older leaves. Deficiency of calcium, on the other hand, shows first in the younger leaves.

## QUESTIONS

**1** Criticise the statement that 'the functions of a leaf include transpiration and respiration'.

**2** In an investigation into water loss in a certain species of plant, eight leaves were removed, grouped into four pairs and smeared with grease as shown in the table. After determining the mass of each leaf, one of each pair was suspended in daylight and the other in darkness, all other conditions being the same. After six hours the mass of each leaf was again determined and its percentage decrease in mass calculated. The results are shown in the table below.

| | Treatment | Percentage loss in mass Light | Dark |
|---|---|---|---|
| A | Greased on both surfaces | 0 | 0 |
| B | Greased on upper surface only | 10 | 2 |
| C | Greased on lower surface only | 2 | 1 |
| D | Left untreated | 12 | 3 |

**a)** What was the purpose of treating the leaves as in **A**?
**b)** Why do leaves **B, C** and **D** lose mass faster in the light than in the dark?
**c)** Give the letter of the treatment which indicates the percentage water loss via the **(i)** lower epidermis, **(ii)** upper epidermis.
**d)** What fraction of lower epidermal transpiration occurs via the cuticle in the light?
**e)** What appear to be the relative numbers of stomata in the upper and lower epidermis? Express as a ratio upper:lower.
**f)** State *three* assumptions you have had to make in answering the above questions.

**3** In an investigation into transpiration, a leaf was suspended by a piece of cotton in a well-lit, draughty place, and weighed every minute for an hour. The change in mass of the leaf is shown in the graph. Can you explain why the gradient of the graph changes?

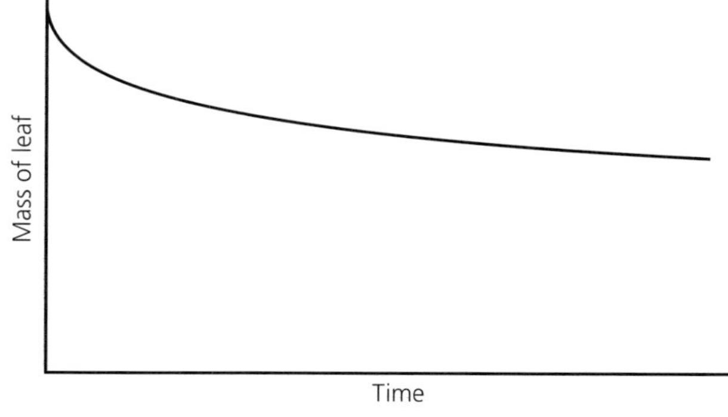

## QUESTIONS

**4** The following results give the distance moved by a bubble along a capillary tube of cross-sectional area 2 $mm^2$ in 4 minutes, when shoots of five species were connected to it.

| Species | Distance moved in 4 mins (mm) | Total leaf area ($cm^2$) |
|---|---|---|
| A | 25 | 50 |
| B | 5 | 40 |
| C | 50 | 75 |
| D | 40 | 30 |
| E | 10 | 50 |

**a)** Which species has the highest rate of transpiration under the conditions of the investigation?
**b)** Calculate the rate of transpiration for species **A**.

**5** Suggest how the spines of a cactus could promote survival in dry habitats.

**6** **a)** Which would you expect to have the greater rate of transpiration under the same environmental conditions – a green leaf or a petal of identical size and shape? Justify your answer.
**b)** Imagine two shoots in identical atmospheric conditions. Shoot A is very young and is importing most of its organic material. Shoot B is mature and is exporting most of the carbohydrate it produces. Which would you expect to be more vulnerable to water loss in water shortage, and why?

**7** Some flowers still wilt when put in water immediately after picking. Suggest why the shoot is unable to absorb water quickly enough.

**8** What would you expect to be the effect (if any) of each of the following treatments on the rate of water uptake by a plant?

**a)** covering the shoot system with a transparent hood
**b)** watering the soil with a dilute salt solution
**c)** lowering the soil temperature
**d)** flooding the soil?

In each case, say whether the effects would be short or long term, and give reasons.

**9** When a plant is transplanted, it is more likely to survive if a transparent plastic hood is placed over the shoot system. Explain why.

**10** One way of estimating the velocity of movement of sugar in the phloem is by measuring the increase in dry mass of a growing fruit such as a pumpkin. Suppose you knew the average increase in dry mass per hour, what *two* other measurements would you need to determine the velocity of flow of the sugar?

EXAM QUESTIONS

## EXAM QUESTIONS

**1** The graph shows the effect of light intensity on the rates of photosynthesis in two different species of plant. Species **A** normally grows in full sunlight and species **B** normally grows in the shade.

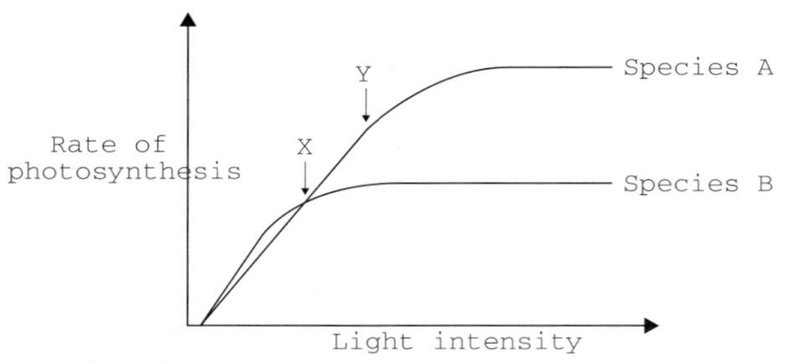


**a)** The rate of photosynthesis in species **A** is limited by light intensity between points **X** and **Y**. What is the evidence from the graph to support this? *(1 mark)*

**b)** Describe and explain how *one* feature of the curve for species **B** shows an adaptation to growing in the shade. *(2 marks)*

**c)** Suggest *one* way in which you would expect the structure of the leaves of species **A** and species **B** to differ. Give a reason for your answer. *(2 marks)*

**AEB**

**2** The diagram shows part of the light-independent pathway of photosynthesis.

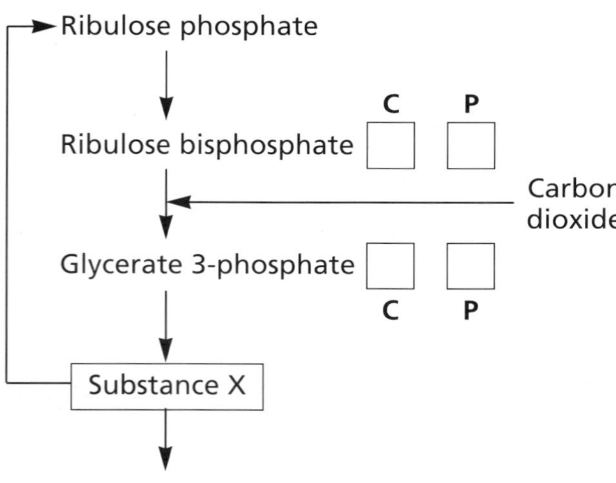


**a)** For the two compounds, ribulose bisphosphate and glycerate 3-phosphate, enter the number of carbon atoms in each molecule in the boxes labelled **C** and the number of phosphate groups in the boxes labelled **P**. *(1 mark)*

**b)** Give the name of substance **X**. *(1 mark)*

**c)** Give *two* specific roles of ATP in the light-independent pathway of photosynthesis. *(2 marks)*

**AEB**

**3** Some of the flowers of the barley cereal (*Hordeum distichon*) are sterile, some are male only, and some produce grain. During the daily flowering period, vast amounts of pollen are shed. The drawing shows a single flower.

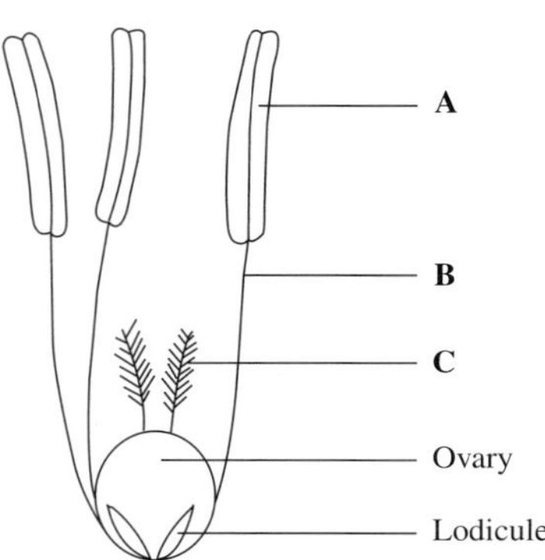


**a) i)** Name the parts of the flower labelled **A**, **B** and **C**.

**ii)** Suggest which part of a generalised flower structure is represented by the lodicules. *(4 marks)*

**b)** Using the information above, give reasons which would support the assumption that barley is

**i)** wind pollinated, *(2 marks)*

**ii)** cross pollinated. *(1 mark)*

**c)** On the outline diagrams below, draw *all* cells and nuclei present at maturity, and label the male and female gametes. *(4 marks)*

A germinated pollen grain

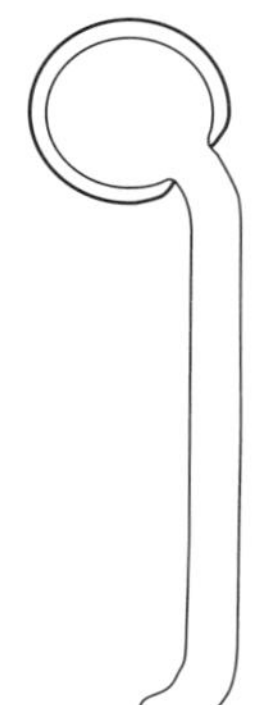

An embryo sac

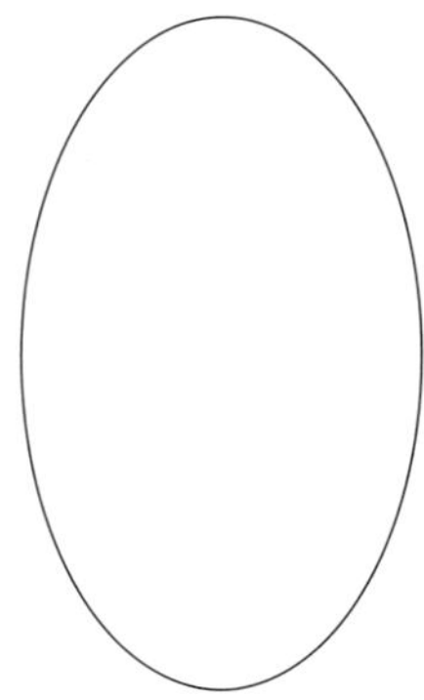

## EXAM QUESTIONS

**d) i)** Name the chief food reserve found in cereal grain.

**ii)** State the rôle of the hormone gibberellin in the mobilisation of this food reserve during germination.

*(3 marks)*
**WJEC**

**4** The diagram below shows vertical sections of a male flower and a female flower of red campion, *Silene dioica* ($2n = 24$). The male and female flowers always occur on separate plants.

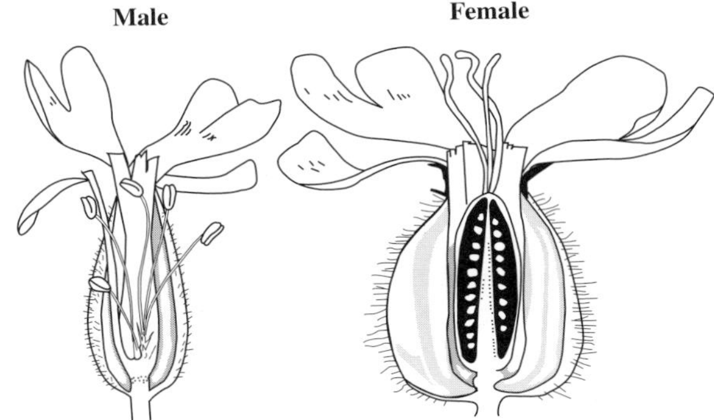


**a)** With reference to the diagram only,

**i)** state *three* ways in which the structure of a male flower *differs* from that of a female flower.

*(3 marks)*

**ii)** state the most likely vector for pollination of the red campion, giving one reason for your answer.

*(1 mark)*

**b)** Suggest the advantages to the red campion of having male and female flowers on separate plants. *(3 marks)*

**c)** State the chromosome number of a single gamete nucleus of a red campion pollen grain. *(1 mark)*
**UCLES**

**5** The diagram shows a section through a leaf, showing the pathways followed by water molecules, as black arrows, and passage of sugar molecules manufactured during photosynthesis, as white arrows.

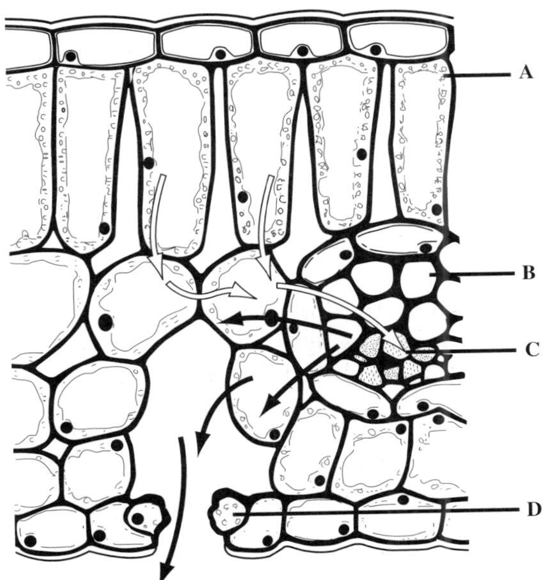


**a)** Identify cells **A** to **D**. *(2 marks)*

**b)** Describe the methods by which water moves in the direction shown by the black arrows in the diagram. *(4 marks)*

**c)** Outline how sucrose is loaded into the cell **C**. *(2 marks)*
**UCLES**

**6** The table below refers to the structure and functions of xylem vessels and phloem sieve tubes in plants.

If the statement is correct, place a tick (✓) in the appropriate box and if the statement is incorrect, place a cross (✗) in the appropriate box.

| Statement | Xylem vessels | Phloem sieve tubes |
|---|---|---|
| Possess living contents | | |
| Provide support | | |
| Composed of cells fused together end to end | | |
| Walls contain lignin | | |

*(4 marks)*
**ULEAC**

## EXAM QUESTIONS

**7** An investigation was carried out into the relationship between the rate of water absorption and the rate of transpiration in sunflower plants at various times of the day. The results are shown in the diagram below.

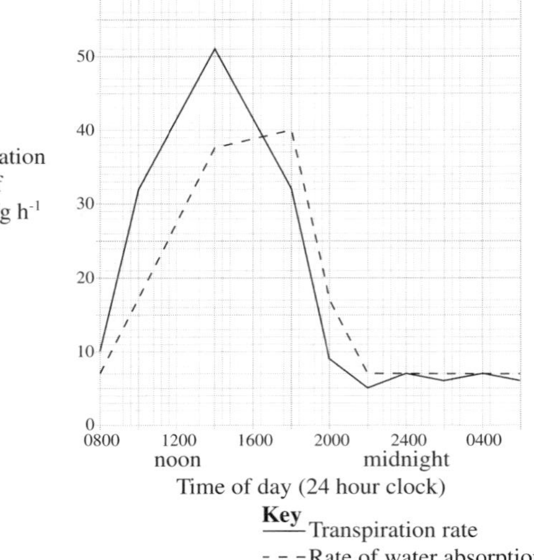


**a) i)** Describe the changes in the rate of transpiration that took place during the experiment. *(3 marks)*

**ii)** Suggest why these changes occurred. *(3 marks)*

**b)** Comment on the relationship between the rate of transpiration and the rate of water absorption during the experiment. *(2 marks)*

**c)** Describe a simple method that you could use in the laboratory to measure the rate of transpiration of a flowering plant. *(4 marks)*

**EDEXCEL**

**8** Air flows in and out of a leaf only via the stomata. The diagram shows a porometer, a device which can be used to investigate this flow.

The porometer cup is attached to the surface of a whole leaf using an air-tight seal. Water is drawn up the graduated scale and the water meniscus set at zero. A stop watch is started and the time taken for the water meniscus to fall a set number of scale units is recorded. The rate at which the water meniscus falls depends on the amount of air drawn through the stomata into the porometer cup.

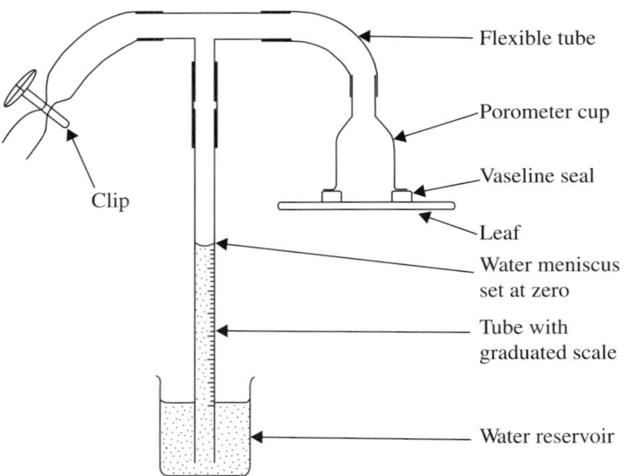


**a)** State *one* function of stomata. *(1 mark)*

**b) i)** Describe how you would use the apparatus to investigate the effect of light intensity on the rate at which air flows through the leaves of a plant. *(4 marks)*

**ii)** The graph below shows the results of this experiment.

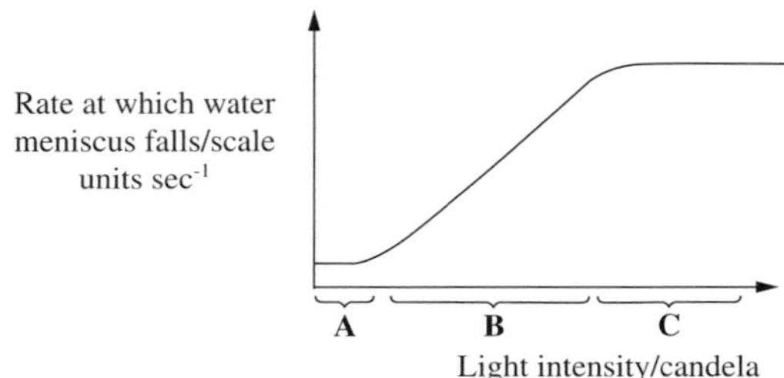


Suggest an explanation for the rate at which the water meniscus falls in regions **A**, **B** and **C**. *(3 marks)*

**iii)** Measurements of intercellular pH in the spongy mesophyll showed a fall in pH in darkness and a rise in daylight. Explain these changes. *(3 marks)*

**c)** Suggest why the behaviour of stomata under the porometer cup may differ from those in the rest of the leaf. *(2 marks)*

**WJEC**

**9 a)** Isolated chloroplasts are able to release oxygen in the presence of an oxidising agent. This is called the Hill reaction.

## EXAM QUESTIONS

The Hill reaction was investigated using the blue dye DCPIP as a substitute for the electron acceptor normally present in the cell. The reaction involved is shown below.

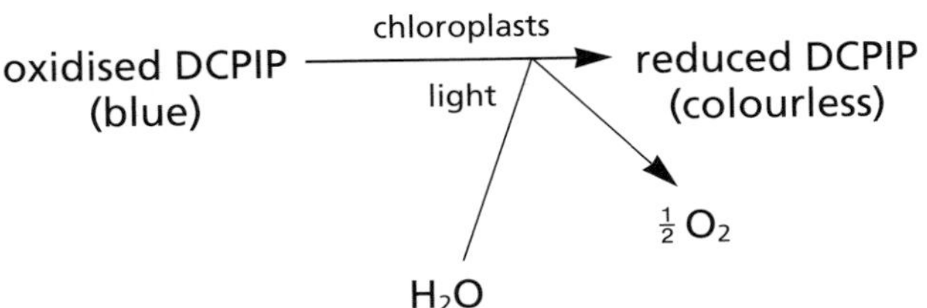


Complete the grid by naming the electron donor and the final electron acceptor in the experiment and in the intact cell. *(4 marks)*

| Situation | Initial Electron Donor | Final Electron Acceptor |
|---|---|---|
| Experiment | | |
| Intact cell | | |

**b)** The progress of the Hill reaction was followed by measuring the absorbance of DCPIP at different time intervals using a colorimeter. This measures the amount of light at a certain wavelength which is absorbed by a sample of solution.

Two light intensities were compared by placing tubes containing 0.5 $cm^3$ chloroplast suspension and 5 $cm^3$ DCPIP solution at 15 and 45 cm from a lamp.

The results are shown in the table.

| Time/min | Absorbance at 600 nm | |
|---|---|---|
| | 15 cm from lamp | 45 cm from lamp |
| 0 | 0.35 | 0.35 |
| 0.25 | 0.2 | 0.31 |
| 0.5 | 0.11 | 0.26 |
| 1 | 0.025 | 0.18 |
| 2.0 | 0 | 0.04 |

*Adapted from Hawcroft and Short (1973) J. Bio. Ed. No 57, 23–26*

**i)** Plot the data on graph paper. *(6 marks)*

**ii)** Which part of which curve shows the most rapid reduction of DCPIP? *(2 marks)*

**iii)** In addition to the experiment described there was a control. Suggest the conditions in the control experiment. *(1 mark)*

**c)** In another experiment the effect of the weedkiller, Diuron, on the Hill reaction was investigated. During the first minute, the rate of change of absorbance in a solution of DCPIP and Diuron was reduced by 95% compared with DCPIP alone. Suggest how Diuron might cause the death of a plant. *(2 marks)*

**WJEC**

**10 a)** The diagram below shows the apparatus used to measure the effect of light intensity on the rate of photosynthesis in *Elodea*.

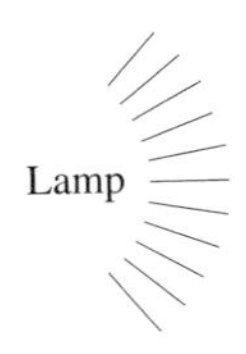


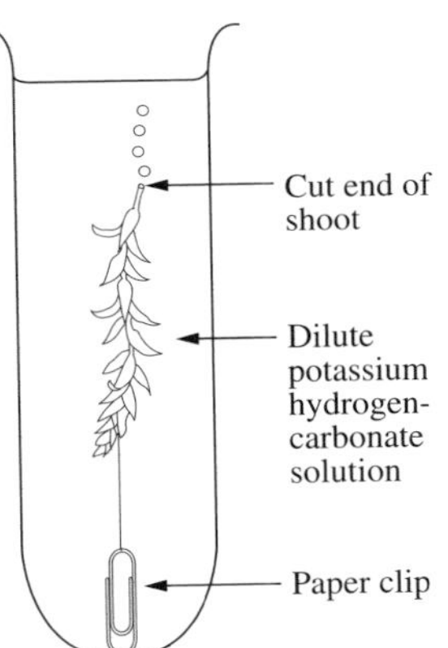


**i)** Explain why it is possible to study the rate of photosynthesis using the above apparatus. *(2 marks)*

**ii)** Explain why potassium hydrogencarbonate was used. *(1 mark)*

**iii)** Give *one* reason why the results using this apparatus may not be accurate. Explain your answer. *(2 marks)*

## EXAM QUESTIONS

**b)** The graph shows the uptake and release of carbon dioxide which takes place in a plant during a 12 hour period.

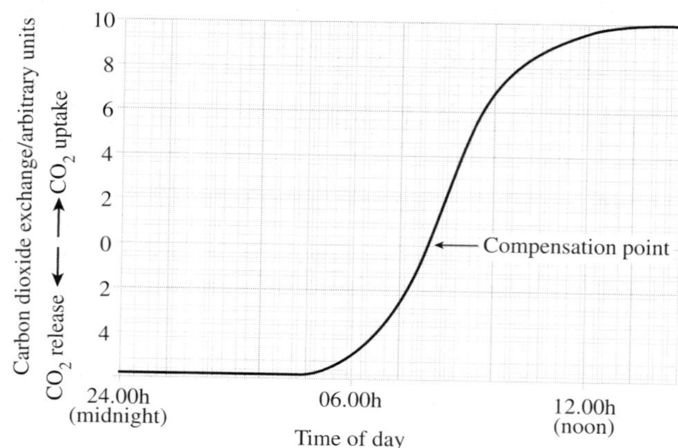


**i)** At the compensation point carbon dioxide is neither taken up nor released. Suggest an explanation for this observation. *(1 mark)*

**ii)** This plant increased its mass only after the compensation point had been reached. Use the information in the graph and your own knowledge to explain the reason for this. *(3 marks)*

**iii)** The compensation period is the length of time between daybreak and the compensation point. In low-growing woodland plants this is shorter than in meadow plants. Suggest why this is an advantage to woodland plants. *(2 marks)*

**WJEC**

**11** The diagram shows the arrangement of tissues in a transverse section through the stem of a lime tree, *Tilia europaea*.

**a) i)** Name the tissues labelled **B**, **D** and **E**. *(3 marks)*

**ii)** Complete the table by naming and stating the function of *each* of the tissues labelled **A** and **C** in the diagram. *(4 marks)*

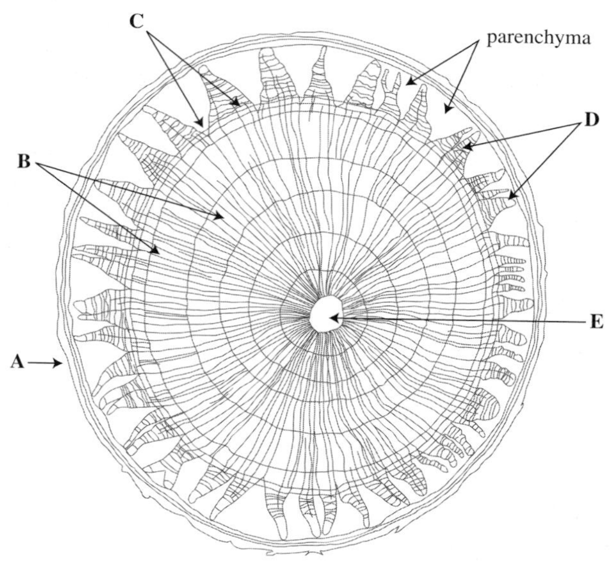


| Structure | Name | Function |
|---|---|---|
| A | | |
| C | | |

**b) i)** How many years old is the tree in the diagram? *(1 mark)*

**ii)** Shade in the whole area on the diagram that shows the second year's growth of the tree. *(1 mark)*

**iii)** As the tree becomes older it may become hollow.

1. Explain why the tree does not die. *(2 marks)*
2. Suggest *one* advantage of hollow branches or trees to other organisms in the community or ecosystem. *(1 mark)*

**c)** Young trees are often attached to stakes with strong plastic ties so that they can withstand winter gales. Suggest why the tree will often die after 3 to 4 years if the tie is not loosened. *(2 marks)*

**WJEC**

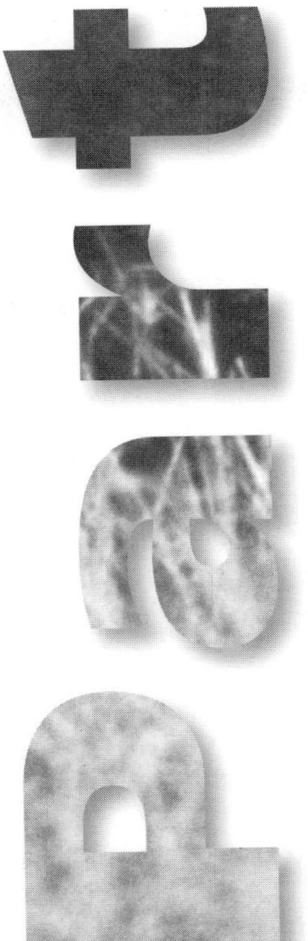

# Part four

# ORGANISMS and THEIR ENVIRONMENT


**CHAPTER 35**
INTRODUCTION to ECOLOGY . . . . . . . . . . 670

**CHAPTER 36**
The PHYSICAL ENVIRONMENT . . . . . . . . . . 679

**CHAPTER 37**
The BIOTIC ENVIRONMENT . . . . . . . . . . 687

**CHAPTER 38**
POPULATIONS . . . . . . . . . . 710

**CHAPTER 39**
COMMUNITIES and ECOSYSTEMS . . . . . . . . . . 725

CHAPTER

# 35 INTRODUCTION to ECOLOGY

## LEARNING OBJECTIVES

By the time you have completed your study of this chapter you should be able to:

- define and distinguish between the terms *environment, habitat* and *niche*.
- define and distinguish between the terms *population, community* and *ecosystem*, giving an example of each.
- distinguish between abiotic and biotic factors, and between conditions and resources, giving examples of each.
- describe the constituents of soil and the role that each plays in the growth of plants.
- explain how soil is formed.

## What is ecology?

The word 'ecology' was first coined by Ernst Haekel in the mid-19th Century to mean the study of organisms in relation to their surroundings (*oikos* is Greek for home). Ecology deals with the ways in which organisms interact with each other and with other aspects of their surroundings. How, for example, are the numbers of individuals in a species regulated? Why is a species rare in one place and common in another, and why was it rarer ten years ago than it is now?

Questions like these are often hard to answer because, as we shall see, most environments are complex and ever-changing. However, certain basic principles are well understood. We know, for instance, why fish die when farm slurry is allowed to run off farmland into lakes. We know why killing all large predators in an area can lead to overgrazing by herbivores. But there are many cases in which the interactions between organisms and their environment are so subtle, with so many variables involved, that ecologists are often unable to explain or make predictions.

### Habitat, environment and niche

The simplest way of describing an organism's surroundings is in terms of its **habitat**, or the place in which it lives. The habitat of a carp, for example, is a lake or pond, and the habitat of the alder tree is the bank of a stream or river.

Most habitats are made up of patchworks of **microhabitats**, each with its own characteristic inhabitants. A leaf miner (the larva of an insect that lives between the upper and lower epidermis of a leaf) and a bark beetle both share the same woodland habitat, but they live in very different microhabitats, each with its own microclimate.

Distinct from an organism's habitat is its **environment**. This includes all the factors in its surroundings that affect its chances of survival and reproduction. A carp may continue to live in the same pond (its habitat), but its environment is ever-changing.

Quite different from an organism's environment is its **niche**. This concept will be dealt with in more detail in Chapter 37, but it is basically an organism's *way of life*, or role in the community. For example, a centipede and a millipede may share the same habitat, but they make their living in very different ways, since a centipede is a carnivore and a millipede is a herbivore.

Ecological phenomena can be studied at a number of different levels of complexity.

- **Populations** A population comprises all the individuals of a given species in a particular habitat, such as the deer on the Scottish island of Rhum, or the trout in a tributary of the river Ribble.
- **Communities** A community comprises all the species in a given area, such as all the inhabitants of a pond. A community is thus a group of populations.
- **Ecosystems** One level up the complexity hierarchy, an ecosystem includes a community together with all the non-living (abiotic) components that interact with the community. A pond ecosystem thus comprises not only all the organisms in the pond, but the water and its dissolved and suspended matter, and the mud at the bottom.
- **Biomes** A biome is a major ecosystem type, extending over a large geographical area – in some cases beyond continental boundaries. The Arctic tundra biome includes those parts of North America and Asia that encircle the Arctic Ocean. Tropical rain forest, savannah and coral reefs are other examples of biomes.

## Autecology and synecology

Instead of studying an entire community or ecosystem, some ecologists focus their attention on a single species. In this kind of study, called **autecology**, other species are included in the study only in so far as they influence the species in question.

Autecological studies often play an important part in the conservation of endangered species. Effective protection of a species requires a detailed knowledge of all the factors that affect its survival and breeding success. Autecological studies are also important in the control of pests, but in this case the knowledge gained is used to reduce the species' reproductive success.

A much broader approach to ecology is **synecology**, in which interactions between many species are studied simultaneously. Synecological studies are important in the conservation and management of ecosystems such as the North Sea, in which fish stocks have become grossly overexploited in recent years.

## Kinds of environmental factors

There are so many environmental factors that ecologists find it convenient to divide them into categories. **Abiotic** factors are physical influences, such as temperature and light intensity. **Biotic** factors include other organisms, for instance predators, parasites and competitors. In plant ecology, soil conditions are referred to as **edaphic** factors, as distinct from **climatic** factors.

The distinction between physical and biotic factors is not quite as clear cut as it might appear. An oak tree for example is a biotic factor for insects feeding on it, but it is also a physical factor in the lives of organisms shaded by it. Also abiotic and biotic factors often interact with each other, so that the effect of one depends on the level of another. Figure 35.1 illustrates this point: the numbers of herons decreased in hard winters, but the actual mechanism was probably not temperature (a physical factor), but the inability to get at the fish beneath the ice. So lack of food (a biotic factor) was probably responsible.

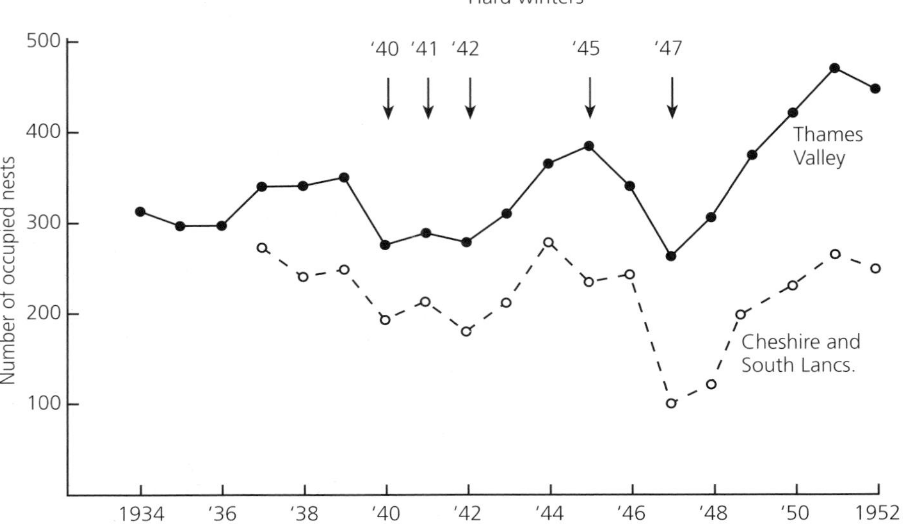


**Figure 35.1**
Fluctuation in two English heron populations over a number of years. Although the populations are hundreds of kilometres apart, the fluctuations show a close relationship (after Lack)

Another way in which physical and biotic factors can interact occurs when one influences the effect of another. Everyday examples that illustrate this are the many imported plants that grow in British gardens. They compete successfully in their countries of origin, but survive the physical conditions in British gardens only if protected from competition from 'weeds'. Other examples of interaction are mentioned in Chapter 36.

Environmental factors can also be classified in a different way. **Resources** are factors that may be competed for, such as food, hiding places, mates and territory in the case of animals, and light, water and mineral ions in plants. **Conditions** are factors that influence organisms without being used up, such as temperature, soil pH and rate of water current flow.

These two classifications are independent, and cut across each other. Light, for example, is a resource for plants but a condition for animals. Some biotic factors, such as competitors, are neither resources nor conditions.

### *Rhythms in the environment*

In all habitats except deep caves and the ocean depths, physical conditions change rhythmically with day and night and with the seasons. On the sea shore they also change with the tides. Because such changes are rhythmic, they can be *predicted*. Most organisms have an 'internal clock' which enables them to change their behaviour before it gets dark, or before the tide goes out. Nocturnal animals, such as cockroaches, become more active as evening approaches even when kept in constant light, temperature and other conditions.

Most organisms use photoperiod to measure time – this is far more reliable than other seasonal cues, such as temperature. In some way (as yet not fully understood) photoperiod is measured using an internal clock which has its own endogenous rhythm. In birds, the clock is located in a part of the brain called the **pineal gland**, located just beneath the skull, and in mammals it is in the hypothalamus, just above the pituitary gland.

Accurate time measurement is crucial to survival. The winter moth (*Operophtera brumata*), for example, lays its eggs in December on the bark of oak, apple and other deciduous trees. The eggs hatch in spring, and the caterpillars march up the twigs just as the buds are opening, and feed on the young leaves. As oak leaves mature they begin to produce large amounts of defensive chemicals called **tannins** which, besides being toxic, also render them indigestible. Winter moth caterpillars only have these few weeks in which to feed and grow. If the eggs are laid too early, the buds will not have opened and the tiny caterpillars will starve. If they are laid too late, the tannin defences of the oak will overtake the older caterpillars, which will also starve.

Caterpillars of the winter moth are important food items for the great tit (*Parus major*) and other birds. Though great tits often lay a second clutch, the first is larger, and egg-laying is timed so that the greatest demands of the young on their parents coincides with the glut of caterpillars. Birds that nest too late are less successful, partly as a direct result of insufficient food, and also because in making more noise, the hungry young are more vulnerable to predators such as weasels.

Great tits and other small birds are food for sparrowhawks, which lay their eggs later than great tits so that the young are at their most demanding just as young great tits are becoming available. Indirectly then, the timing of the sparrowhawk's reproduction is linked to the surge in the output of tannins in the oak and other trees!

# Soil

Soil is the material in which land plants grow and from which their roots obtain water, mineral ions and anchorage. Since plants provide food and shelter for animals, soil plays a central role in the lives of all terrestrial organisms.

## Constituents of soil

Soils vary in their composition and properties, but all consist in varying proportions of mineral particles, humus, living organisms, water, air and soluble mineral ions.

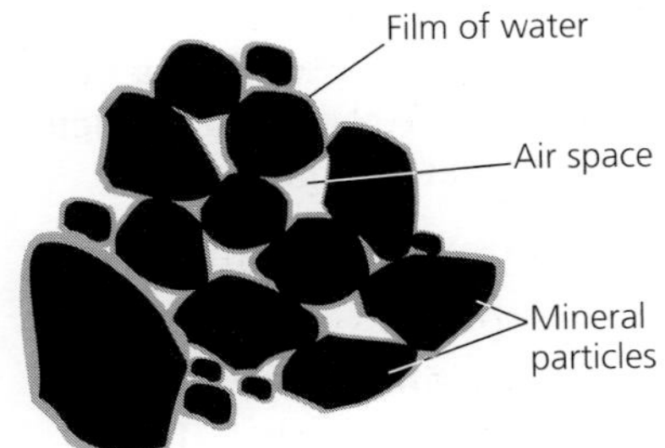


**Figure 35.2**
Soil particles, their film of water and enclosed air spaces. In a well-drained soil, the air spaces are continuous with each other, but after flooding, they may be replaced by water

### *Mineral particles*

The mineral particles of the soil act as its 'skeleton', and are formed by the slow break-up of rock (weathering). Besides providing anchorage for roots, mineral particles also provide support for a network of countless minute air spaces surrounded by water (Figure 35.2).

As far as their effects on soil properties are concerned, the two most important properties of the mineral particles are their size and their chemical composition.

#### *Particle size*

Most soils contain particles of mixed sizes, classified by diameter as follows:

Coarse sand 2.0–0.2 mm

| | | |
|---|---|---|
| Sand | 0.2–0.02 | mm |
| Silt | 0.02–0.002 | mm |
| Clay | <0.002 | mm |

The proportions of the different-sized particles vary with the soil type, though different soils often merge with each other. A soil consisting of roughly equal proportions of all particle sizes is called a **loam**. An average loam consists of 40–70% solid matter, the rest being air and water (Figure 35.3).

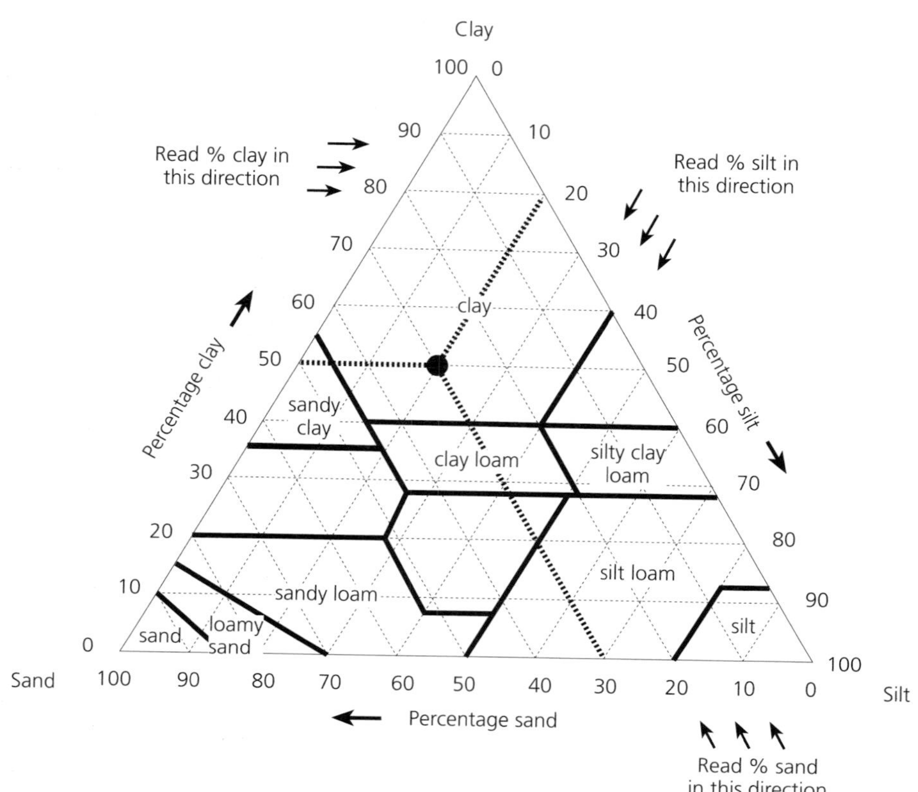


**Figure 35.3▶**
Classification of soils by particle size. For any given soil, its position in the classification is obtained by drawing lines from the sides of the triangle as shown. The hypothetical sample shown consists of 30% sand, 50% clay and 20% silt

Clay particles consist of flat plates of potassium and aluminium silicates (Figure 35.4). They are just below the smallest size that can be resolved by the light microscope and are about the size of virus particles. They are important both chemically and physically:

- their minute size gives them a colossal surface area over which water can be held by attraction (1 gram of the finest clay has a total area of over 100 $m^2$). Clay soils, therefore, have a high water-retaining capacity and drain very slowly, and so can become easily waterlogged.
- they are negatively charged and thus strongly attract metal cations, thereby helping to retain them in the soil.
- in the presence of sufficient calcium, clay particles form a complex with humus, causing the larger particles to aggregate into 'crumbs'. These are typically 3–5 mm in diameter and increase the pore space, thus improving drainage. Soil which forms crumbs is said to have a 'crumb structure'.

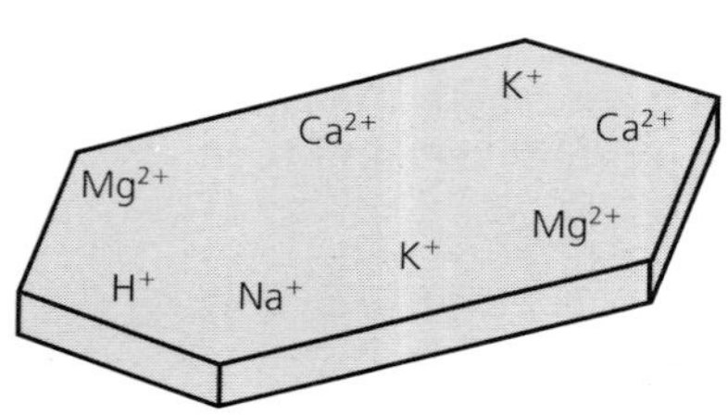


**Figure 35.4**
A clay particle with bound cations

### *Humus*

Humus consists of partly-decayed organic matter and forms a slimy black material which coats the mineral particles and gives the topsoil its characteristic dark colour. Humus helps 'crumb' formation, and greatly increases the ability of the soil to absorb and retain water. Its negative charges attract nutrient cations, helping to retain them in the soil.

Eventually humus is broken down by saprotrophs into carbon dioxide and water, but the rate at which this occurs depends on the conditions. It is much slower, for instance, in acid soils. It is faster in warm conditions, so the humus content of many tropical soils may be low, most of the plant nutrients being stored in the bodies of organisms.

### Living organisms

A fertile soil contains a huge number of organisms of many different kinds (Figure 35.5). Some live in the film of water round the soil particles, for example bacteria, fungi, protoctists, flatworms and nematodes. Though these organisms are considered to be terrestrial, they are technically aquatic. Others, such as mites, live in the air spaces between the particles.

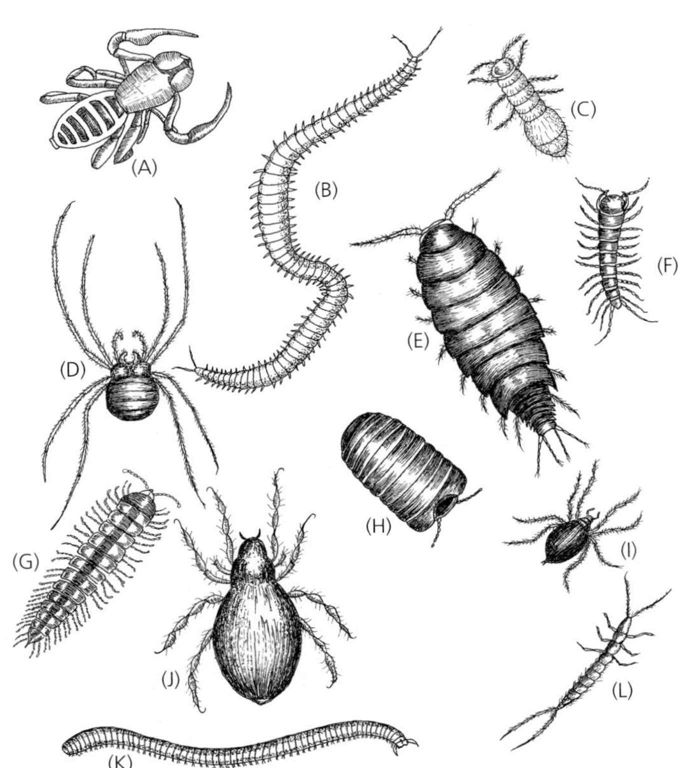


**Figure 35.5**
Various kinds of soil animal (after Cloudsley-Thompson, not to scale) (A) false scorpion, (B) burrowing centipede, (C) springtail, (D) harvestman, (E) woodlouse, (F) centipede, (G) flat-backed millipede, (H) pill millipede, (I) spider, (J) beetle mite, (K) millipede and (L) bristletail

A third category consists of larger, burrowing animals such as earthworms, insects, millipedes, centipedes and, on a much larger scale, mammals such as moles and rabbits. Among the most important of the burrowers are the earthworms (see the information box on page 676).

### Water

In a well-drained soil, water forms a thin layer between the mineral particles. Some of the soil organisms are really aquatic because they are small enough to be able to live entirely within this thin watery film. Because of the soil water, the soil atmosphere is saturated with water vapour, so soil animals do not lose water by evaporation and most of them cannot tolerate dry air for long.

Whereas water is a condition for soil animals, it is a resource for plants. Not all the water in the soil is equally available to plants and from this

## The importance of earthworms in the soil

Of all soil animals, earthworms are perhaps the most important. In a woodland soil, their numbers can be as high as 60–260 $m^{-2}$. It was the 18th Century naturalist Gilbert White who first pointed out the importance of earthworms to soil fertility – a view that was confirmed and developed a century or so later by Charles Darwin.

Earthworms maintain soil fertility in the following ways:

- their burrows aerate the soil and allow rapid drainage after heavy rain.
- they increase the organic matter in the soil. They drag dead leaves down into their burrows, and those which are not eaten break down and enrich the soil. In oak forest, earthworms eat greater amounts of dead leaves than all other soil invertebrates put together. Any worms that die underground also help to enrich the soil.
- besides dead leaves they ingest soil, which is subsequently defaecated. The two species that form casts on the soil surface (*Allolobophora longa* and *A. nocturna*) bring soil from deeper levels to the surface, helping to mix the soil. Charles Darwin calculated that in one particular field, earthworms had been bringing mineral particles to the surface at a rate of 0.22 inches (5.6 mm) a year.

point of view, the soil water can be divided into several parts (Figure 35.6).

- **Gravitational water** is water that flows down through soil without being retained. When all the gravitational water has drained through, a soil is said to be at **field capacity**.
- **Capillary water** is that which is held by surface tension. When an air-water surface is curved, the tension exerts a pull on the water, rendering its pressure negative. The greater the curvature of a curved water surface (meniscus), the more negative its pressure potential. Water in the wider spaces is less tightly held than water in the narrowest spaces since its menisci are less strongly curved. As more and more of the capillary water is absorbed, the remaining water retreats into narrower and narrower spaces and becomes more difficult to absorb. Water in the finest capillary spaces is unavailable to plants.
- **Hygroscopic water** is hydrogen-bonded to mineral particles and is unavailable to plants.
- **Combined water** forms a chemical constituent of certain minerals in the form of hydrates, such as $Fe_2O_3.3H_2O$.

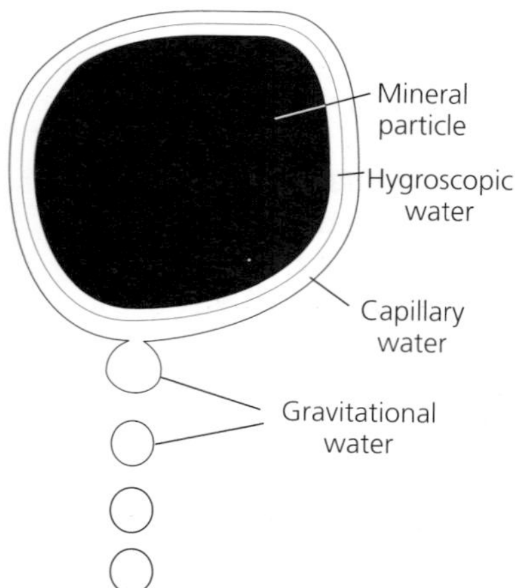


**Figure 35.6**
Different categories of soil water availability

At a certain depth is the **water table**, where the soil is saturated with water. The depth of the water table varies with the kind of soil and with the climate. Water absorbed by roots may be replaced by capillary rise from the water table.

### Air

The air and water content of soils are indivisibly related: the more pore space containing water, the less space is available for air. In a well-drained soil, the air forms a network of interconnected spaces. This allows oxygen to diffuse from the soil surface to the organisms below and $CO_2$ to diffuse in the reverse direction. If the soil is too wet, the air spaces become partly filled with water and no longer connect with each other, causing the soil to become anaerobic. This may be a permanent condition in peat bogs, but in soils which are normally well-aerated, prolonged flooding causes many soil organisms to die.

In a well-drained soil, the soil atmosphere is similar to that of the air above ground, except that it is normally saturated with water vapour and contains a little more $CO_2$ and a little less oxygen.

## *Soluble ions and soil pH*

Nutrient ions include nitrates, phosphates and sulphates of potassium, calcium, magnesium and iron. Some ions are in solution in the soil water, whilst others are held by attraction to negatively-charged clay particles and humus. These may be released into solution as a result of **ion exchange,** in which metal ions are exchanged for hydrogen ions produced by plant roots. As a result of ion exchange, metal ions become more readily available to plants. In very acid soils, the concentration of free ions such as aluminium ($Al^{3+}$) may reach toxic proportions.

Normally, ions are being removed by plants at about the same rate as they are added as a result of decay. This equilibrium is disrupted in agriculture by harvesting, so fertilisers are added to compensate. Soil nutrients can also be washed out, or **leached**, as a result of heavy rain.

The pH of the soil is an important factor influencing the growth of plants and indirectly, therefore, the animals that feed on them. Some plants, for example yew (*Taxus baccata*), are called **calcicoles** because they will only grow in the alkaline soils of chalk or limestone where there is adequate calcium. Others, such as heather (*Calluna vulgaris*), are called **calcifuges** and have a strong preference for acid soils in which the calcium ion concentration may be quite low.

## *Soil formation*

Soil takes thousands of years to form and involves both geological and biological processes. The mineral particles of the soil are produced from rock by various processes, the most important of which are **weathering** and erosion by glaciers. In **alluvial soils** the mineral particles are transported by rivers or by glaciers, and are thus some distance away from the parent rock from which they were formed.

Weathering may be of various kinds. **Physical weathering** results from heating and cooling, especially from the expansion of water in crevices as it freezes. **Chemical weathering** results in the formation of new compounds as a result of the chemical action of water and solutes. Dissolved carbon dioxide forms a weak acid (carbonic acid), which can slowly dissolve basic minerals such as calcium carbonate.

Once fragmentation of the rock is under way it may be further assisted by the action of organisms. Physical weathering is accelerated by the penetration and expansion of roots in crevices. Chemical weathering is assisted by the carbon dioxide produced by organisms such as lichens and mosses. The burrows of earthworms and mammals, such as rabbits, help bring water and air closer to the parent rock where weathering occurs.

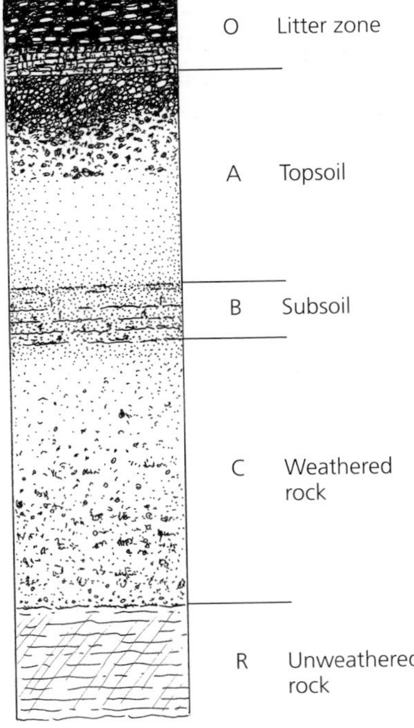


**Figure 35.7**
A generalised soil profile (after Buckman and Brady)

## *Soil profiles*

An undisturbed soil consists of several layers, or **horizons**, which collectively make up the **soil profile** (Figure 35.7). Most soils have three main horizons called A, B and C, which differ in their physical and biological characteristics. Each horizon may be further subdivided, the details varying from one kind of soil to another.

The uppermost layer is the O horizon, or litter zone, which is devoid of mineral particles. It is subdivided into an upper, or $O_1$, layer consisting of identifiable fallen leaves and twigs, and a lower, $O_2$, horizon consisting of largely decomposed and unrecognisable black matter. In temperate soils, the thickness of the $O_1$ horizon may vary with the seasons, being thickest in late autumn and almost gone by late summer.

Below the litter zone is the A horizon, or 'topsoil'. It is the zone from which soluble matter is continually being lost, or **leached,** by slow downward flow of rainwater. It is the darkest horizon because it contains the most humus.

The underlying B horizon, or 'subsoil', is the zone which gains material leached from the A horizon. Here it accumulates and may be precipitated. It is paler and contains less organic matter, and is penetrated only by the deepest tree roots.

Beneath the B horizon is the C horizon. It is devoid of organic matter and consists of the weathered rock or sediment which is the parent material of the soil.

The stratification of soil into horizons develops over thousands of years, and the pattern that develops depends on the kind of vegetation it bears, which in turn depends on climate and geology.

## The importance of ecology today

The origins of ecology go back tens of thousands of years to the times when our ancestors made their living by hunting animals and foraging for plants. Though little direct evidence remains of the plants our distant ancestors ate, studies of modern hunter–gatherer societies suggest that they ate a wide variety. In the 1960s, the anthropologist Richard Lee found that the people of the Kalahari desert in Southern Africa ate 68 species of plant, although some of them only seasonally. The exploitation of many of these plants must have required a detailed knowledge of their distribution and seasonal growth. For hunter–gatherers, then, ecological knowledge was a prerequisite for survival.

Modern ecology sprang from roots that were anything but practical. The industrial revolution generated wealth and, for some, considerable leisure. Charles Darwin was the best known of a large number of Victorian naturalists for whom hunting, fishing and collecting wildlife specimens was a popular pastime.

Ecology has changed in many ways since Darwin's time, and is now a true science. Whereas many early naturalists were content with observation, today's ecologists ask questions and test hypotheses by experiment and by further observation, supported by sophisticated techniques and the statistical evaluation of results.

Ecology has also taken on a new urgency. The scientific and industrial revolutions that generated the necessary leisure time for modern ecology to get started has also led to immense pressure on the Earth's resources. Population and economic growth have led to pollution, overfishing, food shortages and other problems. Dealing with these requires considerable expertise from ecologists specialising in areas such as pest control, pollution and soil conservation.

The ecological wheel has thus turned full circle. Ten thousand years after people began to exchange their dependence on wild plants and animals for farming, we are once again dependent on ecology for our survival.

### QUESTIONS

1 What is the distinction between physical and biotic factors, and why is this distinction not clear-cut?

2 Why do clay soils drain slowly, and why are they hard when dry?

CHAPTER

# 36 The PHYSICAL ENVIRONMENT

## LEARNING OBJECTIVES

By the time you have completed your study of this chapter you should be able to:

- explain three ways in which organisms are affected by their physical environment.
- give examples to show how physical factors can interact with each other.
- give examples of how animals and plants can survive adverse physical conditions.
- describe the physical factors that operate in a habitat such as a pond, rocky shore or woodland.

When we speak of the physical, or **abiotic**, environment we are referring to such non-living factors as temperature and light intensity. Although it is easy to list abiotic factors separately, it is important to realise that in many cases they interact with each other. To make things even more complicated, abiotic factors often interact with biotic ones.

In plant ecology, the distinction is often made between **climatic** factors and **edaphic**, or soil, factors. Like many other distinctions, this one is blurred, because soil conditions are indirectly affected by climate as well as geology.

## Temperature

Temperature has a profound influence on all life because of its effects on enzyme activity – even a moderate rise in temperature results in a marked rise in metabolic rate (except in endotherms).

Though the effect of cold on enzymes is temporary, it may have a lethal effect on the whole organism. The Painted Lady butterfly, for example, migrates to Britain and breeds every summer but it cannot survive the cold of winter. The closely-related peacock and small tortoiseshell butterflies on the other hand, overwinter successfully.

In temperate climates there are large seasonal and daily variations in temperature, which are even greater on land than in water. Since large water bodies absorb heat in the summer and store it in the winter, seas and lakes tend to be warmer in winter and cooler in summer than the land. Land areas near the sea are generally more thermally stable than large continental land masses.

Seasonal temperature changes influence almost every aspect of life. Plants, fungi, bacteria and all animals (except birds and most mammals) become less active in winter. Because photosynthesis slows or even stops altogether, the supply of food for animals is greatly reduced. Birds and mammals are indirectly affected, since the maintenance of body temperature in winter requires increased food intake (Chapter 25).

Organisms survive the winter in various ways. Some, such as robins, blackbirds and most mammals, are able to find enough food to remain active throughout the year. Most however avoid the cold in ways briefly described below.

### *Dormancy*

Many plants and the eggs of many animals survive adverse conditions in a state of greatly reduced activity called **dormancy**. This is not the direct result of low temperatures on metabolism because it requires some environmental factor besides favourable temperatures before activity can be resumed. A common stimulus is exposure to several weeks of winter cold.

Many animals pass the winter in the egg stage. In *Daphnia* (water flea) and aphids (greenfly) reproduction in the summer is by parthenogenesis, in which all individuals are females and eggs develop without fertilisation. In the autumn, males are produced and fertilise the eggs. The resulting fertilised eggs are resistant to cold and other extreme conditions. The following spring, the eggs hatch to give rise to another generation of parthenogenetic females.

### *Hibernation and aestivation*

Animals such as hedgehogs, adders and frogs survive periods of cold in a state of greatly reduced activity called **hibernation**. The dry season equivalent of hibernation is **aestivation**, in which the hazard is dryness rather than cold. Many land snails prepare for aestivation by secreting a thick layer of mucus over the entrance to the shell, which greatly reduces water loss.

### *Diapause*

A special case of hibernation is **diapause**. This is a state of arrested development in many arthropods, and can occur at any stage of the life cycle. In the large and small white butterflies it occurs in the pupal stage. There are several generations a year, but pupae formed in late summer do not develop further until the following spring regardless of how favourable the conditions may be at the time. Only after a period of cold has been endured is diapause broken and development resumed. This is an adaptive response which prevents the young emerging during the winter when conditions are not favourable for survival.

### *Migration*

Many birds migrate to warmer latitudes in the autumn, for example cuckoos and swallows. This behaviour may not just be an escape from cold: by returning to higher latitudes to breed in the spring, more daylight hours are available to forage for food for the chicks. By the time they are nearly full size and are placing the greatest demands on their parents, days are longer than in the tropics so more time is available for feeding.

Some butterflies make extensive migrations. The monarch butterfly (a native of North America) overwinters in Mexico and California and migrates north as far as Canada to breed. The next generation returns south in the autumn.

## *Light*

Light is the energy that drives photosynthesis, so it is the ultimate source of energy for most living things. As a resource for plants it is competed for

– the plants which produce the highest leaves receive the most light.

For many animals light is a source of information. By their ability to refract light onto photosensitive cells many animals use light in vision, which informs them about food, predators and mates. In both animals and plants daylength, or **photoperiod**, is an important seasonal indicator, making it possible to make adaptive changes in advance of unfavourable conditions.

Though visible light is not harmful to animals, it is usually associated with lower relative humidity. For so-called **cryptozoic** animals, avoidance of light is thus a way of preventing dehydration, and may also reduce risks of predation.

### Mineral ions

Mineral ions in the environment can influence organisms in two quite different ways. In low concentrations they are a *resource* for plants, but in higher concentrations they influence the concentration of water, and thus the osmotic *conditions*.

As resources, certain minerals play a key role in the distributions of animals and plants. For example snails and crustaceans, such as the water louse (*Asellus*), are rare or absent from areas in which the water flows through soil low in calcium.

### Salinity

In environments such as inland waters and the open sea the total salt concentration (salinity) averages 35 parts per thousand (35‰); in lakes it is typically less than 1‰. Animals living in these osmotically-stable conditions are usually unable to withstand great changes in the salt concentration of their surroundings and are said to be **stenohaline**.

Estuaries and sea shores on the other hand are osmotically variable. In estuaries, salinity varies with the state of the tide and with distance from the sea. Organisms living on bare rock on a sea shore may be alternately exposed to rain and the desiccating effects of hot sunshine. Estuarine animals have to cope with a medium that can vary between fresh water and sea water. These creatures are able to withstand a wide range of osmotic conditions and are said to be **euryhaline**.

Euryhaline animals survive in one of two ways. Some, such as the short crab (*Carcinus maenas*), rely mainly on the ability to regulate the composition of the blood by some form of active transport. In **osmoconformers**, such as the lugworm (*Arenicola*), the blood varies considerably in salinity, and the animal is dependent on the ability of its cells to tolerate changes in the concentration of the tissue fluid.

Salinity gradients may influence distribution. Figure 36.1 shows the distribution of three species of amphipod crustacean. *Gammarus pulex* is a freshwater species, inhabiting the upper reaches of river estuaries. *G. locusta* cannot survive salinities below about two-thirds that of sea water and occupies the more saline areas. *G. zaddachi* occupies the intermediate regions.

### Oxygen

Oxygen is a resource for animals, but on land and above ground it is present at such a high concentration (20.95%) that it is never competed for and is thus never a limiting factor.

The situation is quite different in many aquatic habitats. Oxygen is sparingly soluble in water and its solubility decreases with rise in temperature. It also diffuses 300 000 times more slowly in solution than in air. In the cold, turbulent waters of shallow streams, the concentration of

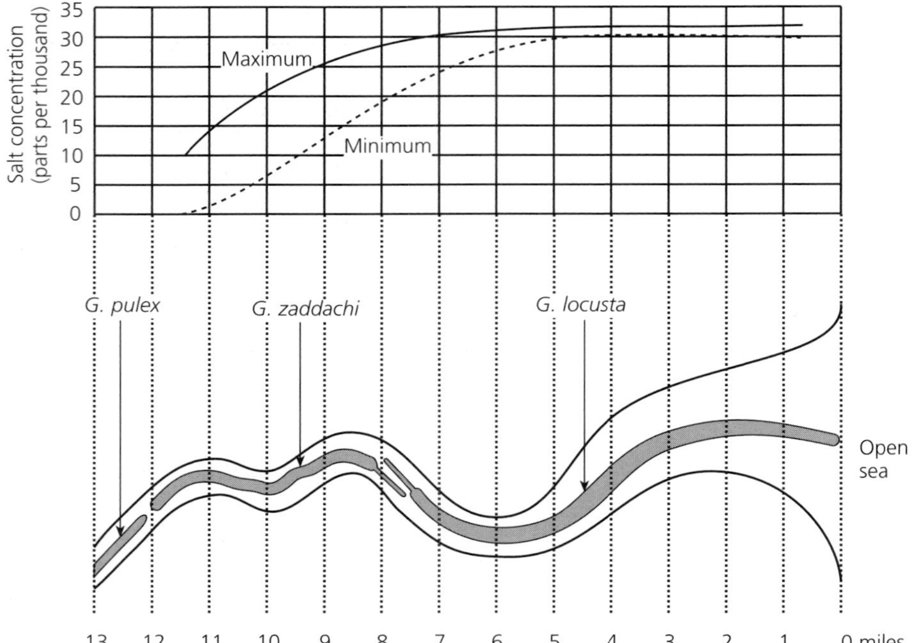


**Figure 36.1**
Distribution of three species of *Gammarus* along British rivers. Relative abundance is indicated by the width of the shaded band (after Spooner)

dissolved oxygen is high enough to support active animals like rainbow trout. In lakes and in slow-flowing rivers and waters, oxygen concentration is usually lower and less active fish such as carp are present.

In stagnant water, such as ponds and canals, oxygen concentrations may fluctuate considerably. Pondweeds may enrich the water with oxygen during the day, but oxygen levels fall at night. In the mud at the bottom of a pond where there is neither turbulence nor photosynthesis, conditions tend to be permanently anaerobic.

## pH of soil and fresh water

Few plants can survive soil pH values below 3 or above 9. Extreme soil pH values can be hazardous to plants for two reasons. First, there is the direct effect of high or low $H^+$ concentration on metabolism itself. The second reason stems from the fact that pH affects the solubility of certain soil ions (Figure 36.2).

Aluminium (as aluminate) becomes much more soluble at low pH values, and may become toxic to plants and animals. Manganese and iron also, although essential nutrients, may reach toxic concentrations at low soil pH values. In very alkaline soils, certain nutrients such as phosphate and manganese become highly insoluble and consequently unavailable to plants.

## Relative humidity

For terrestrial animals and plants, desiccation is always a potential hazard. How real this is depends on three factors: the structure of the outer layer of the body, the surface area/volume ratio of the body and the relative humidity of the air. At any given temperature, the lower the relative humidity, the more rapid the evaporation of water.

Insects and spiders have highly impermeable cuticles and can survive for long periods in air with low relative humidity. Birds and mammals also have water-resistant skins, and their larger size gives them a relatively small surface area through which water can evaporate.

Many land invertebrates, mosses and liverworts have relatively permeable body surfaces and cannot survive in dry air for long.

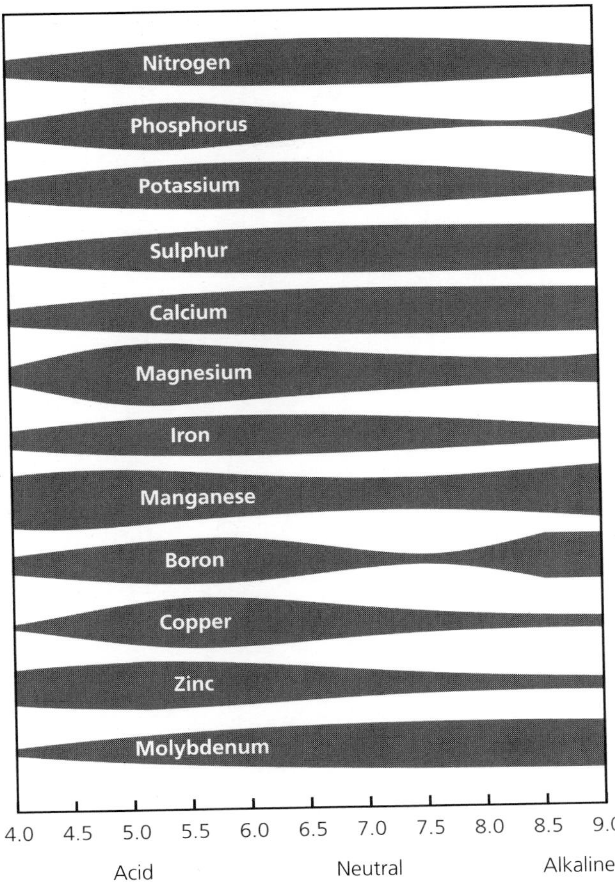


**Figure 36.2►**
Effect of pH on the solubility of soil ions

Millipedes, centipedes, woodlice, earthworms and nematodes are **cryptozoites**, controlling their internal water content by their behaviour and preferring dark places where relative humidities are generally high.

## Water flow

For an organism living in a stream, water flow has costs as well as benefits. It delivers resources such as oxygen and food, but it may also wash the organism away. In powerful currents only encrusting organisms with low profiles, such as some protoctists, can maintain their position. The nymphs of some mayflies have flattened bodies and hooks on the ends of their feet which enable them to hug the rock surface, thus staying in the relative calm but thin 'boundary layer' (Figure 36.3).

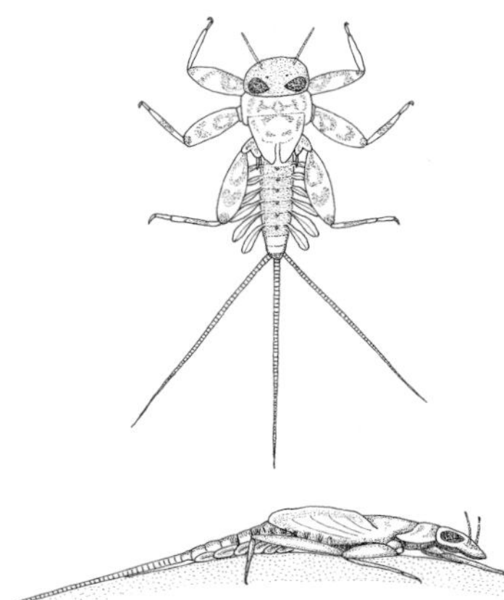

**Figure 36.3**
*Ecdyonurus*, a mayfly nymph adapted to fast-running water (after Mellanby)

## Zonation on rocky shores

Perhaps more than any other habitat, rocky shores show the effects of physical factors. Though the shore inhabitants are considered to be marine, the effect of the tides means that, except for the inhabitants of pools, they all spend periods of **emersion**, in which they are exposed to the air, alternating with periods of **submersion**, in which they are covered by sea water.

Emersion carries the obvious hazard of desiccation, but there are other dangers. Rain can cause osmotic uptake of water. Temperatures may fall near or below freezing in winter, and can reach 40°C or more in the midday sun in summer.

The rhythmic changes in physical conditions are complicated by the fact that the tidal cycle is *just over* 12 hours, so the time of low tide is a

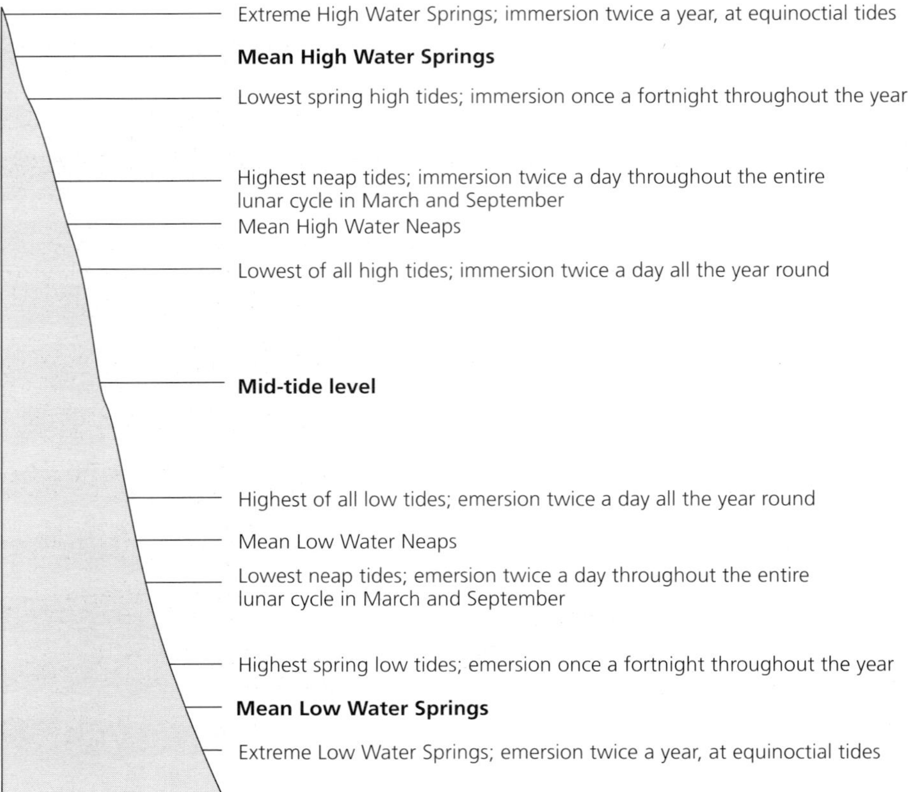


**Figure 36.4**
Ecologically important zones on the shore

little later each day. Dangers of dessication are obviously much more severe when low tide comes at midday than when it occurs at dawn or dusk.

Superimposed on the daily cycle are two other rhythms. Once a fortnight, at around the time of the New Moon and Full Moon, the tides come up higher and retreat lower. These **spring tides** alternate with the **neap tides**, when the tidal range is least, and which occur during the 'half moon' phase of the lunar cycle.

On a much longer time scale is a twice-yearly rhythm. On around March 21 and September 21 the spring tides are even greater than usual. These are called the **equinoctial tides** because they occur at the equinoxes, when all latitudes experience day and night of equal length.

These twice-daily, twice-monthly and twice-yearly cycles have considerable significance for the inhabitants of the shore (Figure 36.4). At the extreme low water level, organisms are uncovered only about two days a year. A little higher up is the mean low water of spring tides where, throughout the year, organisms are exposed once a fortnight at most. Higher still is a level reached by the highest low tides of the year, below which organisms are exposed part of every day – even during neap tides.

Above mid-tide is the lowest high tide level of the year, where organisms are submerged at least once a day throughout the year. Higher up is a level which is submerged at least once a fortnight all year round, and at the very top of the shore, organisms are submerged only at the equinoctial tides. This zone is not clearly defined because it varies greatly with exposure, merging with the splash zone.

Though the gradient in the duration of emersion and submersion is continuous, the distribution of some of the shore organisms may show quite sharp boundaries. Experiments have indicated that competition (Chapter 37) is a probable factor. For example, when brown seaweeds are artificially removed, the red seaweeds which are normally confined to the lower shore grow in abundance. This suggests that competition intensifies the effects of small changes in physical conditions.

## What causes the tides?

Tides are caused by the rotation of the Earth in relation to the gravitational pull of the moon and, to a much lesser extent, the Sun. On opposite sides of the Earth, facing towards and away from the moon, the ocean piles up. Since the Earth is rotating, two **tidal waves** move round the Earth, causing two high tides every 24.8 hours. The reason why the Sun's influence is so small is that although its gravitational pull is much greater than that of the moon, the *difference* in the strength of its pull on the two sides of the Earth is very small (since the Sun is so far away).

The reason why successive tides are 12.4 hours apart (rather than 12 hours) is that the moon orbits the Earth in the same direction as the Earth's rotation, so the Earth has to rotate through slightly more than 360° for any given point on its surface to reach the same position relative to the moon.

What causes the ocean to bulge on the side *away* from the moon, causing two tides a day instead of one? The explanation is as follows. Although we speak of the moon rotating round the Earth, the Earth and moon are actually rotating around their common centre of gravity, which is *to one side* of the centre of the Earth. On the side away from the moon, the tendency of the ocean to continue in a straight line ('centrifugal force') causes it to bulge away from the Earth.

The spring tides occur at the time of the New and Full Moon, which occur twice a month. At these times the gravitational pull of the moon is reinforced by that of the Sun, since Sun, moon and Earth are in line. During neap tides the Sun and moon pull at right angles and tend to cancel each other out.

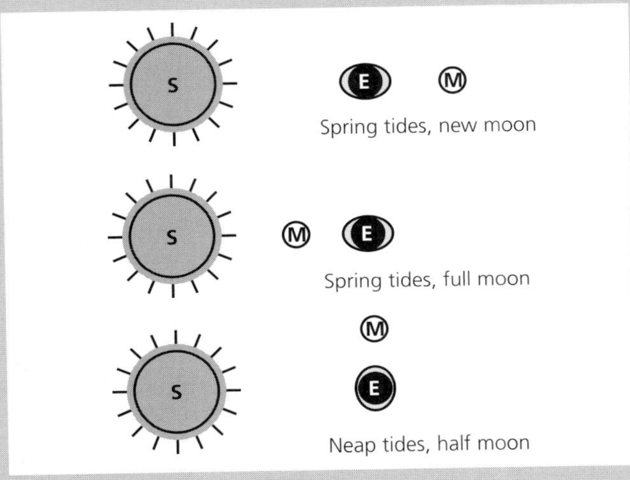


### Tolerance

An organism can only live within a certain range of physical conditions such as temperature and, in the case of plants, soil water content and pH. The range of physical conditions in which an organism can survive is its range of **tolerance**.

The range of conditions an organism's tissues can tolerate in the laboratory is its **physiological tolerance**. In nature an organism must not only withstand physical conditions – it must compete for resources, endure parasites and, in the case of animals, escape from predators. Not surprisingly, the ecological tolerances of most organisms are much narrower than their physiological tolerance.

The idea of a range of tolerance is sometimes expressed by a graph (Figure 36.5) in which the vertical axis represents some index of 'well being' or 'performance' such as rate of growth, number of offspring produced, and so forth.

### Interactions between factors

Though environmental factors have been discussed separately, in reality they often interact with each other. They can do so in different ways.

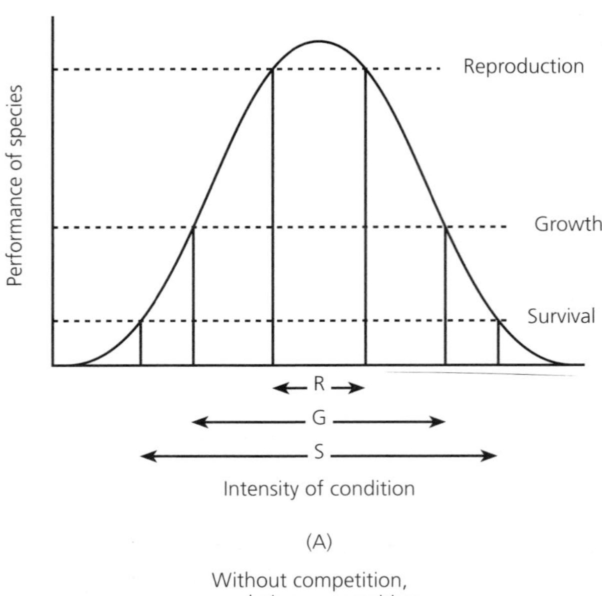


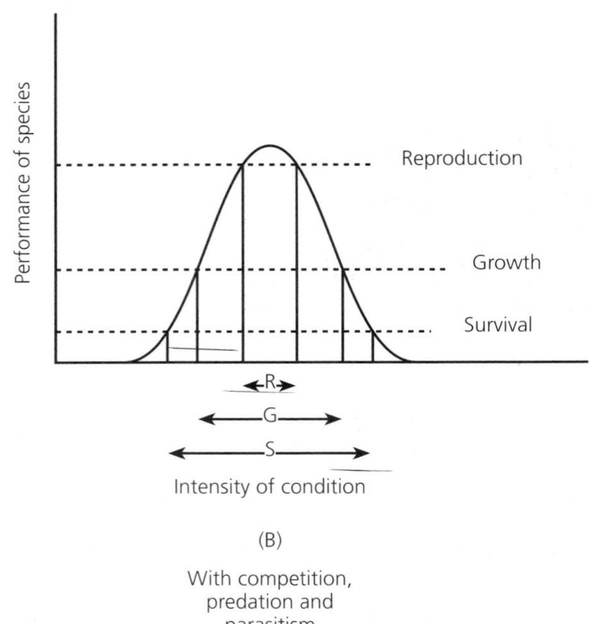


**Figure 36.5**
Tolerance curve showing the effect of a physical factor, such as temperature, on an organism in the absence (A) and presence (B) of adverse biotic factors (modified from Begon, Harper and Townsend)

- One factor directly influences another. For example a rise in the temperature of the air lowers its relative humidity. The solubility of oxygen in water decreases with increasing temperature and increasing salt concentration.
- One factor may influence the effect of another. For example adding nitrate to a soil will only increase plant growth if other minerals are present in sufficient amounts. Thus, addition of nitrate has a much greater effect on plant growth if phosphate is adequate than if phosphate is deficient (Figure 36.6). The mineral that is in shortest supply is said to be the one that *limits* growth. A similar kind of interaction occurs in photosynthesis between light, temperature and $CO_2$.

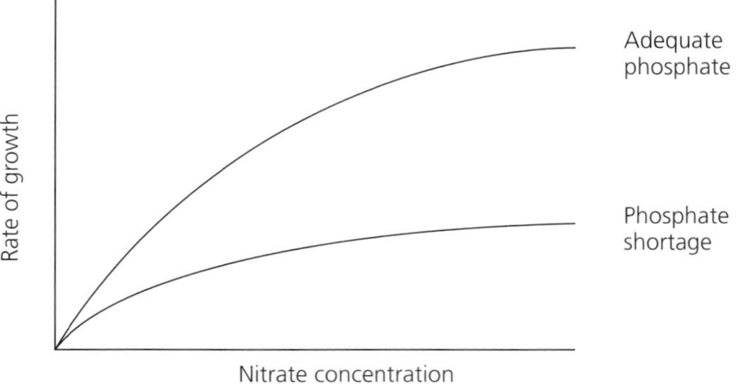


**Figure 36.6**
Interaction between nitrate and phosphate supply on the growth of plants

## QUESTIONS

**1** Why are small organisms more susceptible to adverse physical factors than large ones?

**2** Despite being able to regulate their body temperatures, winter is an adverse time for birds such as robins and blackbirds. Explain why.

CHAPTER

# 37 The BIOTIC ENVIRONMENT

## LEARNING OBJECTIVES

By the time you have completed your study of this chapter you should be able to:

- define competition, predation, parasitism and mutualism, and give an example of each.
- explain why, when there is competition for mates, males tend to compete for females, and why this can lead to sexual dimorphism.
- explain the significance of territory in reproduction, and why there is an optimum size for a feeding territory.
- explain what is meant by social dominance and how it can act to regulate population density.
- describe how intraspecific competition can affect plant growth.
- interpret Gause's experiments on interspecific competition in terms of the Competitive Exclusion Principle.
- explain how the Competitive Exclusion Principle helps to account for species diversity.
- describe some of the ways in which predators and prey influence each other's evolution.
- explain how the malarial parasite and the pork tapeworm are adapted to a parasitic mode of life, and how a knowledge of the life cycles of these two parasites helps us to control them.

When we speak of an organism's biotic environment we mean its relationships with other organisms. It includes the following influences:

- competition, in which organisms utilise a resource that is in limited supply.
- predation, in which one organism, the predator, obtains food from another, the prey, which is killed in the process.
- parasitism, in which one species benefits at the expense of another, called the host, which is not usually killed.
- mutualism, in which both species benefit from the relationship.
- commensalism, in which one species derives benefit but the other is not harmed.

## Competition

A state of competition exists when the demand for a resource exceeds its supply. Thus plants may compete for light, water and minerals, and animals may compete for food, mates or places to breed. **Intraspecific** competition occurs between members of the same species, and **interspecific** competition is between members of different species.

Competition has the following important features.

1 Only if a resource is in limited supply relative to its requirements can it be competed for. Oxygen, for example, is never competed for by land animals since it is so abundant.

2 It operates in a **density-dependent** manner (see Chapter 38). Directly or indirectly, competition always results in a decrease in the reproductive rate of the population, though some individuals (the 'winners') may suffer very little. When competition is for mates or breeding sites, the link with reproduction is clear. If food is being competed for the effect on reproduction is indirect, those that secure more food being more likely to become successful parents.

## Intraspecific competition

Because individuals of the same species resemble each other more closely than members of different species, intraspecific competition is more intense than interspecific competition. Resources that may be competed for by members of the same species include:

- food, e.g. members of a pack of wolves after a kill.
- mates, e.g. red deer and seals.
- breeding territories, e.g. robins and great tits.
- space, e.g. many sessile (attached) animals such as sponges, barnacles and mussels. In plants, competition for space as such is probably uncommon – most overcrowded plants are more likely to be competing for light above ground or for minerals or water below ground.
- light, for example when plants grow close together.
- minerals. This occurs in plants, particularly when other conditions such as light and temperature are favourable to growth. In some situations, such as the upper layers of the sea and lakes in summer, mineral availability usually limits rate of growth of the phytoplankton.

### *Competition for mates*

The ultimate (but of course unconscious) 'aim' of all sexually reproducing organisms is to hand on their genes to the next generation. Except in hermaphrodite (bisexual) species, male and female gametes are produced by different individuals. Though both sexes invest a great deal of energy in reproduction, they do so in very different ways. Males produce vast numbers of minute sperms, while females typically produce much larger, but fewer, eggs.

Since an egg costs more energy to produce than a sperm it represents a bigger investment, so it is in the interests of the female to ensure that her eggs are fertilised by the 'best' sperm. For this reason females tend to be more selective about the male with whom they mate, so in many species males compete for females.

Sperms, on the other hand, are individually cheap to produce, so it is in the interests of the male to ensure that his sperms fertilise as many eggs as possible. Males therefore typically mate with many females, though there are numerous exceptions to this, especially amongst birds.

Competition by males for females takes two forms: fighting and display. In both cases there is strong selection pressure for competitive

**Figure 37.1**
Male deer fighting

ability. This leads to **sexual dimorphism**, in which males differ markedly from females (Chapter 28).

### *Fighting*

In many animals, competition for females is aggressive. Winning a contest requires physical vigour and usually large size. When the winner takes possession of many females, his reproductive success is correspondingly high. In elephant seals for example, only about 10% of the males succeed in winning possession of a 'harem' of females, and these are responsible for almost all the fertilisations in the colony.

The effect of selection for fighting ability is very evident in red deer, in which the stags (males) have antlers and are about half as large again as the hinds (females). Fighting during the autumn 'rut', or breeding season, is usually preceded by roaring, and it seems that in many cases a stag senses the superiority of its opponent and withdraws from the contest without fighting. When threat fails, the males lock antlers and fight by pushing against each other. More often than not the contest is settled without serious injury.

### *Display*

In some animals the female 'judges the quality' of a male's genes by his appearance. While you might think that the gaudiness of a peacock's feathers would not be a good guide to his ability to survive, it seems that indirectly it is. When a bird is in poor health, its plumage is the first thing to suffer. Consequently, when a female is selecting a male to father her offspring, she is able to judge the quality of a male's genes by the state of his plumage, as this is influenced directly by his ability to resist internal parasites. As a result the gaudiest males are more successful at handing on their genes.

There is an obvious disadvantage of bright colours – the male is more conspicuous to predators. However, provided that the advantages of attracting a mate outweigh this, bright colours will still be selected for. Female birds (for whom there is no such conflict of interest) are usually dull and inconspicuous.

## *Competition for territory*

Males do not always compete directly for females. In some species males compete for possession of an area on which to mate. Such a **territory** is occupied and defended against others. In contrast, a **home range** is not defended. Territorial behaviour is very widespread in the animal kingdom, especially amongst vertebrates. It is important to note that it is the area containing the resource that is defended rather than the resource itself.

A territory can exist for varying durations and provide a variety of resources:

- it may be defended all the year round and represent a source of food, for example for the robin.
- it can be a site on which to build a nest and rear young, after which it is abandoned (as in the lesser black-backed gull).
- it may be very transient, used simply as a place for mating, as in red deer and seals.

## *Costs and benefits of territory defence*

Defending a territory is not worthwhile unless the benefits exceed the costs. The benefits depend on how the food is distributed. When it is patchy, animals have to spend too much time and energy finding it for

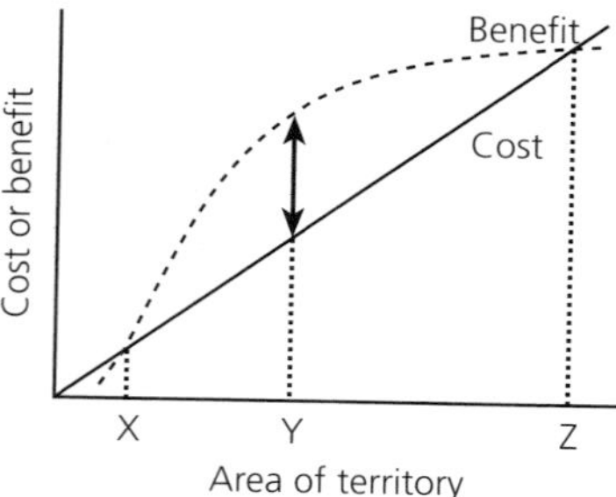


**Figure 37.2**
The costs and benefits of territorial defence. A territory is worth defending if its area lies between X and Z, the optimum being Y (after Krebs and Davies)

defence of the area to be economic. In Tanzania for example, spotted hyenas are territorial in the Ngorongoro crater where the food is abundant all year round, but on the Serengeti plains where food is seasonal, hyenas cover wide areas and are not territorial.

The optimum size for a feeding territory is a compromise between the energy gained from the food in the territory and the cost of its defence (Figure 37.2).

Observations and experiments show that within a given species, territory size tends to be smaller where the food supply is richer (Figure 37.3).

Animals also economise in territorial behaviour by matching the intensity of their defence to the severity of the threat, spending no more energy than they have to. Defence of territory usually occurs at three levels of intensity.

1 A ‘keep out’ signal, such as song in many birds, or scent marking in dogs and many other mammals. In most cases this low level warning is sufficient, and holders of neighbouring territories avoid each other.
2 If this does not work, various forms of threat display may be employed. Lesser black-backed gulls, for instance, raise the head and open the wings slightly, placing these weapons in clear readiness for use.
3 If all else fails, fighting may occur. In most territorial animals the intensity of the defence increases with nearness to the centre of the defender’s territory, so the ‘rightful’ owner rarely loses.

### *The non-territory holder*

When a territory is purely for breeding, non-territory holders cannot reproduce. With feeding territories, non-territory holders are displaced to marginal areas where food is less abundant and where they are consequently less likely to survive. The territory holder is thus at a considerable reproductive advantage over the non-holder. So why do non-territory holders usually give way in territorial disputes?

The answer is probably that even for the non-territory holder, there is *some* possibility of gaining a territory in the future, for instance when a territory holder dies. It may be that the injuries that would be sustained in fighting could reduce the chances of eventual reproduction even more than ‘waiting in the wings’.

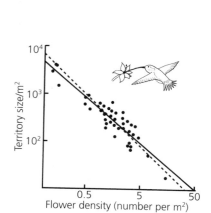

**Figure 37.3**
Evidence that territories in humming birds are of optimum size. Solid line = line of best fit, dotted line = calculated assumption that each territory contains a constant number of flowers. (Redrawn from Krebs and Davies, after Kodric-Brown and Brown)

## *Social dominance*

Territory is only one way in which resources are distributed amongst a population. In many social species resources are shared, but not on a ‘fair’ basis. Certain individuals are of higher *rank* and they consistently have access to the best food and mates. The commonest form of social dominance is a **linear hierarchy**, in which an ‘alpha’ individual dominates all other members of the group, while a ‘beta’ animal dominates all except the alpha, and so on. Social hierarchy is particularly common in birds and mammals, but it also occurs in some reptiles, fish and arthropods.

In most animals social position is determined early in life during brief fights, after which status is seldom contested except by newcomers or as a dominant individual reaches old age. In a newly established group there may be considerable fighting, but once the hierarchy is established there is so little conflict that it is hard to detect. Only the fact that certain individuals consistently defer to others shows that it exists.

The signals by which animals indicate their status are very diverse. In wolves and dogs a dominant animal holds its tail high and its ears erect, while a submissive individual cringes with its tail and ears down (Figure 37.4).

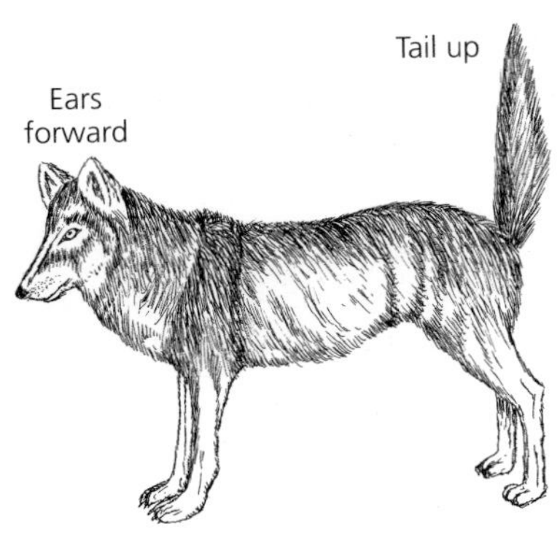

**Figure 37.4**
Dominant and submissive postures in a dog

### Significance of social dominance

The advantages of being dominant are obvious: there is more food and, for males, more matings. But what about the other end of the hierarchy? If being at the bottom means less reproductive success, why doesn't the 'lowest' individual risk all in a fight, and thereby have at least some chance of successful reproduction?

As in the case of non-territory holders, the chances of becoming a parent are probably increased by avoiding conflict and almost certain injury, because dominant individuals may die for other reasons.

## Avoidance of intraspecific competition

One way of avoiding intraspecific competition is by dispersal. In plants this is entirely passive, so they have little or no control over their destination. The larvae of many marine animals can actively move and often have some control over where they settle. Even so, the mortality of larval stages of marine animals is enormous.

Another way in which intraspecific competition is avoided is by having different diets at different stages of the life cycle. In insects with complete metamorphosis, the larva and adult usually have very different diets. Leaf-eating caterpillars, for example, grow into nectar-feeding butterflies.

## Intraspecific competition in plants

In plants, competition for light can affect the average growth in a population and it can also affect the variation in the growth.

An interesting study on the effects of competition on average growth was performed by Palmblad, who sowed seeds of a number of different species of weed at different densities, with each density being replicated in three pots. At the end of the summer the total dry mass of the plants in each pot was determined, as were the number of seeds produced in each pot and the number of plants that had died. The results for one species, Canadian fleabane (*Conyza canadensis*), are shown in Table 37.1 (data are the average for three individual pots).

**Table 37.1**
The results of Palmblad's experiment on competition in Canadian fleabane

| Number of seeds per pot | 1 | 5 | 50 | 100 | 200 |
|---|---|---|---|---|---|
| % germination | 100 | 87 | 56 | 54 | 52 |
| % mortality | 0 | 0 | 1 | 4 | 8 |
| % reproducing | 100 | 87 | 51 | 42 | 36 |
| Total dry mass (g) | 12.7 | 37.24 | 37.75 | 16.66 | 18.32 |
| Mean dry mass per plant (g) | 12.7 | 3.5 | 0.355 | 0.166 | 0.092 |
| Total number of seeds | 55 596 | 59 625 | 40 845 | 35 264 | 38 376 |
| Mean number of seeds per reproducing individual | 55 596 | 13 710 | 1602 | 836 | 534 |
| Number of seeds produced per gram of plant dry mass | 4378 | 3917 | 2301 | 2117 | 2094 |

Notice the following points:

- in this species the percentage germination is significantly reduced at higher densities (though not in some other species in the study).

- a 200-fold increase in sowing density resulted in only a 1.4-fold increase in final total mass of plant matter (18.32/12.7). Most of this is accounted for by a 138-fold decrease in mean mass per plant (from 12.7 to 0.092). Only at higher sowing densities was there any significant mortality.
- whereas the 200-fold increase in sowing density resulted in an increase in total dry mass produced, there was a decrease in total seed output.
- assuming that seed size remained constant as crowding increased, there was a steady decrease in the proportion of organic matter devoted to seed production (from 4378 to 2094 seeds per gram of plant dry matter).

The second kind of effect is illustrated in Figure 37.5, which shows the results of an investigation into the effect of competition on variation in flax. Seeds were sown at different densities, the plants were harvested after different time intervals and the dry masses of individual plants were measured.

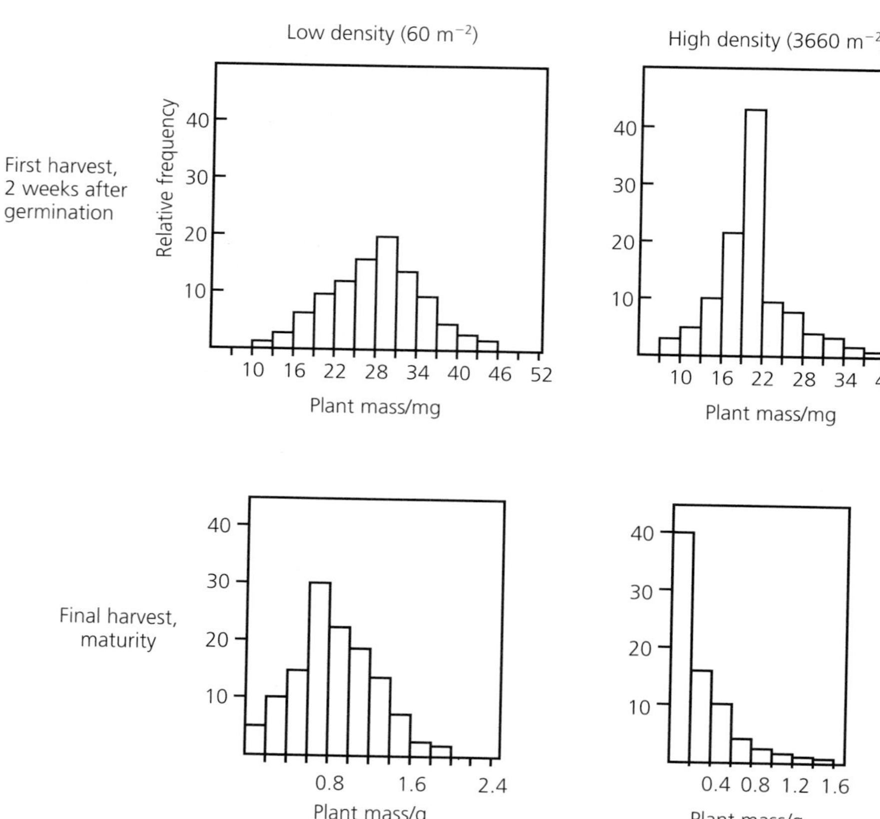


**Figure 37.5**
Effect of competition on variation in flax plants (after Obeid)

At low densities there was considerable variation in plant mass – the distribution was roughly symmetrical, the mean lying approximately in the middle. At higher densities the size distribution at harvest was highly skewed, with proportionately many more undersized plants than at low densities.

The explanation is fairly simple. Germination is normally staggered – some seedlings reaching the light earlier than others. Those that do so first have no competition for light and grow rapidly. Those emerging later are shaded by the early germinators and grow more slowly, leading to greater overshadowing and so on. In a vicious circle, small initial differences in germination time and initial growth rate become greatly magnified.

Early competition for light is not confined to seedlings. For a plant that spends the winter underground as a bulb, corm or rhizome, it is essential that spring growth is as rapid as possible to avoid overtopping by competitors. Herbaceous perennials store considerably more energy than they need for winter survival, with most of the reserves being used for rapid growth even when partly overshadowed by competitors.

## Interspecific competition and the Competitive Exclusion Principle

Since all members of the same population that are of the same age have the same requirements, they would be expected to compete strongly. It is less obvious that members of different species compete with each other. Just because caterpillars of the large and small white butterflies can be seen on the same cabbage, it does not follow that they are competing – there may be more than enough cabbage leaf to go round.

Evidence for the existence of interspecific competition comes from two sources: laboratory experiments and field observations.

### Laboratory experiments on competition

The advantage of laboratory experiments is that variables can be precisely controlled and the numbers of organisms estimated with greater accuracy. Some of the first experiments on competition were carried out by the Russian scientist G.F. Gause (pronounced 'Gauzer'), published in 1934. In one series of experiments he set up mixed and separate cultures of three closely-related species of unicellular protoctists, *Paramecium caudatum*, *P. aurelia* and *P. bursaria*. The paramecia fed on bacteria that in turn fed on the oatmeal or yeast which he regularly added to the cultures. Although Gause had no control over the kinds of bacteria in the cultures, paramecia are filter-feeders and feed indiscriminately on any particles of suitable size. The results are shown in Figure 37.6.

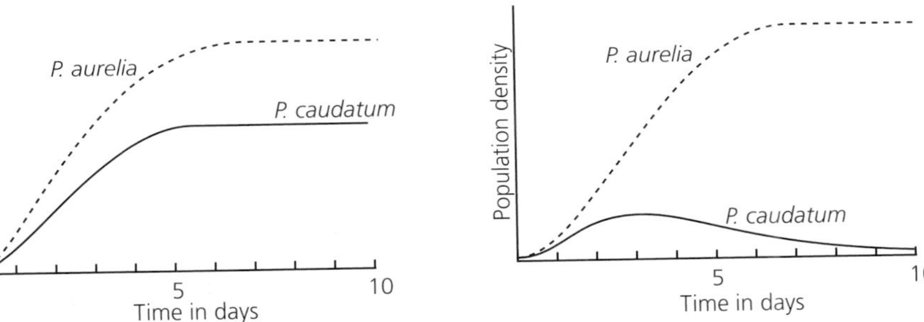

**Figure 37.6**
Competition between *Paramecium* species. Left: single species culture, right: mixed species culture

Although all three species flourished in isolation, in mixed culture *P. caudatum* was eliminated by the smaller *P. aurelia*. That competition was a two-way affair was shown by the slower rate of increase of *P. aurelia* in mixed culture.

A different result was obtained when *P. caudatum* and *P. bursaria* were grown together. Although *P. bursaria* declined, it was not eliminated. It turned out that *P. caudatum* was feeding on the freely-suspended bacteria, while *P. bursaria* was feeding on bacteria on the bottom. The two species were thus coexisting by exploiting different parts of the habitat.

Gause also found that by manipulating the conditions he could influence the outcome. For example, when the water was regularly changed *P. caudatum* won, suggesting that it was less able to tolerate wastes than *P. aurelia*.

Experiments by other scientists on competing species of grain beetle (*Tribolium*) showed that small changes in humidity and temperature could alter the outcome. Other pairs of related species gave similar results: under a given set of conditions, one species always eliminated the other.

Non-competing organisms can also influence the outcome of competition. For example *Tribolium castaneum* is more sensitive to a certain protoctist parasite than a close relative, *T. confusum*. In physical conditions in which *T. castaneum* would normally win, the presence of the parasite causes *T. confusum* to win.

Gause concluded from his results that two species cannot share the same habitat if they have similar requirements. This principle has become known as the **Competitive Exclusion Principle**.

The totality of an organism's requirements constitutes its **niche**, which is often likened to its 'profession', in contrast to its habitat (its 'address'). The Competitive Exclusion Principle, in effect, states that no two species can occupy the same niche indefinitely in the same habitat. A sea anemone and a mussel may live in the same rock pool but they clearly occupy different niches, the anemone feeding on small fish and crustaceans, and the mussel on plankton.

### *Field evidence for competitive exclusion*

There are many examples of animals which at first sight appear to live very similar lives, but which upon closer study can be seen to be utilising different resources.

- The kestrel and sparrowhawk are similarly sized birds of prey, but while the kestrel feeds mainly on voles, mice and other small mammals, the sparrowhawk feeds on small birds which it takes on the wing.
- The larvae of two British butterflies, the orange tip (*Anthocharis cardamines*) and the green-veined white (*Pieris napi*), both feed on Lady's smock (*Cardamine pratensis*). Competition is avoided by specialisation – the orange tip feeding on the seed pods and the green-veined white eating the leaves.
- The field mouse (*Apodemus sylvaticus*) and bank vole (*Clethrionomys glariolus*) share the same habitat, but the field mouse is nocturnal and the bank vole is diurnal (Figure 37.7).

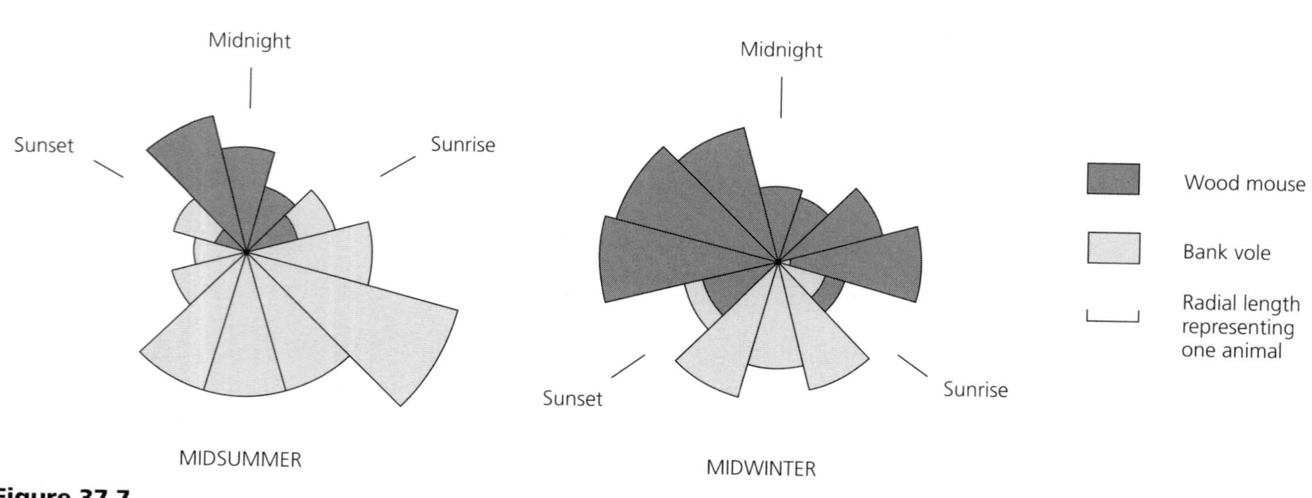


**Figure 37.7**
Activity periods of the field mouse and bank vole (after Brown)

- Competition between various species of birds that feed on mud flats is avoided by differences in beak length (Figure 37.8). The turnstone has a short beak and feeds mainly on animals that live on or very near the surface, such as *Hydrobia* snails. The shelduck has a longer bill and feeds on bivalves living several centimetres down, such as cockles.

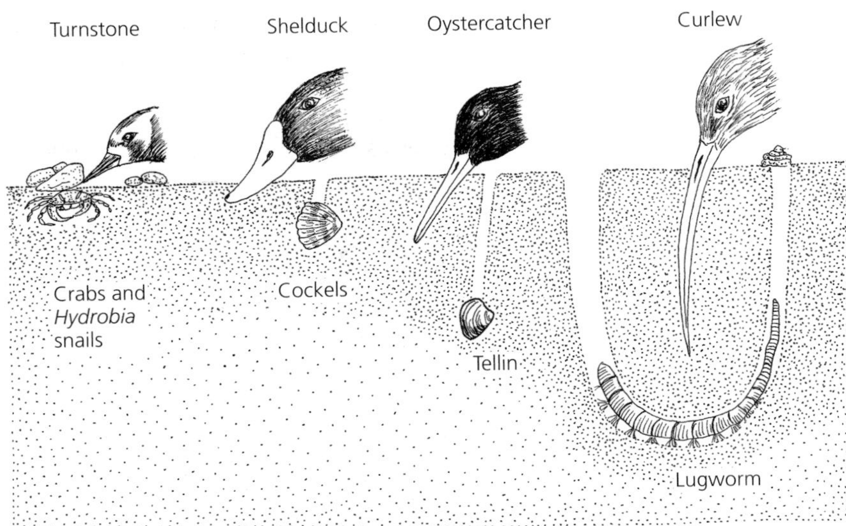


**Figure 37.8**
Niche diversification amongst mudflat birds (after Ford)

Oystercatchers have much longer beaks which enable them to reach tellins. Curlews have even longer beaks and can reach even deeper-lying prey, such as lugworms.

- Some of the strongest evidence for interspecific competition is provided by **character displacement**, exemplified by beak sizes in the Galapagos finches (Chapter 15).
- Forest plants may have differing light requirements. Large trees, such as oak and beech which form the canopy are 'sun plants' and can utilise high light intensities. The shade plants of the understorey, such as ferns, can use dimmer light.
- Many flowers are constructed so that they can only be pollinated by certain kinds of insect (Chapter 30). By concentrating on those flowers whose nectar they are structurally adapted to obtain, insects reduce competition with other species.

The Competitive Exclusion Principle helps to explain why there are so many species (estimated to be 10–30 million). For species sharing the same habitat, the more similar their niches the more intense the competition and the greater the evolutionary pressure for them to diverge.

### *Some problems with competitive exclusion*

If the Competitive Exclusion Principle is valid, then in any given habitat there must be as many niches as there are species. Whereas animals have very diverse feeding habits, plants seem to present problems. All plants require only light, $CO_2$ and certain mineral ions, yet many plant species may share the same habitat. At first sight, there does not seem to be a sufficient number of different ways of obtaining these resources to account for the observed variety of species. But there are several ways out of the problem.

- Though all plants need the same basic mineral ions, the proportions in which they are required vary slightly. Two species could coexist if each were most 'greedy' for a different mineral ion.
- Unlike a controlled laboratory environment, real environments are seldom (if ever) constant. It may be that at any given time of the year, one species population is in the process of gaining the ascendancy over

another, but conditions do not remain favourable for long enough for it to eliminate its competitors.

- Besides varying in time, most natural environments vary in space. Even though in a given area species A may be in the process of eliminating species B, there may be constant immigration by species B from areas in which species B is eliminating species A, and vice versa.
- In real environments, predation may prevent populations reaching levels at which competition becomes important. Most predators show some specialisation, so this would favour greater diversity of species.

#### *Niches change*

An animal's diet may change considerably with its stage of development. This is especially true in species that undergo metamorphosis, as in amphibians and many insects. Even in the adult stage there may be seasonal changes in diet. For instance, badgers eat many young rabbits in spring and summer, but in the autumn they eat more fruits and earthworms.

### *Fundamental and realised niches*

It is a frequently observed fact that in captivity many animals can tolerate a much wider range of physical conditions and can exist on a wider diet than they can in nature. A likely explanation is that in the wild, animals can only survive on food which they can obtain in the face of competition and other adverse biotic influences. A similar effect is seen in plants: most garden plants imported from overseas can only survive if protected from competition by 'weeds'; in their natural habitats in their countries of origin they are competitively superior to other species. Therefore as a result of coexistence with other species, organisms are forced to occupy somewhat narrower niches than they are capable of.

The range of conditions in which an organism can survive under protected conditions constitutes its **fundamental niche**. The narrower, more exacting conditions which it requires in nature constitute its **realised niche**. The difference between fundamental and realised niches is illustrated by the black rat (*Rattus rattus*). In Britain this species is more or less confined to cities, but after it was introduced into New Zealand it rapidly colonised most habitats, from forest to grassland. The house mouse (*Mus musculus*) has similarly spread throughout New Zealand, but in Britain is far less numerous in woods and fields, where it has to compete with field mice (*Apodemus sylvaticus*).

## Predation

A predator feeds on another organism, the prey, killing it in the process. This definition therefore includes a squirrel eating acorns (killing the embryo plant in the process), a water flea feeding on phytoplankton and a weasel feeding on a rabbit. The essential feature of a predator is that it exploits the prey as 'capital'. A parasite, in contrast, feeds on 'interest' (see below).

Predation is also distinct from **grazing**, in which one organism kills *part* of another, leaving the rest of the body to regenerate more tissue. A rabbit is a grazer; it feeds on 'interest' generated by a large number of individual grass plants, which are not killed.

Predators and prey affect each other over two very different time scales.

- Over the short term they may influence each others' numbers in ways explained in more detail in Chapter 38.

- Over the longer term they shape each other's evolution. Predators and prey are in a kind of evolutionary 'arms race'. Predators are under constant selective pressure to improve their prey-catching ability, and consequently prey are under selective pressure to improve their escaping ability. Selection on prey is even more intense than on predators because a prey that fails to escape loses any chance of reproducing, whereas a predator that fails to obtain a meal always has another chance.

### Predation and group living

Both predators and prey may benefit from belonging to a group. For a predator, this can increase food intake in two quite different ways. When prey are sufficiently active to resist and escape, group hunting enables larger and more heavily-armed prey to be killed; a pack of wolves can kill a moose, and two or more lions can tackle a wildebeest with less risk of injury. Second, it may also help in locating food items that are difficult to find but locally abundant (as with some seeds and small insects) because the cost of sharing is small.

For the prey, group membership may provide safety in a number of ways.

- *Greater vigilance*. For animals that rely on speed for escape (such as antelopes and ostriches) early warning is crucial. Living in groups increases the likelihood of early detection of a predator (Figure 37.9). Indirectly, sharing the task of predator-spotting allows more feeding time. An animal feeding on food close to the ground has its head too low to spot an approaching predator. From time to time therefore it must stop feeding and raise its head. A member of a group needs to raise its head less often because the cost of vigilance is shared.

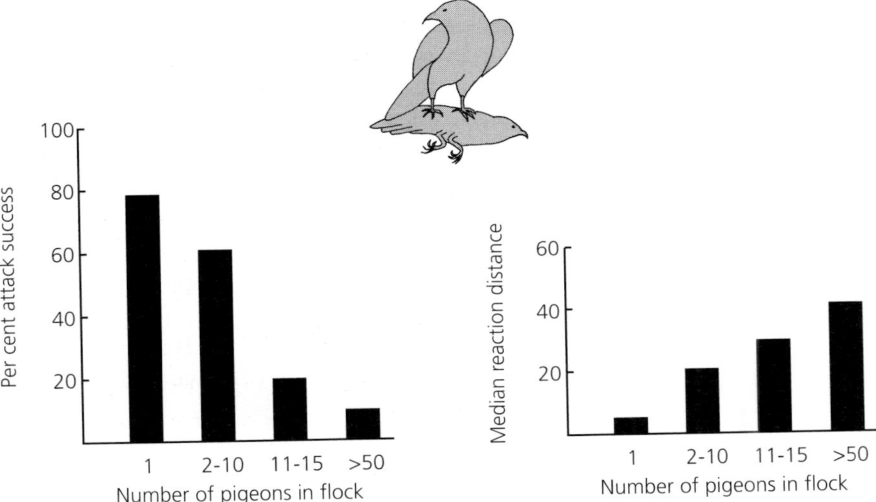


**Figure 37.9**
Effect of flock size on predation in pigeons (redrawn from Krebs and Davies, after Kenward)

- *Increased armament.* Some large herbivores combine to deter predators. A leopard will not attack baboons if there are several males present, but will do so if there is only one male. Many sea birds, such as terns and gannets, nest in large colonies and make mass attacks on a predator.
- *Dilution effect.* In any given predator attack, the larger the group, the less likely any particular prey will fall victim. The presence of other

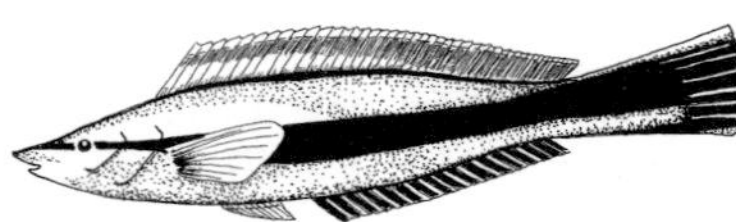

Mimic – the sabretooth blenny

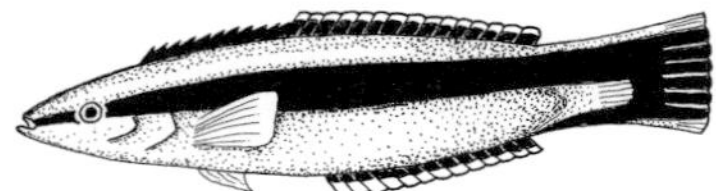

Model – the cleaner wrasse

**Figure 37.10**
Mimicry in sabretooth blenny and cleaner wrasse

prey 'dilutes' the effect of a predator. Of course, larger prey groups are more conspicuous and are more often attacked, but this does not fully counteract the dilution advantage – a herd of 100 antelope are not attacked 100 times as often as a single animal. Even if every predator attack were successful, the chances of an animal being eaten in an attack are smaller if it is a member of a large group.

- *Confusion effect.* Most predators prefer to attack isolated prey because the predator finds it difficult to focus its attention on an individual prey when there are many others, particularly if they are rushing in all directions. Most fish shoals show a 'fountain effect', the shoal splitting into two every time the predator rushes into it.

  Most commercial fish are caught in large numbers precisely because they swim in shoals. Thus, behaviour that is successful against most predators can render them vulnerable to human hunters. It may be that as fishing pressure increases, selection will operate against shoaling in these species.

### *Mimicry*

Some animals avoid being eaten by mimicking a harmful or distasteful animal. Protective mimicry can be of two kinds:

- in **Batesian mimicry**, a palatable animal – the mimic, resembles an unpalatable one – the model. A famous example is the distasteful monarch butterfly, which in North America is mimicked by the palatable viceroy. The effectiveness of Batesian mimicry is self-limiting because the more successful the mimic, the more common it becomes, and the more likely a predator will encounter the mimic rather than the model.
- in **Mullerian mimicry** there is no distinction between model and mimic, with several unpalatable species resembling one another. It is easier for predators to learn to avoid a group of unpalatable species, since a lesson learned after a taste of one species will also apply to other species of similar appearance, e.g. the yellow and black warning coloration of both wasps and hornets.

Some predators also use mimicry. Cleaner fish obtain food by removing parasites from the gills of other fish, which open their mouths for the cleaner. Cleaners are recognised by their 'clients' because of their distinctive colour pattern and mode of swimming. Some predatory fish mimic the cleaner's appearance and movement, and use this deception to get close enough to bite chunks out of the fins and gills of the client fish (Figure 37.10). As with Batesian mimicry, its success is self-limiting, since the deception will only work if the mimic is rarer than the model.

## *Efficient predators*

To make a net 'profit', a predator must gain more energy from its prey than it spends in capturing, subduing and digesting it. Predators that make the biggest energy 'profit' will be the most successful, so it would be expected that there would be strong selection for efficiency. Several studies support this. Figure 37.11 shows the results of one such investigation into predation by shore crabs on mussels.

The crabs easily open very small mussels with their chelae (pincers), but the energy in the flesh is very small. Large mussels, although containing more energy, take much more time and effort to open. Somewhere between the two extremes is an optimum prey size.

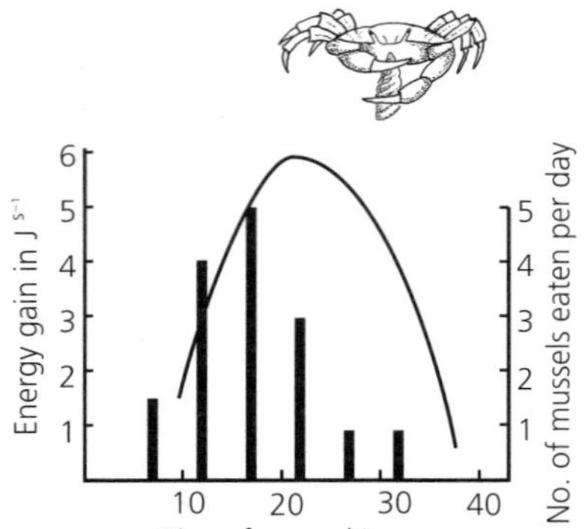

**Figure 37.11**
Efficient predation by the shore crab (*Carcinus maenas*) on mussels (after Einer and Hughes)

The optimum prey size was determined in the following way. First, the energy content of mussels of different sizes was measured. Next, the average time taken by a crab to open mussels of different sizes (and thus of known energy content) was determined. From these measurements the profitability of each mussel size was calculated and plotted. A crab was then presented with a variety of prey sizes, and the size of each prey actually eaten was recorded. The bar graph shows the observed prey preference. The peaks of the two graphs are similar, showing that the crabs were selecting the most profitable prey.

Of course these results do not mean that predators are so clever that they *consciously* choose prey of optimum size. Selection has simply favoured those individuals that select prey of the 'best' size.

## Parasitism

A parasite depends for its food on one organism, the **host**, which is not usually killed (although it may be weakened, rendering it more likely to succumb to other hazards). In general it is in the interest of the parasite to harm its source of food as little as possible.

Just as predators and prey evolve in concert, parasites also come into an evolutionary balance with their hosts. The myxoma virus which causes myxomatosis in rabbits is one example. When the virus was first introduced into Britain in the 1950s it killed over 95% of the rabbits, but over the next few years rabbits became more resistant and the virus became less virulent.

The advantage of resistance in the rabbits is obvious enough, but why should the virus have become less virulent? The answer is that an infected rabbit that remains alive is a source of far more virus particles than if it were to die quickly.

Although parasitism is often considered to be a somewhat 'abnormal' way of life, there are actually more parasitic than free-living species because most free-living organisms are host to several kinds of parasite. To take the red squirrel (*Sciurus vulgaris*) as an example, it may carry three species of lice, three species of flea, two species of tick and four species of nematode worm. Even parasites have their own parasites, called **hyperparasites** or **secondary parasites**, and these may even have **tertiary parasites**.

Animal parasites are divided into two categories: **ectoparasites** which live on the outside of the host, and **endoparasites** which live inside the host's body.

Ectoparasites are less intimately associated with their hosts than endoparasites and in some cases, such as mosquitoes, the relationship is very brief. Many blood-sucking ectoparasites are important as vectors in the transmission of disease. For example the rat flea transmits the bacterium that causes plague, and certain species of *Anopheles* mosquito transmit malaria. Many ectoparasitic insects are wingless, for example lice and fleas. Both have flattened bodies which give some protection against the scratching of a mammalian host, and lice have strong claws which enable them to hold on tightly to the host's body.

Endoparasites typically inhabit internal spaces such as the gut (e.g. tapeworms and some roundworms), bile ducts (e.g. liver fluke) and blood vessels (some flukes and roundworms). Many parasites maintain their position in these places by holding on with hooks, suckers, or both.

Protected and nourished by the host, an endoparasite does not need to expend as much energy producing organs concerned with food-finding and escape, and many endoparasites are structurally simpler than their free-living relatives.

One aspect of an endoparasite's life is particularly precarious – **transmission** to another host. Adult parasites cannot travel from one host to another since they are ill-adapted for survival in the tough outside environment. Many blood parasites use a blood-sucking insect as a vector. Most other endoparasites leave the body as resistant eggs or specialised larval stages. The mortality in transmission is huge and is only offset by the equally high reproductive capacity of the adult.

Besides producing large numbers of offspring, many endoparasites have a complex life cycle often involving a succession of two or even three kinds of host, as in tapeworms.

Most endoparasites are adapted to live in a very limited range of hosts. As with all specialisations, there is a cost attached to this **host-specificity**: the evolutionary future of an endoparasite is dependent upon that of its host. If the host becomes extinct, so does the parasite.

## *The malarial parasite,* Plasmodium

In its effects on human societies malaria is one of the most important diseases, causing high fever, weakness, anaemia and other symptoms. It is actually not one disease but several, caused by four species of single-celled protoctist belonging to the genus *Plasmodium*. The most common form is caused by *P. vivax*, but the most serious infection is due to *P. falciparum* which is potentially fatal because it can attack the brain. It is this form that is responsible for most of the 1–2 million annual deaths, mostly of young children.

**Figure 37.12**
*Anopheles maculipennis*, a *Plasmodium*-carrying mosquito

The disease is spread by mosquitoes belonging to the genus *Anopheles* (Figure 37.12). Mosquitoes breed prolifically in swampy areas, and long before the link with mosquitoes was established malaria was associated with swamps ('mal' and 'aria' are Italian words meaning 'bad air').

### *Life cycle*

The life cycle of the parasite actually involves two hosts – humans and female anopheline mosquitoes. In spreading the disease from one human to another, the female mosquito acts as a **vector** (though from the point of view of the mosquito, which is also parasitised, humans could just as reasonably be regarded as vectors).

The life cycle can be divided into four stages (Figure 37.13):

- asexual reproduction in the human liver.
- asexual reproduction in the human red blood corpuscles.
- sexual reproduction in the mosquito.
- asexual reproduction in the mosquito.

Since the sexually mature stages occur in the mosquito, the mosquito is said to be the **primary host**, and the human the **secondary host**.

Only the female mosquito feeds on blood, which is necessary to provide protein for her developing eggs; males are nectar feeders. When she feeds, she first injects an **anticoagulant** from her salivary glands into the wound. This prevents the blood from clotting and blocking the food canal of her proboscis. If the mosquito is infected with *Plasmodium*, the saliva also contains worm-like **sporozoites**, which are the infective stage of the parasite.

Once in the human host, the sporozoites invade the liver cells. During the next week or so (in which there are no symptoms) the parasite feeds and grows, finally dividing into over a thousand tiny **merozoites**. The now dead liver cell breaks open to release the merozoites into the blood.

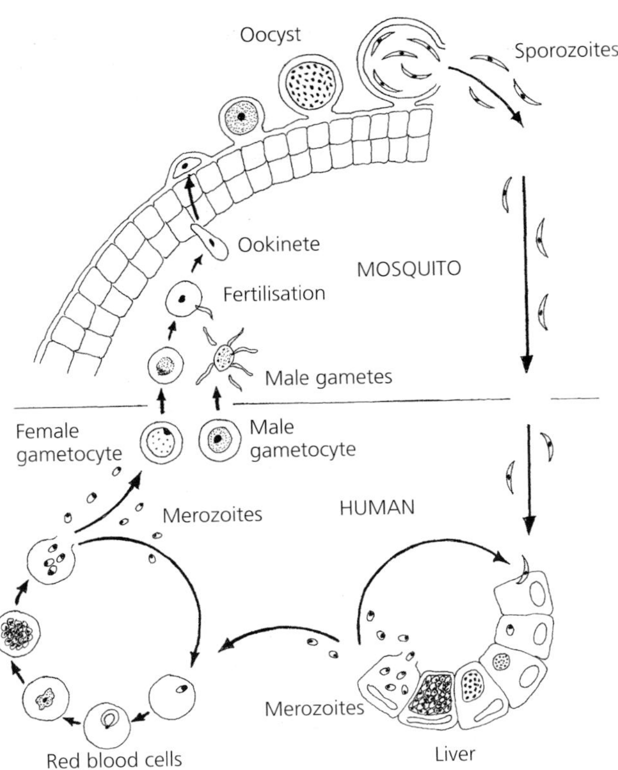


**Figure 37.13**
Life cycle of the malarial parasite

In *P. malariae* the merozoites may reinvade other liver cells, producing further cycles of growth and asexual multiplication. In *P. vivax* some parasites remain dormant in the liver cells and may become active later. This is why some people who have had malaria can suffer a relapse years later in a country free of malaria.

After release into the blood the merozoites enter red corpuscles, feeding on the haemoglobin. After 30–40 hours, depending on the species of *Plasmodium*, the parasite divides into eight or 16 merozoites. These are released into the blood with the disintegration of the red cell. Also released are waste products which cause fever by disrupting the control of body temperature.

A peculiar feature of *P. vivax* and *P. malariae* is that the release of merozoites and the accompanying fever always occurs in the late evening. This is because asexual reproduction in the red corpuscles takes a multiple of 24 hours (48 hours in *P. vivax* and 72 hours in *P. malariae*).

After a number of asexual cycles in the red cells, some of the parasites begin to develop into the sexual stages forming male and female **gametocytes**. Although gametocyte formation takes longer than the asexual cycle (96 hours in *P. vivax*) it still takes a multiple of 24 hours. As a result they also come to maturity in the late evening, just when anopheline mosquitoes are feeding.

How is this extraordinary periodicity achieved? In some way the internal 'clock' of the parasite becomes synchronised with – or **entrained** to – the internal clock of the host. As a time-giving cue, the parasite probably uses the slight daily variations in body temperature of the host.

The gametocytes can develop no further unless they enter a feeding mosquito, and are the only stage that can survive in the gut of a mosquito. Once inside the mosquito, the male gametocytes break free from the red cells and produce up to eight thread-like male gametes. The female gametocyte gives rise to a single female gamete. Fertilisation occurs in the mosquito's gut, after which the zygote develops into a worm-like **ookinete**. This burrows through the gut wall into the surrounding blood space, or haemocoel. The ookinete undergoes meiosis and develops into an **oocyst**. A heavily-infected mosquito may have hundreds of oocysts protruding from the gut, but apparently comes to little harm.

The oocyst grows and, after a certain time (depending on the temperature) it divides into many haploid sporozoites. These make their way to the salivary glands of the mosquito and enter the salivary ducts. They are now ready to develop further, but only if the mosquito injects them into a human before taking a blood meal.

For transmission of the parasite to be successful, the mosquito must take two blood meals – the first from an infected human, and another after an interval long enough for the stages in the mosquito to be completed. Since mosquitoes are ectothermic, this can vary from 8 days to a month or so. Successful transmission is therefore more difficult in cool climates, since the mosquito must survive for longer. Transmission also requires standing water, and in countries with a dry season transmission is impossible for that part of the year.

### *Control measures*

There are two approaches to malaria control: killing the parasite in humans with drugs, and preventing its transmission by the vector.

The role of mosquitoes in the transmission of malaria was proved in 1897 by Ronald Ross, an army surgeon working in India. Although it was many years before all the details of the life cycle were worked out, mosquitoes immediately became the main targets of control. Breeding can be reduced by draining swamps, and numbers of larvae reduced by introducing suitable fish into ponds. Larvae can also be killed by adding a thin layer of oil to the surface of water bodies. This prevents them from opening their breathing tubes at the surface (Chapter 2).

The adults can be killed by spraying dwelling places with insecticides, although resistance to a number of insecticides, such as DDT and malathion, has already evolved.

The first drug to be used against the parasite itself was **quinine**, obtained from extracts of the bark of *Cinchona ledgeriana*, a tree native to the Peruvian Andes. Extracts of the bark were being used two centuries before the parasite causing the disease was discovered. Since then a number of other drugs have been developed, such as **chloroquine**, **mepacrine**, **proquanil** and **artemisine**. Like all forms of chemotherapy, these drugs exploit differences between the metabolism of the parasite and that of the host.

Unfortunately many of these drugs are proving less effective because the parasite has developed resistance and, although new drugs have been developed, resistance to some of these has already appeared.

## *The pork tapeworm,* Taenia solium

Tapeworms make up the class Cestoda within the phylum Platyhelminthes. They are all endoparasites, the life cycle involving two or three hosts. The **primary host**, in which the sexually mature stage occurs, is almost always a vertebrate, the worm living attached to the lining of the intestine. The **secondary hosts**, in which the asexual stages occur, are diverse, although usually fairly specific for a given species of tapeworm. Some species of tapeworm have one secondary host, others have two.

The pork tapeworm is so-called because, although humans are the primary host, the secondary host is normally the pig. Humans normally become infected by eating undercooked pork containing the young stages. The pork tapeworm has always been rare in Jewish and Muslim societies (in which pork is not eaten), and because of stringent control measures it is now rare in most developed countries.

The adult tapeworm is usually 2–4 metres long and consists of two parts (Figure 37.14). A 'head', or **scolex**, anchors the animal to the intestinal lining by means of four suckers and two rings of hooks. The scolex also contains concentrations of receptors and nerve cells which are presumed to be involved in the frequent changes in position that tapeworms are now known to make.

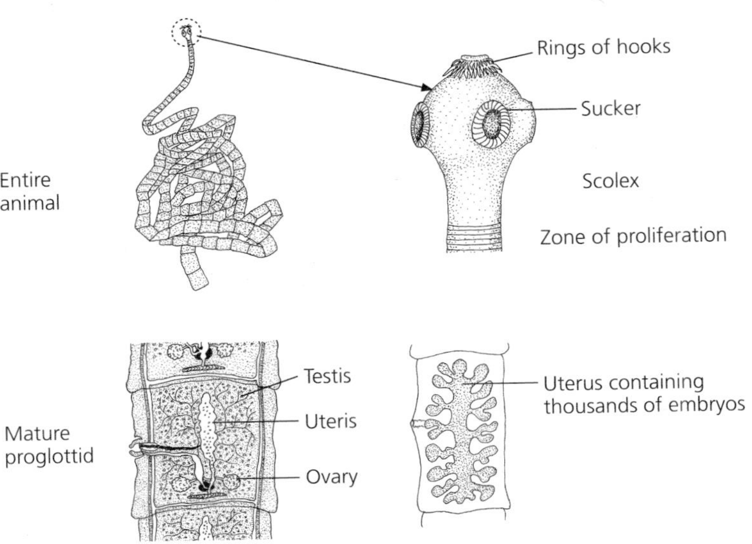


**Figure 37.14**
External features of a pork tapeworm

The rest of the body consists of a long, tape-like **strobila**, divided into up to 1000 flattened 'segments', or **proglottids**. New proglottids are continuously being produced from the zone of proliferation (a region just behind the scolex) at a rate of about 3–4 a day. Mature ones, each about 5–6 mm wide and bearing 30 000–40 000 embryos, are cast off in small groups at the rear.

Tapeworms have no alimentary canal, as predigested food is absorbed from the gut of the host over the general body surface. The surface area for absorption is increased by the flat shape of each proglottid (Figure 37.15).

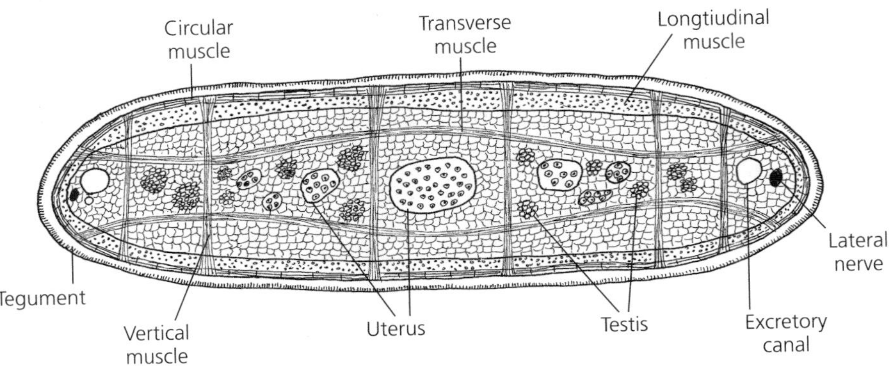


**Figure 37.15**
Transverse section of proglottid of *Taenia* to show flattened shape (after Grove and Newell)

The outer layer of the body is highly specialised for absorption of digested food and resistance to the action of the host's enzymes. The outer layer, which is known to be metabolically very active, is called the **tegument** (Figure 37.16). Cells in its outer layer lack cell boundaries and

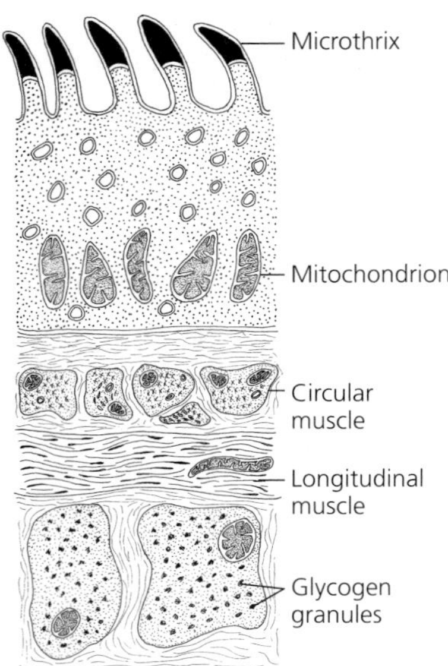


**Figure 37.16**
Section through outer layer of a tapeworm (after Morseth)

bear many minute projections called **microtriches** (singular, microthrix) which in some ways resemble microvilli. Together with the flattened shape of the body, they increase the surface area for the absorption of digested food. The tegument also contains protease inhibitors which protect it from being digested by the host.

There is little oxygen in the intestine, and the animal obtains energy anaerobically by converting glucose into lactic acid and other organic acids. Although there are many mitochondria beneath the microtriches they do not carry out oxidative phosphorylation, but convert pyruvate to acetate, succinate and ATP. The animal generates ATP from glucose very inefficiently, but food is abundant and wastes are easily excreted.

Although the adult tapeworm has no predators and does not need to move to find food, it is more active than one would expect. By means of muscles in the scolex it can periodically release its hold and reattach in a different location. Anti-tapeworm drugs act by paralysing the muscles by which the animals holds on, causing it to be passed out with the faeces.

### *Life cycle*

In the protected environment of the intestine the adult tapeworm can live for many years. Without the need to spend energy meeting the hazards of free-living life, it can devote a high proportion of its resources to reproduction. This is just as well, for the inevitable death of the host means that the young stages must travel through the hostile world outside to reach another host. Although huge numbers of eggs are produced, the transmission of the parasite involves high mortality.

Each proglottid is **hermaphrodite**, containing a complete set of male and female reproductive organs. Although a person may have more than one tapeworm, a given host usually supports only one adult worm (in which case self-fertilisation must occur). Most tapeworms are therefore likely to be homozygous for most of their genes, though when there is more than one worm in a host, cross-fertilisation may occur.

After fertilisation the zygote develops into an embryo called an **onchosphere**, and by the time a proglottid is ready to be cast off, it contains thousands of embryos.

After being cast off, the mature proglottids are egested with the faeces and break open to release over a thousand embryos. Each embryo is protected by a tough outer wall and, although it can survive for long periods on the ground, it normally does not develop further unless it is eaten by a pig (Figure 37.17).

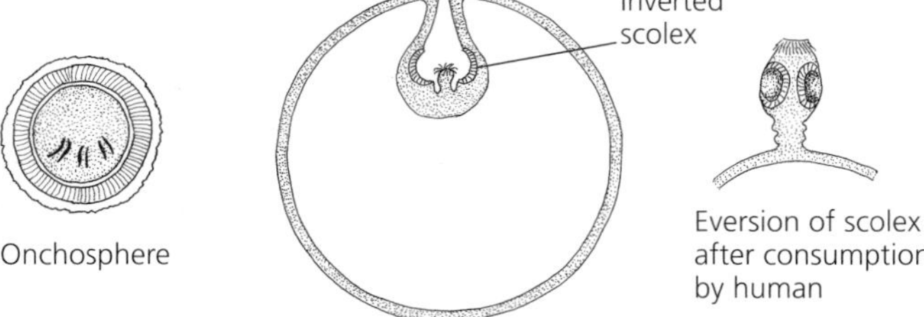


**Figure 37.17**
Bladderworm stage of the pork tapeworm

In this event the protective layer is partly digested away by the pig's enzymes and, stimulated by bile salts, the embryo emerges. By means of its six hooks it bores its way through the wall of the pig's intestine and enters the portal blood vessels which carry it to the liver. From here, the embryos are carried to all parts of the body, but they are especially likely to settle in the muscles. Each embryo now grows into a **cysticercus**, or **bladderworm**, in the wall of which there is an inverted (turned inside out) scolex. It remains in this state until insufficiently-cooked pork is eaten by a human. If this happens the scolex everts, attaches to the intestinal wall and matures within a few weeks, budding off proglottids. The life cycle is summarised in Figure 37.18.

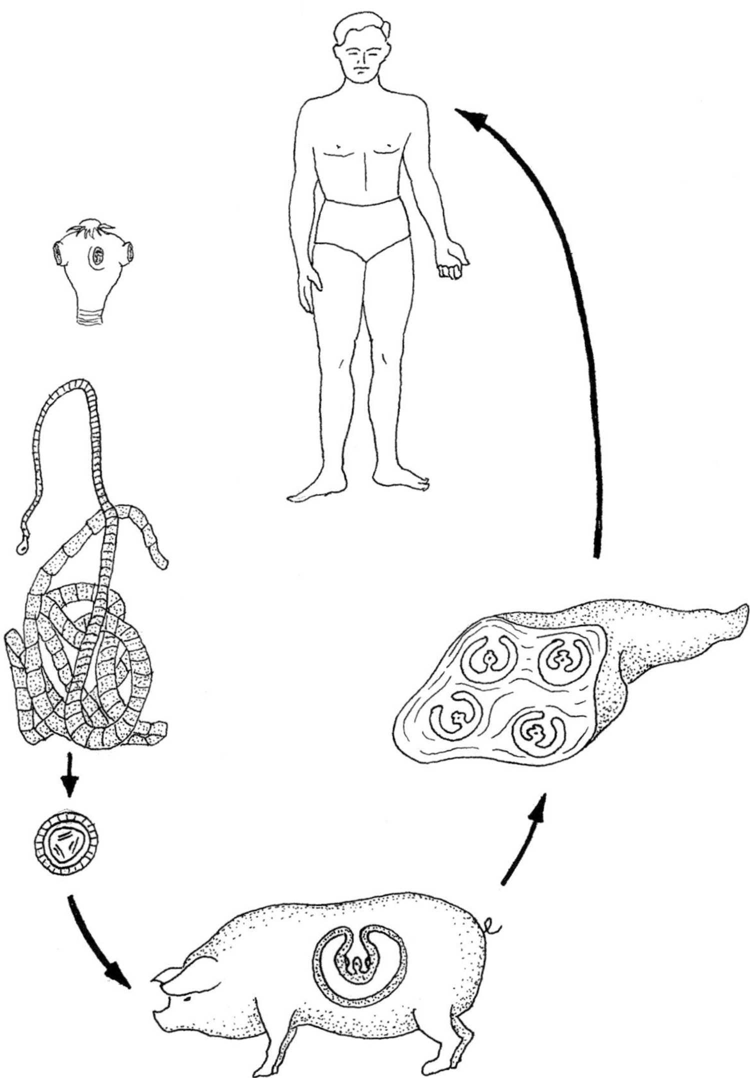

**Figure 37.18**
Life cycle of the pork tapeworm (after Lapage)

Although the pig is the usual secondary host, bladderworms can also develop in humans if onchospheres are accidentally eaten. The onchosphere wall is then digested as would normally occur in a pig, and the result is an infection by cysticerci called **cysticercosis**. The seriousness of the condition depends not only on the number of bladderworms, but on the site of development; a single bladderworm could be fatal if it were to develop in the brain, for example. A massive cysticercosis can occur if the intestine undergoes antiperistalsis (possibly in response to irritation by the worm) causing gravid (egg-laden) proglottids to reach the stomach and release their eggs there.

Prevention of tapeworm infections by breaking the life cycle is, in principle, straightforward, and can be achieved in a number of ways:

- hygienic disposal and treatment of faeces prevents the infection of pigs.
- bladderworms in 'measly' pork are easily visible and with regular inspection of meat in abattoirs, the pork tapeworm is now uncommon.

- thorough cooking of meat prevents infection of humans. Bladderworms are killed by temperatures above 56°C, or below −5°C.

Apart from occasional cases of cysticercotic infections by pork tapeworms, most tapeworms that infect humans do relatively little harm. An exception is the hydatid tapeworm, *Echinococcus granulosus*. The adult is only a few millimetres long and is specific to dogs (or their wild relatives, wolves and coyotes). Although the dog is not harmed, this is not true for the secondary host which is usually a sheep (but may also be a human). In the secondary host the bladderworm undergoes repeated asexual reproduction, budding off 'daughter' cysts and scolices into its cavity. These produce further generations of cysts and scolices. Eventually the cyst may grow large enough to weaken the host to the point where it cannot escape from predators. In this case it is clearly to the *advantage* of the parasite to harm the host (see the information box at the bottom of this page).

## The ultimate parasites

A common feature of many parasites is the loss of structures associated with the normal activities of independent life. Many animal parasites have reduced ability to move and some, such as tapeworms, have no gut.

Viruses can be considered to be the ultimate parasites. Outside the host cell a virus is chemically inactive, with no metabolism of its own. The only way viruses resemble living things is in the fact that they can reproduce. Even the phenomenon of reproduction is restricted to the possession of coded information, since the host cell supplies the necessary raw materials – energy and enzymes. (Retroviruses have one enzyme of their own, but even this is produced by the host cell and is inactive outside it.)

The inability to reproduce outside a host cell is not restricted to viruses. The malarial parasites and some bacteria are also obligate intracellular parasites. Bacteria of the genus *Chlamydia* (common causes of venereal infections) even depend on the host cell to make ATP. Yet even these extreme parasites make proteins using their own ribosomes. Moreover, unlike viruses (which lose their structural identity inside the host cell) they retain their cell membranes inside the host cell and divide by binary fission.

## Parasites that kill

Since a parasite is dependent on its host for food, there is usually strong selection pressure on the parasite to harm the host as little as possible and, in general, the best-adapted parasites have come into an evolutionary accommodation with their hosts. An exception to this 'rule' (where the parasite depends on the secondary host being eaten by the primary host) has already been mentioned.

Other exceptions are tiny wasps called **parasitoids**. The adults are free-living and lay their eggs inside the young stages of other insects. *Apanteles glomeratus*, for example, lays its eggs in the caterpillars of butterflies such as the cabbage white. At first the wasp larvae feed on the fat and other non-essential tissues, but eventually they begin feeding on the vital organs, killing the larva. Insect parasitoids play a major part in the regulation of many insect populations, and some have been used as biological control agents of introduced pests (Chapter 38).

## Are parasites degenerate?

A common feature of many endoparasites is the reduction of those organs associated with free-living life, such as organs of locomotion and sensation. An extreme case is *Sacculina*, a barnacle parasitic in crabs. All traces of barnacle organisation, and even arthropod characters, are absent in the adult which resembles little more than a bag with fine extensions that penetrate throughout the crab's body. Only the larva, which strongly resembles that of barnacles, betrays the animal's relationship with free-living barnacles.

This structural simplicity has led some people to consider parasites to be 'degenerate'. This is, of course, a value-laden term that has no meaning in biology. Although many adult endoparasites are structurally simple, their life cycles are often extremely complex, associated with the demands of transmission from one host to another. Transmission thus demands complexity in time rather than the spatial complexity that is characteristic of free-living animals.



# Mutualism

A surprising number of organisms enter into some form of partnership with a member of another species. A close relationship between members of two species in which both partners benefit is called **mutualism**. Lichens are examples, in which the hyphae of a fungus are interwoven with single-celled 'algae'. A few other examples are described below.

### *Root nodules*

The roots of members of the legume family such as gorse, lupin and clover, have swellings, or **nodules**, containing bacteria belonging to the genus *Rhizobium* (Figure 37.19). The basis of the association is that each partner can do something the other cannot: the bacteria can fix nitrogen and the plant can fix $CO_2$. The bacteria convert nitrogen into ammonia which is used to make amino acids, the organic carbon skeletons being supplied by the plant.

The bacteria live freely in the soil as saprobes, but when soil nitrogen is in short supply they penetrate the root hairs of the legume and stimulate development of a nodule, enabling the legume to compete very effectively with non-legumes.

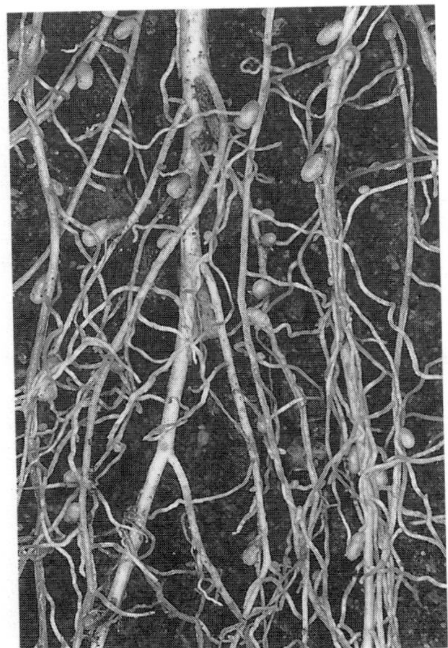

**Figure 37.19**
Root nodules on the roots of white clover

### *Mycorrhiza*

One of the most widespread mutualistic relationships is a **mycorrhiza** (meaning 'fungus-root'), a partnership between a fungus and the roots of a plant. The fungal hyphae form a meshwork in the cortex of the root and extend for up to several centimetres into the soil (Figure 37.20). The hyphae are much better at absorbing minerals (especially phosphate) than the root hairs of the plant, possibly because they are finer than root hairs and extend further into the soil. So effective are the hyphae at absorbing minerals that in heavy fungal infections root hairs may hardly develop, and in many trees the lateral roots are very short. Having absorbed the minerals, the hyphae transport them into the root.

The other side of the coin is what benefit does the fungus get? Radioactive tracer experiments show that, as one might expect of an organism without chlorophyll, the fungus obtains sugar from the plant. The fact that mycorrhizal fungi do not live freely in the soil suggests that, as far as the fungus is concerned, the relationship is **obligate**, or essential for survival.

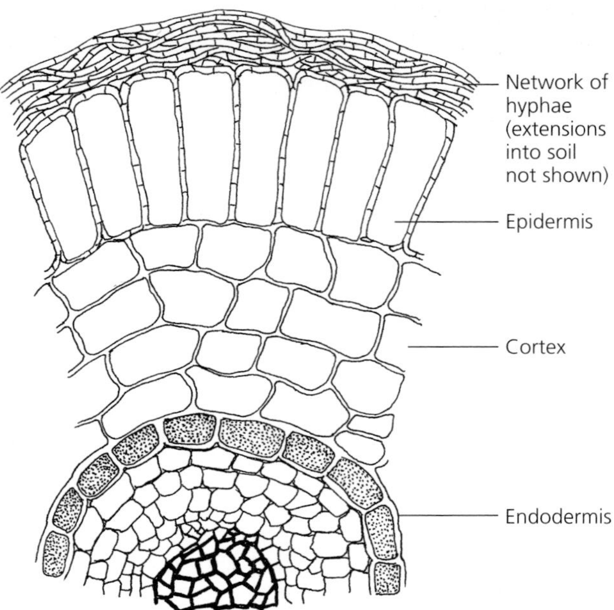


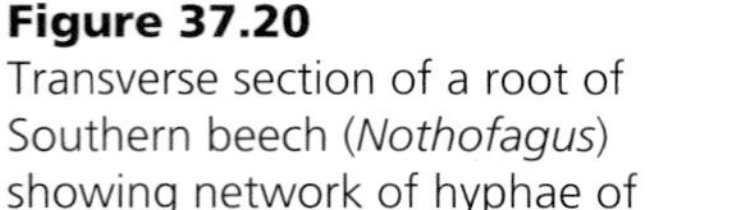

**Figure 37.20**
Transverse section of a root of Southern beech (*Nothofagus*) showing network of hyphae of mycorrhizal fungus (after Stevenson)

Mycorrhizae are more than just widespread – they are the rule rather than the exception. Some of the mushrooms commonly found beneath forest trees are actually the reproductive bodies of mycorrhizal fungi.

### *Gut mutualists*

Very few animals can produce a cellulase enzyme (snails and some marine wood-boring animals are exceptions). The overwhelming majority of herbivores rely on micro-organisms in their gut to do the job for them. Many mammalian herbivores have a large part of the gut specialised for housing these micro-organisms, for example the caecum of rabbits, rats and mice, and the rumen of sheep, cattle and deer. In this protected environment the micro-organisms obtain warmth and the anaerobic conditions that most of them require. The herbivore derives a triple benefit:

- it obtains large amounts of otherwise unavailable carbohydrate.
- the micro-organisms produce essential amino acids, which become available to the animal when the micro-organisms are eventually digested.
- they also produce Vitamin $B_{12}$ (which no animal can produce for itself) and certain other B vitamins.

## *Commensalism*

Commensalism is a relationship between two species in which one benefits and the other neither benefits nor is harmed. Humans have many commensals – many bacteria live harmlessly on our skin and in our colons, although some of the colon bacteria are potentially harmful (*Escherichia coli* for instance can cause infections of the urinary tract, and also food poisoning).

Parasitism, mutualism and commensalism are collectively referred to as **symbiosis**. The essential feature of a symbiotic relationship is its closeness – and often species-specificity – in contrast to the looser relationships of predation and competition. Symbiosis was often used synonymously with mutualism, but in order to avoid confusion 'mutualism' is to be preferred.

## QUESTIONS

**1** True or false? Territorial behaviour
- **a)** decreases the reproductive success of a population.
- **b)** acts as an agent of natural selection.
- **c)** decreases intraspecific competition.

**2** What is the benefit of defending a territory, and why does the benefit not continue to increase with increasing territory size?

**3** Why are dominance hierarchies more common in more complex animals?

**4** The diagram shows the vertical distribution of two species of barnacle, *Semibalanus balanoides* and *Chthamalus stellatus* on rocky shores. *S. balanoides* inhabits most of the intertidal zone, while *C. stellatus* occurs down to mid-tide level but grows more slowly than *Semibalanus*. It was suggested that the reason for the absence of older *Chthamalus* individuals was competition for space by the more rapidly-growing *Semibalanus*.

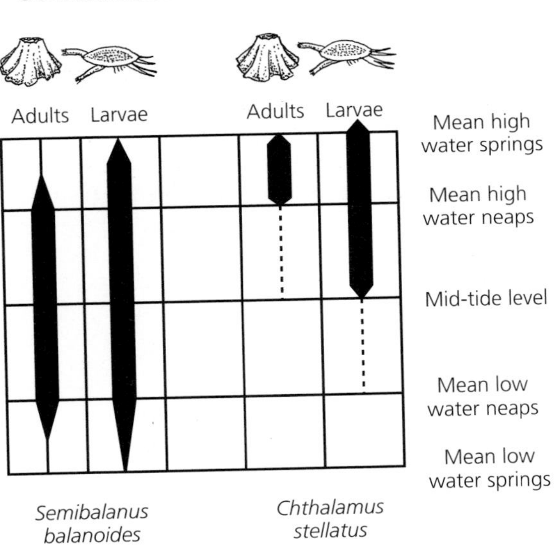


- **a)** How would you test the hypothesis that the inability of *Chthamalus* to reach maturity in the mid- and lower shore is due to competition from *Semibalanus*?
- **b)** Suggest two possible reasons for the inability of *Semibalanus* to grow on the upper shore.

**5** Give an example to illustrate how
- **a)** plants influence the distribution of plants,
- **b)** plants influence the distribution of animals,
- **c)** animals influence the distribution of plants,
- **d)** animals influence the distribution of animals.

**6** Nuthatches are small insectivorous birds of the genus *Sitta*, and are native to Europe and Asia. The diagram shows the heads of two species, *S. neumeyer* and *S. tephronota*. They have very similar beak lengths where they are allopatric (occupy different areas) but are clearly distinct where they are sympatric (ranges overlap).

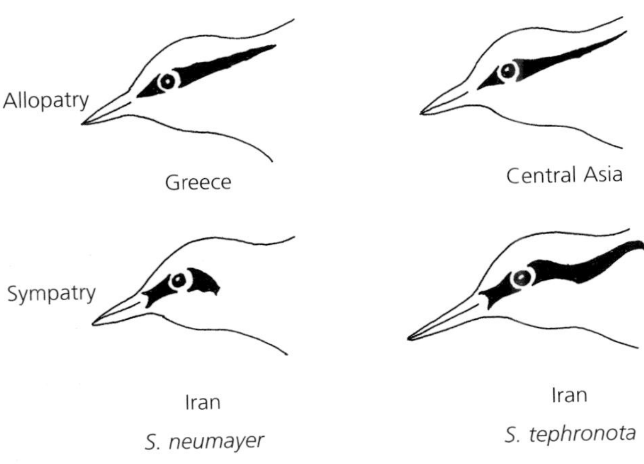


- **a)** What name is given to this phenomenon?
- **b)** Explain how the similarities of **(i)** beak, **(ii)** eye-stripe have probably arisen.

**7**
- **a)** What does the Competitive Exclusion Principle state?
- **b)** Give one piece of field evidence for the existence of competition between species.

CHAPTER 38

# POPULATIONS

## LEARNING OBJECTIVES

By the time you have completed your study of this chapter you should be able to:

- define and give an example of a population.
- state the four processes that can change population density.
- explain the theoretical basis of estimating population densities using quadrats.
- carry out a simple calculation to estimate a total population from mark and recapture data.
- explain precisely what is meant by exponential growth.
- distinguish between relative and absolute population growth rates.
- distinguish between the various phases of sigmoid growth of a population.
- explain what a survivorship curve is and how it can indicate changes in mortality with age.
- interpret age group structures for different populations.
- explain what is meant by density-dependent control, and why populations can only be regulated at a steady level by density-dependent mechanisms.
- interpret graphs of predator–prey cycles of lynx and snowshoe hare.
- describe the various ways in which changes in population densities of prey can influence predation.
- with examples, distinguish between *r* and *K* life cycle strategies, and explain their significance.

## What is population biology?

A population comprises all the organisms of a particular species living in a defined area and sharing a common gene pool; for example, the lugworms in a particular stretch of shore or the house sparrows on the Isle of Wight.

Population ecologists are concerned with problems and phenomena that are not applicable to individuals. Most of all they are concerned with the mechanisms by which the numbers in a population are regulated. Why, for example, are some species common and others rare? Why do some populations remain fairly stable over quite long periods of time, whilst others may fluctuate wildly?

The number of individuals per unit area is the **population density**. At any one time population density is subject to two opposing influences: a tendency to *decrease* due to **mortality** (deaths) and **emigration**, and a tendency to *increase* due to **natality** (births) and **immigration**.

Mortality actually refers to the death *rate*, and is expressed in terms of the number of deaths per thousand per unit time. Similarly natality is expressed in terms of the number of births per thousand per unit time.

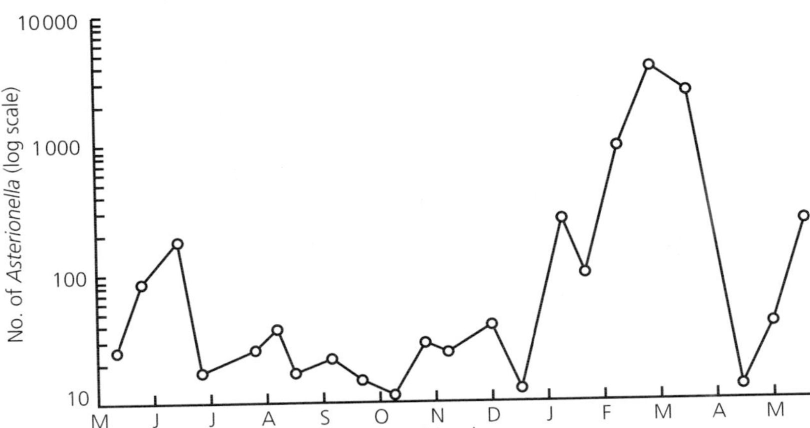


**Figure 38.1**
Fluctuations in numbers of *Asterionella formosa*, a planktonic alga, in surface waters of Lake Rotorua, New Zealand. Note the logarithmic scale, meaning that fluctuations are several thousand-fold. Note also the inverted seasons (data from Flint)

In most populations the balance between natality and mortality is ever-shifting, so that numbers are constantly changing. Fluctuations may be quite small, or they may be so large that only a logarithmic scale can accommodate them (Figure 38.1).

## Estimating population densities

Before ecologists can even attempt to learn how populations are regulated they need reasonably reliable data. Changes in population densities can only be observed if there is some means of estimating numbers.

It is seldom possible to count the entire population – that is, to conduct a **census**. Most often ecologists have to be content with estimates based on **samples**. There are various methods, the suitability of each depending on the species.

### Use of quadrats

With plants and with sessile (attached) and sedentary animals, sampling by the use of **quadrats** is often used. A quadrat is a small part of the habitat whose dimensions are accurately known. The size of a quadrat is thus a known proportion of the total habitat. If the number of individuals in each of a series of quadrats is accurately counted, the mean number per quadrat enables the population density to be estimated.

The size of the quadrat obviously has to be adapted to the situation: when estimating the density of trees it has to be much larger than when estimating the density of barnacles. When quadrats are being used to study the composition of a community, they must obviously be large enough to include most of the species present.

Though quadrats are often square for convenience, they do not have to be any particular shape. In fact in some cases they are not even two dimensional. For example, in estimating the density of cockles on a sandy shore, one would count the number of cockles in a certain area down to a certain depth – i.e. in a certain *volume* of sand.

### Mark, release and recapture

This method, also known as the **Lincoln Index** after the American F.C. Lincoln who introduced it, is most suitable for animals that are fairly mobile but can easily be caught, such as grasshoppers. A reasonably large number are caught, marked and then released and allowed to mix with the rest of population. The following day another sample is caught. Some of these will be marked, and the proportion of these **recaptures** makes it possible to estimate the total number in the population. The accuracy of this estimate depends on the extent to which certain assumptions (listed below) are valid.

Suppose that 100 individuals are marked and released. The larger the total population, the smaller will be the proportion of marked individuals. If, say, the total population is 1000 (a figure that we actually do not know), then about 10% of that population would be marked. The percentage of marked animals in the population can be estimated by taking a second sample a day or so later, a proportion of which would be marked. In the above imaginary case, about 10% would be marked.

It follows that:

$$\frac{\text{number of marked animals released (R)}}{\text{total number of animals in population (T)}}$$

$$= \frac{\text{number of marked animals recaptured (M)}}{\text{total number of animals recaptured (N)}}$$

$$\text{Hence } T = R \times \frac{N}{M}$$

This estimate will be reasonably accurate only if the following conditions and assumptions hold:

- the number released must be large enough for a reasonable number to be recaptured. This number will vary with the species and the situation.
- the marked animals must mix completely with the unmarked ones. Some animals, such as woodlice, live in local aggregations and mixing is unlikely to be complete.
- marked and unmarked individuals must be equally likely to be caught. If some are easier to catch than others, the marked ones are more likely to be recaptured. A higher proportion of marked individuals amongst the recaptures will lead to an underestimate of the total population and vice versa.
- the mark must not reduce an animal's life expectancy in any way. It must be non-toxic and not so conspicuous that it renders it more likely to be seen by predators.
- the mark must persist at least until the day of recapture. In arthropods (which shed their cuticles) some marks will be lost in this way.
- there must be no immigration or emigration.

## Population growth

Most populations fluctuate, but over a long period of time the rises and falls more or less cancel each other out. It follows that over a long period, the tendency for the population to increase must be balanced by its tendency to decrease, i.e.:

$$\text{natality} + \text{immigration} = \text{mortality} + \text{emigration}$$

In a population that is in such a balance, the vast majority of individuals die without reproducing. Most will be weakened by parasites, eaten by predators or be unable to obtain enough food. Many fail to reproduce because they cannot find a mating partner and, in plants, some ovules remain unfertilised.

Because of this, death rates in natural populations are very high. To get an idea of how high mortality normally is, we only have to see what happens when a population grows without any of the usual checks.

### *Exponential growth*

Under favourable physical conditions and in the absence of predators, parasites or competition for food and other resources, a population grows **exponentially**. This means that numbers grow by the same *proportion* each unit time. Doubling every hour, tripling every ten years or increasing by 1.5 times every century, would all be examples of exponential growth.

When numbers are plotted against time, the graph for an exponentially-growing population gets steeper and steeper and, very soon, numbers become too large to be plotted on an ordinary scale. A simple expedient is to plot the **logarithm** of the numbers, thus allowing very large and very small numbers to be plotted on the same graph.

Another advantage of using a log scale is that exponential growth is easy to recognise because it gives a straight line. In other words the logarithm of the numbers of organisms is directly proportional to the time. This is why exponential growth is also called **logarithmic growth** (Figure 38.2).

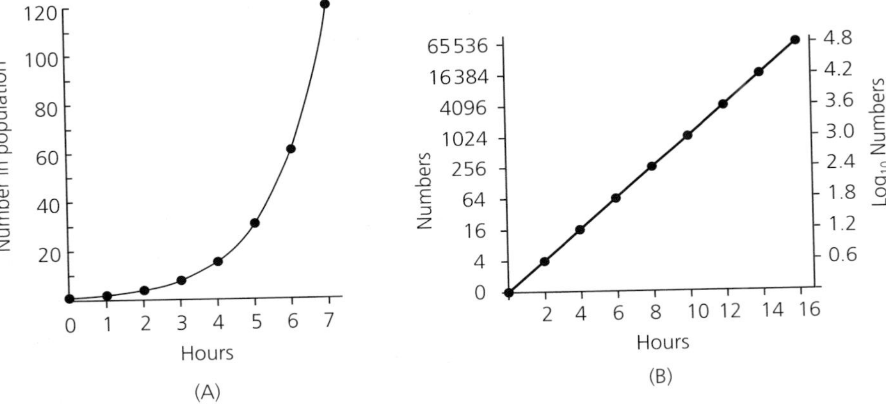


**Figure 38.2**
Graphs showing growth of an imaginary population of bacteria, each bacterium dividing once every hour. Plotted on a linear scale (A). A logarithmic scale accommodates much higher populations (B). Note that both scales in (B) are logarithmic, but are represented in different ways

Though all populations have the capacity for exponential growth, the rate of increase varies greatly. One way of expressing the rate of increase is in terms of the **doubling time**, or time for the population to double. In organisms that reproduce by binary fission this is the same as the generation time, which for some bacteria under ideal conditions may be 20 minutes. For fruit flies at 25°C it is just under two days, and for rabbits it may be about two weeks.

Under ideal conditions the doubling time depends on three features of the organism's biology:

- the **generation time**, or the time it takes for an organism to become a parent. Generation times vary from 20 minutes in some bacteria to about 30 years in elephants.
- the average number of offspring produced by each parent in each reproductive episode; this varies from one in monkeys and apes to hundreds of millions in puffball fungi.
- the number of times an organism reproduces in its life. Some, such as eels, most insects and ephemeral plants, devote all their energy to one gigantic reproductive effort and then die. Others may reproduce more than once.

Of these factors, generation time is by far the most important: a bacterium

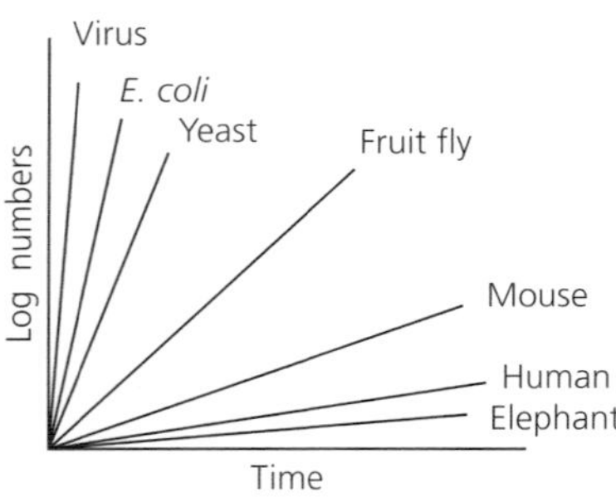


**Figure 38.3▲**
Unrestrained growth of different populations plotted logarithmically

only produces two new individuals per generation, but in the time it takes for an oyster to produce millions of offspring, a bacterium has had time to double its numbers thousands of times. A thousand doublings would give $2^{1000}$ offspring – far more than the estimated number of charged particles in the universe! (Figure 38.3).

In most animals and in many plants, exponential growth is more complicated than in bacteria for two reasons. First, many animals and plants eventually die of old age if they are not eaten by predators. Secondly, in many species successive generations overlap, so a rabbit lucky enough to escape predators could still be reproducing at the same time as its great grandchildren are.

## *Relative and absolute growth rates*

The rate of growth of a population can be expressed in two ways (Figure 38.4).

- **absolute growth rate**, or increase in numbers per unit time, for example 2500 per year, or 300 per week. If the absolute growth rate is constant, it gives a straight line when plotted on a graph with a linear vertical scale (Figure 38.4A).
- **relative growth rate**, or the percentage by which numbers increase per unit time, for example 5% per year. When plotted on a graph, a constant relative growth rate yields an ever-steepening curve if the vertical scale is linear, and a straight line if plotted on a logarithmic vertical scale (Figure 38.4B).

Absolute and relative growth rates are quite independent of each other. For example, an increase of 1 million people in the populations of the United Kingdom and China would represent the same absolute increase, but it would be a much greater relative increase for the U.K.

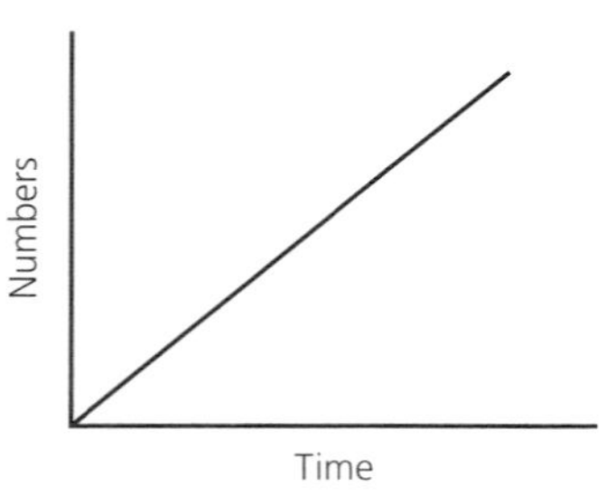


(A) Constant absolute growth rate

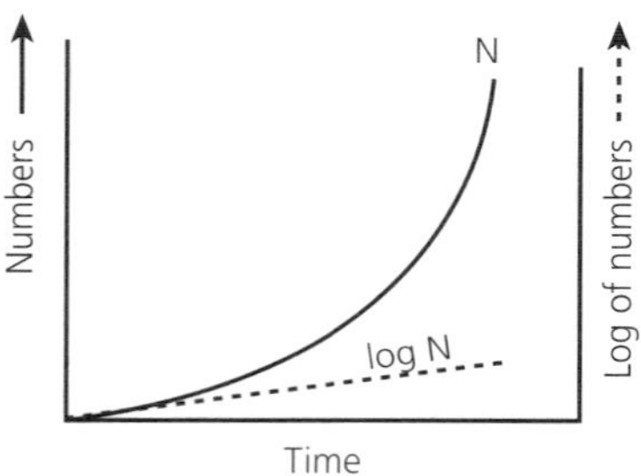


(B) Constant relative growth rate

**Figure 38.4▲**
Absolute and relative growth rates

## *Sigmoid growth*

Under natural conditions, logarithmic growth cannot continue for long – the larger the numbers, the greater the checks on further growth, for example:

- increasing competition for food.
- predators can catch more prey when the latter are more numerous.
- parasites and the diseases they cause can spread more easily from one host to another when the host numbers are large.

Collectively these factors constitute the **environmental resistance**. When numbers have stabilised, the population has reached the habitat's **carrying capacity** and there is an equilibrium between natality and mortality.

Figure 38.5 shows what would be expected to happen if a pair of goldfish were to be introduced into a pond containing plenty of food. The population increases slowly at first, but grows increasingly rapidly before gradually levelling off and stabilising. The graph is described as **sigmoid** because it is 'S'-shaped (*sigma* is the Greek letter 's').

In a natural pond, toxic waste products would be removed by bacteria, and oxygen would be unlikely to be depleted because of production by plants. These two factors would be likely to be important in a single species culture fed artificially.

Let's now take a closer look at sigmoid growth. Bacteria are a convenient example, since they reproduce by dividing into two and do not have overlapping generations. Suppose a million bacterial cells are

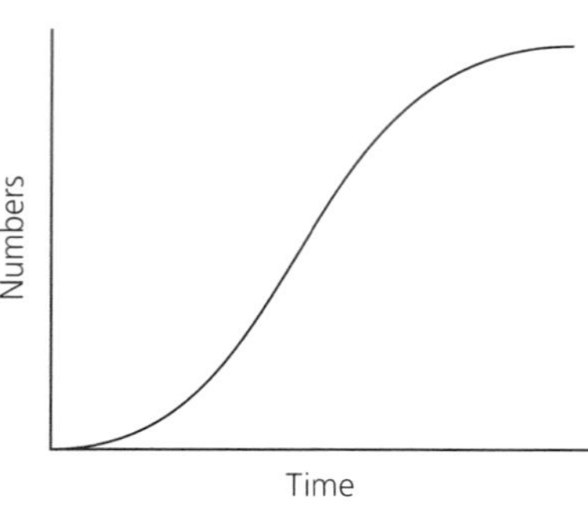


**Figure 38.5▲**
Sigmoid growth of a population of goldfish after introduction into a pond

introduced into a nutrient solution at a constant temperature of 20°C, and that at this temperature each cell can divide every hour. Table 38.1 shows an imaginary set of figures to illustrate the important features of the growth of such a population, and Figure 38.6 shows the figures plotted graphically.

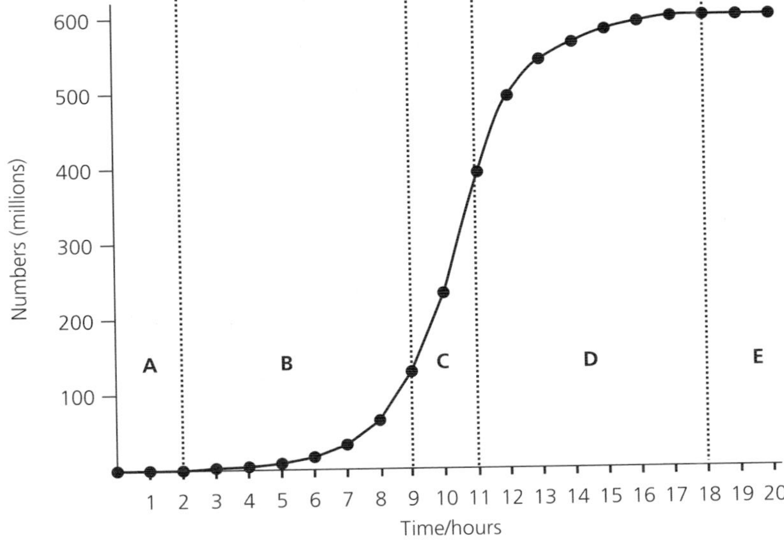


**Figure 38.6▶**
Graph of the growth of an imaginary population of bacteria to show five phases of growth: A, lag phase; B, exponential phase; C, decreasing relative growth rate, increasing absolute growth rate; D, decreasing relative and absolute growth rate; E, stationary phase

**Table 38.1**
Growth of an imaginary population of bacteria in a culture solution

| Growth period | Time (h) | Numbers (millions) | Increase (millions) | % increase |
|---|---|---|---|---|
| A | 0 | 1 | – | – |
| | 1 | 1 | – | – |
| | 2 | 1 | – | – |
| B | 3 | 2 | 1 | 100 |
| | 4 | 4 | 2 | 100 |
| | 5 | 8 | 4 | 100 |
| | 6 | 16 | 8 | 100 |
| | 7 | 32 | 16 | 100 |
| | 8 | 64 | 32 | 100 |
| | 9 | 128 | 64 | 100 |
| C | 10 | 232 | 104 | 81 |
| | 11 | 392 | 160 | 69 |
| D | 12 | 496 | 104 | 26.5 |
| | 13 | 542 | 46 | 20 |
| | 14 | 564 | 22 | 4 |
| | 15 | 582 | 18 | 3.2 |
| | 16 | 592 | 10 | 1.7 |
| | 17 | 598 | 6 | 1.0 |
| | 18 | 600 | 2 | 0.33 |
| E | 19 | 600 | 0 | 0 |
| | 20 | 600 | 0 | 0 |

The graph shows five distinct phases:

**A** If the new medium is chemically different from the previous one, there may be a **lag phase**, during which bacterial metabolism adapts to the new conditions. When, for example, *E. coli* bacteria are transferred from a medium without lactose to one containing lactose, genes for lactose utilisation are activated – a process that takes an hour or two.

**B** A period of exponential growth in which each cell divides at a constant rate. In this hypothetical example the population doubles every hour because every cell divides every hour. Notice that although the relative growth rate is constant, the increasing number of dividing cells means that the absolute growth rate is still increasing.

**C** The relative growth rate is declining because each cell takes longer to grow and divide due to the effects of crowding. Despite this, the absolute growth rate (i.e. the gradient) continues to increase because the increase in generation time is more than offset by the increasing number of dividing cells.

**D** The gradient of the curve begins to decrease because the increase in number of dividing cells is more than offset by their decreased rate of division. Although the numbers are still increasing, the absolute growth rate is declining.

**E** Growth in numbers has ceased. Cells are either dead, or have formed resistant spores.

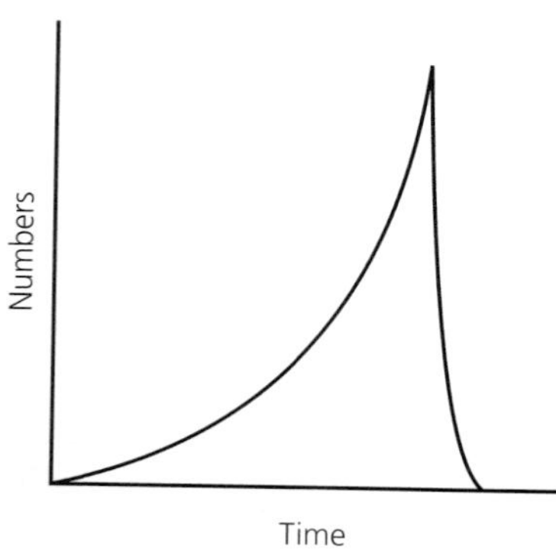


**Figure 38.7**
A 'J' growth curve showing exponential rise in numbers followed by population crash

### *'J' curves*

Not all populations grow in a sigmoid manner. If a population has the capacity to grow faster than its food can regenerate, it may grow exponentially for a while before crashing, producing a 'J'-shaped 'boom and bust' curve (Figure 38.7). This has happened on numerous occasions when herbivores (such as deer) have been introduced into a predator-free area.

## Age-specific death rates – survivorship

Overall mortality figures conceal the important fact that, in most animals and plants, death rates vary with age. This can be shown as a **survivorship curve**. This is a graph of the number of survivors of an original group born, hatched or germinated at approximately the same time. Such a group is called a **cohort**.

Suppose that out of an original cohort of 10 000 individuals, half die every year. 5000 will have survived after 1 year, 2500 after years, 1250 after 3 years, 625 after 4 years, and so on (Figure 38.8).

The number of survivors is normally plotted on a log scale for two reasons. First, it allows large and small numbers to be accommodated on the same scale. Second, it makes it easier to compare death rates at different ages. A straight line indicates that a constant *proportion* is dying each time interval, meaning that the probability of any given individual dying in a given time interval is constant. A steepening curve means that the probability of death is increasing with age, and a decreasing gradient means that life is becoming less dangerous with age.

In most species, death is not equally probable at all ages (Figure 38.9). An oyster, for example, produces millions of eggs which develop into tiny planktonic larvae but very few survive to settle on the bottom, the overwhelming majority being eaten by small fish. Many other organisms have a similarly high infant mortality.

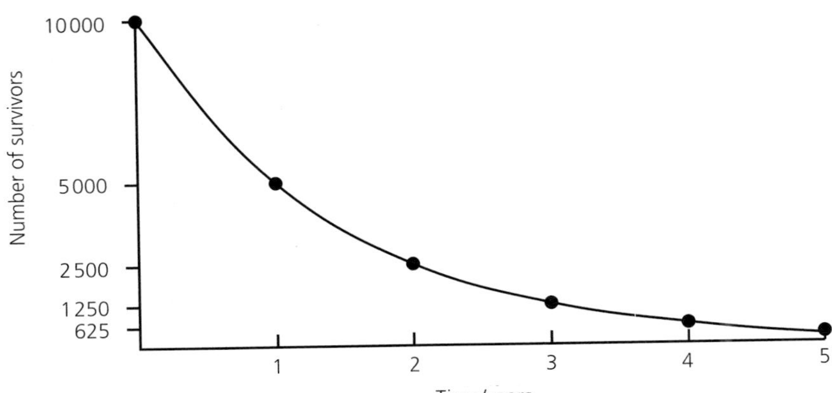


**Figure 38.8**
Graph showing exponential decrease in the number of survivors out of an original cohort of 10 000

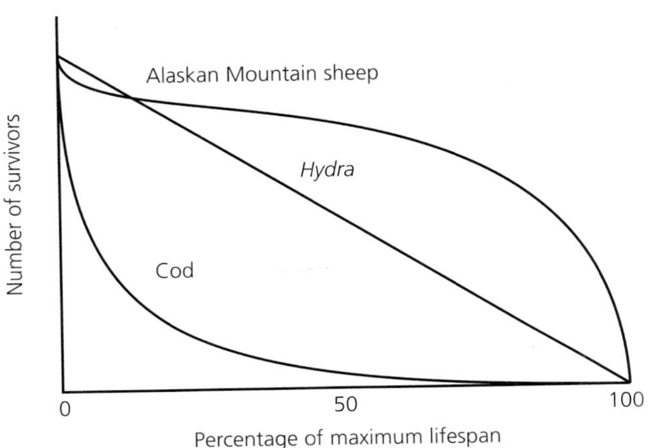


**Figure 38.9**
Three kinds of survivorship curve. In most animals (e.g. cod), mortality is much higher in the young stages. In most mammals, mortality is high in both young and old. In a few animals, such as *Hydra*, mortality is constant throughout life

Humans living in developed countries have a different survivorship in which death rates are quite low until old age. A less common type occurs in organisms in which all stages are equally vulnerable, as in *Hydra*.

A detail that we have ignored until now is this: how are the data for drawing survivorship curves obtained? For each organism, two things need to be known: its age at death and when it died. Age at death can often be determined from clues such as growth rings, as, for example, in fish scales, the horns of sheep, and trees. If the approximate date of death is also known, each dead individual can be assigned to its cohort.

## Age group structure

Changes in mortality with age affect the **age structure**, or the relative proportions of each age group in the population. Figure 38.10 shows the age structure of two human populations in Sweden and India in 1970.

The age structure of a population results from the relationship between natality and mortality during previous years. Because these can vary independently, age structures need to be interpreted with caution. For example, a population with a high proportion of young individuals could result from two quite different causes.

- There may be high infant mortality, few reaching adulthood. If the population size is stable, mortality of each age group can be gauged by the decrease in the size of successive age groups up the pyramid.
- Even if infant mortality is low, a rapidly expanding population results in a high proportion of young individuals.

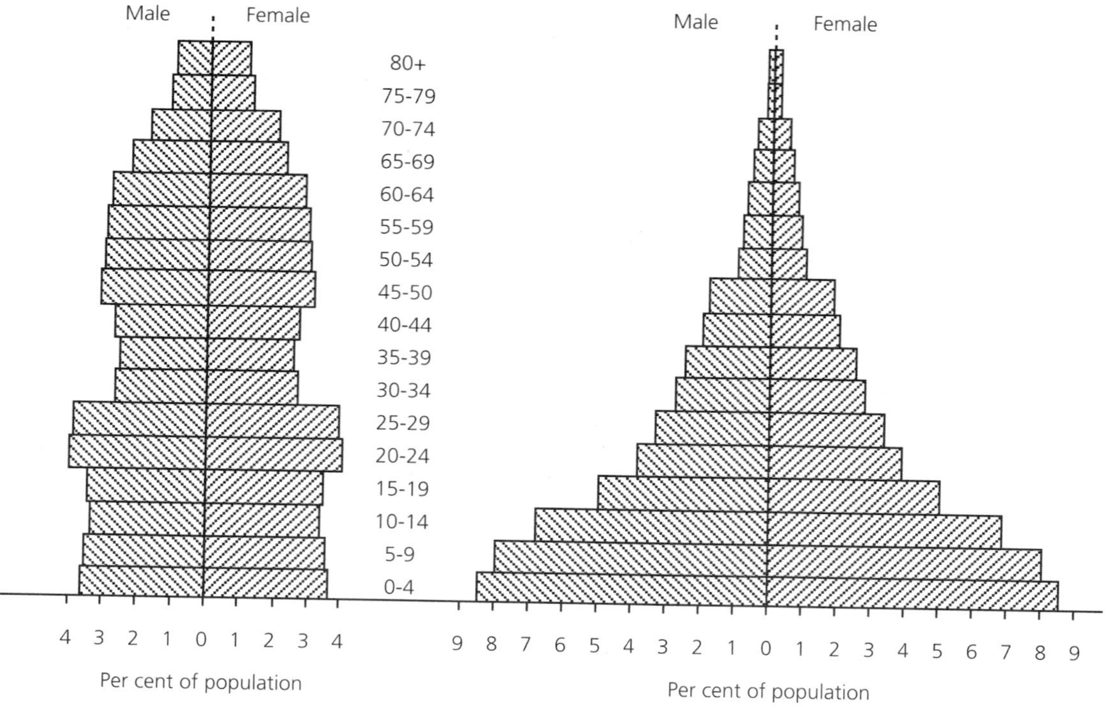


**Figure 38.10**
Age structures for human populations in Sweden and India in 1970 (UN Demographic Yearbook)

## The mathematics of population growth

When a population grows without restraint (i.e. exponentially), the rate of increase is given by

$$\frac{dN}{dt} = rN$$

Where $\frac{dN}{dt}$ = the rate of increase (the gradient of the curve),

$r$ = the intrinsic rate of increase, and
$N$ = the number of organisms.

The intrinsic rate of increase is the number of offspring produced by each parent per unit time. So if a bacterium is dividing once every half hour, it gives rise to four descendants every hour, and $r = 4\ \text{hr}^{-1}$.

The intrinsic rate of increase is only attained if there is no environmental resistance, which is only true when numbers are low. An equation which takes account of environmental resistance is called the **logistic equation:**

$$\frac{dN}{dt} = rN\left(1 - \frac{N}{K}\right)$$

where $K$ = the carrying capacity.

When $N$ is small, $\frac{N}{K}$ is also small, so $\left(1 - \frac{N}{K}\right)$ is nearly equal to 1,

and $\frac{dN}{dt} \approx rN$.

When $N$ is equal to $K$, $\frac{N}{K} = 1$, so

$\left(1 - \frac{N}{K}\right) = 0$, and $\frac{dN}{dt} = 0$, so numbers do not increase.

The term $\frac{N}{K}$ represents the environmental resistance, and the more closely this approaches 1, the lower is the rate of population growth.

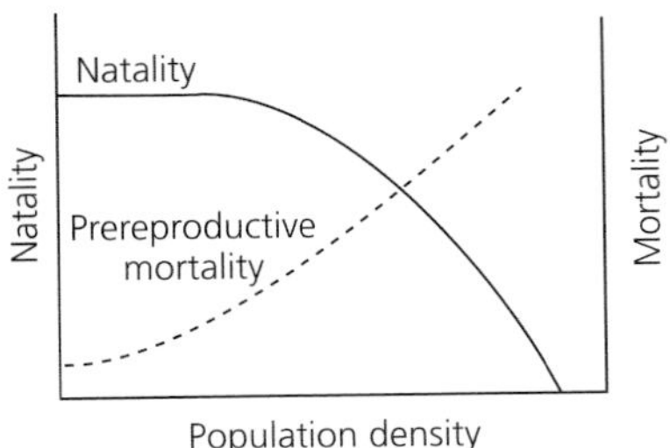


**Figure 38.11**
Density-dependent effects on mortality and natality. As population density increases, the probability that any given individual will die without reproducing increases, and the average number of offspring produced by those that do reproduce, decreases

# Population regulation – density-dependent and density-independent factors

Every environment has a finite carrying capacity, so no population can grow unchecked for long. Sooner or later it either stabilises or falls drastically. Where physical conditions are only favourable for short periods, populations may undergo wild oscillations with short periods of exponential growth followed by a crash. In environments that are permanently hospitable to life, most populations are *regulated* by biological agencies such as competition for resources, predation and parasitism, and undergo only minor fluctuations.

Checks on growth can act either by increasing mortality or decreasing natality – or more often, a combination of the two (Figure 38.11). Since mortality is the number of deaths *per thousand* per unit time, an increase in mortality means an increase in the *proportion* of individuals dying per unit time. In other words, the average life expectancy of an individual must decrease. Similarly, a decrease in natality means that the average reproductive output per individual must decrease. This does not mean that all individuals are so affected; as we shall see, the effects of crowding may, in some cases, fall very unequally on different individuals.

It follows from the foregoing that population control must be **density-dependent**. The essential feature of density-dependent control is that the increase in mortality and/or natality must be greater than the proportion by which the population has increased. Suppose for example a population increases by 10%. A 20% increase in mortality with no change in natality would reduce the numbers, but a 5% increase in mortality would not.

Density-dependent control of numbers thus operates by a kind of **negative feedback** process in which the greater the increase in numbers, the greater the checks on further increase. On the other hand if population density falls below the carrying capacity, the intensity of density-dependent checks is relaxed, allowing the population to rise.

Whereas density-dependent factors are usually of a biological nature such as predation, competition, parasitism and so on, density-independent factors are generally abiotic. There is no special reason, for example, why a storm should kill a higher *proportion* of a dense population of birds than it would of a less dense population.

## Control by predators

Every time a predator kills a prey, the number of prey is reduced. It might therefore seem obvious that predator populations control their prey populations. Although there is clear evidence for this in many cases, it is not always so. Some predators are simply acting as 'executioners' to animals that were doomed anyway. A lesser black-backed gull chick that loses its mother will die even if adult gulls in the colony do not get it first. American muskrats that hold a territory are relatively secure from predation by mink, but for muskrats that do not hold territories, predation is a major cause of death. Although predation may be the immediate cause of death for non-territory holders, the more fundamental cause is competition for territory.

On the other hand there are cases where a predator can exert decisive control over its prey. An example is the small white butterfly (*Pieris rapae*) which was accidentally introduced into New Zealand in about 1930. Within a few years it was devastating cabbage and other brassica crops. In an attempt to control the pest a small parasitoid wasp, *Pteromalus puparum*, was introduced. The wasp lays its eggs in the pupae of the butterfly and the wasp larvae feed on it and eventually kill it. The result was a dramatic decrease in the numbers of the small white, which is now a nuisance rather than a major pest.

## Predator–prey cycles

Some prey appear to be controlled primarily by their food supply, predators having only a subordinate effect. An example is the snowshoe hare and its chief predator, the lynx. In Arctic Canada where these two animals live, there are relatively few other species, so the biotic environment is less complex than in warmer latitudes. Both animals were trapped in large numbers, and records of the Hudson's Bay Company give a picture of population fluctuations over more than a century (Figure 38.12). Both populations undergo regular oscillations every 9–10 years. After a short delay, each hare peak is followed by a lynx peak.

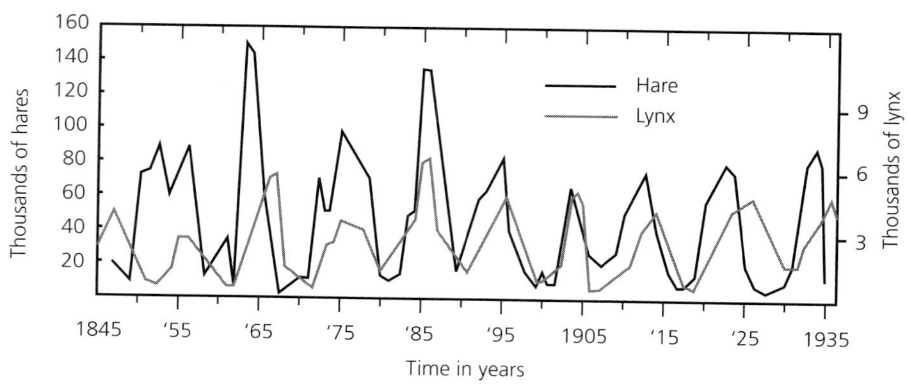


**Figure 38.12**
Fluctuations in numbers of lynx and snowshoe hare in Canada (from Colinvaux, after Machulich)

It seems clear that lynx numbers are fluctuating in response to changes in the supply of prey – with more hares there is more food for the lynx, whose numbers rise. When hares are scarce, lynx numbers decline. This much seems obvious, but is the converse also true? Are the decreases in hare numbers caused by predation by the lynx? There are two reasons for believing that they are not the main factor controlling the hares.

- In parts of Canada where there are no lynx, hare numbers still undergo 10-year cycles.
- Hares have a higher reproductive potential than lynx, with an average of about 10 young per year compared with 2–3 for the lynx. When food is adequate, the hares are always able to outproduce the lynx.

Research has suggested that the regular crashes in hare numbers are due to overgrazing of their food supply. Predation by lynxes while the numbers of hares are low delays the build up of hares. During this time the vegetation makes a full recovery, after which there is a rapid increase in hares.

## Laboratory experiments on predation

Most communities contain more species than the Canadian Arctic example described above. In complex communities, the number of possible interactions between species is simply too great to disentangle individual effects. By setting up artificial predator–prey systems in the laboratory, individual factors can be controlled and studied separately. In a series of classic experiments, C.S. Holling showed that changes in prey

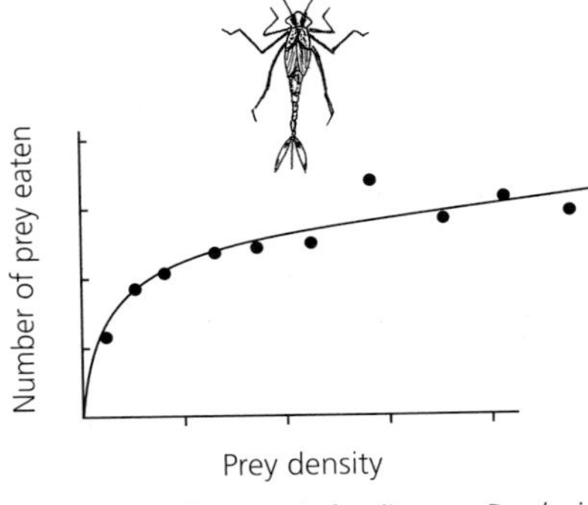


(A) Dragonfly nymph feeding on *Daphnia*

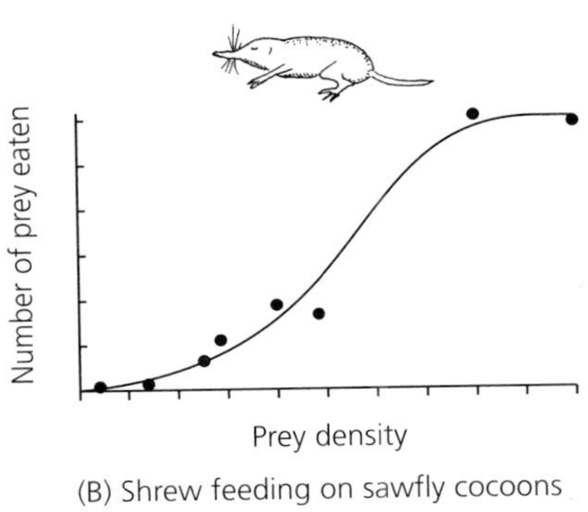


(B) Shrew feeding on sawfly cocoons

**Figure 38.13**
Two different ways in which increasing prey density can influence the rate at which an individual predator consumes prey (see text for explanation)

numbers can influence predators in different ways, which he called **functional** and **numerical** responses.

### *Functional responses*

Here, an increase in prey density causes each *individual* predator to increase its consumption of prey. Two kinds of functional response are shown in Figure 38.13.

One kind (Figure 38.13A) is characteristic of animals which have to search for or lie in wait for their prey, and which either feed on one type of prey or whose prey preference does not depend on how common it is.

The time and energy invested by the predator can be divided into two stages.

1 Searching, which involves all the time and energy spent in locating the prey. In web-weaving spiders and other predators which trap their prey, the equivalent of search time is the time spent waiting for prey to be snared in the trap.

2 Handling, which involves pursuit, capture and eating. Where the prey is large enough to inhibit further search until it has been digested, handling also includes digestion. In animals which build traps that have to be repaired after each capture (e.g. many spiders), repairing the web can be considered to be part of the handling time.

Consider a sea anemone. Rather than searching for prey, it waits until it becomes trapped in its tentacles. An increase in prey density has two effects. First, the time the anemone has to wait for a prey to be caught decreases and is eventually insignificant. Handling time (needed to kill, engulf and digest the prey) remains constant, so above a certain saturation prey density the anemone cannot feed any faster. Second, the predator cannot respond to an increasing density of prey because it is already occupied in handling others. The latter effect causes the graph to flatten out. This means that the proportion of prey eaten *decreases* as prey density increases. A predator showing this kind of response cannot, therefore, by itself regulate a prey population.

A different kind of response is shown in Figure 38.13B. This occurs when a predator changes its preference as the prey becomes more common. The result is an increase in the *proportion* of prey taken as its density increases, indicated by an increase in the gradient of the curve. A predator behaving in this way *could* regulate a prey population. This kind of response is probably most common when the prey is concealed, and the ability of the predators to pick out hidden prey improves with practice and, therefore, with prey abundance. Predators responding in this way tend to be opportunistic, with wider diets, such as foxes and badgers.

### Numerical responses

Here the density of predators increases by immigration and/or increased reproduction. Numerical and functional responses normally occur together. Provided the predators do not interfere with each other, the combined effect is obtained by multiplication. For example, a 5-fold increase in the number of predators (numerical response) and a doubling of their rate of predation (functional response) would result in a 10-fold increase in the rate of predation. If the density of the prey were to rise less than 10 times, the prey population would be brought under control. If it were to increase more than 10 times it would not (though of course it would contribute to regulation by other species). The most effective predators as controlling agents are those which show a strong tendency to shift their prey preference and, at the same time, are able rapidly to build up their numbers either by immigration or reproduction, or both.

### Territory and population regulation

Predation and parasitism act by influencing mortality. In many territorial animals, competition for territory can act as a powerful population regulator by controlling natality. In many territorial birds, such as great tits, only territory-holding males can attract a mate, non-territory-holders living as bachelors. The number of territories available therefore limits the size of the population. Since territory tends to prevent overcrowding, it results in the population size being held at a level that the food supply can support.

## Reproductive strategies – r and K species

The biological 'aim' of every organism is the production of as many surviving offspring as possible (or in some cases to help their relatives reproduce successfully, since they share some of their genes). Reproductive potential is defined as the total number of offspring an organism can produce per year. In many species reproduction is concentrated in a single episode, after which the organism dies – as in the case of many insects. Alternatively it may be repeated, usually seasonally, so that successive generations overlap.

Even in the same taxonomic group, reproductive capacity varies hugely from one species to another. Albatrosses take about 7 years to reach maturity and lay a single egg every two years, but partridges mature in their second year and produce 15 or more eggs in a clutch. Rabbits are sexually mature at 4 months and produce several litters per year with an average of 5–6 offspring in each litter, whereas gorillas do not begin to breed until they are about 10 years old and produce one offspring about every 4 years. How are these differences to be reconciled with the statement that all species are working flat out to reproduce?

The answer is that the important thing is not the number of offspring, but the number of offspring that *survive to reproduce*. Different species achieve this 'aim' in different ways, referred to as **reproductive strategies**. Some devote more energy to producing as many young as possible, but invest little in parental care. Others produce fewer young but spend a good deal of energy on their protection and nourishment.

It is important to appreciate that although the term 'strategy' is used in human affairs to indicate some conscious objective, the term has no such meaning when used in biology. Reproductive strategies, like other aspects of the phenotype, are the product of the genes, which are themselves the product of selection. Genes do not anticipate their use in the future: their existence is the result of their usefulness to previous generations.

Reproductive strategies can be thought of as lying along an imaginary continuum. At one end are the **'*r*'-species**, so-called because they have a high reproductive potential (denoted by the symbol *r* and alternatively known as the **intrinsic rate of increase**). These species mature early and produce many offspring, devoting little resources to their protection.

The *r* strategy is highly effective in environments in which there are short periods of great abundance. Many garden weeds are of this type, germinating and completing their life cycle in the few weeks after the garden has been dug over. *r* species are thus **opportunists**, exploiting environments that can only support them for short periods.

At the other end of the spectrum are the **'*K* species'**, so-called because they are adapted to environments that are ecologically 'full', or near the carrying capacity (symbol *K*). *K*-species tend to be strong competitors; they have longer life cycles and produce fewer but better-protected offspring, and often show parental care. A summary of differences between *r* and *K* strategies is shown in Table 38.2.

**Table 38.2**
Summary of differences between *r* and *K* reproductive strategies

| | *r*-strategy | *K*-strategy |
|---|---|---|
| Habitat | unstable and/or inhospitable | stable and hospitable |
| Body size | small | large |
| Length of life cycle | short | long |
| Generation time | short | long |
| Number of offspring | many | few |
| Parental care | little | considerable |
| Competitive ability | weak | strong |
| Reproductive episodes | single | repeated |
| Dispersal | high | low |
| Population density | highly variable, often over-shooting *K* | less variable, usually near *K* |

*r*-*K* differences can be seen both within and between taxonomic groups. Mammals tend to be nearer the '*K*' end of the spectrum and insects are nearer the '*r*' end, but each group contains its own *r*-*K* variation. Amongst mammals, rodents tend to be near the *r* end, but there is considerable variation even within this group. Guinea pigs for example have small litters of well-developed young, whereas lemmings produce large litters. Whereas most insects lay many eggs with little or no parental care, a female tsetse fly produces about a dozen young that hatch inside the mother and are born as maggots.

## QUESTIONS

**1** A sample of 100 grasshoppers of various ages was taken from a field, and each was marked with non-toxic paint. Twenty were kept in a cage and the other 80 were released back into the field. The following day another 100 grasshoppers were caught, 9 of which were marked. Of the animals that had been marked but had not been released, two no longer bore the mark because they had shed their cuticles. What is the best estimate of the total number of grasshoppers in the field?

**2** In a study of the effect of quadrat size on the number of species recorded, a series of samples was taken using quadrats of different sizes. Which of the following graphs would you expect to show the effect of quadrat size on the number of species recorded?

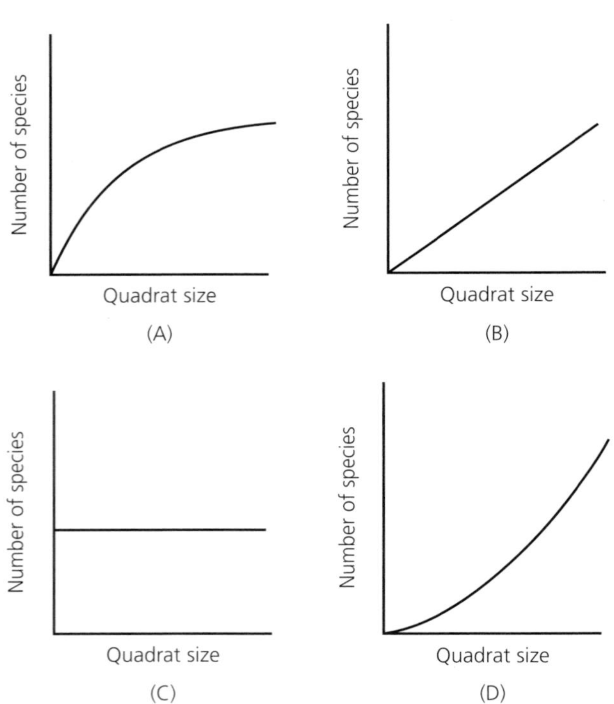


**3** **A**, **B**, **C**, **D** and **E** are five populations whose growth was monitored over equal time intervals. The results are given in the table below. Which one is growing exponentially?

| | Population | | | | |
|---|---|---|---|---|---|
| **Time (years)** | **A** | **B** | **C** | **D** | **E** |
| 1 | 10 | 200 | 5 | 200 | 2 |
| 2 | 20 | 300 | 15 | 400 | 4 |
| 3 | 30 | 450 | 60 | 500 | 16 |
| 4 | 40 | 675 | 300 | 550 | 128 |

**4** In a nutritive medium maintained at 10°C, the number of bacteria increased 256-fold in 16 hours.

**a)** What is the generation time for this bacterium under these conditions?

**b)** Assuming that the rate of growth doubles for every 10°C rise in temperature and that chemical conditions in the medium were similar, by how many times would the number of bacteria have increased at 30°C?

CHAPTER

# 39 COMMUNITIES and ECOSYSTEMS

## LEARNING OBJECTIVES

By the time you have completed your study of this chapter you should be able to:

- define a community and give examples.
- distinguish between the main feeding categories in a community.
- describe what self-sustaining communities have in common.
- explain why the main characteristics of a community are primarily defined by the physical environment.
- give reasons why communities vary in species diversity.
- calculate the index of diversity for a community.
- explain the terms *producer*, *consumer* and *trophic level* in relation to *food chains* and *food webs*.
- describe the various fates of the solar energy incident on a green plant and of the energy entering an animal.
- explain how energy is lost as it passes along a food chain and why few food chains have more than six links.
- explain why biomasses of successive trophic levels are usually of pyramidal shape and why productivities of successive trophic levels are always pyramidal.
- describe the activities of producers, consumers, and micro-organisms in the cycling of carbon and nitrogen in an ecosystem.

## Communities

A community comprises all the organisms in a defined area, and is therefore a group of populations. An oak wood, a pond, a salt marsh and moorland all contain characteristic communities.

Members of a community are not living together simply because they are adapted to similar physical conditions; directly or indirectly they depend on each other. Because of the network of inter-relationships between them, a community is an integrated unit in which a change in any one part can affect many other parts. Like any other level of biological organisation, a community is more than the sum of its constituent parts.

The most important inter-relationships in a community are based on energy, nutrients, shelter and breeding. The trees and other plants of a forest provide food for a great many species of small leaf-eating, sap-sucking, wood-boring and root-eating animals. Long after a tree has died, its timber provides food for fungi and beetle larvae. The shade cast by the trees provides a cool environment for smaller plants such as ferns and mosses, and crevices in the bark provide shelter for a variety of small animals. Trees and shrubs provide safety for a number of birds to build nests.

Forest plants are not only providers – they are also dependent on other members of the community for resources. The supply of mineral ions in the soil is maintained by bacteria, fungi and animals. Animals also act as agents for pollination and seed dispersal.

Superimposed on this web of biological relationships is the effect of the physical environment, which is the ultimate determinant of the community membership. Birch (*Betulus*), for example, is characteristic of cooler climates, whilst alder (*Alnus*) and willow (*Salix*) are adapted to wetter soils than oak or beech. Since many small animals feed on one kind of plant, the animal life of birch, oak and beech woods differs.

## Feeding relationships

The most important kinds of relationships in a community relate to nutrition. On this basis, the organisms of a community can be divided into two groups:

- **autotrophs**, which convert carbon dioxide into organic compounds. Most are **photoautotrophs**, using light from the sun. A few are **chemoautotrophs**, using chemical energy derived from oxidation of inorganic compounds in their environment – for example nitrifying bacteria.
- **heterotrophs**, which obtain organic compounds from *other* organisms ('heterotrophic' means 'other feeders').

Heterotrophs include fungi, animals, most bacteria and many protoctists. They are usually divided into three main categories:

1. **consumers** include most non-parasitic animals. They feed either on other organisms or on parts of them, digesting the food in an internal gut cavity. Food that is alive when eaten is killed. Consumers which feed on detritus (fragments of dead plant and animal matter) are called **detritivores**.
2. **decomposers** include many fungi and bacteria and feed on dead matter. Somewhat arbitrarily they are distinguished from consumers of dead matter by the way they digest it; instead of engulfing their food and digesting it in a gut cavity, they are surrounded by their food, and digest it externally.
3. **parasites** feed on living matter without killing it.

Consumers can be further subdivided. Those feeding directly on plants are called **herbivores** or **primary consumers**, whilst **carnivores** feed on other animals. Carnivores feeding on herbivores are called **primary carnivores** or **secondary consumers**. **Secondary carnivores** or **tertiary consumers** feed on primary carnivores. In some communities, food chains are long enough to accommodate **tertiary carnivores**, which feed on secondary carnivores. Animals feeding on both animal and plant food are called **omnivores**.

Like most classifications and definitions, these feeding categories are not as clear cut as they might appear. Within omnivory, for instance, there is a spectrum of intermediates connecting carnivores and herbivores. At one extreme are mainly carnivorous animals such as foxes, whilst at the other are animals like squirrels that are mainly herbivorous but which also feed on birds' eggs.

Similarly the boundaries of autotrophy are not always clear cut. Insectivorous plants such as the sundew (*Drosera*) not only photosynthesise, but also trap insects on their leaves, digesting them and absorbing the resulting amino acids through their leaf surfaces. While the main benefit is in the form of combined nitrogen (in which boggy soils are usually deficient), the amino acids contain some energy.

Although all communities contain heterotrophs, some lack autotrophs and depend on energy imported from elsewhere. Communities in deep underground streams, for example, depend on organisms and detritus washed down from sunlit areas above. Similarly the ocean floor is a consumer community. Even above ground, communities are not energetically isolated from each other. Although the rocks on the bottom of shallow streams are covered with encrusting algae, much of the animal life depends on detritus washed down from the surrounding land. A stream community is thus a net importer of energy.

### *What determines the membership of a community?*

Ultimately the composition of a community depends on physical conditions such as geology and climate, for it is these factors that determine the kinds of plant that can grow there. The plants in turn have a major influence on the animal life because many animals depend on a particular kind of plant for their food, and many parasites are also highly host-specific.

Physical factors affect both animals and plants directly, but they also influence animals indirectly via their food. The swallowtail butterfly (*Papilio machaon*), for example, is restricted to the Norfolk Broads because the food plant of the larva, milk parsley (*Peucedanum palustre*), is restricted to the swampy habitat of the fens. Most plant galls such as the 'oak apple' are caused by tiny insects that only feed on one kind of plant, and so their distribution is determined by the soil requirements of their food plants.

### Measuring community diversity

To compare the diversity of different communities or of a given community at different times, we need some method of quantifying diversity. This is clearly not simply the number of species; a community of 1000 individuals belonging to 10 species is less diverse than a community of 100 individuals of 100 species. An easily calculated formula that takes into account the number of individuals as well as the number of species is the **Simpson index**, given by the formula:

$$D = \frac{N(N-1)}{\Sigma n(n-1)}$$

Where: $D$ = diversity index
$N$ = total number of organisms present
$n$ = number of organisms per species
$\Sigma$ = 'the sum of'

If $N$ and $n$ are reasonably large, $N - 1$ approximates to $N$ and $n - 1$ to $n$, so the formula approximates to $N^2/\Sigma n^2$. It follows that a high diversity index will be favoured by a large number of species, and/or a small number of individuals per species.

## Ecosystems

An ecosystem is usually defined as a community together with all the non-living components of the habitat. Two processes are central to the functioning of any ecosystem: the flow of energy and the cycling of nutrients.

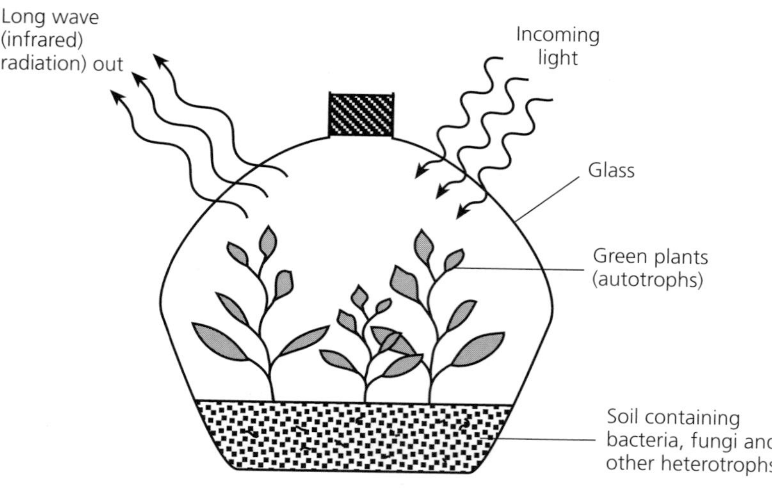


**Figure 39.1**
A simple model ecosystem. In a bottle this size, there would be insufficient plant growth to support any but very small animals

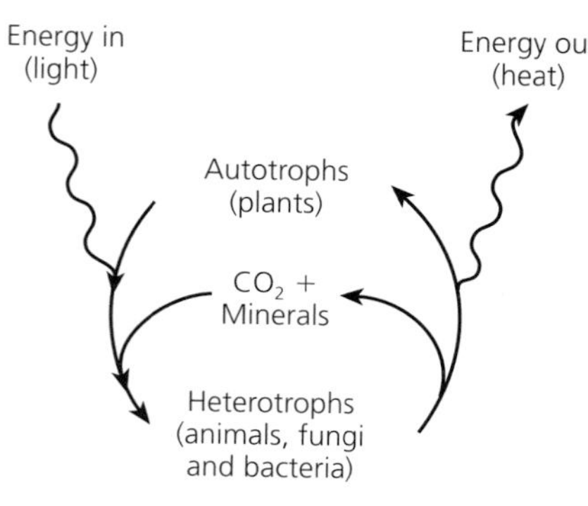


**Figure 39.2**
Interdependence between autotrophs and heterotrophs

Figure 39.1 shows a simple, artificial ecosystem, consisting of a sealed bottle containing some plants, together with soil and its bacteria, fungi and small animals. Since the bottle is gas-tight, the organisms inside are, as far as raw materials are concerned, independent of the world outside. No ecosystem, however, is independent of its surroundings for energy. This enters as light through the transparent walls of the bottle and, after being used to drive living processes, the energy is degraded to heat which leaves the bottle as infrared radiation.

Provided that it receives sufficient light, a simple ecosystem like this can function indefinitely. The food supply of the consumers and decomposers is limited by the rate of growth of the plants. In adequate light, the rate of growth of the plants is limited by the rate at which the consumers and decomposers break down organic matter to simple inorganic nutrients (Figure 39.2).

In this simple system there is a balance between autotrophs and heterotrophs. The plants cannot grow faster than their supply of $CO_2$ and mineral ions allow, and the consumers and decomposers cannot outgrow their food supply. In this situation the total photosynthesis and respiration occurring over each 24 hour period are equal.

Although a balance between autotrophs and heterotrophs is established, it is unlikely that this would exist from the outset when the apparatus was set up. For example there might have been an excess of plant material over consumers and decomposers. In this case the supply of $CO_2$ would be inadequate and leaves would die and fall off faster than new ones would grow. Then as the dead leaves decomposed, the $CO_2$ level would rise and so would photosynthesis, until an equilibrium was established.

Of course ecosystems are not sealed off from each other like the simple model described above. All ecosystems exchange materials and energy with neighbouring ecosystems. Sediment is washed down from land to sea in rivers, and birds flying over sand dunes may deposit faeces. Many insects that live in fresh water as young stages travel from one pond to another as adults, and some animals migrate as adults over long distances. In all of these movements, matter and energy are transferred from one ecosystem to another.

When we consider the biosphere as a whole, exchanges of energy and matter between ecosystems cancel each other out. The Earth is therefore simply a very much larger version of the sealed bottle described above, with energy entering as light and leaving as heat.

## Food chains

A simple way of expressing feeding relationships is by a **food chain**, in which transfer of energy from one organism to the next is represented by an arrow, for example:

oak leaf → winter moth caterpillar → great tit → sparrowhawk

*Chlorella* → *Daphnia* → minnow → pike

In all food chains, the first link is a producer, which provides energy for the other organisms in the chain.

### *Grazing and detritus food chains*

Two kinds of food chain can be distinguished. In a **grazing food chain** such as the two examples above, the primary consumers feed on living plant material. Many plant parts die before being eaten and form the base of **detritus food chains**, in which the primary consumers are detritivores such as earthworms:

dead leaves → earthworm → blackbird → sparrowhawk

Detritus and grazing food chains only differ in the first links, and some organisms may be members of both, for example the sparrowhawk above.

The relative importance of grazing and detritus food chains differs greatly between land and sea. In most terrestrial ecosystems the bulk of the plant matter enters the detritus food chain as dead leaves and roots, only a small proportion entering the grazing food chain. In oceans the reverse is the case. The base of the food chains consists entirely of microscopic floating algae, the **phytoplankton**. These are food for tiny floating animals, the **zooplankton**. Most of the phytoplankton is alive when eaten and thus enters the grazing food chain. A minority die without being eaten and begin to sink, entering the detritus food chain.

## Food webs

A food chain is obviously an oversimplification of feeding relationships. Few organisms feed on only one species (many parasites are exceptions), and few (if any) are eaten by one species. A spider eats many kinds of insects, and an oak tree may be eaten by larvae of over a hundred species of butterflies and moths – just one of the several insect orders present in the tree. In other words food chains are branched and interconnected, forming **food webs** (Figure 39.3).

Figure 39.3 is highly simplified in a number of ways:

- many species are lumped together into single categories, e.g. 'moths', 'aphids', 'wood borers', 'beetles'.
- many animals change their diet as they get older: most caterpillars feed on leaves, but the adult butterflies and moths feed mainly on nectar.
- many adult animals change their diet with the season. Badgers for example eat a lot of young rabbits in spring and summer, but eat earthworms and acorns in the autumn. Great tits eat mainly insects in spring and early summer, but include beech fruits in the autumn.
- in habitats undergoing succession, the species composition slowly changes (see below).

Since every organism in the web is directly or indirectly connected with every other, a change in one population has the potential, in theory, to influence every other species in the community. In communities

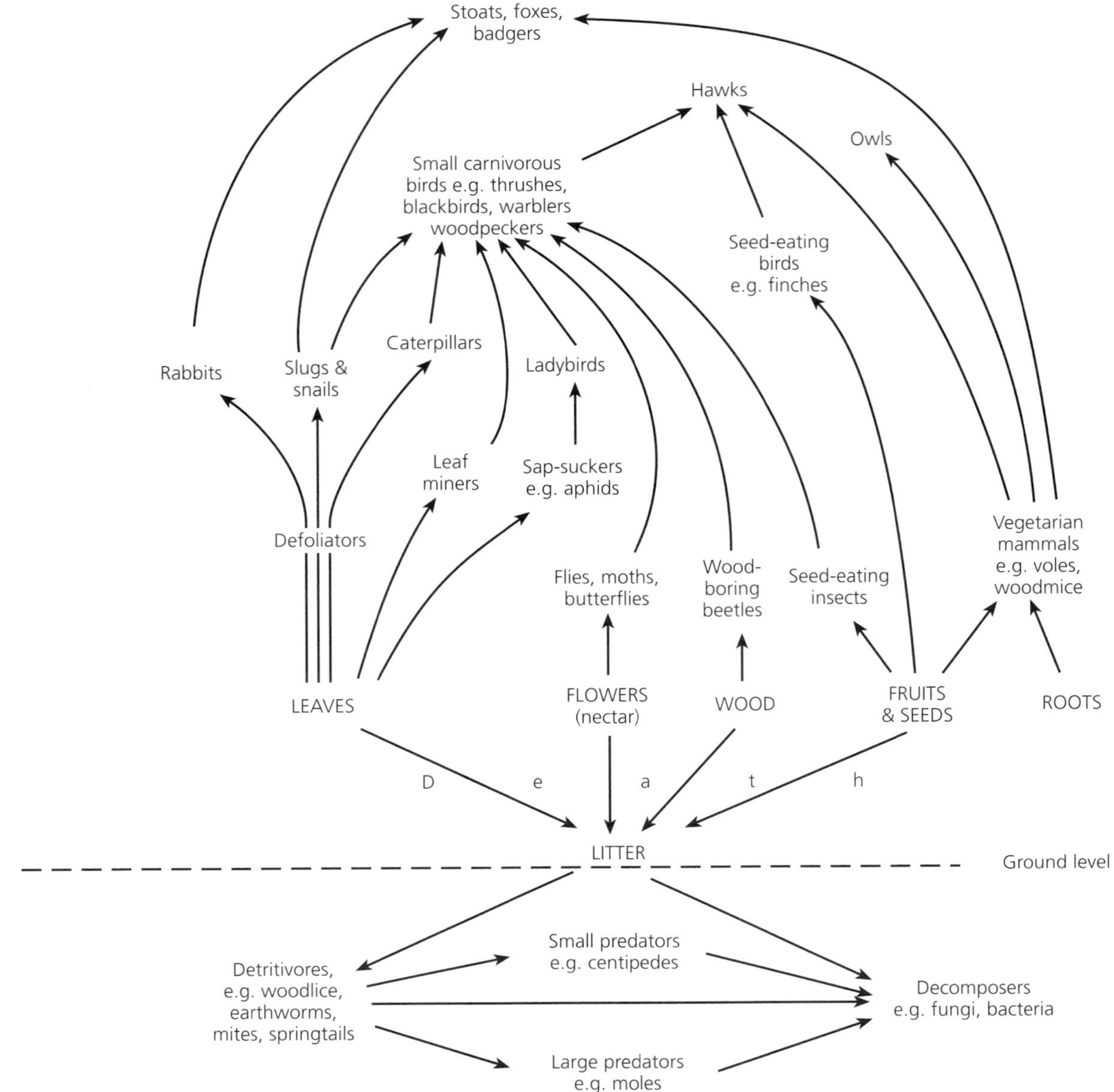


**Figure 39.3**
A woodland food web. In this highly-simplified scheme, numerous relationships have been omitted. For example, the grazing and detritus food chains should be interlinked, as birds feed on earthworms

containing few species, a change in the numbers of one species is likely to have drastic effects on others. An example is the oscillations in the populations of the Canadian snowshoe hare and lynx (Chapter 38). In more complex communities there is much greater scope for animals to switch from one kind of food to another, so a disturbance in one population is more likely to be absorbed by the others. Probably for this reason, complex communities are generally more stable than simple ones.

Working out the dietary habits of animals takes considerable time and effort (see the information box on page 731). A food web may contain thousands of species. Therefore ecologists concerned with the overall pattern of energy flow often simplify the situation for practical purposes by lumping together all producers, all plant-eaters, all primary carnivores, and so on. Each of these major feeding categories constitutes a **trophic level**. Thus all the producers together make up one trophic level, and all the herbivores form another, and so on.

As with many other biological terms, trophic levels are not mutually exclusive groups: many animals belong to two or even more levels simultaneously, and the emphasis on membership of particular levels may change with the seasons. For instance, a badger may predominantly belong to the herbivore level in the autumn when it feeds on acorns, but when it feeds on young rabbits in the spring it spends more of its time as a carnivore.

## What feeds on what?

Construction of a food web requires a detailed knowledge of the feeding habits of many animals. This may take considerable time and still leaves many details uncertain. There are several ways of studying feeding habits.

- Direct observation. Animals can sometimes be seen in the act of feeding, particularly in the case of specialist herbivores that spend all their time on their food plant. This is more difficult with carnivores, which feed intermittently.
- By keeping animals in captivity with a choice of food items. The difficulty is that there is no guarantee that because a predator eats a particular prey item in captivity, it does so in nature.
- A study of gut contents may reveal fragments of undigested food such as parts of exoskeleton and plant cell walls.
- By immunological techniques. A rabbit is injected with an extract of the suspected food and allowed to develop antibodies against it. An extract of gut contents from the animal under study is then mixed with serum of the immunised rabbit. A precipitate indicates the presence of the same antigens in the food as were used to immunise the rabbit, confirming that the suspected material is part of the animal's diet.
- By treating plants with a radioactive tracer such as $^{32}P$ (phosphorus), and monitoring the radioactivity in members of the community at intervals afterwards. Radioactivity appears first in the phloem sap and in insects sucking the sap. Next to become radioactive are the youngest leaves, followed by leaf-eating animals, and then their predators. Members of the detritus food chain take longer to become labelled, since leaves have to senesce and die before they become available as food.

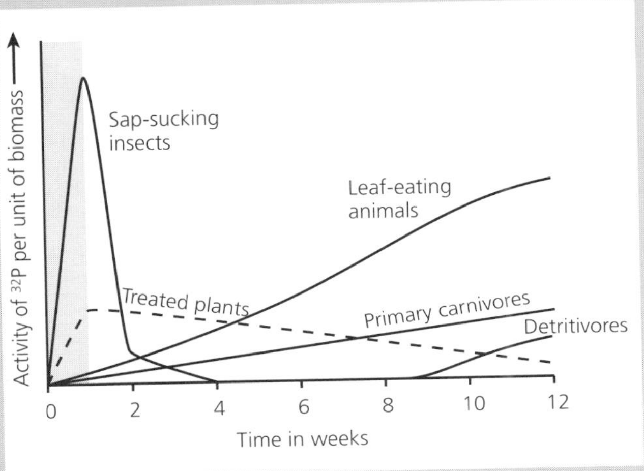


Changes in radioactivity in members of a community after treatment of plants with radioactive phosphorus, $^{32}P$ (modified from a JMB question, June 1978)

A practical consideration with this method is the choice of isotope: this must have a half-life that is sufficiently long for its rate of decay to be insignificant compared with changes in concentration resulting from biological processes. Also the concentrations of isotope used must be too low to be harmful to the organisms over the time scale of the experiment.

## Organisms as energy converters

Few food chains have more than five links, and some have only three. The reason is that of the energy entering an organism, only part is retained as chemical energy in new tissues. This is true of both animals and plants, but since their energy budgets are rather different we will deal with them separately.

## *Energy conversion in plants*

Figure 39.4 shows the typical energy budget of a crop plant under field conditions. The figures are only approximations and the actual efficiency depends on a number of factors, including light intensity, temperature, water supply and mineral supply.

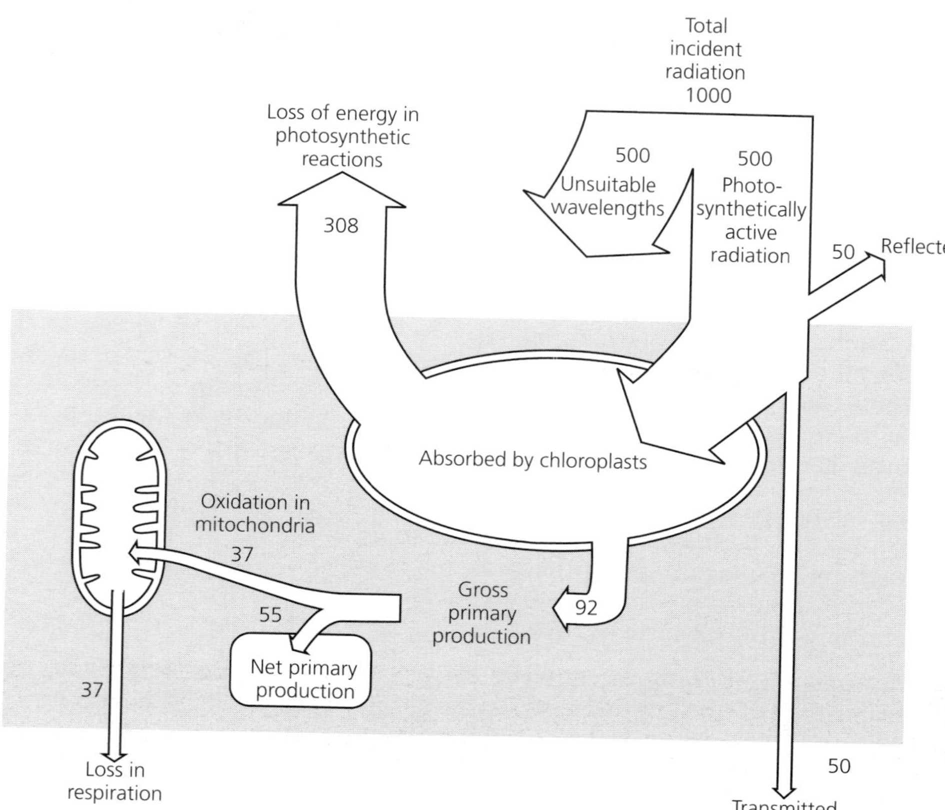


**Figure 39.4**
Energy flow through a crop plant (after Hall)

Of the radiation incident on a leaf, about 50% is of a wavelength suitable for use in photosynthesis. Of this photosynthetically active radiation (PAR), about 80% is absorbed by the chloroplasts, the rest being reflected or transmitted. Of the light absorbed by chlorophyll, only about 20% is converted to chemical energy in the products of photosynthesis; the rest is wasted as heat in photosynthetic reactions. Finally, about a third of the photosynthetic product is used as fuel in respiration; the rest is used for growth and is thus potentially available for consumers.

As a result of these various sources of energy loss, less than 5% of the incident solar energy is converted into plant matter – and this is in a crop plant in which conditions for growth are artificially manipulated in the plant's favour. In many natural communities the efficiency is less than this.

## *Energy conversion in animals*

Figure 39.5 shows the energy budget for a bullock, and gives an idea of how inefficiently farm animals convert grass energy into meat energy. Of the food eaten, about 60% is not even digested and appears as faeces. Of the food that is absorbed, most is used as fuel. Only about 4% of the energy intake is converted into new growth. Thus for every 1000 kJ of energy entering the animal, about 40 kJ is potentially available to a predator such as a human.

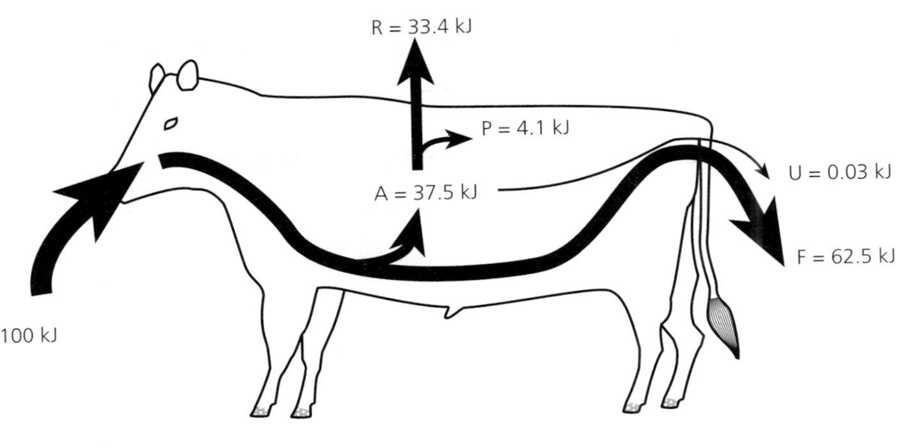


A = assimilated
R = energy lost in respiration
U = energy lost in urea
F = energy lost in faeces
P = energy conserved in production or growth

**Figure 39.5**
Energy budget for a bullock. For every 100 kJ of energy eaten, little more than a third is assimilated, and of this, almost 90% is respired. Of the original food eaten, only 4% is available to humans as meat

Some of this inefficiency is due to the fact that, as endotherms, farm animals spend a high proportion of their energy in maintaining body temperature. As Table 39.1 shows, most ectotherms are more efficient, converting between 10 and 25% of their food energy into new tissue.

**Table 39.1**
Energy budgets for six animals (modified from Advanced Biology Learning Project, C.U.P.)

| Animal | Feeding category | Percentage of total food consumed: Assimilated | Respired | Egested | Net production | % of assimilated food respired |
|---|---|---|---|---|---|---|
| Grasshopper | herbivore | 37 | 24 | 63 | 13 | 65 |
| Caterpillar | herbivore | 41 | 17.5 | 59 | 23.5 | 42.7 |
| Spider | carnivore | 92 | 57 | 8 | 27 | 62 |
| Perch | carnivore | 83.5 | 61 | 16.5 | 22.5 | 73 |
| Elephant | herbivore | 33 | 32 | 66 | 1 | 97 |

For humans, with an increasing population and finite land area, efficiency of food production is becoming increasingly important. A hectare of land devoted to growing potatoes or wheat yields far more energy for humans than a hectare used for beef production, since the cattle waste about 96% of the energy in their food. The implication of this is that from the energy point of view, it makes better sense to use land for growing crops than for meat production.

There is another side to the energy question: we have to invest energy in order to obtain food. In traditional agricultural societies the energy investment was in the form of muscle power. Since the Industrial Revolution, the energy investment in the form of farm machinery, fuel and fertiliser has progressively increased to the point where some food contains less energy than was used to produce it. Much of the energy used in intensive food production comes from an energy subsidy in the form of fossil fuels.

Energy is not however the only consideration in food production. Not only is animal protein generally richer in essential amino acids than plant protein, but Vitamin $B_{12}$ is completely absent from plant food. Also some herbivores can live on vegetation that humans cannot. Therefore by rearing sheep on hills, humans can use energy that would otherwise be unavailable.

## Food pyramids

Because heterotrophs convert their food into flesh very inefficiently, only a small proportion of the energy entering each link is transferred to the next. By the time the energy reaches the top carnivores, a minute proportion of the solar energy trapped by plants remains – the rest has been lost as heat. This loss of energy along a food chain can be described in various forms of **food pyramid**.

### *Pyramids of numbers*

According to an old Chinese saying, 'one hill cannot shelter two tigers'. The relative rarity of carnivores (especially large ones) is part of a broader generalisation: in any ecosystem, herbivores are usually more common than primary carnivores, and primary carnivores are more common than secondary carnivores. If the total numbers of organisms at successive trophic levels in an ecosystem are represented by blocks of proportionate size, they often form a pyramid (Figure 39.6). The lowest layer of the pyramid represents the total number of producers in the ecosystem, the next block represents the numbers of herbivores, and so on.

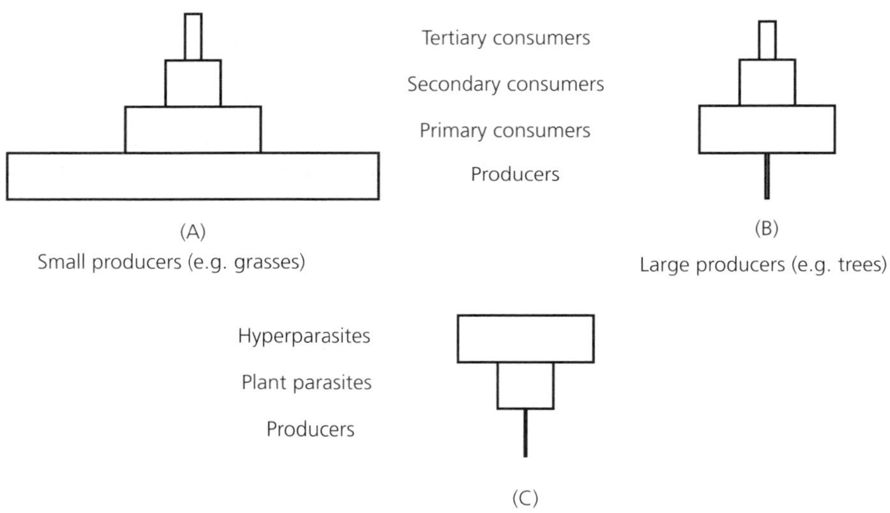


**Figure 39.6**
Pyramids of numbers

The reason for the decrease in numbers of organisms in successively higher trophic levels in Figure 39.6A is that because of energy losses, each organism eats many times its body mass during its life time. Even if predators were the same size as their prey, each predator would eat many prey. More often than not however, predators are larger than their prey, increasing the disparity between numbers of predators and prey.

The size disparity does not always work this way. Some animals are much smaller than their food organism. Thousands of insects for example can feed off a single oak tree, so that part of the pyramid may be **inverted** (Figure 39.6B). A similar effect is created with parasites which are always smaller (but often more numerous) than their hosts (Figure 39.6C). In the extreme example shown below, the small white caterpillar is host to three 'tiers' of parasitoid wasps, each smaller and more numerous than its host:

cabbage white → small braconid wasp → ichneumon wasp → chalcid wasp

Even amongst non-parasitic animals, some predators are smaller than their prey. A stoat for example is less than a quarter the mass of a rabbit.

## *Pyramids of biomass*

The effect of body size is eliminated if total population biomass (in kg per $m^2$) is compared instead of numbers (Figure 39.7). The result is usually a pyramidal shape, indicating that a large total mass of plant material supports a smaller total mass of herbivores, which supports a smaller mass of primary carnivores.

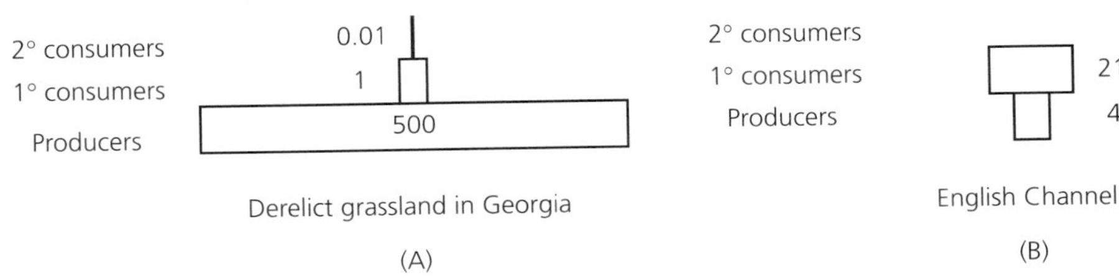


**Figure 39.7**
Pyramids of biomass. Derelict grassland in Georgia (A). An anomalous pyramid, the English Channel in Spring (B). (Block sizes not to scale). Units are g per $m^2$

## *Pyramids of productivity*

Although the biomass in successive trophic levels usually forms a pyramidal shape, it does not have to be so, as Figure 39.7B shows. In a pyramid of biomass each block represents the **standing crop**, which is the quantity of organic matter present *at one time*. Here the producers are single-celled protoctists and the primary consumers are tiny filter-feeding crustaceans. In this example the biomass of a given producer supports a greater biomass of herbivore. How is this possible?

The explanation depends on the fact that it is not the standing crop of the producers that matters, but its **productivity** or **rate of growth**. This is the rate at which new material becomes available to the primary consumers. Under optimal conditions members of the phytoplankton can divide once a day, so every day the producer biomass doubles – or it would if it were not being eaten by animals.

A more revealing kind of food pyramid is a **pyramid of productivity**, or pyramid of **energy flow** (Figure 39.8).

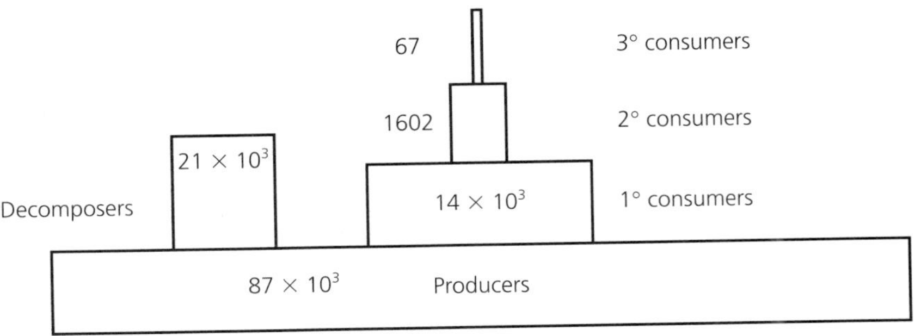


**Figure 39.8**
Pyramid of productivity for a stream at Silver Springs, Florida (after Odum). Note that the decomposers account for a considerable proportion of the energy flow through the ecosystem

Figure 39.9 shows two trophic levels from Figure 39.7B, but with arrows indicating energy flow. The closeness of the arrows is proportional to the energy *flux*, or the amount of energy per $m^2$ flowing through each trophic level per unit time. Although the biomass of zooplankton is greater than that of the phytoplankton, the energy flux is less. This is because some of the energy entering each trophic level leaves as heat rather than remaining as chemical energy available for the next trophic level.

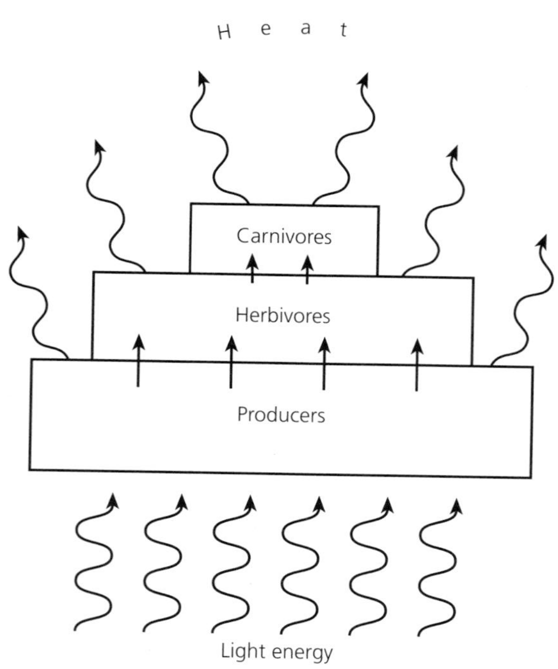


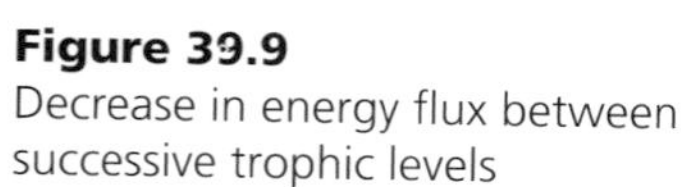
**Figure 39.9**
Decrease in energy flux between successive trophic levels

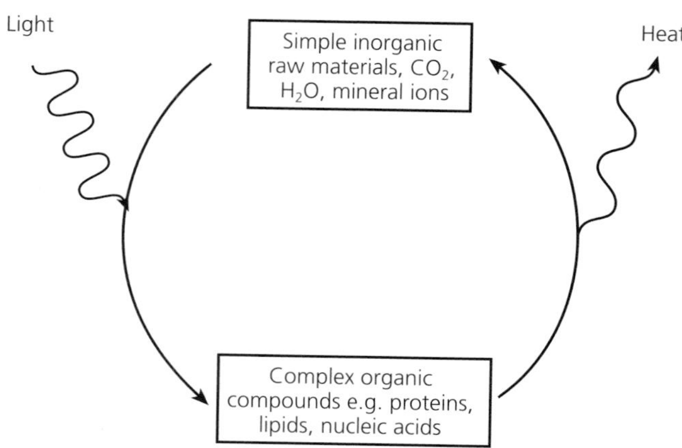


**Figure 39.10**
A generalised nutrient cycle. The nutrient elements are recycled, but the solar energy which drives the cycle is lost as heat

Productivity is expressed as the energy value in the new organic matter produced in each m$^2$ per year. When productivities of successive trophic levels are compared, a pyramidal shape is *always* obtained. This must be so because some energy is always lost in passing from one trophic level to the next.

There is another reason why it is sensible to measure energy rather than biomass. Organisms vary somewhat in their chemical composition, and the various constituents have differing energy values. A starfish, for example, has a limy endoskeleton, which contributes mass but not energy, and animals storing fat are richer in energy than animals without fat stores.

Incidentally, a pyramid consisting of blocks representing the amount of energy present in each trophic level (in kJ m$^{-2}$) could quite legitimately be called a 'pyramid of energy'. However since it does not convey the idea of *rate* (which has a time dimension), the terms 'pyramid of productivity' or 'pyramid of energy flow' are to be preferred to 'pyramid of energy'.

## The cycling of nutrients

Sooner or later all the energy that enters an ecosystem is radiated out into space as heat. Energy is therefore only used once. Matter, on the other hand, is recycled. Each element spends part of its time in complex organic molecules and part in simple, inorganic substances in the abiotic part of an ecosystem (Figure 39.10). There are about 17 elements found in living matter and each has its own cycle.

The details of each cycle vary with the element, but certain features are common to all. Nutrients are absorbed in the inorganic state by autotrophs and passed on to the heterotrophs that feed on them. In a heterotroph, atoms can undergo two possible fates: they may leave the body as inorganic waste, or they may enter another organism as food. In every link in the food chain there are the same two alternative pathways, so sooner or later all elements are returned to the environment and become available for uptake by autotrophs.

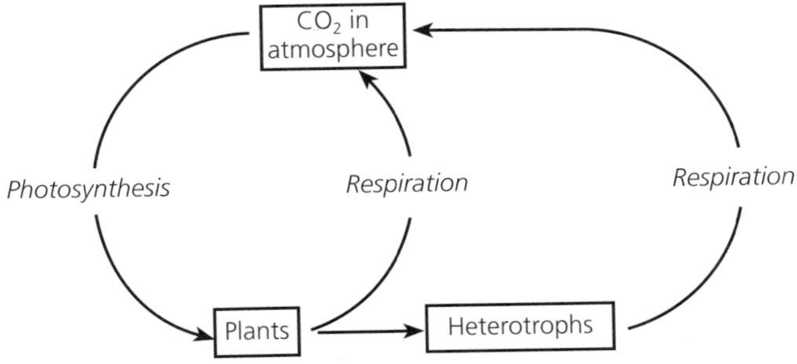


**Figure 39.11**
Overview of the carbon cycle

## *The carbon cycle*

Carbon forms the 'skeleton' of all organic molecules. It enters green plants as $CO_2$ and is converted into carbohydrate in photosynthesis and, on a much smaller scale, by chemoautotrophic bacteria. Carbohydrates are then converted into all the other organic constituents of living matter such as lipids, proteins and nucleic acids. Some of the organic matter made by plants is converted into $CO_2$ in respiration. As organic matter moves along the food chain it is converted back into $CO_2$ in respiration, which re-enters the atmosphere (Figure 39.11).

Figure 39.12 shows the main details of the carbon cycle. Each compartment represents a 'stop' on the cycle. In some compartments carbon is in the oxidised state, such as $CO_2$, carbonate ($CO_3^{2-}$) or hydrogencarbonate ($HCO_3^-$) ions. In organic compounds such as carbohydrates and proteins, carbon is in the reduced state (because at least some of the carbon atoms are directly bonded to hydrogen).

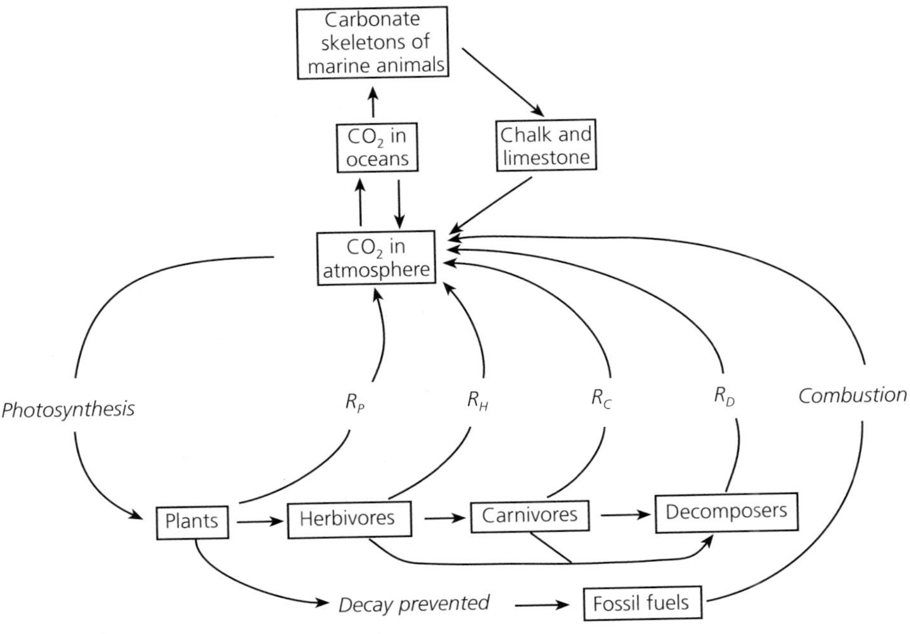


**Figure 39.12**
Details of the carbon cycle. $R_P$, $R_H$, $R_C$ and $R_D$ = respiration by plants, herbivores, carnivores and decomposers, respectively

Until the Industrial Revolution, annual total world photosynthesis more or less equalled respiration and combustion, so the amounts of carbon in each compartment changed little from year to year. In the summer total photosynthesis exceeds total respiration, so $CO_2$ is removed from the environment faster than it is returned, causing its level to fall slightly. In the winter the situation is reversed.

Before the Industrial Revolution these changes more or less cancelled each other out over the year as a whole. More or less – but not exactly; for the following reason. In ecosystems where there is insufficient oxygen for decay (such as swamps) dead bodies of plants and animals accumulate and are slowly converted into coal, oil or natural gas. Although this represents less than 0.001% of the total annual photosynthetic product, when accumulated over millions of years this tiny annual deficit in the return of carbon to the atmosphere has resulted in the laying down of massive deposits of fossil fuel.

Since the Industrial Revolution the balance between the uptake and production of $CO_2$ has been disrupted on a much greater scale, for two reasons: the burning of fossil fuels and deforestation followed by decay and combustion of the timber. Both have contributed to a rise in the level of atmospheric $CO_2$. Possible implications of this are discussed in Chapter 42.

In the cycle we have been concerned with so far, carbon alternates between the oxidised inorganic state and the reduced organic state. As Figure 39.12 shows, there is another cycle in which carbon remains oxidised throughout. In this cycle $CO_2$ is used by planktonic protoctists to make their tiny skeletons of lime ($CaCO_3$). Over tens of millions of years vast numbers of these skeletons are deposited on the ocean floor, forming **chalk**. Over equally long periods this may undergo metamorphosis to become **limestone**. If the ocean floor subsequently rises, chalk and limestone may be exposed to weathering. Some dissolves in rainwater, freeing it to enter the other cycle via photosynthesis.

Of the two cycles, the purely 'inorganic' cycle turns by far the more slowly. This is partly because the amounts of carbon in the chalk and limestone compartment are so enormous, and partly because the rate at which carbon enters and leaves these compartments is relatively low. Because the carbon cycle involves geological as well as biological processes it is called a **biogeochemical cycle**.

## *The nitrogen cycle*

Nitrogen is a constituent of amino acids and nucleotides and, therefore, of proteins and nucleic acids. In all these compounds nitrogen is bonded to hydrogen and is therefore in the reduced state.

The nitrogen cycle differs in an important way from the carbon cycle in that several key stages can only be carried out by bacteria. It actually contains two interlocking loops, one of which 'turns' much more rapidly than the other (Figure 39.13). The rapid cycle involves organisms from all five kingdoms, whilst the slow one involves bacteria only.

Like the carbon cycle, the nitrogen cycle involves oxidation and reduction with associated energy changes. The main cycle can be divided into three phases – occurring in plants, in consumers and decomposers, and in chemosynthetic bacteria, respectively. Interlocked with this cycle are two other natural processes – denitrification and nitrogen fixation. Also affecting the nitrogen cycle are agricultural harvesting, addition of fertilisers and lightning.

### *Reduction by green plants*

Though molecular nitrogen ($N_2$) is superabundant (accounting for 80% of the atmosphere) it is extremely inert and cannot be used by plants. Instead, plants absorb their nitrogen in the form of nitrate ions ($NO_3^-$), which they then reduce to nitrite and then to ammonia:

$$NO_3^- \rightarrow NO_2^- \rightarrow NH_3$$

This is an energetically 'uphill' process, using NADPH derived from photosynthesis. Many plants can absorb ammonia directly from the environment, thus bypassing this stage and saving energy.

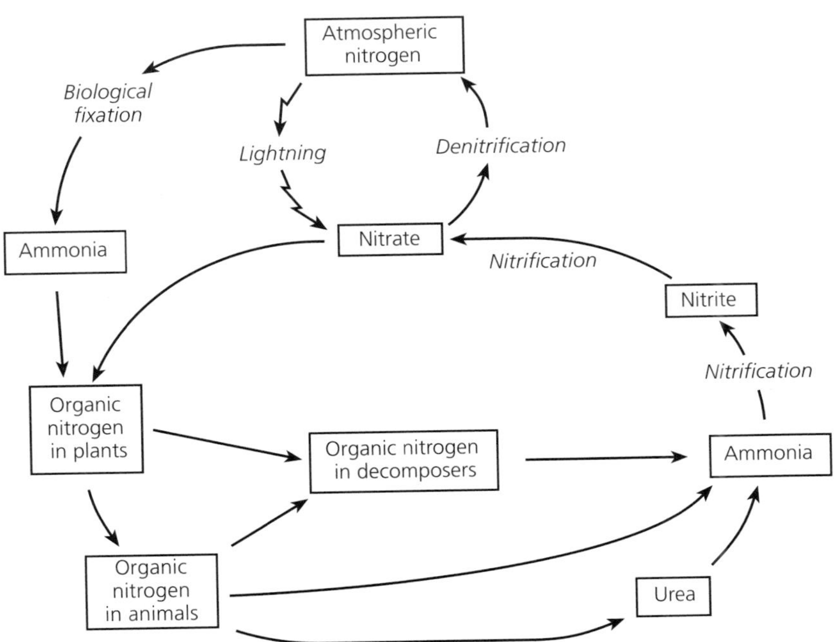


**Figure 39.13**
The nitrogen cycle

Ammonia is then used to make nitrogenous organic compounds such as amino acids and proteins:

$$NH_3 + \text{keto acids} \rightarrow \text{amino acids} \rightarrow \text{proteins}$$

### *Ammonia formation by heterotrophs*

When herbivores feed on plants they digest the plant proteins, converting them to amino acids. Some of the amino acids are used to make herbivore protein. The rest are deaminated to release ammonia:

$$\text{amino acids} \rightarrow \text{keto acids} + NH_3$$

Ammonia then immediately combines with $CO_2$ to form ammonium carbonate:

$$2NH_3 + H_2O + CO_2 \rightarrow (NH_4)_2CO_3$$

A similar process occurs in every heterotroph, with the result that all the organic nitrogen is sooner or later converted to ammonia.

In many animals the ammonia produced in deamination is converted to a less toxic substance such as urea or uric acid. The end result is the same, however, since these excretory products are converted into ammonia by bacteria (which utilise the energy released to drive their metabolism).

### *Nitrification*

The ammonia produced by deamination of amino acids is oxidised by bacteria to nitrates – a process called **nitrification**. It can only be carried out by bacteria and is essentially the reverse of the reduction of nitrate to ammonia by plants. It occurs in two steps, carried out by different kinds of **nitrifying bacteria**. *Nitrosomonas* oxidises ammonium ions to nitrite, and *Nitrobacter* oxidises nitrite to nitrate:

$$2NH_3 + 3O_2 \xrightarrow{\textit{Nitrosomonas}} 2NO_2^- + 2H^+ + 2H_2O$$

$$2NO_2^- + O_2 \xrightarrow{\textit{Nitrobacter}} 2NO_3^-$$

Both reactions release energy which is used to convert $CO_2$ to carbohydrate. These bacteria are thus **chemoautotrophs**. Since nitrification requires oxygen, ammonia may accumulate in anaerobic conditions.

We might add in passing that although nitrifying bacteria are said to be autotrophic, the energy released in nitrification is energy that had originally been expended by green plants in the reduction of nitrate to ammonia. The energy released in the oxidation of ammonia and nitrite therefore represents stored energy originally derived from photosynthesis. In that sense these chemoautotrophs are ultimately as dependent on the solar energy as are the plants, animals and fungi.

### *Denitrification*

Under anaerobic conditions, bacteria such as *Pseudomonas denitrificans* reduce nitrate to nitrogen gas or to nitrous oxide, $N_2O$. This is called **denitrification** and obviously tends to lower soil fertility.

Denitrifying bacteria are **facultatively anaerobic**. This means that they use oxygen if it is available, but if the soil becomes depleted of oxygen they switch to anaerobic respiration (using nitrate as an electron acceptor instead of oxygen).

Anaerobic conditions are particularly likely to occur after prolonged flooding, in soils that are very rich in organic matter, or in sewage. In well-drained soils denitrification is seldom an agricultural problem.

One of the environmental problems of denitrification is that nitrous oxide is contributing to the destruction of ozone (Chapter 42). On the other hand, swampy land provides habitats for specialised and diverse communities of ecological interest.

### *Nitrogen fixation*

Denitrification would eventually lead to serious depletion of nitrate were it not for the opposing process of **nitrogen fixation**, in which nitrogen gas is converted to combined nitrogen. To combine nitrogen with other atoms, the very strong triple bond in the nitrogen molecule must first be broken. The large amount of energy needed can come from two sources: heat generated in combustion or lightning, or the metabolism of living organisms.

#### Non-biological fixation

Non-biological fixation occurs when nitrogen and oxygen are heated to high temperatures – for example in car engines and in lightning flashes they form oxides of nitrogen which react with water to form nitrite and nitrate.

#### Biological fixation

Of the total natural annual nitrogen fixation, 90% occurs by biological mechanisms, the remaining 10% resulting from lightning. The only organisms that can fix nitrogen are prokaryotes – bacteria and cyanobacteria. The process involves a molybdenum-containing **nitrogenase** enzyme and is highly expensive in terms of ATP. Some nitrogen-fixing bacteria live freely in the soil, for example *Azotobacter*, which is aerobic, and *Clostridium*, which is anaerobic. The energy for fixation is derived from organic matter in the soil. The cyanobacteria (blue-green bacteria) produce their own carbohydrate by photosynthesis.

Economically the most important nitrogen fixers are bacteria that live in a mutualistic relationship with plants. By far the most important of these are members of the genus *Rhizobium*, which live with members of the Papilionaceae or legume family, such as clover, peas and beans (see Chapter 37). A number of non-legumes (such as alder) also have nitrogen-fixing nodules.

On a global scale the rate of bacterial nitrogen fixation is very small compared with the rate of nitrate absorption by plants (about $8 \times 10^{12}$ moles N per year, compared with about $500 \times 10^{12}$ moles N per year).

### Human inputs and outputs

In most natural ecosystems nitrogen is recycled with little loss; but in agriculture, harvesting removes large quantities of nitrogen as protein. To some extent this may be replaced by growing clover and other legumes, but considerable quantities of inorganic fertiliser in the form of ammonium nitrate also need to be added.

The disadvantage of inorganic fertiliser is that it leads to a short-term rise in nitrate content, much of which is not used. The surplus may be *leached*, or washed out in run-off, raising the nitrate content of lakes and rivers. The effect is that lakes become enriched with nutrients, or **eutrophic**. As a result plant growth is stimulated, and when the plants die off in the autumn their decay may cause pollution problems (Chapter 42). Ammonium ions are less prone to leaching since their positive charges are attracted by negatively-charged clay particles.

### Interactions between the cycles

Although the carbon and nitrogen cycles are treated separately, all nutrient cycles are, of course, interlocked because many organic molecules contain these and other elements. Nucleotides contain nitrogen and phosphorus in addition to carbon, and some proteins also contain sulphur. A change in the rate of turning of one nutrient cycle must therefore be accompanied by a change in the others.

## Nutrient cycling in oceans and lakes

Although nutrient cycles function in basically the same way in both aquatic ecosystems and on land, there are some important differences. In temperate seas and lakes, the rise in light intensity and water temperature in spring cause a burst in the growth of phytoplankton. At the same time the rise in water temperature reduces its density, so that a layer of warm water forms on top of the cold, denser water below. The boundary between the warm and cold water is called the **thermocline**, and its formation prevents the upper and lower water from mixing. In lakes, the water above the thermocline is the **epilimnion** and the water below is the **hypolimnion.**

Thermal stratification in Lake Windermere in July (redrawn from Macan and Worthington)

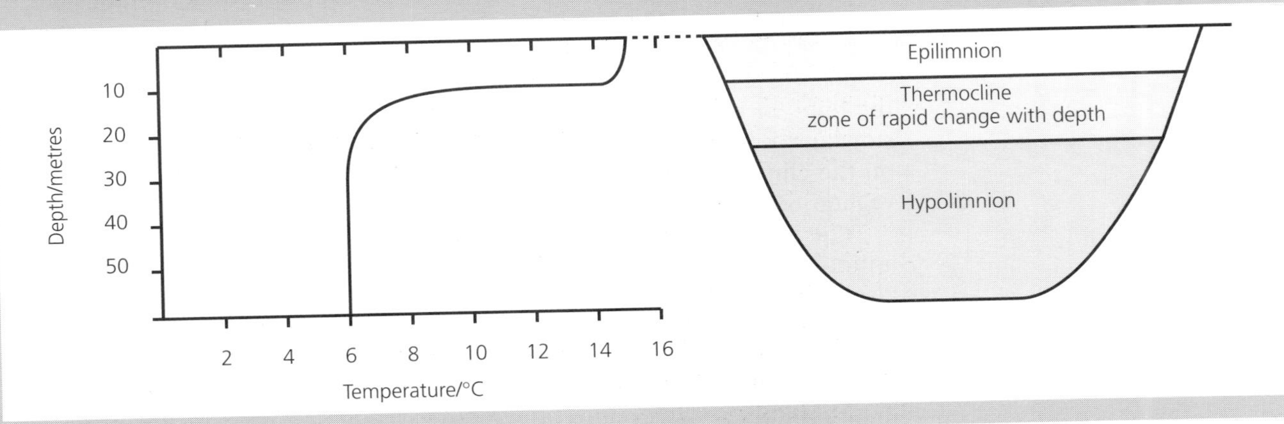

## Nutrient cycling in oceans and lakes *continued*

It is in the surface waters (where the light intensity is greatest) that the uptake of mineral ions by the phytoplankton is fastest. Although much of the nutrients are released back into the water by animals of the grazing food chain, some dead bodies sink down into the colder, darker water below. Since there is little or no upward movement of water in the summer, the nutrients contained in these bodies are out of circulation, and the surface water becomes progressively depleted of minerals.

In the autumn the surface layers cool and the thermocline breaks down. This enables some mixing to occur and there may be a small burst of phytoplankton growth. In winter mixing is more complete, so that by the next spring the surface waters are once again rich in nutrients.

In most tropical oceans the thermocline is a permanent feature, so the illuminated layers are always poor in nutrients and productivity is correspondingly low. In polar seas a thermocline never develops.

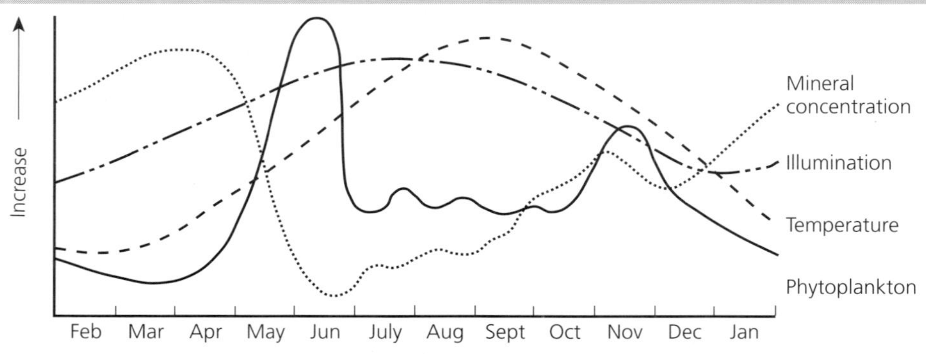


Seasonal changes in biotic and abiotic factors in a lake

## Succession

Succession is a gradual, progressive and predictable change in the composition of a community over a period of time. There are two kinds of succession: primary and secondary.

### *Primary succession*

Primary succession occurs when conditions are initially unfavourable for plant growth, but soil is gradually made fertile by plants. To begin with, physical conditions are harsh and only a few **pioneer species** can survive. Slowly these raise the fertility, and other plants begin to get established and compete with the pioneers, eventually displacing them. The process of competitive replacement continues, with stronger competitors displacing less competitive species, until there is no further change. When the final, or **climax**, community has developed, any further change is very slow and is due to climatic change. The kind of climax vegetation depends on the climate and soil conditions. In most parts of Britain the climax vegetation is deciduous forest.

The increase in the number of plant species also results in greater animal diversity. In addition to providing new food sources, the plants also provide shelter.

Some successions do not reach the climax state because of interference by a biotic factor, such as grazing by rabbits. The final community is called a **deflected climax** and is in equilibrium with a biotic factor rather than with the climate.

### Succession on bare rock

Bare rock dries out very quickly and the only autotrophs that can survive there are lichens. The fungal component of the lichens release $CO_2$, helping to form tiny cracks in the rock surface. Eventually certain species of moss that can survive desiccation, such as *Tortula muralis*, get established, overtopping the lichens and replacing them. As a result of the growth of mosses, enough humus develops to allow slightly larger plants to become established. These outcompete the mosses and eventually replace them. The process continues, with each newly-established species adding to the dead matter and increasing the water-retaining capacity, allowing larger plants to establish themselves.

### Succession in sand dunes

Unlike the situation with bare rock, the changes in time in a sand dune system are reflected by changes in space – the youngest dunes being nearest the sea (Figure 39.14). So walking inland from the high tide mark we encounter progressively later stages in the succession.

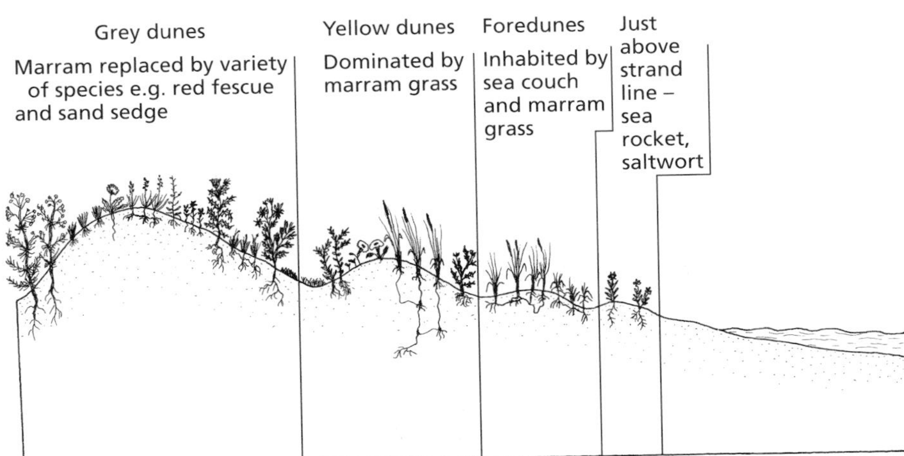


**Figure 39.14**
Succession in sand dunes (after King)

Immediately above the strand line are a few scattered plants such as sea rocket (*Cakile maritima*) and saltwort (*Salsola kali*). They can survive not only salt spray but also occasional immersion by exceptionally high tides, and their succulent leaves and stems enable them to store what little water is available. Though these plants are extremely tolerant of salt and drying, they cannot grow upwards through wind-blown sand so they have little dune-forming ability.

A little higher up are the **foredunes**. Among the pioneer species is sea couch grass (*Agropyron junceiforme*). Though very salt-tolerant, it has limited capacity to grow vertically upwards through sand and cannot form dunes much more than a metre or so high.

The most important sand-binder is marram grass (*Ammophila arenaria*). Like sea couch, its leaves help to trap sand. Although it is unable to tolerate immersion by sea water, its rhizomes have much greater ability to grow upward than sea couch, and as more sand is deposited by the wind, it forms dunes many metres high. Marram grass is characteristic of the **mobile**, or **yellow, dunes**.

As a result of the activity of the pioneer plants, the sand gradually becomes stabilised and some humus begins to form. The slight increase in fertility allows other species to become established. The most important of these are sand fescue grass, *Festuca rubra*, and a sedge, *Carex arenaria*.

These plants begin to form a more continuous mat of vegetation and slowly convert the mobile dunes into more stable **fixed**, or **grey**, **dunes**. At the same time the increasing fertility tips the balance of competitive advantage against marram grass. Though able to survive harsh physical conditions, marram is a poor competitor in a fertile soil and is displaced by other species. As fertility increases, the number of plant species (and of course animal species that feed on them) increases further.

During the process of succession there is not only an increase in species diversity – there is also an increase in total biomass. For this to happen there must be a net import of nutrients. Carbon of course comes from the atmosphere as $CO_2$. Nitrogen and phosphorus are imported by the deposition of bird faeces, and additional nitrogen is fixed by legumes such as bird's foot trefoil (*Lotus corniculata*). In addition, sea spray contains small amounts of all plant nutrients.

### *Succession in lakes*

Most shallow lakes have vegetation round the edge. In the deeper water may be found pondweeds such as *Potamogeton* spp and water lilies. As the plants die, organic matter accumulates and raises the bottom, enabling plants such as reedmace (*Typha latifolia*) to grow. As deposition of dead plant matter continues, sedges make their appearance, followed by alder (*Alnus*) and willow (*Salix* spp). Slowly the lake is filled in, and the climax vegetation is marked by the appearance of trees such as oak and mesophytic herbs.

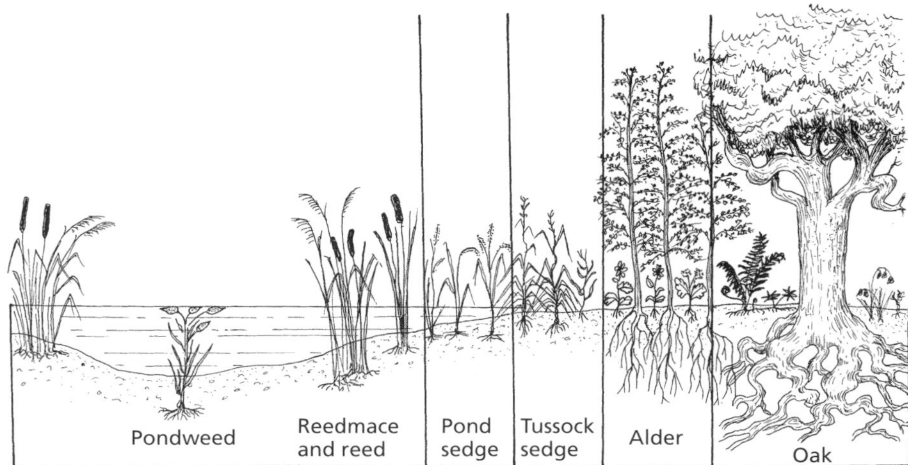


**Figure 39.15**
Succession in lake vegetation (after King)

## Decomposer succession

Humus formation (**humification**) is a complex process involving fungi, bacteria and other micro-organisms. Many small arthropods play an indirect role by chewing up plant matter, as a result of which their faeces have a greatly increased surface area for attack by micro-organisms.

The rate of humification depends on the kind of material and on the temperature. Pine needles take 7–10 years, while leaves of deciduous trees may be broken down in a year or two.

Fungal decay of plant matter follows a regular **succession**, the first species of fungus specialising on the more easily digested starches and sugars, and later ones utilising cellulose, hemicelluloses, proteins and finally tannins.

Succession also occurs on the faeces of rabbits, sheep and other herbivores. First to get established are starch- and sugar-utilising fungi. These are succeeded by fungi that use hemicelluloses, and finally cellulose and lignin-digesting fungi.

### *Types of primary succession*

The entire series of changes in a primary succession is a **sere**. Succession which begins with a dry substratum (such as sand) in which the pioneer plants are xerophytes, is called a **xerosere**, whilst a sere in which the pioneer plants are hydrophytes is a **hydrosere**. A sere beginning with bare rock is called a **lithosere** (*lithos* = rock).

### *Secondary succession*

A quite different kind of succession occurs after vegetation is removed as a result of fire or by cultivation. The early stages of this **secondary succession** can be seen every time a garden is left unweeded. The soil is fully fertile and so it is not long before plants become established. The first to appear are either those which can produce large numbers of wind-dispersed seeds or spores, or those which were already present in the soil as seeds, tubers or other dormant stages.

Since the soil is fully fertile from the beginning, secondary succession is much faster than primary succession. In the tropics where the higher temperatures result in faster growth, it may only take a hundred years for bare ground to change to mature forest.

## QUESTIONS

**1** Explain why
- **a)** productivity in polar seas is much higher than in most tropical seas.
- **b)** there are fewer carnivores than herbivores.
- **c)** an increase in the concentration of atmospheric $CO_2$ can cause an increase in the rate of nitrogen fixation.

**2** Which of the following could *not* affect the circulation of nitrogen in an ecosystem?

- **A** increasing the nitrogen content of the air
- **B** increasing soil water content
- **C** increasing soil nitrate
- **D** decreasing soil temperature
- **E** decreasing soil pH
- **F** adding urea to the soil
- **G** adding phosphate to the soil

**3** At which point in the nitrogen cycle does nitrogen **(a)** join, **(b)** detach from carbon?

**4** Vitamin A is fat-soluble and toxic in large quantities. It is not excreted and in vertebrates is stored in the liver. Suggest why it is dangerous to eat polar bear liver.

**5** Study Table 39.1 and answer the following:
- **a)** Why does the elephant use such a high proportion of its assimilated food for fuel?
- **b)** Suggest why the grasshopper, caterpillar and elephant assimilate a smaller proportion of their ingested food than the spider and perch.

**6** A common problem in many aquaria is green water due to unicellular photosynthetic protoctists. Lakes rarely suffer from green water. Suggest an explanation based on nutrient cycling and energy pyramids.

**7** The table shows some of the differences between two parts of a sand dune system: the foredunes nearest the sea and the fixed dunes furthest away.

Use the information in the table to explain the nature of the changes occurring during a primary succession.

| | Foredunes | Fixed dunes |
|---|---|---|
| Species diversity | low | high |
| Food chains | linear, grazing | branched, detritus and grazing |
| Origin of inorganic nutrients | outside the dunes | inside the dunes |
| Net $CO_2$ uptake | high | low |
| Niche breadth | wide | narrow |

## EXAM QUESTIONS

**1** **a)** Explain what is meant by an ecosystem. *(1 mark)*

**b)** The diagram represents the flow of energy through a grassland ecosystem.

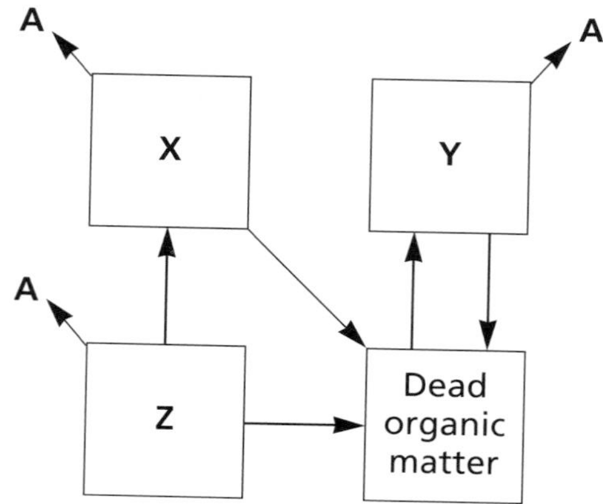


**i)** What biological process is represented by the arrows labelled **A**? *(1 mark)*

**ii)** Give the name of the trophic level of the organisms in boxes **X, Y** and **Z**. *(1 mark)*

**iii)** Briefly describe how carbon in dead organic matter is made available to the organisms in box **Z**. *(2 marks)*

**AEB**

**2** The diagrams show pyramids of numbers representing two different food chains.

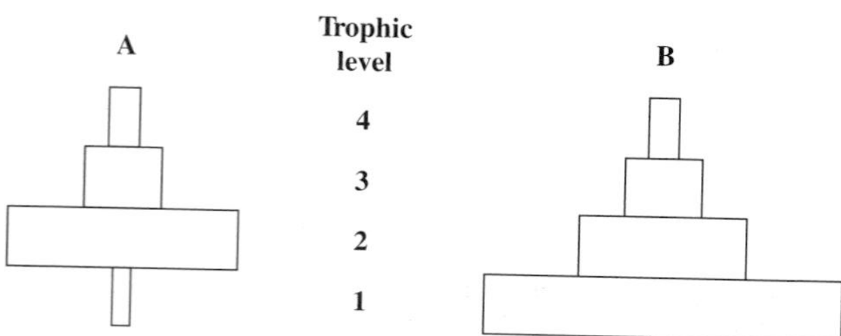


**a)** Give the name of trophic level **3**. *(1 mark)*

**b)** Suggest why the two pyramids of numbers have different shapes. *(1 mark)*

**c)** Redraw pyramid **A** as a pyramid of energy. *(1 mark)*

**d)** Redraw pyramid **B** as it would appear if organisms in trophic level **4** were parasitised by bacteria. *(2 marks)*

**AEB**

**3** In an investigation into the roles of different organisms in the breakdown of leaves, leaf discs of standard size were placed in nylon bags of varying mesh sizes and buried in newly cultivated soil. Every two months the bags were dug up and examined to determine the area of leaf that had disappeared. The table shows the kinds of organisms able to enter through each mesh.

The results of the experiment are shown in the graph.

| Mesh size/mm | Organisms with free entry |
|---|---|
| 7.000 | Micro-organisms, arthropods and earthworms |
| 0.500 | Micro-organisms and arthropods |
| 0.003 | Micro-organisms |

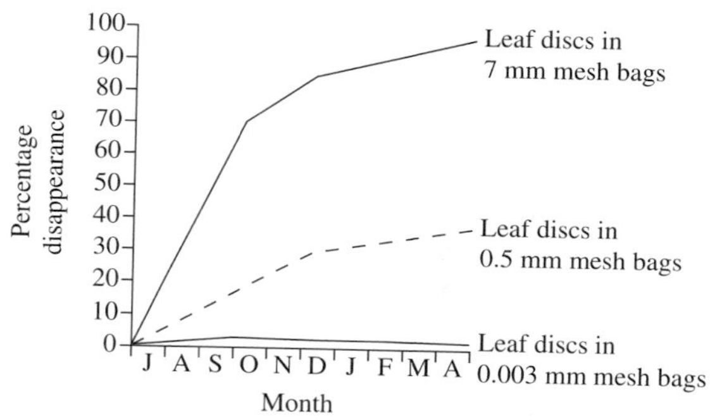


**a)** Describe the relative importance of arthropods, earthworms and micro-organisms in the breakdown of the leaves. *(2 marks)*

**b)** Suggest how the arthropods and earthworms may aid the breakdown of the leaves by micro-organisms. *(2 marks)*

**c)** Describe how micro-organisms break down the leaves. *(3 marks)*

**AEB**

## EXAM QUESTIONS

**4** **a)** Suggest *two* ways in which modern farming might affect the diversity of living organisms. *(2 marks)*

**b)** 'Set-aside' is the common name given to a European Union policy under which farmers receive a subsidy for land taken out of cultivation. A study was carried out to investigate how the amount of time a set-aside field was left uncultivated would affect the species of birds feeding there.

The table below shows the numbers of birds of different species feeding in one field which had been left uncultivated for a year.

| Species | Number of birds of that species feeding in the field |
|---|---|
| Greenfinch | 12 |
| Goldfinch | 8 |
| Wood pigeon | 3 |
| Pheasant | 1 |

**i)** Use the formula  $d = \dfrac{N(N-1)}{\Sigma n(n-1)}$

where d = index of diversity
N = total number of organisms of all species
and n = total number of organisms of a particular species

to calculate the index of diversity for the birds feeding in the field. Show your working. *(2 marks)*

**ii)** Explain why it is more useful in a study of this sort to record diversity rather than the number of species present. *(2 marks)*

**c)** The figure at the top right is a graph showing the relationship between bird species diversity and plant species diversity in this study. The figure below this is a graph showing the relationship between bird species diversity and plant structural diversity for the same study. Structural diversity refers to the different forms of plants such as herbs, shrubs and trees.

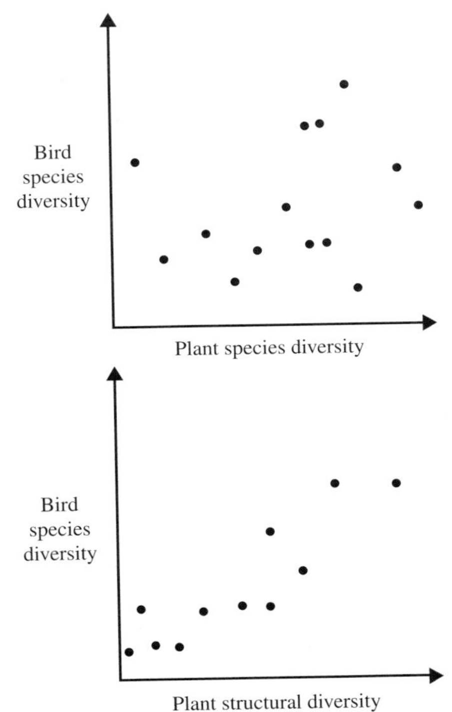


**i)** Explain briefly how you could obtain the data that would enable you to calculate the diversity index for the species of plants growing on a set-aside field. *(3 marks)*

**ii)** Describe the difference in the relationships shown in the two graphs. *(2 marks)*

**iii)** Suggest an explanation for the relationship between bird species diversity and plant structural diversity shown in the second graph. *(2 marks)*

**d)** In another study of fields taken out of cultivation, the figures shown in the table below were obtained.

| Value of index of diversity for bird species | Time in years since cultivation stopped |
|---|---|
| 2.1 | 5 |
| 3.2 | 15 |
| 5.6 | 20 |
| 4.1 | 25 |
| 4.8 | 40 |
| 9.4 | 60 |

## EXAM QUESTIONS

i) Plot these data as a suitable graph on graph paper. *(4 marks)*

ii) Predict what might happen to the bird species diversity in the study summarised in this table over the next 100 years. Explain how you arrived at your answer. *(3 marks)*

**AEB**

**5** The diagram shows the life cycle of the malarial parasite, *Plasmodium falciparum*.

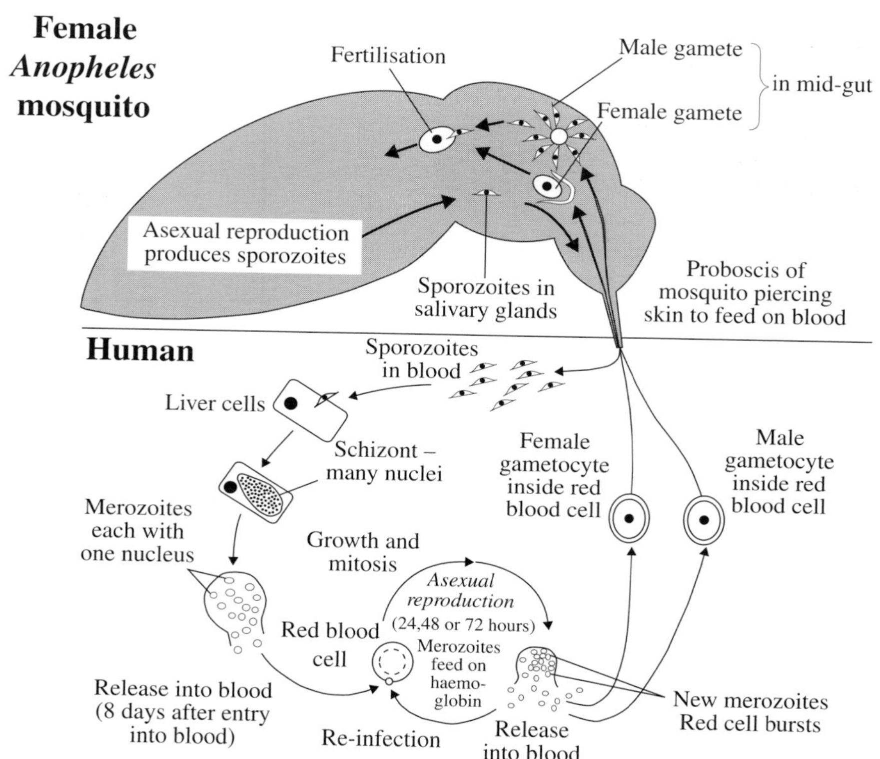


**a) i)** What part does the female mosquito play in the transmission of the parasite? *(1 mark)*

**ii)** From the diagram, what is the minimum time which must elapse after infection before a human could infect another mosquito? *(1 mark)*

**b)** Suggest *one* stage in the life cycle when the parasite would be vulnerable to attack by human antibodies. Give a reason for your answer. *(2 marks)*

**c)** The drugs quinine and chloroquine have been used to treat malaria by killing the parasites in the red blood cells. These drugs are now less effective than they used to be. Suggest why. *(2 marks)*

**d)** Recently, a drug called artemisinin has been developed which has proved effective against the malarial parasite. This drug reacts with iron to release a highly reactive iron oxide which is extremely toxic. Suggest why artemisinin is effective against the malarial parasite. *(3 marks)*

**AEB**

**6** The following procedure can be used for collecting aquatic insect nymphs from a stream.

'An 8 mesh per cm net of 28.5 cm diameter is attached to a metal frame on the end of a long pole. The collector works slowly upstream for a distance of 1 m, lifting stones into the net without removing them from the water. Any animals clinging to the stone are brushed by hand into the net. The net should always be held immediately downstream from the stone with the frame pressed firmly against the stream bed. The contents of the net are washed into a container of clean water and retained.'

## EXAM QUESTIONS

a) i) Give *two* advantages of working slowly upstream. *(2 marks)*

ii) Explain why the net should be pressed firmly against the bed of the stream. *(1 mark)*

b) i) Suggest *two* reasons why this method might *not* yield accurate data about the types and numbers of nymphs in the stream.

ii) Comment, with reasons, on the accuracy of the method for comparing numbers of a single species in the same part of the stream at different times of the year. *(5 marks)*

c) The graph shows seasonal variation in body length of a species of stonefly. Individual nymphs were measured and the vertical lines show the range of body length with the cross bars showing the mean body length for combined samples collected at each visit to the stream.

The visits were at irregular intervals, but always in the first two or three days of each month. The sample sizes are shown by numbers over the collecting months.

i) Predict the mean length of the stonefly nymphs in March when no collection was made.

ii) State the length of the largest nymph collected during the whole period.

iii) Calculate the percentage decrease in sample size between the September and December visits.

iv) Offer an explanation for this decrease. *(4 marks)*

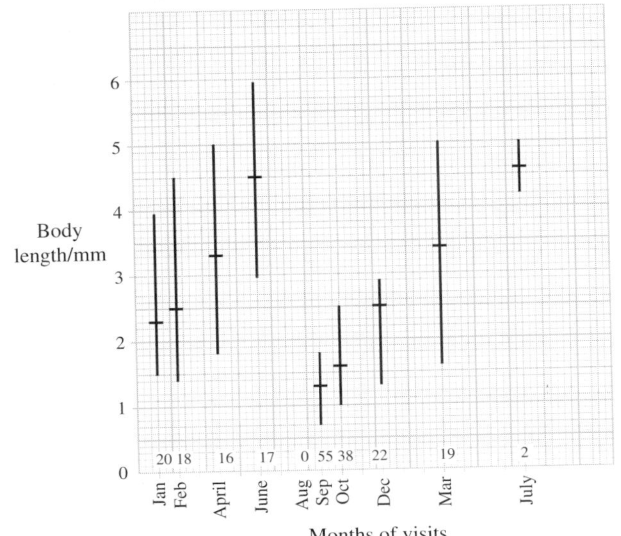


d) Use the graph to deduce

i) the month in which most adult stonefly emerge from the stream,

ii) the month(s) when the eggs are laid,

iii) the total duration (in months) of the nymphal stages. *(3 marks)*

**WJEC**

**7** In an experiment investigating the rate of disappearance of leaf litter, oak leaf discs were placed in nylon mesh bags and buried in newly cultivated pasture.

The table shows the disappearance of oak leaf discs from bags made from 7 mm and 0.5 mm mesh over a period of months.

| Month | Percentage oak leaf area remaining in bags of mesh size: | |
|---|---|---|
| | 7 mm | 0.5 mm |
| June | 100 | 100 |
| August | 81 | 94 |
| October | 30 | 91 |
| December | 13 | 66 |
| February | 9 | 62 |
| April | 6 | 60 |

*Ref: Ecology of Woodland Processes by J.R. Packham and D.J.L. Harding (1982).*

## EXAM QUESTIONS

a) Plot the data on graph paper. *(5 marks)*
   i) Describe the effect of mesh size on the rate of disappearance of leaf litter between June and October. *(2 marks)*
   ii) Suggest an explanation for this. *(2 marks)*
   iii) Explain the variation in the rate of disappearance of litter from the 0.5 mm mesh bags during the period of the experiment. *(2 marks)*

c) i) One estimate of the nitrogen content of leaf litter is 11 g $m^{-2}$ of leaf. Calculate the total amount of nitrogen (g $m^{-2}$) in 50 leaf discs each 1.25 cm in radius contained in the 7 mm mesh bag.

   Assume that the area of each disc is $\pi r^2$ and that $\pi = 3.14$
   Show your working. *(4 marks)*
   ii) If all the leaf discs in the 7 mm mesh bag contained a total of 0.25 g of nitrogen, state the amount of nitrogen that would have been released from the discs during the period June to April. *(1 mark)*
   iii) The nitrogen released as a result of this decomposition is a natural fertiliser, only some of which is available to plants. Suggest *two* ways in which nitrogen is lost. *(2 marks)*

**WJEC**

**8** The diagram below shows the uptake of nutrient elements by plants in soils with different pH values.

a) i) Circle the *one* pH range below in which the majority of nutrients in the soil are most readily available to plants.
   5.5–6.0; 6.0–6.5; 6.5–7.0; 7.0–7.5; 7.5–8.0. *(1 mark)*
   ii) Which *one* element is least available in *both very* acid and *very* alkaline soils? *(1 mark)*
   iii) Describe *fully* the effect of a deficiency of this nutrient in plants. *(2 marks)*

b) i) A farmer planted a crop of corn in a field, pH 5.0. The crop failed. Use the information in the diagram to suggest the *most likely* reason for this. *(1 mark)*
   ii) In this crop only the young seedlings became green before the crop failed. Suggest *one* reason for this. *(1 mark)*
   iii) Suggest how the farmer might improve the field so that he can grow corn. *(1 mark)*

c) The presence of too much of a particular nutrient in the soil may create a low water potential in soil water.
   i) Describe how this might affect the appearance of the plant. *(1 mark)*
   ii) Explain your answer. *(2 marks)*

**WJEC**

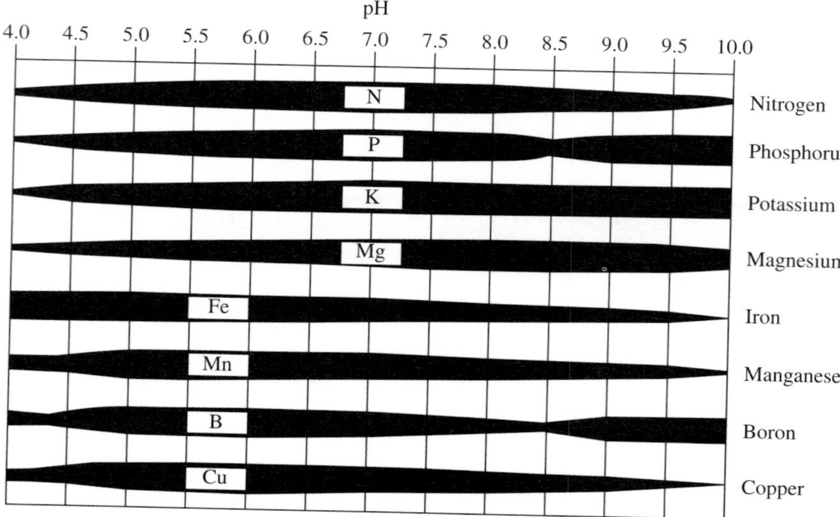

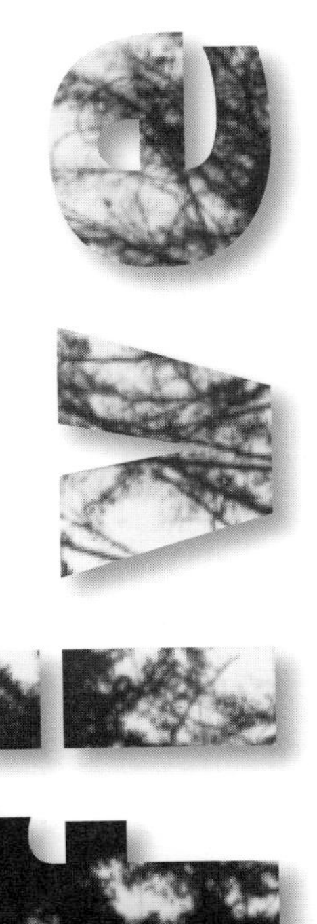

# Part five

# The HUMAN IMPACT


CHAPTER 40
PRIMATE LEGACY . . . . . . . . . . . . . . . . . . . . . . . . . . . . . . . . . . . . . . . . 752

CHAPTER 41
The EVOLUTION of HUMANS . . . . . . . . . . . . . . . . . . . . . . . . . . . . . . 775

CHAPTER 42
The HUMAN PREDICAMENT . . . . . . . . . . . . . . . . . . . . . . . . . . . . . . . 801

CHAPTER

# 40

# PRIMATE LEGACY

## LEARNING OBJECTIVES

By the time you have completed your study of this chapter you should be able to:

- list the features that distinguish primates from other mammals.
- explain the relationship between social life and the evolution of intelligence in primates.
- describe ways in which chimpanzee behaviour foreshadows that of humans.
- state the features that lead us to classify humans with Old World monkeys and apes.
- describe the skeletal features that adapt humans to bipedalism and the manipulation of objects.
- describe how toolmaking by humans has been associated with changes in the teeth, jaws, skull and limb bones.
- describe the features of the vocal tract and brain that make speech possible in humans.
- explain how humans differ from other apes with regard to reproduction and growth pattern.
- explain how molecular taxonomy and chromosomal evidence have led to a reappraisal of the relationships between humans and other apes.

## Primate characteristics

Humans are primates, and as such they are members of an order of mammals that includes the monkeys and apes, together with some lesser known mammals such as lemurs, lorises, bushbabies and tarsiers.

A central theme running through primate organisation is an adaptation to tree-living, or **arboreal** life. Tree-living is precarious, demanding grasping or **prehensile**, limbs that are sufficiently mobile to be able to gain support from points that are both vertically and horizontally apart. In this three-dimensional world each anchorage point is different from the previous one, so locomotion requires moment-by-moment co-ordination of sensory information from the eyes, balance organs and touch receptors. The close-co-ordination between eye and grasping forelimb was a preadaptation for the manipulation of objects and, in our own ancestry, toolmaking.

Though adaptations are often spoken of as if they are separate items in an organism's 'survival kit', they work together as an integrated whole. Nevertheless, for descriptive purposes it is convenient to deal with each system separately. The features discussed in this chapter are most clearly shown by the monkeys and apes. These primates are more **derived**, meaning that they have evolved more from the ancestral type than have the (more primitive) lemurs and their relatives.

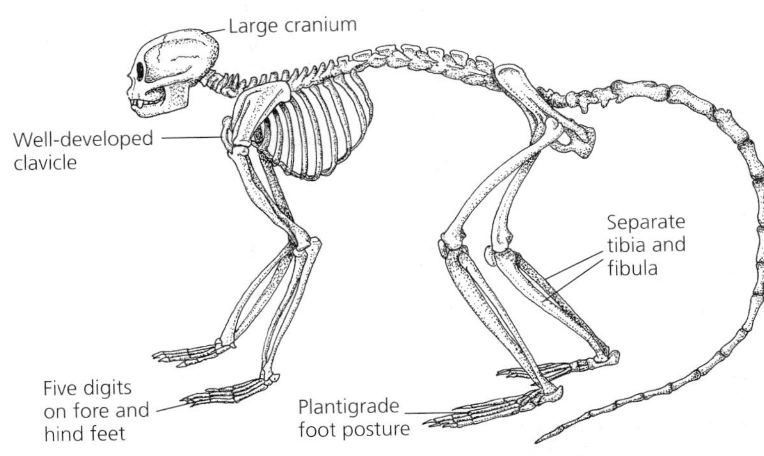


**Figure 40.1**
Skeleton of an old World monkey showing primate features

## *Skeleton*

Unlike most mammals, primates have retained the general-purpose pattern of the early mammals and in this respect are **primitive** (Figure 40.1).

Most generalised of all are the limbs, which closely resemble the **pentadactyl** pattern of land vertebrates (Figure 40.2). The feet lie flat on the ground in the **plantigrade** posture, in contrast to most other mammals which stand with only their toes touching the ground. The five fingers and toes are long and mobile and the first digit of the foot diverges from the other toes, enabling branches to be grasped. To a variable extent the first digit of the hand (the thumb) can press against, or **oppose**, the tips of the fingers, enabling objects to be picked up and manipulated.

Another feature that aids the grasping and manipulating of objects is the fact that the radius and ulna are freely moveable, and in monkeys and apes the forearm can be rotated so that the palm faces upwards. In most other mammals the radius and ulna are bound together by ligaments, with the forefoot permanently in the 'palm-down' position. The tibia and fibula (lower leg bones) of most primates are also separate and can move slightly relative to each other. In most other mammals they are partly fused into a single unit.

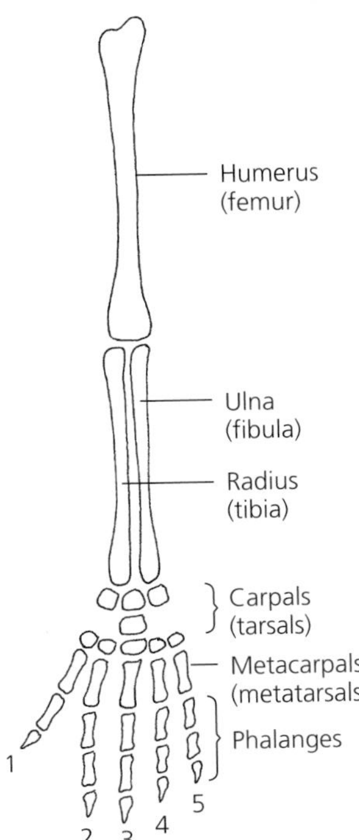


**Figure 40.2**
Basic plan of the pentadactyl limb. Names in brackets refer to the hind limb

## *Vision*

Rapid movement from branch to branch obviously demands good eyesight. Monkey and ape vision differs from that of many other mammals in three respects:

- the eyes face forward, with the two visual fields overlapping so the two eyes see the same object slightly differently (Figure 40.3). By comparing the inputs from the two eyes, the brain obtains a 3-D, or **stereoscopic**, view, enabling the distance of branches to be judged.
- the retina has cones as well as rods, giving highly developed colour vision.
- the retina has a well-developed **fovea**, enabling the perception of *detail* (Chapter 18).

## *Smell*

Arboreal life is primarily a visual world rather than an olfactory one, and in primates the sense of smell is less important than in most other mammals. This is associated with a shortening of the nasal area of the skull – the snout.

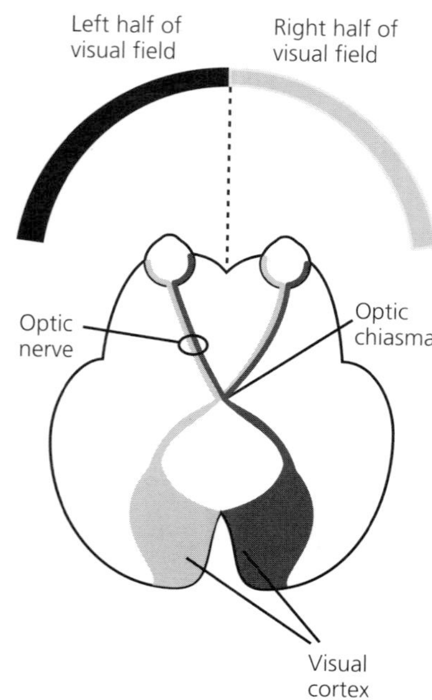


**Figure 40.3**
Stereoscopic vision requires overlapping fields of view

## *Nails*

Non-primate arboreal mammals (such as squirrels) obtain their grip by digging sharp claws into the bark of tree trunks and large branches. Primates have nails which give support to the soft flesh of the tips of the fingers and toes, giving better grip on small branches.

## *Sitting posture*

Nearly all primates sit in an upright posture. Perched horizontally on a branch, an animal cannot turn round quickly without risking a loss of balance. With its neck semivertical, a sitting primate can look round without altering the position of its centre of gravity simply by swivelling its head on its neck. By freeing the hands, upright sitting also enables food to be manipulated easily.

## *Brain*

In relation to body mass, primates have bigger brains than most other mammals. The ratio of brain/body mass in monkeys is two to three times greater than in other mammals (except whales and dolphins), and the great apes have brains that are relatively twice as big as those of most monkeys.

Arboreal life, with an almost infinite variety of foot- and handholds, is anything but predictable. Survival in the trees therefore demands increased emphasis on the more flexible behaviour that comes from **learning**, and the urge to learn is an essential part of a young primate's development. Primates are thus extremely curious and investigate any strange object they encounter.

When a monkey turns an object over in its hands its brain is, at any given instant, analysing information from a whole range of sources.

- Comparison of input from both eyes gives information on its 3-D *shape*.
- Stretch receptors in the ligaments of the finger joints indicate the positions of the fingers, thus helping to build up a 3-D *feel* of the object's shape.
- Touch receptors beneath the skin ridges inform the animal of the surface *texture* of the object (how rough or smooth it is).
- Stretch receptors in the muscles and tendons give information about the tension in the muscles, and therefore about its *weight*.

Though these kinds of information are initially processed in different parts of the brain, they undergo further analysis in which they are compared with each other. The bringing together, or **collating**, of different kinds of information about an object enables the animal to build up a concept of, say, 'stone' or 'stick'. An important implication of this is that an object can be seen as something that exists independently of the rest of the environment, and can therefore be used by the animal as a **tool**. Though humans excel in this respect, many primates can use tools. To date, tool use has been observed in 18 species of monkeys and apes.

Although it may seem self-evident that successful arboreal life goes hand in hand with a large brain, other factors may also be involved. Recent research has shown that primates, especially monkeys and apes, seem to be far brighter than previously realised – considerably more intelligent, it seems, than is needed for routine economic activity such as food-gathering.

As we shall see later, it now seems likely that the most important demand on intelligence stems from the complexity of the **social environment**, in which males manipulate others to gain power, and hence reproductive success.

### Reproduction

The skills needed for safe movement in trees take a long time to learn, and most infant primates have to be carried by the mother for some time after birth. This has influenced reproduction in three ways:

- *single young* – a female can comfortably carry only one young at a time, and twins are rare in most primates. There is a single uterus, formed by fusion of the two uteri characteristic of most other mammals. In nearly all primates there are only two teats, and these are pectoral (on the chest) allowing the mother to hold the infant while it suckles.
- *long gestation* – since the mother needs her hands for climbing, the young must be born at a sufficiently advanced stage of development to be able to cling to her fur.
- *long juvenile dependence* – with the decrease in the number of young, the mother invests more care in each one. There is an increase in the time taken to reach maturity – three years in lemurs, seven in most monkeys, 12 in chimpanzees and about 20 in humans.

## Primate variety

Living primates fall into six fairly distinct groups (Figure 40.4).

1 Lemurs of Madagascar
2 Bushbabies of Africa and lorises of Africa and Asia
3 Tarsiers of South-East Asia
4 New World monkeys of Central and South America
5 Old World monkeys of Africa and Asia
6 Apes (including humans) of Africa and Asia.

**Figure 40.4**
Primate variety

**Figure 40.5**
Nostrils of New World and Old World monkeys

The first two groups are more primitive and make up the suborder **Prosimii**. The monkeys and apes are the more familiar primates and are placed in the suborder **Anthropoidea**. Tarsiers have some of the features of both groups.

## Lemurs and other prosimians

The prosimians are more primitive than the anthropoids in that they retain some mammalian features lost by the anthropoids, and that certain primate characteristics have evolved less far.

- Most prosimians are either nocturnal or show evidence of nocturnal ancestry.
- The sense of smell is more important than in anthropoids, and in most species the snout is relatively long and dog-like.
- The eyes are less forwardly-directed than in monkeys and apes, so the angle of overlap of the two visual fields (the binocular field) is narrower. The photoreceptors are mainly rods and there is no fovea, so detailed examination of objects would be impossible even if there were complete binocular vision. The brain is smaller compared with that of anthropoids.

## Anthropoids

The Anthropoidea are divided into three groups:

- the New World monkeys of Central and South America, for example spider monkeys and howler monkeys;
- the Old World monkeys of Africa and Asia, for example baboons, macaques and langurs;
- the apes, which include gorillas and humans.

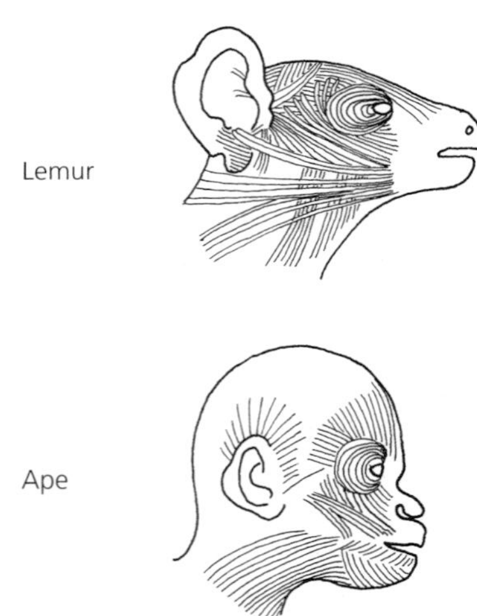


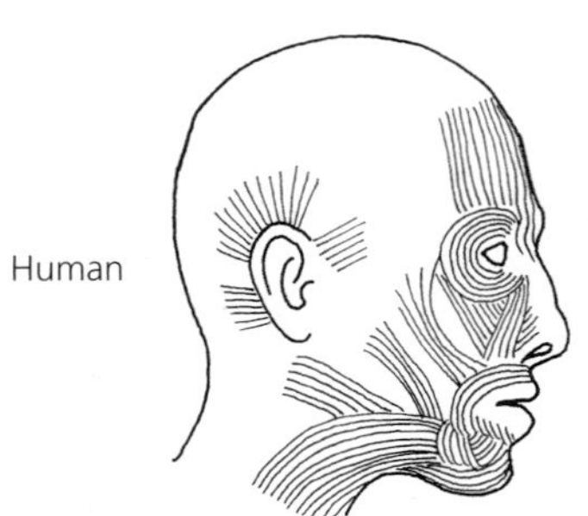


**Figure 40.6**
Facial muscles of lemur, ape and human

### *New World monkeys*

The monkeys of Central and South America differ from the Old World monkeys in a number of ways. The nostrils are wide apart and open sideways, in contrast to those of the Old World monkeys which are close together and open downwards (Figure 40.5).

Although the big toe is fully opposable, the first digit of the hand is scarcely so, and as a result objects are grasped between the second and third fingers or between the fingers and palm. Some New World monkeys have evolved a feature of their own, a **prehensile tail** which is used as a 'fifth limb'. The facial musculature is simple, so there is less communication by expression (Figure 40.6).

### *Old World monkeys*

Old World monkeys differ in a number of ways from those of Central and South America. The thumb shows a considerable degree of opposability. The tail is never prehensile. The skin on the 'bottom' is thickened with fatty, fibrous tissue to form **ischial callosities**. In some species these are surrounded by naked, brightly-coloured skin which becomes enlarged in females just before ovulation. Whereas all the New World monkeys are completely arboreal, many Old World species have become partly terrestrial, for example baboons. Brain and social behaviour are more elaborate than in most New World monkeys.

**Figure 40.7**
Shoulder mobility in monkeys and apes

### Apes

Apes differ from the Old World monkeys chiefly in their locomotory adaptations. Instead of running along small branches as monkeys do, they swing between branches using their arms. This is called **brachiation**, and it enables a large animal to reach fruit at the ends of small branches (except for the gorilla, which is a ground dweller). Brachiation is linked to a number of skeletomuscular adaptations:

- an external tail is absent (the tail vertebrae are fused to form the coccyx).
- the arms are very powerful and are longer than the legs.
- the chest is flatter and wider than in monkeys.
- the hand can be rotated at the wrist (as in a screwdriver action) through 180°. This is about twice as much as in a monkey and enables branches to be grasped from any direction.
- the shoulder joint is even more mobile than in monkeys, but as a result is more easily dislocated (Figure 40.7).

It is worth noting that humans have a number of features associated with brachiation, such as the broad chest and highly mobile shoulder, suggesting that our ancestors too were brachiators.

Besides their locomotory adaptations, apes are considerably more intelligent than monkeys, and have relatively larger brains. They are also much slower breeders than monkeys, with several years between births (five years in chimpanzees).

## Primate social life

In many non-primate species, individuals spend considerable time with each other, living as a group. In monkeys and apes, sociality is almost universal, is often highly complex, and is as characteristic of primates as a grasping hand.

Social life offers a number of significant advantages.

- Predation may be reduced, in three different ways. First, a predator may be deterred from attacking – even a leopard will seldom attack a group of baboons containing several males. Second, with more eyes, a group may spot a predator while it is further away. Increased vigilance also allows each individual to spend more time feeding rather than looking out for danger. Third, predators find it harder to focus on an individual when it is in a group.
- When food is scattered but locally abundant, it is more easily found by a group.
- It removes the necessity for males and females to find each other.
- In primates, with their long period of parental care and high capacity for learning and communication, group life enables individuals to profit from the experience of others. A primate group therefore represents a storehouse of information about the past experiences of many individuals.

Sociality is not without drawbacks. A group is more easily spotted by a predator, and disease can spread more rapidly from one individual to another when they are close together. Although there may be more eyes to find food, there may be more competition when it is located.

Another potential drawback is **inbreeding**, or mating between relatives. Inbreeding has the disadvantage that the probability of relatives sharing the same harmful recessive alleles is greater than would be expected for unrelated individuals, thus making it more likely that the offspring would be homozygous for such alleles. Inbreeding is avoided by the emigration of young adults to other groups. In most Old World monkeys it is usually the males that emigrate, in chimpanzees it is mainly the females, and in gorillas a proportion of both sexes emigrates.

## Communication

Many of the advantages of group life depend on the ability of individuals to communicate with other members of the group. This can have a variety of advantages:

- alarm signals, indicating presence of a predator.
- dominance and submission signals reduce conflict.
- advertisement of fertility in females.

Communication can be achieved by scent, by sound or by visual signals.

### *Vocalisation*

Primates use vocal signals to 'broadcast' simple messages such as 'danger', or 'keep out of my territory' to any other individuals of the same species that are within range. In recent years it has become evident that some signals convey more information than had previously been realised. Vervet monkeys for instance give slightly different alarm calls in response to the presence of leopards, eagles, snakes and baboons (Figure 40.8). That these calls have different meanings was shown by playing recordings of alarm calls that had been given in response to each kind of predator. An 'eagle' alarm caused the monkeys to look up or run into the bushes, a 'snake' alarm caused them to look down into the grass, whilst a 'leopard' alarm provoked a stampede into the trees.

**Figure 40.8**
Vervet monkeys have different alarm signals for different kinds of predator

### *Visual signals*

As predominantly 'visual' animals, monkeys and apes also employ a wide variety of gestures and facial expressions to convey information to other members of the group. This is especially true of apes, whose facial muscles are highly developed and can produce a wide repertoire of expressions (Figure 40.9). Whereas scent and calls are usually directed indiscriminately, facial expressions convey meaning to particular individuals.

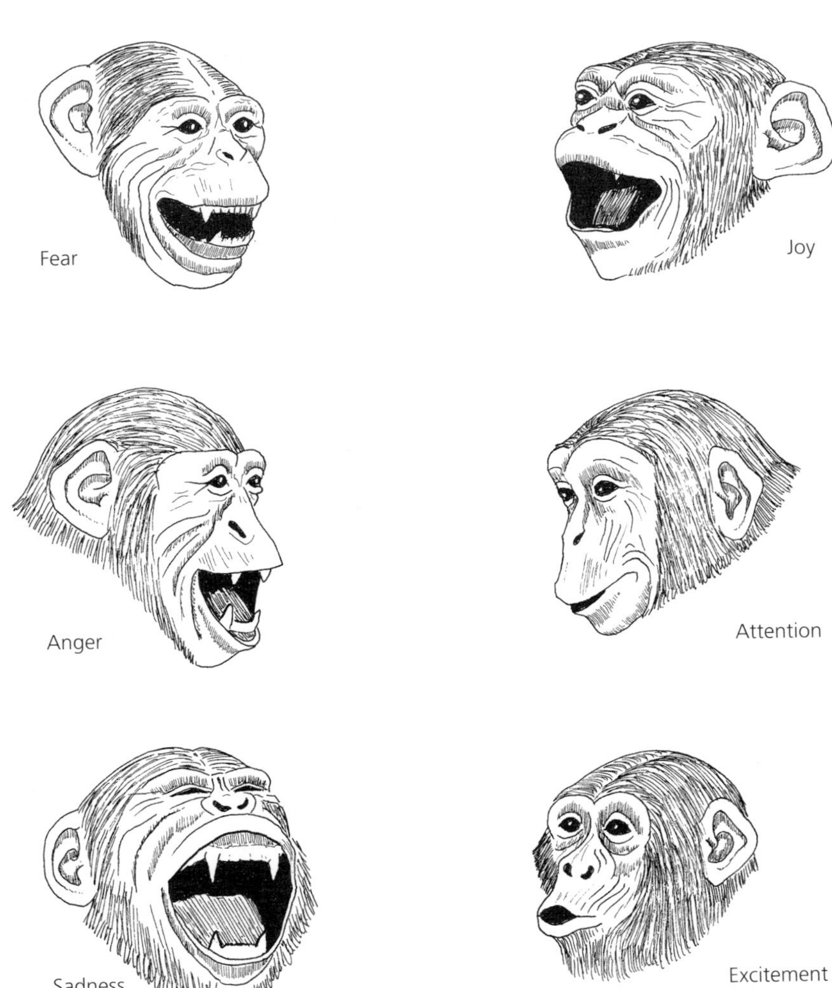


**Figure 40.9▶**
Chimpanzees can express a wide range of emotions by facial expressions

## Social dominance

It is easy to imagine that individuals in a group are co-operating for the benefit of each other. In reality, social life in primates is a constant and ever-shifting balance between *co-operation* and *competition*. Individuals gain mutual benefit in protection from predators, and often join forces in foraging for food. When food is found however, there may be insufficient to satisfy demand, leading to potential competition. Similarly males may compete for access to females.

In many animals that live in reasonably stable groups, overt conflict is reduced by a social hierarchy. Each animal knows its place, sending dominance signals to its subordinates and indicating submission to its superiors (Figure 40.10). Signals differ from species to species. For instance in rhesus macaques teeth baring conveys submission, but in chimpanzees it indicates fear.

**Figure 40.10▲**
In baboons, a 'yawn' indicates threat

In some Old World monkeys and apes, dominance relationships may be much more complex than in the simple system outlined in Chapter 37 and foreshadow some of the complexities of human relationships. For instance, in macaques and chimpanzees two or more males will sometimes form an alliance, enabling them to maintain a position in the hierarchy which none could on its own. In chimpanzees an alpha male may form an alliance with one or more older males against the animal that threatens it most – the beta male.

### Grooming

Most monkeys and apes (and also some prosimians) spend a good deal of time removing ticks and other parasites from the skin of other members of the group. Besides providing an obvious benefit, grooming is an important gesture by which subordinates placate their superiors, since lower individuals usually groom those dominant to them rather than the other way round.

### Friendships and altruism

In baboons and chimpanzees, certain individuals spend more time together and give one another preferential treatment. Baboons will sometimes give way to a 'friend' at a kill, and chimpanzees have been observed to share food with a 'friend'.

In a very rudimentary form this represents a simple kind of the co-operative behaviour that has played such an important part in human cultural evolution (Chapter 41). It is therefore important to try to understand how it could have evolved.

At first sight it might appear that altruism (or helping others) could not evolve by natural selection, since the helper does not seem to benefit. There are actually two possible solutions to the problem.

1 If the animal that is being helped is a relative of the helper, then they share many of their genes in common. By helping a relative, therefore, an animal is actually increasing the chances of its own genes (actually, copies of them carried by the helped individual) being passed on to the next generation. Selection for co-operation between relatives is called **kin selection**, and is most likely to occur in stable groups in which relatives remain reasonably close together.

2 Less easy to explain is the evolution of co-operation between non-relatives. When an animal helps a non-relative, it can only benefit if the help is returned later. Moreover the cost of helping must, on average, be less than the benefit of being helped – in other words there must be a net return on the investment. This form of co-operation takes the form of 'you scratch my back and I'll scratch yours', and is called **reciprocal altruism**. It is really a form of concealed selfishness in which the individual makes a small sacrifice in the short term for the sake of a greater benefit later.

   There are two obvious conditions which are required for reciprocal altruism to evolve: individuals must meet frequently, and they must be able to remember helping and being helped. Both of these conditions are most likely to be met in intelligent, social animals such as Old World monkeys and apes, and it is probably reciprocal altruism that lies behind 'friendships' in baboons and chimpanzees.

### Cultural change in primates

In Old World monkeys, chimpanzees and gorillas, other members of the group are a vital source of information on how to obtain food and escape from predators. The **social** environment of primates therefore plays an essential role in the learning of survival skills.

Knowledge, like genes, is handed down from one generation to the next, though by a very different mechanism. Like genes in the gene pool, the knowledge held by a population can change over time. Moreover, as with genetic information, cultural information can change in different directions in different populations. This process of **cultural evolution** is not unique to humans, as the following examples show.

- One of the best known examples (and a case of cultural evolution that was observed as it happened) was in a group of macaques on Koshima Island off the southern coast of Japan. The monkeys had been fed by scientists at regular intervals with sweet potatoes. In 1953 an 18 month old female called Imo started to wash her potatoes in a nearby stream. During the next ten years the practice of potato-washing spread to other members of the troop, though more slowly to the older ones, some of whom never learned it.
- In different parts of Africa, populations of chimpanzees use slightly different techniques for 'termiting' (Figure 40.11). For example those at Okorobiko in Central Africa strip the bark from the twig before using it, whilst those at Gombe National Park in Tanzania do not. Since chimpanzees learn 'termiting' by watching their mothers, it seems probable that these differences in behaviour have arisen as a result of small changes in behaviour being handed on by copying.

**Figure 40.11**
A chimpanzee 'termiting'

- Chimpanzees at Mahale on the shores of Lake Tanganyika open hard fruits by biting them, whereas those at Gombe bang them against rocks or trees.
- Gombe chimpanzees sponge water up from holes in trees using a crushed leaf, whilst those at Mahale dip their hands into the water and then lick them.

These examples show that monkeys and apes can learn a great deal by imitation and that, once established, a particular habit can be handed down from generation to generation by non-genetic means.

### *Chimpanzee behaviour*

Since the 1960s a number of intensive studies of wild chimpanzees have been made, most notably by Jane Goodall at Gombe in Tanzania. As a result of these and other studies it has become clear that much of what had hitherto been considered to be exclusively human behaviour is also shown by chimpanzees. Some examples are described on page 762.

- *Toolmaking.* Whilst a number of animals use tools, chimpanzees are the only primates that are known to *make* tools by modifying an object to achieve a specific end. Besides using twigs for 'termiting', tools are also used in other ways. For example, leaves are made into sponges by crushing them in the mouth and are then used to draw water from holes in trees.
- *Warfare.* Whilst relationships within a community are peaceful enough apart from the occasional squabble, neighbouring communities may live in a state of mutual and sustained hostility. Parties of males patrol the border of their territory and on occasions raid the territory of a neighbouring community, capturing and killing any they can outnumber. Over a four-year period, Jane Goodall witnessed the wiping out of an entire chimpanzee community in this way.
- *Politics.* Chimpanzees spend a considerable proportion of their time in the pursuit of power and status. As with humans, these aims are achieved by complicated social interactions. Superimposed on the usual dominance hierarchy based on physical strength is a complex web of alliances. As a result, the real power and influence are not always held by the strongest individuals. In obtaining and holding onto power by social manipulation, chimpanzees can thus be said to show *political* behaviour.
- *Self-awareness.* Experiments and observations have shown that humans are not (as previously thought) the only animals to be conscious of their own existence. This was shown in an experiment in which a chimpanzee was first allowed to get used to seeing its reflection in a mirror. The animal was then anaesthetised and a small patch of paint put on its forehead. When it recovered consciousness it was given a mirror, and it tried to pick off the paint. The animal evidently knew that what it was seeing was a reflection of itself, rather than another chimpanzee. Similar experiments with other primates, including gorillas, suggest that the orangutan is the only other primate to have self-awareness.
- *Deception.* Monkeys and apes have a wide repertoire of facial expressions by which they can convey their intentions and mental states. Though this can have its advantages, in the complex world of chimpanzee society it may be advantageous to withhold information from others, or even to actively send false information. This has been clearly shown by Frans de Waal as a result of a long-term study of chimpanzees at Arnhem Zoo in Holland. In one instance he observed a low-ranking male who was showing interest in an oestrous female, concealing his erection from a nearby alpha male. In some cases a low-ranking male would catch the eye of an oestrous female, and then look towards some bushes before quietly disappearing behind them. When the female followed, they copulated out of sight of a higher ranking male. De Waal also describes a case in which human experimenters hid some grapefruit under the ground within sight of a chimpanzee called Dandy, who apparently had not noticed. Hours later when the other chimpanzees were asleep, Dandy returned to the exact spot, dug up the grapefruit and ate them.

These, and many other cases of probable deception, are evidence that chimpanzees can assess the mental states of others, and hence the social consequences of their actions. This enables them to use other individuals as social 'tools'.

Toolmaking, warfare, social manipulation, self-awareness and language (see the information box below) are all attributes that had previously been held to be uniquely human. One by one these supposed barriers between humans and the rest of the animal kingdom have become blurred. It is not unreasonable to suppose that the earliest protohumans were as capable as modern chimpanzees as toolmakers and co-operative hunters, so the question of how these more advanced forms of behaviour began is no longer limited to the field of human evolution. In rudimentary form they probably predated the split between the evolutionary lines leading to humans and other apes.

## Apes and language

Humans are the only animals that can communicate complex information and abstract ideas by sound. Given the ability of chimpanzees to learn quite complex tasks (such as piling boxes on top of each other to reach a banana), it might seem surprising that all attempts to teach four-legged apes to speak have failed, apart from the repetition of a few very simple sounds resembling 'mama' and 'cup'. Is it because four-legged apes lack the brain power to organise sounds into meaningful patterns, or do they simply lack the necessary anatomical features of the air passages?

The answers have emerged from a series of experiments in which apes have been trained to communicate with humans using non-vocal means. In the first experiments, a chimpanzee called Washoe was taught American Sign Language for the deaf (ASL). Her teachers taught her each sign by moving her hands while showing her the object that the sign represented. By rewarding her when she made the correct sign, Washoe was eventually taught to use more than 150 'words' correctly.

Though critics said that Washoe had merely learned that she would be rewarded if she made a certain gesture in a particular situation, she could string signs together into novel sequences. Thus, when shown a swan, she made the signs for 'water' and 'bird'. Washoe had thus used two signs in a way that was more than the sum of its parts, indicating something new.

Other experiments have been carried out using media other than ASL. Some of the most recent involved a chimpanzee called Kanzi. Instead of gestures, Kanzi was shown designs on a board. Each symbol represented a noun, adjective or verb. The teacher touched the symbols in the appropriate sequence. At the same time the teacher spoke the sentence, which usually took the form of a request or a command. In this way, Kanzi learned to use 'words' in a way more like that of a human child, by hearing.

By the time he was ten years old, Kanzi could understand about 200 'words'. Much more impressive than the size of his vocabulary, however, was his ability to comprehend them when strung together in novel sequences. For example, when asked to 'take the vacuum cleaner outdoors', or to 'put the pine needles in the refrigerator', Kanzi had no trouble in carrying out the request.

# A most unusual ape

Though the fossil record is far from complete, we can get some idea of the changes that have occurred by comparing humans with our closest living relatives, the chimpanzees and gorillas.

Of course, humans did not evolve from four-legged apes, at least not from modern ones. It would be more correct to say that humans and four-footed apes evolved from a common ape ancestor. How then is it possible to learn something about our past by comparison with other living apes?

The fact that chimpanzees and gorillas are far more similar to each other anatomically than either is to a human, suggests that they resemble the common ancestor much more closely than humans do. By comparing humans with modern apes we can gain *some* idea of the changes that have occurred in our own ancestry.

## Bipedalism

The most significant development during human evolution was the adoption of two-legged standing and walking. By freeing the hands it provided the precondition for activities demanding mental ability (such as the manufacture of stone tools) millions of years later.

A human, when standing, uses only about seven per cent more energy than when lying down, because the weight of the body is carried passively by bone rather than by active muscle contraction. This energy saving is achieved by concentrating the weight of the body as nearly as possible into a vertical column, so that all parts lie more or less above the point of support (Figure 40.12).

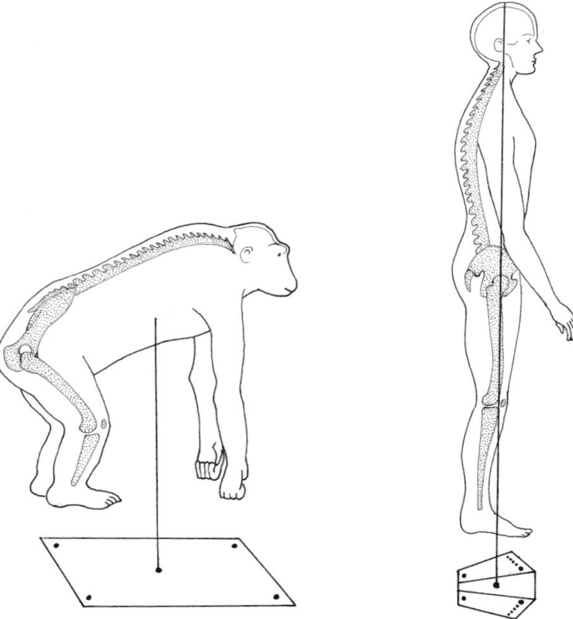

**Figure 40.12**
Area of support and centre of gravity in a chimpanzee and human (after Zihlman and Brunker)

Lacking these adaptations, a non-human ape has to use a lot of muscular energy to prevent itself from falling over, and can thus stand and walk on two legs for short periods only. In a standing human, this principle can be seen in the anatomy of the skull, chest and vertebral column, and – during walking – in the knee joint.

- *Skull.* In a human the foramen magnum (the hole through which the spinal cord leaves the skull) lies nearly in the middle of the skull, so the skull is nicely balanced on the first cervical vertebra (Figure 40.13). Consequently little muscular energy is needed to hold it upright. In the gorilla and chimpanzee the foramen magnum lies further back, causing the head to tend to sag forward. This is counteracted by the large neck muscles that run from the rear of the skull to the neural spines of the cervical vertebrae (Figure 40.14).
- *Vertebral column.* The human vertebral column is 'S'-shaped, enabling the chest to sit almost directly above the joint between the spine and the pelvis (Figure 40.15).

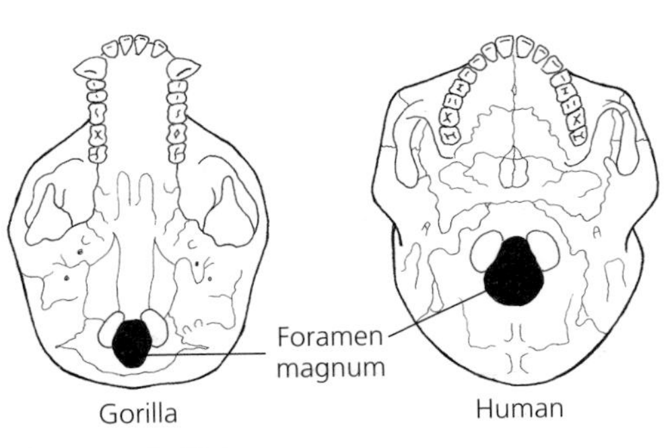


**Figure 40.13▲**
Position of foramen magnum in human and female gorilla

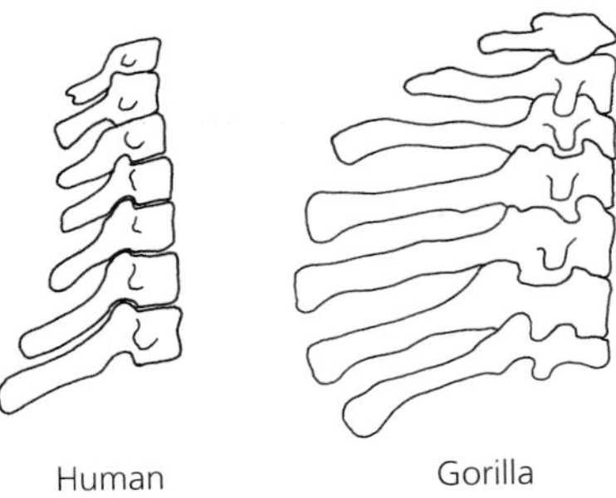


**Figure 40.14▲**
Neural spines of cervical vertebrae of human and gorilla

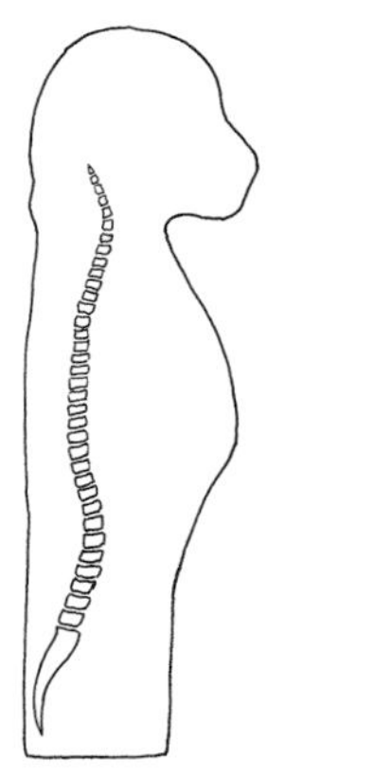

**Figure 40.15▲**
The human vertebral column compared with that of four-footed apes (shown in upright position) (after Schultz)

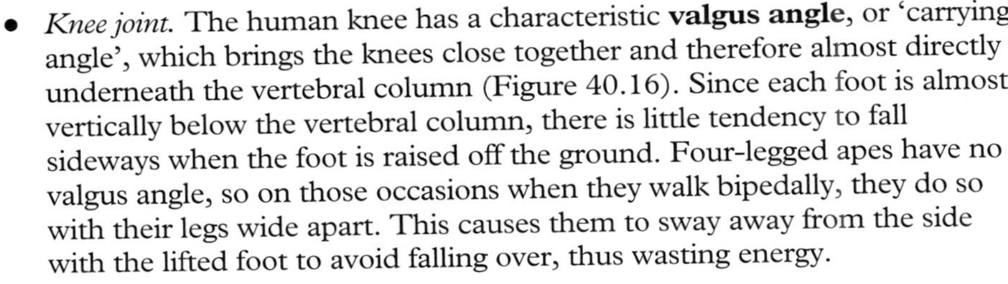

- *Knee joint.* The human knee has a characteristic **valgus angle**, or 'carrying angle', which brings the knees close together and therefore almost directly underneath the vertebral column (Figure 40.16). Since each foot is almost vertically below the vertebral column, there is little tendency to fall sideways when the foot is raised off the ground. Four-legged apes have no valgus angle, so on those occasions when they walk bipedally, they do so with their legs wide apart. This causes them to sway away from the side with the lifted foot to avoid falling over, thus wasting energy.
- *Foot.* In the human foot the big toe points forward, providing the final thrust against the ground just before the foot is lifted. Since it is permanently directed forward, all opposability is lost (Figure 40.17A). The foot is also arched and braced underneath by ligaments, providing a much firmer platform than in four-legged apes (Figure 40.17B).

**Figure 40.16▼**
Knee of human and gorilla compared

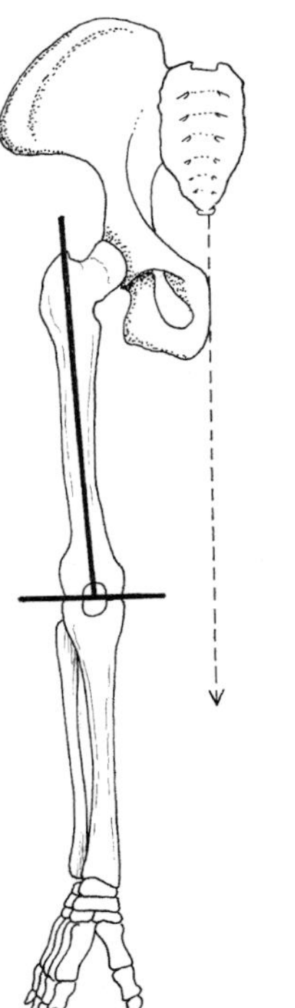


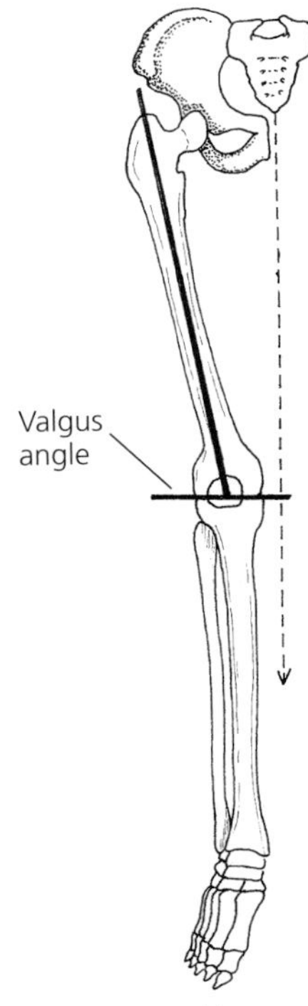

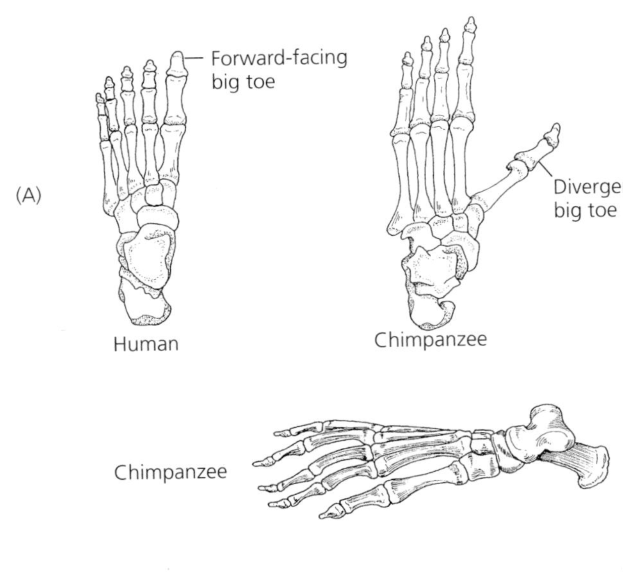

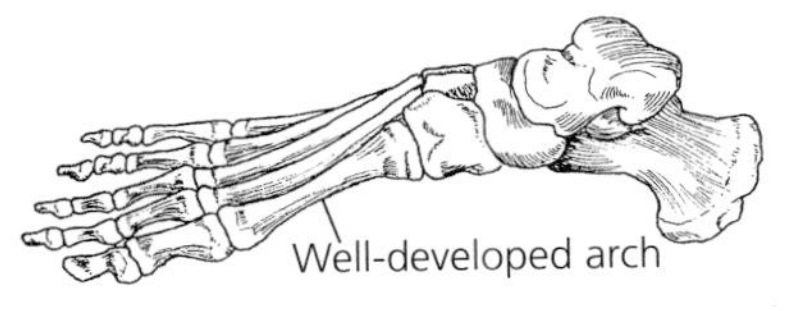


**Figure 40.17►**
Human foot compared with that of a chimpanzee

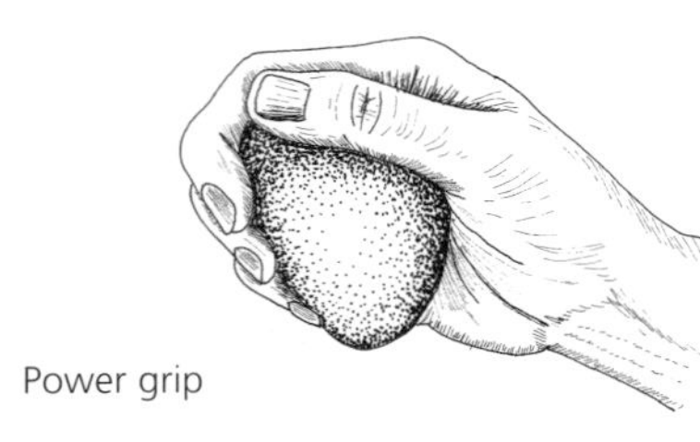


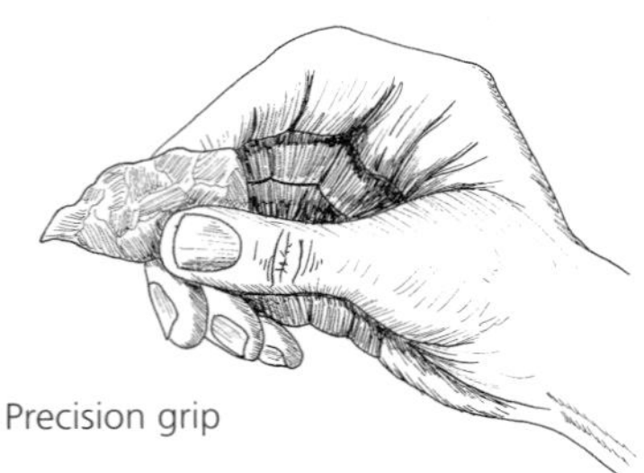


**Figure 40.18▲**
Power grip and precision grip

### *Manipulating objects*

Though monkeys and non-human apes can hold objects in a 'power grip', only humans can achieve the full thumb tip to finger tip 'precision grip' (Figure 40.18), as when holding a pencil or performing other fine movements.

There are two reasons for this: the thumb is longer than in other apes, and it can be more fully rotated at its base (Figure 40.19).

### *Teeth and jaws*

The great apes have much larger teeth and jaws than humans. The action of their powerful jaw muscles sets up vertical stresses in the skull, which are resisted by bony thickenings above the eye sockets called **brow ridges**. In humans, tools enable food to be broken into small pieces before it enters the month, reducing the amount of energy used in chewing. This has been associated with a great reduction in the size of the teeth and jaws in humans, and the loss of brow ridges in modern humans (Figure 40.20).

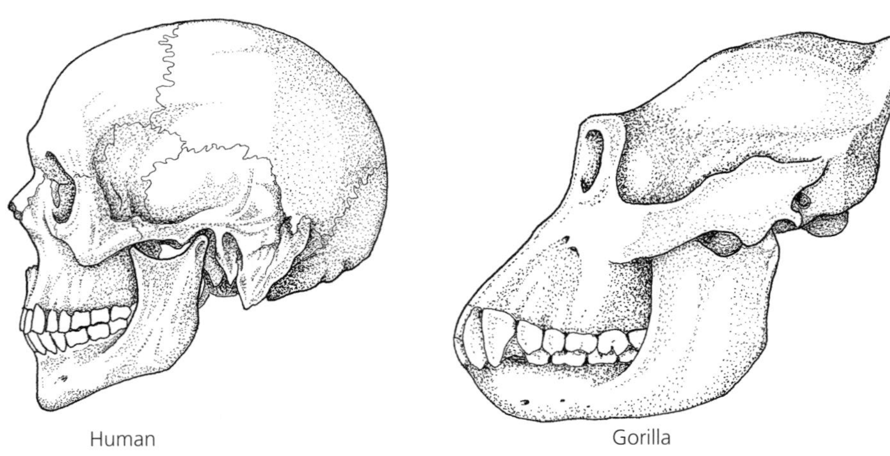


**Figure 40.20►**
Skull of a human and a male gorilla

The use of tools as weapons has also reduced the need for large canines as weapons for defence of the group, and in humans the canines are small in both males and females.

### *Skin*

Desmond Morris called humans 'naked apes', referring to the fact that we have much less hair than other apes. Actually humans have as many hairs per $cm^2$ of skin as chimpanzees – they are just thinner and shorter. Another peculiarity of the human skin is the number of sweat glands, which are much denser than in any other mammal. These two features are probably related, since sweat can evaporate more easily when air can circulate freely.

Why does the human skin have an unusually high cooling capacity? One possible explanation is that it would have enabled our ancestors to forage actively under the African sun when most large carnivores were resting.

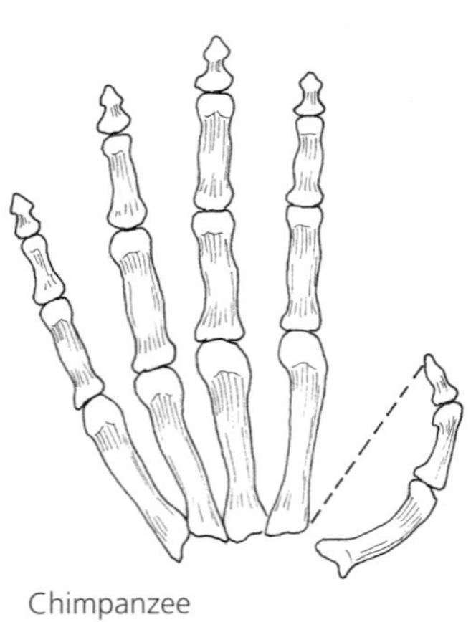


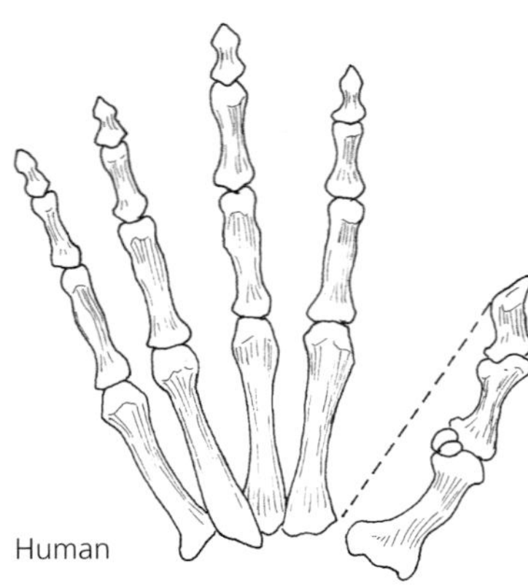


**Figure 40.19▲**
Hand of a human and a chimpanzee showing differences in the length of the thumb

### *Brain*

By far the largest region of the human brain is the **cerebrum**. The human cerebrum is not only relatively and absolutely larger than in other primates, but the outer part, the **cortex** (where nerve cell bodies are concentrated) is also more folded (Figure 40.21). Folding allows the area of the cortex to increase in proportion to the increase in volume (without

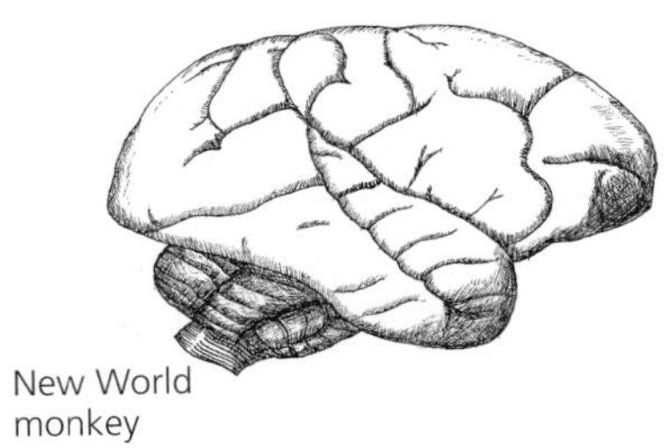


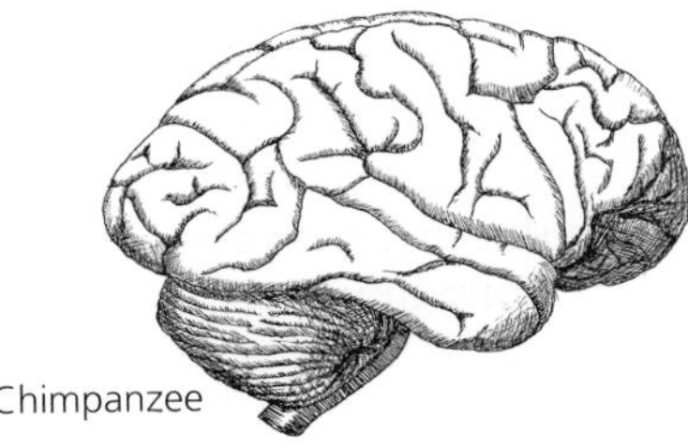


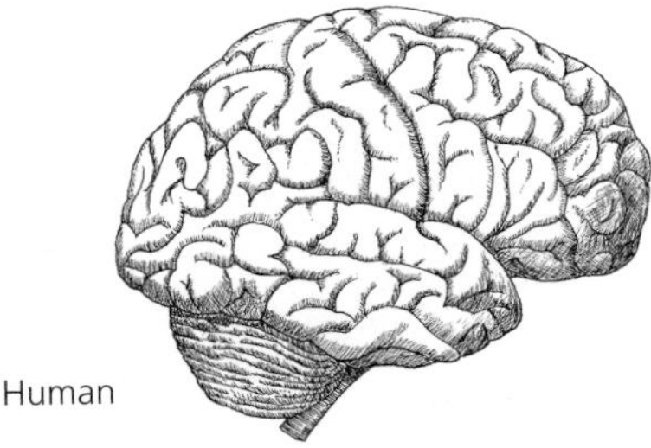


**Figure 40.21▲**
Brains of human, chimpanzee and New World monkey

folding, an eight-fold increase in brain volume would result in only a four-fold increase in cortical area).

## Speech

Although chimpanzees can be taught to communicate simple ideas by sign language, they cannot speak. The reason lies in two features unique to humans: the anatomy of the vocal tract (the route the sound takes from the vocal cords to the lips) and the organisation of the brain.

The lower part of the vocal tract consists of the **pharynx**, which in modern adult humans is much longer than in other mammals (Figure 40.22). This, coupled with the ability of the muscles of the pharynx to vary its diameter, enables modern humans to generate a much greater variety of vowel sounds.

### The brain and language

Speech consists of a succession of **phonetic segments**, each corresponding roughly to a letter of the alphabet. In normal speech phonetic segments are emitted at a rate of between 15 and 25 per second, allowing an average sentence to be transmitted in little over two seconds.

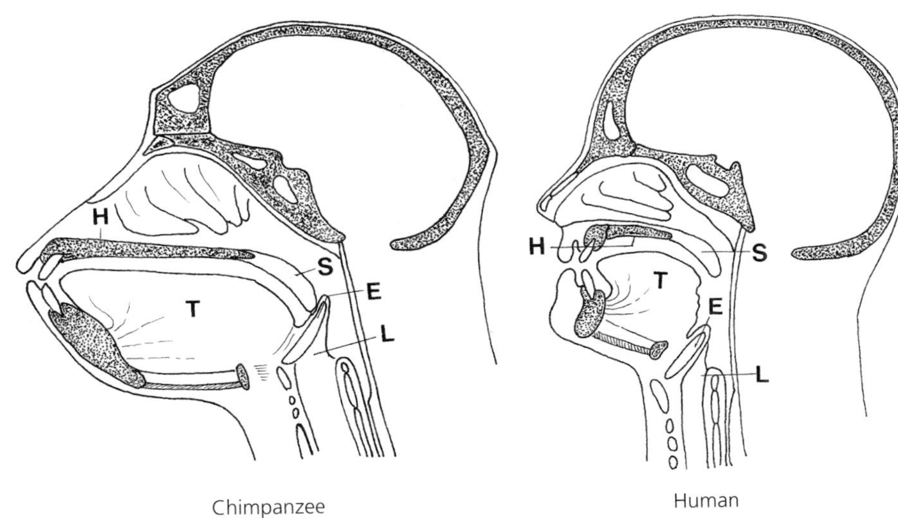


**Figure 40.22►**
Pharynx of human and chimpanzee. E = epiglottis, H = hard palate, L = larynx, S = soft palate and T = tongue

We take for granted that the brain can decode information at such a rapid rate, yet experiments have shown that humans cannot identify non-speech sounds – such as the dots and dashes of Morse code – faster than seven to nine items per second. At this rate it would take several seconds to transmit an average sentence, and by the time the entire sentence had been received, the receiver would have forgotten the beginning. Thus the limits of short-term memory put a premium on the rapid transmission of information, which only speech makes possible.

Another limitation of non-speech communication is that the receiver has to concentrate so hard on decoding the information that he or she cannot absorb the contents of the message. A Morse code expert can receive about fifty words a minute, but at this rate is quite unable to comprehend – let alone remember – the message.

The decoding and comprehension of speech clearly require special brain functions. Two regions of the cerebral cortex are concerned specifically with language, both in the left side (Figure 40.23). **Broca's area** is concerned with the output or motor aspect of communication – the stringing together of sounds into words. Injury to this region causes slow and laboured speech. **Wernicke's area** is concerned with the analysis of incoming sounds, and therefore with understanding.

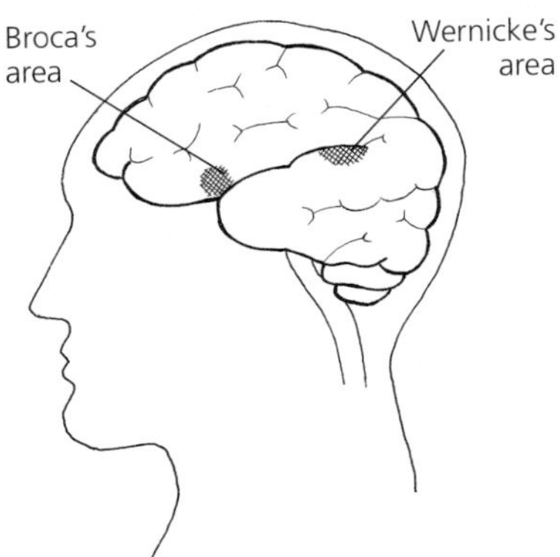


**Figure 40.23▲**
Speech areas of the human brain

### Controlling primitive urges

One part of the brain that has decreased in importance during human evolution is the **limbic system** (Figure 40.24). This region is concerned with the more automatic responses and emotions like pain and fear that are associated with 'basic' functions such as survival and reproduction. When a monkey searches for food, attempts to copulate, runs from a predator, or screams with pain, it is responding to the dictates of its limbic system.

The limbic system forms a major part of the brain in most mammals, but in the higher primates – and most especially in humans – it is much less prominent. Humans are the most socially interdependent of all primates; impulsive urges often have to be controlled in subordination to higher, longer-term priorities. We are not born with self-control – a young child is at the mercy of its limbic system, but gradually the 'higher' centres in the brain take control.

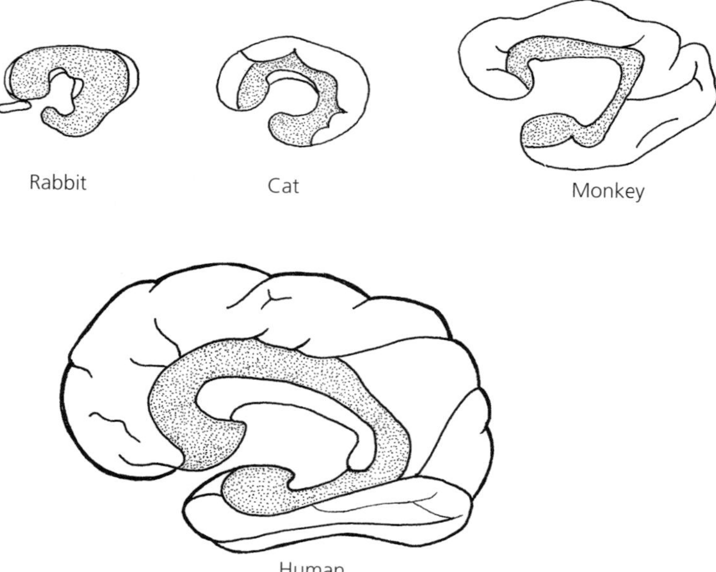


**Figure 40.24**
Relative sizes of the limbic system (the shaded area) in various mammals

## Reproduction

In most mammals ovulation is marked by signals from the female which advertise the fact that she is available for mating and which stimulate sexual arousal in males. During this period of 'heat', or **oestrus**, the female often shows interest in the males and may present her rear for copulation. Humans are also unusual in that there is no oestrus: a female does not show greater interest in mating during the fertile period and males do not know when she is fertile.

From the short-term point of view it might therefore appear that most human sexual activity is wasted. What really matters however is not the rate at which babies can be produced, but the rate at which they can be *reared to maturity*. This requires a long period of parental care and therefore domestic harmony. If males were frantically competing for a female in oestrus, the children would probably suffer.

Another possible advantage of concealed ovulation is that when sexual activity is spread throughout the female's sexual cycle, males are continually interested in the female, thus helping to maintain the stable pair-bonds needed for a long period of child-rearing. A result of the more stable human sexual relationship is that, unlike four-legged apes and Old World monkeys, human fathers form relationships with their offspring and invest more energy in rearing them.

### *Growth and development*

Humans grow and develop differently from other primates in three important ways. Babies are helpless at birth and there is a long period of juvenile dependence, at the end of which there is an adolescent spurt of growth.

#### *Helpless new-born*

Of the total period of human development, only the first 1/27 is spent in the uterus, compared with about 1/16 in other apes. If a human were to spend the same proportion of its total development within the mother as other apes, the gestation period would be *15 months*. In a sense, then, the first six months of life are spent as an 'extra-uterine embryo'. Humans are therefore an exception to the general primate condition of a *long* gestation period.

Because it is born so relatively 'early', the human baby emerges into the world with a brain only a quarter its final size, and is more helpless than any other primate. Three-quarters of the brain's growth occurs in the more stimulating environment of the outside world, so learning can begin sooner.

The explanation for early birth lies in the mechanical problem of getting a large brain through the mother's pelvic girdle. Brain expansion may have gone as far as it can in humans – any further increase would probably be incompatible with efficient walking.

#### *Prolonged juvenile dependence*

During the juvenile period most of the body grows slowly. The brain is an exception, reaching almost adult size when body growth is only 40 per cent complete. This long period of juvenile dependence is associated with a greatly increased capacity for learning. An infant rhesus monkey can learn to solve simple problems within three weeks of birth, but humans learn over a much longer period and eventually achieve far more than any other primate. Humans not only learn skills related directly to survival, but also the customs, beliefs and traditions relating to society – in other words, its *culture*.

#### *Adolescent growth spurt*

During human childhood the emphasis is on the growth of the brain and on learning within the protection afforded by parents. To attain reproductive maturity however, the body has to 'catch up', resulting in the adolescent growth spurt. In other apes this is absent, growth following a smooth trajectory from infancy to adulthood.

## Reclassifying humans: molecular taxonomy

Even in Darwin's time there was no doubt that our closest living relatives were the great apes. Yet there are important differences: humans are relatively hairless, have a huge brain and a number of skeletomuscular features which enable us to walk upright. On these grounds alone biologists felt justified in classifying humans and the great apes in different groups. There was however another factor at work – a deep reluctance to place ourselves too close to the rest of the animal kingdom, as if our subconscious had never fully come to terms with the implications of Darwin's theory of evolution. It is therefore hardly surprising that until

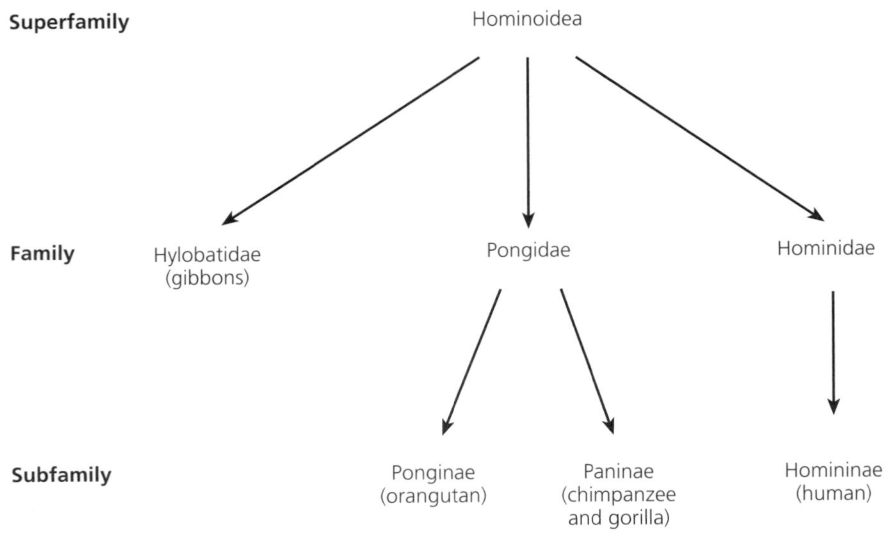

Superfamily — Hominoidea

Family — Hylobatidae (gibbons) — Pongidae — Hominidae

Subfamily — Ponginae (orangutan) — Paninae (chimpanzee and gorilla) — Homininae (human)

**Figure 40.25**
Traditional classification of the apes

recently, biologists placed humans and their great apes in separate families, the Hominidae and Pongidae, respectively (Figure 40.25).

This view has been thrown into upheaval by a relatively new branch of taxonomy. Instead of comparing anatomy (as traditionalists have done) **molecular taxonomists** compare proteins and nucleic acids. In 1967 Vincent Sarich and Allan Wilson came to the revolutionary conclusion that chimpanzees and gorillas are more closely related to humans than to orangutans. Moreover their evidence suggested that the common ancestor of humans and chimpanzees lived as recently as five million years ago – less than half the traditionally held figure (Figure 40.26).

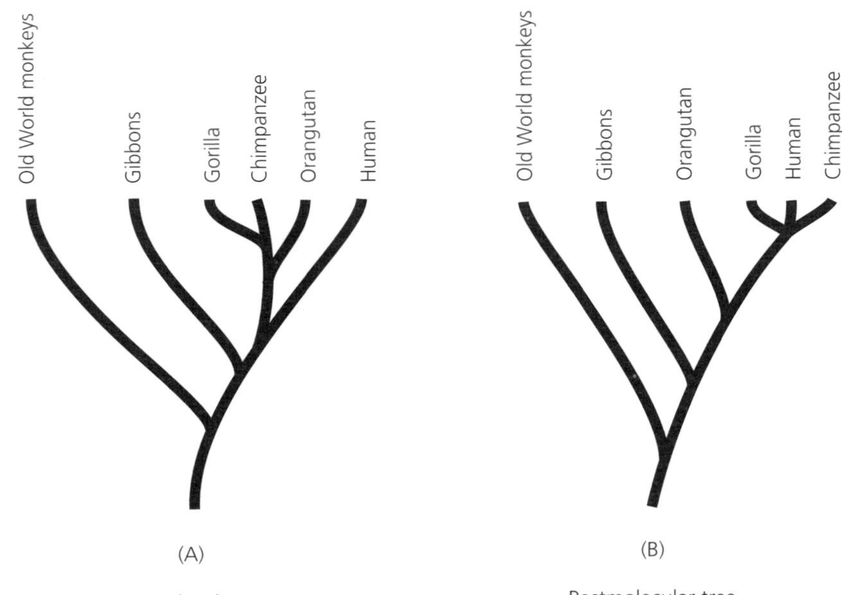


**Figure 40.26**
Pre- and postmolecular evolutionary relationships between humans and other apes

As a result of the revolution in molecular taxonomy, humans and all the great apes (chimpanzees, gorillas and orangutans) are now placed in the same family, the **Hominidae**. Humans are now placed with the other African apes (chimpanzees and gorillas) in the same **subfamily**, the **Homininae**. There has been no change in the status of the superfamily **Hominoidea**, which still includes gibbons (lesser apes) and the great apes. (Notice that superfamilies end in *-oidea*, families in *-idae* and subfamilies in *-inae*.)

An implication of this is that the term 'hominid', meaning a member of the family Hominidae, should be redefined. However, palaeoanthropologists (scientists who study human evolution) continue to use the term to denote bipedal apes on the line leading to humans, or side-branches of that line – an unfortunate ambiguity.

## Evolutionary trees from immunology

A major disadvantage of using anatomical features (as traditional taxonomists have done for much of this century) is that it is to some extent subjective – it is often a matter of opinion whether one difference is more important than another. Molecules, on the other hand, can be compared more objectively because differences can be *quantified*.

For many years the only way of comparing proteins was to use the immune system's ability to recognise 'foreign' substances. In principle, the method is simple. An animal, such as a rabbit, is first injected with human blood serum (plasma minus clotting factors). The plasma proteins act as *antigens*, causing the rabbit to make *antibodies* against them. The rabbit is now *immunised* against the human proteins. When a sample of human serum is mixed with serum of an immunised rabbit, the antibodies in the rabbit serum bind to the human proteins causing a precipitate. The degree of cloudiness can be measured and is used as a standard for comparison with precipitates produced with sera from other species (Figure 40.27).

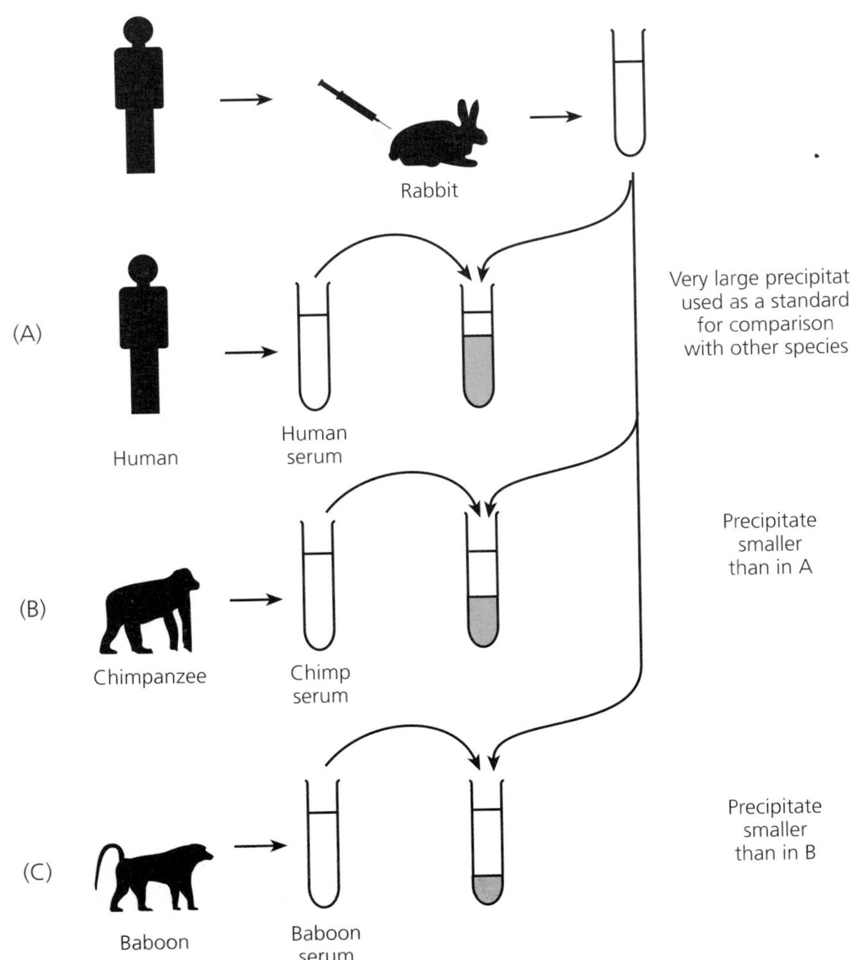


**Figure 40.27**
Using the immune system to compare blood proteins of different species

When chimpanzee serum is mixed with the serum of a rabbit immunised against human plasma, the mixture again turns cloudy, but slightly less so than with human serum. This is because chimpanzee plasma proteins are similar to, but not quite the same as, human plasma proteins, and, consequently they combine less firmly with the rabbit's antibodies (which, remember, were specific for *human* plasma proteins). Baboon serum produces only a slight cloudiness because the proteins in it differ from the human proteins even more than the chimpanzee proteins do. The cloudier the mixture, the more similar are the proteins to human ones.

The basic method of this **serological test** has been used since the turn of the century, but it has been refined greatly since then. Nowadays biologists use purified extracts of particular proteins rather than a mixture of all plasma proteins. The degree of difference between two proteins – say human and chimpanzee albumin – is expressed as the **immunological distance**. Some results of this work are shown in Table 40.1.

**Table 40.1**
Immunological distances between apes

| Species tested (source of antibody) | Animal used to immunise rabbit: Human | Chimpanzee | Gibbon |
|---|---|---|---|
| Human | – | 3.7 | 11.1 |
| Chimpanzee | 5.7 | – | 14.6 |
| Gorilla | 3.7 | 6.8 | 11.7 |
| Orangutan | 8.6 | 9.3 | 11.7 |
| Gibbon | 10.7 | 9.7 | – |
| Old World monkeys (average of six species) | 38.6 | 34.6 | 36.0 |

Notice two important points. First, reading down the middle column rabbit immunised by chimpanzee serum), serum albumin of the chimpanzee is more similar to that of the human (3.7) and gorilla (6.8) than to that of the orangutan (9.3). Second, the albumins of the Old World monkeys are roughly equidistant from those of human, chimpanzee and gibbon. This suggests that after the Old World monkeys and apes diverged, the albumin evolved at roughly the same rate in the three lines leading to human, chimpanzee and gibbons, respectively. This latter point is particularly significant because it implies that this protein evolves *at a constant rate*. A number of other proteins have been shown to behave similarly (though different proteins evolve at different rates).

### Dating the tree

If many proteins evolve at roughly constant rates, they are behaving rather like clocks. It follows that for a given protein, the immunological distance between two species is proportional to how long ago their common ancestor lived. Sarich and Wilson showed that for a number of plasma proteins, the average immunological distance between baboon and human is about six times greater than between chimpanzee and human. The common ancestor of baboon and human should therefore have lived about six times as long ago as the ancestor of chimpanzee and human.

Now it is known from independent fossil evidence that Old World monkeys and apes diverged about 25–30 million years ago. It follows that the most recent ancestor of chimpanzee and human lived about one-sixth as long ago – that is, about four to five million years ago. Although the protein clock lacks a 'dial', it can in this way be *calibrated* using an independent clock – the fossil record.

## Other methods of comparing molecules

The serological method outlined above uses the ability of the immune system to distinguish one protein from another. The basis for the differences between proteins is the amino acid sequence or primary structure of the protein (see Chapter 3).

As the primary structures of more and more proteins were worked out, it became clear that (generally speaking) organisms which on traditional grounds were believed to be closely related, had similar amino acid sequences in their corresponding proteins. The alpha chain of human haemoglobin, for instance, is identical to that of the chimpanzee, and differs from that of the gorilla by only one amino acid, and from that of the less closely-related rhesus monkey by four amino acids. The alpha chain of the much more distantly-related pig differs by 18 amino acids, and similar results have been obtained for other proteins.

The number of amino acid differences between two related proteins gives a measure of the degree of difference between them and can be used as a means of classifying them. Using computers it is possible to derive possible evolutionary relationships, and the resulting 'trees' are usually in fairly close agreement with those obtained using traditional methods.

### *Comparing DNA*

Because most amino acids are specified by more than one DNA triplet (Chapter 11), two species might have identical amino acid sequences yet still differ in their DNA base sequences. Comparisons between the DNA are thus potentially more revealing.

DNA can be compared by **hybridisation** (see figure in the information box on page 170). This technique relies on the fact that, because the hydrogen bonds holding the two strands are weak, they are easily broken by moderate heat. When a sample of DNA is heated to about 85°C, the two strands separate and the DNA is said to 'melt'. If it is then slowly cooled, some double-stranded, or **duplex**, DNA reforms.

To compare human and chimpanzee DNA, a mixture of the two is heated until the DNA melts. The mixture is then slowly cooled and some of the single strands of chimpanzee and human DNA combine to form 'hybrid' duplex DNA. How strongly the chimpanzee and human strands bind together depends on how many complementary base pairs are formed. This in turn depends on how similar the human and chimpanzee base sequences are. The more strongly the two strands are held together, the more energy is needed to separate them, and the higher the temperature at which the hybrid duplexes melt. Chimpanzee–human hybrid DNA melts at only 1°C lower than pure human DNA, and it has been shown that this corresponds to a 97.6 per cent similarity in base sequence. This is a much closer similarity than is found between the DNA of many species that are placed in the same genus – yet the traditional classification places humans and chimpanzees in different families!

## Chromosomes provide further evidence

When the karyotypes (chromosomal characteristics) of humans and great apes are compared, the pattern of banding is seen to be markedly similar – in fact about half the chromosomes are practically identical with respect to banding. Furthermore, where there are differences, it is possible to see how they have probably arisen.

The most obvious difference is in the chromosome number. Since humans are the odd one out with 46 chromosomes, at some stage our ancestors must have 'lost' one pair. Figure 40.28 shows how this probably happened. If the two chimpanzee chromosomes were to break at the points indicated by the dotted lines and then join, the result would be a chromosome with banding identical to that of the human chromosome 2.

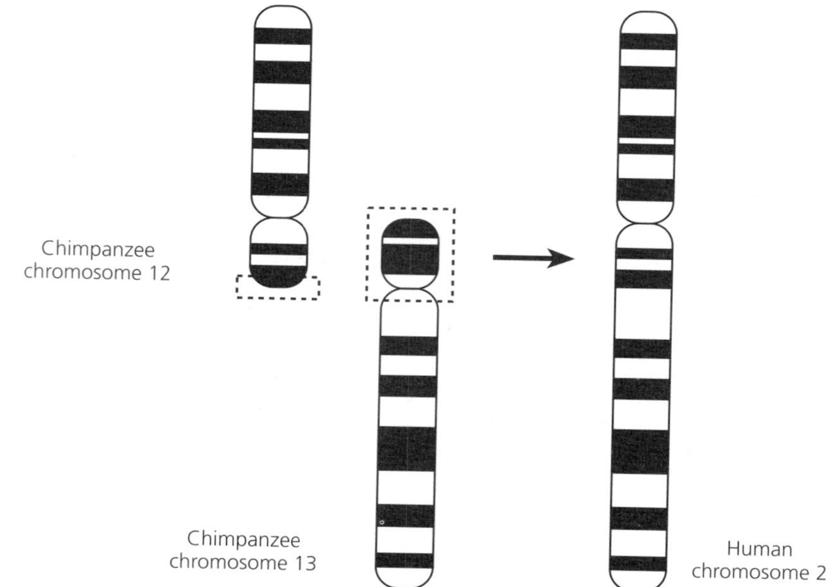


**Figure 40.28**
Fusion of two great ape chromosomes to form human chromosome 2. The enclosed areas have been lost as a result of breakage, followed by joining of the remaining sections

Detailed comparison between the karyotypes of humans and the great apes has enabled biologists to reconstruct most of the chromosomal changes that have occurred in their evolution (Figure 40.29). Most of the differences are between the African apes and humans on the one hand and the orangutan on the other. The chromosomes therefore seem to be telling us the same thing as proteins and DNA – that the orangutan line split away first, and the African apes and humans diverged later.

## QUESTIONS

1 Give *three* features by which you could recognise an animal as:
   **a)** a primate,
   **b)** a prosimian,
   **c)** an anthropoid,
   **d)** a New World monkey,
   **e)** a Old World monkey,
   **f)** a hominoid.

2 Briefly explain the functional significance of the skeletal differences between a human and a great ape with regard to each of the following: skull, lower jaw and teeth, hands, vertebral column, knee and foot.

3 In which ways are humans reproductively different from other apes?

4 What advantage is there in using proteins and DNA in classification, as opposed to anatomy?

CHAPTER

# 41 The EVOLUTION of HUMANS

## LEARNING OBJECTIVES

By the time you have completed your study of this chapter you should be able to:

- explain how fossils are formed and how they are dated by palaeontologists.
- describe factors that may have been involved in the origin of bipedalism.
- describe, with reference to *Australopithecus afarensis*, *Homo habilis*, *H. ergaster*, and *H. sapiens*, the anatomical changes that have occurred in human evolution.
- describe the increasing sophistication of the tools and other artefacts associated with *H. habilis*, *H. ergaster*, *H. neanderthalensis* and *H. sapiens*.
- explain how cultural evolution differs from biological evolution.

## Rocks, fossils and dating

To become fossilised, an organism must first be buried by fine particles. In the sea or in a lake, any corpse that sinks to the bottom may become covered by sediment. On land it is more unusual, since it depends on the animal falling into a bog, or becoming covered by silt washed down by a flood, or by volcanic ash.

Once the corpse is covered by sediment, the process of fossilisation begins. Slowly, over thousands of years, soluble salts penetrate the corpse and become precipitated within it as solid, mineral material. If the sediment is very fine and conditions are anaerobic, soft parts may persist without decay for long enough to become preserved. More often though, only the hard, skeletal parts are fossilised.

Both the bodies and the surrounding sediments are slowly converted into rock under the pressure of the sediments accumulating above. It is in these **sedimentary rocks** that fossils are formed. Because the kind of sediments and their rate of deposition can change over time, sedimentary rocks often form layers, or **strata**. Unless folding of the Earth's crust has occurred, the deeper the stratum, the older it is. Although in any one locality only a limited period of deposition may be represented, there are often sufficient overlaps in different places for a more complete sequence to be worked out (Figure 41.1)

### Geological time

Human evolution extends over several million years, but this is merely an instant compared with the history of the Earth. To illustrate the insignificance of human history compared with geological time, imagine

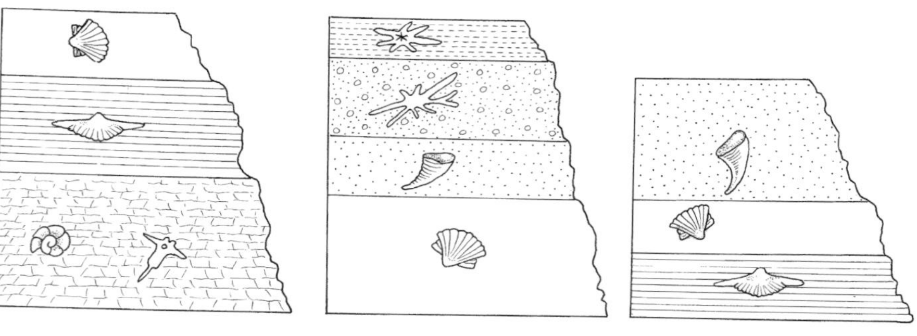

**Figure 41.1**
Overlapping fossil sequences in different localities enable extended sequences to be worked out

the history of the Earth compressed into one year. The origin of the Earth 4600 million years ago would then be at midnight on the night of December 31/January 1. The oldest known fossils (bacteria) would have been formed in early May, but it would not be until November that organisms belonging to eukaryotic groups alive today would have appeared. Primitive jawless vertebrates would make their entrance at the end of November, whilst placental mammals would not show up until December 24. The earliest known protohumans arrive on the scene at about 5.30 p.m. on December 31. Biologically modern humans (100 000 years ago) appear at 11.48 p.m., and farming begins at about a minute to midnight.

Until the middle of this century, geologists had no reliable way of telling how old a rock stratum was. One thing was very obvious though: in passing from lower (and therefore older) to higher strata, the fossil sequence showed a definite pattern. Without knowing the actual ages of the strata, geologists were able to divide up the entire sequence, or 'geological column' as it is called, into three **eras**: the **Palaeozoic** ('Age of Ancient Life'), the **Mesozoic** ('Age of Middle Life') and the **Cenozoic** ('Age of Recent Life'). Each of these eras was divided into **periods**, and the two periods of the Cenozoic were divided into six **epochs**, as shown in Table 41.1.

**Table 41.1**
The latter part of the geological time scale

| Period | Epoch | Duration (million years ago) | Important climatic and biological events |
|---|---|---|---|
| Quaternary | Holocene | 0.01–present | Moderate temperatures; explosive cultural change |
| | Pleistocene | 1.8–0.1 | Repeated glaciations; use of stone tools |
| | Pliocene | 5–1.8 | Human ancestors became bipedal |
| Tertiary | Miocene | 22.5–5 | Extensive drying in Africa; retreat of forests; Asian and African apes diverge |
| | Oligocene | 37–22.5 | Old and New World monkeys diverge |
| | Eocene | 53–37 | Early prosimians |
| | Palaeocene | 65–53 | Earliest primates |

## Dating a fossil

No matter how complete and well-preserved, a fossil is of limited value unless it can be related in time to other fossils. Ideally the palaeontologist would like to know its age in years – this is **chronometric**, or **absolute dating**. In many cases this is not possible, and palaeontologists have to be content with **relative dating**, putting fossils into order of age without knowing the actual age. Before the development of radioisotopic methods (see below), all dating was of the latter kind.

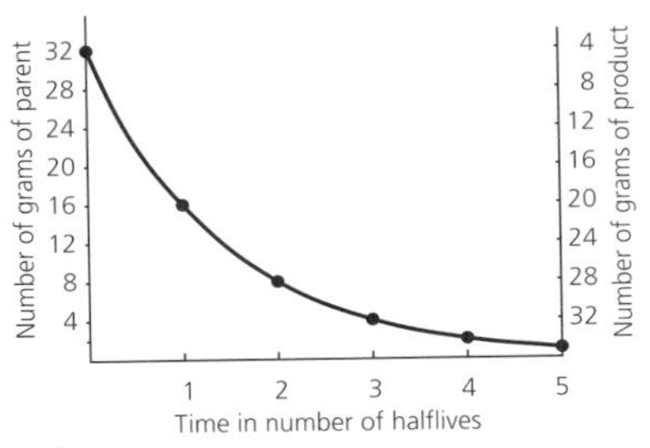


**Figure 41.2**
How the proportions of parent and daughter isotopes change in a sample of radioactive material, starting from 32 grams of parent isotope

Often it is not the object itself that is dated, but the material in which it lies buried. An obvious example is an axe that was made of stone that is far older than the axe itself. In this kind of situation the palaeontologist must be sure that the object and the material in which it is buried have not been moved relative to each other (e.g. by a flood or landslide).

### *Radioisotopic dating*

Every element exists as several chemically-identical forms, or **isotopes**, which differ in their atomic masses. Some isotopes are **radioactive**, each atom eventually 'decaying' into an atom of a different kind. Particles of radiation are given off in the process.

Radioisotopic dating is dependent on the fact that the rate of conversion of 'parent' to 'daughter' isotope is constant for a given isotope. The rate of decay is expressed as the **half-life**, the time for half the isotope to break down. If the half-life of an isotope is one year, then starting with 8 grams, there would be 4 grams after one year, 2 grams after two years, 1 gram after three years, 0.5 grams after four years, and so on (Figure 41.2). Provided that the original concentration of the parent isotope is known, then from the amount remaining (or the amount of daughter product) the number of half-lives that have elapsed can be calculated.

Since the values obtained are based on a sample, they are estimates rather than exact values. This is usually indicated by an uncertainty value, based on the standard deviation of the individual measurements. Thus, 20 000 ± 150 means that the true value probably lies somewhere between 19 850 and 20 150 years.

Several methods of isotopic dating are in common use, two of which (suitable for fossils of very different ages) are described below.

#### **Potassium–argon dating**

Rather than being used to date actual specimens, the potassium–argon method is used to date volcanic rock or ash in or below which a fossil lies. It is thus only useful for fossils in undisturbed sites. When a fossil lies below a layer of ash, the age of the ash gives a minimum age for the fossil. The method was developed in about 1960 and has proved particularly useful for dating protohuman fossils in East Africa, many of which were buried under volcanic ash.

The method depends on the fact that about 0.01 per cent of natural potassium is the isotope $^{40}K$. This decays to argon-40 which remains trapped in the rock. If the rock is heated, the argon escapes and can be collected and measured. Accumulation of argon begins as soon as the rock cools, so its 'age' is really the time since it was last hot (i.e. the time since the last volcanic eruption). The age is calculated from the ratio of $^{40}K$ to $^{40}Ar$. The half-life of $^{40}K$ is so long (about 1260 million years) that it takes about 400 000 years for measurable quantities of argon to accumulate in the rock. This method is therefore only suitable for fossils of at least this age.

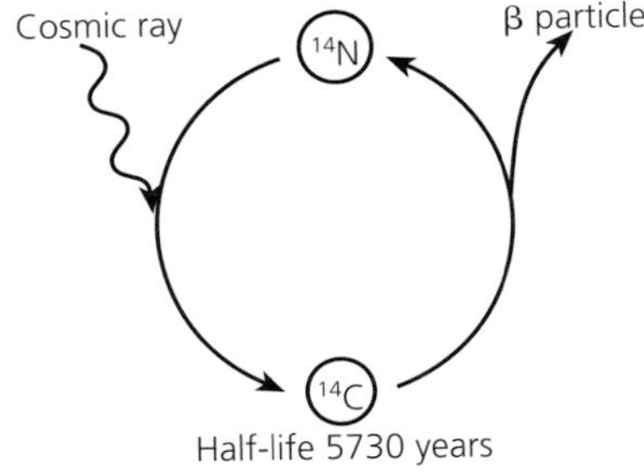


**Figure 41.3**
Formation and decay of carbon-14

#### **Carbon-14 dating**

Since carbon is a constituent of all organic matter, carbon-14 can be used to date actual specimens. The isotope is continuously being produced in the upper atmosphere by the action of cosmic rays on 'ordinary' nitrogen, or $^{14}N$. The $^{14}C$ decays back to $^{14}N$, emitting a particle of radiation in the process (Figure 41.3). The $^{14}C$ atoms produced are oxidised to $CO_2$ which is taken up by plants in photosynthesis. The $^{14}C$ passes along the food chain together with the normal $^{12}C$. The moment the organism dies it ceases to take in carbon, so its $^{14}C$ content begins to decrease. The clock thus starts 'ticking' as soon as the organism dies.

The half-life of $^{14}C$ is short – about 5730 years. After 57 300 years (ten half-lives), the amount remaining is $(\frac{1}{2})^{10}$, or $\frac{1}{1024}$ of the original – which is itself only about one part per million million. This gives such a low level of radioactivity as to be difficult to distinguish from the background radiation. The practical limit for this method is about 40 000 years, after which its reliability decreases greatly.

### *Relative dating: faunal correlation*

Fossils can often be approximately dated from the nature of the animal bones found with them. If for example it is known from independent evidence that a certain species lived during a particular period, then any remains found alongside bones of this species can be judged to be of similar age. In this way fossils can be used to date other fossils. In Africa fossil pigs, elephants and horses have played an important part in dating protohuman remains.

### *Reconstructing fossil sequences*

Even when a newly-discovered fossil has been reliably dated, it is often not possible to find its place in the evolutionary jigsaw because most branches of the evolutionary tree have become extinct. A number of protohuman fossils have turned out to be just such interesting 'side-lines' – that is, cousins of our ancestors rather than the ancestors themselves.

A second problem is that within a single population individuals vary – even within the same age group and sex. Only when scientists have an idea of the *range* of variation within a species can they form an opinion as to whether a fossil is 'new' in the sense that it came from a population not previously known.

Building up a picture of variation within a fossil species is a slow business because few fossils are complete. Most consist of a tooth, a jaw, part of a skull or some limb fragments. With primates the problem is even worse than in most other mammals because of the rarity of primate fossils.

When it comes to naming, palaeontologists face one difficulty with which other biologists do not have to contend. How much does a species have to change before it is sufficiently different from an ancestral type to be considered a new species? Since evolutionary change is continuous, the dividing lines between species are arbitrary.

Faced with these problems it is hardly surprising that many fossils have been named as new species, or even placed in a new genus by their discoverers, only to be later renamed as more fossils are discovered and knowledge increases.

## The fossil evidence

Primates are predominantly forest dwellers and in the acid soils of such habitats, bones tend to soften as their minerals are dissolved. In forests almost the only place where burial by sediments can occur is on river banks.

Fortunately many protohuman remains are found in volcanic ash and in caves where dissolution by acid does not occur. For this reason the fossil record of protohumans which lived in open country is much better known than that of the forest-dwelling apes.

### The australopithecines

The earliest extensive protohuman skeletal remains were found in Ethiopia by Don Johanson and his colleagues in the mid-1970s and have

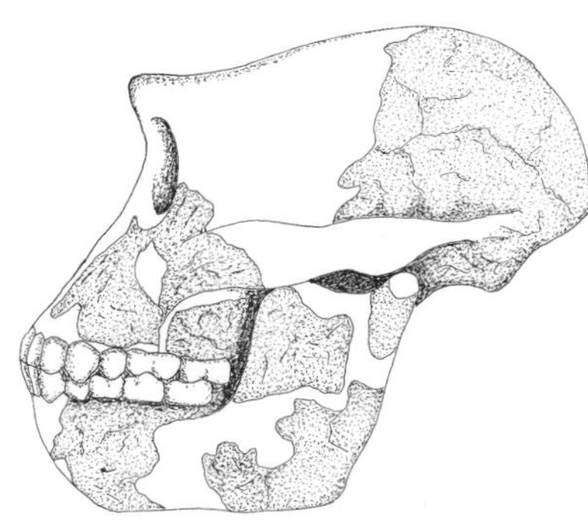

**Figure 41.4**
Skull of *Australopithecus afarensis*

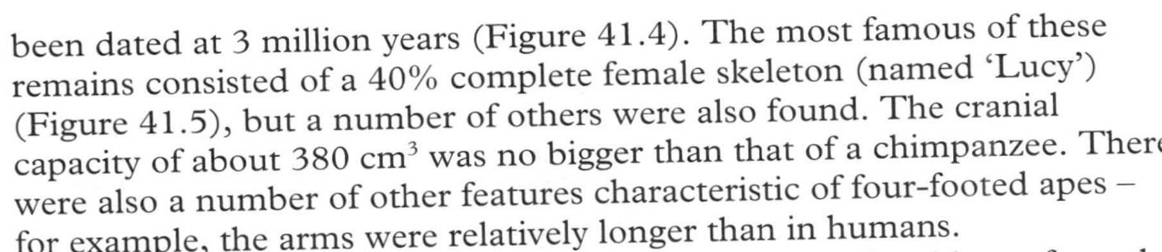

been dated at 3 million years (Figure 41.4). The most famous of these remains consisted of a 40% complete female skeleton (named 'Lucy') (Figure 41.5), but a number of others were also found. The cranial capacity of about 380 $cm^3$ was no bigger than that of a chimpanzee. There were also a number of other features characteristic of four-footed apes – for example, the arms were relatively longer than in humans.

Despite these primitive characters, there was enough evidence from the hip, knee and foot to show that these apes – named *Australopithecus afarensis* – were fully bipedal. Evidence for even earlier bipedalism was found a couple of years later in the form of fossilised footprints in volcanic ash, dated at 3.7 million years.

*Australopithecus afarensis* is the best known of a number of species of small-brained, bipedal, African apes belonging to the genus *Australopithecus* (meaning 'Southern ape'). Some of these australopithecines, such as *A. robustus* and *A. boisei*, had large jaws and molars, were almost certainly herbivorous and left no living descendants. Another species, *A. africanus*, was probably more omnivorous and may have been ancestral to modern humans.

Although there is no good evidence that the australopithecines made stone tools, it is highly probable that they made wooden tools of some kind. Modern chimpanzees make 'termiting' tools from twigs (Chapter 40) and there is no reason to believe that the australopithecines were any less sophisticated in this regard.

## The evolution of bipedalism

The change to bipedalism was probably the single most important event in human evolution because, as Darwin said, it freed the hands for purposes that would make a large brain particularly advantageous. Because it almost certainly provided the impetus for other changes, the problem of how bipedalism got started is one of *the* questions in human evolution.

In a sense, bipedalism is an extension of a tendency shown by most primates towards an upright posture. Bushbabies cling to vertical branches in between leaps, monkeys sit semiupright, apes brachiate with the body suspended vertically, and nearly all primates suckle their young sitting in an upright posture. As the only animal that habitually walks on two legs, humans have simply taken this tendency to its ultimate conclusion.

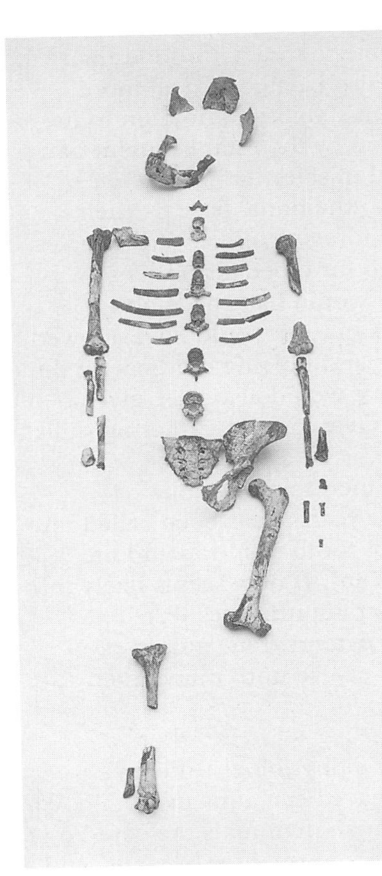

**Figure 41.5**
Photo of the remains of 'Lucy', *Australopithecus afarensis*

### *Costs and benefits of ground living*

The chief benefit of tree life is safety from predators; but on the other hand, leaves and fruits are not rich in protein or energy. Life on the ground offers possibilities of a wider and richer diet, but the risks from predation are greater. Spending an increasing proportion of time on the ground must therefore have been accompanied by some other development that reduced the danger from predation.

### *Bipedalism evolved very quickly*

The fossil record shows that protohumans were fully bipedal over 3.7 million years ago, and the molecular evidence indicates that the ape–human ancestors lived about five to six million years ago. The change from brachiation to bipedalism must therefore have occurred within a period of 2.5 million years at most. Considering the extensive changes in the skeleton and muscles of the hip and leg, this was very rapid evolution and must have been due to very strong selection pressure. A number of advantages have been suggested: the carrying of objects such as infants, food and tools, and the ability to keep cool. Of course these advantages are not mutually exclusive – they may all have played a part.

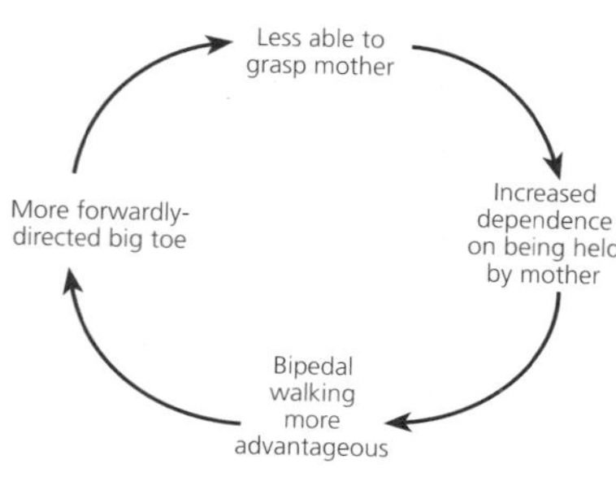


**Figure 41.6**
How two selection pressures can increase each other's effects

### *Carrying infants*

Bipedalism enables an infant to be carried by its mother, and thus to be born at an earlier and more helpless stage of development. Unlike other ape babies, human infants cannot cling to their mothers because there is not enough hair to grasp and because their feet are not prehensile.

Once under way, any tendency towards bipedalism would become self-accelerating (Figure 41.6). A partially-bipedal protohuman would have a big toe tending to face more directly forward, so the foot would be less able to grasp. The infant would therefore be less able to hold on to its mother, who would have to use her arms more to carry it. She would therefore become more dependent on the use of her legs for walking, increasing the advantage in having a forwardly-directed big toe, and so on in a self-accelerating process. This kind of situation, in which two or more changes reinforce one another's effects, is called **positive feedback**. Its essential feature is that, once started, *change becomes self-accelerating*. Therefore it only requires the most minute initial change to produce a very large final result, just as a very small detonator can set off a very large explosion.

### *Food carrying*

Human hunter-gatherers differ from other primates in that they collect food some distance away from where it is eaten, so consumption is delayed; other apes eat it where it is found. Humans also *share* food; though chimpanzees occasionally share it with a 'friend', primates in general never share food except with their offspring.

How might food-sharing be related to upright walking? One hypothesis (put forward by Owen Lovejoy) links bipedalism to prolonged parental care, in the following way. A mother carrying her baby would be less mobile and, with her hands occupied, less able to forage for food than males and non-nursing females. She would also be a good deal more vulnerable to predators. The loss of the infant's ability to hold on to its mother may therefore have been accompanied by the establishment of a home base in which children could be reared in safety.

Many of the females would then be tied to the home base by their children, and so would be dependent on the foraging ability of the males and non-nursing females. The better-adapted for bipedalism the protohumans were, the more effectively they would have been able to gather food and transport it back home. This in turn would have allowed a longer period of dependence by youngsters and nursing mothers on the foragers. More prolonged juvenile dependence would also have given more time for learning, which would have increased food-gathering skills and so on, in another positive feedback cycle.

The notion that bipedalism was linked to increasing juvenile dependence is an attractive one, but in recent years it has been called into question by studies of the correlation between tooth eruption and life cycle data (see the information box on page 786). It now seems likely that as far as growth pattern is concerned, the australopithecines were essentially like four-footed apes, and that the extended juvenile dependence characteristic of humans did not evolve until much later.

### *Standing tall to keep cool*

A quite different hypothesis was suggested by a physiologist, Peter Wheeler. He suggested that bipedalism may have been a means of avoiding excessive heat uptake. In most savannah mammals the core body temperature can rise well above the 37°C that is normal for humans, and in some antelope species it may rise to 45°C. Yet the brain is especially vulnerable to elevated temperatures: a rise of only 4°C produces heat-stroke and can be fatal.

Most mammals of the savannah keep their brain temperatures well below that of the rest of the body interior using a combination of the evaporative surfaces in the nose and a heat exchanging system below the brain. Whereas in primates the carotid arteries go directly to the brain, in most mammals of the savannah they break up into a dense network of branches or **carotid rete**, which rejoin before entering the brain (Figure 41.7). This network of arterial blood vessels passes through a large cavity containing blood returning from the muzzle. This venous blood has been cooled by the evaporation of water from the mucous membranes in the nose, and cools the arterial blood in the carotid rete.

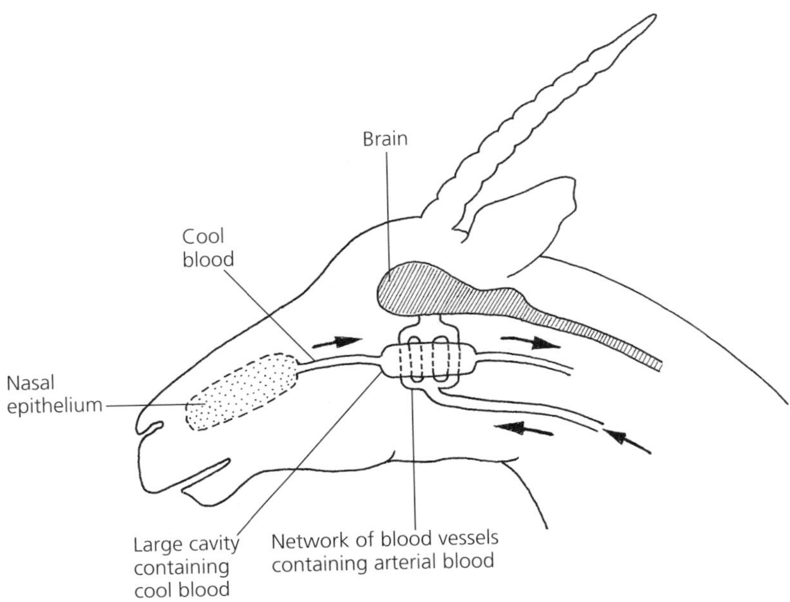


**Figure 41.7**
How an antelope keeps its brain cool in a way that a primate cannot

Primates and other forest dwellers lack this mechanism, and the only way to prevent overheating of the brain is to cool the whole body. When the Sun is directly overhead, an upright human body absorbs only about 40 per cent as much energy as it does in mid-morning or mid-afternoon. Reduced absorption of solar heat is also helped by retention of hair on the top of the head. Also by raising the body above the ground, more of the body is surrounded by cooler, more rapidly-flowing and drier air, speeding up the evaporation of sweat.

Of course, it could be argued that overheating could be avoided by simply staying in the shade of a tree. But this would result in loss of foraging time, and therefore of food supply, so any increase in the ability to be active in the hot Sun would have been advantageous.

If reduced absorption of solar energy was a causative factor, there would have been an additional bonus once the anatomical adaptations associated with bipedalism had evolved; upright walking is more efficient than walking in quadrupedal apes, so less heat would be generated.

### *Other advantages of bipedalism*

Besides the advantages considered above, bipedalism may also have given improved protection against predators in a number of ways. For one thing, if the head is higher up predators can be spotted earlier – vervet monkeys and baboons frequently stand up on their hind legs for a few seconds to look out for possible danger. Secondly, a taller body may seem larger to a predator. Third, an upright posture is more effective for brandishing sticks and throwing stones at predators, as chimpanzees are known to do with leopards.

## Ancient climates

Though piecing fossils together into a continuous sequence is one of the aims of palaeontology, it is really only the beginning. Once some of the pieces are in place, palaeontologists begin to ask questions about how extinct creatures lived and why they evolved in the ways they did.

Answers to these questions require knowledge of ancient environments. One of the most fundamental influences was climate. Palaeontologists are able to deduce a good deal about past climates from evidence provided by the analysis of sediments laid down over long periods. Two of the most fruitful methods of investigating past climates involve the analysis of preserved pollen and oxygen isotopes.

Pollen grains of wind-pollinated plants are produced in huge numbers and are blown over wide areas. Many become trapped and eventually buried in bogs and marshes. Pollen grains are resistant to decay and are often of such characteristic size, shape and surface sculpting that an expert can identify the species of plant. As sediments accumulate over immense periods of time, the pollen grains trapped within them provide a record of the plant species that were abundant at different times. Sediments up to about 40 000 years old can be carbon-dated, so it is possible to construct a record of the changes in the abundance of various species of wind-pollinated plants over long periods. Since each species is associated with particular climatic conditions, preserved pollen can reveal much about past climates. In Europe, for example, birch and willow are associated with much cooler conditions than oak. Although sediments can only be carbon-dated back to about 40 000 years, pollen can be preserved for much longer, and older samples have to be dated using other methods.

A second palaeoclimatological tool uses the fact that there are two stable (non-radioactive) isotopes of oxygen. Besides the normal $^{16}O$, there is a very small proportion of $^{18}O$. The skeletons of microscopic marine organisms contain oxygen, the two isotopes being in the same proportions as in the surrounding sea water.

Since water containing $^{18}O$ is slightly denser, it evaporates more slowly from the ocean surface than water containing the normal $^{16}O$. Ocean water is therefore slightly richer in $^{18}O$ than atmospheric water vapour. Now atmospheric water vapour is the source of rain and snow, the latter accumulating on land in the polar regions. The colder the climate, the more water is locked up in ice sheets, and the higher the proportion of $^{18}O$ in the oceans. As a result, the proportion of $^{18}O$ in the skeletons of marine micro-organisms rises during an ice age. When they die, the skeletons of these organisms fall in a steady rain onto the ocean floor and over long periods of time accumulate as sediments. An analysis of the proportion of $^{18}O$ along a core of ocean sediment can therefore provide a record of changes in global temperatures.

### *Homo habilis*

The earliest known stone tools were found in Ethiopia by Jack Harris and his colleagues, and have been dated at 2.6 million years (Figure 41.8). These tools were of a type first found by Louis and Mary Leakey in the Olduvai Gorge in Tanzania, and are accordingly called **Oldowan** tools. They were also known as **pebble tools** because they were made by striking flakes off one side of a pebble. The flakes themselves were sharp and were probably used for cutting through animal hide and meat. In fact it is not unlikely that the flakes were the tools, and that the pebble cores were discarded.

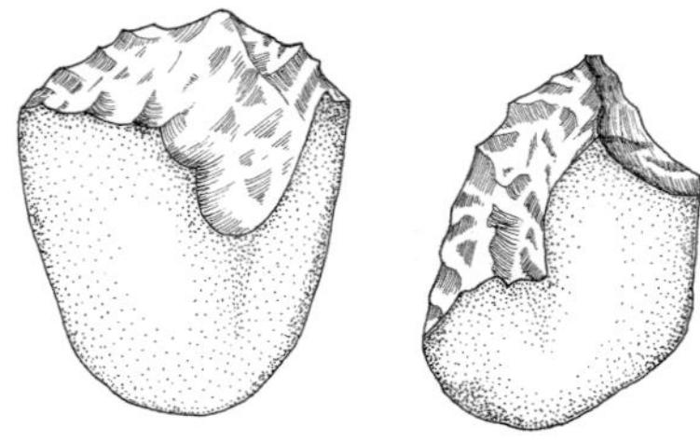

**Figure 41.8**
Oldowan tools used by *Homo habilis*

This idea was put to the test by Lawrence Keeley in 1980. He made flakes for himself and used a microscope to study the wear patterns produced by cutting meat, wood or reeds. He found that each kind of material produced characteristic wear pattern. This enabled him to show that some of the Oldowan flakes had been used to work wood, some to cut meat and some to cut reeds.

Does this mean that pebble tools were not tools at all, but simply the biproducts of flake-making? It seems not, because at some sites cores and flakes are made from different kinds of rock, the cores from lava and the flakes from quartz.

The ability to cut meat into pieces would have enabled large animals to be butchered. The cores may have been used for such purposes as cracking bones to extract the marrow, crushing tough plant food, or perhaps digging up edible bulbs and roots. Whatever their precise roles, stone tools would have opened up new ecological opportunities and a correspondingly wider diet. An important question – still unresolved – is whether these early toolmakers actually killed large game or merely scavenged it.

The first skeletal remains of the creatures that made these tools were found at Olduvai in 1959 by Jonathan Leakey (eldest son of Louis and Mary), and named *Homo habilis*. They consisted of a lower jaw, parts of the braincase and some hand bones. The cranium, though very incomplete, was estimated to have a capacity of about 640 $cm^3$, significantly larger than that of any australopithecine. The view that here at last were the remains of one of the toolmakers was bolstered by the fact that the teeth, though large by modern standards, were distinctly smaller than those of the australopithecines. It seems probable that tools were beginning to do some of the work of teeth.

*H. habilis* has been a centre of controversy ever since. Although further remains were found at Olduvai and later at Koobi Fora on the West shores of Lake Turkana in Kenya, these later finds only seemed to make the picture more complicated.

The most famous of the Koobi Fora fossils was a skull, found by Richard Leakey's team in 1972. It is known by its distinctly unromantic catalogue number, KNM-ER 1470 (short for Kenya National Museum, East Rudolf, Lake Rudolf being the old name of Lake Turkana). It took about two months of painstaking work to collect and reassemble the 300 or so skull fragments to form a largely complete cranium (Figure 41.9A). Dated at about 1.9 million years, its most distinguishing feature was the relatively large cranium with a capacity of about 750 $cm^3$ – only about 250 $cm^3$ less than that of the smallest modern human crania.

Shortly after the discovery of 1470, another skull (KNM-ER 1813, Figure 41.9B) was found that further complicated the picture. Though the cranium was only about 530 $cm^3$, in its small teeth it was more human-like than 1470. For any palaeoanthropologist looking for a neat correlation between the brain expansion and tooth reduction, this was a puzzle.

**Figure 41.9**
Skulls of *H. rudolfensis* (A) and *H. habilis* (B) (previously *H. rudolfensis* was considered to be a variant of *H. habilis*)

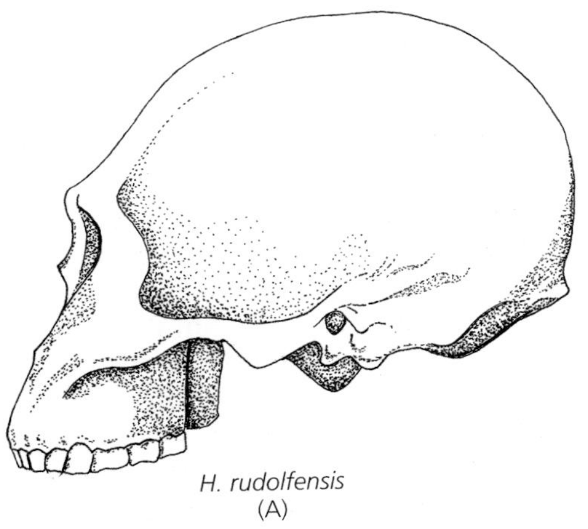

*H. rudolfensis*
(A)

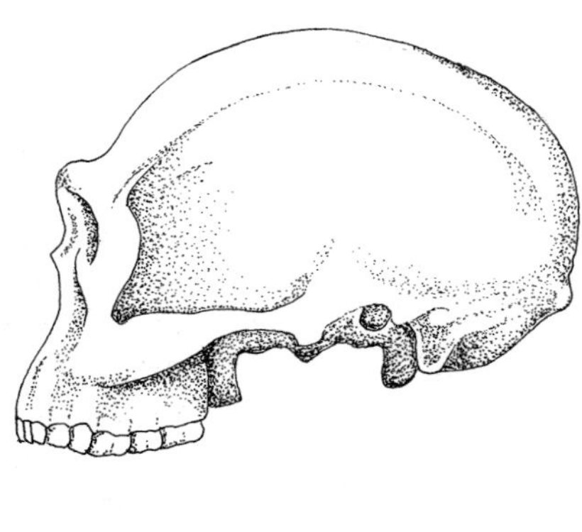

*H. habilis*
(B)

Some palaeoanthropologists consider that these fossils are representatives of a single sexually-dimorphic species, KNM-ER 1470 being a male and KNM-ER 1813 a female. The alternative hypothesis is that these fossils represent more than one species. This view is gaining ground and many consider that KNM-ER 1470 should be placed in the species *H. rudolfensis*.

### Forensic palaeoanthropology

Finding and dating a fossil are just two of the problems with which palaeoanthropologists have to contend. Very few animals remain intact for long after death. Most often carnivores dismember the corpse so bones become scattered and trampled and often crushed by large herbivores. Only when an animal falls into a bog or lake and is rapidly buried is there much chance of an entire skeleton becoming fossilised. Even after burial, sediment may become scoured away again, leaving the bones to be washed away. As a result large collections of bones and teeth may be deposited some distance from where their owners died, and in some cases this can lead palaeontologists to false conclusions about how they got there.

Quite apart from relocation after death, bones may become marked in various ways that can provide clues – and traps – for palaeontologists. Scratch marks on a bone may have been caused by a stone tool, but they could also have been produced by carnivore teeth or gnawing by a porcupine. Distinguishing these different kinds of mark is one of the tasks of **taphonomy**. The taphonomist is rather like a forensic scientist, except that he or she deals with post mortem events that may have occurred millions of years ago.

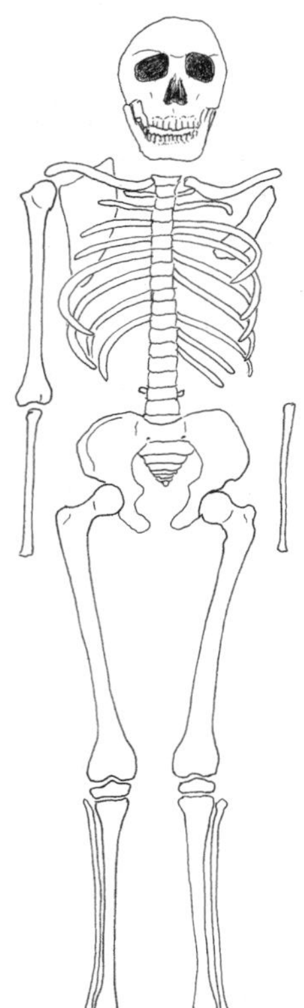

**Figure 41.8**
*H. ergaster*, the 'Turkana Boy'

## *Homo ergaster*

In 1984, Richard Leakey's team made one of the most spectacular discoveries in the history of palaeoanthropology. On the West side of Lake Turkana in Kenya, they found an almost complete skeleton of a male juvenile, dated at about 1.6 million years old (Figure 41.10). Though the 'Turkana Boy' as Leakey called him, was about 9–10 years old, he was already about 1.6 metres (5' 4") tall. This would mean that as an adult, he could have reached almost 1.83 metres, or six feet, which is more than the average for modern males.

Initially, the Turkana Boy and certain other African fossils were considered to be early representatives of *H. erectus*, described below. However, palaeoanthropologists now believe that these African fossils are sufficiently distinct to warrant classification as a new species, *H. ergaster* ('work man'). One of these early African fossils, a female, is 1.7 million years old and is shown in Figure 41.11.

## *Homo erectus*

The first specimens of *H. erectus* were found in Indonesia in the late 19th Century, and early in the 20th Century in China. The ages of the oldest of these Asian fossils have recently been revised to about 1.7 million years. The most recent, dated at about 400 000 years, had cranial capacities estimated to be about 1000 $cm^3$ – just within the range of variation in modern humans (Figure 41.12).

Despite an increased cranial capacity, the skull of *H. erectus* differed in a number of ways from that of modern humans. It was thicker and its roof

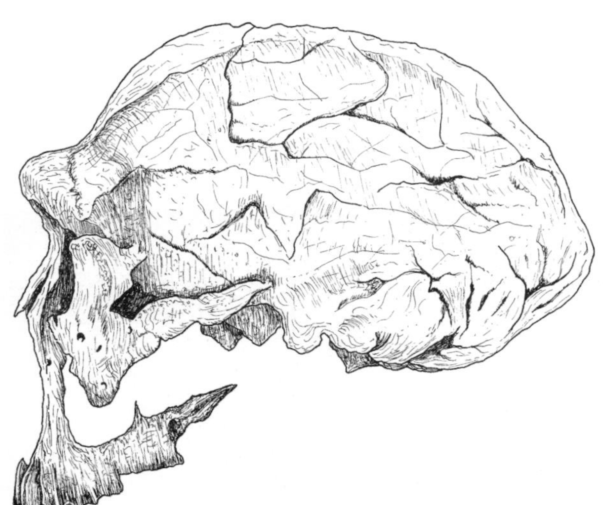

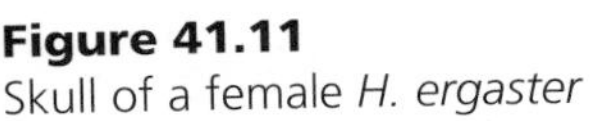

**Figure 41.11**
Skull of a female *H. ergaster*

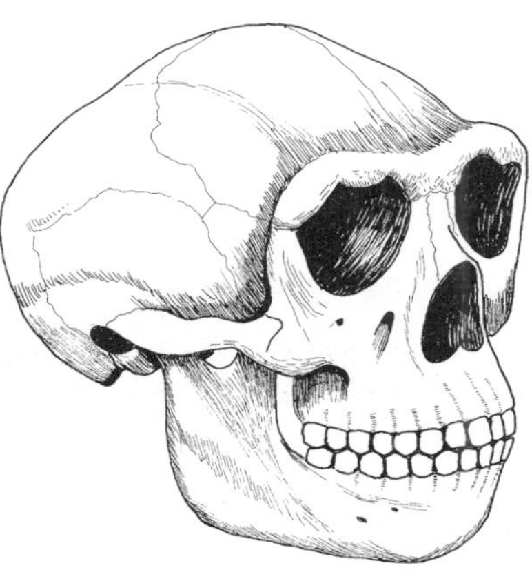

**Figure 41.12**
Reconstruction of skull of *H. erectus* from Zhoukoudian in China

was low with a narrow, sloping forehead with prominent brow ridges. These probably resisted the stresses set up during chewing, since the lower jaw was strongly built – indicative of powerful chewing muscles. The lower jaw was deeper than in modern humans, and there was no trace of a chin. The teeth were smaller than in *H. habilis* but larger than in modern humans.

Despite their relatively large brains, *H. erectus* survived with the use of relatively simple Oldowan type tools. This is taken as evidence that their ancestors left Africa before the development of the more sophisticated, Acheulean tools (see below).

Although there is evidence for the use of fire in southern Africa almost a million years ago by early *H. ergaster*, *H. erectus* is usually credited with being the first to use fire extensively. Evidence for its use at living sites includes the presence of charcoal and charred bones, thermally-altered stones, and some form of hearth such as a ring of stones surrounding charcoal. It is probable that *H. erectus* was limited to maintaining fires that had begun in this way, rather than starting them artificially. Whatever the case, the controlled use of fire would have brought massive benefits.

- It would have provided warmth. During much of the Pleistocene epoch the parts of China where *H. erectus* fossils have been found were enduring the bitter cold of the ice ages, and it seems unlikely that these regions could have been colonised without the use of fire.
- It would provide illumination at night, allowing groups to extend the length of 'home base' activities such as tool-making and butchering carcasses.
- It would have kept away dangerous animals.
- It may have been used to harden wooden spear points.
- Game could be driven towards traps by fire.
- Cooking would have improved the flavour and digestibility of food, and would have made it safer to eat by killing the young stages of parasites.

The use of fire was a landmark in human evolution, since it represented the first use of an 'extrasomatic' energy source – that is, a source of energy outside the body. The increasing ability to supplement human muscle power with other sources of energy was one of the dominant strands in the later stages of human cultural evolution.

Until recently *H. erectus* was considered to be on the direct line of human ancestry, but with the renaming of the African *'erectus'* fossils as *H. ergaster*, *H. erectus* is now considered to be an Asian side branch that left no living descendants.

## Teeth and life history

No matter how many fossils are eventually found, some of the most important evolutionary changes will probably never be fully clarified because they leave no physical record. A case in point is the change from the great ape pattern of growth to the human type. At some stage, there was a trend towards birth at an earlier stage of development and an increase in the period of juvenile dependence.

Until recently it was widely believed that from very early on, protohuman growth followed the human pattern rather than that of four-footed apes. Then in the late 1980s all this changed when Holly Smith, a palaeoanthropologist specialising in the study of teeth, showed that evidence of growth pattern can be obtained from skeletal remains. She made use of the fact that a number of life history and other features of primates tend to be correlated. These include gestation period, age at sexual maturity, litter size, inter-litter interval, body size and longevity. For example, long-lived animals tend to have long gestation periods, few offspring, and long interbirth intervals. By plotting one variable against another for many species, it is possible to obtain the **correlation coefficient**, which is a measure of the extent to which two variables are correlated. If all the points fall on a line, the correlation coefficient is 1.0, meaning that if you know one value you can accurately predict the other one. A correlation of 0.0 means that there is no relationship, so one value tells you nothing about the other.

Smith used this idea to find out whether it would be possible to make predictions about life histories of extinct protohumans (which can't be measured directly) from one that can, such as brain size and body size. A particularly useful variable is the age of eruption of the various teeth, since these can indicate age at death. The strongest correlation – between brain size and the age at which the first molar erupted – was 0.98.

Smith used the brain and body sizes of a number of extinct protohumans to estimate the age of eruption of the first molar. She found that in the australopithecines and *H. habilis* it erupted at just over three years of age – similar to chimpanzees and gorillas. On the other hand, she found that in *H. ergaster* the eruption of the first molar would have occurred at about four and a half years. This is intermediate between the australopithecines and modern humans, in which the first molars erupt at just under six years.

Smith's findings suggest that in human evolution, the increase in the size of the brain and the prolonging of juvenile dependence occurred in parallel. A reasonable hypothesis is that, as explained earlier, helpless young and prolonged parental care can be linked to the problem of getting a large brain through the birth canal.

## *Homo heidelbergensis*

Until recently, the premodern human fossil record was (with the exception of the Neanderthals) rather poor, and was limited to scattered remains from Europe, Africa and Asia. These pre-modern types were, together with the Neanderthals, lumped into a convenient rag bag called 'archaic *H. sapiens*'.

In the early 1990s the picture changed dramatically with the discovery in a cave in Spain of well-preserved remains of at least 30 adult and

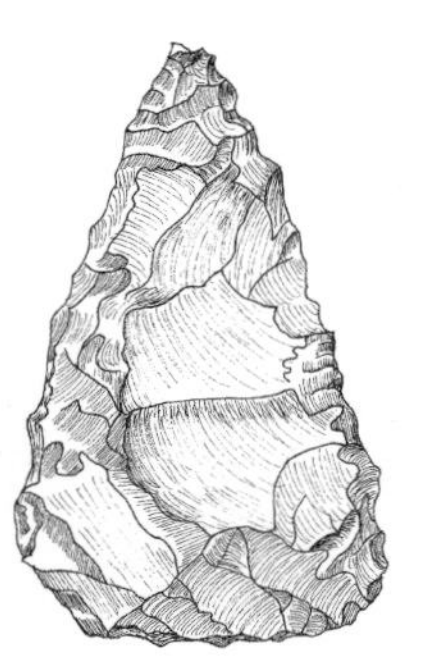
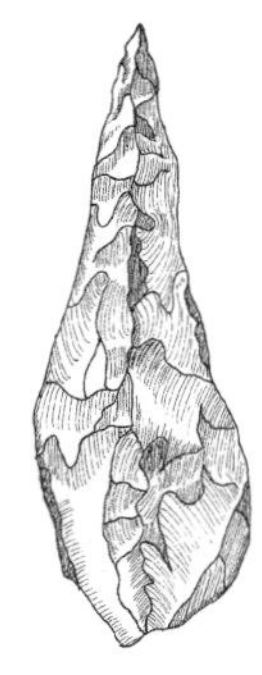

**Figure 41.13**
Acheulean hand axe made by *H. heidelbergensis*

juvenile individuals that were estimated to have died about 300 000 years ago. Though some of their features foreshadowed the Neanderthals, they have been included, together with other non-Neanderthal archaic *sapiens* fossils, in the same taxon, *H. heidelbergensis*. It is now believed that this species most probably gave rise to both *H. neanderthalensis* and *H. sapiens*.

The tools most characteristic of *H. heidelbergensis* were bifacial handaxes, in which the entire stone was shaped into two main faces (Figure 41.13). Tools of this type were characteristic of the **Acheulean** tool industry (so-called because they were first found at the village of St Acheul in Northern France). Together, the Oldowan and Acheulean cultures are known as the **Lower Palaeolithic**, or 'Old Stone Age' (*lithos* is Greek for stone). Acheulean tools first appeared about 1.5 million years ago in East Africa associated with remains of *H. ergaster*, after which their manufacture spread to Europe and India.

## The Neanderthals

The most recent premodern humans were the Neanderthals, so-called because one of the first discoveries in the mid-19th Century was in the Neander valley in Germany ('Tal', then spelled 'Thal', is German for valley). Since then, many Neanderthal remains have been found throughout Europe and also in parts of Western Asia.

The forehead was lower and sloped backward, but despite this, the cranium was slightly larger than that of modern humans, with a mean capacity of about 1500 $cm^3$ compared with 1375 $cm^3$ for modern humans (Figure 41.14). The extra volume was due to the fact that the Neanderthal skull was longer and wider, especially at the rear where there was a characteristic bulge – the so-called 'occipital bun'. Though the chin receded, the jaws were very strong and the teeth were rather larger than in modern humans. The rest of the skeleton indicated that the Neanderthals were thick-set, with prominent ridges for muscle attachment.

**Figure 41.14►**
Neanderthal and modern skulls compared (after Weiner)

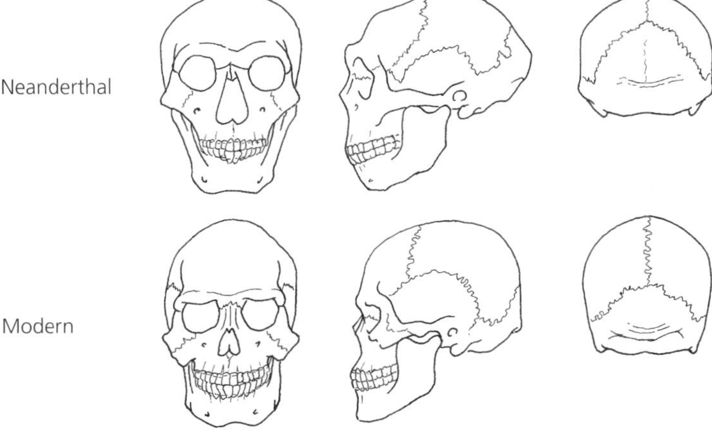


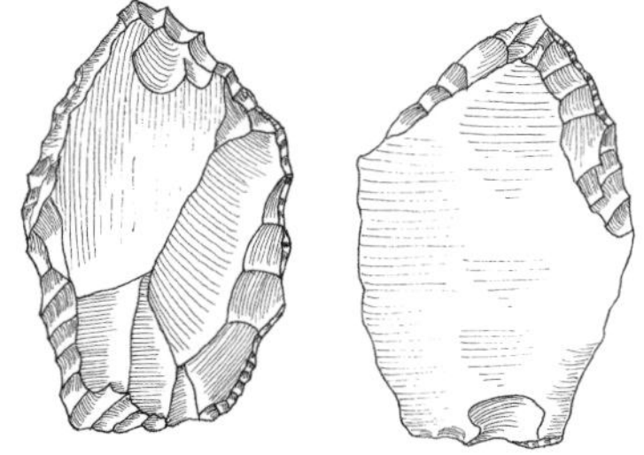

**Figure 41.15**
A mousterian tool (two views) used by Neanderthal people of the Middle Palaeolithic

Evidence of Neanderthal skills lies in their stone tools, which are much more sophisticated and varied than the Acheulean tools of *H. ergaster* and *H. heidelbergensis*. Neanderthal tools are known as **Mousterian**, after Le Moustier in South-western France where they were first found. Unlike the Acheulean tools in which a core was shaped by striking off flakes, Mousterian tools were shaped from the flakes themselves (Figure 41.15).

A core was first shaped by tapping off small flakes, and then a large flake was then struck off; this was then further shaped by striking very small flakes off its edges. Several large flakes could be produced from a single core, which was then discarded. The production of a tool by striking flakes from a prepared core is called the *Levallois* technique, after a suburb of Paris where such tools were first found. Making tools from a prepared core required the ability to envisage the shape of the tool while it was still part of the core, and obviously called for higher mental faculties than the gradual shaping of an Acheulean handaxe.

Mousterian tools were not only more finely made than the Acheulean handaxes, they were far more varied, with over 60 different kinds being recognised. These included blades, chisels, points and scrapers.

Many archaeologists believe that the Neanderthal people buried their dead, and this has been taken to imply some form of belief in life after death. If bodies were simply left, they would soon be dismembered by carnivores. The fact that so many intact skeletons have been found has been taken as evidence that the bodies had been buried.

Neanderthal skeletons are available in larger numbers than those of earlier protohuman types, and it has been possible to gain some idea of death rates at particular ages. Age at death can often be inferred from the eruption of the teeth, and from the extent to which cartilaginous growth centres in bones have been replaced by bone (ossification). Studies of Neanderthal skeletons have shown that less than one in 10 individuals passed the age of 35.

Neanderthal people inhabited the earth between 200 000 and 30 000 years ago, a period known as the **Middle Palaeolithic**. This period included the most recent of the Pleistocene ice ages, and their survival must have demanded not only physical toughness, but also considerable resourcefulness.

**Table 41.2**
Summary of *Homo* species

| | Time B.P. (m.y. = million years) | Cranial capacity | Tool culture | Distribution | First to ... |
|---|---|---|---|---|---|
| *H. habilis* | 1.5–2.0 m.y. | 500–800 | Oldowan | E. Africa | make stone tools |
| *H. ergaster* | 1.7–1.5 m.y. | 850 | Oldowan, Acheulean | E. Africa | |
| *H. erectus* | 1.7 m.y.–200 000 | 980 | Oldowan | Asia | use fire extensively |
| *H. heidelbergensis* | 600 000–250 000 | 1000–1300 | Acheulean | Africa, Europe | |
| *H. neanderthalensis* | 200 000–30 000 | 1200–1750 | Mousterian | Europe, W. Asia | care for aged |
| *H. sapiens* | 100 000–present | 1200–1700 | very diverse | worldwide | make art forms |

### *Relationships to modern humans*

Despite the close similarity between Neanderthal and modern people, Neanderthals were at first considered to belong to a different species, *H. neanderthalensis*. As knowledge increased, many palaeoanthropologists considered that Neanderthals and modern humans should be regarded as two subspecies of *H. sapiens*: *H.s. neanderthalensis*, and *H.s. sapiens*. In recent years, however, there has been an increasing tendency for the pendulum of opinion to swing back again, Neanderthals being accorded full species status.

This shift received powerful support in 1997 when mitochondrial DNA (mtDNA) was extracted from a Neanderthal skeleton and compared with its modern human equivalent (see later in this chapter for an explanation). The results were startling: Neanderthal and modern mtDNA differed much more than would have been expected. In fact, knowing the mutation rate of mtDNA it was possible to calculate that the most recent common ancestor lived about 600 000 years ago! This was three times further back in time than had been supposed, and lent further support to the separate species view.

This latest swing of opinion is only the most recent of many that have occurred since the birth of palaeoanthropology last century, and it is unlikely to be the last. Uncertainly, debate and constant re-examination of the evidence are the very nature of science, and palaeoanthropology would be in an unhealthy state without it. Figure 41.16 shows one possible human evolutionary tree.

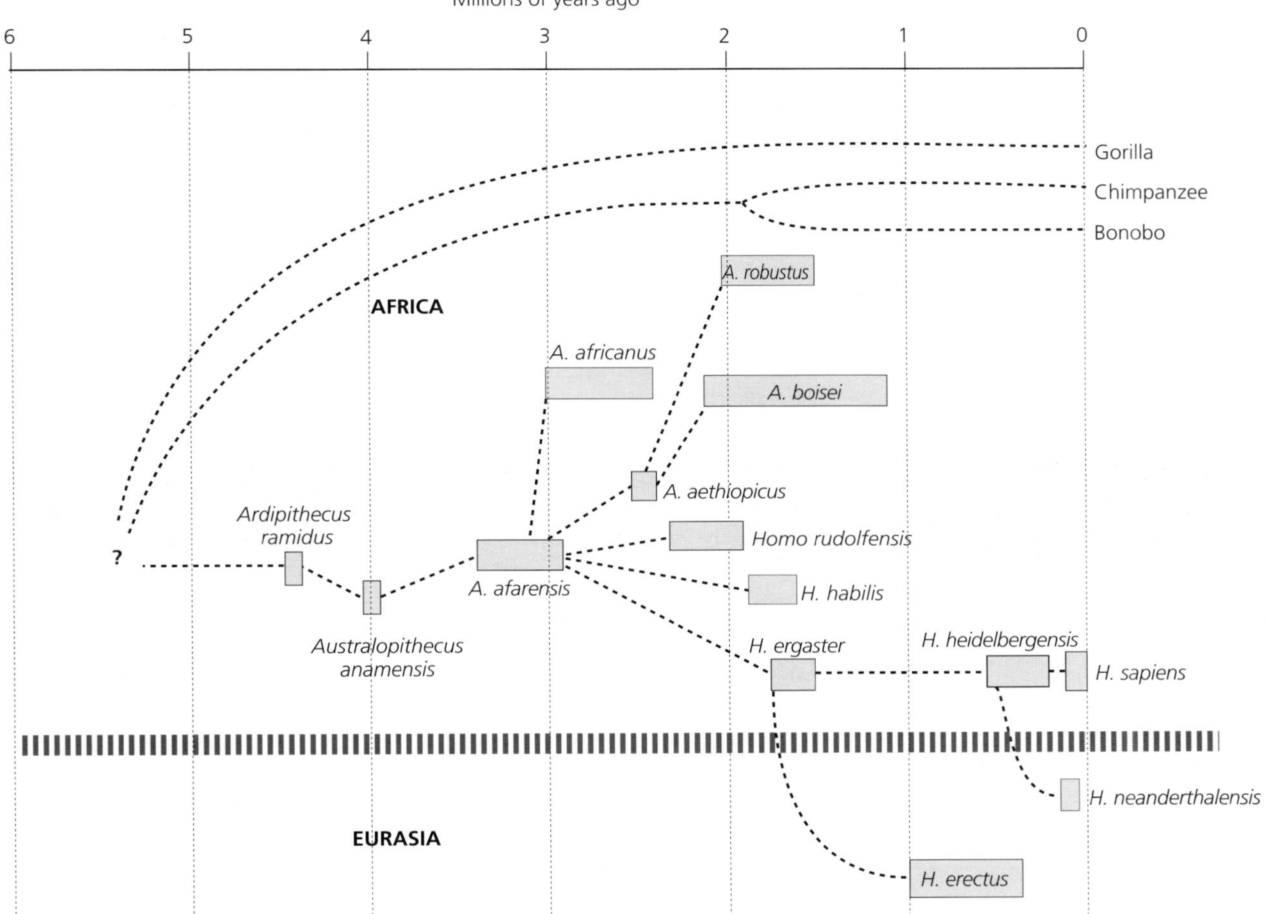


**Figure 41.16**
One interpretation of the human fossil record

## Modern humans

The first prehistoric remains of anatomically modern humans were found in 1868 in a limestone cave in South-western France. The cave was at a place called Cro-Magnon, and these early Europeans became known as Cro-Magnon people. Since then, remains of early modern humans have been found at many other European sites and also in Asia, Africa, Australia and America.

The earliest European remains of modern humans are dated at around 40 000 years B.P. (before the present), but considerably older remains have been found elsewhere. Probably the oldest are some lower jaw and cranial fragments found at the Klasies River Mouth Caves in South Africa. These have been dated at a minimum of 74 000 years, and are possibly as old as 115 000 years old, when Neanderthals were the

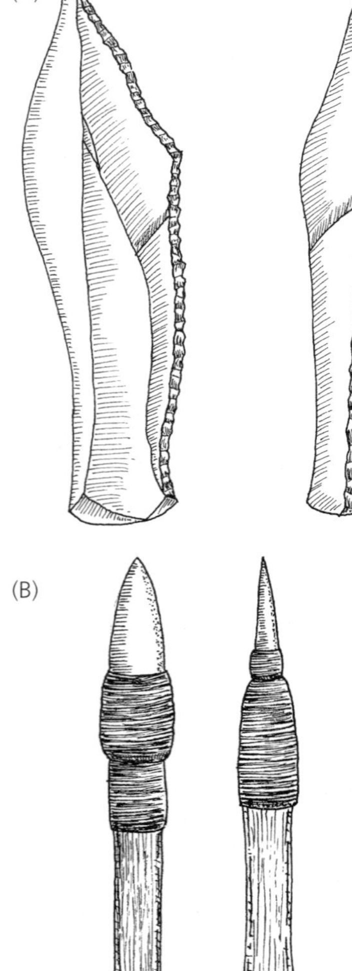


**Figure 41.17**
Tools of the upper Palaeolithic. Stone tools (A) and composite tools (B) made by *H. sapiens*

dominant people in Europe. The implication of these remains is that since Neanderthals were contemporaneous with modern humans, they could not have been their ancestors.

### *Physical characteristics*

The differences between modern and Neanderthal skeletons are really quite small, the most obvious being in the skull (Figure 41.14). In modern humans brow ridges, if present, are small. The large forehead is much steeper and the upper part of the cranium is shorter and higher, and there is no 'bun' at the rear. There is a prominent chin, and the jaws and teeth are less robust, especially the incisors. Another feature suggesting that superior technology was replacing Neanderthal muscle is the more slender limbs and limb girdles of *H. sapiens*.

### *Tool technology*

Since the arrival of modern humans there appears to have been very little evolutionary change – at least as far as skeletal characteristics are concerned. By far the greatest changes since then have been cultural, in the form of improved technology. Tools made by early modern people showed significant advances over those of the Neanderthals, and characterised the **Upper Palaeolithic** phase of human culture (Figure 41.17).

The most important features were as follows:

- stone tools were more finely made, with more cutting edge per kg of stone used.
- they showed a much greater use of materials other than stone, such as bone, antler and ivory. Tough and durable, these materials can not only be carved into various shapes, but can be *perforated*. This was particularly important because it made possible the manufacture of bone needles (and hence stitched clothing) and fish hooks (and therefore a wider diet).
- they combined one kind of material with another to make **composite tools**. For example, spearheads of stone or carved antler were joined to (presumably wooden) handles, probably using thongs of skin (which tighten on drying) or glue from plant gums and resins.
- whereas Neanderthal technology changed little over long periods and showed very little regional variation, the arrival of early modern people in Europe was marked by a great increase in the range and variety of tool types, and an increase in the pace of change. The period between 40 000 and 11 000 years ago in Europe was marked by a succession of tool cultures (referred to as 'industries'), each distinct from its predecessor.

### *Hunting techniques*

The fact that mammoths, woolly rhinoceros, and numerous other large mammals became extinct in Europe in the last 30 000 years shows how efficient modern humans were as hunters. There is evidence that hunting was highly organised and on a large scale. Bones of thousands of horses have been found at the bottom of a cliff in Southern France, suggesting that they had been stampeded over the edge to be butchered at the bottom, and large piles of mammoth bones have been found in what had probably been pits prepared for the purpose.

New weapons were also appearing. By 17 000 B.P., barbed harpoons had been invented, so that for the first time fish could become a significant part of the human diet. By 14 000 B.P., fish hooks had been introduced. The bow and arrow increased a hunter's killing range still further, enabling the energy expended in a slow muscle contraction to be stored in the elasticity of the wood before being explosively released. Furthermore, since arrows are much lighter than spears, several could be carried, allowing another shot if the first one missed. The earliest preserved pieces of bows or arrows are about 10 000 years old, but bony parts of what may have been 20 000-year old arrows have been found in Southern Africa.

### Improvements in the use of fire

Though fire had been in systematic use by early human for hundreds of thousands of years (if not longer) its full potential was first exploited by modern people. There were two key developments. First, they discovered how to start fire by striking sparks off iron pyrite – one of the few minerals that can generate sparks hot enough to light tinder. The earliest known 'firestone' comes from a cave in Belgium and was found alongside remains dated at 15 000 B.P. A groove on one side of the stone suggested that it had been repeatedly struck with another.

Second, it seems that they had learned the technique of creating a draught of air – some ancient hearths in Eastern Europe had channels in the earth leading to them, apparently acting as ducts through which air could be directed. This would have increased the temperature of the fire and, in the much longer term, made it possible to make pottery and later still, smelt metal ores.

### Palaeolithic art

Until the appearance of *H. sapiens*, all energy expended by human ancestors was, as far as we can tell, devoted to matters of immediate survival value. The first evidence for activity of no immediate material benefit was the burial practice of the Neanderthals, but with the arrival of modern humans there was an enormous increase in the importance of satisfying emotional rather than physical needs. This is shown very clearly by the beautiful cave paintings and carved statuettes, beads and other ornaments that have been found with Upper Palaeolithic remains. Most of the animals depicted were horses, deer, oxen, bison and mammoth, all of which were important food for the Cro-Magnon people (Figure 41.18).

### Life expectancy

A considerable number of skeletons of early modern people have been found, and this has made it possible to get some idea of mortality statistics. Women tended to die earlier, presumably because of child bearing. Though men rarely lived beyond 50, it seems that life expectancy for both sexes was considerably longer than for the Neanderthals. While the higher proportion of older people may have been a burden in one sense, their accumulated knowledge (particularly of hazards that occurred rarely) may have more than offset this.

## The origin and spread of modern humans

The origin of modern humans has, until recently, been uncertain. Two hypotheses have been proposed (Figure 41.19).

- The **regional continuity**, or 'candelabra', hypothesis. According to this idea, modern Africans, Asians and Europeans all evolved

**Figure 41.18**
Cave drawings of the Upper Palaeolithic

independently from different populations, so 'racial' characteristics would be very ancient.

- The **replacement** or 'Out of Africa' hypothesis. According to this view all modern humans evolved relatively recently in Africa before spreading to Asia and Europe, replacing the existing, less advanced populations. Local 'races' would thus have a very recent origin.

Though the matter has not yet been fully resolved, genetic and fossil evidence both seem to support the replacement hypothesis.

**Figure 41.19**
Two rival hypotheses for the origin of modern humans

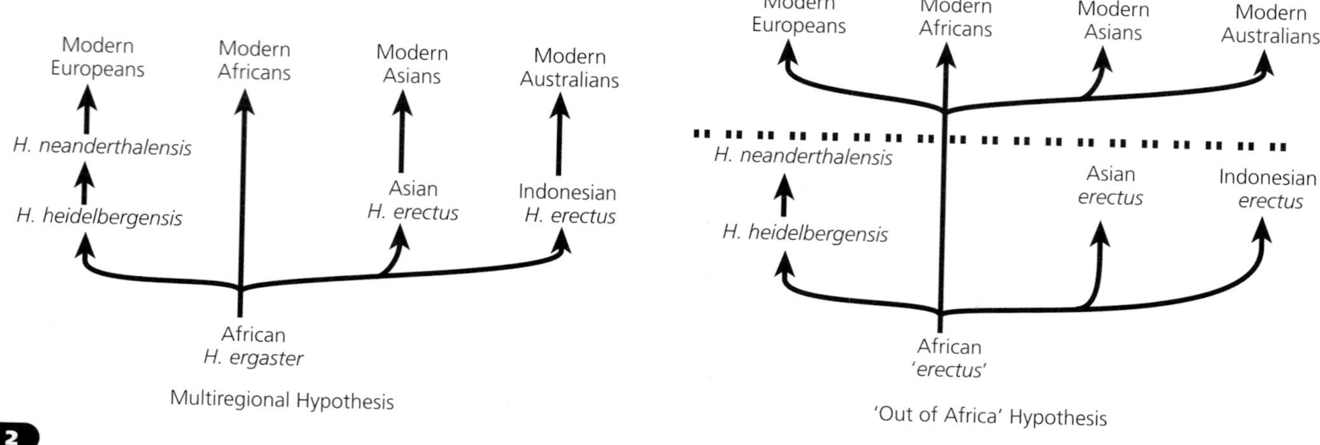

## Evidence from mitochondrial DNA

Like chloroplasts, mitochondria (Chapter 4) are self-replicating with their own DNA. Mitochondrial DNA (mtDNA) has two significant advantages for evolutionary studies:

- it is only inherited via females, because although the middle piece of a sperm contains mitochondria, these do not make any significant contribution to the mitochondria of the zygote since the ovum contains many thousand. This means that all your mtDNA has come from your mother, and all hers came from *her* mother, and so on. No matter how many generations you go back, all your mtDNA comes from only *one* ancestor (the maternal one) in each generation. Going back, say, 10 generations, all your mtDNA comes from only one of the $2^{10}$ ancestors in that generation. This particular female however only contributed $(\frac{1}{2})^{10}$ of your nuclear DNA (there are limits to how many generations you can go back before this relationship begins to break down because there is an increasing likelihood of two ancestors being related).

  As a result of this maternal inheritance, mtDNA can only change by mutation (there can be no recombination between maternal and paternal DNA if there is no paternal contribution).

- mtDNA mutates about 10 times faster than nuclear DNA because it lacks the elaborate 'proof reading' mechanisms of nuclear DNA, with errors in replication going unrectified. Since a given amount of change therefore occurs in about a tenth the time, mtDNA is very useful for the study of short-term evolutionary changes.

### How do biologists compare mtDNA?

To compare the mtDNA of different people, it is not necessary to analyse the entire sequence of 16 569 base pairs. Instead biologists use one of the techniques of genetic engineering. Using restriction enzymes (see Chapter 14), DNA can be broken up into smaller pieces, each enzyme cutting the DNA wherever it encounters a particular base sequence or 'restriction site'. Since each restriction enzyme recognises a different site, different enzymes cut DNA at different points. When a given sample of mtDNA is treated with a particular restriction enzyme, it always yields the same pattern of fragments which can be separated and their sizes determined.

By treating mtDNA from a given individual with different restriction enzymes and then determining the sizes of the fragments produced, the relative positions of the restriction sites can be worked out and represented as a **restriction map**.

### Evolutionary trees from restriction maps

When restriction maps of mtDNA from different human populations were compared, it was found that there were 147 different kinds. These types were not randomly distributed – as might be expected, people from the same ethnic group showed fewer differences than people from different ethnic groups. If the degree of difference between samples of mtDNA from different populations is taken as a measure of their relatedness, then it is possible, using a computer, to construct an evolutionary tree (Figure 41.20).

The tree based on human mtDNA has two main branches, one leading to Africans and the other to Asians, Europeans and native Australians. The split into these two main branches would therefore appear to have occurred in Africa. The descendants of those that left Africa then diversified into Europeans, Asians and Australians.

**Figure 41.20**
Inter-relationships between different populations of modern humans according to evidence from mtDNA

### *Calibrating the tree*

Figure 41.20 only shows the order of branching – it does not tell us anything about when the various splits occurred. As with the evolution of proteins, we need to know the *rate* at which mtDNA evolves. We also need to be sure that it evolves at the same rate in different populations.

As with proteins, the fossil record has enabled a 'dial' to be put on the mtDNA 'clock', and also to show that it 'ticks' (i.e. evolves) at a more or less constant rate. The approximate date of colonisation of Australia and New Guinea is known from fossil evidence to be about 40 000 years ago in both cases. If mtDNA has been evolving at the same rate in the two populations, then each should show about the same amount of variation, and this turns out to be the case.

Knowing the actual rate of change of mtDNA (about three per cent per million years), it is thus possible to estimate the time at which early *H. sapiens* split into the two main lines of descent. African mtDNA differs from that of Europeans, Asians and Australians by about 0.6 per cent, so with a rate of change of three per cent per million years, the split would have occurred about $0.6 \div 3 = 0.2$ million years (or 200 000 years ago) in an African population. So all modern human mtDNA has been derived from a common female ancestor ('Eve') only about 10 000 generations ago.

## Human diversity

Most wide-ranging species show considerable geographical variation, and humans are no exception. Populations native to North America, for instance, have longish noses and straight hair, whilst people indigenous to Southern Africa have short noses and crinkly hair. Most obvious of all, populations native to the tropics have darker skins than those inhabiting temperate zones. What is the biological significance of this variation? Specifically, to what extent are such differences adaptive and how have they come about?

In some cases physical features are clearly adaptive. Two of the more obvious examples are skin colour and body proportions.

### *Skin colour*

The darkness of a person's skin depends mainly on the concentration of the pigment **melanin**, which absorbs ultraviolet (UV) radiation. UV radiation has both harmful and beneficial effects: it causes skin cancer, but enables synthesis of Vitamin D in the skin (Chapter 26). The optimum melanin content is thus a compromise between the need for Vitamin D and protection against skin cancer. In societies living in high latitudes where UV radiation is low, fish may be an important source of Vitamin D. Where fish is not part of the diet (and dietary intake of Vitamin D is therefore low) selection favours pale skin. Where UV radiation is high, dark skin is favoured.

Variation in skin colour at birth is the result of genetic differences, and differences between populations with respect to this are the result of evolutionary change. Tanning, on the other hand, is a purely phenotypic adjustment to environmental changes, with a timescale considerably shorter than the lifespan of an individual.

## *Limb proportions*

Populations also differ with respect to physique. People indigenous to cold regions such as the Inuit (Eskimo) and the Lapps of Northern Scandinavia, tend to be stocky, their lower surface to volume ratio helping to conserve heat. In the tropics, where the ability to lose heat is important, a more slender build is the general rule (Figure 41.21).

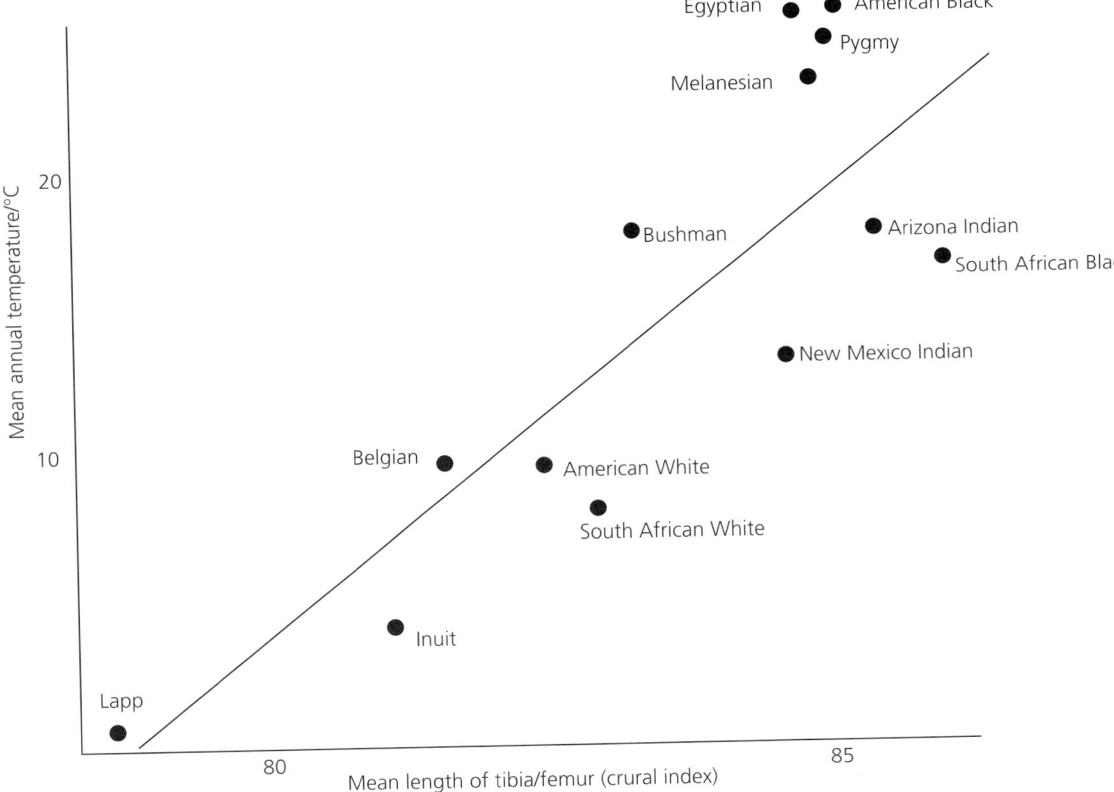


**Figure 41.21**
Adaptive variation in build in different human populations

It seems clear that variation in limb proportions has a genetic basis, since the children born after their parents have moved to a different climate tend to resemble their parents rather than the local people.

## *Do races exist?*

Geographical variation in humans is so marked that it is often possible to tell to what part of the world a person is indigenous. Perhaps for this reason it has been common to regard humanity as being divisible into distinct 'races', and it has been suggested that such 'races' represent the earliest stages in the formation of new species.

Though it may be convenient to refer to aboriginal peoples as Melanesians, Australians, Maori, Hottentots, and so on, these divisions are less clear cut than they appear for the following reasons.

- The features that are used to distinguish 'races' are but a very small fraction of the number of ways humans can differ, many of which are biochemical. Not unnaturally, greater emphasis is given to external features such as skin colour, which make the most immediate impression. Any natural classification should use as much information as is available.
- Whatever characteristic is considered, there are no clear-cut boundaries between populations. Rather, characters and allele frequencies show *gradients* from one region to another. Such a gradient in a characteristic or allele frequency is called a **cline**.
- When different kinds of characteristic are considered, they often vary geographically independently of one another. In other words, in any given population, clines for genes for different characteristics often run in different directions, or one characteristic may show clinal variation whilst another is unchanging over large distances. For instance in passing from Northern Europe to India skin colour becomes markedly darker, whilst facial features change little.
- Genetic differences *within* 'races' are much greater than genetic differences *between* them. If two people are taken at random from the same 'race', they differ by almost as many genes as would two people drawn at random from different 'races'. In other words the *additional* differences due to 'race' are extremely small.

It seems therefore that the idea that human 'races' are biologically distinct from each other has no scientific justification, no matter how obvious the external differences may appear. To maintain that 'races' should be kept 'pure' by preventing intermarriage is therefore nonsense.

## Social 'tools' and brain expansion

The evidence of the australopithecines and *H. habilis* showed that bipedal walking preceded brain expansion, just as Darwin had stated. The earliest *H. habilis* skulls are about 2 million years old, so human ancestors were walking on two legs at least 1.5 million years before the brain began to enlarge. The fact that the first human ancestor known to use stone tools was also the first to undergo significant brain enlargement, is consistent with the view that tool-making provided the impetus for intellectual development and brain expansion.

Attractive though the tool-making hypothesis may appear at first sight, an increasing number of palaeoanthropologists are coming to the view that the driving force for brain enlargement may have been social rather than technological. One reason for this has been the realisation by primatologists that in the laboratory, monkeys (and especially great apes) show far more intelligence than field observations suggest they would need to survive in the wild.

Large brains are expensive to build and to run, so it seems unlikely that monkeys and apes are more intelligent than they need to be. The answer may lie in the extraordinary complexity of social interactions in primate groups. Field observations have shown that chimpanzees and some monkeys form 'friendships', gaining competitive advantage over others in access to food and especially sex. In such an alliance each individual not only knows who is in the same group, but who belongs to another group. Alliances are often temporary, individuals sometimes switching membership. A successful coalition is not necessarily the one containing the most physically dominant individuals. Rather, it wins by precipitating conflict when circumstances are most favourable – when, for example, a key member of the other group is not on hand to give support.

## Social 'tools' and brain expansion *continued*

Social manipulation, or the use of other individuals as 'tools', requires intelligence of an altogether higher order than would be needed to dig up bulbs or hunt for termites. Each individual must have some appreciation of the mental state of every other individual, and be able to foresee the effects of its own actions on others. Those individuals with the greatest political skills would get the most sex, and hence would make the greatest contribution to the next generation.

It may be that social intelligence may lie at the root of that most human characteristic, the capacity for thinking about one's own thinking. By understanding one's own reactions to the actions of others, one is better able to predict the effect of one's actions on other group members, and therefore to make 'political' decisions that yield more power and influence.

## Agriculture – the Neolithic transition

By the end of the most recent ice age (some 12 000 years ago) humans had colonised all tropical and temperate land masses except New Zealand and Madagascar. The total world population at this time is estimated to have been about 10 million people. From the artefacts they left behind, it is clear that they made their living by gathering wild plants and hunting wild animals.

Beginning about 10 000 years ago a dramatic change began, marked by two processes:

- a shift to food production by the domestication of animals and plants.
- a change from a nomadic to a sedentary way of life, with the development of permanent settlements, from small villages to cities.

The domestication of animals and plants marked the change from the Palaeolithic, or Old Stone Age, to the **Neolithic**, or New Stone Age. The shift to agriculture did not occur simultaneously throughout the world, but began independently in several areas. From these small nuclei, farming spread within a few thousand years to most of Eurasia and Africa.

The transition between the Palaeolithic and the Neolithic took a considerable time, and is often referred to as the **Mesolithic**. The use of the term seems to vary with respect to different regions, but there is general agreement that its most distinguishing feature was the extensive use of **microliths**. These were very small (one to three centimetres long) stone flakes, which were set in wood using resin or pitch to produce a longer cutting edge such as would be needed for a sickle (Figure 41.22).

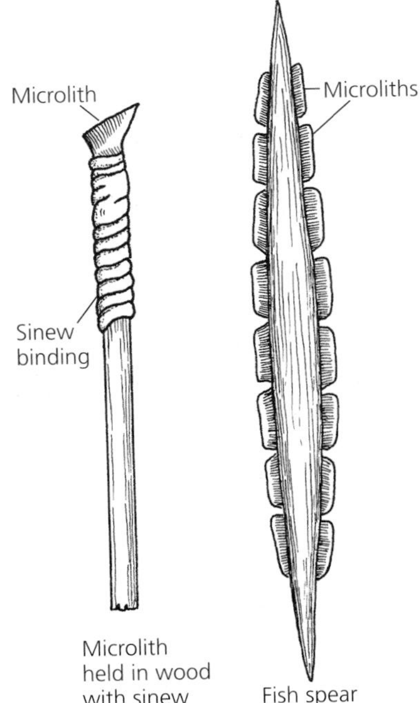


**Figure 41.22**
Microlith tools of the Mesolithic

### *Evidence for domestication*

How do archaeologists decide whether some remains are those of wild or domesticated animals or plants? Obviously this is central to the question of when domestication began. There are two kinds of clue.

- Whereas animals killed by hunting tend to be of various ages, those killed under domestication are likely to be young adults (age can usually be estimated from tooth wear and skeletal characteristics).
- As a result of selection, the characteristics of animals and plants slowly change during domestication. Tusk size in pigs has decreased, and the cross-sectional shape of goats' horns has changed from oval to slightly grooved on one side. In wild barley, the grains are in two rows on the

central axis, whilst in cultivated barley they are in six rows. Evidence of this kind has shown that the harvesting of grain preceded its cultivation by thousands of years. Sickles for cutting grain and quern stones for grinding it predate agriculture by many thousands of years (Figure 41.23).

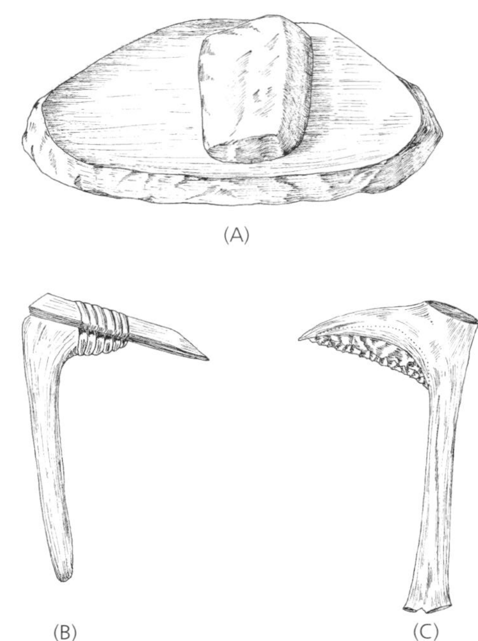


**Figure 41.23**
Some Neolithic tools. Quern stone to grind barley (A), hoe for digging (B) and sickle to harvest grain (C)

### *Origins of domestication*

Between 10 000 and 5000 years ago, agriculture originated independently in several parts of the world. The earliest evidence of domestication comes from a crescent-shaped region stretching between what is now Israel, through Syria and Iraq to Iran. Known as the 'fertile crescent', this area was especially favourable for the development of agriculture, since wild wheat and barley grew there. Sheep and goats were also native to the region. The earliest remains of domesticated animals are those of the dog (Iraq, 12 000 B.P.), whilst sheep and goats were being domesticated about 9000 years ago. Wheat and barley were being cultivated about 10 000 years ago. In the Yangtse Delta region of China, rice, millet, soybean, yams and pigs were being domesticated by about 7000 years ago. In Mexico, maize, beans, cotton, llamas and guinea pigs were under domestication by about 5000 years ago, perhaps earlier. Lima beans and kidney beans were also being domesticated in Peru 8000 years ago. In West Africa yams and oil palms were domesticated, whilst in sub-Saharan Africa, sorghum and millet (both cereals), were grown. From these and probably other centres, domestication of animals and plants spread to other parts of the world.

### *Advantages of domesticating animals*

The change from a nomadic to a settled life is unlikely to have been sudden – we need to envisage a transitional period. In this respect, animal husbandry presents less of a problem than growing crops, since nomadic herdspeople have just such an intermediate way of life. There would have been some obvious advantages in keeping animals alive:

- as a source of milk or hair, an animal may provide a steady income.
- a herd or flock is a 'living larder'; meat 'on the hoof' does not decay.

- when cattle were domesticated (about 8000 years ago) they could be used to carry and lift. The use of non-human muscle power represented another step in humanity's increasing exploitation of energy.
- animals became a source of wealth and could be traded.

## *Advantages of domesticating plants*

The benefits of plant domestication are less obvious than that of animals. Since wild plants seemed to satisfy all nutritional needs with the cost of relatively little labour, why should they have been domesticated with the uncertainties of crop failure? One possibility is that a seminomadic group may have returned annually to an area where wild barley or wheat grew, and perhaps harvested a surplus which may have been used to trade with other groups for scarce resources such as obsidian (volcanic glass) for making stone tools. Dependence on such a regular 'income' might have made it worthwhile staying in places where wild cereals grew most abundantly, and increasing yield by removing other plant species. It may have been some time later that the practice of deliberately sowing seeds developed. Whatever the precise details of the origins of plant domestication, it seems highly probable that it developed in stages, beginning with protection and regular harvesting of wild cereals, followed by deliberate sowing.

## *Agriculture and social complexity*

The Neolithic transition marked the beginning of a massive increase in the tempo of cultural change. Until recently archaeologists believed that the shift from food gathering to food production led to the establishment of permanent settlements. A group that stayed in one place for long enough to harvest crops sown some months earlier would be expected to build larger and more permanent dwellings. Freed from the necessity to carry them around, a settled community would develop a wider range of more specialised tools, including pottery.

These developments were, it was believed, accompanied by important changes in social organisation. The food surpluses that became possible in fertile areas such as river valleys, meant that not everyone needed to be directly concerned with food production. Carpenters, merchants and priests would have traded their services for food produced by farmers. Agriculture was thus seen as a catalyst for the development of social complexity, and the changes it spawned were called the 'Neolithic Revolution'.

This seductively simple and logical idea is now being called into question by recent excavations in the Middle East. One settlement in Syria just over 10 000 years old and consisting of permanent houses, was based entirely on a hunting and gathering economy. Five hundred or so years later it was replaced by a community which had domesticated plants, but continued to depend on the hunting of gazelle for their meat supply. Only later did they domesticate goats and sheep. Hunter gatherers in the Fertile Crescent were not the only people – or indeed the first – to build permanent houses. Excavations in Ukraine in Europe have exposed semipermanent houses built of mammoth bones dating back 30 000 years.

The change from a nomadic to a sedentary way of life can thus no longer be simply attributed to the beginning of agriculture. It is possible that cause and effect may actually be the other way round – the increased social complexity which accompanied the development of settlements may have contributed to the development of agriculture.

Whatever the cause and effect relationships between settlements and agriculture, there can be no doubt that the domestication of animals and plants did contribute directly to further technological developments, including the plough. Planting crops on the most favourable dates requires an accurate knowledge of the seasons, and therefore of the apparent movement of the stars. It was probably this which led to the beginnings of astronomy and to the keeping of a calendar.

## The cultural explosion

The *H. sapiens* populations that replaced the Neanderthals 30 000 years ago were, at least skeletally, indistinguishable from modern humans. The advances since then have been cultural rather than biological. Changes in knowledge, attitudes and customs are examples of **cultural evolution**. This differs in a number of ways from biological evolution.

- Whereas genetic information is stored in DNA and can only be passed from parents to offspring via the gametes, cultural information is carried in the brain. In the last few thousands years it has also been stored outside the body as writing, and very recently, electromagnetically.
- Genetic information only changes *between* generations, at meiosis and fertilisation, but cultural change occurs throughout life.
- Genetic information can only be handed 'downwards' from parents to offspring, but cultural information can be transmitted from any individual to any other, whether related or not, and even from offspring to parents!
- Genetic information changes by mutation and recombination. Because these are *random* processes, any beneficial effect is the result of pure chance; indeed, most genetic changes are at least slightly harmful. Cultural change, on the other hand, is often the result of deliberate intent and is therefore highly *purposeful*.
- Technological change provides the catalyst for further innovation. For example, the technique of baking clay ornaments to make them hard was a precondition for the making of pottery thousands of years later. The (probably accidental) discovery of copper smelting would have increased the likelihood that iron smelting (which requires a higher temperature) would follow. Cultural change can thus work by **positive feedback**, or 'auto-catalysis'. In contrast, genetic changes are quite independent of each other.

### QUESTIONS

**1** What human characters can be regarded as extensions of adaptations to arboreal life?

**2** *Homo habilis* had a larger brain and more slender limbs than the Australopithecines. How might these differences be linked?

**3** How might the exploitation of fire have helped the evolution of the capacity for speech?

**4** Match the following fossils (in italics) to the associated tool cultures:

| | |
|---|---|
| i) Mousterian | A *H. heidelbergensis* |
| ii) Oldowan | B *H. neanderthalensis* |
| iii) Acheulean | C *H. habilis* |

**5** What important advantages does mitochondrial DNA offer compared with nuclear DNA in the study of evolutionary relationships?

**6** What part did the ice ages play in the spread of modern humans?

CHAPTER

# 42

# The HUMAN PREDICAMENT

**LEARNING OBJECTIVES**

**By the time you have completed your study of this chapter you should be able to:**

- **state the causes of the various stages in the growth of the human population.**
- **explain the reasons for the differences in recent population trends between developed and undeveloped countries**
- **explain the causes and environmental effects of the following: discharge of sewage and farm effluent into rivers, use of persistent pesticides, acid rain, crude oil spills, heavy metal pollution, thermal pollution, the rising level of atmospheric $CO_2$ and its consequences, production of CFCs, deforestation.**

## The population explosion

It has been estimated that in the world today there are between 10 million and 100 million species – far more than the number actually named. In all species there is the potential for exponential growth of numbers, but this potential is never realised for long – natality always comes back into equilibrium with mortality because of factors such as predation, parasitism and competition for resources.

The human population has apparently been an exception. Human numbers have been increasing, not only in absolute terms (increase in numbers per year), but until very recently it has also been increasing in relative terms (percentage increase per year).

Population censuses are a recent thing, so we have no precise figures for human populations in ancient times. There are indications, however, that the present increase in human numbers is but the latest (and by far the biggest) of three such surges, each larger than the previous one.

The first began about 2.5 million years ago and coincided with the development and improvement of stone tools, culminating with composite tools of the Upper Palaeolithic (Chapter 41). These would not only have increased the supply of food but also improved defence against large carnivores.

Agriculture triggered a second population surge starting about 10 000 years ago, leading to a further improved food supply. The most recent – and by far the biggest – increase has resulted from a decrease in death rate as a result of the medical revolution of the last two centuries which has greatly reduced the impact of parasites (Table 42.1).

**Table 42.1**
Estimated populations in human evaluation at different times

| Time | Population (millions) |
|---|---|
| 2 000 000 B.P. | 0.125 |
| 300 000 B.P. | 1.0 |
| 25 000 B.P. | 3.3 |
| 10 000 B.P. | 5 |
| 2 000 B.P. | 200 |
| 1650 A.D. | 500 |
| 1850 A.D. | 1000 |
| 1930 A.D. | 2000 |
| 1960 A.D. | 3000 |
| 1975 A.D. | 4000 |
| 1987 A.D. | 5000 |
| 2 000 A.D. | 6000 |

One way of looking at the history of the human population is in terms of changes in the time taken for the population to double. Between 10 000 years ago and the year 1650 the population increased a hundred-fold – from an estimated 5 million to about 500 million. This required an average doubling time of about 1500 years. The next doubling, from 500 million to 1 billion, took 200 years. It took 80 years to double again to 2 billion, and 45 years to double again to 4 billion. Thus human population growth has been more than exponential (which would require a constant doubling time). It took 1–2 million years to reach the first billion, 80 years to add the second, 30 years for the third, 15 years for the 4th, and 12 years to reach the 5th, with a projected time of 13 years to reach the 6 billion mark.

These figures show that the term ‘population explosion’ is an appropriate one. A person born before 1930 will have seen the world’s population triple in his or her lifetime. Until about 1960, each population doubling took half, or less than half, the time of the previous doubling.

Since then however there has been a dramatic slowing of relative growth (though the absolute growth rate is still increasing). This reduction in population growth has been most marked in Western countries, where a sharp decrease in child mortality has reduced the need for large families, and new contraceptive measures have provided the means of achieving them. Whereas in agricultural societies every additional child is potentially another pair of hands, in industrial societies children are, in economic terms, a cost rather than a benefit. This, coupled with the education and emancipation of women, has resulted in some people postponing or even foregoing reproduction altogether in favour of a career. This drastic reduction in population growth following industrialisation is called the **demographic transition** (demography is the study of the growth and structure of human populations).

In many ‘Third World’ countries the history of population growth has been very different. Instead of following industrialisation, the reduction in mortality brought about by modern medicine *preceded* it. As a consequence, there has not only been a rapid increase in numbers, but a change in the age structure of the population, with a great increase in the proportion of people of prereproductive age (e.g. India, Figure 38.10).

This bulge at the base of the age group pyramid has serious implications for future population growth. Even if family size were to be stabilised at two *now*, world population would still approximately double during the next few decades, unless there were an increase in mortality. To bring about sustained zero population growth *now*, the average family size in many ‘Third World’ countries would have to drop to less than two. However, the family size needed for zero population growth would creep upwards to just over two as the age structure of the population gradually stabilised.

## *The effects of the population explosion*

When a population grows without check, one of two things eventually happens. Either it comes into a stable equilibrium as in the logistic growth curve, or it undergoes a crash. The former is what usually occurs when an increase in numbers exerts an immediate density-dependent effect. If however its effects are delayed, there may be a population overshoot, causing a J-shaped curve.

Clearly the human population will stabilise at some point, either as a result of an increase in mortality or a decrease in natality, or a combination of the two. The only points at issue are the relative contribution of changes in natality and mortality, the extent to which these changes will be voluntary, and the time scales over which they will occur.

The effects of human population growth are difficult to predict because the carrying capacity is constantly changing as a result of technological change. For example, had it not been for the development of 'miracle rice' and other new crop plants, world food production would have fallen behind consumption some years ago.

Whilst some people argue that the pace of technological change is so great that it will be equal to any environmental challenge, there are others who argue the opposite. According to this view it is the very speed of change that is the source of danger, because there is insufficient time to evaluate the effects of a technology until we have become dependent on it. Those who take this view would quote the destruction of stratospheric ozone as an example.

In view of these uncertainties, this chapter deals with some of the known effects of population growth where there is little controversy.

## Pollution

All animals produce wastes which are used by other species and thus removed. The rise in human numbers combined with chemical technology has also resulted in the introduction into the environment of new substances, many of which did not exist until recently and cannot be broken down by natural means (such as radioisotopes with very long half-lives).

Defining a pollutant is not as straightforward as one might think. We associate pollution with human activity, but it should be remembered that natural agencies can produce large amounts of highly toxic chemicals. Sulphur dioxide and hydrogen sulphide, for example, are both produced in vast amounts by volcanoes.

Another complication is that some substances, such as copper, are essential to life in trace amounts, but are deadly poisonous at higher concentrations. Nitrate is a plant nutrient, yet is implicated in producing cancer in children when present in drinking water. Carbon dioxide is a nutrient for plants, but it also plays a key role in the heat budget of the Earth, and an increase in its atmospheric concentration may be altering the Earth's climate.

Defining pollution in a way that satisfies everyone is difficult; for political or economic reasons definitions are elastic so that they are not too restrictive on economic activity. One simple but rigid definition would be: 'The introduction, by human agency, of material or energy into an environment which renders it to some degree less able to support human or other life.' On this criterion, a simple test of river water quality would be whether it is safe to drink large quantities of it.

Pollutants are of two general kinds. First, there are substances that occur naturally but which, as a result of human activity, occur in damagingly high concentrations. An example is the raised level of nitrates in drinking water. Substances of the second type are not found in nature, such as polychlorinated biphenyls (PCBs), and are often intrinsically harmful.

Not all pollutants are toxic. Some, such as chlorofluorocarbons (CFCs), are quite harmless in themselves, but become converted into harmful substances after release into the environment.

Pollutants also vary in their persistence. Some are broken down rapidly, so that the problem can be solved simply by preventing release. Others persist indefinitely, such as heavy metals and certain radioisotopes which take hundreds of thousands of years to decay to a small percentage of their original level.

## Clean water

Few people would be willing to drink from a river or lake in their neighbourhood as it may contain any of the following kinds of contaminant:

- pathogens, such as the virus causing poliomyelitis, and the bacteria that cause cholera, food poisoning and typhoid.
- degradable organic materials of biological origin, such as sewage and farm effluent.
- organic chemicals of industrial origin, such as oil, solvents, pesticides, polychlorinated biphenyls and detergents.
- plant nutrients, such as nitrate and phosphate ions.
- other soluble inorganic substances such as heavy metal ions, acids, alkalis and cyanide.
- particulate matter that is fine enough to remain suspended. Even if non-toxic, such materials reduce light transmission and interfere with photosynthesis.
- hot water from cooling systems of power stations. This reduces the oxygen content of water by raising the rate of uptake of oxygen by organisms. It also lowers the solubility of oxygen in water.

### *Sewage and farm effluent*

Faeces often carry the eggs of gut parasites, and in densely-crowded conditions they represent a potential health hazard. Until it was realised that diseases such as typhoid and cholera were spread by sewage-contaminated drinking water, most rivers running through towns were treated as sewers. Even today many coastal towns continue to pipe untreated sewage into the sea.

Quite apart from the threat of disease, depositing sewage into lakes and rivers can have a catastrophic effect on aquatic life, especially fish. This is not because faeces contain toxic materials, but because they are rich in decomposable organic matter. The oxidation of organic matter by saprobic bacteria uses up oxygen. Since oxygen is only sparingly soluble in water, there is usually far more organic matter in the water than the oxygen needed to break it down. Therefore conditions become **anoxic** (without oxygen), killing fish and other aquatic life.

The amount of oxygen needed to break down organic matter is proportional to the amount of organic matter present. This is the basis for the practice of expressing the concentration of organic matter in terms of the amount of oxygen needed to oxidise it. The **biochemical oxygen demand** (**BOD**) is a measure of this, and is expressed in terms of the amount of oxygen removed from a sample of water kept in the dark for 5 days at 20°C. Untreated sewage has a BOD of about 600 g $O_2m^{-3}$, compared with unpolluted river water which has a BOD of less than 5 g $O_2m^{-3}$. Treatment reduces the BOD of sewage to less than 30 g $O_2m^{-3}$.

Although BOD is a measure of the concentration of organic matter (and therefore of the potential for deoxygenation of the water) other factors can influence the level of deoxygenation:

- the rate at which oxygen enters the water from the air, which in turn depends on the surface area to volume ratio of the water and the level of turbulence. For a given concentration of organic matter, fast-running streams are much less likely to become anoxic than deep, slow-moving rivers.

- water temperature, which influences the rate of oxygen consumption by saprobic organisms. At low temperatures, oxygen removal is slow and is more equally counterbalanced by uptake of oxygen from the air.

The effect of organic pollution on aquatic life can be seen in Figure 42.1, which shows some of the chemical and biotic changes below a sewage outlet in a river.

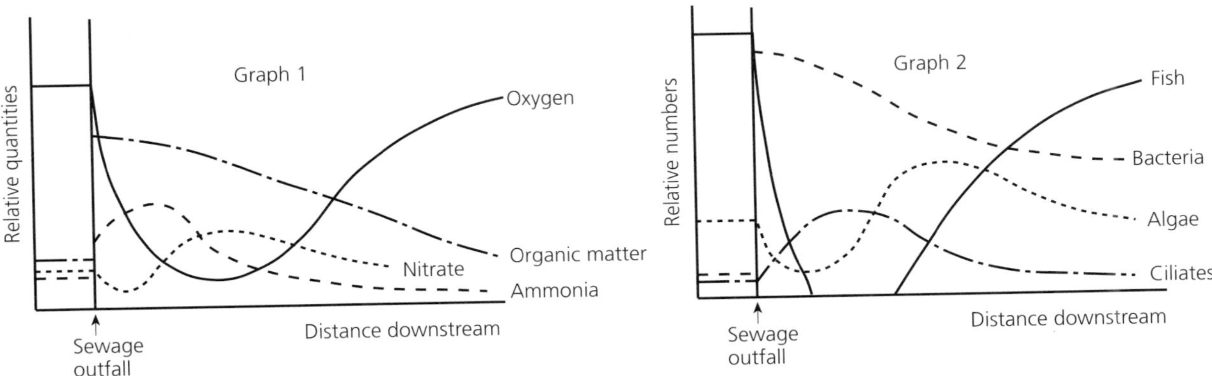


**Figure 42.1**
Chemical and biological changes in river water with increasing distance from a sewage outlet

Immediately below the outfall there is a sharp fall in oxygen due to respiration by saprobes. This provides the ideal conditions for denitrifying bacteria, which cause a decrease in nitrate levels. There is also an increase in ammonia produced by deamination of amino acids by saprobic bacteria, but this is subsequently converted to nitrate by nitrifying bacteria.

The massive rise in bacteria accounts for the increase in the protozoa that feed on them. The so-called 'sewage fungus' actually consists of dense mats of a filamentous bacterium (*Sphaerotilus natans*), within which a variety of protozoa live. The initial reduction in algae can be attributed partly to decreased light penetration due to suspended solids, and also to reduced oxygen, which would be a factor at night (when they are net oxygen consumers). As light penetration improves, the increase in nitrate and phosphate ions stimulates algal growth, with consequent increase in dissolved oxygen levels. One alga, *Cladophora glomerata*, or blanket weed, may form such dense mats that its respiration may cause deoxygenation of the water at night.

There are also changes in the abundance of different animal groups (Figure 42.2). Oligochaete worms, such as *Tubifex*, can tolerate low oxygen levels and thrive immediately below the outfall where there is no competition from other animals. Other animal groups reappear in characteristic sequence reflecting their differing sensitivities to anoxia. Last to reappear are stonefly and mayfly nymphs, and fish.

These changes occur in a regular sequence, both spatially (with increasing distance below an outfall), and temporally (with increasing time after cessation of sewage discharge). So orderly and predictable are the changes that they are used as an indicator of pollution levels, so that in some cases measurements of BOD are unnecessary.

For large cities, improper disposal of sewage is not an option, and sewage is treated in a sewage works. Sewage treatment involves two main stages:

- primary treatment, in which grit, sediment and other non-biodegradable solids are removed by physical means.

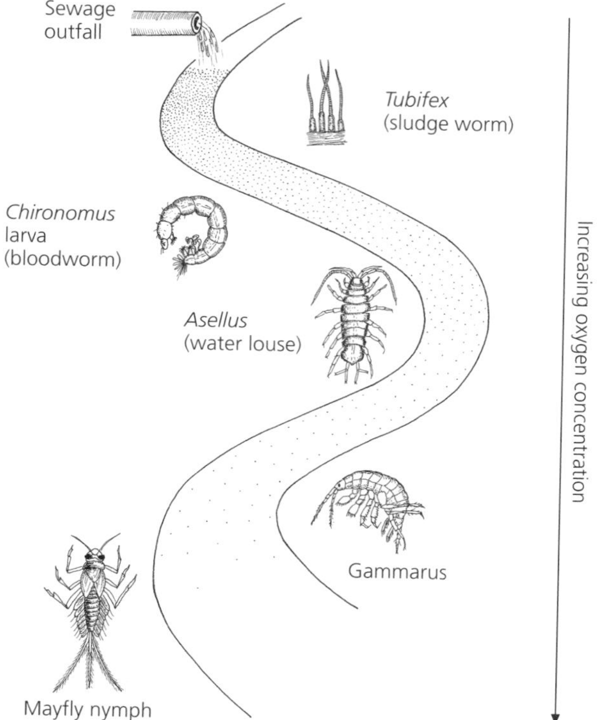


**Figure 42.2**
Some freshwater animals found at different distances below a sewage outlet into a river

- secondary treatment, in which the organic materials are broken down by bacteria. The final product is harmless sludge which, unless it is contaminated by heavy metals and other pollutants, can be used as fertiliser.

Sewage treatment has four advantages:

1 it prevents pathogens (disease-causing organisms) from reinfecting humans.

2 it prevents the killing of fish and other aquatic life.

3 the release of odours associated with decomposition is controlled.

4 sludge can be used as an organic fertiliser by farmers.

## *Eutrophication*

Although associated with pollution, eutrophication is actually a natural process, but one which normally occurs very slowly. When a lake is first formed, its water contains low concentrations of mineral ions and is said to be **oligotrophic** (meaning 'few nutrients'). Phytoplankton is meagre and the water correspondingly clear. Slowly, over thousands of years, mineral ions are washed from rocks into the lake, raising its fertility and causing it to become **eutrophic** (eutrophication literally meaning 'good feeding'). The dense growth of phytoplankton in eutrophic lakes increases the turbidity (cloudiness) of the water. Oligotrophic lakes are 'young' and eutrophic lakes are, under natural circumstances, mature.

Eutrophication can be greatly speeded up by human activity, in which case it is sometimes referred to as 'cultural eutrophication'. Run-off of nitrate and phosphate fertilisers greatly increases the concentration of plant nutrients in lakes, causing a 'bloom', or sudden surge, in plant life.

In itself a bloom is not usually harmful; the problems start in the autumn when the plants die back and decay. If all the oxygen produced by the plants during the bloom could remain in solution, there would be just enough to oxidise the plant matter. Unfortunately, oxygen is so sparingly soluble that most of it passes out of solution into the atmosphere during the bloom. By autumn, then, the amount of oxygen available for the reoxidation of the dead plant material is limited by the small amount that had remained in solution. The water therefore becomes **anoxic**, or deoxygenated, with the attendant effects on fish and other animal life.

A similar result occurs when sewage or farm slurry runs into a lake, with the difference that the first stage – growth of plants – is bypassed so that oxygen depletion is immediate.

## Pesticides

Unlike most other pollutants, pesticides are specifically designed to kill. Their targets are either organisms that compete with humans for food, such as agricultural pests, or animals that play a part in the life cycle of pathogenic organisms.

According to their targets, pesticides are put into broad categories. Thus insecticides are used to kill insects, fungicides kill fungi, and herbicides kill weeds.

Though pesticides have produced great benefits, some of their long-term effects have proved harmful. This is partly because they have not always been used correctly, but often the problems have proved to be more fundamental.

The hazards of indiscriminate use of pesticides are best illustrated by the use of **organochlorine** insecticides such as **DDT**, **aldrin** and **dieldrin**. When first introduced in the 1940s, DDT was highly effective against flies, lice and other insects, and was considered to be almost non-toxic to humans. It rapidly came into widespread use and was of great benefit in the control of lice (which carry typhus) and mosquitoes (some species of which carry malaria).

The public first learned of the dangers of the indiscriminate use of insecticides in 1962 when an American journalist, Rachel Carson, published a book called *Silent Spring*. Amongst the many examples she cited was the case of the grebes of Clear Lake in California. In 1949, in an attempt to get rid of hordes of non-biting but troublesome midges, the lake was sprayed with DDD, a close relative of DDT but less toxic to fish.

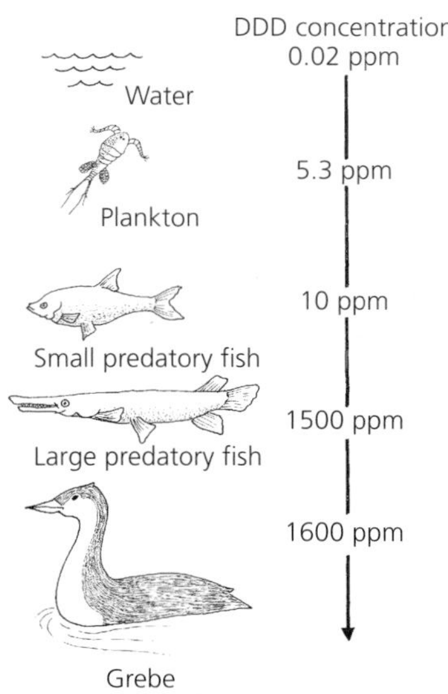


**Figure 42.3**
Food chain concentration of DDD in Clear Lake (after Flint and van den Bosch). ppm = parts per million

The initial results were spectacular, but repeated annual spraying became less and less effective, and was discontinued in 1954 when it was found that grebes were dying in large numbers. The fat of these fish-eating birds was found to contain DDD at concentrations of 1600 ppm, 32 000 times the original concentration (0.02 ppm) of the DDD in the lake water (Figure 42.3).

What had happened was that organochlorines are neither metabolised nor excreted, and since each animal eats many times its own body mass in its lifetime, the DDD becomes concentrated as it passes from one trophic level to the next. Since the grebes are at the end of the chain, they suffer a much higher dose than the midges.

The loss of effectiveness of the insecticide was but one of many cases of the evolution of resistance by pests. In other cases the pests have become even more numerous than they were before treatment because the insecticide has killed insect predators that normally keep the pest in check. Furthermore, predatory insects usually breed more slowly and take longer than the pests to evolve resistance, making the problem worse than ever.

Similar cases of **food chain magnification** have occurred in Britain. During the 1960s it was found that peregrine falcons and other birds of prey were often failing to breed. Tests showed that the birds were

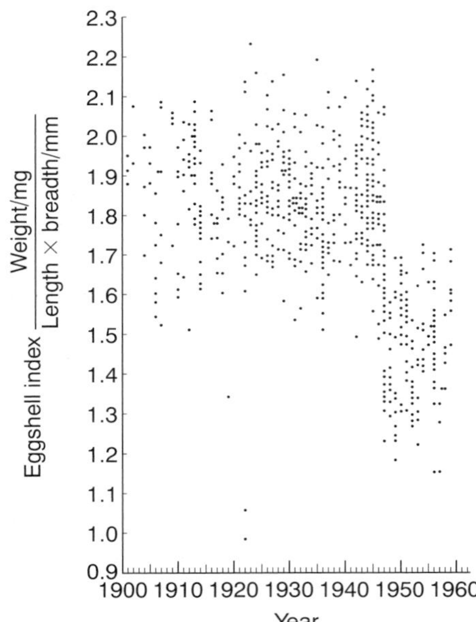


**Figure 42.4**
Change in relative thickness of eggshells of peregrine falcons over 60 years (after Ratcliffe)

concentrating DDT in their tissues, and that their eggs had abnormally thin shells that easily broke (Figure 42.4). DDT concentrations in non-predatory birds showed much lower levels. As top carnivores, peregrines were suffering the highest doses of the insecticide.

Another disadvantage with organochlorines is that they are not **biodegradable** by micro-organisms and are thus highly persistent. As a result of these ecological problems, organochlorines were banned in most countries. They were replaced by **organophosphate** insecticides such as **malathion** and **parathion**. These are related to nerve gases and are even more toxic than organochlorines, but are much less persistent.

## Acid rain

At the moment of its formation rain is pure water, even in industrial areas. As it falls through the air it picks up various materials, such as metal ions derived from dust, and gases such as $CO_2$, $SO_2$ and $NO_2$. $SO_2$ is produced in the burning of coal and oil, and $NO_2$ is formed in internal combustion engines. These latter gases dissolve in water to form nitric and sulphuric acids, and in industrial areas the pH of rain is frequently less than 5 (a pH of 2.4 has been recorded, a hundred times more acidic than pH 4.4).

Acid rain affects life both directly and indirectly. The low pH of the water can itself be harmful to life, but perhaps even more important is the effect on the solubility of certain mineral ions. For example at very low pH values, aluminium may be released from combination with rocks, raising its concentration to toxic levels.

Acid rain has become a serious problem in many countries. It is particularly difficult to deal with because winds do not respect political borders, and one country may suffer pollution generated in a neighbour. For example, much of the $SO_2$ reaching Southern Sweden has been produced in Britain, and Canadians are suffering the effects of $SO_2$ produced in the United States.

The extent to which the pH of lakes is lowered depends on the nature of the underlying rock through which rain water percolates. In limestone areas the effect is relatively small because the acid is neutralised by the basic rock. However in granite areas the pH of rivers is commonly below 5. There are large areas of Norway and Sweden where lakes are devoid of fish.

## Thermal pollution

Large quantities of water are drawn from some rivers and used to cool power stations. The water re-entering the river may be up to 12°C warmer than the river. Warming decreases the solubility of oxygen in water, so if oxygen-saturated water is warmed, some oxygen comes out of solution. If the water is less than saturated, warming initially raises the percentage saturation (though not its oxygen concentration). Only when it reaches 100% saturation is oxygen driven out of solution. For example if water that is 70% saturated is warmed from 15°C to 30°C, it would raise its oxygen-saturation to 92% – insufficient to drive any out of solution. The situation is similar (though the opposite way round) to that of the effect of temperature on the humidity of the air. A fall in temperature raises the *relative* humidity but has no effect on the actual quantity of water vapour in a given volume of air until 'dew point' (100% saturation) is reached. Below this temperature water vapour condenses out of the air, lowering the concentration of water vapour.

Moderate warming of water that is only half saturated thus has no effect on its oxygen content (though it does raise its percentage saturation). Another factor is that as water passes through the cooling towers, it is highly agitated and equilibrates fully with the atmosphere and, even at the warmer temperature, contains more oxygen than most river waters.

In a polluted river, the oxygen saturation rarely rises above 50%, so warming the water to this extent does not *in itself* deoxygenate the water. Indirectly, though, heat does lower dissolved oxygen levels by raising metabolic rates and thus oxygen demand. Possibly for this reason there may be a noticeable change in species composition for some distance below the site of warm water discharge.

## Crude oil

Crude oil consists chiefly of hydrocarbons together with some sulphur-containing compounds. The hydrocarbons are a complex mixture of various molecular weights and volatility, and are of little use until they have been separated by fractional distillation in a refinery. Most of the oil transported from producer to consumer countries is by ocean-going tanker. Both their number and size have grown enormously over the years, and accidental spillages are inevitable.

The volume of water in the ocean is enormous – equivalent to 245 million cubic metres for every human on earth. Crude oil, however, is insoluble and floats on water, remaining concentrated in the form of large **slicks**. Where oil pollution is concerned, it is the *surface* of the ocean, rather than its volume, that matters, and this is only about 10 000 $m^2$ (equivalent to a 100 m $\times$ 100 m square) for every human being on earth.

Oil pollution became headline news in 1967 when the *Torrey Canyon* went aground off the Cornish coast, releasing over 120 000 tonnes of crude oil. Over 80 km of the coast of Cornwall and a stretch of French coast were seriously contaminated. In an attempt to disperse the oil, beaches were sprayed with detergent. Industrial detergents are far more harmful than those used domestically, and the treatment proved to be disastrous; the detergents used were more toxic than the oil, killing crustaceans, molluscs, fish and seaweeds.

More effective methods are now used. One way is to absorb the oil by covering it with a material like straw or wool. Another method involves covering the oil with chemically-treated sand which becomes coated with oil before sinking. Nowadays detergents are not used except on holiday beaches.

The immediate effects of oil spills on marine life are serious. Seaweeds are smothered and die because of lack of light and air. Gills of invertebrates such as bivalves may become clogged, interfering with gas exchange.

The damage to sea birds is particularly serious. With oil-clogged feathers they are unable to fly and quickly lose heat and may die of cold. In attempting to preen themselves, the birds may ingest the oil and become poisoned, developing intestinal ulceration and haemorrhages. Removal of the oil with solvents also removes the natural water-proofing oil from the feathers.

The chief casualties are species that spend much of their time swimming on the surface, such as divers and gulls. The vast majority of affected birds die at sea. After the *Torrey Canyon* disaster about 8000 birds were collected, but only 150 were well enough to be released. Some of these were marked and found dead within a month.

Though the immediate impact of an oil spill on marine life is dramatic, the long-term effects appear to be less serious. Oil is naturally broken down by bacteria, and a few years afterwards, little sign of the disaster remains.

### Heavy metals

Unlike most other forms of pollution, heavy metals are neither produced nor broken down, but are relocated. Human activity merely disperses them more widely from their original concentrated location in mineral deposits in which they are safely locked away.

Lead, cadmium and mercury inactivate enzymes and are probably toxic to humans to at least some degree at all concentrations. Lead and mercury, in particular, are known to affect the functioning of the nervous system. As enzyme cofactors, copper and zinc are essential in the diet at low concentrations but are toxic at higher levels.

The most widely-documented effects of heavy metal poisoning are those of mercury and lead. The best known large scale poisoning due to the release of a heavy metal into the environment was at Minamata Bay in Japan in 1953. Mercury was released by a factory into the bay where it was converted by bacteria into an organic form, **methyl mercury**, which is far more toxic than elemental mercury. The mercury entered the food chain and, via fish, was eaten by fishermen and their families. Many suffered acute nervous disorders including paralysis and, in some cases, death.

The most widespread poisoning in Britain by a heavy metal was due to lead, which was routinely added to petrol in the form of tetraethyl lead to raise its octane rating. After burning in the engine, the lead was emitted in the exhaust fumes as minute particles which settled near roads. In addition, lead was widely used in paints and in domestic water piping (the Latin name for lead is *plumbum* (symbol Pb) from which the term 'plumbing' is derived).

After epidemiological studies showed a correlation between children's blood lead levels and impaired mental performance, lead in plants, domestic piping and petrol was eventually banned.

### The 'greenhouse' effect

Carbon dioxide is present in the atmosphere at a concentration of about 350 parts per million. Besides being a raw material for photosynthesis, it also plays a vital role as a kind of 'heat jacket' round the Earth. Though transparent to visible light, it is opaque to the longer wavelength infrared rays. Sunlight absorbed by the Earth's surface is re-emitted as infrared

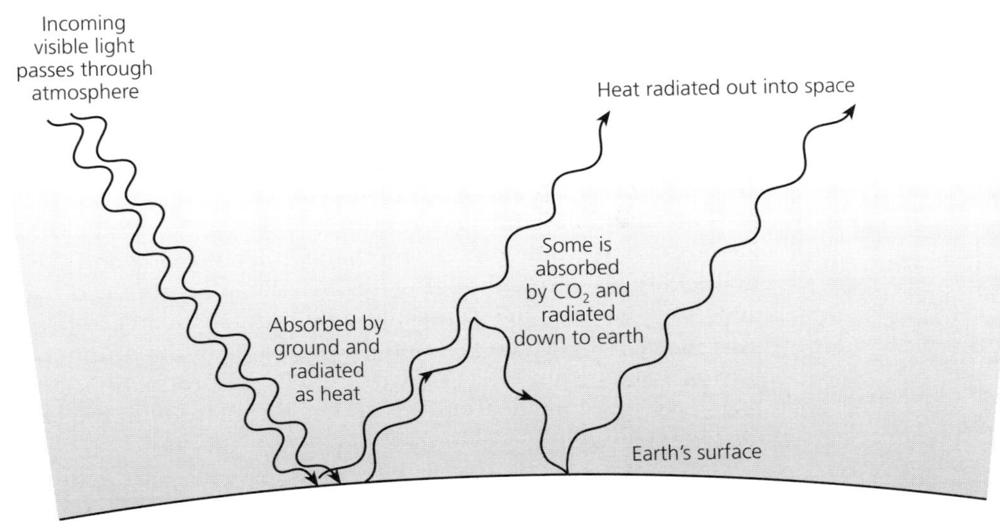


**Figure 42.5**
How carbon dioxide warms the Earth

radiation, which is trapped by atmospheric $CO_2$ instead of radiating out into space. The result is that the average temperature of the earth is 15°C instead of the −20°C it would be without $CO_2$ (Figure 42.5).

Since the Industrial Revolution, the burning of fossil fuels and the oxidation of forest timber (by burning or by decay) has led to a steady increase in atmospheric $CO_2$. As long ago as 1896 the Swedish chemist Svente Arrhenius predicted such a rise and said that it would cause the mean temperature on the Earth to increase. Only recently has the possibility of global warming been taken seriously.

Evidence of a rise in atmospheric $CO_2$ comes from direct measurements made since 1957 at the top of Mauna Loa, a mountain in Hawaii (Figure 42.6). At an altitude of over 4000 metres and in the mid-Pacific, the atmosphere there is relatively uncontaminated by proximity to human habitation. The graph shows not only a steady rise in $CO_2$ from 315 ppm in 1957, but a slight seasonal fluctuation as the balance between respiration and photosynthesis changes (Chapter 39).

Although direct measurements of atmospheric $CO_2$ have only been made on a systematic basis since 1957, analysis of air trapped in ice thousands of years old in Antarctica has shown that before the mid-19th Century, the level of $CO_2$ was about 270 ppm.

The two essential ingredients in the situation are not in dispute: $CO_2$ traps outgoing heat radiation, and $CO_2$ levels are rising. What is uncertain is the extent to which climate will change as a result. Predictions are made more difficult because it appears that about half the net annual production of $CO_2$ is absorbed in the oceans and used in photosynthesis by phytoplankton. Some of the fixed carbon sinks to the ocean floor in the form of dead organisms, delaying or preventing its return to the atmosphere as $CO_2$.

The magnitude of this effect is itself likely to be influenced by warming of the oceans. Since the solubility of $CO_2$ decreases with rise in temperature, the more the seas warm, the less $CO_2$ they can absorb.

Other feedback effects complicate the picture still further. For example a rise in temperature will increase evaporation of water from the oceans. The resulting increase in cloud cover would raise the reflectance of solar energy by the Earth. This would be a *negative* feedback effect and would tend to stabilise world temperature.

On the other hand there may be positive feedback effects. For example, the melting of the permanently frozen ground (permafrost) in North America and Asia, in which there are enormous quantities of dead organic

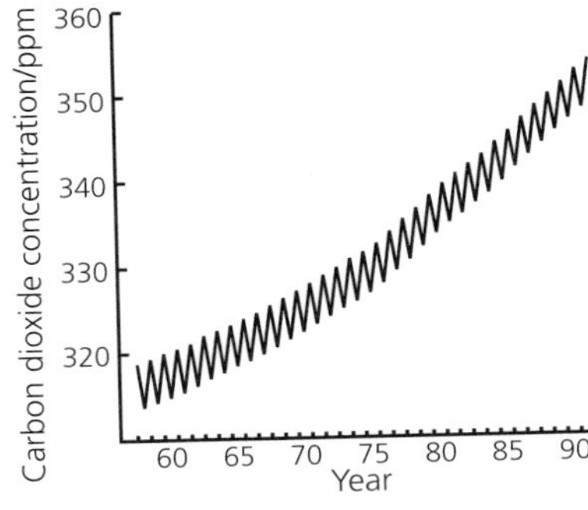


**Figure 42.6**
Changes in $CO_2$ concentration measured at Mauna Loa, Hawaii

**CLEAR THINKING**

It is often said that we need forests because they act as 'sinks' for $CO_2$. As long as a tree is growing, more $CO_2$ enters the tree in photosynthesis than leaves it in respiration. When the tree dies, its carbon is returned to the atmosphere as $CO_2$ by decomposers. In a mature forest the rate of removal of $CO_2$ in photosynthesis is balanced by its addition in respiration. A mature forest is therefore not a sink for $CO_2$.

When a forest is cut down, all the carbon in the trees is converted back to $CO_2$, either by decay or by combustion. The only way this could be prevented would be to treat the timber to stop its decay, and preserve it for furniture or building materials.

Young forests, on the other hand, *are* sinks for carbon, and remain so until their total biomass has stopped increasing. Planting forests, therefore, does help to reduce the atmospheric $CO_2$ content, provided that the trees are not harvested!

matter, would bring about the release of very large amounts of $CO_2$. This would cause further warming and production of more $CO_2$.

Despite these and other uncertainties, most atmospheric scientists agree that if present trends continue, mean annual temperatures are likely to rise by about 2°C in the next century. To put this seemingly small rise into perspective, mean temperatures during the most intense ice age were about 5°C lower than they are today.

The situation is further complicated by the fact that certain other gases, such as water vapour, methane, nitrous oxide and CFCs (see below) are also 'greenhouse' gases. Methane is produced naturally by anaerobic bacteria in swamps, and by mutualistic micro-organisms in the guts of many herbivorous animals. Increases in the numbers of cattle are having a significant effect on world methane production. Nitrous oxide is produced in denitrification, which has increased with the large scale use of nitrate fertilisers.

There are several possible effects of a rise in world temperatures. Some of the most obvious include the following:

- a rise in sea level due to the thermal expansion of the oceans and the melting of terrestrial ice caps (melting of *floating* ice has no effect on sea level).
- a change in the wind patterns and distribution of rainfall with more extremes of weather. In particular the frequency of tropical cyclones (which develop over warm oceans) is likely to increase.
- an increase in the rate of photosynthesis, both as a direct result of increased $CO_2$ and also as a result of warmer conditions.

If photosynthesis increases, could the increased rate of removal of $CO_2$ counteract the effect of fossil fuel burning and deforestation? No, unless extra plant material is allowed to remain as biomass, and is not oxidised by combustion or respiration. Any increased photosynthesis is likely to be in crops, which will be eaten and rapidly oxidised to $CO_2$.

At various times in the geological past, atmospheric $CO_2$ concentrations and global temperatures have been higher than they are now, and species have adapted to them. Is it not likely that similar adaptation will occur in response to human-induced changes that are occurring now? Studies of preserved pollen have shown that many times in the past plants have shifted their distribution in response to changes in climate. Could similar responses occur again?

Species with long-range dispersal mechanisms and short life histories probably would be able to adapt, but it is likely that many trees, which take decades to reach reproductive age, would not.

### *The ozone 'hole'*

The sun is like a gigantic hydrogen bomb except that it 'burns' its fuel relatively slowly. Like a hydrogen bomb it emits the entire spectrum of electromagnetic radiation, from extremely high energy and short wavelength $\gamma$ rays at one extreme, to low energy, long wavelength radio waves at the other. The shortest wavelength radiation is harmful because its particles, or **photons**, have enough energy to break biological molecules.

Just beyond the visible spectrum is ultraviolet radiation, with wavelengths of 200 to 400 nm. Whereas photons of $\gamma$ rays and X-rays have enough energy to indiscriminately smash any molecule, ultraviolet photons have just enough energy to break certain bonds. Most importantly, radiation of around 260 nm is specifically absorbed by DNA, causing mutations (Chapter 13).

UV of wavelengths maximally absorbed by DNA is therefore deadly to life. Fortunately radiation of these wavelengths is all filtered out by ozone in the upper atmosphere. Some UV of slightly longer wavelength (290–340 nm) does get through and causes sunburn and skin cancer.

Ozone is triatomic oxygen ($O_3$). It is a deadly poisonous, pale blue gas that is harmful even in small amounts. At sea level it is formed by electric discharges and, increasingly, by the exhaust gases of motor cars. At a much safer distance, it is also present in the upper atmosphere about 15–50 km above our heads. This region is called the **stratosphere** and is defined as that part of the atmosphere in which temperature increases with altitude, in contrast to the **troposphere** in which it gets colder with increasing altitude.

Ozone is formed by the same radiation that it protects us against, UV. Photons with wavelengths in the region of 190 nm have enough energy to break the bond in diatomic oxygen ($O_2$) to form free oxygen:

$$O_2 \xrightarrow{UV} O + O$$

Free oxygen atoms are extremely reactive and combine with oxygen molecules to form ozone:

$$O_2 + O \rightarrow O_3$$

The reaction yields heat, which is responsible for warming of the stratosphere. Most of the UV of wavelengths shorter than 200 nm is absorbed in ozone formation.

The bonds holding the ozone molecule together are weaker than those in diatomic oxygen and can be broken by longer wavelength (lower energy) UV:

$$O_3 \xrightarrow{UV} O_2 + O$$

The atomic oxygen can then either react with diatomic oxygen to reform ozone, or with ozone to form diatomic oxygen:

$$O_3 + O \rightarrow O_2 + O_2$$

Ozone is thus continuously being destroyed, and until recent decades its destruction and formation were occurring at roughly equal rates.

Since the 1950s, **chlorofluorocarbons** (CFCs) have altered this equilibrium. These gases are unreactive and non-toxic, and were introduced as coolants for refrigerators in 1920 and as aerosol propellants in 1950. Because of their chemical inertness, they are slowly carried up into the stratosphere unchanged where they are broken down by UV radiation, yielding free chlorine atoms. The chlorine then reacts with ozone to form chlorine monoxide, which reacts with atomic oxygen to regenerate a chlorine atom:

$$Cl + O_3 \rightarrow ClO + O_2$$

$$ClO + O \rightarrow Cl + O_2$$

Since the chlorine is regenerated, it takes part in a catalytic cycle. Each chlorine atom breaks down over 100 000 ozone molecules before eventually being destroyed in other reactions (such as combining with another chlorine atom). The overall result, obtained by adding the above two equations, is:

$$O + O_3 \rightarrow O_2 + O_2$$

Since the first concerns were expressed by Sherwood Rowland in 1972,

### Ozone and terrestrial life

Before the evolution of green plants, terrestrial life was impossible because the land was continually being irradiated by UV. In the oceans life was protected because water absorbs UV, and oxygen produced in photosynthesis began to accumulate in the atmosphere. Only then could ozone begin to form and terrestrial life begin to evolve. Life on land thus owes its origin to the phytoplankton.

the potential dangers of CFCs have been confirmed, but the picture has also become more complicated. It was originally supposed that ozone depletion occurred in all parts of the atmosphere exposed to UV, but in October 1980 an area of severe ozone depletion was detected over the Antarctic. This 'hole' has reappeared every spring, only to be dispersed as ozone-low air is carried to lower latitudes. Not only has this ozone 'hole' reappeared every subsequent spring, but it has become bigger, and a similar hole has appeared above the Arctic in the Northern spring.

Why is ozone depletion concentrated in the polar spring? It turns out that much of the chlorine released in the breakdown of CFCs combines with other compounds, thus becoming temporarily unavailable for ozone destruction. In the extremely cold polar air these compounds are broken down in winter, releasing the chlorine. Nothing further happens until the spring sunshine when the UV begins the catalytic breakdown of ozone, forming the 'hole'. Figure 42.7 shows how closely the decline in ozone above Halley Bay in Antarctica correlates with the increase in atmospheric chlorine.

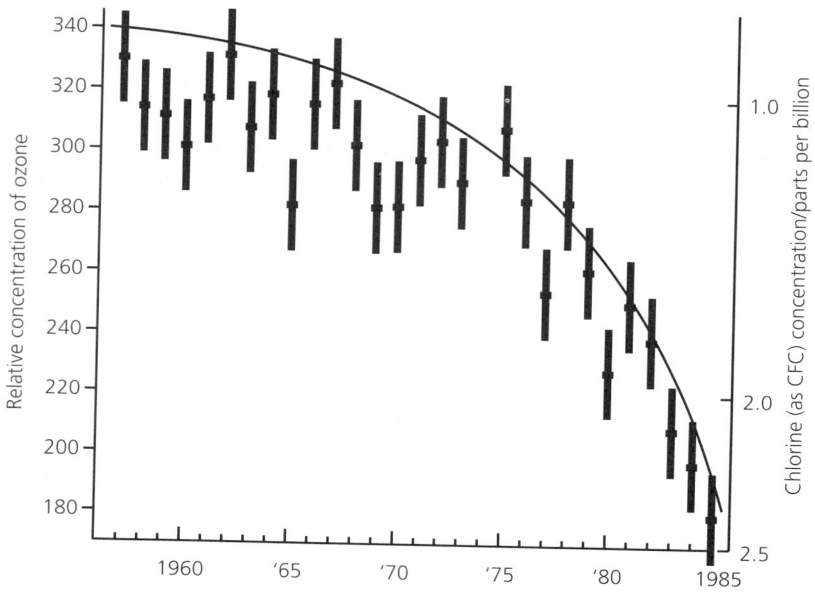


**Figure 42.7**
Close parallel between the increase in CFCs above Halley Bay in Antarctica (smooth curve) and the decrease in the amount of ozone. Vertical bars for ozone measurements are uncertainties. The scale for CFCs is upside-down to facilitate comparison (after Gribbin)

CFCs are not the only 'ozone-eaters'. Nitrous oxide ($N_2O$) produced in denitrification, has the same effect. It is also inert and reaches the stratosphere where it is broken down by UV to nitric oxide, which destroys ozone.

One of the most serious aspects of the CFC problem is that these compounds take between 75 and 110 years to be broken down in the atmosphere, so even if production were completely halted now, the problem would still exist at the end of the 21st Century.

## Deforestation

The world's forests are being cut down at an unprecedented rate; a high proportion of the forests of S.E. Asia have already been felled, and if present rates continue most of the Amazon forest will have disappeared by the middle of the next century.

There are three main motives for cutting down forests:

- the need for warmth; forests are being used for firewood by expanding human populations.
- the need for food; land may be cleared for agriculture.

- to make money; forests in Thailand are being cut down and the timber sold for export, and in parts of Brazil they are being cleared for export beef production.

The effects of deforestation are several:

- the oxidation of the timber, whether by combustion or by decay, contributes to the 'greenhouse' effect.
- on hillsides, it often causes erosion of topsoil and the silting up of waterways.
- forests act as gigantic sponges, absorbing heavy rains and releasing water slowly. On hillsides forests therefore help to prevent flooding and soil erosion; the cause of flooding in Bangladesh may lie thousands of kilometres away in deforested Himalayan slopes.
- after exposure to air and sun, some tropical soils become converted into hard mineral matter called **laterite**, and are completely unsuitable for growth of plants.
- loss of biodiversity. Tropical forests in particular contain huge numbers of species. One species of tree, for example, has been found to have over 1500 species of *beetle*.

### Could deforestation cause the world to run out of oxygen?

**Besides absorbing $CO_2$, trees also produce oxygen, and it has been suggested that deforestation and the burning of fossil fuels could lead to a serious reduction in our oxygen supplies. Is this a realistic threat?**

**It is believed that all the $1.2 \times 10^{15}$ tonnes of oxygen in the atmosphere have been made by plants in photosynthesis. When the first water-splitting bacteria began to produce oxygen, there was little or none in the atmosphere. Slowly, over immense periods of time, oxygen accumulated in the atmosphere, balanced by an accumulation of reduced carbon in the form of fossil fuels and deep ocean sediments.**

**For every molecule of oxygen produced, one molecule of $CO_2$ was reduced to organic compounds. So if all the organic matter were to be burned, all the oxygen would be used up again.**

**Now biomass accounts for a minute fraction (about 0.002%) of the total reduced carbon. Most is in the form of fossil fuel, only a fraction of which is recoverable for burning. If all the forests and recoverable fossil fuel were to be burned, it would only reduce atmospheric oxygen to about 20.05%. This would produce the same effect on the oxygen concentration as climbing to an altitude of 150 metres. There is therefore no possibility that the world could run out of oxygen.**

# Looking back – and forward

Within the space of three million years, an insignificant African ape with a brain no larger than that of a chimpanzee has evolved into a species that has been to the Moon and back. The explosion of human numbers and technology are having effects that some people think are threatening the future of our species. To understand the nature of the human predicament we need to appreciate our own biological nature and evolutionary history.

The dominance of *Homo sapiens* is due to an enormous brain which is larger relative to body mass than in any other species. The evolution of the human brain can be attributed to certain key events in our evolutionary history.

The most fundamental occurred 60 million years or so ago with the evolution of a grasping hand. For these early primates, the close co-ordination between hand and eye was simply a means of holding on to branches. Tens of millions of years later this ability was a preadaptation to the manipulation of objects and, later, the use and manufacture of tools.

Almost as characteristic of primates as a grasping hand is a complex social life, in which each individual knows every other. As explained in Chapter 41 it seems increasingly likely that for a male, reproductive success in a complex social environment requires more than brute strength. In chimpanzees and in some monkeys, males achieve status at least as much by political skills as by brute force, and the requisite social skills require intelligence. It seems likely that the demands of social success remained an important factor in the increase of intelligence throughout much of human evolution.

An event of great significance, was the development of bipedalism over 3.5 million years ago. By freeing the hands it eventually made possible the manufacture of increasingly sophisticated tools. As weapons, these not only increased the effectiveness of humans as predators, but also reduced the threat from large carnivores.

Perhaps equally as important as bipedalism was the evolution of the capacity for speech. As explained in Chapter 40, the transmission of complex ideas would probably have been impossible without speech. Though we will probably never know when our ancestors began to develop spoken language, it must have had a profound impact on learning and probably on the continued evolution of intelligence.

From 2.5 million years ago until about 10 000 years ago stone tools became increasingly sophisticated and probably increasingly important in human survival. The improvement in tool manufacture involved two quite different evolutionary processes:

- the increase in brain size and associated intelligence, resulting from natural selection.
- the acquisition of new techniques and skills, learned from others.

As explained in Chapter 41, the first is an example of **biological evolution** and involves genetic change. The second is **cultural evolution**.

As described in Chapter 40, cultural evolution has occurred to a limited degree in a number of primates besides humans. Only in humans has it accelerated to the point where virtually all change in human populations is now cultural. The increase in the importance of cultural change seems to have occurred in two stages.

- A period between 2 million years and 100 000 years ago – during which cultural change and brain expansion were occurring together, possibly each accelerating the other.
- The last 100 000 years – in which little detectable biological change occurred, but cultural change continued to gather speed.

The pace of cultural evolution surged forward with the development of agriculture, beginning about 10 000 years ago (Chapter 41). Since then improvements in technology have catalysed further improvements, and so on, in a positive feedback cycle.

The explosion of technology has enabled humans to all but eliminate the threat from predators and from many competitor species, and some of the disease-causing parasites. Humans have come close to winning the 'evolutionary arms race' that exists between predators and prey (Chapter 37). This is something that no other species has ever done, and it has resulted in the explosive increase in population and the attendant problems of overcrowding, resource depletion and pollution. Only organisms that can reproduce rapidly have been able to evolve counter-measures to human technology, for example antibiotic-resistant bacteria and insects resistant to insecticides.

The problems presented by the impact of *H. sapiens* are proving politically difficult to deal with for several reasons, some of which are listed below.

- There is no general agreement as to the number of people the Earth can support; opinions vary between 1 billion and 100 billion.
- Though we do not know the Earth's carrying capacity, technology is constantly changing the situation. New varieties of cereals have greatly increased food production in the last 20 years, and have given a 'breathing space'.
- Some of the effects of technology are delayed, and are hence difficult to evaluate until harm has been done. The damage to the ozone layer is a case in point.
- There is a conflict between the short-term needs of the individual and the long-term needs of the group, a conflict encapsulated by Garrett Hardin in his essay *The Tragedy of the Commons*. The problem is essentially this. When a resource is shared by a large group, it is to the disadvantage of the individual to show restraint in exploiting that resource unless everyone using the resource is in some way restrained. Individuals therefore over-exploit resources even though in so doing, the eventual outcome is ruin for all. The history of fisheries exploitation, which has been repeated time and time again, is an illustration.
- Our view of the world is conditioned by the human life span. The idea that present actions may be putting at risk the environment of future generations is therefore not a popular one. Politicians, whose horizons seldom extend beyond the next election, do not find it easy to take the long view if it is likely to cost votes in the short term.

## QUESTIONS

**1** Explain why

**a)** levels of dissolved oxygen in lakes may fall considerably after enrichment with mineral ions, particularly phosphates.

**b)** following repeated doses of insecticide, insect pests may reach higher levels than before treatment with insecticide.

**c)** an agricultural crop is more vulnerable to fungi and insects than the ancestral plant populations.

**d)** to bring about *immediate* zero population growth, the average number of children would have to be reduced to less than 2.0.

**2** True or false:

**a)** If all the world's forests were cut down and the timber treated with preservative to prevent decay, the concentration of $CO_2$ in the atmosphere would rise (ignore the death and decay of leaves, twigs, animals and fungi).

**b)** Deforestation threatens to reduce the supply of oxygen.

**3** Are freedom from disease and hunger compatible in the long term with unlimited freedom to reproduce, consume and pollute?

## EXAM QUESTIONS

**1** An area of rain forest was cleared for agriculture. Two experimental plots of land were marked out. On one, rice was grown continuously. On the other, crops of rice and beans were alternated. No fertiliser was added to either plot. The graph shows the crop yields for the two plots over a period of time.

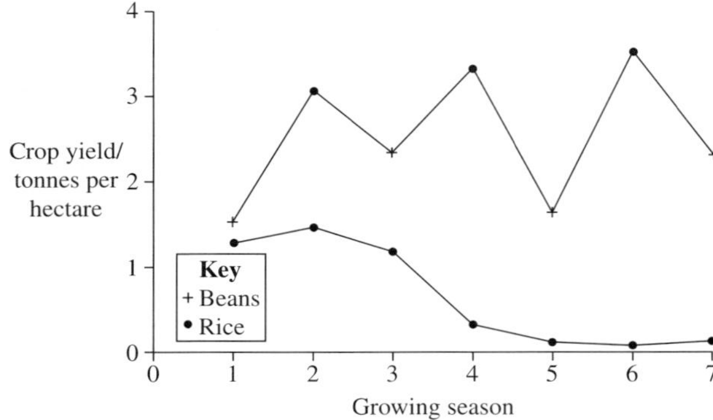


**a)** Explain why:

**i)** the yield fell when rice was grown continuously; *(1 mark)*

**ii)** the crop yield was maintained when rice and beans were alternated. *(2 marks)*

**b)** In another investigation, it was found that the biomass of decomposers fell from 54 g $m^{-2}$ to 3 g $m^{-2}$ when the rain forest was converted to permanent agriculture. Explain how you would expect this fall to influence the recycling of nutrients. *(2 marks)*

**AEB**

**2** **a)** In 1993 a group of scientists found fragments of the fossilised skull of a new hominid near the African village of Aramis. No hip or leg bones were found, yet the scientists believe that the hominid was bipedal.

**i)** Suggest *one* piece of evidence in the fossil fragments which may have led the scientists to believe that the hominid was bipedal. *(1 mark)*

**ii)** Give *two* features of the genus *Homo* which set it apart from the australopithecines. *(2 marks)*

**b)** An important development in the evolution of the genus *Homo* was forming hunter gatherer groups. Each of the following is thought to have exerted a selection pressure on the genus. Suggest the likely outcomes of each of these selection pressures.

**i)** Hunter gatherers must communicate. *(1 mark)*

**ii)** Hunter gatherers must travel considerable distances and carry the kill back to a home base. *(1 mark)*

**iii)** Catching and killing larger, faster animals requires ingenuity. *(1 mark)*

**AEB**

**3** The diagrams below show the skulls of two primates, a New World monkey and a chimpanzee.

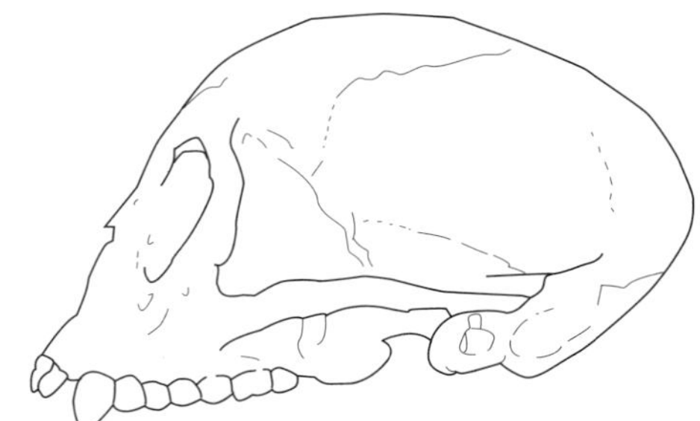

New World monkey

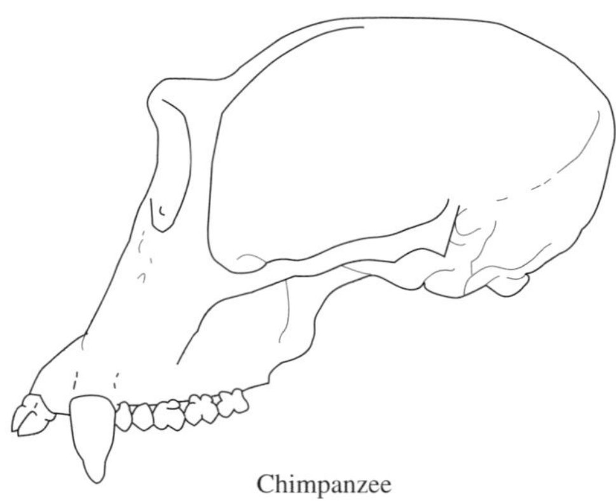

Chimpanzee

**a)** State *two* features, visible in the diagrams, which are characteristic of primates. *(2 marks)*

**b)** Give *two* visible differences, excluding overall size, between the skulls of the New World monkey and the chimpanzee. *(2 marks)*

**EDEXCEL**

## EXAM QUESTIONS

**4** The closeness of the relationship between different hominoids can be estimated by comparing their blood sera using a precipitation test. Antibodies precipitate proteins in the blood serum.

In such a test, human serum is injected into another animal, such as a rabbit. The rabbit responds by producing antibodies. When these antibodies are isolated and mixed with human blood serum, precipitation occurs. If, however, serum from a different hominoid species is used, less precipitation occurs. Species which are closely related to humans produce almost as much precipitation, whereas more distantly related species produce a smaller precipitate.

The results of a blood serum test on a range of hominoids are given below, expressed as percentages.

| | |
|---|---|
| Human | 100% |
| Chimpanzee | 97% |
| Gorilla | 92% |
| Gibbon | 79% |
| Baboon | 75% |

**a)** Antibodies are specific in relation to the molecules with which they interact. In this context, give the meaning of the term *specific*. *(1 mark)*

**b)** What do the results indicate about the similarity to human blood proteins of the proteins in the blood of the baboon? Explain your answer. *(2 marks)*

**c)** The diagrams below show two possible genealogical trees linking four of the above species.

**i)** Which of the trees, **A** or **B**, is best supported by the evidence from the serum precipitation test? Give a reason for your answer. *(2 marks)*

**ii)** On the tree you have chosen in your answer to **i)** add a branch which best shows the relationship of the baboon to the other species as indicated by the precipitation test. *(1 mark)*

**d)** Explain why similar protein structures may indicate a close genetic relationship between two species. *(2 marks)*

**e)** An alternative method of comparing the proteins of related species is to find the amino acid sequences of specific protein molecules such as haemoglobin. Explain why this may be considered a better method than serum precipitation. *(3 marks)*

**EDEXCEL**

**5** *Ectocarpus siliculosis* is a marine alga found growing on the shore and on the hulls of ships.

**a)** *Ectocarpus siliculosis* belongs to Phylum Phaeophyta. State *one* characteristic feature you could use to distinguish this organism from those of other algal Phyla. *(1 mark)*

**b)** Two populations of *Ectocarpus siliculosis*, **A** and **B**, were sampled. Samples of population **A** were collected from an unpolluted shore. Samples of population **B** were collected from the hull of a ship painted with antifouling paint which slowly releases copper into the water.

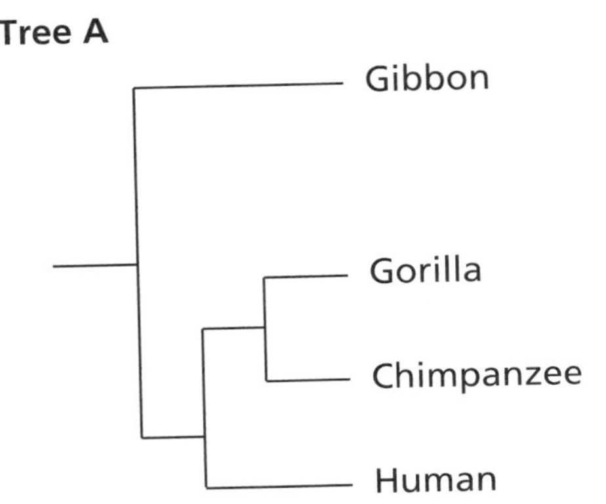


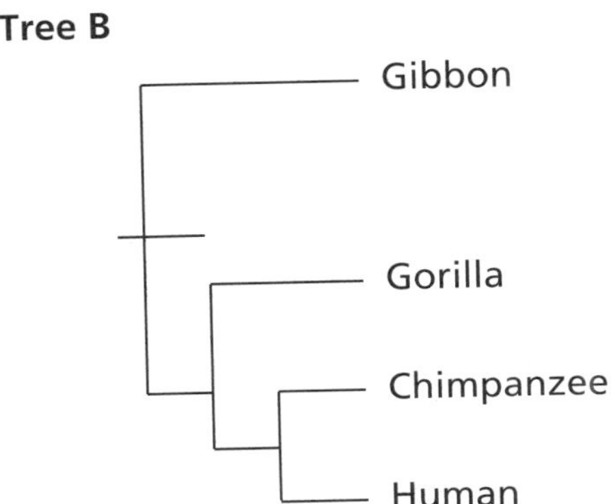

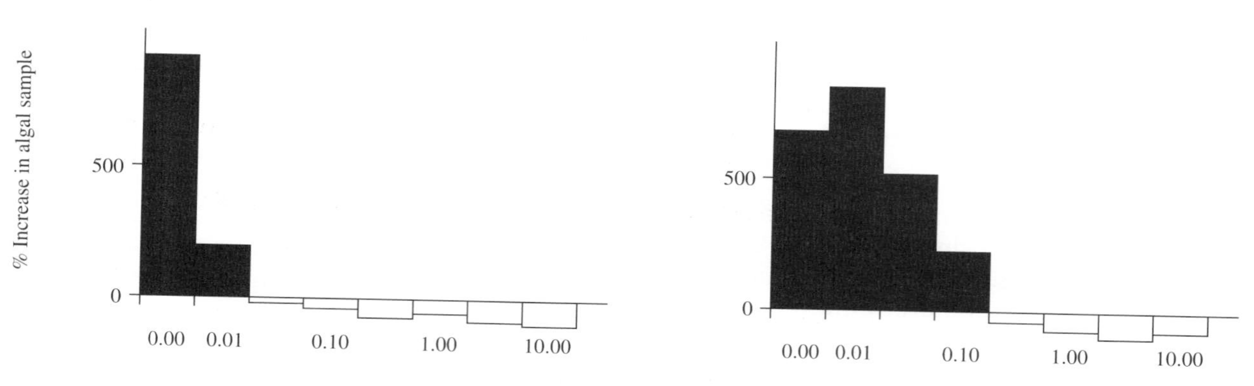


*Ref: Adapted from Studies in Biology No. 130, Edward Arnold (1981).*

The diagram shows the percentage increase in algal samples collected from populations **A** and **B** and cultured, for identical periods of time, in media containing different concentrations of copper.

**i)** Describe *three* ways in which the data for the two populations differ. *(3 marks)*

**ii)** State which population shows the greater tolerance to copper. *(1 mark)*

**iii)** Explain how the copper tolerance observed in this population may have evolved. *(3 marks)*

**c)** Despite thorough regular cleaning and repainting with antifouling paint, the hull of the ship is rapidly recolonized.

**i)** Suggest *one* explanation for this rapid recolonisation. *(1 mark)*

**ii)** Suggest why shipping companies are prepared to spend money to prevent this recolonisation. *(2 marks)*

**d)** Give *one* example of the commercial exploitation of algae by Man. *(1 mark)*

**WJEC**

# Appendix I – Some basic chemical principles

A knowledge of basic chemistry is absolutely essential for those hoping to understand biology – in fact some branches of biology can just as easily be regarded as branches of chemistry, and vice versa. The following short section is intended to help those whose chemistry is a bit rusty.

## Elements, compounds and mixtures

Water is a **compound** because it can be split into simpler substances, in this case hydrogen and oxygen. Hydrogen and oxygen are both **elements** because they cannot be broken down into simpler substances by chemical means.

Compounds differ from mixtures in two important ways:

- a compound has quite different properties from its constituents. Water, for instance, is a liquid at room temperature, while oxygen and hydrogen are both gases.
- whereas the constituents of a mixture can be in any proportions, the elements of a compound are combined in a definite ratio. In water, for example, every gram of hydrogen combines with 8 grams of oxygen.

## Atoms and molecules

All substances consist of tiny particles, which may be of three kinds. **Atoms** are the smallest particles of matter that can exist by themselves. Atoms can combine with each other to form larger particles called **molecules**. A molecule of water, for example, consists of an oxygen atom linked to two hydrogen atoms, and is represented by the formula $H_2O$. Some molecules consist of atoms of only one type – an oxygen molecule, for example, consists of two oxygen atoms (Figure 1).

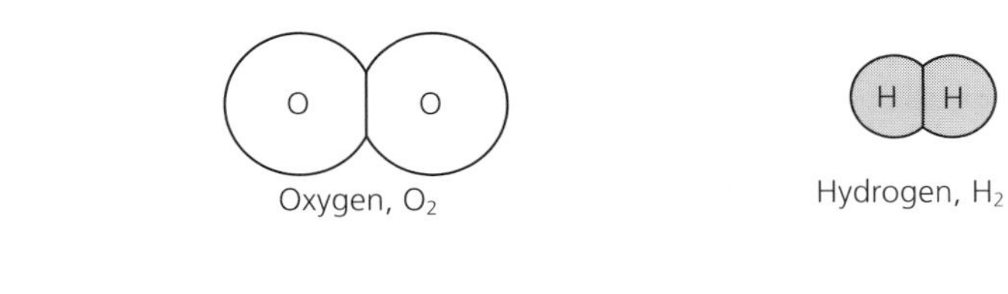


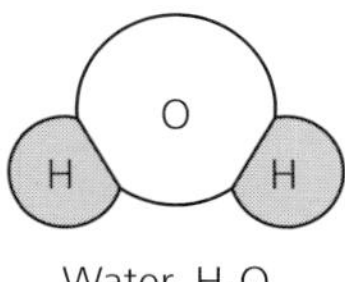


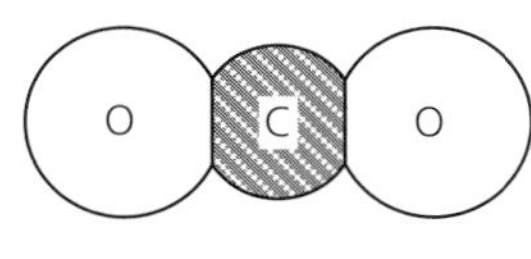


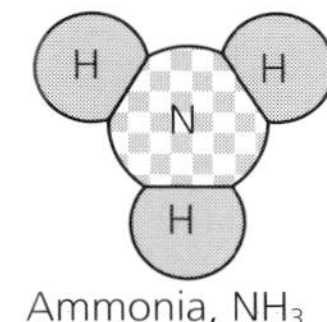


**Figure 1**
Some simple molecules

## Inside the atom

An atom consists of three kinds of even smaller particles. **Neutrons** occupy the central **nucleus** of all atoms except hydrogen, and have no electrical charge. **Protons** are positively charged and are also located in the nucleus. **Electrons** are negatively charged and occupy the outer part of the atom. The mass of an atom is effectively due to its constituent protons and neutrons, as electrons have negligible mass by comparison. In any atom the number of protons and electrons is equal, so there is no net charge.

The number of electrons and protons in an atom is characteristic of an element. Hydrogen, for example, has one proton and one electron, oxygen has eight protons, eight electrons and eight neutrons, and carbon has six protons, six electrons and six neutrons.

In an atom the electrons can be thought of as occupying a series of concentric layers or 'shells'. Each shell can only accommodate up to a certain maximum number of electrons. The inner shell can hold two electrons, the next can hold eight, the next, 18, and the next, 32. Heavier elements have other shells but are not constituents of living matter. Elements in which the outermost shell is full are stable and unreactive, for example helium, neon, argon and other 'noble gases' (Figure 2).

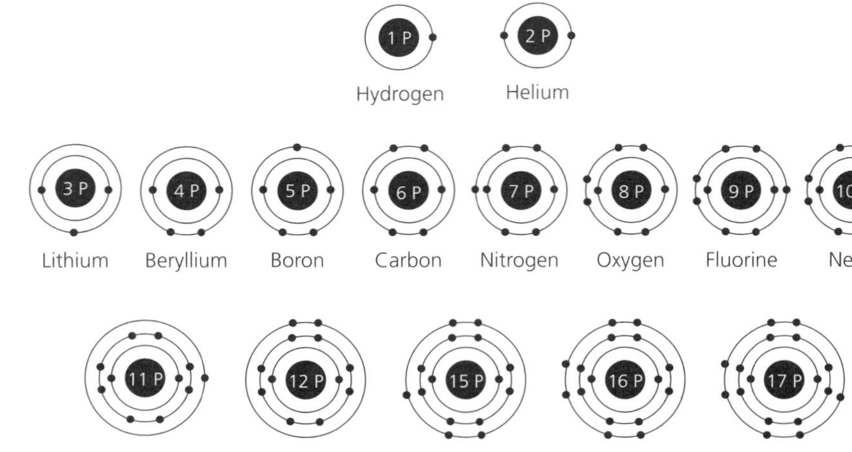


**Figure 2**
Electron configurations of some common elements. 'P' denotes the number of protons, but neutrons are not indicated. In the third row, only those atoms of particular biological importance are shown

Atoms are so small that instead of measuring their masses in conventional units, they are compared with each other in terms of the **atomic mass**. The base for comparison is carbon, which has an atomic mass of 12. Hydrogen has an atomic mass of 1, and oxygen has an atomic mass of 16.

## How atoms combine – chemical bonds

Except for the noble gases (which already have a full outer electron shell), all elements have a tendency to achieve a full outermost shell by reacting with other elements. They can do this in one of two ways:

- by sharing one or more of their outer electrons with other atoms, forming a **covalent bond**.
- by the donation of one or more electrons from one atom to another, forming an **electrovalent bond**.

### *Covalent bonds*

In a covalent bond, electrons are shared between two atoms. As a result each atom achieves the stable number in its outer electron shell. If one pair of electrons is shared, then a **single bond** (represented by a line) is formed (Figure 3).

If two pairs of electrons are shared, a **double bond** (represented by a double line) is formed, as in carbon dioxide (Figure 4).

### *Ions and electrovalent bonds*

When sodium reacts with chlorine, each sodium atom gives up an electron to a chlorine atom (Figure 5). The sodium still has 11 protons but only 10

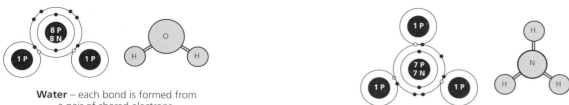


**Water** – each bond is formed from a pair of shared electrons, one of each pair provided by oxygen, one by hydrogen

**Ammonia $NH_3$** – each bond is formed from a pair of shared electrons, one of each pair provided by nitrogen, one by hydrogen

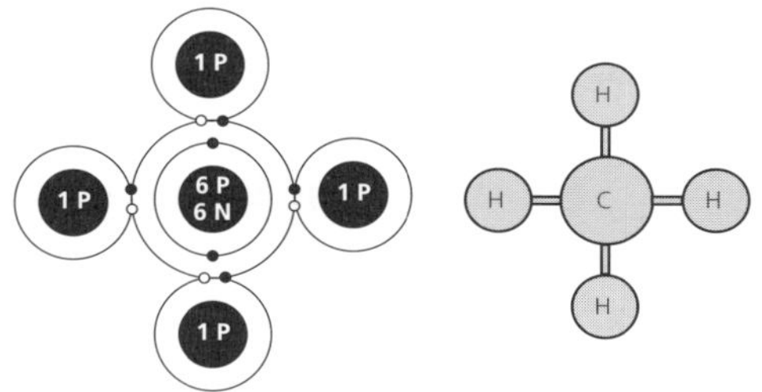


**Methane** – each of the four bonds is formed by a pair of shared electrons, one of each pair provided by carbon, one by hydrogen

**Figure 3▲**
Covalent bonds in some simple molecules

electrons, so it has a positive charge. An electrically-charged particle is called an **ion**. The chlorine now has a surplus electron and is a negatively-charged chloride ion. Positive ions are called **cations** and negative ions are called **anions**.

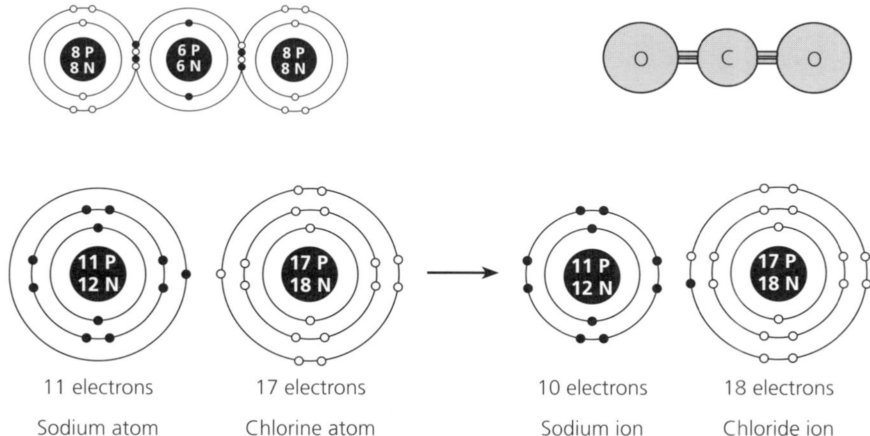


**Figure 4▶**
A double covalent bond

**Figure 5▶**
Formation of an electrovalent bond in sodium chloride

The attraction between the positive sodium ions and the negative chloride ions is called an **electrovalent bond**. In contrast to covalent bonds (which are formed by the sharing of electrons) an electrovalent bond is formed when an atom *donates* one or more electrons to another atom. Electrovalent bonds are not represented by lines because each cation is attracted to all the anions surrounding it in the solution or in the crystal.

Formation of ions is simply redistribution of electrons between the particles – the overall charge remains the same. Many ions are formed from more than one atom, for example sulphate ($SO_4^{2-}$) and phosphate ($PO_4^{3-}$). Although individual ions are charged, in any compound the total number of negative and positive charges is always equal.

## Valency

Each element has a characteristic 'combining power', or **valency**. The valency of an element is the number of its electrons that are involved in the linking with other atoms. The concept of valency can be illustrated by the following examples.

- In reacting to form sodium chloride, each sodium atom gives up one electron to a chlorine atom. Since one electron is transferred, sodium and chlorine both have a valency of 1.
- In a water molecule, the oxygen atom forms two bonds – one to each hydrogen atom. Oxygen therefore has a valency of 2 and hydrogen has a valency of 1. In some compounds oxygen shares its two valency electrons with the same atom, forming a **double bond**.
- In ammonia the nitrogen forms three bonds, each with a different hydrogen atom. Nitrogen is thus trivalent.
- Carbon has a valency of 4, forming single bonds with each of four other atoms (as in methane, $CH_4$), two double bonds (as in carbon dioxide, $CO_2$), or a double bond and two single bonds.

## Isotopes

Although all the atoms of an element are identical chemically, elements can exist in more than one form according to the mass of the atom. These different forms of an element are called **isotopes**. All the isotopes of a given element have the same number of protons and electrons (which determine its chemical properties) but differ in the number of neutrons (which affect the atomic mass). Although the atomic mass of a given isotope is a whole number, this is not so for the element as a whole since it consists of a mixture of isotopes. For example chlorine consists of two stable isotopes with atomic masses of 35 and 37, in proportions that give an average atomic mass of 35.5.

Many isotopes are **radioactive**, giving off radiation, and are termed **radioisotopes**. Whenever an atom emits a particle of radiation, it 'decays' to another isotope – usually of a different element. For example, carbon-14 decays to nitrogen-14. Radioisotopes can thus be used as markers or tracers and are an indispensable item in the cell biologist's tool kit. Several examples are described throughout the book.

## Molecular masses

By adding the atomic masses of their constituent atoms, molecular masses of compounds can be compared. For example water, with two atoms of hydrogen and one of oxygen, has a molecular mass of $(1 \times 2) + 16 = 18$, and carbon dioxide has a molecular mass of $12 + (2 \times 16) = 44$. Glucose ($C_6H_{12}O_6$) has a molecular mass of $(12 \times 6) + (1 \times 12) + (16 \times 6) = 180$. A glucose molecule is thus 180 times bigger than an atom of the most common isotope of hydrogen.

## Chemical equations

A chemical equation is a shorthand way of summarising a chemical reaction. For example the burning of glucose can be summarised by the equation:

$$\underset{\text{glucose}}{C_6H_{12}O_6} + \underset{\text{oxygen}}{6O_2} \rightarrow \underset{\text{carbon dioxide}}{6CO_2} + \underset{\text{water}}{6H_2O}$$

The above equation tells us that when glucose is burned, each glucose molecule reacts with six molecules of oxygen and produces six molecules of carbon dioxide and six molecules of water.

Notice that in a chemical reaction, no atoms are destroyed and no new ones created – existing atoms are simply *rearranged*. There must therefore be the same number of each kind of atom on the left and right of the equation. In other words, the equation must *balance*.

## Moles

When measuring quantities of chemicals, biochemists frequently use a quantity called a **mole**. A mole is the molecular mass of a substance expressed in grams. A mole of glucose ($C_6H_{12}O_6$) is therefore equal to $(12 \times 6) + (1 \times 12) + (16 \times 6) = 180$ grams. Its great practical use lies in the fact that *a mole of any substance contains the same number of molecules* (Figure 6).

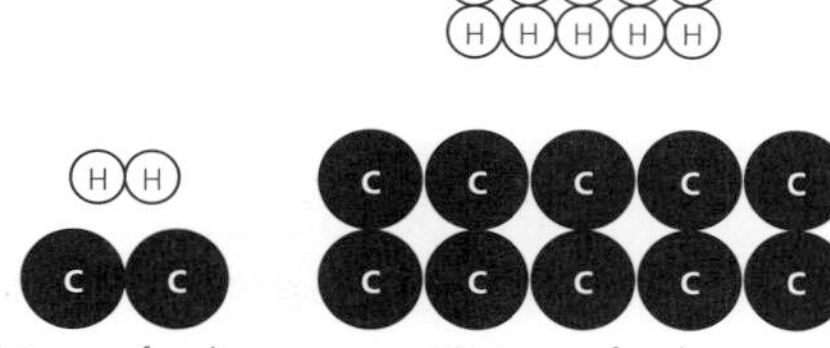
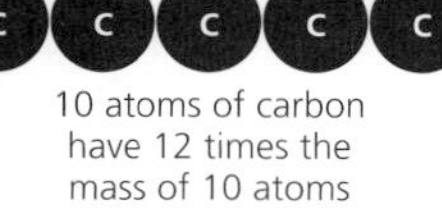
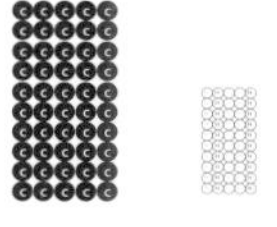

H
C
1 atom of carbon has 12 times the mass of 1 atom of hydrogen
H H
C C
2 atoms of carbon have 12 times the mass of 2 atoms of hydrogen
H H H H H
H H H H H
C C C C C
C C C C C
10 atoms of carbon have 12 times the mass of 10 atoms of hydrogen
N carbon atoms have 12 times the mass of N hydrogen atoms.
The number of atoms in any given mass of hydrogen is therefore the same as the number of atoms in 12 times the mass of carbon

**Figure 6**
The mole concept

By expressing quantities in moles, it is possible to calculate amounts of chemicals involved in chemical reactions. Take, for instance, the burning of glucose, summarised by the equation given previously. The equation tells us that one mole of glucose (180 grams) combines with six moles of oxygen ($6 \times 32 = 192$ grams), producing six moles of carbon dioxide ($6 \times 44 = 264$ grams) and six moles of water ($6 \times 18 = 108$ grams).

Another useful thing about moles is that for gases at the same temperature and pressure, equal numbers of moles of all gases occupy equal volumes. Thus 44 grams of carbon dioxide occupy the same volume as 32 grams of oxygen. Conversely, at the same temperature and pressure equal volumes of all gases contain the same number of moles (and the same number of molecules).

## Acids and bases

The acidic taste of such substances as vinegar and acetic acid is due to the presence of hydrogen ions, or **protons** ($H^+$). An **acid** is therefore a substance which can provide hydrogen ions. For example, hydrochloric acid dissociates into a proton and a chloride ion:

$$HCl \rightarrow H^+ + Cl^-$$

Hydrochloric acid is a **strong acid** because it dissociates completely into ions. **Weak acids**, such as ethanoic acid, dissociate incompletely, and some protons and ethanoate ions recombine to form ethanoic acid molecules. The result is an equilibrium in which both processes occur simultaneously:

$$CH_3{-}COOH \rightleftharpoons CH_3{-}COO^- + H^+$$

An important thing about acids is that they combine with **bases**. A base is a substance that can neutralise an acid by combining with hydrogen ions. For example, sodium hydroxide reacts with hydrochloric acid to form sodium chloride and water:

$$NaOH + HCl \rightarrow NaCl + H_2O$$

If we write the substances in terms of ions, we see that only the $H^+$ and $OH^-$ ions take part:

$$Na^+ + OH^- + H^+ + Cl^- \rightarrow Na^+ + Cl^- + H_2O$$

Since the $Na^+$ and $Cl^-$ ions are on both sides of the equation they can be deleted. The essential reaction is between the $H^+$ ions and the $OH^-$ ions, and the reaction can be represented by the equation:

$$H^+ + OH^- \rightarrow H_2O$$

Like acids, bases vary in strength. Strong bases, such as sodium hydroxide, dissociate completely to give $OH^-$ ions, while weak bases, such as ammonium hydroxide, do so only partly.

## The pH scale

Living things can only tolerate a limited range of hydrogen ion concentrations in their surroundings. The hydrogen ion concentration of a solution is expressed as its **pH value**. Pure water (which is neutral) has a pH of 7. Acidic solutions have pH values of less than 7, and alkaline solutions have pH values of greater than 7.

One important thing about the pH scale is that it is logarithmic (pH stands for the **p**ower of the reciprocal of the **h**ydrogen ion concentration). A solution of pH 1 has 10 times the $H^+$ ion concentration of a solution of pH 2, and 100 times the $H^+$ concentration of a solution of pH 3, and so on.

## Oxidation and reduction

Some of the most important chemical reactions in living cells involve the transfer of electrons between atoms, ions or molecules. Oxidation is the *loss* of electrons, while reduction is the *gain* of electrons. Any electron transfer involves both processes, the electron donor being oxidised and the electron acceptor being reduced. For example, when chlorine reacts with iron to form ferric chloride, electrons are transferred from iron to chlorine. As a result, the iron atoms are oxidised to $Fe^{3+}$ ions and the chlorine atoms are reduced to chloride ions, $Cl^-$.

$$2Fe + 3Cl_2 \rightarrow 2Fe^{3+} + 6Cl^-$$

In the above reaction, chlorine is acting as an electron acceptor, or **oxidising agent**, and iron is acting as an electron donor, or **reducing agent**. Oxidising and reducing agents vary in their 'strength': a strong oxidising agent, for example, has a powerful tendency to take electrons from other substances, but may itself lose electrons to an even more powerful oxidising agent.

The definition of oxidation given in junior courses is 'the addition of oxygen'. Fundamentally this also involves a transfer of electrons, albeit a partial one. When methane is burned, for example, it forms $CO_2$ and water. In a methane molecule the carbon and hydrogen share electrons more or less equally, but in $CO_2$ the oxygen takes more than its 'fair' share of the electrons (it is a strong electron acceptor). Because the carbon atom loses some of its electron share, it has been oxidised.

# Appendix II – Use of statistics in biology

The use of statistics is an essential tool in most scientific work. Statistics are used to clarify data and to help their interpretation. They can be used in two ways:

- **descriptive statistics** are concerned with graphical representation of data.
- **inferential statistics** use mathematical techniques in order to draw conclusions from data.

Any scientific study involves the recording of data in the form of either measurements or counts, or both. 'Data', incidentally, is the plural of *datum*, although many people use the term incorrectly in the singular sense (e.g. 'data is . . .')

Data are of two kinds:

- **continuous data** consist of *measurements*. The widths of leaves, are continuous data, because each datum can take any number of values between the narrowest and the widest.
- **discrete data** consist of *counts* because there are no intermediates between adjacent values (a nest cannot contain 2.5 eggs!)

## Graphs

It is often much easier to interpret data if they are represented in the form of a **graph**. There are a number of different kinds, each suitable for a particular kind of situation.

### Line graphs

Line graphs are used for continuous data and show how one quantity is affected by another. An example could be the effect of temperature on the rate of heartbeat of *Daphnia* (Figure 1). Heart rate is the **dependent variable** because it depends on temperature, which is the **independent**

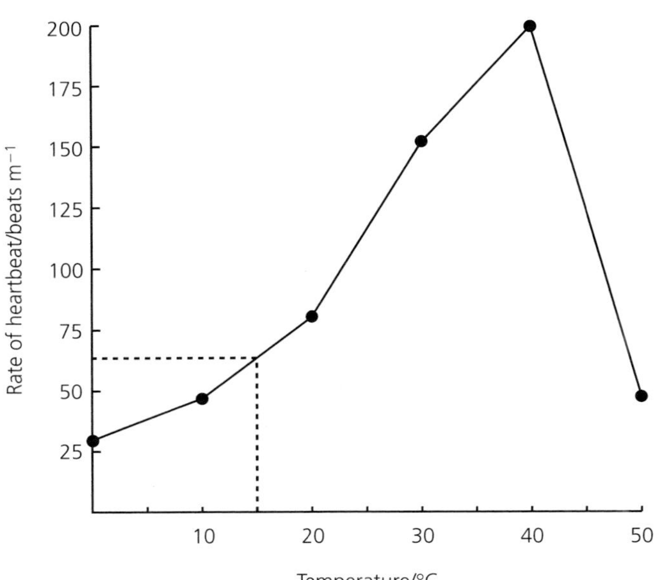


**Figure 1**
A line graph showing the effect of temperature on the rate of heartbeat in *Daphnia*

**variable**. The values of the independent variable are decided upon in advance, whilst those of the dependent variable are obtained by experiment or observation.

When plotting a line graph you should bear the following points in mind.

- Give it an informative *title*. 'A line graph showing the effect of temperature on the rate of heart beat in *Daphnia*' says a lot more than 'Heartbeat graph'.
- The axes should be the right way round, with the independent variable on the horizontal ($x$) axis and the dependent variable on the vertical ($y$) axis.
- Label the axes and give the units of measurement, e.g. 'Temperature in °C, 'Rate of heart beat (beats per minute)'.
- Scales should be chosen to use as much of the available space as possible (but avoid using awkward scales such as 2 squares to represent 5 units).
- If certain values of the independent variable have not been used, the intervals on the horizontal scale should take account of this. If, for instance, temperatures used are 10, 20, 25, 30, 35, 40 and 45, then the interval between 10 and 20 should be double that of the others – otherwise the shape of the graph will be distorted.
- Points should be large enough for their position to be clearly visible. Use crosses, circles or squares rather than dots.
- Do not extrapolate the graph (i.e. do not extend it beyond the data), unless you use a dotted line.
- When you draw two or more graphs on the same paper, remember to provide a key to distinguish them.

One problem which is often encountered is whether to join the points up or to draw the line of best fit. When measurements are suspected to be subject to appreciable error – either in the measurements themselves or in the experimental technique – it is best to draw the line of best fit.

Since temperature and heart rate are both measurements, we can use the line to estimate heart rate at temperatures that were not used. For example, by drawing the dotted lines as shown we can say that under the conditions of the experiment the heart rate at 15°C would have been about 63 beats $min^{-1}$.

## Bar charts

By connecting points on a line graph we are able to estimate values between points. A bar graph is used when there are no intermediates, either because a variable is discrete or because it cannot be quantified.

Figure 2A shows the average wing diameter for five species of British butterfly. Although averages are continuous data because they can take any values, species are not quantities, so the horizontal axis does not represent a scale.

In Figure 2B the $x$ axis expresses a continuous variable (time), but the number of eggs can only take discrete values.

Figure 2C shows the variation in clutch size in partridges. The columns represent frequencies of clutches with increasing numbers of eggs. Although this is a frequency distribution, it is not a histogram (see below) since egg numbers cannot take on intermediate values.

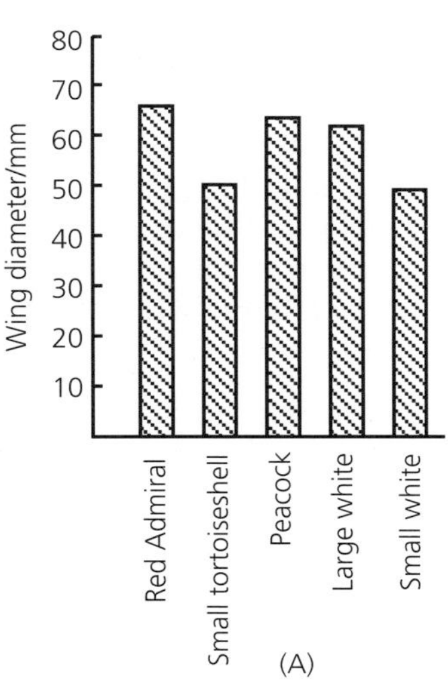


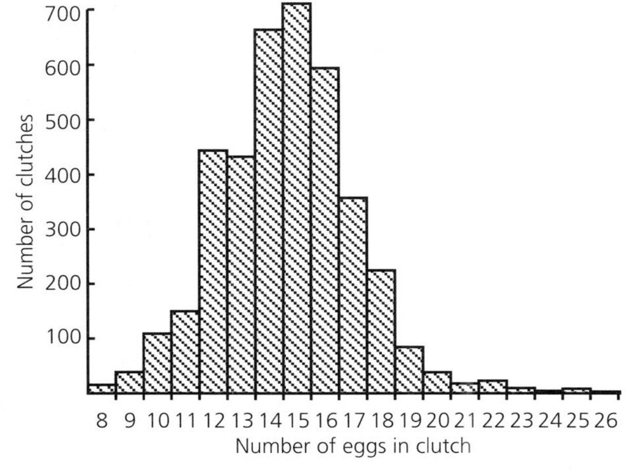


**Figure 2▲**
Three examples of data illustrated by a bar chart: (A) variation in wingspan of some male butterflies; (B) number of eggs laid by a fruit fly; (C) variation in clutch size in partridges

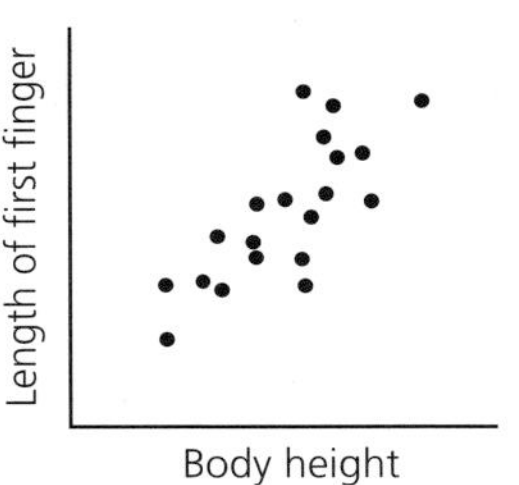


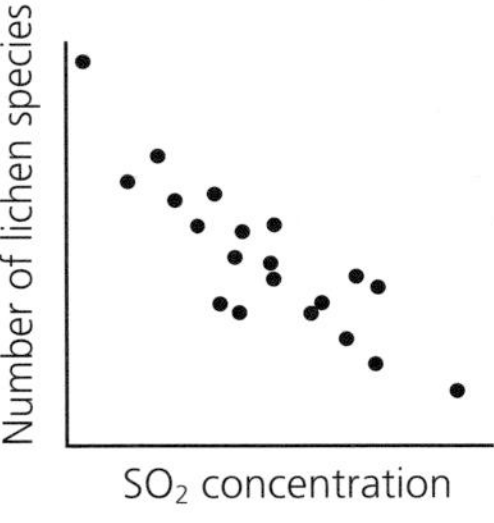


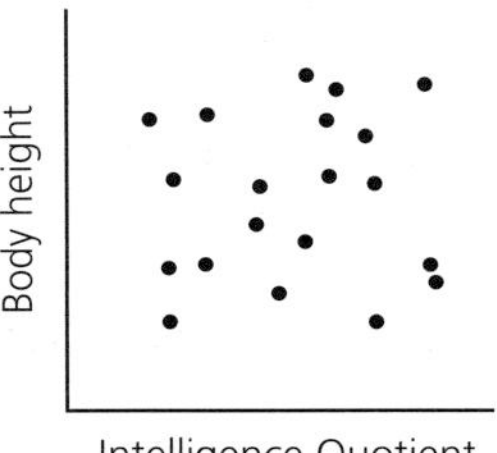


**Figure 3▲**
Imaginary scattergrams illustrating different kinds of correlation

## Scattergrams

A scattergram is used to determine whether two variables are correlated – in other words, whether the value of one is any indication of the value of the other. Suppose for example you want to know whether there is any relationship between the size of islands and the number of species of birds, you would measure the area of a large number of islands and for each, record the number of species. Since the hypothesis is that island size affects number of species (rather than the other way round), it would be logical to regard area as the independent variable and plot it on the horizontal axis. Each island would be represented by a point on the graph, but the points *are not joined up*. Note that it is quite possible for two islands to have the same number of species (same *y* value) but different areas (*x* value).

Where there is no reason to regard either variable as dependent on the other, the axes could be either way round (e.g. finger length vs blood pressure).

Whether variables are correlated depends upon the way the points are clustered (Figure 3). In the island example above, there is a tendency for larger islands to have more species. The two variables are said to be **positively correlated**. If the spread of dots shows the reverse pattern, the data show **negative correlation**, as for example the atmospheric sulphur dioxide level and the number of lichen species in an area.

The stronger the correlation, the more the dots tend to be clustered along a line. When a correlation is absolute, all the points lie along a line so that for every value of one variable we can precisely predict the value of the other. If the line is straight the relationship is said to be **linear**, but many relationships are non-linear. If there is no correlation the points are randomly scattered, e.g. body height and Intelligence Quotient.

In many cases a correlation may not be strong enough to show up on a graph, in which case statistical methods must be used to determine whether there is any correlation. Correlation between two variables may be quantified as the **correlation coefficient**, in which a value of 0 indicates no correlation, +1 indicates absolute position correlation and −1 indicates absolute negative correlation.

When two factors are correlated it may indicate that one is a cause of the other, as for instance in the relationship between smoking and cardiovascular disease and lung cancer. In many cases, there may be no

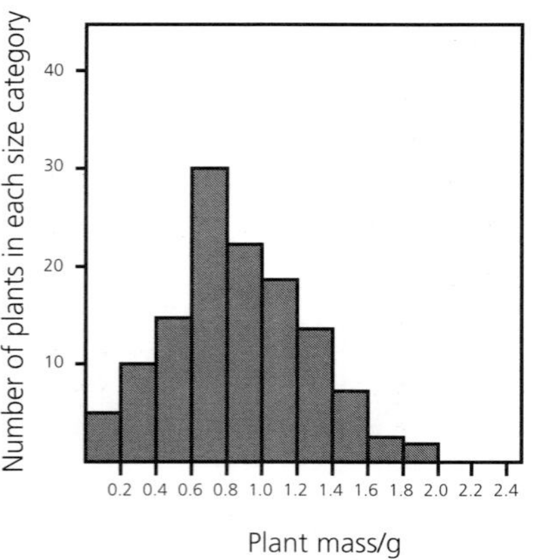


**Figure 4▲**
A histogram showing variation in size of flax plants

causal connecton, or alternatively both variables may be related to the same causative agent.

## Histograms

A histogram is used to show the relative frequency of different measurements *and is used only with continuous data*. When plotting a histogram, measurements are first divided up into classes. The most suitable size of each class is decided by the investigator and will depend on certain factors including the number of measurements (Figure 4).

Note the following points:

- the figures on the *x* axis denote the boundaries between categories.
- the width of each column is proportional to the size of the interval. If the area under the entire graph is taken as 1, the area of each column indicates the proportion of measurements falling into that interval.
- the columns must touch.

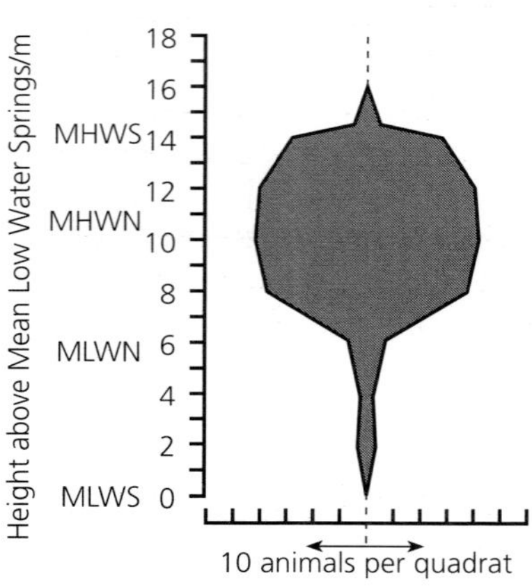


**Figure 5▲**
A kite diagram showing the variation in abundance of the limpet *Patella vulgata* along a transect

## Kite diagrams

A kite diagram is used to show how the abundance of an organism changes along a chosen line or transect. The dependent variable (some measure of abundance) is plotted as two symmetrical lines, one on each side of the axis showing the independent variable (distance along the transect). Figure 5 shows the distribution of the limpet *Patella vulgata* along a transect up a rocky shore.

## Pie charts

A pie chart is used to indicate the percentages of various constituents of a whole, for example the proportions of various nutrients in a particular kind of food (Figure 6). In an apple, for instance, water forms about 84%, and is represented by a segment with an angle of 84% of 360°, or 302°.

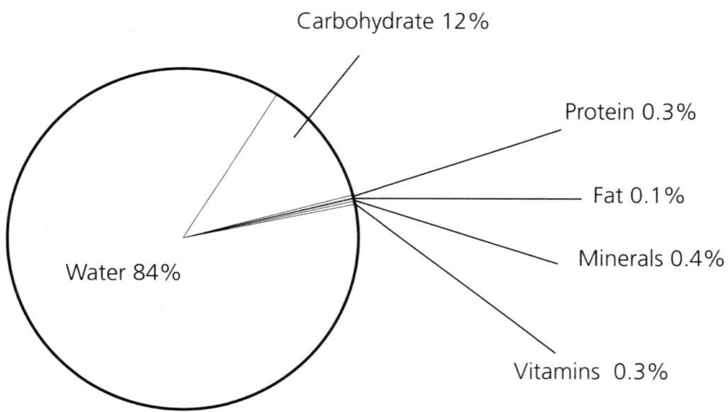


**Figure 6►**
A pie chart showing the approximate nutrient composition of an apple

# Facts from figures: statistical tests

Having obtained the results of an investigation, a scientist is faced with the prospect of interpreting them. In some cases the results may be sufficiently clear-cut to require no further treatment, but generally some form of statistical analysis is necessary.

**Table 1**
Approximate probabilities of various outcomes when tossing a coin 10 times

| Head:tails | Probability |
|---|---|
| 10:0 | 0.001 |
| 9:1 | 0.01 |
| 8:2 | 0.05 |
| 7:3 | 0.12 |
| 6:4 | 0.2 |
| 5:5 | 0.25 |

## The need for statistical analysis

Suppose you are investigating the response of woodlice to light. You place, say, 10 woodlice in a 'choice chamber', one side of which is dark and the other light. After 10 minutes you count the number on each side. Suppose that all 10 are on the dark side. The probability that all 10 would move to the dark side by chance is very small $(\frac{1}{2})^{10}$, so you would be justified in concluding that woodlice (or at least, these particular woodlice) prefer dark to light. A result of 9 dark:1 light would have been almost as clear-cut. Problems arise with results like 8:2, 7:3 or 6:4. The probabilities that these results could occur by chance are the same as those for coin tossing and are given in Table 1.

It is important to appreciate that probabilities are affected by absolute numbers as well as ratios. For example, a six heads and four tails result would often be obtained by chance, a 60:40 result would be very rare, and 600:400 would be close to impossible.

If only chance is operating, the most probable result is 5:5. This is called the **expected value**. In statistics, 'expected' has a different meaning from the usual one. It means the most frequent result when a large number of trials are performed and when chance alone is operating. If, for example, a coin is thrown ten times by 100 people, five heads and five tails would be the most common result. The probability of getting tails in any given toss is therefore $\frac{5}{10} = 0.5$. Even though the probability of getting 5:5 by chance is only about 0.25, every other result is even less likely. The greater the deviation from the 'expected' value of 5:5, the less probable it is.

What we have to decide is how great a deviation from the expected value we can reasonably accept as being due to chance. At what point do we suspect that the woodlice are not moving randomly, but have a preference for darkness?

### The Null Hypothesis

The first step in a statistical analysis is to erect a **Null Hypothesis ($H_0$)**. In the case of the woodlice it could be stated thus:

"*Woodlice show no preference for dark or light conditions, any deviation from the expected value being due to chance alone*"

Having erected a Null Hypothesis, the investigator must then test it statistically. It is important to realise that such a test cannot prove that chance is or is not the only factor at work – it can only tell us the probability that any observed results could have occurred by chance.

### Significance

Though it is not impossible that one could toss a coin 100 times and get all heads, it is so unlikely that it would be much more reasonable to conclude that there was something wrong with the coin. Table 1 shows that there is a continuous spectrum of probabilities between the least and most probable. At what point are we justified in concluding that chance is not responsible for a result? The investigator has to choose a probability below which we regard a deviation as **significant** and can thus justifiably reject a Null Hypothesis. In most situations a probability of 0.05 is used. A calculated probability of 0.05 means that we can be 95% confident that the result is not due to chance. If the calculated probability is less than 0.05, we say that the result is significant.

### What kind of test?

The first and most important step in a statistical analysis is to decide on the kind of test to use. Two of the most commonly used tests in biology are the $\chi^2$ (Chi-squared) and the $t$ test.

## The $\chi^2$ test

The $\chi^2$ (the Greek letter, pronounced 'ki') test is used where discrete data (counts) can be classified in one of a number of distinct categories or classes, and where we have a prior expectation as to how many there should be in each group if chance alone is operating. Thus in the woodlouse example mentioned previously, each woodlouse is either on the light side of a choice chamber or it is on the dark side, and the total **observed** number on each side is then compared with the expected numbers.

Consider the following example. A fruit fly heterozygous for ebony body and vestigial wings was mated with an ebony-bodied, vestigial-winged fly, and the offspring were as follows:

| | |
|---|---|
| grey-bodied, long-winged | 92 |
| grey-bodied, vestigial-winged | 86 |
| ebony-bodied, long-winged | 97 |
| ebony-bodied, vestigial-winged | 101 |
| Total | 376 |

Within the limits of chance, is this result consistent with a 1:1:1:1 ratio?

Having erected the Null Hypothesis that the fruit fly numbers do not differ significantly from a 1:1:1:1 ratio, the $\chi^2$ test is used to test the Null Hypothesis. First, the value of $\chi^2$ is calculated, and then a table of $\chi^2$ is used to determine the probability that a $\chi^2$ value as high as this or higher could be obtained by chance.

$\chi^2$ is calculated as follows:

1 Calculate the expected values ($E$). This is the total divided by the number of possible kinds of fly in terms of body colour and winglength $= \frac{376}{4} = 94$.

2 Calculate the differences between the observed ($O$) and the expected values.

3 Since some of the deviations will be negative, they are then squared so all values are positive.

| | Observed | Expected | $O - E$ | $(O - E)^2$ |
|---|---|---|---|---|
| grey-bodied, long-winged | 92 | 94 | −2 | 4 |
| grey-bodied, vestigial-winged | 86 | 94 | −8 | 64 |
| ebony-bodied, long-winged | 97 | 94 | +3 | 9 |
| ebony-bodied, vestigial-winged | 101 | 94 | +7 | 49 |

4 Obtain the value of $\chi^2$ using the formula

$$\chi^2 = \Sigma \frac{(O - E)^2}{E}$$

where $O$ = observed value, $E$ = expected value and $\Sigma$ = 'the sum of'.

In this example the data fall into four categories, so there are four values for $\frac{(O - E)^2}{E}$. These are then added together to obtain $\chi^2$:

$$\frac{4}{94} + \frac{64}{94} + \frac{9}{94} + \frac{49}{94} = \frac{126}{94} = 1.34$$

### Degrees of freedom

The probability that any given value of $\chi^2$ could be exceeded by chance depends on the number of **degrees of freedom**, which is always one less than the number of classes. Thus in the fruit fly example there are four classes so there are three degrees of freedom.

The number of degrees of freedom is the maximum number of classes whose numbers could change independently without altering the total. In the fruit fly example there are 376 flies. If the numbers of flies in three classes are known, then the number in the fourth category is fixed.

### Using the table of $\chi^2$

Table 2 shows part of the table of $\chi^2$. Looking along the line for three degrees of freedom we see that 1.34 corresponds to a probability of between 0.8 and 0.5. In other words, if chance alone were operating, we would expect a $\chi^2$ value of 1.34 or higher with a probability of between 0.5 and 0.8. To put this another way, if this experiment were done a very large number of times, a $\chi^2$ result as high as or higher than this would be obtained in 50–80% of the experiments.

**Table 2**
Table of $\chi^2$ (d.f. = degrees of freedom)

| | Probability | | | | | | | | | |
|---|---|---|---|---|---|---|---|---|---|---|
| **d.f.** | 0.98 | 0.95 | 0.8 | 0.5 | 0.2 | 0.1 | 0.05 | 0.02 | 0.01 | 0.001 |
| **1** | 0.001 | 0.004 | 0.064 | 0.055 | 1.64 | 2.71 | 3.84 | 5.41 | 6.64 | 10.83 |
| **2** | 0.04 | 0.103 | 0.466 | 1.386 | 3.22 | 4.61 | 5.99 | 5.41 | 9.21 | 13.82 |
| **3** | 0.185 | 0.352 | 1.005 | 2.366 | 4.64 | 6.25 | 7.82 | 7.82 | 11.35 | 16.27 |
| **4** | 0.429 | 0.711 | 1.649 | 3.357 | 5.99 | 7.78 | 9.49 | 9.49 | 13.28 | 18.47 |
| **5** | 0.752 | 0.145 | 2.343 | 4.351 | 7.29 | 9.24 | 11.07 | 13.39 | 15.09 | 20.52 |
| | ← Do not reject $H_0$ | | | | | | | Reject $H_0$ → | | |

This probability is considerably higher than the 0.05 which we regard as significant, so we cannot reject the Null Hypothesis. In other words, we have no reason to suppose that the result does not conform to a 1:1:1:1 ratio.

*Point to note:* This does not mean that the deviation is necessarily due to chance. It simply means that if there is something other than chance at work, *the test has failed to reveal it*. Strictly speaking, therefore, we cannot accept a Null Hypothesis, for to do so would be to conclude that chance has been entirely responsible, and we cannot be sure of that. Rather than accepting $H_0$, it is therefore better to 'not reject' it.

## Analysis of samples

The $\chi^2$ test is used in situations where an observed set of data fit with a theoretical expectation. Another common – but completely different – situation in biology is where, for example, we have two sets of measurements and want to know if they differ significantly. For example, we might want to know whether female Red Admiral butterflies have a larger wingspan than males. In this situation, we would take a sample and measure the wingspan of each individual, noting the sex of each. We would then compare the average male wingspan with the average female wingspan, and use a ***t* test** to see whether there was a significant difference between them. However, before dealing with the *t* test itself, we must deal with some preliminary groundwork.

## Samples

The number of observations a scientist can make is limited, so we usually have to be content with a **sample**. The population from which a sample is drawn is called the **parent population**. Although the parent population may really exist (such as the carp in a lake), it can also be imaginary. A series of measurements of transpiration rate, for example, is a sample of an infinite number of measurements that, in theory, could have been taken.

If we take repeated samples from the same population, they will vary slightly due to chance – the larger the samples, the less they vary. Each individual sample therefore provides an **estimate** of the properties of the parent population, with larger samples giving more reliable estimates than small samples. There are two particularly important things we need to know about a sample:

1 a measure of the **central tendency**, or a figure that is most representative of the sample as a whole.
2 a measure of the spread, or **dispersion**, of the data.

## Measures of central tendency

There are three of these: mean, mode and median (Figure 7).

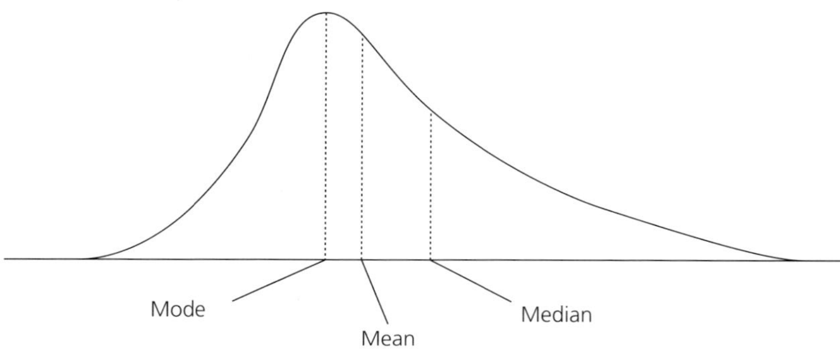


**Figure 7**
Measures of central tendency: mean, mode and median

- The **mean**, or 'average'. This is the sum of the data divided by the number of data, and is given by $\frac{\Sigma x}{N}$ where $x$ = an individual measurement and

  $N$ = total number of data. It divides the area under the curve into two equal halves. The mean of a population is given by the symbol $\boldsymbol{\mu}$, and the mean of a sample is given by $\bar{x}$.
- The **mode**. This is the most numerous value, and is represented by the peak of the frequency distribution. A sample may have more than one peak, for example, if body size were plotted against frequency. The two peaks in the distribution could represent males and females respectively. In this case, the distribution would be **bimodal**.
- The **median**. This is the value midway between the highest and the lowest.

In a symmetrical distribution, the mean, mode and median are the same, but in a **skewed distribution**, as in Figure 7, they are not.

### *Measures of dispersion: standard deviation*

A sample mean provides an estimate of the mean of the population from which the sample has been drawn. How reliable is this estimate? The reliability depends on the sample size and the variability of the data – i.e. their spread around the mean. The **standard deviation** is one of the most commonly used statistics because it takes into account *all* the data (the *range* only takes into account the largest and smallest). The standard deviation of a sample ($s$) is given by the formula:

$$s = \sqrt{\frac{\Sigma(x - \bar{x})^2}{N - 1}}$$

where $x$ = an individual measurement,
$\bar{x}$ = the mean of all the data,
$N$ = the number of data

The population standard deviation is denoted by the symbol $\sigma$.

Each individual measurement differs from the mean by $(x - \bar{x})$. Since the data lie on either side of the mean, some values of $(x - \bar{x})$ are negative. By squaring each value of $(x - \bar{x})$, all are made positive. The values of $(x - \bar{x})^2$ are added together to give $\Sigma(x - \bar{x})^2$. This is then divided by $N - 1$, and the square root then calculated to give the standard deviation. For large samples, $N$ can be used instead of $N - 1$ (since $N$ approximates to $N - 1$).

Calculating standard deviations used to be rather laborious, but scientific calculators have keys which enable data to be entered and the values of $N$, $\bar{x}$ and $s$ read out directly.

Why is standard deviation important? As we shall see later, biologists often have to compare samples in order to find out whether the parent populations differ. Just how much do two sample means have to differ before one is justified in concluding that the parent populations differ? This depends on how reliably the sample means reflect the population means. The reliability of a sample mean is greater if:

- it is large
- the standard deviation is small.

Figure 8 illustrates this point.

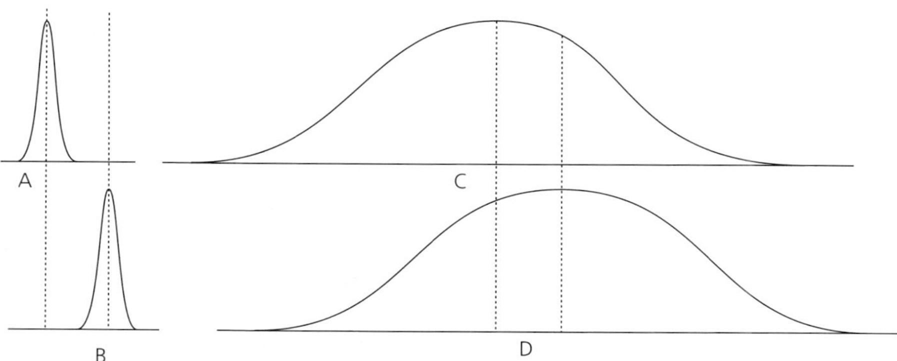


**Figure 8**
Importance of standard deviation in comparing means of samples. The means of samples A and B differ as much as the means of C and D, yet we can be much more confident that the differences between A and B reflect differences between the parent populations

## The normal distribution

Many biological variables (such as body height and blood pressure) have an approximately **normal distribution**. The curve of a normal distribution is bell-shaped, indicating that most of the data are clustered about the mean, with fewer and fewer data at an increasing distance from the mean (Figure 9).

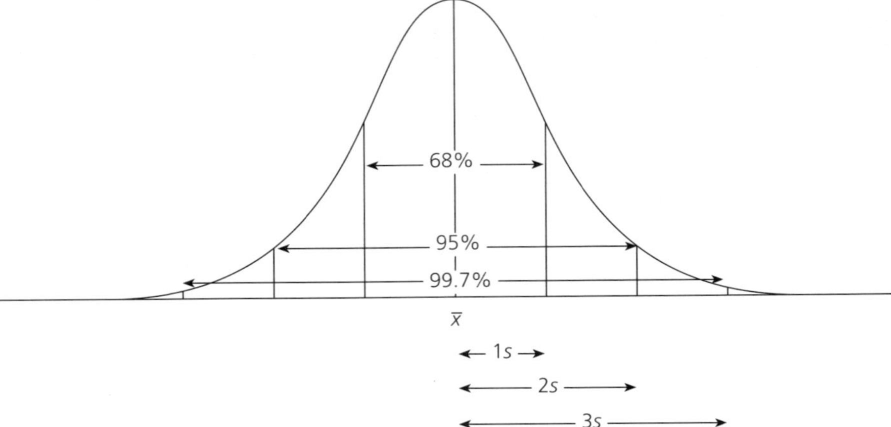


**Figure 9**
The normal distribution

A normal distribution is not simply a bell-shaped curve – it has certain well-defined properties.

- It is symmetrical about the mean – in other words the mean, median and mode are the same.
- The point at which the curve changes from convex to concave corresponds to one standard deviation.
- There is a precise and very useful relationship between deviation from the mean and probability of occurrence. For example, 68% of the data differ from the mean by no more than one standard deviation (34% above, 34% below). 95% of the data fall within two standard deviations from the mean, and 99.7% of the data differ from the mean by no more than three standard deviations.

In the normal distribution, any given number of standard deviations above or below the mean includes a defined proportion of the population. Table 3 shows how the number of standard deviations is related to probability (conventional statistical tables show standard deviations to two decimal places).

### *z* scores

Suppose that a beetle is 20 mm long and that the mean for the population is 17 mm. The beetle is therefore 3 mm longer than the mean. How unusual is this particular beetle? Without some indication of the variability of the population, we cannot tell. We need to know the standard deviation for body length in this particular population. If we know that the standard deviation is 2 mm, then we know that the beetle is $\frac{3}{2} = 1.5$ standard deviations above the mean.

For a normally-distributed variable, a deviation from the mean expressed in terms of the number of deviations is called a ***z* score**, and is given by:

$$z = \frac{x - \bar{x}}{s}$$

In this particular example, the $z$ score is $\dfrac{20 - 17}{2} = 1.5$.

Provided that beetle length is approximately normally distributed, Table 3

**Table 3**
Some probability values from the normal distribution

| *z* score (number of standard deviations above the mean) | Probability that *z* will exceed the mean by that value or less |
|---|---|
| 0.1 | 0.0398 |
| 0.2 | 0.0793 |
| 0.3 | 0.1179 |
| 0.4 | 0.1554 |
| 0.5 | 0.1915 |
| 0.6 | 0.2257 |
| 0.7 | 0.2580 |
| 0.8 | 0.2881 |
| 0.9 | 0.3159 |
| 1.0 | 0.3413 |
| 1.1 | 0.3643 |
| 1.2 | 0.3849 |
| 1.3 | 0.4032 |
| 1.4 | 0.4192 |
| 1.5 | 0.4332 |
| 1.6 | 0.4452 |
| 1.7 | 0.4554 |
| 1.8 | 0.4641 |
| 1.9 | 0.4713 |
| 2.0 | 0.4772 |
| 2.1 | 0.4821 |
| 2.2 | 0.4861 |
| 2.3 | 0.4893 |
| 2.4 | 0.4918 |
| 2.5 | 0.4938 |
| 2.6 | 0.4953 |
| 2.7 | 0.4865 |
| 2.8 | 0.4974 |
| 2.9 | 0.4981 |
| 3.0 | 0.4986 |

enables us to find out how unusual it is. For each *z* score there is a corresponding probability, which is the proportion of the data that exceed the mean by that *z* score or *less*. In this case 0.4332 (43.32%) of the beetles would be expected to lie between the mean and 1.5 standard deviations above the mean.

Figure 10 illustrates the situation for this imaginary beetle. The shaded area represents the proportion of the beetles with a body length between the mean and a *z* score of 1.5 standard deviations. In most cases we want to know what proportion of the data would be expected to exceed the mean by a given amount or *more*, so we subtract the probability given in the table from 0.5. In this case only $0.5 - 0.4332 = 0.0668$, or about 6.7% of the population would be as long as 20 mm or longer.

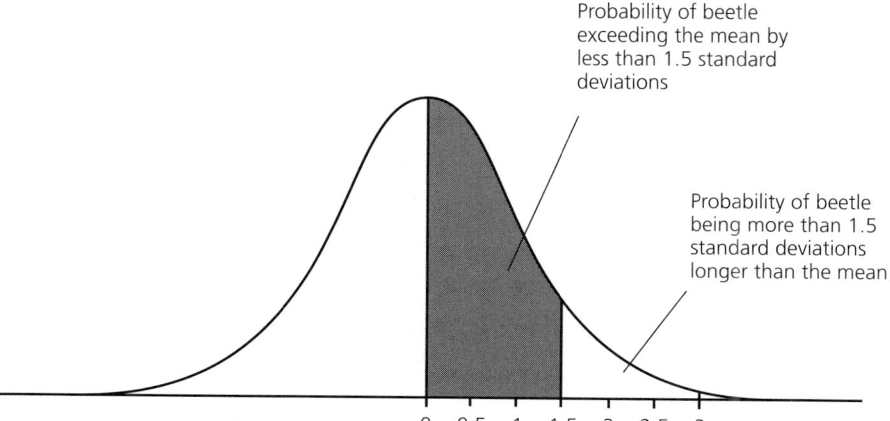


**Figure 10**
Proportion of a population of beetles with a body length exceeding the mean by 1.5 standard deviations or less

## Comparing the means of two samples – standard error of the difference

Do female shore crabs tend to be larger than males? Does a particular drug tend to reduce blood pressure? Does a plant grow faster when the atmosphere is enriched with $CO_2$? These are typical of the kind of questions that many biologists face.

To solve questions like these we have to compare samples – a sample of male crabs with a sample of females, for instance. The difficulty is that because of chance, the means of any two samples would be expected to differ – *even if we compared two samples of the same sex.*

A difference between a sample mean and the mean of the population from which it is drawn is due to **sampling error**. This is not of course an error in the usual sense, but an effect that is purely due to chance. The question is, how much difference between two sample means can we attribute to sampling error before we are entitled to suspect that females are generally larger than males? In other words, just how much can two sample means differ before we can reasonably conclude that they were taken from populations that, as far as size is concerned, are different?

To begin with, imagine that we take two samples from the *same* population – say, two samples of females. Suppose also that we find the mean of each, and then subtract the mean of the first sample from the mean of the second sample. This would give the difference between the two means $\bar{x}_1 - \bar{x}_2$. Suppose we repeated the process many times, obtaining the differences between many such pairs of means. We would then have a large number of values for $\bar{x}_1 - \bar{x}_2$.

Next, suppose we were to find the *mean* of all these values of $\bar{x}_1 - \bar{x}_2$. If the number of such differences between sample means was very large, we would find two things:

- the mean difference between sample means would be close to zero. This is because negative differences (when $\bar{x}_1 < \bar{x}_2$) would roughly cancel out positive differences (when $\bar{x}_1 > \bar{x}_2$). For an *infinite* number of paired samples, the mean difference between sample means would be zero.

- small differences between sample means would be more common than larger ones. In fact the greater the difference between means, the less common they would be.

An important thing about a large group of differences between sample means is that *they are normally distributed.* This is true *even if the variable in question is not normally distributed.* For this to be true, however, the samples must be large enough – the two samples must add up to more than about 30.

Now a large number of differences between sample means not only has its own mean (zero) but it also has a standard deviation. The standard deviation of the difference between sample means is called the **standard error of the difference**, S.E.$_{\bar{x}_1-\bar{x}_2}$. A standard error is simply a standard deviation as applied to means rather than individual data.

Now, recall that it is possible to find out how unusual an individual measurement is by expressing its deviation from the mean in terms of a $z$ score. A difference between two means can also be expressed in terms of the number of standard errors from the mean. To do this we need to know the standard error of the difference.

How do we find out the standard error of the difference? Amazingly, it is not necessary to take huge numbers of pairs of samples. S.E.$_{\bar{x}_1-\bar{x}_2}$ can actually be estimated *from a single pair of samples*, using the formula

$$\text{S.E.}_{\bar{x}_1 - \bar{x}_2} = \sqrt{\frac{s_1^2}{N_1} + \frac{s_2^2}{N_2}}$$

You will see from this formula that the standard error of the difference is larger if the samples have large standard deviations ($s_1$ and $s_2$), and if the sample sizes ($N_1$ and $N_2$) are small. This underlines the point illustrated in Figure 8.

**Example**
A sample of 40 mussels from the exposed side of a bay has a mean length of 65 mm and a standard deviation of 5 mm. A sample of 60 mussels from the sheltered side of the bay has a mean length of 67 mm and a standard deviation of 6 mm. Are we justified in concluding that the mussels on the sheltered side of the bay are larger than those on the exposed side?

The first step is to erect a Null Hypothesis:

*'The mussels on the two sides of the bay do not differ in mean size'.*

The Null Hypothesis relates to the parent populations from which the samples were drawn. $H_0$ is, in effect, saying that the mussels on the two sides of the bay belong to the same population, and that the difference between sample means is due to chance.

Difference between means = 3 mm. To find out how many standard errors this represents, we first calculate the standard error of the difference:

$$\text{S.E.}_{(\bar{x}_1 - \bar{x}_2)} = \sqrt{\frac{25}{40} + \frac{36}{60}} = 1.1$$

A difference of 3 mm is $\frac{3}{1.1} = 2.7$ standard errors. If the mussels on the sheltered side of the bay are not larger than those on the exposed side, then the most probable difference between two sample means would be zero. To find out the probability that for two samples drawn from the same population, one mean could exceed the other by 2.7 standard errors or more, we use the table for the normal distribution (Table 3 and Figure 11).

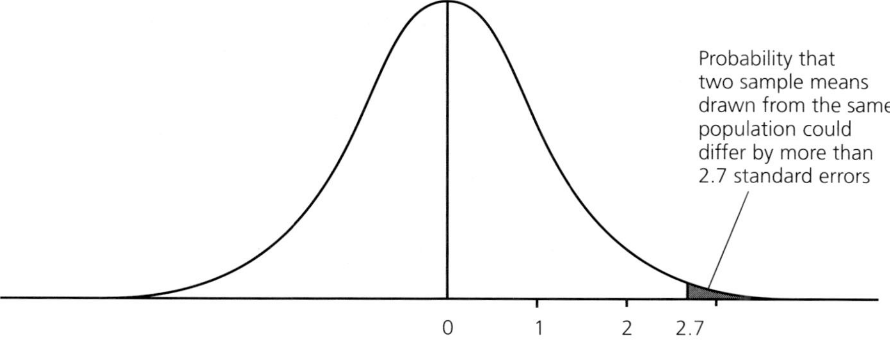


**Figure 11**
Use of the normal curve in comparing sample means

Table 3 shows the probability that a given difference between sample means would exceed the mean difference (zero) by 2.7 standard errors *or less* is 0.4865. This means that the probability that a given difference between means would exceed zero by 2.7 standard errors *or more* is $0.5 - 0.4865 = 0.0135$. In other words, if the parent populations are the same, only in 1.35% of cases would the mean of a sample from the sheltered side exceed the mean of a sample from the exposed side. Since a probability of 0.0135 is considerably less than 0.05, this deviation is highly significant.

## The *t* test

In an earlier section it was stated that differences between means are normally distributed *even if the individual data are not*. This is only true for reasonably large samples, in which $N_1 + N_2$ is more than about 30. For smaller samples, differences between sample means follow a ***t* distribution**. Strictly speaking there are many *t* distributions, one for each number of *degrees of freedom*. The smaller the number of degrees of freedom, the wider the *t* distribution, indicative of the decreasing uncertainty associated with smaller samples (Figure 12).

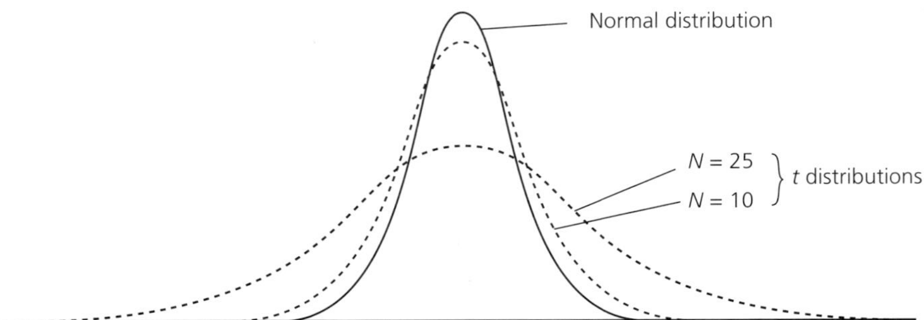


**Figure 12**
Comparison between a *t* distribution and the normal distribution

The number of degrees of freedom is one less than the number of data in the two samples ($N_1 + N_2 - 2$). If we had sampled 14 mussels from the exposed side and 16 from the sheltered side the number of degrees of freedom would be 28.

Just as a *z* score is a deviation from the mean divided by the standard deviation $\left(\frac{x - \bar{x}}{s}\right)$, a *t* value for a difference between means is obtained by dividing the difference by its standard error:

$$t = \frac{\bar{x}_1 - \bar{x}_2}{\sqrt{\frac{s_1^2}{N_1} + \frac{s_2^2}{N_2}}}$$

### *Using the t table*

Having calculated *t*, we use the *t* table to find the probability that a *t* value as high as, or higher than, the calculated value could have been obtained by chance (Table 4).

The table is set out in horizontal lines, each line corresponding to a particular number of *degrees of freedom*. The probability values are set out in two lines at the top of the table. The line used depends on whether we are doing a **one-tailed** or a **two-tailed test**, which in turn depends on the kind of question asked. In the case of the mussels this could take two forms.

- Do the mussels differ in size on the sheltered and exposed sides of the bay? In this case there is no prior expectation as to which way round any difference may lie, so we are concerned with both sides of the *t* distribution. Accordingly, we perform a **two-tailed test**, using the upper line of probabilities (*2p*).
- Are those on the sheltered side larger than those on the exposed side? In this situation we would only be interested in one side of the *t* distribution and have carried out a **one-tailed test**, using the lower line of probabilities (*p*).

**Table 4**
Table for values of $t$

| $2p$ | 0.2 | 0.1 | 0.05 | 0.02 | 0.01 | 0.001 |
|---|---|---|---|---|---|---|
| $p$ | 0.1 | 0.05 | 0.02 | 0.01 | 0.001 | 0.0005 |
| **d.f.** | | | | | | |
| **1** | 3.08 | 6.3 | 12.7 | 31.82 | 63.66 | 636.6 |
| **2** | 1.89 | 2.92 | 4.3 | 6.96 | 9.92 | 31.59 |
| **3** | 1.64 | 2.35 | 3.18 | 4.54 | 5.84 | 12.94 |
| **4** | 1.53 | 2.13 | 2.78 | 3.75 | 4.6 | 8.6 |
| **5** | 1.48 | 2.02 | 2.57 | 3.37 | 4.03 | 6.86 |
| **6** | 1.44 | 1.94 | 2.45 | 3.14 | 3.71 | 5.96 |
| **7** | 1.42 | 1.89 | 2.37 | 2.99 | 3.50 | 5.40 |
| **8** | 1.40 | 1.86 | 2.31 | 2.90 | 3.36 | 5.04 |
| **9** | 1.38 | 1.83 | 2.26 | 2.82 | 3.2 | 4.78 |
| **10** | 1.37 | 1.81 | 2.23 | 2.76 | 3.17 | 4.59 |
| **12** | 1.36 | 1.78 | 2.18 | 2.68 | 3.05 | 4.32 |
| **15** | 1.34 | 1.75 | 2.13 | 2.60 | 2.95 | 4.07 |
| **20** | 1.32 | 1.72 | 2.09 | 2.53 | 2.84 | 3.85 |
| **30** | 1.31 | 1.70 | 2.04 | 2.46 | 2.75 | 3.65 |
| **60** | 1.30 | 1.67 | 2.00 | 2.39 | 2.67 | 3.46 |
| **–** | 1.28 | 1.64 | 1.96 | 2.33 | 2.58 | 3.29 |

## *Another look at significance*

Having calculated the probability that two sample means could differ purely by chance, what do we do with this information? Suppose the probability turned out to be 0.001. With a probability of one in a thousand, we would feel confident that chance was not responsible and that the two populations really were different. But what about a probability of 0.1, or 0.2? Where do we draw the line between 'virtually impossible' and 'quite likely'?

What we have to do is to decide on a probability below which it is *reasonable* to conclude that chance alone is not responsible. Most often, a probability of 0.05 is used. In this case a probability of less than 0.05 is said to be significant.

There is nothing sacred about this value of 0.05. In some instances a higher level of certainty might be needed. If, for example, a new drug is being tested for safety, a probability of 0.0001 might be considered necessary to indicate significance. Whatever level of probability is chosen, the calculated value will be more or less than this.

Notice that by deciding on the level of probability *before* doing the test, the question of whether or not to reject $H_0$ is taken out of our hands. If we chose the level of significance after carrying out the test, there would be a danger that we might choose a level that would enable us to get the 'desired' result.

### Example

Suppose that for our samples of 16 male and 14 female butterflies, $t$ is 2.0. Reading along the line for 30 degrees of freedom (the nearest value in the table to 28), we find that the value of $t$ nearest to 2.0 is 2.04. Significance for a one-tailed test is indicated by a $t$ value of 1.70 or more, corresponding to a probability of less than 0.05. Since 2.0 lies to the right of 1.70, the probability of exceeding 2.0 by chance is less than 0.05. Accordingly we could conclude that the difference between two means was significant, and so would reject the Null Hypothesis.

If we had simply asked whether males and females differed in size, a two-tailed test would have been required. In this case the critical value of $t$ would have been 2.04. Since 2.0 lies to the left of 2.04, we would have concluded that the difference was *not significant*. We would therefore not have been justified in rejecting $H_0$.

### *Accuracy, reliability and validity*

These terms are often used interchangeably, but in science they have quite distinct meanings. Suppose, for example, that a student is asking the question: 'What effect does light intensity have on leaf size in sycamore trees?' She measures the area of six leaves from a sycamore tree growing in a sunny situation, and six leaves from a sycamore tree growing in a more shaded place, and calculates the mean for each group. Though her individual measurements may be extremely *accurate*, her samples are so small that she cannot be sure that each is a *reliable* estimate of the area of the leaves she did not measure.

Moreover, there may be other reasons why the leaves on the two trees differ in size. For instance, leaf size may change with the age of the tree or it may be influenced by the nutrient levels in the soil. In either case she would not be measuring the effects of what she thought she was measuring, so her results would not be *valid*.

# ANSWERS to REVIEW QUESTIONS

## Chapter 1

1 a) Muscles that move the external ear and muscles that raise the hairs in the skin.
b) Stamens, petals and nectaries of a dandelion.

2 Houseflies produce far more offspring than humans, greatly increasing the chances that some offspring will have some resistance. Houseflies have a much shorter life cycle than humans so selection has more 'biological time' in which to act.

3 B

4 a) Lamarck would have said that the musical skill developed by the parents as a result of practice would have been handed on to the children.
b) A modern biologist would say that musical talent has at least some genetic basis, and that children of talented musicians are thus more likely to inherit genes that favour the development of musical talent than children of parents lacking in musical talents.

## Chapter 2

1 B

2 i) C ii) A iii) D iv) B

3 499

4 $C_{15}H_{26}O_{13}$

5 A monosaccharide has the same number of oxygen atoms as carbons, a disaccharide has one fewer oxygen than carbons (since a water molecule is removed in its formation) and a trisaccharide has two fewer oxygen atoms than carbons. Hence $C_7H_{14}O_7$ could be a monosaccharide, $C_{11}H_{20}O_{10}$ could be a disaccharide and $C_9H_{14}O_7$ could be a trisaccharide.

6 i) a fat = D
ii) a saturated fatty acid = A
iii) an unsaturated fatty acid = C
iv) a carbohydrate = E
v) an amino acid = B

7 Four: gly-gly, ala-ala, gly-ala, ala-gly (the last two are different).

8 $\{2 \times (141 - 1)\} + \{2 \times (146 - 1)\} = 570$

9

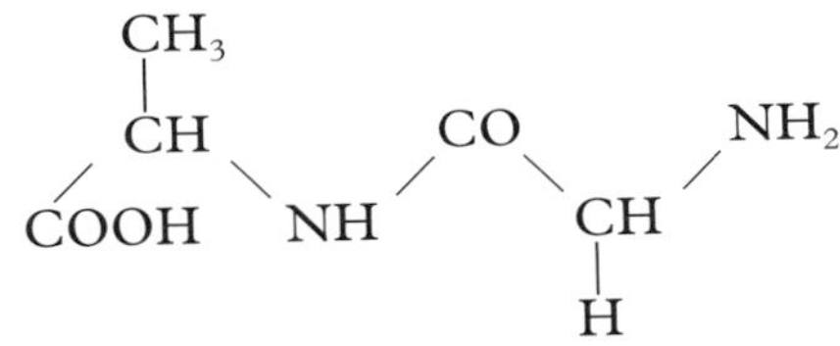


## Chapter 3

1 A chloroplast, nucleus, cell wall and a skin cell can be seen with a school microscope;
a mitochondrion and a Golgi body can only be seen with the best light microscopes;
a lysosome, haemoglobin molecule and a cell membrane can only be seen with the electron microscope.

2 Since the high power objective is 10 times as powerful, the field of view must be a tenth of 3 mm, or 300 µm.

3 C, E, B, D, A, F

4 i) A ii) C iii) C iv) C v) C, D vi) C, D vii) C viii) A ix) C x) C, D xi) C, E xii) C
N. B. Mature mammalian red blood cells lack all the above except a plasma membrane

5 i) D ii) L iii) K iv) F v) H vi) C vii) E viii) I ix) J x) B xi) A xii) G

## Chapter 4

1 a) T b) F c) F d) F e) F f) T g) F

2 See text.

3 See text.

4 See text.

5 a) A, C, E, F
b) A, F

## Chapter 5

1 Combustion occurs at high temperatures, respiration at moderate temperatures.
In combustion all the energy is released as heat (and some light), in respiration about half the energy is conserved as chemical energy in the form of ATP.

2 a) ATP b) starch

3 starch > glucose > pyruvate > ATP

4 Since fat contains less oxygen than carbohydrate, it undergoes more oxidation, releasing more energy.

5 Ethanol ($C_2H_6O$), since it contains proportionately less oxygen.

6 Lactate is produced by addition of hydrogen to pyruvate. When subsequently removed as NADH, the latter provides additional energy on oxidation.

7 C

8 B

## Chapter 6

1 All are false.

2 a) B (diffusion has nothing to do with need but results from a difference in concentration).
b) B. A rise in temperature raises the respiratory

rate and hence steepens the concentration gradient (the direct effect of temperature on diffusion is very small).

3 B

4 C

5 Immediately after mixing, the water potential of the red cells and plasma are identical. But, since glycerol is more concentrated in the plasma than in the red cells, it diffuses into the red cells, lowering their water potential and causing water to follow by osmosis until the red cells burst. Urea has smaller molecules than glycerol and enters the red cells more quickly. Glucose molecules are too large to enter red cells by simple diffusion.

## Chapter 7

1 A = centromere; B = chromatid; C = chromosome; D = homologous pair.

2 There should be six daughter chromosomes moving to each pole, the chromosomes should be single-stranded, and each daughter chromosome should resemble the one from which it separated – the one moving to the other pole.

3 Independent assortment.

4 a) Interphase, b) Anaphase

5 a) A or E (chromosome(s) single-stranded, one of each kind).
   b) C if the cell is not going to divide again, B if it is.
   c) D or F (chromosome(s) double-stranded, one of each kind).
   d) B (diploid, chromosomes double-stranded).

6 The products are usually genetically different.

7 A is incorrect: the chromatids do not cross over and then break.

8 a) For each chromosome pair there are two kinds of gamete, so for eight chromosomes (four pairs) there are $2^4 = 16$ possible combinations.
   b) For each chromosome pair there are three possible combinations, so for four pairs there are $3^4 = 81$ kinds of zygote.

## Chapter 8

1 C

2 a) Bill must be heterozygous.
   b) 0.25 (the same as for the first child).
   c) Bill's father must be heterozygous since he received a recessive allele from Bill's paternal grandfather.

3 a) 1 b) 2 c) 8

4 All are true.

5 a) 2 b) 8 c) 4 d) 32

6 1 and 2, 3 and 5, 4 and 6, since these combinations never occur together.

7 i) C ii) A iii) D iv) B and E

In B, the recombinants are more numerous than the parental types. In E, the alleles *F* and *f* are not segregating in a 1:1 ratio (35 + 5 = 40% *F*, 15 + 45 = 60% *f*)

8 a) i) *G i* and *g I*
      ii) *G I* and *g i*
   b) i) *G H i* (no crossover) > *GHi* (crossover between *H*/*h* and *I*/*i*) > *Ghi* (crossover between *H*/*h* and *I*/*i*) > *GhI* (double crossover)
      ii) A) *g h i* B) *g H i*

9 4 (one in each chromatid of each member of a homologous pair).

10 a) 0 (his father's mother has no Y-chromosome)
   b) 1

11 a) F b) T c) T d) F

## Chapter 9

1 If both man and woman are homozygous the children would be normal, since each parent would provide the normal allele the other lacks.

2 a) Since two reactions (and therefore two enzymes) are involved in the conversion of tyrosine to melanin, a mutation in either of the two genes concerned could result in albinism. If each parent was homozygous recessive for a different gene, they would complement each other and the child would be heterozygous for both loci and thus phenotypically normal.
   b) A recessive allele in one of the parents could have mutated to the normal allele.

3 a) Three kinds. If defective and normal β chains combine equally readily with α chains, then, $\frac{1}{2} \times \frac{1}{2} = \frac{1}{4}$ of the resulting haemoglobin molecules would have both β chains defective, $\frac{1}{4}$ would have both β chains normal and $\frac{1}{2}$ would have one of each kind.
   b) Since there are three kinds of haemoglobin molecule in a heterozygote, they would stack less readily than in homozygotes, in which all the haemoglobin molecules are the same.

4 $\frac{1}{2} \times \frac{1}{2} \times \frac{1}{2} \times \frac{1}{2} = \frac{1}{16}$ would be normal.

## Chapter 10

1 C T A G T A T T A

2 The bases.

3 C (adenine can vary quite independently of guanine. Cytosine and thymine are similarly independent of each other.)

4 A

5 C

6 B. The sequence of bases in one strand defines the sequence in the other.

7 30% (20% must be adenine, so the remaining 60% must consist of equal proportions of guanine and cytosine).

8 a) Each chromatid (consisting of a double-helical molecule of DNA) would consist of one radioactive and one non-radioactive strand.
b) Half the chromosomes would be radioactive and half non-radioactive.

9 a) 20
b) 80 (40 chromosomes, each consisting of 2 chromatids).
c) 40 (20 chromosomes, each consisting of 2 chromatids).

10 a) In M13 phage adenine and thymine are not present in equal amounts and neither are guanine and cytosine.
b) The DNA must be single-stranded.
c) Heating the DNA should result in no change in density or in UV absorption (such as occurs when double-stranded DNA 'melts').

## Chapter 11

1 a) Two, one for the α chains and one for the β chains.
**b)** For the α chain: $141 \times 3 = 423$, plus 3 for start and 3 for stop = 429 bases.
For the β chain: $146 \times 3 = 438$, plus 3 for start and 3 for stop = 444 bases.
For the whole protein it would need $429 + 444 = 873$ bases.

2 Five. $2^5$ would give 32 possible sequences (four would only give $2^4 = 16$).

3 leu-asn-pro-gly

4 A Double-stranded DNA
B Double-stranded RNA
C Double-stranded DNA
D Single-stranded DNA
E Single-stranded RNA

5 a) mRNA, tRNA, rRNA
b) mRNA, tRNA, rRNA
c) DNA and rRNA
d) DNA, tRNA and rRNA
e) mRNA
f) DNA, mRNA, tRNA, rRNA

6 a) F b) T c) T

7 A

8 i) N ii) O iii) O iv) N v) I vi) O vii) I viii) I ix) I x) N xi) I xii) O

9 1 D 2 B 3 A 4 E 5 D 6 C

## Chapter 12

1 i) D ii) I iii) F iv) A v) J vi) C vii) B viii) E ix) H x) G

2 Most mutations are harmful in an unchanging environment. An organism that has successfully reproduced obviously has 'good' alleles, so any *random* change is likely to prove harmful in the same environment.

3 Mutation has no relation to *need*. A change in the environment (e.g. radiation or an increase in temperature) may increase the *rate* of mutation, but has no effect on the *kind* of mutation that occurs.

4 Organisms that only reproduce asexually.

5 In haploid organisms all alleles are expressed, so disadvantageous alleles cannot be protected by dominance. Any allele that is not of immediate utility is thus likely to be eliminated by selection. In diploid species, recessive alleles are protected in heterozygotes by dominant alleles.

6 a) A b) E c) B d) D e) C

7 a) Chromosomes do not have 'partners' with which to pair at meiosis.
b) 14

## Chapter 13

1 a) Restriction endonuclease
b) DNA ligase
c) Reverse transcriptase
d) DNA polymerase

2 To ensure that the bases at the ends of the human DNA are complementary with the bases at the cut ends of the plasmid DNA.

3 It is single-stranded.

4 Eukaryote genes contain introns which could not be 'edited out' in a prokaryote.

5 a) The enzyme recognising four bases, since a short sequence would occur by chance more often than a longer sequence.
b) The probability of a specific sequence of four bases occurring would be $(\frac{1}{4})^4 = \frac{1}{256}$. The probability of a specific sequence of eight bases occurring would be $(\frac{1}{4})^8$, or $(\frac{1}{4})^4 \times (\frac{1}{4})^4$. Since 8-base sequences would be generated $(\frac{1}{4})^4$ as frequently as 4-base sequences, the ratio of fragment lengths would be 256:1.

## Chapter 14

1 a) Z
b) Y (higher frequency of homozygotes than predicted by the Hardy–Weinberg equation).
c) X (higher frequency of heterozygotes than predicted by the Hardy–Weinberg equation).

2 a) Frequency of Rh-negative phenotype $= 1 - 0.85 = 0.15$.
Frequency of Rh-negative allele = $\sqrt{0.15} = 0.3873$
b) $2\{0.3873 \times (1 - 0.3873)\} = 0.4746$
c) At-risk pregnancies occur when mother is Rh-negative and child is Rh-positive. This occurs in two kinds of mating:
all cases when the father is an Rh-positive homozygote and
half the cases when he is heterozygous.
Frequency of first kind of mating $= 0.15 \times q^2$

Frequency of second kind of mating
$= 0.15 \times 2pq$
But, only half the second kind of mating produces an Rh-positive child.
Therefore, frequency of at-risk pregnancies
$= (0.15 \times q^2) + \frac{1}{2}(0.15 \times 2pq)$
$= 0.15 \times (q^2 + pq)$
$= 0.0919$
Hence 9.19% of pregnancies are at risk.

3 First cousins share a higher proportion of their genes in common than unrelated people, and so are more likely to be heterozygous for the same harmful genes. Their children are thus more likely to be homozygous for harmful recessive alleles.

4 a) Less food is available for each nestling.
b) With smaller clutches each nestling has more food, so they are larger and more fully developed when they leave the nest.
c) Small clutches will lead to fewer offspring with a higher survival rate. Large clutches result in more offspring but with reduced life expectancy. The optimum clutch size is a compromise between these two opposing factors.
d) Stabilising selection.

5 a) See text.
b) Interspecific mating is disadvantageous, so in areas where the two species are sympatric, individuals that can discriminate between their own and another species would be at an advantage and genes responsible for this discrimination would be selected for. In areas where the two species are allopatric there would be no selection pressure.

## Chapter 15

1 Kingdom
2 Genus *Rattus*
3 An elephant's tusk and a tiger's incisor are homologous (they are both teeth). Spines of hedgehog and sea urchin are analogous.
4 They have been modified under similar selection pressures (digging).
5 Differences can be quantified and are thus less subjective.
6 *Ranunculus ficaria* (in handwriting, with both words underlined).
7 They are not immediately obvious, especially when (as is often the case) only a corpse is available.
8 a) Ferns have prominent veins in the leaves and a well-developed root system; mosses do not.
b) Flowering plants have flowers and the seeds are hidden and protected within a fruit. Conifers have cones, the scales of which bear unprotected seeds.
c) Cartilaginous fish have externally visible gill slits, the upper and lower lobes of the tail fin are different, and the mouth is beneath the head. Bony fish have gill slits protected by an operculum (gill cover), the two lobes of the tail fin are usually alike, and the mouth is usually terminal.
d) Reptiles have leathery skins (dry except in aquatic species). Amphibians have soft, usually moist skin.
e) Cnidarians are soft-bodied and the mouth is surrounded by tentacles. Echinoderms have spiny skins and lack tentacles, though they have five rows of tube feet radiating from the mouth.
f) Annelids have soft, moist skins; terrestrial fly larvae have dry cuticles with spiracles.

9 a) Arthropoda
b) Mollusca
c) Echinodermata
d) Cnidaria
e) Platyhelminthes

## Chapter 16

1 a) Since its length has doubled, its volume (and hence its mass) would have increased by $2^3$ or 8 times, to 84 grams.
b) Its surface would have increased by $2^2 = 4$ times.
c) That the body shape remains constant.

2 D

## Chapter 17

1 a) It would fall since the $K^+$ concentration gradient would be reduced.
b) It would increase, since the $Na^+$ gradient would be steeper.
c) No immediate effect, but it would gradually diminish with unopposed leakage of $K^+$ ions down its concentration gradient.

2 a) No effect, though it would lower the threshold for a stimulus.
b) No effect.
c) No immediate effect, though with repeated stimulation it would diminish as the $Na^+$ gradient runs down.

3 The time taken for the excitation to pass between two recording electrodes and the distance between them.

4 The most excitable fibres propagate impulses more quickly, and they produce larger action potentials.

5 The length of axon involved at any instant is the distance the impulse moves in the duration of the action potential (0.5 ms in this case). At $100 \text{ m s}^{-1}$ the impulse travels 1 m in 10 ms. In

0.5 ms ($\frac{1}{20}$ of this time) it travels $\frac{1}{20}$ of a metre, = 50 mm, or 50 internodes.

## Chapter 18

1 A receptor detects *stimuli*; taste is a sensation.
2 The relaxation of muscle does not generate a force. By relaxing, a muscle may permit movement by ceasing to oppose another force – in this case the tension in the choroid.
3 Under water the eye does not refract the light sufficiently (because water is denser than air and has a higher refractive index), so light would not be refracted enough. To see clearly the eye would need to be abnormally powerful, as in short-sight.
4 Figure 18.25A shows that rods are unstimulated by red light, and so are fully dark-adapted.
5 In dim light small differences in light intensity result in large differences in impulse frequency, giving greater contrast between different parts of the image. In bright light the reverse is true.

## Chapter 19

1 a) This promotes the transmission of tension between muscle and tendon fibres.
b) This minimises the delay between the arrival of an action potential and the contraction of the parts of the muscle fibre furthest from the motor end plate.
c) Since the myofibrils do not have to pass around nuclei or mitochondria, tension is developed parallel to the long axis of the muscle.
2 a) C b) B
3 In a narrow fibre the distance between blood and mitochondria is less, speeding up the diffusion of oxygen.
4 i) C ii) B iii) A
5 On average each neuron supplies $4 \times 10^6/8000 = 500$ muscle fibres in the case of the quadriceps, and $2 \times 10^3/850 = 2.3$ fibres in the case of the laryngeal muscle. This enables the laryngeal muscle to make the much finer grades of response needed to effect tiny changes of pitch in speech.
6 A = glycogen; B = glucose; C = pyruvate; D = lactate; E = carbon dioxide; F = oxygen; G = water; H = ADP; I = ATP.

## Chapter 20

1

| | Nervous | Hormonal |
|---|---|---|
| **Speed of travel** | rapid | slow |
| **Distribution of signal** | fixed pathways | throughout body |
| **Duration of signal** | short | long |
| **How strength is varied** | by frequency | by concentration |

2 Postganglionic neurons of adrenal medulla; posterior pituitary neurons secreting ADH and oxytocin.
3 a) Pancreas
b) Thyroid/parathyroid and adrenal cortex/adrenal medulla
4 To permit the strength of the signal to be reduced as well as increased.
5 See information box on page 376.

## Chapter 21

1 A; stability of body temperature in a worm living at the bottom of the ocean is due to stability of the environment rather than regulation by the animal.
2 Osmoregulation by contractile vacuole in *Paramecium* and other freshwater protozoa; the stomatal closure in a flowering plant in response to water shortage.
3 Only after a change has been detected can the necessary corrective measures be initiated.

## Chapter 22

1 The plasma would contain the higher *concentration* of oxygen, but the *amount* would be much smaller than in an equal volume of whole blood. In the latter case most of the oxygen would combine with haemoglobin, leaving less to dissolve in the plasma.
2 a) The curve for 40°C would lie to the right of that for 37°C.
b) Active muscle is slightly warmer than resting muscle, so it would promote the release of oxygen.
3

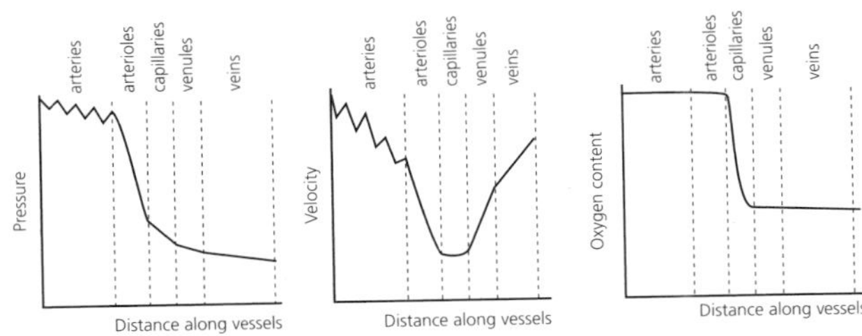


4 Elastic tissue in large arteries absorbs the energy of the pulse; as the pulse diminishes along the arterial tree so does the proportion of elastic tissue. Muscle regulates the distribution of blood, which can only be varied by controlling the resistance to flow along smaller branches.
5 D (pulmonary artery).
6 a) Cross striations and rapid, powerful contractions.
b) Myogenic activity and immunity to fatigue.
7 The heart (arteries are dilated by blood pressure).
8 B, since the flow through the lungs is always

equal to the flow through the *entire* systemic system.

9 B

10 B; although the pressure falls slightly, vessel diameter increases many times from venules to main veins.

11 It reduces the resistance to flow of tissue fluid through the pores between adjacent capillary wall cells, increasing the rate of tissue fluid formation and thus of materials such as glucose (which cannot leave the blood by diffusion through the plasma membranes of capillary wall cells).

## Chapter 23

1 a) i) $0.5 \times 15 = 7.5$ dm$^3$ min$^{-1}$
ii) $2.5 \times 30 = 75$ dm$^3$ min$^{-1}$.
b) Ventilation increases in proportion to oxygen uptake and $CO_2$ production.
c) 10 times.

2 a) No, because the mitochondria cannot use the oxygen delivered to them.
b) No, because there is insufficient haemoglobin to carry oxygen from the lungs.
c) Yes. Oedema would increase the length of the diffusion path. Raising the oxygen concentration in the alveoli would help restore the steepness of the oxygen concentration gradient.
d) Yes. In emphysema there is a reduction in the cross-sectional area of the diffusion path: administering oxygen would steepen the gradient and thus raise the rate of diffusion of oxygen per unit area of membrane.

## Chapter 24

1 A = bladder; B = renal artery; C = Bowman's capsule; D = end of proximal tubule.

2 a) A very slight dilution (same as the very slight dilution of the blood plasma).
b) Very little; the production of an increased volume of more dilute urine results from decreased reabsorption from the collecting ducts.

3 a) Increased concentration of urea.
b) No effect, since increase in insulin output would prevent a large increase in blood glucose.

4 Cooling reduces active transport, and hence the reabsorption of sodium and glucose. The former reduces the reabsorption of water and hence increases urine output. The latter results in some glucose not being reabsorbed, so it appears in the urine; this effect also reduces water reabsorption by lowering the solute potential of the tubular fluid.

5 a) 0.12 g min$^{-1}$ (using formula given in the information box on page 437).
b) Creatinine. The U/P ratio for inulin is 120, meaning that it is 120 times as concentrated in the urine as in the plasma. For a substance to have a higher U/P ratio than inulin, more must be leaving the nephron than is entering – i.e. it must be secreted into the nephron. Only creatinine has a higher U/P ratio than 120.
c) Urea has a U/P ratio of 60, or half that of inulin. Hence half the $Na^+$ ions must be reabsorbed.

6 Mannitol lowers the solute potential of the tubular fluid and reduces the amount of water reabsorbed.

7 See text, and also Chapter 22 'Giving blood'.

## Chapter 25

1 Water has a much higher specific heat capacity than air and conducts heat better.

2 Sweat glands, smooth muscle in dermal arterioles, and the hair erector muscles.

3 a) A cooling of the blood would be detected by the hypothalamus, provoking vasoconstriction in the dermal arterioles and a decrease in sweating. In warm air this would warm the skin by reducing heat transfer from the core to the skin.
b) In cool air temperatures the same responses would occur but in this case the reduction in heat transfer to the skin would *cool* the skin.
c) That information from receptors in the skin can be overridden by information from receptors deep in the body.

4 a) Blood from the most metabolically active parts of the body enters the right side of the heart. Because of evaporation into the air in the bronchial tree, blood flowing through the lungs is cooled slightly.
b) It would increase; more heat would enter the blood passing through the muscles and increased ventilation would cause greater cooling as the blood flows through the lungs.

## Chapter 26

1 i) C ii) F iii) E iv) D v) B vi) G vi) A

2 a) That something else besides proteins, fats, carbohydrates, minerals and water is needed for healthy growth.
b) Milk contains something other than proteins, fats, carbohydrates, minerals and water that is essential for healthy growth.
c) The rats must have been able to store some of the nutrients in the milk.
d) Use of two groups of rats; several rats in each group; reversal of the diet so that each rat was given both diets.

3 a) Maize is deficient in certain essential amino

acids, and peas are deficient in *different* essential amino acids. Neither peas nor maize given alone supplies all the essential amino acids but when given in combination each makes up for the deficiencies of the other.
  b) The loss in mass would be due in part to metabolised body protein, the nitrogen of which would be converted to urea and excreted. Since Group A lost the greatest mass they would have produced the most urea.

4 a) A and B
  b) A and D
  c) A and E
  d) A and C
  e) A and F

5 a) At low concentrations the uptake of galactose is much faster. At higher concentrations the galactose transport mechanism becomes saturated, further increase in concentration having little effect on rate of uptake.
  b) Xylose, since it does not show the saturation effect characteristic of a carrier mechanism.
  c) It would resemble that for galactose, since facilitated diffusion involves a carrier.
  d) i) It would be reduced, since transport by carrier involves conformational change in the carrier protein, which is temperature-sensitive.
    ii) No effect, since the energy for transport comes from the concentration difference rather than ATP.

## Chapter 27

1 1) F 2) K 3) I 4) B 5) R 6) A 7) O 8) C 9) H 10) D 11) N 12) E 13) G 14) J 15) T 16) S 17) L 18) P 19) M 20) Q

2 An antigen is a substance, not an organism. A disease-causing organism, or pathogen, carries antigens (substances which can provoke the production of antibodies) on its surface.

3 Inbred strains are genetically very similar to each other.

4 Phagocytes engulf bacteria much more readily if the bacteria are coated with antibody; antibodies are produced by plasma cells which develop from B cells.

5 The clonal selection hypothesis is essentially Darwinian in that it depends on the selection of variants produced *before* the need arises, by a *random* process, unrelated to need. Those that 'work' are then selected for. The instructional hypothesis is Lamarckian because it proposed that each antibody is produced in response to exposure to an antigen – i.e. *after* the encounter with the antigen, and thus occurs in response to need.

## Chapter 28

1 a) Although sperms are still being produced they cannot reach the urethra and be ejaculated.
  b) Sexual desire is influenced by testosterone, which continues to leave the testis via the blood.
  c) Since his semen is free of sperm it is free of DNA.

2 Under the influence of gonadotrophic hormones from the pituitary of the mature rat, the testis would mature and begin producing sperm and testosterone.

3 The need for incubation of the eggs (when feeding cannot occur) means that in most bird species the female cannot rear the young on her own. Desertion by the male would thus put his own genes (actually copies of them, in his offspring) at risk.

4 Early in pregnancy the ovary is the only source of the progesterone and oestrogen necessary for maintenance of pregnancy. After 4 months the placenta produces large amounts of these hormones.

5 i) Glucose cannot be transported by simple diffusion (otherwise D and L forms would cross at equal rates). Though some form of specific carrier mechanism must be involved, facilitated diffusion (which is non-active) could play a part.
  ii) Transport of amino acids must be active since it occurs against a concentration gradient.

## Chapter 29

1 Plants absorb thinly dispersed, very dilute nutrients. A finely dissected shape enables a large surface area to be exposed to the surroundings.

2 Specialised supporting tissues enable body shape to be maintained in the face of gravity and wind. Development of specialised transport tissues is associated with interdependence of the root and shoot. Storage tissue is associated with survival over adverse conditions – seasons are much more extreme on land.

3 A = sieve tube element; B = collenchyma cell; C = parenchyma cell; D = fibre; E = vessel member.

## Chapter 30

1 See text.

2 A

3 A

4 a) By working from older, functionally female flowers to younger, functionally male flowers it reduces the chances of self-pollination.
  b) The older flowers may produce more nectar than younger ones.

5 Only bee genes that are advantageous to the bees can be selected for: bees are selfish; they are not doing flowers a favour, but visit flowers for nectar and pollen.
6 There would be less incentive for an insect to visit other flowers, so pollination would be less likely.
7 Insects visiting non-nectar flowers do so for the large amounts of pollen they produce. A flower that is functionally female would not be visited.
8 An insect visiting a bilaterally symmetrical flower takes up the same position each time, so the anthers and stigma can be located where they are most likely to touch the insect's body, reducing pollen wastage.
9 a) inside a pollen grain, b) inside the embryo sac.
10 Zygote is to embryo as ovule is to *seed* and ovary is to *fruit*.
11 21 (endosperm is triploid).

## Chapter 31

1 a) Protein
b) Energy reserves run out before the plumule reaches the light.
c) Cotyledons
d) It must be very low.
2 a) Place in an oven at 105°C until there is no further change in mass.
b) Since drying kills an organism, its dry mass can only be measured once. A change in dry mass is estimated by taking samples of known and similar live mass, and measuring their dry masses at known intervals of time.
c) That if initial live masses are similar, then so are the dry masses. This can be checked by measuring live and dry masses of a number of samples at the same time. Reliability of estimates can be increased by taking larger samples.
d) C; osmotic uptake of water.
e) i) B ii) E iii) D
f) Cellulose
g) The transport mechanism itself involves structures which must remain in place during the transport process.
3 False. A fat-storing seed may convert a given mass of fat into a larger mass of sugar, though the latter *always has a lower energy value*.
4 i) B ii) C iii) D iv) A
5 a) Elongation of a stem is partly due to water uptake by vacuolating cells, which depends on their having a lower (more negative) water potential than the neighbouring xylem due to the active uptake of ions. When the stomata close, water continues to enter the roots, raising the water potential in the xylem. This makes it easier for the expanding cells to vacuolate by absorbing water.
b) During the day when transpiration is rapid, water potential in the xylem is low, making it harder for vacuolating cells to absorb water. Following the rise in water potential in the xylem after stomatal closure at night the rate of vacuolation by the cells increases.
c) Growth slows down, probably because the ion uptake necessary to maintain a low water potential in vacuolating cells fails to keep pace with water uptake. As a result the solute potential of the expanding cells rises and rate of water uptake falls.

## Chapter 32

1 C and D
2

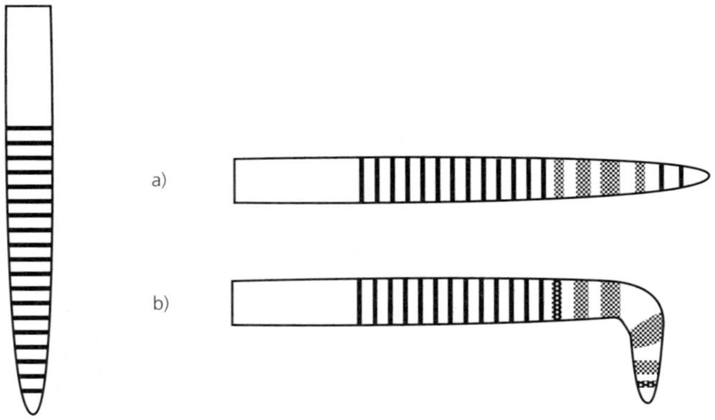

a)

b)

3 D, C, A, B
4 More pronounced apical dominance and stronger negative gravitropism in poplar would explain the difference in shape.
5 Young shoots produce auxin, which promotes the formation of lateral roots.
6 External stimuli are not necessary for the *production* of growth substances: tropisms result from a change in their lateral distribution.
7 The validity of Went's *Avena* curvature test depended on growth rate being proportional to auxin concentration, but the graph shows that between $10^{-3}$ and $10^{-1}$, growth rate is proportional to the *log* of auxin concentration.

## Chapter 33

1 A = carbon dioxide; B = water; C = oxygen; D = carbohydrate; E = light; F = chlorophyll; G = chloroplast.
2 Mass of $CO_2$ absorbed per unit time per unit leaf area or gram of plant material.
Mass of oxygen produced per unit time per unit leaf area or gram of plant material.
Increase in dry mass per unit time per unit leaf area or gram of plant material.

3 a) light intensity
b) temperature
c) carbon dioxide

4 2.0 + 0.5 = 2.5 mg $CO_2$ per hour

5 a) See text.
b) An entire plant since it contains the greatest proportion of respiring to photosynthesising tissue.
c) It would raise it. A rise in temperature increases respiration but in dim light, has no effect on photosynthesis. Hence a brighter light would be needed to balance an increased respiratory rate.

6 B. At low to moderate intensities, red is more effective than green, but in very bright light the plant is light-saturated in red light. In green light, light saturation is reached at higher intensities since a lower proportion of green light is used.

7 a) Respiration and photosynthesis.
b) They are equal since $CO_2$ concentration does not change.
c) As $CO_2$ concentration falls the rate of photosynthesis falls.
d)

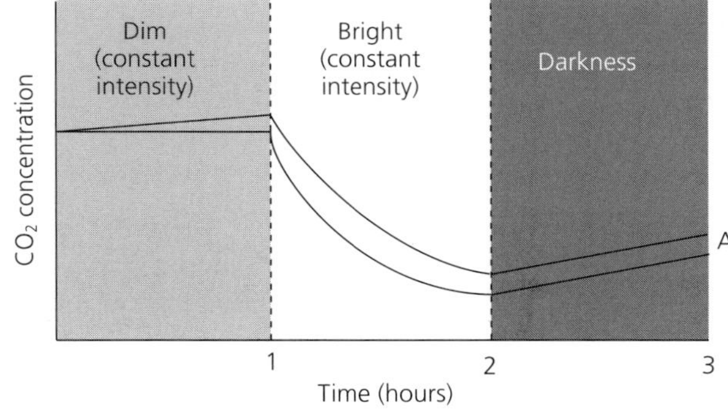


8 a) A = $H_2O$ B = $O_2$ C = $NADP^+$
D = NADPH E = GP F = GALP
G = RuBP H = $CO_2$
b) X = granum, Y = stroma.
c) C and G.

9 a) The total amount of respiration or photosynthesis occurring in 24 hours.
b) Total photosynthesis is only slightly greater than total respiration, so the plant would only just be gaining in dry mass.

10 Phototropism would be normal since it is most sensitive to blue light. Phytochrome-mediated responses would not occur since neither form of the pigment absorbs blue light to any great extent. Photosynthesis would not be directly affected since blue light is absorbed by both PS I and PS II. Indirectly, however, it would, since the straightening of the plumule and hypocotyl and leaf expansion in seedlings are mediated by phytochrome. Light-dependent germination would not occur. Seasonal responses involving photoperiodism would probably be severely affected.

11 Flooding causes the soil to become anaerobic, inhibiting respiration and the active uptake of ions.

12 a) Metal ions are positively charged and are attracted into the negatively charged interior of the root cells. If the electrical attraction exceeds the effect of the concentration difference, metal ions could diffuse into the root against a concentration gradient.
b) Nitrate ions are negatively charged and so must enter against electrical and concentration gradients. Entry of negative ions is therefore active.

## Chapter 34

1 Transpiration is an inevitable process and so cannot be regarded as a function. Respiration occurs in mitochondria which occur in living cells throughout the plant. It would therefore be inappropriate to regard a leaf as a source of respiratory energy for the rest of the plant.

2 a) To check that greasing does prevent gases leaving the leaf.
b) Stomata are open in the light so transpiration is faster.
c) i) B ii) C
d) Loss in mass due to lower stomatal transpiration = 8% (obtained by subtracting 2 from 10).
e) 8:1 (2 − 1 = 1% loss due to upper stomata).
f) That all loss in mass is due to water; that stomata on upper and lower surfaces are of equal size; that stomata close completely in the dark.

3 The initial rapid loss occurs via open stomata. As the leaf becomes dehydrated the stomata close, reducing the rate of water loss.

4 a) D
b) $\frac{25 \times 2}{4 \times 50} = 0.25 \text{ mm cm}^{-2}$

5 Grazing is a particularly serious threat in dry soils for two reasons. Growth is slow (stomatal closure reduces $CO_2$ uptake), and tissue damage breaks the cuticle and increases water loss. Spines render a plant less palatable.

6 a) The leaf. A petal only needs enough stomata to cater for respiratory gas exchange, whereas a Leaf has to absorb $CO_2$ for photosynthesis. Since oxygen is 600 times more concentrated than $CO_2$, fewer stomata are needed.
b) Since shoot A undergoes mainly respiratory gas exchange, it can 'afford' to have its stomata nearly closed.

7 Air locks form in the xylem vessels, so the only route for water movement is through the cell walls.

8 a) Decrease, since with no air currents, humidity round the shoot would increase.
b) Decrease, since the salt would lower the solute potential of the soil, decreasing the steepness of the osmotic gradient between soil and root. With continued salt uptake the gradient would become steeper again, and water uptake would increase.
c) Decrease, since salt uptake would decrease, reducing the osmotic gradient between soil and root.
d) Initially no effect, but anaerobic conditions would inhibit salt uptake and thus water uptake.

9 Placing a transparent plastic hood over the shoot system restores the balance between water uptake and water loss.

10 Concentration of sugar in the phloem sap and total cross-sectional area of the sieve tubes. The calculated value would be an under-estimate because some of the sugar arriving in the fruit would be used in respiration.

## Chapter 35

1 See text.

2 A large number of very fine pore spaces (as in clay) has a much greater surface area than a small number of large spaces (as in sand), so it offers much more resistance to water flow. As a soil dries out the water menisci retreat between the mineral particles. Since the pore spaces in clay are much narrower than in sand, the menisci are much more curved, and pull the mineral particles together much more strongly than in sand. Also, the clay particles interlock over a much greater surface than sand particles do.

## Chapter 36

1 Small organisms have a greater surface area to volume ratio and the distance between the interior of the body and the environment is less. With steeper gradients and shorter distances between the environment and the outside, small organisms have to work harder to maintain differences between the inside and outside of the body.

2 Food, and therefore energy, for maintaining body temperature is harder to obtain.

## Chapter 37

1 a) False. While it decreases the reproductive success of non-territory holders it increases the reproductive success of territory holders.
b) True. Non-territory holders are less likely to reproduce.
c) False. While it decreases intraspecific competition once territories have been established, there is still the initial competition to gain territories.

2 There is better access to resources such as food, breeding sites and mates. While the cost of territory defence continues to rise with increase in territory size, the benefits do not since there is a limit to the resources that each individual can use.

3 To form a dominance hierarchy each individual must know every other individual in the group, which requires a more complex brain.

4 a) Remove *Semibalanus* individuals and see if there is any effect on the numbers of *Chthamalus* reaching maturity.
b) Inability to tolerate prolonged exposure to air, and inability to compete with *Chthamalus*.

5 a) Mistletoe growing on apple, ferns growing in shade of trees.
b) Many insects have highly specific food requirements and live on their food plants, e.g. cabbages are food for larvae of the large white butterfly.
c) Plants of grazed pasture are dependent on herbivores to keep down potential competitors. When myxomatosis entered Britain in the 1950s, the number of species of herbaceous plants in some rabbit-grazed land decreased markedly.
d) Any animal parasite.

6 a) Character displacement.
b) i) When species are sympatric there is some competition for food, so selection would favour differences in diet and thus of beak length.
ii) If the eye-stripe plays a part in species recognition in breeding behaviour, then in sympatric populations selection would favour those individuals of *S. neumeyer* that differ most from *S. tephronota*, leading to divergence. In allopatric populations there would be no such selection pressures.

7 See text.

## Chapter 38

1 Since two out of 20 (10%) of the marked (but not released) animals lost their mark through ecdysis, a similar proportion of the released animals probably lost their mark. If a similar proportion of the animals released lost the mark, then about 10 would have been recaptured (one having lost its mark).

$$\text{Hence: } \frac{80}{\text{total population}} = \frac{10}{100}$$

$$\text{Total population} = 800$$

2 A
3 B
4 a) A 256-fold increase represents eight doublings, or eight cell divisions. Hence each division takes 2 hours.
b) A 20°C rise in temperature would cause a four-fold increase in growth rate, giving time for $4 \times 8 = 32$ divisions. The number of bacteria would thus increase $2^{32} = 4.3 \times 10^9$ fold, or 4.3 billion times.

## Chapter 39

1 See text
2 A. The nitrogen content of the air is far above a level that could limit the rate of nitrogen fixation. All the other factors could affect the nitrogen cycle (though in some cases the effect is indirect).
3 a) When ammonia combines with organic compounds to form amino acids.
b) When amino acids are deaminated to form ammonia.
4 Polar bears are 'top carnivores', being at the end of a long food chain. Each link concentrates Vitamin A, so by the time it reaches polar bears it has reached high levels.
5 a) Respiratory energy is used for growth and for maintenance. An elephant grows very slowly and during this long period of growth it uses a large amount of energy for maintenance.
b) Few herbivores can digest cellulose completely, even with the help of mutualistic micro-organisms.
6 In a lake the population density of fish is limited (at least in part) by the primary productivity in the lake. Filter-feeding crustaceans such as *Daphnia* filter out planktonic algae so the water does not turn green. In an aquarium the density of fish is much higher than can be supported by production in the aquarium and they have to be artificially fed. This boosts the algal population in two ways: the fish eat all the *Daphnia* and excrete enough wastes to fertilise the water and promote algal growth.
7 Species diversity is low in the fore-dunes because very few species can survive the adverse conditions there. With increase in fertility the number of species increases, as does the complexity of feeding relationships. Biomass is low in the fore-dunes so there is little release of nutrients by decay, most minerals being imported from outside in the form of organic matter (such as faeces, dead leaves etc.). As biomass increases there is a correspondingly high net uptake of $CO_2$. As species diversity increases, possibilities for alternative food species, and thus niche diversity, increase.

## Chapter 40

See text.

## Chapter 41

1 A grasping hand with an opposable thumb; a large brain; upright stance (primates sit upright); single offspring; a long period of juvenile dependence.
2 The development of more sophisticated tools, made possible by an increase in brain size, would have reduced the amount of work done by 'muscle power', which would have enabled the limbs to become more slender.
3 See text.
4 i) B ii) C iii) A
5 See text.
6 During an ice age sea levels fall, with the result that some land masses become linked. It was almost certainly by such means that people reached America from Asia. It would also have made the colonisation of Australia easier by reducing the amount of open sea that would have to be crossed.

## Chapter 42

1 a) See text.
b) See text.
c) Plants in an agricultural crop are genetically far less diverse than wild populations, so if one plant is susceptible the rest of the crop is also likely to be.
d) See text.
2 Both are false (see information box on page 815).
3 A topic for discussion (and too contentious to discuss here!)

# INDEX


Abiotic factors, 679*ff*
Abscisic acid (ABA), 598
  bud dormancy, 604
  stomata, 643
Abscission, 598, 604
Absolute growth rate, 567
Absorption of food, 479–81
Accommodation, 337
Accuracy, 842
*Acetabularia* (role of the nucleus), 174
Acetylcholine, 318
Acetylcholinesterase, 318
Acetyl coenzyme A, 83
Acid rain, 808
Acidosis, 386
Acrosome, 514
Actin, 116, 350
Actinomorphic flower, 550
Actinozoa, *284*
Action potential
  nerve, 312
  muscle, 354
Action spectrum,
  photosynthesis, *613*
  phototropism, 593
  phytochrome-mediated responses, 602
Activation energy, 66
Active immunity, 498
Active site (enzyme), 67
Active transport, 5, 102, 479–80, 637
Actomyosin, 351
Adaptation, receptor, 332
Adaptive radiation, 243–4
Addison's disease, 381
Adenine, 163
Adenylate cyclase, 375
Adipose tissue, 300, 455
Adolescent growth spurt, 769
ADP, 80
Adrenal cortex, 381
Adrenal gland, 380
Adrenal medulla, 323, 381
Adrenaline, 323, 376, 381
Adrenergic neuron, 321
Adrenocorticotrophic hormone (ACTH), 381
Aerobic respiration, 77
Aestivation, 680
*Agaricus* (mushroom), *275*
Age group structure (population), 717
Agglutination, *491*, 492
Agriculture, origin, 797–800
*Agrobacterium*, 218
AIDS, 494
Air
  alveolar, 421
  atmospheric, 421
  in leaf, 627
  in soil, 676
Air and water breathing, 416
Albinism, 156, *157*, 225–6
  Himalayan, 193
Aldose sugars, 34
Aldosterone, 381
Aleurone layer, 563
Algae, 278–9
Alimentary canal, 470*ff*
Alkaptonuria, 153
Allantois, 517
Allard, Henry, 599
Alleles, 125
  frequency in populations, 223
  multiple, 147
  segregation, 124
Allergen, 498
Allopatric speciation, 239–40
Allopatry, 237
Allopolyploidy, 201–2
Allosteric enzymes, 73, 92
Alpha helix, 25
Alternation of generations, 279, 358
Altitude effects
  breathing, 425
  transpiration, 645
Altruism, 760
Alveolar air, 421
Alveoli, 418
Amino acid, 20
  absorption (mammals), 480
  deamination (mammals), 483
  essential, 465
  transamination, 465
  turnover, 482–3
  utilisation, 482
Amino group, 20
Aminopeptidase, 471
Ammonia, 738–9
Ammonia excretion, 434
Ammonotely, 434
Amnion, 516, *517*
Amniotic cavity, 516
*Amoeba*, *276*
AMP, 80
  cyclic, 375
Amphibians, 293
Amphipathic molecule, 31
Amylase, 473, 477
Amylopectin, 37
Amyloplast, 42, 594
Amylose, 37
Anabolism, 3, 73
Anaemia
  iron-deficiency, 467
  sickle-cell, 28, 193, 230
Anaerobes
  facultative, 271
  obligate, 271
Anaerobic respiration, 88
Analogous features, 8, 263
Anaphase
  meiosis, 119, 120
  mitosis, 115
Ancient climates, 782
Androecium, 546
Anemophily, 551
Aneuploidy, 198
Angina, 404
Angiospermophyta, 281
Angiotensin, 445
Anion, 822
Annelid worms, *286*
Annual rings, 577
Annuals, 559
Annulus, 541
*Anopheles* mosquito, 700
Anther, 546
Antheridium
  fern, 542
  moss, 539
Anthropoids, 756
Antibiotic resistance, 212
Antibody, 490, *491*
Anticoagulant, 700
Anticodon, 181
Antidiuretic hormone (ADH), 377, 442
Antigen, 490
Antipodal cells, 554
Anus, 481
Aorta, 398
Aortic bodies, 424

*Apanteles*, 706
Apes, 757
 chromosomal evolution, 774
 language, 763
Apical dominance, 596
Apical growth, 530
Apical meristem, 568
Apicomplexa, 278
Apocrine glands, 455
Apodeme, 368
Apomixis, 556
Apoplast, 649
Apoptosis, 52
Appendix, 481
Aqueous humour, 335
Arachnida, 290
Arachnoid layer, 326
Archegonium,
 moss, 540
 fern, 542
Areolar tissue, 299
Art, earliest, 791
Artefact, 48
Arteries, 403
Arterioles, 403
Arthropods, 287
Artefact, 48
Ascospore, 274
Ascus, 274
Asexual reproduction
 animals, 501
 plants, *see* vegetative propagation
Aster, 114
Asthma, 418
Astigmatism, 338
Atherosclerosis, 404
Atoms, 821
ATP , 54, 80–1
 control of production, 92
ATP synthase, 85
ATPase, 104
Atrioventricular node, 400
Atrio-ventricular valve, 398
Atrium, 397
Attenuated pathogen, 498
Autecology, 671
Autocatalysis, 474
Autoimmunity, 490, 498
Autonomic system, 321
Autophagosome, 53
Autopolyploidy, 200
Autoradiography, 53, 623
Autotroph, 4, 271, 461, 530, 726
Autralopithecines, 778–9
Auxin analogues as weedkillers, 597
Auxin, 556, 591, 592, 596
 cambial activity, 596
 cell walls, 493
 fruit development, 596
 phototropism, 592
 root development, 596
*Avena* curvature test, 591
A-V node, 400
Avery, Oswald, 161
Axon, 305
Axon hillock, 320
*Azotobacter*, 740

B lymphocyte, 492, *495*
β-oxidation, 91
Bacillus, 269
*Bacillus thuringiensis*, 220, 273
Backcross, 126
Bacteria, 269–72
 aerobic and anaerobic, 271
 blue-green, 269
 disease-causing, 272
 endospores, 272
 flagella, 270
 industry, 273
 mutualistic, 272
 photosynthetic, 616
 plasmids, 212
 reproduction, 271
 saprobic, 272
 shapes, 269
Bacterial transformation, 161, 211
Bacteriophage, 61, 161, 213
Balanced diet, 470
Bark, 575
Baroreceptors, 409
Barr body, 143
Basal body, 53
Basal metabolic rate (BMR), 455
Base substitution, 193
Basement membrane, 298
Basidiomycota, 275
Basidiospore, 276
Basidium, 275
Beadle, George, 153–4
Benedict's test, 32
Beri beri, 468
Bernard, Claude, 384
Biennials, 559
Bilayer, 31
Bile, 476
Bimodal distribution, 834
Binominal nomenclature, 265
Bioassay, 591
Biochemical oxygen demand (BOD), 804
Bioelements, 16
Biogeochemical cycle, 738
Biolistics, 220
Biological clocks, 603
Biome, 671
Biotic factors, 687
Bipedalism
 advantages, 779–81
 evolution, 779–81
 skeletal adaptations, 764
 thermoregulation, 780
Birds, 293
Birth, 519
*Biston betularia*, 232
Bisphosphate vs diphosphate, 82
Bivalent, 119
Bladder, *432*
Bladderworm, 704
Blastocoel, 515
Blastocyst, 515
Blending inheritance, 122
Blind spot, 339, *340*
Block mutation, 196
Blood,
 composition, *390*
 clotting, 396
 groups, 499
 loss, adjustment to, 410
 oxygen transport, 391–5
 pressure regulation, 445
Blood brain barrier, 407
Blood systems, 388*ff*
 closed, 389
 open, 389
Body temperature, 447*ff*
Bohr effect, 394
Bonds (chemical), 822–4
Bone marrow, 489
Bone, 360
Bonner, James, 600
Bony fish, 293
Borthwick, Harry, 587
Boundary layer, 645
Boveri, Theodor, 110
Bowman's capsule, 432
Boysen-Jensen, P, 590
Brachiation, 757
Bracing of bones, 367
Brain, 323–4
Breath holding, 428
Breathing centre, 423
Breathing Distress Syndrome, 420
Breathing in mammals, 422–4
Broca's area, 467
Bronchial air, 421
Bronchiole, 418
Bronchus, 417

Brown algae, 278
Brow ridges, 766
Brown, H T, 628
Brown, Robert, 109
Brown fat, 86, 457
Brunner's glands, 476
Brush border , 478
Bryophyta, 279
Buchner, E, 65
Bud, 531
Bud dormancy, 604
Bud scale, *582*
Buffers, 23
Bulbs, 581
Bundle of His, 401
Bundle sheath, 634
Burnet, McFarlane, 492
Buttercup flower, 546

C3 and C4 photosynthesis, 634–5
Caecum, 481
Calcicole, 677
Calciferol, 469
Calcifuge, 677
Calcitonin, 378, 466
Calcium,
  blood clotting, 396
  diet, 466
  gravitropism, 596
  membrane permeability, 636
  middle lamella, 636
  muscle contraction, 354
  regulation of in blood, 469
  synaptic transmission, 318
  uptake from gut, 469
Callose, 660
Callus, 598
Calorimeter, 463, *464*
Calvin cycle, 622
Calvin, M, 622
Calyx, 546
CAM plants, 635
Cambium
  cork, 568, 579
  vascular, 575, *576*, *578*
Canaliculi
  bone, 362
  liver, 482
Cancer
  as mutation, 196
  defence against, 497–8
Capacitation (sperm), 514
Capillaries, 406
Capillarity 19,
  in soil, 676
  rise of sap, 652
Capsid, 61
Capsomere, 61
Capsule (bacteria) 60
Capsule, moss, 540
Capture–recapture method, 711
Carbamino haemoglobin, 396
Carbohydrase, 471
Carbohydrates, 31–8
Carbon cycle, 737–8
Carbon dioxide
  effect on breathing, 424–5
  effect on climate, 810–11
  effect on oxygen transport, 394
  transport in blood, 395
  uptake by leaf, 627
Carbon monoxide poisoning, 393
Carbon-14, 623, 777
Carbonic anhydrase, 390, 395, 475
Carboxyl group, 20
Carboxyhaemoglobin, 393
Carboxypeptidase, 471
Cardiac cycle, 401
Cardiac infarction, 404
Cardiac muscle, 399–*400*
Cardiac output, 402
Cardioaccelerator centre, *409*
Cardioinhibitor centre, *409*
Carnivore, 726
Carotene, 468, 610
Carotenoids, 610
Carotid bodies, 424
Carotid sinus, 409
Carpal, 364
Carpel, 545, 546
Carrying capacity, 714
Cartilage, 300
Cartilaginous fish, 293
Cascade effect, 375, *376*
Casein, 475
Casparian strip, 655
Catabolism, 3, 73
Catalysts, 64
Cation, 822
cDNA (complementary DNA), 211
Cell, 3, 40–60
  cycle, 112
  enlargement, *569*, *570*
  mediated immunity, 492
  plant and animal compared, 59
  plate, 116
  prokaryotic compared with eukaryotic, 59, *60*
  sap, 99, 533
  theory, 41
Cell wall
  bacteria, 60
  plants, 42, 57
Cellulase, 708
Cellulose, 37, 57
  digestion, 708, 744
Census (population), 711
Centipedes, 289
Central Dogma, 188
Central nervous system, 304
Centrifugation, 50, 166–7
Centriole, 56
Centromere, 110
Centrosome, 56
Cephalopods, 291
Cerebellum, *324*, 325
Cerebral cortex, 766
Cerebrospinal fluid, 326
Cerebrum, 325, 766
Cervix, 506, 519
Cestoda, *286*
Chaeta, 286
Character displacement, 243, 695
Chargaff, Erwin, 164
Chase, Martha, 161
Chelicera, 290
Chemical equations, 822
Chemiosmosis, 85, 620
Chemoautotroph, 740
Chemoreceptors, 329
Chemotropism, 589
Chewing, 473
Chi squared test, 832
Chiasma, 119
Chilopoda, 289
Chimpanzee
  behaviour, 761–3
  chromosomes, 774
  DNA, 773
Chitin, 37, 272, 368
*Chlamydomonas*, *279*
Chlorenchyma, 533
Chloride shift, 395
Chlorofluorocarbons (CFCs), 813
Chlorophylls, 610
Chlorophyta, 279
Chloroplast, *55*, 533
  pigments, 609, 612
Chlorosis, 636
Cholecalciferol, 469
Cholecystokinin-pancreozymin, 478
Cholesterol, 405, 477
Cholinergic neuron, 320
Cholinesterase, *see* acetylcholinesterase
Chondrin, 300
Chondroblast, 300
Chondrocyte, 300

Chordata, 291
Chorionic gonadotrophin, 516
Choroid, 334
Chromatid, 112
Chromatin, 112
Chromatography, *611*
Chromoplast, 42, 533
Chromosome, 6, 42, 110
  deletion, 196
  duplication, 196
  fusion, 197
  homologue, 111
  inversion, 197
  mapping, 137
  translocation, 197
Chylomicron, 480
Chyme, 475
Chymotrypsin, 477
Ciliary body, 335
Ciliates, 277
Cilia, 55–6, 348, 350
  oviduct, 517
  air passages, 418
  *Paramecium*, *350*
Circadian clock, 603
Circulation, single and double, 397
Circumcision, 503
Cisterna, *51*
Citric acid cycle, 83
Clade, 264
Class (taxonomy) 258
Classification, 258*ff*
  artificial, 266
  natural, 261
  phylogenetic, 261
Clathrin, 107
Clay, 674
Cleavage, 515
Cleidoic egg, 434
Cleistogamic flowers, 552
Climate, ancient, 782
Climatic factor, 671
Climax community, 742
Cline, 796
Clinostat, *see* Klinostat
Clitoris, 506
Clonal selection hypothesis, 492
Clone, 492
Cloning vector, 211
Closed blood system, 286, *292*
*Clostridium botulinum*, 271
Club mosses, 281
Cnidaria, 282
Coated pits, 107
Co-evolution, 549
Co-operative binding, 393
Coccus, 269
Codominant alleles, 147
Codon, 178
Coelenterata, 282
Coelom, 286–*7*
  and earthworm locomotion, 370
Coelomic fluid, 370
Coenocyte, 41, 273
Coenzyme, 70
Coenzyme A, 83
Cofactor, 70
Cohesion–tension hypothesis, 651–4
Cohort, 258
Coleoptile, 563
Coleorhiza, 563
Collagen, *25*, 299, 361, 454
Collecting ducts, *433*
Collenchyma, 534
Colon, 481
Colour blindness, 141, 142
Colour vision, 344–5
Commensalism, 708
Community, 671, 725
Companion cell, 536, 658
Compensation point (light) 632
Competition
  interspecific, 693–6
  intraspecific, 688–93
Competitive exclusion, 693
Competitive inhibition (enzymes), 71
Complement, 488
Complementary genes, 148
Compounds, 821
Concentration gradient, 96
Condensation reactions, 22
Conditioned reflex, 310
Cones (eye), 343
Conidiospore, 274
Conidium, 274
Coniferophyta, 281
Conjugated protein, 28
Conjugation (bacteria), 271
Conjunctiva, 334
Connective tissues, 299
Consumers, 726
Continuous variation, 123, 149
Contraceptive (pill), 512
Contractile vacuole, 277, *383*
Convection, 453
Convergent evolution, 261
Convoluted tubule, 433
Copulation, 512
Cork cambium, 568, 579
Corm, 582
Cornea, 334
Corolla, 546
Coronary artery, 399
Corpus callosum, 326
Corpus luteum, 507, 511
Correlation coefficient, 786
Cortex
  kidney, *432*
  plant, 571, 575
Cortical reaction, 515
Cortisol, 387, 484
Cotransport, 103–4
Cotyledon, 556, 562
Countercurrent multiplier, 441
Counterflow
  gills, 415
  heat 456
  kidney, 442
Courtship, 512
Covalent bond, 822
Cowper's gland, 503
Cranial nerve, 304
Crassulacean acid metabolism, 635
Creatine phosphate, 358
Creationism, 8
Cretinism, 380
Crick, Francis, 164, 180
Crista (mitochondrion), 54
Critical daylength, 599
Crossing over, 117, 134
Crossover value (COV), 137
Crude oil, 809
Crustacea, 288
Crypt of Lieberkühn, 478
Cryptorchid, 503
Cryptozoic animals, 681
Ctenidium, 290
Cultural evolution, 761, 800, 816–7
Culture solutions, 635
Cuticle (arthropods), 368
Cuticle (plant), 280, 532, 642
Cutin, 532
Cyanide, 86
Cyanobacteria, 740
Cyanogenesis, 234
Cyclic AMP, 375
Cyclic GMP, 342
Cysticercosis, 705
Cysticercus, 704
Cytochrome c evolution, 263
Cytochrome, 85
Cytokinesis, 115
Cytokinins, 598
Cytopharynx, 278
Cytoplasm, 42
Cytosine, 163
Cytosis, 106

Cytoskeleton, 56
Cytosol, 51
Cytotoxic T cell, 495

Darwin, Charles, *7*, 9–11
Darwin's finches, 243
Dating of fossils, 776–8
Day-neutral plants, 600
DCMU (dichloro-phenyl-methyl urea), 622
DDT, 807
Dead space, 421
Deadnettle, *547*
Deamination, 431, 483–4
Decomposers, 726, 744
Dedifferentiation, 575
Deficiency disease, 467
Deficiency disease (plants), 635
Deflected climax, 742
Deforestation, 814–5
Deglutition (*see* swallowing)
Degrees of freedom, 833
Dehydrogenase, 81
Deme, 223
Demographic transition, 802
Denaturation
  DNA, 170
  proteins, 28, 69
Dendrite, 6, 305
Dendrochronology, 577
Denitrification, 88, 740
Density-dependent factors, 719
Density-gradient centrifugation, 50
Deoxyribonuclease, 477
Deoxyribonucleic acid, *see* DNA
Derived characters, 752
Desmosome *58*, 298, 400, 454
Detritivores, 726
Deuteromycota, 276
Dextrin, 473
Diabetes insipidus, 443
Diabetes mellitus, 385
Diapause, 680
Diaphototropism, 589
Diaphragm, 422
Diastole, 401
Dichogamy, 551
Dichotomous key, 267
Dicotyledon, 282
Dictyosome *see* Golgi body
Differential permeability, 97
Differentiation, 297
  plants, 568, 569–72
Diffusion, 95–7
Digestive enzymes, 471
Digit, 364
Dihybrid cross, 127
Dikaryon, 538
Dioecious plants, 552
Dipeptidase, 479
Diploblastic animals, 283
Diploid, 111
Diplontic life cycle, 537
Diplopoda, 289
Disaccharides, 32, 35
Discontinuous variation, 123
Dispersal (seeds and fruits), 558–9
Dispersion (data), 835
Display, 689
Disruptive selection, 233
Distal tubule, *433*, 440
Disulphide bond (proteins), 27
Diversity index, 727
DNA (deoxyribonucleic acid), 120, 160–71
  annealing, 209
  denaturation, 170
  fingerprint, 217
  genetic material, 171
  hybridisation, 170, 773
  library, 212
  ligase, 209
  mitochondrial, 793–4
  polymerase, 167, *169*
  repair, 171
  replication, 166–9
  restriction maps, 793
  structure, 163–6
  synthesis, 112
  tandem repeats, 217
  transcription, 175
DNA polymerase, 167
Domain (taxonomy), 268
Dominant trait, 124
Dormancy, 680
  buds, 604
  seeds, 586–8
Double circulation, 397, *398*
Double fertilisation, 555
Down's syndrome, 199
*Drosophila*, 134
Dry mass, 567
Dual innervation, 322
Ductus arteriosus, 520
Ductus venosus, 521
Duodenum, 476

Earthworm, 286
  movement, *370*
Eccrine glands, 455
Ecdysis, 368
Echinodermata, 291
Ecology, 760*ff*
Ecosystem, 671, 728
Ectoderm, 283, 287
Ectoparasite, 699
Ectoplasm, 276
Ectothermy, 448
Edaphic factor, 671
Effector, 303
Egestion, 431
Ejaculation, 513
Elastase, 477
Elastic cartilage, 301
Elastin, 299, 454
Electromagnetic spectrum, *334*, 609
Electron, 821
Electron microscope, 47
Electron transport
  chloroplasts, 618–9
  mitochondria, 84
Electrophoresis, 24
Electroporation, 218
Electrovalent bond, 822
Elements, 821
Embryo (flowering plant) 555, *556*
Embryo sac, 554
Emerson, R, 618
Emphysema, 420
Endemic species, 243
Endergonic reaction, 79
Endocrine system, 372*ff*
Endocytosis, 106
Endoderm, 283, 287
Endodermis, 575, 655
Endogenous rhythm, 455
Endometrium, 506
Endonuclease, 208
Endoparasite, 699
Endopeptidase, 471
Endoplasm, 276
Endoplasmic reticulum, 50, *51*
Endoskeleton, 359
Endosperm, 555, 563
Endospore, 272
Endosymbiosis, 55
Endothermy, 448
Energy conversion
  animals 732–3
  plants, 732
Energy substrate, 77
Engelmann, T, 615
Enhancement effect, 618
Enteroceptors, 329
Enteron, 283
Enteropeptidase, 477
Entomophily, 550

Entrainment, 701
Entropy, 5
Environment
  abiotic, 671–2
  biotic, 671–2
  social, 761
Environmental resistance, 714
Enzyme–product complex, 68
Enzyme–substrate complex, 67
Enzymes, 64–75
  activation energy, 66
  immobilised, 74
  mode of action, 67
  naming, 65–6
  specificity, 65
  substrate, 65
  turnover number, 68
Enzymes, effect of
  cofactors, 70
  inhibitors, 71
  pH, 70
  temperature, 69
Ephemeral plants, 559
Epicotyl, 565
Epidermis
  mammal, 453
  flowering plant, 572, 642–3, 646
Epigeal germination, 565
Epiglottis, 474
Epilimnion, 741
Epistasis, 148
Epithelia, *298*
Escombe, F, 628
Erythrocytes, *see* red blood corpuscles
Erythropoietin, 425
Ethene, 598–9
Etiolation, 565
*Euglena, 269*
Eukaryote, 47, 268
Eukaryotes, 268
Euploidy, 198
Euryhaline fish, 681
Eutherian mammals, 293
Eutrophic lakes, 741, 806
Evolution, 7–11,
  cultural and biological, 816
  evidence, 7–8
Excretion, 5, 431*ff*
Excretory products, 431
Exergonic reaction, 66, 79
Exine, 552
Exocrine glands, 372
Exocytosis, 107
Exodermis, 574
Exon, 177
Exopeptidase, 471
Exoskeleton, 287, 359, 368
Expected value, 831
Expiration, 422
Exponential growth, 713
Extensor muscle, 365
Exteroceptors, 329
Extra-cellular fluid, 383
Eye (mammal), 333*ff*

$F_1$ generation, 124
$F_2$ generation, 124
Facilitated diffusion, 104
Facilitation, 320
Factor VIII, 396
Facultative anaerobe, 271
FAD (flavine adenine dinucleotide), 81
Fallopian tube, 505
Family (taxonomy), 258
Fats, 29, 30
  energy yield, 77
  conversion to carbohydrate, 565
  production from carbohydrate, 384
Fatty acids, 29
Feeding, 4
Field capacity (soil), 676
Fermentation, 88–90
Fertilisation
  flowering plant, 555
  mammal, 502, 513
Feulgen stain, 160
Fibre, dietary, 465–6
Fibres, 533
Fibrin, 396
Fibrinogen, 396
Fibroblast, 299
Fick's laws of diffusion, 97, 419
Fighting, 689
Fimbria, 270, 506
Fire, 785, 791
Fish
  circulation, 377, 398
  gas exchange, 414
Flagellum
  eukaryote, 55–*6*
  prokaryote, 270
Flavoprotein, 81
Flexor muscle, 365
'Florigen', 602
Flower
  dicotyledon, 281, 546–9
  monocotyledon, 281, 548
Flowering, control of, 599–604
Flukes, *286*
FMN (hydrogen carrier), 80
Foetal
  circulation, 520
  haemoglobin, 395
  membranes, 517
Foetus, 516
Follicle (ovary), 505
Follicle-stimulating hormone (FSH), 509
Food chain, 729, 731
Food pyramid, 734
Food vacuole, 106
Food web, 729
Foramen ovale, 520
Forebrain, 325
Foreskin, 503
Fossils, 775
Fossil fuels, 737
Founder effect, 236
Fovea, 338, 344–5
Frame shift, 194
Franklin, R, 164
Fructose, 34
Fruit, 281, 544
  development, 596
  ripening, 598
  types, 557
FSH, *see* follicle stimulating hormone
Fucoxanthin, 278
*Fucus*, 278
Functional response (ecology), 721
Fungi, 272–6

G phase, 112
Galapagos Islands, *9*
Gall bladder, 476
Gallstones, 477
Gametogenesis, 502
Gametocytes (*Plasmodium*), 701
Gametophyte, 279
  fern, 542
  flowering plant, 553–4
  moss, 539
Ganglion, 308
Gap junction, 58, *59*, 400
Garner, W, 599
Gas exchange, 77, 412
Gas exchange organ, 413
  fish, 414
  insects, 428
  mammal, 418
Gastric glands, 474
Gastric juice, 474
Gastric secretion, control of, 475
Gastrin, 475
Gastropoda, 291
Gastrulation, 516

Gause, G F, 693
Gene, 125, 179
  cloning, 207
  flow, 224
  linkage, 133–7
  mapping, 137–9
  mutation, 191–6
  operator, 187
  pool, 222
  probe, 214
  regulation, 184–7
  regulatory, 186
  structural, 186
  technology, 206*ff*
  therapy, 220
Generation time, 271, 713
Generative nucleus, 553
Generator potential, 331
Genetic bottleneck, 236
Genetic code, 178–80
Genetic drift, 224, 235
Genetic engineering, 206*ff*
Genetic fingerprinting, 217
Genetic load, 228
Genotype, 125
Genus, 258
Geographical isolation, 239
Geological time, 775–6
Geotropism, *see* gravitropism
Germ and soma, 12
Germ layers, 516
Germ line therapy, 220
Germination, 563
  cold-requirement
  epigeal, 565
  garden pea, 564
  hypogeal, 565
  maize, 566
  sunflower, 565
Gestation, 516
Giant axon, 310
Gibberellins, 597
Gills, 413, *415*
Girdle scar, *582*
Glans, 503
Glial cells, 304
Global warming, 810–11
Glomerular filtrate, 434
Glucagon, 385
Gluconeogenesis, 385, 484
Glucose, 32–3
Glucose isomerase, 273
Glucose regulation in blood, 384
Glyceraldehyde, 34
Glycerol, 29
Glycocalyx, 49
Glycogen, 37, 358
Glycogenesis, 482
Glycogenolysis, 482
Glycolysis, 77, 82
Glycoproteins, 29
Glycosidic bond, 35
Goblet cell, 418
Goitre, 380
Golgi body, *51*
Gonadotrophic hormones, 509
Gonadotrophin releasing hormone, 509
Graafian follicle, 507
Grades of organisation, 264
Gram stain, 269
Granulocyte, 489
Granulosa cells, 505
Granum, *55*
Graphs, 827
  bar, 828
  histograms, 830
  kite, 830
  pie, 830
  scattergrams, 828
Gravitropism, 564, 593–6
Green algae, 279
Greenhouse effect, 810
Grey matter, *308*, *325*
Griffith, Frederick, 161
Grooming, 760
Growth rings and climate, 577
Growth, 6
  curves, 567
  human, 769
Guanine, 163
Guard cell, microfibrils, 644
Guard cells, 643
Gut, structure of, 471, 472
Guttation, 650
Gynoecium, 546

Habitat, 670
Haem, 391
Haemocoel, 287, 290
Haemocyanin, 391
Haemoglobin, 26, 390, 391–5
  foetal, 395
Haemolytic disease, 500
Haemophilia, 141, 396
Half-life
  hormone, 376
  isotopic decay, 777
Hämmerling, Joachim, 174
Hamner, Karl, 600
Hanson, Jean, 351
Haplodiplontic life cycle, 538
Haploid, 111
Haplontic life cycle, 537
Hardy–Weinberg principle , 224–6
Haversian canal, 361
Heart, 398*ff*
  A-V node, 400
  cycle, 401
  disease, 404
  muscle, 399–400
  output, 402
  pacemaker, 400
  regulation of output, 402
  S-A node, 400
  systole, 401
Heartwood, 579
Heat loss (mammal) 451
Heat stroke, 459
Heavy metals, 810
Helper T cells, 495
Hemicellulose, 58
Hendricks, Sterling, 587
Henle, F J, 433
Hepatic portal vein, 480, *483*
Hepaticae, 279
Hepatocyte, 482
Herbaceous plants, 580
Hermaphrodite, 704
Hershey, Alfred, 161
Hertwig, Oscar, 109
Heterogametic sex, 140
Heterospory, 281
Heterotroph, 4, 271, 295, 461, 726
Heterozygous, 125
Hexose sugars, 32
Hibernation, 680
High altitude breathing, 425
High density lipoproteins (LDLs), 405
Hill reaction, 616
Hindbrain, 325
Hirudinea, 287
Histamine, 488
Histone, 171
HIV, 494
Holozoic nutrition, 462
Home range, 689
Homeostasis, 6
  leaf, 644
  mammal, 382*ff*,
Homeothermy in animals, 447
Hominidae, 770
Hominiae, 770
Hominoidea, 770
*Homo*
  *erectus*, 784
  *ergaster*, 784

*habilis*, 782
*heidelbergensis*, 786–7
*neanderthalensis*, 787
*rudolfensis*, 783
*Homo sapiens*,
diversity, 794
origin, 791–2
Homogametic sex, 140
Homology, 8, 262
Homologues (chromosomes), 111
Homospory, 545
Homozygous, 125
Hooke, R, 40
Horizon (soil), 677
Hormones, 372*ff*
'Hormone' weedkillers, 597
Host (of parasite), 699
specificity, 700
Human diversity, 794–5
Human lymphocyte antigen (HLA), *496*
Humerus, 364
Humus, 674
Humoral-mediated immunity, 492
Huxley, 351
Hyaluronidase, 514
Hybridoma, 497
*Hydra*, 283
Hydrogen bond
DNA, 165
protein, 25, 27
water, 17,
Hydrogencarbonate indicator, 78
Hydrolase, 66
Hydrolysis, 22
Hydrophilic group, 18
Hydrophytes, *649*
Hydrosere, 745
Hydrostatic support, 572, *573*
Hydrozoa, *284*
Hyperglycaemia, 384
Hypermetropia (*see* long sight)
Hypoglycaemia, 384
Hyperparasite, 699
Hypersensitivity, 498
Hypertension, 406
Hyperthyroidism, 380
Hypertonic solution, 99
Hyperventilation, 428
Hypha, 273
Hypocotyl, 565
Hypogeal germination, 565
Hypolimnion, 741
Hypophyseal portal system, 377
Hypothalamus, 325, 377, 450, 458, 508
Hypothermia, 459
Hypothesis, 7
Hypothyroidism, 380
Hypotonic solution, 99

IAA (indole-3–acetic acid), 591
Ileum, 477
Imbibition, 564
Immunity, 486*ff*
acquired, 486
active, 498
cell-mediated, 492
humoral, 492
passive, 498
Immunisation, 498
Immunoglobulin, 491
Immunological distance, 772
memory, 488, *496*
Implantation, 515
Inbreeding, 226, 758
Incompatibility in plants, 552
Incomplete dominance, 146
Independent segregation, 117
Indole-3–acetic acid (IAA), 591
Induced fit (enzyme), *67*
Indusium, 541
Industrial melanism, 232
Industrial revolution, 737
Inflammation, 488
Inflorescence, 549
Inhibin, 508, 509
Inhibitors (enzyme), 71
Inhibitory synapses, 318
Initials, 577
Innate resistance, 486
Iodine
test for starch, 37
in thyroxine, 378
Insecticides, 807–8
Insects, 288–9
gas exchange, 428
Inspiration, 422
Instructional hypothesis (immunity), 492
Insulin, 385–6
Integration, 320
Integuments, 545, 553
Intercalated discs, 400
Intercostal muscles, 422
Interlukins, 495
Intermediate filaments, 57, 298
Intermediate neuron, 307
Internal environment, 384
Internode, 531
Interphase, 112
Interstitial cell stimulating hormone
*see* luteinising hormone
Intervertebral disc, 363
Intestinal enzymes, 479
Intine, 552
Intrafusal fibres, 359
Intrinsic factor, 475
Intrinsic rate of increase, 723
Intron, 177, 215
Inulin, 37
Ion channels, 104
exchange, 677
Ionic bond, 822
Ionising radiation, 195
Ions, uptake by roots, 637, *638*
Iris (of eye), 335
Iron, dietary, 467
Irritability, 303
Ischaemia, 404
Ischial callosities, 756
Islets of Langerhans, 385
Isoelectric point, 23
Isolating mechanisms, 237–8
Isomerase, 66
Isomers, 16
Isotonic solution, 99
Isotopes, 824

J curve, 716
Jacob, François, 186
Jeffreys, Alec, 217
Jejunum, 477
Jellyfish, 284
Johansen, Donald, 778
Joints, 364
Juxtaglomerular apparatus, 445

*K*-species, 723
Kanzi (chimpanzee), 761
Karyotype, 111
Keratin, 25, 57, 454
Ketone bodies, 464
Ketose sugars, 34
Kettlewell, Bernard, 232
Keys, 266
Kidney, 432
Kin selection, 760
Kinetochore, 114
Kingdom, 258
Klinefelter's syndrome, 199
Klinostat, 594
Krebs cycle, 82, 83
Kupffer cell, 482
Kymograph, 355

Labour, 519
Lactase, 479

Lactation, 522
Lacteal, 478
Lactate fermentation, 90
Lactose, 35
  intolerance, 187
Lacuna (bone), 362
Lag phase, 716
Lakes
  acidification, 808
  eutrophication, 741
  nutrient cycling, 741
Lamarck, Jean Baptiste, 9
Larva, 289
Larynx, 417
Latent period, 356
Lateral meristem, 568
Laterite (soil) 815
Lateral roots, 575
Leaching, 677
Leaf
  compound, 531
  fall, 604
  mosaic, 631
  photosynthetic adaptations, 626–31
  simple, 531
  support, 630
Leakey
  Jonathan, 782
  Louis, 782
  Mary, 782
  Richard, 783, 784
Leeches, *286*
Lemur, 756
Lens (eye), 335
Lenticel, *579*, *581*, 642
Leucocyte, 488–9
Levers, 365–7
Leydig cells, 508
Lichens, 707
Life cycle
  fern, 541, *543*
  flowering plants, 537*ff*
  moss, 539
Ligament, 364
Ligase, 66
Light as environmental factor, 680–1*ff*
  seed germination, 587
Lignin, 280, 532
Limbic system, 768
Limiting factors
  mineral nutrition, 686
  photosynthesis, 615
Lincoln Index, 711
Linkage, 133
Linkage group, 139
Linnaeus, Carolus, 258, 546–7
Lipase, 477
Lipids, 29
Lipoproteins, 405
Lithosere, 745
Liver, 476, 482
Liver fluke, *285*, 286
Liverworts, 279
Loam, 674
Lock and key (enzymes) 67
Locus (gene), 138
Logarithmic growth, 713
Long sight, 338
Long-day plants, 599
Loop of Henle, *433*, 441
Low density lipoproteins (LDLs), 405
Lower critical temperature, 458
'Lucy' (fossil), 779
Lugworm (oxygen transport), *394*
Lung, 413
  surfactant, 420
  volumes, 426
Luteinising hormone, 509, *510*, *511*, 512
Lyase, 66
Lycopodophyta, 281
Lycopods, 544
Lymph node, 409, *489*
Lymph, 408
Lymphatic system, 408
Lymphocytes, 489, 492, 495
Lymphoid tissue, 489
Lymphokines, 495
Lynx, 720
Lysosome, 52, *53*
Lysozyme, 487

M-phase, 112
Macleod, Colin, 161
Macronutrients (plants), 635
Macrophage, 489
Macula densa, 445
Magnesium as plant nutrient, 636
Major histocompatibility complex (MHC), *496*
Maize grain, 566
Malaria parasite, 700
  control, 702
Malnutrition, 470
Malonate (inhibitor), 71
Malpighi, Marcello, 658
Malpighian layer, 453
Malpighian corpuscle, 533
Maltase, 479
Maltose, 35
Malthus, Thomas, 10
Mantle, 290
Mark, release and recapture, 711
Marram grass, 648, 743
Marsupials, 293
Maskell, E J, 658
Mason, T G, 658
Mast cell, 299, 488
Mastication (see chewing)
McCarty, Maclyn, 161
Mechanoreceptors, 329
Median, 834
Medulla (renal) *432*
Medulla oblongata, 325
Medusa, 283
Mean, 834
Meganucleus, 277
Megasporangium, 545, 554
Megaspore, 281, 545, 554
Megasporophyll, 544
Meiosis, 112, 116–20
Melanocyte, 453
Melatonin, 325
Membrane potential (roots), 637
Membrane structure, 48
  fluid mosaic model, 48–9
Memory cells, 495
Mendel, Gregor, 122
Mendel's laws, 133
Meninges, 326–*7*
Meningitis, 327
Menopause, 512
Menstrual cycle, *510*
Menstruation, 510
Meristem
  apical, 530
  lateral, 531
Merozoites 700
Meselsohn, Matthew, 166
Mesentery, 473
Mesoderm, 284
Mesogloea, 283
Mesolithic, 797
Mesophyll, 626
Mesophyte, 647
Mesosome, 60
Messenger RNA, 175–6
Metabolic pathway, 71
Metabolic water, 444
Metabolism, 3
Metacarpals, 364
Metachronal rhythm, 277
Metalloproteins, 28
Metamorphosis, 289
Metaphase
  meiosis, 119, 120
  mitosis, 115

Metaphloem, 571, *572*
Metatherian mammals, 293
Metaxylem, 571, *572*
Methane (*as* greenhouse gas), 812
Micelle, 480
Microfibrils, 37, 352
Microfilament, *57*
Microhabitat, 670
Microlith tools, 797
Micronucleus, 277
Micronutrients, 636
Micropyle, 554
Microscope
  electron, 47
  optical 40–1, 45–6
Microsporangium, 545
Microspore, 281, 544
Microsporophyll, 545–6
Microtriches, 704
Microtubules, 57
Microvilli, *49*
  intestine, 478
Micturition (*see* urination)
Miescher, Friedrich, 160
Midbrain, 325
Middle lamella, 58, 116
Migration, 680
Milieu interieur, 384
Milk, 522
Millipedes, 289
Mimicry, 698
Mineral deficiency symptoms (plants), 635
Mineral ions, 19
  dietary, 466–7
  soil, 677
  translocation in plants, 657, 662
  uptake by plants, 637, *638*
Minimal medium, 154
Minute volume, 427
Mitchell, Peter, 85
Mitochondrion, *54*, 87
Mitochondrial DNA (mtDNA), 793–4
Mitosis, 111, 113–5
Mode, 834
Molarity, 98
Mole (chem), 825
Molecular taxonomy, 263–4, 770–3
Molluscs, *290*
Molybdenum, 636
Monera, 268
Monkeys, 756
Monoclonal antibody, *497*
Monocotyledon, 282
Monoecious plants, 552
Monocyte, 489
Monod, Jaques, 186
Monoecious, 552
Monohybrid cross, 123
Monosaccharides, 32
Monosomy, 198
Morgan, Thomas, 142
Morula, 515
Mosquito, 700
Mosses, 279–80
Motor end plate *see* neuromuscular junction
Motor neuron, 305
Motor unit, 353
Movement, 6
Mucilage, 574
Mucin, 473
*Mucor*, 272, 274
Mucosa, 471
Mucus, 418
Mule, 238
Multicellularity, 297
Multiple alleles, 147
Musci, 279–80
Muscle
  antagonism, 365
  cardiac, 399–400
  red and white fibres, 357
  skeletal, 349
  smooth, 301
  spindle, 359
  striated, 302
  tissue, 301
  twitch, 356
Muscularis mucosae, 471
Mutagens, 195
Mutation
  gene, 190–6
  chromosomal, 196–203
  randomness, 191–2
  somatic, 192
Mutualism, 707
Myasthenia gravis, 498
Mycelium, 273
Mycology, 272
Mycorrhiza, 274, 638, 707
Myelin sheath, 305, *306*, 315
Myofibril, 349
Myogenic contraction, 400
Myoglobin, 25, 393, 357
Myometrium, 506
Myopia (*see* short sight)
Myosin, 116, 350
  walking, 352
Myxoedema, 380
Myxomatosis, 699
NAD (nicotinamide adenine dinucleotide),
NADP (nicotinamide adenine dinucleotide phosphate), 617
Nastic movements, 589
Nasties, *see* nastic movements
Natural classification, 261
Natural selection, 10
Neanderthals, 787
Nectar, 546
Negative feedback, 73
  breathing regulation, 421
  ecology, 719
  glucose regulation, 386
  intestine, 478
  thermoregulation, 459
  thyroid regulation, 379
Nematocyst, 283
Nematoda, *286*
Neolithic, 797
Nephron, 432
Nervous system, 304*ff*
Neural plate, *340*
Neural tube, 339
Neurogenic contraction, 400
Neuroglia, 304
Neuromuscular junction, 320
Neuron, 304–7, *306*
  relay, 307
  efferent, 305
  afferent, 307
  adrenergic, 321
  cholinergic, 320
Neurosecretory cells, 377
*Neurospora*, 153, 274–5
Neurotransmitter, 320
Niche, 694–6
Nicotine, 319
Nicotinic acid, 81, 468
Night blindness, 468
Nitrate, 635, 738–9
Nitrate reductase, 636
Nitrification, 739
Nitrifying bacteria, 739
Nitrite, 635, 738–9
*Nitrobacter*, 739
Nitrogen, plant growth, 635–6
  cycle, 738–41
  fixation, 740
Nitrogenase, 740
*Nitrosomonas*, 739
Nociceptors, 329
Node of Ranvier,
Node (stem), 531
Non-disjunction, 199
Non-endospermic seeds, 555
Noradrenaline, 381

Normal distribution, 835–7
Notochord, 292
Nucellus, 545, 553
Nuclear envelope, 47, 50
Nucleic acids, 160
Nucleoid, 60
Nucleolus, 42
Nucleoside, 163
Nucleosome, 171
Nucleotide, 39, 163
Nucleus
  atomic, 821
  brain, 324
  cell, *42*, 50, 109
Null hypothesis, 831
Numerical response, 722
Nutrient cycles, 736–742
Nutrition, 4
Nutritional mutants, 154
Nymph, 289

*Obelia*, 284
Obesity, 470
Obligate relationship, 707
Oedema, 408
Oesophagus, 474
Oestradiol, 508
Oestrogens, 508
Oestrous cycle, 509
Oestrus, 510
Oil immersion microscopy, 46
Oligochaete worms, 287
Oligosaccharides, 35
Oligotrophic lakes, 806
Omnivore, 726
Oncogene, 196
One gene–one polypeptide, 152–4
One-tailed and two-tailed tests, 840
Oogenesis, 506
Oogonium, 506
Ookinete (*Plasmodium*), 701
Open systems, 102
Open blood system, 389
Operculum (fish), 415
Operon, 186–7
Opsin, 340
Opsonin, 491
Oral groove, 278
Order (taxonomy), 258
Organ
  animal, 283, 298
  plant, 536, 572, 631
Organelle, 3, 47
Organic compounds, 14
Organochlorines, 807
Organophosphates, 808
Organs, plant, 536, 572
  animal, 298
Orgasm, 512
Ornithine cycle, 484
Osmoconformer, 681
Osmoreceptors, 442
Osmoregulation
  mammal, 439*ff*
  leaf, 644,
Osmosis, 97–100
Osmotic potential, *see* solute potential
Osteocyte, 361
Osteomalacia, 469
Outbreeding, 227
Ovary
  mammal, 505
  plant, 281, 544
Oviduct, 505
Ovule, 546, 554
Oxidation, 826
Oxidative phosphorylation, 85
Oxidoreductase, 66
Oxygen
  atmospheric, 421
  measurement of uptake, 78, 427
  transport in blood, 391–5
Oxyntic cell, 475
Oxyhaemoglobin dissociation, 391, *392*
Oxytocin, 378, 519
Ozone, protection from UV, 812

P-protein, 659
Pacemaker, 400
Pacinian corpuscle, 330
Palaeolithic
  Lower, 787
  Middle, 788
  Upper, 790
Palisade mesophyll, 626
Palynology, 553
Pancreas, *477*
Pancreatic juice, 477
Pancreozymin, 478
Pappus, 558
*Paramecium*, *4*, 277, 350
Parapatric speciation, 241
Paraquat, 622
Parasites, 462
Parasitism, 699*ff*
Parasitoid, 706
Parasympathetic nervous system, 321
Parathyroid glands, 466
Parenchyma, 533
Parental type, 135
Parietal cell, *see* oxyntic cell
Partial pressure, 392
Parthenogenesis, 557
Passive immunity, 498
Pasteur, L, 65
Pathogen, 271
Pectin, 58
Pedicel, 546
Pedigrees, 144–6
Pellagra, 468
Pelycopoda, 291
Penicillin, 274
*Penicillium*, 274, *276*
Penis, 503
Pentadactyl limb, 363, 753
Pentose, 32
Peppered moth, 232
Pepsin, 474
Pepsinogen, 474
Peptidases, 471
Peptide bond, 21
Peptidoglycan, 60
Perennation, 580
Perennials, 580
  herbaceous, 560
  woody, 560
Perianth, 548
Pericardium, 398
Pericarp, 556
Perichondrium, 300
Pericycle, 575
Periosteum, 362
Peripheral nervous system, 304
Peristalsis, 472
Perithecium, 275
Peritoneum, 473
Permease, 104
Peroxidase, 54
Peroxisome, 54
Pesticides, 807
Petal, 546
pH
  environmental factor, 677, 682, *683*
  scale, 826
Phagocytosis, 106, 277, 488
Phalanges, 364
Pharynx, 474
Phasic receptors, 332
Phellogen, 579
Phenotype, 125
Phenylketonuria (PKU), 156
Pheromones, 510, 453
Phloem, 280, 536
  leaf, 626
  loading, 661

structure, 657–8
transport mechanism, 659
Phosphate, dietary, 467
Phosphoenolpyruvate (PEP), 634
Phosphofructokinase, 92
Phospholipid, 31
Phosphoproteins, 29
Phosphorus as plant nutrient, 636
Photoautotroph, 726
Photoblastic seeds, 587
Photolysis of water, 617
Photomorphogenesis, *588*
Photonasty, 589
Photoperiod, detection of, 601
Photoperiodism, 599–603
Photophosphorylation, 619
Photoreceptors, 329
Photorespiration, 633
Photosynthesis, 530, 606*ff*
action spectrum, 613
effect of $CO_2$, 633
effect of light intensity, 632
effect of temperature, 633
limiting factors, 615
measuring rate, 609
need for chlorophyll, 608
need for $CO_2$, 607
need for light, 607
need for water, 608
net and gross, 609
oxygen production, 608
photostage, 615
synthetic stage, 615
water supply, 633
Z-scheme, 618, *619*
Photosynthesis–transpiration compromise, 642
Photosynthetic unit, *614*
Photosystems I and II, 618–19
Phototropism, 589–92
action spectrum, 592, *593*
detection of light, 592
Phragmoplast, 116
Phrenic nerve, 423
Phycobilin, 278
Phycoerythrin, 278
Phylogeny, 260–1
Phylum, 258
Phytochrome, 587, 602
Phytoplankton, 729
Pia mater, 326
Pilus, 270
Pineal gland, 325
Pinocytosis, 107
Pioneer species, 742
Pith, 571, 533
Pituitary, 325, 377
Placenta, 516,
exchange of substances, 518
*Planaria*, 285
Plant growth regulators, 586
Plantigrade foot posture, 753
Plants, organisation, 530
Plants, way of life, 530
Plasma, 390
Plasma cells, 491, *495*
Plasma membrane, 42
Plasmagel, 276
Plasmalemma, *see* plasma membrane
Plasmasol, 276
Plasmid, 60, 212, 271
Plasmodesma, 533
*Plasmodium*, 700
Plasmolysis, 99–100
Plastid, 42, 54, 533
Platelets, 390, 397
Platyhelminthes, 284–5
Pleated sheet, 25
Pleiotropy, 150
Pleural cavity, 417
Pleural membrane, 417
Plumule, 556, 562
Pluripotent, 490
Podocyte, *436–7*
Poikilothermy, 448
Polar body, 506
Polar groups, 17
Pollen grain, 545
Pollen mother cell, 552
Pollen, 552
tube, 281, 554
Pollination, 549
cross, 549
insect, 550
self, 549
wind, 551
Pollution, 803*ff*
Polychaete worms, 287
Polygamy, 509
Polymerase chain reaction (PCR), 216
Polymorphism, 234
Polyp, 283
Polypeptide, 21
Polyphyletic group, 261
Polyploidy, 200–2
Polysaccharides, 32, 36–38
Pons, 325
Population, 671, 710*ff*
explosion, humans, 801
growth, 712, 718
Porifera, 282
Pork tapeworm, 702–6
Positive feedback, 314, 800, 789
Post-translational processing, 183–4
Potassium as plant nutrient, 636
Potometer, 646
Predation, 696*ff*
Predator efficiency, 698
Predator–prey cycle, 720
Pregnancy, 519
Prehensile tail, 756
Preprocessing of lymphocytes, 493
Presentation time, 594
Pressure potential, 98
Pressure-flow hypothesis, 660
Primary cell wall, 569
growth, 568
oocyte, 506
spermatocyte, 504
Primate, 752*ff*
social life, 757
variety, 755
vocal communication, 758
Primitive characters, 753
Primordial germ cells, 504
Probability, 131
Procambium, 569, *570*
Productivity, 735
Proembryo, 555
Proenzyme, 72
Progesterone, 508
Proglottid, 703
Prokaryote, 47, 59, *60*, 268
Prolactin, 522
Promoter, 176
Pronucleus, 515
Propagule, 558
Prophase
meiosis, 119, 120
mitosis, 113
Proplastids, 533
Proprioceptor, 329
Prosimians, 756
Prostaglandins, 464–5, 520
Prostate gland, 503
Prosthetic group, 28, 70
Protandry, 551
Protease, 463
Proteins, 20–29
carrier, 103
conjugated, 28
denaturation, 28
diet, 465
digestion, 471
energy source, 463
fibrous, 24
globular, 24

Proteins (*cont'd*)
  membrane, 48
  primary structure, 25
  quaternary structure, 26
  secondary structure, 25
  synthesis, 173*ff*
  tertiary structure, 25
Prothallus, fern, 542
Protoctista, 269
Protogyny, 551
Proton, 821
Proton motive force, 85
Proton pumping, 82, 83–4, *621*
Protophloem, 569, *572*
Prototherian mammals, 293
Protoxylem, 569, *572*
Proximal tubule, *433*, *438*
Pseudopodium 106, 276
Puberty, 507–8
Pulmonary circulation, 398
Pulvinus, 589
Punnett square, 125
Pupa, 289
Pupil reflex, *336*
Pure line, 123
Purines, 164
Purkinje tissue, 401
Pylorus, 475
Pyramid
  biomass, 735
  energy flow, 735
  numbers, 734
  productivity, 735
  renal, *432*
Pyrimidine, 164

Quadrat, 711
Quern stone, 798
Quiescence, 586
Quinine, 702

*r*-species, 723
Radiation, 195
Radicle, 556, 562
Radioisotopes
  dating, 824
  ecology, 731
Radius, 364
Radula, 290
Ragworm, *286*
Ray initials, *578*
Rays, *578*
Receptacle, 546
Receptor, 329*ff*
  potential, 331
  tonic and phasic, 332
Recessive, 124
Reciprocal cross, 124
Recombinant DNA, 206
Recombinants, 127, 135
Rectum, 481
Red algae, 278
Red blood corpuscles, 390
Redi, F, 7
Reduction, 826
Reflex action, 308–10
Reflex arc, 308
Refractory period, 312
Relative humidity, 644, 682
Relaxin, 519
Releasing hormones, 377
Relative growth rate, 567
Reliability, 842
Renal artery, 432
Renal vein, 432
Renin, 445
Rennin, 475
Replica plating, 191–2
Reproduction, 6
  asexual, 501
  sexual, 501
Reproductive strategies, 722
Reptiles, 293
Residual volume, 427
Resolving power, 45
Resources vs conditions, 672
Respiration, 3, 54, 77, 82–8
  aerobic, 77
  anaerobic, 88
  control of, 92
Respiratory centre, *see* breathing centre
Respiratory chain, 84
Respiratory distress syndrome, *see* Breathing distress syndrome
Respiratory organ, *see* gas exchange organ
Respiratory quotient, 91–2
Respirometer, 78
Resting potential, 311
Restriction enzyme, 208
Restriction maps and evolution, 792
Retina, 338–46
Retinal, 340
Retinal convergence, 343
Retrovirus, 63, 211
Reverse transcriptase, 63, 211
Rf value, 611
Rhesus system, 499
*Rhizobium*, 707
Rhizoid, 280
Rhizomes, 580
Rhizopoda, 276
*Rhizopus*, 41, 272–4
Rhodophyta, 278
Rhodopsin, 340–2
Riboflavine, 81, 468
Ribonuclease, 183, 477
Ribonucleic acid, *see* RNA
Ribose, 32, *33*
Ribosomal RNA, 180
Ribosomes, 51, *182*
Ribozyme, 177
Ribulose bisphosphate, 622
Ribulose bisphosphate carboxylase, 633
Rickets, 469
Rigor mortis, 353
Ring species, 244
RNA polymerase, 176
RNA, 42, 173, 175
  messenger, 175
  processing, 177
  ribosomal, 182
  transfer, 180, *181*
Rods, 340
Root cap, 574
Root cap, role in gravitropism, 595
Root hairs, 532, 574
  ion uptake, 637
Root
  adventitious, 532
  internal anatomy, *655*
  mineral uptake, 637
  nodules, 707
  pressure, 650, *651*, 654
  system, 532
  water uptake, 656
Rough endoplasmic reticulum, 51
Rubisco, 633–4
Runners, 580

S phase, 112
Saliva, 473
Salivary amylase, 473
Saltatory propagation, 315–16
Samples, 711, 834
Sampling error, 838
Sand dune succession, 743
Sap, ascent of, 650–
Saprotroph, 271, 462
Sapwood, 579
Sarcolemma, 349
Sarcomere, 349
Sarcoplasmic reticulum, 354
Sarich, Vincent, 770
Saturated fatty acid, 29, *30*
Scaling, 296, 369

Scanning electron microscope, 47
Scapula, 364
Schleiden, M, 41
Schwann, T, 41
Schwann cell, 305
Scientific method, 6
Sclera, 334
Sclereids, 534
Sclerenchyma, 533
Scolex, 703
Scorpions, 290
Scrotum, 502
Scutellum, 563
Scyphozoa, *284*
Sea urchin, 291
Sebaceous glands, 455
Second messenger, 375
Secondary cell wall, 569
  growth, 568, 575, *576*
  oocyte, 507
  pump, 106
  spermatocyte, 504
  succession, 745
Secretin, 478
Sedimentary rocks, 775
Seed, 281
  dispersal, 557–9
  dormancy, 586–8
  germination, 563
  endospermic, 555
  maize, 563
  non-endospermic, 555
  pea, 562
  structure, 562
  sunflower, 562–3
Segregation, 116, 124
*Selaginella*, life cycle, *544*
Selection, 224
  directional, 231
  disruptive, 233
  stabilising, 229
Selective permeability, 97
Semen, 503
Seminal fluid, 503
Seminal vesicle, 503
Seminiferous tubule, 502
Sensitivity, 6, 303
Sensory neuron, 307
Sepal, 546
Sere, 745
Serological test, 772
Sertoli cell, 505
Serum, 396
Set point, 386, 450
Seta, *see* chaeta
Sewage, 804
Sex determination, 140
Sex hormones, 507
Sex-linkage, 141
Sex pilus, 270
Sexual dimorphism, 513
Shade leaves, 629, *630*
Shepherd's purse (embryo), 555, 556
Shoot system, 531
Short sight, 338
Short-day plants, 599
Sickle cell anaemia, 28
Sickle cell haemoglobin, 230
Sieve tube element, 536, 658
Sigmoid growth curve, 567, 714
Significance, 831, 841
Simple reflex, 308
Simpson Index, 727
Single circulation, 397, *398*
Sinoatrial node, 400
Sinus venosus, 397
Sinusoids (liver), 482
Size and strength, 364
Skeletal muscle, 349
Skeleton, 359*ff*
  appendicular, 363
  axial, 363
  fast and slow twitch, 357
  functions, 360
  hydrostatic, 301, 359, 369, *370*
  jointed 359
Skewed distribution, 834
Skin (mammal), 451, 453–5
  colour in humans, 796
  in defence, 487
Skull, 363
Sleeping sickness, 277
Sliding filament hypothesis, 351
Smith, Holly, 786
Smooth endoplasmic reticulum, 51
Smooth muscle, 301
Snowshoe hare, 720
Social dominance, 690, 759
Sodium pump, 313
Sodium regulation, 444
Sodium, dietary, 467
Soil atmosphere, 676
  constituents, 673*ff*
  formation, 677
  profile, 677
  water, 676
Solute potential, 98
somatic cell therapy, 220
Somatic doubling, 200
Somatic mutation, 192
Sorus, 541
Sources and sinks, 659
Southern blotting, 217
Spatial summation, 320, 343
Speciation, 237*ff*
Species, 258
Speech, 767
Spermatid, 504
Spermatogenesis, *504*
Spermatogonium, 504
Spermatozoa, 503
Sphincter, 472
Spiders, 290
Spinal cord, *305*, *308*
Spinal nerve, 304
Spindle (cell division), 114
  muscle, 359
Spiracle, 288, 428
Spirilla, 269
*Spirogyra*, 279
Spirometer, 425, *426–7*
Spleen, 489
Sponges, 282
Spontaneous generation, 6
Sporangiophore, 273
Sporangium, 273, 541
Spore, 116, 273, 540
Sporophyll, 545
Sporophyte, 279, 540, 541
Sporozoite, 700
Stahl, F, 166
Stalked particles, 54, 85
Stamen, 545
Standard deviation, 835
Standard error of the difference, 838
Standing crop, 735
Starch, 37
Statoliths, 594
Stem cell, 490
Stem tubers, 581
Stenohaline, 681
Stereoisomers, 34
Stereoscopic vision, 346–7, 753
Steroids, 31
Stickleback, 238
Stigma, 546
Stimulus, 303, 328
  modality, 329
Stoma, 18, 280
Stomach, 474
Stomata, diffusion through, 628
Stomatal movement, 643
Stratum (rock), 775
Stretch reflex, 359
Strobila, 703
Style, 546
Suberin, 533, 579
Submucosa, 471

Substomatal air space, 628
Succession
  primary, 742
  secondary, 745
Sucrase, 479
Sucrose, 35
Sulphur as plant nutrient, 636
Summation, 320, 343
Sumner, J, 65
Sun and shade leaves, 629, *630*
Surface tension, 18
Surfactant, 420
Survivorship, 716
Svedberg unit, 50
Swallowing, 474
Sweat gland, 455
Sweat, 452, 487
Swim bladder, 293
Symbiosis, 708
Sympathetic nervous system, 321
Sympatric speciation, 242
Sympatry, 237
Symplast, 649
Synapse, 316–21
Synapsis, 119
Synaptic
  cleft, 318
  inhibition, 318
  synaptic vesicle, 318
  synaptic knob, 317
Syncytium, 41
Synecology, 671
Synergids, 554
Synovial fluid, 364
Systole, 401
Szent-Györgi, Albert, 351

$T_2$ phage, 161
T-DNA, 218–9
T lymphocyte, 492, *496*
T-system, 354
t test, 840
*Taenia solium*, 702–6
Tagma, 287
Talmage, David, 492
Tandem repeat, 217
Tannin, 43
Tanning, 368, 673
Tapetum (anther), 552
Tapeworm, *286*, 702–6
  life cycle, 704
  structure, 702–4
Target cell, 373, 374
Tatum, Edward, 153–4
Taxon, 258
Taxonomy, 261
Tegument, 703
Teleology, 12
Telophase
  meiosis, 119, 120
  mitosis, 115
Temperature effects
  ecological factor, 679
  enzymes, 69
  photosynthesis, 633
  transpiration, 645
Temperature coefficient ($Q_{10}$), 69
Template strand, 176
Temporal summation, 320
Tendon, 364
Terminal transferase, 212
Terminator sequence, 176
Territory, 689
Testa, 556, 563
Testcross, 126
Testis, 502
Testosterone, 508
Tetanus, 356
Tetrad (pollen), 552
Tetraploidy, 200
Tetrasomy, 198
Tetrose, 32
Thalamus, 325
Thermal pollution, 809
Thermocline, 741
Thermodynamics, Second Law, 5
Thermonasty, 589
Thermoneutral zone, 458
Thermoreceptors, 329, 450
Thermoregulation, 447–60
Thiamine, 468
Thigmotropism, 589
Thirst, 443
Threshold stimulus
  muscle, 356
  nerve, 312
Thrombin, 396
Thrombocytes, *see* platelets
Thromboplastin, 396
Thrombosis, 397, 404
Thylakoid, *55*
Thymine dimer, 195
Thymine, 163
Thymus, 489
Thyroid gland, 378
Thyroid-stimulating hormone, 379
Thyrotrophin-releasing factor, 380
Thyroxine, 378, 457
Ticks, 290
Tidal volume, 426
Tides, 684
Tight junction, *58*, 106
Tissue fluid, 383, 407–8
Tissue, 297
Tissues
  mammal, 297–302
  plant, 533
Tolerance, 685, *686*
Tonic receptors, 332
Tonoplast, 43, 99
Tools, 754
  Acheulean, 787
  composite, 790
  Mousterian, 787
  Oldowan, 782
Torpor, 449
Torsion, 291
Toxoid, 499
Trachea, 288, 417, 428
Tracheid, 534
Tracheole, 428
Tracheophyta, 280
Transamination, 465
Transcellular transport, 105
Transcription (DNA), 175–6
Transducin, 342
Transfer cell, *661*
Transfer RNA, 180–1
Transferase, 66
Transgenesis, 206
Translation of genetic information, 180
Translocation, 657–62
Transmission electron microscope, 47
Transpiration, 641*ff*
  effects of, 644
  factors affecting, 644–6
  leaf structure and, 646
  measurement, 646
  pull, 651, *652*
Transverse tubules, 354
Trematoda (flukes), *286*
Tribe, 258
Tricarboxylic acid cycle, 83
Trichromatic theory (vision), 344
Triglycerides, 29
Tri-iodothyronine, 378
Triose sugar, 32
Triplet code, 178
Triploblastic animals, 284
Triploidy, 201
Trisomy, 198
Trochophore larva, 287
Trophic hormones, 377
Trophic level, 730
Trophoblast, 515
Tropism, 589–96
Tropomyosin, 352

Troponin, 352
*Trypanosoma*, 277
Trypsin, 477
Tube foot, 291
Tubulin, 57, 112
Tunica adventitia, 403
  intima, 402
  media, 402
Turbellaria, *286*
Turgor, 100
Turner's syndrome, 199
Twins, 519

Ulna, 364
Ultrafiltration (kidney) 434
Ultraviolet radiation, 163, 195
*Ulva*, 279
Umbilical cord, 517
Undulipodia, 56
Unsaturated fatty acid, 29, *30*
Uracil, 164
Urea formation, 483–4
Urea, 431, 444
Ureotely, 434
Ureter, 432
Urethra, 432
Uric acid, 431, 434
Uricotely, 434
Urination, *432*
Uterus, 505

Vaccination, 498
Vacuole 43, 99, 533
Vagina, 506
Vagus nerve, 322, 409
Validity, 842
Valve (heart), 398
Van Niel, C B, 616
Vas deferens, 502
Vasa recta, 442
Vascular bundle, 571
Vasectomy, 505
Vasomotor centre, 405
Vector, 700
Vegetative propagation, 580
Veins, 405
Vena cava, 398
Ventilation, 77, 414
Ventricle
  brain, 324
  heart, 397
Vernalisation, 603–4
Vertebrae, 363
Vertebrates, 292
Vessel (xylem), 534
  formation, 535
  member, 534
Vestigial structure, 8
Vibrio, 269
Villus, 478
  chorionic, 518
  trophoblastic, 515
Virion, 61
Viruses, 60–3
  as mutagens, 195
Vital capacity (lungs), 426
Vitalism, 65
Vitamin $B_{12}$, 468, 475
Vitamin K, 396, 481
Vitamins, 467–9
Vitreous humour, 335
*Volvox*, 279

Water potential, 97
Water stress, 643
Water table, 676
Water vascular system, 291
Water, 17–19
  loss in plants, 642–7
  transport in plants, 649–54
  uptake in plants, 654–7
Watson, James, 164
Weathering (rock), 673
Weismann, August, 12
Went, Frits, 591
Went-Cholodny hypothesis, 596
Wernicke's area, 767
Wheat, polyploidy, 203
White blood cells, 489
White matter, 308
Wilson, Allan, 770
Winogradsky, S, 616
Winter buds, *582*
Womb, *see* uterus
Wood, *578*
Woodland food web, *730*
Woody plants, 582

X chromosome, 140
X-rays
  crystallography, 165
  mutagen, 195
Xanthophylls, 610
Xeromorphic characters, 647
Xerophyte, 647, *648*
Xerosere, 745
Xylem, 280, 534
  leaf, 626
  primary, 568–71
  secondary, 576–7

Y chromosome, 140
Yeast, 90, 276,
Yellow body, 507

Zigzag dance (stickleback), 238
Zona pellucida, 505
Zonation on rocky shores, 683
Zooplankton, 729
Zygomorphic flowers, 550
Zygomycota, 274
Zygote, *111–12*, 501
Zymogen, 72, 471